Deep Marine Systems

Deep Marine Systems: Processes, Deposits, Environments, Tectonics and Sedimentation

Kevin T. Pickering & Richard N. Hiscott
With contribution from Thomas Heard

This work is a co-publication between the American Geophysical Union and Wiley

WILEY

This edition first published 2016 © 2016 by Kevin T. Pickering & Richard N. Hiscott

This work is a co-publication between the American Geophysical Union and Wiley. This book has been published in full colour with the aid of generous financial support by Nexen Petroleum UK Limited, a wholly owned subsidiary of CNOOC Limited.

Registered office: John Wiley & Sons, Ltd, The Atrium, Southern Gate, Chichester, West Sussex, PO19 8SQ, UK

Editorial offices: 9600 Garsington Road, Oxford, OX4 2DQ, UK
The Atrium, Southern Gate, Chichester, West Sussex, PO19 8SQ, UK
111 River Street, Hoboken, NJ 07030-5774, USA

For details of our global editorial offices, for customer services and for information about how to apply for permission to reuse the copyright material in this book please see our website at www.wiley.com/wiley-blackwell.

Library of Congress Cataloging-in-Publication Data

Pickering, K. T. (Kevin T.)
  Deep marine systems : processes, deposits, environments, tectonics and sedimentation / by Kevin T.
Pickering & Richard N. Hiscott ; with contribution from Thomas Heard.
    pages cm
  Includes bibliographical references and index.
  ISBN 978-1-118-86549-1 (cloth) – ISBN 978-1-4051-2578-9 (pbk.) 1. Marine sediments. 2. Plate
tectonics. I. Hiscott, Richard N. II. Title.
  GC380.15.P54 2015
  551.46'86 – dc23
                                                          2015007967

A catalogue record for this book is available from the British Library.

Wiley also publishes its books in a variety of electronic formats. Some content that appears in print may not be available in electronic books.

Cover image: Palaeogene deep-marine sediment gravity-flow deposits, Andaman Islands, India (beds young to right): © Kevin T. Pickering

Set in 9/11pt Minion by SPi Global, Chennai, India

Printed and bound in Singapore by Markono Print Media Pte Ltd

1 2016

Contents

Preface

It is now 25 years since, along with Francis Hein as a co-author, we published *Deep Marine Environments: Clastic Sedimentation and Tectonics* with Unwin-Hyman (Pickering *et al.* 1989). During this time, there have been enormous advances in our understanding of modern and ancient deep-water environments, the physical processes that operate and the resulting deposits. Technology has allowed a much greater interrogation of deep-water environments, for example through improvements in the use of side-scan sonar, the acquisition and processing of seismic-reflection data, together with considerable advances in both industrial and academic deep-water drilling. Experimental, theoretical and observational studies have led to a significantly improved understanding of the fluid dynamics and depositional processes in deep-water settings.

A greater understanding of the architectural elements that form the building blocks of deep-water systems, and the application of sequence stratigraphy to such deposits, have all led to an appreciation of the dynamic nature of sedimentation in response to allocyclic (extra-basinal) and autocyclic (intra-basinal) controls, such as changes in local, regional or global relative base level (eustatic, tectonic, etc.). Experimental modelling of stratal geometry is increasingly used in an attempt to constrain the range of what is physically possible (e.g., using large flume tanks such as EUROTANK based at Utrecht University, The Netherlands; St Anthony Falls Laboratory, University of Minesota/USGS; the STEP Basin, University of Texas at Austin). Research in the past two decades has led to a greater understanding of the actual or potential importance of climate change and weather patterns at a whole range of temporal scales in driving environmental change and deep-water deposition, for example Milankovitch cyclicity, sunspot cycles, extreme weather, and non-linear ocean-atmosphere dynamics (*cf.* Dansgaard–Oeschger cycles and Heinrich events). Many of these processes still remain relatively poorly understood, but they are increasingly being invoked as possible explanations for stratigraphic patterns observed in deep-water sedimentary environments. The possible importance of internal tides in the ocean as a process in sediment transport and deposition was not even considered when we wrote our first book. These are just some of the recent major advances in deep-water sedimentology that make this book very different to the 1989 book. It is, however, beyond the scope of this current book to consider the physics of such processes. Instead, we discuss how they may explain physical observations from deep-water successions.

As with the first book, this second volume grew out of our desire to integrate process-based, environmental and large-scale plate-tectonic aspects of both modern and ancient deep-marine sedimentation into a unified and comprehensive text. Our research and experience over the last 25 years persuaded us to write a new book that includes the most significant advances in deep-water sedimentology. Edited volumes that cover parts of this rapidly expanding area of knowledge are available, but in our humble opinion they are unable to provide the broad ranging and consistent conceptual approach of an integrated text. We have endeavoured to produce an up-to-date, discursive and well-balanced text on deep-marine sedimentology. We have focussed on clastic sedimentology, and in particular siliciclastic sediments, although much of the text is equally applicable to carbonate, volcaniclastic or lithologically mixed environments. Aspects of basin tectonics and structural controls on the development of deep-marine basins are addressed, but the detailed geochemical and petrographic aspects of deep-marine environments are beyond the scope of this book.

There are many ways in which to tackle a book of this type. As with Pickering *et al.* (1989), we have divided the text into three parts. In the first part, we discuss the fundamental building blocks of deep-marine environments and the nature of sediment transport and deposition in the deep sea. After considering the quantitative and semi-quantitative aspects of deep-water sedimentation, we provide a systematic classification of the range of sediment deposits (i.e., facies). Unlike the 1989 book, we have added chapters on trace fossils in deep water (ichnology), time–space integration (incorporating sequence stratigraphy), and statistical properties of sediment gravity-flow (SGF) deposits, focussing on the vertical stacking of facies to form sequences and a discussion on how such sequences can be recognised using objective, statistical criteria.

The second part of the book is concerned with specific deep-marine environments. Individual chapters focus on: sediment drifts and abyssal sediment waves, modern submarine fans and related depositional systems, and ancient submarine fans and related depositional systems. Unlike the approach followed in our 1989 book, we have subsumed any description and discussions of submarine canyons, submarine channels and sheet-like systems within two chapters, one on modern and the other on ancient systems. This is done in appreciation of the intimate linkage between the up-dip and down-dip parts of deep-marine depositional systems, and is in line with the approach used by sequence stratigraphers. In recent years, there have been several integrated outcrop–subsurface studies of deep-marine systems that have greatly improved our understanding of such environments, for example the Ainsa Project in the Spanish Pyrenees (based at University College London, UK); the SLOPE Project in the Karoo, South Africa (originally based at Liverpool University, and now co-located at Manchester and Leeds University, UK); the Clare Basin drilling project in western Ireland (based at University College Dublin, Ireland), and a drilling programme in the Taranaki Basin of North Island, New Zealand (based at the University of Auckland, New Zealand).

Our approach has been to review the important ideas and models associated with the environments, with only a few key modern and ancient case studies that have been carefully chosen to reflect the major developments in understanding that are associated with the particular environment.

The third part of the book, tectonics and sedimentation, integrates the various deep-marine environments discussed in the preceding part of the text into a plate-tectonic framework. We have adopted a traditional four-fold division of plate settings into evolving and mature extensional systems, subduction margins, foreland basins, and oblique (strike)-slip margins. These chapters draw extensively on the large data base and publications resulting from scientific ocean drilling, such as the Ocean Drilling Program (ODP), the Integrated Ocean Drilling Program and International Ocean Discovery Program (both abbreviated to IODP), especially programmes such as the Nankai Seismogenic Zone Experiment (NanTroSEIZE).

One of our principal aims has been to produce a truly integrated book with a standardised nomenclature and conceptual approach to each subject. In many cases this has required that we re-classify the facies of authors whose work we have cited, or that we use different names for the underlying flow processes in order to accord with the process models used in Chapter 1 of the book. We apologise to any authors who are unhappy with this recasting of certain aspects of their work, but there is no alternative if the objective is to create a text with consistent terminology from one chapter to the next. It will be apparent to many readers of this book that we draw upon our own experience with many examples taken from those research areas most familiar and accessible to us. Any uncredited field photographs in the book are from our personal collections.

The trace-fossil chapter (Chapter 3) required specialist knowledge, so we invited Thomas Heard to contribute a chapter on ichnology. Creating the first draft for the remaining chapters was shared between the co-authors in the following way: Chapters 1, 5, 6, 7 and 9 were led by Richard Hiscott, and Chapters 2, 4, 8, 10, 11 and 12 by Kevin Pickering. Large parts of Chapter 5 were extracted from the PhD thesis of former Hiscott student Sherif Awadallah, who graciously gave his permission to use this material. The first drafts were then reviewed and modified by the other co-author, going through many iterations until we were satisfied with the content. Many colleagues kindly read the manuscript at various stages of preparation; their constructive comments and advice have substantially improved the scientific content and readability of the text. Early drafts of chapters were used in a taught Masters degree programme by Kathleen Marsaglia (University of California, Northridge) and the student feedback was used to rewrite and improve portions of the text.

We are very grateful for the help and professionalism of the staff at Wiley, without whose commitment and encouragement this book would not have been possible, particularly Ian Francis and Kelvin Matthews. Ian is particularly thanked for his involvement in the book at the stage of the transformation of a contract and book plan into preliminary chapters. Despite our busy schedules and delays in getting chapters written, Ian maintained his faith in us to deliver and gently encouraged us to keep working toward a final product.

We are indebted to our many research students who have worked with us over our careers in the quest to understand a range of issues related to deep-water sedimentology and associated transport processes. KTP thanks his students: Keith Myers, Richard Blewett, Sarah Davies, Tommy McCann, Julian Clark, Vincent Hilton, John Millington, Nick Drinkwater, Sarah Gabbott, Clare Stephens, Clair Souter, Jane Alexander, Jordi Corregidor, Christine Street, David Hodgson, Susan Hipperson, Sarah Boulton, Martin Gibson, Nicole Bayliss, Thomas Heard, Clare Sutcliffe, Kanchan Das Gupta, Gayle Hough, Richard Ford, Rachel Quarmby, Veronica Bray, Bethan Harris, Edward Armstrong, Pierre Warburton, James Scotchman, Blanca Cantalejo and Nikki Dakin. RNH thanks those of his students involved in deep-marine or gravity-flow research: Scott Gardiner, David Mosher, Paul Myrow, Tim England, Louise Quinn, Cuiyan Ma, Chengsheng (Colin) Chen, Abdelmagid Mahgoub, Sherif Awadallah, Martin Guerrero Suastegui, Renee Ferguson and Bursin Işler. Finally, we thank our respective spouses and families for their multifarious support over the many years when a substantial part of the family 'leisure' time was taken up by 'the book'; thanks Louise Pickering and Paula Flynn.

Kevin T. Pickering
Richard N. Hiscott
July 2015

About the companion website

This book is accompanied by a companion website:

      www.wiley.com/go/pickering/marinesystems

The website includes:

- Powerpoints of all figures from the book for downloading
- PDFs of tables from the book
- Complete set of high-resolution core photographs from the UCL Ainsa Project

PART 1

Process and product

CHAPTER ONE
Physical and biological processes

(a) Experimentally produced turbidity current. Courtesy Jeff Peakall. (b) Upper part of sediment slide deposits (Facies F2.1) draped by siltstone turbidites (Facies D2.2 and D2.3) in deep-marine volcaniclastics, Miocene Misaki Formation, Miura Peninsula, southeast Japan.

Deep Marine Systems: Processes, Deposits, Environments, Tectonics and Sedimentation, First Edition. Kevin T. Pickering and Richard N. Hiscott.
© 2016 Kevin T. Pickering and Richard N. Hiscott. Published 2016 by John Wiley & Sons, Ltd.
Companion Website: www.wiley.com/go/pickering/marinesystems

1.1   Introduction

This chapter has two main functions. First, there is an introduction to the main processes responsible for the physical transport and deposition of sediments derived from land areas and carried into the deep sea. Second, the origin of pelagic sediments (oozes, chalks, cherts) and organic-rich muds (e.g., black shales and sapropels with >2% organic matter) is explained. For these sediments, transport of material from an adjacent land mass is either not required, for example in the case of accumulation of biogenic skeletons, or is far less important than the chemistry of the seawater at the site of deposition. The biogenic process of bioturbation is considered in Chapter 3.

The three main processes responsible for transporting and depositing particulate sediments seaward of the edges of the world's continental shelves are (i) bottom-hugging sediment gravity flows (e.g., turbidity currents and debris flows), (ii) thermohaline bottom-currents that form the deep circulation in the oceans and (iii) surface wind-driven currents or river plumes that carry suspended sediment off continental shelves. Tidal currents, sea-surface waves and internal waves at density interfaces in the oceans appear to be only locally important as transport agents on the upper parts of slopes and in the heads of some submarine canyons.

In order to appreciate how sands and gravels encountered in deep-marine petroleum reservoirs are deposited, it is essential to understand the dynamics of sediment gravity flows. A *sediment gravity flow* (SGF; Middleton & Hampton 1973) is a bottom-hugging density underflow carrying suspended mineral and rock particles, mixed together with ambient fluid (most commonly seawater). A SGF is a special type of *particulate gravity current* (McCaffrey *et al.* 2001) – in other flows belonging to this broad category the particles can be snow and ice (e.g., in a powder snow avalanche), or the fluid phase can be hot volcanic gases (e.g., in a pyroclastic surge). In engineering practice, such mobile solid and fluid mixtures are called *granular flows*, *slurry flows* or *powder flows*, depending on the size of the particles, whether the fluid phase is a liquid or a gas and whether cohesive forces are significant. In this book, we will use the more geologically relevant term 'sediment gravity flow', but anyone undertaking a literature search needs to be aware of the alternative terminology used in other disciplines.

The particles in SGFs spend most of their time in suspension rather than in contact with the seafloor. In the more dilute SGFs, particles in suspension eventually settle to the seafloor where they accumulate, either with or without a phase of traction transport. This is called *selective deposition* (Ricci Lucchi 1995; *incremental deposition* of Talling *et al.* 2012), because particles are deposited one by one according to size, shape, density or some other intrinsic property. In concentrated dispersions or cohesive debris flows, the particles are not fully free to move independently of one another and therefore accumulate by massive deposition (Ricci Lucchi 1995; *en masse* deposition of Talling *et al.* 2012). The distinction between *selective deposition* and *massive deposition* (or en masse deposition) is a useful one, because the former deposits are commonly laminated and the latter commonly structureless, poorly organised, plastically deformed, or contain evidence of intense particle interaction and/or pore-fluid escape. The ultimate end-member example of en masse deposition is *coherent sliding* in which masses of semi-consolidated material move downslope while retaining some of the organisation and stratigraphy of the original failed successions. Sediment slides come to rest as deformed, folded and/or sheared units.

Let us start by considering a typical event responsible for basinward sediment transport along a continental margin. The transport can be divided into four phases: (i) a phase of flow initiation; (ii) a period during the early history of the flow when characteristics of the transporting current change rather quickly to a quasi-stable equilibrium state; (iii) a phase of long-distance transport to the base of the continental slope or beyond and (iv) a final depositional phase. In many cases, the concentration of solid particles changes systematically along the flow path. Particle concentration is an important variable because mixtures of sediment and water can only become fully turbulent if the concentration is low. Without turbulence, it is difficult to suspend and transport mineral-density particles for long distances, and tractional sedimentary structures like current-ripple cross-lamination cannot form. Figure 1.1 shows how a range of deep-marine transport processes can be assigned to one or more of these four stages of flow evolution, and shows how the flow concentration might change through time. For example, sediment suspended by a storm on a continental shelf or by tidal currents in the head of a submarine canyon forms low-concentration suspensions that might continue to move downslope as turbidity currents, transferring particulate sediment tens to hundreds of kilometres farther seaward. Other initiation processes, like the disintegration of sediment slides on steep slopes, can generate more concentrated SGFs such as submarine debris flows. These debris flows may never become more dilute or develop turbulence, and therefore are less likely to transport their sediment load far into the deep-marine basin.

There is a fundamental difference in the way that non-turbulent, highly concentrated SGFs (e.g., debris flows) deposit their sediment load and evolve as compared to turbulent and water-rich flows like turbidity currents. As a highly concentrated SGF decelerates, the internal resistance to flow (e.g., friction between adjacent particles, or electrostatic cohesive forces between clay minerals) eventually exceeds the gradually decreasing gravitational driving force. When this happens, the flowing mass ceases to travel basinward and is deposited. Because the shear stress responsible for internal deformation and therefore 'flow' increases downward toward the base of the moving mass (due to the cumulative weight of the overlying material), it is the basal part of the mass which last stops deforming. This final immobilisation of the basal part of the flow only occurs after higher levels have ceased to deform (Fig. 1.2). In the case of a slowly decelerating, texturally homogeneous debris flow, the top of the flow will stop deforming first, and then internal deformation will cease in a sequential manner from the top of the flow (where the shear stress is lowest) downward. If we equate the cessation of internal deformation to the phenomenon called 'deposition', then such flows might be said to deposit (and acquire their textures and fabrics) 'from the top downward'. In the case of decelerating debris flows, this gradually thickening zone with little or no internal deformation at the top of the flow has been called a 'rigid plug' (Johnson 1970). In contrast, most decelerating turbulent SGFs deposit their load selectively, grain by grain, during a period when solid particles rain downward to the base of the flow. This can be thought of as deposition 'from the base upward', with the result that many deposits of this type show graded bedding because of a progressive decline in flow energy during the depositional phase. Unlike the non-turbulent flows, these low-concentration SGFs become increasingly dilute as they lose their sediment load (Fig. 1.1), eventually losing their identity when the flow density decreases to a value close to that of clear seawater.

Although the way in which deposition proceeds can help explain the development of texture and the internal organisation of many deep-sea facies, the distinction between deposits formed 'from the top downward' and those formed 'from the base upward' is sometimes blurred. This happens in some SGFs of intermediate concentration or

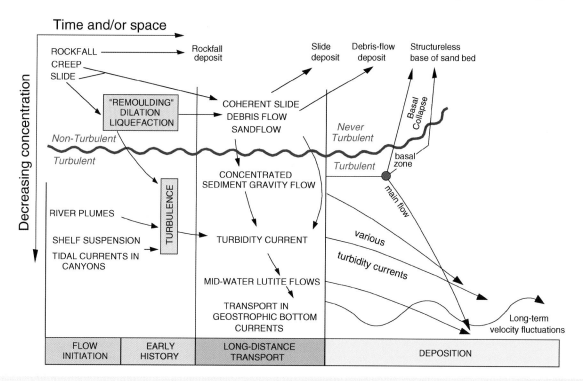

Fig. 1.1 Simplified conceptual overview of the evolution of sediment gravity flows and other deep-marine transport processes as a function of concentration. The horizontal axis is time and/or space, but no units are implied because the evolution of some flows is much longer than for others. For example, turbidity currents might flow for only hours to days, whereas contour-following geostrophic bottom-currents (i.e., thermohaline currents) have velocity fluctuations lasting thousands of years. Notice that non-turbulent flows tend to deposit en masse, so that the deposit is simply the original flow, arrested in place when driving forces are no longer adequate to keep the material moving. In contrast, turbulent flows lose their sediment load by settling, and therefore become increasingly less concentrated during the depositional phase. Modified from Middleton and Hampton (1973) and Walker (1978).

clay content (e.g., Baas *et al.* 2011), because the downward rain of solid particles during times of rapid deceleration suppresses turbulence in the near-bed region of the flow. Such flows are strongly stratified in terms of their density and concentration, so that the basal and upper parts of the flow behave differently. If the average concentration is low (<5% by volume), particles are lost by settling from the upper fully turbulent part of the flow and the flow becomes progressively more dilute. Near the base of the flow, however, frictional resistance can eventually in some cases immobilise a basal sheared layer, and the properties of the resulting deposit will be more akin to those of the more highly concentrated SGFs. If the average concentration is high (>5% by volume) and a significant amount of clay is present, rapid deceleration leads to the accumulation of a bipartite bed with an upper division of *fluid mud* containing variable quantities of silt and sand, as well as clay (Baas *et al.* 2011). Above this bipartite deposit, the upper part of the still-moving viscous flow may be immobilised, because of insufficient shear stress, to form a variably thick semi-rigid plug that thickens downward with decreasing flow velocity, leading to a very complicated final deposit.

With this brief introduction behind us, let us now probe more deeply into the processes responsible for deep-marine sedimentation. What is the best point of departure for a systematic assessment of sedimentary processes in the deep sea? We take our guidance from Figure 1.1, and begin with shelf-edge processes that can either initiate SGFs or independently move sediment into deeper water. In deep parts of the ocean basins, thermohaline currents locally are important agents of sediment transport, but mostly these areas receive their

sediment from infrequent SGFs, separated in time by long periods of relative quiescence. SGFs are responsible for a wide spectrum of modern and ancient facies, so their classification and description form a significant part of this chapter. Each type of SGF creates unique sedimentary textures and structures, which are described after each flow process is explained. The chapter concludes with other issues that are important in understanding the deep-marine environment, like the accumulation of pelagic sediments and deep-sea bioturbation.

1.2 Shelf-edge processes

1.2.1 High-level escape of mud from the shelf

Suspended sediment concentrations in shelf areas may be quite high due to the input of mud-laden river water, or stirring of the bottom by waves (Geyer *et al.* 2004), tidal currents or internal waves at density interfaces (Cacchione & Southard 1974). This suspended sediment may be advected off the shelf by ambient currents, possibly wind-driven, or by transport in cascading cold water that may flow off the shelf in the winter months (Postma 1969; McCave 1972; McGrail & Carnes 1983; Wilson & Roberts 1995; Ivanov *et al.* 2004). Suspensions of fine-grained sediment may also leave the shelf as dilute turbidity currents (lutite flows), moving along the bottom onto the lower slope and rise, or along density interfaces in the ocean water (Fig. 1.3) (Postma 1969; McCave 1972; Gorsline *et al.* 1984). These dilute suspensions may move down a smooth upper slope as

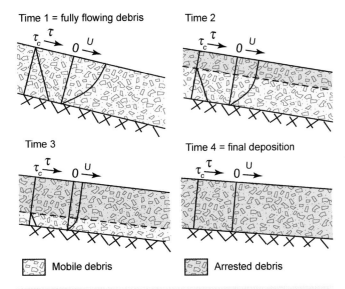

Time 1 = fully flowing debris   Time 2

Time 3   Time 4 = final deposition

▨ Mobile debris   ▨ Arrested debris

Fig. 1.2 Four snapshots during the deceleration and eventual deposition (Time 4) of a non-turbulent debris flow, showing how textures, fabrics and internal structures of the eventual deposit are locked into place by the progressive downward thickening of a 'rigid plug' (arrested debris). These are streamwise vertical cross-sections through the flow. In the 'rigid plug', there is little to no internal deformation because gravity-induced shear stress (τ) is less than the critical shear stress (τ_c) needed to overcome resisting forces (due to internal grain friction and electrostatic cohesive forces). Decreasing slope explains the decreasing shear stress. Profiles of shear stress (τ) and velocity (U) are shown in each case. In the 'rigid plug', the change in velocity with depth is zero, although the velocity of the plug itself is positive up until Time 4. In SGFs of this type, the material is effectively deposited 'from the top downward', and the base of the flow is the last part to be deformed by shearing (e.g., Time 3).

bypasses the slope, leading to maximum rates of accumulation on the continental rise (Nelson & Stanley 1984).

On narrow shelves, plumes of suspended sediment from river deltas can extend beyond the shelf-slope break (Emery & Milliman 1978; Thornton 1981, 1984), directly contributing fine-grained sediments to slope and rise areas. In polar areas, sediment-laden spring meltwater may actually flow from the land across the surface of floating sea ice and deposit its load directly onto the continental slope (Reimnitz & Bruder 1972).

Mud that leaves the shelf either by 'high-level' escape in river plumes, by dilute turbidity-current flow, or by movement along density interfaces in the water column over the slope eventually settles to the seafloor to form the bulk of what are called *hemipelagic deposits*. Deposition rates are on the order of 10–60 cm kyr^{-1} (Krissek 1984; Nelson & Stanley 1984). As is true for strictly pelagic sediments, the finest particles in the high-level suspensions are probably carried to the bottom as aggregates in the form of faecal pellets (Calvert 1966; Schrader 1971; Dunbar & Berger 1981). In regions of higher mud concentration, for example off river mouths, a significant quantity of mud forms aggregates called *floccules* (or flocs). The 'stickiness' that holds silt- and clay-sized particles together in flocs is provided by electrostatic attraction, organic matter and bacteria (Gibbs & Konwar 1986; Curran *et al.* 2002; Geyer *et al.* 2004). Flocs from a number of environments settle at speeds very close to 1 mm s^{-1} (~100 m day^{-1}), which is much faster than the settling rates of the same material once disaggregated (Gibbs 1985a, b; Hill 1998; Geyer *et al.* 2004). Even in mud-laden turbidity currents, the percentage of sediment deposited as flocs may commonly exceed 75% (Curran *et al.* 2004). The diameters of flocs, based mainly on measurements on continental shelves, are >100–200 μm, although floc densities are low and decrease with increasing floc size (Hill & McCave 2001).

Fine-grained suspensions that move seaward across the edges of continental shelves may vary seasonally in (i) grain size and (ii) content of suspended organics. In anaerobic/dysaerobic basins, this fine-scale seasonal cyclicity can be preserved in the sediment record; on oxygenated basin slopes all such lamination would be destroyed by burrowers. Dimberline and Woodcock (1987) and Tyler and Woodcock (1987) convincingly argue that submillimetre-thick interlaminations of silt and organics in the Silurian Welsh Basin are a result of alternations of (i) spring algal blooms with (ii)

unconfined sheet flows, or may be confined by gullies or canyons in which suspension is augmented by weak tidally forced flows (Shepard *et al.* 1979; Gorsline *et al.* 1984). There is evidence that most mud transported off the shelf by dilute turbidity currents (lutite flows)

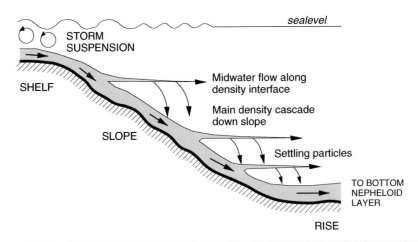

Fig. 1.3 Schematic representation of lutite flows cascading downslope. The increase in length of the arrows indicates an increase in concentration toward the base of the continental slope. The *nepheloid layer* is a part of the water column near continental margins where the suspended sediment concentration is particularly high because persistent currents prevent deposition of fine-grained suspended load. Redrawn from McCave (1972).

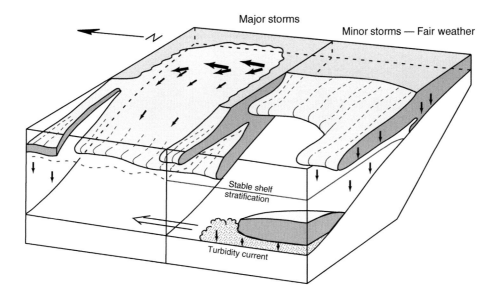

Fig. 1.4 Model for hemipelagites (Silurian Bailey Hill Formation) and turbidite sandstones (Brimmon Wood Member) in the Welsh Basin. Suspended silts, fine sands and organics (blue 'tongues') were advected off the shelf by waves and currents (large arrows and smaller downslope-oriented arrows), forming dilute bottom- and mid-water flows. The particulate materials then settled vertically from these flows (short vertical arrows). Annual seasonal layering in the hemipelagites was preserved under anaerobic/dysaerobic conditions. Redrawn from Dimberline and Woodcock (1987).

increased winter discharge of silt into the basin. The assumption of annual cyclicity leads to reasonable sediment accumulation rates of 60–150 cm kyr^{-1} (Dimberline & Woodcock 1987). A general depositional model (Fig. 1.4) involves bottom-hugging and midwater dilute flows advected off the shelf during fair-weather periods (hemipelagic laminated silts/organics, depending on the season), and during storms (silty/sandy graded beds with irregular order of internal structures).

1.2.2 Currents in submarine canyons

Current-meter data have been collected in submarine canyons to depths of over 4000 m (Shepard *et al.* 1979). Generally, the currents alternate directions, flowing up and down canyons with periodicities from 15 min to 24 h. The longest recorded unidirectional flows are five days down-canyon, off the Var River, France (Gennesseaux *et al.* 1971), and three days down-canyon in the Hudson Canyon off New York (Cacchione *et al.* 1978), in both cases with variable speeds.

Progressive vector plots of measured current data from many canyons tend to show a net down-canyon flow, although the results from canyons off the Eastern Seaboard of the United States show approximately equal durations of up- and down-canyon flow (Fig. 1.5). The time periods over which current speeds change vary considerably. The periodicity of most currents approximates to semi-diurnal tidal cycles at depths greater than about 200 m (e.g., Shepard *et al.* 1974). In canyons associated with small tidal ranges, such as off the west coast of Mexico, the length of a canyon-current cycle only approaches the tidal frequency at much greater depths. Shepard *et al.* (1979) summarised the relationship between the average cycle period of the up- and down-canyon alternating flows, the depth where data were recorded along each canyon axis and the local tidal range. In general, small tidal ranges and shallow depths tend to be associated with short average cycles, whereas large tidal ranges and/or deep water tend to be associated with long average cycles.

Although most currents flow up or down canyon, in some cases such as in Hueneme Canyon off the Santa Clara Delta, California, there is a considerable spread of flow directions. Hudson Canyon current data bear little or no relationship to the canyon orientation compared to the good agreement shown for Carmel Canyon off California. Currents that flow at an angle to the 'normal' up- or down-canyon direction are referred to as cross-canyon flows (Shepard & Marshall 1978). Cross-canyon flows are most common in wide canyons, for example in the Kaulakahi Channel off northwest Kauai, although the relatively narrow Hudson Canyon is the site of numerous cross-canyon flows. Strong cross-canyon flows tend to occur at low tide, possibly related to strong wind-driven currents becoming effective at slack low tide. In the Santa Barbara Channel, west of Santa Cruz Island, California, the cross-canyon bottom flows are mainly toward the east, similar to the direction of the surface currents in this area (Shepard & Marshall 1978). The origin of cross-canyon flows is poorly understood. One hypothesis of Shepard *et al.* (1979) is that these currents meander in wide canyons, in a similar manner to the way in which a small subaerial stream meanders in a wide valley.

Data from relatively shallow current-meter stations suggest a correlation between wind speed and the magnitude and direction of currents within canyons (Fig. 1.6). Pressure waves, preceding a storm, may be responsible for at least some, or part, of these current patterns. In other cases, however, there appears to be no correlation; for example, during a storm in La Jolla Canyon with 65 km h^{-1} onshore winds, maximum current speeds increased as wind speeds rose, although the up- and down-canyon periodicity did not vary until finally a large down-canyon surge up to 50 cm s^{-1} was recorded (Shepard & Marshall 1973a, b). Unfortunately, the current-meter was damaged during this surge and therefore any additional increases in speed that may have occurred went unrecorded – the meters were retrieved 0.5 km down-canyon, partially buried by sediments and kelp. Also, during this storm and probably during the current surge, a trough with walls 0.5 m high was excavated into the silty sand of the canyon

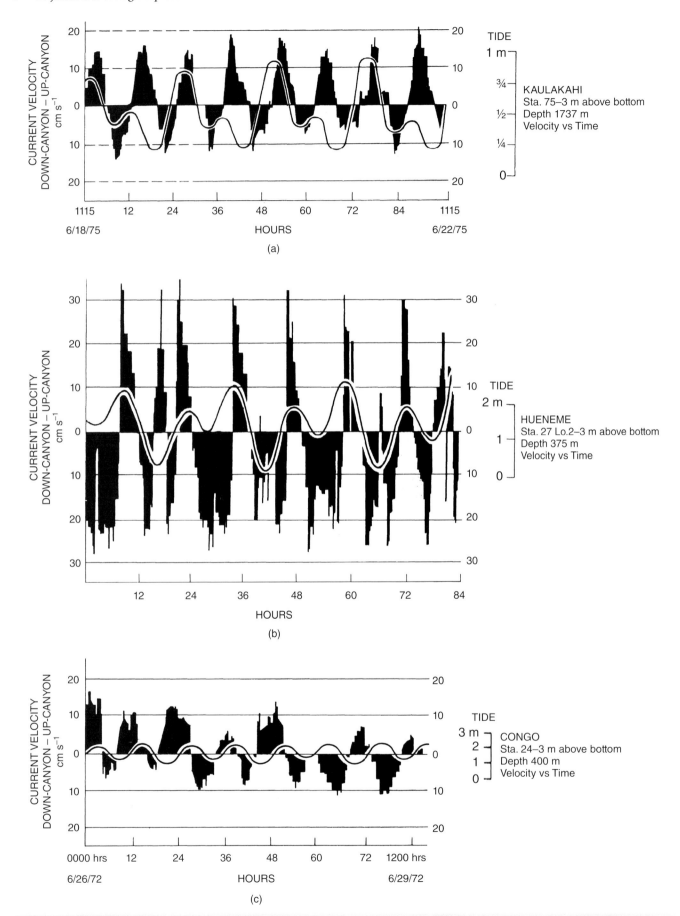

Fig. 1.5 Diagrams to show the periodicity of oscillating up- and down-canyon currents. Tide relationship obtained from the predicted tide at the nearest reference station. (a) Kaulakahi Canyon between Kauai and Niihau islands, Hawaii; (b) Hueneme Canyon, California; (c) Congo Canyon, west Africa. Redrawn from Shepard and Marshall (1978).

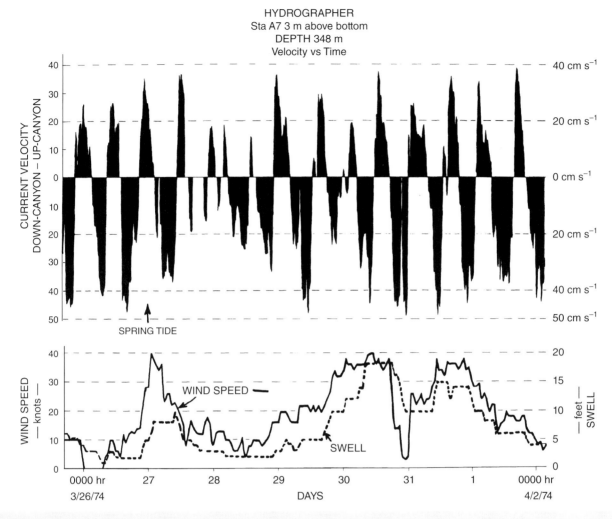

Fig. 1.6 Relationship of wind speed and swell height to the magnitude of up- and down-canyon currents during a storm period in Hydrographer Canyon, off Massachusetts. The slowest currents occurred during periods of reduced wind speeds and reduced swell. Redrawn from Shepard and Marshall (1978).

floor. Shepard and Marshall (1973a, b) ascribed the current surge and its associated erosional features to the passage of a storm-generated turbidity current flowing down-canyon. Similar down-canyon currents, with velocities up to 190 cm s⁻¹, have been reported from the head of Scripps Canyon during an onshore storm (Inman *et al.* 1976), and from other canyons (Gennesseaux *et al.* 1971; Reimnitz 1971; Shepard *et al.* 1975).

The measured current velocities are, at times, sufficient to transport sand. Shanmugam (2003) advocates that care be taken in interpreting the origin of tractional structures, such as current-ripple lamination, in canyon deposits, because some of this lamination might have been produced by tidal currents rather than turbidity currents.

The 'ambient' or 'normal' contemporary sedimentation within canyons appears to be mainly the deposition of finer grained suspended matter, presumably entrained by the periodic up- and down-canyon currents (Drake *et al.* 1978). In one recent monitoring study, however, an energetic SGF transported sand and gravel hundreds of metres down Monterey Canyon (Paull *et al.* 2003). The deposit from this event is at least 125 cm of porous sand and gravel, locally with a mud matrix and high water content. While sediment

transport by energetic SGFs appears to be unimportant in canyons at the present time, the ancient record suggests that this was not always the case (Chapter 8). The more energetic events might simply occur with such low frequency that they are rarely recorded in modern studies, or they might be largely restricted to times of lower sea level.

1.2.3 Internal waves

Internal waves form along density interfaces in stratified water masses. The surface of density change may be the temperature-dependent pycnocline or a contact between relatively fresh surface water (e.g., near a river delta) and underlying seawater. In continental-margin settings, the waves are associated with internal astronomical tides (diurnal or semi-diurnal) and are generated near the edge of the shelf (Wright 1995). The wave period depends on the vertical density gradient, and ranges from about 20 min at open-ocean thermoclines to somewhat less than 5 min where a freshwater plume overrides seawater. Amplitudes are of the order of 10 m. Many internal waves are solitons or groups of solitons with particularly large amplitudes and energies.

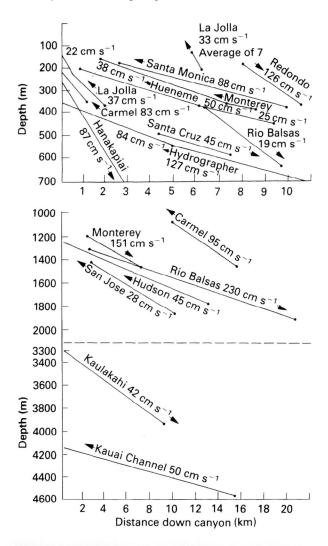

Fig. 1.7 Graphs to show the most likely direction and approximate speed of internal waves up and (less frequently) down the axes of various submarine canyons. The speed of wave advance is approximate because of errors in matching wave crests between current-meter stations, particularly in cases where the current appears to be up-canyon. Redrawn from Shepard and Marshall (1978).

Strong near-bed currents can develop where internal waves shoal as the pycnocline and seabed converge on a sloping shelf (Cacchione & Southard 1974; Wright 1995). The shoaling and breaking of solitons can generate intense turbulence (Kao *et al.* 1985) capable of suspending sediment that can subsequently move downslope under the influence of gravity. Internal waves are inferred to be active in submarine canyons. Similarities in time–velocity patterns for up- and down-canyon flows, when phase-shifted by the cycle length, point to the likely up-canyon advance of internal waves at depths shallower than 1000 m in many canyons, and at depths from 1000–2000 m in a few canyons (Fig. 1.7; Shepard & Marshall 1978). In other canyons, internal waves advance seaward. Shepard *et al.* (1979) ascribe the down-canyon advance of internal waves in Santa Cruz, Santa Barbara and Rio Balsas canyons to the introduction of moving water masses into the heads of the canyons.

1.2.4   Sediment slides and mass transport complexes (MTCs)

The downslope component of gravity can cause sediment masses previously deposited on a delta front or on the upper continental slope to move into deeper water, either by increments or during a single episode. Slow downslope movement without slip along a single detachment surface, that is without failure, is referred to as *creep*. No structures in ancient successions have been unambiguously attributed to creep, although features in seismic profiles have been interpreted to have been formed by this mechanism (Hill *et al.* 1982; Mulder & Cochonat 1996). More rapid downslope movements immediately following failure events generate sediment slides (Fig. 1.8) and submarine debris flows. Sediment slides can result in little-deformed to intensely folded, faulted and brecciated masses (Barnes & Lewis 1991) that have translated downslope from the original site of deposition. The head of the slide mass tends to display extensional features, whereas the toe suffers compression, folding and thrusting (Fig. 1.8). If the primary bedding is entirely destroyed by internal mixing, with soft muds and/or water mixed into the slide, then a slide may transform into a cohesive debris flow (Fig. 1.9).

According to the ISSMGE Technical Committee on landslides, submarine mass movements can be divided into slides (translational or rotational), topples, spreads, falls and flows (Locat 2001; *cf.* seismic examples described by Moscardelli *et al.* 2006; Moscardelli & Wood 2007). The various types of mass movements generate a spectrum of deposits which, when intimately associated with one another, are best referred to simply as *mass transport complexes* (MTCs). As defined by Pickering and Corregidor (2005):

> Mass transport complexes include chaotic deposits, typically with visco-plastically deformed rafts of disrupted bedding, cobble-pebble conglomerates, pebbly mudstones, mud-flake breccias, and pebbly sandstones. These deposits represent a range of processes, including slides, slumps, turbidity currents and debris flows.

We recommend that where a single event is believed to be responsible, the term '*mass transport deposit*' (MTD) be used, with MTC being reserved for stacked multiple events – accepting that in many cases such a distinction can prove difficult.

MTCs are common surficial sediments in the modern oceans and in ancient deep-water settings (e.g., Schipp *et al.* 2013; Shanmugam 2015). For example, Embley (1980) claims that 'at least 40% of the continental rise of eastern North America … is covered by a veneer of mass-flow deposits [slides] including debris flows'. On marine slopes, there is probably a complete gradation between coherent slides and thoroughly mixed cohesive debris flows. When sliding occurs, the movement occurs along a sharp or diffuse basal failure surface at some depth below the seafloor (Fig. 1.10). Along this surface, the shear stress produced by the sum of gravitational acceleration and cyclic accelerations due to seismic shocks (Morgenstern 1967) and passing surface waves (Lu *et al.* 1991) or internal waves exceeds the internal shear strength of the sediment. The shear strength depends on a variety of sediment properties like water content, texture, pore pressures and organic content.

Fine-grained sediment has a variety of geotechnical properties (Bennett & Nelson 1983) that are useful indicators of its physical state and that help determine under what conditions the sediment will fail and generate a slide. Conditions that favour initiation of

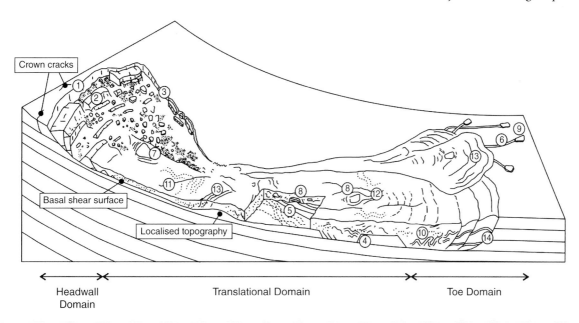

Fig. 1.8 Schematic representation of a sediment slide. Circled numbers show (1) headwall scarp, (2) extensional ridges and blocks, (3) lateral margins, (4) basal shear surface ramps and flats, (5) basal shear surface grooves, (6) basal shear surface striations, (7) remnant blocks, (8) translated blocks, (9) outrunner blocks, (10) folds, (11) longitudinal shears (= first-order flow fabric), (12) second-order flow fabric, (13) pressure ridges, (14) fold and thrust systems. From Bull *et al.* (2009).

slides in muddy terrigenous sediments are a function of (i) bottom slope (Moore 1961), (ii) sedimentation rates (Hein & Gorsline 1981) and (iii) the response of the sediment to cyclic stress produced by earthquake shaking (Morgenstern 1967). Sedimentation rates on basin-margin slopes vary widely, but Hein and Gorsline (1981) conclude that a rate of 30 mg cm^{-2} yr^{-1} must be attained before slope failures become common. For example, in the Santa Barbara Basin, California Borderland, sedimentation rates exceed 50 mg cm^{-2} yr^{-1}, and debris flows are widespread on slopes of <1° (Hein & Gorsline 1981). In many areas, sediment slides preferentially occur on very low slopes (Fig. 1.11), suggesting the involvement of water-rich, rapidly deposited sediments.

The sedimentation rate effectively determines the water content and shear strength, *S*, of the sediment, although shear strength is also a function of other variables such as content of organic matter (Keller 1982), generation of gas in the sediment by decay of organics or by gas-hydrate decomposition (Carpenter 1981), and binding by bacteria and fungi (Meadows *et al.* 1994). According to Keller (1982):

> cohesive sediments with greater than about 4–5% organic carbon [have] … (1) unusually high water content, (2) very high liquid and plastic limits, (3) unusually low wet bulk density, (4) high undisturbed shear strength, (5) high sensitivity, (6) high degrees of apparent over consolidation, and (7) high potential for failure [by liquefaction] in situations of excess pore pressure.

The stability of sediments on a sloping bottom has traditionally been analysed using a static infinite slope model (Moore 1961; Morgenstern 1967). Consider a blanket of sediment on an inclined surface. Beneath the seabed, the shear stress increases downward in a linear fashion because of the cumulative weight of sediment. The sedimentary blanket will remain stable as long as the strength of the sediment increases downward at a faster rate than the rate of increase

in the shear stress. Strength is generated by internal friction, electrostatic cohesive forces and organic binding. In many natural situations, persistent high water contents (leading to elevated pore-fluid pressures) and gas evolution because of organic-matter decomposition prevent effective consolidation, and a depth is reached where the shear stress along an inclined bedding surface exceeds resisting forces. Slippage and therefore failure along this bedding surface is then inevitable. Of course the failure surface must, at its down-slope end, cross bedding and rise upward to the seabed in order for the translating mass to glide freely into deeper water. Booth *et al.* (1985) developed the concept of a safety factor, *SF*, which is the ratio of resisting forces to shearing forces. For ψ = excess pore-fluid pressure, Z = depth measured vertically below the seabed, $\gamma' = \rho_s\, g'$, ρ_s = sediment density, reduced gravity $g' = g(\rho_s - \rho)/\rho_s$, ρ = density of seawater, g = gravitational acceleration, ϕ = angle of internal friction (a characteristic of the material), and α = slope angle,

$$ SF = \left[1 - \frac{\psi}{\gamma' Z \cos^2\alpha} \right] \left[\frac{\tan \phi}{\tan \alpha} \right] \tag{1.1} $$

Case studies (Athanasiou-Grivas 1978) show that the probability of failure is low for *SF* >1.3, and that failure is a virtual certainty for *SF* <0.9. Notice that steadily increasing sediment density (because of normal consolidation) increases *SF*, whereas excess pore-fluid pressure has the opposite effect. Booth *et al.* (1985) provide a nomogram to determine *SF* under undrained conditions given sedimentation rate, coefficient of consolidation, sediment thickness, slope angle and angle of internal friction. Excess pore pressure is obtained from consolidation theory (Gibson 1958), under the assumption that these excess pressures are entirely the result of trapping of pore-water in compacting, fine-grained sediment of low permeability.

The infinite slope model can be extended to the case of superimposed ground accelerations due to earthquakes (Morgenstern 1967; Hampton *et al.* 1978). Horizontal peak accelerations, like earthquake

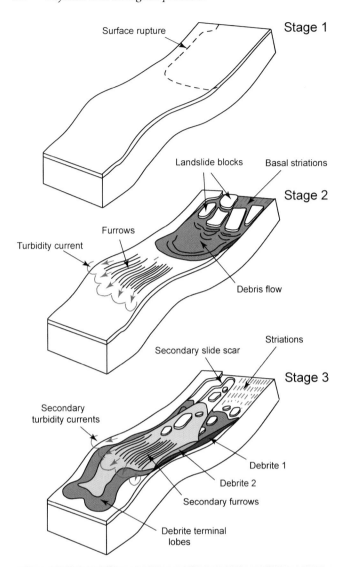

Fig. 1.9 Conceptual model of submarine slide evolution (Gee *et al.* 2006). Stage 1 shows seafloor rupture. Stage 2 shows tabular blocks, basal striations, debris flow and a turbidity current generated in the headwall area. Downslope of the headwall area, turbidity currents erode furrows in the seafloor. Stage 3 shows the development of secondary slide events within the headwall, triggering secondary debris flows and turbidity currents.

intensities, decrease away from the epicentre (Fig. 1.12). An earthquake safety factor, *ESF*, can be expressed as (Booth *et al.* 1985):

$$ESF = \frac{SF\gamma' \tan\alpha}{\gamma a_{\mathrm{X}} + \gamma' \tan\alpha} \qquad (1.2)$$

where $\gamma = \rho g$. and a_{x} = horizontal acceleration coefficient expressed in terms of gravity (e.g., 0.1 g).

Figure 1.13 is taken from Booth *et al.* (1985), and allows estimation of the earthquake-induced horizontal ground acceleration required to reduce *ESF* to 1.0 for a wide range of slopes and safety factors, and a reasonable range of specific weights. The increase in excess pore pressures caused by ground shaking (Egan & Sangrey 1978) must be taken into account in estimating the safety factor. Clearly, 'even small earthquake-induced accelerations are very detrimental to the stability of a submarine slope' (Morgenstern 1967).

1.3   Deep, thermohaline, clear-water currents

Large parts of the deep ocean basins, especially the Atlantic Ocean, are characterised by geostrophic currents moving at mean speeds of 10–30 cm s^{-1} (McCave *et al.* 1980; Hollister & McCave 1984), with short 'gusts' reaching about 70 cm s^{-1} (Richardson *et al.* 1981). Deep circulation in the oceans is the result of thermohaline effects. In the North Atlantic Ocean, for example, dense cold water sinks off the coast of Greenland and in the Norwegian Sea and moves southward as a bottom current (Worthington 1976); in the South Atlantic, ice formation in the Weddell Sea causes an increase in salinity and hence density, the dense seawater sinks and flows northward along the bottom (Stommel & Arons 1961; Pond & Picard 1978: p.134). Other regions of the world's oceans that are characterised by spreading cold bottom water are outlined by Mantyla and Reid (1983). The deep ocean bottom currents are deflected to the right in the northern hemisphere and to the left in the southern hemisphere by the Coriolis effect, with the result that they are banked up against the continental slope and rise on the western sides of ocean basins, effectively flowing parallel to the bathymetric contours. Two examples are the *Western Boundary Undercurrent (WBU)*, which sweeps along the continental rise of eastern North America (Fig. 1.14) at depths of about 2000–3000 m and at peak velocities of about 25–70 cm s^{-1} (Stow & Lovell 1979), and the *Deep Western Boundary Current* (DWBC), which occupies the same region at depths of 4000–5000 m (Richardson *et al.* 1981). The WBU is derived from the Norwegian Sea, whereas the DWBC appears to be formed of Antarctic bottom water (Hogg 1983). These currents carry a dilute suspended load – generally <0.1–0.2 g m^{-3} – that forms the thick bottom *nepheloid* layer (Ewing & Thorndike 1965; Biscaye & Eittreim 1977). Concentrations may briefly reach values much higher, up to at least 12 g m^{-3} (Biscaye *et al.* 1980; Gardner *et al.* 1985). Most of the fine-grained suspended material is winnowed from the seafloor; the rest is probably added to the current by cascades of cold shelf water or lutite flows originating at the edge of the continental shelf (Postma 1969).

The WBU and DWBC are capable of long-distance transport of fine-grained sediments. According to Heezen and Hollister (1971), distinctive red mud derived from the weathering of Carboniferous and Triassic bedrock in the Gulf of St Lawrence area (eastern Canada, 45° N latitude) has been transported at least as far south as the Blake Plateau (30°N), a distance of about 2000 km. On the Newfoundland Rise (Carter & Schafer 1983), the high-velocity core flow of the WBU ($U \leq 35$ cm s^{-1}) intersects the bottom at depths of 2600–2800 m, and is capable of transporting sediment grains, of approximate diameter 0.1 mm, 1–15% of the time. The seabed beneath this core zone is sandy. Finer grains are effectively maintained in suspension as a nepheloid layer up to 800 m thick.

On the continental rise off Nova Scotia, the high-velocity core flow of the DWBC ($U \leq 70$ cm s^{-1}) is at depths of 4500–5000 m (Richardson *et al.* 1981; Bulfinch & Ledbetter 1984). Characteristics and effects of the DWBC were studied in great detail during the multidisciplinary 'High Energy Benthic Boundary Layer Experiment' (HEBBLE; for an excellent summary of findings, see Nowell & Hollister 1985; McCave & Hollister 1985; Hollister & Nowell 1991a,b). The bottom beneath the DWBC consists of coarse silts moulded into longitudinal ripples (Bulfinch & Ledbetter 1984; Swift *et al.* 1985;

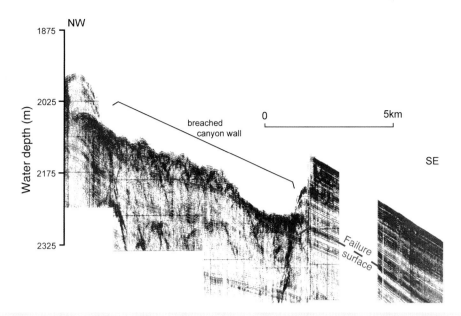

Fig. 1.10 Dip seismic-reflection profile through a failed part of the wall of Munson Canyon, US Atlantic coast. The failure surface is overlain by a chaotic MTC to the left, and is ~150 m below the seabed to the right. The apparent downward step in the failure surface at the edge of the depression is an artifact ('pullup') created by the differing acoustic travel-time in water and sediment. Modified from O'Leary (1993).

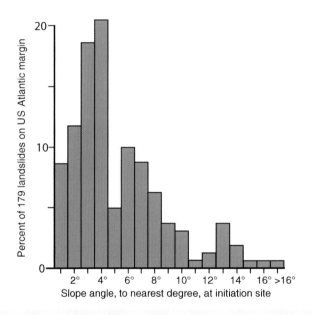

Fig. 1.11 Frequency distribution of submarine slides (MTCs) on the US Atlantic margin as a function of seabed slope at the site of initiation. Redrawn from Booth *et al.* (1993).

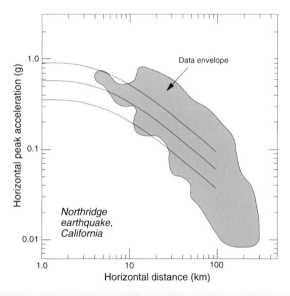

Fig. 1.12 Graph of the horizontal-component peak acceleration versus distance from the epicentre of the 1994 Northridge earthquake, California. The green envelope encloses >150 data points plotted by Mueller (1994). The curves represent the average peak acceleration (solid line) and the ±1 standard deviation accelerations (dashed lines) expected for a magnitude 6.7 earthquake. Redrawn from Mueller (1994).

Tucholke *et al.* 1985). The silt size fractions most affected by the core flow span 5–8 ϕ (where $\phi = -\log_2$[size in millimetres]). Net accumulation rates are not high (5.5 cm kyr^{-1}), but instantaneous rates can be, due to alternation of periods of rapid erosion and rapid deposition from a highly concentrated nepheloid layer (Hollister & McCave 1984). Temporal variations in the bottom flow are very complicated, and may involve significant variations in flow speed and reversals of flow direction. Times of strongest and most variable flow are called 'deep-sea storms' by Hollister and McCave (1984). The

'storms' last from a few days to several weeks, are characterised by current speeds in excess of 20 cm s^{-1} and result in high concentrations of suspended sediment. The 'deep-sea storms' result from an interplay between the deep circulation and wind-driven currents in the surface layer of the ocean created by atmospheric storms passing overhead (Faugères & Mulder 2011). Based on a five-year record in the HEBBLE

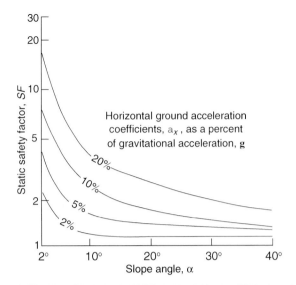

Fig. 1.13 Horizontal ground acceleration, a_x, required to reduce the static safety factor to a value of 1.0 for given slope angles and sediment density in the range $1.5–2.0\,\mathrm{g\,cm^{-1}}$ (simplified from Booth *et al.* (1985). For example, if $SF = 2.0$ and $\alpha = 10°$, ground accelerations of about 0.07 g or greater will reduce safety factor to 1.0 or less, and failure will be likely. On the same slope with $SF = 4.0$, accelerations of at least 0.2 g would be needed to cause failure. Note that SF is itself a function of bottom slope and excess pore pressure (Eq. 1.1).

area off the coast of Nova Scotia, about three such 'storms' occur each year, and occupy about 35% of the time (Hollister & McCave 1984).

Contour current deposits, or *contourites* (Hollister & Heezen 1972), may be treated as two end members: (i) muddy contourites and (ii) sandy contourites (see also Section 6.3). Muddy contourites are fine grained; mainly homogeneous and structureless; thoroughly bioturbated (McCave *et al.* 2002); and only rarely show irregular layering, lamination and lensing. They are poorly sorted silt- and clay-size sediments with up to 15% sand. They range from finer-grained homogeneous mud to coarser-grained mottled silt and mud, and their composition is most commonly mixed biogenic and terrigenous grains. According to Hollister and McCave (1984), short-term depositional rates of mud can be extremely high, about 17 cm yr^{-1}, followed by rapid biological reworking.

Sandy contourites comprise thin irregular layers (<5 cm) that are either structureless and thoroughly bioturbated, or may possess some primary parallel or cross-lamination which may be accentuated by heavy minerals or foraminiferal tests (Bouma & Hollister 1973; Stow & Faugères 2008). Grading may be normal or inverse, and bed contacts may be sharp or gradational. Grain size ranges from coarse silt to, rarely, medium sand, with poor to moderate sorting. The sandy facies is produced by winnowing of fines by stronger flows (Driscoll *et al.* 1985), and physical sedimentary structures only seem to be preserved where the currents are particularly focused and strong, as is the case where Mediterranean water flows out of the Strait of Gibralter into the Gulf of Cadiz (Stow & Faugères 2008), or where tidal or wind-driven currents are forced through constricted straits so that high velocities are maintained to hundreds of metres water depth (Colella & d'Alessandro 1988; Ikehara 1989).

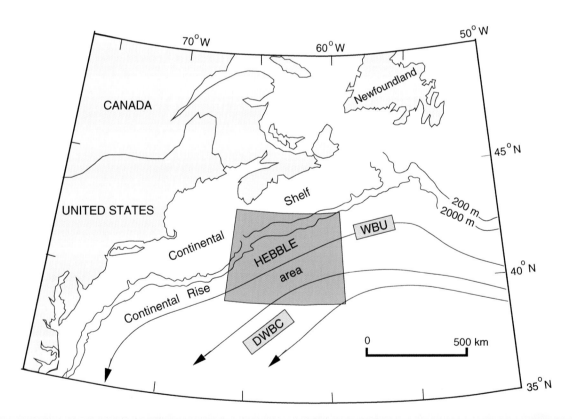

Fig. 1.14 Approximate tracks of Western Boundary Undercurrent (WBU) and Deep Western Boundary Current (DWBC) along the eastern continental margin of North America. The HEBBLE area was the site of detailed long-term measurements of bottom currents, and is an acronym for High Energy Benthic Boundary Layer Experiment. Redrawn from Hollister and McCave (1984).

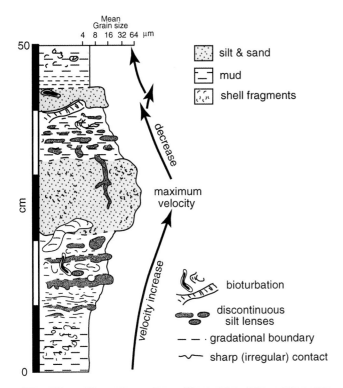

Fig. 1.15 Schematic model showing upward coarsening and fining of contourite facies from the Faro Drift, offshore southern Portugal. Redrawn from Gonthier *et al.* (1984).

Muddy and sandy contourites commonly occur together in upward-coarsening to upward-fining sequences (Faugères *et al.* 1984; Stow & Piper 1984b). A complete sequence shows inverse grading from a fine homogeneous mud, through a mottled silt and mud, to a fine-grained sandy contourite facies, and then normal grading back to a muddy contourite (Fig. 1.15). The changes in grain size, sedimentary structures and composition probably are related to long-term (1–30 kyr) fluctuations in the mean current velocity (Stow

& Piper 1984b). Stow and Faugères (2008) note that the time of maximum velocity (Fig. 1.15) might result in a hiatus and/or erosion if the currents become so strong that no accumulation is possible. In such cases, top cut-out motifs may occur if erosion cuts away the top of the idealised doubly graded profile, or base cut-out motifs may occur above the erosional unconformity.

In particular successions, it may be difficult to distinguish between mud turbidites and muddy contourites (Bouma 1972; Stow 1979). Also, the reworking of sand turbidites can result in bottom-current-modified turbidite sands, believed to be common on continental slopes and rises. In the central parts of ocean basins, bottom currents are known to construct large sediment drifts (Chapter 6) of almost pure biogenic material (McCave *et al.* 1980; Stow & Holbrook 1984). Such biogenic contourites may be indistinguishable from true pelagites.

Since the early 1990s, a number of authors have maintained that rythmically interbedded sands or large and well-sorted sand lenses in oilfields were emplaced, or largely reworked, by bottom currents (Mutti 1992: p. 19; Shanmugam *et al.* 1993a, b, 1995; Shanmugam 2008). Some of the supposed bottom-current deposits are coarse grained. The authors of this book doubt that such sands were deposited or largely reworked by bottom currents, because the required flow velocities and variability would be significantly greater than any known from the modern oceans. Features like sharp-topped ripple lenses and climbing ripple lamination are not diagnostic of bottom-current transport, as some have claimed (Shanmugam *et al.* 1995; Jordan *et al.* 1994; Shanmugam 2008). Sharp-topped ripples can form when turbidity currents bypass a part of the seafloor, perhaps because of flow unsteadiness (e.g., the 'depletive waxing flow' of Kneller 1995). Alternatively, the flow could leave sharp-topped ripples if it were deficient in silt sizes. Climbing ripples always require rapid deposition from suspension during ripple migration (Allen 1971; Jobe *et al.* 2012), consistent with deposition from turbidity currents (Fig. 1.16). The main reason, however, that the authors of this book dismiss the notion that sandy contourite deposits replete with stratification are the norm (Shanmugam 2008), is the meagre evidence for such stratified sands in modern bottom-current deposits because of widespread and thorough bioturbation except in very rare situations (Stow & Faugères 2008). We take the view

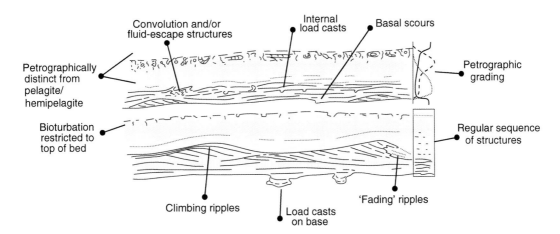

Fig. 1.16 Diagnostic criteria for the recognition of fine-grained turbidites. Rapid deposition from a decelerating SGF produces both wet-sediment deformation structures like load casts, and climbing ripples. Accumulation rates under thermohaline currents are much lower, preventing the development of these types of sedimentary structures. The co-occurrence of several of these structures is sufficient to rule out deposition by clear-water bottom currents (contour currents). Redrawn from Piper and Stow (1991).

that 'actualism' trumps arguments based on the presence of surficial bedforms in areas where bottom currents are active today. Apparently the stratification produced by the migration of these bedforms does not survive into the geological record because of intense sediment disturbance by burrowers under conditions of slow sediment accumulation. See Section 6.5 for additional discussion of this issue.

1.4 Density currents and sediment gravity flows

(1) A density current results when a more dense fluid or mobile plastic material moves beneath a less dense material under the influence of gravity. *Fluids* are materials like water and air that deform continuously when subjected to even the smallest shear stress. *Plastics* resist deformation until a critical level of shear stress is reached, after which they deform continuously unless the shear stress later declines below the critical value.

(2) The density contrast with the ambient fluid might result from compositional differences (e.g., oil flowing beneath water), from temperature differences (e.g., cold air entering a warm room), from the presence of suspended material (e.g., particulate gravity currents), or where there are strong salinity contrasts (e.g., laboratory saline currents and natural saline underflows as in the modern Black Sea – Di Iorio *et al.* 1999; Hiscott *et al.* 2013).

Dilute density currents consisting mainly of water or air are turbulent on even low slopes, unless they are extremely thin.

Natural turbulent density currents are widespread in the oceans and the atmosphere (Simpson 1982, 1997); they occur as powder snow avalanches (Hopfinger 1983), characterise many volcanic eruptions (Cas & Wright 1987), and carry suspended sediment from land areas into lakes and ocean basins. In the laboratory, turbulent density currents have been formed of both suspensions (e.g., Middleton 1966a, b) and saline solutions (e.g., Hallworth *et al.* 1996 – Fig. 1.17; Gladstone *et al.* 2004 – Fig. 1.18).

In sediment gravity flows (SGFs), particles and water move down slopes because the mixtures have a density greater than that of the

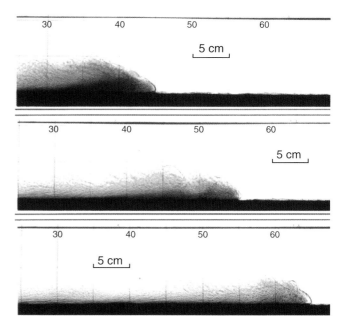

Fig. 1.17 Sequential development of a turbulent density current formed by release of an alkaline saline solution containing a pH indicator into an acidified freshwater ambient environment. The pH indicator stains the saline current purple, changing to red once mixing with the overlying ambient fluid results in a neutral pH. Mixing is strongly developed in the upper part of the flow and in the wake behind the head of the current. There, the visual contrast between red and colourless regions provides a detailed image of the shapes of the turbulent eddies. See Hallworth *et al.* (1996) for details.

ambient fluid, normally seawater. Initially, gravity acts solely on the solid particles in the mixture, inducing downslope flow; the admixed water is a passive partner in this process. Put another way, gravity pulls the grains, and the grains pull the water. If sufficient potential energy is converted into kinetic energy in the evolving flow, then the

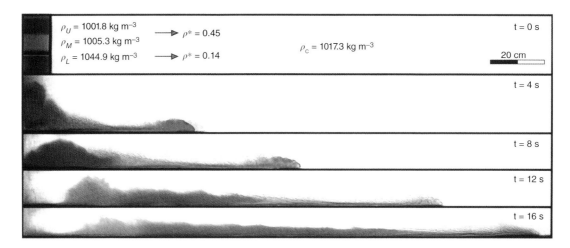

Fig. 1.18 Sequential development of a three-layered saline, turbulent density current in which the lower layer has the greatest density. ρ^* = density contrast between layers; ρ_c = average flow density; t = time after flow initiation. The three starting colours (red, yellow, blue) were created by artificial dyes, whereas transitional colours show the extent of fluid mixing during evolution of the flow (e.g., red+yellow = orange; blue+yellow = green). See Gladstone *et al.* (2004) for details.

flow may become turbulent and the eddies in the fluid phase then become fundamental to the maintenance of the suspension. The flow will continue to move if the following conditions are satisfied: (i) the shear stress generated by the downslope gravity component acting on the excess density of the mixture exceeds frictional resistance to flow; and (ii) the grains are inhibited from settling by one of several support mechanisms. Only a few support mechanisms are believed to be responsible for maintaining sediment in suspension on seafloor slopes of a few degrees or less.

(1) *Turbulence* characterises low-viscosity fluids in which inertial forces dominate viscous forces. Turbulence is the superimposition of swirling eddies and seemingly random velocity fluctuations on the average downstream velocity. The upward components of the velocity fluctuations diffuse sedimentary particles into the flow according to their settling velocity, so that the finest particles are evenly distributed throughout the flow, even though they are more dense than the turbulent fluid.

(2) *Buoyancy* is the support provided to an object by a dense surrounding fluid phase. If the surrounding fluid has the same density as the object, then the object has no immersed weight and seems to float aimlessly in the fluid. If the fluid is more dense, then buoyancy is positive and the object floats on its surface (e.g., dry wood floating in water). If the fluid is less dense than the object but more dense than water, then the downward gravitational forces on the object are reduced, and are equivalent to the gravitational forces on a much smaller object in clear water. Buoyancy permits relatively dense sediment–water mixtures like debris flows to carry large clasts even at low velocities.

(3) *Grain collisions and near-collisions* (also called *grain interaction*) transfer some of the downstream momentum of moving particles to an upwardly oriented dispersive pressure (Bagnold 1956) as faster moving grains ricochet off more slowly moving grains beneath them. The moving mass of grains dilates (expands), thus increasing the vertical spacing between the particles and reducing grain-to-grain friction. This support mechanism only operates at high particle concentrations and cannot alone maintain a SGF on low slopes.

(4) *Excess pore pressure* and *pore-fluid escape* result when a dispersion of grains settles too quickly to allow the interstitial pore-fluid to escape upwards. Instead, low permeability impedes the upward flow of escaping pore-water, causing fluid pressures in the pore spaces to significantly exceed the expected hydrostatic pressure. This excess pore pressure keeps the grains separated (as if separated by an inflated pillow) so that friction is reduced and the grains can continue to move relative to one another. In local areas, the pressured pore-fluid can escape rapidly along preferred channel-ways, potentially elutriating fine matrix material and forming porous fluid-escape pillars.

(5) *Matrix strength* is a property of concentrated mixtures of fine-grained or poorly sorted sediment and water. Small shear stresses do not cause such mixtures to flow because internal deformation is resisted by friction between adjacent grains (*frictional strength*) and electrostatic attraction between clay and silt particles (*cohesive strength*). This is different to the behaviour of Newtonian fluids, where the applied shear stress 'τ', is proportional to the fluid viscosity 'μ' x velocity gradient 'du/dy'. An example is water, which deforms (i.e., flows) no matter how small the applied shear stress. For materials with matrix strength, some critical shear stress must be applied before they will move – likewise, these materials will cease to move even on low slopes if the downslope component of gravity is not sufficient to generate the required shear stress along the basal surface of the flow. Once a material of this type is moving, matrix strength is believed to play a roll in reducing the tendency for large clasts to settle, because in order to do so they have to push cohesive and/or granular material out of the way, and overcome a certain amount of residual frictional and cohesive strength.

The relative importance of the various support mechanisms in the principal sediment gravity flows is summarised in Figure 1.19. The names of the flows in this figure are explained in Section 1.4.1 but for the moment it is probably sufficient to know that concentration increases to the right, from turbidity currents to cohesive flows.

A variety of late-stage depositional processes can leave their imprint on a deposit. Many of these depositional processes are not unique to a particular transport mechanism. For this reason, a clear distinction must be made between long-distance transport agents and local depositional mechanisms in explaining the origin of various deep-sea deposits. For example, Middleton and Hampton (1973, 1976) recognised grain-flow deposits as an end-member facies in the spectrum of SGF deposits. *Grain flows* derive their particle support entirely from the dilation induced by grain collisions and near collisions (Bagnold 1956). A familiar example of grain flow is the avalanching of sand down the front of a dune. However, pure grain flows cannot move on the gentle slopes that characterise ocean-basin margins, and instead require slopes of more than ~13° (Straub 2001). Because of this minimum slope requirement, many beds that have in the literature been referred to as 'grain-flow deposits' must instead have been deposited from decelerating *concentrated density flows* or *inflated sandflows* (Section 1.4.1) which can travel on slopes <1–2°. In concentrated density flows, turbulent suspension provides the long-distance particle support, but during rapid deceleration and grain settling, turbulence is increasingly replaced near the base of the flow by grain interaction effects resulting from particle collisions and near collisions above a bed under shear (Bagnold 1956; Rees 1968). Hence, the final deposit mainly or entirely records the effects of grain collisions and elevated particle concentrations (Fig. 1.19), leading to poor organisation, possible inverse-to-normal grading, and possible poorly developed lamination. This is in spite of the fact that long-distance transport might have been provided by a large turbulent flow. In inflated sandflows, a high concentration of clasts ranging in size from coarse silt to gravel guarantees that grain interaction effects are strongly recorded in the deposits (Fig. 1.19).

1.4.1 Classification

There has been a recent dramatic improvement in the understanding of characteristics and behaviour of SGFs (e.g., Talling *et al.* 2012). This has resulted from (i) improved hindcast analysis of recent SGFs in the oceans (a method for testing a mathematical model, where known or estimated inputs for past events are input into the model to see how well the output matches the known result); (ii) a post-1995 revitalisation of research into the flow dynamics and depositional mechanisms of such currents and (iii) completion of a number of relevant experiments on concentrated flows. New classifications of flow processes have resulted, with an interesting cross-fertilisation between those who study deep-marine SGFs and those who study the dispersal of volcaniclastic materials (e.g., Pierson & Costa 1987; Gladstone *et al.* 2004).

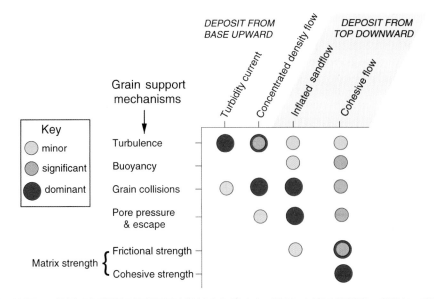

Fig. 1.19   Relative importance of particle-support mechanisms for the four varieties of SGF recognised in this book, and defined in Section 1.4.1. Where orange and red symbols are superimposed, the support varies from significant to dominant. See also Figure 1.22.

The most widely accepted approach to classification has been to attempt, even though difficult, to infer the likely dominant transport mechanism for the entire flow event from both the deposits and theory (Middleton & Hampton 1973; Lowe 1982; Middleton 1993; Hiscott *et al.* 1997a; Mulder & Alexander 2001; Talling *et al.* 2012). This approach has been challenged by G. Shanmugam and colleagues in numerous publications (Shanmugam 1996, 1997, 2000, 2002, 2003; Shanmugam & Moiola 1995; Shanmugam *et al.* 1994, 1995, 1997) who prefer to use different names for component parts of a single decelerating flow (e.g., the upper part is a turbidity current and the lower part a sandy debris flow) based on changing characteristics of the deposits (e.g., normally graded versus ungraded sand in a single event deposit). G. Shanmugam and colleagues advocate that nearly all traction-generated lamination (e.g., planar lamination, ripple lamination, climbing-ripple lamination) is formed by bottom currents, not SGFs. In addition, they insist that 'process terms [should] refer only to depositional mechanisms, not transport mechanisms' (Shanmugam 2000: p. 302). If pushed to the extreme, this approach might lead to the classification of all avalanche-generated foresets on current ripples and dunes as the deposits of grain flows! This is not our preferred approach. Instead, we seek to interpret long-distance transport processes and processes operating during the final stages of deposition from experimental datasets and inferences based on field examples of deep-water deposits.

Many of the classification pitfalls identified by G. Shanmugam and colleagues result from the fact that most large, natural, non-cohesive SGFs are vertically stratified in terms of their properties, so that the near-bed conditions are considerably different to conditions in the main part of the flow. Also, the final deposit is commonly much thinner than the full flow thickness, so that the imprint left by the near-bed processes is enhanced (Fig. 1.20). We see no merit, however, in mentally slicing such flows into separate components, and in assigning different names to these component parts of what is actually a single SGF. Instead, if a single SGF has a spectrum of support mechanisms from top to base, or from front to back, then in this book we will explicitly deal with this as a stratified or hybrid flow, not two or more different flows. As a cautionary note, however,

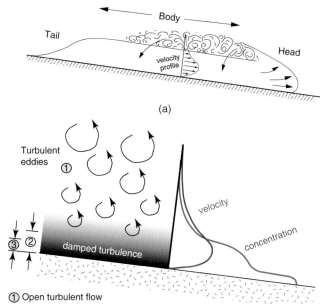

① Open turbulent flow
② High fallout, high concentration; suppressed turbulence
③ Maximum deposit thickness with all deposition below this level

(b)

Fig. 1.20   (a) Simplified depiction of a turbidity current or concentrated density flow divided into head, body and tail regions. (b) Conceptual view of the vertical stratification of velocity and concentration expected in the body of such a flow. Rapid particle fallout during deposition can increase the near-bed concentration to the point that turbulence is damped, grain collisions become common, and deposits become poorly organised. The eventual deposit is much thinner than the flow that created it.

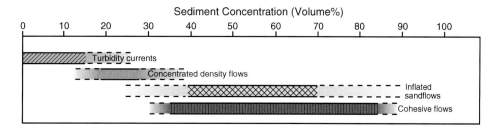

Fig. 1.21 Approximate solids concentrations typical of SGFs, modified from Mulder and Alexander (2001). Dashed lines show the possible extensions of sediment concentrations to lower and higher values than those deemed to be typical. Note the overlaps between different flow types, which result from the effects of flow stratification and different textures of the sediment load.

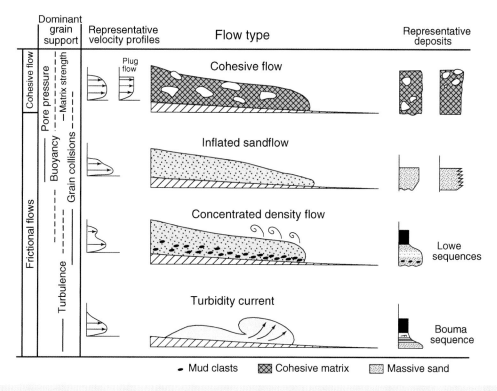

Fig. 1.22 Summary of flow characteristics, typical deposits, and grain-support mechanisms for cohesive and frictional (non-cohesive) SGFs, modified from Mulder and Alexander (2001).

G. Shanmugam is correct in his view that there is a formidable challenge in correctly interpreting the long-distance transport mechanism of many deep-marine sandy and gravelly deposits, because of the strong imprint of late-stage depositional effects.

We have elected to use an existing classification as the foundation for this and subsequent chapters, with some modification of terminology. The classification of Mulder and Alexander (2001) is based on sound theoretical understanding of flow rheology, supported by diverse research results from natural settings. The essence of this classification is outlined below. The reader is directed to the original paper for more extensive background and explanation.

SGFs are subdivided according to their rheological behaviour into predominantly *cohesive flows* and *frictional flows* (the latter called 'granular flows' by some researchers). Approximate ranges for the volume concentrations of solids are shown in Figure 1.21. Cohesive flows have matrix strength resulting from electrostatic attraction between fine particles in the mud fraction. They are differentiated from all the other flows discussed in this section because they have a pseudoplastic rheology and, hence, do not tend to become diluted by either particle loss (via deposition) or entrainment of ambient water. In effect, cohesive flows tend to 'hold together'. In contrast, frictional (non-cohesive) flows are made up of discrete particles dispersed in water. The behaviour of frictional flows is related directly to the relative proportion of grains and water. In general, frictional flows are characterised by selective deposition, and cohesive flows by en masse deposition.

We recognise four types of sediment gravity flow: *cohesive flows, inflated sandflows, concentrated density flows* and *turbidity currents* (Fig. 1.22). The last three are frictional flows. Cohesive flows can be further subdivided into *debris flows* and *mudflows*. The solid fraction in mudflows consists of <5% gravel by volume and mud : sand >1 : 1. Mudflows transport little or no coarse sediment except for isolated large blocks. Debris flows consist of more poorly sorted sediment (>5% gravel with a variable sand proportion) and may transport

Table 1.1   Published names for SGFs compared to the classification used in this book. Two names set in bold are used by Mulder and Alexander (2001) but are not used in this book

This book	Approximately equivalent terms
(Cohesive) debris flow & mudflow	Debris flow & mudflow
Inflated sandflow	Liquefied flow (Middleton & Hampton 1976)
	Density-modified grain flow (Lowe 1976a)
	Cohesionless debris flow (Postma 1986)
	Sandflow (Nemec *et al.* 1988; Nemec 1990)
	Sandy debris flow (Shanmugam 1996)
	Hyperconcentrated density flow (Mulder & Alexander 2001)
Concentrated density flow	High-concentration turbidity current (Lowe 1982)
Turbidity current	Low-concentration turbidity current (Middleton & Hampton 1973)
	Turbidity flow (Mulder & Alexander 2001)

boulder-sized clasts of soft sediment or rock and very large rafts or olistoliths. In cases of very coarse-grained material and so little mud that cohesion is insignificant, the alternative terms *inflated sand/gravel flows*, or *inflated gravel flows* can be used.

The four flow types recognised here replace a number of other terms that Mulder and Alexander (2001) argue do not correctly accord with the physical behaviour of natural flows. Approximately equivalent terms are listed in Table 1.1. We also show the terms of Mulder and Alexander (2001) that appear in their original classification but that we have replaced in this book. For example, we have avoided their use of the term 'hyperconcentrated' because of its potentially ambiguous meaning (see below). We support the suggestion of Shanmugam (2000) that the general term 'sediment gravity flow' should be used whenever the transport and depositional processes are unconstrained.

We have elected to replace two of the names proposed by Mulder and Alexander (2001), but otherwise retain the essence of their classification. Other authors (including Mulder & Alexander 2001) use the term 'turbidity flow', but we avoid this for essentially the same reason that we would not refer to a debris flow as a 'debris current'. The terms 'turbidity current' and 'debris flow' have clear precedence in literature extending back to the early part of the twentieth century. We also avoid the name 'hyperconcentrated density flow' (Mulder & Alexander 2001) because the name 'hyperconcentrated flow' has been used in variable ways to describe the more fluid pulses of stream flow associated with certain volcanic debris-flow events (Beverage & Culbertson 1964; Pierson & Costa 1987). To avoid confusion with this type of stream flow, we instead use the term 'inflated sandflow' (with variants of 'inflated sand/gravel flow' and 'inflated gravel flow'). This term builds on the use of the term 'sandflow' by Stanley *et al.* (1978), Nemec *et al.* (1988) and Nemec (1990) for laminar sand-laden flows characterised by strong grain interactions and liquefaction. The adjective 'inflated' is used because the term 'sandflow' alone invites confusion with the rather precisely defined process called 'grain flow' (Bagnold 1956), in which particles are mostly in collisional contact (or near contact) with one another, as is the case when grains avalanche down the face of a dune. Unlike grain flow, the process envisaged by Mulder and Alexander (2001), Nemec *et al.* (1988) and the authors of this book involves particles which are more dispersed than in a grain flow, kept apart by both grain collisions and elevated pore-fluid pressures and capable of moving on very low slopes.

In describing processes and deposits in this chapter, we have incorporated observations from the literature using the equivalencies provided above (plus others outlined by Mulder & Alexander 2001). Original authors might not support the licence we have taken with

their work, but otherwise the chapter (and reader) would be burdened by an unworkable number of overlapping and different terminologies. In a textbook, rather than review article, we are comfortable with ensuring consistent and simple terminology for the reader.

All four flow types recognised in this chapter are capable of long-distance transport of particulate sediments into the deep sea on relatively gentle slopes (<5°). Cohesive-flow deposits and inferred inflated sandflow deposits are common on slopes less than 1° (Prior & Coleman 1982; Damuth & Flood 1984; Simm & Kidd 1984; Thornton 1984; Nelson *et al.* 1992; Aksu & Hiscott 1992; Masson *et al.* 1993; Schwab *et al.* 1996); the flows are capable of travelling for hundreds of kilometres from upper continental slopes to abyssal plains (Embley 1980). Turbidity currents can flow long distances on flat basin floors or even upslope (Komar 1977; Elmore *et al.* 1979; Hiscott & Pickering 1984; Pickering & Hiscott 1985; Underwood & Norville 1986; Lucchi & Camerlenghi 1993). For example, Pleistocene turbidity currents carried distinctive coal fragments at least 1800 km from the eastern Canadian continental margin to the Sohm Abyssal Plain (Hacquebard *et al.* 1981), and Chough and Hesse (1976) suggest that turbidity currents flow for 4000 km in the Northwest Atlantic Mid-Ocean Channel. Concentrated density currents travel onto the middle parts of even large submarine fans where the gradients are extremely low (Pirmez *et al.* 1997).

Mulder and Alexander (2001) carefully distinguish between dominant particle support mechanisms and depositional mechanisms (Fig. 1.23). The utility of this distinction can be demonstrated with reference to the contentious issue of the origin of so-called 'massive sands' (Stow & Johansson 2000). In the case of inflated sand-flow deposits, en masse deposition by frictional 'freezing' leads to a chaotic grain fabric (Hiscott & Middleton 1980), whereas macroscopically similar, unstratified deposits of concentrated density flows have a strong *a*-axis fabric with a high imbrication angle throughout much of the deposit (Hiscott & Middleton 1980). This results because concentrated density flows deposit their load from suspension, and the strong shear near the aggrading bed strongly aligns the particles. Similarly, structureless T_a divisions of turbidity-current deposits can form whenever the suspension fallout rate is too high to allow sufficient grain traction to form lamination (Lowe 1988; Arnott & Hand 1989; Allen 1991; Hiscott *et al.* 1997a), but the grain fabric of such deposits is characteristically well organised with *a*-axes of elongate grains parallel to flow and imbricated upflow ~10–15°. As advocated by Hiscott *et al.* (1997a), careful grain-fabric studies can be a powerful aid in distinguishing the transport mechanisms for structureless (also referred to as 'massive') sands.

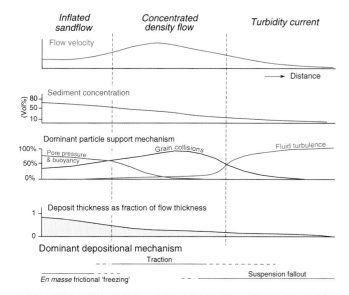

Fig. 1.23 The gradual transition in grain-support mechanisms between inflated sandflows, concentrated density currents and turbidity currents. Cross-overs in the relative importance of these mechanisms are used to define these SGFs. As is the case for cohesive flows, the thickness of the deposit of an inflated sandflow is similar to the thickness of the flow itself. Turbidites, in contrast, are much thinner than the associated turbidity current. Modified from Mulder and Alexander (2001).

1.4.2   Transformations between flow types

It has been proposed that density currents might become stratified, with discontinuities in concentration (Fig. 1.24), or might change their character dramatically while moving downslope as a result of *flow transformations*. Four varieties of flow transformation are recognised:

(1) As defined by Fisher (1983), *body transformations* involve down-current changes between turbulent and laminar flow, or between a coherent slide and a debris flow as seawater is incorporated (e.g., as interpreted for ancient examples by McCave & Jones 1988; Jones *et al.* 1992; Talling *et al.* 2004; Pickering & Corregidor 2005; Strachan 2008; Haughton *et al.* 2009; Talling *et al.* 2010).

(2) *Gravity transformations* involve gravitational segregation into a lower, laminar, highly concentrated part and an upper, turbulent, less concentrated part (e.g., as studied experimentally by Postma *et al.* 1988).

(3) *Surface transformations* occur where the front or top of a highly concentrated flow is eroded by shear beneath the overlying ambient fluid, creating a more dilute, turbulent daughter current (e.g., as described from experiments by Hampton 1972 and Talling *et al.* 2002, and inferred for an ancient deposit by Strachan 2008).

(4) *Fluidisation transformations* occur mainly above highly concentrated pyroclastic flows as a dilute cloud of elutriated material forms a secondary, less concentrated, and turbulent flow (elutriation = the upward flushing out of fine particles by escaping fluids).

The stratified flows produced by gravity, surface and fluidisation transformations might deposit composite beds with abrupt grain-size or textural breaks (Gladstone & Sparks 2002; Talling *et al.* 2004; Pickering & Corregidor 2005; Amy & Talling 2006; Strachan 2008; Haughton *et al.* 2009; Talling *et al.* 2010; Fig. 1.25), or the two parts of the flow might ultimately take different paths or travel different distances, thus forming spatially separate deposits.

It is very difficult in ancient deposits to determine whether a composite bed formed by a flow transformation, or whether there might have been two (or more) separate flows that contributed to what seems to be a single deposit. Amy and Talling (2006) discuss this dilemma in relation to intimately interbedded turbidites and debrites in the Apennines, with the debrites having much more limited extent than the possibly co-genetic turbidites (Fig. 1.26). Haughton *et al.*

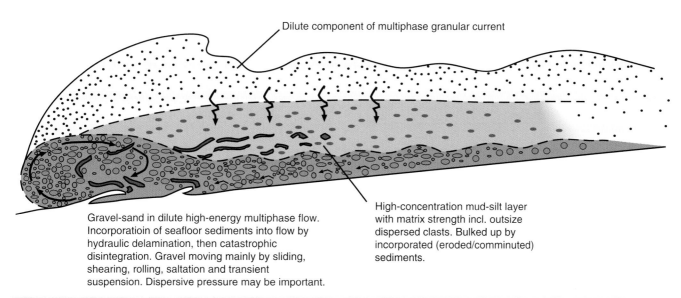

Gravel-sand in dilute high-energy multiphase flow. Incorporatioin of seafloor sediments into flow by hydraulic delamination, then catastrophic disintegration. Gravel moving mainly by sliding, shearing, rolling, saltation and transient suspension. Dispersive pressure may be important.

High-concentration mud-silt layer with matrix strength incl. outsize dispersed clasts. Bulked up by incorporated (eroded/comminuted) sediments.

Fig. 1.24 Multiphase hybrid SGF proposed by Pickering and Corregidor (2005) to explain certain disorganised beds in the Ainsa Basin, southern Pyrenees.

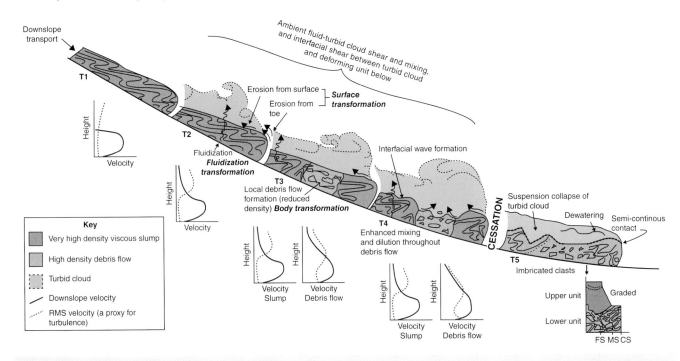

Fig. 1.25   Schematic drawing of the Little Manly Slump flow transformation model from Strachan (2008). T1 to T4 represent progressive development of transformation with time. Vertical velocity profiles for T1 to T4 are indicated and, where two are present, indicate the differences between slump and debris flow vertical velocity profiles. T5 shows deposition of units following slump cessation, together with a log showing a vertical profile. Surface and body transformations are inferred during T2 and T3. (FS = fine sand; MS = medium sand; CS = coarse sand.)

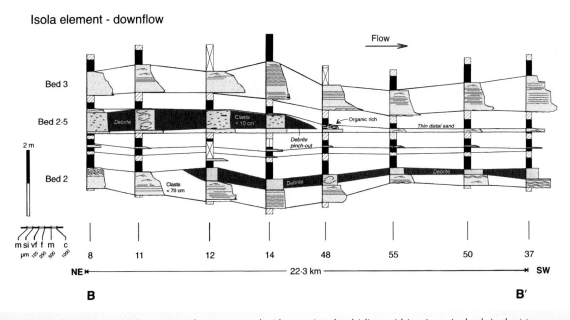

Fig. 1.26   Limited distribution of debrite units when compared with associated turbidites within tri-partite beds in the Marnoso arenacea, Italy. From Amy and Talling (2006).

(2003) concluded that 'linked' debrites in Jurassic deep-water deposits in the North Sea (Fig. 1.27) formed from separate flows that were triggered at the same time as sand-load concentrated density flows, but eventually came to rest on top of the freshly accumulated sand. In this case, two separate flows with different rheology were apparently generated by the same failure event, leading to bi-partite beds with sharp textural discontinuities.

Talling *et al.* (2007) have studied an enormous sediment slide and its runout deposits (debris flow and turbidite) from the Agadir Basin, and the Seine and Madeira abyssal plains, located offshore northwest

LITHOLOGY

■ mudrock

▓ sandy mudrock

▒ clay-prone sandstone

□ sandstone

SEDIMENTARY STRUCTURES

quartz granules

consolidation lamination

banding

vertical dewatering sheets

sheared dewatering sheets

parallel lamination

mudclasts

swirled fabrics

carbonaceous matter

sand clasts

sand injections

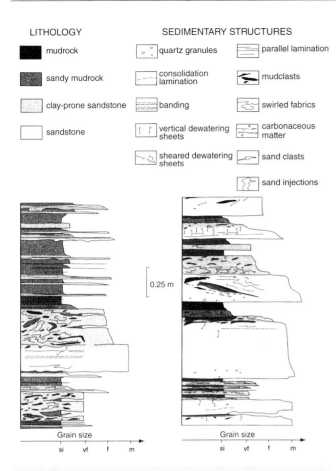

0.25 m

Grain size

si   vf   f   m

Grain size

si   vf   f   m

Fig. 1.27 Examples of linked debrites and Facies Class B sandstones (see Section 2.4) from the fringe of a Jurassic sand-prone fan in the subsurface of the North Sea. From Haughton *et al.* (2003).

Africa. Beyond the base of the continental slope, a mostly >1 m thick bed consists of a lower stratified sand and a middle muddy sand that is interpreted as a co-genetic debrite. The debrite is overlain by the graded mud top of the turbidite. The total volume of this linked turbidite and muddy debrite is ∼125 km³ assuming a bulk density of 1.8 g cm⁻³, which is about ten times the annual discharge of all rivers to the world ocean. Their work shows that extensive debrites can form downflow from abrupt slope breaks and areas of significant seafloor erosion. The 'linked' debrite forms the centre of the deposit and is encased within turbidite sandstone and mudstone.

Based on a detailed study of 1–10 m-thick muddy SGF deposits from the Madeira Abyssal Plain, McCave and Jones (1988) and Jones *et al.* (1992) proposed their deposition from dense ($\Delta\rho = 5$–100 kg m⁻³), non-turbulent flows. The deposits are ungraded, structureless mud. The proposed density contrast with seawater is equivalent to a volume concentration of less than ∼6 volume%; non-turbulent mud-laden flows of this character have been produced experimentally by Baas and Best (2002) at velocities of ∼0.33 m s⁻¹. McCave and Jones (1988) and Jones *et al.* (1992) believe that a fully turbulent flow decelerated, went through a body transformation to form an essentially laminar and viscous SGF. With turbulence severely damped, the suspension consolidated into a cohesive layer with inter-particle forces preventing the differential settling of coarser grains. Consideration by Masson (1994) of the small cross-sections of the channels through which the muddy SGFs passed en route to

the Madeira Abyssal Plain, and the lack of prominent levées along these channels (suggesting little overspill), supports the notion of high flow densities.

Gravity flows that undergo transformations along their path have been termed *composite flows* by Haughton *et al.* (2009), with the deposits referred to as *hybrid event beds*. These authors propose a classification (Fig. 1.28a) that includes such composite flows and their deposits. Contrary to the better-understood case of gradual flow dilution and deposition of progressively more organised and finer-grained deposits in the downdip direction (Fig. 1.28b1), composite flows increase in concentration distally and transform, in part, into mudflows or debris flows (Fig. 1.28b2). Three processes can force an increase in concentration and suppression of turbulence (Haughton *et al.* 2010; Fig. 1.29): (i) deceleration of a clay-rich flow so that viscous effects become dominant; (ii) segregation of clay components into trailing and lateral parts of a flow so that turbulence intensity is damped; and (iii) addition of clay to the suspension by disintegration of soft clasts eroded along the travel path – a process referred to as 'bulking'. The first of these processes has been studied experimentally by Sumner *et al.* (2009) and Baas *et al.* (2011). With rapid deceleration of a flow carrying ∼10% suspended clay, silt and sand, Baas *et al.* (2011) hypothesise that a basal relatively clean sand division can accumulate rapidly because of a sharp drop in the total transport capacity for sand. This drop results from a decline in turbulence support that occurs more quickly than the parallel increase in cohesive support.

An alternative way to produce a hybrid event bed is for synchronously triggered debris flows and concentrated density flows to deposit one after the other, with the latter flow outrunning its more viscous partner (Fig. 1.29d).

1.5 Turbidity currents and turbidites

1.5.1 Definition and equations of flow

Turbidity currents are density currents in which the denser fluid is a grain suspension, with particles supported largely by the upward velocity fluctuations associated with turbulent eddies (Bagnold 1966; Leeder 1983). Entrained sediment diffuses throughout the flow thickness, but the highest particle concentrations are at the base of the flow (Stacey & Bowen 1988; Middleton 1993; Felix 2001). As with open-channel flows in flumes and rivers (Rouse 1937), the coarsest size fractions are also carried toward the lower part of the flow. Depending on the manner in which the flow was initiated, the turbidity current may be (i) relatively short, quickly passing an observation point on the seafloor (surge-type flow); or (ii) relatively long with steady discharge due to prolonged input from a long-lived source (steady and uniform discharge – generally river-fed as 'hyperpycnal' flows) (Mulder & Syvitski 1995; Mulder *et al.* 2001, 2003; Alexander & Mulder 2002; Felix *et al.* 2006). *Hyperpycnal flow* is the term used to describe river discharge which, because of a high suspended load, is more dense than seawater so travels down the delta front as an underflow.

Talling *et al.* (2012) provide an historical reminder that Ph. Kuenen and co-workers coined the term 'turbidity current' because the density currents they were investigating were 'turbid' (murky because of suspended solids) rather than because they were turbulent. However, it is now rather common practice to associate this term with turbulent flows. We therefore follow Mulder and Alexander (2001) in restricting

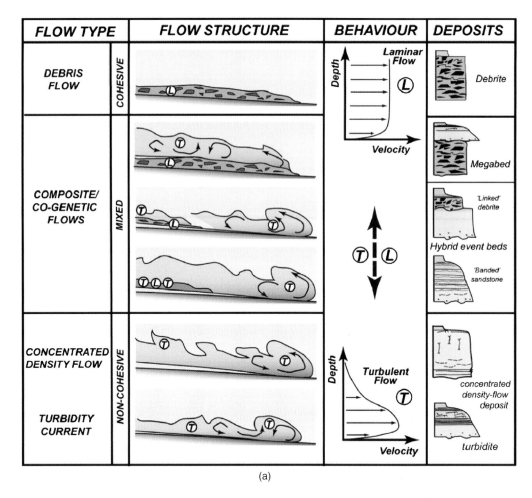

(a)

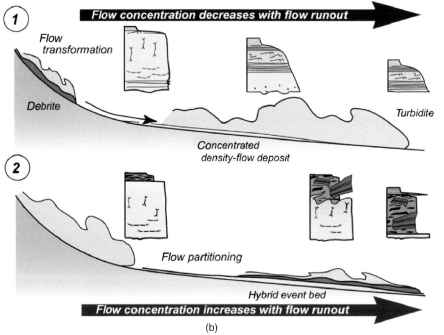

(b)

Fig. 1.28 (a) Haughton *et al.* (2009) classification scheme for event beds emplaced by subaqueous sediment gravity flows. (b1) Debrites, concentrated density-flow deposits and turbidites dominate the record of many deep-water systems and record increasing downdip dilution of the flow (debrites passing to concentrated density-flow deposits and eventually turbidites). (b2) In some systems there is instead a downdip progression from non-cohesive flows (depositing concentrated density-flow deposits and turbidites) to flows transformed into components with radically different rheology, with the deposits of the cohesive flow components increasingly dominant distally. From Haughton *et al.* (2009) who used 'high- and low-density turbidity currents' instead of 'concentrated density-flows and turbidity currents'.

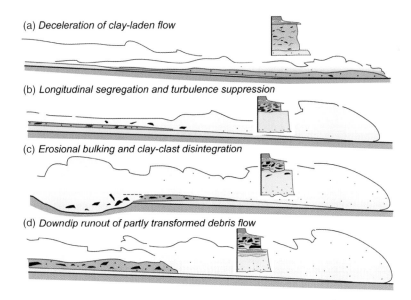

Fig. 1.29   Summary of depositional origin for hybrid event beds as a result of (a) loss of turbulence and deceleration of a clay-rich flow, (b) longitudinal segregation of clays and clay flakes to suppress turbulence in the rear and margins of an otherwise turbulent flow, (c) bulking and disintegration of clay clasts to release clay near-bed in an otherwise turbulent flow and (d) downdip runout of a flow that was either synchronously triggered with a concentrated density flow, or that partially transformed to generate a forerunning concentrated density flow. From Haughton *et al.* (2010).

the term 'turbidity current' to flows in which turbulence is the dominant support mechanisms (Figs. 1.19 and 1.23). The concentration of solids in the lower part of such flows can exceed the Bagnold (1962) limit of 9% by volume (Fig. 1.21), beyond which grain-to-grain interaction begins to occur. Natural flows with a mixture of sediment sizes are stratified according to density and other properties, so that two flows of identical average density might have quite different near-bed concentrations. Since it is the near-bed concentration that controls settling rates, the intensity of turbulence and the formation of sedimentary structures, it is unavoidable that the deposits of some turbidity currents might resemble those of the more dilute concentrated density flows. For example, basal structureless divisions in some turbidites likely imply more about the rate of deposition from suspension than the absolute concentration (Lowe 1988; Arnott & Hand 1989; Allen 1991). The rate of suspension fallout will vary with the rate of flow deceleration, so that two turbidity currents with exactly the same average concentration and turbulence intensity might deposit structureless sand at one locality (rapid deposition) and laminated sand at a different locality (protracted deposition).

Concentration has a bearing on flow velocity and therefore competence on basin margin slopes. On gentle slopes, coarse sands and gravels are likely to be transported by inflated sandflows or concentrated density flows rather than turbidity currents. Concentration influences flow density and viscosity, but has little effect on the flow mechanics of a turbidity current until particle concentrations become so high that (i) interparticle collisions become an important component of grain support even well above the bed (Fig. 1.23), or (ii) turbulence, particularly near the bed, becomes damped (Fig. 1.30).

Modelling of deposition from surge-type turbidity currents is difficult because equilibrium velocities and sediment concentrations may never be attained, and because of scaling problems in small-scale experiments (Middleton 1993). Experimental modelling of density surges by Laval *et al.* (1988) shows that the velocity of the surge is effectively proportional to the square root of the initial volume, and that surge velocity increases with increasing initial density of the flow, being proportional to the square root of the ratio of the excess density to the density of the ambient fluid. These results are corroborated by the numerical simulations of Zeng and Lowe (1997a, b).

Dade *et al.* (1994) and Dade and Huppert (1995) developed mathematical models to explain flow evolution and deposition from turbidity currents. On slopes, the primary cause of decreasing flow concentration is entrainment of ambient seawater (Dade *et al.* 1994). On horizontal surfaces like basin plains or the very low slopes on distal submarine fans, flows gradually become less concentrated as their suspended load is deposited (Dade & Huppert 1995); seawater entrainment is much less significant than on slopes (Stacey & Bowen 1988). For deep-water gravity currents travelling across horizontal surfaces (flow thickness <0.075 water depth), the box model of Dade and Huppert (1995) predicts a maximum deposit thickness of about three times the average deposit thickness, located 1/5 of the way between the most proximal deposit and the most distal deposit. The run-out distance, measured from the point of arrival onto the basin plain, is proportional to the initial reduced gravity and volume of the surge, and is inversely proportional to the average particle settling velocity. Sorting is predicted to improve in the downcurrent direction for mixed-load currents.

On uniform slopes, the model of Dade *et al.* (1994) predicts deposit thickness and texture, evolution of flow velocity and Froude number, and depositional runout distance. Critical parameters that control evolving flow properties are the rates of sediment loss through deposition, and dilution of the suspension through seawater entrainment. When applied to a hypothetical natural surge travelling about 300 km, this model predicts growth in flow thickness from 100 m to >1 km, and in length from 1 km to >10 km, solely because of seawater entrainment.

Sophisticated numerical simulations, tested against natural data from a fjord setting (Bute Inlet), have been formulated by Zeng and

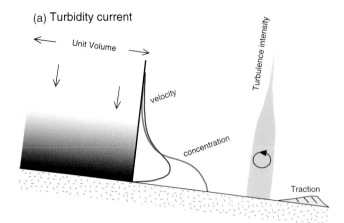

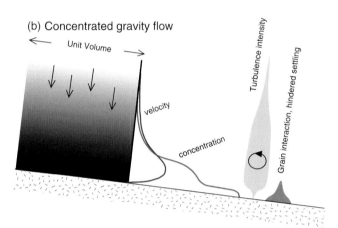

Fig. 1.30 Contrasting near-bed grain support between a turbidity current and a concentrated density flow. Although both types of SGF are turbulent away from the sedimentation surface, the latter has strong grain interaction at the aggrading bed, and turbulence is sufficiently damped to prevent the development of tractional sedimentary structures.

Lowe (1997a, b). The simulated flows have prominent grain-size stratification and relatively low concentration. The simulated deposits show good distribution grading like that observed in natural turbidites. For initial boundary conditions of higher concentration, the predicted grading is poorer and limited to the coarsest parts of the size spectrum (coarse-tail grading). These simulated flows compare best with the smallest of the natural flows that have been observed in Bute Inlet.

For continuously-fed flows or unusually large-scale surge-type flows on a constant slope, a condition of effective flow steadiness (i.e., $\partial U_B/\partial t = 0$; U_B = body velocity) may exist in the long body of the current. The maximum velocity is in the lower part of the body of the current (Fig. 1.31) and is ~1.6 times the mean velocity (Felix 2004). The finest grain sizes have an essentially uniform concentration throughout the flow, whereas the coarsest grain sizes are most concentrated near the base of the flow (Rouse 1937; Hiscott 1994a). Because the time-averaged velocity is lower near the base of the flow than in the vicinity of the velocity maximum, coarse size fractions lag behind fine fractions; this differential transport rate may lead to basal

inverse grading in the deposits of turbidity currents (Hand & Ellison 1985; Hand 1997).

The flow steadiness described above is at best an average condition, because velocity fluctuations are superimposed on the body of the flow by the passing of interfacial waves along internal density boundaries and along the top of the current (Simpson 1997; Baas & Best 2002), and by longer-period pulsing described from natural currents by Best et al. (2005) but still poorly understood.

Experiments of Gladstone et al. (2004) provide insight into how vertical and streamwise stratification might affect grading and textures of deposits. These authors demonstrate that strong vertical grain-size stratification inevitably leads to downflow lateral grading in the flow (Fig. 1.18). In most cases, the lower part of the flow will contain coarser sediment. If it also has a higher concentration, then the coarse fractions may outpace the upper part of the same flow with its finer sediment load (opposite to the predictions of Hand 1997). Sequential deposition as the flow passes will produce a graded bed, with the smoothness of the grading (or the degree of step-wise grading) depending on the initial density structure of the flow.

A comprehensive mathematical model for flow evolution and deposition from turbidity currents is presented by Pratson et al. (2000). In this book, we provide simpler equations for homogeneous, non-erosional and non-depositing flows, primarily to clarify what factors control velocity and thickness of turbidity currents. For more sophisticated treatments, the reader is referred to Pratson et al. (2000), Stacey and Bowen (1988) and other references in preceding paragraphs.

As shown by Middleton (1966b), the velocity of the body of a turbidity current is given by a Chezy-type equation,

$$U_B{}^2 = \left[\frac{8g}{f_o + f_i}\right]\left[\frac{\Delta\rho}{\rho + \Delta\rho}\right] d_B \tan\alpha \qquad (1.3)$$

where $\Delta\rho$ = density difference between the flow and seawater, ρ = density of seawater, d_B = body thickness, α = bottom slope in degrees, f_o = dimensionless Darcy–Weisbach friction coefficient for bed friction, and f_i = dimensionless friction coefficient for interfacial friction at the top of the flow. Friction coefficients have been determined empirically for rivers and in flumes, but turbidity currents differ from rivers in that there is also friction between the flow and the overlying water. According to Middleton and Southard (1984), $f_o + f_i$ for large natural turbidity currents is likely to be about 0.01. For subcritical flow conditions (Froude number, $F < 1.0$), $f_o \gg f_i$, but for supercritical flows ($F > 1.0$), $f_i > f_o$ due to intense mixing at the upper interface of the flow.

The velocity of the head of the current, U_H, does not appear to depend significantly on the bottom slope for turbidity currents moving over low slopes; Middleton (1966a) gives:

$$U_H{}^2 = 0.56g\, d_H\left[\frac{\Delta\rho}{\rho + \Delta\rho}\right] \qquad (1.4)$$

On steeper slopes, from perhaps 2–10°, a more general result (Hay 1983a) includes a dependence of head velocity on the bottom slope:

$$U_H{}^2 = g\, d_H\left[\frac{\Delta\rho}{\rho + \Delta\rho}\right](0.50\cos\alpha + t\sin\alpha) \qquad (1.5)$$

Experiments with small surges suggest that the empirical factor t is in the range 1.6–4.0 (Hay 1983a). Note that for $\alpha = 1°$ and $t = 3.3$, Equation 1.5 reduces to Equation 1.4, but for $\alpha = 10°$, the numerical

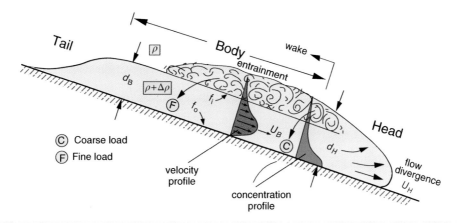

Fig. 1.31 Idealised streamwise cross-section of a turbidity current, divided into head, body and tail regions. Settling from the wake behind the head produces a lateral size grading in the flow.

constant in Equation 1.4 would be 1.1, not 0.56. The ratio of the head velocity to the body velocity (U_H/U_B) is approximately 1.0 on gentle slopes but <1.0 on steeper slopes (Fig. 1.32). The head does not grow unchecked as suspension from the body of the current overtakes it. Instead, an equilibrium is developed. The rate of body flow into the head region is balanced by loss of suspension from a region of intense turbulence and flow separation at the back of the head (Simpson 1982, 1997; Fig. 1.31). This ejected material settles back into the flow top according to fall velocity, with the coarsest grains returning to the body nearest the head and the finest grains returning far behind the head. This process amplifies the lateral size grading that might result from density stratification in the flow (Gladstone *et al.* 2004). The result is a coarser suspension near the head and a finer suspension near the tail (Walker 1965).

Equations 1.3 through 1.5 can only provide guidance as to the behaviour of a turbidity current at a single location along its flow path, because flow concentrations and thicknesses evolve downcurrent as seawater is entrained into the suspension and as deposition removes suspended load. A special case in which flow thickness is partly regulated over long flow distances occurs in deep-sea levéed channels.

Here, dramatic flow thickening is counteracted by overspill across the levées. Such channelised flows may also have their sediment concentration regulated by flow shortening through time, as the rear part of the body (travelling on steeper slopes) progressively catches up with the frontal part of the body (travelling on gentler slopes) (Hiscott *et al.* 1997b). As a result, turbidity currents are able to maintain their fundamental properties and travel long distances through such levéed channels (e.g., Chough & Hesse 1976).

1.5.2 Natural variations and triggering processes

Initially, the theory of flow and deposition of turbidity-currents was based mainly on observations from simple, small-scale experiments (Middleton 1993), leading to predicted facies characteristics simpler than those encountered in nature. Natural currents, however, are strongly stratified (Gladstone *et al.* 2004) and typically encounter irregular seabed topography, opposing slopes, or are confined within meandering channels, all of which modify flow behaviour and direction. The result may be structurally complex beds or beds showing step-wise grading and a complex repetition of structures (e.g., Pickering & Hiscott 1985; Marjanac 1990; Pickering *et al.* 1992, 1993a; Edwards 1993; Edwards *et al.* 1994; Haughton 1994; Gladstone *et al.* 2004). The influences of basin topography and density layering of the ambient fluid (Figs 1.17 and 1.18) have only recently been studied experimentally (e.g., Kneller *et al.* 1991; Alexander & Morris 1994; Kneller 1995; Rimoldi *et al.* 1996; Gladstone *et al.* 2004). It will be challenging to extrapolate these results to large natural scales, but it is critical that complex topography and special oceanographic conditions be taken into account when interpreting ancient successions.

Natural marine turbidity currents have been studied only superficially because of the inherent difficulties in devising and deploying monitoring systems. More has been published on turbidity currents in lakes and reservoirs (see review in Middleton 1993). Human-made marine turbidity currents formed by the dumping of mine tailings have been studied by Normark and Dickson (1976), Hay (1987a, b; see also Hay *et al.* 1982), and Normark (1989). Nascent turbidity currents in Scripps Submarine Canyon have been monitored up to the point at which current-meters were lost, giving minimum flow speeds of 1.9 m s^{-1} (Inman *et al.* 1976). Natural delta-fed turbidity currents in a protected fjord were monitored (Fig. 1.33) and their deposits studied and later numerically modelled by Zeng *et al.* (1991, 1997a, b). Larger

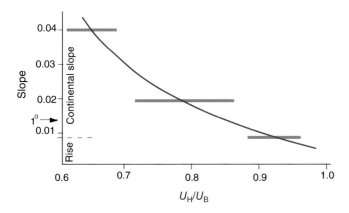

Fig. 1.32 Ratio of head velocity to body velocity plotted against bottom slope for experimental turbidity currents (5-m flume) of Middleton (1966a). Horizontal bars show extent of scatter in the data. Ranges of bottom slope for continental margins are superimposed.

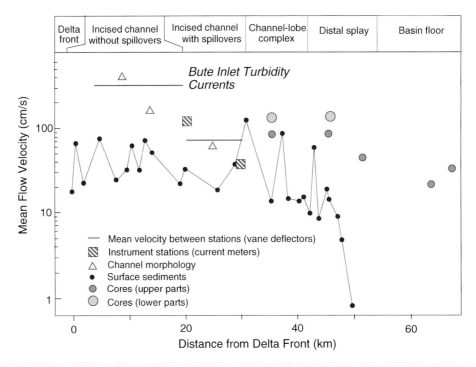

Fig. 1.33 Measured flow velocities of turbidity currents in Bute Inlet compared with velocity estimates inferred from the texture of turbidites along the flow path. Redrawn from Zeng *et al.* (1991).

natural flows have only been studied indirectly, by examining records of submarine cable breaks or displacements of instrument packages (Heezen & Ewing 1952; Heezen *et al.* 1954; Gennesseaux *et al.* 1980; Normark *et al.* 1993b), variation in levée heights of submarine channels at channel bends (Komar 1969; Pirmez 1994), and the height of flow deposits and deposit characteristics (Bowen *et al.* 1984; Piper & Savoye 1993). Quantitative results of some of these studies are summarised in Table 1.2.

The best documented large natural flow was generated by a 7.2 magnitude earthquake in the Laurentian Channel, eastern Canada, in 1929. The so-called '1929 Grand Banks turbidity current' broke a succession of submarine cables and deposited a graded fine sand to silt layer about 1 m thick (Fig. 1.34) in a water depth of about 5200 m (Heezen & Ewing 1952; Heezen *et al.* 1954; Fruth 1965). According to Piper *et al.* (1988), the turbidity current carried about 200 km³ of sediment. Calculated maximum velocity and flow depth are 19 m s⁻¹ and about 400 m. In order to carry sufficient sediment through the valleys of the Laurentian Fan to account for the volume of the turbidite on the Sohm Abyssal Plain, the 1929 turbidity current must have taken 2–3 hours to flow past a point along one of the fan valleys. Piper *et al.* (1988) and Hughes Clarke *et al.* (1990) initially attributed the formation of large gravel bedforms with wavelengths of 10–70 m in the Laurentian Fan valleys (Fig. 1.35) to the 1929 turbidity current. Now, the gravel transport and reworking into large bedforms is interpreted to have taken place during a Pleistocene glacial-lake outburst event which formed a powerful hyperpycnal gravity flow (D.J.W. Piper, in Wynn *et al.* 2002b).

Triggering mechanisms for natural turbidity currents include the following processes (Normark & Piper 1991; Van den Berg *et al.* 2002): (i) hyperpycnal flow from rivers and glacial meltwater (Heezen *et al.* 1964); (ii) sand liquefaction (Seed & Lee 1966; Andresen & Bjerrum

1967) in canyon heads, triggered by storms, earthquakes or local failures, followed by water entrainment and acceleration; (iii) breach failure which results from gradual back-sapping (retrogression) of an over-steepened slope in cohesionless material after an initial failure (Van den Berg *et al.* 2002; Mastbergen & Van den Berg 2003); (iv) erosion of the front of a moving debris flow (Hampton 1972; Talling *et al.* 2002) or thickening and dilution of a debris flow as it undergoes an hydraulic jump (Weirich 1988); (v) dilution of sediment slides (Ricci Lucchi 1975b; Cita *et al.* 1984; Hughes Clarke *et al.* 1990); (vi) suspension of sediments in canyon heads by edge waves associated with storms (Inman *et al.* 1976, Fukushima *et al.* 1985); (vii) ignitive flow of shelf suspensions that descend a basin slope and (viii) ignitive flow of ash from pyroclastic falls. *Hyperpycnal flow* is the term used to describe river discharge which, because of a high suspended load, is more dense than seawater so travels along the seabed as an underflow. The term *ignition* (Parker 1982) refers to a state in which an accelerating density current entrains more sediment along its path, which causes it to continue to accelerate and grow in size. According to Normark and Piper (1991), ignition is favoured by initial volume concentrations of particles of about 0.01, initial velocities of about 1 m s⁻¹ (fine-grained sand) to 1.2 m s⁻¹ (medium-grained sand), and slopes exceeding 3° (Fig. 1.36). Pratson *et al.* (2000) instead explain that ignition is most effective on slopes of ∼2–2.5°; below this range the turbidity current will decelerate and die, whereas on higher slopes the quantity of entrained sediment is less and the turbidity current smaller.

Earthquakes are commonly called upon as prospective triggers for turbidity currents (Hiscott *et al.* 1993; Beattie & Dade 1996). Earthquake recurrence intervals follow a power law, as do many turbidite bed thicknesses (Hiscott *et al.* 1993; Rothman *et al.* 1994; Drummond & Wilkinson 1996). According to Kuribayashi and Tatsuoka

Table 1.2 Estimates of mean flow parameters for natural turbidity currents, using a seawater density of 1030 kg m⁻³ (unless given). Highest velocities at top of table

Location	M=measured C=calculated H=hindcast	Basis of calculations	Density contrast ($kg\ m^{-3}$)	Body velocity ($m\ s^{-1}$)	Comments	Source
Laurentian Fan, 1929 event	C, H	Cable breaks, seabed features, theory	~95	19	Gravel waves now believed older (Wynn et al. 2002b)	Piper et al. (1988)
Monterey Fan, California	C	Levee height differences, theory		20–4		Komar (1969)
Var middle and lower valley	C, H	Sand distribution and size, theory	~95–25	~10–2	Proximal to distal ranges given	Piper & Savoye (1993)
Var Canyon and Valley	C, H	Cable breaks, seabed features, theory	25–8	~8–3	Turbidity current 'plume' only; concentration assumed	Mulder et al. (1997)
Labrador Sea (NAMOC)	C, H	Grain size	87	8	Base of in-channel flows	Klaucke et al. (1997)
Bute Inlet, Canada	M, C	Current-meters, grain-size of deposits	26–2	3.3–0.5	Sea-water density 1023 kg m⁻³	Zeng et al. (1991)
Scripps Canyon, California	M	Current-meters		1.9		Inman et al. (1976)
Rupert Inlet, Canada	C	Channel geometry, tailings discharge rate	100	~1.3	Surge-type turbidity currents	Hay (1987b)
Zaire submarine valley	M	Current meters, sediment traps		1.2	Coarse sand 40 m above floor; velocity for 150 m elevation	Khripounoff et al. (2003)
Labrador Sea (NAMOC)	C	Channel geometry	12–1	0.86–0.05	Proximal to distal ranges given	Klaucke et al. (1997)
Navy Fan, California	C, H	Grain size, channel geometry, theory		0.8–0.1		Bowen et al. (1984)
Labrador Sea (NAMOC)	C, H	Grain size	4	0.45	Top of in-channel flows	Klaucke et al. (1997)
Rupert Inlet, Canada	C	Channel geometry, tailings discharge rate	30	0.4	Continuous-flow of mine-tailings discharge	Hay (1987)
Reserve Fan, Lake Superior	M	Current meters and water samples	0.05	0.12–0.08		Normark (1989)

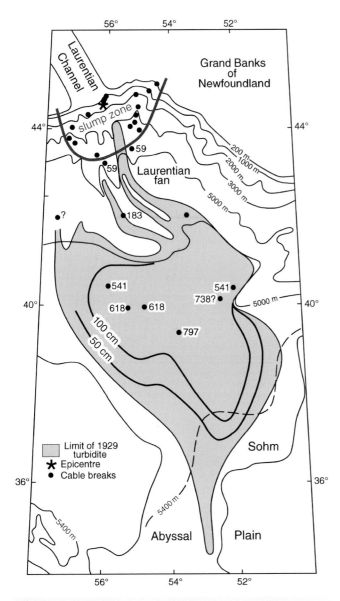

Fig. 1.34 Distribution of the sand turbidite generated by the 1929 Grand Banks earthquake. Numbers beside cable break positions are the time of each break, in minutes, after the earthquake. Bathymetric contours are in metres. Turbidite thickness is contoured in centimetres. Redrawn from Piper *et al.* (1988).

(1977) and Keefer (1984), only earthquakes with magnitudes greater than about 5.0 can cause significant sand liquefaction and, as the distance from the epicentre increases, so does the minimum magnitude required for liquefaction. Their data indicate, for example, that a magnitude 7.0 shock can liquefy sediment as far as about 100 km from the epicentre. Using historical records for the period 1905–64, Duda (1965) determined a recurrence interval, for those earthquakes stronger than magnitude 7, of 0.063 yr (23 days) for the entire length of the circum-Pacific convergent margins, about 40 500 km. For each 100 km segment of trench, therefore, the average recurrence interval is about 25 yr, in general agreement with more recent data around Japan (Mogi 1990). These frequent earthquakes would be capable of liquefying sediment and triggering sediment gravity flows over a wide area (Keefer 1984).

Hyperpycnal flows originate where sediment-laden river water attains a greater density than seawater, so that it is able to descend directly to the seabed at the river mouth and flow seaward as a turbidity current. Mulder and Syvitski (1995) explained the requirements for hyperpycnal flow and demonstrated the likelihood of such flows in modern oceans, particularly seaward of certain Asian rivers with high sediment loads. They concluded that sediment concentrations of at least 40 kg m^{-3} are needed to initiate hyperpycnal flow. In a groundbreaking re-evaluation of these ideas, Parsons *et al.* (2001) showed experimentally that sediment concentrations could instead be as low as ~1 kg m^{-3} because of *finger convection* between the warm, fresh, sediment-laden overflow exiting a river mouth, and an underlying colder, saline ambient water mass. The convection triggers instability and the removal of large amounts of suspension by a bottom-hugging hyperpycnal flow. The key to this process is the presence of a downward-increasing dominant gradient in salinity, and an upward-increasing gradient in temperature. Instead of only 9 of 147 rivers predicted to generate hyperpycnal flows annually (Mulder and Syvitski 1995), Parsons *et al.* (2001) predict that 61 of the rivers should show this behaviour. These particular 61 rivers 'produce 53% of the world's oceanic sediment load and are therefore responsible for a significant portion of the [modern] sediment record' (Parsons *et al.* 2001: p. 477).

The reader might take issue with calling the experimental flows of Parsons *et al.* (2001) 'hyperpycnal', because they did not exit directly from the proxy river channel, but rather developed through convection-driven collapse of a surface hypopycnal plume. However, this collapse occurred (and in nature is predicted to occur) very close to the river mouth, so that the resulting SGF effectively has the same close link with river input that characterises true hyperpycnal flows. The situation is not unlike the seaward gravity-driven flow of mud-laden suspensions produced where delta-derived mud is kept in suspension, and concentrated, by wave stirring through a process called 'frontal trapping' (Geyer *et al.* 2004). High mud concentrations in the wave boundary layer are able to initiate what Geyer *et al.* (2004) also call hyperpycnal flows, because suspended mud builds up near the seabed adjacent to the river mouth, and because the wave-induced mixing of seawater with fresh river water ensures that the suspension eventually exceeds the density of fully marine water. In some cases, the mud concentration in the zone of 'frontal trapping' can reach that of so-called 'fluid mud' (>10 000 mg litre^{-1}; Geyer *et al.* 2004).

A somewhat different approach was taken by Felix *et al.* (2006), who looked more carefully at the effects of density stratification and estuarine mixing, and concluded that hyperpycnal flows can be generated even when the density difference between the ocean and the particulate suspension is only a few tenths of 1 kg m^{-3}, much lower than advocated by Mulder and Syvitski (1995) and lower than proposed by Parsons *et al.* (2001).

In a study offshore of the Var River, France, Mulder *et al.* (1997) demonstrated that day-long hyperpycnal flows would occur each 5–21 years, and Mulder *et al.* (2001) showed that deposits from such flows are inverse-to-normally graded (Fig. 1.37). The number of turbidites in the geological record that were emplaced by hyperpycnal flows is still unknown, but the dominance of normal rather than inverse-to-normal grading in ancient deposits suggests that the deposits of hyperpycnal flows might not be as common as predicted by Parsons *et al.* (2001) and Felix *et al.* (2006). There are ancient examples, however, that possess the attributes expected for deposits from hyperpycnal flows (Soyinka & Slatt 2008). Additional data from modern natural settings will be needed to assess the true importance of hyperpycnal flow as an initiating mechanism for turbidity currents.

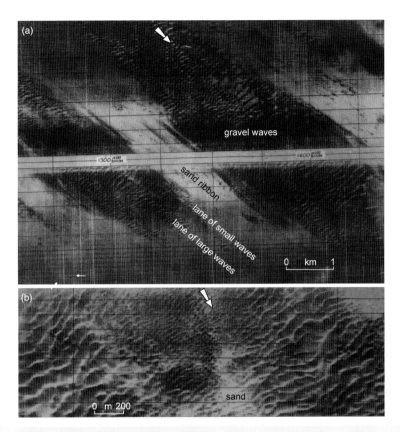

Fig. 1.35 Giant gravel waves in the Eastern Valley of the Laurentian Fan, likely formed under a hyperpycnal flow produced by an outburst flood during decay of continental ice sheets. (a) Fields of gravel-rich waves in lanes of larger and smaller waves, buried by sand ribbons. (b) Detail of rather sinuous gravel-rich waves of various sizes, locally covered by sand patches, using 1-km swath system. From Wynn *et al.* (2002b). Copies of original graphics courtesy of D.J.W. Piper, Geological Survey of Canada.

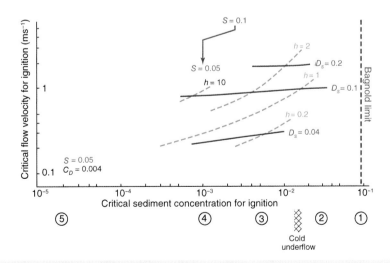

Fig. 1.36 Optimum conditions for ignition of turbidity currents, from Normark and Piper (1991). The plot shows critical velocity and critical concentration for various flow thicknesses (*h*, metres) and representative grain sizes (D_s, mm) on a slope of $S = 0.05$, assuming a drag coefficient of $C_D = 0.004$ (based on four-equation model results of Parker *et al.* (1986)). Also shown is the approximate offset in the concentration scale resulting from an increase in *S* to 0.1. Circled numbers are measured suspended-sediment concentrations from a number of settings: 1 = Huanghe River, China, in flood (Wright *et al.* 1988); 2 = Sustina River, Alaska, in flood (Hoskin & Burrell 1972); 3 = maximum measured discharge into Glacier Bay, Alaska (Hoskin & Burrell 1972); 4 = ignitive condition for sand in the head of La Jolla Canyon (Fukushima *et al.* 1985); 5 = nearshore sediment concentration after flood of Santa Clara River (Drake *et al.* 1972). The 'cold underflow' is the approximate sediment concentration in cold river water required to produce a marine underflow (Gilbert 1983).

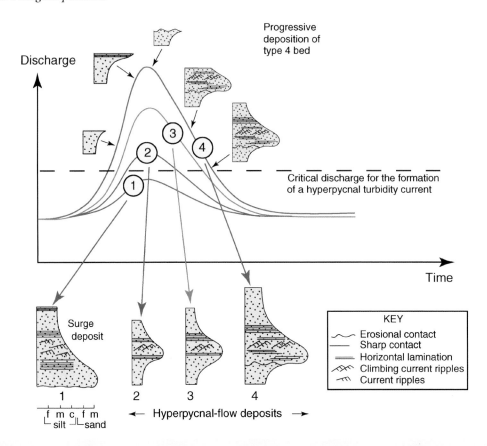

Fig. 1.37 Comparison of normally graded surge-generated turbidites and inverse-to-normally graded hyperpycnal-flow deposits produced by the rising and then falling flood stage of a river. The grain size scale below column 1 has divisions of fine (f), medium (m) and coarse (c) silt, and fine (f) and medium (m) sand. Redrawn from Mulder *et al.* (2001).

Table 1.3   Typical gradients of submarine fans and basin-margin slopes

Location	Bottom gradient	Source
Upper passive-margin slope	3–6°	Heezen *et al.* (1959)
Lower passive-margin slope	1.5–3°	Heezen *et al.* (1959)
Passive-margin rise	0.1–1°	Heezen *et al.* (1959)
Transform-margin scarps	10–30°	Aksu *et al.* (2000)
Forearc basin flanks	6–12°	Tappin *et al.* (2007)
Accretionary prism lower slope	>8°	Tappin *et al.* (2007)
Submarine canyons	1–3°	Nelson & Kulm (1973)
Upper fan channels	0.2–0.5°	Barnes & Normark (1984)
Suprafan lobes	0.1–0.4°	Barnes & Normark (1984)
Lower fan	0.1–0.2°	Barnes & Normark (1984)
Carbonate bank slopes	4–40°	Mullins & Neumann (1979)

1.5.3   Supercritical flow of turbidity currents

A fundamental distinction can be made between turbidity currents that are subcritical (Froude Number, F <1.0) and those that are supercritical (F > 1.0), where

$$F^2 = \frac{U_B^2}{\left(\dfrac{\Delta\rho}{\rho + \Delta\rho}\right) g d_B} \tag{1.6}$$

For dilute flows with $\Delta\rho/(\rho+\Delta\rho) \approx \Delta\rho/\rho$, Equation 1.6 reduces to

$$F^2 = \frac{U_B^2}{RCg d_B} \tag{1.7}$$

where R is the submerged weight of the grains (~ 1.65 g cm^{-3} for quartz) and C is the volume fraction of grains. This formulation for the Froude number is only valid for dilute turbidity currents, but is useful when trying to understand the relationship between flow concentration and the transition to supercritical flow. Clearly, the most dilute currents ($C \ll 1.0$) will be supercritical even at low velocities (but see Huang *et al.* (2009) for caution regarding determination of the critical densimetric Froude number for transitions from supercritical to subcritical flow).

For a reasonable friction factor, $f = 0.02$, Komar (1971) concluded that turbidity currents would be supercritical on slopes >0.5°, a value exceeded on many basin-margin slopes and on the upper parts of submarine fans (Table 1.3; *cf.* Section 4.13). The transition to subcritical flow occurs as the bottom gradient declines, and may involve a hydraulic jump with intense turbulence and flow homogenisation (Middleton 1970; Komar 1971), as well as creation of an upcurrent-migrating bore in confined mini-basins (Toniolo *et al.* 2006). Hydraulic jumps have considerable effect on the formation of widespread scours and lenticular deposits of all scales (Mutti & Normark 1987, 1991; Garcia & Parker 1989; Alexander & Morris 1994; Vicente Bravo & Robles 1995).

1.5.4 Autosuspension in turbidity currents

It is frequently claimed that turbidity currents can effectively carry sand-sized detritus across basin-margin slopes without leaving a deposit: there may even be net erosion in these areas (i.e., flow *ignition*). Weaver (1994) demonstrated slope bypassing by large mud-rich turbidity currents entering the Madeira Abyssal Plain (*cf.* Stevenson *et al.* 2013) by examining their reworked coccolith assemblages. Such 'bypassing' requires that the turbidity current be at the least self-sustaining on the slope, or while flowing through slope channels. This process of 'self maintenance' (Southard & Mackintosh 1981) was named *autosuspension* by Bagnold (1962), and is summarised in Figure 1.38. In words, the excess density of the grain suspension, combined with the downslope component of gravitational acceleration, induces basinward flow. The turbulence generated by the flow maintains the grains in suspension (Middleton 1976; Leeder 1983; Eggenhuisen & McCaffrey 2012), and the suspension maintains its density contrast with the overlying seawater, allowing continued flow, turbulence generation and effective grain suspension. According to Middleton (1966c), Allen (1982: II, p. 399) and Pantin (1979), true autosuspension is probably not common in nature, except for thick turbidity currents carrying fine particles on steep slopes. For all other cases, some of the suspended load settles through the flow and is deposited.

Southard and Mackintosh (1981) and Middleton and Southard (1984) claim that Bagnold's (1962) mathematical formulation of the autosuspension criterion fails to find support in experiments. This shortcoming is ascribed to flaws in Bagnold's (1962) energy-balance equation, resulting in a 'fallacious system of energy bookkeeping' (Paola & Southard 1983). As outlined by Middleton and Southard (1984), Bagnold's (1962) equations do not account for the fact that only a small percentage of a flow's power, about 2%, is available to suspend sediment, the rest being expended to overcome frictional resistance at flow boundaries and to produce turbulence and heat, most of which does not contribute to grain suspension. This over-estimation of available power by Bagnold (1962) was remedied by Pantin (1979), who introduced an efficiency factor, e, into the formulation of an autosuspension criterion. This criterion is:

$$e\alpha U_s > w \qquad (1.8)$$

where α = slope angle, U_s = transport velocity of the suspended sediment and w = grain settling velocity. In a worked example, Pantin

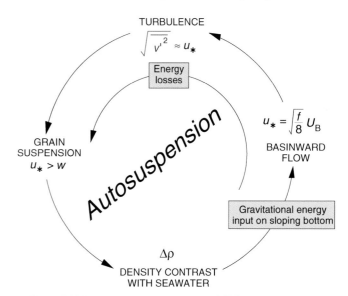

Fig. 1.38 Conceptual diagram to explain autosuspension. If gravitational energy input = energy losses, the flow will be self maintaining, and grains with settling velocity w will be kept in suspension by vertical velocity fluctuations of average strength $\sqrt{v'^2}$, approximated by u_*.

(1979) sets $e = 0.01$. In general, Pantin (1979) shows that flow density is the main control on whether autosuspension will occur. Below a critical density, which varies with slope, flow thickness, grain size and drag coefficient, a turbidity current will 'subside' and deposit its sediment. Above the critical density, the flow will 'explode' (i.e., ignite) and will achieve autosuspension. This latter condition is only predicted for sediment finer than fine sand. Autosuspension has been achieved in small-scale laboratory experiments by Pantin (2001). Application of the theory of Pantin (1979) and Parker (1982) to conditions for initiation of erosive turbidity currents in Scripps Submarine Canyon, California, suggests that ignition of turbidity currents will occur for initial down-channel velocities in excess of about 0.5 m s^{-1} (Fukushima *et al.* 1985), in general agreement with field observations on the conditions required for initiation of flows in this canyon by Inman *et al.* (1976).

1.5.5 Effects of obstacles in the flow path

In natural settings, turbidity currents must flow over or around obstacles. In extreme cases, the obstacle is insurmountable and the turbidity current is reflected back on itself, leading to peculiar deposits with internal flow reversals and grain-size breaks created by successive passes of the depositing flow over the same site (Pickering & Hiscott 1985; Haughton 1994). More commonly, the flow is deflected, or perhaps is forced to decelerate in order to climb over the obstacle. The deviation from the original flow path is greatest when the height of the obstacle is comparable to the flow thickness. Laboratory-scale experiments have been used to study the effects of different obstacle heights and geometries (Kneller *et al.* 1991; Alexander & Morris 1994; Kneller 1995; Morris & Alexander 2003). Deposits are predictably thin over the top of obstacles, but can also thicken abruptly in front of barriers oriented oblique to flow because of hydraulic jumps created by the flow perturbation at the obstruction (Fig. 1.39).

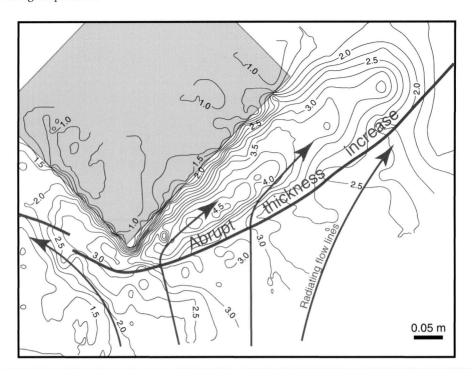

Fig. 1.39 Thickness of the sedimentary deposit in an experimental tank in the presence of an obstacle oriented oblique to the radiating flow. Contours are in millimetres of thickness. The spreading turbidity current (purple vectors) came from the bottom of this map. The obstacle (green) is a wedge with leading edge 2.4 cm high. Initial suspension density was 1.222 g cm^{-3} (10%), maximum head velocity was 24 cm s^{-1}, and grain size was 80 μm. The obstacle induced a hydraulic jump. Downflow of this jump, flow expansion caused a velocity drop and enhanced deposition (beyond the red line marked 'abrupt thickness increase'). Flow vectors (purple) changed sharply at the position of the jump. Redrawn from Morris and Alexander (2003).

1.5.6 Turbidites

Deposits of turbidity currents are called *turbidites*. They show evidence of grain-by-grain deposition from evolved currents (i.e., those in which particles are sufficiently mobile that they can become size-sorted either vertically or laterally in the flow). Such evidence includes size grading, presence of lamination, or organised grain fabric with moderate to low angles of imbrication (Hiscott & Middleton 1980; Arnott & Hand 1989). As explained by Mulder and Alexander (2001), the imprint of fluid turbulence and grain-by-grain behaviour should be stronger in these deposits than the imprint of grain-to-grain interactions (Fig. 1.22). The character of turbidites depends as much on the mode of deposition as on the long-distance transport mechanism (Fig. 1.1). Depositional process is mainly a function of (i) flow concentration near the sediment bed and (ii) rate of deposition. Rate of deposition is dependent on the rate of decrease of both flow *competence* – the coarsest particle that can be transported, and flow *capacity* – the sediment discharge, in units of volume or mass per unit time, integrated over the cross-section of the flow. Competence and capacity both decrease as mean velocity decreases (Hiscott 1994a). Velocity may decrease for any of the following reasons: (i) decreasing bottom slope; (ii) flow divergence; (iii) increased bed friction; (iv) increased particle interaction (intergranular friction); (v) decreasing flow density due to deposition; or (vi) deflection of slow, mud-rich flows by contour currents or by the Coriolis effect so that they are constrained to move roughly parallel to the slope contours rather than down the slope (Hill 1984a). Van Andel and Komar (1969) provide mathematical expressions for the momentum losses in a turbidity current due to grain-to-grain friction, bottom and interfacial friction,

and sediment loss due to deposition. Flows with a high proportion of suspended mud can flow for a greater distance than mud-poor flows without suffering a crippling degree of sediment loss, even on quite gentle slopes (Salaheldin *et al.* 2000). Such flows are therefore more 'efficient' in moving both their mud- and sand-size loads into the basin. The relative 'efficiencies' of turbidity currents have been used by some workers to characterise types of submarine fans (Chapter 7).

The slowing of a turbidity current with velocity u can be expressed mathematically as $du/dt < 0$, where t is time and $du/dt = \partial u/\partial t + u \cdot \partial u/\partial x$. The slowing can be one of, or a combination of, (i) a temporal deceleration at a fixed observation point (non-steady flow with $\partial u/\partial t < 0$, called *waning flow*), or (ii) a spatial deceleration along the flow path (non-uniform flow with $u \cdot \partial u/\partial x < 0$, called *depletive flow* by Kneller 1995 and Kneller & Branney 1995). Deposition is assured by a combination of depletive and waning flow, but other combinations are possible (Fig. 1.40). Five combinations of spatial and temporal deceleration and acceleration (the latter called *accumulative flow* and *waxing flow* by Kneller 1995) may generate deposits with distinctive grading profiles or structural sequences (Fig. 1.41). Accumulative and/or waxing flows can cause erosion and sediment entrainment (e.g., flow ignition), whereas flows with no net acceleration or deceleration (i.e., $\partial u/\partial t + u \cdot \partial u/\partial x = 0$) will bypass, leaving little if any record. The determination of a velocity/time/distance history for a turbidity current (Fig. 1.42) is an instructive way to understand the progress of deposition.

Komar (1985) attempted to use the median size of eleven samples through a mainly laminated and cross-laminated Miocene turbidite to infer the velocity history of the waning turbidity current. He found serious discrepancies between the velocity estimates based on grain

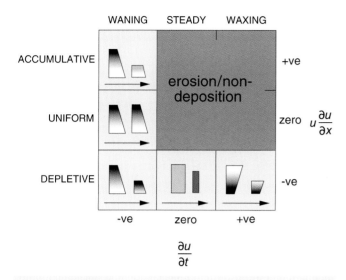

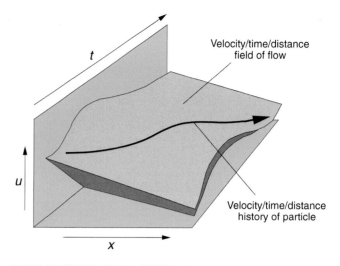

Fig. 1.40 Definition diagram for spatially and temporally accelerating and decelerating turbidity currents. The former accumulate or deplete, whereas the latter wax and wane. Arrows point downflow. Predicted grading profiles are shown, with white = sand and black = mud. Redrawn from Kneller (1995).

Fig. 1.42 Graphical representation of the velocity history of a turbidity current (arrow) as it changes with time (t) and distance along the flow path (x). Redrawn from Kneller (1995).

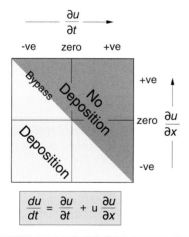

Fig. 1.41 Subdivision of acceleration space into fields characterised by deposition, non-deposition (or erosion) and bypass. Redrawn from Kneller (1995).

size and an estimate based on the transition from upper to lower flow regime conditions. Hiscott (1994a) showed that this discrepancy might be the result of an unjustified assumption that turbidity currents deposit their suspended load because they become incompetent to carry some of that load. Instead, Hiscott (1994a) argued that the fundamental control on deposition is the progressive loss of the *capacity* to maintain a particular sediment discharge as turbulence intensity decreases (see also Leeder *et al.* 2005). Even at velocities considerably higher than those needed to suspend individual particles of a particular size, a turbidity current that is at its full capacity must lose some of this load through deposition if the flow velocity decreases.

As an example, consider a turbidity current carrying a maximum particle size of 1 mm. Using the suspension criterion of $u_* > w$, a flow with shear velocity greater than about 12 cm s^{-1} would be competent

to suspend the coarsest grains (Blatt *et al.* 1980: p. 65). Nevertheless, a flow at full capacity (i.e., holding as much suspended load as turbulence intensity allows), carrying a maximum size of 1 mm, and decelerating from shear velocities of 50 → 20 cm s^{-1} (mean velocities about ~15 → 5.5 m s^{-1}) would drop about 85% of its suspended load (including essentially all of the 1 mm particles). This deposition would not occur immediately, because of a lag between the time of the onset of grain settling and the time at which a particle arrives at the bed. Nevertheless, a large fraction of the original suspended load could be deposited by such a flow, even though it would be at all times competent to carry even the largest grains in suspension. What controls deposition is the natural limit on the *amount* of suspended load that can be carried (i.e., sediment discharge), rather than the limit on the size of the largest particle that can be suspended in a clear flow of the same velocity. Clearly, flow velocities calculated using competence considerations can only provide minimum estimates – a turbidity current might be travelling much faster than such an estimate based solely on competence, yet still deposit a wide range of suspended particles.

Experimental monitoring of deposition from turbidity currents has only rarely been attempted. These laboratory studies are inadequate analogues for deposition from much larger natural flows, but provide some insight into depositional processes. Middleton (1967) monitored deposition of graded beds from both low- and high-concentration flows (i.e., from turbidity currents and concentrated density flows). Low-concentration flows deposited beds with good distribution grading; because of the nature of the experiments, no traction transport took place, and no lamination nor bedforms were generated. Lüthi (1981) studied deposition from unconfined turbidity currents that were free to expand laterally after leaving a narrow entry slot. With increasing distance from the entry slot, velocity and grain size both decreased, and sedimentary structures changed from parallel lamination, to climbing-ripple lamination, to fine parallel lamination. These structures are the same as in vertical sections through ancient turbidites, and correspond respectively to the Bouma (1962) divisions T_b, T_c and T_d.

By far the most detailed observations of the internal characteristics of turbidites come from study of ancient sediments. The Bouma

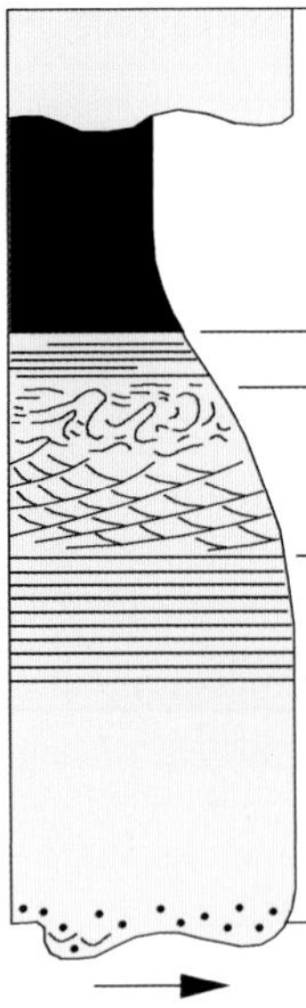

GRAIN SIZE	BOUMA (1962) DIVISIONS	INTERPRETATION
Mud	*e* Laminated to homogeneous mud	Deposition from low-density tail of turbidity current + settling of pelagic or hemipelagic particles
Silt	*d* Upper mud/silt laminae	Shear sorting of grains and flocs
Sand	*c* Ripples, climbing ripples, wavy/convolute laminae	Lower part of lower flow regime of Simons *et al.* (1965)
Sand	*b* Plane laminae	Upper flow regime plane bed
Coarse Sand	*a* Structureless or graded sand to granule	Rapid deposition with no traction transport, possible quick (liquefied) bed

Fig. 1.43 Ideal sequence of sedimentary structures in a turbidite bed (from Bouma 1962, with interpretation from Harms & Fahnestock 1965; Walker 1965; Middleton 1967; Walton 1967; Stow & Bowen 1980).

(1962) sequence (Fig. 1.43) represents a summary of the transitions observed in 1061 beds in the Grès de Peïra-Cava in the Maritime Alps of southern France. In a complete Bouma-type turbidite, the T_a division was probably deposited during a phase of rapid fallout from suspension, without traction. Other divisions have been interpreted in relation to the progression of bedforms observed during deceleration of flow in flumes (Harms & Fahnestock 1965; Walker 1965), but with omission of dunes. Most natural turbidites do not contain all five Bouma divisions, but instead are missing one or more basal division. Beds lacking upper divisions and dominated by structureless sand are more likely the deposits of concentrated density flows (see Section 1.6). The Bouma sequence is a good model for medium-grained deposits of many surge-type turbidity currents (Mulder & Alexander 2001), but is too general for fine-grained turbidites that consist entirely of Bouma's T_d and T_e divisions.

For a completely contrary opinion on the genesis of the Bouma (1962) divisions T_b–T_d, readers are directed to Shanmugam (2000, 2008), who attributes the laminated nature of these divisions to reworking of the tops of sand beds by deep-water bottom currents generated by thermohaline circulation, ocean tides, shoaling internal waves or atmospheric winds. In his view, the original sand beds are more likely to have been deposited by sandy debris flows (our inflated sandflows) than by turbidity currents. Shanmugam (2000) only ascribes the normally graded parts of a basal T_a division to direct accumulation beneath a turbidity current. As one argument, Shanmugam (2000) points to the uncommon occurrence of complete Bouma sequences in nature, and extends this argument further to suggest that if turbidity currents are responsible for the lamination, then there should be documented examples of beds possessing all 16 divisions of the Lowe (1982), Bouma (1962) and Stow and Shanmugam (1980; Fig. 1.47) idealised structural sequences, which those authors have attributed to accumulation beneath decelerating turbulent density currents. To quote: 'The absence of a complete turbidite bed with 16 divisions in the geologic record suggests that the ideal turbidite facies models are wrong' (Shanmugam 2000: p. 316). We dismiss this suggestion because it presupposes that the deceleration of a SGF takes place at a single point, so that the

entire range of sediment grain sizes and the entire spectrum of sedimentary structures might be found in a single core or outcrop. Instead, let us consider a SGF that descends the upper slope as an inflated sandflow, transforming through water entrainment and dilution into first a concentrated density flow and then a turbidity current. The first deposits would consist of the coarsest grain-size fractions, either structureless because of the high rate of fallout, or with structures indicative of particle interaction (e.g., spaced stratification of Hiscott 1994b). This material would be left behind, perhaps covered by a thin veneer of fine fallout of sediment from a residual dilute cloud (likely later eroded by subsequent flows), and the remainder of the flow would bypass this area and move farther downslope to deposit finer sediment with different sedimentary structures and different structural divisions. Hence, Shanmugam (2000) might be theoretically correct that a single SGF could deposit all 16 structural divisions if it had a sufficiently broad range of grain sizes in suspension, but these divisions would logically be strung out for tens to hundreds (or even thousands) of kilometres along the track of the evolving flow, and so would never be found at a single locality. More specific arguments against the interpretation of the laminated Bouma divisions as bottom-current deposits are found in Section 6.3 of this book.

1.5.7 Cross-stratification in turbidites

The general absence of dune-scale cross-stratification in Bouma-type turbidites can be explained in at least five ways.

- Walker (1965) suggested that deposition from turbidity currents is too rapid for bedforms to equilibrate with the decelerating flow. The result is that dunes, which require a significant time to develop because of their size, only begin to form by the time that velocity decreases to values consistent with ripple, not dune, stability. Allen (1969) used a similar argument (Fig. 1.44), that is, for grain sizes available in most turbidity currents 'the range of flow power appropriate to these forms was traversed too quickly, being narrow, to permit their growth' (Allen 1982: II, p. 414).

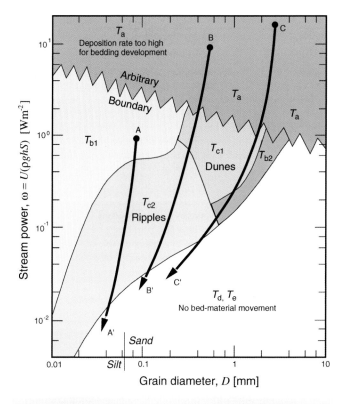

Fig. 1.44 Explanation of Allen (1969, 1982) for the absence of dune-scale cross-stratification in most turbidites. Once the rate of grain fallout declines sufficiently for tractional bedforms to develop, the stability field for dunes has either been bypassed, or there is not enough time for dunes to grow before current ripples become the stable bedform.

This is exacerbated by the fact that high rates of sediment fall-out during the early phases of deposition prevent the formation of lamination (see also Arnott & Hand 1989).

- Walker (1965) suggested that many turbidity currents may be too thin to allow formation of dunes, which only form if the ratio of flow depth to bedform height is about 5:1. For natural flows with velocity maxima below the flow top, and perhaps with strong internal density stratification, Walker (1965) suggested that flow thicknesses of 5–10 m might be necessary for the growth of dunes 50 cm high.

- For the appropriate flow conditions, most turbidity currents are depositing sediment too fine for production of dunes (Fig. 1.44 path AA′; Walton 1967; Allen 1982: II, p. 414). Experiments show that dunes are absent in sediment finer than 0.1 mm (Allen 1982: I, p. 339). This explanation gains strong support from the observation that coarse-grained bioclastic turbidites differ from generally finer grained terrigenous turbidites in that they may contain dune cross-stratification in association with the Bouma T_b division (Hubert 1966a; Thompson & Thomasson 1969; Allen 1970).

- Cross-stratification might be absent due to the occurrence of a hydraulic jump between deposition of the Bouma T_b and T_c divisions. Although upper flow regime planar lamination does not necessarily indicate supercritical flow, it may do so. Even long-wavelength antidune lamination (Hand *et al.* 1972) may be so subdued as to appear flat. Downstream of a hydraulic jump,

flow depth increases and velocity decreases sharply. This rapid velocity drop could cause complete omission of the dune stability field. Upstream migration of the hydraulic jump would result in superposition of ripples on a previously flat sediment bed. This mechanism is restricted to submarine slopes or the upper parts of submarine fans, where hydraulic jumps are believed to occur (Komar 1971).

- High suspended sediment concentrations in depositing currents might extend the conditions for stability of upper flow regime plane beds to lower velocities, so that the eventual transition to the lower flow regime would occur within the stability range for ripples, not dunes (Lowe 1988).

There are no data on the mechanics of deposition from large natural turbidity currents, but it is not improbable that all five explanations for lack of dunes in turbidites are valid in particular cases. It is interesting that increasingly numerous side-scan sonar surveys of submarine fan channels and channel-termination areas have documented the presence of large sand and gravel bedforms (Fig. 1.35), both transverse and parallel to flow (Piper *et al.* 1988; Malinverno *et al.* 1988; Hughes Clarke *et al.* 1990; Wynn *et al.* 2002b). To date, there is no evidence that these bedforms contain cross-stratification like that recognised in outcrops or cores (Piper & Kontopoulos 1994).

1.5.8 Antidunes in turbidites

It is unclear whether supercritical turbidity currents ($F > 1.0$) deposit any diagnostic sedimentary structures. In rivers and flumes, supercritical flows mould antidunes on the bed. These antidunes may migrate either upcurrent or downcurrent to produce internal lamination with low angles of dip (Middleton 1965). Skipper (1971) and Skipper and Bhattacharjee (1978) described cross-bed sets with wavy upper profiles at the base of some thick turbidites in the Ordovician Cloridorme Formation of Quebec, Canada, and interpreted these bedforms as short-wavelength antidunes (wavelength ~65 cm), based primarily on the observation that the foreset dip directions opposed the local palaeoflow as deduced from flutes. The hydrodynamic interpretation of these bedforms proved to be difficult (Hand *et al.* 1972), because antidune wavelength beneath turbidity currents should be of the order of 12 times the flow depth. Pickering and Hiscott (1985) have re-interpreted this part of the Cloridorme Formation as a basin-floor sequence deposited in a constricted foreland basin, in which large turbidity currents were repeatedly reflected and deflected from marginal slopes. Many individual graded beds have sole markings and divisions of cross-stratification or ripple lamination indicating flow reversals during deposition. Grain fabric data obtained by Pickering and Hiscott (1985) from the bedforms described by Skipper (1971) indicate that the depositing current flowed in a direction opposite to that indicated by most flutes in the sequence. These fabric results are supported by experiments of Yagishita and Taira (1989). The bedforms are therefore not antidunes, but instead record the migration of dunes under subcritical flow conditions (Fig. 1.45).

Prave and Duke (1990) and Yagishita (1994) described wavy bedforms with lengths of about 1 m from turbidites, and interpreted these as antidunes. The lamination within the structures described by Yagishita (1994) is *spaced stratification*, which we argue below is produced beneath vigorous unsteady flows. Both of these interpretations have the same problem as the original Skipper (1971) interpretation, in that the bedform wavelength is unreasonably short for antidunes beneath turbidity currents, unless the turbidity currents were strongly stratified. Although the jury is still out on this issue, we suspect that

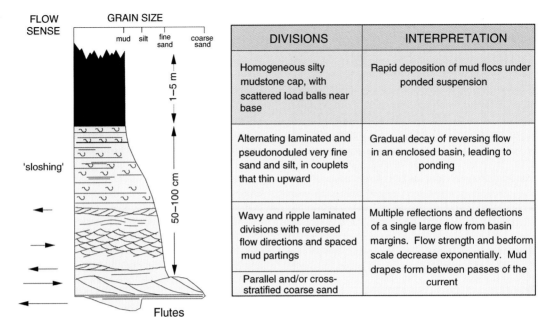

Fig. 1.45   Generalised sequence of internal structures produced beneath large reflected turbidity currents (Pickering & Hiscott 1985). The thick mud cap is deposited after the flow becomes ponded within a confined basin. Single beds of this type may be >10 m thick.

true antidune lamination is either exceedingly rare or is not discernable in deposits because of its long wavelength and low amplitude in turbidites.

Some exceedingly large mud waves on the levées of the Monterey Fan (Fig. 1.46; wave height 2–37 m, wavelength 0.3–2.1 km) have been interpreted as antidunes formed beneath low-concentration, low-velocity turbidity currents ($u \sim 10\,\mathrm{cm\ s^{-1}}$) $\sim$100–800 m thick (Normark *et al.* 1980). Other levée systems and some unconfined slopes are characterised by long-wavelength mud waves that are interpreted to have been initiated as antidunes (Ercilla *et al.* 2002, Normark *et al.* 2002). The layering in these mud waves would appear flat in both deep-sea cores and outcrops. Younger fine-grained turbidites and hemipelagic sediments may drape the wave forms and maintain their morphology, well after conditions of supercritical flow cease to occur.

1.5.9   Turbidites from low-concentration flows

The deposits of low-concentration, mud-load turbidity currents have been described by Piper (1978), Stow (1979), Stow and Bowen (1980), Stow and Shanmugam (1980) and Stow and Piper (1984b). The Bouma (1962) divisions for sandy turbidites are too general for mud turbidites; many beds contain only Bouma's T_d and T_e divisions. Several schemes to further subdivide mud turbidites have been proposed, and are outlined in Figure 1.47. Stow and Shanmugam (1980) recognised nine divisions, numbered T_0 to T_8. The physical structures in these nine divisions are believed to result from suspension fall-out and traction (T_0–T_2), shear sorting of silt grains and clay floccules in the bottom boundary layer (T_3–T_5) and suspension fall-out with no traction (T_6–T_8). Division T_0 corresponds to Bouma's division T_c. As with Bouma-type turbidites, complete structure sequences are unusual; top-absent, base-absent and middle-absent sequences are common (Stow & Piper 1984b). The very thin, regular laminations of division T_3 have been attributed to shear sorting near the base of the

flow and then alternating deposition of silt grains and clay particles by settling through the viscous sublayer (Stow & Bowen 1980; Kranck 1984). According to this model, as the silt grains and clay floccules fall toward the bed, the increased shear in the boundary layer causes the clay flocs to break up (Fig. 1.48a). The silt grains then settle through the viscous sublayer to form a silt lamina (Fig. 1.48b). As more sediment is supplied to the top of the boundary layer, the mud concentration builds up, and some reflocculation may occur (Fig. 1.48c). At some critical concentration, the clays are able to form sufficiently large aggregates that they escape disaggregation by shear stresses and are able to settle rapidly through the viscous sublayer to form a mud lamina (Fig. 1.48d). The cycle of silt and mud deposition is then repeated for successively finer grain sizes.

A second interpretation for fine-scale lamination in fine-grained turbidites has been proposed by Hesse and Chough (1980). The silt-rich laminae are believed to form during periods when bursts and sweeps repeatedly disturb the viscous sublayer of the flow, resulting in transport of clay-sized particles away from the boundary. The composite nature of individual silt laminae is used as support for this model. Clay-rich laminae are attributed to periods when bursts and sweeps are suppressed, perhaps by passage of large eddies. Alternatively, flow pulsing (Best *et al.* 2005) might account for the fluctuations in the viscous sublayer.

Stow *et al.* (1990) has proposed the term *hemiturbidite* for subtly banded, largely bioturbated muds on the distal Bengal Fan. He interprets these muds to represent deposition during the waning stages of turbidity current flow, probably over time periods from a few weeks to a few months for a 1 m-thick unit. *Flow lofting* (that is, detachment of the final dilute suspension from the seafloor because of positive buoyancy) produces the final cloud of suspended clay and mud that settles to produce the uppermost mud layer. On the Amazon Fan, Damuth and Kumar (1975a) described grey 'hemipelagic' muds that were interpreted to have been derived from dilute, thick, mud-laden turbidity currents spreading over the entire

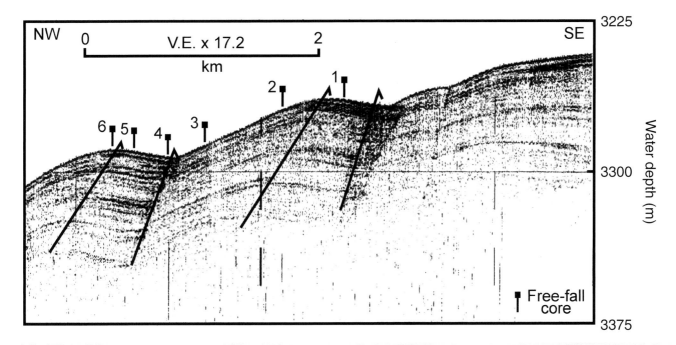

Fig. 1.46 3.5-kHz profile of sediment waves on western levée of Monterey Fan valley. The levée crest is 20 km to the right. Arrows show migration of sediment-wave crests and troughs. From Normark *et al.* (2002).

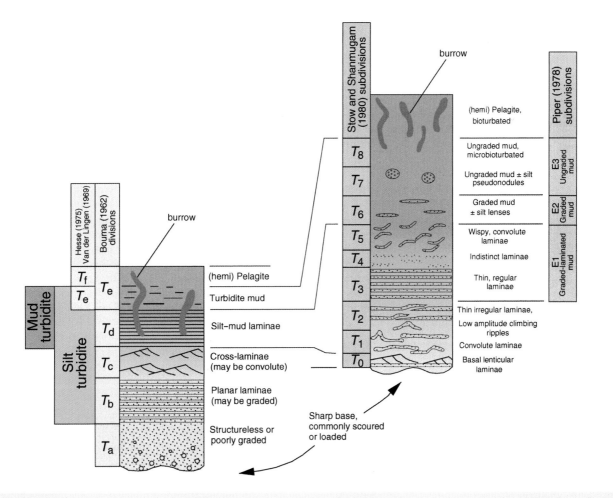

Fig. 1.47 Summary of schemes for subdivision of fine-grained turbidites, based on Hesse (1975), Piper (1978), van der Lingen (1969) and Stow and Shanmugam (1980).

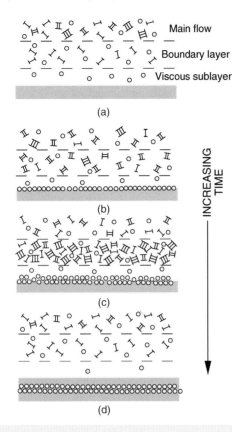

Fig. 1.48 (a–d)Schematic representation of the four stages of silt and mud deposition through the boundary layer of a turbidity current to form silt and mud laminae. From Stow and Bowen (1980).

fan. These muds are probably analogous to the hemiturbidites of Stow *et al.* (1990).

1.5.10   Downcurrent grain size – bed thickness trends in turbidites

Turbidites derived from a single source with a consistent range of grain sizes have been shown to exhibit a well-defined relationship between bed thickness and grain size (Fig. 1.49) (Sadler 1982). Sadler's curves indicate a downslope trend from (i) proximal, relatively thin, medium-grained beds, to (ii) intermediate, relatively thick, somewhat coarser grained beds, to (iii) progressively finer-grained and thinner beds with increasing distal transport. The 'distal limb' of this trend is relatively easy to explain. Sadler (1982) calculated that 'for a given flow density and slope and with a uniform grain-size distribution, bed thickness becomes proportional to the square root of bed shear stress cubed'. Hence the decrease of both competence and capacity as turbidity currents decelerate leads to a related decrease of both grain size and bed thickness. The 'proximal limb' of the trend involves a temporary downslope increase in both grain size and bed thickness, and requires a special explanation. Sadler points to experiments of Kuenen (1951) that show a longitudinal decline of density and competence from the head to the tail of a turbidity current. As a flow that is in a state of autosuspension decelerates, deposition will begin beneath the tail of the current where competence and flow density are relatively low. 'The most proximal part of the deposit will therefore thicken and coarsen downcurrent until the point at which all of the flow, except a small region near the head, will be depositing sediment' (Sadler 1982: p. 48).

Dade and Huppert (1995) also predicted that deposit thickness first increases and then decreases for beds laid down on basin plains. They give a different explanation than Sadler (1982) for this geometry: 'At early times, the rapidly moving body of the turbidity current passes over a point very quickly and so little mass is laid down despite high rates of deposition … At large times, duration of passage of the surge over a fixed point is much longer, but the slowly moving current has relatively little to deposit.'

Dade *et al.* (1994) developed a mathematical model that relates deposit thickness for turbidites laid down on slopes to flow velocity,

$$BT = \frac{U_{B} w_{s} \cos\alpha}{kF^{2} g'_{b}} \tag{1.9}$$

where w_{s} = average settling velocity, BT = bed thickness, $g'_{b} = g\phi_{b}[(\rho_{clasts} - \rho)/\rho]$, ρ = seawater density, ϕ_{b} = volume fraction of solids in the bed, α = slope gradient, U_{B} = surge velocity, k = (surge height)/(surge length) and F = Froude number (nearly constant except near the distal point of deposition). In this case, BT is proportional to the square root of bed shear stress (because $\tau_{o} \propto [U_{B}^{2}]$), not $\tau_{o}^{3/2}$ (Sadler 1982).

1.5.11   Time scales for turbidite deposition

Time scales for deposition can be looked at in three ways:

(1) the time taken to deposit a turbidite at one location on the seafloor;
(2) the total time that it takes for a turbidity current to travel from the place where it was initiated to the place downslope where it loses its identity;
(3) recurrence intervals (or frequencies) of turbidite emplacement.

The time taken to deposit a turbidite at one location on the seafloor is similar to, but potentially less than, the time taken for the turbidity current responsible for the deposit to pass a particular cross-section. If the head and frontal part of a turbidity current are erosional or non-depositional, then the deposit might form only during a short part of the total transit time. At the extreme, it might take several weeks for the cloud of fine-grained suspension to dissipate after passage of a large turbidity current, particularly in confined basin plains (Pickering & Hiscott 1985). This recovery phase for the return of the local water column to its pre-flow condition is not generally considered when evaluating the time to deposit turbidites at a single site. Instead, accumulation times for sand/silt divisions, commonly Bouma divisions T_{a} through T_{d}, are assessed. Arnott and Hand (1989) studied the development of sedimentary structures and fabric under experimental conditions of a steady rain of suspended sediment to the bed. They determined a linear relationship between imbrication angle in degrees, β, and the bed aggradation rate in cm min^{-1}, x; that is

$$\beta = 10.7 + 1.36x \tag{1.10}$$

For some Ordovician turbidites from the Appalachians, this relationship gives aggradation rates of 7–12 cm min^{-1}, and a requirement of 1.25 h to deposit a 9.1 m-thick sand bed described by Hiscott and Middleton (1980). Allen (1991) used aggradation rates published for

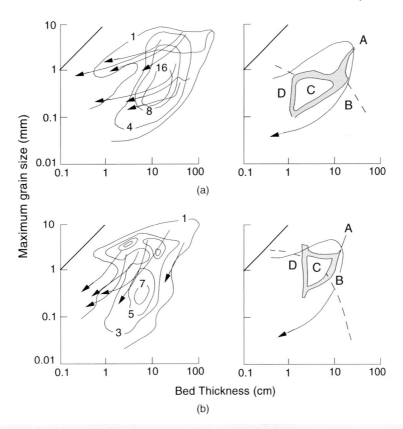

Fig. 1.49 Maximum grain size plotted against bed thickness for Lower Carboniferous turbidites of (a) the Rhenaer Kalk (498 beds), and (b) the Posidonienkalk (235 beds), from Sadler (1982). The left-hand diagrams show contours of density of data points representing total bed thickness and corresponding maximum grain size. Arrows are selected vertical grading curves using the x-axis scale as a 'distance to top of bed'. The right-hand diagrams show fields for Bouma divisions T_a through T_d found at the base of the beds for which maximum size was determined. Green zones = overlap of fields for divisions T_b, T_c and T_d. Dashed line = lower limit of field for division T_a, which has extensive overlap with the other fields. Arrow = modal horizontal grading curve, based on contour pattern in left-hand diagram.

the transition from divisions T_a to T_b (Arnott & Hand 1989) and the transition from division T_b to climbing ripple lamination of division T_c (Allen 1971) to estimate a few tens of minutes for deposition of the 27 cm-thick turbidite described by Komar (1985). Piper *et al.* (1988), using deposit volume and channel discharge estimates for the 1929 Grand Banks event, concluded that the turbidity current must have taken 2–3 h to pass a particular channel cross-section on its way to the Sohm Abyssal Plain. Jobe *et al.* (2012) calculated sedimentation rate and accumulation time for 44 climbing-ripple cross-laminated intervals from three field areas using TDURE (a mathematical model developed by Baas *et al.* (2000)). For T_c divisions and T_{bc} beds averaging 26 cm and 37 cm thick, respectively, average climbing-ripple cross-lamination and whole-bed sedimentation rates were 0.15 mm s^{-1} and 0.26 mm s^{-1} and average accumulation times were 27 min and 35 min, respectively.

For relatively compact surge-type flows, the time required for a turbidity current to travel from the place where it was initiated to the place downslope where it loses its identity because nearly all sediment has been deposited is approximately the total path length divided by the time-averaged head velocity. Time scales for large natural currents are a few days (e.g., Dade & Huppert 1995 for the Black Shell turbidite). Timing of cable breaks indicates that the 1929 Grand Banks turbidity current took about one day to reach about 34°N on the Sohm Abyssal Plain (Doxsee 1948; Piper *et al.* 1988).

Turbidity currents transiting the ~800 km-long Amazon Channel crossing the Amazon Submarine Fan (Chapter 7) at average speeds of about 1 m s^{-1} (Pirmez 1994) would take about 9 days to complete their journey. A slow-moving mud-load turbidity current crossing the Navy Fan in the California borderland flowed for about 6 days (Bowen *et al.* 1984). Some turbidity currents are believed to be semi-continuous events, either during extended periods of peak river discharge for hyperpycnal flows, or when industries discharge suspensions over a period of weeks to months (Hay 1987b).

The frequency of turbidity currents can be estimated by dividing the age of a stratigraphic column by the number of turbidites it contains. This procedure requires that there has been no seafloor erosion. Only those turbidity currents which reached the site and deposited sediment there can be included, so this procedure underestimates the total number of turbidity currents generated by upslope failures, many of which might have followed a path leading away from the sampling site. In successions containing beds of different thicknesses, the frequencies of deposits of different thickness can be evaluated separately, although the small number of some beds can make frequency calculations untrustworthy.

There is a wide reported range of recurrence intervals, dependent both on variable rates of sediment supply and triggering mechanisms for initial failures. For example, thin Pleistocene turbidites forming colour-banded muds in the levées of the Amazon Fan, which accumulated at rates as high as 25 m kyr^{-1}, are potentially near-annual in their frequency (Hiscott *et al.* 1997b). Piper and Normark (1983) estimated recurrence frequencies of 1–1000 yr for turbidity currents reaching different parts of the Navy submarine fan. Simm *et al.* (1991) determined a frequency of about 25 kyr for large fine-grained turbidity currents reaching the Madeira Abyssal Plain during the past 127 kyr. For seven bed-thickness classes from 1–3000 cm in an Oligocene forearc-basin succession from the western Pacific, Hiscott *et al.* (1993) determined turbidity-current recurrence intervals of 3–1000 yr.

1.6   Concentrated density flows and their deposits

Experimental and theoretical understanding of concentrated density flows lags far behind the understanding of turbidity currents. The high concentrations (~15–40% by volume, Fig. 1.21) that lead to distinctive deposits might be limited to the lower parts of stratified flows which, farther above the bed, are less dense and more turbulent. The features seen in the deposits of these flows certainly demonstrate significant grain-to-grain interaction and imply a suppression of turbulence near the sediment bed (Fig. 1.20). According to Mulder and Alexander (2001) and others, depositional processes include: (i) rapid en masse deposition of a quick bed (Middleton 1967) due to an increase of intergranular friction; (ii) generation of inverse grading at the base of the flow because of grain collisions and dispersive pressure; (iii) formation of *spaced stratification* (formerly called *traction carpets*) under unsteady flows (Hiscott & Middleton 1979, 1980; Hiscott 1994b; Sohn 1997); (iv) grain-by-grain deposition with little subsequent traction transport (Arnott & Hand 1989); (v) alternate deposition of bedload and suspended load (Walker 1975a); (vi) deposition of mudclasts well above the base of the bed and (vii) expulsion of pore-fluids when the primary, unstable, open-grain packing collapses, generating fluid-escape structures (Lowe & LoPiccolo 1974).

Because the transition between a turbidity current and a concentrated density flow depends on the relative importance of fluid turbulence and other mechanisms in providing particle support, it is possible for a single current to undergo one or more transition between these flow types during its history. This might occur if an undulatory seabed, or variable amount of flow confinement, leads to alternating increases and decreases in speed along the flow path (Kneller & Branney 1995; Fig. 1.42). Sharp decreases in speed would induce higher rates of particle fallout and intensified grain-to-grain interactions in the lower part of the flow. If these alternative support mechanisms dominate the depositional phase, then the deposits will reveal structures characteristic of deposition from a concentrated density flow rather than a turbidity current. Hence, it should always be remembered that these two flow types might be transitional when making interpretations about depositional environment and deposit distribution.

1.6.1   Deposits from concentrated density flows

In experiments on deposition from high-concentration flows, Middleton (1967) demonstrated rapid, en masse deposition leading to the formation of a 'quick bed' that was easily deformed by Helmholtz waves at the top of the deposit. The deposit was characterised by coarse-tail grading. Field studies have revealed sedimentary structures not present in the Bouma (1962) sequence for turbidites. One of these was initially called traction-carpet stratification, but has since been renamed *spaced stratification* by Hiscott (1994b). The term *traction carpet* (Dzulynski & Sanders 1962: p. 88; Lowe 1982 division S2) was used to describe generally 5–10 cm-thick, inversely graded stratification found in some concentrated density-flow deposits (Fig. 1.50). As originally envisioned, clasts in such carpets were believed to be sheared as a dense dispersion just above the bed, maintained by dispersive pressure (Bagnold 1956), until a critical thickness was reached that prompted en masse deposition. Within each such deposit, intense grain interaction was believed to be responsible for both a basal inverse grading, and a strong alignment of grains with imbrication angles commonly >20° (Hiscott & Middleton 1979).

Hiscott (1994b) identified serious flaws in the physical model which had been used to explain shearing of a *traction carpet* beneath a sediment gravity flow. Specifically, shear stress should be constant throughout such a traction carpet (cf. Sumer *et al.* 1996; Pugh & Wilson 1999), so there is no reason to expect deposition to take place in a series of steps as each carpet individually 'freezes' from the top downward, as was proposed by Hiscott and Middleton (1979) and Lowe (1982). Legros (2002) subsequently discounted dispersive pressure as a viable mechanism to form the observed inverse grading. In addition, shear-cell experiments (Savage & Sayed 1984) suggest that even large shear stresses cannot shear a mobile bed to depths greater than about ten grain diameters. Hiscott (1994b) proposed a more descriptive term for the inversely graded deposits: *spaced stratification*. He interpreted this type of stratification to result from strongly fluctuating hydrodynamic conditions and vigorous burst–sweep cycles beneath large turbidity currents as they rain sediment onto the seabed. At the base of each stratum, inverse grading is attributed to a 'kinetic sieve' mechanism (Middleton 1970; Savage & Lun 1988) whereby small particles in a thin agitated dispersion are able to filter through the voids between larger particles, eventually dominating the lower part of the sheared layer. The conclusion that traction carpets are not the explanation for spaced stratification is supported by Sohn (1997), who demonstrated that deposition from traction carpets must occur by progressive aggradation rather than frictional 'freezing' of a sheared layer. This style of deposition will leave no layering in the deposit, and in sandy sediments is unlikely to produce inverse grading (Sohn 1997).

Sedimentary structure sequences for the deposits of concentrated density flows have been presented by Aalto (1976), Hiscott and Middleton (1979), Hein (1982), Lowe (1982) and Mulder and Alexander (2001). The analysis of Hiscott and Middleton (1979) is based on a transition matrix (9 states, Table 1.4) for 214 medium- to coarse-grained sandstone beds (mean thickness 380 cm) from inferred submarine-fan deposits (Hiscott 1980) of the Tourelle Formation of Québec, Canada. The preferred transitions, inferred from Markov Chain Analysis, are presented in Figure 1.51 with process interpretations. These transitions differ somewhat from those presented by Hiscott and Middleton (1979), due to a change in the method of calculation of the independent trials matrix (modified procedure includes iterative fitting of row and column totals; Powers & Easterling 1982). Except for some basal scour-and-fill structures, the first deposits accumulated by rapid en masse deposition, producing either a division of structureless coarse- to medium-grained sand or, in the case of highly unsteady flow, a division of spaced stratification. There is no evidence of grain-by-grain traction on the bed during the early stages of deposition, perhaps because the development of planar stratification is suppressed if the rate of bed aggradation

Fig. 1.50 *Spaced stratification* within a graded bed of coarse to medium sandstone. The base and top of the bed are not shown. Each 5–10 cm-thick stratification band is inversely graded at its base and then essentially structureless and ungraded above. The sandstone has a strong grain fabric with high grain imbrication angle. Scale divisions are 10 cm.

Table 1.4 Transition matrix for 214 thick sandstone beds, Tourelle Formation

'State'	Number of upward transitions								
	1	**2**	**3**	**4**	**5**	**6**	**7**	**8**	**9**
1 Basal scour	0	155	20	2	0	0	0	0	37
2 Structureless ('Massive') or graded	135	0	10	13	21	10	9	12	23
3 Internal scours	4	27	0	0	0	0	0	0	1
4 Planar lamination	7	1	0	0	5	1	2	0	0
5 Current-ripple lamination	25	0	0	0	0	0	0	3	0
6 Convolution	12	0	0	0	0	0	1	0	0
7 Cross-stratification	9	0	0	2	1	0	0	0	0
8 Muddy lamination	16	0	0	0	0	0	0	0	0
9 Spaced stratification	6	50	2	0	1	2	0	1	0

exceeds $\sim 4\,cm\,min^{-1}$ (Arnott & Hand 1989). At the top of most beds, deceleration of the more dilute tail of the current produced a sequence of structures like the Bouma sequence. In other examples, the dilute tail of the current reworked the sandy top of the initial deposit into large ripples or dunes, and then presumably continued down the fan surface to deposit its fine load at more distal sites (cf. Lowe 1982).

Hein (1982) documented sequences of sedimentary structures in channelised conglomerates, pebbly sandstones and coarse sandstones that are similar to those recognised by Walker (1975a) for conglomerates and Hiscott and Middleton (1979) for sandstones. Important additional divisions recognised by Hein (1982) are those bearing the imprint of syn- and post-depositional fluid escape; dish structures are particularly abundant in these rocks. Also, the pebbly sandstones of Hein (1982) contain more cross-stratification than was observed in finer deposits by Hiscott and Middleton (1979). Hein (1982) interpreted some of this cross-stratification as the product of reworking of the tops of the deposits by dilute flows that spilled into the submarine

channel from other, nearby channels (see Section 7.4.5, flow stripping process of Piper & Normark 1983).

Soh (1987) studied the grain fabric of thickly bedded graded conglomerates from the Miocene–Pliocene of central Japan and concluded that the main transport process just before deposition changed from grain shearing, dominated by particle collisions near the base of flows, to viscous current drag near flow tops, where concentrations were lower.

A general theme that runs through these descriptions is an early phase of deposition dominated by grain-to-grain interactions and pore-fluid escape, followed by a transition to more turbulent conditions and slower, selective deposition. The implication is that these flows were density stratified with turbulence suppression limited to the lower parts of the flows. It is unclear whether there was a sharp interface between the different parts of these flows, whether the pronounced flow stratification was long-lived, or whether it was mainly an outcome of intense particle fallout as deposition started.

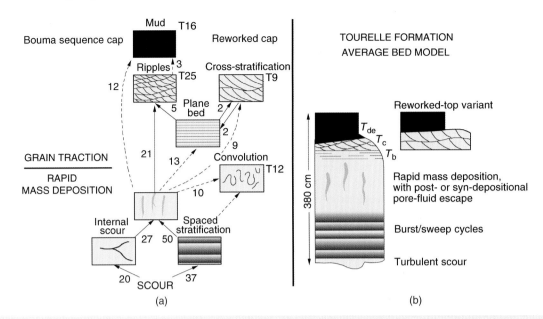

Fig. 1.51 (a) Preferred transitions between nine 'states' in 214 thick sandstone beds (average thickness 380 cm) deposited from concentrated density flows, Tourelle Formation, Québec, Canada. Numbers indicate the number of transitions between states; T25 signifies that a rippled division is the highest division in 25 beds. Complete beds (tops preserved) only occur 62 times – the other 152 beds had their tops bevelled during amalgamation. Structural divisions were produced either by rapid mass deposition ('freezing') or by selective grain-by-grain deposition with traction transport. In some cases, post-depositional reworking of the top of the bed is indicated by medium-scale cross-stratification. Arrow weight indicates the statistical significance of transitions (based on method of Powers & Easterling 1982: p. 922) with the levels of significance being >98% (solid arrow) and >93% (dashed arrow). No arrows are shown if there is only one transition between states, regardless of significance level. (b) Generalised bed model based on the transition diagram, with interpretation. Where appropriate, structural divisions have Bouma (1962) T_{a-e} labels.

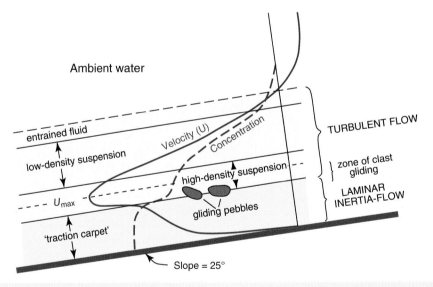

Fig. 1.52 Mechanism proposed by Postma *et al.* (1988) for the transport of large mud clasts within a strongly stratified SGF. At 25°, the slope for the experimental demonstration of this effect is unreasonably high for natural SGFs.

1.6.2 Large mud clasts in concentrated density-flow deposits

Many concentrated density-flow deposits contain large mud clasts, either dispersed well above the base of the bed, or concentrated into clusters or trains (Walker 1985; Postma *et al.* 1988; Shanmugam *et al.*

1995). Such clasts are commonly described to be 'floating' in the bed, and for this reason proposals have been made to explain how shale clasts can be suspended by sediment gravity flows, or by stratified two-layer flows. Postma *et al.* (1988) employed flume experiments on high slopes of 25° to show how large clasts can be carried along the interface between a highly concentrated, laminar, basal inertia-flow, and a less concentrated, fully turbulent overriding flow (Fig. 1.52). They proposed that larger natural two-layer flows could operate in the same way on much lower slopes after undergoing a 'gravity transformation' (Fisher 1983). Shanmugam *et al.* (1995) instead interpreted 'floating' clasts in structureless sands as a strong indicator that the deposits were emplaced by inflated sandflows (his sandy debris flows), not turbidity currents or concentrated density flows. Hiscott *et al.* (1997a) disputed this generalisation, noting that concentrated density flows are deposited from the base upward through progressive aggradation (Kneller & Branney 1995), so that part way through the depositional phase the seafloor corresponds to some level well above the base of the eventual deposit. If a mud clast is rolling or sliding along the bed and then ceases to move, it will subsequently be buried by sand yet will appear to be suspended in sand. Outsized mud clasts rolling and sliding beneath a concentrated density flow might be expected to reach their final resting place well after deposition of some thickness of sand, because they would travel much more slowly than the mean velocity of the current (Hand & Ellison 1985).

Recent interpretations of 'hybrid event beds' (Haughton *et al.* 2009) explain large floating clasts in deposits having a graded sandy basal division as the product of flow transformation ('gravity transformation') from a turbulent density flow to a debris flow. The basal part of such deposits records accumulation from a turbulent flow, so is organised and graded (Fig. 1.28a). The clast-rich interval is more muddy and disorganised, because it was deposited from a cohesive late-stage phase of the flow capable of carrying large clasts. Alternatively, the large clasts might have been carried by a co-genetic debris flow that was triggered at the same time as a concentrated density flow (Fig. 1.29), leading to stacked deposits comprising a single event bed with large mud clasts in the upper part (Haughton *et al.* 2010).

1.7 Inflated sandflows and their deposits

Inflated sandflows are envisaged to range in density from ~40–70% by volume (Fig. 1.21), most of which is coarse silt or larger in size, and the bulk of which is sand. Variable amounts of gravel can also be present. Cohesion is minor (Figs 1.19 & 1.22), and voids between particles are well connected. Mulder and Alexander (2001) explain that this facilitates the ingesting of seawater and therefore the transition of many inflated sandflows to concentrated density flows. Inflated sandflows lack significant cohesive strength, but do possess frictional strength because of grain-to-grain interlocking when concentrations become very high, for example as velocity decreases and deposition starts. As a result, deposition from inflated sandflows takes place by frictional 'freezing'.

Even though cohesion is not a significant contributor to particle support, a small amount of interstitial mud, or very poor sorting, are required to permit these sandflows to operate on low slopes. Mud, or the very small pore throats that characterise poorly sorted sand, will dramatically reduce the permeability of the mobile sediment and therefore slow the dissipation of excess pore-fluid pressures that contribute to particle support (Fig. 1.19). Without the elevated pore-fluid pressures or cohesion provided by small amounts of clay, inflated sandflows would depend entirely on grain interaction for maintenance of the suspension, and would be pure grain flows. Grain flows, as noted earlier, cannot operate on slopes less than ~13°. As pointed out by Mulder and Alexander (2001), as little as 2% by volume clay will permit flow on low slopes if the remaining sediment is fine-grained sand. A maximum of ~20–25% by volume clay is required at water contents of 25–40% by volume if the rest of the

sediment is coarse-grained sand ± gravel (Hampton 1975; Marr *et al.* 2001). In natural poorly sorted mixtures of sand, the fine-grained sand between any larger sand grains will ensure that no more than ~2% by volume clay is required to significantly reduce the rate of pore-fluid escape.

When clay content is low and the sandy sediment is poorly sorted, inflated sandflows owe their mobility to the extended persistence of elevated pore-fluid pressures, and to grain-to-grain interactions. When mud forms ~10–25% of the detrital fraction, the flows may resemble the 'slurry flows' described by Lowe *et al.* (2003). As an explanation for the long runout distances seen in the more muddy sandflows, Mulder and Alexander (2001) point to the increased coherence of the flow mass that results from the presence of cohesive particles (i.e., clays). 'Coherence' is the ability of a mixture to hold together and support sediment (Marr *et al.* 2001). For flows with low coherence (i.e., low clay contents), the rheological strength of the material is unable to fully resist the dynamic stresses generated by the flow (Marr *et al.* 2001), leading to uninhibited internal deformation.

Inflated sandflows are more susceptible to 'shear mixing' with ambient seawater than clay-rich cohesive flows, because of their lower coherence and connected voids (Marr *et al.* 2001; Talling *et al.* 2002). Shear mixing is the erosion of the front and top of a SGF because of its movement beneath ambient fluid. This erosion produces a secondary dilute turbidity current (Hampton 1972).

1.7.1 Deposits of inflated sandflows

The high particle concentrations do not permit selective deposition and the formation of lamination by traction transport. Grain sorting cannot occur by the differential settling of coarser and finer grains (Mulder & Alexander 2001); hence, deposits lack normal grading or only show poor coarse-tail grading (Marr *et al.* 2001). They may be inversely graded, however, because of grain-to-grain interactions or because the base of the flow undergoes greatest and most prolonged shear, leading to reduced competence (Hampton 1975). The dissipation of elevated pore-fluid pressures often produces fluid-escape structures like dish structures and fluid-escape pillars. In more muddy deposits, a minor but significant amount of cohesion facilitates the development and preservation of a variety of shear laminae and banding, and deformation structures (Lowe *et al.* 2003). If a turbidity current is generated by shear mixing at the upper interface of the inflated sandflow, an organised turbidite might directly overlie the primary deposit, but more likely will accumulate more distally on lower slopes (Marr *et al.* 2001).

Although frictional 'freezing' is the cause of deposition, Mulder and Alexander (2001) conclude that the entire flow does not deposit en masse. They point to the experiments of Major (1997) on subaerial frictional flows, in which deposits formed by the successive accretion of thin layers which ultimately formed a thicker deposit. However, thick subaqueous flows might behave differently, and 'freeze' from the top downward as shear stress drops below the frictional strength of the material. Because the shear stress is lowest toward the flow top, the first grain interlocking and arrested internal motion might take place there, leading to the formation of a semi-rigid plug which would thicken downward as deceleration continues.

Branney and Kokelaar (2002) produced a set of synthetic bed models for deposition from mainly cohesionless gravity flows (Fig. 1.53). The main variables that distinguish the flows responsible for these deposits are flow concentration, shear rate and deposition rate (Fig. 1.54). For relatively low concentrations in turbidity currents, high shear rates produce tractional structures whereas low shear

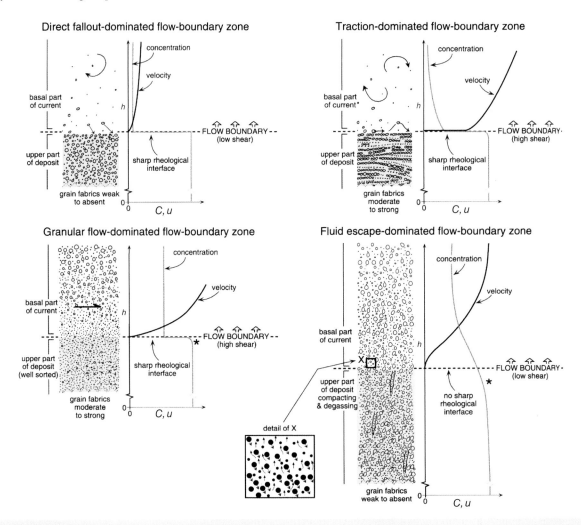

Fig. 1.53 Relationship between deposit characteristics, shear rate and flow concentration in sandy gravity-flow deposits. From Branney and Kokelaar (2002).

rates with fine-gained suspensions can lead to little traction and structureless deposits. For relatively high concentrations, high shear rates at the accumulating bed produce a sharp interface and strong grain fabrics, whereas low shear rates across a diffuse depositional boundary (perhaps because of strong pore-fluid escape and elutriation) lead to weak fabrics in deposits with signs of syndepositional liquefaction and fluid escape.

1.8 Cohesive flows and their deposits

1.8.1 Definitions and equations of flow

Cohesive flows are characterised by strong coherence, so that they hold together while flowing along the seabed (Marr *et al.* 2001). The coherence is created by cohesive forces between electrostatically charged clay minerals. Mulder and Alexander (2001) distinguish between *mudflows* and *debris flows*. Their mudflow deposits have less than 5% by volume gravel and more mud than sand. Their debris-flow deposits consist of more poorly sorted sediment (>5% by volume gravel with variable sand proportion) and may include boulder-sized

clasts of soft sediment or rock and very large sediment rafts. The name 'debris flow' commonly is used for both a flow process and a deposit. This usage generally does not create problems because the context makes the meaning clear. For clarity, however, we will henceforth use the term 'debris flow' for the process only, and *debrite* for the deposit.

Cohesive flows are only well described from modern subaerial settings (Johnson 1970, 1984; Pierson 1981; Takahashi 1981; Middleton & Wilcock 1994). According to Takahashi (1981: p. 58), cohesive flows are flows 'in which the grains are dispersed in a water or clay slurry with the concentration a little thinner than in a stable sediment accumulation … [and] in which all particles as well as the interstitial fluid are moved by gravity'. Subaerial cohesive flows are capable of transporting boulders up to 2.7 million kg (Takahashi 1981), have bulk densities in the range of 2.0–2.5 g cm^{-3}, and may move at speeds of 20 m s^{-1}. Catastrophic submarine cohesive flows may carry (or push) enormous slabs weighing up to about 2300 million kg (immersed weight, Marjanac 1985).

Experiments and theory (Hampton 1975, 1979; Rodine & Johnson 1976) suggest that only a small amount, about 5%, of interstitial matrix – mud + water slurry – is required to allow flow on surprisingly gentle slopes. The matrix serves several functions. It (i) lubricates the

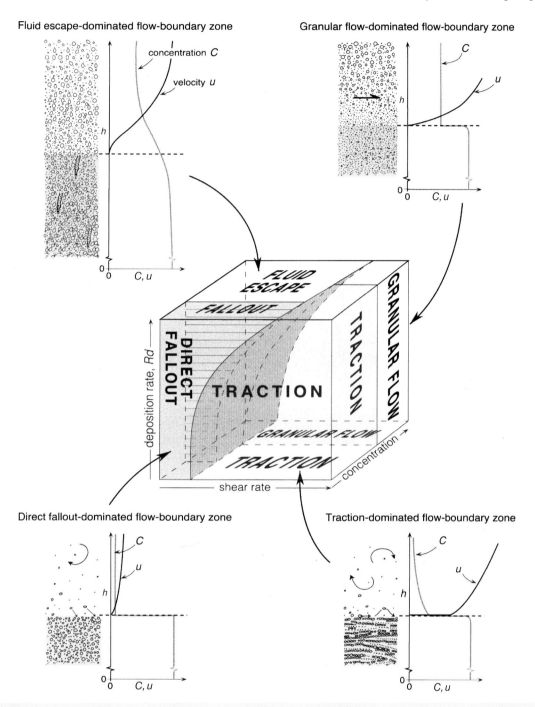

Fig. 1.54 Process cube to explain deposits from sandload SGFs. High concentrations (the two upper scenarios) form typical deposits of inflated sandflows and some concentrated density flows. From Branney and Kokelaar (2002).

larger clasts so that they are able to slide past one another, (ii) provides buoyant support for clasts of only slightly higher density and (iii) commonly exhibits elevated pore pressures that increase buoyancy and the lower frictional resistance to flow (Pierson 1981; Ilstad *et al.* 2004a).

In terms of rheology, cohesive flows resemble wet concrete and exhibit *strength*, which can be divided into two components: cohesive strength due to electrostatic attractions between clay-size particles, and frictional strength due to interlocking and surface contacts between clasts and between the flow and its bed. According to Pierson (1981), frictional strength far exceeds cohesive strength as a mechanism of clast support. Strength allows cohesive-flow deposits to stand up in relief above its surroundings, with steep meniscus-like margins called *snouts*. Strength, and buoyancy of clasts in the matrix, permits the flow to carry clasts above the bed that are more dense than the bulk density of the flow itself.

Materials with strength will not deform until a critical yield strength is exceeded. They therefore behave as *plastic* materials. Once deformation (flow) begins, laminar flow generally prevails in those parts of the cohesive flow where the critical yield strength is less than the shear stress. Elsewhere, friction and/or cohesion resist deformation. Particle support during flow comes from a combination of (i) frictional resistance to settling through finer matrix, similar in explanation to the kinetic sieve process described by Middleton (1970); (ii) matrix cohesion or strength that is not exceeded by the downward-directed force exerted by dispersed clasts; (iii) buoyancy (partial support only); (iv) elevated pore pressures in the cohesive matrix and (v) dispersive pressure (Bagnold 1956, 1973; Pierson 1981). Lowe (1982) believes that, in many cases, the largest clasts are not actually suspended, but remain in contact through rolling, sliding and intermittent bouncing downslope.

Cohesive flows move as a result of deformation in a basal zone of high shear stress. Lower shear stresses high in the flow, or at the flow margins, do not always exceed the yield strength of the material, so that the upper part of the flow may be rafted along as a semi-rigid plug (Johnson 1970). As total shear stress decreases (i.e., as bottom slope decreases), or as intergranular friction increases, the semi-rigid plug thickens by downward growth until the entire mass ceases to move ('freezes') when bed shear stress declines to a value lower than the yield strength. Likewise, low shear stresses at flow margins result in marginal 'freezing' and construction of levées. Johnson (1970) favoured either a Bingham plastic or Coulomb viscous rheological model for debris flow (see Iverson 1997 for other models). A mathematical expression for Bingham plastics is:

$$\tau = k + \mu \left[\frac{du}{dy} \right]; \quad \tau_{\text{crit}} = k \quad (1.11)$$

where k = strength of the cohesive flow = critical shear stress for movement (τ_{crit}), μ = dynamic viscosity of the sediment-water mixture after movement begins, and du/dy = the rate of change of velocity at any level, y, in the flow. Note that for $k = 0$, this equation reduces to Newton's law of viscosity. A similar expression for Coulomb viscous materials is:

$$\tau = C + \sigma_n \tan \phi + \mu \left[\frac{du}{dy} \right]; \quad \tau_{\text{crit}} = C + \sigma_n \tan \phi \quad (1.12)$$

where C = the cohesive strength component, due to electrostatic attractions between clay particles; $\sigma_n \tan \phi$ = the frictional strength component, due to intergranular friction; σ_n = normal stress; and ϕ = angle of internal friction. In contrast, Takahashi (1981) favours a 'dilatant-fluid' rheological model based on dispersive pressure (Bagnold 1956). Locat (1997) has developed a bilinear rheological model with two viscosity terms. At high strain rates, the debris behaves as a Bingham material {Eq. 1.11} with low viscosity, and at low strain rates it behaves as a Newtonian fluid with a high viscosity. The bilinear model provides a good fit to experimental data (Imran *et al.* 2001).

Although cohesion is a primary support mechanism in both mudflows and debris flows, natural flows exhibit a wide variation in the relative importance of grain-support mechanisms. In particular, elevated pore-fluid pressures are important in explaining the mobility of many cohesive flows (Pierson 1981; Pierson & Costa 1987; Ilstad *et al.* 2004a). Once mud-rich sediments fail and the electrostatic bonds responsible for cohesion are broken, the material tends to remain mobile throughout its downslope flow, even on low slopes. This is a reflection of the *thixotropic behaviour* of clays and fine silts, in which strength is dependent on the recent deformation history of the material.

1.8.2 Turbulence of cohesive flows

Most cohesive flows are laminar, with no fluid mixing across streamlines. Large flows may be turbulent (Enos 1977; Middleton & Southard 1984), but should not be classified as turbidity currents or concentrated density flows because they lack the strong vertical concentration gradients of more dilute currents (e.g., Fig. 1.30), and because they re-acquire laminar behaviour and develop a semi-rigid plug with deceleration. Even in laminar flows devoid of eddies, secondary circulation due to clast rotations and encounters, or due to flow meandering, may cause churning and internal mixing.

For Bingham plastics, the criterion for turbulence is based on both the *Reynolds Number, R.* and the *Bingham Number, B*, where

$$R = \frac{Ud\rho}{\mu} \quad (1.13)$$

and

$$B = \frac{\tau_{\text{crit}} d}{\mu U} \quad (1.14)$$

In these equations, U = mean flow velocity, d = flow thickness, ρ = flow density, and μ = dynamic viscosity after flow begins. Experimental data (Fig. 1.55), originally plotted by Hampton (1972), indicate that for large values of either R or B, a conservative criterion for turbulence is:

$$R \geq 1000B \quad (1.15)$$

which is equivalent to

$$\frac{\rho U^2}{\tau_{\text{crit}}} \geq 1000 \quad (1.16)$$

This last dimensionless product was named the *Hampton Number* by Hiscott and Middleton (1979), who used reasonable values for

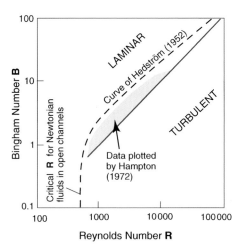

Fig. 1.55 Relation of Bingham Number to critical Reynolds Number for turbulence in a Bingham plastic. Experimental data are for pipe flow but scales have been adjusted to the correct values for 2D SGFs using the thickness of the flow as the length scale. From Hiscott and Middleton (1979), based on Hampton (1972).

flow strengths and densities to show that even large, fast debris flows probably would not be turbulent.

Basal scour beneath fully freighted laminar cohesive flows generally is insignificant (Takahashi 1981), possibly because in cases where hydroplaning results this allows the flow to ride along on top of a thin sheet of overpressured water or wet, low-viscosity, mud (Mohrig *et al.* 1998; Ilstad *et al.* 2004b). Also, the lack of turbulence decreases the chance of erosion because fluid scour is inoperative. Nevertheless, underlying sediments may be plucked up and incorporated into the flow because of high shear stresses along the basal interface (Hiscott & James 1985; Dakin *et al.* 2012). Where large cohesive flows impinge upon the base of submarine slopes, they can gouge and disrupt the underlying strata (e.g., Hiscott & Aksu 1994) and produce 10–30 m-deep erosional channels, grooves and scour pits, perhaps facilitated by partial liquefaction of the rapidly loaded substratum of unconsolidated or weakly lithified sediments (Dakin *et al.* 2012). Cohesive-flow deposits also may occur in channels cut by other processes, as these flows seek out bathymetric lows.

1.8.3 Competence of cohesive flows

The competence – largest clast that can be carried (D_{max}) – of cohesive flows was determined by Hampton (1975) to be:

$$D_{max} = \frac{8.8k}{g(\rho_{clast} - \rho_{matrix})} \tag{1.17}$$

where k = strength, g = gravitational acceleration and ρ = density. Using kaolinite-water slurries as matrix, Hampton (1975) determined that (i) competence decreases approximately exponentially with increasing weight percent water, from about 20 cm at 40% water to about 2 cm at 60% water (approximate lower limit of cohesive flows, Fig. 1.21), (ii) competence of slurries is less after shear than before shear – by about 0.5 mm at 60% water, (iii) competence decreases with flow duration for durations less than about one hour and (iv) competence is independent of flow velocity.

For debris flows with more than 20 volume% coarse sand and gravel, the competence increases dramatically with increasing concentration of coarse clasts (Hampton 1979). This is because the large clasts load the matrix and produce elevated pore-fluid pressures that counteract the tendency of the clasts to settle. These excess pore pressures (above hydrostatic) also reduce the strength of the flow by reducing normal stress (Eq. 1.12). At volume concentrations above 50%, clast collisions and near collisions provide additional support (Rodine & Johnson 1976), and competence continues to increase dramatically above that predicted by Equation 1.17 (Hampton 1979). At these concentrations, grain-to-grain interactions may be the dominant support mechanism (Pierson 1981; Takahashi 1981), and the more sand-laden flows are transitional into inflated sandflows.

Natural subaerial analogues to submarine cohesive flows commonly flow in an unsteady manner as a series of advancing 'waves' that may overtake one another or be separated in time by more fluid flows (Fig. 1.56). This process affects the streamwise texture of the flow and the eventual vertical profile of the deposit.

1.8.4 Deposits of cohesive flows, including debrites

Because of the mode of deposition, cohesive-flow deposits are poorly sorted, lack distinct internal layering but may have crude stratification due to non-uniform migration through the flow of the base of the semi-rigid plug during deceleration (Hampton 1975; Thornton 1984; Aksu 1984), have a poorly developed clast fabric (Lindsay 1968; Aksu 1984; Hiscott & James 1985), irregular mounded tops and tapered flow margins or snouts. Grading is generally poor, but both normal and inverse grading may occur (Naylor 1980; Aksu 1984; Shultz 1984). Inverse grading may develop because the base of the flow undergoes the greatest and most prolonged shear, leading to reduced competence (Hampton 1975). Individual cohesive-flow events may deposit separate tongues or lobes of material that have quite different textural characteristics (for a subaerial example, see Johnson 1984: pp. 266–74), making it difficult to distinguish separate flows in the geological record (Fig. 1.57).

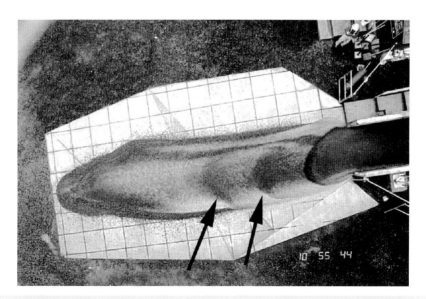

Fig. 1.56 Experimental subaerial debris flow with a series of advancing surface waves (arrows). The grid squares have sides of 1 m, and a human figure in the top right also provides scale. From Major (1997: his fig. 7E).

Fig. 1.57 Amalgamated experimental debris flow deposits separated by a cryptic surface (arrow) that could easily be missed in a natural exposure. Leftmost scale divisions are in centimetres. Photograph courtesy of J. J. Major. A different view of the arrowed amalgamation surface is shown in Major (1997: his fig. 10A).

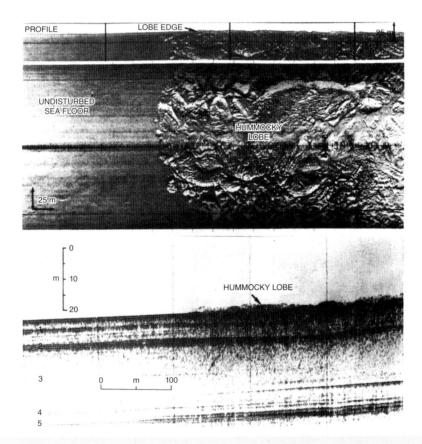

Fig. 1.58 100 kHz side-scan sonar image (and relief profile), and matching sub-bottom profile of hummocky debrite lobe in Bute Inlet, Canada. From Prior *et al.* (1984).

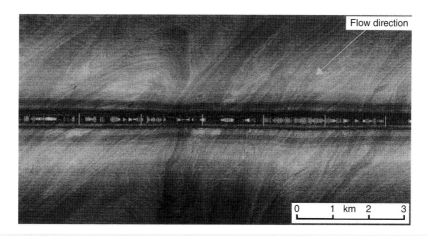

Fig. 1.59 TOBI 30 kHz sonograph of 'woodgrain' texture, consisting of flow-parallel banding, pressure ridges and longitudinal shears, on the surface of the Saharan Debris Flow deposits at ~4350 m water depth. Light tones are high backscatter and the vehicle track is along the centre-line of the image. From Weaver *et al.* (1995).

Fig. 1.60 SeaMarc1A mosaic showing shrub-shaped SGF deposits at the terminations of dendritic channel network on the Mississippi Fan, and interpreted by Schwab *et al.* (1996) and Talling *et al.* (2010) as the deposits of mudflows. From Paskevich *et al.* (2001) and http://pubs.usgs.gov/of/2000/of00-352/htmldocs/images.htm.

Increasingly, submarine cohesive flows are being imaged by high-resolution side-scan sonar systems. Some flows have irregular blocky surfaces (Fig. 1.58) (Prior *et al.* 1984), whereas others show prominent woodgrain-like flow lines like the surfaces of some glaciers, indicating ductile flow (Fig. 1.59) (Masson *et al.* 1993; Weaver *et al.* 1995). These flow lines resemble the compression ridges produced in small-scale experiments by Marr *et al.* (2001). Low-viscosity flows that transited the Mississippi Fan channel system form shrub-like digitate deposits (Fig. 1.60) just beyond the channel mouths and over the tops of low distal levées (Twichell *et al.* 1992; Schwab *et al.* 1996; Talling *et al.* 2010). These digitate deposits have been interpreted by Talling *et al.* (2010) as ~1 m-thick mudflow deposits carrying large clay and fine-sandstone slabs, passing downward into clean graded sand deposited during a turbulent phase of the same flow event. According to Talling *et al.* (2010), a body transformation occurred along the flow path as clay concentration increased in the upper part of the flow, leading to damping of turbulence and development of cohesion. Similar finger-like sandy deposits (inferred to be hyperconcentrated-flow deposits – the inflated sandflow deposits of this chapter) are present in channel-mouth lobes of the Monterey Fan channel (Klaucke *et al.* 2004). Extensively cored 'mass-transport complexes' on the Amazon Fan, with the acoustic characteristics of mudflows, predominantly consist of contorted and disoriented blocks of silt and mud (Piper *et al.* 1997). The blocky flows and the woodgrain-banded flows presumably possess different rheology and viscosities.

Shultz (1984) attributed grading style and volume concentration of matrix in debrites to the relative importance of cohesion, clast interaction (dispersive pressure) and fluid behaviour. The result is a continuum of deposit characteristics (Fig. 1.61) intermediate in character between four distinct end-member facies: *Dmm* = 'massive' (structureless) matrix-supported debrite, *Dmg* = graded matrix-supported debrite, *Dci* = inversely graded clast-supported debrite, and *Dcm* = 'massive' (structureless) clast-supported debrite (Shultz 1984).

The larger clasts – cobbles and boulders – of some cohesive-flow deposits are concentrated near the base or in the middle of the deposit. Some large boulders may be grounded on the underlying substrate (Masson *et al.* 1993), whilst others project from the tops of the beds (Fig. 1.22), allowing a quantitative assessment of rheological strength using a Bingham plastic model (Johnson 1970: p. 487).

Bingham plastic strength can also be calculated from the shape of debris snouts and from the thickness, T_{crit}, at which flow ceased (Johnson 1970: p. 488), according to the following equation:

$$T_{crit} = \frac{k}{\gamma_d \sin \alpha} \tag{1.18}$$

where γ_d = specific weight of the cohesive flow and α = slope angle. Note that $\gamma_d = g' \rho_{flow}$ where $g' = g(\rho_{flow} - \rho_{water})/\rho_{flow}$.

Calculations of cohesive-flow strength based on either the extent of clast projection or snout shape are given by Hiscott and James (1985) and Kessler and Moorhouse (1984), who calculated strengths in the range 10^2 to 1.0 Pa (1 Pa $= 10^5$ dynes cm^{-2} = 1 N m^{-2}) for Cambro-Ordovician and Jurassic deposits, respectively. These estimates are in the same range as values calculated for subaerial flows by Johnson (1970).

1.8.5 Submarine versus subaerial cohesive flows

All evidence suggests that submarine cohesive flows and their deposits differ only in minor ways from subaerial equivalents, although associated facies and processes of post-depositional modification are clearly different. Shear stress is dependent on the density difference between the flow and the ambient fluid, so one might predict that somewhat higher slopes would be needed to permit subaqueous versus subaerial cohesive flow, given similar yield strengths. Recall, however, that submarine cohesive flows have been documented from very low slopes. Hydroplaning might help explain this high degree of mobility (Mohrig *et al.* 1998). In general, submarine flows probably have lower yield strengths than subaerial flows as a result of entrainment of seawater and wet mud, lack of downward percolation of water from the flow itself into the substrate, and elevated pore-fluid pressures (Pierson 1981) due to greater amounts of interstitial fluid. This suggestion is supported by the observation that, for deposits of equal thickness, subaqueous cohesive-flow deposits contain smaller boulders than subaerial deposits (Nemec *et al.* 1980; Gloppen & Steel 1981; Nemec & Steel 1984), that is, subaqueous cohesive flows are weaker than subaerial flows.

1.9 Accumulation of biogenic skeletons and organic matter

Far from continental sources of detritus, four main types of pelagic input dominate the world oceans: *biogenic silica* in high-productivity zones along the equator (radiolarians) or at about 60° latitude (diatoms); *biogenic carbonate* (foraminifera, nannofossils, pteropods) in other areas shallower than the carbonate compensation depth (CCD); ice-rafted particles along the trackways of drifting icebergs; and red clays where no biogenic or glaciogenic particles are supplied (Figs 1.62 and 1.63). Under special circumstances, pelagic sediments may accumulate near continents, but only where terrigenous input is relatively very low (e.g., Gulf of California; Calvert 1966). Unconsolidated biogenic pelagic sediments are called *ooze*. If sufficiently buried and lithified, siliceous ooze becomes diatomite or radiolarite, and eventually *chert*, whereas calcareous ooze becomes *chalk* and eventually aphanitic limestone.

It is appropriate here to differentiate between the terms *pelagic* and *hemipelagic* as they apply to sedimentary deposits. In this book, pelagic grains are defined as those grains that initially enter the marine hydrosphere beyond the shelf-slope break, or are created by

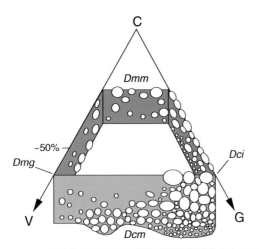

Fig. 1.61 Schematic diagram of relationships among debrite types. See text for abbreviations. Redrawn from Shultz (1984).

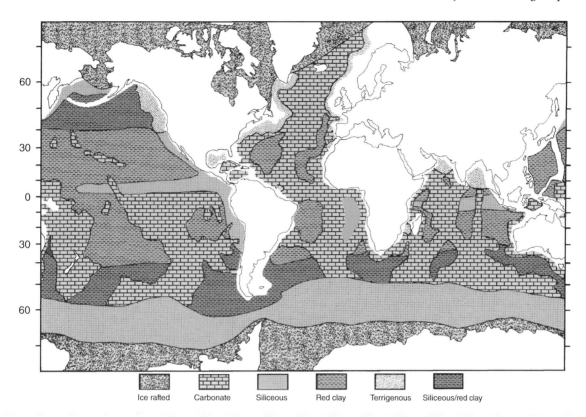

Fig. 1.62 World distribution of dominant sediment types in the oceans. Redrawn from Jenkyns (1986).

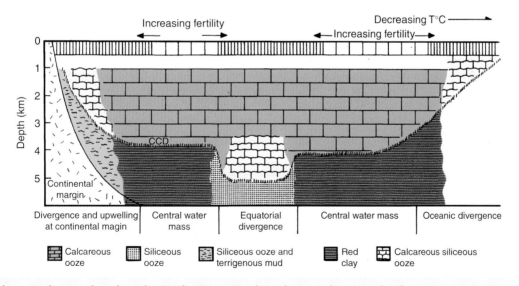

Fig. 1.63 Schematic diagram, based on the Pacific Ocean, to show the main location of sediment types relative to the CCD and near-surface organic productivity. Redrawn from Ramsay (1977).

organisms in this open-ocean region, and that subsequently settle to the seafloor. Pelagic particles, therefore, include biogenic siliceous skeletons (diatoms, radiolaria), calcareous skeletons (foraminifera, nannofossils), wind-blown dust (e.g., Staukel *et al.* 2011; Wan *et al.* 2012; Xu *et al.* 2012) or volcanic ejecta that lands on the sea surface of the open ocean, and debris liberated by melting icebergs. Hemipelagic sediments contain a pelagic component, generally 5–50% but locally

as much as 75%, with the remainder consisting of terrigenous mud that initially enters the ocean at the coast, either by coastal erosion or through river systems (deltas, estuaries etc.). This terrigenous component is advected off the shelf by storms and ocean currents.

The tests of dead planktonic organisms, with their empty chambers, settle more slowly than would be anticipated based on diameter alone. For example, planktonic foraminifera settle at the same rate as quartz particles with a diameter approximately 2.4 times less (Berger & Piper 1972). Because of advection by ocean currents, such slow settling might be expected to result in geographic mismatches between areas of plankton productivity and areas of their seafloor accumulation. Serious mismatches are, in fact, not observed in the modern oceans for a number of reasons (Berger & Piper 1972): (i) surface currents mostly transport settling tests parallel with the pelagic facies belts; (ii) deep currents commonly return settling tests closer to the place where they originated in surface waters and (ii) many tests travel to the seafloor in faecal pellets, and therefore settle (as a group) far more rapidly than they would as single tests.

Organic matter can be derived from land masses as *terrestrial organics*, or from the preservation of algal and other *marine organic compounds*. The two main processes which may lead to burial of organic matter are high rates of accumulation (leaving inadequate time or oxidising agents to decompose the material at the seafloor), and accumulation under anoxic to dysaerobic seabed conditions. High rates of accumulation, particularly of marine organic matter, are favoured by oceanographic upwelling of nutrient-rich waters along continental margins or at open-ocean divergence zones. Fully or partly enclosed bodies of water may have their own particular chemistry (e.g., stratification and poorly oxygenated bottom water) which can lead to accumulation of organic-rich sediments. Examples are *sapropels* in the Mediterranean Sea (Kidd *et al.* 1978; Rossignol-Strick 1985; Emeis & Weissert 2009; Möbius *et al.* 2010) and Japan Sea (Ishiwatari *et al.* 1994; Stax & Stein 1994) which formed periodically in the Neogene and Quaternary, and Albian *black shales* which have widespread distribution in the world's oceans (Jenkyns 1980). Sapropel is defined as mud with >2% total organic matter (Kidd *et al.* 1978). Decomposition of this organic matter in the sediment uses up the available oxygen and makes the pore-water anoxic. If the bottom water is also anoxic, then neither deposit feeders nor filter feeders can live in the sediment, and it is entirely undisturbed by burrowers, leading in many cases to the finely laminated deposits typical of ancient black shales.

The maximum depth for accumulation of deep-marine biogenic carbonate particles is dependent upon the position of the *carbonate compensation depth* (CCD), and the shallower *aragonite compensation depth* (ACD) for pteropods. These critical depths, below which calcareous remains do not occur, are functions of ocean circulation, latitude, seawater chemistry and geologic time (Fig. 1.64). The net result of the spatial and temporal variation of these controls is the generation of vertically and laterally changing facies, facies associations and sequences. Such changes result in geographically and stratigraphically distinct units or 'packets' of sediments, each with their own environmental interpretation.

Highstands of sea level are associated with an amelioration in climate, increased biological diversification, reduced mid-water oxygen concentration, an open-ocean shoaling of the CCD (Fig. 1.64), a proliferation of deep-marine 'condensed successions' of pelagic sediment and the accumulation of muddy abandonment facies over many deep-marine clastic systems. At the present time, during the Holocene sea-level highstand, most of the world's deep-marine clastic systems are essentially dormant and being mantled by pelagic or hemipelagic sediments. The surfaces of many submarine fans are veneered by about 1 m of light-brown, pelagic, foraminiferal ooze or marl, representing approximately the last 11 kyr; for example on the Amazon Fan (Damuth & Kumar 1975a; Damuth & Flood 1985), and the Mississippi Fan (Bouma *et al.* 1986).

Pelagic depositional rates are typically 1–60 mm kyr^{-1} (Fig. 1.65). However, these rates may reach about 100 mm kyr^{-1} on outer shelves and upper slopes in areas of upwelling. Ocean currents control the mixing of cold and warm water masses, biological productivity, sites of upwelling and the distribution of various chemogenic sediments such as phosphorites. Oceanic circulation, interacting with water depth and basin physiography, oxygen minima or anoxia, and biological productivity, govern the potential distribution of open-ocean (e.g., biogenic) sediments.

Although modern ocean circulation patterns are well known, and despite their profound impact on the distribution of fine-grained

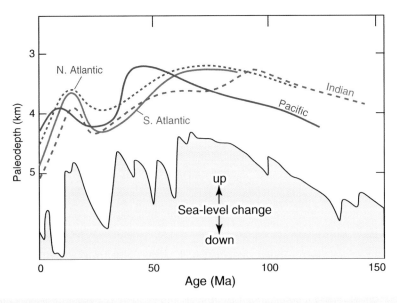

Fig. 1.64 Fluctuations in the depth of the CCD since 150 Ma in the Atlantic, Pacific and Indian oceans and the variation in global sea level. Redrawn from Kennett (1982).

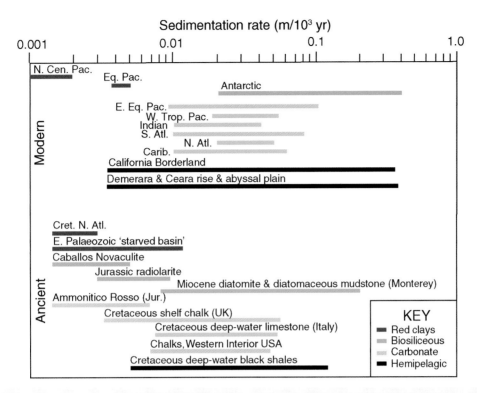

Fig. 1.65 Rates of sediment accumulation of some modern and ancient pelagic and hemipelagic sediments. Note that estimated rates for ancient sediments are ~60–70% of modern rates. Redrawn from Scholle and Ekdale (1983).

and biogenic facies, there are few data on ancient ocean current patterns, except constraints provided by DSDP and ODP drilling. For example, a comparison of Palaeocene and Oligocene world oceans, based partly on DSDP data (Fig. 1.66) shows the important role that the distribution of continents plays in controlling ocean circulation. In the Palaeocene, prior to the collision of India with the Asian Plate, there was an equatorial circum-global Pacific-Tethys current, and clockwise subpolar gyres inferred for the Southern Pacific and Atlantic. By the Oligocene, the remnant of Tethys was a relatively small fragmented ocean basin, and much of the present-day ocean circulation was established, including the Circum-Antarctic Current.

Results from various DSDP sites around the Central Atlantic show that even in the Early Cretaceous, a proto-Gulf Stream was established (Fig. 1.67), with pelagites and hemipelagites comprising varve-type laminations, graded claystones and limestone–shale couplets (Robertson 1984; Sheridan, Gradstein *et al.* 1983). The fine-grained varve-type lamination, formed by fluctuating proportions of terrigenous plant material, marine plankton and clastics, may reflect short-periodicity (tens to hundreds of years) climatic changes. The graded claystones and black shales represent fine-grained turbidite redeposition from within or near the oxygen minimum zone on the upper continental slope. The limestone–shale couplets suggest climatic variation on time scales of 20 000–60 000 years, with the organic-rich shales formed during wetter periods when abundant plant material entered the sea (Robertson 1984). The abundance of radiolaria in the pelagic chalks suggests upwelling to produce fertile surface waters, possibly due to the inflow of nutrient-rich waters from western Tethys (Fig. 1.67). This is just one example

of the importance of palaeoceanography in interpreting ancient deep-marine mixed siliciclastic and carbonate sediments, including organic-rich shales.

1.9.1 Environmental information from biogenic skeletons

A number of fossil groups can provide information on past values of seawater salinity (e.g., dinoflagellate cysts, benthic foraminifera, ostracods) and temperature (e.g., planktonic foraminifera). These constraints are largely based on species proportions (through statistically based transfer functions) and morphological variations including test abnormalities. Here, we emphasise two criteria from benthic foraminiferal studies that provide constraints on palaeo water depth and water-column oxygenation.

Pioneering work in the California Borderland and Neogene successions of the western United States established the technique of determining palaeo water depth from the assemblage of the benthic foraminfera, in particular (Natland 1933; Bandy 1953; Ingle 1975). Ingle (1975, 1980) explains procedures and limitations, and provides some statistical data for offshore California. For example, planktonic foraminifera constitute 0–10%, 10–50%, 20–80% and <30% of total fauna for inner shelf, outer shelf, upper slope (<1000 m) and basin, respectively, whereas radiolarians increase dramatically at the shelf edge from <5 to ~500–1000 specimens per gram. Only the known upper depth limits of benthic foraminiferal species can be used to establish the palaeo water depth, because post-mortem downslope

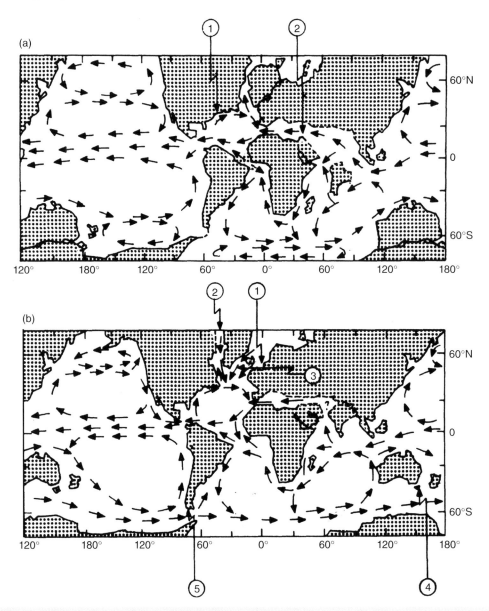

Fig. 1.66 Schematic reconstruction of the Palaeocene and Oligocene distribution of continents and ocean surface-circulation patterns. Numbers refer to following: (a) Palaeocene; 1, Proto Gulf Stream; 2, Tethys Current; (b) Oligocene; 1, Norwegian-Greenland Sea; 2, North Labrador Passage; 3, Greenland-Iceland-Faroe Ridge; 4, South Tasman Rise; 5, Drake Passage. From Leggett (1985) after Haq (1981).

transport contaminates samples with the skeletons of upslope fauna. For example, 50–100% of benthic foraminiferal tests found beyond the shelf edge offshore California have been displaced from upslope. When applied to ancient successions, great care must be taken because the modern upper depth limits depend on the stratification of temperature, nutrients, oxygen concentration, etc. in the modern ocean. As a result, Ingle (1980) advises that modern upper depth limits from offshore California can be used for cool periods in the Palaeogene and Neogene, but that for warmer climatic periods (e.g., late Palaeocene – early middle Eocene, latest Oligocene, middle Miocene) the upper depth limits from more tropical areas in the eastern Pacific and Gulf of Mexico are more reliable. The water-depth ranges which can be discriminated using benthic foraminifera become wider with increasing depth (e.g., upper bathyal = 150–500 m, whereas upper middle bathyal = 500–1500 m).

The use of benthic foraminifera to estimate palaeo water depth is most reliable in the Neogene and Palaeogene, where species are likely to have had similar behaviour to modern relatives. However, this approach has also been used successfully in Upper Cretaceous rocks (Sliter 1973; England & Hiscott 1992).

Benthic foraminifera can also be used to quantify past concentrations of dissolved oxygen in bottom waters, unless conditions were too anoxic for life. Kaiho (1991) defined a *benthic foraminiferal oxygen index* (BFOI), which displays a linear correlation to the oxygen content of bottom waters at least in the range of values below $3 \, \mathrm{ml \, l^{-1}}$ (Kaiho 1994). The BFOI uses the relative proportions of three benthic foraminiferal morphogroups to estimate past bottom water oxygen content. Kaiho adapted the morphogroup concepts of Bernhard (1986), and Corliss and Chen (1988), to define these groups. 'Dysoxic' species have flattened, tapered, or elongate tests

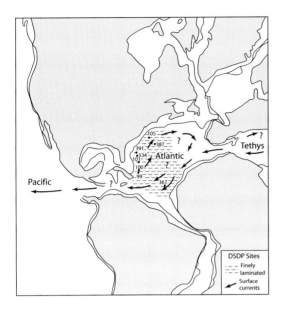

Fig. 1.67 Reconstruction of the central Atlantic Ocean in the Early Cretaceous, showing location of some DSDP sites, the inferred surface circulation and probable distribution of varve-type laminated sediments. Redrawn from Robertson (1984).

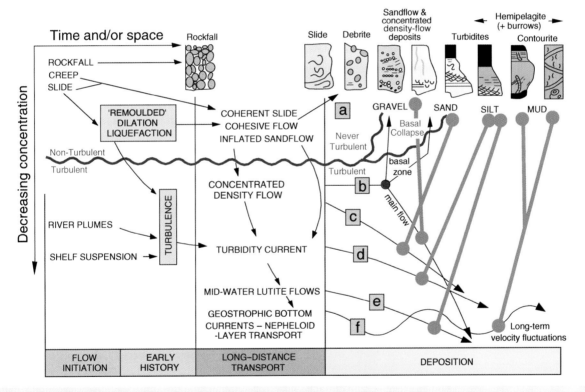

Fig. 1.68 Summary of the evolution of sediment gravity flows and other deep-marine transport processes as a function of particle concentration. Compare with Figure 1.1. Deposits are arranged across the top of the diagram at the end of each evolutionary pathway. The box labelled with the letter a relates to dense, cohesive SGFs that never became turbulent. Box b corresponds to concentrated density flows which, during their depositional phase, become strongly stratified into a non-turbulent lower part and a thicker turbulent upper part. The lower part of this flow eventually deposits en masse rather than by suspension settling. Boxes with letters c–e correspond to turbulent SGFs of variable concentration – these lose their sediment load by selective (grain-by-grain) deposition, forming graded beds with tractional structures. Eventually, all sediment is dropped and concentration approaches zero. Grey 'dumbells' link each evolving flow with its deposit. The box and flow labelled f is a thermohaline contour current with low concentration and very long-period velocity fluctuations (e.g., thousands of years). Clearly, the time scale for these currents and their fluctuations is much longer than the time scales for other flows in the diagram, which might be no longer than hours to days. Modified from Middleton and Hampton (1973) and Walker (1978).

which typically display a thin, porous wall and lack ornamentation. 'Oxic' species display a variety of test forms, including spherical, plano-convex, biconvex, and rounded trochospiral. This group includes species that have large, ornamented tests and thick walls. Finally, the 'intermediate' or 'suboxic' group contains taxa that are typically larger than the 'dysoxic' forms, and display some ornamentation. This group includes tapered and cylindrical forms, planispiral forms, and those trochospiral forms that have small, thin test walls. The BFOI has been used successfully by Kaminski *et al.* (2002) to assess past environmental conditions in the salinity-stratified water column of the Marmara Sea, Turkey, where dysoxic conditions have existed for the last 10,000 years.

Summary

The main mechanisms for long-distance lateral transport of sediment in the deep sea are: (i) turbidity currents; (ii) concentrated density flows; (iii) cohesive flows; (iv) deep, thermohaline, clear-water currents that commonly flow parallel to bathymetric contours; (v) movement of dilute mud suspensions in water masses that drift or cascade off the shelf and (vi) mass movement in the form of sediment slides. The summary Figure 1.68 is a more developed version of Figure 1.1, and relates the various transport processes to their deposits. The deposits of cohesive flows are poorly sorted, generally lack stratification, have a poorly developed clast fabric, and stand up above their surroundings as irregular mounds with a tapered marginal snout. Deposits of concentrated density flows (and

less common inflated sandflows) show evidence for grain-to-grain interaction (as inverse grading) and pore-fluid escape (as dish structures or pillars); however, these processes are restricted to the time of deposition and reveal little about the long-distance transport mechanism.

Fine-grained sediment can be deposited by turbidity currents, by bottom currents (contour currents), or by settling from dilute suspensions at the top of, or within, the water column (pelagites and hemipelagites). The deposits have some diagnostic characteristics, but are in many cases difficult to interpret because of a superposition of more than one process, or because of post-depositional bioturbation.

Sediment slides can involve all size grades of sediment, but are most common in poorly consolidated, water-rich muds found in areas with high depositional rates. The susceptibility of sediment masses to sliding, even on very low slopes, is sharply increased by cyclic vibrations generated by earthquakes.

Biogenic sediments and organic-rich muds are not shown in Figure 1.68 because they do not form beneath discrete flow events. Biogenic sediments accumulate where biological productivity is high in surface waters, and where post-mortem dissolution of hard parts is prevented or reduced by favourable shallow depositional depths (carbonate) or high accumulation rates (silica). Organic-rich sediments accumulate in areas of high productivity (e.g., upwelling zones) where accumulation rates are too high to allow significant seabed oxidation of organic matter, or where bottom waters are anoxic because oxidising organic matter consumes dissolved oxygen faster than it can be supplied by sluggish marine currents.

CHAPTER TWO
Sediments (facies)

(a)

(b)

(a) Armoured mud ball of gravel and pebble clasts, Ainsa I Fan, Ainsa Quarry, Middle Eocene Ainsa Basin, Spanish Pyrenees. Camera lens cap scale. (b) Erosive base to debris flow deposit (Facies A1.4), Middle Eocene Morillo I Fan, Ainsa Basin, Spanish Pyrenees. Compass-clinometer scale (centre of sandstone). *Cf.* Dakin *et al.* (2013).

Deep Marine Systems: Processes, Deposits, Environments, Tectonics and Sedimentation, First Edition. Kevin T. Pickering and Richard N. Hiscott.
© 2016 Kevin T. Pickering and Richard N. Hiscott. Published 2016 by John Wiley & Sons, Ltd.
Companion Website: www.wiley.com/go/pickering/marinesystems

2.1 Introduction

In this chapter, we first briefly describe five classification schemes for deep-water siliciclastic sediments, then comment briefly on seismic facies, followed by a detailed overview of the facies scheme which is used throughout this book. The five schemes and the most recent comprehensive literature reference for each are the:

- Mutti and Ricci Lucchi descriptive scheme (Mutti 1992);
- Mutti genetic scheme (Mutti 1992);
- Walker descriptive scheme (Walker 1992);
- Ghibaudo coded scheme (Ghibaudo 1992);
- Pickering *et al.* (1986a, 1989) hierarchical and descriptive scheme (this book).

Strictly process-oriented classifications of sediment gravity flows (SGFs), such as the Mulder and Alexander (2001) scheme explained in Section 1.4.1, are not considered because they are not designed to be used in the field.

2.2 Facies classifications

The seminal classification is that of Emiliano Mutti and Franco Ricci Lucchi (Mutti & Ricci Lucchi 1972, 1974, 1975; Mutti 1977, 1992). This descriptive classification (Fig. 2.1) is primarily restricted

to gravels, sands, and silts deposited by sediment gravity flows. Therefore, it does not provide a guide for the classification of muds and biogenic sediments. Each deposit of a sediment gravity flow is not normally assigned to a facies. Rather, groups of broadly similar beds typically form each Mutti and Ricci Lucchi facies. Hence, this scheme is relatively coarse and is best used for the description of medium- to large-scale bed clusters in outcrops.

The Mutti (1979, 1992) genetic scheme (Fig. 2.2) can be used as a theoretical basis for understanding the evolution of single masses of failed sediment as they transit a basin slope and flow out onto the adjacent basin floor. Some genetic facies may logically progress from one into the next (e.g., F7 → F8 → F9), and it is these groups that form what Mutti *et al.* (1994) call 'elementary facies tracts'. Note that such downflow transitions are generally inferred, not observed. Also, sand-load turbidity currents and concentrated density flows may rarely if ever form from cohesive debris flows, and many debris flows may never undergo a transition to another flow type. Hence, this genetic scheme is best viewed as a set of possible flow transformations and evolutionary pathways based on a wealth of field experience. The scheme can be useful in cases where a process slant is desired, for example on a field trip or in overviews of basin infilling.

The Walker (1978, 1992) scheme is simplified from the Mutti and Ricci Lucchi (1972) scheme, but uses descriptive names rather than alphanumeric codes to classify facies and facies associations (Table 2.1). Hence, it is particularly easy to use for overview

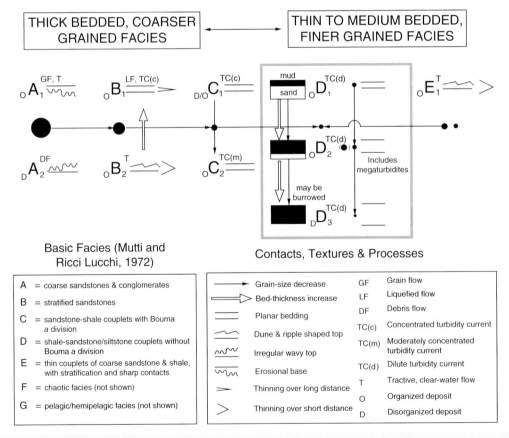

Fig. 2.1 Summary of relative grain sizes (diameters of filled circles), geometry of basal and upper contacts, primary depositional structures, and emplacement processes (superscripts to the right) for deep-marine facies A–E of Mutti and Ricci Lucchi (1972). Subscripts to the left indicate whether the deposits are organised (O) or disorganised (D); subscripts to the right are subfacies numbers.

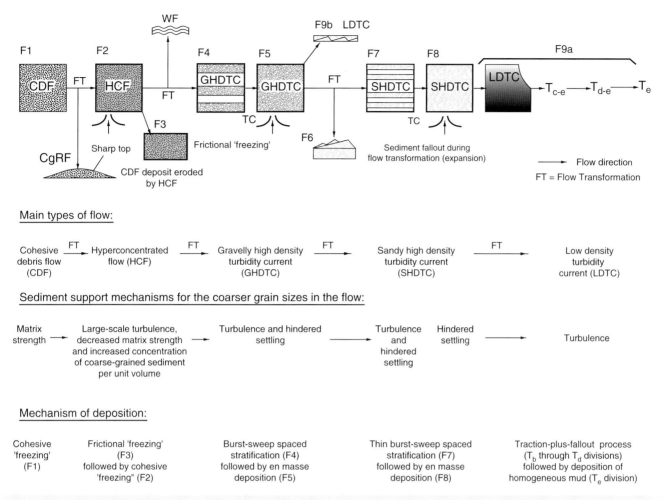

Fig. 2.2 Gravity-flow facies F1–F9 and WF of Mutti (1992) with a summary of transport and depositional processes associated with cohesive debris flows to low-density turbidity currents. F1 = beds with muddy matrix, plastic deformation, dispersed large clasts (some projecting from bed top). F2 = beds with basal scours, dispersed large mudstone clasts, crude grading of muddy sand/gravel mixtures. F3 = clast-supported gravels, mostly unstratified, commonly inversely graded. F4 = thick coarse-grained sands with spaced stratification (Hiscott 1994b). F5 = thick coarse-grained sands with fluid-escape structures and poor sorting. F6 = coarse-grained planar and cross-stratified sands, ungraded (same as Mutti 1977 Facies E). F7 = thin, coarse-grained sands with horizontal, commonly very thin, laminae akin to spaced stratification (Hiscott 1994a). F8 = deposits of structureless medium- to fine-grained sand, possibly graded (equivalent to Bouma 1962 T_a division). F9 = deposits of thoroughly current-laminated very fine sand and silt capped by mudstone (equivalent to Bouma 1962 T_b through T_e divisions). WF = <20 cm-thick deposits of very poorly sorted coarse sand to pebbles with faint wavy laminae generally overlying F2 and underlying F4. CgRF = conglomerate remnant facies with convex top. TC = traction carpets. FT = flow transformation. Note that some facies in this figure constitute 100% of an event-deposit, whereas others form divisions within the succession of deposits left by a single passing gravity flow. In this book we use a different classification of flow types (see Table 1.1 for equivalencies).

descriptions both in the field and in published reports. Muds, biogenic sediments and chemogenic sediments are not classified.

The Ghibaudo (1992) scheme (Fig. 2.3) is styled after schemes previously published for fluvial and glacial deposits, in which a set of simple codes can be used in the field to describe the essential characteristics of single flow units. Like the Mutti and Ricci Lucchi (1972) scheme, it focusses on the deposits of sediment gravity flows, and cannot deal adequately with muds, biogenic sediments or chemogenic sediments. Depending on the level of detail desired, more complete or less complete coding can be used during field description.

The scheme of Pickering *et al.* (1986a, modified in 1989 and for this book) permits classification of all sediment grades from gravel to mud. It is hierarchical, allowing variable levels of detailed classification down to the level of individual flow units, and it can be easily

modified or augmented with new facies. This scheme also facilitates the addition of 'subfacies' if there is a requirement to define even more refined distinctions, for example in detailed statistical analyses of beds showing different combinations of the Bouma divisions (see Section 5.2). In this classification, unlike the descriptive classification of Mutti and Ricci Lucchi (1972), the deposit of each sediment gravity flow is typically assigned a specific facies name. The main drawback of this scheme is that it uses a set of alphanumeric codes that cannot easily be memorised without dedicated effort. However, this scheme is not designed primarily for field description, so facies codes need not be memorised.

The Pickering *et al.* scheme is used throughout this book, and is explained and illustrated in detail in this chapter. Before doing this,

Table 2.1 Outline of the Walker (1978, 1992) facies scheme

Facies association *Facies*	Key features
Classic turbidites	All contain Bouma divisions
Thick-bedded turbidites (≥1 m)	
Thin-bedded turbidites	
— standard variety	*Single ripple train, few rip-up clasts*
— CCC turbidites	*Climbing ripples, convolutions, abundant* *rip-up clasts*
Massive sandstones	Erosion and amalgamation, subtle to absent graded bedding, fluid-escape structures
Pebbly sandstones	Normal grading, horizontal stratification, rare cross-bedding, imbrication, ubiquitous amalgamation
Conglomerates	
Graded-stratified conglomerate	*No inverse grading, imbrication,* *horizontal and cross-stratification*
Graded conglomerate	*No inverse grading or stratification,* *imbrication present*
Inverse-to-normally graded conglomerate	*No stratification, imbrication present*
Disorganised conglomerate	*No grading, stratification, nor imbrication*
Pebbly mudstones, debris flows, slumps and slides	
Pebbly mudstones	*Pebbles and distorted sedimentary clasts*
Debris flow deposits	
Slumped shales and mudstones	*Many soft-sediment folds*
Slumped thin-bedded turbidites	
Slumps with angular stratified blocks	

however, we will comment briefly on the use of seismic profiles to deduce facies characteristics.

2.2.1   Seismic facies

Only high-resolution seismic systems provide acoustic images that can be compared with features seen at outcrop. These include 3.5 kHz sub-bottom profiles, and records obtained using surface or deep-towed boomers or low-energy sparkers (e.g., HUNTEC deep-towed system). The vertical resolution of these systems is of the order of a few tens of centimetres (boomer) to about 1 m. Airgun or water-gun profiles, particularly those obtained during surveys for the purpose of hydrocarbon exploration, typically have a resolution no better than several tens of metres (McQuillin *et al.* 1984), and cannot readily be used to recognise depositional facies comparable to those determined at outcrop.

Seismic facies from 3.5 kHz profiles have been clearly described and interpreted by Damuth (1980). In general, the greater the acoustic penetration, and the more regular the sub-bottom reflections, the more fine grained and evenly bedded is the sediment. In contrast, sand and gravel tend to reflect a large amount of acoustic energy and absorb much of the remainder. Therefore, internal reflections cannot be identified and underlying sediments are masked. Interpretation of 30 to 35 Hz seismic data shows excellent examples of mass transport deposits with associated megascours at the base of some MTCs as described by Moscardelli *et al.* (2006) and Moscardelli and Wood (2007).

Boomer systems have been widely used on continental shelves where extensive ground truth is available from piston- and vibro-cores (King 1981; Piper & Fader 1990). Sand deposits return a diffuse and commonly incoherent reflection with a high-amplitude surface return and inhibit deep penetration of acoustic energy. Muds with laminae or beds of silt return moderate amplitude, parallel reflections, with thin sand beds returning high-amplitude reflections. Mud appears relatively transparent, giving low-amplitude reflections. Acoustic penetration is greatest in muddy sections. Debrites return incoherent reflections and commonly have an irregular surface. Muddy debrites have low surface backscatter and have little effect on attenuating the depth of penetration of acoustic energy; sandy debrites and deposits of inflated sand flows have high surface backscatter and inhibit deep penetration of acoustic energy.

On the deep-water fans of Santa Monica Basin, California Borderland, Normark *et al.* (1998) and Piper *et al.* (1999) determined that low-amplitude reflections in deep-towed boomer data indicate principally mud. Moderate- to high-amplitude parallel reflections result from mud interbedded with silt and thin sand beds. Higher amplitude returns correlate with a higher proportion of thicker sand beds. Thick-bedded sands return incoherent reflections with high backscatter that limit penetration of acoustic energy.

2.2.2   The Pickering *et al.* classification scheme

We use the term 'facies' to mean a body of sedimentary rock, or sediments, with specific physical, chemical and biological characteristics. For the deposits of sediment gravity flows, the scale of individual facies matches the scale of single flow deposits. The chief attributes used to define the different facies are bedding style and thickness, sedimentary structures, composition and texture. Below, individual facies are described and illustrated with photographs.

For the sake of brevity, we have mainly used the terminology for modern, unconsolidated sediments throughout this chapter. The

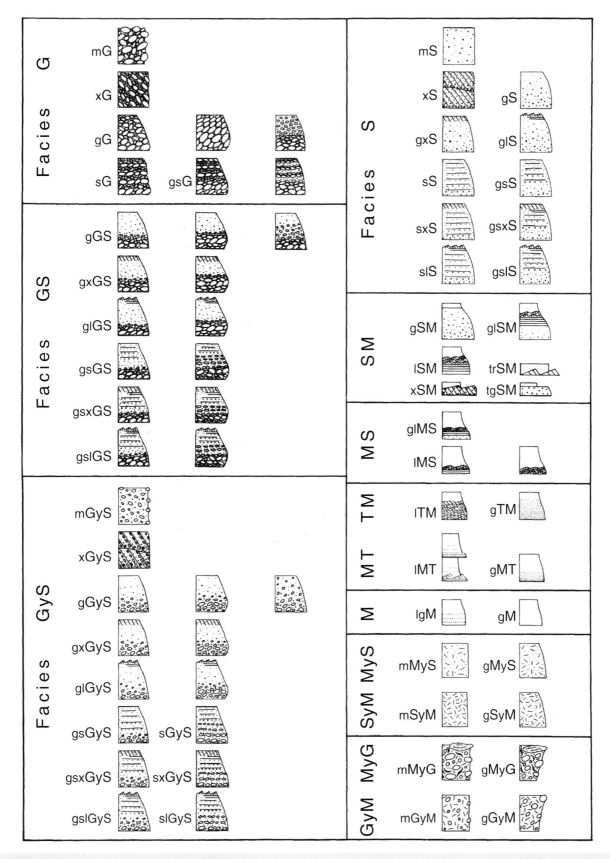

Fig. 2.3 Facies classification of Ghibaudo (1992) employing strings of upper and lower case letters as codes for gravity-flow deposits. G = gravel; GS = gravel–sand couplet; S = sand, M = mud and T = silt, either alone or as part of a sand–mud (SM), silt–mud (TM), mud–sand (MS) or mud–silt (MT) couplet; Gy = gravelly; My = muddy; g = graded, s = plane-stratified, x = cross-stratified, l = laminated, r = ripple-laminated, m = massive; t = thin-bedded.

Table 2.2 Facies classes, facies groups and individual facies used in this book

Class	
Group	
Facies	See text for detailed descriptions

A Gravels, muddy gravels, gravelly muds, pebbly sands, ≥5% gravel
    A1 Disorganised gravels, muddy gravels, gravelly muds and pebbly sands
        A1.1 Disorganised gravel
        A1.2 Disorganised muddy gravel
        A1.3 Disorganised gravelly mud
        A1.4 Disorganised pebbly sand
    A2 Organised gravels and pebbly sands
        A2.1 Stratified gravel
        A2.2 Inversely graded gravel
        A2.3 Normally graded gravel
        A2.4 Graded-stratified gravel
        A2.5 Stratified pebbly sand
        A2.6 Inversely graded pebbly sand
        A2.7 Normally graded pebbly sand
        A2.8 Graded-stratified pebbly sand

B Sands, ≥80% sand grade, <5% pebble grade
    B1 Disorganised sands
        B1.1 Thick/medium-bedded, disorganised sands
        B1.2 Thin-bedded, coarse-grained sands
    B2 Organised sands
        B2.1 Parallel-stratified sands
        B2.2 Cross-stratified sands

C Sand-mud couplets and muddy sands, 20–80% sand grade, <80% mud grade (mostly silt)
    C1 Disorganised muddy sands
        C1.1 Poorly sorted muddy sands
        C1.2 Mottled muddy sands
    C2 Organised sand-mud couplets
        C2.1 Very thick/thick-bedded sand-mud couplets
        C2.2 Medium-bedded sand-mud couplets
        C2.3 Thin-bedded sand-mud couplets
        C2.4 Very thick/thick-bedded, mud-dominated, sand-mud couplets
        C2.5 Medium to very thick-bedded, mud-dominated slurried sand-mud couplets

D Silts, silty muds and silt-mud couplets, >80% mud, ≥40% silt, 0–20% sand
    D1 Disorganised silts and silty muds
        D1.1 Structureless silts
        D1.2 Muddy silts
        D1.3 Mottled silt and mud
    D2 Organised silts and muddy silts
        D2.1 Graded-stratified silt
        D2.2 Thick irregular silt and mud laminae
        D2.3 Thin regular silt and mud laminae

E ≥95% mud grade, <40% silt grade, <5% sand and coarser grade, <25% biogenics
    E1 Disorganised muds and clays
        E1.1 Structureless muds
        E1.2 Varicoloured muds
        E1.3 Mottled muds
    E2 Organised muds
        E2.1 Graded muds
        E2.2 Laminated muds and clays

F Chaotic deposits
    F1 Exotic clasts
        F1.1 Rubble
        F1.2 Dropstones and isolated ejecta
    F2 Contorted/disturbed strata
        F2.1 Coherent folded and contorted strata
        F2.2 Brecciated and balled strata

G Biogenic oozes (>75% biogenics), muddy oozes (50–75% biogenics), biogenic muds (25–50% biogenics) and chemogenic sediments, <5% terrigenous sand and gravel
    G1 Biogenic oozes and muddy oozes
        G1.1 Biogenic ooze
        G1.2 Muddy ooze
    G2 Biogenic muds
        G2.1 Biogenic mud
    G3 Chemogenic sediments

---

terms gravel, sand, mud, silt and clay are, therefore, used to include the ancient lithified rock types conglomerate, sandstone, mudstone, siltstone and claystone. Bed thicknesses are defined according to Ingram (1954):

- laminae, <1 cm;
- very thin beds, 1–3 cm;
- thin beds, 3–10 cm;
- medium beds, 10–30 cm;
- thick beds, 30–100 cm;
- very thick beds, >100 cm thick.

Our classification is hierarchical (Table 2.2, Fig. 2.4). *Facies classes* are divided into two or more *facies groups*, which are each further subdivided into constituent *facies*. The seven facies classes are defined largely on: (i) texture of the gravelly, sandy or silty divisions of the beds (Fig. 2.4); (ii) relative thickness of mud interbeds or caps; and also on (iii) internal organisation for Facies Class F, and on composition for Facies Class G (Table 2.3). For the second-order classification, Facies Classes A–E can be divided into disorganised and organised facies groups (A1, A2, etc.), that is groups with beds that lack clear stratification or grading and groups with beds that show clearly defined sedimentary structures. Facies Class F is mainly disorganised and can be divided into two groups, characterised respectively by (i) exotic clasts and (ii) contorted strata. A third group was included in this class by Stow (1985, 1986), but the scheme was revised to include these deposits in Facies Class A (Pickering *et al.* 1986a, this book). Facies Class G is divided into biogenic oozes, muddy oozes, biogenic muds and chemogenic sediments.

For the purposes of large-scale mapping and reconnaissance fieldwork, subdivision of strata into facies classes or facies groups may be appropriate. For much more detailed analysis, recognition of the more specific individual facies will be necessary. In erecting models to describe and define the various deep-water sedimentary

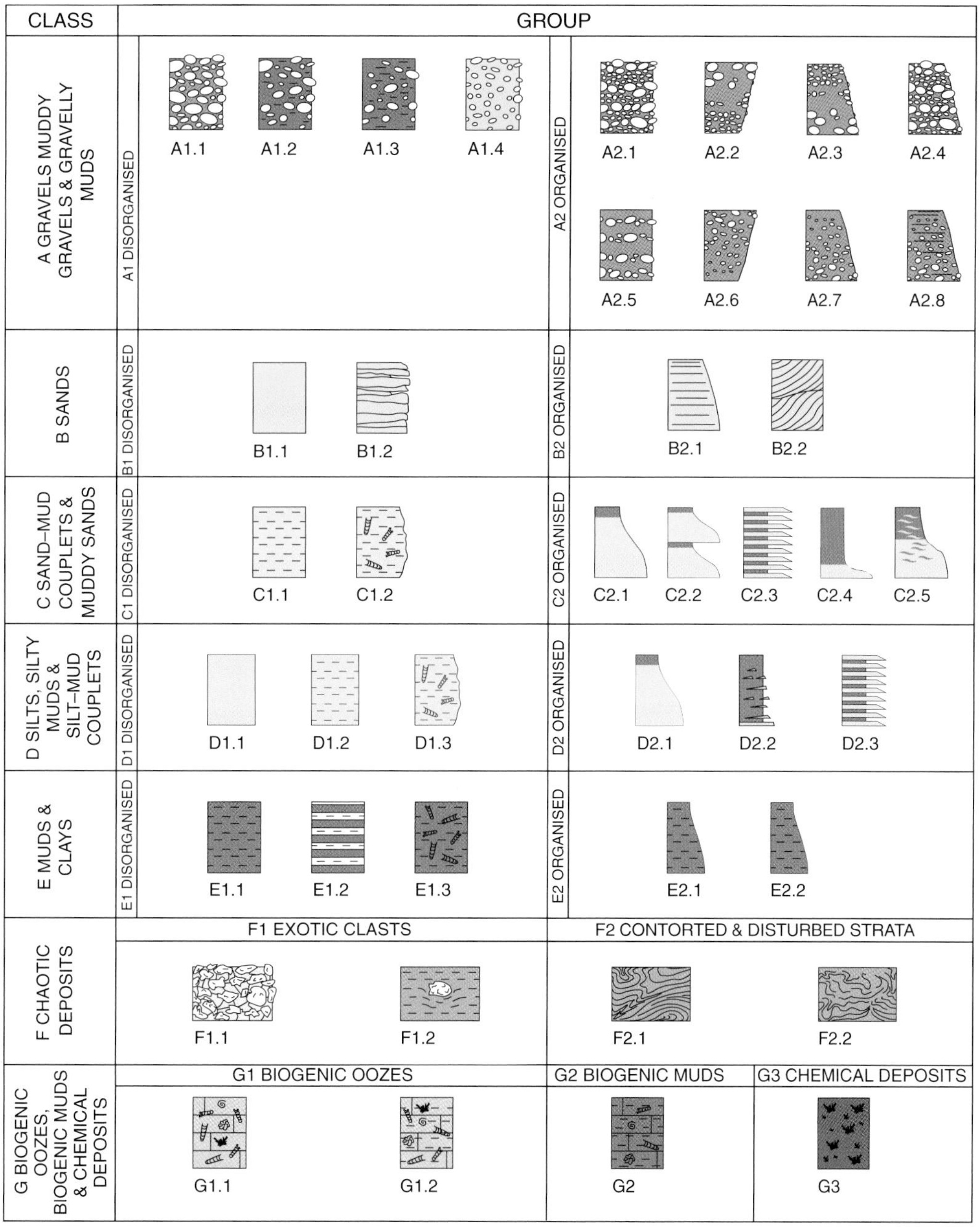

Fig. 2.4 Summary of facies scheme used in this book. Modified from Pickering *et al*. (1995a).

environments, it is often useful to lump facies together, and it is at this level of description that our facies classes and groups become particularly useful.

We have retained the general outlines of the Mutti and Ricci Lucchi scheme. The main differences are: (i) the abolition of their Facies E that is now distributed among the other facies; (ii) the restriction of Facies Class D (their Facies D) to silt and silt-mud units, rather than including sands; (iii) the addition of a much-needed new Facies Class E for muds and (iv) the definition of three tiers of classification rather than two, allowing for a greater number of facies within a scheme that is still manageable.

We do not use the term 'facies association' in our classification because facies associations represent the temporal and spatial arrangement of facies in the sedimentary record. Instead, our facies classes and groups are composed of texturally similar facies taken out of context from vertically or laterally associated deposits with different inferred origins.

2.3 Facies Class A: Gravels, muddy gravels, gravelly muds, pebbly sands, ≥5% gravel grade

Facies Class A includes the coarsest grained deep-water clastic sediments, with ≥5% pebble-grade or coarser material. This facies class includes clast-supported gravels, gravels with a supporting sand matrix, muddy gravels and gravelly muds. The latter two lithologies

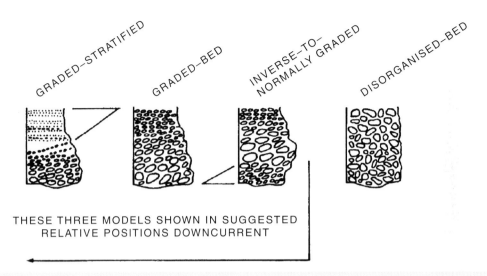

GRADED-STRATIFIED   GRADED-BED   INVERSE-TO-NORMALLY GRADED   DISORGANISED-BED

THESE THREE MODELS SHOWN IN SUGGESTED
RELATIVE POSITIONS DOWNCURRENT

Fig. 2.5   Descriptive facies for resedimented conglomerates proposed by Walker (1975a, 1976, 1977, 1978).

may contain more mud than gravel, but their transport process may be identical to that of deposits of this class with a smaller amount (<10%) of mud matrix; these mud-rich deposits were included in Facies Class F by Stow (1985, 1986).

Deep-water gravels and pebbly sands are commonly termed 'resedimented' to set them apart from fluvial and shallow-marine deposits; they are believed to have accumulated first in shallow water and subsequently to have been transported into deeper water (Walker 1975a). In many cases, features of resedimented gravels taken in isolation from other attributes of a succession are not sufficient to prove deep-water deposition.

Walker (1975a, 1976, 1977, 1978) established a set of descriptive facies for resedimented conglomerates (Fig. 2.5, Table 2.1). He believed that the depositional process and the rate of deposition determined the degree of organisation in the final conglomerate deposit. His fundamental dichotomy between 'organised' and 'disorganised' conglomerates forms the basis of most other classifications (e.g., Kelling & Holroyd 1978; Piper *et al.* 1978), including ours. Walker (1975a) speculated that conglomerate facies are arranged spatially such that disorganised beds are most proximal, inverse-to-normally graded beds are intermediate in position, and graded and graded-stratified beds are most distal. Surlyk (1984) carefully documented the spatial distribution of conglomerate facies for 15 km away from a steep, faulted basin margin in east Greenland, and found no such spatial pattern. Instead, conglomerate facies appear to have no simple relationship with proximality.

2.3.1   Facies Group A1: Disorganised gravels, muddy gravels, gravelly muds and pebbly sands

Gravels in this group may be supported by clast contacts, by a sand matrix, or by a mud matrix. Bed thickness varies, although these deposits tend to occur in medium to thick and very thick beds. The shape of beds generally reflects the topography over which the sediment gravity flows travelled. Four facies are recognised: A1.1, disorganised gravels; A1.2, disorganised muddy gravels; A1.3, disorganised gravelly muds; and A1.4, disorganised pebbly sands.

2.3.1.1   Facies A1.1: Disorganised gravel (Fig. 2.6)

In general, this facies is more thickly bedded than other gravel facies. Exceptionally, single beds may be several tens of metres thick. More commonly, thicknesses are 0.5–5 m. In some cases, beds or layers may be thin to very thin stringers of gravel as little as one pebble thick. Beds may be flat-based to deeply scoured. Upper surface geometry may be irregular, wavy, or with individual clasts projecting out of the bed. Laterally, there may be gradual dilution of the clasts by sand, such that ill-defined stringers of clasts give a crude stratification and the gravel body has a pod-like shape. Alternatively, primary flow margins may be abrupt snouts with a shape like the margin of a water droplet on a smooth horizontal surface.

Clast size ranges from fine pebble to boulder grade, and beds are characteristically poorly sorted (Fig. 2.6). Clast shape is dependent upon composition and inherited shape. Consequently, disorganised gravels have been described with both angular and well-rounded clasts. Clasts lack a well-ordered fabric, although concentrated, elongate clasts may exhibit a poorly defined parallel alignment with bedding, or a slight imbrication (Hiscott & James 1985). Variously oriented, plastically deformed, mudstone intraclasts are common. Facies A1.1 may pass laterally and/or vertically into Facies A1.4.

Transport process: concentrated density flows, inflated sand/gravel flows and debris flows.
Depositional process: 'freezing' on decreasing bottom slopes due to intergranular friction and cohesion.
Selected references: Hendry (1973), Marschalko (1975), Carter and Norris (1977), Long (1977), Stanley *et al.* (1978), Surlyk (1978, 1984), Winn and Dott (1978), Johnson and Walker (1979), Nemec *et al.* (1980), Hein and Walker (1982), Okada and Tandon (1984), Hiscott and James (1985), Fullthorpe and Melillo (1988), Bøe (1994), Bernhardt *et al.* (2011), Dykstra *et al.* (2012), Bayliss & Pickering (2015a, b).

2.3.1.2   Facies A1.2: Disorganised muddy gravel (Fig. 2.7)

Facies A1.2 is matrix-supported, structureless muddy gravel with 10–50% mud- or clay-grade matrix (Fig. 2.7). Units are medium

(a)

(b)

(c)

(d)

Fig. 2.6 Facies A1.1: disorganised gravel. (a) Ordovician fill of slope channel at Grosses Roches, Gaspé Peninsula, Quebec. (b) Middle Eocene Gerbe I System, Ainsa Basin, Spanish Pyrenees. (c) Middle Eocene Charo Canyon fill, Ainsa Basin, Spanish Pyrenees. Hammer ~35 cm long. Note, near-vertical erosional edge of these gravel deposits cut into thin-bedded facies. (d) Middle Eocene Morillo I fan System, Ainsa Basin, Spanish Pyrenees. Note that Facies A1.1 occurs as a decimetre-thick layer of cobbles and gravel ~25 cm above the human scale.

bedded to very thick bedded. Bed shape may appear tabular in small outcrops, but many beds taper to a blunt snout. The bases of beds generally show little erosion into underlying units, but the tops of beds are commonly irregular and hummocky. Large cobbles and boulders may be evenly dispersed throughout the bed, may project above the top of the unit, or may define a coarse-tail grading. Enormous blocks or *olistoliths* may be contained in 100–200 m-thick megabeds containing component divisions identical to this facies, for example a block 300 × 150 × 30 m in a 170 m-thick megabed in Yugoslavia (Marjanac 1985), and platform carbonate slabs 50 m thick and about 1 km long in 'megaturbidites' that reach thicknesses of about 200 m in Spain (Labaume *et al.* 1983a, b).

Clasts tend to show a polymodal grain size distribution, and have a poorly ordered fabric. If a fabric exists, it occurs as a poorly defined parallel to subparallel lamination and/or a crude alignment of platy or rod-shaped clasts in bedding. Clast composition may be igneous,

sedimentary, metamorphic, lithified biogenic material, or unlithified sediment. This facies is commonly associated with other Class A deposits or with deposits of Class F.

Transport process: cohesive debris flows. Enormous slabs may slide into place on a cushion of overpressured or liquefied mud (Labaume *et al.* 1983a, b).

Depositional process: 'freezing' on decreasing bottom slopes due to intergranular friction and cohesion.

Selected references: Jeffery (1922), Crowell (1957), Johnson (1970), Hampton (1972), Mutti and Ricci Lucchi (1972), Rodine and Johnson (1976), Embley (1976), Enos (1977), Kurtz and Anderson (1979), Winn and Dott (1979), Page and Suppe (1981), Lowe (1982), Naylor (1982), Labaume *et al.* (1983a), Middleton and Southard (1984), Hiscott and James (1985), Marjanac (1985), Pickering & Corregidor (2005), Dykstra *et al.* (2012).

202,99M

(a)   (b)

Fig. 2.7 Facies A1.2: disorganised muddy gravel. (a) Plio–Pleistocene Hata Formation, Ashigara Group, central Japan. Scale divisions 10 cm. (b) Well A3 core photograph, Middle Eocene Ainsa System, Ainsa Basin, Spanish Pyrenees. Core width ~6.5 cm. Depth marker in well is metres below ground level.

2.3.1.3 Facies A1.3: Disorganised gravelly mud (Fig. 2.8)

Facies A1.3 includes (i) the pebbly mudstones and olistostromes described from many rock successions, particularly at ancient active continental margins, and (ii) ice-rafted deposits with high concentrations of gravel, including many so-called *Heinrich* layers in the North Atlantic Ocean (Broecker *et al.* 1992). Characteristics are much like those of Facies A1.2, except that the deposits contain 50–95% mud- or clay-grade sediment (Fig. 2.8). Beds range from decimetres to metres in thickness; exceptional beds may be tens of metres thick. Most beds are laterally discontinuous and highly irregular in shape, and display marked variations in the degree of internal organisation, matrix content and bed shape over very short lateral distances. Grading is absent in pebbly mudstones and olistostromes, but may be present in otherwise disorganised ice-rafted units.

Clast compositions are like those for Facies A1.2, with the common addition of abundant silt-mud chips and slabs. In ancient examples, the ductility contrast between the matrix and clasts may result in a tectonically sheared matrix surrounding relatively undeformed clasts.

Transport process: cohesive mud flows (debris flows), and settling from melting icebergs.
Depositional process: for mud flows, 'freezing' on decreasing bottom slopes as shear stress at the base of the flow becomes less than the cohesive strength. For ice-rafted sediments, grain-by-grain deposition.
Selected references: as for Facies A1.2, plus Kuhn and Meischner (1988), Cremer (1989), Eyles (1990), Wang and Hesse (1996), Strand *et al.* (1995), Normark *et al.* (1997), Cornamusini (2004), Bernhardt *et al.* (2011), Bayliss & Pickering (2015a, b).

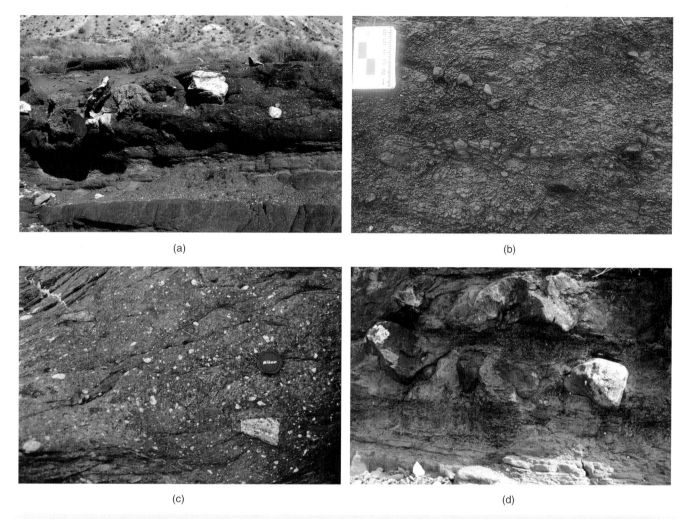

Fig. 2.8 Facies A1.3: disorganised gravelly mud. (a) Miocene (Tortonian) Tabernas Basin, southeast Spain. Camera lens cap for scale. (b) Pliocene–Pleistocene Hata Formation, Ashigara Group, central Japan. Notebook 19 cm high. (c) Upper Precambrian Kongsfjord Formation, Finnmark, Arctic Norway. Camera lens cap for scale. (d) Eocene Richmond Formation, Jamaica. Camera lens cap for scale.

2.3.1.4 Facies A1.4: Disorganised pebbly sand (Fig. 2.9)

Facies A1.4 is distinguished by the dispersion of larger clasts in a sand matrix. Mud-grade sediment can account for up to a few percent. Bed shape and thickness are similar to Facies A1.1. Where clasts are widely dispersed, the definition of bedding surfaces is poor. Scouring, loading and large sole marks are well documented from pebbly sands. Clasts of fine to coarse pebble grade appear to be most common, with dispersed cobbles and boulders being less common. Clast concentration is variable, with irregular patches and stringers of more concentrated clasts occurring down to one pebble in thickness; mudstone clasts are also common to Facies A1.4, and where concentrations are very high, and the clasts are angular, the beds are best termed mud-flake breccias.

Grading, stratification and preferred orientation of the larger clasts are generally absent. Larger clasts may be concentrated toward the base of a bed and then pass abruptly up into clast-poor pebbly sand. Alternatively, there may be a gradual upward decrease in the size of the 'floating' clasts to give the appearance of a coarse-tail grading.

Transport process: concentrated density flows, inflated sand flows.
Depositional process: rapid collective grain deposition of a pebble–sand mixture due to increased intergranular friction as the flows decelerate.
Selected references: Dzułynski *et al.* (1959), Bartow (1966), Ricci Lucchi (1969), Walker and Mutti (1973), Carter and Lindqvist (1975), Lowe (1976a), Piper *et al.* (1978), Surlyk (1978), Walker (1978), Winn and Dott (1978), Hein (1982), Hein and Walker (1982), Cornamusini (2004), Dziadzio *et al.* (2006), Janocko *et al.* (2012b), Stright *et al.* (2014), Pickering *et al.* (2015).

2.3.2 Facies Group A2: Organised gravels and pebbly sands

Organised gravels and pebbly sands are described with close reference to the evolutionary scheme of Lowe (1982). Our recognition of eight facies (Table 2.2) may appear cumbersome, but the facies are easy to recognise and in predominantly conglomeratic successions these divisions allow more flexibility than any simpler classification.

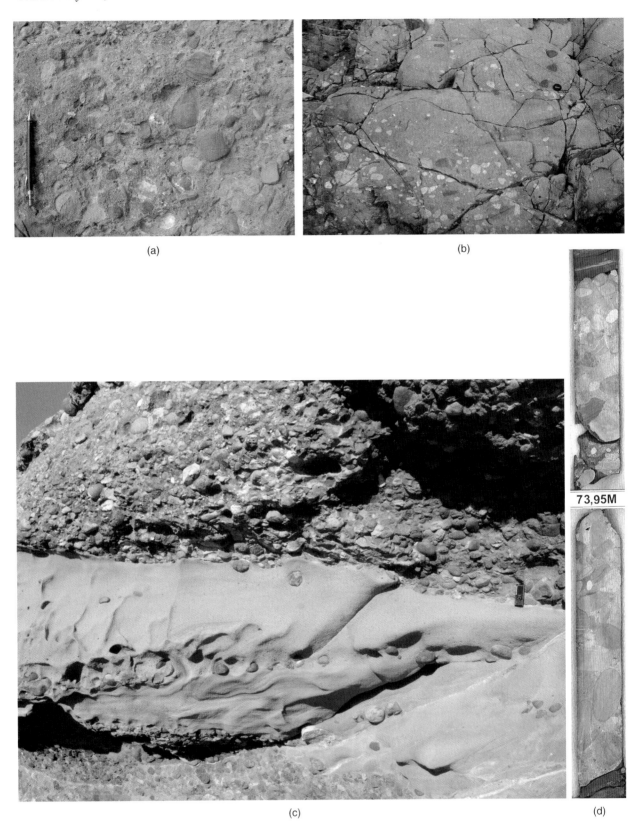

(a)

(b)

(c)

(d)

Fig. 2.9   Facies A1.4: disorganised pebbly sand. (a) Middle Eocene, Gerbe I System, Ainsa Basin, Spanish Pyrenees. (b) Upper Precambrian, Rybachi-Sredni Peninsula, northern Russia. Camera lens cap for scale. (c) Late Cretaceous–Palaeocene Carmelo (Pigeon Point) Formation, Point Lobos, California (*cf.* Anderson *et al.* 2006). Note the crude stratification defined by pebble stringers. Mobile telephone for scale. (d) Well A3 core photograph, Middle Eocene Ainsa System, Ainsa Basin, Spanish Pyrenees. Core width ~6.5 cm. Depth marker in well is metres below ground level.

(a)   (b)

Fig. 2.10 Facies A2.1: stratified gravel. (a) Cambrian–Ordovician Cap Enragé Formation, Quebec Appalachians. Top is to the left. (b) Late Cretaceous–Palaeocene Carmelo (Pigeon Point) Formation, Point Lobos, California (*cf.* Anderson *et al.* 2006). Note the low-angle stratification defined by pebble-rich layers.

2.3.2.1 Facies A2.1: Stratified gravel (Fig. 2.10)

Stratification in resedimented gravels is most commonly reported in fine pebble gravels (Figs 2.10) and pebbly sands. The best ancient examples of stratified, coarse grained, clast-supported gravels known to us have been described by Winn and Dott (1977, 1979) from southern Chile. These beds are lenticular and wedge-shaped with inclined strata up to 12 m thick and with dune-shaped bodies up to 4 m thick. Individual strata range from a single pebble thickness to over 1 m thick. Scours and erosional sole marks have been described from these ancient gravels.

Imbrication is well developed: Winn and Dott (1977, 1979) report *ab* planes of clasts dipping upstream with *a*-axes parallel to the flow direction. Large-scale stratification, especially where stratal boundaries are not sharp, may be difficult to distinguish from individual, stacked, graded and structureless gravel beds.

Modern gravel waves of 2–5 m height and 50–100 m wavelength have been described from water depths of 2000–4500 m on the Laurentian Fan (Piper *et al.* 1988). Similar gravel waves occur on the Var Fan offshore from Nice, France (Malinverno *et al.* 1988). The Laurentian Fan gravel waves were molded by a late Pleistocene glacial outburst event (D.J.W. Piper, in Wynn *et al.* 2002b). Although one might expect these gravel waves to contain stratified or cross-stratified sediment, limited submersible observations instead indicate structureless to graded gravel (Hughes Clarke *et al.* 1990). To date, no modern deep-water bedforms have been shown to contain internal stratification like that of Facies A2.1. Ito and Saito (2006), however, have described Pliocene–Pleistocene examples from the Boso Peninsula, southeast Japan.

Transport process: concentrated density flows.
Depositional process: grain-by-grain deposition from suspension and then traction transport as bed-load.
Selected references: Winn and Dott (1977, 1979), Hein (1982), Piper *et al.* (1988), Sohn (1997), Wynn *et al.* (2002a), Ito and Saito (2006), Dykstra *et al.* (2012), Ito *et al.* (2014).

2.3.2.2 Facies A2.2: Inversely graded gravel (Fig. 2.11)

Inversely graded beds of gravel make up a significant proportion of many resedimented, coarse-grained successions. Beds are commonly lenticular, with basal erosion, lateral thinning and variations in clast concentration causing complex bed shapes. Inversely graded beds reach a maximum thickness of several metres, but most commonly are 0.5–4 m thick. Poor sorting and large clast sizes are typical of this facies. Clast imbrication is well developed.

The entire bed may be inversely graded (Figs 2.11), or the gravel overlying the basal inversely graded part may be structureless or normally graded. In thicker inversely graded beds, the lowest 5–20% of the bed commonly contains finer clasts (generally fine to coarse pebbles) than the immediately overlying part. The transition between these two parts of the bed is commonly abrupt; the coarser clasts are concentrated at a certain distance above the base of the bed.

The tops of inversely graded gravel deposits may be abrupt, showing a sharp break between gravel and sand, or there may be an upward increase in sand content such that the uppermost part of a bed has a bimodal size distribution.

Transport process: concentrated density flows undergoing late-stage flow transformations.
Depositional process: rapid deposition of a concentrated traction carpet/dispersion near the bed due to increased intergranular friction. The strong imbrication is caused by intense grain interaction. The inverse grading may have the same cause, or might result from coarser subpopulations lagging behind finer subpopulations along the transport path (Hand & Ellison 1985).
Selected references: Davies and Walker (1974), Surlyk (1978, 1984), Howell and Link (1979), Johnson and Walker (1979), Winn and Dott (1979), Nemec *et al.* (1980), Watson (1981), Hein (1982), Okada and Tandon (1984), Soh (1987), Surpless *et al.* (2009).

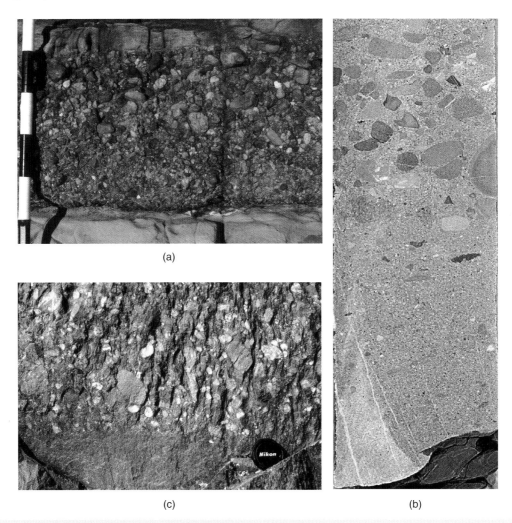

(a)

(c)

(b)

Fig. 2.11 Facies A2.2: inversely graded gravel. (a) Cretaceous Nakaminato Formation, Ibaraki Prefecture, Japan. Scale divisions 10 cm. (b) Well A3, Middle Eocene Ainsa System, Ainsa Basin, Spanish Pyrenees. Core width ~6.5 cm. (c) Lower Silurian (Llandovery) Caban Coch conglomerates, Elan Valley, Welsh Basin, UK. Camera lens cap scale.

2.3.2.3 Facies A2.3: Normally graded gravel (Fig. 2.12)

Normally graded gravels tend to be finer grained than inversely graded or disorganised beds within any single succession. Bed thicknesses, however, are similar to those of the other gravel facies, ranging from about 0.5 to several metres (Fig. 2.12). Beds show marked thickness changes as a result of localised deep scours and more gradual, large-scale down-cutting into pebbly sands. Normally graded gravel appears to be one of the most abundant of the clast-supported facies.

Normal grading occurs in several modes. Abruptly graded beds mainly show a coarse-tail grading where the coarsest clasts are only present in the lowest part of the bed and rapidly give way upward to fine pebbles. Gradual distribution grading from cobbles to granule sand is less common, but can occur in very thick beds. Imbrication appears to be less well developed in this facies than in the inversely graded gravel facies (Facies A2.2).

Transport process: concentrated density flows.
Depositional process: grain-by-grain deposition from suspension.
   The clasts undergo little or no traction transport after reaching the bed, probably because of relatively rapid deposition.

Selected references: Marschalko (1964), Hendry (1972, 1978), Mutti and Ricci Lucchi (1972), Davies and Walker (1974), Walker (1977), Winn and Dott (1978), Nemec *et al.* (1980), Hein (1982), Hein and Walker (1982), Surlyk (1984), Siedlecka *et al.* (1994), Surpless *et al.* (2009).

2.3.2.4 Facies A2.4: Graded-stratified gravel (Fig. 2.13)

Walker (1975a, 1976) established a 'graded-stratified bed model' in which pebbly sands with parallel, inclined and cross-stratification overlie graded gravel (Fig. 2.13). Beds of graded-stratified gravel are generally thinner bedded and finer grained than other clast-supported gravels. Bed shape is less variable, with sharp planar bases, although some scouring characterised by trough-shaped scour-and-fill stratification is common.

Clast size may decrease progressively toward the top of the bed. Alternatively, the lower clast-supported part of the bed may show 'delayed grading'. Graded-stratified gravels are considered to be transitional between clast-supported cobble/pebble gravels and graded-stratified pebbly sands (Facies A2.8).

Fig. 2.12 Facies A2.3: normally graded gravel. (a) Middle Eocene Morillo System, Ainsa Basin, Spanish Pyrenees. Compass-clinometer for scale. (b) Cretaceous Nakaminato Formation, Ibaraki Prefecture, Japan. Scale divisions 10 cm. Note dish structures in underlying sandstone. (c) Middle Eocene Gerbe System, Ainsa Basin, Spanish Pyrenees. 15-cm scale.

Transport process: concentrated density flows.

Depositional process: grain-by-grain deposition from suspension followed by traction during deposition of the upper part of the bed only.

Selected references: Hubert *et al.* (1970), Hendry (1972), Mutti and Ricci Lucchi (1972), Davies and Walker (1974), Rocheleau and Lajoie (1974), Walker (1975a, 1976, 1977, 1978), Aalto (1976), Surlyk (1978, 1984), Johnson and Walker (1979), Hein (1982), Hein and Walker (1982), Vicente Bravo and Robles (1995), Cornamusini (2004), Pickering *et al.* (2015).

2.3.2.5   Facies A2.5: Stratified pebbly sand (Fig. 2.14)

Beds of stratified pebbly sand show an alternation of pebble- and sand-rich layers (Fig. 2.14). Commonly, individual strata have gradational contacts with both normal and inverse grading. Strata may pinch and swell and split into irregular stringers and lenses. Stratification may also occur on a finer scale with alternations of (a) granule sand with a few pebbles and (b) coarse- to medium-grained sand.

Bed shape is extremely variable, with typical bed thicknesses from 0.5–3 m. Defining individual flow deposits may be difficult because a pebbly sandstone unit may be composite. Generally, entire beds are poorly sorted; coarse pebbles and, rarely, cobbles may be present as irregular stringers and scour fills.

Transport process: concentrated density flows.

Depositional process: grain-by-grain deposition from suspension followed by traction transport as bed load.

Selected references: Hendry (1973, 1978), Hein (1982), Hein and Walker (1982), Lowe (1982), Surlyk (1984).

2.3.2.6   Facies A2.6: Inversely graded pebbly sand (Fig. 2.15)

Pebbly sands with well-developed inverse grading throughout are analogous to the inversely graded gravels of Facies A2.2, and appear to be relatively rare. It is more usual to find a pebble-free zone several cm thick that passes rapidly up into structureless or normally graded pebbly sand. Alternatively, there may be alternations at a scale of 5–10 cm of (i) pebble-poor zones and (ii) pebbly sand, giving an indistinct stratification. In such cases, it is very difficult to define individual sedimentation units.

Facies A2.6 tends to occur in planar-based beds that are generally thinner than beds of other pebbly-sand facies. Well-developed, multiple, inversely graded layers suggest a transition into thicker bedded granule sands with 'spaced stratification' (Facies B2.1).

Transport process: concentrated density flows.

Depositional process: the inverse grading might be formed by a 'kinetic sieve' process, whereby coarser grains rise to the top of a sheared dispersion (Middleton 1970). Alternatively, the inverse grading might result from coarser subpopulations lagging behind finer subpopulations along the transport path (Hand & Ellison 1985).

Selected reference: Lowe (1982).

2.3.2.7   Facies A2.7: Normally graded pebbly sand (Fig. 2.16)

Normally graded pebbly sands are very common in deep-water clastic successions. Generally, this facies is thicker bedded than the stratified and inversely graded pebbly sands. Common scour structures tend to give most beds an irregular appearance. Bed contacts may be diffuse where amalgamation occurs and clast concentrations are low.

Facies A2.7 typically displays well-defined normal grading that is most commonly coarse-tail grading, although distribution grading also occurs. There are many reported examples of beds 2–3 m thick showing normal grading from base to top.

Transport process: concentrated density flows.

Depositional process: grain-by-grain deposition from suspension, with rapid burial and no significant traction transport on the bed.

Selected references: Hubert *et al.* (1970), Aalto (1976), Long (1977), Walker (1977, 1978), Stanley *et al.* (1978), Watson (1981), Hein (1982), Surpless *et al.* (2009).

(a)

(b)

(c)

Fig. 2.13 Facies A2.4: graded-stratified gravel. (a) Middle Eocene Morillo System, Ainsa Basin, Spanish Pyrenees. Compass-clinometer for scale. (b) Lower Ordovician succession at Grosse Roches, Gaspé Peninsula, Quebec. Gerard Middleton provides scale. (c) Upper Precambrian Kongsfjord Formation, Finnmark, Arctic Norway. Camera lens cap for scale.

2.3.2.8 Facies A2.8: Graded-stratified pebbly sand (Fig. 2.17)

Facies A2.8 is similar to Facies A2.7 in terms of bed thickness, bed shape and clast size. Lateral transitions between these two facies are common. Basal scouring is common, and upper bed contacts are sometimes poorly defined where stringers of pebbles occur high in the bed.

Beds of this facies show an overall grading from base to top, although layers of coarser clasts are repeated upward throughout beds (Fig. 2.17). However, clasts coarser than very fine pebbles and granules appear to be confined to the lower graded portions of beds. Stratification may be parallel, oblique, multiple sets, draping scours, or megaripple cross-bedding. Facies A2.8 is considered to be transitional between graded-stratified gravels of Facies A2.4 and the sands of Facies Class B.

Transport process: concentrated density flows, becoming more dilute with time at a single locality (i.e., waning conditions of Kneller 1995).

Fig. 2.14 Facies A2.5: stratified pebbly sand. (a) Oligocene Annot Sandstone Formation, Contes sub-basin, southeast France. Mobile phone left of centre for scale ~10 cm long. (b) Close-up of left-hand picture to show detail of stratification in Facies A2.4. Mobile phone at lower left for scale ~10 cm long. (c) Eocene Richmond Formation, Wagwater Group, Jamaica. Camera lens cap for sale.

Depositional process: grain-by-grain deposition from suspension. Initially, deposition is so rapid that no subsequent traction transport takes place. At higher levels in the deposit, grains are transported as bedload to form stratification before being buried.

Selected references: Hubert *et al.* (1970), Rocheleau and Lajoie (1974), Aalto (1976), Walker (1978), Hein (1982), Hein and Walker (1982), Surlyk (1984), Pickering *et al.* (2015).

2.4 Facies Class B: Sands, >80% sand grade, <5% pebble grade

This class comprises sand beds with <20% mud and silt matrix and <5% pebble-grade material. Facies Class B is divided into an organised and a disorganised facies group, based on the presence or absence of well-defined sedimentary structures. Bed thickness and

Fig. 2.15 Facies A2.6: inversely graded pebbly sand. Eocene, ODP Site 1276A, Newfoundland Basin (210-1276A-9R2, 55–74 cm). Above the large pebbles, the 50 cm-thick deposit becomes normally graded. This SGF deposit is at the base of ~2 m of amalgamated sand beds. Scale bar 10 cm.

Fig. 2.16 Facies A2.7: normally graded pebbly sand. Pliocene Kiyomusi Formation, Boso Peninsula, Japan. Scale divisions 10 cm.

shape are highly variable. Most beds in this facies class cannot be described successfully using the Bouma (1962) scheme for classic turbidites.

Deposits with characteristics common to our Facies Class B are well documented; for example, some of the 'arenaceous facies' of Mutti and Ricci Lucchi (1972), the structureless beds of Sanders (1965), some 'fluxoturbidites' of Stanley and Unrug (1972), Dzułynski *et al.* (1959) and Kuenen (1964).

2.4.1   Facies Group B1: Disorganised sands

Disorganised or structureless sands, comparable to Walker and Mutti's (1973) Facies B1 and B2 massive sands with or without dish structure, and Mutti and Ricci Lucchi's (1972) Facies B sands, are recorded from many deep-water successions (Stauffer 1967; Carter & Lindqvist 1975; Keith & Friedman 1977; Piper *et al.* 1978; Hiscott & Middleton 1979; Cas 1979; Hiscott 1980; Lowe 1982; Normark *et al.* 1997; Plink-Björklund & Steel 2004; Lien *et al.* 2006; Janocko *et al.* 2012b). Facies Group B1 consists of two facies.

2.4.1.1   Facies B1.1: Thick/medium-bedded, disorganised sands (Fig. 2.18)

Facies B1.1 consists of laterally continuous, parallel-sided to highly irregular, medium to thick beds (Fig. 2.18). Sole marks tend to be rare. Grading is absent or poorly developed as a coarse-tail grading with small pebbles and granules concentrated in a thin basal layer. The most obvious internal sedimentary feature may be fluid-escape structures. These tend to occur in the upper half of beds and include subvertical sheet structures (Laird 1970), dish structures (Fig. 2.18) and fluidisation pipes and pillars (Fig. 2.18) (Wentworth 1967; Lowe & LoPicollo 1974; Lowe 1975; Mount 1993; Nichols *et al.* 1994). Dish structures seem to be characteristic of the better sorted sands of this facies, in which upward percolation of escaping pore-water is able to form relatively impermeable 'consolidation laminations' (Lowe & LoPicollo 1974) that, when breached, develop a characteristic dish shape. In more poorly sorted sands, pore-fluid escape is not general, but is localised along sheets and pillars (e.g., Hiscott & Middleton 1979) with clay-plugged wall regions (Mount 1993).

In some cases, liquefaction was so pervasive in what were then shallowly buried beds of this facies that bed tops developed either diapir-like upward protuberances (Fig. 2.18), pinch-and-swell geometry as a result of intrastratal flow of sand, or abundant pseudo-clasts generated by wholesale downward collapse of overlying units (Fig. 2.18). It is probably similar severe disruption that leads to outcrop-scale, or larger, sand injection complexes (e.g., Hiscott 1979; Nichols 1995; Dixon *et al.* 1995).

Transport process: concentrated density flows, inflated sandflows (see Table 1.1 equivalencies).
Depositional process: deposition from a rapidly decelerating flow, so that tractional structures are unable to form (Arnott & Hand 1989; Kneller & Branney 1995), or rapid mass deposition due to intergranular friction in a concentrated dispersion. In either case, the resultant open grain packing may collapse during or after deposition of the entire bed, resulting in escape of surplus pore-fluids and formation of fluid-escape structures and/or other deformation structures.
Selected references: Stauffer (1967), Middleton (1970), Carter and Lindqvist (1975), Aalto (1976), Lowe (1976a,b), Keith and

(a) (b)

Fig. 2.17 Facies A2.8: graded-stratified pebbly sand. (a) Lower Ordovician Tourelle Formation, Gaspé Peninsula, Quebec. Arrow 10 cm long. (b) Silurian Milliners Arm Formation, New World Island, Newfoundland, Canada. Note load and flame structures created during deposition with residual shear by the overriding flow, suggesting palaeoflow towards right. Camera lens cap scale.

Friedman (1977), Piper *et al* (1978), Cas (1979) (a2 division), Hiscott and Middleton (1979), Jordan (1981), Hein (1982), Lowe (1982) (S3 division), Surlyk (1984), Kneller and Branney (1995), Hiscott *et al*. (1997a), Normark *et al*. (1997), Badescu *et al*. (2000), Plink-Bjorklund and Steel (2004), Cornamusini (2004).

2.4.1.2 Facies B1.2: Thin-bedded, coarse-grained sands (Fig. 2.19)

Facies B1.2 is distinguished from the other facies in the class by its thin-bedded nature and very coarse grain size (Fig. 2.19). Pebbles are rare. Angular silt and mud clasts may occur. The beds are internally structureless.

Beds are typically irregular with common wedge-shaped or pinch-and-swell geometry. Tops are sharp. This facies is associated with Facies B2.2 in many cases, and in some respects resembles Facies E of Mutti and Ricci Lucchi (1972, 1975), but is generally coarser grained and lacks internal structures. There is no grading, and rarely small pebbles may occur within beds as stringers. In some beds, pebbles protrude above the bed top into overlying facies.

Transport process: bed-load transport beneath turbulent SGFs or strong bottom currents. Transport may be short, and the major process may be winnowing out of finer grain sizes.
Depositional process: grain-by-grain deposition of bed load.
Selected references: Mutti and Ricci Lucchi (1972, 1975), Mutti (1977).

2.4.2 Facies Group B2: Organised sands

This group includes any sands showing well-defined sedimentary structures that are clearly not part of the Bouma sequence for sand-mud turbidites. Various fluid-escape structures may overprint, the original current-generated structures. Our Facies Group B2 contains deposits that show many of the features of Mutti and Ricci

Lucchi's (1972) subfacies B1 and B2. We recognise two facies in this group.

2.4.2.1 Facies B2.1: Parallel-stratified sands (Fig. 2.20)

Hiscott and Middleton (1979, 1980) were the first to document this facies in detail. Much of the stratification in this facies is defined by bands, several centimetres to 10 cm thick, each showing basal inverse grading (Fig. 2.20). These bands are present as an internal structure within thick to very thick beds that are overall normally graded. From the base to the top of each stratification band, Hiscott and Middleton (1979) recognised the following sequence of structures and textures: (i) a basal horizontal or near-horizontal erosional surface; (ii) a subdivision of inversely graded sand, typically grading from 2–3 ϕ to approximately 1 ϕ and (iii) a subdivision of structureless or massive −1 to 1 ϕ sand, commonly with well-developed grain imbrication.

Hiscott (1994a) renamed this structure 'spaced stratification' to replace the popular genetic name 'traction-carpet stratification' (Section 1.6.1). Sand between divisions of spaced stratification is structureless and may contain fluid-escape structures like Facies B1.1. Bed tops grade into silt and may have an upper division of ripple lamination.

Transport process: concentrated density flows.
Depositional process: repeated burst-sweep cycles, with the downward sweep of a large turbulent eddy forming both the lower erosional surface and the inverse grading in a sheared grain layer ≤ 1 cm thick, followed by sufficiently rapid bed aggradation to deposit several centimetres of sand lacking lamination (Hiscott 1994b). Thicker structureless divisions record longer periods of rapid grain-by-grain fallout without traction (like Facies B1.1).
Selected references: Chipping (1972), Mutti and Ricci Lucchi (1975) (subfacies B1), Hendry (1972), Hiscott and Middleton (1979, 1980), Lowe (1982), Hiscott (1994b), Sohn (1997), O'Brien *et al*. (2001), Plink-Bjorklund and Steel (2004), Cantero *et al*. (2012); Stevenson *et al*. (2014).

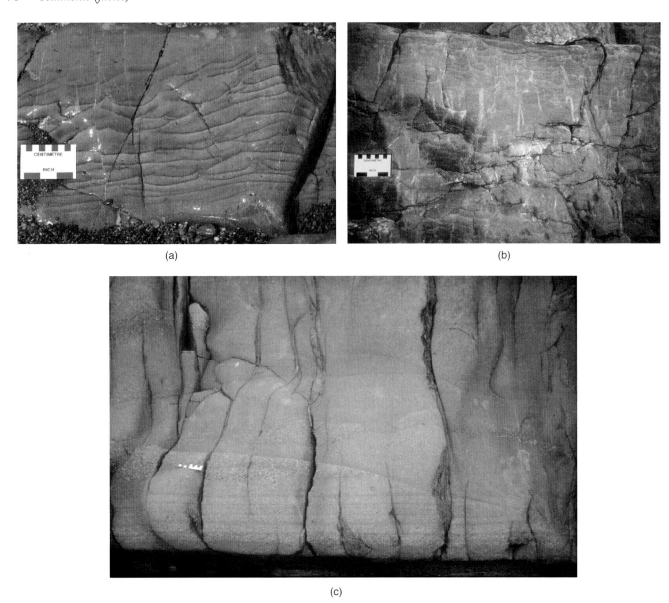

Fig. 2.18 Facies B1.1: thick/medium-bedded, disorganised sands. All scales in centimetres. (a), (b) Cambrian–Ordovician Cap Enragé Formation, Quebec Appalachians. Paler sediment between dish structures, and paler sediment in fluid-escape pillars, is deficient in matrix because of flushing by escaping pore-water. (c) Lower Silurian (Llandovery) Aberystwyth Grits, Welsh Basin. Amalgamated bed including at least four scour-based event-deposits of structureless sand above a basal unit of Facies B2.1 parallel-stratified sand.

2.4.2.2 Facies B2.2: Cross-stratified sands (Fig. 2.21)

The granule grade to coarse-grained sands of Facies B2.2 are much better sorted than sands of other facies. The thin-bedded varieties of this facies are distinctive in having a very coarse grain size for their thickness.

The beds consist of cross-stratification in sets that are typically from 10–25 cm thick (Fig. 2.21). They may occur as solitary sets or co-sets that are tabular to trough shaped. Internally, the lamination may have a low dip, or may rest at or near the angle of repose. Beds are commonly irregular, with lensing, splitting and amalgamation; basal contacts may be erosive and the tops are sharply defined. Foresets are defined by alternations of coarser and finer grained layers with concentrations of coarser grained material toward the toes of the foresets. Over-steepened and recumbently folded cross-stratification (similar to Type 1 of Allen & Banks 1972) may occur in Facies B2.2. Beds rarely contain abundant mud chips.

Transport process: bed-load transport beneath turbidity currents or strong bottom currents in confined channels.
Depositional process: avalanching (grain flow) or intermittent suspension transport of grain dispersions over the crests of medium- to large-scale bed-forms, or into scours.
Selected references: Hubert (1966b), Scott (1966), Piper (1970), Mutti and Ricci Lucchi (1972), Mutti (1977), Hiscott and Middleton (1979), Hein (1982), Hein and Walker (1982), Pickering (1982a), Lowe (1982), Vicente Bravo and Robles (1995), Lien *et al.* (2006),

Fig. 2.19 Facies B1.2: thin-bedded, coarse-grained sands. (a) Middle to Upper Ordovician (Caradoc–Ashgill) Point Leamington Formation, north-central Newfoundland, Canada. (b) Nummulite-rich bed, Well L2, fan lateral-margin and basin-slope deposits, Middle Eocene Ainsa System, Ainsa Basin, Spanish Pyrenees. Core width ~6.5 cm.

Pickering *et al.* (2015). NB: lenticular deposits with relatively thin-bedded 'dune' forms were defined as Facies E by Mutti and Ricci Lucchi (1972, 1975) and Mutti (1977); we instead assign such beds to this facies.

2.5 Facies Class C: Sand–mud couplets and muddy sands, 20–80% sand grade, <80% mud grade (mostly silt)

Most beds of Facies Class C can be described using the Bouma (1962) sequence for classic turbidites. Bed shape is variable and cannot be used to differentiate the constituent facies. In general, however, beds are sheet-like. In many cases, the deposits of single sediment gravity flows are graded, with most of the mud present as a cap on the top of one or more sandy division.

Deposits of this class characterise deep-water successions (e.g., Kuenen & Migliorini 1950; Dzułynski *et al.* 1959; Bouma 1962, 1964; Bouma & Brouwer 1964; Dzułynski & Walton 1965; Macdonald *et al.* 2011a). The depositional processes for sand beds in this class are well understood because of considerable experimental and theoretical work (Section 1.5).

This facies class is similar to, although not strictly analogous to, Facies C1 and C2 of Mutti and Ricci Lucchi (1972, 1975) and Walker and Mutti (1973). Facies Class C beds are the 'classic' or 'classical' turbidites of Walker (1976, 1978, 1992). The terms 'proximal' and 'distal' as qualifying terms for the beds of this class (Walker 1967b, 1970) are not employed in our classification to distinguish thicker from thinner beds, because thin beds need not be distal relative to the point of sediment supply (Nelson *et al.* 1975).

Facies Class C is divided into an organised and a disorganised group. Facies distinctions are made on the basis of textural homogeneity for the disorganised group, and bed thickness for the organised group. Bed thickness is a useful environmental indicator and permits easy recognition in the field. Bed thickness is broadly related to grain size (Sadler 1982; Middleton & Neal 1989; Dade *et al.* 1994) and to the sequence of internal sedimentary structures, such that the facies represent a gradational spectrum from the coarsest grained and thickest to the finest grained and thinnest beds.

2.5.1 Facies Group C1: Disorganised muddy sands

2.5.1.1 Facies C1.1: Poorly sorted muddy sands (Fig. 2.22)

Facies C1.1 is characterised by a high content of silt- and clay-grade sediment (up to 80%) in poorly sorted beds (Fig. 2.22). Maximum grain sizes are typically in the range of coarse- to fine-grained sand. Normal distribution grading may occur and, in some cases, a coarse-tail grading may be present in very coarse- to coarse-grained sand in the lower part of the bed. The uppermost part of the bed is silty mud and the lower part is muddy sand. Bounding surfaces are generally clearly defined, with the bases showing a range of sole markings. The tops are planar or gradational.

Internal sedimentary structures are mainly absent, but indistinct parallel lamination may occur in the lowest few centimetres, as well as convolute lamination associated with pseudonodules of silty mud. These liquefaction structures can give the beds a swirled appearance, and are partly responsible for the fact that they have been termed 'slurry beds' by some workers (Morris 1971; Hiscott & Middleton 1979; Strong & Walker 1981).

Silty mud clasts or 'chips' occur in varying proportions (Fig. 2.22). In some beds, large rafts up to several metres long are found 'suspended' within the deposit. These rafts may be much longer than the beds are thick (Hiscott 1980).

Fig. 2.20 Facies B2.1: parallel-stratified sands. (a) Fine-grained sandstone example, Permian Karoo System, South Africa. (b), (c) and (d) are from the Lower Ordovician Tourelle Formation, Quebec Appalachians. All scales in centimetres. (b) 165 cm-thick bed, graded overall but consisting of inversely graded spaced stratification 'bands'. (c) Lower part of graded bed showing stratified division. (d) Detail of part of a division of spaced stratification with arrows at the base of each stratum. Each stratum is inversely graded in its basal 1–1.5 cm from fine sand to medium or coarse sand. (e) Middle Eocene, Ainsa System, Ainsa Basin, Spanish Pyrenees, Well A1. Core width ~6.5 cm.

Transport process: sand/mud-load cohesive flows.

Depositional process: rapid mass deposition due to increased intergranular friction or cohesion. The deposit remains sufficiently plastic for gravitationally induced loading to occur.

Selected references: Wood and Smith (1959), Burne (1970), Morris (1971), Skipper and Middleton (1975), Mutti *et al.* (1978), Hiscott and Middleton (1979), Hiscott (1980), Pickering (1981b), Strong and Walker (1981), Pickering and Hiscott (1985), Myrow and Hiscott (1991), Normark *et al.* (1997), Lowe and Guy (2000), Lowe *et al.* (2003), Pyles and Jennette (2009), Brunt *et al.* (2012), Talling *et al.* (2013), Terlaky & Arnott (2014).

2.5.1.2   Facies C1.2: Mottled muddy sands (Fig. 2.23)

Mottled muddy sands occur mostly as very thin to medium beds (<1 cm to ~20 cm), with an irregular to sheet-like shape. Both tops and bases of beds may be sharp or gradational and are variable from one part to another of the same bed. Cross-lamination and parallel lamination are rare (Fig. 2.23), but irregular concentrations (layers or lenses) of coarser material are common. An indistinct normal or inverse grading can be present.

Bioturbation is commonly pervasive and may mask any primary physical structures. The grain size is mostly fine sand (gradational to the mottled silts of Facies D1.3), and beds are poorly to moderately

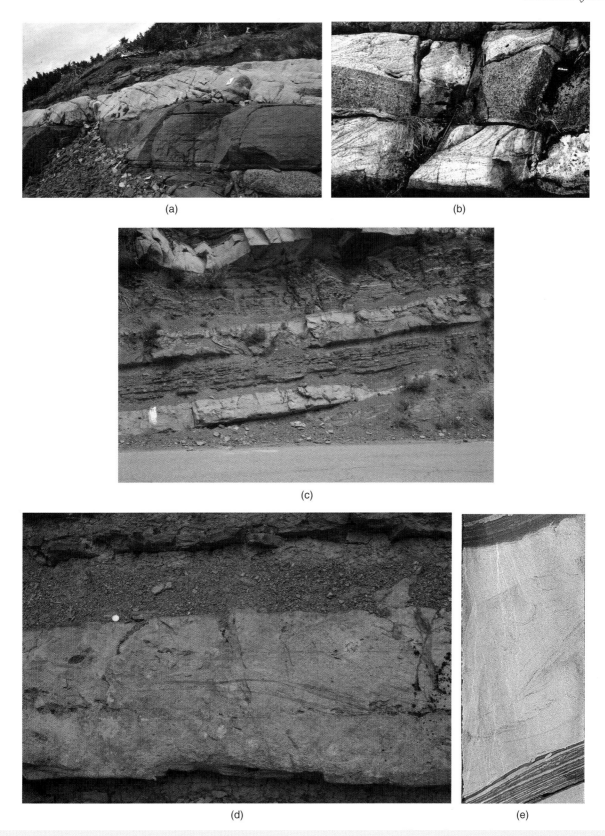

Fig. 2.21 Facies B2.2: large-scale cross-stratified sands. (a) Cambrian Blow-me-down Brook Formation, Woods Island, Bay of Islands, western Newfoundland, Canada. Notebook (centre right) is ~20 cm high. (b) Upper Precambrian Kongsfjord Formation, Finnmark, Arctic Norway. (c) Bedform (~25 cm thick) of very coarse-grained, cross-stratified, sandstone, Oligocene Grès d'Annot Formation, Peira Cava, southeast France. (d) Close-up of part of Facies B2.2 shown in part (c) to show multi-storey sets; 1 Euro coin for scale. (e) Cross-stratified sands in core. Middle Eocene, Ainsa System, Ainsa Basin, Spanish Pyrenees. Core width ~6.5 cm.

(a)   (b)

Fig. 2.22 Facies C1.1: poorly sorted muddy sand. (a) Cretaceous (Albian), ODP Site 1276A, Newfoundland Basin (210-1276A-69R2, 43–65 cm). Note floating and flattened shale clasts and sharp bed top. Scale bar 10 cm. (b) Close-up of part of a 2–3 m-thick SGF deposit, Cretaceous Nakaminato Formation, Ibaraki Prefecture, Japan. Scale divisions 10 cm.

well sorted. The grains are typically of both terrigenous and biogenic origin, depending on the original sediment source.

Transport process: winnowing of fines, and short, repeated bedload transport of sands by strong bottom currents.
Depositional process: grain-by-grain deposition of coarse load, and thorough post-depositional mixing of sand and mud by burrowers.
Selected references: McCave *et al.* (1980), Stow (1982), Stow and Piper (1984a,b), Gonthier *et al.* (1984), Shipboard Scientific Party (2004a).

2.5.2   Facies Group C2: Organised sand–mud couplets

Facies Group C2 consists of moderately well sorted to poorly sorted sand–mud couplets showing partial or complete Bouma sequences (Fig. 2.24). Beds tend to show marked normal grading (Kuenen 1953; Ksiazkiewicz 1954). Both tool and scour marks are common sole markings. The bases of beds may show deep scour structures, load structures, or may be smooth and planar. The tops of beds are generally smooth to planar if the upper part of the bed contains substantial amounts of mud. Bioturbation may occur within or throughout the bed, but is more common toward the tops of beds. Liquefaction structures, including convolution and fluid-escape structures, are common.

The Bouma (1962) sequence of internal structures, T_{abcde}, may be complete, but more commonly beds show base-, top- or middle-absent sequences. Climbing-ripple lamination may be present in fine- to very fine-grained sands/sandstones (*cf.* Jobe *et al.* 2012).

We recognise five facies in this group, the first three based entirely on bed thickness (Table 2.4):

- Facies C2.1, very thick- and thick-bedded sand–mud couplets;
- Facies C2.2, medium-bedded sand–mud couplets;
- Facies C2.3, thin-bedded sand–mud couplets;

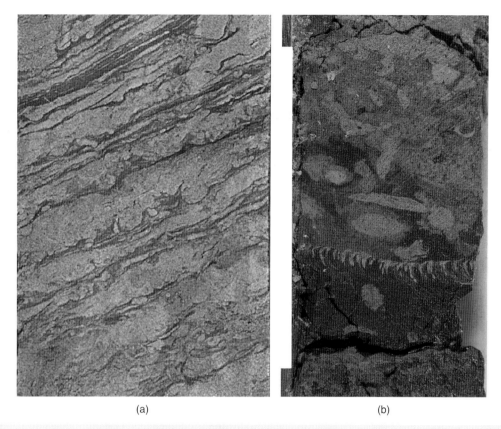

(a)   (b)

Fig. 2.23 Facies C1.2: mottled muddy sands in core. (a) Middle Eocene Ainsa System, Ainsa Basin, Spanish Pyrenees. (b) Middle Miocene, ODP Site 792E, Izu-Bonin forearc basin (126-792E-21R2, 6–18 cm). Scale bar 10 cm.

- Facies C2.4, thick- to very thick-bedded, mud-dominated sand–mud couplets, and
- Facies C2.5, medium to very thick-bedded, mud-dominated slurried sand–mud couplets.

Beds in Facies C2.1 generally begin with Bouma division T_a, (Fig. 2.25a) those in Facies C2.2 with division T_b (Fig. 2.25b), and those in Facies C2.3 with division T_c (Figs 2.25c, d, e). A distinctive form of Facies C2.3 is very thin beds, commonly <3 cm thick, with sand:mud >1.0, in which there are low amplitude ripples (<2 cm high) with relatively long wavelengths up to decimetres. These ripples may occur with stoss-side erosion and only lee-side preservation, followed abruptly by a silt/mud drape giving the beds a 'form surface'. Some of Mutti's (1977) Facies E examples are included in this category. Another unusual variant of Facies C2.3 is documented from ODP Leg 169 on a sedimented oceanic spreading centre (the Bent Hill massive sulphide deposits, Middle Valley at the northern end of the Juan de Fuca Ridge) as sulphide-banded, cross-laminated sandstone in which the development of Cu-rich sulfide mimics sedimentary structures and replaces the SGF sandstone deposits (Zierenberg *et al.* 2000).

Facies C2.1 is deposited from flows transitional between concentrated density flows and turbidity currents. Facies C2.2 through C2.3 are deposited from turbidity currents of progressively lower concentration and velocity.

The predominant depositional process is grain-by-grain deposition from suspension, followed either by burial (Bouma division T_a; Arnott & Hand 1989) or by tractional transport as bed load (Bouma

divisions T_b, T_c). The muddy upper divisions are deposited in the same manner as mud turbidites of Facies Class D.

Facies C2.4 comprises thick- to very thick-bedded, mud-dominated sand–mud couplets, generally 1–15 m thick, that show internal evidence of flow reversals during deposition (Fig. 2.26), and that commonly have a mud cap that accounts for about 80% of the thickness of the deposit. The sandy basal divisions are graded, although the grading may be step-wise with mud breaks between divisions with opposed senses of flow (Hiscott & Pickering 1984; Pickering & Hiscott 1985). Internal structures include megaripple form sets, ripple and climbing-ripple lamination, wavy and parallel lamination, and pseudonodules. Pickering and Hiscott (1985) interpret these unusual turbidites as the deposits of large-volume, high-concentration turbidity currents confined within small basins, such that multiple deflections and reflections of the initial current occur during deposition of the sand-silt load. The thick silty mud caps are deposited by rapid settling of flocs formed in the highly concentrated mud cloud that becomes ponded above the basin floor after cessation of flow. In order to emphasise the importance of reflections and ponding in deposition of Facies C2.4, Pickering and Hiscott (1985) called the emplacing flows *contained turbidity currents*.

Selected references: literature on Facies C2.1, C2.2 and C2.3 is abundant. Instead of providing a long list, we refer the reader to references in the books edited by Doyle and Pilkey (1979), Siemers *et al.* (1981) and Tillman and Ali (1982). Examples of Facies C2.4 are described by van Andel and Komar (1969), Ricci

(a) (b)

(c)

Fig. 2.24 (a) Facies association of mainly Facies Class C beds (Class C2.1 with minor C2.2). Upper Precambrian Kongsfjord Formation, Finnmark, Arctic Norway. (b) Facies association of mainly Facies Class C beds (Facies C2.2). Lower Silurian (Llandovery), Lake District, England. (c) Facies association of mainly Facies Class C beds (Facies Class C2.2 and C2.3), Upper Carboniferous (Namurian) Ross Formation, County Clare, western Ireland. Compass-clinometer scale right of centre.

Lucchi and Valmori (1980), Ricci Lucchi (1981), Hiscott and Pickering (1984), Pickering and Hiscott (1985), Pickering *et al.* (1992, 1993a, b), Edwards (1993), Edwards *et al.* (1994), and Haughton (1994).

Facies C2.5 (a new facies designation here: Fig. 2.27) comprises medium to very thick-bedded, mud-dominated, slurried sand–mud couplets, generally tens of centimetres to several metres thick. Characteristic features include abundant evidence of wet-sediment deformation, pseudonodules, shear laminae, contorted and discontinuous laminae, plastically folded and disrupted laminae. The basal, more sandy, part of the bed may be planar laminated or cross-laminated, but is generally thin compared to the total deposit thickness. The extent of internal bed contortion and the high mud content have resulted in such beds being called 'slurry beds' by Lowe and Guy (2000) and Lowe *et al.* (2003). (The term 'slurry bed' has also been applied to disorganised muddy sands with neither structure nor grading–Facies C1.1 in our classification). Lowe and Guy (2000) identified seven sedimentary structure division types in Britannia oilfield slurry beds: (M1) current-structured and structureless divisions, which tend to mark the bases of beds; (M2) mixed slurried and banded divisions; (M3) wispy laminated divisions;

(M4) dish-structured divisions; (M5) fine-grained, micro-banded to flat-laminated divisions; (M6) foundered and mixed layers that were originally laminated to micro-banded; and (M7) vertically water escape-structured divisions. Lowe *et al.* (2003) proposed that the flows commonly began deposition as what we here call concentrated density flows, characterised by turbulent grain suspension and bed-load transport and deposition or direct suspension sedimentation (M1). Mud was transported within the flows in large part as particles or floccules that behaved hydrodynamically as silt- to sand-sized grains. As flows waned, both mud and mineral grains settled, increasing near-bed grain concentration and flow density. Mud grains settling into the near-bed layers were disaggregated by shear and abrasion against rigid mineral grains, resulting in increasing clay surface area, flow cohesion, shear resistance and viscosity. Eventually, suppression of turbulence in the near-bed layers transformed them into cohesion-dominated shear layers or viscous sublayers. Banding in M2 is thought to reflect the formation, evolution and deposition of such cohesion-dominated shear layers. More rapid fallout from suspension in less muddy flows resulted in the development of thin, short-lived viscous sublayers to form wispy laminated divisions (M3) and, in the least muddy flows with the highest suspended-load fallout rates, direct suspension sedimentation formed dish-structured M4 divisions. Lowe *et al.* (2003) showed that these divisions are stacked to form a range of bed types including: (i) dish-structured beds; (ii) dish-structured and wispy laminated beds; (iii) banded and wispy laminated and/or dish-structured beds; (iv) predominantly banded beds and (v) thickly banded and mixed slurried beds (Fig. 2.27). These different bed types form mainly in response to the varying mud contents and suspended-load fallout rates in the depositing flows. Blackbourn and Thomson (2000) proposed a variation on this scenario for the origin of banding. They suggested that, during deposition, coarse biotite and 'other light components (clays, carbonaceous fragments, etc.)' settled to the bases of the flows, where they were retained in near-bed layers through the same kinetic sieving process discussed by Lowe *et al.* (2003). As clay and mica contents increased in the near-bed layers, clay flocculation formed gel-like aggregates that increased viscosity, dampened turbulence and ultimately triggered sedimentation of both the mica-rich near-bed layer and a largely mica-free zone at the base of the overlying flow, forming a dark-light band couplet. The darker colour of the dark bands is attributed to later pressure solution associated with the 'biotite' grains and not to higher mud contents.

Selected references: Lowe and Guy (2000), Lowe *et al.* (2003), Romans *et al.* (2011), Piper *et al.* (2012), Paull *et al.* (2014).

2.6 Facies Class D: Silts, silty muds, and silt–mud couplets, >80% mud, ≥40% silt, 0–20% sand

Facies Class D contains those sediments that are dominantly silt and clay grade. Key synthesis papers include those by Mutti (1977), Nelson *et al.* (1978), Piper (1978), Stow (1979, 1981), Stow and Lovell (1979), Lundegard *et al.* (1980), Kelts and Arthur (1981), Kelts and Niemitz (1982), Stanley and Maldonado (1981), Gorsline (1984), Stow and Piper (1984, b), Piper and Deptuck (1997), and Bouma and Stone (2000, and references therein). The deposits range from sediments with over 90% medium- to coarse-grained silt, to those with about 40% silt, much of which may be very fine grained. The coarser silts are commonly in distinct beds or laminae interstratified with mud and clay. This class encompasses a wide range of sedimentary characteristics: (i) beds >1 m thick to laminae <1 mm thick; (ii) parallel-sided, lenticular or highly irregular layers; (iii) structureless or thinly laminated sediment with a variety of other small-scale current-generated structures; (iv) poorly developed to absent normal or inverse grading, and graded-laminated units and (v) layers of coarse-grained silt with a relatively high proportion of fine-grained sand and sections with only 10% fine silt laminae in mud.

Sediments in this class include those transported by most of the main processes outlined in Chapter 1. In particular, they may form from the tail of concentrated density flows, the body of turbidity currents, or may be deposited from suspension in deep-water bottom currents. Silts and clays can also be transported by surface currents and winds, to settle through the water column and contribute to the hemipelagic deposits of Facies Class G. Facies Class D deposits are commonly the materials that form coherent and incoherent slides, giving rise to some of the facies of Class F.

Within Class D, we recognise two main facies groups, disorganised (D1) and organised (D2), both of which contain several facies. As with Facies Group C2, the subdivision of Facies Group D2 is based on layer thickness.

2.6.1 Facies Group D1: Disorganised silts and silty muds

Facies Group D1 contains all those silts, muddy silts and irregularly interlayered silts and muds that show little regular or consistent organisation. They may, however, show poor indistinct grading and irregular layering and lensing.

2.6.1.1 Facies D1.1: Structureless silts (Fig. 2.28)

Facies D1.1 commonly occurs as medium- to thick-bedded, parallel-sided, essentially structureless silts. There may be a poorly defined normal and, rarely, an inverse grading at the base of the bed. Commonly, the sediments are fine to coarse silt size, and are commonly sandy with floating mud clasts. The silts may range from poorly sorted to well sorted.

Transport process: silt-laden concentrated density flows, or highly fluid silty cohesive flows.
Depositional process: rapid mass deposition from a concentrated dispersion, due to a combination of increased cohesion and intergranular friction.
Selected references: Piper (1973, 1978), Jipa and Kidd (1974), Stanley and Maldonado (1981), Stow (1984), Stow *et al.* (1986), Edwards *et al.* (1995), Normark *et al.* (1997), Pyles and Jennette (2009), Oliveira *et al.* (2011).

2.6.1.2 Facies D1.2: Muddy silts (Fig. 2.29)

Facies D1.2 occurs as thin- to thick-bedded, poorly sorted, essentially structureless muddy silts. Grading is absent unless as an ill-defined normal grading. Typically, the base of the bed is sharp, possibly resting on a scoured surface, whereas the upper surface is gradational into finer grained facies. Bioturbation is commonly localised in the upper part of Facies D1.2 beds.

Transport process: mud-laden concentrated density flows. Some sediment creep or sliding may contribute to transport.
Depositional process: rapid deposition of silt grains and mud flocs from suspension with no size sorting either in the viscous sublayer or on the bed.
Selected references: Piper (1978), Chough and Hesse (1980), Stow (1984), Wetzel (1984), Bouma *et al.* (1986), Stow *et al.* (1986), Stow *et al.* (1990), Awadallah *et al.* (2001), Rebesco and Camerelenghi (2008), Surpless *et al.* (2009).

2.6.1.3 Facies D1.3: Mottled silt and mud (Fig. 2.30)

Typically, Facies D1.3 consists of very thin beds, laminae, lenses and mottles of silt in mud. Bed shape is characteristically irregular and both bases and tops of the beds vary from sharp to gradational. Normal and inverse grading may occur on the scale of individual laminae and over intervals up to several tens of centimetres thick, although the grading is mostly irregular in nature. Sorting is poor to moderate. Bioturbation is extensive. With increasing grain size, this facies grades into Facies C1.2 (mottled muddy sands); fine grained examples grade into Facies E1.3 (mottled muds).

Transport process: long-lived bottom currents.

(a) (b) (c)

(d) (e)

Fig. 2.25 *(continued)*

(f)

Fig. 2.25 Facies C2.1, 2.2 and 2.3. (a) Graded Facies C2.1 sand–mud couplet without tractional structures, Lower Miocene, ODP Site 793, Izu-Bonin forearc basin (126-793B-19R3, 70–95 cm). Scale 10 cm. (b) Bouma T_{bcd} bed of Facies C2.2, Lower Oligocene, ODP Site 792, Izu-Bonin forearc basin (126-792E-56R4, 13–34 cm). Scale 10 cm. (c) Rippled lenses of very fine sand, ODP Site 942, Amazon Fan (155-942A-6H5, 93–109 cm). (d, e) Variant of Facies C2.3, thin-bedded sand–mud couplets as very coarse-grained beds, Middle Eocene Ainsa System, Ainsa Basin, Spanish Pyrenees. (f) Climbing ripple-lamination or ripple drift, Middle–Upper Ordovician Cloridorme Formation, Gaspé Peninsula, Quebec, Canada. Scale divisions 10 cm.

Depositional process: grain-by-grain deposition from suspension, with subsequent pervasive bioturbation destroying most of the original physical sedimentary structures.

Selected references: Piper and Brisco (1975), Stow (1982), Faugères *et al.* (1984), Gonthier *et al.* (1984), Stow and Piper (1984a), Bouma *et al.* (1986), Chough and Hesse (1985), Strand *et al.* (1995), Normark *et al.* (1997), Rebesco and Camerelenghi (2008).

2.6.2 Facies Group D2: Organised silts and muddy silts

The facies of Group D2 consist of silts and silty muds either as discrete beds or as interlaminated units of mud and silt. This group also includes fissile, organic-rich muds with silt lenses or laminae. Stratification, grading and a predictable sequence of sedimentary structures are common attributes of Facies Group D2.

2.6.2.1 Facies D2.1: Graded-stratified silt (Fig. 2.31)

Facies D2.1 is thin to medium bedded (<30 cm), rarely thick bedded. Soles are typically sharp and scoured, whereas bed tops tend to be gradational. Normal distribution grading prevails. Internal sedimentary structures can be described using the Bouma (1962) turbidite model (Fig. 2.31). In many cases, Facies D2.1 occurs as beds that are thoroughly laminated. Deposits are of silt grade, grading upward into clay-grade sediment. This facies, to a certain extent, overlaps in character with sandy Facies C2.2 and C2.3.

Transport process: turbidity currents.
Depositional process: grain-by-grain deposition from suspension, followed by traction transport along the bottom to produce

lamination. Clay-grade tops are deposited from suspension as flocs, with no subsequent traction transport.

Selected references: Dewey (1962), Piper (1973, 1978), Jipa and Kidd (1974), Pickering (1982a, 1984a), Stow and Piper (1984), Rigsby *et al.* (1994), Strand *et al.* (1995), Normark *et al.* (1997), Pyles and Jennette (2009), Ghadeer and Macquaker (2011), Romans *et al.* (2011), Expedition 317 Scientists (2010), Expedition 339 Scientists (2012).

2.6.2.2 Facies D2.2: Thick irregular silt and mud laminae (Fig. 2.32)

The sediments of Facies D2.2 are typified by medium to thick lenticular silt laminae in mud (Fig. 2.32), and/or thin, irregular, convolute and sub-horizontal silt laminae. In some cases, extreme loading of the silt laminae into the underlying muds produces deep, irregular load structures with intervening mud flame structures protruding upward into the silt layers, or detached load balls (pseudonodules). Typically, silt : mud ratios exceed 2 : 1. Facies D2.2 commonly contains thick silt laminae with sharp, commonly rippled tops and scoured, sharp bases. An internal micro-lamination and slight normal grading may be present through individual laminae. Groups of laminae may be arranged in normally graded, laminated units showing partial structural sequences. The sediment grade is typically medium to coarse silt interlaminated with fine silt and mud. Facies D2.2 is gradational with Facies D2.1 and D2.3. Facies D2.2 is equivalent to the Piper (1978) E1 and Stow and Shanmugam (1980) T_{0-2} divisions.

Transport process: turbidity currents, or relatively weak bottom currents.

Fig. 2.26 Facies C2.4: thick- to very thick-bedded, mud-dominated sand–mud couplets. (a) Facies C2.4 with lighter-coloured, thick, structureless, silty mudstone caps interbedded with mainly Facies Class C turbidites, Upper Cretaceous, Flysch di Monte Cassio, belonging to the so-called 'Helminthoid Flysch'. Note river and village for scale. (b) Basal part of Facies C2.4 bed, Cloridorme Formation, Middle–Upper Ordovician (Caradoc–Ashgill), Gaspé Peninsula, Quebec, Canada. Note flow reversals shown by bi-directional cross-stratification at base of bed, (see arrows); 5 cm-scale. (c) Middle part (below mud cap) of Facies C2.4 bed, Cloridorme Formation, Middle–Upper Ordovician (Caradoc–Ashgill), Gaspé Peninsula, Quebec, Canada. Gradually thinning of laminated to muddy/pseudonoduled couplets is attributed to the seiche created when a large-volume gravity flow was ponded in a basinal depression. (d) Inverted ~12 m-thick Facies C2.4 bed near Cotefablo, Middle Eocene Jaca Basin, Spanish Pyrenees. Person (~2 m high) has top of head coincident with the base of sandstone part of bed, with prominent grey-coloured structureless upper part to right. Note well-developed slaty cleavage in silty part of bed.

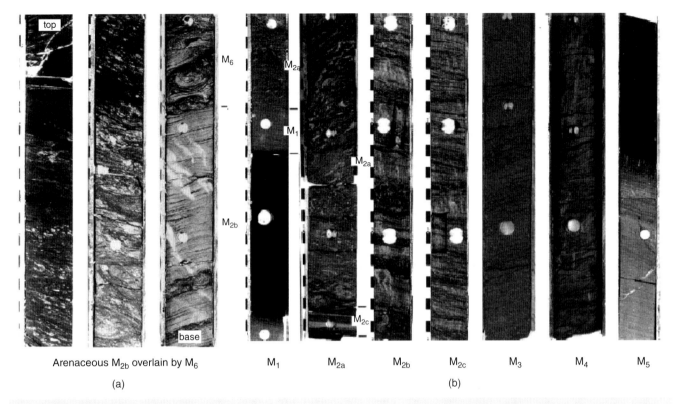

Arenaceous M₂ᵦ overlain by M₆

(a)

M₁   M₂ₐ   M₂ᵦ   M₂c   M₃   M₄   M₅

(b)

Fig. 2.27 Facies C2.5: medium to very thick-bedded, mud-dominated, slurried sand–mud couplets. (a) Sedimentary structure divisions M6 and M2b of slurry-flow beds. The M6 division shows a mixed and deformed unit of very fine-grained sand and silt that passes upwards into deformed mudstone. From Lowe *et al.* (2003). (b) Sedimentary structure divisions M1, M2, M3, M4 and M5 of slurry-flow beds. M1 shown here is a thin flat-laminated layer that marks the base of slurry bed 50, well 16/26-B1. It immediately overlies a mudstone unit and is overlain by an M2a mixed slurried division. M2a (base of bed 46, well 16/26-B8) shows a thin M2c mesobanded division overlain by a very thick mixed slurried to megabanded M2a unit. M2b and M2c (bed 46, well 16/26-24) consist of macrobanded and mesobanded sandstone respectively. The M3 division of wispy laminations (beds 34/36, well 16/26-B8) shows characteristic fine subvertical water-escape channels (light streaks). Dish-structured M4 division (bed 62, well 16/26-B5) shows moderately curved dishes cross-cut by large, vertical and subvertical water-escape channels (light). Core at far right (top of bed 86, well 16/26-B10) shows an M4 division of faint flat dish structures overlain by an M5 division of fine, flat sandy microbanding that grades upwards into laminated silt and mud without cross-laminations or other current structures. Scale units shown on left side of core segments are 0.1 foot (30 mm) long. From Lowe *et al.* (2003).

Depositional process: fairly rapid grain-by-grain deposition from suspension (turbidites only), followed by traction transport of the silt load. Lack of bioturbation indicates rapid accumulation or absence of metazoans (Precambrian cases).

Selected references: Piper (1972a, 1972b, 1973, 1978), Stow (1976, 1979, 1981, 1984), Nelson (1976), Nelson *et al.* (1978), Lundegard *et al.* (1980), Stow and Shanmugam (1980), Kelts and Arthur (1981), Kelts & Niemitz (1982), Pickering (1982b, 1983a, 1984b), Stow and Piper (1984), Gardiner and Hiscott (1988), Normark *et al.* (1997), Dolan *et al.* (1990), Wignall and Pickering (1993), Rebesco and Camerelenghi (2008).

2.6.2.3 Facies D2.3: Thin regular silt and mud laminae (Fig. 2.33)

Facies D2.3 occurs as thin to medium beds containing horizontal silt laminae in mud, with some slightly lenticular, indistinct and wispy silt laminae (Fig. 2.33). These laminae are commonly grouped into graded-laminated units in which successive silt laminae become finer grained upward over intervals of about 2–10 cm. These units show regular sequences or partial sequences of structures. Silt:mud ratios range from about 2:1 to about 1:2. The silt laminae show sharp to gradational tops and bases. Grain sizes are mostly medium to fine silt and clay. Facies D2.3 is gradational with Facies D2.2 and E2.1. This facies is equivalent to the Piper (1978) E1, E2 and the Stow and Shanmugam (1980) T_{2-6} divisions.

Transport process: mainly turbidity currents, but possibly weak bottom currents.

Depositional process: for turbidites, slow uniform deposition from suspension, with shear sorting of silt grains and clay flocs in the viscous sublayer of the turbidity current as described by Stow and Bowen (1980). Examples deposited by bottom currents require an environment hostile to burrowers.

Selected references: as for Facies D2.2, and Stow and Bowen (1980), Rigsby *et al.* (1994), Normark *et al.* (1997), Expedition 339 Scientists (2012).

Fig. 2.28 Facies D1.1: structureless silts. Structureless, closely spaced, thin coarse silt beds from levée deposits of the Yellow Channel–Levée System, Amazon Fan (ODP Leg 155: Interval 155-937C-6H-3, 36–58 cm). After Normark *et al.* (1997).

Fig. 2.29 Facies D1.2: muddy silts interbedded with Facies B1.2 and C2.3 sandstones, Miocene Monterey Formation, California, USA (pencil 14 cm long). After Surpless *et al.* (2009).

Fig. 2.30 Facies D1.3: mottled silt and mud. Cretaceous (Cenomanian), ODP Site 1276A, Newfoundland Basin (210-1276A-32R1, 48–62 cm). Scale bar 10 cm.

2.7 Facies Class E: ≥95% mud grade, <40% silt grade, <5% sand and coarser grade, <25% biogenics

Facies Class E includes some of the finest deep-sea sediments, both silty clays and clays. Layers vary from very thick beds (several metres thick) and essentially unbedded thick sections, to very thin beds and laminae. Layers are mostly parallel-sided, although broad basal scours may occur. Contacts may be sites of bioturbation. Individual beds can be structureless, graded, irregularly or finely laminated, and can show varying degrees of bioturbation.

Modern representatives of this facies class are widespread, and include the pelagic (red) clays of abyssal oceanic depths, thick proximal mud turbidites, thin basinal turbidites, muddy contourites, and much of the hemipelagic cover of continental slopes and rises. These deposits may undergo later creep, or may contribute to submarine slides or debris flows.

There have been few syntheses of data from deep-sea muds, so that little is known in detail of environments and processes of deposition. Efforts in this direction have been made by Piper (1978), Stow and Lovell (1979), Stanley and Maldonado (1981), Stanley (1981), Gorsline (1980, 1984), Hoffert (1980) and Stow and Piper (1984b).

As with other facies classes, there appears to be a natural distinction between a disorganised (E1) and an organised (E2) facies group. Many of the attributes of the fine-grained facies in this class are best studied with the aid of X-radiographs of thin slabs (rocks or cores), scanning electron microscopy, or with the aid of peels (rocks).

2.7.1 Facies Group E1: Disorganised muds and clays

Facies Group E1 comprises muds, silty muds and clays, typically in thick, uniform, essentially structureless sections. Sedimentary features are generally subtle. Origin is in many cases enigmatic. Three facies are recognised.

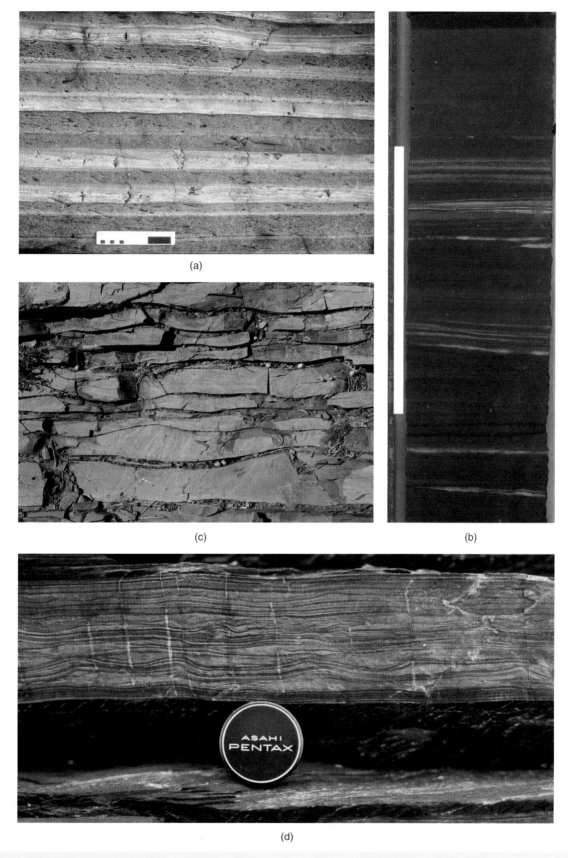

Fig. 2.31 Facies D2.1: graded-stratified silt. (a) Lower Silurian (Llandovery) Aberystwyth Grits, Welsh Basin; 15 cm scale bar. (b) Cretaceous (Cenomanian–Albian), ODP Site 1276A, Newfoundland Basin (210-1276A-37R3, 95–114 cm). Scale bar 10 cm. The highest essentially structureless bed above the scale bar is Facies D1.1. (c) Permian Karoo System, South Africa. Coin scale ~2 cm diameter. (d) Upper Precambrian Kongsfjord Formation, Finnmark, Arctic Norway. Camera lens cap for scale.

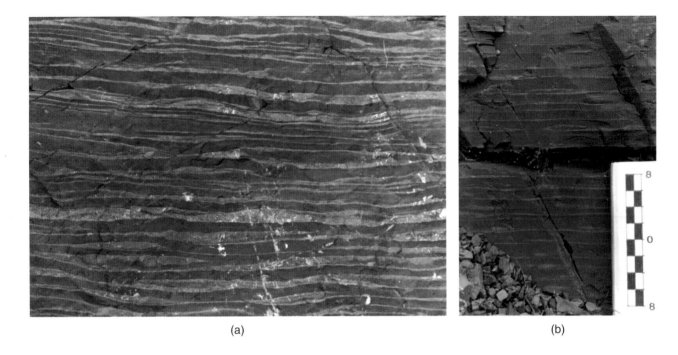

(a)

(b)

(c)

Fig. 2.32 Facies D2.2: thick irregular silt and mud laminae. (a) and (b) Haydrynian (upper Proterozoic) Conception Group, False Cape, Avalon Peninsula, Newfoundland. Thickness 18 cm at left, scale in centimetres at right. (c) Upper Jurassic (Kimmeridge) Boulder Beds, northeast Scotland. This has been referred to as 'tiger-stripe' facies.

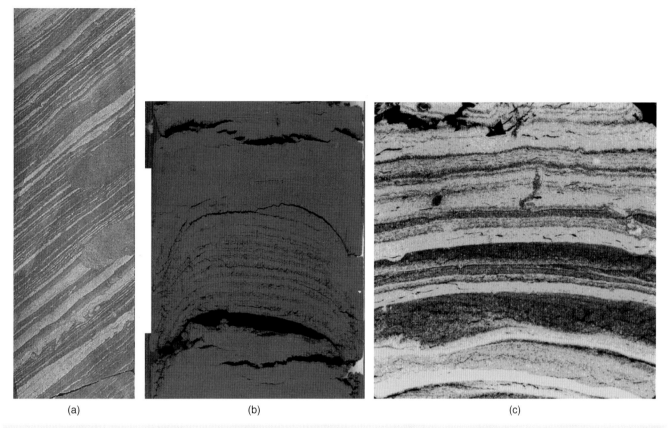

Fig. 2.33 Facies D2.3: thin regular silt and mud laminae in core. (a) Middle Eocene Ainsa System, Ainsa Basin, Spanish Pyrenees. Inclination of beds due to coring through dipping section. (b) ODP Site 931, Amazon Fan (155-931B-4H6, 116–121 cm). Scale bar 10 cm. (c) Photographic print from thin section of silt (dark) and mud (pale) laminae like those in the middle image, ODP Site 931, Amazon Fan (155-931A-4H2, 78–80 cm). Thickness shown 1.5 cm.

2.7.1.1   Facies E1.1: Structureless muds (Fig. 2.34)

Structureless muds commonly occur in thick sections (one to tens of metres thick); bedding is poorly defined or absent. This facies appears to be common in both modern and ancient successions. While there is a notable absence of structures, both primary and secondary, an indistinct textural, compositional, or colour banding or lamination, and zones of burrow mottling, may be locally developed. The muds are either clay grade, or mixed silt and clay grades; sand-sized material is less common. Composition may be remarkably uniform, being dominantly terrigenous. Facies E1.1 tends to occur in association with biogenic muds (G2), mud turbidites (E2.1), slides (F2) and muddy debris flow deposits (A1.3).

Transport process: largely unknown, but is likely to include thick, mud-laden turbidity currents and lateral transfer of hemipelagic material by deep-ocean currents or by sliding.
Depositional process: probably rapid deposition of flocs, based partly on lack of significant planktonic biogenics. Rapid deposition may result from ponding of mud-rich turbidity currents in confined basins (Pickering and Hiscott 1985). McCave and Jones (1988) and Jones *et al.* (1992) have proposed that mud-laden turbidity currents collapse in distal basin plains to form viscous mudflows responsible for the deposition of structureless muds.
Selected references: Piper (1978), Stanley and Maldonado (1981), Stanley (1981), Kelts and Arthur (1981), Stow (1984), Pickering and Hiscott (1985), Stow *et al.* (1986), McCave and Jones (1988), Hiscott *et al.* (1989), Jones *et al.* (1992), Rigsby *et al.* (1994), Rothwell *et al.* (1994), Flood *et al.* (1997), Normark *et al.* (1997), Oliveira *et al.* (2011).

2.7.1.2   Facies E1.2: Varicoloured muds (Fig. 2.35)

Many mud-dominated successions contain not only the classic 'pelagic red clay', but also interbedded muds of various colours (red, green, brown, grey, etc.), mostly lacking sedimentary structures; these are assigned to Facies E1.2. Such successions may be up to tens of metres thick. Individual beds or layers are defined on the basis of colour changes. Typically, layers are parallel-sided with either abrupt or gradational bases and tops. Primary physical structures are mostly absent, whereas secondary structures such as bioturbation, mottling and burrowing tend to be common. Grain sizes are mainly silt grade to clay grade. Facies E1.2 is predominantly terrigenous in composition ($\leq$25% biogenic material), typically with a significant volcanigenic fraction and authigenic minerals. Ferromanganese nodules and crusts are common and enrichment in trace metals may occur. Subtle differences in chemical composition, oxidation states and the amount of organic carbon control the colour differences. High organic carbon content can give a black mud (Facies E2.2). Facies E1.2 is commonly associated with biogenic oozes and muddy oozes (G1) and fine-grained turbidites (D2 and E2).

(a)   (b)

Fig. 2.34 Facies E1.1: structureless muds. Clayey nannofossil ooze, interpreted as turbidite mud (T_e), Upper Pliocene–Pleistocene ODP Leg 135 (Lau Basin, southwest Pacific) Site 835 (Interval 135-835A-1H-4, 20–45 cm). (a) Core photograph and (b) X-radiograph of same interval. Core width ~6 cm. ODP, Site 835 (*cf.* Rothwell *et al.* 1994).

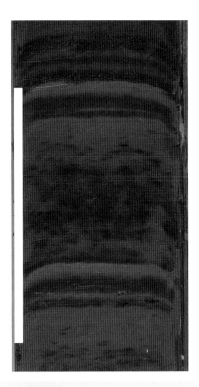

Fig. 2.35 Facies E1.2: varicoloured muds. ODP Site 934, Amazon Fan (155-934A-1H3, 7–22 cm). The colour banding is accentuated by FeS staining. Scale bar 10 cm.

Transport process: lateral transport of hemipelagic material by ocean currents and/or aeolian action; in situ biogenic production.

Depositional process: settling of individual particles or particle aggregates (faecal pellets).

Selected references: Arthur and Natland (1979), Hoffert (1980), Arthur *et al.* (1984), Jenkyns (1986), Normark *et al.* (1997), O'Brien *et al.* (2001).

2.7.1.3   Facies E1.3: Mottled muds (Fig. 2.36)

Facies E1.3 consists of relatively uniform, thin to thick intervals of mud that are poorly bedded. Relict primary sedimentary structures include wavy, indistinct, or fine parallel layering (Fig. 5.16). Indistinct burrows (mottling) are ubiquitous. Clay-grade material is dominant, but relatively silty intervals (silt mottles, pockets and blebs) may be common. Where silt content becomes substantial, Facies E1.3 grades into Facies D1.3. Compositionally, this facies shows considerable variation, but typically consists of mixed terrigenous and biogenic components, with or without volcanigenic debris. The facies is commonly associated with probable contourite deposits (D1.3 and C1.2), biogenic muds (G2) and fine-grained turbidites (D2, E2).

Transport process: transport as suspended load in bottom currents, commonly contour currents. Silt grains may have moved as bed load beneath the same currents.

Depositional process: settling of particles or particle aggregates from suspension. Extensive post-depositional bioturbation.

Selected references: Stow (1982), Faugères *et al.* (1984), Gonthier *et al.* (1984), Chough and Hesse (1985), Strand et al. (1995), Normark *et al.* (1997), O'Brien *et al.* (2001).

2.7.2   Facies Group E2: Organised muds

Muds and clays that show some internal organisation are assigned to Facies Group E2. There is a complete gradation with the facies of Group E1. Facies include muds, silty muds and clays in thin isolated beds or thicker layers. There are two distinctly different facies: graded muds, and finely layered or laminated muds.

2.7.2.1   Facies E2.1: Graded muds (Fig. 2.37)

Facies E2.1 is widespread, especially in deep basinal settings, although it can occur more proximally. It can occur as single, isolated, graded beds >1 m thick, or as thick, repetitive successions of graded muds of variable thickness. Individual beds may show broad scoured bases; thicker beds can be lenticular. Most beds show normal grading. Bioturbation becomes increasingly more important toward the top of many graded mud beds.

Colour or compositional grading is generally more marked than grain-size grading, although beds of dominantly terrigenous composition tend to be siltier toward the base, with increasing clay content upward. These beds may have thin silt laminae at the base. Wherever biogenic material is present, the overall texture tends to be slightly coarser. Facies E2.1 is gradational with silt-mud turbidites (D2.3 and D2.1), and commonly occurs in association with Facies Classes F and G.

Transport process: turbidity currents.

Depositional process: grain-by-grain or floc settling during flow deceleration. No significant traction precedes burial. Post-depositional bioturbation.

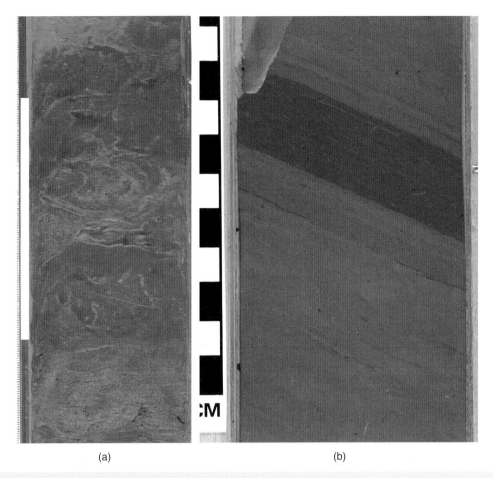

(a)   (b)

Fig. 2.36 Facies E1.3: mottled muds. (a) ODP Site 941, Amazon Fan (155-941B-1H4, 12–31 cm). Scale bar 10 cm. (b) Well A6, Middle Eocene Ainsa System, Ainsa Basin, Spanish Pyrenees. Scale bar in centimetres.

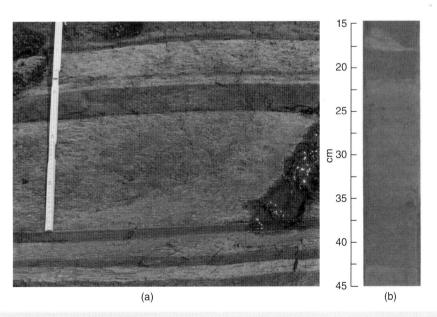

(a)   (b)

Fig. 2.37 Facies E2.1: graded muds. (a) Pale (green shades) and dark grey banded muds, Cap des Rosiers Formation, Cap Ste-Anne, Quebec. The pale units have the sharpest bases so are inferred to be the SGF deposits, capped by dark grey, presumably hemipelagic muds. Scale ~45 cm long. (b) Silty mud, poorly sorted and carbonaceous near base grading up into fine mud, DSDP Leg 96, Mississippi Fan, Gulf of Mexico (Sample 615-22-2, 15–45 cm). After Stow *et al.* (1986).

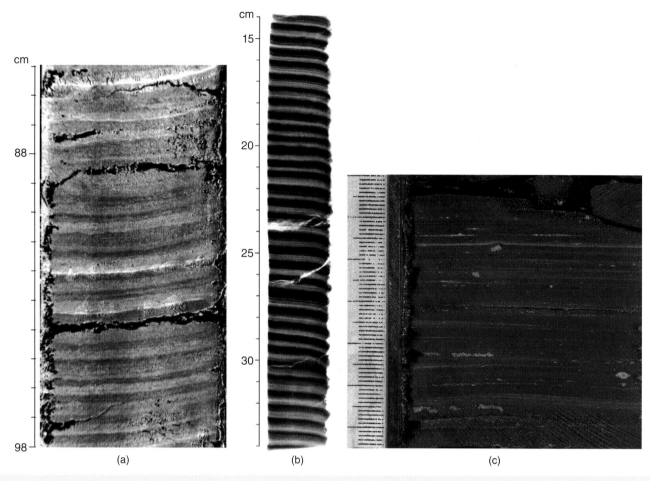

Fig. 2.38 Facies E2.2: laminated muds and clays from ODP Leg 169S, Saanich Inlet, off Vancouver Island, British Columbia (see Bornhold *et al.* 1998). (a) Alternations of paler laminae (diatom ooze) and darker laminae of diatomaceous mud with rare very light-coloured sub-laminae of silty muds (Core interval 169S-1034B-3H-6, 85–98 cm). (b) X-ray radiograph of alternating laminated sediments showing cm-scale pale (diatom-rich) and dark layers (containing terrigenous sediment) interpreted as annual 'varves', within which there are up to 10 intra-annual sub-laminae (Core interval 169S-1034B-4H-3, 14–34 cm). (c) Cretaceous (Albian) ODP Site 1276A, Newfoundland Basin (210-1276A-72R5, 103–111 cm). Pyrite nodules attest to the sulfidic nature of these organic-rich laminated muds. Scale bar 6.5 cm long.

Selected references: Piper (1978), Kelts and Arthur (1981), Stow (1984), Stow *et al.* (1986), Bouma *et al.* (1986), Weedon and McCave (1991), O'Brien *et al.* (2001), Lucchi *et al.* (2002), Bernhardt *et al.* (2011).

2.7.2.2 Facies E2.2: Laminated muds and clays (Fig. 2.38)

Facies E2.2 occurs in thin to thick mud-dominated sections (up to tens of metres thick) and typically constitutes 10–60% of sections in which it occurs. Individual beds or intervals range from 1 cm to decimetres in thickness, commonly with fine parallel lamination or distinct 'varves' (Fig. 2.38). The laminae or 'varves' show slight colour, compositional and/or textural variations. Bioturbation is generally absent, although small-scale local burrows may be evident. Clay-grade and fine silt-grade sediments dominate this facies and, in some cases, there are thin silt laminae.

Composition is mixed terrigenous and biogenic, with 1–10% organic carbon (more rarely >20%) and common but minor iron sulfides. Facies E2.2 includes sapropels, the precursors of the marine 'black shales' of the ancient geological record. This facies is commonly associated with oozes and biogenic muds (Facies Class G) and with fine-grained turbidites (E2.1 and D2).

Transport process: in situ production of organic matter in the overlying water column, transport of hemipelagic material by ocean currents, periodic introduction of mud as suspended load in turbidity currents.

Depositional process: grain-by-grain or aggregate settling from the transporting currents. Little bed-load traction of the silt fraction. 'Varves' are related to periodic fluctuations of terrigenous input. Anoxic bottom waters favour the preservation of organic matter.

Selected references: Arthur *et al.* (1984), Stow and Dean (1984), Dimberline and Woodcock (1987), Tyler and Woodcock (1987), Normark *et al.* (1997), Stow *et al.* (1990), Gersonde *et al.* (1999), O'Brien *et al.* (2001), Lucchi *et al.* (2002), Oliveira *et al.* (2011), Brunt *et al.* (2012).

(a)

(b) (c)

Fig. 2.39 Facies F1.1: rubble. (a) Deep-towed camera photograph of rock talus with attached coral on the volcanic slope of Mata Fitu submarine volcano in the NE Lau Basin ~150 km SW of Samoa. Water depth ~2440 m. The two green dots just above coral are lasers with separation of 15 cm. NOAA- Earth Ocean-Interactions and WHOI-MISO TowCam (http://www.whoi.edu/page.do?pid=17619). (b) Oligocene, ODP Site 793, Izu-Bonin forearc basin (126-793B-87R2, 67–84 cm). Primary voids between volcanic clasts are filled with diagenetic cement. (c) Rubble as unsorted talus covering a larger block, Logatchev hydrothermal field (14°45′N, Mid-Atlantic Ridge). From Petersen *et al.* (2009).

2.8   Facies Class F: Chaotic deposits

Facies Class F consists of more or less chaotic mixtures of sediments, including other deep-water facies emplaced by large-scale downslope mass movements or by ice rafting. We exclude deposits resulting from in situ liquefaction. The thickness and shape of the deposits in this facies class ranges from single, isolated clasts (pebbles, cobbles, boulders) to whole sections of continental slope up to hundreds of metres thick. There are two facies groups: F1, exotic clasts; and F2, contorted/ disturbed strata.

Lateral and vertical transitions between facies can occur over very short distances, for example from the central to marginal parts of submarine slides. Where vertical or lateral facies changes are abrupt and abundant, a researcher may choose to describe a succession only in terms of facies groups.

2.8.1   Facies Group F1: Exotic clasts

Facies Group F1 comprises blocks or clasts, either in complete isolation from, or associated with, other similar clasts. Generally, any interstitial 'matrix' is very poorly sorted, and shows a bimodal to polymodal grain-size distribution, although there may be overall systematic textural and compositional variations laterally. The 'matrix', or sediments enveloping the exotic (F1) clasts, is commonly much finer grained than the clasts. There is never any conclusive evidence that the 'matrix' is contemporaneous with clast deposition. Rather, textural relationships suggest that the 'matrix' percolated into the interstices or draped the clasts after deposition.

In many ancient examples, Group F1 deposits have been affected by subsequent tectonism. In some cases, shaly 'matrix' may have acquired a fissility due to compaction alone, but in other examples it may be pervasively sheared during tectonic deformation, even though the exotic clasts are relatively undeformed (Abbate *et al.* 1970).

Within this group, two facies are recognised: F1.1, rubble; and F1.2, dropstones and isolated ejecta.

2.8.1.1   Facies F1.1: Rubble (Fig. 2.39)

Facies F1.1 consists of a chaotic jumble of mainly angular to subangular cobbles and boulders, generally as a talus or scree fringing relatively steep submarine cliffs and slopes (Fig. 2.39). In ancient examples, the matrix tends to resemble a rock flour devoid of any current-related sedimentary structures; infilling of residual hollows between the larger clasts is indicated by sediment drape. In many modern examples, the rubble is either devoid of interstitial sediments or contains limited and patchy matrix.

The clasts produce mainly compressional deformation and rupturing of underlying sediments in the form of depressed bedding, small-scale syn-sedimentary faults, buckling and attenuation of beds surrounding the clasts. Wherever large blocks occur, detailed sedimentary logging and mapping may be required to reveal sediment drape and the 'exotic' nature of the blocks; faunal data may be necessary to establish age relationships of the rubble constituents in this facies.

Clast compositions include igneous, metamorphic and sedimentary rock types. The degree of lithification of the rubble clasts varies considerably. Also, clasts may have experienced internal wet-sediment deformation. Isolated, displaced clasts, or exotic blocks, are well documented from ancient successions, for example, limestone blocks up to 250 m in maximum dimension from the

Mesozoic Laganegro Basin, southern Italian Apennines (Wood 1981), isolated blocks to 50 × 50 × 100 m in the Longobucco Group of southern Italy (Teale 1985), and the so-called 'fallen stack' of Devonian Caithness Flagstone (45 × 27 × ? m) in the Upper Jurassic of northeast Scotland (Bailey & Weir 1932; Pickering 1984b). Facies F1.1 tends to be associated in particular with deposits of Facies Groups A1 and F2. Extremely large banks of accumulated slide blocks have been documented from offshore volcanic islands with individual blocks having dimensions up to hundreds of metres by several kilometres, for example, the Hawaiian Ridge by Lipman *et al.* (1988), Moore *et al.* (1989) and Jacobs (1995), and from offshore Montserrat by Watt *et al.* (2012).

Transport process: submarine rockfalls, avalanching and sliding along overpressured glide planes, debris flow.
Depositional process: cessation of movement because of basal friction. Some large blocks travelling in debris flows may have become 'grounded', even though the rest of the flow continued into the basin, leaving little or no depositional record in the vicinity of the block.
Selected references: Bailey and Weir (1932), Flores (1955), Abbate *et al.* (1970). Hsü (1974), Surlyk (1978), Pickering (1984b), Teale (1985), Normark *et al.* (1993a), Petersen *et al.* (2009).

2.8.1.2   Facies F1.2: Dropstones and isolated ejecta (Fig. 2.40)

Dropstones and isolated ejecta of Facies F1.2 are generally substantially larger in size than their 'matrix', or host sediment, occurring

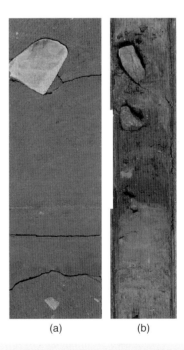

(a)   (b)

Fig. 2.40 Facies F1.2: dropstones and isolated ejecta. (a) Dropstone in background mudstone from ODP Leg 119 Site 739, continental shelf of Prydz Bay, East Antarctica, Core 119-739C-34R-3, 68–87 cm (see Hambrey *et al.* 1991). (b) ODP Site 645, Baffin Bay, Canadian Arctic (105-645F-3H4, 31–61 cm). Isolated dropstones occur in mud above a pale-coloured ice-rafted unit rich in detrital carbonate. Scale bar 10 cm.

either in isolation (Fig. 2.40) as groups, or in lumps of till. In high latitudes, however, large quantities of gravel-, sand-, silt-, and even clay-size material can be derived from melting icebergs (Srivastava *et al.* 1987; Dowdeswell *et al.* 1994), and a true gravel facies (Facies A1.3) is developed. The composition, shape and other attributes of the clasts are mainly a function of the source area. For example, dropstones may show glacial striations, polishing, or faceted faces. Such features are supportive of a glaciomarine interpretation, but are not diagnostic of that environment as glacial clasts may be reworked in other environments.

Transport process: encased within, or carried on top of, floating ice. Very rarely, the rafting agent may be seaweed or floating trees. Volcanic bombs are ejected during eruption, and follow ballistic trajectories to the sea surface.

Depositional process: grain-by-grain settling. For ice-rafted material, the release is triggered by (a) melting, or (b) the sudden overturning of sediment-laden icebergs.

Selected references: Boltunov (1970), Ovenshine (1970), Anderson *et al.* (1979), Edwards (1986), Srivastava *et al.* (1987), Cremer (1989), Eyles and Eyles (1989), Dowdeswell *et al.* (1994), Krissek (1989, 1995), Breza and Wise (1992), McKelvey *et al.* (1995), Rea *et al.* (1995), Thiede and Myhre (1996), Cofaigh and Dowdeswell (2001), Barker and Camerlenghi (2002), Tarlao *et al.* (2005), Schultz *et al.* (2005), Benediktsson (2013), Ojakangas *et al.* (2014). These references include ODP Leg 145 papers that deal with ice-rafted sand deposits.

2.8.2 Facies Group F2: Contorted/disturbed strata

Penecontemporaneously deformed packets of beds and layers resulting from lateral translation of parts or the entire sediment packet along discrete shear or glide surfaces are assigned to Facies Group F2. The thickness of the packets is typically on the order of metres, although the range is from centimetres to hundreds of metres. Bounding surfaces range from smooth, planar and parallel-sided, to highly irregular with deep erosional scours at the base of the packets. Internally, bedding varies from virtually undisturbed to highly disturbed, and it is this criterion that separates Facies F2.1, coherent folded and contorted strata, from Facies F2.2, dislocated and brecciated strata.

Selected references: Modern and ancient examples of this facies group are described in books edited by Doyle and Pilkey (1979), Watkins *et al.* (1979), Saxov and Nieuwenhuis (1982), Jones and Preston (1987), Schwab *et al.* (1993), Lykousis *et al.* (2007), Mosher *et al.* (2010), and Yamada *et al.* (2012). As sediment slides can contain a range of facies listed under Facies Group F2, the reader is referred to: Moore (1961), Laird (1968), Lewis (1971), Ricci Lucchi (1975a), Roberts *et al.* (1976), Woodcock (1976a, b, 1979a, b), Clari and Ghibaudo (1979), Doyle and Pilkey (1979), Watkins *et al.* (1979), Pickering (1982b, 1984a, 1987a), Saxov and Nieuwenhuis (1982), Barnes and Lewis (1991), Normark *et al.* (1997), Piper *et al.* (1997), Lucente and Pini (2003), Masson *et al.* (2006), Vanneste *et al.* (2006), Surpless *et al.* (2009), Oliveira *et al.* (2011), Watt *et al.* (2012).

2.8.2.1 Facies F2.1: Coherent folded and contorted strata (Fig. 2.41)

Facies F2.1 comprises folded and contorted, essentially coherent to semi-coherent strata in irregularly shaped layers or horizons from centimetres to tens of metres thick (Fig. 2.41). Discrete internal glide or shear surfaces may be visible and define the bounding surfaces of the deposit (Fig. 2.41). Upper and lower bounding surfaces vary from smooth and planar to highly irregular; there are commonly dramatic changes in layer thickness along strike.

Internal structure, bed thickness and grain sizes of this facies are highly variable. There may be a consistent sense of overturning in any folded interval, and locally such folds may be of relatively constant wavelength and amplitude. Typically, it is the coarser grained beds and laminae that have preserved their lateral continuity in this facies; mud-rich sediments tend to deform plastically and may inject microfaults (Pickering 1983b, 1987a). Facies F2.1 commonly passes over short lateral and vertical distances into Facies F2.2.

In ancient rocks, the scale of many Facies F2.1 layers generally precludes an appreciation of their vertical and especially lateral dimensions. Exceptional outcrops can show the internal complexity of large-scale sediment slides (e.g., 10^2 thickness and 10^4 m width 'Casaglia Monte della Colonna sediment slide' in the Miocene Marnoso arenacea Formation, Italian Apennines; Lucente and Pini 2003). In modern continental-margin and other deep marine environments, this facies is recognised on the basis of deep and shallow seismic profiles, long-range sidescan sonar, and deep-sea drilling. In seismic profiles, Facies F2.1 horizons typically appear as acoustically unstratified sediments and/or chaotic acoustic horizons. Sidescan images show irregular hummocky topography. In cores, this facies is suggested by common reversals of bedding dip, and by stretched, attenuated, sheared, faulted (typically microfaulted), folded and overturned strata.

Transport process: slides and rotational slumps, either due to depositional overloading of weak sediments, or to cyclic or single shocks (earthquakes, tsunami).

Depositional process: cessation of movement on decreasing bottom slopes because gravity forces no longer exceed or balance basal and internal friction.

2.8.2.2 Facies F2.2: Brecciated and balled strata (Fig. 2.42)

Facies F2.2 is gradational from F2.1, and is typified by highly brecciated and balled strata in a chaotic jumble of fragments, in layers of varying thickness but generally thinner than those of Facies F2.1. Layer shape is highly variable from parallel-sided to very irregular and lens-shaped. Typically, the layers consist of a relatively fine-grained 'matrix' with fragments of original bedding and laminae in angular to well rounded pieces (Fig. 2.42). Facies F2.2 is an intraformational deposit with relatively uniform composition for any given layer. A crude ghost stratigraphy may be present and commonly many of the fragments are folded/contorted as isolated prolapse structures and single folds on a scale of decimetres.

The layers of balled strata, so named because of the roundness of the fragments, are the most common variant of Facies F2.2. Typically, bounding surfaces are smooth, although layer thickness is variable along strike. Relict bedding is visible as plastically deformed bundles of laminae/beds, or as 'wisps' within an almost 'homogenised' layer. Layers of balled strata may pass over short lateral distances through a zone of pervasively microfaulted sediment into Facies F2.1.

The brecciated variants of Facies F2.2 are characterised by abundant, poorly imbricated, unordered, elongate, angular to subangular intraformational fragments, commonly of fine-grained sands, silts,

(a)　　　　　　　　　　　　　(b)

(c)　　　　　　　　　　　　　(d)

(e)

Fig. 2.41 Facies F2.1: coherent folded and contorted strata. (a) Middle Eocene Ainsa System, Ainsa Basin, Spanish Pyrenees. (b) 'Black Flysch' (Deva Formation), Arminza, northern Spain. Scale in middle fold axis 1 m long. (c) Pliocene–Pleistocene, Miura Peninsula, southeast Japan. (d) Middle Ordovician Cloridorme Formation, Gaspé Peninsula, Quebec. Scale divisions 10 cm. (e) Ross Formation ('Ross Slide'), Upper Carboniferous (Namurian), Bridges of Ross, County Clare, western Ireland. Human scale in centre of image.

(a)

(b)

Fig. 2.42 Facies F2.2: brecciated and balled strata. (a) Middle to Upper Ordovician (Caradoc–Ashgill) Point Leamington Formation, north-central Newfoundland, Canada. Hammer for scale. (b) Small-scale syn-sedimentary faults within Facies F2.1, Upper Jurassic (Kimmeridgian), northeast Scotland.

muds and clays, in layers on the order of decimetres thick. Upper and lower surfaces tend to be irregular. Undercutting of partially eroded sediments may occur. Upper surfaces may be draped by overlying sediments. Commonly, brecciated strata are associated with fluid-escape structures.

Transport process: gravity-induced sliding, during which internal deformation and brecciation occurs.
Depositional process: as for Facies F2.1.

2.9 Facies Class G: Biogenic oozes (>75% biogenics), muddy oozes (50–75% biogenics), biogenic muds (25–50% biogenics) and chemogenic sediments, <5% terrigenous sand and gravel

Biogenic and chemogenic sediments are ubiquitous throughout the world's oceans, and are widely recognised in ancient deposits. Most facies of Class G result either from slow settling of material through the water column in the absence of substantial bottom currents, or from direct chemical precipitation. Many of the processes that are proposed for the accumulation of some of these sediments, however, require advective currents to furnish (i) suspended sediments, (ii) nutrients and (iii) other chemicals to the water column. Three distinctive facies groups can be distinguished: G1, biogenic oozes and muddy oozes; G2, biogenic muds; and G3, chemogenic sediments.

The chief distinguishing features of Groups G1 and G2 include: (i) evidence for low to very low rates of sediment accumulation and continuous bioturbation (except in anoxic basins; Byers 1977); (ii) an absence of primary sedimentary structures or other evidence of sustained bottom currents; (iii) an essentially uniform composition within any given succession except for regular cyclicity related to climatic or other controls; (iv) a variable biogenic component, mainly of planktonic tests; (v) a very fine-grained and commonly far-transported terrigenous component and (vi) commonly a significant authigenic fraction.

The true biopelagic sediments (Facies Group G1) accumulate in the open ocean, and primarily consist of whole or disarticulated skeletons of planktonic organisms, together with some minor amounts of very fine silt and clay, much of which reaches the open ocean by aeolian transport. The actual proportion of terrigenous material may be increased by preferential dissolution of the biogenic fractions. Total dissolution results in the varicoloured pelagic clays of Facies E1.2 (≤25% biogenics; generally <1%).

Biogenic muds (Facies Group G2) accumulate on continental margins and in other settings near terrigenous sediment sources. They consist of indigenous biogenic material (>25–50%) and silt- and clay-grade terrigenous detritus. Lower biogenic contents result in transitions with Facies E1.1 and E1.3. Both these latter facies and Group G2 deposits can be broadly referred to as hemipelagites.

Chemogenic sediments (Facies Group G3) are composed almost entirely of authigenic minerals such as ferromanganese nodules and phosphorites. These deposits are very complex, and a satisfactory discussion is far beyond the scope of this book. The interested reader is referred to comprehensive books on these intriguing deposits by Horn (1972), Glasby (1977), Bentor (1980), Baturin (1982) and selected chapters in Hüneke and Mulder (2011).

2.9.1 Facies Group G1: Biogenic oozes and muddy oozes

2.9.1.1 Facies G1.1: Biogenic oozes (Fig. 2.43)

Biogenic oozes are most typical of the open ocean basins far from terrigenous sources. They have been the subject of much study over the past 145 years; hence, their sedimentary characteristics are well known. Some important syntheses are to be found in Arrhenius (1963), Hsu and Jenkyns (1974), Cook and Enos (1977) and Jenkyns (1986).

Oozes are composed predominantly of the tests of planktonic organisms (>75%), either calcareous (coccoliths, foraminifera, pteropods and nannoplankton) or siliceous (radiolaria, silicoflagellates), or a mixture of both (Berger 1974). Also included in Facies G1.1 are 'diatomites' that contain the remains of diatoms, a type of hard-shelled algae, and that typically have a particle size ranging from <1 μm to >1 mm, but typically 10–200 μm. These major components are soluble to a significant degree in seawater. The other components (Lisitzin 1972) may include very fine-grained terrigenous material (principally quartz, feldspars and clay minerals), volcanigenic minerals (e.g., palagonite and derived clay minerals), authigenic minerals (e.g., phosphates, barite, zeolites, ferromanganese nodules and coatings) and rare extraterrestrial material (tektites and iridium-rich dust). Under normal oxic conditions, the organic carbon content is extremely low, but under anoxic conditions pelagic black shales can contain >20% organic carbon (Isaacs 1981; Arthur *et al.* 1984). Biogenic opal can also be a constituent or major component of this facies (e.g., Anderson & Ravelo 2001; Hillenbrand & Fütterer 2001). Green clays in the South China Sea have been interpreted as due to diagenetic alteration of organic-rich claystones (Tamburini *et al.* 2003).

Rates of accumulation are low, commonly from <1 mm kyr^{-1} to 10 mm kyr^{-1}, although this can be an order of magnitude higher under zones of upwelling. The sediments are typically thoroughly homogenised by bioturbation, and devoid of any primary current-formed structures. A variety of burrow types may be preserved (Fig. 2.43), dependent on varying environmental factors such as water depth, grain size, rate of sediment accumulation and redox conditions (Seilacher 1967; Werner & Wetzel 1982). Some of the main diagnostic trace fossils are *Zoophycus*, *Chondrites*, *Planolites*, *Scolicia*, *Trichichnus*, *Teichichnus* and *Lophoctenium* (see Chapter 3).

The grain size of biogenic oozes is largely dependent on the composition of the biogenic components. Coccolith plates are very small (clay size), whereas some foraminifera-rich or diatom-rich oozes may have a mean grain size that is in the silt range. The terrigenous fraction is largely clay grade. Full grain-size analyses of pelagic oozes are rarely conducted because data on the hydraulic equivalence of biogenic particles (Berthois & le Calves 1960; Maiklem 1968; Berger & Piper 1972; Braithwaite 1973) are too meagre to allow useful interpretation. The characteristics, composition and distribution of the different types of biogenic oozes are dependent on (i) the water depth and the location of the carbonate compensation depth (CCD); (ii) the source and type of terrigenous/volcanic material, together with the processes that supply this material; (iii) surface water productivity and the type of biogenic materials produced; (iv) surface currents and bottom circulation patterns; (v) climate and basin physiography and (vi) the physiochemical conditions (Lisitzin 1972; Berger 1974).

Biogenic components settle relatively slowly through the water column, are buried very slowly, and therefore are exposed for relatively

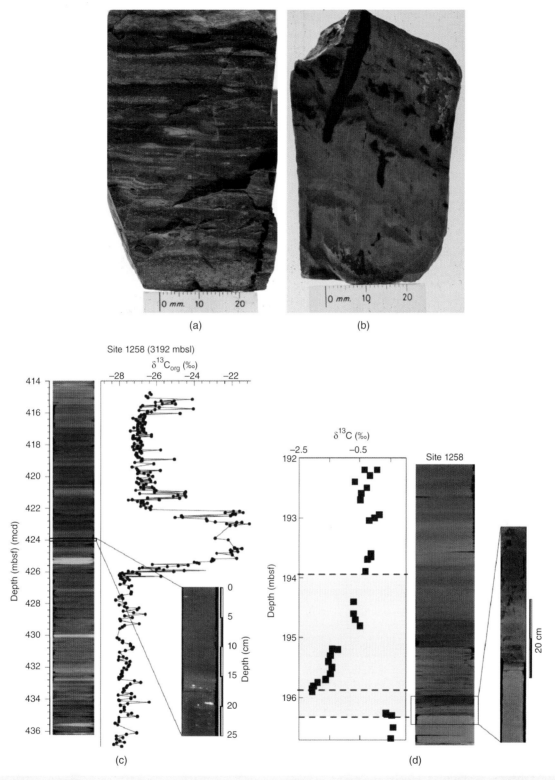

Fig. 2.43 Facies G1.1: biogenic ooze. Siliceous bioturbated radiolarian (biogenic) mudstones (chert), occurring as both red (a) and grey deposits (b), Middle to Upper Ordovician (Caradoc–Ashgill) Point Leamington Formation, north-central Newfoundland, Canada. Examples of biogenic oozes have been interpreted as associated with major oceanographic events, for example from ODP Leg 207 in a transect at Demerara Rise, western tropical Atlantic (*cf.* Mosher *et al.* 2007). (c) $\delta^{13}C_{org}$ record from ODP Site 1258 showing the excursion considered to represent the Late Cretaceous OAE 2 (Oceanic Anoxic Event 2; *cf.* Erbacher *et al.* 2005). The adjacent photo is a highly vertically compressed photomosaic of recovered cores. Depths are in metres below the seafloor. The inset plate is a sample core image within OAE 2, displayed with the correct aspect ratio. (d) $\delta^{13}C$ record at ODP Site 1258 through the Palaeocene/Eocene Thermal Maximum (PETM) (after Nuñes & Norris 2006). The adjacent core photo is highly vertically compressed. The inset core photograph is of the boundary and is shown with the correct aspect ratio.

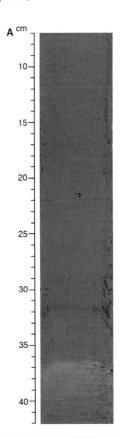

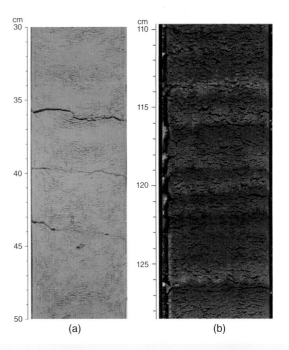

Fig. 2.45 (a) Facies G2.1: biogenic mud. Radiolarian ooze (interval 191-1179C-20H-6, 30–50 cm). Northwestern Pacific Ocean, ODP Site 1179. After Kanazawa *et al.* (2001). (b) Holocene diatomaceous ooze, IODP Expedition 318 Site U1357. Note the distinct seasonal laminations.

Fig. 2.44 Facies G1.2: muddy ooze. Structureless mud-bearing diatom ooze (top part) with one lamina in interval 178-1099A-2H-3, 31–32 cm, of *Chaetoceros* spp. spores, and the top part of a turbidite in interval 178-1099A-2H-3, 7–42 cm. Palmer Deep, ODP Leg 178 Site 1099. After Barker *et al.* (1999).

Transport process: introduction of clays by aeolian action and/or as suspended load of very dilute ocean currents.

Depositional process: grain-by-grain or aggregate settling through the water column.

Selected references: as for Facies G1.1.

2.9.2   Facies Group G2: Biogenic mud

2.9.2.1   Facies G2.1: Biogenic mud (Fig. 2.45)

Both this facies and its facies group were named 'hemipelagite' in an earlier published version of our facies scheme (Pickering *et al.* 1986a), but inspection of the attributes and interpretations of Facies Class E indicates that many Class E muds also resemble hemipelagic deposits described in the literature. For this reason, the genetic term 'hemipelagite' does not appear here as a facies name, and Facies G2.1 is restricted to 25–50% biogenic content from its previous 5–75% range (Pickering *et al.* 1986a). No requirement is placed on silt content, although most examples of this facies contain mixtures of silt and clay in the terrigenous fraction. Sand content may range up to 15%, but is mainly of biogenic origin (i.e., tests of planktonic organisms).

These sediments are poorly sorted and show no systematic grading. In published descriptions of sedimentary successions, they are commonly referred to with some Facies Class E deposits as 'background sedimentation', 'normal', 'ubiquitous' or 'interbedded' facies. At high latitudes (e.g., in the Arctic Ocean and Baffin Bay), ice rafting may be a major contribution to fine-grained muds and biogenic muds.

Hemipelagites (Facies G2.1, E1.1, E1.3 + other Class E deposits) are generally not described in detail although they constitute the greater part of many successions. Particularly useful detailed descriptions of

long time periods, either in the water column or on the seafloor. The actual processes of settling as single grains or as larger flocs and pellets are discussed by Gorsline (1984) and McCave (1984).

Depositional process: grain-by-grain or aggregate settling through the water column.

Selected references: Arrhenius (1963), Lisitzin (1972), Berger (1974), Broecker (1974), Hsü and Jenkyns (1974), Jenkyns (1986), Werner and Wetzel (1982), Arthur *et al.* (1984), Gorsline (1984), Isaacs (1984), McCave (1984), Bouma *et al.* (1986), Stow *et al.* (1990), O'Connell (1990), Rothwell *et al.* (1994), Bahk *et al.* (2000), Whitcar and Elvert (2000), Nederbragt and Thurow (2001), Exon *et al.* (2004), Lyle and Wilson (2004).

2.9.1.2   Facies G1.2: Muddy ooze (Fig. 2.44)

There is a continuum of facies from biogenic ooze with >75% biogenic material to pelagic clay with <25% biogenics (Facies E1.2). Muddy biogenic oozes are a relatively common intermediate sediment type, with attributes that are transitional between an ooze and a clay. They differ from true hemipelagic sediments in possessing little terrigenous silt, and in being an open-ocean rather than continental-margin facies. Layers of this facies are devoid of physical sedimentary structures, but are commonly pervasively bioturbated.

modern hemipelagites include those of Stanley *et al.* (1972), Moore (1974), Rupke (1975), Kolla *et al.* (1980a), Stanley and Maldonado (1981), Hill (1984b), Isaacs (1984) and Thornton (1984). A number of papers describe possible hemipelagites from ancient slope and basinal environments (Hesse 1975; Piper *et al.* 1976; Ingersoll 1987a,b; Hicks 1981; Pickering 1982b; Wang & Hesse 1996). Such facies are commonly homogeneous and structureless, with poorly defined bedding or no bedding. There are no primary current-formed structures such as lamination or erosional contacts, although a depositional lamination may be preserved under anoxic conditions (e.g., Isaacs 1984; Thornton 1984) or in Precambrian strata deposited before the advent of vigorous burrowers.

Under normal oxic conditions, bioturbation is ubiquitous and pervasive, commonly resulting in completely homogenised sediment with a mottled appearance. Burrow traces may be preserved, with the same major trace fossil assemblages as for Facies G1.1. Iron sulfide filaments (mycelia) and mottles are also common features of hemipelagic deposits.

The biogenic input to muds and biogenic muds may be calcareous or siliceous. It is commonly of mixed planktonic and in situ bathyal benthic species. The terrigenous components are mostly uniform in any given area of an ocean and display very gradual compositional trends toward the source area(s). The nature of hemipelagites varies considerably throughout the geologic record and in modern oceans, depending on the tectonic, climatic, source area and oceanic physiochemical conditions. There may be far-transported, wind-blown and ice-rafted terrigenous detritus, constituting substantial proportions of the total sediment volume in some cases.

Transport process: introduction of terrigenous materials by aeolian action, ice rafting, suspension transport in mid- and bottom-water currents from river deltas and other coastal areas.

Depositional process: grain-by-grain or aggregate settling through the water column.

Selected references: Stanley *et al.* (1972), Hesse (1975), Rupke (1975), Piper *et al.* (1976), Stanley and Maldonado (1981), Pickering (1982b), Gorsline (1984), Gorsline *et al.* (1984), Isaacs (1984), Thornton (1984), O'Connell (1990), Wang and Hesse (1996).

2.10 Injectites (clastic dykes and sills) (Figs 2.46–2.50)

Fluidisation structures, such as clastic dykes and sills, are a common feature of deep-water systems, and tend to be associated with liquefaction and brittle deformation, such as sedimentary faults. Post-depositional sediment deformation can involve any of the facies described above and range from small and mesoscopic scale (Fig. 2.46), to larger scale features (Fig. 2.47), and even km-scale features, for example mud volcanoes (see review of their formation processes by Dimitrov 2002). Mud volcanoes are described from many depositional systems worldwide, with particularly well-studied examples including those described from the Barbados accretionary prism and surrounding area (Langseth *et al.* 1988; Henry *et al.* 1990; Lance *et al.* 1998; Sullivan *et al.* 2004), the Banda arc (Barber *et al.* 1986; van Weering *et al.* 1989), and the Nankai accretionary prism and forearc (Pickering *et al.* 1993b; Kuramoto *et al.* 2001; Ashi *et al.* 2008). Fluidisation and liquefaction structures, such as small-scale dish-and-pillar (pipe) structures, can be thought of as amongst the earliest-formed injection structures that are commonly

contemporaneous with sediment deposition, as seen in the consistent sense of downcurrent overturning (Fig. 2.46b and c). Injectites also form synchronous with seismic activity, for example as observed in their occurrence along fault zones (Figs 2.42b and 2.46d).

There is a considerable observational, experimental and theoretical literature on these features and mechanisms by which they form. As this topic is beyond the scope of this book, more information and examples of liquefaction, fluidisation and injection structures can be found in: Waterston (1950), Lowe (1975, 1976b), Lowe and LoPiccolo (1974), Hiscott (1979), Archer (1984), Nichols *et al.* (1994), Lonergan *et al.* (2000), Boehm and Moore (2002), Jonk *et al.* (2003), Briedis *et al.* (2007), Hamberg *et al.* (2007), Hubbard *et al.* (2007), Lonergan *et al.* (2007), Bouroullec and Pyles (2010), Gamberi & Rovere (2010, 2011), Vigorito and Hurst (2010), Hurst *et al.* (2011) and Jackson and Sømme (2011). A key requirement is that the liquefied material be trapped within an impermeable host succession so that it cannot dewater and regain its strength, at least not until it has been injected or contorted by overburden pressures. In the case of injection (i.e., dykes and sills), the liquefied material moves down a pressure gradient toward the seafloor.

In many cases, injection occurs before the host succession is fully compacted, creating the opportunity for later compaction-driven folding of near-vertical dykes. Geometric analysis of the folds can be used to back-calculate the burial depth at the time of the injection (Hiscott 1979). The same technique can be used wherever injections penetrate uncompacted sediment, even if those injections are of igneous origin (e.g., Karner & Shillington 2005).

A particularly impressive outcrop example of sediment injection structures is in the Upper Cretaceous to Lower Palaeocene complex (PGIC), Coastal Ranges, California (Fig. 2.47). This is a complex of dykes and sills with a possible decompacted stratigraphic thickness of 1600 m to 1200 m today (Braccini *et al.* 2008; Hurst *et al.* 2011). The Cretaceous Moreno Shale Formation is the fine-grained host rock for the intrusions that generally dip at low angles to the east (Jolly & Lonergan 2002). The source of the clastic dykes and sills is the underlying Dosados sandstone member ~80 m stratigraphically deeper. Isotopic analysis of calcite deposited during the expulsion of the cold seep fluids dates the injections as having reached the surface over a 2 Myr interval ~62 Ma (Minisini & Schwartz 2007). Their age is based on marine fossils contained within the sandstone injections that were extruded onto the palaeo-seafloor.

Clastic injections can be associated with hydrocarbons. For example, in the Santa Cruz area, California, the Upper Miocene deep-marine Santa Cruz Mudstone (bio-siliceous unit of diatomaceous mudstones and opal-CT porcelanites that accumulated contemporaneously with the Monterey Formation elsewhere in the same basin) contains a dyke and sill complex that hosts hydrocarbons (Thompson *et al.* 1999; 2007; Boehm & Moore 2002) (Fig. 2.48).

Figure 2.49 summarises the geometry of sandstone intrusions and remobilised sandstones in a review by Hurst *et al.* (2011). They recognised four architectural elements: parent units, dykes, sills and 'extrudites'. Parent units are depositional sandstones that show features formed both by depositional processes and post-depositional sand and fluid mobilisation, and form an interconnected system of sandstones together with sandstone intrusions. Sandstone dykes are discordant, locally tabular bodies that may include mudstone clasts, organic matter and diagenetic cements) that over the majority of their length cross-cut sedimentary bedding at various angles. Sandstone sills are essentially tabular bodies of injected material whose boundaries are approximately concordant with that of bedding in the host strata but locally may be discordant with bedding along

(a)

(b)

(c)

Fig. 2.46   Meso- and small-scale sandstone dykes and sills. (a) Sandstone clastic dyke, Upper Precambrian Konsgfjord Formation, Finnmark, Arctic Norway. Lower thick sandstone bed shows local normal bedding orientation and was donor bed for the dyke that intruded up to just below human scale. (b) Pipe or pillar structure in fluidised bed, Upper Precambrian Konsgfjord Formation, Finnmark, Arctic Norway. The consistent sense of overturning of the pipes/pillars is approximately downcurrent as deduced from sole marks and current ripples in surrounding beds. (c) Top surface of sandstone bed that shows pervasive dish-and-pillar structure, Upper Precambrian Konsgfjord Formation, Finnmark, Arctic Norway. Penknife for scale. Note the preferred orientation of the sandstone injections that has the same orientation as the strike of palaecurrents in surrounding beds. (d; facing page) Clastic dyke of scoriaceous and pumiceous material cutting through volcaniclastic deposits and offset by layer-parallel faults (left of scale), Miocene Misaki Formation, Miura Peninsula, Japan; 15-cm bar scale. (e) Top view of sandstone injections (dykes) with a preferred orientation immediately below an MTD (fallen block), Ainsa I Fan, Middle Eocene Ainsa Basin, Spanish Pyrenees. The emplacement of the MTD may have been responsible for the injections; 10 cm scale bar.

(d)

(e)

Fig. 2.46 *(continued)*

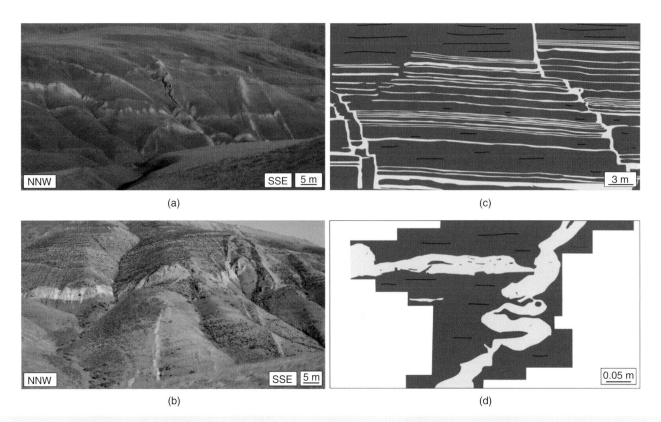

(a)   (c)

(b)   (d)

Fig. 2.47 Large-scale sandstone dykes and sills from the Upper Cretaceous–Lower Palaeocene Marca Canyon, Panoche Hills, California. In these plates, the sandstones are lighter grey and the mudstones dark grey. See Hurst *et al.* (2011).

Fig. 2.48 Part of large clastic dyke–sill complex in the Miocene Monterey Formation correlative stratigraphic interval, Yellow Bank Beach, near Santa Cruz, California. Note near-vertical, dark-coloured, hydrocarbon-bearing dyke in centre. The yellow/orange-coloured lower half of the outcrop left of centre is a clastic sill that fed the dyke seen in the centre. From Boehm and Moore (2002).

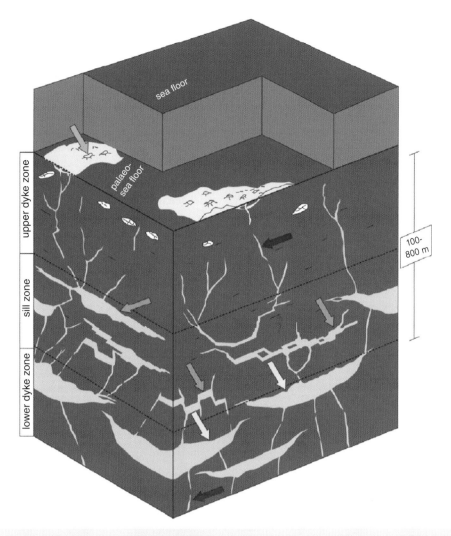

Fig. 2.49 Synoptic diagram of a sand injectite complex with a tripartite architecture based on outcrop and subsurface observations in host rock (dark grey). Remobilised parent sandstone units (yellow arrows); sandstone dykes (red arrows) and sills (blue arrows); irregular sandstone intrusions (orange arrow); sandstone extrudites, for example, sand volcanoes (green arrow). From Hurst *et al.* (2011).

Fig. 2.50 Sand volcano, Ross Formation, Namurian Clare Basin, western Ireland. Pen for scale ~14 cm long.

University College London (UCL) Ainsa Project
Ainsa II Fan channel facies-association
(Facies A1.1, B1.1, C1.2 (minor C1.1))

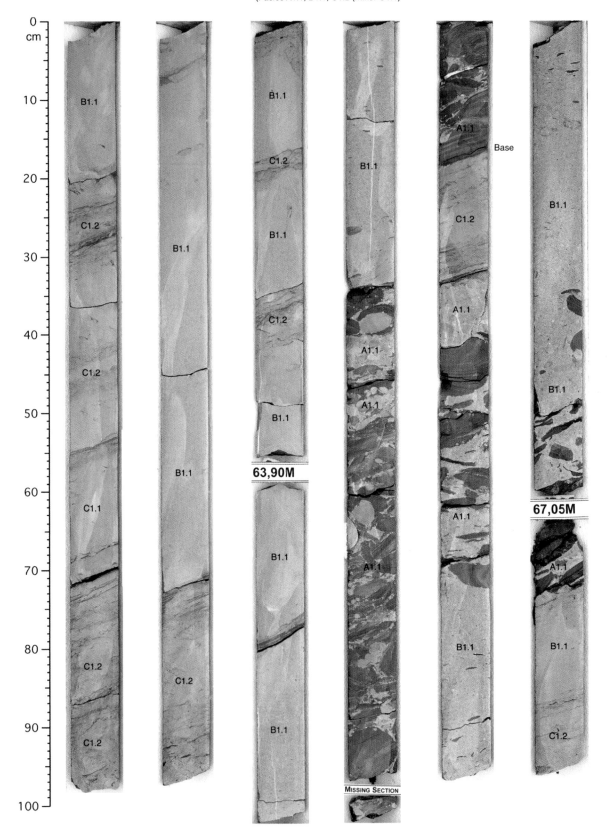

Fig. 2.51 Example of channel facies association, Ainsa II Fan, Middle Eocene Ainsa System, Ainsa Basin, Spanish Pyrenees. 'Base' refers to the inferred erosional base of one of the nested (laterally offset-stacked) submarine channels observed in the nearby outcrop (*cf.* Pickering & Corregidor 2005; Pickering & Bayliss 2009). Core depths in metres below ground level, younging from right to left.

their upper and lower margins. Irregular sandstone intrusions have discordant margins and abrupt changes in thickness. Extrusive sandstones form by venting of sand onto the seafloor and can be connected to and fed by underlying dykes (Hurst *et al.* 2011). Figure 2.50 shows a typical sand volcano in the Namurian Clare Basin, western Ireland.

2.11 Facies associations

The facies groups and classes presented in this chapter demonstrate and emphasise the common attributes and, in many cases, the common origins of suites of deep-marine deposits. There is no stratigraphic connotation to the facies groups and classes, with the result that this facies classification takes no account of the facies associations found in natural deposits, which instead reflect spatial and/or temporal shifts in the environmental conditions in a part of a deep-marine basin, for example the progradation of a depositional lobe, or the lateral migration or back-filling of a submarine channel (e.g., Fig. 2.51). Subsequent chapters of this book consider facies distributions in the real world, and therefore emphasise vertical and lateral facies associations, and how to interpret these associations in terms of changing seabed morphology, transport processes, palaeoceanography conditions, and tectonic factors of the receiving basin and hinterland.

CHAPTER THREE
Deep-water ichnology

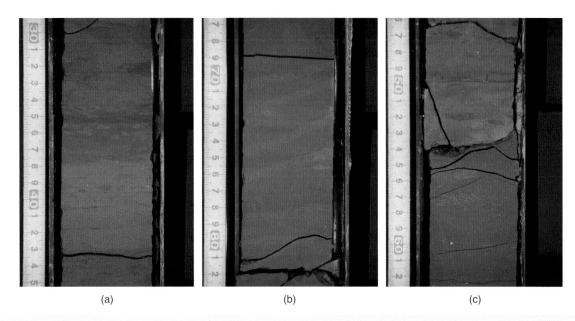

(a)　　　　　　　　　　　(b)　　　　　　　　　　　(c)

Trace fossils from IODP Expedition 322, Kashinosaki Knoll, northeastern Shikoku Basin, Japan (NanTroSEIZE subduction inputs).
(a) *Palaeophycus, Thalassinoides, Planolites* and *Chondrites*. C0011B-6R-2, 30–45 cm (b) *Zoophycos*. Core interval C0011B-6R-1, 67–83 cm. (c) *Chondrites, Planolites* and *Thalassinoides*. Core interval C0011B-12R-1, 46–62 cm.

3.1　Introduction

Ichnology represents an important and accessible tool to sedimentologists because it can reveal palaeoenvironmental information beyond that gained from the study of sedimentary structures alone. The results can easily be incorporated into any type of sedimentological study. Ichnology can help sedimentologists reconstruct palaeoenvironments, stratigraphers correlate sedimentary strata, palaeontologists determine the nature of fossil communities, geochemists determine the effect of organisms on sediment composition (Ekdale *et al.* 1984), and petroleum geologists characterise reservoirs and improve reservoir models. In deep-water environments, trace fossils are particularly valuable because they commonly represent the only preserved macro fossils. Preservation of trace fossils, particularly on the soles of sediment gravity flow (SGF) deposits, can be of exceptional quality, which can aid detailed interpretations of palaeoenvironmental conditions before, during and after the deposition of each SGF deposit. This chapter introduces some general principles of ichnology with an emphasis on deep-water trace fossils, to highlight some of the important applications of ichnology in sedimentary studies. Two case studies are presented to illustrate the methods and uses of ichnological studies in palaeoenvironmental analysis and reservoir characterisation. The first case study is based on Heard and Pickering (2008) who examined the ichnology of the Middle Eocene Ainsa–Jaca basins, Spanish Pyrenees. The second case study examines the results of a core-based ichnofabric study from the Ainsa turbidite system (Heard *et al.* 2014).

The strength of ichnology stems from the fact that the activity of organisms is dictated by a number of dynamic controlling factors, such as substrate consistency (e.g., soft versus hard), nutrient supply, level of oxygenation, hydrodynamic energy, rate of sediment accumulation, salinity and chemical toxicity (Frey *et al.* 1990; Gingras *et al.* 2008). The study of ichnology, therefore, can reveal important information about the depositional environment, including: (i) the primary consistency of the sedimentary substrate (i.e., *looseground, soupground, softground, stiffground, firmground* or *hardground*) (Ekdale & Bromley 1991; Lewis & Ekdale 1992; Wetzel & Uchman 1998b, 2012); (ii) interstitial oxygen levels in organic-rich sedimentary units (Ekdale & Mason 1988; Savrda & Bottjer 1986, 1987, 1989, 1994; Wignall 1991); (iii) trophic state of the water

Deep Marine Systems: Processes, Deposits, Environments, Tectonics and Sedimentation, First Edition. Kevin T. Pickering and Richard N. Hiscott.
© 2016 Kevin T. Pickering and Richard N. Hiscott. Published 2016 by John Wiley & Sons, Ltd.
Companion Website: www.wiley.com/go/pickering/marinesystems

(eutrophic versus oligotrophic conditions) (Tunis & Uchman 1996b; Wetzel & Uchman 1998a); (iv) intensity of bioturbation in stratigraphic sequences (Bottjer & Droser 1991; Droser & Bottjer 1986, 1988, 1989); (v) palaeoflow direction (Monaco 2008); (vi) tiering patterns (Bromley & Ekdale 1986; Rajchel & Uchman 1998); (vii) subsequent taphonomic history of one or more phases of biogenic activity (Taylor & Goldring 1993); (viii) depth of erosion (Wetzel & Aigner 1986; Savrda & Bottjer 1994); (ix) discontinuity surfaces (Ghibaudo *et al.* 1996; Savrda *et al.* 2001a,b; Hubbard & Shultz 2001; Knaust 2009); (x) sedimentary cycles (Erba & Premoli Silva 1994; Heard *et al.* 2008); and (xi) processes of deposition (Frey & Goldring 1992). Ichnology can also reveal important information about biologically induced petrophysical heterogeneity for use in reservoir characterisation studies (e.g., Pemberton & Gingras 2005; Tonkin *et al.* 2010).

The study and application of ichnology is well developed for shallow- and marginal-marine sedimentary systems, but less so for deep-water systems. This is, in part, due to the fact that research on shallow- and marginal-marine systems developed earlier than it did for deep-water systems. Despite considerable research into the characterisation of deep-water systems in the past decades, very few of these studies have attempted to fully integrate ichnology and sedimentology. There are, however, a number of detailed ichnological studies on deep-water systems based on the taxonomy of deep-water trace fossils (e.g., Häntzschel 1975; Crimes 1977; Książkiewicz 1977; Crimes *et al.* 1981; Uchman 1995, 1999; Wetzel & Uchman 1998a, Tchoumatchenco & Uchman 1999; Orr 2001; Uchman *et al.* 2004; Miller III 2007; Knaust 2012; Knaust & Bromley 2012). Such publications have established a detailed catalogue of taxonomic descriptions of deep-water trace fossils which are now being integrated into sedimentological studies (e.g., Uchman 2001; Heard & Pickering 2008; Hubbard & Shultz 2008; Monaco *et al.* 2010; Phillips

et al. 2011; Cummings & Hodgson 2011; Hubbard *et al.* 2012; Callow *et al.* 2013).

3.2 General principles of ichnology

Ichnology is the study of biogenic structures formed by the activity of an organism or organisms which modify the substrate (Bromley 1996). Biogenic structures include those produced in both unconsolidated substrates (biogenic sedimentary structures) and rigid substrates (bio-erosion structures). In general, the most common biogenic sedimentary structures studied in ichnology are bioturbation structures, which are formed by the activity of organisms (Frey & Pemberton 1985). Two types of biogenic sedimentary structures are typically distinguished: trace fossils and bio-deformational structures (Schäfer 1956) (Fig. 3.1). Trace fossils have a definite shape and sharp outlines, allowing ichnotaxonomic classification. Typically, they are classified in terms of preservation (toponomy), behaviour (ethology) and type (taxonomy) (Seilacher 1953a, b).

Bio-deformational structures lack a sharp outline and are not characterised by a distinct shape which would enable classification. Typically they are formed in soft to soupy sediment and destroy formerly existing physical structures (Wetzel 1983). They have been interpreted to be common in sediments with a high organic matter content where behavioural specialisation which would lead to the formation of trace fossils is superfluous (Wetzel 1981, 1983).

3.2.1 Preservational classification of trace fossils

The study of the processes which lead to the burial and preservation of fossils is known as *taphonomy* (Rindsberg 2012). These

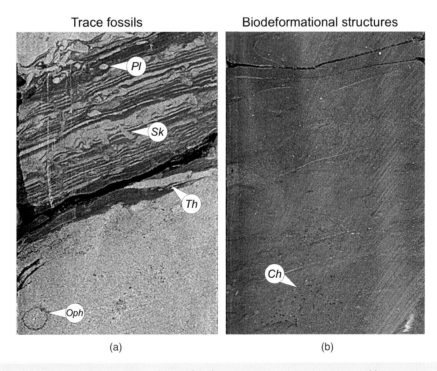

| Trace fossils | Biodeformational structures |

Fig. 3.1 Examples of bio-deformational structures and trace fossils in core (Ainsa core L1, core width ~6 cm). (a) Trace fossils associated with fine- to medium-grained sandstones and laminated siltstone–mudstone intervals (*Th* = *Thalassinoides*; *Oph* = *Ophiomorpha*; *Sk* = *Skolithos*; *Pl* = *Planolites*). (b) Bio-deformational structures overprinted by later trace fossils (*Ch* = *Chondrites*).

processes provide the ichnologist with important information on the trace-making organism(s) as well as palaeoenvironmental conditions at the time the traces were formed. In deep-water systems, trace fossils are commonly exceptionally preserved on the soles of sandstone and siltstone beds deposited by SGFs. Consequently, the activity of the organisms at the time of deposition can commonly be determined. According to Seilacher (2007), preservation occurs as a result of the exhumation of structures created by shock erosion followed by casting by the deposit of a subsequent SGF. Shock erosion occurs as a result of the sudden acceleration of water ahead of the approaching SGF, causing the removal of unconsolidated surface mud and the exposure of the burrow systems. If the SGF is in its depositional phase, these burrow systems are then cast at the base of the bed (Seilacher 2007). Consequently, deep-water systems provide an excellent natural laboratory for the ichnologist and sedimentologist to study biogenic activity on the seafloor because the trace-fossil record provides a frozen snapshot of the activity of organisms at the time of deposition. As the preservation of trace fossils on the soles of SGF deposits is related to a combination of erosion and deposition, the nature of the SGF has a significant impact on the preservation potential of burrows. Typically, trace fossils on the soles of SGF deposits are best preserved in association with low-concentration gravity flows (e.g., turbidity currents) in their depositional phase. This is because burrows uncovered by shock erosion can immediately be cast by fallout from suspension. Preservation potential of trace fossils is reduced when associated with concentrated density flows (Section 1.4.1) because these more energetic flows erode the seafloor to a greater depth than many endichnial open burrow networks, resulting in their destruction.

Where trace fossils are preserved in deep-water systems, the morphological expression of the trace (toponomy) is dependent on whether the trace fossil was preserved within a mudstone or sandstone, or at the boundary between the two (Bromley 1996). Stratinomic classification schemes to describe this morphological expression have been devised by a number of authors including Seilacher (1964a, b), Simpson (1957), Martinsson (1965, 1970) and Chamberlain (1971). The most common preservational classification schemes adopted by ichnologists are those by Martinsson (1965, 1970) and Seilacher (1964a,b) (Fig. 3.2). The scheme proposed by Seilacher (1964a, b) comprises both descriptive and genetic terms, with descriptive terms based on the relationship of the trace fossil to the casting medium, and genetic terms based on the relationship of the structure to the contemporary surface. There are two main subdivisions of descriptive terms: full-relief and semi-relief (Fig. 3.2). full-relief trace fossils are preserved within the casting medium, and semi-relief trace fossils are preserved at lithological interfaces. Semi-relief structures can in turn be divided into those formed on the upper surface (epi-relief) and those on the lower surface of the casting medium (hypo-relief). The descriptive terms 'convex' or 'concave' can be used to distinguish semi-relief trace fossils preserved as ridges or grooves. Three genetic terms were also proposed by Seilacher (1964a,b): exogenic, endogenic and pseudoexogenic. Exogenic traces are surface traces, endogenic traces are formed within the host bed and pseudoexogenic traces are formed in a homogeneous medium such as mud and subsequently uncovered by erosion and cast with sand. In the scheme established by Martinsson (1965, 1970), biogenic structures on the lower surface of the casting medium are termed *hypichnia*, whilst corresponding structures on the upper surface are called *epichnia*. Structures within the casting medium are termed *endichnia*, and those mostly outside the medium *exichnia* (Fig. 3.2).

3.2.2 Ethological classification of trace fossils

Trace fossils reflect the behavioural patterns of trace-making organisms. The classification of trace fossils according to these behavioural patterns was originally devised by Seilacher (1953a,b, 1964b). This classification scheme has been widely adopted in part because trace-making organisms are rarely identified in the fossil record, and therefore in the absence of any knowledge on the trace-making organism, understanding the ethological function of a trace fossil is highly significant. Initially Seilacher (1953a,b, 1964b) recognised five ethological groups. This classification scheme has subsequently been expanded (e.g., Ekdale 1985; Frey *et al.* 1987; Bromley 1996), with the more common groups described below (Table 3.1):

*Edifices constructed above substrate (**aedifichnia**)*: Structures built of sediment more or less cemented by the architect, for example, termite colonies (Bromley 1996).

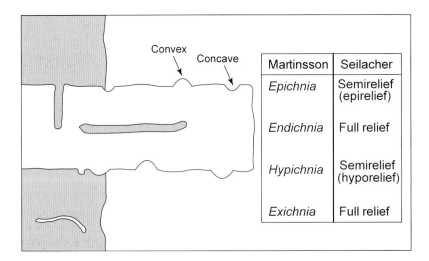

Martinsson	Seilacher
Epichnia	Semirelief (epirelief)
Endichnia	Full relief
Hypichnia	Semirelief (hyporelief)
Exichnia	Full relief

Fig. 3.2 Stratinomic classification of trace fossils in relation to the casting medium (e.g., sandstone). Redrawn from Seilacher (1964a), Martinsson (1965) and Bromley (1996).

Table 3.1 Ethological classification of trace fossils including a list of common deep-water trace fossils within each ethological group

Ethological classification	Description	Deep-water examples
Aedifichnia	Constructed above substrate	
Agrichnia	Traps and gardening	*Paleodictyon, Cosmorhaphe, Helminthorhaphe, Desmograpton, Megagrapton, Chondrites*
Calichnia	Breeding	*Hormosiroidea*
Cubichnia	Resting	*Lockeia, Bergaueria*
Domichnia	Dwelling	*Ophiomorpha, Arenicolites, Skolithos*
Equilibrichnia	Equilibrium	*Diplocraterion, Teichichnus*
Fodinichnia	Feeding	*Thalassinoides, Phycosiphon, Zoophycos, Planolites*
Fugichnia	Escape	Escape structures
Pascichnia	Grazing	*Planolites, Nereites, Scolicia, Helminthopsis, Halopoa*
Praedichnia	Predation	
Repichnia	Crawling	Arthropod tracks

*Traps and gardening traces (**agrichnia**)*: Regularly patterned burrows produced for bacterial farming or as traps to capture meiofauna or microorganisms. Many graphoglyptids are examples of *agrichnia* (Ekdale *et al.* 1984).

*Structures made for breeding purposes (**calichnia**)*: Structures made exclusively for raising larvae or juveniles (Bromley 1996).

*Resting traces (**cubichnia**)*: Structures made by organisms which settle onto or dig into the substrate surface. These traces in many cases closely mirror the ventral anatomy of the trace maker (Frey & Pemberton 1985).

*Dwelling traces (**domichnia**)*: Permanent or semi-permanent domiciles. Mostly for the hemi-sessile suspension feeders or, in some cases, carnivores (Pemberton *et al.* 2001). The burrow walls may be strengthened and most dwelling structures are later passively filled with sediments.

*Equilibrium traces (**equilibrichnia**)*: Traces formed by organisms responding to aggrading or degrading substrates.

*Feeding traces (**fodinichnia**)*: Characterised by a combination of deposit feeding and dwelling. The structure has a degree of permanence but its morphology reflects exploitation of the substrate for food (Bromley 1996).

*Escape traces (**fugichnia**)*: Panic escape by organisms due to burial by rapidly deposited sediment such as SGF deposition.

*Grazing traces (**pascichnia**)*: Where a trackway or locomotion trace clearly indicates exploitation of a particular region of the substrate for food, such as meandering or spiral traces. Grazing traces commonly follow surfaces parallel with the seafloor and generally reflect very efficient coverage of space (Bromley 1996).

*Predation traces (**praedichnia**)*: Traces resulting from predation typically are common in hard substrates as round drill holes in shells and as shell damage by predators (Ekdale 1985). Soft substrate disturbance due to predation is not easily recognisable in the fossil record.

*Crawling traces (**repichnia**)*: Structures reflecting directed locomotion and lacking the systematic probing patterns characteristic of grazing and feeding traces (Frey & Pemberton

1985). Secondary activities may be involved such as feeding.

3.2.3 Taxonomic classification of common deep-water trace fossils

The basis of the taxonomic classification of trace fossils is the morphology of a trace as an expression of the behaviour of an organism (Bromley 1996). In general, the main features used to describe the morphology of a trace include the general form of the trace, the nature of the wall and lining, branching, the fill and the presence or absence of *spreiten*. The two fundamental ranks of ichnotaxa are the ichnogenus and ichnospecies (abbreviated as igen. and isp.).

The classification of deep-water trace fossils above the ichnogeneric level is generally based on a system developed by Książkiewicz (1977) and later modified by Wetzel & Uchman (1998a). This simple, informal, non-interpretative classification of deep-water trace fossils is based solely on morphological criteria. Nine morphological groups are recognised: (i) circular and elliptical; (ii) simple and branched; (iii) rosetted; (iv) *spreite*; (v) winding; (vi) spiral; (vii) meandering; (viii) branched winding and meandering and (ix) nets. A brief description of the most common deep-water trace fossils is given below. Detailed taxonomic descriptions of deep-water trace fossils are beyond the scope of this chapter and readers are referred to more detailed descriptions in publications by Häntzschel (1975), Książkiewicz (1977), Uchman (1995, 1999), Wetzel & Uchman (1998a) and references therein. Both 3D outcrop (bedding plane) and 2D (core) descriptions are given.

3.2.3.1 Simple branched structures

Skolithos (Haldeman 1840)

Diagnosis: Unbranched, vertical to steeply inclined, straight to slightly curved, cylindrical to subcylindrical, lined or unlined

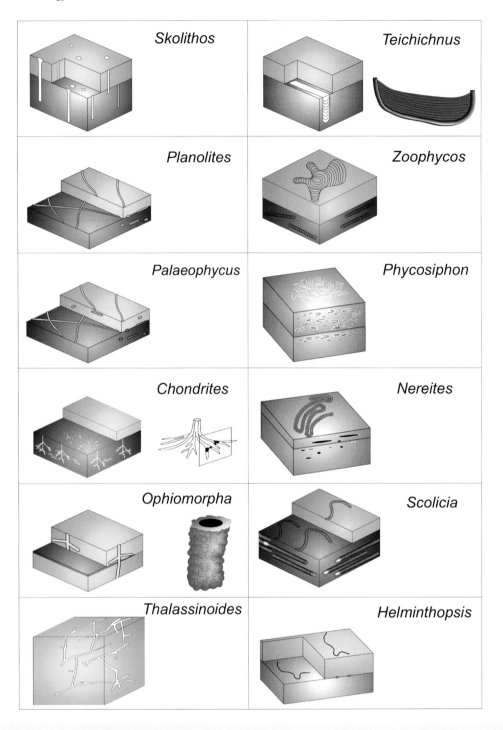

Fig. 3.3   Block diagrams of common trace fossils in deep-water environments.

structures with or without funnel-shaped top (after Alpert 1974; Schlirf 2000) (Figs 3.3, 3.6d).

Remarks: *Skolithos* is a facies-crossing trace fossil, found in virtually every environment. In both shallow- and deep-water environments, it is commonly indicative of relatively high levels of wave or current energy. It has been interpreted as a dwelling burrow (domichnia) produced by annelids or phoronids (Schlirf & Uchman 2005). In deep-water environments, it is typically a post-depositional trace fossil.

In core: *Skolithos* is typically expressed as a vertical to subvertical, lined or unlined burrow.

Halopoa (Torell 1870)

Diagnosis: Generally horizontal trace fossil covered with longitudinal irregular ridges or wrinkles, which are composed of several imperfectly overlapping cylindrical probes (after Wetzel & Uchman 1998a) (Figs 3.6h,j).

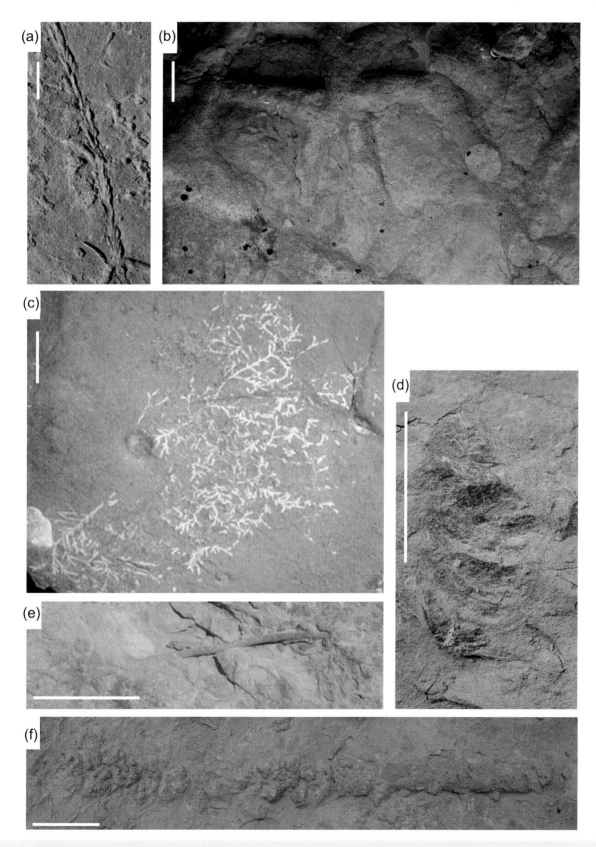

Fig. 3.4 Outcrop images of trace fossils from the Middle Eocene Ainsa–Jaca basins, Spanish Pyrenees. (a) *Halopoa storeana* isp. n. (Uchman 2001). Hypichnial full-relief. Scale bar: 2 cm. See Uchman (2001) for discussion. (b) *Thalassinoides suevicus* (Rieth 1932). Hypichnial full-relief. Scale bar: 2 cm. See Howard and Frey (1984) for discussion. (c) *Chondrites intricatus* Brongniart 1823. Endichnial full-relief. Scale bar: 2 cm. See Uchman (1998) and Fu (1991) for discussion. (d) *Teichichnus* isp. Epichnial full-relief. Scale bar: 2 cm. See Seilacher (1955) for discussion. (e) *Planolites* ispp. Endichnial full-relief. Scale bar: 2 cm. See Pemberton and Frey (1982) for discussion. (f) *Ophiomorpha rudis* (Książkiewicz 1977). Epichnial full-relief. Scale bar: 2 cm. See Uchman (2001) for discussion.

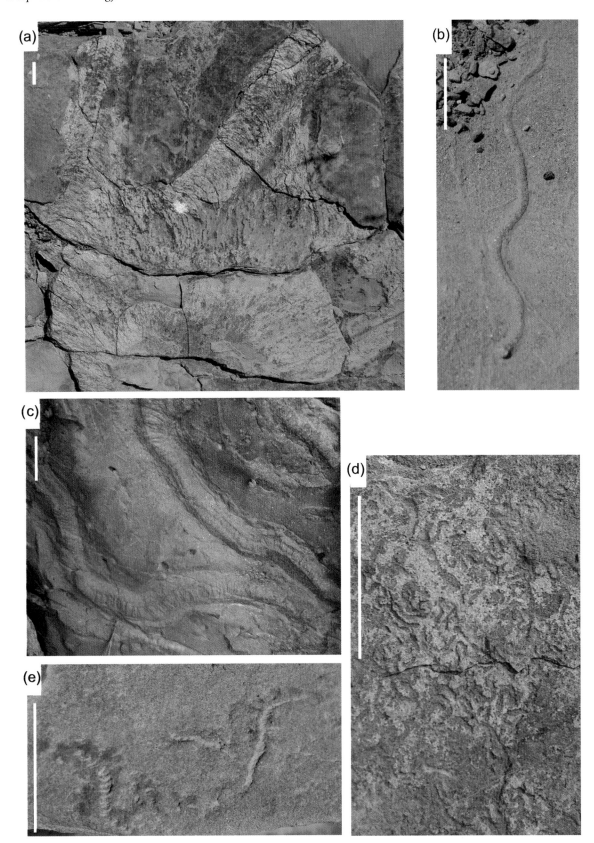

Fig. 3.5   Outcrop images of trace fossils from the Ainsa–Jaca basins. (a) *Zoophycos insignis* (Squinabol 1890). Endichnial full-relief. Scale bar: 2 cm. See Uchman (1999) for discussion. (b) *Helminthopsis* ispp. Hypichnial full-relief. Scale bar: 2 cm. See Wetzel and Bromley (1996) for discussion of ichnogenus. (c) *Scolicia plana* (Książkiewicz 1970) Endichnial full-relief. Scale bar: 2 cm. See Uchman (1998) for discussion. (d) *Phycosiphon incertum* (Fischer-Ooster 1858). Endichnial full-relief. Scale bar: 2 cm. See Wetzel and Bromley (1994). (e) *Nereites missouriensis* (Weller 1899) Endichnial full-relief. Scale bar: 2 cm. See Uchman (1995) for discussion.

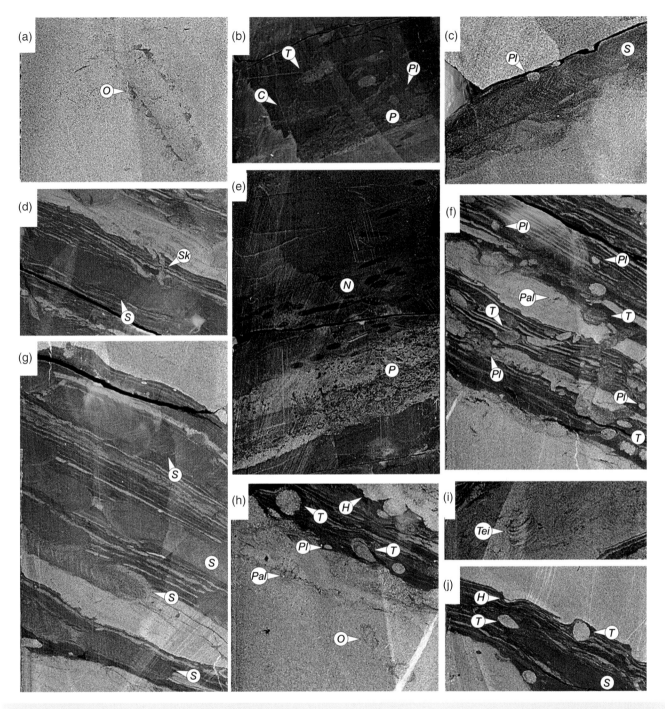

Fig. 3.6 Images of common trace fossils in Ainsa Basin cores. The core is ~6 cm wide. (a) *Ophiomorpha* (*O*) in coarse-grained sandstone. (b) *Phycosiphon* (*P*), *Planolites* (*Pl*), *Thalassinoides* (*T*) and *Chondrites* (*C*) in mudstones. (c) *Planolites* (*Pl*) and *Scolicia* (*S*) in medium-grained sandstones and siltstones. (d) *Skolithos* (*Sk*) and *Scolicia* (*S*) in fine-grained sandstones, siltstones and mudstones. (e) *Nereites* (*N*) and *Phycosiphon* (*P*) in fine-grained sandstones and mudstones. (f) *Planolites* (*Pl*), *Thalassinoides* (*T*) and *Palaeophycus* (*Pal*) in fine-grained sandstones and siltstone–mudstone couplets. (g) *Scolicia* (*S*) in fine-grained sandstones and siltstone–mudstone couplets. (h) *Thalassinoides* (*T*), *Halopoa* (*H*), *Planolites* (*Pl*), *Ophiomorpha* (*O*) and *Palaeophycus* (*Pal*) in medium-grained sandstones and siltstone–mudstone couplets. (i) *Teichichnus* (*Tei*) in medium-grained sandstones. (j) *Thalassinoides* (*T*), *Halopoa* (*H*) and *Scolicia* (*S*) in medium-grained sandstones and siltstone–mudstone couplets.

Remarks: *Halopoa* is a facies-crossing trace fossil occurring in both shallow- and deep-water environments. It is typically preserved as hypichnial full-relief burrows occurring at the base of thin- to medium-bedded sandstones. The *Halopoa* producers were probably deposit-feeders (pascichnion) which reworked the sandstone/mudstone interface, backfilling their burrows. In deep-water environments, it is typically a post-depositional trace fossil.

In core: *Halopoa* is expressed as cylindrical burrows, 7–12 mm wide, preserved in full-relief with furrows and wrinkles on the exterior of the burrows. In general, identifying *Halopoa* in core is challenging because the furrows and wrinkles which characterise the surface of the cylindrical burrows are difficult to observe.

Planolites (Nicholson 1873)

Diagnosis: Unlined, rarely lined, rarely branched, straight to tortuous, smooth to irregularly walled or annulated trace fossils, circular to elliptical in cross-section, of variable dimensions and configurations. Fill is essentially structureless, differing in lithology from the host rock (after Pemberton & Frey 1982; Wetzel & Uchman 1998a) (Figs 3.3, 3.4e, 3.6b,h,c,f).

Remarks: *Planolites* is an extremely facies-crossing ichnogenus occurring in virtually all environments from freshwater to deep-marine. It is preserved as endichnia, hypichnial ridges and epichnial grooves. *Planolites*-forming organisms were probably polyphyletic vermiform deposit-feeders (Pemberton & Frey 1982).

In core: Cylindrical tunnels with a fill which is essentially structureless and differing in lithology from the host rock. Typically, the diameter of *Planolites* tunnels is 3–7 mm.

Palaeophycus (Hall 1847)

Diagnosis: Branched or unbranched, smooth or ornamented, lined, essentially cylindrical, predominantly horizontal trace fossils of variable diameter: fill typically structureless, of the same lithology as the host rock (after Pemberton & Frey 1982) (Figs 3.3, 3.6h).

Remarks: *Palaeophycus* is distinguished from the morphologically similar *Planolites* by wall linings and the character of the burrow fills, with the burrows passively filled by percolation of sediment into open, lined burrows. *Palaeophycus* is produced probably by polychaetes (Pemberton & Frey 1982). As with most passively filled lined burrows, *Palaeophycus* is interpreted as a dwelling structure. It is a facies-crossing ichnogenus occurring in virtually all environments from freshwater to deep-marine.

In core: Cylindrical tunnels with distinct lining. The sediment fill is the same lithology as the host rock.

Chondrites (Sternberg 1833)

Diagnosis: Regularly branching three-dimensional burrow system consisting of a mastershaft open to the surface which ramifies at depth to form a dendritic network (after Fu 1991; Uchman 1998) (Figs 3.3, 3.4c, 3.6b).

Remarks: *Chondrites* is a feeding system of unknown infaunal deposit feeders. *Chondrites* occurs in variable substrates including the Bouma T_b division of SGF deposits, but most commonly in fine-grained siliciclastic and calcareous rocks. It occurs in normal marine to dysaerobic conditions, and is interpreted as being indicative of dysaerobic conditions where bioturbation intensity and/or ichnodiversity are low. It is commonly the last trace to be formed as

the trace-making organism exploited declining levels of oxygen in the substrate.

In core: *Chondrites* forms clusters of small tunnels in core, with each tunnel system being ~1–3 mm in diameter. It may occur in abundance in both totally homogenised sediments, and intervals characterised by low intensity bioturbation.

Ophiomorpha (Lundgren 1891)

Diagnosis: Simple to complex burrow systems lined at least partially with agglutinated pelletoidal sediment. The burrow fill is typically structureless and of the same lithology as the host rock. Branching is irregular (after Howard & Frey 1984; Uchman 1995, 2009) (Figs 3.3, 3.4f, 3.6a,h).

Remarks: *Ophiomorpha* represents dwelling burrows (domichnia) of organisms and occurs in both shallow- and deep-water environments. *Ophiomorpha* is produced mainly by shrimps comparable to the recent *Callianassa major* (Uchman 1995). The trace-making organism penetrates deep into the sediment (up to 2.5 m vertical depth) maintaining an open connection to the seafloor.

In core: *Ophiomorpha* is easily recognisable in core due to its distinctive knobbly exterior. *Ophiomorpha* typically occurs within sandstones emplaced by turbidity currents and concentrated density flows and ranges between 6–25 mm in diameter.

Thalassinoides (Lundgren 1891)

Diagnosis: Three-dimensional burrow systems consisting predominantly of smooth-walled, typically unlined, essentially cylindrical components (after Howard & Frey 1984; Uchman 1995) (Figs 3.3, 3.4b, 3.6b,h,f,j).

Remarks: *Thalassinoides* is characteristic of fine-grained, coherent substrates, in which wall reinforcement is unnecessary. *Thalassinoides* is mainly produced by deposit-feeding crustaceans and is typically a post-depositional trace. It is a facies-crossing form, most typical of shallow-marine environments, but is also common in deep-marine environments.

In core: The three-dimensional burrow system of *Thalassinoides* is rarely preserved in core, but evidence of branching may be preserved on the rough outer surface of the core. The cylindrical burrows are typically 5–20 mm in diameter, with the fill differing in lithology from the host rock.

Teichichnus (Seilacher 1955)

Diagnosis: Long, *spreiten* structure consisting of a pile of vertically stacked gutter-shaped laminae (after Seilacher 1955) (Fig. 3.3, 3.4d, 3.6i).

Remarks: *Teichichnus* is interpreted to have been formed by a deposit-feeding, worm-like organism and can be classified as a fodinichnion (Pickerill *et al.* 1984). It is formed by the upward migration of a horizontal to subhorizontal tunnel produced by an organism moving back and forth in the same vertical plane probing the sediment for food. The trace-making organism migrates upward in its burrow to keep up with sedimentation. It occurs in both shallow- and deep-water environments.

In core: *Teichichnus* is recognised in core by its distinctive vertical to sub-vertical stack of gutter-shaped near-horizontal laminae. In transverse sections it is characterised by gently curved *spreiten* which are typically concave-up. In longitudinal profile it may form an elongate J-shaped structure comprising elongate wavy laminae.

3.2.3.2 Radial structures

Zoophycos (Massalongo 1855)

Diagnosis: *Spreiten* structures comprised of numerous small, U- or J-shaped protrusive burrows of variable length and orientation which surround a central shaft in distinct levels or coilings (after Wetzel 1991; Uchman 1999) (Figs 3.3, 3.5a).

Remarks: *Zoophycos* is generally assumed to be the trace of an unknown deposit-feeding organism. It is associated with the *Zoophycos* Ichnofacies.

In core: *Zoophycos* occurs as a horizontal to low angle burrow with distinct *spreite* laminae. Each burrow ranges between 2–7 mm in thickness.

Phycosiphon (Fischer-Ooster 1858)

Diagnosis: Extensive small-scale *spreite* trace fossil comprising repeated narrow, U-shaped lobes enclosing *spreite* of millimetre to centimetre scale, branching regularly or irregularly from an axial *spreite* of similar width (after Uchman 1999) (Figs 3.3, 3.5d, 3.6b,e).

Remarks: This trace fossil is interpreted to have been produced by a post-depositional opportunistic deposit feeder (see Section 3.3). It typically occurs in silty mudstone and fine-grained sandstones from shallow- to deep-water environments. *Phycosiphon* represents the grazing trails of worm-like organisms. *Phycosiphon* and *Phycosiphon*-like trace fossils ('Phycosiphoniform' burrows) were recently reviewed in detail by Bednarz and McIlroy (2009).

In core: In core *Phycosiphon* is characterised by a dark backfilled centre surrounded by a thin halo of pale sediment. It commonly occurs in small dense patches which penetrate vertically as well as horizontally.

3.2.3.3 Winding and meandering structures

Nereites (Macleay 1839)

Diagnosis: Winding to regularly meandering, more or less horizontal trails, consisting of a median backfilled tunnel enveloped by a zone of reworked sediment (after Uchman 1995) (Figs 3.3, 3.5e, 3.6e).

Remarks: *Nereites* is interpreted as a grazing trace (pascichnion) and is probably produced by a worm-like sediment feeder (Mangano *et al.* 2000). It is typical of the *Nereites* Ichnofacies and is most commonly associated with SGF sequences characterised by medium- to thin-bedded sandstones, siltstones and silty mudstones. Morphologically, *Nereites* is similar to *Phycosiphon* but has a larger burrow diameter and lacks the vertical loops (Callow *et al.* 2013).

In core: In core *Nereites* has a dark centre and pale mantle, representing the median backfilled tunnel enveloped by a zone of reworked sediment. Individual tunnel zones range between 1–4 mm in diameter and are typically horizontal in orientation.

Scolicia (De Quatrefagues 1849)

Diagnosis: Simple, winding, meandering to coiling bilobate or trilobate backfilled trace fossils with two parallel, locally discontinuous, sediment strings along their underside. The underside between the strings is flat or slightly convex up. Backfill laminae are composite and may be biserial on the upper side (after Uchman 1998) (Figs 3.3, 3.5c, 3.6d,g,c).

Remarks: It is most likely that Cretaceous and younger *Scolicia* are produced by echinoids (Uchman 1995). *Scolicia* is a facies-crossing

trace, and is common in silty mudstones and commonly forms thick vertical intervals of bioturbated sediment up to 1 m thick.

In core: *Scolicia* occurs as large (9–20 mm thick), essentially horizontal burrows filled with meniscate laminae. In cross-section, *Scolicia* is approximately oval in outline with a central ridge at the base.

Helminthopsis (Heer 1877)

Diagnosis: Simple, unbranched, elongate, cylindrical tube with curves, windings, or irregular open meanders (after Wetzel & Bromley 1996) (Figs 3.3, 3.5b).

Remarks: *Helminthopsis* is a facies-crossing trace fossil which was probably produced by polychaetes or priapulids (Fillion & Pickerill 1990). It is typically preserved as hypichnial semi-relief ridges and is particularly common in deep-water environments.

In core: *Helminthopsis* is hard to positively identify in core, forming hypichnial semi-relief cylindrical tunnels typically 1–4 mm in diameter.

3.2.3.4 Graphoglyptids

Graphoglyptids (Füchs 1895) are relatively small, patterned (mainly meander, star and net-shaped) pre-depositional, infaunal, dominantly horizontal open-burrow networks, which are some of the most characteristic trace fossils of deep-water systems younger than the Cambrian (Seilacher 1977, 2007; Miller 1991; Orr *et al.* 2003; Fürsich *et al.* 2007). Typically, these open-burrow networks were formed several millimetres to tens of millimetres below the seafloor in fine-grained, soft mud and were exposed by incoming erosional gravity currents and immediately cast by suspension fallout. Due to the delicate nature and shallow position of these burrows on the seafloor, preservation is unlikely where erosion cuts particularly deep into the substrate (see Section 3.2.1). The specialised farming behaviour of graphoglyptids as well as the small size of the tunnel networks are interpreted as an adaptation to nutrient- and oxygen-poor (oligotrophic) conditions. The tunnels are interpreted to have been used to cultivate bacteria or other chemo-autotrophs in the mucus linings of tunnel walls which were later harvested by the organism (Seilacher 2007). Delicate, rarely preserved vertical shafts to the surface enhanced ventilation to aerate the bacterial gardens. Due to the horizontal nature of these trace fossils, positive identification of graphoglyptids in core is rare because they typically appear as small depressions or cylindrical structures at the base of sandstone beds.

Modern graphoglyptid-like structures are known from bathyal to abyssal settings only (e.g., Ekdale 1980; Wetzel 1983), and Mesozoic–Cenozoic fossil graphoglyptids are usually considered diagnostic elements of deep-water settings. However, there are a number of shallow-water examples of graphoglyptids including examples from shallow-water channel-levée systems of the upper Middle Eocene CCa member, Cerro Colorado Formation, Fuegian Andes (Olivero *et al.* 2010) as well as several examples from the Upper Triassic and Jurassic sedimentary basins of Iran (Fürsich *et al.* 2007). Common graphoglyptids include radial structures such as *Lorenzinia*, winding and meandering structures such as *Cosmorhaphe* and *Helminthorhaphe*, spiral structures including *Spirohaphe* and networks including *Megagrapton* and *Paleodictyon*.

Lorenzinia (Gabelli 1900)

Diagnosis: Simple, short, smooth hypichnial ridges, arranged in one or two circular rows, radiating from an oval or circular central area.

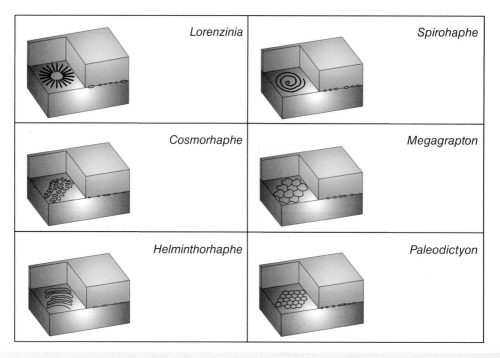

Fig. 3.7 Block diagrams of common graphoglyptids in deep- water environments.

The ridges are very similar or different in length and are regularly or irregularly distributed (after Uchman 1995) (Fig. 3.7, 3.8a).

Remarks: In general, *Lorenzinia* is a 3D burrow system of various radial elements forming a wreath joined by a central ring. It has been interpreted as the product of holothurians, crabs, annelids, or sipunculids (Uchman 1998, and references within).

Cosmorhaphe (Füchs 1895)

Diagnosis: Unbranched graphoglyptid trace fossil with two orders of meanders or undulations (after Seilacher 1977) (Figs 3.7, 3.8d).

Helminthorhaphe (Seilacher 1977)

Diagnosis: Non-branching trace fossil of small string diameter with only one order of smooth systematic meanders of very high amplitude (Uchman 1998) (Figs 3.7, 3.8b).

Spirohaphe (Füchs 1895)

Diagnosis: Thin, spirally coiled trace fossil, supposedly multi-floored (after Uchman 1998) (Figs 3.7, 3.8c).

Megagrapton (Książkiewicz 1968)

Diagnosis: Trace fossil commonly preserved as hypichnial irregular nets (after Uchman 1998) (Fig. 3.7, 3.8e).

Paleodictyon (Meneghini 1850)

Diagnosis: Three-dimensional burrow system consisting of horizontal net composed of regular to irregular hexagonal meshes and vertical outlets. It is the net which tends to be preserved (after Uchman 1995) (Figs 3.7, 3.8f).

3.3   Colonisation of SGF deposits: Opportunistic and equilibrium ecology

The colonisation of SGF deposits by trace-making organisms is characterised by two population strategies: opportunistic ('r-selected') and equilibrium ('K-selected') (Ekdale 1985). Opportunistic ('r-selected') species include those organisms which exhibit high reproduction rates, rapid growth rates, broad environmental tolerances and generalised feeding habits. In contrast, equilibrium ('K-selected') species are those which are typically slow to colonise new environments, but over the long run are better at adapting to environmental changes compared to the more rapidly colonising opportunists (Pianka 1970; Ekdale 1985; Vossler & Pemberton 1988; Uchman 1995).

R-selected opportunistic trace-making organisms are pioneers and will bioturbate a deposit fairly rapidly to exploit whatever the sediment has to offer and then move on to another site. They are common in environments where ecological conditions fluctuate greatly and may become inhospitable to most life. K-selected equilibrium species generally exhibit lower reproductive and growth rates than r-selected trace-making organisms, and their environmental tolerances are much narrower (Ekdale 1985; Bromley 1996). They are commonly specialised feeders, which are adapted to specialised niches in stable, predictable and largely unchanging environments and are exemplified by graphoglyptids. Typically, they are members of high-diversity, persistent, climax communities in which individual species abundances and population densities are generally low (Ekdale 1985). In deep-water clastic systems, r-selected opportunistic trace-making organisms commonly produce post-depositional trace fossils (*sensu* Kern 1980) because they colonise deposits rapidly (Uchman 1995; Tunis & Uchman 1996b). Trace fossils such as *Ophiomorpha* represent examples of post-depositional opportunistic taxa (Fig. 3.9). These trace fossils are produced by trace-making organisms which

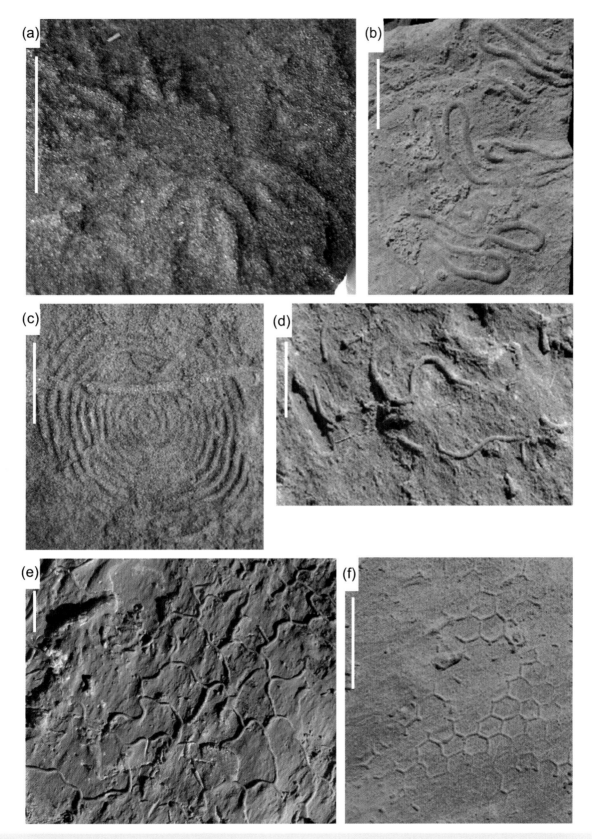

Fig. 3.8 Outcrop images of trace fossils from the Ainsa–Jaca basins. (a) *Lorenzinia nowaki* Książkiewicz 1970. Hypichnial semi-relief. Scale bar: 2 cm. See Uchman (1998) for discussion. (b) *Helminthorhaphe japonica* (Tanaka 1970) Hypichnial semi-relief. Scale bar: 2 cm. See Uchman (1998) for discussion. (c) *Spirohaphe involuta* (De Stefani 1895). Hypichnial full-relief. Scale bar: 2 cm. See Seilacher (1977) for discussion. (d) *Cosmorhaphe sinuosa* (Azpeitia Moros 1933) Hypichnial semi-relief. Scale bar: 2 cm. See Seilacher (1977) for discussion. (e) *Megagrapton submontanum* (Azpeitia Moros 1933). Hypichnial semi-relief. Scale bar: 10 cm. See Uchman (1998). (f) *Paleodictyon strozzii* (Meneghini 1850). Hypichnial semi-relief. Scale bar: 1 cm. See Uchman (1995) for discussion.

Trace fossils within (*endichnia*) and at top (*epichnia*) of SGF deposit

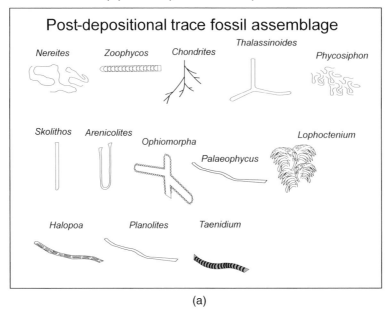

(a)

Trace fossils on base of SGF deposit (*hypichnia*)

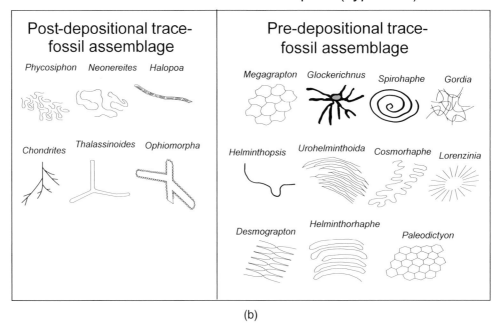

(b)

Fig. 3.9   Examples of deep-water trace-fossil assemblages. (a) Post-depositional deposit assemblages within and at the top of SGF deposits. (b) Post-depositional and pre-depositional assemblages at the base of SGF deposits.

burrowed deep into the SGF deposit to exploit buried nutrients which cannot be reached by smaller organisms. In contrast, K-selected equilibrium trace-making organisms typically form pre-depositional trace fossils (*sensu* Kern 1980) because they are connected with long intervals of time between deposition of SGF deposits (Uchman 1995; Tunis & Uchman 1996b). These structures are preserved as semi-reliefs on the soles of sandstone SGF deposits and represent trace fossils produced on or very close to the sediment–water interface. Common examples include graphoglyptides (Fig. 3.9). Typically the soles of SGF deposits reveal two suites, a pre-depositional suite and also a post-depositional suite (Figs. 3.9, 3.10). The post-depositional suite results from colonisation by organisms borrowing down to the base of the bed from its top, as well as by organisms which survived burial by the newly deposited sediment (Kern 1980).

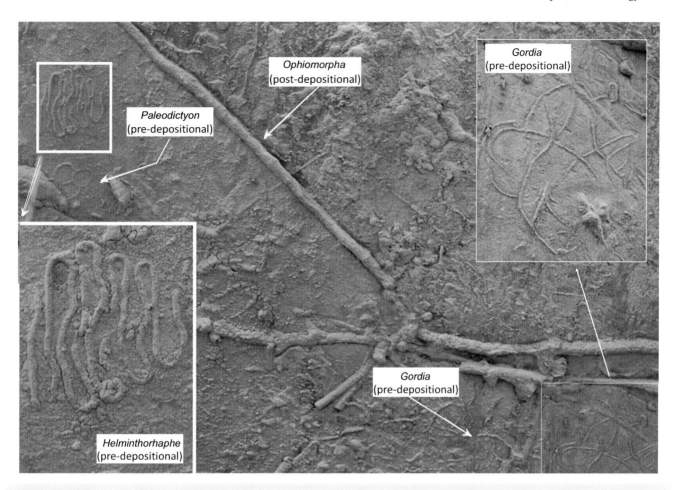

Fig. 3.10 Diverse trace-fossil assemblage on the sole of a SGF sandstone bed, Middle Eocene Ainsa Basin, Spanish Pyrenees. The pre-depositional trace-fossil assemblage (*Paleodictyon, Helminthorhaphe, Gordia*) is preserved by the depositing SGF and is later cross-cut by a post-depositional trace-fossil assemblage (*Ophiomorpha*).

3.4 Ichnofacies

The archetypal ichnofacies concept was originally developed by Seilacher (1953a, b, 1964a, 1967) to describe certain assemblages of trace fossils which are characteristic of particular palaeoenvironmental conditions recurrent through time and space. The concept was soon adopted by sedimentologists and ichnologists as a means of discerning relative palaeobathymetry (e.g., Picket *et al.* 1971). Today the ichnofacies concept is regarded as an important tool for discerning relative palaeobathymetry, but it is also recognised that palaeobathymetry is only one aspect of the ichnofacies concept. Subsequent work has shown that the fundamental controlling factors are not water depth or distance from the shore, but substrate consistency, food supply, hydrodynamic energy, rate of deposition, grain size, salinity, oxygen levels, temperature and toxic substances (e.g., Frey & Seilacher 1980; Frey *et al.* 1990; Pemberton *et al.* 1992; Pemberton *et al.* 2001). It is the response of organisms to these environmental conditions which defines an ichnofacies.

Initially, Seilacher (1967) established six ichnofacies, named after characteristic ichnogenera, which included four softground marine ichnofacies (*Skolithos, Cruziana, Zoophycos* and *Nereites*), one substrate controlled marine ichnofacies (*Glossifungites*) and one softground continental ichnofacies (*Scoyenia*). Three ichnofacies were

added to the original classification, with one more softground marine ichnofacies (*Psilonichnus*) (Frey & Pemberton 1987) and two for borings (*Teredolites and Trypanites*) (Bromley *et al.* 1984; Frey & Seilacher 1980). Due to the broad nature of these original ichnofacies, more recently there has been a proliferation of new ichnofacies. Typically, these new ichnofacies are either described with the use of sub-ichnofacies or as epithets such as 'proximal', 'distal', 'stressed' or 'unstressed', to Seilacher's original ichnofacies (see Seilacher 1974; Bromley 1996; Pemberton *et al.* 2001; McIlroy 2004, 2008). Deep-water environments are characterised by the *Nereites* and *Zoophycos* Ichnofacies. The original marine ichnofacies of Seilacher (1953a, b, 1964a) are briefly described below, with a more detailed description of the deep-water *Nereites* and *Zoophycos* Ichnofacies.

Glossifungites Ichnofacies

This ichnofacies develops in firm but unlithified substrates: that is, dewatered muds. In deep-water environments, firm substrates are exposed on the seafloor by two processes: (i) Erosion of semi-consolidated mudstones and siltstones by bypassing SGFs or large-scale sediment slides/slumps (e.g., Hubbard & Shultz 2008; Knaust 2009). (ii) Sediment starvation (i.e., low sedimentation rates) along with bottom-current winnowing (e.g., Savrda *et al.* 2001a,b; Knaust 2009). The *Glossifungites* Ichnofacies consists of a mixture of burrows and borings. Burrows generally consist of permanent vertical to steeply inclined domiciles, which generally exhibit no evidence of burrow wall reinforcement but may exhibit scratched surfaces (Uchman *et al.* 2000; Pemberton *et al.* 2001). Most trace fossils are suspension feeders. Characteristic trace fossils include *Arenicolites*, *Rhizocorallium*, *Diplocraterion*, *Gastrochaenolites*, *Thalassinoides* and *Spongeliomorpha* (Pemberton *et al.* 2001). In deep-water systems, firmgrounds may be associated with regionally extensive, sequence-stratigraphically significant discontinuity surfaces associated with erosional processes (e.g., Hubbard & Shultz 2008; Hubbard *et al.* 2012; Savrda *et al.* 2001a, b). They may also be associated with localised erosional surfaces at the base of, or within, individual channels. In the latter case, the stratigraphic significance of the *Glossifungites* Ichnofacies is probably minimal (Hubbard *et al.* 2012). Examples of local firmgrounds have been observed in the Ainsa Basin, Spanish Pyrenees (Heard & Pickering 2008), as well as the Rosario Formation, Baja California, Mexico (Callow *et al.* 2013). In the Ainsa Basin, firmgrounds at or near the base of individual channels are dominated by either *Thalassinoides* isp. or *Arenicolites* isp. while the firmgrounds recognised by Callow *et al.* (2013) are characterised by 'U'-shaped aff. *Ilmenichnus* trace fossils. According to Hubbard *et al.* (2012), passively filled firmground burrows with a fill which is coarser grained than overlying deposits may represent the only evidence of bypassing SGFs in the absence of coarse-grained lag deposits.

Psilonichnus Ichnofacies

Associated with supra-littoral/upper littoral, moderate- to low-energy marine and/or aeolian conditions typical of beach to backshore to dune environments (Pemberton *et al.* 2001).

Skolithos Ichnofacies

Represents lower littoral to infra-littoral, moderate- to high-energy conditions; generally corresponding to the beach foreshore and shoreface (Frey & Pemberton 1985) and is dominated by vertical shafts such as *Skolithos*, *Diplocraterion* and *Arenicolites*.

Cruziana Ichnofacies

Associated with infra-littoral and shallow circa-littoral marine substrates between wave base and storm wave base (Frey & Pemberton 1985; Pemberton *et al.* 2001). Typically it is associated with a moderate- to relatively low-energy environment in which the preservation potential of upper tiers is high resulting in a high ichnodiversity, high intensity trace-fossil assemblage (Bromley 1996; Pemberton *et al.* 2001).

Zoophycos Ichnofacies

This ichnofacies typically occurs from below storm wave base to bathyal water depths in quiet water conditions, or protected epeiric sites which are more or less deficient in oxygen. Offshore sites are in areas free of SGFs or significant bottom currents (Seilacher 1967; Frey & Seilacher 1980; Frey & Pemberton 1985; Pemberton *et al.* 2001). This ichnofacies is characterised by a high intensity, low ichnodiversity trace-fossil assemblage dominated by grazing traces and shallow feeding structures (Frey & Pemberton 1985). Horizontally to gently inclined spreiten structures are commonly present (Pemberton *et al.* 2001). Common trace fossils include *Zoophycos*, *Thalassinoides*, *Phycosiphon* and *Chondrites* (Bromley 1996; Uchman & Wetzel 2011). Bio-deformational structures are also common,

resulting in sediments of the *Zoophycos* Ichnofacies commonly being totally homogenised as a result of bioturbation.

Nereites Ichnofacies

This ichnofacies is most common in lower slope and basin floor settings in which generally low-energy environments are subject to periodic SGFs. The *Nereites* Ichnofacies is characterised by graphoglyptids and meandering trace fossils which are preserved as semi-relief trace fossils because of burial by incoming SGFs (Seilacher 1967, 2007). Environments beyond the influence of SGFs, such as parts of the slope and abyssal plain, typically do not exhibit the *Nereites* Ichnofacies and are more commonly associated with the *Zoophycos* Ichnofacies (Uchman & Wetzel 2011, 2012).

The *Nereites* Ichnofacies comprises a number of ichnotaxa which have been classically regarded as 'shallow-water' trace fossils, such as *Ophiomorpha* and *Thalassinoides*. These trace fossils are common in deep-water systems throughout the world and are now regarded as normal components of deep-water trace-fossil assemblages (e.g., Crimes *et al.* 1981; Uchman 1995, 2001; Heard & Pickering 2008; Phillips *et al.* 2011). In a study by Uchman (1995) in the Marnoso-arenacea Formation, Italy, a large range of *Ophiomorpha* burrow dimensions were recorded, which was interpreted to indicate a full age spectra of trace-making organisms living in the deep water. It was therefore interpreted that rather than being transported from shallow-water environments by SGFs (*cf.* Föllmi & Grimm 1990), the trace-making organisms lived, reproduced and died in the deep sea. This interpretation is supported by the fact that ichnotaxa such as *Ophiomorpha* and *Thalassinoides* have been observed in every sub-environment of a submarine fan from slope canyon to the distal basin floor (e.g., Uchman 1995, 2001; Heard & Pickering 2008).

The *Nereites* Ichnofacies has been divided into a number of sub-ichnofacies: the *Ophiomorpha rudis*, *Nereites* and *Paleodictyon* sub-ichnofacies (Seilacher 1974; Uchman 2001). The *Ophiomorpha rudis* sub-ichnofacies occurs in thick-bedded sandstones typical of proximal axial environments such as channel axis and proximal lobe environments (Uchman 2001, 2009; Heard & Pickering 2008). This sub-ichnofacies is typified by trace fossils such as *Ophiomorpha rudis*, *Ophiomorpha annulata* and *Scolicia strozzii* (Uchman 2001, 2009). Seilacher (1974) proposed the *Paleodictyon* sub-ichnofacies and the *Nereites* sub-ichnofacies within the *Nereites* Ichnofacies, which tend to occur in sandier and muddier parts of deep-water SGF systems, respectively. The *Nereites* sub-ichnofacies is dominated by post-depositional backfilled burrow systems made by sediment feeders (e.g., *Nereites*, *Phycosiphon*, *Zoophycos*) and typically occurs in distal or proximal off-axis environments such as the basin-plain, inter-channel or slope. The *Paleodictyon* sub-ichnofacies is characterised by an abundance of open tunnels preserved on the soles of SGF deposits (e.g., *Paleodictyon*) and is more common in environments such as the channel margin, lobe or fan fringe (López-Cabrera *et al.* 2008; Heard & Pickering 2008; Phillips *et al.* 2011; Uchman 2001, 2007, 2009). A number of recent studies have illustrated the limitations of using the sub-ichnofacies of the *Nereites* Ichnofacies to describe the ichnological variations in deep-water submarine fan and related sub-environments (e.g., Uchman 2001; Heard & Pickering 2008; Monaco *et al.* 2010; Cummings & Hodgson 2011). For example, Heard and Pickering (2008) observed that each of the sixteen submarine fan and related environments recognised in the Ainsa–Jaca basins are characterised by distinct trace-fossil assemblages. In the deep-marine Basque Basin, northern Spain, Cummings and Hodgson (2011) noted that a single stratigraphic succession could contain all three sub-ichnofacies of the *Nereites* Ichnofacies and therefore suggested the current sub-ichnofacies only provide an informal way of catagorising trace-fossil assemblages based on a general position within a submarine fan system (e.g., proximal to distal, or axial to off axis). In a study by Monaco *et al.* (2010) from the Oligocene to Miocene foreland basins of the northern Apennines, Italy, trace fossils were grouped into ichnocoenoses (trace-fossil assemblages formed at an ecologically synchronous time by a single endobenthic community) rather than sub-ichnofacies of the *Nereites* Ichnofacies. It is clear, therefore, that in detailed studies of submarine-fan and related environments, the use of sub-ichnofacies may be of limited value.

3.5   Ichnofabrics

An alternative or complementary method of ichnological analysis was developed by Ekdale and Bromley (1983) based on the description of the fabric of a sediment created by the action of organisms, and is known as ichnofabric analysis. The ichnofabric concept was developed in order to understand the imprint left by organisms in cases where individually distinct and identifiable trace fossils are absent (Ekdale *et al.* 2012). The concept was developed following a number of significant studies on the ecology of trace-fossil associations in modern environments (e.g., Schäfer 1956; Reineck 1973; Berger *et al.* 1979; Wetzel 1981; 1983). It is now one of the most dynamically developing fields of ichnology (e.g., Taylor & Gawthorpe 1993; Taylor *et al.* 2003; McIlroy 2004, 2008; Knaust 2009; Callow & McIlroy 2011; Callow *et al.* 2013) partly due to its applicability in core-based studies where the positive identification of trace fossils in vertical section can be challenging (see Section 3.6). There is no formal agreement on naming conventions for ichnofabrics, but ichnofabrics are typically named after the dominant or most characteristic ichnogenera. Because ichnofabric analysis is done at a small bed-by-bed scale, a single outcrop or core can yield a large inventory of different ichnofabrics. As a consequence, McIlroy (2007) suggested grouping similar ichnofabrics into ichnofabric associations. This level of detail may prove particularly useful in palaeoenvironmental studies (*cf.* Callow *et al.* 2013).

Typically, ichnofabric analysis involves detailed observations on the intensity and type of bioturbation, relative abundance of ichnotaxa, tiering and the sedimentary fabric. A detailed description and methodology for ichnofabric analysis has been outlined by Taylor and Goldring (1993) and McIlroy (2004), a summary of which is outlined in the following text.

Intensity of bioturbation

The intensity of bioturbation is a major component of ichnofabric analysis and is important because it reflects the duration of the colonisation events, which, in turn, is related to rates of sedimentation/erosion (Taylor & Gawthorpe 1993). The bioturbation intensity also reflects bottom-water oxygen conditions (e.g., Ekdale 1985; Ekdale & Mason 1988). There are two main schemes for the assessment of bioturbation intensity in current use: the descriptive bioturbation indices of Taylor & Goldring (1993) and the semi-quantitative ichnofabric index ('ii') scheme of Droser and Bottjer (1986, 1991). The ichnofabric index scheme is based on the degree of biogenic disruption of the primary sedimentary fabric (Fig. 3.11). Indices range from *ii* = 1 for non-bioturbated sediment, to *ii* = 6 which reflects total homogenisation. For totally homogenised sediments, a secondary index can be used to describe discrete trace fossils superimposed on a background of homogenised sediment (*ii* = 2/6 to *ii* = 4/6). The alternative scheme introduced by Taylor & Goldring (1993) is based on a measurement of the percentage area bioturbated with each grade of bioturbation clearly defined in terms of burrow density, amount of burrow overlap and the sharpness of the original sedimentary fabric. This scheme is based on an earlier scheme developed by Reineck (1963) and has become popular amongst ichnologists due to its precise and descriptive nature. However, its use can be highly complex and time consuming. Both schemes are particularly well suited to core analysis, whereas at outcrop on horizontal bedding planes, both schemes have certain limitations. Consequently, an alternative, outcrop-based scheme was developed by Miller and Smail (1997) based on a bedding plane bioturbation index.

Diversity

Ichnodiversity can be regarded as a proxy for organism diversity rather than a direct measure. This is because a single trace-making organism can produce more than one type of trace fossil, whilst different trace-making organisms can form one type of trace fossil. Ichnodiversity can be particularly powerful in palaeoenvironmental analysis; by measuring ichnological diversity, it is possible to recognise changes in environmental conditions. For example, in deep-water systems, environments associated with a high frequency of erosive SGFs may be characterised by a low ichnodiversity trace-fossil assemblage. In contrast, environments with a low frequency of erosive SGFs are usually represented by a higher ichnodiversity trace-fossil assemblage because long-term stable conditions may become established between the arrival of successive SGFs. Stressed environments can also be recognised by changes in the dimensions of burrows (McIlroy 2004).

Infaunal tiering

Ecologic conditions change with depth below the sediment–water interface; sediment strength increases, porosity and permeability decrease, organic matter becomes decomposed, and oxygen content of the pore-water decreases (Wetzel & Uchman 1997; Wetzel & Uchman 1998a). As a result, organisms and their traces below the sediment–water interface are subdivided into a vertical sequence known as tiering (Ausich & Bottjer 1982). Each tier consists of co-occurring trace fossils which cross-cut one another at a distinct depth. The vertical zonation of the trace fossils and associated tiers is determined by the depth the trace fossils penetrate below the sediment–water interface. The preservation of a tiering profile is reliant upon rapid killing off of the community (e.g., burial under a SGF deposit), because with continued incremental deposition, deeper burrows tend to overprint shallower ones. Consequently, deep-water systems are well suited to studying tiering because of the way SGFs deposits preserve the tiering structure of the bioturbate zone as a 'frozen profile' (Wetzel & Uchman 1998a). In SGF deposits, two zones of bioturbation can be distinguished, the spotty layer (*sensu* Uchman 1999) and the elite layer (Ekdale & Bromley 1991; Uchman 1999), which correspond roughly to the mixed layer and transitional layer observed in Recent deep-water deposits (Ekdale & Berger 1978; Bromley 1996) (Fig. 3.12). The *spotty layer* represents the uppermost zone of bioturbation and is characterised by a homogenised background associated with biodeformational structures along with sharp to diffuse trace fossils. The deeper tier *elite layer* is associated with distinct trace fossils which either overprint an intensely bioturbated background in the upper part (upper elite layer) or primary sedimentary structures in the lower part (lower elite layer).

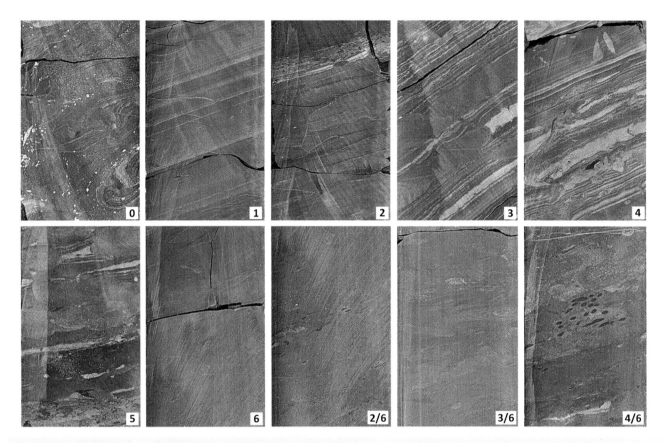

Fig. 3.11 Examples of ichnofabric indices (*sensu* Droser & Bottjer 1986), Ainsa 6 core, Ainsa Basin, Spanish Pyrenees (see text for details). The core is ~6 cm wide. From Heard *et al.* (2008).

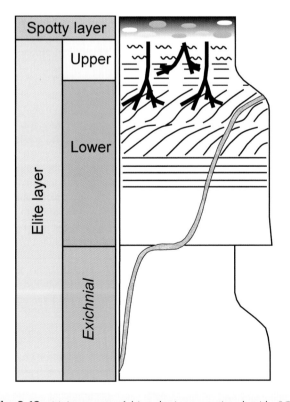

Fig. 3.12 Main zones of bioturbation associated with SGF deposits (see text for details). From Uchman (1999).

In thin-bedded turbidite successions or continuously accumulating deposits, tiering can be very complex due to the occurrence of multiple episodes of bioturbation as deeper burrows overprint shallower ones. In many cases, elite trace fossils, which are commonly late-stage bioturbators and may be visually striking due to a prominent fill/burrow lining, may visually dominate an ichnofabric (Bromley & Ekdale 1986; Bromley 1996; McIlroy 2004). Tiering is significantly influenced by variations in environmental conditions. The tier structure and variations in tiering can therefore provide considerable palaeoenvironmental information such as sedimentation rates, benthic food content, pore- and bottom-water oxygenation, substrate consistency and evidence for surfaces of erosion or non-deposition (Wetzel & Aigner 1986; Wetzel & Uchman 1998a).

3.6 Trace fossils in core

At outcrop, where trace fossils are typically studied in three dimensions as semi-relief structures on bedding planes, the positive identification of trace fossils can be relatively straightforward. However, in cores, trace fossils are typically studied in two dimensions as full-relief structures on slabbed core sections. Consequently, information about the morphology of a trace fossil may be limited in a core slab resulting in great challenges when trying to positively identify trace fossils. Typically, most trace fossils in cores are identified to the ichnogenus level, with some trace fossils, such as graphoglyptids, almost impossible to recognise, much less identify. The study of trace fossils in cores can be enhanced by using high-resolution X-ray images to give a sense of the structures in three dimensions. Other techniques include using large thin slices of a core viewed in transmitted light (Garton & McIlroy 2006) or constructing three-dimensional computer reconstructions based on regularly spaced images produced by serial grinding (Bednarz & McIlroy 2009). Despite the limitations of identifying trace fossils in cores, due to the lack of weathering and the vertical continuity of the observed section, core-based ichnological studies do have many advantages over outcrop studies. For example, burrow boundaries and walls commonly can be viewed in detail in cores, meaning many trace fossils which are classified on the nature of burrow boundaries or walls (e.g., *Ophiomorpha*) can generally be identified to the ichnospecies level. Fine-grained intervals which at outcrop may be intensely weathered are typically well preserved in cores, so that detailed studies on fabrics can be made. As a consequence, many studies in core now utilise the ichnofabric approach (e.g., Taylor & Goldring 1993; Taylor *et al.* 2003; Knaust 2009). In contrast, most outcrop studies are focused on the identification of individual ichnotaxa, and therefore tend to concentrate on the ichnofacies approach with the hope of using variations in trace-fossil assemblages and bioturbation intensity to assist in the interpretation of different environments (e.g., Uchman 2001; López-Cabrera *et al.* 2008; Heard & Pickering 2008). Comparing the results of outcrop and core studies can be challenging because the nature of the information derivable from outcrops and cores is very different. However, a number of recent outcrop studies have also adopted the ichnofabric approach, recording identical information to those obtained from cores. These outcrop studies have used high quality exposures comparable with the quality available in cores (e.g., Wetzel & Uchman 1998a; Phillips *et al.* 2011; Callow *et al.* 2013). The results provide an important source of information to guide future core-based studies. There are also many examples of detailed core-based studies which have successfully applied the ichnofacies concept (e.g., Pemberton 2001). Both the ichnofabric and ichnofacies approaches take into consideration primary sedimentary structures and other aspects of the sediment, and because of the strengths and weaknesses of both methodologies, future studies should aim to integrate both approaches to produce a more complete analysis.

3.7 Case study I: Trace fossils as diagnostic indicators of deep-marine environments, Middle Eocene Ainsa–Jaca basins, Spanish Pyrenees

3.7.1 Introduction

This case study is based on an outcrop study of the Middle Eocene Ainsa–Jaca basins, Spanish Pyrenees by Heard and Pickering (2008). It builds upon an earlier ichnology study of the Ainsa–Jaca basins by Uchman (2001). Heard and Pickering (2008) aimed to integrate sedimentology and ichnology at outcrop to better understand the potential uses of trace fossils in deep-water palaeoenvironmental studies.

3.7.2 Study area: Ainsa–Jaca basins

The Ainsa–Jaca basins represent the deep-marine part of the Middle Eocene South Pyrenean foreland basin (Fig. 4.33) (Pickering & Corregidor 2000, 2005; Remacha *et al.* 2003). The infill of the proximal deep-marine Ainsa Basin consists of eight coarse clastic depositional complexes or systems, deposited in a variety of settings. These proximal clastic depositional systems correlate with unconfined sheet

systems in the more distal Jaca basins (Mutti *et al.* 1985; Das Gupta & Pickering 2008). A total of sixteen submarine-fan and related environments have been recognised within the Ainsa–Jaca basins, each with distinct trace-fossil assemblages. The most common environments are basin-slope, canyon-fill, channel-axis, channel off-axis, channel-margin, outermost channel-to-levée-overbank, proximal interfan, channel-lobe transition, lobe, lobe-fringe, fan-fringe and distal basin floor (Bayliss & Pickering 2009). For details, see Section 4.8.

3.7.3 Trace-fossil distributions

Uchman (2001) and Heard and Pickering (2008) recognised that a distinct trace-fossil assemblage, ichnodiversity and intensity of bioturbation characterised each submarine-fan and related environment (Figs 3.13, 3.14). Proximal and axial parts of the sandy systems, such as canyon-fill, channel-axis, and intrachannel off-axis environments, are characterised by low-ichnodiversity, low-bioturbation-intensity trace-fossil assemblages dominated by post-depositional trace fossils such as *Ophiomorpha* (Figs 3.4f, 3.6a) and *Thalassinoides* (Fig. 3.4b, 3.6h). A similar trace-fossil assemblage is also present in the channel-lobe transition in the Jaca Basin. Off-axis environments such as channel-margin and outermost channel-to-levée-overbank exhibit a general increase in ichnodiversity and bioturbation intensity compared to more proximal and axial environments. Maximum bioturbation intensity occurs in the commonly totally bioturbated sediments of the proximal interfan, and greatest ichnodiversity is found in the channel-margin and thin-bedded sandstones at the top of sandbodies. Off-axis environments, in particular channel-margins, are characterised by an increase in pre-depositional trace fossils, including graphoglyptids. The basin-slope is characterised by relatively low-ichnodiversity, high-bioturbation-intensity trace-fossil assemblages. In the distal Jaca Basin, the greatest average bioturbation intensity and ichnodiversity are in the lobe fringe, with the lobe also showing a relatively high ichnodiversity and bioturbation intensity. A general trend of decreasing bioturbation intensity and ichnodiversity is developed from the fan fringe to distal basin floor. A similar trend is also shown by the number of pre- and post-depositional trace fossils and graphoglyptids with a decrease in pre- and post-depositional trace fossils as well as graphoglyptids from the fan fringe to distal basin floor (Figs 3.13, 3.14).

3.7.4 Interpretation

Heard and Pickering (2008) interpreted the low-ichnodiversity, low-bioturbation-intensity, dominantly post-depositional trace-fossil assemblages of the proximal, axial parts of the sandy systems as a consequence of a number of factors. First, in these environments, SGFs are generally highly erosive and, therefore, tend to remove the fertile top layer of the seafloor, resulting in the preservation of only the trace fossils created by the deepest burrowers. Consequently, such environments are associated with low numbers of pre-depositional trace fossils due to their poor preservation potential. Second, large, robust, deeply burrowing trace-making organisms such as those associated with *Ophiomorpha* are well adapted to surviving burial by newly deposited thick sand beds (Wetzel & Uchman 2001). The traces of smaller organisms which may have survived burial, or colonised the deposits after migrating from outside the area affected by the newly deposited sand bed, may be destroyed by these large, robust trace making organisms. Third, the coarse-grained nature of the deposits in proximal axial environments may exclude burrowers adapted only to

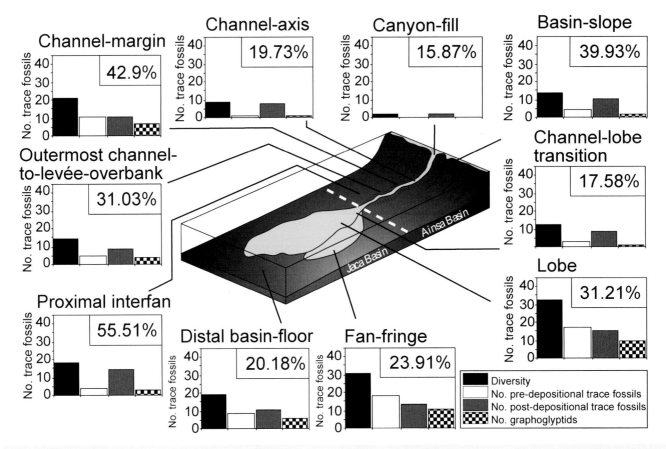

Fig. 3.13 Summary diagram of average bioturbation, ichnodiversity, number of pre- and post-depositional trace fossils and number of graphoglyptids for the most characteristic environments of the deep-water complex of the Ainsa–Jaca basins. Redrawn after Heard and Pickering (2008).

fine-grained sediments (Tchoumatchenco & Uchman 1999) and may also prevent the preservation of the activities of smaller organisms.

The high-ichnodiversity, high-bioturbation-intensity trace-fossil assemblages of proximal off-axis and distal-fan environments, which are characterised by an abundance of pre-depositional trace fossils and graphoglyptids, are interpreted to result from a number of factors, including: (i) the increased frequency of taphonomically beneficial SGFs which had sufficient erosive capability to expose and cast the pre-depositional trace-fossil assemblage; (ii) decreased rates of sediment accumulation and a decrease in bed thickness and grain size of the deposits so that organisms had greater time to colonise and bioturbate the newly deposited sediments. The reduced bed thicknesses compared to more proximal and axial environments may also have enabled a greater variety of organisms to survive burial compared to more proximal and axial environments.

3.8   Case study II: Subsurface ichnological characterisation of the Middle Eocene Ainsa deep-marine system, Spanish Pyrenees

3.8.1   Introduction

In the first case study, we show how trace fossils can be powerful discriminators of deep-marine clastic submarine-fan and related

environments in an outcrop-based study. Many sedimentology studies, however, particularly those in the oil and gas industry, are based purely on subsurface data. This second case study, based on the Ainsa channel system in the Ainsa Basin, demonstrates the uses and value of trace fossils and ichnofabric analysis in core description projects. Heard *et al.* (2014) studied ichnology and sedimentology using cores from six wells which were drilled at 400–500 m spacing, to subsurface depths ~250 m each.

3.8.2   Trace-fossil distributions and ichnofabrics in the Ainsa System, Ainsa Basin, Spanish Pyrenees

Trace fossils recorded by Heard *et al.* (2014) include *Skolithos, Halopoa, Planolites, Palaeophycus, Chondrites, Trichichnus, Ophiomorpha, Thalassinoides, Zoophycos, Phycosiphon, Nereites, Scolicia* and *Teichichnus*. The reader is referred to Section 3.2.3 and Figures 3.3 and 3.6 for descriptions of these trace fossils. Through an integrated semi-quantitative ichnological analysis based on bioturbation intensity and ichnodiversity, infaunal tiering, colonisation styles, and a detailed sedimentological analysis, nine recurrent ichnofabrics are recognised in the Ainsa System (Table 3.2, Fig. 3.15).

Observations from the core study of Heard *et al.* (2014) have many similarities to those from the outcrop study (Heard & Pickering 2008). For example, a general increase in bioturbation intensity from axial to off-axis environments was observed in the cores

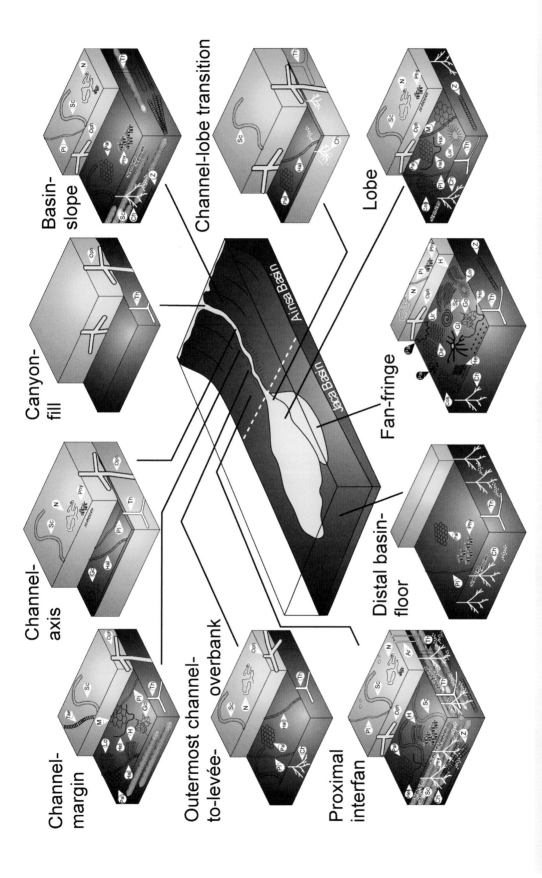

Fig. 3.14 Summary of characteristic trace-fossil assemblages in submarine fan and related environments of the Ainsa–Jaca basins. (L = *Lockeia*; Sk = *Skolithos*; Ar = *Arenicolites*; Nu = *Nummulites*-lined burrow; H = *Halopoa*; Pl = *Planolites*; Ch = *Chondrites*; Oph = *Ophiomorpha*; Th = *Thalassinoides*; S = *Saerichnites*; Lor = *Lorenzinia*; Gl = *Glockerichnus*; Z = *Zoophycos*; Phy = *Phycosiphon*; Lo = *Lophectenium*; N = *Nereites*; Sc = *Scolicia*; Tae = *Taenidium*; Pr = *Protovirgularia*; Cos = *Cosmorhaphe*; He = *Helicolithus*; H = *Helminthorhaphe*; Hel = *Helminthopsis*; Sp = *Spirohaphe*; Ur = *Urohelminthoida*; M = *Megagrapton*; De = *Desmograpton*; Pa = *Palaeomeandron*; Pal = *Paleodictyon*). Redrawn from Heard & Pickering (2008).

Table 3.2 Summary of the ichnofabrics observed in core in the Ainsa System. See Chapter 2 for facies descriptions

Ichnofabric (IF)	Ichnology	Sedimentology	Facies*	Environment
***Ophiomorpha* IF**	Vertical to horizontal orientated *Ophiomorpha rudis*. Overall low bioturbation intensity	Medium- to thick-bedded amalgamated sandstones	A1.4, C2.1 & B2.1	Channel axis
***Thalassinoides* IF**	Sharp-walled, undeformed, horizontal to vertical orientated *Thalassinoides*	commonly situated below erosional surfaces.	B1.1, B2.1, C1.2, C2.1,	Channel off-axis environments
***Ophiomorpha-Thalassinoides* IF**	Low-moderate bioturbation intensity, low ichnodiversity trace-fossil assemblage dominated by *Ophiomorpha*, *Thalassinoides* and *Planolites*.	Thin- to thick-bedded *amalgamated* to non amalgamated sandstones	A2.7, B1.1, C1.2, C2.1, C2.2 & D2.3	Channel off-axis
***Thalassinoides-Phycosiphon* IF**	Moderate bioturbation intensity, moderate to high ichnodiversity trace-fossil assemblage dominated by biodeformational structures as well as *Planolites*, *Thalassinoides*, *Phycosiphon*, *Scolicia* and *Chondrites*	Thin- to medium-bedded sandstones and interbedded siltstone–mudstone laminae	B2.1, C1.2, C2.1, C2.2, D1.3, D2.3 & E1.3	Channel margin
***Scolicia* IF**	Moderate bioturbation intensity, moderate to high ichnodiversity trace-fossil assemblage dominated by *Scolicia*, *Planolites*, *Phycosiphon*, and *Thalassinoides*			
***Planolites* IF**	Moderate to high bioturbation intensity, high ichnodiversity trace-fossil assemblage dominated by *Planolites*, *Phycosiphon*, *Scolicia* and *Thalassinoides*.	Thin-bedded sandstones and siltstone–mudstone laminae	C2.3, D1.3, D2.3, E1.3 E2.2	Overbank-levee
***Planolites-Chondrites* W**	Moderate to high bioturbation intensity, high ichnodiversity trace-fossil assemblage dominated by *Planolites*, *Phycosiphon*, *Chondrites* and *Thalassinoides*.			
***Phycosiphon-Planolites* IF**	Moderate bioturbation intensity, high ichnodiversity trace- fossil assemblage dominated by biodeformational structures as well as *Phycosiphon*, *Planolites*, *Scolicia* and *Thalassinoides*.	Thin-bedded sandstones and siltstone–mudstone laminae	Cl.2, C2.1, D 1.3, D 2.3, E1.3 & E2.2	Channel abandonment
***Biodeformational-Phycosiphon* IF**	High bioturbation intensity, moderate ichnodiversity trace-fossil assemblage dominated by biodeformational structures as well as *Phycosiphon*, *Planolites*, *Thalassinoides* and *Chondrites*.	Siltstone–mudstone laminae with rare thin-bedded sandstones	C1.2, C2.3, D1.3 & E1.3	Interfan

(Fig. 3.15). Channel-axis environments are characterised by the low-ichnodiversity *Ophiomorpha* ichnofabric and the *Thalassinoides* ichnofabric (Fig. 3.15, Table 3.2). The channel margin is characterised by a high-bioturbation-intensity and high-ichnodiversity trace-fossil assemblage dominated by the *Thalassinoides–Phycosiphon* ichnofabric (Fig. 3.15, Table 3.2). Individual channel-margin elements in the Ainsa II and III fans are also associated with the *Scolicia* ichnofabric. Levée-overbank environments are characterised by high-bioturbation-intensity and high-ichnodiversity trace-fossil assemblages, with levée-overbank deposits of the Ainsa I Fan associated with the *Planolites–Chondrites* ichnofabric and those in the Ainsa II and III fans, the *Planolites* ichnofabric. Interfan deposits between the Ainsa I and Ainsa II and III fans, as well as underlying the Ainsa System, are characterised by a high-intensity, high-diversity trace-fossil assemblage and are associated with the biodeformational *Phycosiphon* ichnofabric (Fig. 3.15). The abandonment deposits at the top of the Ainsa II and III fans are characterised by a moderately low bioturbation intensity but high-diversity trace-fossil assemblage

associated with the *Phycosiphon–Planolites* ichnofabric (Fig. 3.15, Table 3.2).

3.8.3 Interpretation

The increase in ichnodiversity and bioturbation intensity from channel axis to off-axis environments such as the channel margin and levée-overbank is consistent with the outcrop study of the Ainsa–Jaca basins by Heard and Pickering (2008). The reader is, therefore, referred to Section 3.7.4 for a detailed interpretation of the ichnological variations from axis to off-axis environments. In the core study, Heard *et al.* (2014) interpreted a number of intervals in channel-axis and channel off-axis environments as firmgrounds (*Glossifungites* Ichnofacies) which were not recognised at outcrop. These intervals are characterised by the *Thalassinoides* ichnofabric and are interpreted to have been formed by the local exhumation of the substrate by erosive SGFs resulting in the exposure of semi-consolidated deposits on the seafloor (see Section 3.4).

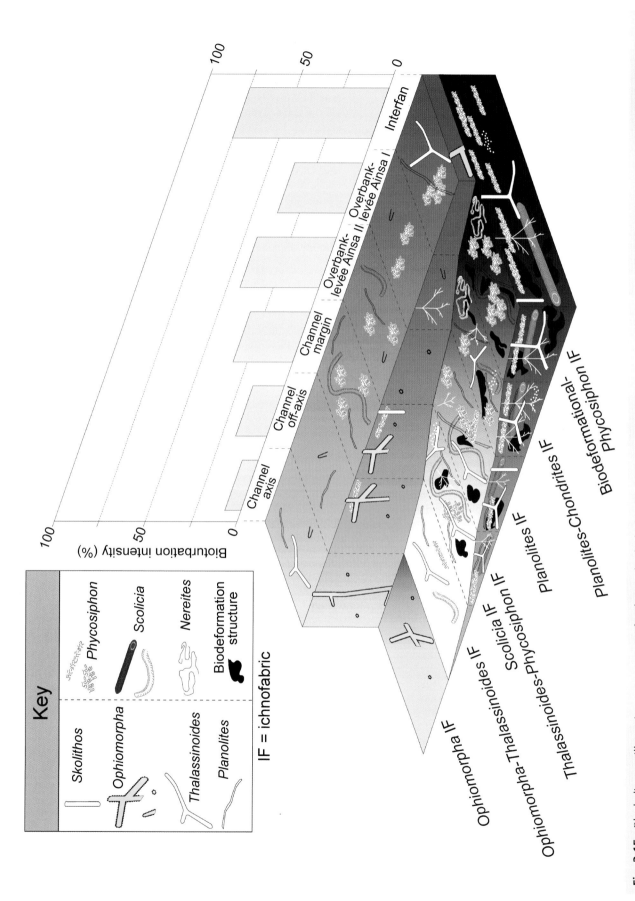

Fig. 3.15 Block diagram illustrating common trace fossils in individual ichnofabrics from channel-axis to external levée-overbank and interfan. The average bioturbation intensity associated with each ichnofabric is illustrated in the graph, whilst the relative proportion of sandstone (yellow) versus marlstone (grey) associated with each ichnofabric is also provided.

3.9 Summary of ichnology studies in deep-water systems

The general trend of increasing bioturbation intensity and ichnodiversity observed in the Ainsa–Jaca basins from channel-axis to off-axis environments in confined submarine fan and related environments, as well as increasing ichnodiversity from channel-lobe transition to fan-fringe environments, is consistent with a number of other studies. For example, a similar distribution of trace fossils has been observed in the Eocene Jaizkibel Fan of northern Spain (Crimes 1977), the Cretaceous–Eocene Gurnigel and Schlieren Flysch of Switzerland (Crimes *et al.* 1981), the upper Palaeocene to Lower Eocene flysch del Grivó in the Julian Pre-Alps of Italy and Slovenia (Tunis & Uchman 1992, 1996a), the Miocene Marnoso-arenacea Formation of the Northern Apennines (Uchman 1995), the Eocene SGF deposits of the Istria Peninsula (Tunis & Uchman 1996b), the Miocene Cingöz Formation of the Adana Basin (Uchman & Demircan 1999), the Eocene Greifensteiner Schichten of Austria (Uchman 1999), the Eocene Tarcau sandstone of the Eastern Carpathians, Romania (Buatois *et al.* 2001) the early-Middle Eocene Kusuri Formation of the Sinop-Boyabat Basin, Turkey (Uchman *et al.* 2004), a Campanian submarine fan in the Norwegian Sea (Knaust 2009), Oligocene to Miocene foreland basins of the Northern Apennines, Central Italy (Monaco *et al.* 2010), the deep-marine Basque Basin, northern Spain (Cummings & Hodgson 2011), the Grès d'Annot Basin, southeast France (Phillips *et al.* 2011) and the Rosario Formation, Baja California, Mexico (Callow *et al.* 2013).

In general, these studies have shown that channel-axis environments are characterised by a low-bioturbation-intensity and low-ichnodiversity trace-fossil assemblage dominated by *Ophiomorpha*. For example, in the Grès d'Annot Basin, the channel axis is dominated by *Ophiomorpha rudis* (the *Ophiomorpha rudis* ichnofabric of Phillips *et al.* 2011). In channel-axis deposits of the Rosario Formation, Callow *et al.* (2013) recognised *Ophiomorpha*-rich ichnofabrics including the *Ophiomorpha*/Phycosiphoniform ichnofabric as well as the Phycosiphoniform/*Chondrites* ichnofabric. In the foreland basins of the Northern Apennines, Monaco *et al.* (2010) observed that channel-axis environments are dominated by *Ophiomorpha rudis* and *Scolicia strozzii*. The high-bioturbation-intensity and high-ichnodiversity trace-fossil assemblages which characterise channel off-axis environments in the Ainsa Basin have also been well documented in other basins worldwide (e.g., Wetzel & Uchman 2001; Uchman *et al.* 2004; Knaust 2009; Monaco *et al.* 2010; Cummings & Hodgson 2011; Phillips *et al.* 2011; Callow *et al.* 2013). For example, in the Campanian submarine fan documented by Knaust (2009), a high-bioturbation-intensity, moderate-ichnodiversity ichnofabric dominated by *Scolicia* characterises heterolithic sediments interpreted as proximal overbank deposits. In the Kusuri Formation, Sinop-Boyabat Basin, Turkey, the highest ichnodiversity is associated with overbank deposits (Uchman *et al.* 2004), whilst in the foreland basins of the Northern Apennines, the highest ichnodiversity and bioturbation-intensity trace-fossil assemblages are seen in overbank deposits (Monaco *et al.* 2010). In the Rosario Formation, off-axis environments such as the outer channel/terrace and inner levée exhibit the highest ichnodiversity, bioturbation intensity and deepest tiering (Callow *et al.* 2013). In the levée deposits of the Rosario Formation, a general trend of decreasing bioturbation intensity, and frequency of vertical and horizontal burrows, extends from inner external levée deposits to outer external levée deposits (Kane *et al.* 2007; Callow *et al.* 2013). In terms of ethological variations in proximal submarine fan and related environments, Cummings

and Hodgson (2011) observed a trend of decreasing domichnia and increasing fodinichnia from channel-axis to channel off-axis environments in the Basque Basin. A distinct increase in graphoglyptids characterises off-axis environments, particularly the channel margin and levée-overbank, combined with a decrease in domichnia and increase in pascichnia. These observations are consistent with those from the Ainsa–Jaca basins (Uchman 2001; Heard & Pickering 2008).

In unconfined submarine-fan and related environments, a number of studies have shown that the highest ichnodiversity and bioturbation intensity occurs in lobe and lobe-fringe environments due to diverse graphoglyptid-enriched trace-fossil assemblages (e.g., Uchman 2001; Heard & Pickering 2008; Monaco 2010; Cummings & Hodgson 2011). In the foreland basins of the northern Apennines, Monaco *et al.* (2010) observed, from proximal lobes to distal detached lobes, a change from ichnocoenosis dominated by post-depositional opportunistic ichnotaxa such as *Ophiomorpha rudis* and *Scolicia strozzii*, with only very rare pre-depositional trace fossils, to high-ichnodiversity, high-bioturbation-intensity ichnocoenosis dominated by *Nereites*, *Phycosiphon*, *Halopoa* and graphoglyptids. In the Basque Basin, Northern Spain, Cummings and Hodgson (2011) also observed a proximal-to-distal trend of decreasing domichnia and increasing fodinichnia. This observation is consistent with those from the Ainsa–Jaca basins (Uchman 2001; Heard & Pickering 2008). Cummings and Hodgson (2011) also observed a high abundance of *Scolicia* in channel-lobe transition deposits, as well as fan-fringe environments. This observation contrasts to observations from the Ainsa–Jaca basins (Uchman 2001; Heard & Pickering 2008) where *Scolicia* is rare in fan-fringe environments.

The trace-fossil distributions documented in these studies should not be strictly applied to all submarine fans, as distributions may vary between basins; for example, in the Upper Jurassic to Lower Cretaceous SGF deposits of southwest Bulgaria, most ichnotaxa occur in the proximal facies (Tchoumatchenco & Uchman 2001), whilst McCann and Pickerill (1988) concluded that proximal and distal lobe-fringe facies in the Cretaceous SGF sandstones of the Kodiak Formation of Alaska contain a lower ichnodiversity and low-density trace-fossil assemblage compared with channel-levée or inter-channel environments.

3.10 Concluding remarks

Today, ichnology is recognised as an important tool available to sedimentologists studying deep-water systems. This chapter has outlined some of the major concepts developed over the past half-century which have advanced ichnology to a point where it has become increasingly relevant to more than just sedimentological and palaeontological studies. Ichnology is now important in sequence stratigraphy, reservoir characterisation, petroleum exploration, petrophysics, geochemistry, zoology and ecology. The power of ichnology stems from the fact that the activity of organisms is influenced by dynamic controlling factors. These controlling factors can vary significantly between depositional environments, which in turn have a major impact on the preserved trace-fossil assemblages.

In this chapter, some of the applications of ichnology in palaeoenvironmental studies have been illustrated through two case studies from the Ainsa–Jaca basins. These studies utilised both the ichnofacies and ichnofabric methodologies to characterise variations in trace-fossil assemblages in submarine-fan and related environments. At outcrop, proximal, channel-dominated environments display a clear trend of increasing bioturbation intensity, ichnodiversity and the number of pre-depositional trace fossils from channel-axis to off-axis environments (Figs 3.13–3.15). This trend is also reflected in the ichnofabrics observed in cores. These variations in the trace-fossil assemblages are directly related to the influence of dynamic controlling factors controlling the activity of organisms. For example, the preservation potential of pre-depositional trace fossils is partially controlled by the nature of the SGFs. In channel-axis environments, the preservation potential of trace fossils is low due to the high frequency of concentrated density flows (Section 1.4.1) associated with significant tractional erosion. The coarse-grained nature of the deposits, which may exclude burrowers adapted only to fine-grained sediments, and the high frequency of thick sand beds which result in only large, robust, deeply burrowing organisms occurring in large numbers, also have a significant impact on the preserved trace-fossil assemblages in proximal axial environments. In contrast, slower rates of sediment accumulation, an increased frequency of taphonomically beneficial SGFs and a reduction in grain size and bed thickness of the deposits may all contribute to an increase in average bioturbation, ichnodiversity and number of pre-depositional trace fossils in proximal off-axis and distal environments.

At outcrop, where trace fossils are studied in three dimensions on bedding planes, emphasis is put on variations in trace-fossil assemblages and bioturbation intensity in different environments. In contrast, in core-based studies where the positive identification of individual ichnotaxa can be difficult, emphasis is put on describing the texture of the sediment created by the action of organisms as well as analysis of ichnodiversity and bioturbation intensity.

Several researchers have attempted to address the resulting discrepancy in datasets by describing in detail the sedimentary fabrics in outcrop-based studies, either in naturally polished sections, such as stream or wave-cut outcrops, or by polishing portions of the outcrop (e.g., Wetzel & Uchman 2001; Phillips *et al.* 2011; Callow *et al.* 2013). The results of such studies should provide an important comparison to core-based studies.

Our understanding of the uses of ichnology in deep-water systems has rapidly advanced in the past decade due to the better integration of sedimentology and ichnology (e.g., Uchman 2001; Heard & Pickering 2008; Phillips *et al.* 2011; Callow *et al.* 2013; Heard *et al.* 2014). Outcrop-based ichnological studies have benefited from the considerable volume of research on the sedimentological characterisation of deep-water systems, as well as detailed morphological and taxonomic studies of trace fossils (e.g., Häntzschel 1975; Książkiewicz 1977; Uchman 1998). Such studies have led to the expansion of the ichnofacies concept and continued development of ichnofabric analysis. Today, ichnology is integral to most core-based studies such as Integrated Ocean Drilling Program (IODP) core analysis as well as petroleum-industry core evaluation. With petroleum exploration expanding into ultra-deep water, and deeply buried, poor seismically imaged reservoirs now being actively explored and developed, considerable well costs necessitate the use of all available data, including ichnology, in reservoir characterisation. Continued exploration and development of deep-water systems in the petroleum industry and the integration of ichnology into sedimentological, sequence stratigraphic and palaeontological studies, suggest the continued rapid advancement of ichnology and its applications.

CHAPTER FOUR
Time–space integration

Candidate maximum flooding surface (prominent organic-rich metre-scale dark interval of mudstones in middle foreground immediately above vehicles), Middle Eocene Guaso System, Ainsa Basin, Spanish Pyrenees. The prominent sandbody is interpreted as submarine channel (prodelta) deposits (see Sutcliffe & Pickering 2009).

4.1   Introduction

The understanding of deep-water depositional systems has advanced significantly in recent years. The impetus was the publication of a proposed global sea level or eustatic record by Haq *et al.* (1987) (Fig. 4.1), followed by its application to seismic stratigraphy and lithostratigraphy. Since the publication of the Haq *et al.* (1987) eustatic sea-level chart, there have been several attempts to test and improve on this work (e.g., presenting Palaeozoic sea-level changes by Haq & Schutter 2008), but also criticism (e.g., Miall 1992b). Amongst the more notable recent studies that attempt to reconstruct global sea-level changes in deep time is that of Kenneth Miller and his co-workers (e.g., Kominz *et al.* 2008; Miller *et al.* 1991, 1996,

1998, 2003, 2005a, b, 2011). They have employed back-stripping techniques and comparative studies (Fig. 4.2). A particular problem with evaluating the Haq *et al.* (1987) and updated eustatic sea-level charts is the absence of any integrated, published and widely available datasets that were used to construct them.

In the past, much of our understanding of deep-water sediment transport processes and depositional systems came from studies of outcrops, modern fan systems, and 2D reflection seismic data (Bouma 1962; Mutti & Ricci Lucchi 1972; Normark 1970, 1978; Walker 1978; Posamentier *et al.* 1991; Weimer 1991; Mutti & Normark 1991). More recently, our understanding of these systems has improved because of: (i) interest from the hydrocarbons industry in deep-water exploration and development (e.g., Pirmez *et al.* 2000),

Deep Marine Systems: Processes, Deposits, Environments, Tectonics and Sedimentation, First Edition. Kevin T. Pickering and Richard N. Hiscott.
© 2016 Kevin T. Pickering and Richard N. Hiscott. Published 2016 by John Wiley & Sons, Ltd.
Companion Website: www.wiley.com/go/pickering/marinesystems

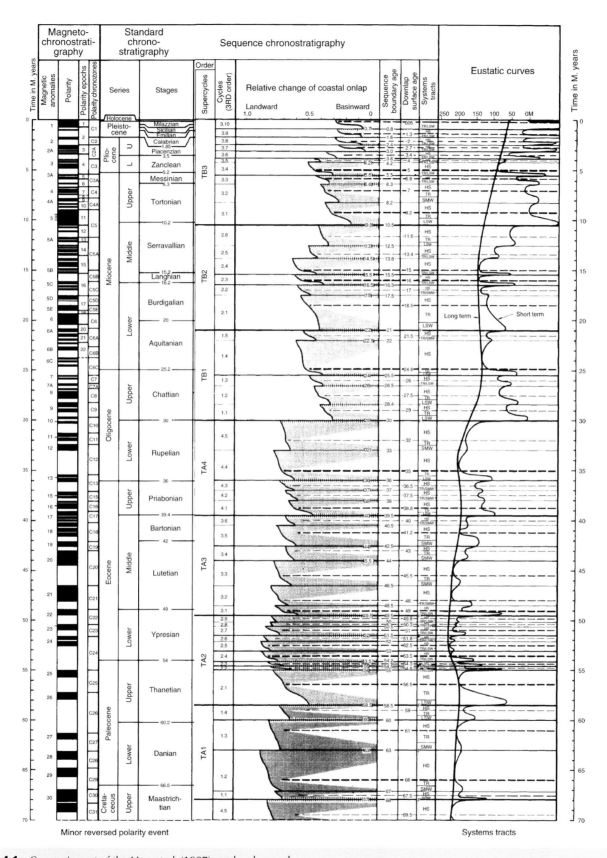

Fig. 4.1 Cenozoic part of the Haq *et al.* (1987) sea-level record.

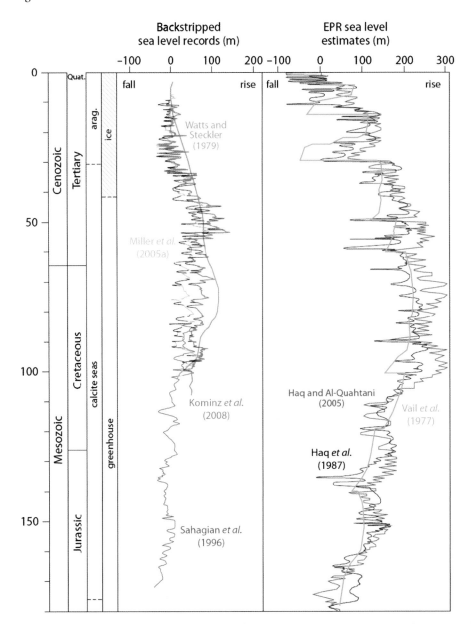

Fig. 4.2 Comparison of back-stripped sea-level records from New Jersey (blue, Miller *et al.* 2005a; brown, Kominz *et al.* 2008), Russian platform (pink, Sahagian *et al.* 1996), and Scotian margin (grey, Steckler & Watts 1978) with Exxon Production Research Company (EPR; green = Vail *et al.* 1977a; black = Haq *et al.* 1987; maroon = Haq & Al-Qahtani 2005). Note the much higher amplitude of the EPR estimates. From Miller *et al.* (2011).

and the advent of widely available high-quality 3D seismic data from many deep-water environments (e.g., Beaubouef & Friedman 2000; Posamentier *et al.* 2000); (ii) recent drilling and coring of both near-surface and reservoir-level deep-water systems (e.g., Twichell *et al.* 1992); (iii) increasing utilisation of deep-tow side-scan sonar and other imaging devices (e.g., Twichell *et al.* 1992; Kenyon & Millington 1995) and (iv) proliferation of integrated outcrop-subsurface studies, particularly in the upper Miocene Mount Messenger Formation, North Island, New Zealand (Browne & Slatt 1997, 2002; Haines *et al.* 2004; Hulme *et al.* 2004), Permian Karoo System, South Africa (Johnson *et al.* 2001; Grecula *et al.* 2003a, b; Andersson *et al.* 2004; Sixsmith *et al.* 2004), Permian Brushy Canyon Formation, Guadalupe Mountains, west Texas (Batzle & Gardner 2000; Carr & Gardner 2000; Gardner *et al.* 2000), and Middle Eocene Ainsa Basin, Spanish

Pyrenees (Pickering & Corregidor 2005; Das Gupta & Pickering 2008; Heard & Pickering 2008; Heard *et al.* 2008; Pickering & Bayliss 2009). Three-dimensional seismic data provide an unparalleled perspective of deep-water depositional systems, where vertical resolution reaches 2–3 m. Seismic time slices, horizon-datum time slices and interval attributes provide images of deep-water systems in map or plan view that can be analysed from a geomorphologic perspective. Geomorphologic analyses lead to the identification of depositional elements, which, when integrated with seismic profiles, can give significant stratigraphic insight; for example, as outlined in the case studies (exercises) in sequence stratigraphy in Abreu *et al.* (2010). Finally, calibration by correlation with borehole data, including logs, conventional core and biostratigraphic samples, can provide the interpreter with an improved understanding of the geology of deep-water systems. What determines the architecture of deep-water deposits are the controlling parameters of flow process, flow volume, sand-to-mud ratio, slope length, slope gradient and rugosity of the seafloor, not the age of the deposits. The application of a sequence stratigraphic approach to understanding deep-water systems is an attempt to make predictive sense of very complex natural systems that are controlled by many forcing variables, such as tectonics, global climate, regional weather patterns and catastrophic events such as volcanic eruptions and bolide impacts.

Sequence stratigraphy 'is the study of rock relationships within a chronostratigraphic framework of repetitive, genetically-linked strata bounded by surfaces of erosion or non-deposition, or their correlative conformities' (Van Wagoner *et al.* 1988). These bounding surfaces are called *sequence boundaries*. The methodological approach of sequence stratigraphy is used to subdivide the sedimentary record into packages that are defined by bounding unconformities and prominent internal surfaces, and that result from changes in eustatic sea level, tectonic uplift and subsidence, and rates of sedimentation (sediment flux). Sequence stratigraphic analyses are made from seismic cross-sections, well logs and outcrop studies of sedimentary rocks to infer changes of relative base level, sediment calibre (bulk mean grain size) and sediment flux. Sequence stratigraphy, therefore, permits predictions about the lateral continuity and extent of lithofacies.

The much wider range of sedimentary features (including well-developed condensed horizons) and biological indicators (fossils and trace fossils) make sequence stratigraphic interpretations much easier in shallow-marine compared to deep-water settings. For example, at least in shallow marine environments, 'the most readily identifiable surface is the transgressive surface … The second most easily recognised surface is the surface of maximum flooding …' Haq *et al.* (1987). The deep-water extensions of these and other fundamental surfaces are not easy to recognise because of many factors, including erosion and non-deposition in areas of sediment bypass between the shelf and the adjacent basin, so that key seismic reflections cannot be traced into deep water with confidence, or at all.

Most of the emphasis in sequence-stratigraphic models has been placed on shallow-marine successions and their development. Our approach will be to explain briefly sequence-stratigraphic concepts for the entire continental to deep-marine profile, and then to highlight and discuss issues of particular relevance to deep-water systems. In many cases, fundamental surfaces and styles of deposition are defined exclusively in the shallow-marine realm, and can only be extrapolated into deep water based on broad assumptions about linkages between shelf and basinal depositional processes.

Accommodation space (now commonly referred to simply as 'accommodation') is defined as the space available for sediments to accumulate within a sedimentary basin, or more particularly at a site within a basin. This space is a function of: (i) eustatic sea-level change; (ii) tectonic uplift and subsidence and (iii) sediment flux. The rate of change of accommodation space, A, with time, t, is expressed as $\delta A/\delta t$. An increase in accommodation space $= +\delta A/\delta t$. A decrease in accommodation space $= -\delta A/\delta t$. The first derivative with respect to time (hence $\delta/\delta t$) of the eustatic sea-level curve (rate of change) plus the first derivative of the tectonic subsidence curve (rate of change) equals the rate of change in relative sea level. In deep-water environments, even substantial changes in accommodation space vertically above the basin floor, or above deeper parts of the adjacent basin floor, will have no direct effect on depositional patterns in these areas because the percentage change in the total water depth will be small. Subtle changes in accommodation space in adjacent shelf

and coastal environments, however, may have a profound effect on sedimentation in the deep basin, especially in cases where the shelf is narrow to non-existent. This is because even small reductions in accommodation space in near-shore and shallow-marine settings will tend to drive depositional systems basinward with a geometry characterised by progradational stacking patterns. Where $-\delta A/\delta t$ is fast, as in a relatively rapid eustatic sea-level fall, or fast tectonic uplift rates of coastal and shallow-marine environments, and where sediment flux is high, shelf-edge deltas will tend to develop (e.g., Muto & Steel 2002), creating wedges of gravitationally unstable clastic sediment poised for failure and redeposition into deep water.

On any relative sea-level curve, an *inflection point* is identified where the curve changes from a concave-upward to convex-upward shape, or vice versa. The inflection point on a sinusoidal sea-level curve occurs when the rate of sea-level fall or rise is at its most rapid. A sequence boundary will usually start to develop prior to the inflection point of a relative sea-level fall. A condensed section will usually accumulate prior to or at the inflection point of a relative sea-level rise (generally a time of maximum rate of transgression). The critical issue in gaining an understanding of deep-water systems comparable to that available for shallow-water settings is the recognition of sequence boundaries (see below), particularly as few basins – at least those that outcrop – preserve the linked non-marine, shallow-marine, slope and deep basin environments.

A fundamental concept in sequence stratigraphy involves what is called the base level. *Base level* is the surface to which sediment can aggrade if there is sufficient sediment supply. Accumulation either fills a depression up to base level or erosion bevels an elevated surface down to base level. While many studies tend to link changing base level to changing relative sea level, this assumption has been challenged for many situations (*cf.* Christie-Blick & Driscoll 1995). For example, base level can also be related to the water surface of lakes, local equilibrium surfaces associated with river systems, or the spillover points of intraslope basins that limit the ponding of through-going turbidity currents. Base level is linked to the concept of an equilibrium point, which is the site along a depositional profile where the rate of eustatic change equals the rate of subsidence or uplift. The equilibrium point separates zones of synchronous rising and falling relative sea level (Posamentier *et al.* 1988). Any understanding of changing relative base level must be conditioned by an appreciation of the typical rates of change and the amplitude of such changes (Table 4.1). For example, John *et al.* (2011) analysed ODP Leg 194 sediment cores from the Marion Plateau, offshore northeast Australia, to define the mechanisms and timing of sequence formation on mixed carbonate–siliciclastic (calciclastic) margins, and to estimate the amplitude of Miocene eustatic sea-level changes. They demonstrated that sequences on the Marion Plateau are controlled by glacio-eustasy since sequence boundaries are marked by increases in $\delta^{18}O$ (deep-sea Miocene isotope events Mi1b, Mbi-3, Mi2, Mi2a, Mi3a, Mi3, Mi4 and Mi5, respectively), which reflect increased ice volume primarily on Antarctica. Their back-stripping estimates suggest that sea level fell by 26–28 m at 16.5 Ma, 26–29 m at 15.4 Ma, 29–38 m at 14.7 Ma and 53–81 m at 13.9 Ma. Combining back-stripping with $\delta^{18}O$ estimates yields sea-level fall amplitudes of 27 ± 1 m at 16.5 Ma, 27 ± 1 m at 15.4 Ma, 33 ± 3 m at 14.7 Ma and 59 ± 6 m at 13.9 Ma. Using a similar approach, John *et al.* (2011) further estimated eustatic rises of 19 ± 1 m between 16.5 and 15.4 Ma, 23 ± 3 m between 15.4 and 14.7 Ma, and 33 ± 3 m between 14.7 and 13.9 Ma. They suggested that sea level fell by 53–69 m between 16.5 and 13.9 Ma, implying that at least 90% of the East Antarctic ice sheet was formed during the middle Miocene.

Table 4.1 Cycle hierarchies for driving depositional systems. Modified From Tilman and Weber (1987)

Tectono-eustatic/ eustatic cycle/ order	Shallow-marine sequence stratigraphic unit	Duration Myr	Relative sea-level amplitude (m)	Relative sea-level rise/ fall rate (cm/kyr)
First		>100		<1
Second	Supersequence	10–100	50–100	1–3
Third	Depositional sequence	1–10	50–100	1–10
	Composite sequence			
Fourth	High-energy sequence, parasequence and cycle set	0.1–1	1–150	40–500
Fifth	Parasequence high-frequency cycle	0.01–1	1–150	60–700

The assumptions made in the Exxon approach to sequence stratigraphy are that: (i) the rate of seafloor subsidence at any single location on a profile is constant; (ii) total subsidence increases in a basinward direction; (iii) deposition along a divergent continental margin results from a constant rate of sediment supply (flux); (iv) marine proximal-to-distal environments are characterised by a shelf, slope and basin and (v) the eustatic sea-level trend is curvilinear, approaching sinusoidal (at least resolvable as a series of sine curves) (Posamentier *et al.* 1988). These assumptions impact directly on the development of depositional sequences and, therefore, the predicted architecture of deep-marine systems.

A *depositional sequence* is defined by Brown and Fisher (1977) as a three-dimensional assemblage of lithofacies, genetically linked by active (modern) or inferred (ancient) processes and environments (river, delta, barrier island, etc.). A *systems tract* comprises genetically associated stratigraphic units that are deposited during a specific segment of the eustatic curve (i.e., eustatic lowstand = lowstand wedge; eustatic rise = transgressive systems tract; rapid eustatic fall = lowstand fan, etc.) (Brown & Fisher 1977, Posamentier *et al.* 1990, Posamentier & Allen 2000). These units are represented in the rock record as three-dimensional facies assemblages. They are defined on the basis of bounding surfaces, position within a sequence and parasequence stacking pattern (Van Wagoner *et al.* 1988). A *parasequence* is a relatively conformable succession of genetically related beds or bedsets, commonly with an upward-coarsening trend, bounded by marine flooding surfaces and their correlative surfaces (Van Wagoner *et al.* 1988). An unresolved issue is how shallow-marine parasequence stacking is manifested in the downdip contemporaneous deep-water systems. A major problem in deep-water systems, especially uplifted ancient deposits, is that age dating is commonly inadequate for recognising high-frequency cycles that are typical of parasequences (10^4–10^5 years).

Coastal onlap is the landward limit on the basin margin of sediment accumulation, whether it is marine or non-marine. Inflection points that develop during falling sea level (F points) are characterised by a basinward shift in coastal onlap. With a constant rate of sediment supply, the rate of progradation is inversely proportional to the rate of aggradation and the rate of subsidence. Clearly the magnitude of the basinward or landward shift in the coastal onlap, together with the shelf physiography (shelf width, depth and any shelf basins that can act as storage for large volumes of sediment), will have a profound effect on the propensity for shelf bypass and redeposition of detritus into deeper water.

Deep-water systems are commonly interpreted as parts of lowstand systems tracts. A *Lowstand Systems Tract (LST)* is the set of depositional systems active during a time of relatively low sea level (base level) following the initiation of the development of a sequence boundary on the shelf and in non-marine settings. If a distinct shelf-slope break exists and relative sea level (base level) has fallen sufficiently, the lowstand systems tract may include two distinct parts: the 'lowstand fan' and the 'lowstand wedge'. Following the relative fall in sea level (base level) that produces the sequence boundary, relative sea level (base level) begins to stabilise and eventually begins to rise, but at a very slow rate. This slow rate of creation of accommodation coupled with relatively high supply of sediment results in the progradational stacking typical of the lowstand wedge (Fig. 4.3). Whilst the notion of a lowstand fan and lowstand wedge are idealised abstractions, it is clear that these represent only parts of a continuum of deep-water environments that are preferentially created during a relative fall in base level. As such, they have little utility as truly discrete and clearly identifiable deep-water facies associations. We prefer to envisage each as a type of submarine fan, with the so-called 'wedge' being typified by backfilling processes in submarine channels and associated canyons as relative base level rises. In reality, given sufficient sediment flux, the position of submarine fans within a basin will be heavily influenced by factors such as gradient and topography (including the position of slope basins in relation to sediment routing). Thus, the models shown in Figure 4.3 do not exhaust the possibilities for the location of submarine fans within a basin.

The *Transgressive Systems Tract (TST)* consists of a retrogradational (landward-stepping) set of parasequences. It is underlain by the transgressive surface and overlain by the maximum flooding surface. As relative sea level (base level) continues to rise, accommodation space in nearshore settings is produced at a faster rate than it can fill with sediments, and a retrogradational set of parasequences forms. At each flooding surface in the transgressive systems tract (i.e., at the top of each parasequence), the short-term accelerated rate of water-depth increase caused by shifting patterns of sediment supply or high-frequency sea-level (base-level) fluctuations adds to the long-term (lower frequency) rate of relative sea-level (base-level) rise to produce an unusually rapid landward shift in the shoreline and a highly pronounced flooding surface. In deep water it is tempting, but perhaps unjustified, to speculate that the parasequences developed in nearshore areas could match one-for-one with stacked submarine-fan lobe deposits that show an overall fining upwards, and with the timing of aggressive backfilling of submarine channels.

The *Highstand Systems Tract (HST)* consists of an aggradational to progradational set of parasequences that overlies the maximum flooding surface and that is overlain by the next sequence boundary. As the parasequences pass from aggradational to progradational stacking, the flooding surfaces are increasingly subdued at the expense of overall shallowing. During the development of the highstand systems tract, the rate of relative sea-level (base-level) rise begins to slow and relative sea level (base level) eventually begins to fall prior to the development of the next sequence boundary. Throughout the

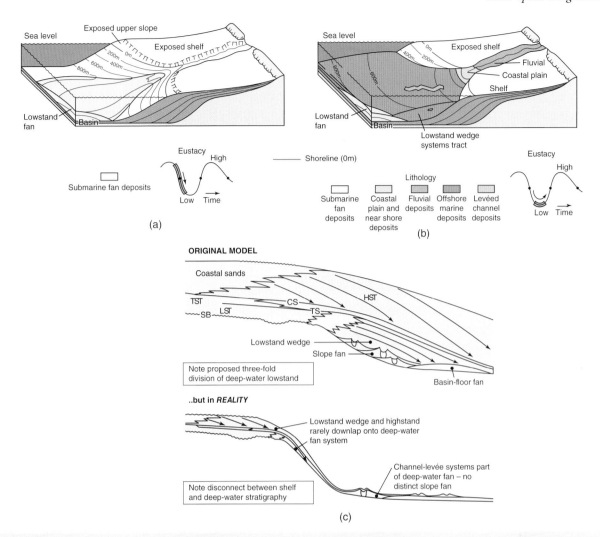

Fig. 4.3 (a) Components of the lowstand systems tract and lowstand fan. (b) The lowstand systems tract, lowstand wedge. From (Posamentier *et al.* (1988). (c) Comparison of the original depositional model of Vail (1987) and Posamentier and Vail (1988a), and more recent 'in reality' perceptions about the position of siliciclastic systems in deep-water basins from the SEPM Stratigraphy Web (http://www.sepmstrata.org/page.aspx?pageid=41). HST = highstand system tract; LST = lowstand system tract, and TST = transgressive system tract.

time of the highstand systems tract however, accommodation space is created or destroyed at a relatively slow rate. Coupled with the increased supply of sediment to the shelf as estuaries are filled, progradational stacking is increasingly favoured over aggradational stacking. As relative sea level (base level) begins to fall, a new sequence boundary begins to form; this sequence boundary will begin to erode into the underlying highstand systems tract. In deep water, the early HST is likely to be associated with abandonment of coarser clastic systems like submarine fans, and draping by hemipelagic and pelagic facies associations (e.g., Facies Class E). During the late HST, progradation of deltas across the shelf and onset of inner-shelf erosion can reactivate the delivery of coarser detritus to slope and basinal environments.

A *Type 1 Sequence Boundary (SB1)* is an unconformity (and correlative conformity in deep water) that is associated with stream rejuvenation and fluvial incision, sedimentary bypass and subaerial exposure of the entire shelf, and an abrupt basinward shift of facies and the position of coastal onlap. It is interpreted to form when the rate

of fall in base level exceeds the rate of basin subsidence at the shelf break (i.e., the seaward edge of the shelf), producing a relative fall in sea level (base level) at that position. A *Type 2 Sequence Boundary (SB2)* consists of an unconformity marked by subaerial exposure and a downward shift in coastal onlap to a point landward of the shelf break; however, it lacks subaerial erosion associated with stream rejuvenation and there is a less pronounced basinward shift in facies. Seaward of the inner shelf, a SB2 becomes a conformable surface. This type of sequence boundary is interpreted to form when the rate of fall in base level is less than the rate of basin subsidence at the shelf break, so that no relative fall in sea level (base level) occurs at that location. Instead, the lowstand shoreline of a SB2 is situated in the middle of the former (highstand) shelf. It seems reasonable to predict that many deep-water clastic systems that abruptly overlie fine-grained basinal deposits are associated with a SB1, rather than a SB2 (e.g., Fig. 4.4). However, the magnitude and longevity of any relative fall in base level, together with the available supply of sediment of a particular calibre, and the basin and slope gradients (including topography), will

(a)

(b)   (c)

Fig. 4.4   (a) Candidate Type I sequence boundary and palaeo-valley fill, Namurian Tulig cyclothem, Clare Basin, western Ireland. Above the unconformity (arrow), there are non-marine fluvial (deltaic?) sediments, including coal horizons. Below the unconformity, there is an interval of offshore marine prodelta fine-grained deposits, below which several packets of sandstones are seen. Cumulative cut-down visible in this image is ~4 m, but total observed cut down is ~15 m. (b) Flute casts in fluvial sandstones at the base of the palaeo-valley fill shown in (a) (c) Regional marine flooding surface at top of palaeo-valley-fill sandstones shown in (a). Surface characterised by abundant *Zoophycos* and phosphate concretions.

determine any vertical sequences and their thicknesses. At present, there is insufficient stratigraphic control from a variety of depositional systems, where good age and spatial correlation can be established between shallow-marine and deep-marine systems, to develop sophisticated models linking events on the shelf with their associated deep-water response.

Another important component of any sequence stratigraphic model is the nature of marine flooding surfaces in deep water. A *marine flooding surface* records the first significant flooding across a shelf, heralding the onset of transgression across the so-called *transgressive surface* (TS). The marine flooding surface may coincide with the previous sequence boundary if no lowstand facies accumulated above the sequence boundary. A TS is commonly a surface marked by hardgrounds cemented by carbonate and overlain by widespread winnowed, sorted and often conglomeratic sediment, or lag. This coarse veneer is left behind by the process of shoreface erosion, which takes place as the shoreline sweeps landward; this process is

called *ravinement* (Swift 1968) and results in the decapitation of the upper shoreface and subaerial parts of the coastal system. When the shoreline reaches its maximum landward position, the transgression is complete and the contemporary depositional surface is called the *maximum flooding surface* (MFS). Glauconitic and phosphatic sediments are commonly associated with widespread condensed sections well offshore at the time of maximum flooding.

Many marine flooding events, especially on the deeper parts of continental shelves, slopes and ponded basins, tend to be associated with the accumulation of organic-rich shales. Such deposits record prolonged periods of the slow accumulation of hemipelagic and pelagic deposits (Facies Group E and G), hence the term *condensed intervals*: they also commonly show high natural gamma-ray values. Such intervals have been used as proxies for palaeoceanographic conditions which can be correlated over large distances in the subsurface, for example as described for the Upper Carboniferous cyclothemic black shales from the mid-continent of North

American (Samson *et al.* 2006). However, other researchers have argued that 'hot' shales (with high natural gamma-ray values) are primarily due to globally high temperature excursions that created extensive algal blooms and anoxic oceanic conditions (so-called *oceanic anoxic events*, abbreviated to OAEs; *cf.* Arthur & Schlanger 1979; Schlanger *et al.* 1987) in abyssal water depths; for example, for the Eocene shales in deep-marine Gulf of Mexico wells (Sercombe & Radford 2007). Sercombe and Radford (2007) dismissed the possibility that these hot shales are linked to a maximum flooding surface as they would have accumulated where the influence of eustatic sea-level changes in deep water was insignificant. Nevertheless, any geological time intervals of elevated global temperatures (greenhouse periods), including brief periods of extreme global warmth (with OAEs), could be explained by a combination of high eustatic sea level, the extensive flooding of continental shelves and reduced coarse clastic sediment delivery to deep water, thus favouring the accumulation of 'hot' organic-rich shales. For more information on OAEs, the reader is referred to the review by Jenkyns (2010) and references therein.

Where a TS extends over LST valley fill, it commonly can be recognised in electrical logs by a small local increase in the resistivity curve in response to carbonate cementation along a hardground surface. The TS is then overlain by a minor low in resistivity (a shale 'kick'). In deep water, surfaces (intervals) dating from a time of maximum flooding of the shelf are commonly characterised by hemipelagic and pelagic facies and facies associations (Facies Class E). Such condensed intervals can provide the best means for dating ancient deep-marine depositional systems as they may be enriched in pelagic fossils (e.g., microfossils or ammonites) and offer the best chances of finding any preserved volcanic ash layers.

Einsele (1985) considered the response of sediments to sea-level changes in differing storm-dominated margins and epeiric seas, particularly the Mesozoic epicontinental, mud-dominated seas in the slowly subsiding basins of Germany. Such basins contrast markedly with the rapidly subsiding shelf-margin seas subject to rapid changes in sea level that are glacio-eustatically driven, upon which much of the Exxon philosophy is based. He pointed out that the base level to which sediments aggrade is the storm wave base rather than sea level *sensu stricto*. Einsele (1985) believes that sediment accumulation patterns are particular to depositional sites and conditions. Perhaps the principal differences between the Exxon model and this perspective is that, for Einsele (1985): (i) regressions are perceived as gradual rather than abrupt; (ii) purely aggradation depositional units reach their maximum thickness where basin subsidence is most rapid, for example in the centre of basins and (iii) the base level to which sediments may aggrade is the storm wave base.

Although philosophically similar in many respects to the Exxon approach, Galloway (1989a, b) has put forward the concept of a genetic stratigraphic sequence. A *genetic stratigraphic sequence* is the sedimentary product of a depositional episode, each sequence comprising (i) a progradational facies-association; (ii) an aggradational facies-association and (iii) a retrogradational or transgressive facies-association. Genetic sequences are bounded by a sedimentary veneer or surface that records the depositional hiatus that occurs over much of the transgressed shelf and adjacent slope during maximum marine flooding.

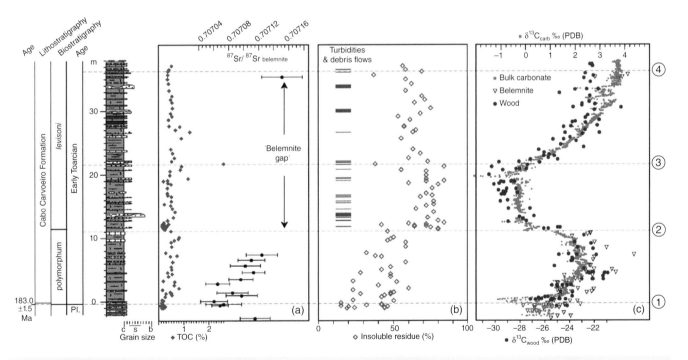

Fig. 4.5 The Early Toarcian section from Ponta do Trovão and Praia do Abalo, Peniche, Portugal, currently the sole candidate for the Global Stratotype Section and Point (GSSP). The graphic log and sample heights are based on the section measured only in the cliff exposures at this locality. (Ammonite biostratigraphy from Mouterde (1955).) (a) Total Organic Carbon (TOC) data from hand specimens; Sr-isotope data from belemnites, normalised to a value of 0.710250 for NIST 987. (b) Stratigraphic distribution of SGF deposits, including debrites, shown together with insoluble residue from bulk rock samples. (c) High-resolution C-isotope data from bulk sediment, fossil wood and belemnites. Horizons labelled 1, 2, 3 and 4 represent key levels correlatable between Yorkshire (England) and Peniche (Portugal). The single black shale at 21.5 m in the section with a TOC value of 2.6% has a $\delta^{13}C_{org}$ value of −29.5‰. From Hesselbo *et al.* (2007). See text for discussion.

Hesselbo *et al.* (2007) caution against adopting simple sequence-stratigraphic interpretations that link slope instability and sediment supply to global sea-level fall. In a study of an Upper Pliensbachian to Lower Toarcian, fully marine hemipelagic succession of coccolith-bearing marls, limestones and redeposited sediments (the Lemede and Cabo Carvoeiro formations, exposed in cliffs and foreshore around Peniche, western Portugal), they show that the lowest occurrence of SGF deposits, including debrites, in the Peniche succession coincides with the base of a negative carbon-isotope excursion and their highest occurrence coincides with its top (Fig. 4.5). This succession is interpreted to have accumulated on a northwest-facing carbonate ramp in the extensional Lusitanian Basin. The inference made by Hesselbo *et al.* (2007) is that redeposition processes into deep water were most active during an interval of extreme global warming and high global sea level. In fact, such an observation is predicted for carbonate-dominated submarine fans, because increased storminess and carbonate production are likely at highstands (see Section 4.10), and there may be greater slope instability due to the release of gas hydrates in a warmer ocean. As a cautionary note, the global nature of this particular carbon-isotope excursion has been challenged by McArthur (2007).

4.2 Submarine fan growth phases and sequence stratigraphy

4.2.1 Early models for fan development and relative base-level change

Mutti (1985) developed a model for the evolution of submarine fan systems during rises and falls of sea level that incorporates the features of high- and low-efficiency submarine fans into a single scheme (see also Chapter 7). Mutti recognises three types of deep-water system, differing in the distribution of sand deposits and in relative scale. The three types may develop independently of one another, or may develop in succession in large delta-fed systems as sea level varies. The discussion that follows pertains to large delta-fed systems, and is therefore most applicable to submarine fans in orogenic settings with fast rates of sediment supply. When sea level falls, large volumes of delta-front and delta-slope sediment become unstable and generate a series of large-volume, high-momentum sediment gravity

flows (SGFs) through retrogressive slope failure. These SGFs do not deposit in proximal erosional channels, but rather carry their load far out into the basin to form extensive sheet-like sand bodies, or packets, 3–15 m thick, that grade distally into finer grained lobe-fringe deposits. Individual sand beds have mostly flat bases. In vertical sections, the sandbodies alternate with lobe-fringe deposits to form units several hundreds of metres thick that are classified as Type I systems by Mutti (1985), and that correspond to the *high-efficiency fans* of Mutti (1979). As the volume of individual SGFs decreases, either due to slowly rising sea levels or smaller slope failures on a slope of reduced gradient, sands are not carried as effectively into the basin, and instead accumulate at the distal ends of distributary channels and on small lobes immediately downfan from the channels. The result is a *low-efficiency fan* of the type described by Normark (1970), with sand-rich suprafan lobes. These are classified as Type II systems by Mutti (1985). Continued sea-level rise results in mud deposition on the submarine fan surface, but in the vicinity of a large river delta eventual progradation of the delta to the shelf edge during the highstand leads to deposition on the submarine fan surface of mud-rich channel–levée complexes in which sands are restricted to the proximal parts of small channels. The bulk of the sediment is mud that represents overbank deposition from the channels. These muds undergo slumping and sliding from levée crests, producing characteristic multiple angular unconformities. These mud-rich deposits are assigned to Type III systems by Mutti (1985).

The full evolutionary sequence from Type I to Type III is only developed for delta-fed submarine fans with fast rates of sediment accumulation, such that large volumes of unconsolidated sediment can accumulate on the shelf during highstands of sea level to guarantee a source for subsequent large lowstand failures. Where shelf sedimentation rates are slower, lowstand Type II systems alternate vertically with mud blankets of intervening highstands. A fundamental conceptual innovation in the Mutti (1985) model was the suggestion that sand packets in a vertical succession of ancient submarine fan deposits may not represent deposits of suprafan lobes like those described by Normark (1970), but instead may represent extensive sand sheets produced during a lowstand of sea level, perhaps covering all of what might previously have been considered as the middle fan and lower fan. Viewed in this way, mudstone units between sandstone packets were never lateral equivalents of sandy facies, but instead were deposited uniformly over the submarine fan surface during a time of higher sea level. Likewise, upper-fan

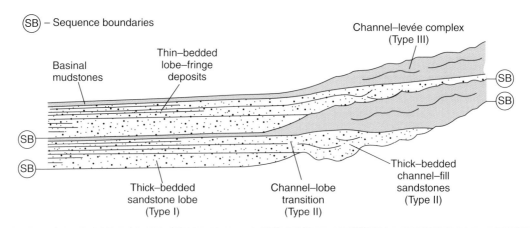

Fig. 4.6 Changing character of sequence boundaries of Mutti (1985) from the basin margin into the deeper parts of the basin. In this hypothetical example, the sequences are complete and show an evolution from Type I to Type II to Type III systems. See text for explanation.

channel–levée complexes are considered in Mutti's (1985) scheme to be younger than, not equivalent to, middle- and lower-fan deposits (see also Mitchum 1985).

According to Mutti (1985), channel–levée complexes and upslope deltaic sediments of the highstand provide the bulk of material that fails during the next drop in sea level, producing Type I deposits of a younger sequence. The result is that sequence boundaries in the basin are marked by the sudden introduction of sheet-like sand bodies, whereas sequence boundaries on the slope and at the base-of-slope are marked by erosional unconformities (Fig. 4.6). On a smaller scale, Mutti (1985) proposes that minor short-term fluctuations of sea level are responsible for 3–15 m-thick alternations of thick-bedded (Facies C2.1 and C2.2) and thin-bedded (Facies C2.3 and D2.1) lobe and lobe-fringe deposits so commonly seen in ancient submarine fan deposits (Fig. 4.7). The wireline log response that characterises such packets of thinner and thicker bedded strata is like that illustrated by Hsü (1977) for the Repetto Formation of the Ventura Basin.

The general model of Mutti (1985) is a useful guide to facies development in small delta-fed systems of tectonically active areas (e.g., Morgan & Campion 1987), but its applicability to large passive-margin fans like the Indus Fan and Amazon Fan has been questioned by Kolla and Coumes (1987) and Hiscott *et al.* (1997c). On the Indus Fan, lowstand deposits are of Type III (i.e., channel–levée complexes), not Type I, and the deposition of channel–levée complexes continues from the time of a low sea-level stand throughout the subsequent period of rising sea level. Kolla and Coumes (1987) attribute the apparent lack of any Type I deposits in modern, large, passive-margin fans to generally fine sediment size and great distances to the uplifted source areas. On the Amazon Fan, Type I and Type III sand-prone

sheets and channel–levée complexes are contemporaneous, and tens of 5–25 m-thick sandstone packets like those in ancient successions accumulated during single lowstands of sea level (Hiscott *et al.* 1997c).

Having considered the larger scale stratigraphic architecture of submarine fans as a response to changing relative base level (both tectonically and eustatically driven), smaller scale depositional environments, such as discrete submarine channels, may also record such controls. Figure 4.8 shows a sequence-stratigraphic explanation for a fining-upward trend in confined settings such as those in an erosional or erosional-depositional submarine channel or a submarine canyon.

Posamentier and Kolla (2003) recognise a vertical sequence similar to that described and interpreted to have been driven by autocyclic processes by Pirmez *et al.* (1997 – see below), but in contrast they interpret this sequence to have resulted from changing relative sea level and the type of dominant mass-flow process associated with submarine channels (Figs 4.9, 4.10). This genetic sequence is based on a number of studies but in particular offshore Indonesia. Their succession comprises cohesive-flow deposits at the base (corresponding to the initial period of relative sea-level fall), overlain by so-called frontal-splay-dominated, and then levéed–channel-dominated sections (corresponding to the subsequent period of early and late relative sea-level lowstand, respectively). The succession is capped by deposition of cohesive-flow and condensed-section deposits (corresponding to periods of rapid sea-level rise and highstand, respectively).

IODP Leg 308 in the Gulf of Mexico (Flemings, Behrmann, John & Expedition 308 Scientists 2005) concluded that the evolution of Brazos–Trinity Basin IV sedimentation (~200 km south of Galverston, Texas in ~1400 m of water, the terminal basin of a chain of five mini-basins that are separated by interbasinal highs) resulted

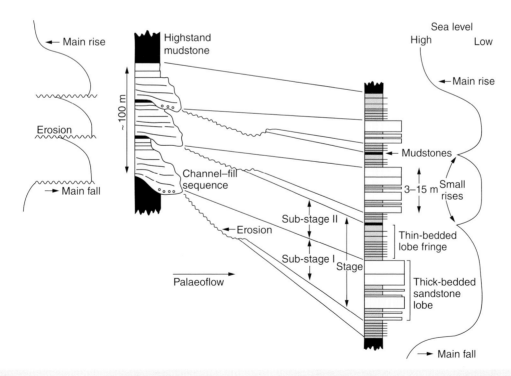

Fig. 4.7 Sea-level changes and resulting facies cyclicity in a submarine fan system comprising channel-fill and lobe elements. Sub-stages result from minor oscillations in sea level on a major fall/rise excursion. Falls in sea level can result in proximal unconformities and distal sandstone packets (Facies Class B and C) 3–15 m thick. Minor rises can result in proximal backfilling of channels and distal deposition of thin-bedded lobe-fringe deposits (Facies Class D). From Mutti (1985). See text for explanation.

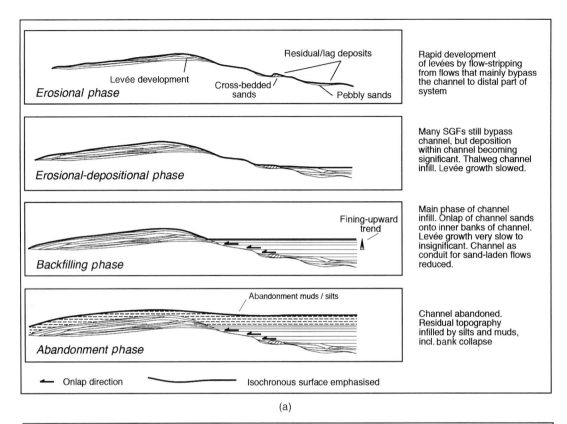

(a)

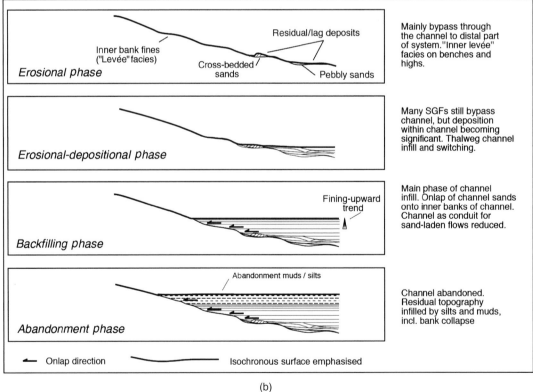

(b)

Fig. 4.8 Channel-fill models to show three principal phases in the history of a channel. Changes in the channel phases are related to changes in relative base level (accommodation space, eustasy and tectonic uplift/subsidence). (a) Channel–levée–overbank complexes. (b) Canyon and non-levéed channels. Modified from Pickering *et al.* (1995b).

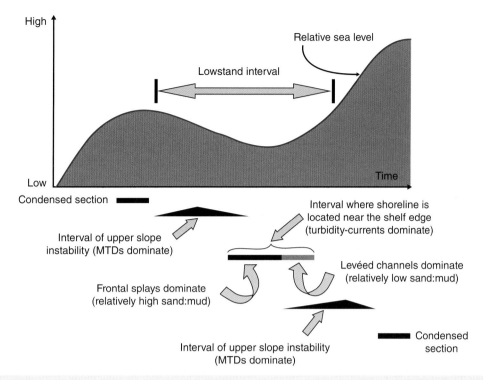

Fig. 4.9 Schematic depiction of the relationship between relative sea level and type of dominant mass-flow process. The succession comprises cohesive-flow deposits at the base (corresponding to the initial period of relative sea-level fall), overlain by frontal-splay-dominated and then levéed-channel dominated sections (corresponding to the subsequent period of early and late relative sea-level lowstand, respectively). The succession is capped by deposition of cohesive-flow and condensed-section deposits (corresponding to periods of rapid sea-level rise and highstand, respectively). From Posamentier and Kolla (2003).

from a complex interaction between fluvio-deltaic dynamics, sea-level changes, and the interaction between SGFs (e.g., turbidity currents) and submarine topography. During a eustatic lowstand in sea level corresponding to MIS 6, the basin received significant input of terrigenous sediments, but the complete absence of sand and silt was interpreted to indicate that either SGFs were filling basins updip (e.g., fill-and-spill model, Section 4.6.2) or that deltaic systems were trapping sands in areas adjacent to the Brazos–Trinity slope at the time. During the stepwise sea-level fall between MIS 5e and MIS 2, the basin received up to 175 m of SGF deposits including sediment slides/slumps, and cohesive-flow deposits. A pause in SGF deposition occurred from MIS 5a to ~ MIS 4. This period included both a relative rise and a relative fall in sea level. The absence of SGFs entering the Brazos–Trinity Basin IV during a relative sea-level fall must have been the result of factors other than base-level changes, perhaps including lateral shifts in the sediment source on the shelf or trapping of sediments in Basins I or II updip preventing the spillover of turbidity currents into Brazos–Trinity Basin IV.

Attempts to incorporate growth phases of submarine fans into sequence stratigraphic models have produced mixed results. Without doubt, lower base level promotes the delivery of greater amounts of sediment, and a greater proportion of sand, to deep-water systems. As an example, the use of sequence-stratigraphic principles, particularly the concept of a sea-level lowstand fan, led to the discovery of a major submarine fan system off the Pearl River, China (Pang *et al.* 2007). In some cases, however, high rates of sediment supply can mask the effects of a higher base level so that some submarine fans remain active even at the time of maximum flooding (Kolla & Macurda 1988). The inverse of the statement that sand

supply increases at lowstands is, unfortunately, not true, and the lack of appreciation of this fact has resulted in many misinterpretations of facies successions in ancient deposits. Specifically, we believe that it is not true as implied Mutti (1985; Fig. 4.7) and as interpreted for many ancient successions (e.g., King *et al.* 1994) that an alternation of fine-grained intervals and sand-prone intervals (called 'packets') at a scale of tens of metres implies high-frequency changes in base level. Instead, it has been shown that alternating sand packets and fine-grained facies typically accumulate during the same part of a sea-level cycle as a result of autocyclic processes (e.g., channel and lobe switching; Hiscott *et al.* 1997c; Nelson & Damuth 2003). Deposits of single sea-level lowstands in even smaller submarine fans can be hundreds of metres thick (Normark *et al.* 1998), so do not match the 10–20 m-thick sand packets so common in the geological record.

A likely reason for the misinterpretation of alternating sandy and muddy packets as a record of base-level fluctuations is their superficial similarity to shallow-marine parasequences. In the case of parasequences, it is generally believed that they result from short-term coastal progradation followed by inundation and flooding. Without direct proof or a chronological tie to nearby shelf deposits, however, there is no justification for assuming that the top of each sandy packet in a deep-water succession corresponds to a rise in base level on the shelf. Growing experience from the coring of Pleistocene–Holocene deep-water systems suggests that this is rarely, if ever, the case unless the deep-water sedimentation rate is very low.

In many cases, stratal geometries and facies architecture of deep-marine systems cannot simply be attributed to a falling regional (or global) sea level, but instead can reflect the complex interplay

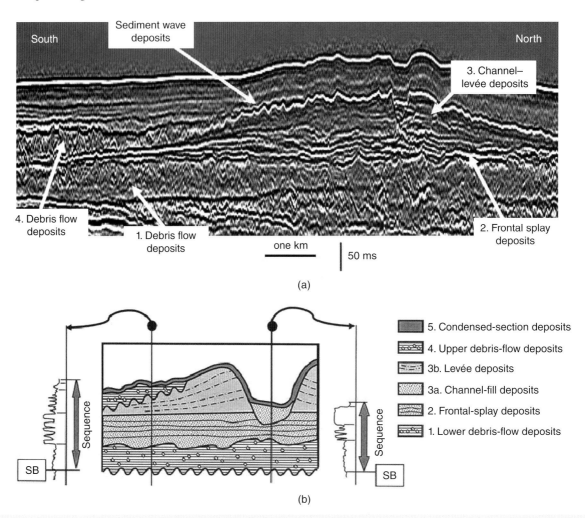

Fig. 4.10   (a) Seismic reflection profile from offshore Indonesia illustrating the stratigraphic succession of a deep-water sequence. Cohesive-flow deposits (1) are overlain by frontal-splay deposits (i.e., sandy lobes) (2), channel–levée deposits (3), and again cohesive-flow deposits (4). The entire succession is then inferred to be mantled by a thin veneer of condensed-section deposits (5). (b) Schematic depiction of an idealised deep-water depositional sequence, with two hypothetical log profiles. From Posamentier and Kolla (2003).

of several factors like global climate change, tectonic uplift and subsidence, changes in sediment supply, deep-water current activity and weather (*cf.* Püspöki *et al.* 2009; Sattar *et al.* 2012). As emphasised by Shannon *et al.* (2005), sequence stratigraphic analysis in deep-water settings must recognise and evaluate along-slope, downslope and vertical components of sediment supply. In their study, Shannon *et al.* (2005) looked at examples from the deep-water Cenozoic basins of the northwest European Atlantic passive margin, which contain a set of seismic megasequences bounded by unconformable to conformable surfaces, correlative from the shelf via slope to deep basin environments. The deep-water successions contain regional, locally diachronous, submarine unconformities that were created in response to changes in deep-water current circulation and slope processes, rather than subaerial erosion. These bound aggradational contourite drifts which show upslope accretion against basin margins. The deep-water successions interfinger at two stratigraphic levels with seaward-prograding shelf-slope wedges that record differential tectonic movements, involving coeval uplift and subsidence. The youngest wedges formed in the last 4 Ma and record seaward progradation of the margin by up to 100 km, overprinting high-amplitude variations in sea level and coeval with changes in

deep-ocean currents. Sediment supply by fluvial sources gave way in the Pleistocene to direct glacial sediment supply to the slope. This example by Shannon *et al.* (2005) demonstrates the inherent complexity of controls on deep-water systems and stresses the importance of looking to underlying guiding principles to explain the temporal and spatial distribution of deep-water systems.

A commonly held belief that is widespread in the sequence-stratigraphic literature is that maximum lowstands produce a different type of submarine fan than times of rising sea level; the deposits are referred to as *basin-floor fans* and *slope fans*, respectively (Vail 1987; Posamentier *et al.* 1988; Posamentier *et al.* 1991). Mutti (1985, 1992) interpreted these to consist, respectively, of sand-prone sheets (his Type I system) succeeded by channel–levée complexes (his Type III system), with an intermediate type of deposit (Type II) interpreted to accumulate as lowstand channels are backfilled during the initial rise in base level (Fig. 4.6).

The original concept of basin-floor and slope fans came from studies of modern fans, with a strong input from early ideas concerning the development of the Amazon submarine fan (J.E. Damuth, pers. comm. 1995). Scientific drilling of the Amazon Fan instead clearly shows that distal sand sheets and channel–levée units form at the same time (Shipboard Scientific Party 1995a), with avulsion events promoting the deposition of sand-prone sheet-like deposits that are progressively buried by levées constructed by the same channels that feed sand to the sheets (Fig. 7.10 – HARP model of Flood *et al.* 1991). Hence, the concept of the sequential development of basin-floor fans and slope fans seems to lack merit (Nelson & Damuth 2003).

What does seem clear is that the net delivery rate of terrigenous detritus to submarine fans decreases at highstands. For many fans, this leads to highstand abandonment and deposition of hemipelagic deposits rich in biogenic material (Damuth & Kumar 1975a). These highstand deposits, however, might be very thin (tens of centimetres) compared with hundreds of metres of lowstand SGF deposits arranged into sand-prone and mud-prone packets formed by autocyclic processes (Hiscott *et al.* 1997c).

As a summary of our views on submarine fan growth and sequence stratigraphy, the authors believe that the simple, all-purpose submarine-fan models presented by Mutti and Ricci Lucchi (1972) and Walker (1978), served the geoscience community well during the 1970s and early 1980s but are now understood to be too simple. The research community now appreciates the variability in size, shape, physiography and facies composition of modern submarine fans, and much current research is focused on evaluating the relative importance of the controlling variables such as tectonics and climate change. Classifications of submarine fans based on shape (e.g., radial versus elongate, as in Pickering 1982c; Stow 1985) have proved inadequate because shape is a dependent variable, reflecting the tectonic setting of the basin. In an attempt to integrate the complexity of sediment sources into submarine fan classifications, and to differentiate between fans according to grain-size variability, Reading and Richards (1994) proposed a substantially more elaborate classification (Table 7.1), but even this scheme fails to address the sequence stratigraphic (dynamic) nature of deep-marine clastic systems. Mutti and Normark (1987), and later Pickering *et al.* (1995b), took a different approach, similar to the architectural analysis suggested for fluvial systems (cf. Miall 1986). Mutti and Normark (1987) proposed a hierarchy of depositional scales, and discussed some of the major lithofacies assemblages or 'elements' that constitute deep-water systems (Section 7.4.3). The approaches of Mutti and Normark (1987) and Reading and Richards (1994) represent contrasting (and complementary) attempts to incorporate new data into submarine-fan classification schemes, focusing primarily on the identification and description of morphological components (architectural elements) and on the range of fan sizes and compositions, respectively. These newer schemes attempt to unravel the inherent complexity in deep-marine clastic systems, in order both to better describe and predict depositional patterns, but perhaps more importantly to relate cause and effect, magnitude and frequency of natural processes. As correctly pointed out by Middleton (2003), the architectural-element approach permits an almost limitless variety in the facies composition of submarine fans (see also Fig. 7.49), so that a universally applicable model can never be realised.

4.2.2 California Borderland submarine fans and base-level change

Undoubtedly, lowstand fans are only a component of any sequence stratigraphic understanding of deep-water systems. In a study using an extensive grid of high-resolution and deep-penetration seismic-reflection data, associated with deep-marine canyon–channel systems contributing to the southeast Gulf of Santa Catalina and San Diego Trough since 40 ka, Covault *et al.* (2007) show that a comparable volume of coarse-grained deposits were deposited in the California Borderland deep-water basins regardless of the height of sea level. A regional seismic-reflection horizon (40 ka) was correlated across the study area using radiocarbon age dates from the Mohole borehole and US Geological Survey piston cores. The study focussed on the submarine fans fed by the Oceanside, Carlsbad and La Jolla Canyons, all of which head within the length of the Oceanside littoral cell. The Oceanside canyon–channel system was active from 45–13 ka, and the Carlsbad system was active from 50 ka (or earlier) to 10 ka. The La Jolla System was active over two periods, from 50 ka (or earlier) to 40 ka, and from 13 ka to the present. One or more of these canyon–channel systems has been active regardless of the sea level. During sea-level fluctuations, shelf width between the canyon head and the littoral zone is inferred to be the primary control on canyon–channel-system activity. Highstand submarine fan deposition occurs when a majority of the sediment within the Oceanside littoral cell is intercepted by one of the canyon heads, currently La Jolla Canyon. Since 40 ka, the computed sedimentation rate on the La Jolla highstand fan has been >2 times the combined rates on the Oceanside and Carlsbad lowstand fans. Figure 4.11 shows the model proposed by Covault *et al.* (2007). These California Borderland fans are mainly fed by longshore currents feeding sediment directly into canyon heads very close to the shoreline. The strength of longshore currents and, therefore, their importance as a sediment delivery process to deep water, depends upon the wind-shear across the basin, or fetch, and also the physiography of the basin and the orientation, width and bathymetry of its connection to the open ocean. Thus, in open ocean-facing continental margins or restricted basins with strong ocean gyres, such as the Gulf of Mexico, longshore currents may be responsible for highstand fans. However, in many ancient deep-water systems this may not be the case, and glacio-eustasy may still be the dominant influence on the volume of sediment (and its calibre) reaching deep water.

A contrasting view of coarse-grained sediment supply to small siliciclastic deep-marine systems offshore from southern California (Fig. 4.12) was presented by Normark *et al.* (2006), who tracked sediment from river source through the turbidity-current initiation process to ultimate deposition, and evaluated the impact of changing sea level and tectonics. The Santa Monica Basin is almost a closed system for terrigenous sediment input, and is supplied principally from the Santa Clara River. Hueneme Fan is supplied directly by the river, whereas the smaller Mugu and Dume fans are fed by southward longshore drift. To map the Upper Quaternary fill of the Santa Monica Basin, they used a dense grid of high-resolution Huntec deep-tow seismic-reflection profiles tied to radiocarbon ages for ODP Site 1015 back to 32 ka. Over the last glacial cycle, sedimentation rates in the distal part of Santa Monica Basin averaged 2–3 mm yr^{-1}, with increases at times of extreme relative sea-level lowstand. Coarser-grained mid-fan lobes prograded into the basin from the Hueneme, Mugu and Dume Fans at times of rapid sea-level fall. These pulses of coarse-grained sediment resulted from river channel incision and delta cannibalisation. During the extreme lowstand of the Last Glacial Maximum, sediment delivery was concentrated on Hueneme Fan, with mean depositional rates of up to 13 mm yr^{-1} on the middle and upper fan. During the marine isotope stage (MIS) 2 transgression, enhanced rates of sedimentation of >4 mm yr^{-1} occurred on the Mugu and Dume Fans, as a result of distributary switching and southward littoral drift providing sediment to these submarine fan systems. Longer-term sediment delivery to Santa Monica Basin was controlled by tectonics, so that prior to MIS 10 the

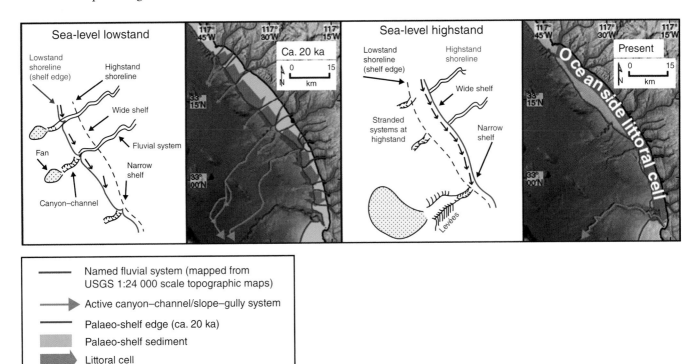

Fig. 4.11 Summary diagrams and palaeogeographic reconstructions illustrating the influence of shelf width between canyon heads and the littoral zone on canyon–channel system activity. From Covault *et al.* (2007).

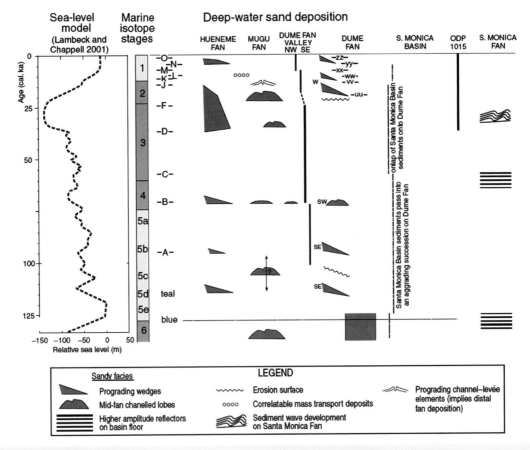

Fig. 4.12 Age model for key reflectors showing timing of principal coarse-sediment supply to the Hueneme, Mugu and Dume fans. Sea-level curve from Lambeck and Chappell (2001). From Normark *et al.* (2006).

Anacapa Ridge blocked the southward discharge of the Santa Clara River into the Santa Monica Basin.

From their study of the deep-marine systems in Santa Monica Basin, Normark *et al.* (2006) concluded that the style and distribution of SGF sedimentation is a consequence of two main factors that are both controlled by sea level: the rate of supply of coarse-grained sediment (greatest during incision and cannibalisation) and the style of initiation of SGFs, influencing whether sand is deposited principally on the middle fan or transported to the basin plain. These two factors appear more important than the absolute position of sea level.

4.2.3 Recent studies of ancient submarine fans and inferred base-level changes

A study, based on multiple well logs, of the sequence stratigraphy of the Lance–Fox Hills–Lewis shelf margin in southern Wyoming (Figs 4.13a,b) suggests that high rates of sediment supply were critical in causing progradation of this moderately wide Maastrichtian shelf margin, at a minimum rate of 47.8×10^{-6} km yr^{-1}, and the generation of large, sand-rich submarine fans during every shoreline regression across the shelf (Carvajal & Steel 2006). In this system, the submarine fans developed from shelf-margin clinoforms that show systematically rising shelf-edge trajectories (a proxy for rising relative sea level), as well as from those that show flat trajectories (a proxy for stable to falling relative sea level). Carvajal and Steel (2006) argue that the latter, producing more sediment bypass, resulted in bigger and thicker submarine fans, whereas the former produced somewhat smaller and thinner fans. They termed the former 'highstand fans' and suggested caution be used in using the lowstand model for systems dominated by high rates of sediment supply (sediment flux). Carvajal and Steel (2006) describe at least 15 deltaic regressive shelf transits during a total time interval of <~1.8 Myr, and argue that many of these shelf transits occurred while relative sea level was rising, that is, that the basinal siliciclastic systems represent highstand deposits. However, it could be argued that the deep-water systems might in fact have accumulated during interludes of falling sea level within a time period where sea level was relatively high (e.g., ~21 kyr precession cycles within larger scale ~100 kyr eccentricity cycles). Thus, whilst the submarine fans appear to have evolved during an overall, longer-term rising sea level, the case remains unproven that these are 'highstand' fans *sensu stricto*.

In conclusion, attempts to incorporate growth stages of submarine fans into the various systems tracts of sequence-stratigraphic models have been premature and have resulted in deeply rooted misconceptions (Nelson & Damuth 2003). Instead, it should be realised that base-level variation is only one of several controls on the facies successions of deep-water systems. In general, low base level favours the accumulation of thick SGF successions, but clustering of sand-prone and mud-prone deposits on the scale of <20–50 m is more likely the result of channel switching than a response to subtle and rapid base-level variations.

Sea-level shifts from the innermost shelf out to the shelf edge may produce bayhead, inner-shelf, mid-shelf and shelf-margin deltas. Porbski and Steel (2006) suggest that these delta types are distinguishable in the ancient record and that such distinction has advantages as compared to the conventional, entirely process-based classification of deltas. Bayhead and inner-shelf deltas tend to form thin clinoforms (metres to 10s of metres amplitude, respectively), and as they aggrade with rising relative base level they generate a 'tail' of thick paralic deposits. Mid-shelf deltas produce clinoforms as high as the mid-shelf water depth, tend to follow a subhorizontal

trajectory, generate little or no paralic tail, and are commonly thinned by transgressive ravinement. Shelf-edge deltas contemporaneous with a stable-to-falling relative base level usually have no paralic tail, create by far the highest clinoforms, and can have a thick succession of sandy SGF deposits on the delta fronts. If base level falls below the shelf margin, a shelf-edge delta becomes incised by its own channels and large volumes of sand can be delivered onto the slope and the basin floor.

Porbski and Steel (2006) point out that many deltas require a strong fluvial drive to attain a shelf transit, and as they approach the outer shelf they tend to become wave dominated. Tidal influence can increase on the outermost shelf if relative sea level is falling, if the shelf-break is poorly developed and if basinal water depth is shallow. During transgression, the system tends to be tidally and/or wave influenced. Deltas that transit back and forth on the shelf on short time scales (10–100 kyr) and that are driven largely by sea-level fluctuations are called accommodation-driven deltas by Porbski and Steel (2006). Deltas that can reach the shelf edge without sea-level fall are termed supply-driven deltas. These highstand deltas deposit thick, sandy, stacked parasequences during their shelf transit, and they tend to have an extensive muddy delta front on reaching the shelf-edge area. Conceptually, such deltas are not normally incised at the shelf edge, and they produce a progradational, shelf-edge attached, sandy slope apron (Exxon: *shelf-margin systems tract*) rather than basin-floor fans, except in cases of extremely high supply. Sequence boundaries are best developed on accommodation-driven deltas, and are likely to be represented on a variety of time scales (third, fourth and fifth order of Table 4.1). Sequence boundaries in supply-dominated deltas may be identifiable only at lower-order time scales, or they may be non-existent.

Undoubtedly, we are only just beginning to understand and appreciate the complexity of sedimentary responses to changing global climate at Milankovitch frequencies. The over-simplistic notion of eustatic lowstands always being contemporaneous with increased sediment flux to deep-marine environments and *vice versa* during highstands is becoming increasingly challenged. For example, Ridente *et al.* (2009) used a combination of sequence stratigraphy and multiproxy chronostratigraphic data, including $\delta^{18}O$ records, to unravel a record of Milankovitch cyclicity through the past ~400 ka on the Adriatic margin in water depths up to ~260 m (i.e., shelf and slope). Chronostratigraphic data and stable isotope records from planktonic and benthic foraminifera were obtained from borehole PRAD1-2, a 71 m-long continuous core drilled on the western Adriatic slope. The cored interval includes the distal parts of stacked progradational wedges that are the main building blocks of 100 kyr depositional sequences. These sequences and their internal progradational units reflect a cyclic interplay between eustatic sea level and oceanographic circulation, with linked feedbacks on fluctuating sediment supply and regional long-shore sediment dispersal. Calibration of the age of slope units with isotope stratigraphy, and direct slope-shelf correlation, suggest that two types of progradational clinoforms alternate in phase with the pattern of composite climate and eustatic sea-level changes derived from $\delta^{18}O$ records. Ridente *et al.* (2009) noted that changes in clinoform geometry of the kind observed in the Adriatic are consistent with a mechanism of switching supply from dominant advection on a flooded shelf during highstands, to overall sediment starvation on a narrowed shelf during lowstands. Seismic and core data suggest that this mechanism is related to ~100 and 20 kyr cyclicity reflecting rearrangement of the oceanographic setting and sediment pathways during the last four glacial–interglacial cycles. They further noted that the development

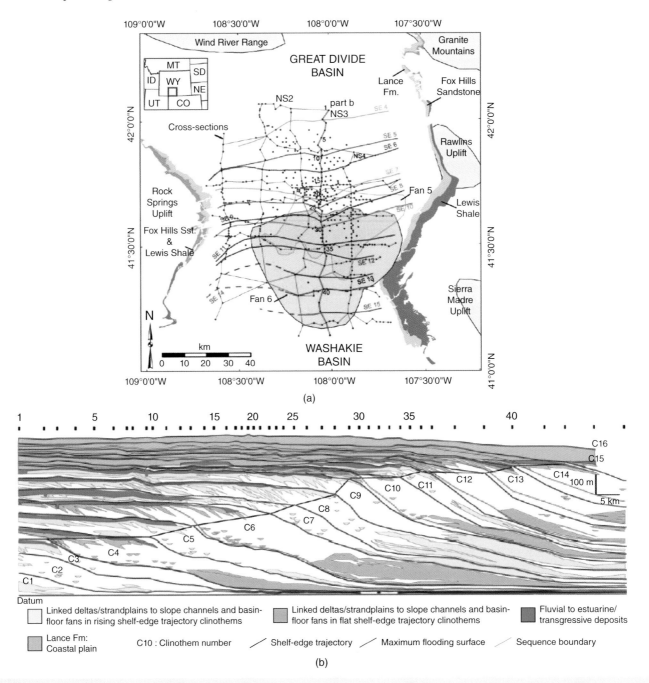

Fig. 4.13 (a) Location of Maastrichtian Lance–Fox Hills–Lewis shelf margin in southern Wyoming, well data base, shelf-edge positions (interpreted for the time of maximum flooding at beginning of each cycle), and two submarine fans. Fan 5 was deposited during a rising shelf-edge trajectory, whereas larger fan 6 was generated during the development of a flat shelf-edge trajectory. (b) NS3 cross-section (orange line with well number labels in (a) showing clinothems 1–16 from shelf to slope, to basin floor with considerable vertical exaggeration). Note that submarine fan maximum thickness does not necessarily coincide with this cross section, because fan depocentres shifted through time. Trajectory quantification was done using NS2, NS3 and NS4 cross-sections. From Carvajal and Steel (2006).

of depositional sequences on the Adriatic margin is largely dominated by eustatic sea-level changes. Conventional sequence stratigraphic models, however, do not predict the considerable thicknesses of the HSTs and the seaward thinning of the LSTs. Classic sequence stratigraphy also holds that accommodation exerts the main control on sequence architecture, that the filling of the accommodation occurs from land, and that deposition shifts seaward across the shelf-slope profile only after all the more landward accommodation has been progressively filled. This Adriatic-margin example, however, accounts for rapid and significant changes in sediment flux driven by changes in the hydrological–oceanographic system, and therefore emphasises the importance of sediment bypass and redistribution in shaping sequences and margin architecture during short-term eustatic sea-level change.

4.3 Tectono-thermal/glacio-eustatic controls at evolving passive continental margins

Present-day passive continental margins (Chapter 9) provide an ideal natural laboratory in which to appreciate both the magnitude and amplitude of changing relative base level over a complete range of temporal scales. The Brazilian continental margin (Fig. 4.14), with its current economic importance as a major hydrocarbon province, provides a good example.

At evolving passive continental margins, first- and second-order cycles (Table 4.1) are generated by isostatically driven subsidence due to thinning of the crust and cooling of the continental lithosphere and its replacement by higher density mantle. Since extension and thinning tend to involve the entire lithosphere (typically ~125 km), and not just the uppermost 30 km (i.e., the crust), higher density lithospheric mantle is displaced by lower density asthenosphere during stretching, partially cancelling out the subsidence due to crustal thinning. The thermal recovery time of the lithosphere is long compared to most rifting events (100 Myr versus 10 Myr); therefore, isotherms within the stretched lithosphere are squeezed together and syn-rift basins experience anomalously high heat flow. Following rifting, thermal gradients slowly return to normal and a post-rift section as thick as, or thicker than, the syn-rift succession is deposited. In many passive margins, the cessation of the last phase of active extension and the onset of thermal subsidence essentially coincide with the initiation of sea-floor spreading. Apparent flexural rigidity is time

dependent, with the lithosphere appearing more rigid when rapidly loaded and relaxing over a period of several million years. It is these plate-tectonic processes that create the overall transgressive event(s) that are associated with evolving passive margins. Over the full period of development of the passive margin, there is a substantial deepening from terrestrial (including lacustrine) to marine deep-water basinal sediments. Within the Brazilian margin, these events are characterised by, from oldest to youngest, the: (i) continental pre-rift megasequence; (ii) continental rift megasequence; (iii) transitional evaporitic megasequence; (iv) shallow carbonate platform megasequence and (v) marine transgressive megasequence (Fig. 4.14; Bruhn 1998). Once seafloor spreading and the creation of oceanic crust is underway, the rate of sediment supply eventually exceeds the rate of creation of accommodation space so that shelf-slope progradation occurs to create the marine regressive megasequence (Fig. 4.14). Higher frequency, smaller amplitude, changes in relative base level serve to generate third- and higher-order cycles.

At Neogene passive continental margins, third- and higher-order cycles can readily be explained by glacio-eustatic processes, for example as documented along the northern Gulf of Mexico (Fig. 4.15) (Prather *et al.* 1998).

An instructive example of how changing relative base level conspired to determine the overall depositional (and erosional) architecture of a continental margin can be found in the results of IODP Expedition 318 – Wilkes Land glacial history and the Cenozoic East Antarctic Ice Sheet evolution from Wilkes Land margin sediments, Antarctica (Figs 4.16, 4.17a, 4.17b, 4.18) (Expedition 318 Scientists

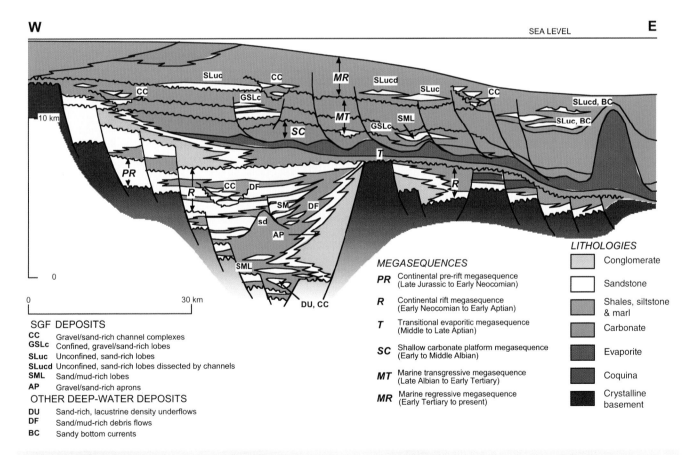

Fig. 4.14 Simplified stratigraphy of the Brazilian continental margin to show the six depositional megasequences and their duration. Redrawn From Bruhn (1998).

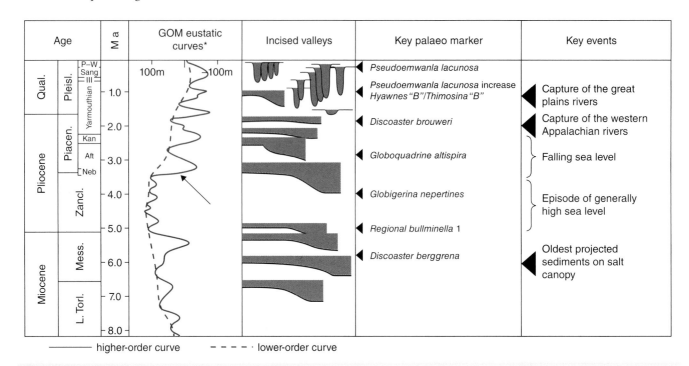

Fig. 4.15 The Neogene Gulf of Mexico (GOM) mini-basins show a first-order control by glacio-eustatic sea-level variations. Correlation chart for the GOM Neogene showing eustatic curves, times of shelf incision (from Prather *et al.* 1998). Asymmetrical shaped symbols in the 'Incised Valleys' column represent large canyons and collapsed shelf edges, and symmetrical symbols represent smaller shelf-edge gorge features (both from mapping at Shell Offshore and Shell E & P Technology). The eustatic sea-level curve is modelled from Haq, Hardenbol and Vail (1987) and Styzen *et al.* (1994). The transgressive–regressive cycles and biostratigraphic zones from the GOM are plotted against the absolute ages for the stage and epoch boundaries of Harland *et al.* (1990).

2010). Base level was controlled by a combination of tectonics, climate change and eustatic sea level. Expedition 318 occupied seven sites (Fig. 4.16) in water depths of ~400 and 4000 mbsl (metres below sea level). Figure 4.18 summarises the tectono-stratigraphic history of the Wilkes Land Antarctic margin since 53 Ma (Expedition 318 Scientists 2010). These cartoons show the tectonic history of the so-called Australo–Antarctic Gulf (at 53 Ma), the onset of the second phase of rifting between Australia and Antarctica (Close *et al.* 2009), the continuously subsiding margins and deepening, to create the present ocean/continent configuration. Tectonic and climatic change turned the initially shallow, broad subtropical Antarctic Wilkes Land offshore shelf into a deeply subsided basin with a narrow ice-infested margin (Fig. 4.18) with thick Oligocene and notably Neogene deposits, including SGFs, contourites and MTDs. Expedition 318 Scientists (2010) interpreted the results as suggesting an ice-free ('Greenhouse') Antarctica until ~33 Ma. An important aspect of this study is that the stratigraphic context of contourite drifts is placed within the evolution of a passive continental margin.

4.4 Eustatic sea-level changes at active plate margins

Glacio-eustasy has been widely inferred, and in many cases shown, to be a primary control on sedimentation in the open ocean and along passive continental margins (Mitchum 1977; Mitchum *et al.* 1977a, b; Vail *et al.* 1977a, b; Pitman 1978; Haq *et al.* 1987, 1988; Van Wagoner 1987, 1990; Posamentier 1988; Posamentier & Vail 1988a; Vail & Posamentier 1988; Greenlee & Moore 1988). More recently, the role of glacio-eustacy has been documented for siliciclastic-dominated deep-marine successions at active convergent plate margins, including foreland basins (Pickering *et al.* 1999; Plint 2009).

In order to test the relative importance of glacio-eustasy at tectonically active plate margins during times of substantial polar ice, a detailed study of the (~1180–600 ka) Pliocene–Pleistocene deep-marine parts of the Kazusa Group (Fig. 4.19), a forearc basin fill, onland southeast Japan, was undertaken by Pickering *et al.* (1999). A high-resolution $\delta^{18}O$ and $\delta^{13}C$ record from planktonic foraminifera (*Globorotalia inflata*) was obtained. This was combined with a high-resolution study of the magnetic susceptibility, total organic carbon (TOC) and %$CaCO_3$ in order to evaluate the sedimentary response to any glacio-eustatic changes (Fig. 4.20). The chemical and magnetic data reveal globally recognised glacial–interglacial cycles, with sandy intervals correlating with inferred glacials, suggesting that relative sea-level changes during glacial–interglacial cycles exerted the primary control on sediment accumulation in fifth-order cycles in the deep-marine forearc basin. Each sandy interval is ~100–150 m thick (Fig. 4.20) and consists of several sandstone packets alternating with more shaly intervals. Cross-spectral analysis of $\delta^{18}O$ and $\delta^{13}C$ data from hemipelagic and pelagic mudstones reveals a Milankovitch control both at the precession and eccentricity modes, with a shift in their relative importance at ~900 ka (the middle Pleistocene Revolution, MPR) (Fig. 4.21). This study shows that at times when there was substantial polar ice, the main control on sediment accumulation at active plate margins can be glacio-eustatic, and supports the sequence-stratigraphic paradigm developed for passive continental margins that high-frequency global sea-level changes can exert a primary control on deep-marine siliciclastic deposition, but at a scale

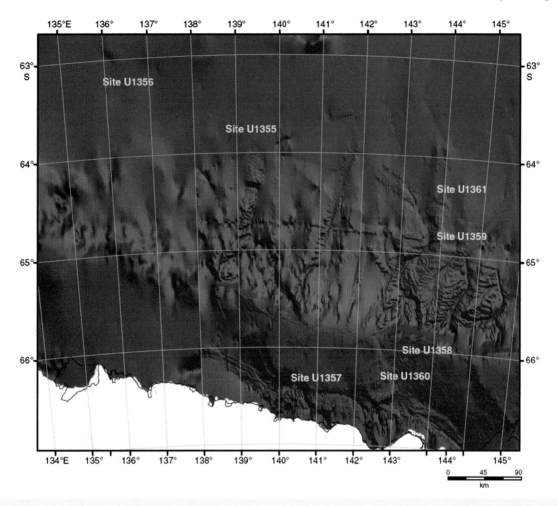

Fig. 4.16 Detailed bathymetry for part of the area drilled during IODP Expedition 318. The irregular morphology of the continental shelf is characterised by several >1000 m-deep inner-shelf basins (Site U1357), erosional troughs extending from these basins to the shelf edge (Site U1358), and shallow banks adjacent to the troughs (Site U1360). The slope is incised by many canyons that on the continental rise evolve to channel–levée complexes (Sites U1356 (not shown), U1359 and U1361). From Expedition 318 Scientists (2010).

(here ~100–150 m; Fig. 4.20) greater than the scale of individual sandstone packets.

The Neogene Niigata-Shin'etsu Basin in the Northern Fossa Magna, a backarc setting of the Japan arc, provides a contrasting example of the interplay between tectonics, climate and sedimentation in deep-marine systems at active plate margins. The basin was created as a rift basin during the Middle Miocene and converted to a compressional basin in latest Late Miocene time due to basin inversion (Takano 2002). The basin was filled with a thick succession of submarine-fan deposits. The various fan types which evolved are related to the basin tectonics. During the post-rift phase, essentially radial sandy fans predominated because there was no distinct topographical control, whereas confined trough-fill deep-marine systems characterised the basin-inversion and compressional phases when syn-depositional folding restricted the spatial distribution of deep-water clastic systems (Takano 2002). In the Pliocene–Pleistocene, submarine-fan systems apparently tended to develop mainly during relative highstands in sea level. The Pliocene to Lower Pleistocene sediments of the basin are divided into two third-order depositional sequences, Kkb-III-A and Kkb-III-B, in ascending order, which formed in response to relative sea-level changes (Takano 2002). An assessment of the temporal and spatial

distribution of the depositional systems appears to show no distinct differences in sedimentation patterns of the submarine fans during LST (Lowstand Systems Tracts), TST (Transgressive Systems Tracts) and HST (Highstand Systems Tracts) in third-order sequences. Estimates of the sedimentation rates also support the interpretation that these submarine-fan systems tended to develop predominantly in the late stage of TSTs and the early stage of HSTs. Although submarine fans also developed during lowstands, the TST and HST fans tended to be larger and coarser grained than those associated with the lowstands. This resulted from particular climatic conditions in the Japan Sea at this time. During relative highstands of sea level, a warm ocean current flowed into the Japan Sea through the Tsushima Strait, resulting in the warming of seawater. A dry and cold monsoon from the northwestern continent induced a large amount of cloud formation due to evaporation from the warm seawater, causing enhanced precipitation. The increased runoff and sediment discharge generated an increased flux of coarse detritus into the basin. In contrast, during relative lowstands of sea level, a warm ocean current could not flow into the Japan Sea because the Tsushima Strait became shallow or subaerially exposed, resulting in the cooling of seawater. These conditions caused dry weather because of the substantially reduced evaporation, and the sediment-supply rate correspondingly

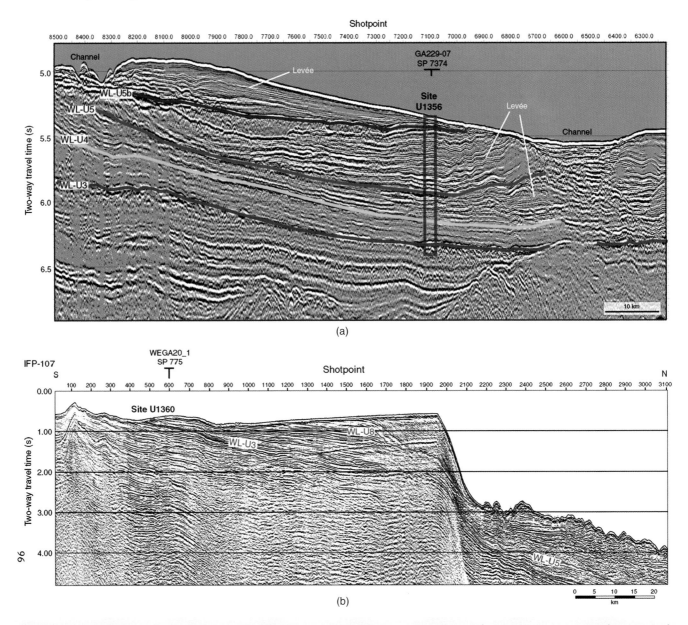

Fig. 4.17 (a) Multichannel seismic reflection profile across Site U1356 showing regional unconformities L-U3, WL-U4 and WLU5. Red rectangle = approximate penetration achieved at Site U1356. (b) Multichannel seismic reflection Profile IFP 107 across Site U1360. Profile shows the main regional unconformities defined in the Wilkes Land continental shelf. Drilling at Site U1360 targeted unconformity WL-U3. Red rectangle = approximate penetration achieved at Site U1360. From Expedition 318 Scientists (2010).

decreased. Additionally, it is possible that phases of tectonic uplift in the hinterland might have been coincident with the highstand stages, resulting in high sediment-supply potential during these highstand phases. Since the basin originated as a rift basin, there was insufficient shelf area for sediment accumulation at the basin margin, so changes in rates of sediment delivery from the provenance area influenced the basin-floor sedimentation more strongly than the effects of relative base-level changes.

IODP Expedition 317 was devoted to understanding the relative importance of global sea level (eustasy) versus local tectonic and sedimentary processes in controlling continental margin sedimentary cycles in the Canterbury Basin, on the eastern margin of the South Island of New Zealand (Expedition 317 Scientists 2010). Despite its

proximity to a major plate boundary, represented by the Alpine Fault, the onshore Canterbury Basin has been an area of relative tectonic stability since Late Cretaceous rifting (Lu *et al.* 2003); thus, it is perhaps best considered a passive margin but with a major strike-slip plate boundary on its western side. ODP Expedition 317 recovered Eocene to Recent sediments, including a late Miocene to Recent record when global sea-level change was dominated by glacio-eustasy. The high rates of Neogene sediment supply and accumulation resulted in the preservation of a high-frequency (0.1–0.5 Myr) record of depositional cyclicity. Additionally, the Canterbury Basin is adjacent to an uplifting mountain chain, the Southern Alps, and strong ocean currents. Currents have locally built large, elongate sediment drifts within the prograding Neogene section. Although Expedition 317

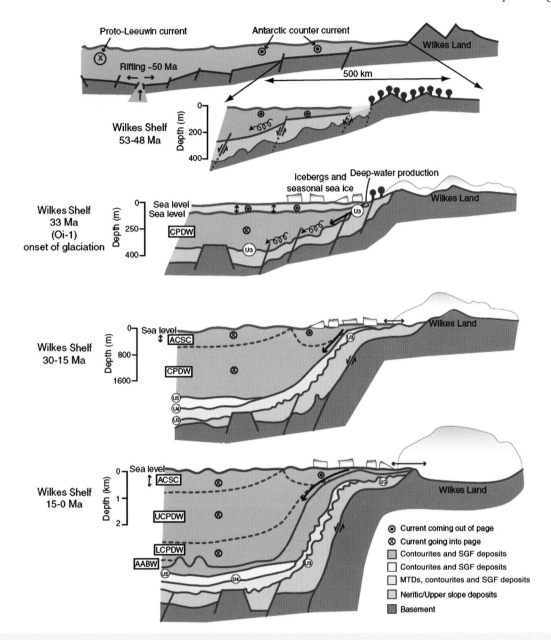

Fig. 4.18 Conceptual illustration of tectonic, geological, sedimentological and climatic evolution of the Wilkes Land margin since the middle early Eocene (~54 Ma). U3, U4 and U5 refer to seismic unconformities WL-U3, WL-U4 and WL-U5. CPDW = Circumpolar Deep-water, ACSC = Antarctic Circumpolar Surface Water, UCPDW = Upper Circumpolar Deep-water, LCPDW = Lower Circumpolar Deep-water, AABW = Antarctic Bottom Water. From Expedition 318 Scientists (2010). Note that this interpretation of the stratigraphy favours greenhouse conditions until ~33 Ma.

did not drill into one of these elongate drifts, currents are inferred to have strongly influenced deposition across the basin, including in locations lacking prominent mounded drifts. Upper Miocene to Recent sedimentary successions were cored in a transect of three sites on the continental shelf (landward to basinward, Sites U1353, U1354 and U1351) and one on the continental slope (Site U1352). Continental slope Site U1352 represents a complete section from modern slope terrigenous sediment to hard Eocene limestone, with all the intervening lithologic, biostratigraphic, physical, geochemical and microbiological transitions. The site also provides a record of ocean circulation and fronts since ~35 Ma. The early Oligocene

(~30 Ma) 'Marshall Paraconformity' was the deepest drilling target of Expedition 317 and is hypothesised to represent intensified current erosion or non-deposition associated with the initiation of thermohaline circulation following the separation of Australian and Antarctica (*cf.* Lu *et al.* 2003; Lu & Fulthorpe 2004).

Whilst many studies of relatively coarse-grained sediments in modern and ancient deep-marine canyon-fan systems have identified a strong eustatic/climatic (including glacio-eustatic) control (e.g., Hiscott 2001; Babonneau *et al.* 2002; Barker & Camerlenghi 2002; Anka *et al.* 2004; Ridente *et al.* 2009; Armitage *et al.* 2010; Backert *et al.* 2010; Bourget *et al.* 2010), others have emphasised regional and/or

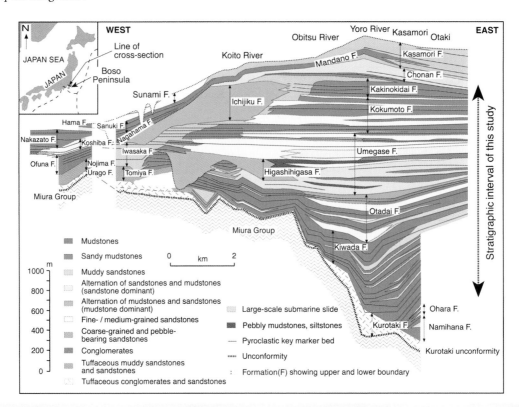

Fig. 4.19 Location map and stratigraphy of the Kazusa Group, southeast Japan. Inset map indicates Boso Peninsula and line of stratigraphic cross-section; also, dashed lines show plate boundaries defined by subduction zones. From Pickering *et al.* (1999).

local tectonics as the dominant driver (e.g., Nelson *et al.* 1999; Babic & Zupanic 2008; Anka *et al.* 2009; Winsemann & Seyfried 2009; Athmer *et al.* 2011). Laminated fine-grained sediments, such as organic-rich sediments and biogenic opal, are generally interpreted as recording a strong climatic signal (e.g., Hillenbrand & Fütterer 2001; Lyle & Wilson 2006; Kroon *et al.* 2007; Mosher *et al.* 2007).

The Permian deep-water succession of the Karoo Basin, South Africa, accumulated during an ice-house period in Earth history in a basin that has been variously interpreted as a backarc basin, thermal-sag or foreland basin (e.g., Tankard *et al.* 2009). Tankard *et al.* (2009) divided the subsidence history of the Karoo Basin into 'pre-foreland (thermal-sag) phase' (Dwyka, Ecca and Lower Beaufort Groups) and a 'backarc foreland-basin phase' as a response to the uplift of the Cape Fold Belt during the Early Triassic (Upper Beaufort Group); an interpretation supported by sediment provenance analyses showing that the Cape Fold Belt was not a sedimentary source during deposition of the Ecca Group (Johnson 1991; Andersson *et al.* 2004; Van Lente 2004). The Upper Ecca Group accumulated during the thermal-sag phase in a marine continental interior basin with palaeoenvironmental conditions that evolved from deep-marine starved to shallow-marine environments (Flint *et al.* 2011). The Ecca Group in the Laingsburg depocentre (the southern and eastern part of the Karoo Basin) shows an ~1300 m-thick succession that records the progradation of a marine basin margin with deposits ranging from deep-water basin plain (Vischkuil and Laingsburg Formations) through submarine slope (Fort Brown Formation) to shallow-marine deposits (Waterford Formation) (Fig. 4.22). Flint *et al.* (2011) placed this succession in a sequence-stratigraphic framework that comprises

three major cycles (their *composite sequence sets*), made up of *composite sequences* and *sequence sets*, which are composed of *elementary depositional sequences*. Each depositional sequence comprises one sandstone/siltstone-prone unit overlain by a regionally mapped mudstone-prone unit (Fig. 4.22). They interpreted the sandstone/siltstone-prone units as lowstand systems tract deposits, and the mudstone-prone units as the transgressive and highstand systems tract deposits (Flint *et al.* 2011).

Figueiredo *et al.* (2010) document the stratigraphy of a mudstone-dominated submarine-slope succession from the Karoo Basin (Fig. 4.22), and interpret the organised stratigraphic stacking as likely caused by glacio-eustasy during the Late Permian icehouse period. The stratigraphy consists of a 470 m-thick, claystone-dominated succession containing five sandstone-prone units (Units D/E, E, F, G and H). Units D/E to Unit F show an overall trend of thickening upward and basinward stepping. This stacking pattern is reversed from the top of Unit F to the base of Unit H, above which basinward stepping is again observed. Different architectural styles of sandstone-prone deposits occupy predictable stratigraphic positions within the basinward-stepping section, starting with what are interpreted to be intraslope lobes through channel–levée complexes to entrenched slope valleys. Sandstone percentage is highest in the intraslope lobes and lowest in the slope valley fills, reflecting a change from depositional to bypass processes. The landward-stepping stratigraphy is dominated by claystone units with thin distal fringes of distributive deposits. The upper basinward-stepping succession (Unit H) is interpreted as a distributive system possibly linked to a shelf-edge delta. Across-strike complexity in the distribution of sandstone-prone units

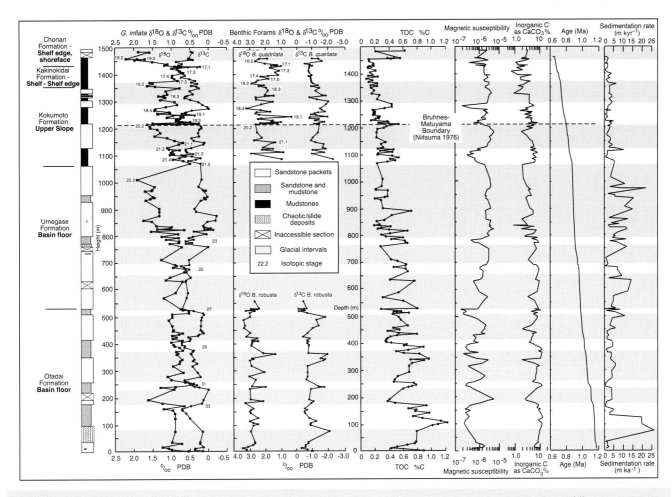

Fig. 4.20 Lithology, $\delta^{18}O$, $\delta^{13}C$, TOC, magnetic susceptibility and inorganic carbon as percent carbonate for the middle Kazusa Group, Boso Peninsula, southeast Japan. In the stratigraphic interval where the legend has been placed, there is a change in benthic foraminifera species and the abundance is low within the very sandy sections. Numbers adjacent to the $\delta^{18}O$ curve refer to marine isotope stages. The position of the boundaries between glacials (shown in blue) and inter-glacials is based primarily upon the isotopic data, but modified to respect the results from the other proxy data (TOC, magnetic susceptibility and inorganic carbon as %$CaCO_3$). From Pickering *et al.* (1999).

is interpreted to have been controlled by cross-slope topography driven by differential compaction processes. Hemipelagic claystones separating the sandstone-prone units represent shutdown of sand delivery to the whole slope and are interpreted to have formed during relative sea-level highstands. Eleven depositional sequences are identified, nine of which are arranged into three composite sequences (Units E, F and G) that together form a composite sequence set (Fig. 4.22).

The Laingsburg Formation in the Karoo Basin (Fig. 4.22), includes a series of sandstone-dominated deep-water clastic deposits (Units A and B of the Laingsburg Formation and Unit C to Unit G of the overlying Fort Brown Formation) separated vertically by regional mudstones. The succession records basin-floor to upper-slope deposition during the Permian icehouse climate (Di Celma *et al.* 2011). Unit C provides nearly continuous exposures over tens of kilometres, and the presence of regionally persistent mudstone markers (lower and upper C mudstones) allows the distribution of sedimentary facies and architectural elements of a deep-water slope system to be documented for more than 30 km downslope and 20 km across-slope. The spatial and temporal distribution of architectural elements and interpreted

depositional environments (external levées, channel belts confined by a combination of basal erosion and overbank aggradation along the margins, and distributive frontal splays) reveals distinct changes in the sedimentation pattern and stratal architecture of the deep-water system through time. Unit C evolved in a stepwise manner from a weakly incised, levée-confined channel belt and its downdip distributive frontal splays (C1), through a more entrenched and sinuous set of channel–levée complexes that fed submarine fans farther into the basin (C2), to a regionally back-stepping package of thin-bedded deposits of a distal distributive system (C3). Unit C is interpreted as a lowstand sequence set, composed of three depositional sequences, each of which includes a sandstone-dominated lowstand systems tract (C1, C2 and C3) and a mudstone-prone interval that is taken to represent the transgressive and highstand systems tracts. The overlying combined transgressive and highstand sequence set is marked by the 30 m-thick C–D mudstone interposed between Unit C and Unit D. Unit C and the C-D mudstone together form the Unit C composite sequence. At the scale of the Unit C lowstand sequence set, the evolutionary trend from C1 to C3 reflects an overall basinward then landward stepping of the system across the depositional profile, which

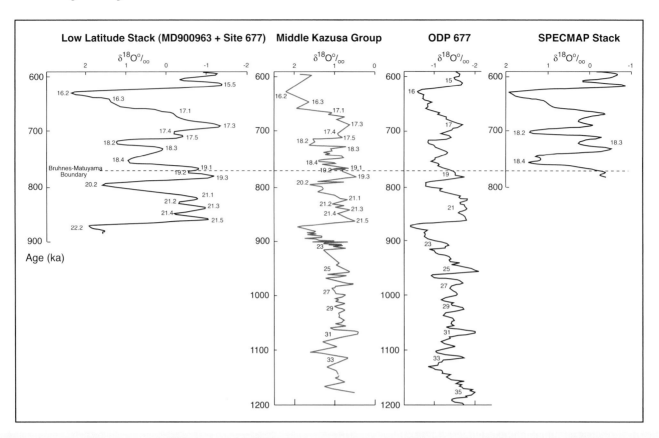

Fig. 4.21   Low Latitude Stack (MD900963 + Site 677), ODP Site 677 and SPECMAP Stack (Imbrie & Imbrie 1980; Imbrie *et al.* 1984; Bassinot *et al.* 1994; Shackleton *et al.* 1990) correlation with Kazusa Group results, shown in red. Even-numbered stages and sub-stages represent glacial and/or colder (stadial) intervals and odd-numbered stages interglacial and/or warmer (inter-stadial) intervals. Nannofossil datum levels (Mita 1993 pers. comm.) and palaeomagnetic boundaries (Okada 1995 pers. comm.) in the middle Kazusa Group are placed on the revised chronology used by Pickering *et al.* (1999).

is interpreted as the product of an overall waxing then waning of flow energy (volume and efficiency).

4.5   Changing relative base level and sediment delivery processes

An important, but unresolved, uncertainty in the application of sequence stratigraphic principles to understanding the linkage between terrestrial, shallow- and deep-water systems is the identification of the operative sediment delivery process(es). What criteria can be used by the geologist in the field? Intuitively, it seems reasonable to expect that each change in base level will change the relative importance of candidate sediment-delivery processes (e.g., storm-induced failures or off-shelf advection, tidal currents, longshore drift, direct river input, various slope failure processes), as well as the sediment flux.

In general, a relative fall in base level (e.g., a fall in sea level) leads to early lowstand systems associated with increasingly active gullying of a basin slope and coarse sediment delivery to deep-water basins, including accelerated erosion and sediment bypass within submarine canyons, for example in the western Gulf of Lion (Baztan *et al.* 2005). During lowstands, storm wave base shifts towards the shelf break, resulting in a greater propensity for cyclic storm pounding both to

cause sediment failure, and suspend and advect sediment basinwards. A reduction in accommodation space on the shelf, and the associated enhanced sediment delivery rates to the shelf edge, tend to cause the active development of shelf-edge deltas (Muto & Steel 2002; Dixon *et al.* 2012) which, in turn, funnel coarse clastics onto the slope and into deep water. As base level continues to fall, shelf-edge deltas are increasingly able to provide a direct transport link between the fluvial and deep-water systems. Such conditions favour riverine input during flood events as (probably fine-grained) hyperpycnal flows, along with other SGFs (Normark & Piper 1991; Plink-Björklund & Steel 2004). Also, during falls in relative base level, there is a seaward shift in the so-called 'avulsion nodes' within channelised systems in deep water (i.e., sites where channels have a tendency to abruptly shift laterally). Slope failure events, whether slumps, slides or cohesive flows, become more significant with falling relative base level, because the shelf edge becomes less stable under the enhanced influence of storms and greater sediment flux. Also, any relative fall in base level will reduce the hydrostatic pressure and can lead to an increase in the dissociation of gas hydrates that further reduce the critical shear stress required to mobilise slope sediments (*cf.* arguments that link greenhouse conditions with the catastrophic release of gas hydrates from continental margins, e.g., MacDonald 1990; Mascarelli 2009; Maslin *et al.* 2010, and references therein).

In any depositional system, it is commonly difficult to constrain the actual triggering mechanism for re-sedimentation processes;

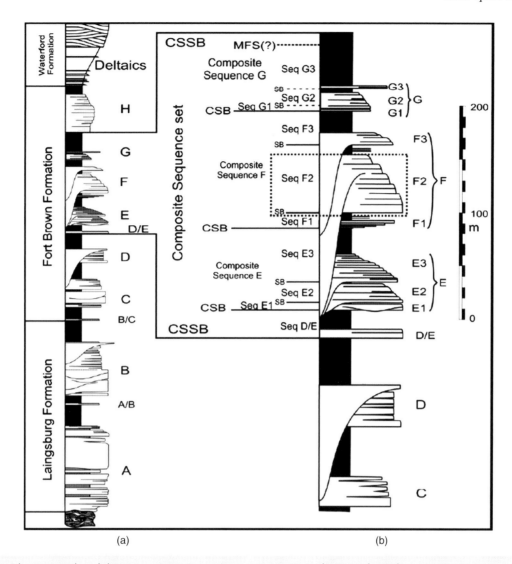

Fig. 4.22 (a) Lithostratigraphy of the Upper Ecca Group (Upper Permian) in the Laingsburg depocentre, Karoo Basin, South Africa. (b) Schematic lithostratigraphy and hierarchical sequence stratigraphy of the uppermost so-called 'Composite Sequence Set'. From Figueiredo *et al.* (2013).

for example earthquake triggering, extreme flood events (including hyperpycnal flow), or other processes (*cf.* discussion of triggers in Carter *et al.* 2012, 2014; Collela *et al.* 2013). Criteria for differentiating the deposits emplaced by these processes, however, are now being developed. A case study from Lake Tahoe, western USA, where a variety of re-sedimentation processes are now known to have contributed to the stratigraphy of the deep lake sediments, illustrates a potentially useful methodological approach. Lake Tahoe is 34 km from north to south and reaches depths of 505 m. It is a tectonically active basin on the edge of the US Basin-and-Range province, and has a relatively flat basin floor with locally steep basin slopes, including vertical scarps up to 40 m high. Debris aprons occur at the base of the lake margins on the western, southern and eastern sides (Fig. 4.23). These debris aprons have been interpreted as the terminal moraines of the ~160 ka Tahoe Glaciation, generated on the western and southern sides of the lake, and that are now found on the floor of the lake (Gardner *et al.* 2000). The McKinney Bay debris avalanche is inferred to have been sourced from the western margin of Lake Tahoe as a

three-stage process: (i) an initial stage when large blocks were carried out into the basin; (ii) flow reflection against the eastern margin and back towards the west, north and south and (iii) burial of the debris avalanche deposit by post-failure sedimentation (Gardner *et al.* 2000). Earlier interpretations of these debris aprons include a failure origin related to glacial-age slumping of ice-cemented lake-margin sediment (Hyne 1969; Hyne *et al.* 1972, 1973) with the mass wasting of the western margin caused by a collapse of ice dams near the outlet of Lake Tahoe that catastrophically lowered the lake level (Birkeland 1964).

An extensive programme of seismic work and coring by the US Geological Survey has led to an improved understanding of the sediment delivery processes to the deep lake floor. Cores LT99-9 (south) and LT99-4 (north) recovered a record dating back ~7.5 cal ka in cores ~1.5 m long. Whilst it might seem intuitive that such a tectonically active region would generate most turbidity currents by seismic events, this turns out not to be the case, at least since 7.5 cal ka (Figs 4.24, 4.25). Instead, most SGF deposits in these

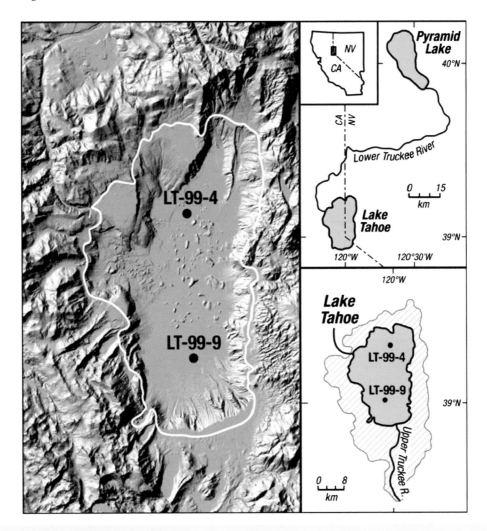

Fig. 4.23 USGS digital elevation model showing Lake Tahoe bathymetry, watershed relief and location of the two cores studied by Osleger *et al.* (2009). Outline of lake shoreline shown in white. Note the large blocks of redeposited material on the lake floor – the McKinney Bay debris avalanche. Upper right: Lake Tahoe drains to Pyramid Lake in western Nevada through the Lower Truckee River. Lower right: Map showing relatively small size of Lake Tahoe watershed (diagonal lines) relative to the lake. From Osleger *et al.* (2009).

cores, interpreted as turbidites, appear to have been emplaced by flows generated during extreme riverine flood events at 5.6, 4.6, 4.5, 4.25 and 3.1 ka (Osleger *et al.* 2009). The deposits of these (hyperpycnal) flows are characterised by sediment up to coarse silt grade showing basal inverse, then normal grading, very similar to the marine hyperpycnal flow deposits described by Mulder *et al.* (2001) (Fig. 1.37). Analysis of the organic fraction shows a terrigenous signal in the C/N ratios and $\delta^{13}C$ used to discriminate lacustrine algae, C_3 and C_2 land plants. The sediments contain a significant land-derived isotopically light soil component (Osleger *et al.* 2009).

The importance of internal tides (found nearly everywhere in the oceans) in suspending, moving and ultimately depositing fine-grained sediments remains poorly understood. We consider such tides here as they have the potential to lead to deposition of fine-grained deposits in deep-marine systems that may be unrelated to relative base-level changes, though the intensity of the tidal currents will be affected by such changes. Internal tides also are susceptible to forcing on astronomical time scales (e.g., de Boer & Alexandre 2012). *Internal tides* are created by barotropic tides over topography, typically extending to water depths of several hundred

metres, but in some cases reach depths greater than 1 km. Barotropic tidal currents are the periodic water motions accompanying the tidal changes in sea level, basically as large-scale waves with a wavelength of ~6000 km move around the ocean basins. Internal tides can be important in the delivery of fine-grained sediment to the deep sea; for example, in the Mozambique Channel, in the narrowest passage between Mozambique and Madagascar (Manders *et al.* 2004; da Silva *et al.* 2009). Manders *et al.* (2004), using deployed current meters, found that internal tidal currents were everywhere strongest near the surface (~4 cm s^{-1} at 250 m depth, and up to 12 cm s^{-1} near the pycnocline in the generation area), decreasing to <3 cm s^{-1} at 600 m depth, and increasing slightly near the bottom. Da Silva *et al.* (2009) studied the Sofala shelf in the Mozambique Channel, from which they identified two distinct types of internal wave train travelling oceanward away from the shelf break. They suggested that these result from direct generation at the shelf break, and from 'local' generation ~80 km from the shelf break, respectively, because of the surfacing of internal tidal rays at the thermocline. They also investigated seasonal differences in the wave patterns, which penetrate more extensively into the channel during the southern summer and appear slightly

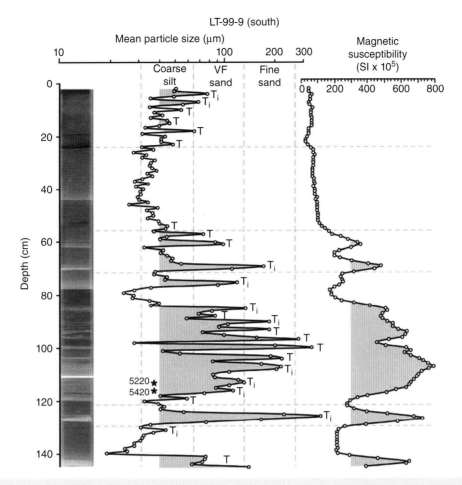

Fig. 4.24 X-radiograph, log plot of mean particle size, magnetic susceptibility and AMS dates of core LT-99-9 (south), Lake Tahoe. The particle-size and magnetic susceptibility curves are built from 145 data points. The shaded areas in the particle-size plot are based on an arbitrary size cut-off of 40 μm. SGF deposits interpreted as turbidites are marked by the letter T, with those having inversely graded bases marked by T_i. From Osleger *et al.* (2009).

farther to the south during the southern winter, from which they surmised that the local generation process is more likely to occur during the winter when the stratification is reduced. Using a simple two-layer calculation, da Silva *et al.* (2009) show that the long-wave phase speed is about the same at both times of year (1.4 m s⁻¹), with the weaker thermocline in October being compensated for by a deeper upper layer.

Internal tides are also well documented from the Makassar shelf (Hatayama *et al.* 1996; Nummedal *et al.* 2001; Ray *et al.* 2005; Robertson & Ffield 2008). In this area, the tidal range is only ~2 m, and rainfall and runoff are high. These are amongst the calmest seas in the world with maximum wave heights of <1 m (Seasat data). The thermocline moves vertically by ~40 m on a semi-diurnal basis, and this movement generates strong internal tides and tidal currents that persist into the adjacent deep-water areas. Current meters have recorded current speeds of nearly 0.5 m s⁻¹ at 965 m water depth (e.g., Nummedal 2001). The sedimentary response to such strong and deep internal tides is the development of muddy sediment waves showing aggradational stacking and updip accretion. Such sediment waves are not contourites (Chapter 6), despite their morphologic and grain-size similarities. 3D seismic reflection data show an extensive field of large, low-amplitude bedforms within, between, and in a band shelfward of the canyon heads. The bedforms are generally asymmetric with

internal onlap geometries that demonstrate accretion upslope and laterally away from the canyon floor towards the inter-canyon ridges. The bedforms have amplitudes on the order of 10 m and wavelengths on the order of 1 km. Piston cores collected from the bedforms contain dominantly clayey mud with widely spaced thin sand beds (Nummedal 2001).

The origin of large-volume SGF deposits, such as Facies C2.4 'megaturbidites', is generally ascribed to catastrophic slope failures that result in the downslope transport of substantial amounts of sediment from relatively shallow- to deep-water environments. Whilst a seismic trigger (Mutti *et al.* 1984) or the release of gas hydrates (clathrates) (Bugge *et al.* 1987; Nisbet 1992) is commonly inferred, Rothwell *et al.* (1998) instead proposed that eustatic sea-level changes might set up the conditions for failure. They used radiocarbon dating in five widely spaced cores to constrain the date of emplacement of a large-volume (~500 km³) bed in the Balearic Basin of the western Mediterranean to ~22 000 calendar years before present. This is the main sedimentation event in the Balearic Basin in the past 100 kyr, and was deposited when eustatic sea level was at its lowest level during the Last Glacial Maximum. Rothwell *et al.* (1998) hypothesise that the failure resulted from clathrate destabilisation caused by a eustatic lowstand, but accept that 'other triggering mechanisms, such as seismic shock, cannot be ruled out'.

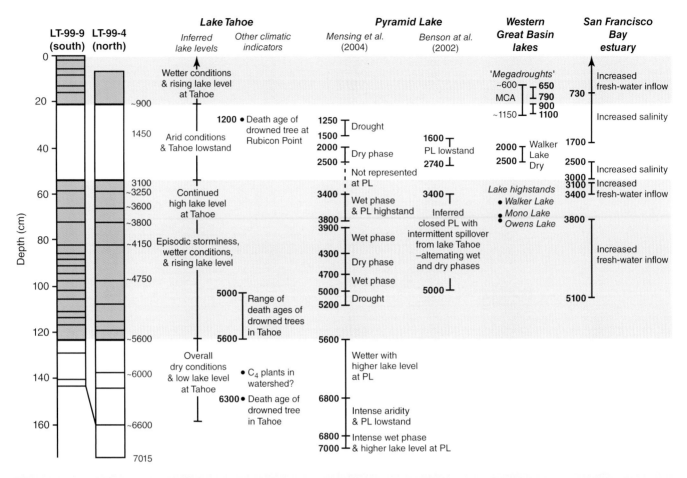

Fig. 4.25 SGF deposit frequency, inferred Tahoe lake levels and correlative palaeoclimatic events. The columns on the left show correlative intervals of low and high SGF deposit frequency; the stratigraphic position of individual SGF deposits (shown by the horizontal lines in each core) is defined by the maximum mean grain size. AMS and ash dates are shown without the ● symbol, whereas interpolated ages of specific SGF deposits that are correlative between cores are shown with the ● symbol. PL = Pyramid Lake; MCA = Medieval Climate Anomaly. From Osleger *et al.* (2009).

4.6 Autocyclic processes

4.6.1 Autocyclicity in submarine channels

In recent years, there has been increasing recognition of the potential importance of autocyclic processes in controlling erosion and deposition in the deep sea. In the 1970s and 1980s in particular, sequence-stratigraphic paradigms were relied upon perhaps too heavily to interpret most facies changes in deep water as the result of base-level (i.e., sea-level) changes. Whilst observational, experimental and theoretical work suggest that autocyclicity should occur, it is difficult to unequivocally reject all but autocyclic controls in a dynamic environment of, for example, changing global climate, regional weather patterns, tectonics and other processes. This section, however, explores examples of possible autocyclicity.

In contrast to the large-scale evolution of continental margins, smaller-scale submarine environments may show a predictable stratigraphic architecture that records changes in relative base level. For example, in a study of the Amazon Fan channel–levée–overbank and related deposits, Pirmez *et al.* (1997) documented an evolutionary sequence of deposits associated with channel development (Figs 4.26, 4.27) (see also Section 7.2.3).

On the middle Amazon Fan, sheet-like sand-prone units are identified in seismic profiles as high-amplitude reflection packets (HARPs). HARP deposition coincides in time with the initiation of a new channel segment after channel bifurcation. Toward the lower fan, HARP units tend to stack directly above each other, as overbank deposits thin downfan; these HARP units probably contain deposits formed basinward of the mouths of channels (Fig. 7.10). The SGF deposits within the HARP intervals include sandbodies 5–25 m thick formed of sand beds 0.1–4 m thick (Fig. 7.11). These sandbodies correlate with episodes of channel bifurcation on the middle fan. Most beds thicker than ~1 m contain mud clasts, interpreted to result from upslope levée erosion and channel entrenchment after channel bifurcation. On the lower fan, bed clusters and sets of amalgamated beds form sandbodies as thick as 50 m, with individual beds commonly greater than 3 m thick and containing abundant mud clasts. Most bed clusters show no apparent trends in bed thickness (Table 5.2). A key point here is that whilst these sand-bed clusters could be interpreted as a response to changing relative base level (i.e., falling, then rising with back fill), the chronological data gathered from the cores prove that they formed during a single lowstand of sea level, without any changes in relative base level (Hiscott *et al.* 1997c). They can be attributed

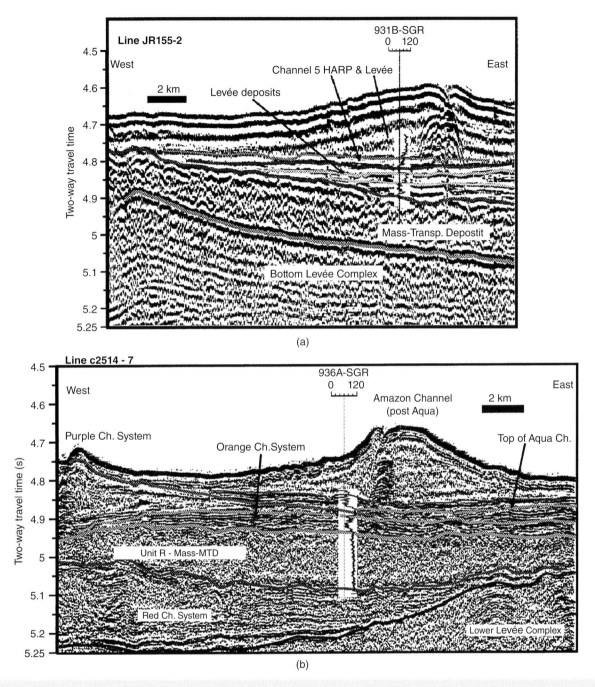

Fig. 4.26 Interpreted seismic lines across ODP Site 931 (a) and Site 936 (b) on the Amazon Fan. Integration of gamma-ray trace with seismic is based on a synthetic seismogram match, and travel time is derived from a sonic log. From Pirmez *et al.* (1997).

entirely to channel avulsion processes and gradual abandonment through a period of fixed base level (compare with fill-and-spill model, Section 4.6.2).

Channel bifurcation on the Amazon Fan results in the abrupt supply of sand to topographic depressions (i.e., inter-channel lows) between channel–levée systems (Fig. 7.10). The sand units correspond to HARPs seen on the seismic reflection data, above an angular and erosional unconformity marking the onset of development of the channel–levée system. In the vicinity of the bifurcation points, the HARPs are composed of single sand beds up to 12 m thick and amalgamated beds >30 m thick (Fig. 7.11).

A channel bifurcation results in the formation of a knick-point that probably leads to flow acceleration, enhancing the erosive power of turbidity currents. Channel entrenchment and changes in channel sinuosity across, and upslope of, the bifurcation site likely result in an increased supply of sand and clasts to the flows. Downslope of the bifurcation site, flows are relatively free to spread laterally, bounded only by the topographic highs of adjacent channel–levée units (Fig. 7.10). The sand mobilisation to form HARP successions is likely episodic (forming 'packets') because entrenchment probably proceeds by piecemeal headward erosion in channel segments undergoing gradient changes. Periodic shifts in the position of focused sand

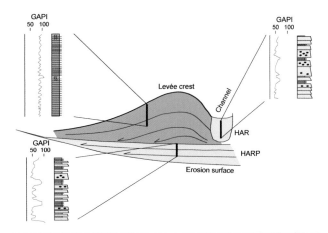

Fig. 4.27 Summary of geometry, stratigraphy, lithofacies and gamma-ray signature of components of Amazon Fan channel–levée systems, including a sheet-like sandy succession formed by redistribution of upfan channel-floor sands into an inter-channel low following avulsion (high-amplitude reflection packet – HARP), downlapping mud/silt deposits of an aggrading levée, and aggrading channel-floor sands of the new channel segment (high-amplitude refections – HAR). Each column represents ~30–50 m of section. Model based on data from ODP Sites 931, 935, 936, 944 and 946. From Pirmez *et al.* (1997).

deposition in the inter-channel low also can lead to clustering of the thicker beds in the HARP unit.

In summary, in the middle fan, HARP units associated with individual bifurcations form 10–30 m-thick sand-bed clusters that are sharply separated above and below from thin-bedded overbank deposits. On the lower fan, similar clusters appear to be stacked upon each other with little, if any, intervening overbank deposits, forming sand-prone units up to 100 m thick. Where such sandy successions are overlain by an aggrading channel-axis deposit (HAR), total sand thickness may be even greater. In the Amazon Fan, the sand-prone units (HARPs and HARs) are encased in thin-bedded silts and silty clays that would provide seals for stratigraphic traps if the sheet-like deposits eventually become a hydrocarbon reservoir. The sharp-based clusters are interpreted to result from upstream avulsion and abrupt lateral switching of the local fan depocentre.

In subaerial settings, autocyclic cut-and-fill (incision-backfilling) processes are recognised to cause switches between channel and more sheet-like geometries in both field and experimental studies (e.g., Whipple *et al.* 1998; Hamilton *et al.* 2013, and references therein). If such a process occurred in deep-water settings, it seems reasonable to expect that there might be evidence of a single or discrete upstream sediment input point (observed or inferred from a radial palaeo-flow), compensational stacking patterns within the less-confined, more sheet-like, elements of the channel-prone units, caused by less predictable aggradational stacking of 'lobe-like' intrachannel elements (*cf.* compensational stacking patterns documented from the Palaeocene Heimdal Formation, North Sea, by Fitzsimmons *et al.* 2005).

Laboratory flume-tank research has shown that autocyclicity can be important (e.g., Lancien *et al.* 2004; Hamilton *et al.* 2013). From 30 laboratory experiments reproducing submarine straight canyons as well as meandering channels, Lancien *et al.* (2004) simulated a submarine ramp draped by a sediment blanket. They modelled sustained density-current flow by a continuous brine stream injected at the top of the ramp. In these experiments, the slope of the plane, the

input flow rate and the brine density were controlled. Using an optical acquisition technique, it was possible to measure instantaneously the topography of the sediment surface at successive times during the erosional-channel formation and frontal lobe deposition. Following an initial stage in which the density current spread over the bed, the channel inception phase suddenly began, followed by a positive feedback mechanism facilitating further erosion. A phase of regressive erosion then occurred, and in some cases a steady state was obtained. By computing the difference between successive maps, elaborate time-varying maps of sedimentation and erosion rates in the system were created. Stacking of these maps produced a 3D cube of sedimentation rates showing autocyclic phases of incision, bypass and sedimentation. Cross-sections through this 3D cube show morphologies very similar to those observed in seismic datasets across submarine channel systems (Lancien *et al.* 2004). Whilst informative, such experiments can only show what is possible, and not necessarily what actually drives erosion and deposition in actual channelised deep-marine systems, simply because such experiments do not vary the sediment flux, sediment calibre (e.g., flow type) and simulated water-level variations. Autocyclic controls, however, might be expected to be more important during times of strong sediment input, and where gradients are steeper, as in the submarine part of fan deltas.

Even if the initiation mechanisms for deep-water SGFs are allocyclic (such as eustatic sea-level changes), Skene *et al.* (2002) argue that levées should not be particularly sensitive indicators of the external controls. They propose that levée architecture probably responds primarily to some common process inherent in overbank flow, so levée architecture should solely reflect autocyclic controls. Skene *et al.* (2002) then propose the corollary that:

> as sequence stratigraphic models rely on allocyclic forcing to create surfaces of time-stratigraphic significance that can be widely correlated (Van Wagoner et al. 1990), the lack of a strong allocyclic signal in levée architecture means that placing levée architecture into a sequence stratigraphic framework is unlikely to produce predictive models for levée deposition, except to the extent that allocyclic channels influence channel size.

In predominantly aggradational channel systems, vertical transitions from sheet-like to channelised deposits have been explained by autocyclic progradation of levéed channels over their own sandy, channel-terminus lobes (Fig. 7.10). Conversely, the vertical transition to more sheet-like deposits has been explained by channel avulsion, with unconfined, overbank deposits of successor channels progressively mantling the abandoned channel (Fig. 7.9 near Site 935). Such autocyclic changes, however, cannot explain repetitive switches between incised channels and more sheet-like deposits. A thinning-upward trend may be explained by autocyclic control, in which intra-channel deposition was not keeping pace with the increasing height of the levée crest above the channel floor, leading to increasing confinement of more of the flows.

What controls vertical transitions from sheet-like to channelised deposits? In a study of this problem, McCaffrey *et al.* (2002) describe repeated vertical alternations in sandbody geometry between incised channels and more sheet-like sandy elements from the Eocene (Priabonian) Grès de Champsaur deep-water siliciclastic system, deposited in a distal setting in the Alpine foreland basin of southeast France. The channels form symmetric, lenticular, erosional features, ~900–1000 m wide (measured between the lateral limits of incision) and ~65–115 m deep, being traceable axially for up to 5 km. The

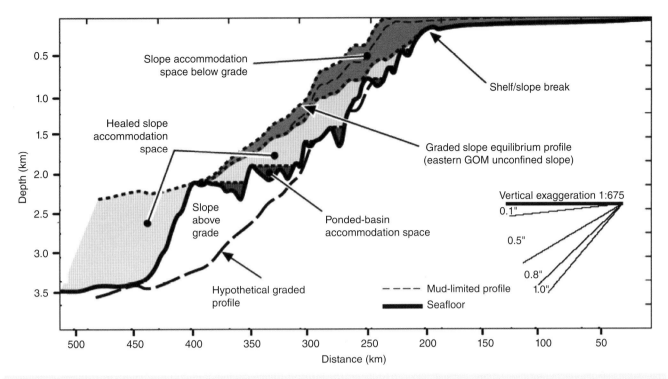

Fig. 4.28 Seafloor profile across the central Gulf of Mexico to show the distribution of accommodation space on a typical above-grade slope profile: (1) ponded-basin accommodation space (darkest-coloured legend); (2) slope accommodation space (light grey legend), and (3) healed-slope accommodation space (intermediate grey-coloured legend) (modified from Prather *et al.* 1998). The graded slope profile comes from the present-day unconfined slope of the eastern GOM where it dips south at ~0.88°. From Prather (2000).

channel fill is capped by a laterally continuous sandy sheet-like unit, lying above fine-grained deposits beyond the channel margins. No intrachannel elements can be traced into the substrate deposits, suggesting that, prior to infill, the channels were open seafloor conduits of essentially the same dimensions as the preserved channel deposits. The channels are vertically stacked, although axial erosion juxtaposes younger channel-axis deposits against the fill of older channels and their channel-capping sheet-like sandstones. The main channel-fill facies comprises coarse-grained, amalgamated sandstones that are typically parallel- or cross-stratified (Facies Group B2). Subsidiary facies of finer grained sandstone–mudstone couplets (Facies Groups C2 and D2) and clast-bearing muddy debrites (Facies Group A1) are commonly preserved as erosional remnants, suggesting a complex channel history of aggradation and erosion. The repeated cycles of channel incision, infill and transition to sheet-like sandstone development suggest repetitive incision and healing of the palaeo-seafloor. McCaffrey *et al.* (2002) argue that autocyclic processes alone cannot account for the Grès de Champsaur channels and, instead, proposed a process of incision linked to the development of relatively steep axial gradients (parallel to the mean dispersal direction), then a return to more sheet-like deposition with the re-establishment of lower axial gradients, with the repetitive switch between erosional channels and more sheet-like sandstones driven by changes in sediment input rate against a background of ongoing seafloor tilting.

4.6.2 Fill-and-spill model for slope basins

Slopes classified on the basis of topography are differentiated into: (i) *above-grade slopes* with either intraslope basins, or stepped depositional profiles and (ii) *graded slopes* that lack significant topography.

Implicit in the discussion of depositional processes that follows is that three main types of accommodation space are present on most above-grade slopes: (i) ponded-basin, (ii) healed-slope and (iii) slope (Fig. 4.28) with lesser amounts of accommodation in incised submarine canyons and gorges. Accommodation space is defined in this setting as the spaces between various graded depositional profiles. Locally active erosional and depositional processes, in what Ross (1989) refers to as the slope readjustment model, maintain these graded profiles across continental slopes (Ross 1989; Thorne & Swift 1991; Ross, W.C. *et al.* 1995). Pyles *et al.* (2011) defined stratigraphic grade as "the similarity of the morphology of successive slope-to-basin profiles in a genetically related depositional system." They proposed that stratigraphic grade is a predictor of stratal (reservoir) architecture and used four methods to collectively define stratigraphic grade: (i) regional stacking patterns of fourth-order stratigraphic surfaces, (ii) the relationship between the trajectory of the shelf edge and the trajectory of the depocenter for fourth-order stratigraphic units, (iii) morphology of the slope-to-basin profiles of fourth-order stratigraphic surfaces, and (iv) the similarity of the morphologies of slope-to-basin profiles within a system. Using this methodology, Pyles *et al.* (2011) concluded that several characteristics of stratal (reservoir) architecture of fourth-order cycles are related to stratigraphic grade: (i) longitudinal distribution of sandstone in fourth-order cycles; (ii) location of maximum sandstone relative to the depocenter of fourth-order cycles; (iii) lengths of fourth-order submarine fans, and (iv) longitudinal and vertical changes in associations of architectural elements.

Within the context of equilibrium slope profiles, one of the more recent influential applications of sequence stratigraphic principles to deep-water depositional systems has been the 'fill-and-spill' model outlined in Figure 4.29 (Prather *et al.* 1998; Prather 2000; *cf.*

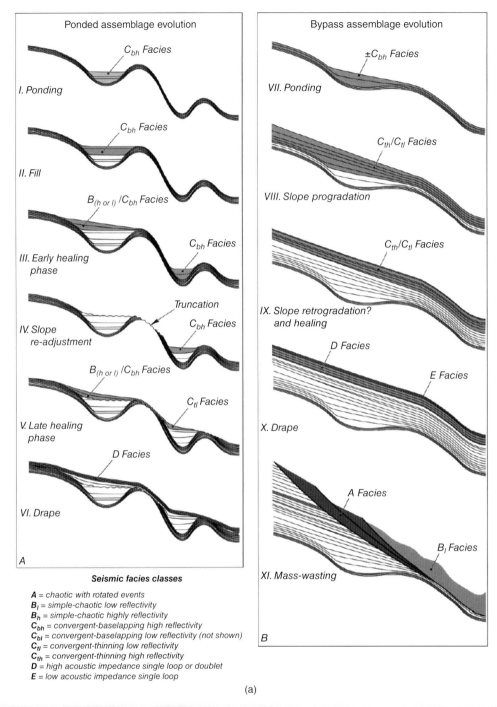

Ponded assemblage evolution

I. Ponding

C_{bh} Facies

II. Fill

C_{bh} Facies

III. Early healing phase

$B_{(h\,or\,l)}$ /C_{bh} Facies

C_{bh} Facies

IV. Slope re-adjustment

Truncation

C_{bh} Facies

V. Late healing phase

$B_{(h\,or\,l)}$ /C_{bh} Facies

C_{tl} Facies

VI. Drape

D Facies

A

Bypass assemblage evolution

VII. Ponding

±C_{bh} Facies

VIII. Slope progradation

C_{th}/C_{tl} Facies

IX. Slope retrogradation? and healing

C_{th}/C_{tl} Facies

X. Drape

D Facies

E Facies

XI. Mass-wasting

A Facies

B_l Facies

B

Seismic facies classes

A = chaotic with rotated events
B_l = simple-chaotic low reflectivity
B_h = simple-chaotic highly reflectivity
C_{bh} = convergent-baselapping high reflectivity
C_{bl} = convergent-baselapping low reflectivity (not shown)
C_{tl} = convergent-thinning low reflectivity
C_{th} = convergent-thinning high reflectivity
D = high acoustic impedance single loop or doublet
E = low acoustic impedance single loop

(a)

Fig. 4.29 (a)   Idealised ponded depositional sequence (A). Capture of submarine fans (I) occurs in ponded accommodation space created by salt withdrawal; fans eventually fill the accommodation space (II). Healing of the slope occurs after the ponded-basin fills and gravity flows spill downslope as the sill separating the upslope basin from downslope basin is overtopped (III). A localised truncation surface forms from erosion of the upslope basin as the equilibrium profile adjusts to the elevation of the downslope basin (IV). Backfill of the space above the truncation surface occurs as the downslope basin fills and the slope between the two basins aggrades to a local equilibrium profile (V). Muddy gravity flows and/or hemipelagic deposits drape the basins after the slope grades to the equilibrium profile or there is a decrease in sediment influx resulting from either a rise in eustatic sea level or slope-system avulsion (VI). Idealised bypass depositional sequence (B). Captured submarine fans fill ponded accommodation space where the rate of local basin subsidence due to salt withdrawal exceeds the rate of sediment influx (VII). Progradation of the slope occurs once the ponded accommodation space fills (VIII). Retrogradational parasequence sets suggest slope progradation is followed locally by 'healing phase' deposits reducing the local slope gradient (IX). Muddy SGF deposits and/or hemipelagic deposits drape the slope after the depositional surface grades to the equilibrium profile and/or there is a drop in the rate of sediment influx due to a rise in eustatic sea level or slope-system avulsion (X). Aerially extensive mass wasting occurs as the regional slope steepens beyond the angle of repose for rapidly deposited muds during slope progradation and/or basinward tilting (XI). See Prather *et al.* (1998) for further explanations of seismic facies classes. Compare this hypothetical model with a real example from the Brazos–Trinity depositional system in the western Gulf of Mexico (Fig. 4.29b). Redrawn from Prather *et al.* (1998, 2000).

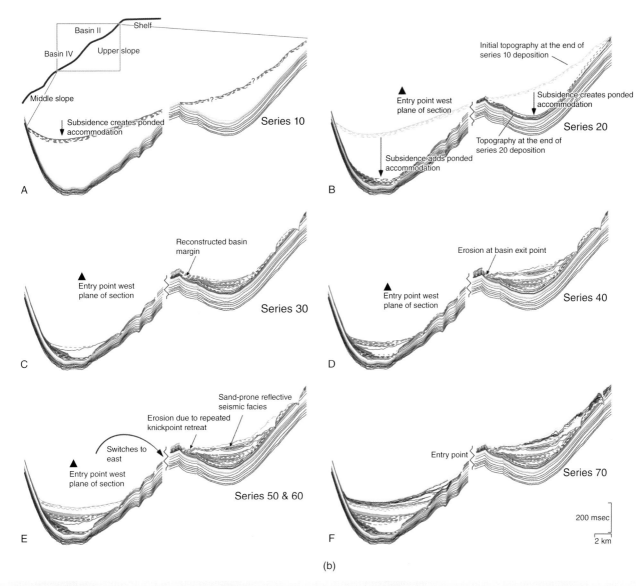

(b)

Fig. 4.29 (b) Stratigraphic evolution of basins II and IV in the Brazos–Trinity depositional system, located on the upper slope, offshore Texas, USA, western Gulf of Mexico. The Brazos–Trinity depositional system consists of four linked late Pleistocene intra-slope basins (I–IV). (A) Initially basin II and IV were part of an unconfined stepped slope. Salt withdrawal created shallow ponded accommodation (<60 m) in Basin IV. Inset profile shows hypothetical position of the Brazos–Trinity basin relative to the shelf and slope. Black line work shows the hypothetical slope configuration at the end of Series 10 deposition. (B) The oldest deposits of the linked Brazos–Trinity basins occur in Basin IV before formation of Basin II. Salt withdrawal creates ponded accommodation in Basin II and increases ponded accommodation in Basin IV. (C) Deposition began in Basin II as SGFs pond. Black line work shows the hypothetical slope configuration at the beginning of Series 20 deposition. Subsidence at the time of Series 20 deposition brought the slope very near to its present structural configuration. (D) First evidence of significant bypass occurred during deposition of Series 40 in Basin IV and corresponds to erosional truncation and early pass of the Basin II. (E) Basin II and IV filled contemporaneously during deposition of Series 50/60. (F) The shallowest apron in Basin IV developed during significant bypass through a levéed channel in Basin II. From Prather *et al.* (2012a).

Beaubouef & Friedmann 2000; Beaubouef *et al.* 2003). This model considers ponded accommodation space and the way in which this space is created and destroyed on a basin slope, with concepts derived mainly from the Gulf of Mexico (GOM). Ponded-basin accommodation space lies within three-dimensionally closed topographic lows on continental slopes. Ponded-basin accommodation space typically forms within intraslope basins as the result of localised salt (e.g., Fig. 9.9) or shale withdrawal, but may also be related to structural movements unrelated to withdrawal processes. In areas of salt withdrawal, ponded-basin accommodation space is usually circular to semi-circular and increases into the distal part of the basin. In areas of shale withdrawal, ponded-basin space commonly takes the form of long, linear to arcuate, doubly-plunging synclines located within out-board deep-water thrust belts or the hanging-walls of faults that cut the seafloor at the time of deposition. Fill-and-spill processes dominate deposition on such slopes: the ponded intraslope basins eventually form low gradient stepped-equilibrium profiles (*cf.* Satterfield & Behrens 1990; Winker 1993; Prather *et al.* 1998; Prather *et al.* 2012a, b).

Seismic facies in GOM intra-slope basins reflect the interplay of a variety of deep-water depositional processes and the evolution of accommodation space on the slope (Prather *et al.* 1998). This interplay of processes results in a transition from an early, sand-prone ponded-basin-fill succession (*ponded facies assemblage*) to a later shale-prone, slope-bypass succession (*bypass facies assemblage*) (Fig. 4.29a). Convergent-base-lapping facies (Prather *et al.* 1998; Prather 2003) in combination with localised chaotic and draping facies dominate the ponded facies assemblage. Convergent-thinning facies with widespread chaotic and draping facies dominate the bypass facies assemblage. These units represent filling of different types of slope accommodation space. Stratigraphically, the transition from ponded to bypass facies assemblages can be sharp or gradational over hundreds of metres. Transitions occurred across the central GOM during the late Pliocene between 2.0 and 1.8 Ma, and in the early Pleistocene between 1.2 and 1.0 Ma. Nearly synchronous transitions throughout basins in the upper to middle slope suggest that increased sediment supply resulting from a second-order sea-level fall (Table 4.1), and the capture of large drainage areas by the Mississippi River during the Pleistocene, are the primary controls on development of this large-scale stratigraphic architecture. Figure 4.29b is an example to show how the Brazos–Trinity depositional system, western Gulf of Mexico, filled linked slope basins (Prather *et al.* 2012a; *cf.* Prather *et al.* 2012b for an example of a submarine apron perched on the upper slope off the Niger Delta).

Mobile substrates with relatively large amounts of ponded-basin and healed-slope accommodation space across the mid-slope characterise above-grade slopes (Prather 2000). There are many above-grade slopes around the world but the Gulf of Mexico (GOM) is amongst the best calibrated. The GOM is characterised by high rates of intra-slope basin subsidence (locally $>10\,\mathrm{km\,Myr^{-1}}$) and has been influenced by meltwater spikes. Ponded-basin fill-and-spill processes dominate early phases of deposition on above-grade slopes and precede progradation of unconfined slopes (Prather *et al.* 1998; Prather 2000). A stepped equilibrium profile across a series of ponded basins forms a base for late unconfined slope progradation. Deposition in the unconfined slope occurs in both slope and healed-slope accommodation space. These spaces are filled as progradational delta fronts build beyond critical angles, and later collapse. Sediments tend to be mud-prone because the slope angles are too high to allow sand-prone submarine-fan deposition. Sands are generally associated with channels with some localised relatively unconfined sand sheets.

Computer-assisted stratigraphic modelling is an important step in devising, visualising and testing depositional process models. Numerical simulation of sedimentary basin stratigraphy is useful for developing and quantifying concepts of basin evolution, predicting facies distribution and architecture, testing exploration scenarios rapidly in frontier basins, constraining interpretations of subsurface data and performing sensitivity tests to evaluate the fundamental controls on an observed basin stratigraphy (Aigner *et al.* 1987; Lawrence *et al.* 1987, 1989, 1990; Lamb *et al.* 2004; Violet *et al.* 2005; Toniolo *et al.* 2006).

Prather (2000) used an in-house proprietary stratigraphic forward modelling application (STRATAGEM) to visualise and distil a GOM-based depositional process model that might be applicable to other slope and base-of-slope systems. Forward models show that the introduction of meltwater spikes does not radically change the depositional architecture of the slope but the spikes do force gravity-flow deposition to continue into eustatic sea-level rises. These models also show that rapid salt withdrawal beneath intra-slope basins controls the distribution of accommodation space across the slope and thus its stratigraphic architecture.

From this study of GOM mini-basins, Prather (2000) concluded that: (i) rapid salt withdrawal beneath intra-slopes basins is a major control on the distribution of accommodation space across the slope and thus its stratigraphic architecture; (ii) basins with more episodic sediment fluxes are more likely to have thick ponded-basin successions with numerous reservoir/seal pairs than slope systems with more constant sediment fluxes; (iii) periods of sustained high sediment flux force progradation of unconfined slope deposits over earlier ponded-basin successions; (iv) periods of little sediment flux (condensed zone deposition) at the end of each depositional cycle allow the local depositional surface to drop below the stepped equilibrium profile, re-initiating the formation of ponded-basin accommodation space; (v) large-amplitude and high-frequency variations in sediment supply promote formation of additional ponded accommodation space throughout the history of a subsiding intraslope basin increasing the likelihood of stacked reservoir/seal pairs; (vi) without the climatically induced high-frequency changes in sediment flux, the rate of basin subsidence might not be sufficient to build and maintain ponded-basin accommodation space for long periods; (vii) without ponded-basin accommodation space, the sand content of the entire slope system drops because there is little room on the slope to accommodate gravity-flow sand sheets; (viii) above-grade slope systems without episodic sediment fluxes are less likely to have the thick ponded-basin successions, with numerous reservoir/seal pairs, than slope systems with highly variable sediment fluxes like the GOM and (ix) lack of thick ponded basin successions is more likely in slope systems whose sediments are sourced from non-glaciated continents or regions without a strong high-frequency (climatic) sediment-supply driver. These fill-and-spill models have been applied to many topographically complex slope mini-basins worldwide (e.g., Smith 2004; Gee *et al.* 2007; Valle & Gamberini 2010; Bourget *et al.* 2011).

4.6.3   Autocyclicity in fan deltas

Although we are primarily concerned with deep-water environments in this book, it is instructive to briefly consider fan deltas as they record fan-building processes generally with high sediment accumulation rates, particularly in tectonically active settings. Flume and experimental modelling by van Dijk *et al.* (2009) has demonstrated the potential importance of autocyclic processes of sediment storage and release. Van Dijk *et al.* (2009) undertook analogue experiments on fan deltas with constant extrinsic variables (discharge, sediment supply, sea level and basin relief) to demonstrate that fan-delta evolution consists of prominent cyclic alternations of channelised flow and sheet flow. The channelised flow was initiated by slope-induced scouring followed by headward erosion to form a channel that connected with the valley, while the removed sediment was deposited in a rapidly prograding delta lobe. The resulting decrease in channel gradient caused a reduction in flow strength, mouth-bar formation, flow bifurcation and progressive backfilling of the channel. In most experiments, in the final stage of channel filling, sheet flow briefly coexisted with channelised flow (semi-confined flow). Subsequent autocyclic incisions were very similar in morphology and gradient, although they eroded deeper into the delta plain and, as a result, took more time to backfill. The duration of the semi-confined flow increased with successive cycles. During the period of sheet flow, the delta plain aggraded up to the 'critical' gradient required for the initiation of autocyclic incision. This critical gradient was found to be dependent on the sediment transport capacity, defined by the input conditions. These autogenic cycles of erosion and aggradation confirm earlier findings that storage and release of sediment and associated slope variation can play an important role in fan-delta evolution. The erosional surfaces produced by the autocyclic incisions were well preserved by the back filling process in the deposits of the fan deltas. These erosional surfaces can easily be misinterpreted as bounding surfaces produced by variations in climate, sea level or tectonics. As many submarine fans are connected updip to fan-delta, or comparable, sediment sources, it seems reasonable to assume that any updip autocyclicity may find expression in the deep-water siliciclastic environments. Discriminating between allocyclic and autocyclic processes, therefore, requires a good 3D stratigraphic framework, excellent age dating, the correlation of depositional units and/or unconformities and their correlative conformities from basin to shelf, and additional data such as stable-isotope analyses to tie any local environmental events to potentially globally-synchronous oceanographic/climatic changes.

4.7   Palaeo-seismicity and the stratigraphic record

In seismically active areas, the deep-water basin succession may preserve a record of palaeo-seismicity. As historic records of earthquakes are relatively short, stratigraphic data might be used to reconstruct their frequency (perhaps more easily for large earthquakes), with the potential to infer a recurrence interval. However, since SGFs are triggered by several processes, it is challenging to ascribe individual deposits to a precise triggering mechanism. We do not even know if every large earthquake generates a SGF, or under what circumstances this may occur. Offshore coring from the Boxing Day 2004 magnitude 9 earthquake off Sumatra has identified a SGF deposit that is interpreted to represent redeposition from this event (Patton *et al.* 2013).

Despite the obvious problems, several studies have shown that seismic activity can be directly linked to the generation of specific SGFs, for example the Grand Banks earthquake in 1929 (Piper & Aksu 1987). In the rock record, the concept of earthquakes linked to Facies C2.4 megaturbidites, leading to the introduction of the term *seismo-turbidite* by Emiliano Mutti (1984) and co-workers (Labaume *et al.* 1983a, b, 1985, 1987), has been widely applied and extended to studies in modern and ancient deep-water basins that include beds of all thicknesses and grain sizes (Hiscott & Pickering 1984; Pickering & Hiscott 1985; Slazca & Walton 1992; Hiscott *et al.* 1993; Gorsline *et al.* 2000; Nakajima 2000; Nakajima & Kanai 2000; Nilsen 2000; Shiki 2000; Iwai *et al.* 2004; Anastasakis & Piper 2006). Whilst intuitively reasonable, these interpretations are not the only possible explanation for many of the beds.

Masson *et al.* (2011a, b) showed that SGF deposits, which they interpreted as turbidites, emplaced at ~6600 and ~8300 cal. yr BP in the Tagus Abyssal Plain off Portugal, correlate with erosional hiatuses in two submarine canyons on the continental margin. The SGF deposits were sourced from simultaneous submarine failure events in both canyons, from which they inferred that the regional trigger was an earthquake. An earthquake recurrence interval for the continental margin of ~4000 years was estimated by extrapolation to deeper SGF deposits in the basin succession. However, there are unresolved mismatches between known earthquake magnitudes and the scale of the inferred contemporaneous sedimentary event. For example, the 1755 AD earthquake which caused widespread devastation in southwestern Iberia had a magnitude >8.5 but the associated SGF deposit in the abyssal plain is typically ~5 cm thick, whereas older SGF deposits can be up to >1 m thick. Given the large magnitude of the 1755 AD earthquake, any differences in SGF deposit thicknesses are unlikely to have a simple relationship to the relative size of triggering earthquakes. Instead, Masson *et al.* (2011a) suggested that the offshore location of the 1755 AD earthquake, coupled with low sediment accumulation rates during the Holocene, might have limited the size of the associated slope failure and SGF deposit.

In 1996, a project to investigate seismic activity and the generation of SGFs along the Cascadia margin was initiated (Goldfinger 2012) (Fig. 4.30). This margin has one of the best records of palaeo-seismic activity and good long-distance correlation of SGF deposits. These correlations have been possible because of the presence of the regionally datable Mazama ash and several distinctive SGF deposits. To date, almost all the cores taken in Cascadia Basin contain this unique ash layer from the eruption of Mount Mazama, dated to 6845 ± 50 radiocarbon yr BP (Bacon 1983). The ash was distributed to the Cascadia canyon/channel system via the drainage basins of major rivers. As only channel cores contain the ash, this suggests that air fall was not significant. Griggs and Kulm (1970) used the Mazama ash to calculate that the mean recurrence interval for the post-Mazama SGFs in the Cascadia Channel is 410–510 years.

Using the Mazama ash, the stratigraphy in Cascadia Basin was elucidated through the recognition and correlation of SGF deposits, in particular the so-called T5, T7, T11 and T16 events. A notable feature is the lack of significant changes in the character of the individual SGF deposits (e.g., grading style) from proximal (Juan de Fuca) to distal (Cascadia) channel sites. The transport distance is ~300 km between these two sites, and the SGFs would have passed through the confluence with Willapa Channel, which drains most of the Washington margin with multiple canyon systems. Correlations of individual event-deposits from site to site are aided by high-resolution physical-property data such as gamma-ray, density, P-wave velocity and magnetic susceptibility. The magnetic and density 'fingerprints' of each SGF deposit reflect their grain-size distributions. A typical signature comprises 1–3 coarse fining-upward sandy pulses (partial Bouma sequences), capped by a fining-upward silty cap, indicating the final waning of the gravity flow (Fig. 4.31). Goldfinger and his co-workers (Goldfinger *et al.* 2000, 2003, 2007, 2008, 2009, 2012, 2013) found that individual events can be correlated not only within channels, but also between separate channels that show no physical connection (e.g., Fig. 4.32). Some correlated events are as much as 500 km apart, but share basic characteristics such as event size, number of coarse sandy pulses, and even subtle details of the shape of the physical property signatures (proxies for grain-size distribution). These observations all support earthquake triggering. One possible explanation that Goldfinger and co-workers argue is that multiple coarse pulses in what seem to be the deposits of single events might reflect the rupture of multiple fault segments or asperities during each earthquake, giving insight into the source-time function of the earthquakes.

4.8   Deconvolving tectonic and climatic controls on depositional sequences in tectonically active basins: Case study from the Eocene, Spanish Pyrenees

A fundamental aspect of understanding any deep-water depositional system is the deconvolution of tectonic and global climatic controls at a range of temporal and physical scales. In many cases, this might

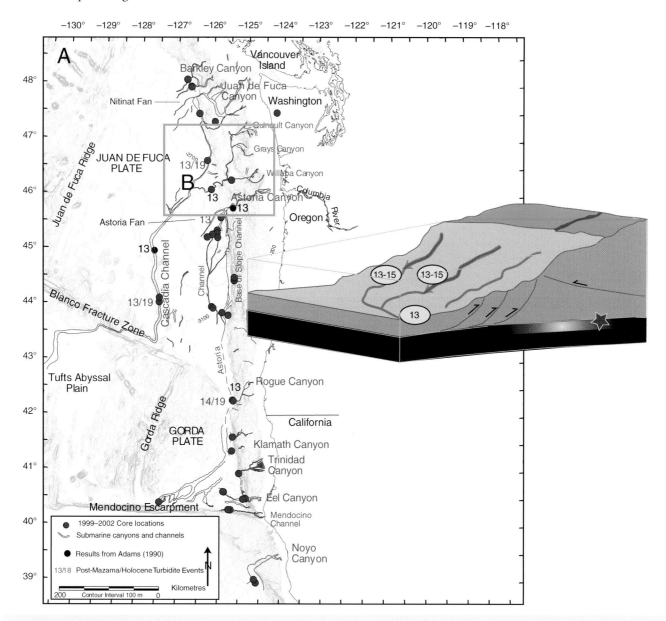

Fig. 4.30   Cascadia margin submarine canyons, channels and 1999–2002 core locations: major canyon systems outlined. Major canyon systems and the number of post-Mazama and/or Holocene turbidites are shown in red. Mazama ash is not present in Barkley Canyon cores or in the cores south of Rogue Canyon. **INSET** Synchronicity test at a channel confluence as applied where Washington channels merge into the Cascadia Deep Sea Channel. The number of events downstream should be the sum of events in the tributaries, unless the SGFs were triggered simultaneously by an earthquake. This site is marked 'B' on the Juan de Fuca plate. From Goldfinger *et al.* (2003). See also Goldfinger *et al.* (2008, 2013).

not be possible, for example because of a lack of good age constraints and a robust age model, a poor understanding of the plate-tectonic setting, and insufficient outcrops (or core) through suitable lithologies for detailed analysis, including time-series analysis. Additionally, even in sedimentary basins where one can be reasonably confident about deconvolving tectonic and climatic controls, caution should be exercised about extrapolating the results to other basins. This section is relatively lengthy because the case study involves a basin that is visited by large numbers of researchers, students and industry geologists, geophysicists and engineers, and arguably is amongst the best natural laboratories worldwide for studying the interaction of tectonics and climate, at a range of temporal scales, on deep-marine

deposition. It is also a deep-marine basin where it appears that specific and different controls on deposition can be identified and at a range of time scales.

Let us consider syn-sedimentary thrusting in orogenic belts. Using a modelling approach to analyse the behavior of individual thrust units associated with asymmetric doubly vergent thrust wedges, Naylor and Sinclair (2007) have proposed that rates of surface uplift, frontal accretion and exhumation should be punctuated on a time scale linked to thrust sheet geometry and convergence rates. This time scale ranges from 0.1–5 Myr for various settings, and should be calculated before external forcing factors, such as climate, are invoked. They propose, for example, that for the central Southern Pyrenees,

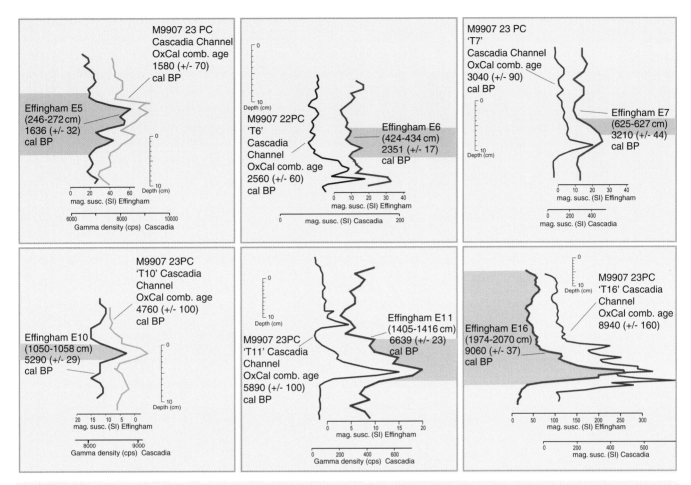

Fig. 4.31 Correlations between Cascadia Channel core M9907-23PC and core MD02-2494 from Effingham inlet, western Vancouver Island. Each graph shows the magnetic susceptibility record (blue) from an Effingham inlet (inner basin) SGF deposit, and a magnetic susceptibility or gamma density record from the 1999 cores in Cascadia Channel in purple. These events are interpreted as seismites based on wall rock signature from the adjacent Fiord walls (shown in grey), and by comparison to the historical SGF deposit triggered by the 1946 Vancouver Island earthquake. The records show a striking similarity in general size, number of sandy pulses (magnetic and density peaks) and, in some instances, detailed trends. Radiocarbon ages are also first-order compatible, but have separations of 100–200 years in some cases. Offshore ages are the OxCal combined ages with 2σ ranges. The combined age data and stratigraphic correlation suggest that the Effingham SGF deposits and Cascadia Basin SGF deposit signatures record the same earthquakes. From Goldfinger *et al.* (2012).

a lower limit on the time scale of internal variability ascribed to punctuated thrust activity is ~4 Myr; for southwestern Taiwan a time scale for the tectonic signal is ~0.15 Myr, and for the Himalayan foothills of central Nepal ~1.2 Myr. Incorporating the arguments of Naylor and Sinclair (2007), together with a considerable data base, this section summarises the results of research in the Eocene of the Spanish Pyrenees, where it appears possible to identify both tectonic and climatic controls on deposition. The central southern Pyrenees accreted the Montsec and Sierras Marginales thrust sheets during Eocene time with a regional convergence rate of ~6 km/Myr (Vergés *et al.* 1995). The thrust sheets have a high aspect ratio associated with strong, competent limestones detached on a weak Triassic evaporate layer, each thrust sheet with a length of ~24 km. This results in a minimum limit on the time scale of internal (tectonic) variability of ~4 Myr, a result that is consistent with the recognition of a 4–5 Myr tectonic driver to explain the two successions in the deep-marine Ainsa Basin that are separated by an angular unconformity, with the time-averaged driver for the individual sandbodies being an order of magnitude faster and, therefore, better explained by climatic or other

processes. This case study illustrates the way in which the tectonic and climatic drivers on deposition can be deconvolved.

The accumulation of lower to Middle Eocene (Ypresian/Cuisian and Lutetian) deep-marine sediments in the Ainsa Basin, Spanish Pyrenees (Fig. 4.33), was broadly contemporaneous with maximum rates of tectonic subsidence in the 'South Pyrenean foreland basin', and coincident with maximum rates of shortening and thrust-front advance during the Lutetian (Muñoz 1992; Vergés *et al.* 1995, 1998). The foreland basin evolved with mainly non-marine and marginal-marine environments in the eastern sectors, whilst farther west there was an overall change from fluvio-deltaic to deep-marine systems (Fig. 4.33). Micropalaeontology suggests that water depths in the Ainsa Basin were in the range 400–800 m (Pickering & Corregidor 2005).

Within the Eocene parts of this basin, both during the foreland basin and thrust-top (piggy-back) stage, an estimated cumulative thickness of 4 km of deep-marine deposits accumulated as the Hecho Supergroup (Mutti 1977, 1983/1984, 1985; Mutti & Johns 1979, Mutti *et al.* 1984; Pickering & Corregidor 2005; Pickering & Bayliss 2009).

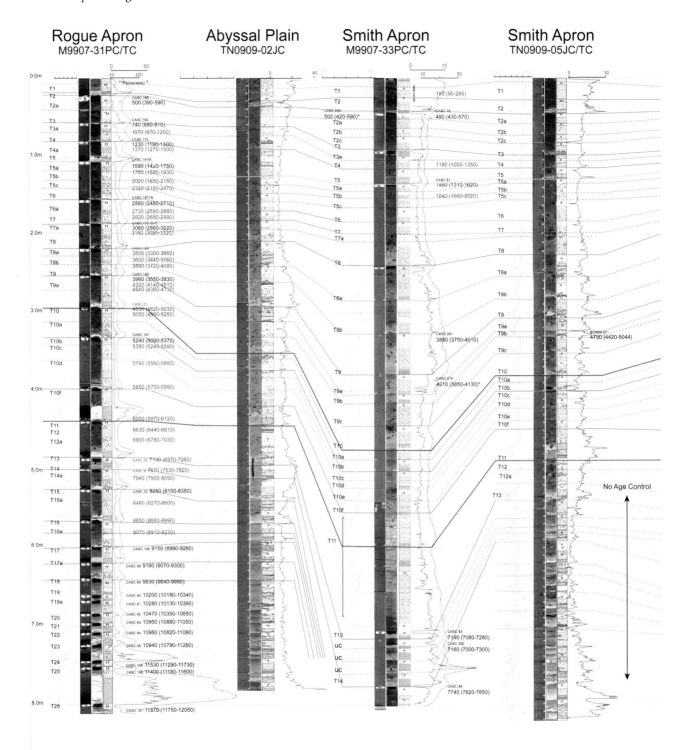

Fig. 4.32 Correlation of SGF deposit core records along ~130 km of the southern Cascadia margin from Rogue Apron to the Trinidad Plunge Pool. Three SGFs (interpreted as turbidites) T5, T10 and T11 are colour coded to match the corresponding seismic reflectors that were recognised. Depth of key reflectors is compatible with the trends in the seismic section (though definitive correlation is not possible) after velocity correction: deepening at Smith, then shallowing slightly before deepening significantly near Trinidad Plunge Pool. Many of the Rogue mud turbidites appear to thicken and coarsen southward, as do most of the margin-wide turbidites, with the exception of T4, T6, T7 and T8, which thin southward (interpreted T numbers assigned to beds after final correlation). From Goldfinger *et al.* (2013).

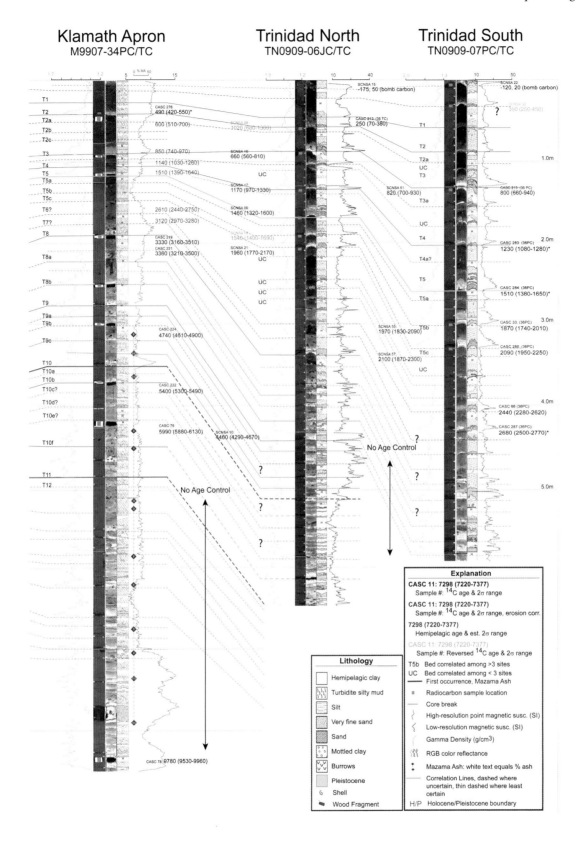

Fig. 4.32   *(continued)*

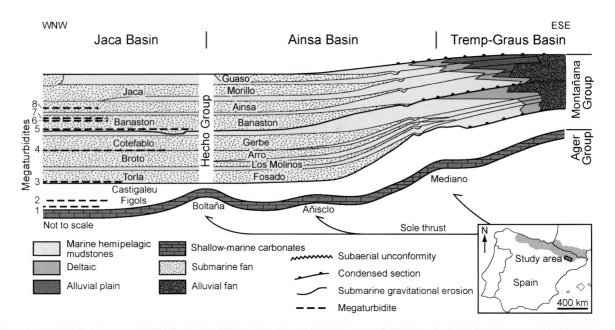

Fig. 4.33 Stratigraphy of the Hecho and Montañana groups within the South Pyrenean foreland Tremp–Graus, Ainsa and Jaca basins. Sediments were supplied axially from alluvial fans within the Tremp–Graus Basin. The Mediano Anticline represents the syn-depositional growth structure separating the shallow-marine and sub-aerial environments of the Tremp–Graus Basin from the deep-marine Ainsa Basin. Within the Ainsa Basin, channelised submarine fans were confined between the Mediano and Boltaña anticlines. West of the Boltaña Anticline, sediments of the Jaca Basin consist of submarine lobes and contain basin-wide Facies C2.4 megaturbidite event beds (MT-1 to MT-8). Correlations are based upon various published research and unpublished conference field guides (Nijman & Nio 1975; Mutti 1983/1984; Mutti *et al.* 1985; Muñoz *et al.* 1998; Nijman 1998; Remacha *et al.* 1998; Remacha *et al.* 2003). Inset: study area (Ainsa Basin) in the south-central Pyrenees, northern Spain. From Scotchman *et al.* (2015a).

The early/Middle Eocene deep-marine siliciclastic sediments in the Ainsa Basin were deposited above a foundered carbonate platform (Barnolas & Teixell 1994), and broadly coincided with maximum rates of tectonic subsidence, shortening and thrust-front advance in the South Pyrenean basins (Verges *et al.* 1995, 1998; *cf.* 2–4 Myr phases of thrusting in coeval and younger stratigraphic sections by Burbank *et al.* 1992). The north–south-trending structures of the Ainsa Basin experienced clockwise vertical axis rotations that vary from 70° in the east (Mediano Anticline) to 55° in the west (Boltaña Anticline): clockwise vertical axis rotations of 60°–45° occurred from early Lutetian to late Bartonian when the folds and thrusts of the 'Ainsa Oblique Zone' developed, with a second rotation event of at least 10° which took place since the Priabonian (Muñoz *et al.* 2013). Such large rotations of the Gavarine thrust sheet from early Lutetian to late Bartonian suggest that the Ainsa Basin was already a thrust-top basin and that subsidence cycles were more likely driven by tectonic mechanisms more complex than simple repeated phases of flexural loading (e.g., 'seesaw tectonics' as discussed below).

The Ainsa Basin deep-marine deposits can be divided into two angular-unconformity-bound units (Muñoz *et al.* 1994, 1998; Fernandez *et al.* 2004; Pickering & Bayliss 2009; Scotchman *et al.* 2015a). The younger unit is both structurally less deformed and shows a west-southwest shift in depositional axis (Pickering & Corregidor 2005; Pickering & Bayliss 2009) (Fig. 4.34) and a first-order tectonic control on accommodation and deposition (but not necessarily sediment delivery to the basin). These two 'tectono-stratigraphic' units (the Lower and Upper Hecho Group, respectively) contain eight coarse clastic systems (Fig. 4.34), each of the order of 100–200 m thick, and vertically separated from one another by up to several tens of metres of mainly marls with lesser amounts of thin- to very

thin-bedded sandstone SGF deposits. Each system typically contains 2–6 individual mappable sandbodies from 30–100 m thick (amounting to at least 25 throughout the basin), locally stratigraphically separated from one another by tens of metres of mainly thin- and very thin-bedded sandstones with subordinate marls; in other parts of the basin, they may be amalgamated. The deep-marine deposits are overlain by up to several hundred metres of slope, outer shelf and prodelta deposits, then ~0.5 km of fluvio-deltaic and related sediments fed mainly from the south (Dreyer *et al.* 1999).

The eight coarse clastic depositional systems in the Ainsa Basin (Fig. 4.33), incorporating ~25 sandbodies or channelised submarine fans (Fig. 4.34a), were deposited in various deep-marine settings, which from the oldest are: (1) Fosado (2 sandbodies) = lower-slope erosional channels; (2) Los Molinos (3 sandbodies) = lower-slope erosional channels; (3) Arro (3 sandbodies) = canyon/base-of-slope channel system; (4) Gerbe (2 sandbodies) = canyon/lower-slope erosional channels (Millington & Clark 1995a, b; Clark & Pickering 1996a); (5) Banastón (6 sandbodies) = base-of-slope erosional channel and proximal basin-floor confined/channel system, but previously interpreted as a canyon system (Clark & Pickering 1996a; Bayliss & Pickering 2015a); (6) Ainsa (3 sandbodies) = lower-slope erosional channels (e.g., Ainsa 1 in quarry outcrops ~1 km south of Ainsa) and proximal basin-floor channelised fans (e.g., Ainsa I immediately west of Labuerda village, and Ainsa II and III in the vicinity of the town of Boltaña) (Pickering & Corregidor 2005; Pickering *et al.* 2015); (7) Morillo (3 sandbodies) = lower-slope to base-of-slope erosional confined channels/channel system (*cf.* similar interpretation by Clark & Pickering 1996a; Bayliss & Pickering 2015b) and (8) Guaso (2 sandbodies) = 'structurally-confined clastic ramp' (Sutcliffe & Pickering 2009; Scotchman *et al.* 2015b). For the Banastón, Ainsa and Morillo systems, the base-of-slope is marked by a change in mean palaeoflow from a more westerly to more north-northwesterly direction, and appears to have coincided broadly with the present-day Rio Cinca and Mediano reservoir. The base-of-slope and proximal basin-floor environments are characterised by abundant sediment slides and slumps (Facies Class F deposits) and debrites (Facies Class A deposits). The Ainsa Basin sandbodies have been referred to simply as slope channels (e.g., Benevelli *et al.* 2003; Fernandez *et al.* 2004; Falivene *et al.* 2006a, b; Bakke *et al.* 2008), but this interpretation ignores their complexity that includes all depositional elements of submarine fans (such as lobe deposits overlain and cut into by scour-and-fills and/or small channels and then larger channels (e.g., Ainsa II Fan panel in Fig. 8.28), and fails to recognise the down-system change in depositional style from erosional channels enclosed in mainly type I MTDs/MTCs to less-confined erosional–depositional channelised sandbodies separated by mostly undeformed, bedded, fine-grained and thin-bedded sandy SGF deposits and marlstones (Fig. 4.34). Whilst the Ainsa Basin floor probably had a gradient not dissimilar to that of modern continental slopes (rather than that of flat abyssal plains), it seems more sensible to consider the Ainsa Basin, relative to the surrounding and confining environments, as a lower-slope, base-of-slope and proximal basin-floor system (*cf.* interpretations by Pickering & Bayliss 2009), with the Jaca Basin as the more distal basin-floor system. Similarly, modern submarine fans have segments dominated by erosional channels, passing distally into erosional–depositional channels and then more distal lobe and related deposits (see Chapter 7).

Dating of the Ainsa Basin, based mainly on foraminifera, suggests that the deep-marine strata of the basin are Middle/Late Ypresian and Lutetian in age (Pickering & Corregidor 2005; Das Gupta & Pickering 2008; Heard *et al.* 2008); with the overlying deltaic sediments being Bartonian (Dreyer *et al.* 1999). The deep-marine deposits span ~10 Myr (time scale of Gradstein *et al.* 2004); consistent with dating of the more distal deposits in the Jaca–Pamplona Basin (Payros *et al.* 1999).

In general, the base of each sandy channelised submarine fan is defined by a chaotic deposit or type II mass-transport deposit/complex (MTD/MTC), typically including pebbly mudstones (see Pickering & Corregidor 2005 for definitions of MTCs; an 'MTD' being the deposit from a single event). A predictable, idealised, genetic vertical sequence that tends to fine upwards has been recognised and interpreted, using sequence stratigraphic principles, to have resulted from a relative base-level change, interpreted as due to tectonic cycles by Pickering and Corregidor (2005) but a eustatic sea-level control by Pickering and Bayliss (2009) (Figs 4.35 to 4.37). Some of these sequences are capped by erosive pebbly type II MTDs/MTCs, probably reflecting more widespread shelf/slope failure in the coastal areas towards the latter stages of a sea-level lowstand and/or limited coarse clastic recharge to coastal environments and reworking into deep water during a lowstand. Thus, the type II MTDs/MTCs are the least predicatble component of these idealised vertical sequences. Palaeoflow in the southeasterly lower-slope erosional channels is generally towards ~290° compared with the more northerly proximal and axial basin-floor where palaeoflow tends to be towards ~320°. The widths of the sandy channel fills typically vary between ~400–800 m, but if the channel margin heterolithics, levée and overbank fine-grained and thin-bedded sandy SGF deposits and marls are included, the widths are ~2.5–4 km. At any time, it appears that only one channel was active in any sandy submarine fan, probably in the deepest axial part of the basin. Every sandy fan and any channels (typically 10–30 m thick) show a lateral offset

stacking towards the west-southwest, away from the eastern basin slope created by the growing Mediano Anticline (Fig. 4.34a, b).

Although each sandbody and associated, enveloping, thin-bedded deposits have previously been interpreted as accumulating due to (i) tectonic pulses (Pickering & Corregidor 2005), or (ii) Milankovitch-controlled cyclicity (~400-kyr-long eccentricity mode) (Pickering & Bayliss 2009), the principal results provide a generic depositional model for other deep-marine clastic systems irrespective of whether they are tectonically or eustatically (climatically) driven. The generic vertical sequence is typified by: (i) not only do MTDs/MTCs constitute a major component of the stratigraphy of the Ainsa deep-marine basinal sediments, but also particular types of MTC tend to occupy (although not exclusively) a characteristic stratigraphic position in relation to the evolution of individual fans; (ii) the stepwise foreland migration of individual fans away from the deformation front indicates a primary tectonic control; (iii) the stepwise foreland migration of individual low-sinuosity channels (likely on time scales of tens of thousands of years) was probably driven by the dynamics of the base-of-slope fold-and-thrust wedge, tectonically driven by incremental thrust-tip propagation and/or gravity-driven periodic mass wastage of the slope to create base-of-slope, mounded, type I MTDs/MTCs; (iv) the deep-marine expression of the inferred coarse-clastic sediment pulses began with large-scale basin-slope collapse as sediment slides and cohesive flows (type I MTDs/MTCs) that formed much of the seafloor topography for each fan and contributed to their lateral confinement; (v) the uppermost slope and any shelf edge, including the narrow shelf, then collapsed, redepositing unconsolidated sands and gravels into deep water (type II MTDs/MTCs); (vi) the basal coarse clastics with type II MTDs/MTCs as a significant component facies, are overlain by an interval of mainly channelised and amalgamated sandy deposits, in which major erosional events, including re-incision processes associated with channel development, are characterised by pebbly mudstones and sandstones rich in angular, locally derived intraclasts (type III MTDs/MTCs) and (vii) the channelised sands pass up into several tens of metres of less confined, non-amalgamated, medium- and thin-bedded, fine-grained sands and marls.

The deposits of depositional stages (iv)–(vii), above, represent the most active period of submarine fan growth, initially by the development of erosional channels, followed by several cycles of sediment bypass, and back fill, giving way to non-channelised, fine-grained sandy deposition. This progression is interpreted to be a response to the flushing out of coarse clastic materials from coastal and near-coastal fluvial systems. During the last stage of active fan growth ((vii) above), sediment accumulation rates probably remained high, and the degraded submarine slope was regraded and healed by fine-grained depositional events. The large amount of woody material and the high non-marine palynomorph signal in these sandy deposits suggest direct river input probably as hyperpycnal turbidity currents for the silty marls. In the upper few metres of each sandy submarine fan, a thinning-and-fining-upward trend accompanies a return to background marl deposition, representing submarine fan abandonment. Many sequences are overlain by intraformational sediment slides (typically type I MTDs/MTCs but rarely type II MTDs/MTCs) that attest to the increasing seafloor gradients associated with the regrading and healing stage in slope development. These organised, predictable vertical sedimentary sequences provide a testable generic model for the development of submarine fans where fan growth is strongly influenced by tectonic processes.

The Ainsa Basin was bounded to the east by the syn-depositional Mediano Anticline and thrust front associated with the lateral-ramp

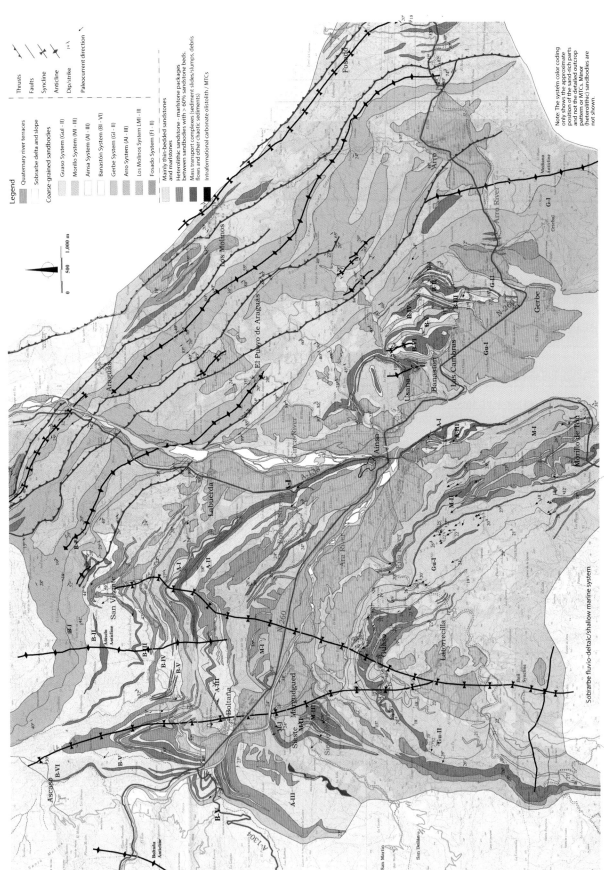

Fig. 4.34 (a) (caption on p. 179)

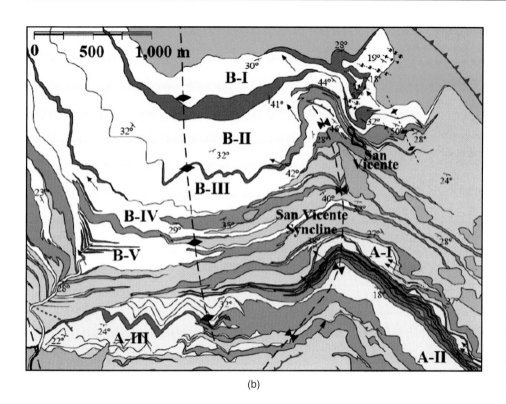

(b)

Fig. 4.34 (a) Facies-association map of the deep-marine systems in the Ainsa Basin, with enlarged area (b) from the NW quadrant of this map to more clearly show the lateral offset stacking as mapped in the Banastón System. From Pickering and Cantalejo (2015), modified from Pickering and Bayliss (2009).

margin of the Pyrenean thrust units (Cotiella nappe), and to the west by a seafloor high, now represented by the Boltaña Anticline (Mutti *et al.* 1985; Holl & Anastasio 1993; Poblet *et al.* 1998; Dreyer *et al.* 1999; Pickering & Corregidor 2005). The sediments within the basin were deformed by both syn-sedimentary and post-depositional tectonics. The Boltaña Anticline broadly coincides with the change from more channelised (Ainsa Basin) to less confined and more sheet-like sandy deposits (Jaca Basin). Although it is uncertain when the Ainsa Basin became a thrust-top basin, it was located within the flexural wavelength of the orogenic load and would have responded to the foreland-basin dynamics that drove subsidence.

Intrabasinal and basin-margin tectonics controlled the position of at least the younger sandy systems (Upper Hecho Group) and their constituent fans (Pickering & Bayliss 2009; see thematic set of papers on the Upper Hecho Group in the journal *Earth-Science Reviews* 2015), in a process of 'seesaw tectonics', by: (i) westward lateral offset stacking of sandy channelised fans due to encroachment of the eastern side of the basin, represented by the Mediano Anticline, and (ii) eastward (orogenward) relocation of the depositional axis of each sandy system, induced by phases of relative uplift of the Boltaña Anticline (Figs 4.34, 4.38). Pickering and Bayliss (2009) also inferred a similar process for the older systems (Lower Hecho Group), something that cannot be verified because of a lack of sufficient suitable outcrops. During basin infill, uplift of the Boltaña Anticline led to increasing basin narrowing and depositional confinement for the youngest deep-marine Morillo and Guaso sandy systems. Unlike the earlier depositional systems, the overlying Sobrarbe fluvio-deltaic system was supplied from a more southerly entry point into the Ainsa Basin. All the deep-marine sandy systems in the Ainsa Basin were fed from a southeastern point source, via canyons and erosional lower-slope channels that cut into the growing lateral-ramp zone, including the Mediano Anticline. The fundamental driver for these tectonic processes must have been controlled

by linked and partitioned movement on the thrusts associated with the syn-depositional growth of the Mediano and Boltaña anticlines (Bentham *et al.* 1992). Where outcrop mapping is good, each system tends to show overall decreasing confinement with age, probably reflecting periods of relative tectonic quiescence after phases of uplift-related movement on the basin-bounding anticlines. Although differential compaction could be invoked to explain at least some of the offset (compensational) stacking pattern for the sandy submarine fans, the consistent westward relocation of successive submarine fans in any system, together with a demonstrable westward progradation of the lower basin-slope sediment entry points between systems, argue against compaction as a primary driver. Compaction of the marls over the sandy submarine fans, however, would have acted to encourage offset stacking.

The recognition of Milankovitch periodicities in laminated deep-marine fine-grained siliciclastic successions, as in the Ainsa Basin, remains poorly documented, especially in such tectonically active basins. Controversy remains about the role of climate, tectonics and/or autocyclicity in driving deposition in such basins. Although there is an important tectonic control on the distribution of the sandbodies in the Ainsa Basin, Heard *et al.* (2008) and Cantalejo and Pickering (2014, 2015) used trace fossils to infer that Milankovitch periodicity exerted a control on the intertonguing pelagic, hemipelagic and fine-grained, thin-bedded sandy and silty turbiditic sediments. They applied a quantitative analysis of trace-fossil abundance and intensity in an essentially continuous 230 m-long core (Well A6), comprising very thin- and thin-bedded siliciclastic SGF deposits (Figs 4.39, 4.40). After removing the sediment slides and cohesive flow deposits, as these represent geologically instantaneous events approximately two orders of magnitude thicker than the typical laminated SGF deposits in the core, spectral analysis was applied to the bioturbation intensity data, revealing a cyclicity at the 41 kyr and ~112 kyr (possibly an average of the 95 and 125 kyr)

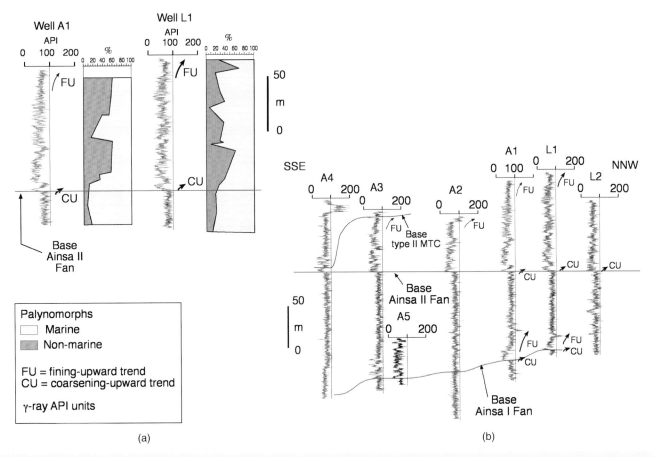

Fig. 4.35 (a) Total gamma-ray logs for seven drill sites (API units) ~400 m apart in the Ainsa fan systems, Ainsa Basin, Spanish Pyrenees. Note coarsening-upward trend (CU) at base of parts of the Ainsa II Fan, and fining-upward trend (FU) of Ainsa fans in their uppermost few metres by gradual abandonment. Intrafan sequences are not shown. (b) Comparison of gamma-ray logs for Wells A1 and L1 to show that the dramatic increase in land-derived (relative to marine) palynomorphs occurs with a significant time lag after the onset of submarine fan sedimentation. This increase is interpreted to record direct fluvial input to the deep-marine basin after the shelf had collapsed. See Fig. 4.37 for sequence-stratigraphic explanation. From Pickering and Corregidor (2005).

Milankovitch frequencies (Fig. 4.41). Heard *et al.* (2008) proposed that this drove environmental changes in bottom-water conditions in the Ainsa Basin, most likely leading to critically varying levels of oxygenation (stratification) of bottom waters. Cantalejo and Pickering (2014, 2015) revisited the trace-fossil data from the same core and confirmed, but with much higher confidence levels, that the Milankovitch frequencies are indeed present.

More recently, using a multi-proxy approach, including spectral gamma-ray logging (SGR), geochemical analyses (e.g., $CaCO_3$ content) and XRF multi-element scanning, Cantalejo and Pickering (2014, 2015), and Scotchman *et al.* (2015b), identified Milankovitch (and possible sub-Milankovitch) millennial-scale climate variability in Ainsa Basin outcrops and split-cores (Well A6) (Figs. 4.42a,b, 4.43a,b). Four commonly used analytical approaches were used to test for cyclicity in the Ainsa A6 well core and all yielded similar results, although smoother results were achieved with the REDFIT method which has the advantage of being used for discontinuous proxy data (Fig. 4.42a), more typical of many geological situations. Any higher-frequency, sub-Milankovitch, millennial-scale climatic oscillations within these fine-grained intervals would probably have to be explained by nonlinear ocean–atmosphere dynamics (*cf.* Quaternary Bond cycles of ~1500 years and Dansgaard–Oeschger cycles

of ~1470 years). The presence of sub-Milankovitch millennial-scale oscillations in the greenhouse climate of the Middle Eocene, however, would appear inconsistent with an origin that is dependent upon Quaternary-like conditions as is the case for the Bond and Dansgaard–Oeschger cycles. If present, sub-Milankovitch millennial-scale oscillations would suggest that there may be pervasive millennial-scale climatic variability throughout geologic time that were driven by an external mechanism such as solar forcing, rather than Quaternary-specific oceanic and cryospheric conditions. Such high-frequency sub-Milankovitch millennial-scale oscillations may be reflected at outcrop in the distinctively colour-banded Facies Class D, E and G deposits of the Ainsa Basin (Fig. 4.44a,b).

A further study by Cantalejo and Pickering (2015) in the Ainsa Basin presented time-series analysis of outcrop spectral gamma-ray and sandstone turbidite intensity data from eight fine-grained, essentially interfan sections located between coarse-grained sandbodies (submarine fans) throughout the Ainsa Basin (Fig. 4.43b). This study confirmed that the interfan deposits were orbitally driven and that Milankovitch cyclicity can be recognised throughout the entire stratigraphic evolution of the Ainsa Basin during a time when the basin was tectonically active. They therefore concluded that orbital parameters most likely paced the cyclic delivery of the fine-grained

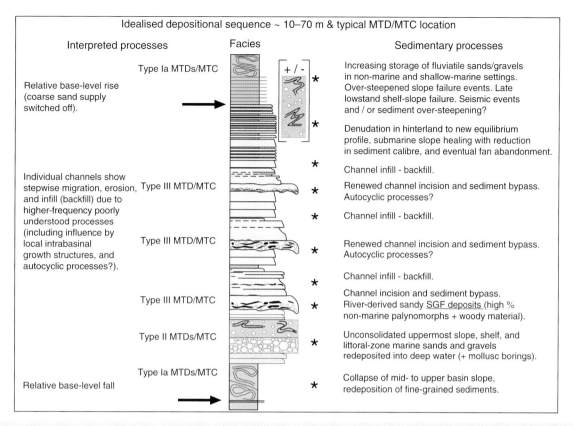

Fig. 4.36 Idealised vertical clastic sequence in the deep-marine Ainsa Basin, probably created by a combination of eustatic, tectonic and autocyclic processes. Note that whilst this sedimentary motif represents a complete sedimentary sequence in a single sandbody (fan), non-deposition and erosion, together with basinal location (axial, lateral, etc.) result in some sequences being base- or top-absent, creating somewhat different depositional patterns. The sequence-stratigraphic implication of each major MTD/MTC type is indicated, although type II MTDs can occur at any position in this sequence. Thicknesses of type Ia MTDs/MTCs below sandbodies are typically tens of metres. From Pickering and Bayliss (2009), modified from Pickering and Corregidor (2005).

sediments mainly by river- and delta-supplied turbidity currents from the non-marine and shallow-marine Tremp Basin immediately to the east. Sediment accumulation rates determined from spectral analysis show a gradual decrease throughout the deep-marine stratigraphy from ~57 to ~24 cm/kyr (Cantalejo & Pickering 2015).

The Middle Eocene, when the Ainsa Basin sediments accumulated, was a time of progressive cooling and deterioration of global climate and marks the transition from the warm and ice-free world of the Palaeocene thermal maximum (PETM at ~56 Ma) and early Eocene to the so-called 'icehouse conditions' of the Oligocene (~34 Ma) (Miller *et al.* 1987, 2011; Zachos *et al.* 2001, 2008; Coxall *et al.* 2005; Liu *et al.* 2009) – albeit with a transient warming in the late Middle Eocene at ~41.5 Ma that appears to have lasted ~600 kyr (Bohaty & Zachos 2003). Transient ice caps may have formed at time scales of ~400 kyr (possibly linked to thermal expansion–contraction of ocean waters with sea-level change measured in metres rather than tens of metres or more; *cf.* Burgess *et al.* 2008). Whilst the dominant control for the delivery of coarser clastic sediments to the deep-marine basin (as sandbodies at least tens of metres thick) may have been tectonically and/or eustatically driven, higher-frequency Milankovitch modes appear present in the finer-grained, thinner-bedded sections and may have been linked to changing storminess and continental (riverine) runoff in the absence of any significant eustatic sea-level changes (Cantalejo & Pickering 2014, 2015). These interpretations

are consistent with Milankovitch cycles being widely recognised in other Eocene siliciclastic, mixed siliciclastic-carbonate and carbonate successions of broadly comparable age to the Ainsa Basin (e.g., immediately east of the Ainsa Basin in the Montañana Group, Tremp Basin, by Weltje *et al.* (1996); the Montserrat alluvial fan/fan-delta complex of the southeast Ebro Basin by Gómez-Paccard *et al.* (2011); a study of the early to Middle Eocene in the Pyrenean continental to deep-marine record by Payros *et al.* (2009); immediately west of the Ainsa Basin in carbonates of the Jaca Basin by Huyghe *et al.* (2012); the Laramide intermontane Bighorn Basin, USA, by Abels *et al.* (2013), and the Green River Formation, USA (Machlus *et al.* 2008; Aswasereelert *et al.* 2013). Palaeogeographic reconstructions place the Pyrenees (Ainsa Basin) at a latitude ~35° N during the Eocene (Hay *et al.* 1999), a latitude that is known to be sensitive to astronomically induced climate changes.

Sequence boundaries in shallow-water sediments at ODP Leg 189 Site 1171 (South Tasman Rise) compare well with other stratigraphic records (New Jersey and northwestern Europe) and with times of increasing $\delta^{18}O$ in deep-sea records, indicating that significant (>10 m) eustatic changes occurred during the early to Middle Eocene (51–42 Ma), that is, at the same time as deep-marine deposition in the Ainsa Basin (Pekar *et al.* 2005). Pekar *et al.* (2005) identified sequence boundaries on the South Tasman Rise and dated them using lithology, bio- and magneto-stratigraphy, water-depth changes,

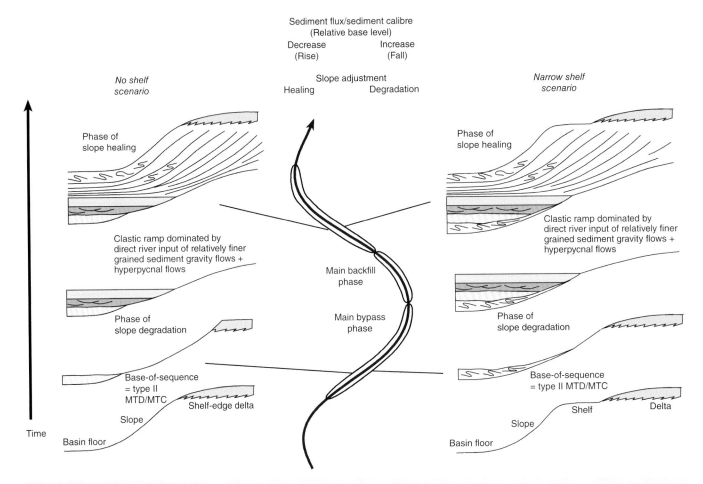

Fig. 4.37 Sequence-stratigraphic interpretation of clastic sequences in the deep-marine Ainsa Basin for a scenario with a shelf-edge delta (effectively no shelf), and with a shelf. The evolution of these sequences is shown within the context of changing sediment supply (flux) and sediment calibre (bulk mean grain size), and inferred changes in relative base level (tectonically or eustatically driven). Type Ia MTCs occur throughout the basin but are most common both immediately below, and at tens of metres above, the coarser clastic intervals. Slope and delta-front failure events are shown as most likely during both the initial collapse of the upper slope (as relative base level falls), and during the time interval of most rapid slope healing associated with regrading processes that increase the slope gradients (as relative base level rises). From Pickering and Corregidor (2005).

$CaCO_3$ content and physical properties (e.g., photo-spectrometry). The sequence boundaries are characterised by a sharp bioturbated surface, low $CaCO_3$ content and an abrupt increase in glauconite above the surface. Foraminiferal biofacies and planktonic/benthic foraminiferal ratios were used to estimate water-depth changes. Ages of six sequence boundaries (50.9, 49.2, 48.5–47.8, 47.1, 44.5 and 42.6 Ma) from ODP Site 1171 correlate well to the timings of $\delta^{18}O$ increases and sequence boundaries identified from other Eocene studies. The synchronous nature of sequence-boundary development from widely separated sites and the similarity of this timing to $\delta^{18}O$ increases indicates a global control that must be glacio-eustatic. This is despite the widespread belief that this part of the Eocene was a time free from continental ice sheets. The apparent operation of a glacio-eustatic control is supported, however, by previous modelling studies that suggest a decrease in atmospheric CO_2 levels below a threshold that would support the development of an Antarctic ice sheet at ~51 Ma (Pekar *et al.* 2005). Estimates of sea-level amplitudes range from ~20 m for the early Eocene (51–49 Ma) to ~25–45 m for the Middle Eocene (48–42 Ma) using constraints established for the

Oligocene $\delta^{18}O$ record. These results open up the distinct possibility that (glacio-) eustasy might have been the primary driver in switching on and off the supply of coarser sand to the deep-water systems of the Ainsa Basin, although tectonics cannot be ruled out. Additionally, other studies suggest that ice may have begun to form on Antarctica by the Middle Eocene (Tripati *et al.* 2006), but the extent of this ice and the influence of associated sea-level drawdown on sedimentary systems remain poorly understood.

High-frequency tectonic pulses that mimic Milankovitch periodicities are described by Beekman *et al.* (1995) and Vakarelov *et al.* (2006), but none have been demonstrated to operate with such regularity over a 10^5 yr time frame, nor with the Milankovitch ratios described by Cantalejo and Pickering (2014, 2015). Also, during icehouse conditions, such tectonic pulses tend to be subordinated/masked due to high-amplitude eustatic sea-level changes caused by glacio-eustasy (Vakarelov *et al.* 2006) and/or the thermal expansion and contraction of seawater. During greenhouse conditions, such as those postulated by some researchers for the Middle Eocene (Zachos *et al.* 2001, 2008; DeConto & Pollard 2003; Coxall *et al.* 2005), such tectonic pulses

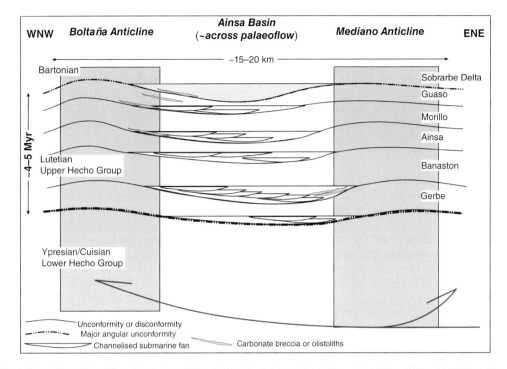

Fig. 4.38 Seesaw tectonic explanation for the sandbodies (fans) of the Upper Hecho Group, which may have been combined with phases of subsidence driven by flexural loading. Yellow represents discrete, mappable, channelised sandy submarine fans, and the blue represents inter-fan fine-grained, marlstone-rich sediments. The green is the overlying Sobrarbe deltaic deposits. Each system is underlain by an unconformity and correlative conformity. Modified From Pickering and Bayliss (2009). See text for explanation.

may have been significant for the Ainsa Basin. Presently, there is no clear mechanism to explain high-frequency, regular, cyclical tectonic processes in actual sedimentary basins where one might reasonably expect tectonic processes to be more erratic (irregular) and of variable duration. The good correlation between the predicted frequencies for Milankovitch cycles during the Eocene (Berger *et al.* 1992) and the frequencies calculated in the time-series analysis by Cantalejo and Pickering (2014, 2015) suggests that the principal driver of the finer-grained and thinner-bedded siliciclastics in the Ainsa A6 well was climatic rather than tectonic.

Any packaging of sediments within the Ainsa Basin could be construed as due to entirely autocyclic processes (*cf.* Lancien *et al.* 2004; Dennielou *et al.* 2006; Kane & Hodgson 2011). However, as Cantalejo and Pickering (2014, 2015) present strong evidence for Milankovitch processes in the sediments within the Ainsa Basin, one would have to argue that this is purely a coincidence and that autocyclicity just happened to operate at similar frequencies and with Milankovitch ratios. The presence of basin-wide decimetre-scale dark-coloured mudstone bands (Sutcliffe & Pickering 2009; see Fig. 4.44) suggests that extrabasinal processes were operating to influence environmental change within the Ainsa Basin. Although the principal control on the accumulation of the large sandbodies (sandy submarine fans) within the Ainsa Basin remains unresolved, it is likely to be due to the interaction of tectonics, climate and even autocyclic processes.

4.9   Problems in determining controls on sediment delivery

It is widely believed that cyclic sedimentation in tectonically active basins is caused by external forcing such as eustatic sea-level

fluctuations, episodic tectonic events, or varying sediment supply due to climate change, although local tectonics has been inferred to create sub-Milankovitch-scale, ~10-kyr, depositional units (e.g., Ito *et al.* 1999). Kim and Paola (2007), however, created cyclic sedimentation in an experiment under conditions of constant fault slip rate and sediment discharge with no base-level fluctuation. The experiment, designed to study sedimentation in a simplified extensional relay ramp system, led to reorganisation of the experimental fluvial channel network in a cyclic manner causing local variation in sediment supply to the hanging-wall basin, where subsidence was maximised. Scaling the experimental results to field length and time scales suggests that comparable autogenic cycles could produce 10–20 m-thick sets of strata on time scales of 10^5 yr, comparable to observed cases attributed to allogenic (allocyclic) effects. Although the experiment was designed to replicate a fluvial system, it seems reasonable to assume that, under suitable circumstances, such autogenic processes may deliver sediment to deep-water environments at time scales comparable to orbital (Milankovitch) cycles. However, this prediction remains untested.

A further problem involved in the interpretation of controls on deep-marine systems, particularly at the time scale appropriate for stacking of parasequences (Table 4.1), involves consideration of the effects, if any, of longer-term modulations of the Milankovitch cycles. Whilst the significance of the precession (~19 kyr and 23 kyr), obliquity (~41 kyr) and eccentricity cycles (~95 kyr and 125 kyr) are well known, the possible influence on sedimentation of the longer-term modulated cycles remains effectively unexplored; for example, the computed ~405 kyr and ~800 kyr eccentricity, 1.215 Myr modulation in the amplitude of obliquity, 2.475 Myr modulation in the amplitude of precession (e.g., Berger & Loutre 1991; Laskar 1999; Rial 1999;

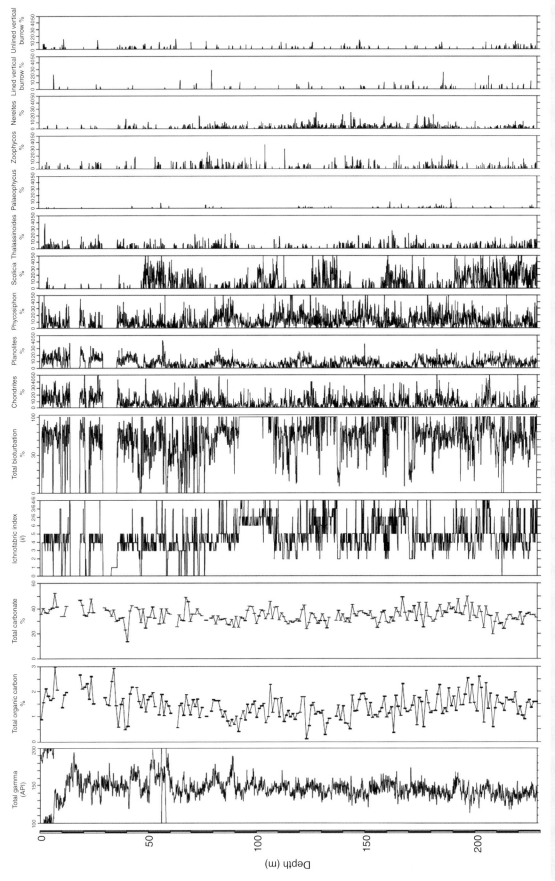

Fig. 4.39 Graphs showing downhole data from Ainsa Basin Well A6, including natural gamma in American Petroleum Institute units (API), Total Carbonate = TC (%), Total Organic Carbon = TOC (%), ichnofabric indices (ii), total bioturbation (%), and bioturbation intensities of common trace fossils. From Heard *et al.* (2008).

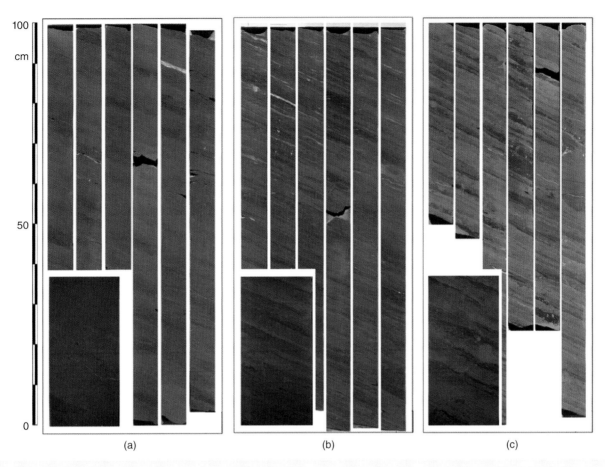

Fig. 4.40 Examples of bioturbation intensity from Well A6, including inset close-ups of the core (~6 cm wide). (a) Totally homogenised sediment. (b) Non-homogenised sediment. (c) Non-homogenised sand-rich sediment (from the upper unit of the core). From Heard *et al.* (2008).

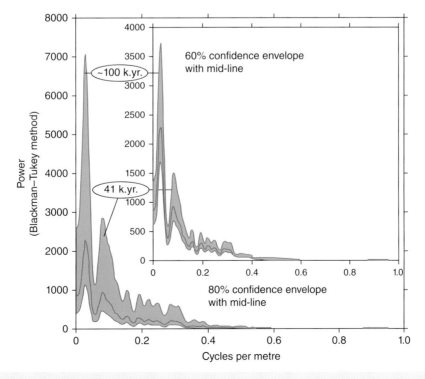

Fig. 4.41 Spectral analysis of intensity of bioturbation in Ainsa Well A6, for core from 71.2–230 m subsurface. This is a Blackman–Tukey spectrum using a Bartlett window. The bandwidth is 0.0327781. The error estimations are shown. If the smaller peak represents the 41 kyr cycle, then the large peak would be ~112 kyr. From Heard *et al.* (2008).

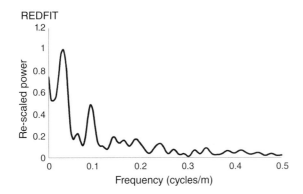

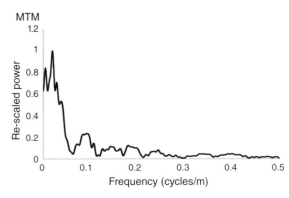

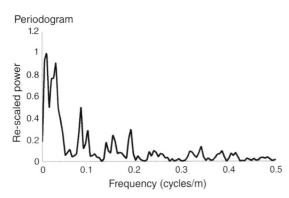

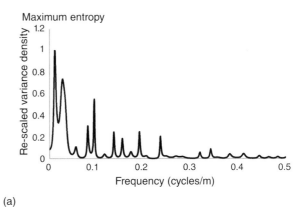

(a)

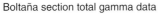

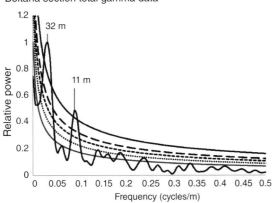

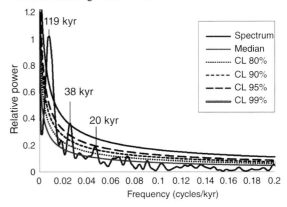

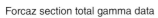

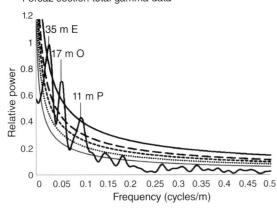

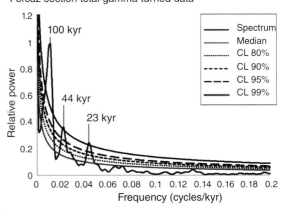

(b)

Fig. 4.42 (a) Cyclostratigraphic analysis undertaken on total spectral gamma-ray records of an ~165 m-long fine-grained sandy turbidite succession in the Banastón and Ainsa Systems, Ainsa Basin. For comparative purposes, spectral estimation has been determined by using four different methodologies: (i) REDFIT; (ii) multi-taper method (MTM); (iii) maximum entropy, and (iv) periodogram modified with a Barlett window. REDFIT uses 6 WOSA (Welch's overlapped segmented average) segments, MTM using four tapers and maximum entropy uses M=N/3. In the WOSA spectrum, the time series is segmented and the resulting periodograms are averaged to reduce their variance, at the expense of frequency resolution. Spectral computations used REDFIT (Schulz & Mudelsee 2002) and Analyseries (Paillard *et al.* 1996) software packages. The four analytical approaches all yield similar results, although smoother results are achieved with the REDFIT method which has the advantage of being used for discontinuous proxy data (more typical of many geological situations). (b) Cyclostratigraphic analysis of orbitally-tuned total spectral gamma-ray records in the Upper Hecho Group (Ainsa Basin). The Boltaña section is ~165 m long and is in interfan deposits of the Banastón and Ainsa Systems. The Forcaz section is ~185 m long and is in interfan deposits of the Ainsa and Morillo Systems. Both successions reveal Milankovitch frequencies with good confidence levels >99%. The data have been orbitally tuned to eccentricity using polarity reversals as anchor points from palaeomagnetic studies undertaken across the Ainsa Basin. Precession cycles are strengthened after tuning. Spectral estimation used REDFIT software (Schulz & Mudelsee 2002). Courtesy of UCL Deep Water Research Group, Blanca Cantalejo.

Palike *et al.* 2001) etc. Without a detailed understanding of palaeo-geography, and basin–hinterland tectonics, it will be very difficult to deconvolve tectonic and climatic signals. A remaining enigma is that ice ages appear to be dominated by the eccentricity cycle. The change in annual average solar insolation is small (~0.5%), but this cycle records by far the largest climate change. Two possible explanations are that: (i) The eccentricity cycle modulates the effects of precession (no change in insolation when e = 0), and/or (ii) Some process or processes amplify the temperature change, for example as a positive feedback loop.

Several researchers have argued that the relative importance of tectonic control in the stratigraphic record should be more clearly expressed during greenhouse periods of Earth history due to the lack of overmasking effects of high-frequency and high-amplitude sea-level changes typical for icehouse periods. The recognition of the importance of tectonics, especially at temporal scales comparable to the duration of typical transgressive–regressive cycles (parasequence stacking), has been frustrated by poor temporal resolution. An outcrop and subsurface-based study by Vakarelov *et al.* (2006) of a Cenomanian shallow-marine siliciclastic succession, Cretaceous Western Interior, USA, constrained by bentonite-based geochronol-ogy and detailed biostratigraphy, examined a 2.2 Myr interval where they identified four tectonically-driven erosional surfaces that dominate the preserved stratigraphy. Biostratigraphic correlation to a sea-level curve for the Cenomanian – where coeval high-frequency low-amplitude eustatic cycles are believed to exist – led to Vakarelov *et al.* (2006) proposing that minor tectonic pulses locally dominated over any effects of eustasy and exert the principal control. They therefore concluded that, at least in this basin, tectonic processes occurred at frequencies and at time scales comparable to the eustatic transgressive–regressive cycles more usually ascribed to Milankovitch ~400 kyr cycles.

Most researchers assume that the sediment flux to a basin varies with periodicities caused by climate change or vertical movements (tectonics) in the source area and, therefore, should have a direct control on the high-resolution stratigraphic record. Castelltort and Driessche (2003) applied a numerical model to investigate plausible time scales for point-sourced clastic sediment supply to basins. Their approach was based on the *sedimentary system concept*, which simplifies natural systems into three zones of dominant processes, the *erosion, transfer* and *sedimentation* sub-systems.

In a transfer sub-system, rivers convey sediments from the upstream source area to the depositional sub-system. This process may be considered to a first-order behaviour as diffusive (Paola *et al.* 1992; Humphrey & Heller 1995; Dade & Friend 1998; Métivier 1999; Métivier & Gaudemer, 1999; Allen & Densmore 2000). In this case,

the response time, *T*, of the transfer sub-system is in the form:

$$T = L^2/K \tag{4.1}$$

with *L* as the length of the sub-system and *K* as its coefficient of diffusivity.

The larger the transfer sub-system, the longer its response time. The more diffusive a transfer sub-system is, the shorter its response time. In natural systems, Dade and Friend (1998) calculated river diffusivities by using the water flux per unit width and a sediment mobility parameter, which incorporates the effects of bedload and suspended load in transport. They found response times of: 85 kyr for the Brahmaputra River; 74 kyr for the North Platte River; 65 kyr for the Mississippi River; 21 kyr for the Indus River; 5.5 kyr for the Cheyenne River and 2.4 kyr for the Savannah River (i.e., in the range of 1 kyr to tens of thousands of years). Métivier (1999) and Métivier and Gaudemer (1999) show that the diffusivity coefficient of a river approaching equilibrium conditions scales with its output sediment flux *Qst*, width *W*, and mean slope, $\partial z/\partial x$:

$$K = \frac{Qst}{[W(\partial z/\partial x)]} \tag{4.2}$$

With the above two equations, Métivier (1999) and Métivier and Gaudemer (1999) derived first-order response times in the range of 10^5–10^6 years for some large Asian river flood plains, which accounts for their strong buffering action for high-frequency sediment input variations (Métivier 1999; Métivier & Gaudemer 1999).

Castelltort and Driessche (2003) applied a diffusive model to several rivers, from long- (>1000 km) to intermediate-length (>300 km) rivers, where the transfer sub-system acts as a buffer for short period sediment pulses (10^4–10^5 year time scales). This suggests that high-frequency stratigraphic cycles in clastic successions fed by large drainage systems are unlikely to simply record sediment supply cycles with periodicities of tens to hundreds of thousands of years. Using the sedimentary system concept, however, it appears that the sediment flux is a derivative of tectonic and climatic changes in the source area and that this released sediment mass must be transported from its source to the depositional site, which is unlikely to be instantaneous.

The sedimentary system concept, therefore, implicitly suggests that the first-order controls on the time scales of variation in sed-iment flux to a basin are the response times of both the erosion and transfer sub-systems. The analysis provided by Castelltort and Driessche (2003) using a simple diffusive model for fluvial entities suggests that intermediate and large transfer sub-systems (>300 km) will buffer high-frequency (≤100 kyr) sediment input disturbances coming from the erosion sub-system. This is in agreement with the

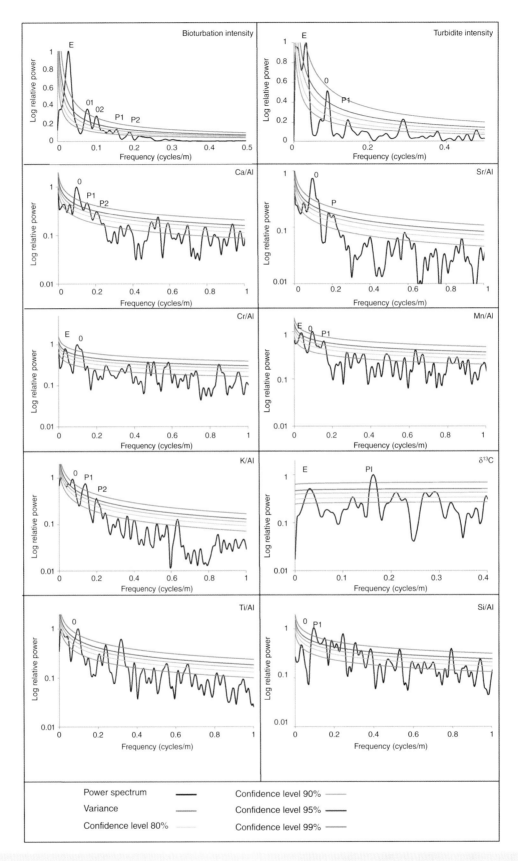

Fig. 4.43 (a) (caption on p. 190)

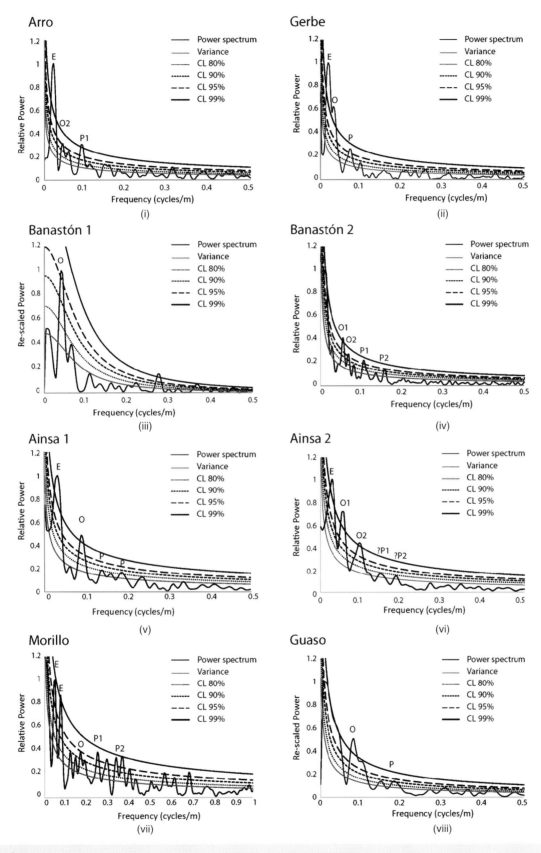

Fig. 4.43   (b) (caption on p. 190)

Fig. 4.43   See page 188–189. (a) Time-series analysis from the Ainsa Well A6 of multi-element XRF data using a 4 cm measurement interval; bioturbation and sandstone turbidite intensity are at a 10 cm spacing and δ^{13}C at 1 m interval. Turbidite intensity was calculated from ~230–80 m core depth by counting the number of sandy turbidites ≥ 0.2 mm thick in successive 10 cm-thick vertical core intervals. Bioturbation and turbidite intensity cover the same stratigraphic interval. Spectra have been re-scaled in order to compare spectra from different variables (*cf.* Weedon 2005). The most prominent peaks in these data appear to be associated to eccentricity (E) and obliquity (O), while frequencies associated with precession (P) are less prominent for most elements. From Cantalejo and Pickering (2014). (b) Time-series analysis of total-gamma data from stratigraphic intervals throughout the Middle Eocene Ainsa Basin using REDFIT. REDFIT spectral results used between 2 and 6 WOSA (Welch-Overlapped-Segment-Averaging Procedure of Welch 1967) segments and a minimum of 5 degrees of freedom. 0FAC (oversampling factor) maintained at 4 and Nsim (number of simulations) established at 1000. REDFIT results plotted using a power regression method, with the exception of the Banastón section 1, were the AR1 (1st order autoregressive process) of Mann and Lees (1996) and show a lower summative square error and a better curve fit. For comparative purposes, the graphs have been re-scaled. Analysis used the software developed by Schulz and Mudelsee (2002). From Cantalejo and Pickering (2015).

(a)   (b)

Fig. 4.44   Candidate millennial-scale sub-Milankovitch cycles in the Middle Eocene Ainsa Basin, Spanish Pyrenees. (a) Interfan sediments between the Banastón and Ainsa Systems. Human scale right of centre. (b) Guaso System. Human scale right of centre. The colour banding reflects grain-size differences. These cycles appear to represent 1.5–2 kyr duration as they occur in successions where sediment accumulation rates are ~30 cm kyr^{-1}, giving average precession cycles encompassing ~6 m of vertical stratigraphy.

buffering action evidenced for large Asian floodplains facing potential high-frequency climate-induced sediment flux variations during the last 2 Ma (Métivier 1999; Métivier & Gaudemer 1999). The transfer sub-system, therefore, plays an important role in the final stratigraphic record where allogenic forcing may have occurred. In sedimentary systems with short (perhaps <300 km) to negligible transfer sub-systems, such as catchment–fan systems, high-frequency variations of sediment flux to the depositional sub-system may occur in equilibrium with climate changes in the source area (e.g., for the Eocene Ainsa Basin – see Section 4.8 above). In such systems, if the influences of basin factors, in conjunction with the sediment flux, can be deconvolved, the stratigraphic record may provide valuable information about short-term climatic and tectonic changes in the source area. In sedimentary systems fed by intermediate to large transfer sub-systems, for example, for large deltas, high-frequency (100 kyr) cycles of sediment flux may not occur in equilibrium with allogenic changes in the source area. In such cases, high-frequency stratigraphic cycles are unlikely be an equilibrium response to such allogenic forcing. In such successions, only over longer time periods, that is, >hundreds of thousands of years, is it likely that variations in sediment supply will be in equilibrium with climate or tectonic changes in the source area – and therefore be recorded in the stratigraphy (*cf.*

Sloss 1979; Raymo *et al.* 1988; Galloway & Williams 1991; Liu & Galloway 1997; Peizhen *et al.* 2001). The high-resolution stratigraphic record of basins fed by intermediate to large transfer sub-systems can provide information about high-frequency effects caused by basin tectonics or eustasy, but not about high-frequency climatic or tectonic changes in the upstream zones. The conclusions of Castelltort and Driessche (2003) do not preclude stratigraphic responses of alluvial basins to rapidly changing sediment supply or diffusivity (Paola *et al.* 1992). A weakness of the analysis by Castelltort and Driessche (2003) is that it was based on modern river data and on assumptions, such as the square root relationship between drainage area and river width, which may be different in ancient sedimentary systems (Paola 2000). Also, they approximated the transfer of sediments using linear diffusion, something that may not be accurate.

Clearly, from the foregoing discussion, there is little agreement amongst different researchers about the dominance of various cyclic or pseudo-cyclic drivers in sediment delivery to a basin, be they climatic, tectonic or somewhat random (autocyclic). Thus, many challenging problems remain in determining controls on sediment delivery from shallow- to deep-marine settings, including the application of the most appropriate quantitative approach to characterise sediment delivery from fluvial to deep-water settings.

4.10   Carbonate versus siliciclastic systems

There is a widespread consensus amongst sequence stratigraphers that, in a glacio-eustatically driven world, lowstands in sea level are associated with increased sediment flux to deep-water environments. The frequency of gravity-flow events is also believed to increase at these times. In general, many studies of modern and ancient submarine-fan systems have supported this hypothesis (e.g., Mississippi Fan DSDP and Amazon Fan ODP legs).

Carbonate deep-marine systems commonly appear to respond in a manner opposite to siliciclastic systems (Table 4.2). During eustatic highstands, carbonate production on the shelf is significantly more vigorous and, therefore, a greater volume of carbonate material is carried by more frequent flows to the deep sea, where it accumulates as calcareous SGF deposits (Mullins 1983; Tucker & Wright 1990; Grammer & Ginsburg 1992; Schlager *et al.* 1994; Bracchert *et al.* 2003). This process is called *highstand shedding* (Schlager *et al.* 1994). Also, the wider range of grain types in carbonate SGFs may permit them to more effectively transport coarse, sand-grade sediment to more distal sites in a marine basin compared to siliciclastic counterparts of equivalent initial volume (Hodson & Alexander 2010).

In a study of ODP Site 626 cores from the Straits of Florida, Fulthorpe and Melillo (1987) argue that the thick, carbonate SGF deposits that accumulated during a relatively brief period in the middle Miocene were caused by overloading-induced instability of the adjacent slope with bank-derived sediment during sea-level highstands. Individual Facies A1.2 flow deposits are up to 19 m thick and contain chalk and limestone clasts in a matrix of muddy carbonate sand. Correlation with flow deposits of the Great Abaco Member of the Blake Ridge Formation (in the Blake–Bahama Basin), sediment-slide and slump deposits north of Little Bahama Bank, and a middle Miocene slide scar on the western Florida margin, however, suggests some uncertainty as slope failure might have been triggered by tectonic activity (Fulthorpe & Melillo 1987).

In a comparative study of the distribution of calcareous SGF deposits from the Bahama Bank (an aseismic area) and the Nicaraguan Rise (with many earthquakes), Andresen *et al.* (2003) show that both deep-water areas have a similar number of events with the timing of calcareous SGF deposits being the same in both areas, and with greater frequency associated with warmer, wetter periods and eustatic sea-level highstands. These enhanced rates of carbonate delivery are attributed to a more active nepheloid layer, rather than increased storminess during warmer and wetter periods, but it is also possible that an increased frequency (and possible average intensity) of tropical storms must play at least some part in increasing the frequency and magnitude of carbonate-load SGFs.

Table 4.2   Carbonate versus siliciclastic systems

System type	Siliciclastic	Carbonate
Most active growth	Lowstand in sea level	Highstand in sea level
Major sediment delivery	Shelf-edge deltas	Storms
Minor sediment delivery	Redeposition from slope failure	Redeposition from slope failure
	Riverine input	Redeposition from forereefs
Common associated deposits	Mudstones	Organic-rich marls, oozes, cherts
Fan radius	Variable	Typically <10s km

Roth and Reijmer (2004) have argued that Bahamian surface currents are directly linked to the North Atlantic atmospheric circulation, and thus periods with high current speeds are proposed to coincide with strong atmospheric circulation and vice versa. A marine sediment core from the leeward margin of Great Bahama Bank (GBB) was subjected to a multi-proxy study. The aragonite-dominated core MD992201 contains a record of the past 7230 years at a decadal time resolution and has sediment accumulation rates reaching 13.8 m kyr^{-1}. Aragonite mass accumulation rates, age differences between planktonic foraminifera and aragonitic grains, and palaeo-temperature distribution reveal changes in aragonite production rates and palaeocurrent strengths. Aragonite precipitation rates on GBB are controlled by exchange of carbonate ions and CO_2 loss due to temperature–salinity conditions and biological activity. These are all dependent on the current strength. Palaeocurrent strengths on GBB show high current velocities during the periods 6000–5100 yr BP, 3500–2700 yr BP and 1600–700 yr BP; lower current speeds existed during the time intervals 5100–3500 yr BP, 2700–1600 yr BP, and 700–100 yr BP.

The development of the sedimentary architecture at the leeward toe-of-slope of the Great Bahama Bank (Ocean Drilling Project Leg 166, Bahama Transect) since ~6 Ma has been shown to be a response to sea-level fluctuations and other major palaeooceanographic and climatic changes (Anselmetti *et al.* 2000; Reijmer *et al.* 2002). A major sequence boundary close to the Miocene/Pliocene boundary (~5.6–5.4 Ma) is interpreted to reflect a major sea-level fall followed by a sea-level rise, which elsewhere led to the re-flooding of the Mediterranean Sea at the end of the Messinian and increasing sea-surface temperatures on the Great Bahama Bank. Distinct erosional horizons also developed during the Pliocene (~4.6 and 3.3–3.6 Ma; Reijmer *et al.* 2002: their fig. 2). These are interpreted to reflect sea-level change, and the intensification of the Gulf Stream that occurred when the Isthmus of Panama was uplifted to a critical threshold. The Gulf Stream brings warm, saline and nutrient-poor waters to the Bahamas. Starting at the early-late Pliocene boundary (~3.6 Ma), this palaeooceanographic reorganisation and the enhanced sea-level fluctuations that were associated with the late Pliocene intensification in Northern Hemisphere glaciation (since 3.2 Ma) led to a gradual change from a ramp-type to a flat-topped-type bank morphology. There was a parallel change from a skeletal-dominated to a non-skeletal-dominated (mainly peloidal) sedimentary system. Increased sea-level fluctuations during the latter half of the Pleistocene led to intensified highstand shedding into the surrounding basins.

In the siliciclastic system of one of the Northern Slope mini-basins (offshore Texas–Louisiana shelf), Mallarino *et al.* (2005) found that deposition of sandy and muddy SGF deposits occurred mostly during lowstands at the subglacial and glacial scale, during Marine Isotopic Stage (MIS) 5d and 5b, late MIS 3 and MIS 2. The greatest volumes of sand were delivered to mini-basin 4 during MIS 2. In contrast, in the carbonate system of two mini-basins northeast of Pedro Bank, deposition of SGF deposits occurred mostly during highstands of MIS 5, early MIS 3 and MIS 1. It might therefore be reasonable to assume that in a mixed carbonate-siliciclastic system, siliciclastic SGF deposits and calcareous SGF deposits would alternate between times of sea-level lowstands and highstands, respectively. The Pandora Trough is such a mixed system in which two cores were acquired by Mallarino *et al.* (2005). Sediments in the lower 10 m of these cores represent mostly MIS 2 and consist of numerous quartz-rich sandy SGF deposits up to 20 cm thick interbedded with dark grey mud. The uppermost 2 m, corresponding to MIS 1, are made of light grey mud with one 30 cm-thick calcareous SGF deposit layer. Thus, as predicted for a mixed carbonate–siliciclastic system, siliciclastic sandy SGF deposits dominated during the MIS 2 lowstand whereas calcareous SGF deposits occurred at the end of the following transgression as soon as the adjacent carbonate platform tops were re-flooded (Ashmore, Eastern Fields and Portlock Reefs; Mallarino *et al.* 2005).

In a study of a progradational–retrogradational carbonate platform of late Palaeocene age from the Galala Mountains, Eastern Desert, Egypt, Scheibner *et al.* (2003) studied the changing facies associations from the platform margin in the north to the hemipelagic basin in the south, to explain the development of the carbonate platform as a consequence of eustatic sea-level fall and tectonic uplift at ~59 Ma (calcareous nannofossil biozone NP5). From this time onwards, the facies distribution along the platform-to-basin transect can be subdivided into five facies belts comprising nine different facies associations. Patch reefs and reef debris were deposited at the platform margin and horizontally bedded limestones accumulated on the upper slope. Sediment slides and cohesive-flow deposits occupy the lower slope. In the sub-horizontal toe-of-slope, mass-flow deposits pass into calcareous SGF deposits. Farther southwards in the basin, only hemipelagic marls were deposited. Between 59 and 56.2 Ma (NP5–NP8), the overall carbonate platform system prograded in several pulses. Distinct changes in facies associations from 56.2–55.5 Ma (NP9) resulted from rotational block movements that induced increased subsidence at the platform margin and a coeval uplift in the toe-of-slope areas. This resulted in the retrogradation of the carbonate platform. The patch-reef and reef-debris facies associations were replaced by the larger foraminifera shoal association. The retrogradation is also documented by a significant decrease in sediment-slide and cohesive-flow deposits on the slope and calcareous SGF deposits at the toe-of-slope.

The late Quaternary is associated with large glacial/interglacial climatic variations with tens to ~120 m, high-amplitude sea-level changes at Milankovitch frequencies during which carbonate platform tops were cyclically subaerially exposed and re-flooded. Jorry *et al.* (2010) studied the accumulation of calcareous SGF deposits, aragonite onset/abrupt increase in fine-grained sediments, and their timing in deep-marine basins adjacent to carbonate platforms. In particular Jorry *et al.* (2010) looked for a correlation between the age of the first SGF event and aragonite onset/abrupt increase, and the time of re-flooding of carbonate platform tops during deglaciations. Three basins were studied adjacent to isolated platforms in the Bahamas, the Northern Nicaragua Rise and the Gulf of Papua, selected as representative of pure carbonate versus mixed systems, in quiescent versus tectonically active settings. These platforms possess different carbonate bank-top morphology, ranging from atoll to relatively deeply and narrowly flooded flat-top banks. Despite these differences, each case shows a clear relationship between the timing of platform-top re-flooding and the initiation of significant carbonate export by SGFs and low-density plumes into the surrounding basins. The concept of a 're-flooding window' was introduced by Jorry *et al.* (2010) to characterise the prolific period of time during which bank and atoll-tops were flooded to produce a substantial export of bank-derived aragonite and of calcareous SGF deposits into adjacent deep-marine basins. Jorry *et al.* (2010) concluded that the main re-flooding windows occurred principally during the latter part of the sea-level rise at each glacial termination, those periods being marked by some of the highest rates of sea-level rise. Analysis of a long piston core from the earthquake-prone Walton Basin (Northern Nicaragua Rise) suggests that sea-level rise, not seismicity, controlled

the generation of SGFs through the last four glacial/interglacial transitions. The study by Jorry *et al.* (2010) shows that the most significant export of sediments in oceanic basins surrounding isolated carbonate platforms is driven by the combination of local and regional factors (e.g., the bathymetry on platform top) and external forcing (insolation variations, triggering climate changes, resulting in sea-level fluctuations) at glacial terminations. Their findings demonstrate a strong link between climate, sea level and the accumulation of SGF deposits (mainly calcareous SGF deposits) in several carbonate basins during the last four glacial–interglacial cycles. All three case studies show that the initiation of calcareous SGF deposit deposition and the onset of enhanced aragonite production and basinward transport coincide with abrupt rises of sea level at glacial terminations.

Another aspect of carbonate SGF deposits is the way that they record hydrodynamic sorting in the depositing current. In a study of Triassic calcareous SGF deposits, in a 100 m-long core and from nearby outcrops of the basinal Buchenstein Formation, composition and thickness variations show that the volume of calcareous SGF deposits, relative to background (hemipelagic) sediment, changed from 15% in the lower part to 60% in the upper part of the formation, reflecting a steady progradation of nearby platforms (Maurer *et al.* 2001). The composition of the sand fraction of 214 SGF deposits, interpreted as turbidites, reveals that micritic peloids (average 23%) and lithoclasts (16%) are by far the most abundant constituents. They are interpreted as two different varieties of *in situ* precipitated micrite (automicrite), which probably formed under the influence of microbes and constituted the principal building material of the adjacent platforms. Platform-derived skeletal fragments account for only 0.5% of the grains. Variations in SGF deposit composition were quantified using Spearman's rank correlation and cluster analysis. The most significant compositional variations in the cored succession seem to be related to hydrodynamic sorting in the turbidity currents and to the gradual shift from distal to more proximal SGF deposits as the platforms prograded basinward. Cluster analysis of the 214 samples shows a major subdivision into micrite- and sparite-dominated SGF deposits, the former more distal and the latter more proximal based on associated outcrop studies. Clusters associated with micrite-dominated SGF deposits are enriched in radiolaria and thin-shelled bivalves, whereas the clusters related to sparite-dominated SGF deposits show an abundance of lithoclasts. This subdivision is interpreted by Maurer *et al.* (2001) to result from sorting effects in a turbidity current, with carbonate mud and more delicate particles travelling farther into the basin. There is no indication that the composition of the SGF deposits fluctuated significantly under the influence of sea-level variations. This is not surprising because the dominant automicrite facies of the platforms only migrated laterally, but shows no significant textural change during sea-level cycles.

Payros and Pujaltea (2008) provide an overview of modern and ancient calciclastic submarine fans. They conclude that calciclastic submarine fans range in length from just a few to >100 km. They distinguish three different types: (i) coarse-grained, small-sized (<10 km) fans, characterised by an abundance of calcirudites, a scarcity of mud, relatively long levéed channels and small radial lobes; (ii) medium-grained, medium-sized fans (10–35 km long) typified by an abundance of calcarenites and lesser amounts of calcirudites and mud, with a tributary network of slope gullies merging to form a levéed channel that opens to the main depositional site, characterised by extensive lobes and/or sheets, which eventually pass into basinal deposits through a narrow fan-fringe area (*cf.* Puga-Bernabéu *et al.* 2009); and (iii) fine-grained, large-sized fans (>50 km and generally close to 100 km long) rich in calcarenites and mud, poor in calcirudites, having wide and long slope-channels that feed very extensive calciturbiditic sheets. In terms of grain-size distribution, the three fan types compare well with sand/gravel-rich, mud/sand-rich and mud-rich siliciclastic submarine fans, respectively (Table 7.1). However, they show notable differences in terms of size and sedimentary architecture, a reflection of the different behaviour of their respective SGFs. Most calciclastic submarine fans were formed on low-angle slopes and were sourced from distally steepened carbonate ramps subjected to high-energy currents. Under these conditions, shallow-water, loose, grainy sediments are transferred to the ramp slope and eventually funnelled onto the submarine fan by SGFs. These conditions seem to be more easily met on leeward margins in which the formation of reefs is hampered by cool waters, nutrient enrichment or oligophoty.

Another circumstance that contributes to the transfer of shallow-water sediments to the distal ramp slope is a low sea level, forcing the so-called carbonate factory closer to the slope break and destabilising sediments that become subaerially exposed and drained of their pore-water, removing former buoyant support of the sediment mass. The most important factor controlling the development of calciclastic submarine fans, however, appears to be the existence of an efficient funnelling mechanism forcing SGFs to merge downslope and build a point-sourced sedimentary accumulation. In most cases this requires a major slope depression associated with tectonic structures, an inherited topography, or large-scale mass failures.

4.11 Computer simulations of deep-water stratigraphy

One of the exciting aspects of time–space integration of deep-water systems has been the development of increasingly sophisticated computer models. In deep-water reservoir modelling, a quantitative appreciation of the spatial distribution of architectural elements is important for estimates of pore-volume distribution and the connectivity of reservoir sandbodies. In particular, this is critical for rock- and fluid-volume estimates, reservoir-performance predictions and development-well planning (Reza *et al.* 2006). Optimising reservoir management requires the integration of realistic geological and engineering attributes. Despite the over-simplifications and assumptions that go into such models, one of their main functions is to help constrain thinking by showing how stratal geometries may be influenced by various controlling factors. This aspect of deep-water systems is beyond the scope of this book; therefore, this short section only serves as a brief introduction to this topic.

Several stochastic modelling techniques have been developed for building deep-water reservoir models and can be divided into three categories: (i) cell-based approaches that primarily implement two-point geostatistics (Deutsch & Journel 1998) and multipoint geostatistical concepts (Strebelle *et al.* 2002); (ii) object-based or Boolean approaches that tend to build more geologically realistic reservoir models using nonlinear features (Haldorsen & Lake 1984; Haldorsen & Chang 1986; Jones 2001; Deutsch & Tran 2002) and for which geological objects are conditioned to hard data (e.g., wells) and also honour stratigraphic relationships and interpretations; (iii) stochastic surface-based techniques (Xie *et al.* 2000; Pyrcz *et al.* 2005) that capture the tendency of the deposits of individual flow events to show compensational stacking patterns within lobe deposits (*cf.* Fig. 7.39).

An integrated stochastic reservoir-modelling approach was developed by Reza *et al.* (2006) (ModDRE – Modeling Deep-water Reservoir Elements) to account for geomorphic and stratigraphic controls in order to generate a deep-water reservoir architecture. Using the Fortran programming language, ModDRE attempts to mimic geologically realistic results of process-based techniques but incorporates stochastic modelling throughout the process. Information on stratal-package evolution can be integrated into the reservoir-modelling process. A slope-area analytical approach is implemented to account for topographical constraints on channel and sheet-form reservoir architectures and their distribution. Inferred sediment-source statistics and architectural-element variability (from seismic, outcrop and stratigraphic studies) associated with relative base-level changes can also be used to constrain the reservoir-element statistics. Based on these geomorphic and stratigraphic constraints, reservoir elements, such as channels and lobes, are built into the model sequentially (in stratigraphic order).

4.12 Laboratory simulations of deep-water stratigraphy

Experimental modelling of stratal geometry is being increasingly used in an attempt to interpret stratal packages and their geometry, and to constrain the range of what is physically possible (e.g., using large flume tanks such as EUROTANK based at Utrecht University – e.g., Postma *et al.* 2008 – The Netherlands; St. Anthony Falls Laboratory, University of Minesota/USGS; and the STEP Basin, University of Texas at Austin).

From flume experiments, Postma *et al.* (2008) concluded that: (i) For large temporal and spatial scales, stratigraphic styles (preservation of strata) in analogue flume models can best be simulated by nonlinear diffusion. (ii) The rate of infill in accommodation is typically nonlinear, and decreasing towards grade with steepening of the slope. Thus, the time that is needed to fill a unit width of river-shelf accommodation up to its grade (equilibrium) is not a simple calculation of the volume of sediment that is available against the available space, because before the accommodation is filled much of the sediment supplied will bypass the system into the downslope systems, such as SGF slope-apron fans. And (iii) if a nonlinearity exponent of $m = 1.5$ is assumed (which is not an unreasonable value for coarse-grained real-world prototypes), then errors in rate of infill in accommodation between flume model and real-world prototype usually are not worse than 10%, and only exceptionally reach 30%. Such results open up possibilities to calibrate numerical models of sedimentary system evolution by analogue experiments.

Paola *et al.* (2009) concluded that the growth of quantitative analysis and prediction in Earth-surface science has been accompanied by growth in experimental stratigraphy and geomorphology. Experimenters have grown increasingly bold in targeting landscape elements from channel reaches up to the scale of entire erosional networks and depositional basins, often using very small facilities. The experiments produce spatial structure and kinematics that, although imperfect, compare well with natural systems despite differences of spatial scale, time scale, material properties and number of active processes. Experiments have been particularly useful in studying a wide range of forms of self-organised (autogenic) complexity that occur in morphodynamic systems. Autogenic dynamics creates much of the spatial structure we see in the landscape and in preserved strata, and is strongly associated with sediment storage and release. The observed consistency between experimental and field systems, despite large differences in governing dimensionless numbers, is what we mean by 'unreasonable effectiveness'. We suggest that unreasonable experimental effectiveness arises from natural scale independence. We generalise existing ideas to relate internal similarity – in which a small part of a system is similar to the larger system, to external similarity – in which a small copy of a system is similar to the larger system. We propose that internal similarity implies external similarity, though not the converse. The external similarity of landscape experiments to natural landscapes suggests that natural scale independence may be even more characteristic of morphodynamics than it is of better-studied cases such as turbulence. We urge a shift in emphasis in experimental stratigraphy and geomorphology away from classical dynamical scaling and towards a quantitative understanding of the origins and limits of scale independence. Other research areas with strong growth potential in experimental surface dynamics include physical–biotic interactions, cohesive effects, stochastic processes, the interplay of structural and geomorphic self-organisation, extraction of quantitative process information from landscape and stratigraphic records and closer interaction between experimentation and theory.

Van Dijk *et al.* (2009) concluded that fan deltas are excellent recorders of fan-building processes because of their high sedimentation accumulation rate, particularly in tectonically active settings. Although previous research focuses mainly on allogenic controls, there is clear evidence for autogenically produced storage and release of sediment by flume and numerical modelling that demands further definition of characteristics and significance of autogenically forced facies and stratigraphy. Analogue experiments were performed on fan deltas with constant extrinsic variables (discharge, sediment supply, sea level and basin relief) to demonstrate that fan-delta evolution consists of prominent cyclic alternations of channelised flow and sheet flow. The channelised flow is initiated by slope-induced scouring and subsequent headward erosion to form a channel that connected with the valley, while the removed sediment is deposited in a rapidly prograding delta lobe. The resulting decrease in channel gradient causes a reduction in flow strength, mouth-bar formation, flow bifurcation and progressive backfilling of the channel. In the final stage of channel filling, sheet flow coexists for a while with channelised flow (semi-confined flow), although in cycle 1 this phase of semi-confined flow was absent. Subsequent autocyclic incisions are very similar in morphology and gradient. However, they erode deeper into the delta plain and, as a result, take more time to backfill. The duration of the semi-confined flow increases with each subsequent cycle. During the period of sheet flow, the delta plain aggrades up to the 'critical' gradient required for the initiation of autocyclic incision. This critical gradient is dependent on the sediment transport capacity, defined by the input conditions. These autogenic cycles of erosion and aggradation confirm earlier findings that storage and release of sediment and associated slope variation play an important role in fan-delta evolution. The erosional surfaces produced by the autocyclic incisions are well preserved by the backfilling process in the deposits of the fan deltas. These erosional surfaces can easily be misinterpreted as climatically-, eustatically- or tectonically-generated bounding surfaces.

4.13   Supercritical versus subcritical fans

An exciting new approach to understanding the stratigraphic development of submarine fans, including their internal architecture and stacking patterns, is to consider the relative importance of deposition from supercritical and subcritical flows. Most submarine slopes, particularly with gradients >0.5° (Section 1.5.3), are likely to contain evidence of supercritical flow. Only in the last few years has there been a recognition of the importance of up-system-migrating bedforms (e.g., from Squamish prodelta slope, British Columbia, by Hughes Clarke *et al.* 2012a) linked to what are referred to as 'cyclic steps', where flows transition between subcritical and supercritical flow (Cartigny *et al.* 2011, 2014; Postma *et al.* 2014; also, see review by Talling 2015). These crescentic bedforms, which have a 30–40 m wavelength, heights of ~2 m, downhill-facing lee slopes of over 40° and uphill facing stoss sides of up to ~10°, even show a strong tidal cyclicity linked with river discharge into deep water during prolonged flow events (Hughes Clarke *et al.* 2012b). Locally, slopes on the Squamish prodelta can reach >45° (Hughes Clarke *et al.* 2011). Such concepts have been applied to understanding alluvial fans with incision–back filling cycles and lobe formation during fan growth with supercritical distributaries (Hamilton *et al.* 2013). Autogenic cycles of channelisation, terminal deposit formation, channel backfilling and channel abandonment have been observed in the formation of alluvial fans and deltas. The main distinctive characteristics of supercritical cycles in alluvial fans are revealed by how the flow interacts with the terminal deposit. At the channel-to-deposit transition of an alluvial fan, the flow undergoes a weak hydraulic jump, resulting in rapid sedimentation, de-channelisation and lateral expansion of the flow, and deposition of any remaining sediment on top of the channel fill and floodplain. The deposits created by this process often cap the channel and propagate up-system, erasing the morphology of the excavated channel (Hamilton *et al.* 2013).

The evolution, architecture and stratigraphy of submarine fans dominated by supercritical flows have been discussed by Hoyal *et al.* (2011, 2014), Demko *et al.* (2014), and Hamilton *et al.* (2015). Hoyal *et al.* (2014) concluded that sandy submarine fans from steep and tectonically active margins are typically small radius (<10 km), and could be dominated by deposits emplaced at hydraulic jumps, thereby limiting channel extension or forcing short avulsion lengths. They also noted that typical geometries include large-scale backset bedding as a lobe propagates up-system behind a migrating hydraulic jump. In cross-section, this process is inferred to produce a simple mounded lobe superimposed on a more narrow channel geometry, creating what has been referred to as a 'steershead'. Supercritical bedforms and bedform sequences that Hoyal *et al.* (2014) observed from several outcrops worldwide led them to call these 'supercritical fans'.

It should be noted, however, that experiments with coarse-grained sediment, unlike fine-grained material, suggest that if the deposition rate is sufficiently high, a flow either transits the length of the low-slope fan as a supercritical flow and shoots off the fan–canyon break without responding to it, or dissipates as a supercritical flow before exiting the fan (Kostic & Parker 2007).

Demko *et al.* (2014) described the deposits of migrating, aggrading bedforms emplaced beneath supercritical sediment gravity flows (supercritical dunes, antidunes and cyclic steps), which they recognised in submarine-fan and fan-delta successions in the Ainsa, Ebro, and Tabernas Basins in Spain. They define 'morphodynamic successions' (sets of genetically related strata that record the evolution of flow conditions deduced from these units) in mouth-bar, delta front, clinoform gully and perched and terminal lobe depositional settings. The morphodynamic successions contain laminae, lamina sets, bed geometries and morphologies that are interpreted to be the result of bedform configurations formed under stable and evolving supercritical sediment gravity flow conditions. The sedimentary structures that they describe include decimetre-scale, sigmoid, humpback trough cross-stratification, decimetre- to metre-scale asymptotic and downlapping upcurrent-dipping foresets, brinkpoint and topset preservation, grain and outsized clast imbrication, and associated scour and fluidisation structures, all of which have been replicated in flume experiments. Demko *et al.* (2014) interpret the sigmoid, humpback cross-stratification as having formed by the migration of dunes. Upcurrent-dipping asymptotic foreset strata are interpreted to have formed beneath long-wavelength antidunes, while upcurrent-dipping, downlapping foreset strata are interpreted to have formed beneath traction carpets encroaching into the scour pit formed by an upcurrent-migrating cyclic step. The preservation of brinkpoint and topset strata is related to bedform climb angle, and the ratio of apparent downstream migration to bed aggradation rate (Demko *et al.* 2014). Grain and clast fabric, including fluidisation features, reflect local flow conditions on the lee and foreset sides of bedforms. Ancient bedforms linked to supercritical flows have been shown by Pickering *et al.* (2015).

In contrast to the sedimentary structures interpreted to reflect deposition from a waning turbidity current to develop the classic Bouma sequence (Section 1.5.6), Postma and Cartigny (2014) propose a suite of facies linked to large-scale flow dynamics (Figs 4.45 and 4.46).

The application of these concepts to a better understanding of submarine fans (e.g., conditions for the development of 'supercritical fans' versus 'subcritical fans') has yet to be fully realised. These facies can still be described with the classification scheme of Section 2.2.2, but would have been formed during hydraulic jumps, and potentially while flows were crossing cyclic steps.

4.14   Hierarchical classification of depositional units

A hierarchical scheme for the systematic description of the deep-water systems starts with the smallest lithologic unit, a *bed*, which is either the basal coarse divisions (gravel + sand + silt)

emplaced by a single flow event, or might encompass the deposits of more than one flow event in the case of an amalgamated bed (Fig. 5.1). Groups of similar beds are defined as *bedsets*, for example beds of similar thickness, grain-size distribution and sedimentary structures. For deep-water facies that occur as sand–mud or silt–mud couplets because of eventual upward passage into an interbed of mud or mudstone (e.g., Facies Groups C2 and D2; Fig. 2.4), we allow the mud caps to form part of the bedset. Groups of bedsets constitute a *sandy storey*. At a higher order, the following depositional elements are recognised: *channel fill* (or inferred channel fill), *channel complex* (or inferred channel complex) and *channel complex set* (or inferred channel complex set).

A comparable terminology is adopted for *mass transport deposits* that build *mass transport storeys*, *mass transport elements* and *mass transport complexes*. The terminology for mass transport deposits follows that of Pickering and Corregidor (2005) for mass transport complexes, but with the caveat that we define single depositional events from sediment slides, slumps and cohesive flows as mass transport deposits (MTDs).

Storeys are fourth order and were originally defined by Friend (1979) as continuous ribbons of sediment within a channel (Fig. 4.47). Storeys can follow fining-upward or coarsening-upward trends (Campion *et al.* 2005). Commonly, they appear to show a coarsening-upward trend in axial environments and fine upwards in off-axis/marginal environments. Storeys may have an erosional base or may be purely depositional.

Outcrop-based studies of Permian deep-water deposits in the Karoo Basin, South Africa, led Flint *et al.* (2008) to define slope channel-fills (storey sets), channel complexes and complex sets that are organised into ten depositional sequences that stack into three composite sequences. Their composite sequences (one or more related sequences) are mappable for tens of kilometres down dip and across strike. In axial positions, sequence boundaries represent regional changes in slope-channel complex-set stacking patterns, fill style, and hence connectivity across regional erosion surfaces. Complex sets are identified on the basis of one or more related complex-fill style and stacking pattern (Flint *et al.* 2008). Sprague *et al.* (2008) extended the deep-water hierarchical approach of Flint *et al.* (2008) from complex

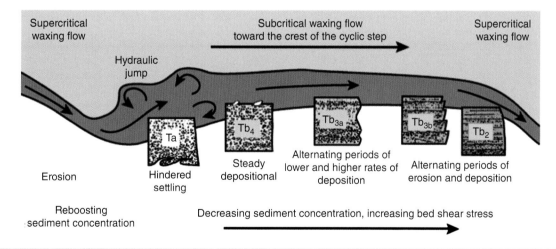

Fig. 4.45 Facies produced by deposition on a cyclic step. Flow along the lee side (i.e., downcurrent side) of the bedform is supercritical and strongly erosive. Flow along the stoss side (i.e., upcurrent side) is subcritical and depositional. Note the change in facies along the stoss side of the bedform, believed to vary with the sediment concentration and the increasing shear stress toward the crest. T_a and T_{b4} are both structureless, but only T_a is characterised by coarse-tail normal grading. From Postma and Cartigny (2014).

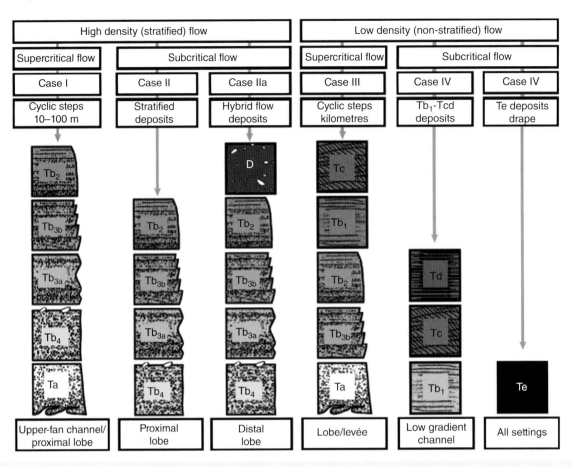

Fig. 4.46 Six suites of facies characteristics (facies-associations) related to case I–IV flow dynamics and depositional environment. D = debrite. Case 1 = supercritical basal layer; case 2 = subcritical basal layer; case 3 = supercritical flow; case 4 = subcritical flow. From Postma and Cartigny (2014).

scale (slope-channel complexes, levée complexes and frontal-lobe complexes resolvable on 30 Hz seismic data) down to well-log and core scale (individual storeys and storey sets). Identification of storeys and storey sets is easier away from axial areas of bed amalgamation. Storey sets stack to form complexes, which are organised into complex sets that exhibit characteristic stacking patterns, thus controlling sandbody connectivity at the limit of seismic resolution. These hierarchical reservoir building blocks are independent of depositional environment (Sprague *et al.* 2008).

Figures 4.48 and 4.49 show how this scheme was adopted with a sequence stratigaphic approach by Figueiredo *et al.* (2013) for an upper-slope submarine channelised system, Karoo Basin, South Africa. Figure 4.49 shows their model for the sequence stratigraphic evolution of the LST of the F2 elementary depositional sequence (located in Fig. 4.22).

4.15   Concluding comments

Whilst global climate change (including glacio-eustasy) can be shown to be important in the delivery of coarse clastics to many deep-marine basins, there are examples where tectonic processes appear to have been more important. Interestingly, in some studies in continental basins associated with active tectonics, it has been argued that climate change is more important in driving erosion and sediment

delivery than tectonic uplift (e.g., Garcia *et al.* 2011). An integral part of any analysis is an understanding of the time scales over which environmental change occurs, and the rates and magnitudes of those changes. Also, global climate change does not necessarily have to be associated with glacio-eustasy. As alternatives, thermal expansion and contraction of the oceans (steric changes), changes in the hydrological cycle, changing storminess, precipitation and continental runoff may all be important factors in sediment flux. Different Milankovitch frequencies (precession, obliquity, short and longer-term eccentricity, together with modulating effects of these), will conspire in deep time to create different stratigraphic signals. Likewise, tectonic processes operate at a range of time scales, rates and magnitudes. Plate-tectonic processes, such as the creation and destruction of oceanic gateways will also be important in driving global climate change at time scales from millions to tens of millions of years (e.g., Smith & Pickering 2003). Also, siliciclastic and carbonate systems appear to respond differently to global climate change and base-level changes.

An example of the importance of oceanic gateways can be found in the Cretaceous–Holocene history of the Tasmanian gateway (Exon *et al.* 2004). The southwestern South Tasman Rise finally separated from Antarctica at the time of the Eocene–Oligocene boundary (~33.5 Ma). The Tasman Rise then subsided, and the continental margin of Tasmania collapsed. The Tasmanian Gateway opened to deep water, disrupting oceanic circulation at high southern latitudes

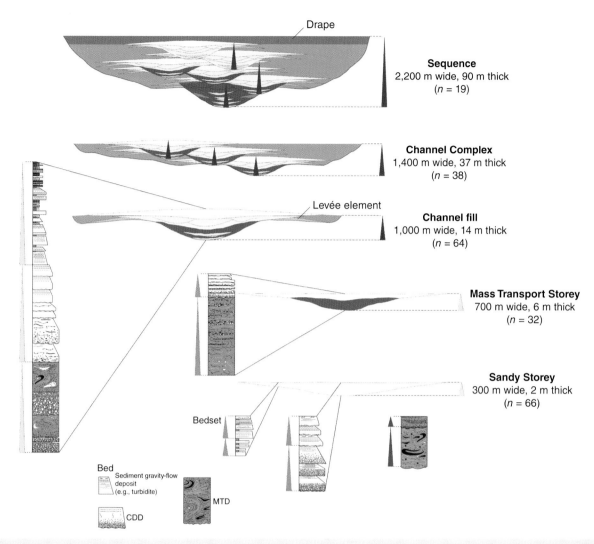

Fig. 4.47   Hierarchical classification of the fundamental building blocks for describing the architecture of deep-water deposits based on Sprague *et al.* (2002, 2005), with examples from the Mid Eocene Ainsa Basin, Spanish Pyrenees. See also the hierarchical terminology of Flint *et al.* (2008), Sprague *et al.* (2008) and Figueiredo *et al.* (2013). From Pickering and Cantalejo (2015).

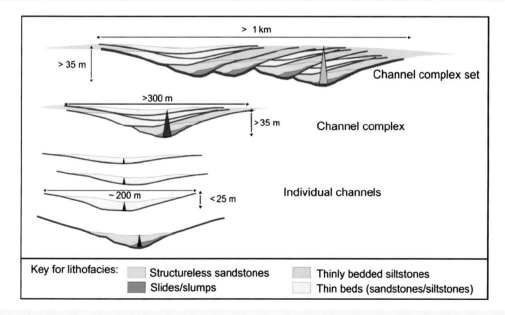

Fig. 4.48   Hierarchy of the upper slope submarine channelised system, Karoo Basin, South Africa. The scale of the smaller single channel elements makes them difficult to resolve in seismic data, even in 3D high resolution data. From Figueiredo *et al.* (2013).

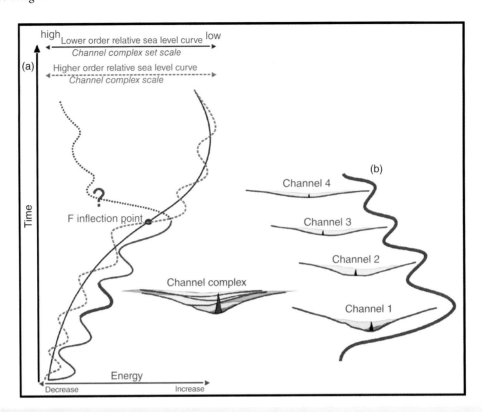

Fig. 4.49 Model for the sequence stratigraphic evolution of the LST of the F2 depositional sequence located in Fig. 4.22. (a) The red curve shows the simplified inferred energy profile during F2 development. At a sequence scale, the lower portion shows an overall increase in energy representing a degradation phase, but punctuated by higher-order decreases in energy, which at the channel complex scale represent local aggradation. The highlighted decreasing energy step represents the channel complex enlarged in (b) to show higher-order variations within the channel complex scale responsible for the single channels. Cartoons depict the channel complex and constituent channels within the context of the energy profile. The lower portion of the energy profile curve is a solid line because it was inferred from the geological record. Whilst the upper part of the curve shows an overall decrease in energy, the data preclude a definitive interpretation; therefore, this segment of the curve is dotted. Curves for relative sea-level variation are interpreted from the geological record. From Figueiredo *et al.* (2013).

and leading to one of the major climatic shifts of the Cenozoic. A substantial reduction in siliciclastic supply ensued, together with the flow of warm currents from northern latitudes that favoured the accumulation of carbonate. In eastern sites, Oligocene bathyal carbonates directly overlie an unconformity caused by the onset of the Antarctic Circumpolar Current (ACC), but a similar facies change occurred later in the west. Siliceous biogenic sediments characterised the Antarctic margin, now isolated from warm water by the ACC. Steady northward movement kept the Tasmanian region north of the southern-hemisphere Polar Front throughout the Neogene, and pelagic carbonates accumulated. Thus, the stratigraphy of the Antarctic continental margins, including the nature of deep-marine sediments, was strongly influenced by the opening of an oceanic gateway.

Another very important aspect of sediment delivery to deep water is the growth of mountain belts. Champagnac *et al.* (2012) have shown that climatic conditions, not just tectonic forcing, are a key factor in determining the shape, size and relief of 69 mountain belts worldwide. Climate variables in this study included latitude (as a surrogate for mean annual temperature and insolation, but most importantly for the likelihood of glaciation) and mean annual precipitation. To quantify tectonics they used shortening rates across each range. As a measure of topography, the mean and maximum elevations and relief were calculated over different length scales. Champagnac

et al. (2012) showed that climatic factors (with negative correlation) and tectonic factors (with positive correlation) together explain substantial fractions (>25%, but <50%) of mean and maximum elevations of mountain ranges, but that shortening rates alone account for smaller portions (<25%) of the variance in most measures of topography and relief. Relief was determined to be insensitive to mean annual precipitation, but does depend on latitude, especially for the relief calculated over small (~1 km) length scales, which Champagnac *et al.* (2012) attributed to glacial erosion. Larger-scale relief (averaged over length scales of ~10 km), however, correlates positively with the tectonic shortening rate. Moreover, the ratio between small-scale and large-scale relief, as well as the relative relief (the relief normalised by the mean elevation of the region) has its strongest positive correlation with latitude.

The failure of present-day shortening rates to account for more than 25% of most measures of relief raises the question: Is active tectonics overrated in attempts to account for present-day relief and exhumation rates of high terrain? (Champagnac *et al.* 2012). This fundamental question has profound importance for deposition in both marine and non-marine deep-water environments.

We caution against simple interpretations that always favour climate (including sea-level) change over tectonics, or the converse, and stress that it is important to understand each depositional system and basin within the context of its hinterland tectonics and climate/weather. Only then, and with good age control, is it possible to interpret the likely cause and effect of sediment flux and deposition in any basin. It is also likely that, as our knowledge of deep-water systems improves, we will find that rarely does a single control operate but rather the stratigraphic architecture results from a combination of climate, tectonics and autocyclicity, and that in any sedimentary basin these variables change in relative importance throughout basin evolution.

For further reading, excellent examples of the stratigraphic architecture of both modern and ancient deep-marine systems, and the controls on their deposition, can be found in the SEPM (Society for Sedimentary Geology) open access Special Publication No. 99 (*Application of the principles of seismic geomorphology to continental slope and base-of-slope systems: case studies from seafloor and near-seafloor analogues*) edited by Prather *et al.* (2012b).

CHAPTER FIVE

Statistical properties of sediment gravity flow (SGF) deposits

Silurian (Llandovery) Aberystwyth Grits, west Wales. UK. Geological hammer for scale.

5.1 Introduction

Perhaps the earliest important statistical tabulation of the characteristics of sediment gravity flows was the gathering, by Bouma (1962), of quantitative data on the succession of sedimentary structures in individual flow deposits of turbidity currents, leading to widespread acceptance and use of the Bouma sequence (Section 1.5.6). It was later recognised that gravity-flow deposits beginning with Bouma division T_c, for example, are deposited farther along a depositional system than, or lateral to (in levée–overbank systems), gravity-flow deposits beginning with divisions T_a or T_b (Mutti & Ricci Lucchi 1972, 1975). Furthermore, the succession of sedimentary structures deduced by Bouma (1962) guided the hydrodynamic interpretation of turbidites as deposits from gradually waning turbulent flows (Walker 1965).

This early example proved that statistical data are crucial for deciphering flow processes and palaeoenvironments.

In recent years, emphasis has shifted to parameters extracted from frequency distributions of deposit thicknesses, or from vertical trends in deposit thicknesses or grain sizes. A prime example is the use of asymmetric trends in bed thickness (so-called 'upward thinning' and 'upward thickening' cycles or trends) to infer palaeoenvironment in submarine-fan systems (Section 7.4.3). Decisions as to whether deposit-thickness populations are log-normal or fit power-laws, for example, have been used to guide the interpretation of depositional processes (Malinverno 1997; Rothman & Grotzinger 1994, 1997; Carlson & Grotzinger 2001; Sinclair & Cowie 2003. Felletti & Bersezio 2010a, Pantopoulos *et al.* 2013). Even the shapes of individual flow deposits can be quantified and have been used

Deep Marine Systems: Processes, Deposits, Environments, Tectonics and Sedimentation, First Edition. Kevin T. Pickering and Richard N. Hiscott.
© 2016 Kevin T. Pickering and Richard N. Hiscott. Published 2016 by John Wiley & Sons, Ltd.
Companion Website: www.wiley.com/go/pickering/marinesystems

to infer flow rheology or flow confinement (the 'aspect ratio' of Aksu & Hiscott 1992; Talling *et al.* 2007; Felletti & Bersezio 2010a; Malgesini *et al.* 2015). In this chapter, we focus our attention mainly on statistics derived from frequency distributions of deposit thicknesses and grain sizes. We also consider the environmental significance of asymmetric trends in deposit thicknesses that have been widely used since about 1970 to assign turbidite successions to channelised or unchannelised parts of ancient submarine fans.

A few definitions are required at the outset. We use the term *bed* for a lithologically distinct stratum of, for example, shale or sandstone/conglomerate. This is the traditional usage of field geologists (Fig. 5.1), and does not distinguish whether a bed is formed of one or more flow deposits (the latter would be an *amalgamated bed*). To explicitly identify single flow deposits, we use the term *SGF deposit*. A SGF deposit may include both basal coarse-grained divisions of sand and gravel, and upper muddy divisions, so that a single flow deposit might actually span two distinct 'beds' in the parlance of field geologists. If the objective of a researcher is to look only at the coarser sand/gravel divisions of a SGF deposit, then we advocate the more specific term *SGF coarse divisions*. Although this nomenclature might seem awkward at first, it does ensure the unambiguous designation of all or part of a single flow deposit. In the literature, some authors use 'bed' as defined above, but others use 'bed' for the entire deposit of a single flow, encompassing both coarse-grained divisions and potentially thick shale caps. We wish to avoid such double meanings.

The thickness of a SGF deposit at a given point within a sedimentary basin is determined by the shape of the deposit and the distance to the source, with deposit shape depending upon factors such as initial volume of sediment, grain-size(s) and flow concentration (Middleton & Neal 1989; Rothman & Grotzinger 1996; Malinverno 1997; Carlson & Grotzinger 2001). The frequency distributions for the thicknesses of SGF deposits and SGF coarse divisions at any sampling site are dependent upon: (i) the range of settling velocities of the grains

that accumulate because of declining competence and capacity of the flows (Hiscott 1994a; Leeder *et al.* 2005); (ii) the rate of temporal deceleration of the depositing flows and (iii) the concentrations of the flows. A frequency distribution of SGF deposit thicknesses thus contains information on flow hydrodynamics, seabed physiography and the texture of the sediment supplied by the source. All three types of information are important for reconstructing the evolution of ancient sedimentary basins.

Predicting the frequency distribution of the thicknesses of SGF coarse divisions within deep-water reservoir units has obvious economic implications for hydrocarbon exploration, particularly if thicknesses can be related to deposit volumes. Such frequency distributions are an important component of reservoir models (e.g., Flint & Bryant 1993 and references therein). If characteristic frequency distributions of SGF deposit thicknesses (or coarse-division thicknesses) exist for different submarine fan sub-environments then interpretations of cores and borehole images (e.g., Formation MicroImager [FMI] data) might be improved. Several statistical distributions have been suggested for the thicknesses of turbidites and the deposits of other sediment gravity flows (SGFs). Here, we consider log-normal, power-law, exponential and gamma distributions (Figs 5.2 and 5.3).

In recent years, the power-law distribution has been suggested as the most appropriate distribution for deposit thicknesses in many turbidite successions. A power-law relationship takes the form $N = a\,T^{-\beta}$, where N = number of flow deposits thicker than T, T = SGF deposit thickness, β = the power-law scaling exponent typically in the range of 1 to 5, and a = a constant. Departures from a power-law distribution have been used to suggest erosion or amalgamation of SGF deposits. It has been suggested that environments dominated by different processes might be recognised by assessing the degree to which these processes have caused departures from a power-law distribution (e.g., Rothman & Grotzinger 1996; Felletti & Bersezio 2010a). When plotted on log-log graph paper, a single straight line defines a

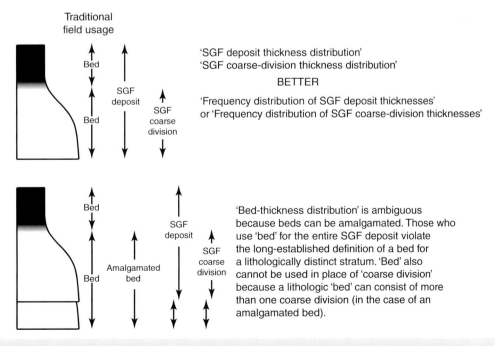

Fig. 5.1 Definition diagram for *beds*, *SGF deposits* and *SGF coarse divisions*. The upper diagram presents the case of a single flow deposit. The lower diagram depicts an amalgamated bed consisting of two flow deposits. Notice that the application of these definitions becomes more complex in amalgamated beds (lower diagram), but allows greater precision than alternative terminologies.

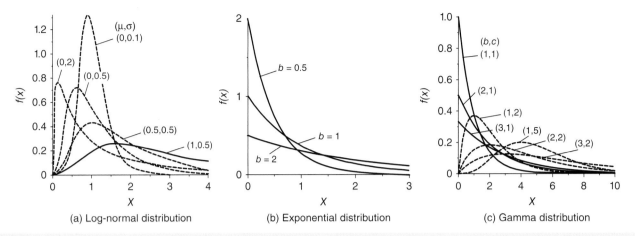

Fig. 5.2 Probability density functions (pdfs) of three distributions that have been suggested to match SGF deposit thicknesses. The power-law distribution is shown in Fig. 5.3c. The parameters for each distribution are shown on the diagram. The log-normal distribution (a) is skewed to the right (i.e., it has an extended tail to the right of the mode). The probability density function, *f(x)*, starts at 0, increases until the mode and decreases after the mode. Two parameters called scale (μ) and shape (σ) control this distribution. If μ is constant, the skewness of the pdf increases as the σ value increases. If σ > 1, the pdf rises sharply for very small values of *x*, follows the *y* axis, reaches the mode early, and then decreases sharply like the exponential distribution. The exponential distribution (b) has only one parameter: the scale parameter *b*, where *b* = 1/λ. As *b* increases (λ decreases) the distribution is stretched out to the right; if *b* decreases, the distribution is squeezed towards the origin. The pdf starts at *x* = 0 at a level where *f(x)* = λ and decreases exponentially as *x* increases. It has a concave-upward shape. As *x* approaches infinity, the pdf approaches 0. The gamma distribution (c) has two parameters: the scale parameter (*b*) and the shape parameter (*c*). At a constant *c*, an increase in *b* tends to compress the distribution towards the *x*-axis, while an increase in *c* tends to make the distribution more symmetrical. From Awadallah (2002).

power-law distribution. Deposit-thickness data, however, commonly plot as 2–3 straight-line segments with different slopes (i.e., with different values of β; Fig. 5.3). Such segmentation of a power-law distribution has been ascribed to variable depositional processes, or to spatial differences in the point of flow initiation or 'spreadability' of the flows generated by different depositional processes, such as levée spillover flows versus un-confined flows (Malinverno 1997; Pirmez *et al.* 1997). Large volume flows tend to spread over a relatively wide area and, therefore, are predicted to have large β values compared with smaller volume flows (e.g., those confined to channels, levées and seafloor and slope depressions) that might have small β values (Malinverno 1997; Rothman & Grotzinger 1997).

Some workers have suggested that different statistical distributions might be characteristic of different deep-marine environments and processes; for example, log-normal or negative exponential frequency distributions versus power-law cumulative frequency distributions. For example, Drummond (1999) proposed that the turbidite bed-thickness frequency in the Upper Devonian Brallier Formation of the Central Appalachian Basin displays a negative exponential distribution.

In a study of well-exposed deep-water units of the Oligocene Cengio and Bric la Croce–Castelnuovo Turbidite Systems (Tertiary Piedmont Basin, Italy), Felletti and Bersezio (2010a) attempted to quantify the degree of confinement of several deep-water systems using SGF coarse-division thickness distributions. In this basin, the cumulative frequency distribution of coarse-division thicknesses follows a segmented power-law relationship. The deviation from the trend observed for the thickest beds was found to be associated with a deficiency of beds in a certain thickness range. This behaviour suggested to Felletti and Bersezio (2010a) the existence of a thickness threshold that segregates (i) the deposits of the fully confined flows that could deposit thick aggrading beds, with minor downcurrent flow transitions, from (ii) the rest of the SGF coarse-division

thickness population. This threshold decreases from the lowermost (and most confined) to the uppermost units of the Cengio Turbidite System (CTS). No thresholds were identified for the uppermost Bric la Croce–Castelnuovo Turbidite Systems (BCTS) that spread over a wider basin than the CTS, suggesting unconfined deposition. The BCTS was presumably larger than the area that could be bathed by one turbidity current. Thus, Felletti and Bersezio (2010a) used SGF coarse-division thickness statistics to provide some general guidance about interpretation of depositional setting, the degree of confinement, erosion/amalgamation and bypass of sediment gravity flows.

Rothman and Grotzinger (1996) and Carlson (1998) showed that the thickness distributions of debris-flow deposits and the coarse divisions of turbidites in the Kingston Peak Formation are different. In an earlier study, Rothman *et al.* (1994) used thickness measurements of the coarse divisions of Kingston Peak turbidites to obtain a good fit to a power-law; in the 1994 study, debris flow deposits were identified in the section but omitted from the cumulative distribution. The debris-flow deposits were subsequently considered by Rothman and Grotzinger (1996), with the conclusion that the scaling exponent for the debris-flow deposits is significantly different from that of the turbidites. Rothman and Grotzinger (1996) inferred that rheological properties are important controls on SGF coarse-division thicknesses (*cf.* Talling 2001). The thickness distribution of the coarse divisions of slurry beds within the Marnoso arenacea (Italian Apennines), comprising both a slurried division and an underlying division of structureless sand, deviates systematically from a log-normal distribution, and more closely follows a normal distribution. The differences between the thickness distributions for the coarse divisions of turbidites and debrites (or slurry-flow deposits) emphasises the importance of flow rheology in determining deposit shape and the thickness distribution (Rothman & Grotzinger 1996).

The coarse divisions of sandy and silty SGFs interbedded with shales (including muddy turbidite tops) are particularly amenable

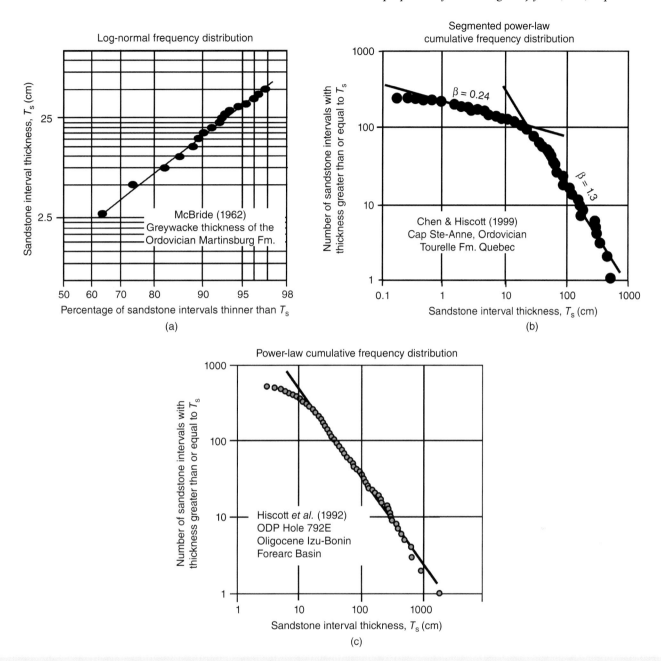

Fig. 5.3 Examples of the three contrasting frequency distributions for the thicknesses of the coarse divisions of turbidites (from Talling 2001). (a) Log-normal frequency distribution (from McBride 1962). In this graph, the x-axis is a probability scale whereas the y-axis is logarithmic. When plotted in this format (or with the x and y variables reversed as is more customary), log-normal populations plot as straight lines. (b) Segmented power-law distribution of cumulative frequency with two different power-law exponents, β (from Chen & Hiscott 1999a). (c) Single power-law distribution of cumulative thickness (from Hiscott et al. 1992). Note that the power-law cumulative frequency distributions significantly over-predict the number of very thin deposits as compared to actual field data.

to study using high-resolution micro-resistivity logs (e.g., Formation MicroScanner (FMS) or Formation MicroImager (FMI), both registered trade names of Schlumberger Corporation). This is because of the sharp contrast in resistivity between the coarser basal divisions of SGFs and the associated shales. There are limitations to such data, however, because the resolution of the micro-resistivity logs does not allow single flow deposits to be confidently discerned in amalgamated beds, unless there are strong grain-size changes at the amalgamation surfaces, and does not allow the precise top of each flow

deposit to be recognised in muddy intervals. Therefore, analysis of thickness information is restricted to SGF coarse divisions, and an assumption must be made that sandstone beds do not contain subtle amalgamation surfaces. For example, Hiscott et al. (1992, 1993), Pirmez et al. (1997) and Awadallah et al. (2001) used FMS images to document facies architecture and thickness distributions for the coarse divisions of SGFs from the Izu–Bonin forearc, the Amazon Fan and Woodlark Basin, respectively. The micro-resistivity images permitted intervals of low core recovery in Ocean Drilling Program

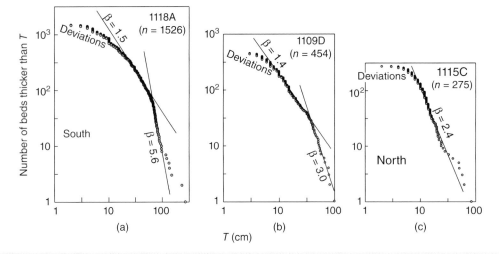

Fig. 5.4   Logarithmic graphs of the number of sand and silt beds thicker than *T*, versus *T*, for data derived from FMS images, Woodlark Basin, offshore Papua New Guinea. These sand and silt beds are the coarse divisions of turbidites. As predicted for power-law distributions, the data plot as straight-line segments with a slope that is defined by the characteristic exponent of the distribution. (a) Hole 1118A; (b) Hole 1109D and (c) Hole 1115C. See text for explanation. From Awadallah *et al.* (2001).

(ODP) boreholes to be described bed-by-bed, just like the intervals that were successfully cored and described at sea. As an illustration, FMS images were obtained from three boreholes along a N–S transect in the western Woodlark Basin, offshore Papua New Guinea (Huchon *et al.* 2001; Awadallah *et al.* 2001). This is a lower to upper Pliocene, northward-thinning, turbidite succession that accumulated on rifted continental crust. The source of the sand, silt and clay detritus was to the north and west, resulting in oblique to longitudinal gravity-flow supply into the propagating rift. In all three boreholes, the number of sand and silt SGF deposits decreases exponentially with increasing thickness of their coarse divisions; there is also a significant 'tail' of deposits with relatively thick coarse divisions (Fig. 5.4). Log-log graphs showing the number of SGF deposits with coarse divisions thicker than *T*, versus *T*, consist of one or two straight-line segments, except for deviations at low values of *T* near the resolution limit of the FMS tool. Each straight-line segment suggests a power-law relationship with a different scaling exponent, β, in the range 1.5 to 5.5. Notice, however, that the relative size of the subpopulations conforming to each power law can be very different. For example, in Figure 5.4b the number of flow deposits consistent with β = 1.4 is ~300 − 30 = 270, whereas the number of flow deposits conforming to a relationship with β = 3.0 is ~30.

In summary, thickness distributions of coarse divisions of Pliocene turbidites from the western Woodlark Basin are broadly exponential (many thin deposits and few thick deposits) but also can be approximated by power-law distributions with either one or two linear segments (i.e., one or two β values for different ranges in the thickness of coarse divisions). These patterns can be interpreted to result from partial confinement of the less voluminous flows in the rift basin, perhaps because of subtle seabed topography and/or channels (Awadallah *et al.* 2001). Broad conformity with a power-law model is consistent with triggering of turbidity currents by earthquakes in the tectonically active rift. Frequencies of turbidites reach ~930 flow deposits per million years at ODP Sites 1109 and 1118 in the period 3.6–3.45 Ma (beginning somewhat earlier at Site 1109). Subsequently, frequencies drop at Sites 1109 and 1115 but remain unchanged at Site 1118 until ~2.6 Ma. This apparent higher rate of arrival of turbidity currents closest to the basin axis might indicate (i) greater trapping

of bottom-hugging gravity flows in the morphologically deeper axial trough or (ii) preferential deposition of only mud from distal turbidity currents once they had travelled beyond the axial zone of the rift.

At times, the rate of deposition of sand and silt turbidites at ODP Sites 1118 and 1109 was ~1000–1500 beds per million years, whereas ODP Site 1115 never received more than ~450 beds per million years. ODP Site 1118, nearest the axis of the rift basin, had the longest history of high-frequency turbidite deposition (~1500 beds per million years, equivalent to a recurrence interval of 670 years). Turbidite recurrence intervals are approximately one order of magnitude longer than the recurrence intervals of magnitude 7.0 and larger earthquakes in this region, suggesting that factors other than the availability of seismic triggers are required to generate turbidity currents. The lower frequency of turbidites with sand and silt divisions north of the axis of the rift, when compared to the turbidite frequency of similar deposits at axial ODP Site 1118, is ascribed to a combination of lateral and distal fining, so that many flows responsible for graded beds of sand and silt at ODP Site 1118 only deposited mud at ODP Sites 1109 and 1115. Presumably the mud was subsequently homogenised by burrowing in areas of slow accumulation (Awadallah *et al.* 2001).

In a study of the Miocene Marnoso arenacea, Italian Apennines, Talling (2001) proposed that turbidite thicknesses (i.e., SGF-deposit thicknesses, Fig. 5.1) and the thicknesses of turbidite coarse divisions have frequency distributions comprising the sum of a series of log-normal frequency distributions. These strata were deposited predominantly in a basin-plain setting, and amalgamated beds are relatively rare. SGF deposits truncated by erosion were excluded from this analysis. Each of the separate log-normal frequency distributions characterises deposit thicknesses, or coarse-division thicknesses, for a given basal grain-size or basal Bouma division. The median thickness of each log-normal population is positively correlated with the average basal grain size of that population (Talling 2001). The deposit-thickness population created by summing all the separate log-normal subpopulations has a segmented power-law trend. The power-law exponent corresponding to the basal grain-size and median thickness is different for turbidites with a basal T_a or T_b division and those with only T_c, T_d and T_e divisions. These two types of turbidite correspond to 'thick bedded' and 'thin bedded' categories of previous workers, respectively; following the classification of Chapter 1, some of the 'thick-bedded' deposits might be the deposits of concentrated density flows rather than turbidity currents. The bimodal thickness distribution of 'thin-bedded' and 'thick-bedded' deposits is reflected by a change in the power-law exponent. This change in the β value is proposed to be related to: (i) a transition from viscous to inertial settling of sediment grains; and (ii) hindered settling at high sediment concentrations. The analysis by Talling (2001) supports the view that T_a and T_b divisions were deposited from flows with relatively high concentration and that the flows carrying distinct grain-size modes undergo different depositional processes. According to Talling (2001), the occurrence of both log-normal and segmented power-law frequency distributions can be explained in a holistic fashion. Power-law frequency distributions of SGF thicknesses had previously been linked to power-law distributions of earthquake magnitudes or volumes of submarine slope failures. The log-normal population observed by Talling (2001) for a given grain-size class supports an alternative view: that SGF thicknesses are determined by the multiplicative contribution of several randomly distributed parameters, in addition to the settling velocity of the grain size present. Talling's (2001) study led him to propose that a log-normal frequency distribution of the thicknesses of coarse divisions might be most appropriate in reservoir modelling, with grain size providing a guide to the median thickness of each log-normal subpopulation. Grain size also influences permeability and porosity, so that it might be possible to define internally consistent sets of values for (x,y,z) = (grain size, median thickness of coarse divisions, permeability) as input into a robust reservoir model for the deposits of turbidity currents and concentrated density flows.

5.2 Cloridorme Formation, Middle Ordovician, Québec

Awadallah (2002) measured >27 000 beds in the lower Cloridorme Formation, Québec Appalachians, that were deposited in a deep-marine setting similar to modern basin plains and distal submarine-fan systems (Hiscott *et al.* 1986). The distributions of the thicknesses of SGF coarse-divisions vary from log-normal for sandstones of Facies Classes B and C, to exponential for siltstones of Facies Class D (Awadallah 2002; Awadallah & Hiscott 2004). Facies of the Cloridorme Formation and their interpretation are summarised in Table 5.1. Because the study of Awadallah (2002) is based on such a large database, we provide a thorough overview of the results as a demonstration of the potential utility of SGF deposit-thickness statistics. Our discussion starts with shales and progressively proceeds to coarser deposits.

Bed-thickness measurements for 13 479 shales belonging to Facies Class E were binned into thickness classes at 5 cm increments, and then plotted as histograms (Fig. 5.5a). Shale bed thicknesses vary from 1–510 cm, resulting in >100 thickness classes. Figure 5.5a shows a very high peak in the 5–10 cm class and a long tail of thicker beds. Many shale beds >140 cm thick represent the mud caps of megaturbidites of Facies C2.4 and C2.5. The shape of the curve is most similar to a log-normal distribution. The exponential and gamma distributions are less suitable as confirmed by quantile-quantile (q-q) graphs (Chambers *et al.* 1983), where most of the points (up to the 99th percentile) are better fitted by a log-normal distribution (Fig. 5.5b).

A quantile-quantile graph, or q-q graph, is a useful way to judge how well a particular distribution function fits a set of field data (Sylvester 2007). A quantile derives its name from the fraction (or percent) of data points below its value; for example, the 0.3 (or 30%) quantile is the value below which 30% of the data fall and above which 70% of the data fall. A q-q graph is a probability graph or scattergram with the quantiles of the observed or measured population on the *x*-axis, and the expected values for one or more candidate distribution (e.g., log-normal, exponential, power-law) on the *y*-axis. To construct a q-q graph, the data are sorted from smallest to largest. If the data best fit a normal distribution, then a plot of the measured quantile values against the expected values for a normal population with the same mean and standard deviation should define a straight line. The expected normal population values are calculated by taking the *z*-values of $(I - 0.5)/n$ where I is the rank in increasing order. Curvature of the plot indicates departures from a normal distribution. A q-q graph is also useful for detecting outliers, which plot as points that are far away from the overall pattern of the majority of the data.

A graph of probability versus log thickness (Fig. 5.5c) shows a large percentage of the shale beds falling along straight-line segments,

Table 5.1 Percentages of deep-marine facies classes in each of nine time-slices (and sub-slices) at three geographically separated composite sections from the lower Cloridorme Formation (Awadallah 2002). From the most proximal RE-PCDR sections, the SYE-SH sections are ~14 km westward (downcurrent) and the PF sections ~25 km westward. Notice the downcurrent thickness changes of each synchronously-deposited slice, and the upward increase in the percentages of Facies Classes B and C (bold italics) as sand-prone depositional lobes encroached from the east. Shading shows mound- and wedge-shaped lobe and lobe-fringe deposits. Localities of composite sections are: PF = Pointe à la Frégate; SYE-SH = St-Yvon (east) – St-Hélier; RE-PCDR = Ruisseau à l'Échalote – Pointe des Canes de Roches (see maps in Awadallah & Hiscott 2004)

Time-slice/ Sub-slice	PF (west)						SYE-SH (central)						RE-PCDR (east)				
	Total thickness (m)	covered (m)	B%	C%	D%	E%	Total thickness (m)	covered (m)	B%	C%	D%	E%	Total thickness (m)	B%	C%	D%	E%
SL 7	**87.20**	12.50	*1.80*	22.40	19.70	56.10	**104.40**	0.00	*6.90*	20.70	15.80	56.60					
SL6-2	**77.27**	0.00	0.00	14.70	16.90	68.40	**103.10**	0.56	*4.00*	15.70	25.30	57.30					
SL6-1	**65.75**	0.00	*0.40*	10.60	16.50	72.50	**97.13**	0.00	*2.70*	12.70	17.80	64.70	**89.70**	*7.60*	21.10	12.80	58.50
SL5	**43.00**	0.00	0.00	12.50	14.80	72.70	**41.50**	0.00	0.00	10.20	11.10	78.70	**38.70**	*1.40*	9.00	26.60	63.00
SL4-2	**73.70**	0.00	0.00	10.50	17.70	71.80	**77.20**	0.00	0.00	8.90	18.90	72.20	**80.10**	*2.00*	11.60	19.20	67.20
SL4-1	**58.50**	0.00	0.00	9.00	14.80	76.20	**57.40**	0.00	*0.50*	14.20	20.20	65.10	**55.90**	*0.80*	17.50	13.70	68.00
SL3	**74.50**	0.00	0.00	11.40	10.80	77.80	**75.80**	0.00	0.00	8.50	19.40	72.10	**65.30**	0.00	7.20	20.50	72.30
SL2	**57.80**	0.00	0.00	11.60	9.70	78.70	**64.20**	0.00	0.00	11.50	6.10	82.40	**58.60**	0.00	6.40	13.40	80.20
SL1	**36.14**	0.00	0.00	17.10	7.60	75.30	**32.90**	5.10	0.00	16.80	3.00	80.20	**32.70**	0.00	13.70	7.20	79.10

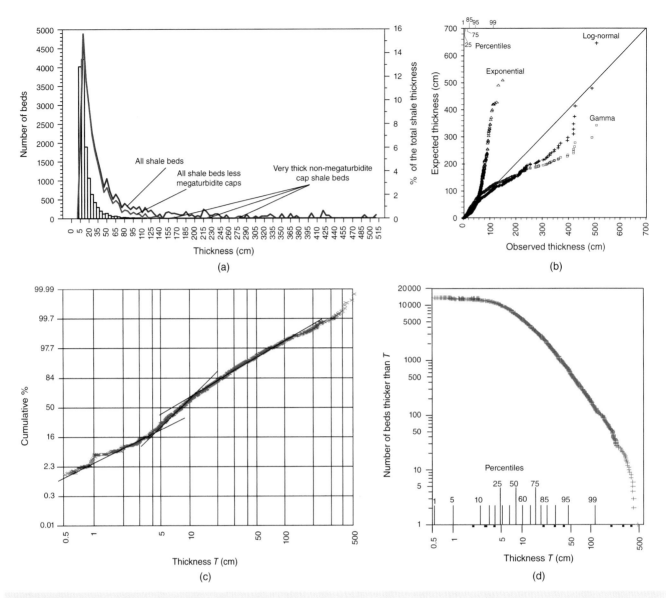

Fig. 5.5 (a) Thickness distribution of the shale beds of Facies Class E, lower Cloridorme Formation. The blue line represents the percent each class interval represents of the thickness of the shale. The rugosity of this line is somewhat decreased when the thicknesses of megaturbidite mud caps are removed and the data re-plotted. (b) q-q graph of the actual bed thicknesses of the shale beds versus the expected thicknesses for a log-normal distribution (crosses), an exponential distribution (triangles) and a gamma distribution (squares). The log-normal distribution best fits the data, because beds that are as thick as 120 cm plot along the $y = x$ line. However, this better fit relative to the other two distributions only pertains to an additional 5% of the population. (c) A graph of probability versus the logarithm of bed thickness can be divided into three or more line segments (three are shown), suggesting the overlap of two or three log-normal distributions. The first is for beds <4 cm thick (16% of the population), the second for beds 4–10 cm thick (~45% of the population) and the third for beds 10–120 cm thick (~40% of the population). (d) Graph of $N > T$ versus T, showing a line with two or more approximately straight segments. Note the inflection at ~4 cm, similar in value to the thickness at which there is a prominent break in the slope on the probability versus log thickness graph of part (c). From Awadallah (2002).

suggesting distinct log-normal populations for various bed-thickness populations. The first is for shale beds <4 cm thick (16% of the population), the second for beds 4–10 cm thick (~45% of the population) and the third segment is for beds that are 10–120 cm thick (~40% of the population). Erosion or winnowing might have affected the thickness of the thinner beds, but there is no clear evidence for this. Instead, Awadallah (2002) concluded that beds <4 cm thick were probably deposited by processes distinct from those that emplaced the thicker beds. For the 4–120 cm population, processes such as

winnowing, erosion and factors related to the size of a flow or its dynamics might have controlled the shale-bed thicknesses. The tail of the thicker beds (>120 cm) has a steeper gradient than the line segments that characterise the rest of the population. The increase in gradient for beds >300 cm thick is due to the presence of 21 beds that represent a very small percentage of the total bed population of 13 479 beds. These 21 beds represent the thick mud caps of megaturbidites (*cf.* Pickering & Hiscott 1985), and have a different origin to other shale beds.

On a graph of $N > T$, versus T (Fig. 5.5d), a power-law scaling is suggested by the linear trend of the central part of the shale dataset. Ninety-eight percent (1st to 99th percentile) of the beds can be fitted by two straight lines. The most prominent trend is for beds ranging from ~17–200 cm thick (~22% of the population of shale beds), with a β value of 1.6. However, this long linear trend is actually of secondary importance because ~77% of the population comprises beds <17 cm thick (with those beds ~4–17 cm thick having a β value of ~0.9). Because power-law graphs are presented in log-log space, what appear to be small portions of the graphed data can in fact include a majority of the measured beds. The illusion that the central

part of a power-law graph is of greatest significance needs to be resisted during the interpretation process.

Awadallah (2002) argued that an alternative way to assess a power-law distribution is to focus on the subpopulation that includes most of the beds. The segment of the line in Figure 5.5d with a β value of ~0.9 represents this subpopulation (i.e., beds 4–17 cm thick representing ~55% of the population). Extending this line in the direction of thinner beds predicts a greater number of beds than the 13 000 actually measured, suggesting that thinner beds are under-represented because of erosion, or difficulty in distinguishing the very thinnest shale layers (beds and laminae). As a means of

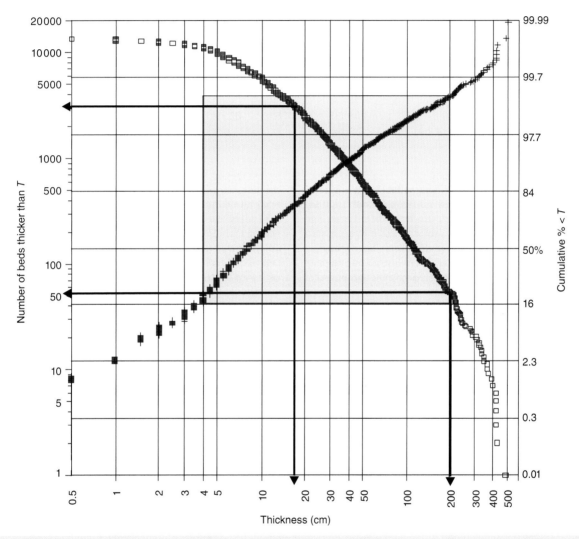

Fig. 5.6 Comparison between a probability-log graph (crosses) and a log-log graph (squares) for thicknesses of the shale beds of Facies Class E, Cloridorme Formation. The probability-log scale tends to compress the data for the beds that are in the middle of the population (16–84%) and expands the graph for the very thin beds and laminae, and for the very thick beds (tails of the population). The log-log scale expands or spreads the very thick beds and compresses the thin beds into a small part of the graph. The bed-thickness data plot as segmented lines in both presentation formats. Some of the segments are less obvious because of distortion introduced by the different nonlinear scales. The log-normal distribution is perceived to be a better fit to the data than the power-law distribution because a greater proportion of the population can be fitted by a more-or-less single straight line or line segments of nearly constant slope (~82% in the yellow area from 16 to ~99 cumulative %, corresponding to beds that are 4–200 cm thick). The most obvious trend that can be fitted by a single best-fit line on the log-log graph of $N > T$ versus T is the trend for beds 17–200 cm thick. These beds represent only ~23% of the shale beds. Other less obvious linear trends on the log-log graph might be more representative (e.g., beds ~4 to ~20 cm thick, accounting for ~60% of the beds). This is because more than 60% of the beds of Facies Class E are thin to very thin. From Awadallah (2002).

testing whether a power-law or log-normal distribution best charac-terises the thicknesses of shale beds and laminae of Facies Class E, Awadallah (2002) re-plotted the data on a single graph with two dif-ferently scaled *y*-axes (Fig. 5.6). He concluded that the log-normal distribution better fits the data, as a single essentially straight line encompasses *a greater proportion of the beds* than the linear trend dis-played on the graph of *N > T* versus *T*.

A subset of 137 beds >100 cm thick represent thick mud caps to megaturbidites of Facies C2.4 and C2.5; these were investigated separately. Both the log-normal and gamma distributions fit these data equally well, while beds >110 cm thick can be fitted to a single log-normal distribution. Seismic activity has been suggested as a triggering mechanism for flows that deposit such megaturbidites (Labaume *et al.* 1987; Mutti *et al.* 1988), while others have suggested that their initiation is controlled by fluctuations in sea level (Weaver *et al.* 1998). For these 137 beds, the β value is ~2, a value that is within the range of values noted for turbidites in seismically active areas (Hiscott *et al.* 1992; Awadallah *et al.* 2001). The graph of probability versus log thickness and a q-q graph of these data show that the log-normal distribution provides a good fit for 80% of these thick mud caps.

Based on the results of the Cloridorme study (Awadallah 2002), it seems reasonable to conclude that shale bed thickness is controlled by many interacting parameters, such as erosion, volume of sediment carried in a turbidity current and the spreadability (degree of con-finement). The observed log-normal population is probably a com-bination of several subpopulations (*cf.* Talling 2001). Some of these subpopulations approximate power-law distributions with a scaling parameter ranging from 1–2, but like many other SGF successions, the fit to a power law only applies to a subset of beds that represent a small part of the population. Thinner beds that form a greater fraction of the population might also follow a power-law distribution but have a power-law scaling parameter closer to zero than the parameter for the thicker beds.

Awadallah (2002) also measured the thicknesses of 11 249 Facies Class D siltstone beds, which he interpreted as the basal coarse divi-sions of turbidites. The bed-thickness data for these coarse divisions were grouped into 5-cm class intervals as the range of bed thicknesses is ~95 cm. The siltstone bed-thickness distribution shows a promi-nent peak in the 0–5 cm class interval with a long tail towards the thicker beds. More than 90% of the beds are <10 cm thick (Fig. 5.7a). The shape of this distribution most closely resembles an exponential or gamma distribution, with a q-q graph favouring a gamma distri-bution (Fig. 5.7b). However, coarse divisions >30 cm thick (0.5% of the total) better fit a log-normal distribution. On a graph of prob-ability versus log thickness, the data are best fitted by two linear trends (Fig. 5.7c). This suggests the presence of two statistical sub-populations for the thicknesses of turbidite coarse divisions – these might have been deposited by different processes or under different flow conditions. Approximately 60% of the coarse divisions belong to the subpopulation of beds <4 cm thick. Coarse divisions of Facies Class D that are >70 cm thick deviate from the log-normal trend valid for siltstones 4–70 cm thick. These >70 cm-thick deposits tend to occur within sandstone packets or might have mistakenly been considered the deposits of single events (i.e., they might be subtly amalgamated beds).

When the thicknesses of the coarse divisions of Facies Class D turbidites are plotted on a *N > T* versus *T* graph (Fig. 5.7c), three separate line segments are observed, namely for deposits <4 cm thick (~60% of the population), deposits 4–12 cm thick (32% of the population), and deposits >12 cm thick (~7% of the population). The

segment that represents deposits 4–12 cm thick is not readily apparent on the *N > T* versus *T* graph due to the compression of the central part of the probability scale.

Awadallah (2002) also demonstrated that particular facies within a facies class can have different distribution parameters. For example, when Facies D2.2 and D2.3 are removed from the entire Facies Class D population, leaving 9013 Facies D2.1 coarse divisions, there is a decrease in the value of the exponential distribution parameter (λ). Whilst the log-normal and gamma distributions might be regarded as a good fit for the thicknesses of Facies D2.1 coarse divisions, a log-normal distribution provides a better fit for coarse divisions >30 cm thick (only 3% of the population; Fig. 5.8b).

A comparison of all three siltstone Facies D2.1, D2.2 and D2.3, on a graph of probability versus log thickness (Fig. 5.9a) shows that all three populations have broadly similar slopes and that the plot for Facies D2.2 is essentially coincident with the plot for Facies D2.1. A possible explanation for this coincidence is that the facies have a common origin, for example most Facies D2.2 turbidites might have originated as Facies D2.1 deposits that underwent syn-depositional modification and post-depositional deformation. The graph of prob-ability versus log thickness for most of Facies D2.3 coarse divisions has a similar slope to the slopes for Facies D2.1 and D2.2, but because Facies D2.3 deposits are thinner bedded, they plot in a different part of the graph. It might be that the thinner deposits of Facies D2.3 accu-mulated from turbidity currents with relatively small sediment loads, or that they accumulated in particular locations or under conditions that were atypical for most Facies Class D turbidites.

We now consider the power-law scaling of the thicknesses of Facies D2.1 coarse divisions (Fig. 5.8c). Only deposits >~12 cm thick can be fitted by a linear trend (β = 2.7). However, these deposits represent only ~8% of the population. Most of the remaining Facies D2.1 deposits do not display an obvious linear trend. By expanding (or stretching) the remaining 92% of the population, two additional linear trends can be fitted to the data, one for deposits 3–12 cm thick (52% of the facies), and another for deposits <3 cm thick (40% of the facies). Facies D2.1 coarse divisions 3–12 cm thick have a power-law exponent β = 1.34, whereas deposits <3 cm thick are consistent with β = 0.3. The inclusion of Facies D2.2 and D2.3 in the global plot for Facies Class D (Fig. 5.7c) does not appear to significantly modify the shape of the graph, because most of the D2.3 deposits are very thin and plot in a small area of the graph. These very thin deposits do, however, better define the segmented shape of the plot on an expanded scale, because they are <3 cm thick and are numerous.

In summary, q-q graphs suggest that the thicknesses of the coarse divisions of siltstone turbidites of Facies Class D are best fitted by an exponential or log-normal distribution. A log-normal model accommodates 3–10% more of the deposits than an exponential model. Any departure from a single log-normal distribution (i.e., a segmented probability distribution) might reflect variations in flow characteristics for thinner and thicker Facies Class D deposits, which, in turn, might have been controlled by seafloor topography.

Awadallah (2002) grouped the thicknesses of 1842 Facies Class C coarse divisions into 5-cm class intervals (Fig. 5.10a). The resul-tant histogram and q-q graphs (Fig. 5.10a, b) suggest a log-normal distribution. Thin- and medium-thickness deposits are resolved as prominent modes. On a graph of probability versus log thickness (Fig. 5.10c), the coarse divisions of Facies Class C plot as a slightly curved (bowed) line that can be fitted by two or three discrete line seg-ments. The segmented line suggests several log-normally distributed subpopulations. The Facies Class C deposits indeed naturally fall into two subpopulations: ~1400 deposits with mudstone tops found

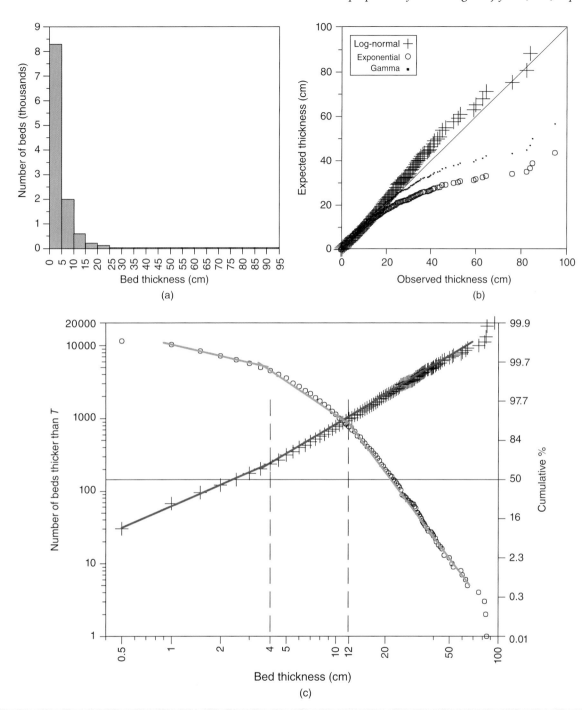

Fig. 5.7 (a) Histograms of the thicknesses of coarse divisions for Facies Class D suggesting that the exponential distribution might be the best fit to the data. (b) q-q graph of the data for Class D; deposits <~20 cm thick most closely follow a gamma distribution whereas the log-normal distribution is a better fit for coarse divisions >20 cm thick. (c) Probability-log graph (crosses) suggesting that the thicknesses of the coarse divisions of Facies Class D might be fitted by the log-normal distribution with two linear trends for coarse divisions thinner and thicker than ~4 cm. Log-log scale (circles) shows that deposits >12 cm thick follow a power-law distribution with a unique exponent. Other less well-defined linear trends could be fitted for deposits <12 cm thick (e.g., deposits 4–12 cm thick). From Awadallah (2002).

outside of amalgamated sandstone 'packets', and ~450 deposits found within amalgamated units. Deposits 40–60 cm thick represent >80% of both of these subpopulations. Very few of the deposits within amalgamated units are >50 cm thick. Facies Class C coarse divisions that are >230 cm thick plot along a line with a different slope. This departure results from the occurrence of four

megaturbidites of Facies C2.5. The probability scale causes the tails of the thickness population to be graphically stretched so that the very thin deposits or very thick deposits are emphasised. In reality, coarse divisions 60–230 cm thick, forming the basal parts of megaturbidites described by Pickering and Hiscott (1985), only constitute ~5% of the Facies Class C population.

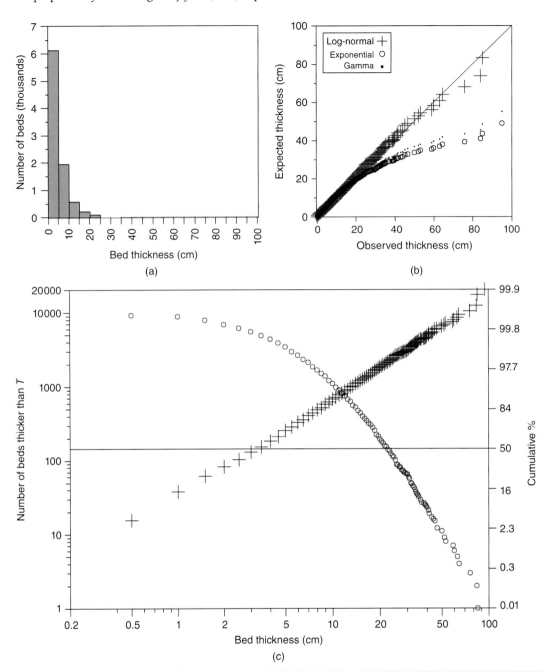

Fig. 5.8   (a) Histogram of bed-thickness data for coarse divisions of Facies D2.1, Cloridorme Formation, Québec. The removal of Facies D2.2 and D2.3 from the Class D dataset does not change the shape of the histogram (*cf.* Fig. 5.7a). The scale parameter, λ = (1/b) is decreased, however, causing the histogram to be less peaked. (b) The q-q graph shows that the removal of Facies D2.2 and D2.3 makes the log-normal distribution a better fit for the deposits that are at the tails of the distribution. (c) On the log-log graph (circles), the shape does not change significantly from that of Fig. 5.7c because the thin deposits of Facies D2.3 that were removed from the dataset affect more-so the shape of the probability plot (crosses), but less-so the log-log plot because these coarse-division thicknesses occupy a small area of the graph. From Awadallah (2002).

Power-law scaling is similar, regardless of whether all Facies Class C coarse divisions are considered together, or whether the amalgamated and non-amalgamated subpopulations are considered separately. Deposits that are 20–100 cm thick (~37% of the population), have a β value of 2.0. Awadallah (2002) proposed that Facies Class C turbidites in the Cloridorme Formation with coarse divisions >20 cm thick were deposited from very large flows. Approximately 25 beds

(<2% of the entire population of Facies Class C beds) that are >100 cm thick deviate from the β = 2 linear trend and instead have a β value of 3. These beds are mainly megaturbidites (Facies C2.4 and C2.5). Deposits 7–20 cm thick, accounting for 40% of the Facies Class C population, do not show a linear tend on a $N > T$ versus T graph.

In amalgamated sandstone packets, SGF deposits of Facies Class C that are >~10 cm thick plot as a line comprising two segments.

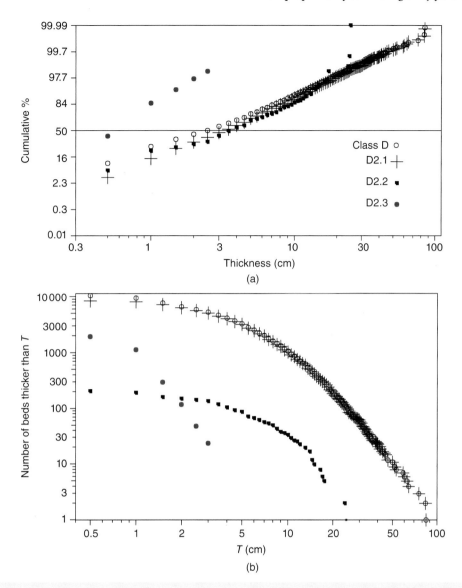

Fig. 5.9 Comparison between the different siltstone facies of Facies Class D, Cloridorme Formation, Québec. (a) Facies D2.1 and Facies D2.2 plot close to each other, reflecting a similar range in the thicknesses of their coarse divisions and possibly a similar origin. Facies D2.3 deposits (red dots) plot away from the other facies because of their thinness, suggesting that the flows that deposited this facies were also different. (b) Log-log graph of the same data. Note that, in this graph, Facies D2.3 also plots in a different area with a different slope, suggesting distinct depositional processes that created thin beds with a narrow range in thickness. Note also that although there are >1500 siltstone laminae of Facies D2.3, their effect on the shape of the plot for all of the siltstone beds of Facies Class D is minor because very thin beds and laminae plot within a narrow thickness range. Facies D2.2 beds plot in a different part of the graph because they are fewer than beds of Facies D2.1 or Facies D2.3. Their inclusion in the global plot for all Facies Class D deposits has little effect because of their small number; also, the D2.2 thicknesses plot in the same range of *T* where there are many more deposits of Facies D2.1. Essentially, Facies D2.1 swamps the global plot. From Awadallah (2002).

Deposits 10–50 cm thick (~60% of the population of amalgamated beds) have a β value of 1.3 (Fig. 5.10c, Amalgamated Class C line), whereas individual flow deposits >50 cm thick (~7% of the population) have a considerably greater β value of ~3.9. The increase in the slope of the line segment for those amalgamated deposits >50 cm thick suggests that these deposits are under-represented, perhaps because of erosion. Thinner amalgamated sandstones are, in contrast, relatively over-represented, perhaps being erosional remnants of originally thicker SGF deposits.

In summary, the log-normal distribution appears to best fit the thickness population of the basal sandy divisions of Facies Class C

turbidites. The thicknesses of amalgamated (top-cut-out) deposits form segments on a probability graph and a graph of *N* > *T* versus *T* with a break in slope (i.e., a change in β value) at a coarse division thickness of ~50 cm.

Awadallah (2002) also investigated different facies of Facies Class C in an attempt to understand if graded or ungraded deposits or deposits with clasts, for example, follow different statistical distributions or scaling values, and what information might be obtained from such variations. This investigation follows the lead of Rothman and Grotzinger (1995), who suggested that debris-flow deposits from the Kingston Peak Formation, California have a smaller power-law

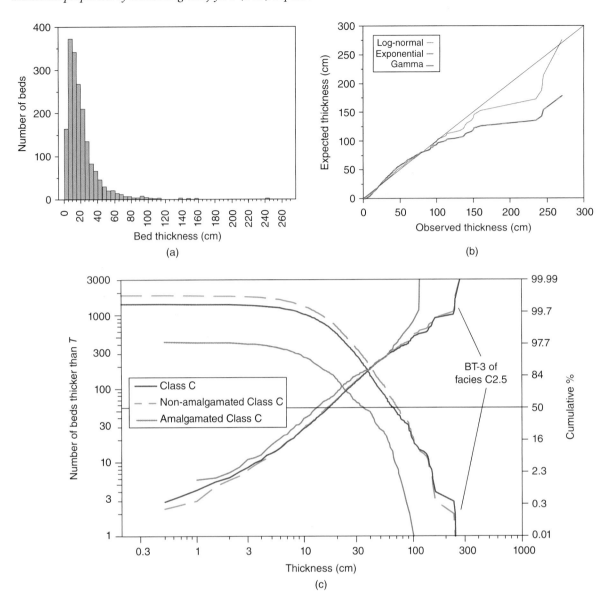

Fig. 5.10 Facies Class C thickness graphs, Cloridorme Formation, Québec. (a) A histogram incorporating all of the thicknesses of Facies Class C coarse divisions suggests a log-normal distribution. This is supported by the q-q graph (b). Megaturbidite bed BT-3 occurs in four different locations and is the cause of the nonlinear departure of the q-q graphs for bed-thicknesses >2 m. BT-3 is also an anomaly on the log-log (power-law) graph (c). Other thick deposits (mostly megaturbidites >100 cm thick) form a linear trend in this graph. There is another linear trend in the global dataset for deposits <7 cm thick. The thicknesses of coarse divisions in amalgamated sandstone 'packets' are considered separately (Amalgamated Class C line). They show linear trends for individual flow deposits 9–50 cm thick and >50 cm thick. No deposits in amalgamated units are thicker than 100 cm. From Awadallah (2002).

scaling value (β = 0.49) compared to associated turbidites (β = 1.4), an outcome they attributed to different flow rheologies.

Muddy sandstones (Facies C1.1 – also known as 'slurry' deposits) were interpreted by Awadallah (2002) as debrites. Their coarse divisions have variable thicknesses that plot as several linear trends on a graph of probability versus log thickness. On a log-log graph, they show a power-law scaling parameter of β = 1.7 which is within the range of β values of other Facies Class C sandstone facies interpreted as turbidites. Awadallah (2002) therefore concluded that turbidites and debrites of the Cloridorme Formation cannot be differentiated based on their β scaling parameter.

SGF deposits of Facies Class B are uncommon in the part of the Cloridorme Formation studied by Awadallah (2002); only 270 examples were encountered, representing <1% of all beds measured. Only six disorganised flow deposits of Facies Group B1 were observed, and because of their limited number they were not included in the analysis of the thickness distribution. The remaining 264 deposits belong to two organised sandstone facies. Facies B2.1 is represented by 56 individual flow deposits, 36 of which have their tops cut away because they occur in amalgamated units. Facies B2.2 is more common (208 deposits, with only 11 in amalgamated units). The size of this population was considered sufficiently

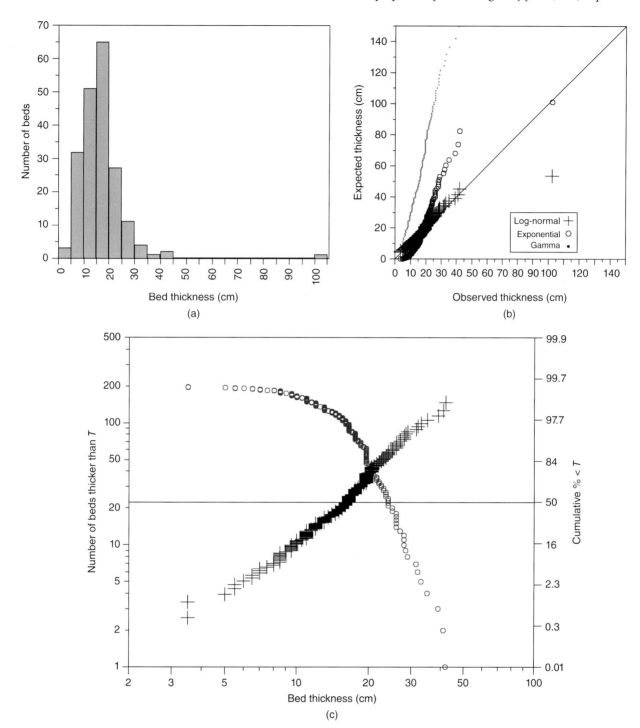

Fig. 5.11 Thickness distributions for the coarse divisions of Facies B2.2. A log-normal distribution best fits the data (a,b). (c) The segmented nature of the plots on both the probability-log graph (crosses) and log-log graph (circles) suggests several subpopulations. From Awadallah (2002).

large for statistical analysis. The coarse divisions of Facies B2.2 are bi-partite, poorly graded and locally display planar laminations or mud clasts. Awadallah (2002) interpreted each bed to have accumulated from a single concentrated gravity flow that became stratified before it reached the site of deposition. It is possible that the flows experienced a gravity transformation similar to that proposed for

other bi-partite or tri-partite beds by Talling *et al.* (2004) and Amy and Talling (2006).

The log-normal distribution best describes the 197 non-amalgamated (i.e., top-preserved) coarse divisions of Facies B2.2 (Fig. 5.11a, b). On a graph of probability versus log thickness, and a log-log graph of *N* > *T* versus *T*, the data form segmented linear

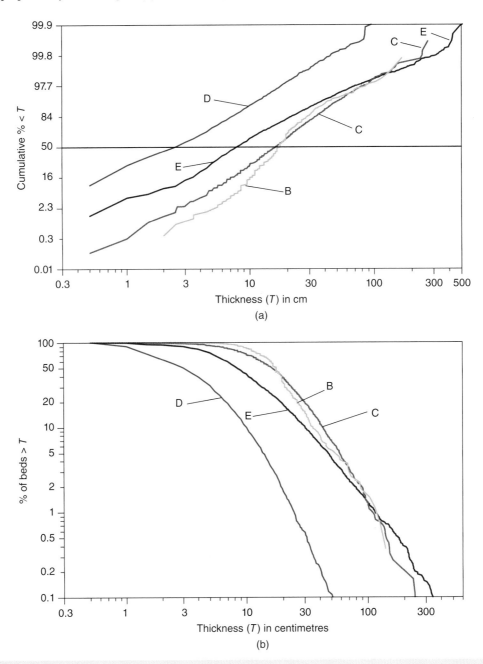

Fig. 5.12 Comparison between the thickness distributions for the coarse divisions of four facies classes (B through E) measured by Awadallah (2002) in the Cloridorme Formation. The same datasets are presented using a probability-log thickness scale (a) and a log-log scale (b). In order to fairly compare facies classes represented by quite different numbers of beds, the y-axis was transformed to a percentage scale. From Awadallah (2002).

trends. The first segment is for coarse divisions that are 8–16 cm thick (80 deposits, or 38%) and has a slope defined by a $\beta = 0.9$. The other segment is for deposits that are 16–42 cm thick (103 deposits, or 50%), with a $\beta = 4.1$. The power-law scaling exponent for the thicker deposits is quite different to exponents for Facies Class C sandstones with a similar thickness range. The steep slope means that most deposits fall within a fairly narrow range of thickness.

Different facies classes were compared to investigate if deposits that have different texture preferentially follow a log-normal or a power-law distribution. The four facies classes, E, D, C and B, were

plotted on a probability-log scale (Fig. 5.12a). Not one of the four classes shows a simple linear trend – instead, the plots are stepped or bowed. The degree of departure from a linear trend is least for Facies Class D and for Facies Class C coarse divisions <~70 cm thick; Facies Classes E and B show the most pronounced departures from a linear trend. These departures suggest more than a single population. For example, the bowing in the trend of Facies Class E in the 1–7 cm range can be traced to the fact that many of these deposits are caps of thin and very thin siltstone and sandstone turbidites (e.g., Facies D2.3). The bowing in the trend of Facies Class E for beds >100 cm thick reflects the fact that these are the mud caps of megaturbidites that formed by ponding of large flows in a confined basinal setting (Pickering & Hiscott 1985). Distinct subpopulations in the tails of the larger population to which they belong are better displayed on a graph of probability versus log thickness (e.g., megaturbidite caps in the Facies Class E population). For Facies Class B, there is a change in the slope of both graphs for beds that are <20 cm thick. This deviation is caused by the 197 non-amalgamated deposits of Facies B2.2 (73% of the depositional units of Facies Class B). About 75% of these 197 coarse divisions are <20 cm thick.

Each facies class plots in different *x–y* space depending on the thickness of the deposits in the class. The median thickness ranges from 2.5 cm for Facies Class D to 16–17 cm for Facies Classes B and C. This suggests that grain size might, on average, vary directly with the thickness of the SGF coarse divisions (*cf.* Talling 2001). Other studies have shown that some coarse-grained deep-water sandstones are characteristically thin (e.g., Facies E of Mutti & Ricci Lucchi 1972). Thin deposits that are coarse in grain size are rare in the Cloridorme Formation. The thicknesses and grain sizes of coarse divisions are probably both dependent on many factors that include flow size and velocity, particularly in settings where flows are largely depositional rather than erosional or bypassing (*cf.* Mutti 1977). The relationship between grain size and thickness does not apply to the shale layers because they are the upper muddy divisions of sediment gravity flows, so do not record the coarsest load of each SGF.

With regard to power-law scaling, the four facies classes all plot as segmented lines or show (i) a linear trend for the thicker deposits, and (ii) a curved trend for the thinner deposits (Fig. 5.12b). The linear trends represent varying proportions of the population for each facies class. For example, the linear trend for Facies Class D is for coarse divisions that are $> \sim 12$ cm thick; these represent ~6% of the population and are mostly from subfacies that occur as very thick beds, many found within sandstone 'packets'. The deposition of these thicker deposits is attributed by Awadallah (2002) either to periods of increased sediment discharge, or to interaction of flows with bottom topography. The linear trends for other facies classes probably reflect the summation of several log-normally distributed subpopulations (*cf.* Talling 2001). The region on the graph where a facies class plots is interpreted in terms of both the grain size and the flow size. Sandstones of Facies Classes C and B plot in the same region, while shale (Class E) plots to the right of the numerous Class D siltstones. Perhaps the sediment gravity flows that were responsible for many of the siltstones deposited thicker mud caps. A similar argument is used to explain the observation that megaturbidite caps plot to the right of Facies Class C sandstones.

Non-amalgamated sandstone facies that display different sequences of sedimentary structures (Facies Class C), and some of the siltstone subfacies (Facies Class D), were examined separately by Awadallah (2002) using percentage data (Fig. 5.13). Subfacies of Facies D2.1 that begin with Bouma T_{bc} divisions and Bouma T_{cd} divisions have a median thickness greater than subfacies with only the Bouma T_c division. This suggests that flows that deposited the more complete Bouma sequences might have been relatively larger. For the sandstone facies, the coarse divisions of Facies C2.2 with a basal T_b division have a greater median thickness than deposits of Facies C2.1 with its structureless basal division (T_a) or Facies C2.3 with its basal division of ripples or climbing ripples (T_c). This suggests that flows that deposited Facies C2.2 might have been bigger than those that deposited Facies C2.1 and C2.3, regardless of the grain size. This is somewhat different to the observations of Talling (2001) who noted that deposits with a basal T_a or T_b division are usually

thicker than those with a basal T_c division. He hypothesised that more-concentrated flows deposit beds that have a basal T_a or T_b division whereas less-concentrated flows deposit thinner beds. In the dataset from the Cloridorme Formation, this hypothesis is only supported for beds of Facies C2.2 and Facies C2.3, whereas other factors might control bed-thickness distributions of Facies C2.1.

The lower coarse divisions of megaturbidites (Facies C2.4) represent a special case that is related to the large size of the depositing flows. Less than 1% of the coarse divisions of megaturbidites are >100 cm thick, while ~4% are 90–130 cm thick and >50% are 30–90 cm thick. These subpopulations plot with distinct linear trends in Fig. 5.13b.

Facies or subfacies that have similar sedimentary structures but differ in texture (Facies C2.2, Facies C2.3 and selected subfacies of Facies D2.1) all show essentially linear and parallel trends (Fig. 5.13a), but plot in a different region of the graph of probability versus log thickness because of their different grain sizes. On a log-log graph (Fig. 5.13b), these deposits have similar trends except for offsets caused by differing texture.

Awadallah (2002) used all of his SGF thickness data and SGF coarse-division thickness data to evaluate whether the type of statistical distribution changes laterally or vertically within the Cloridorme Formation. He measured the thicknesses of SGF coarse divisions in nine time-slices that have their bases and tops defined by precisely correlated megaturbidites (Table 5.1, Fig. 5.14). The correlations are corroborated by the tracing of several geochemically finger-printed tuffs (Awadallah & Hiscott 2004). These synchronous depositional slices range in thickness from ~30–100 m. Downslope (westward) changes in slice thicknesses are interpreted to be the result of seabed compensation effects on a relatively flat basin plain for the older part of the succession (slices SL1–SL3 everywhere, and SL1–SL5 toward the west; Fig. 5.15), and mounding or westward tapering of sand-prone lobe and fan-fringe deposits higher in the succession (shaded Class B and Class C abundances in Table 5.1; also Fig. 5.16).

A general conclusion for all time-slices is that Facies Class C coarse-division thicknesses are best fitted by a log-normal or gamma distribution, while Facies Class D thicknesses are best fitted by an exponential distribution or a gamma distribution. Because Class D flow-deposits are more numerous than Class C deposits (by an order of magnitude in some time-slices), the combined datasets for the thicknesses of both sand-prone and silt-prone coarse divisions plot very much like a gamma distribution on a q-q graph.

The change in depositional environment from a basin-plain to a distal-fan setting (the latter with its sand-prone lobes) is not evident through comparison of histograms of coarse-division thicknesses below and above the change in setting. Although there is an increase in the number of Facies Class C deposits at this transition, there is an even greater increase in the number of Class D siltstone-based SGF deposits, masking any differences caused by the thicknesses of the sandier Class C and Class B deposits.

In the basin-plain part of the succession (time-slices 1–3), >95% of the data for coarse-division thicknesses fall along a single straight line on log-probability paper, suggesting a log-normal or gamma thickness distribution. On log-log graphs of the percent $>T$ versus T, three segments can be fitted. This contradicts claims that the basin-plain deposits in the Cloridorme Formation follow a single linear trend (Carlson & Grotzinger 2001). The first segment represents coarse divisions that are <5 cm thick; these account for ~80% of each population. The second segment is for deposits 5–40 cm thick (~20% of the population) and a third segment is for deposits >40 cm thick. The coarse divisions <5 cm thick are mostly parts of

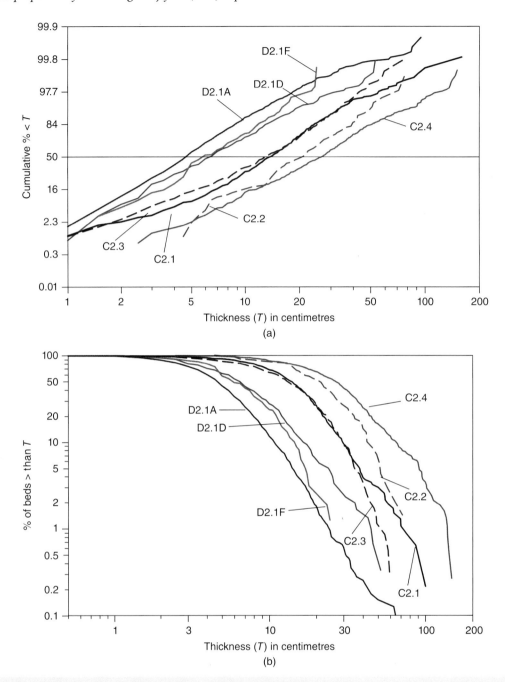

Fig. 5.13 Comparison between the thickness data of facies and subfacies that have similar sequences of sedimentary structures but that differ in texture, or that have similar texture but differ in the sequence of sedimentary structures. Both a probability-log scale (a) and a log-log scale (b) are used. In order to fairly compare facies and subfacies represented by quite different numbers of beds, the y-axis was transformed to a percentage scale. D2.1A contains only a T_c division; D2.1D begins with T_{bc}; D2.1F with T_{cd}. From Awadallah (2002). See text for discussion.

the thin siltstone turbidites that dominate the basin-plain deposits, whereas thicker coarse divisions are from sand-based turbidites, concentrated density-flow deposits or megaturbidites (*cf.* Pickering & Hiscott 1985). The coarse divisions of megaturbidites are thicker farther west, which might demonstrate their ponding in that direction.

Through the lateral transition from (i) the encroaching deposits of a distal submarine fan or fan fringe to (ii) the basin-plain (e.g., east to west across time-slice 6-1, Table 5.1), the probability-log

graphs mostly show linear trends for the thicknesses of SGF coarse divisions, but a strong departure from this simple trend in the most proximal area, for beds >70 cm thick (Fig. 5.14, Sub-slice 6-1, lower right). The graphs of percent $>T$ versus T, used to evaluate power-law scaling, are less steep for data from the more proximal sections because of a greater proportion of thicker event-deposits in the better developed lobe facies that characterise these sections. There are also significant changes in the slopes of the power-law plots

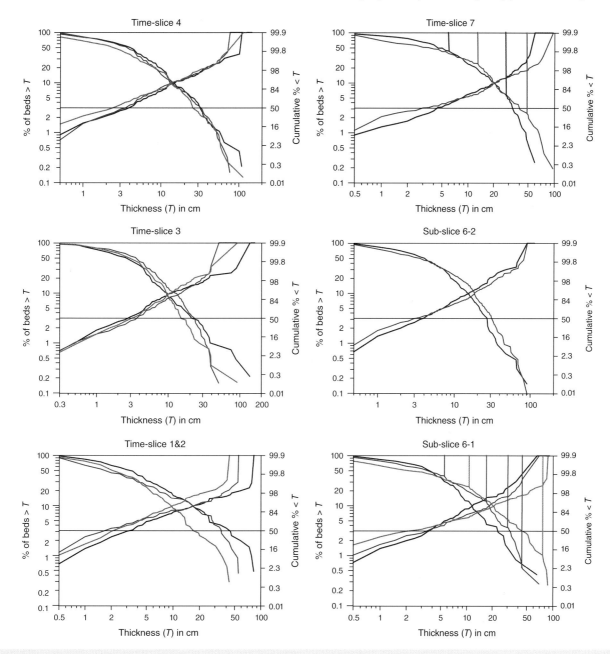

Fig. 5.14 Comparison of thickness distributions for sandstone and siltstone coarse divisions differentiated according to Cloridorme Formation outcrop areas RE-PCDR (green, most proximal), SYE-SH (red, central) and PF (blue, most distal). Each graph considers a different stratigraphic time-slice, bounded above and below by widely traceable megaturbidites. Vertical lines on the plots for Sub-slice 6-1 and Slice 7 mark sharp changes in the slopes of the graphs with corresponding colours. From Awadallah (2002).

with increasing deposit thickness in the proximal sections (Fig. 5.14, Sub-slice 6-1, data for area RE-PCDR, with slope breaks marked by green vertical lines). These discontinuities in power-law scaling are attributed to a combination of factors, including irregularities in coarse-division thicknesses due to amalgamation and the preferential deposition of deposits within certain thickness ranges (i.e., as beds of similar thicknesses in packets). In downcurrent (western) areas where amalgamation is not common, irregularities in the thickness plots are ascribed to preferential deposition as a consequence of bottom topography (i.e., compensation effects).

In the field example from the Cloridorme Formation, even a very large dataset of >27 000 SGF deposits does not provide evidence for

predictable changes in the thickness distributions of SGF deposits as they are traced from a fan fringe to the adjacent basin-plain. In both the proximal and distal parts of the succession, finer-grained turbidites of Facies Class D dominate the thickness populations and therefore control the overall nature of the statistical distributions. One outcome, however, is that more proximal and sand-prone facies associations show greater irregularity in their coarse-division thicknesses because of local erosion and enhanced deposition. They also show a wider range of coarse-division thicknesses than their more distal equivalents.

Numerical models have been employed by other researchers to predict the frequency distribution of preserved bed thicknesses

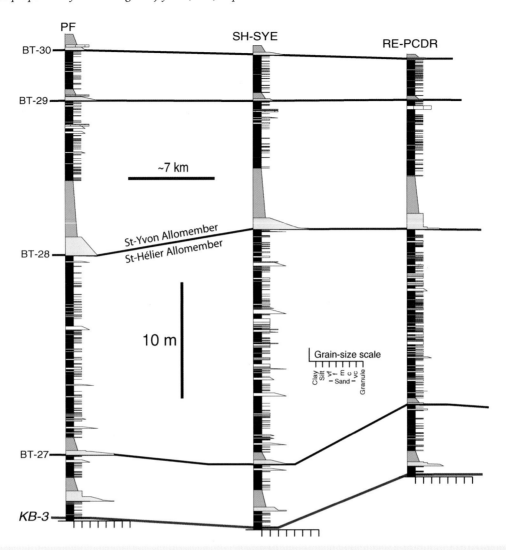

Fig. 5.15   Correlation of megaturbidites for ~25 km in the lower part of Awadallah's (2002) time-slice 3, basin-plain facies-association of the lower Cloridorme Formation. The megaturbidites have yellow sandy basal divisions and green silty mudstone tops. The downcurrent direction is from right to left. See Table 5.1 for location names. Modified after Awadallah and Hiscott (2004).

and coarse-division thicknesses that would result from alternating episodes of deposition and erosion (Kolmogorov 1951; Mizutani & Hattori 1972; Dacey 1979; Muto 1995). These models suggest that, if both erosional and depositional thicknesses are taken from the same type of frequency distribution (normal, geometric or negative exponential), the resulting thickness population is a truncated version of that distribution. Hence, if both erosional and depositional thicknesses are taken from log-normal populations, a truncated log-normal distribution of preserved deposit thicknesses would be expected. The modelling studies of Dacey (1979) and Muto (1995) show a similarity between the thick-bedded tail of such a truncated log-normal distribution and the shapes of negative exponential or geometric frequency distributions.

The Cloridorme Formation results presented by Awadallah (2002) suggest that a comprehensive understanding of the depositional conditions and the factors that control the thicknesses of beds is required before attempting to infer submarine-fan sub-environments, reservoir potential, sandbody geometry and triggering mechanisms from the character of bed-thickness distributions. The only exception

that has shown some promise to date is the potential that power-law graphs can record the degree of confinement of the flows that deposited certain facies or changing degrees of confinement through time (e.g. Malinverno 1997; Felletti & Bersezio 2010a). Contributions from more than one entry point can compromise such analysis (Chakraborty *et al.* 2002).

5.3   Vertical trends

A powerful concept that developed during the 1970s was that thinning-upward and thickening-upward trends in submarine-fan successions could be used to specify the original depositional setting (e.g., fan channels versus fan lobes). It is now widely accepted that this hypothesis is flawed, both because the mechanisms of trend generation are quite hypothetical, and because almost none of the trends proposed in the literature have been rigorously defined using statistical techniques – instead, most were subjectively identified, so there is little agreement among researchers as to their validity. This general

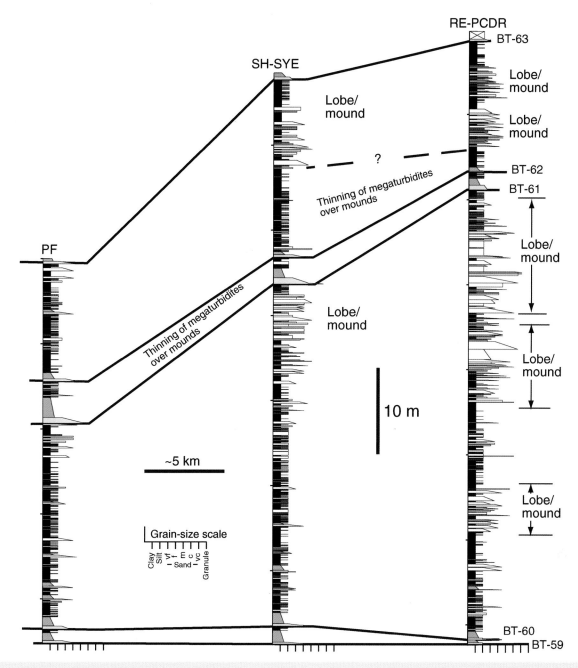

Fig. 5.16 Correlation of megaturbidites for ~25 km in the upper part of Awadallah's (2002) time sub-slice 6-1, lobe-fringe facies-association of the lower Cloridorme Formation. The megaturbidites have yellow sandy basal divisions and green silty mudstone tops. The downcurrent direction is from right to left. See Table 5.1 for location names. Modified after Awadallah and Hiscott (2004).

lack of statistical rigour when interpreting field examples is surprising, since tests formulated to statistically demonstrate the presence (or absence) of asymmetric trends in deep-marine deposits are available (e.g., Waldron 1987; Chen & Hiscott 1999a).

Trends (previously referred to as 'sequences'), rhythms or cycles, define an essentially progressive change in bed thickness and/or grain size, both for vertically-stacked beds or packets of beds. For consistency, the term *trend* is adopted in this book. Asymmetric trends were first recognised in flysch successions by Vassoevich (1948), although vertical trends were recorded a half-century earlier (Bertrand 1897). Research from 1960 to 1980 emphasised the importance of asymmetric thinning- or thickening-upward trends in deep-water successions (e.g., Nederlof 1959; Ksiazkiewicz 1960; Kelling 1961; Dzulynski & Walton 1965; Kimura 1966; Sestini 1970; Sagri 1972; Mutti & Ricci Lucchi 1972, 1975; Mutti 1974, 1977; Ricci Lucchi 1975b; Shanmugam 1980; Ghibaudo 1980). The objective recognition of trends at a variety of scales is important in understanding intrabasinal and extrabasinal controls on sedimentation.

By convention, thickening- and/or coarsening-upward trends are designated as 'negative', and thinning- and/or fining-upward trends as 'positive'. This terminology is based, to a large extent, upon the trend classification of Ricci Lucchi (1975a). As suggested by Ricci

Lucchi (1975a) and Shanmugam (1980), the largest scale trends are defined as *first-order trends*, and generally involve hundreds to thousands of metres of succession, commonly crossing the contacts of geological formations. These trends involve an upward change in grain size, bed thickness and sediment composition. Overall progradation or retreat of major related depositional environments may generate first-order trends. Examples include: (i) the ~600 m-thick Namurian Edale Shales → Mam Tor Sandstones Shale Grit → Grindslow Shales → Kinderscout Grit thickening- and coarsening-upward trend, northern England, interpreted as an upward-shoaling succession from deep-water basinal shales to delta-top sandstones (Allen 1960; Walker 1966a; Collinson 1969, 1970; McCabe 1978); (ii) the progradational suites of basin plain → outer fan → middle fan → inner fan turbidites and hemipelagites in the northern Apennines, Italy (Ricci Lucchi 1975b); (iii) the pelagic starved-basin to outer fan deposits of the Middle Ordovician Blockhouse Formation, Tennessee (Shanmugam 1980); (iv) the overall 3200 m-thick, thinning- and fining-upward trend of the inner → middle → outer fan turbidite system of the upper Precambrian Kongsfjord Formation, northern Norway (Pickering 1981a, 1985); and (v) the Carboniferous (Mississippian) progradational trends described from the Ouachita Mountains, Oklahoma and Arkansas (Niem 1976).

First-order trends reflect major changes in sediment input to a basin, probably governed by long-term changes in sea level, overall subsidence or uplift of a basin, major progradation of linked clastic systems (e.g., delta–slope–fan systems), large-scale strike-slip displacements, and gradual peneplanation or rejuvenation of a source area. First-order trends can describe the overall history of related depositional systems and, with good chrono- and litho-stratigraphic control, it may be possible to elucidate the specific factors generating such trends.

Second-order trends are defined on a scale of tens to hundreds of metres. They are identified as progressive changes in the thickness of beds or packets of beds, all within the same depositional system. Examples of second-order trends are documented from the upper Precambrian Kongsfjord Formation, northern Norway, where hundreds of metres of fan-fringe sediments show an increased or a decreased proportion of outer-fan lobe deposits (Pickering 1981b). Second-order trends may develop because of extrabasinal or intrabasinal processes governing the distribution of sediments within a clastic system.

Second-order trends showing a thickening- and coarsening-upward character develop on oceanic plates as they approach subduction zones and receive increasing amounts of terrigenous sediments. This vertical change is from open-ocean pelagites → hemipelagites and thin-bedded, fine-grained turbidites → thicker bedded and coarser grained trench/slope turbidites (e.g., Schweller & Kuhn 1978; Pickering *et al.* 1993a, b; Underwood *et al.* 1993).

Third-order trends occur on a scale of metres to tens of metres, and rarely at slightly larger or smaller scales. Third-order trends are the most commonly reported, but most subjective and controversial. Trends on this scale are believed to reflect intrabasinal controls. Examples include claims of thinning-and-fining-upward submarine channel fills (e.g., Mutti & Ricci Lucchi 1972, 1975; Ricci Lucchi 1975a, b; Mutti 1977, 1979; Walker 1978; Normark 1978; Hendry 1978; Pickering 1982a), and thickening-and-coarsening-upward fan-fringe and lobe deposits (e.g., Mutti & Ricci Lucchi 1972, 1975; Ricci Lucchi 1975a, b; Mutti 1977; Normark 1978; Walker 1978; Ghibaudo 1980; Pickering 1981b).

The basis for recognition of third-order trends is almost always entirely subjective, and in many cases random alternations of bed thickness (and/or grain size) can generate asymmetric trends that are as convincing as those that have been subjectively identified in field studies (Hiscott 1981; Chen & Hiscott 1999a). This is particularly true for inferred submarine-fan deposits, so that claims that third-order thinning- or thickening-upward trends are useful tools for environmental assessment (e.g., Ricci Lucchi 1975b) should be viewed with caution.

Thinning- and fining-upward trends appear to be more plentiful than thickening- and coarsening-upward trends at the second-order and third-order levels. If sea-level (i.e., relative base-level) fluctuations are partly responsible for many trends (Mutti 1985), then the difference in abundance of these asymmetric trends may indicate that relative sea-level falls are more rapid than rises. An alternative, perhaps more plausible explanation is that falls in relative base level lead to an increased frequency of erosive events, whereas rises, on average, are associated with less erosive processes. Thus, trends associated with a rise in relative base level have the greatest preservation potential. A good example of the latter case is a ~200 m-thick thinning-and-fining-upward trend (Fig. 5.17) generated by a rise in sea level from the Pleistocene to Holocene in a middle-upper fan channel of the Mississippi Fan (Pickering *et al.* 1986b).

Statistical tests for the presence of asymmetric trends can be designed to evaluate (i) whether trends of either decreasing or increasing SGF thickness (or grain size, or SGF coarse-division thickness) predominate in a selected section, and (ii) whether a small segment of such an interval that appears to exhibit thickening or thinning upward in the field does in fact possess a statistically verifiable trend. Query (ii) is not as interesting as query (a), because even undisputed asymmetric cycles can be produced with low probability by random variations in bed thickness in long stratigraphic sections.

Waldron (1987) outlined how a signs-of-differences test proposed by Moore and Wallis (1943) can be used to address query (i), that is, whether a long stratigraphic section has trends with consistent asymmetry. The test compares the total number of decreases in bed thickness observed in the section, m (i.e., the number of negative differences in thickness from each bed to the next overlying bed) to the mean number of decreases, μ_m, expected in a section with N beds, where:

$$\mu_m = (N - 1)/2 \qquad \text{for } N > 12 \qquad (5.1)$$

standard deviation of m, σ_m is given by:

$$\sigma_m = [(N + 1)/12]^{0.5} \qquad (5.2)$$

The test statistic, z, is given by:

$$z = (m' - \mu_m)/\sigma_m \qquad (5.3)$$

where $m' = (m - 0.5)$ if $m > \mu_m$, and $m' = (m + 0.5)$ if $m < \mu_m$. In a random succession, z is approximately normally distributed with a mean of zero and standard deviation of 1.0. The null hypothesis that trends with consistent asymmetry are not present can be assessed at various significance levels. If the null hypothesis is rejected, and $z > 0$, the trends thin upward; if $z < 0$, the trends thicken upward. Waldron (1987) also indicates how data can be smoothed to filter out the effect of random fluctuations superimposed on truly cyclic, asymmetric trends. Note that a significant result using this signs-of-difference test might result from the sections showing a steady trend of bed thickness or grain size with no cyclicity. Conversely, a mixture of positive and

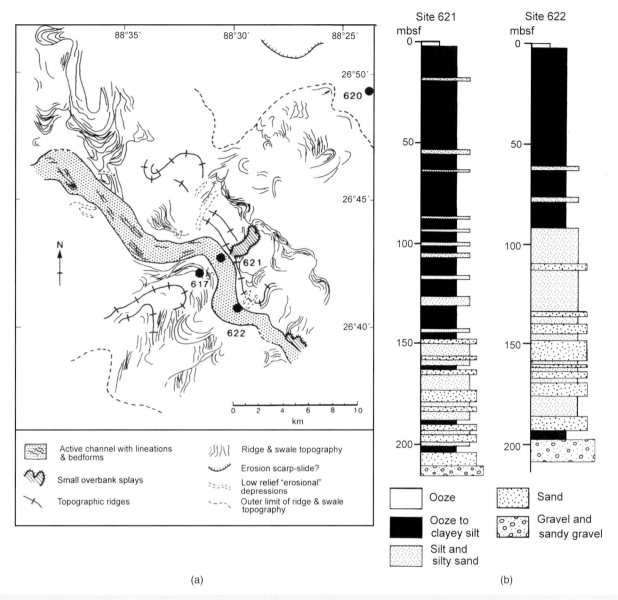

Fig. 5.17 Location of DSDP Leg 96 Sites 621 and 622 (a), and their upward fining lithostratigraphy (b), from the meandering channel of a channel–levée–overbank system on the Mississippi Fan. mbsf = metres below seafloor. From Pickering *et al.* (1986).

negative asymmetric trends could cancel one another out giving a non-significant value of *z* (see also Chakraborty *et al.* 2002).

There is an important conceptual difference between asymmetric trends that show a gradual and progressive trend in bed thickness (or grain size) and those that show step-like trends. Consider two thinning-upward trends of ten beds, both above a fine-grained interval, one progressive and the other involving three steps (Fig. 5.18). Excluding the possibility of ties, for the gradual trend (Fig. 5.18a), the probability, P_G, of observing such an ordered trend is given by:

$$P_G = \sum_{i=1}^{i=9} P(t_i + 1 < t_i) = 1/9! = 2.7 \times 10^{-6} \qquad (5.4)$$

where t_i is the thickness of the *i*th bed. The calculated value of P_G was obtained by considering the probability that ten bed thicknesses taken

from the population of beds would occur in only this one regular arrangement of consistently decreasing thickness from base to top.

If (i) the trend shows a step-wise decrease in bed thickness above a thin-bedded interval, (ii) beds tend to occur in 'packets' of either thick, medium, or thin beds (Fig. 5.18b), and (iii) the three types of packet are equally abundant in a section, then the probability, P_{SW}, that a step-like trend will occur above a thin-bedded interval is given by:

$$P_{SW} = (1/2)^3 = 0.125 \qquad (5.5)$$

Clearly, $P_{SW} \gg P_G$ in this case, and the step-like trend is far less likely to represent a statistically significant trend because it can be generated quite readily through random stacking.

In Figure 5.18b, the packets of quite different bed thickness really represent three different 'states' of the deep-water system,

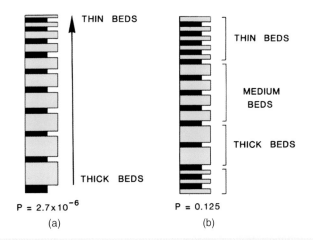

Fig. 5.18 Idealised thinning-upward trends: (a) gradual, and (b) step-like. See text for explanation. From Pickering *et al.* (1989).

each perhaps corresponding to a different environment (e.g., lobe axis versus lobe fringe), and each with its own distribution of bed thicknesses. In our experience, many deep-water fan deposits are characterised by step-like trends (*cf.* Ricci Lucchi 1975b), and alternations of 'packets' with substantially different bed thicknesses (e.g., Fig. 5.19).

Within packets with step-like thickness profiles, the thickness of a bed is not independent of the thickness of associated beds, so the test of Waldron (1987) and other trend tests described below are inappropriate. Instead, we suggest a subdivision of successions with step-like changes in bed thickness or grain size into 'states' (facies) based on these characteristics, and then application of Markov

Chain analysis to identify preferred transitions between these states. Procedures are given by Powers and Easterling (1982) and Harper (1984). This approach was used successfully by Hiscott (1980) to identify a tendency for thinning- and fining-upward in part of the Ordovician Tourelle Formation, Québec, Canada, but unsuccessfully by Hein (1979) to identify asymmetric trends in the less organised submarine channel fills of the Cambro-Ordovician Cap Enragé Formation, Québec.

Chen and Hiscott (1999a) completed a thorough study of statistical tests for the existence of third-order trends in a number of SGF successions, and applied three powerful correlation tests (Kendall's, Spearman's and Pearson's correlation tests) and four tests for randomness (rank difference test, turning points test, median crossing test, and length-of-runs test) to 28 bed-by-bed sections with a wide coverage in geological time, tectonic settings, facies characteristics and depositional environments. They considered 20 field sections with a total thickness of 2513 m and 5916 beds, measured bed-by-bed in the Italian Apennines, the California Great Valley, Barbados, the Gulf Islands of British Columbia and the Gaspé Peninsula of Québec. Eight additional sections were compiled from the Pleistocene Amazon Fan (ODP Leg 155), the Kongsfjord Formation of northern Norway and the Pennsylvanian Jackfork Group of Arkansas (Table 5.2). The localities where sections were measured show either a channel-filling geometry or a relatively unconfined (non-channelised) lobe geometry. Sections in the Italian Apennines were included because they formed the basis of Ricci Lucchi's (1975a) hypothesis that asymmetric cycles correspond to submarine fan channel and lobe deposits. Sections in the California Great Valley were claimed, by Ingersoll (1981), to consist of unambiguous asymmetric cycles. Sections in the Gaspé Peninsula of Québec convinced Hiscott (1980) that upward thinning and thickening cycles are not common in submarine fan deposits.

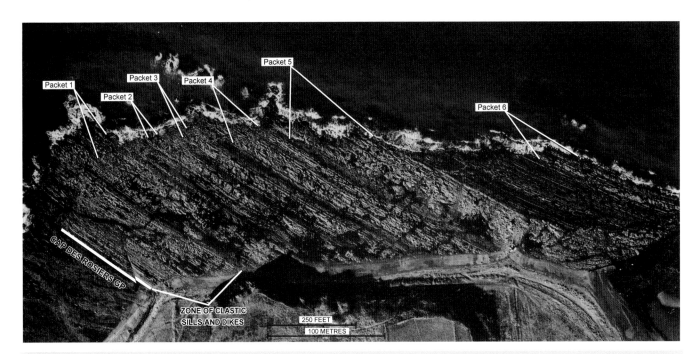

Fig. 5.19 Alternating sandstone packets (numbered) and shaly lobe-fringe and inter-channel deposits in the Ordovician Tourelle Formation of Québec, Canada. Note that packet 5 has an erosional base. The depth of this channel is 9 m. From Hiscott and Devries (1995).

Table 5.2 The number and percentages of sandstone packets with asymmetric thinning- and thickening-upward trends (at a significance level α = 10%) in individual sections or locations, and the cumulative binomial probability of rejecting the null hypothesis *j* or more times (*j* = the number of identified asymmetric cycles) in *n* random trials (*n* = the number of sandstone packets). The null hypothesis is that cycles were generated at random. From Chen and Hiscott (1999a)

Location	No. of sections	Thickness (m)	No. of sandstone packets	No. that thicken up	No. that thin up	Asymmetric cycles (%)	Cumulative binomial probability
Italy							
Santerno Valley	4	445.3	30	2	0	9.1%	0.69
Savio Valley	2	49.5	3	1	0		
California							
Monticello Dam	1	383.7	23	1	3	17.4%	0.19
Cache Creek	1	278.0	22	1	2	13.6%	0.38
Québec							
Cap Ste-Anne	2	222.3	9	0	2	22.2%	0.23
Petite-Vallée	1	452.5	67	3	4	10.5%	0.53
Barbados	2	109.7	10	0	2	20%	0.26
Amazon Fan Sites 931, 944, 946	3	335.1	17	0	2	11.9%	0.52
British Columbia	7	572.9	26	1	3	15.4%	0.26
Northern Norway	3	248.0	37	2	2	10.8%	0.55
Arkansas (DeGray Lake)	2	212.9	42	1	2	7.1%	0.82
Total	28	3309.8	286	12	22	11.9%	0.17
Number and percentage expected from 286 samples of a random population for α = 10%				~ 14	~ 14	10%	

Chen and Hiscott (1999a) provide equations for the computation of test statistics, reproduced in part below. The field variable for these tests could be bed thickness, some measure of grain size of the SGF, SGF thickness or SGF coarse-division thickness (Fig. 5.1). For consistency, we primarily use the last of these options in the next sections. We encourage continuing efforts to objectively document third-order trends in other deep-water successions, because such trends may provide the most sensitive geological indicators of short-term fluctuations in sea level (Mutti 1985), processes of channel and lobe switching (Chapter 7), climatic changes in a source area, pulses of tectonic activity and retrogressive slope failure (Pickering 1979).

5.3.1 Tests for randomness

Three of the four tests for randomness preferred by Chen and Hiscott (1999a) address query (i)–do trends of either decreasing or increasing SGF thickness (or grain size, or SGF coarse-division thickness) predominate in a selected section? Equations for the *turning points test* are taken from Kendall (1976). For this test, SGF coarse-division thickness values, x_i, in a succession are replaced by 1 if either of the following statements is true: x_i is greater than both x_{i-1} and x_{i+1}; or x_i is less than both x_{i-1} and x_{i+1}. Otherwise, x_i is replaced by 0. A single 1 represents a turning point. The number of turning points, *TP*, is approximately normally distributed, with:

$$\text{mean} = \mu_{TP} = \frac{2(n-2)}{3} \qquad (5.6)$$

and

$$\text{Standard deviation} = \sigma_{TP} = \sqrt{\frac{16n - 29}{90}} \qquad (5.7)$$

The Z statistic, assumed to be an observation from a standard normal distribution, is given by:

$$Z_{TP} = \frac{TP - \mu_{TP}}{\sigma_{TP}} \qquad (5.8)$$

If $|TP - \mu_{TP}|$ is sufficiently large at a selected significance level, then the null hypothesis that the series of bed thicknesses was generated by a purely random process would be rejected. Because the number of turning points in a succession is always one less than the number of runs up and down, this method is conceptually identical to the runs-up-and-down test of Murray *et al.* (1996).

The *median crossing test* is explained by Fisz (1963), and is based on a binary series. For SGF coarse-division thicknesses x_i and median thickness x_m, if $x_i < x_m$, x_i is assigned a new value of 0; if $x_i > x_m$, x_i is replaced by the value 1. If $x_i = x_m$, no value is assigned to x_i (i.e., this SGF coarse division is ignored because it is neither larger or smaller than the median), and *n* is reduced by 1. If the original succession was generated by a purely random process, then *M*, the total number of times either 0 is followed by 1 or 1 is followed by 0, is approximately normally distributed, with:

$$\text{mean} = \mu_M = \frac{n-1}{2} \qquad (5.9)$$

and

$$\text{standard deviation} = \sigma_M = \sqrt{\frac{n-1}{4}} \qquad (5.10)$$

The Z statistic is:

$$Z_M = \frac{M - \mu_M}{\sigma_M} \qquad (5.11)$$

If Z_M lies within a selected confidence interval, then the hypothesis that the series of SGF coarse-division thicknesses resulted from a random process cannot be rejected at that significant level, α. This test is similar to the runs-about-median method (Wald and Wolfowitz 1944), because *M* is always one less than the number of runs on either side of the median.

The *length-of-runs test* was proposed by Gold (1929). A run length, *s*, is defined as the number of consecutive thicknesses either all above or all below the median value of SGF coarse-division thickness. If M_s denotes the total number of runs of length s, then for a random process the expected value of M_s is:

$$E(M_s) = \frac{n + 3 - s}{2^{s+1}} \qquad (5.12)$$

and the sum:

$$L = \sum_{s=1}^{s'} \frac{[M_s - E(M_s)]^2}{E(M_s)} \qquad (5.13)$$

is distributed as chi-square with $(s' - 1)$ degrees of freedom, where s' is the maximum run length in the succession. This is a one-tailed test. If L falls below a pre-selected critical value (determined by the significance level, α), then the hypothesis of randomness cannot be rejected at that significance level.

Chen and Hiscott (1999a) used these randomness tests to investigate segments of longer sections, where each segment that they considered is bounded by a statistically significant change in bed thickness or grain size (Fig. 5.20). Hence, the segments are akin to 'packets' described in many deep-water successions. A *packet* (or 'cluster') is defined as a group of beds sharing similar facies characteristics and overall grain size distinct from strata immediately above and below (e.g., Sullwold 1960; Ojakangas 1968; Corbett 1972; Nilsen & Simoni 1973; Hiscott 1980; Pickering *et al.* 1989). A sandstone packet may be the fill of a channel or the deposit of a lobe, whereas a thin-bedded siltstone–mudstone packet may represent deposits of a levée or an inter-channel or interlobe environment. The basal and upper boundaries of a packet are either a sharp change (i.e., a step or discontinuity) or a progressive change of bed thickness and/or grain size.

The weakness of the three randomness tests described above is that they do not evaluate whether the cycles thin upward, thicken upward or are symmetrical (only the first two would constitute 'trends'). To assess asymmetry, trend tests (below) are required. For example, long sections might simply be organised into alternating packets of thinner/finer and thicker/coarser event-beds (Chen & Hiscott 1999b), with no inherent order within each of the packets. Also, runs tests are very sensitive to noise in the data, so that they can fail to recognise an asymmetric trend with superimposed noise in the form of very thin and/or thick SGF event-beds. Procedures designed to overcome this shortcoming, for example the smoothing of noisy data by taking moving averages, are not entirely satisfactory because they corrupt the primary data by introducing dependence between successive measurements (see the debate of this issue in Heller & Dickinson 1985; Waldron 1986; Lowey 1992; Murray *et al.* 1996).

5.3.2   Correlation tests to identify asymmetric trends

If asymmetric cycles are suspected, then trend tests can be used to objectively determine if a series of SGF event-deposits actually thins/fines or thickens/coarsens upward (say, in a sandstone packet, or between segments of a section that are bounded by sharp changes in bed characteristics). Chen and Hiscott (1999a) recommended using all three correlations tests described in this section, because in some cases they found that only two of three tests were able to identify asymmetric cyclicity. In *Kendall's rank correlation test* (also called Kendall's τ test), the rank (in thickness) of each event-bed (or SGF coarse division) is compared with the rank of all overlying event-beds within a sandstone packet (Kendall 1969, 1976; Kendall & Gibbons

1990). A negative score (the target bed is thicker than an overlying bed) or a positive score (the target bed is thinner than an overlying bed) is obtained for each comparison. Kendall's τ is calculated by:

$$\tau = \frac{p - Q}{n(n - 1)/2} \qquad (5.14)$$

or, if ties exist, by:

$$\tau = \frac{p - Q}{\sqrt{p + Q + Ex}\sqrt{p + Q + Ey}} \qquad (5.15)$$

Where p is the total number of positive scores, Q the total number of negative scores and n the number of beds. Ey is the sum of 'extra y': when the rank of the thickness of the event-bed (the x variable) is tied with that of an overlying event-bed, an 'extra y' is counted. Ex is the sum of 'extra x'; for field datasets from turbidite successions, $Ex = 0$, because event-bed numbers (the y variable) cannot be tied. τ varies from -1 to $+1$, with consistent thickening upward giving $\tau = +1$, and consistent thinning upward giving $\tau = -1$. No asymmetric trend gives $\tau = 0$. τ is approximately normally distributed, with:

$$\text{mean} = \mu_\tau = 0 \qquad (5.16)$$

and

$$\text{standard deviation} = \sigma_\tau = \sqrt{\frac{2(2n + 5)}{9n(n - 1)}} \qquad (5.17)$$

The test statistic is:

$$Z = \frac{\tau - \mu_\tau}{\sigma_\tau} = \frac{\tau}{\sqrt{\dfrac{2(2n + 5)}{9n(n - 1)}}} \qquad (5.18)$$

For a random succession of SGF coarse-division thicknesses and for sample size as small as 4, τ approximates a normal distribution (Daniel 1978). The null hypothesis of no trends can be evaluated at any appropriate significance level. For SGF successions, the test is two-tailed. For example, if a test is undertaken at the significance level $\alpha = 10\%$, and if the Z value falls into the 5% area under the standard normal distribution curve either at the right or left tail, then the null hypothesis that there is no trend is rejected.

Spearman's rank correlation test involves calculating Spearman's ρ (Siegel 1956; Kendall 1969; Daniel 1978; Press *et al.* 1986; Rock 1988; Kendall & Gibbons 1990). $R(x_i)$ is defined as the rank of x_i among the other x's, with $R(x_i) = 1$ if x_i is the smallest observed value of x. $R(y_i)$ is similarly defined as the rank of y_i among the other y's. If ties occur among the x's or y's, each tied value is assigned the mean of the rank positions for which it is tied. For series of SGF event beds, the variable y is the event-bed number, which is ranked in increasing order from bottom to top: 1, 2, 3, 4 …; and the variable x is the thickness of interest (e.g., SGF deposit thickness, or SGF coarse-division thickness; Fig. 5.1) – the thickness is replaced by its rank in the packet. When there are no ties, Spearman's rank correlation coefficient ρ is defined as:

$$\rho = 1 - \frac{6\Sigma d_i^2}{n^3 - n} \qquad (5.19)$$

where:

$$\Sigma d_i^2 = \sum_{i=1}^{n} [R(x_i) - R(y_i)]^2 \qquad (5.20)$$

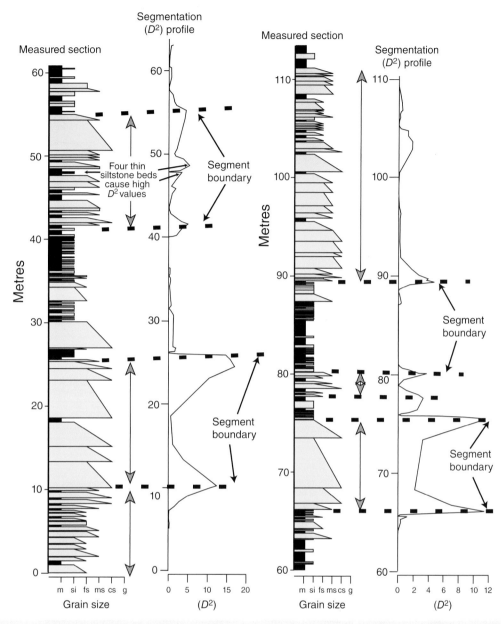

Fig. 5.20 An example of packet segmentation from the Cap Ste-Anne section, Tourelle Formation, Gaspé Peninsula, Québec. The turbidite section is segmented into thick-bedded and thin-bedded sandstone packets and shale–siltstone intervals using the split-moving window technique (Webster 1980). The peak values of D^2 indicate packet boundaries. Packets that were statistically tested by Chen and Hiscott (1999a) are marked by orange arrows. The top of the uppermost packet could not be picked by the split-moving window technique because of the lack of sufficiently thick overlying section. From Chen and Hiscott (1999a).

and n = sample size. When ties are present, the test statistic is slightly more complicated:

$$\rho = \frac{\Sigma x^2 + \Sigma y^2 - \Sigma d_i^2}{2\sqrt{\Sigma x^2 \Sigma y^2}} \tag{5.21}$$

where:

$$\Sigma x^2 = \frac{n^3 - n}{12} - \Sigma T_x \tag{5.22}$$

and

$$\Sigma y^2 = \frac{n^3 - n}{12} - \Sigma T_y \tag{5.23}$$

$$T = \frac{m^3 - m}{12} \tag{5.24}$$

where m = the number of observations tied at a given rank. All summations are from 1 to n. Spearman's ρ ranges from -1 to $+1$, with $\rho = 1$ when event-bed thickness (x) and event-bed number (y) have identical rankings (a perfect upward-thickening trend), and $\rho = -1$

when the two series of values have reversed directions of increasing rank (a perfect upward-thinning trend). The significance of ρ is tested by computing:

$$t = \rho\sqrt{\frac{n-2}{1-\rho^2}} \qquad (5.25)$$

which is distributed approximately as a student's t distribution with $n-2$ degrees of freedom. To test for asymmetric upward thinning and thickening trends, a two-tailed test is preferred.

In *Pearson's correlation test*, Pearson's γ is a measurement of the linear relationship between two variables, x and y (Hoel 1971; Downie & Heath 1983; Press *et al.* 1986). It is given by the formula

$$\gamma = \frac{n\Sigma xy - (\Sigma x)(\Sigma y)}{\sqrt{[n\Sigma x^2 - (\Sigma x)^2][n\Sigma y^2 - (\Sigma y)^2]}} \qquad (5.26)$$

In calculating γ, y (event-bed number) and x (event-bed thickness) are actual values, instead of their rank. The value of γ varies from $+1$ through 0 to -1, with $\gamma = 1$ indicating a perfect direct correlation between the two variables (a perfect upward thickening trend), $\gamma = -1$ a perfect inverse linear correlation between the two variables (a perfect upward-thinning trend), and $\gamma = 0$ no linear correlation (not an asymmetric trend). In computing the significance of γ, the follow statistic is used:

$$t = \gamma\sqrt{\frac{n-2}{1-\gamma^2}} \qquad (5.27)$$

which is distributed approximately as a student's t distribution with $n-2$ degrees of freedom (Press *et al.* 1986). The test is two-tailed.

An analysis of sample size requirements indicates that packets comprising as few as six beds can be tested for asymmetric trends by the three correlation tests at the 5% significance level (Fig. 5.21). When tests use a 10% significance level, packets with as few as four beds can be successfully tested.

The power of these correlation tests is the probability of rejecting a false null hypothesis of no order, calculated as $1-\beta$, where $\beta =$ Type II error. In order to evaluate the power of the three methods for discovering the presence of asymmetric trends, Chen and Hiscott (1999a) generated and tested several artificial sandstone packets that were designed to include most patterns of bed thickness seen in turbidite successions: simple asymmetric trends; asymmetric trends with superimposed noise; symmetric trends; stepwise cycles; clusters of thick and thin beds; and random distributions created by shuffling of data. 'Noise' is defined here as the presence of very thin and/or very thick event-beds superimposed on an otherwise increasing or decreasing trend in SGF event-bed thickness. 'Noise' can also be random positive or negative deviations from an otherwise smooth trend or pattern.

All artificial sandstone packets with clear trends passed two, or more commonly three, of the correlation tests at higher than the 90–95% confidence level ($\alpha < 0.10$ or $\alpha < 0.05$, respectively). These are packets 1, 2, 3, 4 and 7 (Fig. 5.22). Kendall's rank correlation test identified both clear and complex upward thickening or thinning trends at less than the 10% significance level. Spearman's rank correlation test was found to be slightly more sensitive to noise than Kendall's test, and Pearson's linear correlation test is particularly affected by noise in the form of very thick and very thin event-beds (e.g., artificial packet 5, Fig. 5.22). The three correlation tests usually

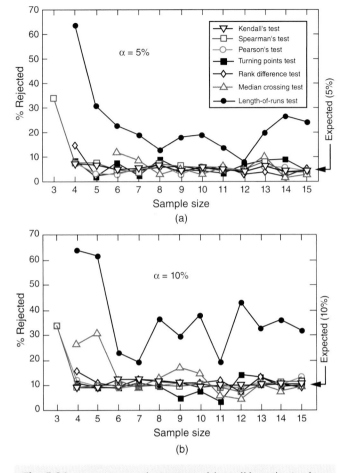

Fig. 5.21 Percentages of rejections of the null hypothesis of 'no trend' for 3 to 15 randomly shuffled thickness values at (a) the 5% significance level, and (b) the 10% significance level. Key is the same for both graphs. Almost all tests perform well until the sample size decreases below 4–6 measurements. From Chen and Hiscott (1999a).

fail to identify symmetric trends (e.g., packet 9), stepwise patterns (e.g., packet 10), and clustering of thin and/or thick beds (e.g., packets 11 and possibly 8). For such patterns, the null hypothesis is more likely to be rejected by tests for randomness.

One test for randomness is designed for use on segments of an otherwise long section, using the same intervals that can be tested using the three trend tests. This is the *rank difference test* (Meacham 1968), for which event-bed thickness (or SGF coarse-division thickness) is replaced by its rank (R_i) in a sandstone packet, with the thinnest event-bed (R_1) being assigned rank 1. The R statistic is calculated by:

$$R = \sum_{i=2}^{n}|R_i - R_{i-1}| \qquad (5.28)$$

where $n =$ the number of event-beds and $R =$ sum of the absolute values of the rank differences between successive event-beds in a single sandstone packet. For a random succession of thicknesses, R is normally distributed, with:

$$\text{mean} = \mu_R = \frac{(n+1)(n-1)}{3} \qquad (5.29)$$

and

$$\text{Standard deviation} = \sigma_R = \sqrt{\frac{(n-2)(n+1)(4n-7)}{90}} \qquad (5.30)$$

A Z statistic is computed as follows:

$$Z_R = \frac{R - \mu_R}{\sigma_R} \qquad (5.31)$$

If Z_R lies within a certain confidence interval, then the null hypothesis that the cycle results from a random process cannot be rejected at the corresponding significance level, α. For example, if we test a series of event-bed thicknesses at α = 0.1, and if the value of Z_R falls within the central 90% of a standard normal distribution, then the null hypothesis that the series of thickness is random is accepted.

5.3.3 Realisation that asymmetric trends can be formed, at low probability, by random processes

There is danger in using the discovery of a few statistically significant asymmetric third-order trends to support an interpretation that a certain type of asymmetry is characteristic of a sedimentary succession, or that the perceived type of trend constrains the likely depositional environment. This is because a small number of perfectly good trends are predicted to result from truly random processes, but at low probability. In order to evaluate the importance of this effect, Chen and Hiscott (1999a) conducted an extensive set of Monte Carlo simulations using the field data corresponding to sections in Table 5.3. The results confirm that asymmetric thickness trends have essentially no statistical significance in deep-water successions, and therefore cannot provide a key criterion for identification of sub-environments in these depositional systems. It follows that models for deep-marine systems based on the widely publicised hypothesis that asymmetric cycles are common (e.g., Mutti & Ricci Lucchi 1972; Ricci Lucchi 1975b; Walker 1978; Shanmugam & Moiola 1988; Mutti 1992) must be treated with great caution or abandoned. Instead, other criteria such as specific facies characteristics, large-scale geometry and degree of sand-bed clustering (Chen & Hiscott 1999b; Felletti 2004; Felletti & Bersezio 2010b) might provide the best tools for discrimination of sub-environments.

As a starting point, Chen and Hiscott (1999a) objectively split the field sections into packets at points of quick vertical facies change. These discontinuities in event-bed thickness and facies were picked using the *split-moving window method* (Webster 1973, 1980) and *maximum likelihood estimation* (Radhakrishnan *et al.* 1991). An example is given in Figure 5.20. The segmentation of the sections is based on the assumption that sharply bounded segments or event-bed packets with different means and variances of event-bed thickness, grain size and/or other measurements reflect different facies associations, and therefore different depositional environments in deep-water systems. Any attempt to recognise trends that bridge segments with distinctly dissimilar properties would violate a critical assumption of any statistical test that all observations come from a single population. In this way, Chen and Hiscott (1999a) selected 286 sandstone packets from the 28 turbidite sections. Then, these packets were evaluated using the three correlation tests (Kendall's, Spearman's and Pearson's) coded in a Fortran-77 program called ASYMRAN.FOR. Only 34 (11.9%) of the sandstone packets passed tests designed to

Table 5.3 List of turbidite sections tested for asymmetric trends by Chen and Hiscott (1999a). See Table 5.2 for number of sections, cumulative thicknesses measured and number of event-beds

Location	Unit (Fm = Formation)	Proposed Environments	Number of sections	Thickness (m)	Number of beds	Measurer or reference
Italy						
Santerno Valley	Miocene Marnoso arenacea	Lobes and basin-plain sand sheets (Ricci Lucchi 1975b; Cattaneo and Ricci Lucchi 1995)	4	445.3	738	C. Chen
Savio Valley		Channel fill (Ricci Lucchi 1975b)	2	49.5	75	C. Chen
California						
Monticello Dam	Cretaceous Venado Fm	Channel fill (Ingersoll 1978a)	1	383.7	379	C. Chen
Cache Creek	Cretaceous Sites Fm	Lobes (Ingersoll 1978a)	1	278.0	565	C. Chen
Québec						
Cap Ste-Anne	Ordovician Tourelle Fm	Lobes (Hiscott 1980)	2	222.3	523	S. Awadallah
Petite-Vallée	Ordovician Cloridorme Fm	Lobes (Hiscott *et al.* 1986)	1	452.5	2893	S. Awadallah
Barbados	Eocene Scotland Fm	Channel fill (Larue and Speed 1983)	2	109.7	183	R. N. Hiscott
Amazon Fan	ODP Sites 931, 944, 946	Lobes related to channel bifurcation	3	335.1	485	Pirmez *et al.* (1997)
British Columbia	Cretaceous Nanaimo Group	Channel fill (England and Hiscott 1992)	7	572.9	560	R.N. Hiscott
Northern Norway	Precambrian Kongsfjord Fm	Channel-lobe transition zone (Drinkwater 1995)	3	248.0	825	N.J. Drinkwater
Arkansas (DeGray Lake)	Pennsylvanian Jackfork Group	Crevasse-splay lobes and levées (Bouma *et al.* 1995)	2	212.9	2514	M.B. DeVries
Totals			28	3309.8	9740	

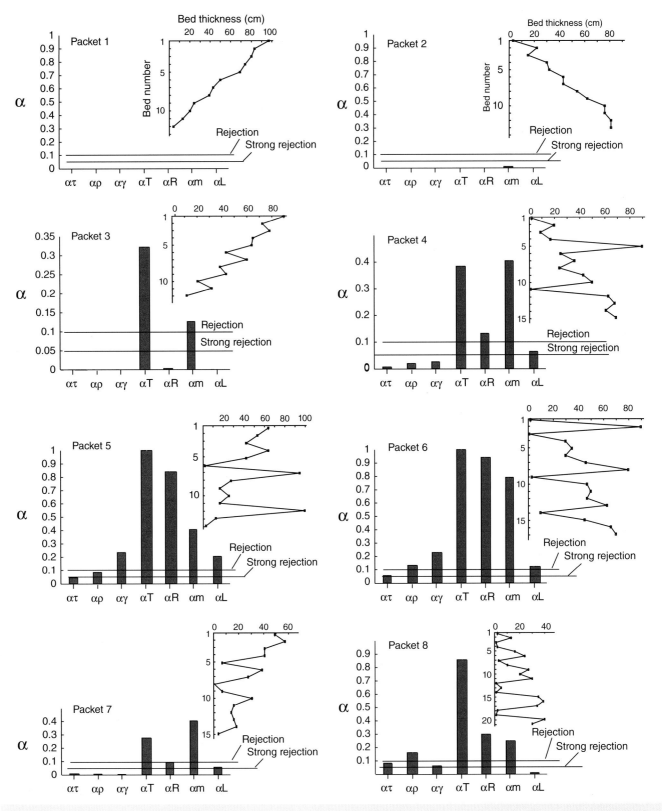

Fig. 5.22 Graphs, this page and next, showing results of experiments on the power of the seven statistical techniques in testing for asymmetric and other trends in 16 artificially generated and random packets. In the bar graphs, the height of the bars indicates, from left to right, the significance level, α, at which each packet passed Kendall's (ατ), Spearman's (αρ), Pearson's (αγ), the turning points (αT), the rank difference (αR), the median crossing (αm) and the length-of-runs (αL) tests. Note that if a series of event-beds passed a test at α ≤ 5%, the null hypothesis was strongly rejected; if at α ≤ 10%, the null hypothesis was rejected; if at α > 10%, then the null hypothesis was accepted. Packets 1 to 11 were artificially generated: 1 and 2 show nearly perfect asymmetric trends; 3 through 8 have overall asymmetric trends but with 'noise' in the form of very thick and/or very thin event-beds; 9 displays a symmetric pattern; 10 a stepwise pattern; and 11 a grouped pattern of thin and thick event-beds. Packets 12 through 16 (facing page) were created by a random-shuffle program. From Chen and Hiscott (1999a).

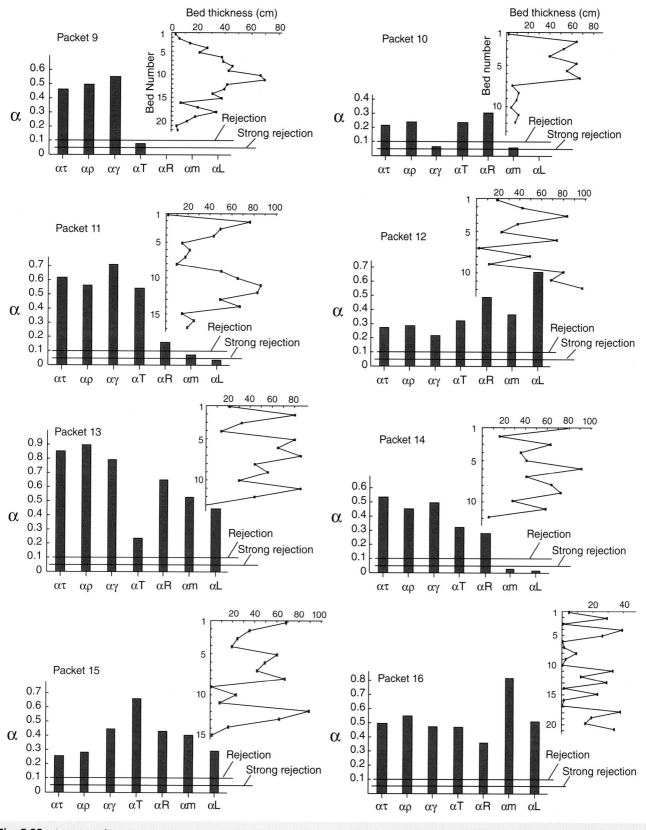

Fig. 5.22 (*continued*)

Table 5.4 The number (percentage) of the 286 sandstone packets of Chen and Hiscott (1999a) for which the null hypothesis of 'no trend' (i.e., no correlation) was rejected at a significance level α = 10 %. For each combination of correlation tests, the results of tests for randomness are provided as additional constraints for deciding the likelihood that the packets are truly asymmetric

Combinations of tests passed	Number (percentage) of packets for which the null hypotheses were rejected at α ≤ 10%, and conclusion	
3 correlation tests		
4 tests for randomness	2 (0.7%)	asymmetric trend
3 tests for randomness	4 (1.4%)	asymmetric trend
2 tests for randomness	5 (1.7%)	asymmetric trend
1 test for randomness	5 (1.7%)	asymmetric trend
no test for randomness	4 (1.4%)	asymmetric trend
2 correlation tests		
3 tests for randomness	1 (0.3%)	other nonrandom pattern
3 tests for randomness	1 (0.3%)	asymmetric trend
2 tests for randomness	3 (1.0%)	asymmetric trend
1 test for randomness	3 (1.0%)	asymmetric trend
no test for randomness	2 (0.7%)	asymmetric trend
Only Kendall's test		
no test for randomness	3 (1.0%)	asymmetric trend
Only Spearman's test		
1 test for randomness	2 (0.7%)	asymmetric trend
no test for randomness	1 (0.3%)	random
Only Pearson's test		
1 test for randomness	2 (0.7%)	random
no test for randomness	3 (1.0%)	random
No correlation test		
4 tests for randomness	3 (1.0%)	other nonrandom pattern
3 tests for randomness	11 (3.8%)	other nonrandom pattern
2 tests for randomness	39 (13.6%)	other nonrandom pattern
1 test for randomness	62 (21.7%)	random
no test for randomness	130 (45.5%)	random
Total of 34 packets with asymmetric trends recognised in 286 sandstone packets (11.9%)		

identify asymmetry at the 10% significance level (Table 5.4). Packets in Figure 5.23, selected from these 286 analysed sandstone packets, are examples of asymmetric cycles recognised by the three correlation tests. Figure 5.24 shows examples of non-asymmetric or random cycles; these can pass the three correlation tests only by using an uncomfortably large significance level in the range 10–32%. Using α ≤ 32%, as Lowey (1992) did, attributes order to many random packets; that is, the Type I error is unacceptably large.

Thirty-three of the 286 packets passed Kendall's test at α ≤ 10%; 30 passed Spearman's test at the same significance level; and 31 passed Pearson's test at this level. Consider the general case of 286 random trials with a predetermined success rate is 10%, for example 286 attempts (with replacement) to draw a black ball from a jar of 900 white and 100 black balls. The cumulative binomial probability (Kreyszig 1967: p. 775; Simpson *et al.* 1960: p. 137) of 33 or more successes is 0.22 (compare Kendall's test result); of 30 or more successes is 0.43 (compare Spearman's test result); and of 31 or more successes is 0.35 (compare Pearson's test result). The successful drawing of a black ball is analogous to the successful recognition of an asymmetric trend *even when there is no correlation between event-bed thickness and event-bed number*, but where the chance of success (i.e., Type I error) is still 10%. Probabilities in the range 0.22 to 0.43 are sufficiently high to suggest that the number

of asymmetric trends in SGF coarse-division thickness identified by the three correlation tests in the 286 sandstone packets can easily be attributed to random processes.

In the 34 statistically identified asymmetric packets, the ratio of upward thinning to upward thickening packets is 22 : 12 (=1.83). If the odds are 50 : 50 for upward thinning versus upward thickening in a random population, the probability of obtaining 22 or more upward-thinning cycles out of a total of 34 asymmetric cycles is 0.06, which is marginally significant because for only one less success (i.e., 21 or more times out of 34) the probability is 0.11. Hence, the 22 : 12 result might indeed indicate the operation of some deterministic process or processes, capable of generating a very small surplus of upward-thinning trends, superimposed on what are otherwise largely random events. This does not contradict the general conclusion that upward thinning and thickening trends are not important in turbidite successions.

Using a different and complementary approach, Chen and Hiscott (1999a) confirmed with Monte Carlo simulation that the number of asymmetric trends identified in the original set of sandstone packets is indistinguishable from the number expected to result from random processes. The SGF coarse-division thicknesses measured in each of the 286 original sandstone packets were randomised by Chen and Hiscott (1999a) by sequentially shuffling them 20 000 times to generate one shuffled sandstone packet. Then, the 286 randomly shuffled packets were tested for asymmetric trends and randomness. The number of packets for which the null hypothesis of individual tests was rejected at the significance levels α = 5% and 10%, respectively, was counted. This process was repeated 100 times so that 100 sets of 286 randomly shuffled packets were generated and tested. The mean, maximum and minimum number of packets that passed individual tests at α = 5% and 10% for the 100 sets of 286 randomly shuffled packets were obtained.

The percentages of asymmetric trends identified by the three correlation tests in the original set of 286 sandstone packets (from field data) are almost identical to the mean values of the percentages of asymmetric cycles identified in the 100 sets of 286 randomly shuffled packets (Fig. 5.25)!

5.3.4 Asymmetric trends in the grain size of SGF deposits

Chen and Hiscott (1999a) also evaluated the importance of asymmetric grain-size trends from their field localities. In order to test for grain-size trends, a grain-size 'score' was calculated for each bed, based on the φ scale, where the score = −4 × (thickness proportion of pebble and granule divisions) + 0 × (thickness proportion of coarse sand divisions) + 1.5 × (thickness proportion of medium sand divisions) + 3 × (thickness proportion of fine sand divisions) + 7 × (thickness proportion of silt divisions). The numerical constants are midpoint φ sizes for each grain-size class recognised in the field. Hence, the grain-size score is an estimate of average grain size for each SGF coarse division. Values of this score could be calculated for the measured sections in California, Italy, Barbados and British Columbia, because individual gravelly and sandy SGF deposits in these sections had been divided during field description into divisions according to grain size and structures.

The grain-size scores of 86 sandstone packets at these localities were statistically tested for trends by using the Fortran-77 program ASYMRAN.FOR. Based on comparisons between the different tests, Chen and Hiscott (1999a) recommend that a packet be considered to coarsen or fine upward only if it passes Kendall's test and at least

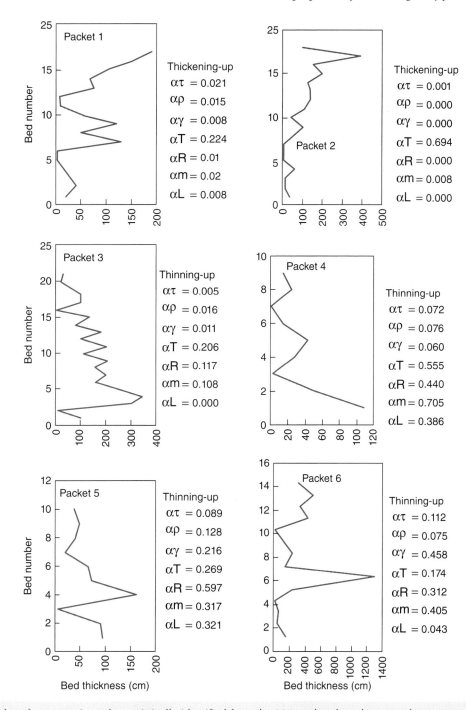

Fig. 5.23 Examples of asymmetric cycles statistically identified from the 286 analysed sandstone packets. ατ, αρ, αγ, αT, αR, αm and αL are significance levels for the Kendall's, Spearman's and Pearson's correlation tests, the turning point test, the rank difference test, the median crossing test and the length-of-runs test, respectively. If the calculated significance level is >0.10 (i.e., >10%), the null hypothesis that there is 'no trend' cannot be rejected. Note that packets 1, 2, 3 and 4 passed all three correlation tests at a significance level α < 10%; packet 5 passed Kendall's correlation test at α < 10%, Spearman's correlation test at α < 15% and Pearson's correlation test at α = 21.6%; packet 6 passed Spearman's test at α < 10% and Kendall's test at α < 15%, but Pearson's test at α = 45.8%. From Chen and Hiscott (1999a).

one of the other two correlation tests at a significance level α ≤ 10%. On the basis of this criterion, 25 (29%) of the 86 sandstone packets coarsen or fine upward (Table 5.5). Figure 5.26 shows examples of some of these packets. The geographic distribution of the 25 coarsening-and-fining-upward packets is very uneven. For example,

in the sections measured in the northern Italian Apennines, only 3 (10%) of 30 sandstone packets show grain-size trends, whereas 10 (50%) of 20 sandstone and conglomerate packets at the Monticello Dam, California, have upward fining or coarsening character. Eight of the latter fine upward and two coarsen upward.

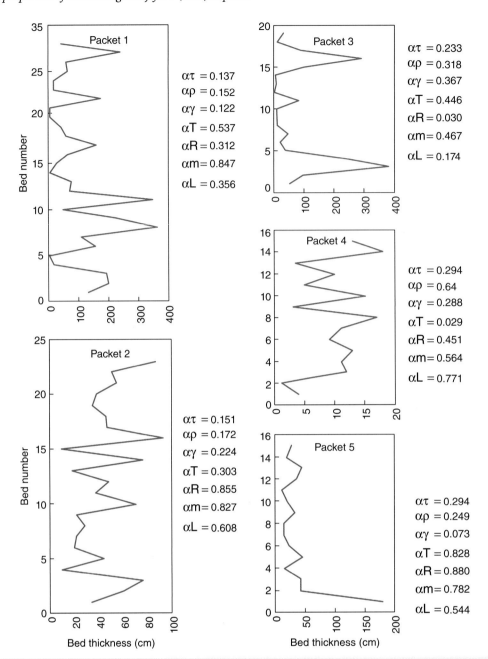

Fig. 5.24 Examples of packets that passed three or two correlation tests (Kendall's, Spearman's and Pearson's correlation tests) at significance levels between 10% and 32%. ατ, αρ, αγ, αT, αR, αm and αL are explained in the caption to Figure 5.23. Using a limiting value of α = 10%, none of these series of deposits can be considered to show asymmetric trends (i.e., the null hypothesis of 'no trend' cannot be rejected). Note that packet 5 does not show an asymmetric trend, but it passed Pearson's test at α < 10%. This indicates that Pearson's test is not as powerful as Kendall's and Spearman's correlation tests. From Chen and Hiscott (1999a).

Proportionally more sandstone packets show significant grain-size trends than trends in SGF coarse-division thicknesses. Fourteen of the 25 asymmetric grain-size cycles are not accompanied by a parallel trend in SGF coarse-division thickness; one upward-fining packet shows an opposite trend in SGF coarse-division thickness (upward thickening); and the remaining ten fining-and-coarsening-upward packets are consistent with upward thinning and thickening trends. These examples suggest that grain size might not correlate

positively (or at all) with SGF coarse-division thickness in turbidite sandstone packets.

All the turbidite sections with a significant number of asymmetric grain-size trends also have more upward-thinning trends than upward-thickening trends, and consist mainly of coarse-grained sediments, especially pebbly sand. On the basis of coarse texture, strong amalgamation, ubiquitous basal erosion and the map-scale distribution of these turbidite bodies, a channel setting is inferred for the

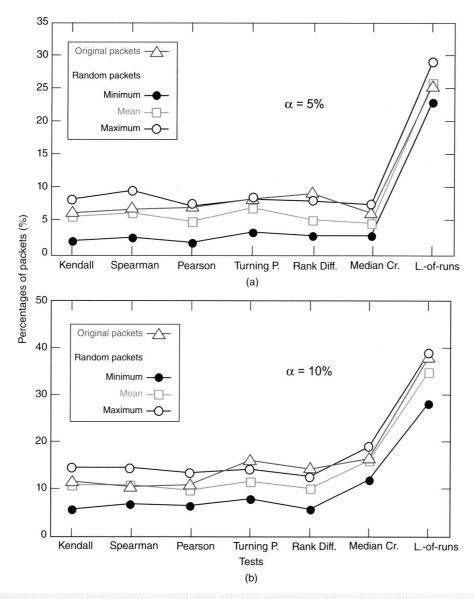

Fig. 5.25 Graphs of percentages of the original 286 packets for which the null hypothesis of individual tests for bed-thickness trends was rejected at (a) α = 5% and (b) α = 10%, compared with the minimum, mean and maximum percentages for 100 sets of 286 random packets shuffled from the original packets. From Chen and Hiscott (1999a).

Table 5.5 The number and percentage of sandstone packets with asymmetric trends in grain size at various turbidite sections

Type of grain-size trend	Associated trend in event-bed thickness	Monticello Dam	Cache Creek	Italian Apennines	Barbados	British Columbia
Upward coarsening	upward thickening	0	1	2	0	0
	upward thinning	0	0	0	0	0
	no asymmetric trend	2	0	0	1	1
Upward fining	upward thickening	1	0	0	0	0
	upward thinning	3	2	0	2	1
	no asymmetric trend	4	1	1	2	2
Total number of packets with asymmetric grain-size trends		10	3	3	5	4
Number of packets tested		20	14	30	10	12
Percentage of packets with grain-size trends		50.0%	21.4%	10.0%	50.0%	33.3%

Packet A from Monticello Dam section (California)

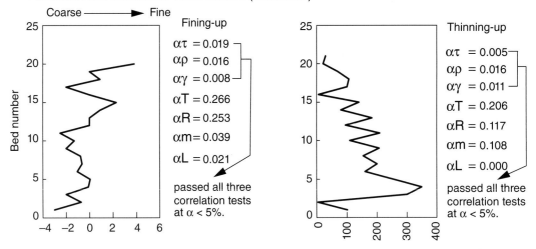

Packet B from Monticello Dam section (California)

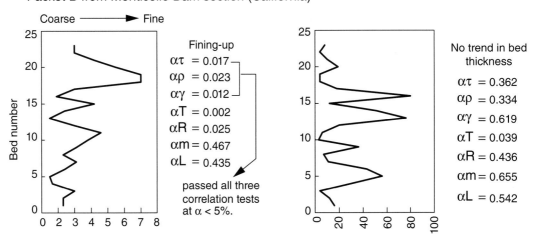

Packet C from Cache Creek section (California)

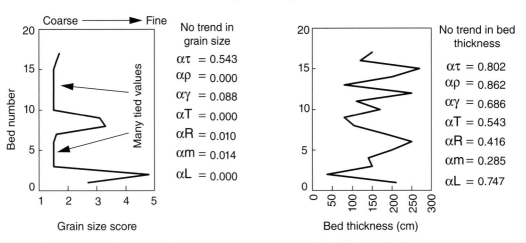

Fig. 5.26 Examples of grain-size trends, accompanied by the profiles of SGF coarse-division thicknesses of the corresponding sandstone packets. Packet A constitutes an upward-fining and -thinning trend; B is an upward-fining packet without an asymmetric trend in bed thickness; C contains many tied values and did not pass Kendall's test ($\alpha\tau = 54.3\%$). $\alpha\tau$, $\alpha\rho$, $\alpha\gamma$, αT, αR, αm and αL are explained in the caption to Fig. 5.23. From Chen and Hiscott (1999a).

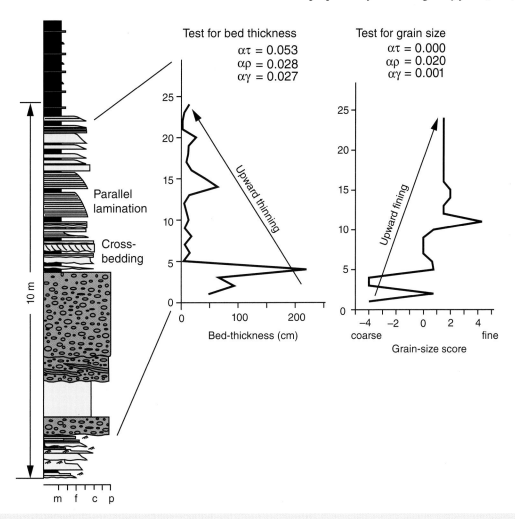

Fig. 5.27 A thinning-upward and fining-upward packet statistically identified ($\alpha < 10\%$) in the Monticello Dam section, Great Valley, California. This packet is interpreted as the fill of an abandoned channel. From Chen and Hiscott (1999a).

lower part of the Monticello Dam section (Fig. 5.27), for Barbados sections and for British Columbia sections. Intervening mud- and silt-rich packets in these sections might be overbank (levée) deposits. These sections resemble the basal parts of channel–levée systems cored on Amazon Fan (Flood *et al.* 1995).

In an active channel or 'nest' of stacked channels (e.g., Clark & Pickering 1996a: p. 183, 206), erosional and depositional phases may alternate. Backfilling of channels can lead to infilling of the lower part of a channel by a channel-mouth lobe (e.g., Normark *et al.* 1998). Subsequent rejuvenation of the channel can erode and incise downward into the previous channel or lobe deposits. Because coarse sediments are deposited in the axial area or thalweg of a channel, and finer sediments at the mouth of the channel, cycles of incision and backfilling ('incision-backfill cycles') might produce upward-fining cycles, although not necessary upward-thinning trends (Fig. 5.28). A few upward-fining sandstone packets (e.g., Fig. 5.27) are possibly the result of filling of an abandoned channel ('channel-abandonment cycle'). This type of cycle was previously hypothesised by Mutti and Ghibaudo (1972), and was confirmed by coring of an abandoned meander loop on the Amazon Fan at ODP Site 934 (Fig. 5.17). Some upward-fining and upward-coarsening packets that have been interpreted by Chen and Hiscott (1999a) as lobe deposits at the

middle and upper parts of the Monticello Dam section, and some upward-thinning packets identified in their Cap Ste-Anne sections and Amazon Fan sections, might have been formed by 'compensation' (Mutti & Sonnino 1981) as lobes migrate and shingle against one another (Swart 1990; Jordan *et al.* 1991; Bouma *et al.* 1995).

Except for preferential upward fining in some channel deposits, Chen and Hiscott (1999a) concluded that SGF coarse-division thicknesses and event-by-event grain sizes show no preferred asymmetric vertical patterns. Those thinning- and thickening-upward packets that can be recognised are not systematically accompanied by upward fining and coarsening trends. These results, however, do *not* exclude the possibility that other criteria might be useful for the distinction of submarine fan sub-environments, like careful facies analysis, or mapping of sand-body geometry on scales of hundreds of metres, or statistical tests for other types of organisation. Mutti and Normark (1987: their figures 7 and 13) provide facies criteria to distinguish sub-environments without reference to asymmetric cyclicity (Figs 7.40 & 7.41). Chen and Hiscott (1999a) and Felletti and Bersezio (2010a) show how the Hurst statistic (Hurst 1951, 1956) can be used to determine the degree of facies clustering as an indicator of depositional setting in submarine fan deposits (see Section 7.4.3.1).

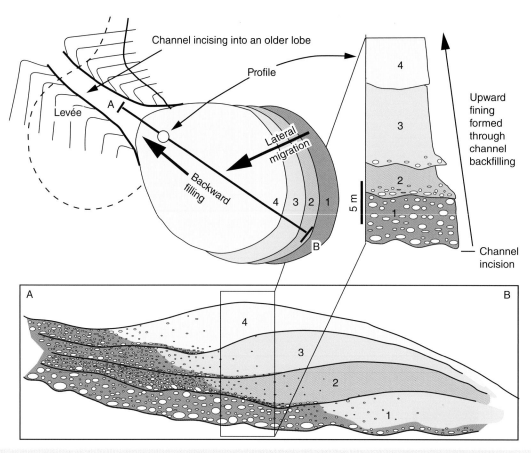

Fig. 5.28 Hypothetical explanation for the genesis of upward-fining packets at channel terminations, as a consequence of channel incision (basal contact of unit 1) and backfilling (units 1 through 4). From Chen and Hiscott (1999a).

PART 2

Systems

CHAPTER SIX

Sediment drifts and abyssal sediment waves

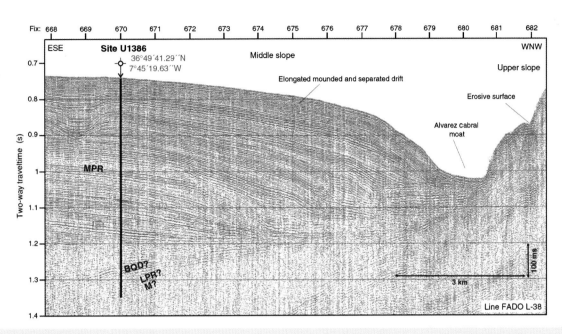

Sparker seismic reflection profile (Line FADO L-38) across the Faro Drift on the middle slope showing location of IODP Site U1386. An important change in the overall architectural stacking of the mounded contourite deposits from a more aggrading depositional sequence to a clear progradational body is associated to the mid-Pleistocene revolution (MPR) discontinuity. BQD = base Quaternary discontinuity, LPR = intra-lower Pliocene discontinuity, M = late Miocene discontinuity. IODP Expedition 339 Scientists (2012).

6.1 Introduction

Sediment drifts are large mounds or ridges of muddy sediments that accumulate beneath persistent thermohaline currents like those that sweep the deep slope and continental rise of the western side of the Atlantic Ocean (Fig. 6.1; Stow & Holbrook 1984; McCave & Tucholke 1986; Kidd & Hill 1987; Rebesco & Camerelenghi 2008; Rebesco *et al.* 2014). These sediment drifts have lengths up to hundreds of kilometres, widths of tens of kilometres and relief of 200–2000 m (Johnson & Schneider 1969). *Abyssal sediment waves* are smaller in scale, and may occur in groups (or trains) on a relatively flat seabed, or as an integral component of a sediment drift. It has been known since the 1960s (Heezen *et al.* 1966) that sediment drifts develop beneath geostrophic bottom currents that, in general, flow parallel with the bathymetric contours as a consequence of there being a balance between the Coriolis Effect tending to deflect the current up a submarine slope, and a buoyancy force inducing the relatively dense bottom water to descend that slope. These geostrophic

currents are called *contour currents*, and the sediments that form the drifts and abyssal sediment waves have been called *contourites* (Hollister & Heezen 1972). Contourites have been identified from all the world's oceans (Heezen 1977; Carter & McCave 1994; Roveri 2002; Stow *et al.* 2002b; Stow & Faugères 2008; Expedition 317 Scientists 2010; Expedition 339 Scientists 2012; Expedition 342 Scientists 2012; Münoza *et al.* 2012; Uenzelmann-Neben & Gohl 2012; Gong *et al.* 2015). The largest database exists for the western North Atlantic Ocean, but large sediment drifts are also present in other regions, for example the southwest Pacific Ocean (Carter & McCave 2002).

The pair of terms, 'contour currents' and 'contourites', imply a relationship between seabed bathymetric contours and flow vectors, which is in most cases impossible to prove in ancient deposits. Nevertheless, on our spinning Earth it is inevitable that major bottom-water circulation in ocean basins of the past would have been deflected by the Coriolis Effect to run parallel with the bathymetric contours of basin-bounding slopes. For this reason we use both terms liberally in this chapter and throughout the book. Readers who prefer to be

Deep Marine Systems: Processes, Deposits, Environments, Tectonics and Sedimentation, First Edition. Kevin T. Pickering and Richard N. Hiscott.
© 2016 Kevin T. Pickering and Richard N. Hiscott. Published 2016 by John Wiley & Sons, Ltd.
Companion Website: www.wiley.com/go/pickering/marinesystems

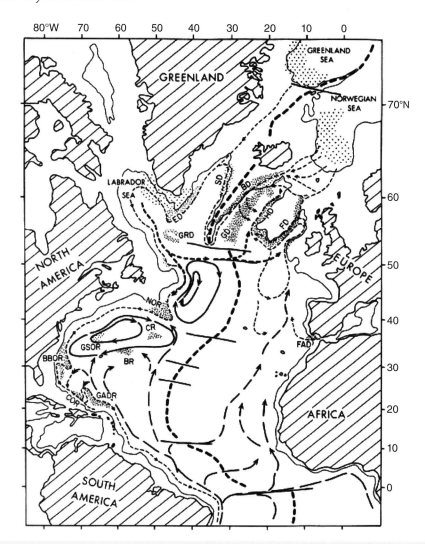

Fig. 6.1 North Atlantic present-day deep-water circulation and major sediment drifts (close stipple). FAD, Faro Drift; FD, Feni Drift; HD, Hatton Drift; GD, Gardar Drift; BD, Bjorn Drift; SD, Snorri Drift; ED, Eirik Ridge; GRD, Gloria Drift; NOR, Newfoundland Outer Ridge; CR, Corner Rise; BR, Bermuda Rise; GSOR, Gulf Stream Outer Ridge; BBOR, Blake–Bahama Outer Ridge; COR, Caicos Outer Ridge, and areas of bottom-water production shown in wide stipple. Redrawn from Stow and Holbrook (1984).

more cautious in cases where the orientation of basin slopes is poorly known, or where palaeocurrent information is scarce, might substitute the terms 'deep-water bottom currents' and 'bottom-current deposits' as advocated by Shanmugam (2008). Of course, 'deep-water bottom currents' are not limited to the thermohaline circulation targeted by this chapter. As itemised by Shanmugam (2008) they also include tidal currents and currents produced by shoaling internal waves (both primarily in submarine canyons), and the lower levels of particularly thick wind-driven currents and tidal currents where they are funnelled through narrow straits (e.g., Colella & d'Alessandro 1988; Ikehara 1989).

Beneath thermohaline currents, the primary control on the overall shape of a sediment drift is the form of the pre-existing seafloor and the influence of this topography on the thermohaline circulation. The speed of contour currents is increased in zones of steeper bottom

gradient and diminished in zones of reduced gradient (McCave & Tucholke 1986; *cf.* McCave *et al.* 1980). Sedimentation preferentially occurs in the more tranquil, lower gradient zones. The morphology of the sediment drifts depends on the location of the axis of maximum flow in relation to the pre-existing seafloor topography (e.g., Vandorpe *et al.* 2014) and on the distribution of suspended sediment matter within the flow (McCave & Tucholke 1986). Depending on the rate of change of the seafloor gradient, double drifts or 'plastered' drifts occur at the base-of-slope (Fig. 6.2). If the rate of change of the seafloor gradient is very abrupt, the sediment drifts may become separated from the slope by a prominent 'moat'. In areas of more complex seafloor morphology (e.g., in areas of combined or reversing flows around a corner or a ridge) 'detached' drifts may form.

In order to quickly orient the reader as to the scale of sediment bodies considered in this chapter, we include at this point seismic

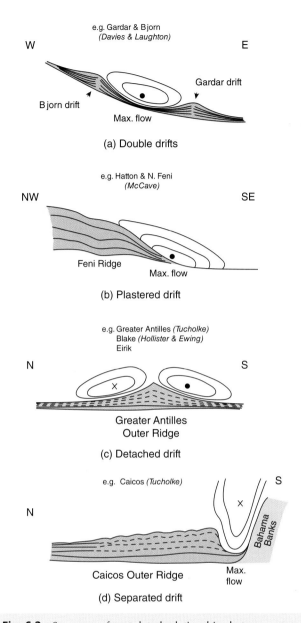

Fig. 6.2 Summary of postulated relationships between current pathways and sediment-drift formation. Dots represent current flow toward the reader; X is current flow away from the reader. Redrawn from McCave & Tucholke (1986).

abyssal sediment waves and their origin. Then, the facies of sandy and muddy contourites will be described, including criteria to distinguish bottom-current deposits from turbidites and other deep-water tractional bedforms. We will then critique claims of bottom-current reworking in turbidite successions, and evaluate criteria for the recognition of ancient sediment drifts.

6.2 Distribution and character of contourites and sediment drifts, North Atlantic Ocean

Sediment drifts do not occur in all areas of bottom-current influence. Emery and Uchupi (1984) have mapped the area of bottom-current deposits in the Atlantic Ocean (Fig. 6.7). This essentially coincides with the location of prominent *nepheloid layers* (Section 1.3), which are thick clouds of suspended sediment near continental margins (Fig. 6.8), maintained by turbulence in the thermohaline circulation and fed by dilute turbidity currents and lutite flows descending the adjacent continental slopes. In much of this area, bottom currents modify the texture of seabed sediments (McCave & Tucholke 1986; McCave *et al.* 2002), but accumulation rates are very slow. The burrowed sediments that result are difficult to distinguish from hemipelagic sediments formed where bottom currents are less active, because all primary textures are destroyed by organisms (Stow *et al.* 2002a).

Contour currents seek the western sides of ocean basins. The water masses transported by these currents acquire a higher density than surrounding water masses as a result of chilling, and freshwater extraction by ice formation at high latitudes (Arctic and Antarctic). The major areas of modern bottom-water formation are the Norwegian Sea and Weddell Sea (McCave & Tucholke 1986; Zenk 2008; Fig. 6.1). Because the Atlantic Ocean deepens toward the equator, the denser water created at high latitudes sinks and flows down a bathymetric gradient toward the equator. Coriolis Effect turns these slow-moving currents (mostly <50 cm s^{-1}) to the right in the northern hemisphere, and the left in the southern hemisphere – these deflections result in flow along the western sides of the ocean basins or, rarely, the western sides of other features like the mid-Atlantic Ridge (Fig. 6.1). As a consequence of several factors – density contrasts with other water masses, volumes of each water mass and flow velocities that dictate the amount of Coriolis deflection – the deep thermohaline currents flow most strongly in a corridor at water depths of ~2500–3500 m, coincident in most places with the deep continental slope and continental rise.

In the western North Atlantic Ocean, the primary thermohaline current is called the *Western Boundary Undercurrent* (WBUC). It transports various water masses, primarily North Atlantic Deep Water, Norwegian Sea Overflow Water and far-travelled but dense Antarctic Bottom Water. Along the path of this current, particularly downflow from irregularities in the seafloor or the continental margin, sediment mounds accumulate that are the sediment drifts considered in this chapter (Fig. 6.1). The dimensions and thicknesses of a number of large sediment drifts in the North Atlantic Ocean are given in Table 6.1.

At sharp changes in the shape of the continental margin, prominent 'spurs' are developed (e.g., Sackville Spur (Fig. 6.3) and Eirik Ridge (Fig. 6.1)). These are streamlined shapes moulded by the WBUC, and

profiles of sediment drifts near Orphan Knoll and Flemish Cap, offshore Newfoundland (Fig. 6.3). These drifts display a series of 'mound' and 'moat' topographies (Kennard *et al.* 1990). In zones of maximum current flow, near an abrupt change in bottom slope, a moat occurs adjacent to the slope, with a mound farther offshore (Fig. 6.4). Double mounds form where the maximum current velocities are concentrated along the mid-slope region in areas of more gentle, uniform gradients (Fig. 6.5). In zones of complex flow, with strong flow asymmetry, as in the Sackville Spur (Figs 6.3 and 6.6), local deposition favours the growth of a large sediment drift.

In this chapter, we will first describe the variety of bottom-current deposits in the North Atlantic Ocean, including a discussion of

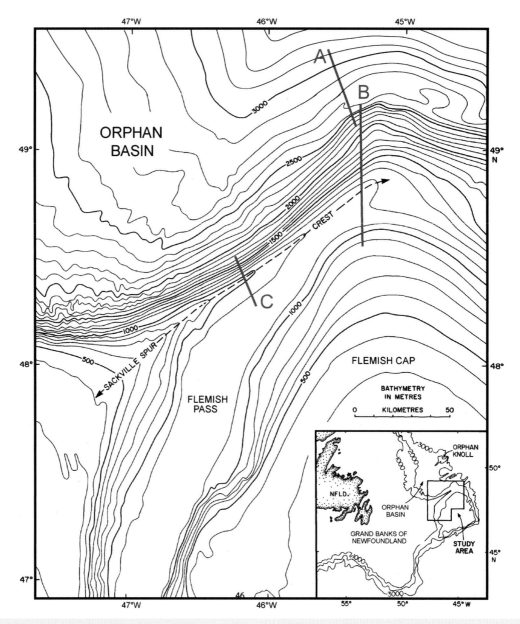

Fig. 6.3   A segment of the Newfoundland slope and rise, off eastern Canada, characterised by current-influenced deposition. The Sackville Spur is a triangular-shaped sediment drift northeast of the Grand Banks of Newfoundland and northwest of Flemish Cap. In this map view, deep parts of the Labrador Current and the Western Boundary Undercurrent sweep from west to east along the upper slope, and lower slope and rise, respectively. Contours in metres. A, B and C indicate cross-sections shown in Figs 6.4, 6.5 and 6.6, respectively. Modified from Kennard *et al.* (1990).

are analogous to snow drifts that form in the lee of obstacles. Masson *et al.* (2002) classify the moulded seabed into four categories: *broad sheeted drifts*, *elongate drifts*, *sediment waves* and *thin contourite sheets*. This classification has similarities to that of Myers (1986) and Myers and Piper (1988), who recognise:

- chaotic basin-fill facies, with variable amplitude, irregular to hummocky, discontinuous reflectors – hyperbolic reflectors are present throughout, but most common near the upper bounding surface (similar to broad sheeted drifts);

- mounded stratified facies, dominated by well-stratified reflectors that undulate gently or build asymmetric mounds with wavelengths of tens of kilometres (similar to elongate drifts);
- mounded-chaotic facies, with locally subparallel to slightly irregular reflectors of a more continuous nature, exhibiting an overall mound-like geometry, with wavelengths of tens of kilometres (another type of elongated drift);
- sediment-wave facies, consisting of symmetric to asymmetric, stratified mounds, with relief of 20–250 m and variable wavelengths, commonly 2–10 km;

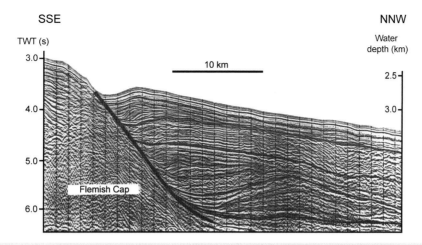

Fig. 6.4 Seismic section A, north of Flemish Cap (see Fig. 6.3 for location). This attached drift has accreted to the continental margin since at least the Miocene (chronology from Kennard *et al.* 1990, their figure 6a). Flow direction of the Western Boundary Undercurrent is out of the page toward the reader. Vertical exaggeration is ~7:1. From Mitchum (1985).

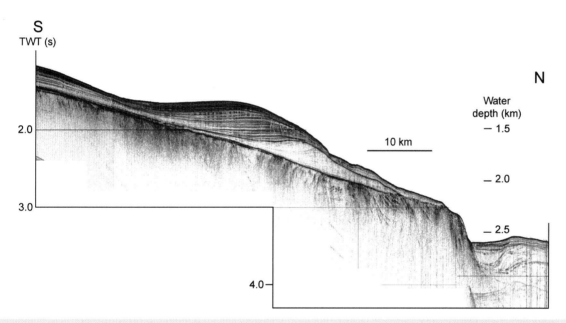

Fig. 6.5 Seismic section B, showing a double mound north of Flemish Cap (see Fig. 6.3 for location). The more northern drift lies along strike from the deposit in Fig. 6.4 and has similar character, with a moat against the adjacent slope. Flow direction of the Western Boundary Undercurrent is out of the page toward the reader. The shallower mound rests on a pronounced unconformity. The better stratified upper part of its sediments are of Pliocene to Recent age based on similar profiles in Kennard *et al.* (1990, their figure 6a). This is a single-channel high-resolution profile acquired by the Geological Survey of Canada (Atlantic) and provided to the authors by D.C. Mosher, D.J.W. Piper and D.C. Campbell. Vertical exaggeration is ~16:1.

- wavy-stratified facies, characterised by discontinuous, wavy and overlapping, moderate-amplitude reflectors (similar to thin contourite sheets).

We will frame our discussion of bottom-current deposits using the classification of Masson *et al.* (2002), which includes both drifts and smaller-scale deposits. However, the geometric diversity of the larger sediment drifts is better captured by the summary sketches of McCave and Tucholke (1986) shown in Figure 6.2. The reader should become familiar with both classifications. Additional summary diagrams and a variety of case studies are presented in Stow *et al.* (2002b).

6.2.1 Broad sheeted drifts

Based on the examples from Rockall Trough, *broad sheeted drifts* blanket basin floors with thicknesses of many tens of metres. They have extremely low slopes to either side of the drift crest (~0.5°). High-resolution profiles show asymmetric growth, with

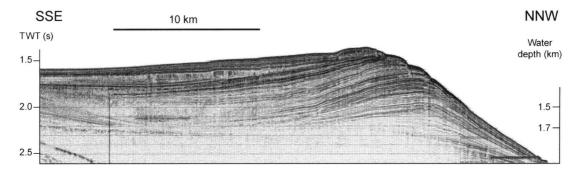

Fig. 6.6 Seismic section C, Sackville Spur (see Fig. 6.3 for location). At these relatively shallow water depths, sediment transport and accumulation can be attributed to deep portions of the surface Labrador Current (Kennard *et al.* 1990). Flow direction is out of the page toward the reader, with the flow focused to the north-northwest of the crest of Sackville Spur, so along the right-hand side of this profile. This is a single-channel high-resolution profile acquired by the Geological Survey of Canada (Atlantic) and provided to the authors by D.C. Mosher, D.J.W. Piper and D.C. Campbell. Vertical exaggeration is ~7:1.

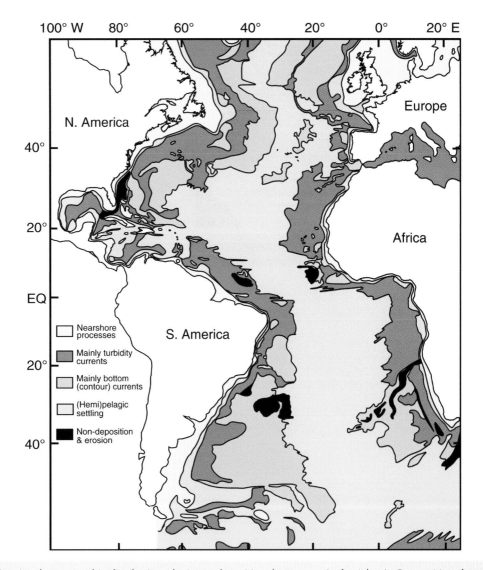

Fig. 6.7 Map showing the geographic distribution of primary depositional processes in the Atlantic Ocean. Nearshore processes are not treated in this book. Adjacent to continents, *hemipelagic settling* deposits terrigenous muds, whereas in the central part of the ocean basin *pelagic settling* deposits biogenic oozes and biogenic muds. Redrawn from Emery and Uchupi (1984).

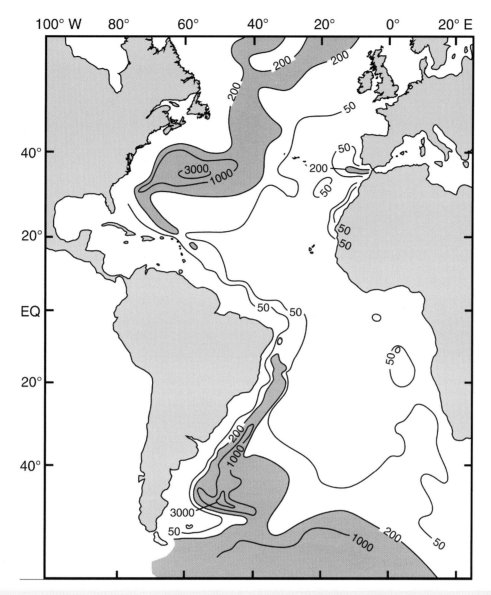

Fig. 6.8 Map showing the mass of particles suspended in the deep-water nepheloid layer of the Atlantic Ocean, contoured in units of µg cm^{-2}. This suspended material is mainly deeper than 3 km, and is concentrated on the western side of the ocean basin (purple tone) where contour currents are most vigorous. Redrawn from Emery and Uchupi (1984).

accumulation higher on one side of the crest than the other. Fields of sediment waves may cover parts of the surface of such drifts.

6.2.2 Elongate drifts

Elongate drifts occur adjacent to the base of, and parallel to, relatively steep slopes, usually separated from the slope itself by a distinct moat. In cross-section, elongate drifts are strongly asymmetric, with a short, steep flank facing upslope (towards the moat) and a much more gentle flank facing downslope (Fig. 6.4). Sedimentation rates are highest on the more gentle downslope flank. In the Rockall Trough, sidescan sonar and bottom samples suggest that coarse-grained sediment is present beneath the upslope flank and moat, and sand/silt contourites typify the gentler, downslope flank of the drift. The Blake–Bahama Outer Ridge is an elongate drift based on this classification, with a depression or moat on its landward

side (Fig. 6.9). Because of multiple crests, the flow direction of the WBUC is highly variable around the Blake–Bahama Outer Ridge, following the intricate contours around its flanks. This complexity is quite typical of the paths of geostrophic bottom currents in areas of complex topography (e.g., Masson *et al.* 2002).

6.2.3 Sediment waves

Sediment waves produced by thermohaline currents can reach wavelengths and heights of 10 km and 150 m, respectively (Wynn & Stow 2002). Crestlines tend to be oblique to the local slope, and the waves migrate upcurrent and upslope. In plan view (Fig. 6.10), waveforms have bifurcating crests. In cross-section, the waves can be ~symmetrical, asymmetrical with deposition on both steep and gentle sides, or more strongly asymmetrical with episodic erosion on one side of the sediment wave (Fig. 6.11). These geometries

Table 6.1 Dimensions and attributes of North Atlantic sediment drifts, from Kidd and Hill (1987; see original tables for full list of references)

Drift name	Approx. length (km)	Approx width (km)	Thickness (m)	Attributes of Sediment Waves			Current velocity (cm/s)
				Wavelengths (km)	Wave heights (m)	Migration sense	
Bahama Outer Ridge	600	400	600	2–3	20–100	Upslope, upcurrent	10–20
Bermuda Rise	700	90	1000	4–6	20–30	Upslope, upcurrent	4–15
Blake Outer Ridge	600	400	600	2–6	20–100	Upslope, upcurrent	5–20
Bjorn Drift	830	100	600	no data	no data	no data	7–20
Caicos Outer Ridge	333	100	1425	no data	no data	Upslope, upcurrent	5–15
Corner Rise	700	150	750	no data	no data	Upslope, upcurrent	5–15
Eirik Drift	355	230	no data	~2	~50	Upslope, upcurrent	18–20
Feni Drift	600	100	1500–1700	0.5–4	23–50	Upslope, downcurrent	5–15
Gardar Drift	1000	130	1300–1600	1.4–1.5	28–35	None	7–12
Gloria Drift	375	330	900–1400	~2	~50	Upslope, downcurrent	3–7
Greater Antilles Outer Ridge	1800	220	700	no data	no data	Upslope, upcurrent	3–17
Gulf Stream Drift	120	70	750	no data	no data	Westward	9–10
Hatteras Outer Ridge	500–550	50	1300	no data	no data	Upslope, upcurrent	9–10
Hatton Drift	65	50	700	no waves	no waves	None	6–24
Hudson Drift	30	5	40	no waves	no waves	None	10–20
Isengard Drift	480	72	no data	0.3–0.5	<20	Upslope, downcurrent	5–20
Newfoundland Outer Ridge	500	200	400	no data	no data	no data	5–35
Snorri Drift	250	100	300–500	1–2	20–30	no data	not reported

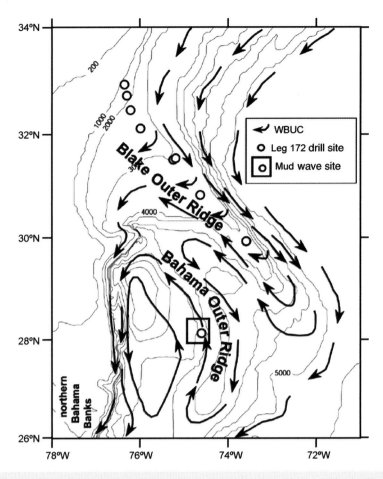

Fig. 6.9 Map of the Blake–Bahama Outer Ridge showing the stream lines of bottom flow. Isobaths are in metres. Bottom flow directions at the sediment wave site (box) are towards the north-northwest. Redrawn from Flood and Giosan (2002).

Fig. 6.10 30 kHz sidescan sonar image from the flank of a broad sheeted drift in the Rockall Trough area, from Masson *et al.* (2002). High-backscatter stripes (white) correspond to the lee (downcurrent) wave slopes. Low backscatter stripes (black) correspond to upcurrent wave slopes and areas of thicker accumulation of Holocene sediment. Locally converging crests indicate bifurcation. The seabed shoals from ~1200 m to ~1100 m from lower left to upper right, and the bedforms are migrating upslope.

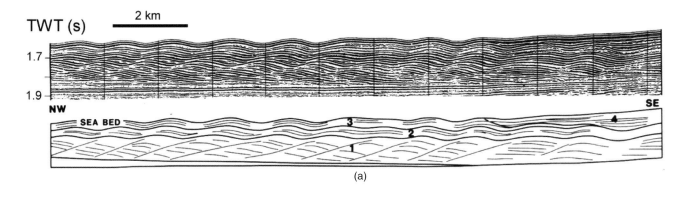

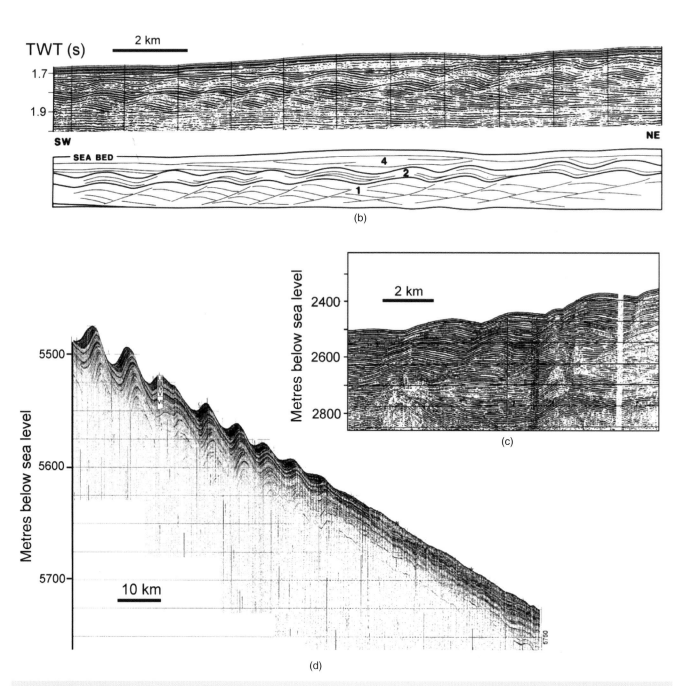

Fig. 6.11 Seismic profiles and line drawings of profiles across sediment waves and climbing bedforms. (a, b) Profiles in Rockall Trough (Richards *et al.* 1987) with 1 = climbing-dune unit, 2 = transitional dune unit, 3 = draping sinusoidal dune unit, 4 = mass-transport unit. (c) Airgun profile of sediment waves on the Var sedimentary ridge (Migeon *et al.* 2000). (d) Parasound profile across the western border of a sediment-wave field on the western flank of the Zapiola Drift, Argentine Basin (von Lom-Keil *et al.* 2002).

are analoguous to the types of climbing ripple lamination described for centimetre-height current ripples by Jopling and Walker (1968). Sediment waves are very long-lived phenomena, and can represent millions of years of accumulation ($>10^5$–10^6 years) (Roberts & Kidd 1979; Tucholke 1979).

Sediment is not actually transported from one side of each sediment wave to the other in order to account for the waveform migration. Instead, the migration is the geometric consequence of a higher depositional rate on one side of the waveform than on the other side, as demonstrated by coring across a sediment wave at ODP Site 1062 (Fig. 6.12; Flood & Giosan 2002). This differential deposition has been attributed to the influence of large scale internal 'lee waves', caused by perturbations on the seafloor (Flood 1978). Flood (1978, 1988) proposed that the internal waves develop on a density gradient rather than on a sharp interface. He determined that the inverse of the gradient Froude number, κ, is given by:

$$k = Nh/U \qquad (6.1)$$

where N is the stability, h is the wave height and U is the free-stream velocity. κ ranges from 0.4 to 1.5 for sediment waves on the Blake–Bahama Outer Ridge. Beneath the internal lee waves, the bottom current decelerates on the upflow flank of the waveform and accelerates on the downflow flank. This results in different accumulation rates from one flank to the other, and enhancement of the waviness on the seabed (Fig. 6.13). Once established, the sediment waves may act as perturbations to subsequent flow, and 'seed' additional lee waves.

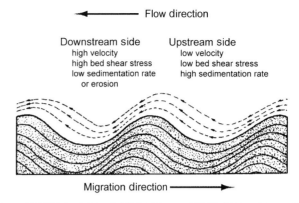

Fig. 6.13 The lee-wave model of sediment wave evolution showing the interaction of a flowing stratified water column passing over the sediment wave profile. The flow slows as water climbs the upstream face of the wave and accelerates on the downstream face, resulting in enhanced deposition rates on the upstream side. Redrawn from Flood (1988).

Although fine-grained sediment waves produced by thermohaline currents differ little in their geometry and seismic response from sediment waves on submarine-fan levées, they are entirely different in facies (*cf.* distinguishing criteria proposed by Faugères *et al.* 1999). The submarine-fan sediment waves are formed of thin bedded turbidites (Normark *et al.* 2002), whereas the abyssal sediment waves are bioturbated and generally rich in biogenic particles. During equilibrium conditions of growth, the abyssal waves accumulate at a much slower rate than the sediment waves associated with submarine fans. For example, Flood and Giosan (2002) calculate sedimentation rates of ~20 cm kyr^{-1} and migration rates of ~0.4 m kyr^{-1} for sediment waves on the Blake–Bahama Outer Ridge. In contrast, channel levées on the Amazon Fan, which are locally ornamented by sediment waves, accumulated at rates of ~10–25 m kyr^{-1} (Shipboard Scientific Party 1995a).

6.2.4 Thin contourite sheets

Thin contourite sheets cover the continental slope northwest of the UK to water depths of 1000 m or more, and consist of muddy sand to sandy gravel (Masson *et al.* 2002). In side-scan sonar images, the upper part of the slope shows high backscatter and is underlain by a gravelly sand <20 cm thick. The lowest part of the slope exhibits low backscatter and is underlain by a 10–25 cm-thick Holocene layer of silt or fine sand. The sediment is well or very well sorted and gravel clasts are rare or absent. The thin sheet-like nature of the sand body, its surface covering of ripples, and good sorting indicate that this is a thin, sandy contourite sheet.

6.2.5 Other abyssal current-generated structures

In addition to sediment waves and current ripples, there are a number of other seabed features associated with active bottom currents. Smaller features show a decreasing hierarchy in size from mud waves → furrows → transverse and longitudinal ripples, and other small current structures that can be observed on bottom photographs (McCave & Tucholke 1986) (Table 6.2). Furrows on the Blake–Bahama Outer Ridge are (i) oriented at about 25–35° to the strike of sediment-wave crestlines; (ii) spaced 20–125 m apart; and (iii) traceable along their length for distances up to 5 km. In plan view, furrows are very

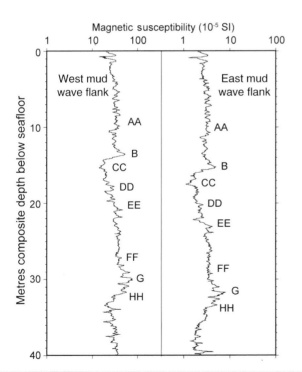

Fig. 6.12 High-resolution down-core records of magnetic susceptibility showing the correlation between the west and east flanks of the sediment wave located in Fig. 6.9. Features marked B and G are correlated based on colour lightness, while features marked AA, CC, DD, EE, FF and HH are correlated based on magnetic susceptibility (mcd = metres composite depth). From Keigwin *et al.* (1998) and Flood and Giosan (2002).

Table 6.2 Indicators of relative (increasing from 1 to 8) current strength from seafloor photographs (From McCave & Tucholke 1986)

Currents		Degree of bedform development	Bottom water
Tranquil	1	Flocs of organic/mineral detritus; undisturbed animal tracks	Clear
	2	Subtle smoothing; rare flocs	
	3	Weak lineation; appearance of poorly developed tool marks; flocs of detritus in lee of obstacles	
	4	Small crags with tails of mineral detritus	
	5	Appearance of barchan ripples of silt/sand; crags and tails; weak scour crescents in front of obstacles	
	6	Mounds and tails; longitudinal ripples; common cornices; crags and tails widespread	
	7	Well developed crags and tails very common; well developed scour crescents around obstacles; some erosional plucking of seabed and of existing bedforms	
Strong	8	Strong and widespread development of erosional plucking; tool marks; scours around obstacles; cohesive sediment exposed and unconsolidated silt/sand absent except in protected areas	Very cloudy

Note: Scales not linear: the lowest speeds (scale 1) are <5 cm/s, whereas the fastest (scale 8) are probably >40 cm/s

straight and show 'tuning fork' junctions opening into the current. In cross-section, furrows are 14 m wide, 0.75–2 m deep, and steep-sided with flat floors (Hollister *et al.* 1976a). Bottom current measurements (20 m above the bottom) show maximum current velocities of 8–10 cm s^{-1} oriented parallel to the strike of the furrows. The thickness of the benthic boundary layer at this site is approximately half the spacing of the furrows, suggesting that the furrows are related to secondary helical circulation patterns within the benthic boundary layer. These small-scale furrows were interpreted by Hollister *et al.* (1976a) to be the cause of the fine-scale hyperbolic echos commonly seen on 3.5 kHz and 12 kHz records.

At other sites between the transition from the Blake–Bahama Outer Ridge to the abyssal plain, larger furrows (20 m deep × 50–150 m wide × 50–200 m spacing) are clearly erosional. As with the smaller-scale mudwave furrows, these larger furrows also show 'tuning-fork' patterns in plan view, which open in an upcurrent direction. Bottom current measurements (20 m above the seafloor) reached a maximum of 8 cm s^{-1}, compared with a maximum of 4 cm s^{-1} above the adjacent abyssal plain. Composite vertical deep-tow temperature profiles in the erosional furrow area show that the boundary layer averages about 60 m thick (Fig. 6.14). The well-mixed boundary layer is believed to result from interactions between deep, steady geostrophic currents and the bottom topography (Hollister *et al.* 1976a). This results in the development of secondary circulations, which maintain the erosional furrow topography.

Bottom photographs in areas swept by currents show rippled silts and sands, current lineations and current tails, and gravel lags (Fig. 6.15). Muds are deposited at sites marginal to the axis of the boundary flows (McCave 1982; Schafer & Asprey 1982; Carter & Schafer 1983). Deposits at current-swept and marginal sites correspond, respectively, to the 'sandy' and 'muddy' contourites of Stow (1982) and Stow and Lovell (1979).

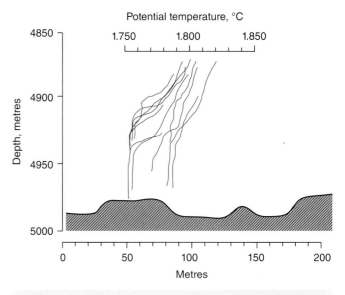

Fig. 6.14 Composite of vertical deep-tow temperature profiles showing a well-mixed bottom boundary layer over large furrows near the Blake–Bahama Outer Ridge–Abyssal Plain transition. Note the relationship between the well-mixed layer thickness and furrow spacing. No vertical exaggeration. Redrawn from Hollister *et al.* (1976).

Larger scale erosional features have been linked to strengthening contour currents during periods of global cooling, lowered sea level, and an increase in thermohaline circulation, for example, features observed from a middle slope setting in the eastern Gulf of Cadiz close to the Strait of Gibraltar, containing two prominent terraces and two large erosional channels up to 3–6 km wide (Hernández-Molina *et al.* 2014). In the same area, pockmarks have been recognised that are sub-circular to irregularly elongated or lobate in shape and 60–919 m in diameter (León *et al.* 2014). Pockmark formation is believed to have resulted from gas and/or sediment pore-water venting from overpressured shallow gas reservoirs trapped in coarse-grained contourites or levée deposits and Pleistocene palaeochannel fills, with venting likely triggered by hydraulic pumping associated with topographically forced internal waves (León *et al.* 2014).

6.3 Facies of muddy and sandy contourites

There is some confusion in the literature about the characteristic facies that form sediment drifts and related waveforms. In the 1970s, it was observed that the seabed beneath the core of the WBUC consisted of current ripples formed of well-sorted sands with heavy mineral concentrations and transported biogenic grains (e.g., foraminifera: Heezen & Hollister 1971; Hollister & Heezen 1972). What was not considered at that time was the almost universally slow depositional rate of these sediments, which invariably leads to thorough

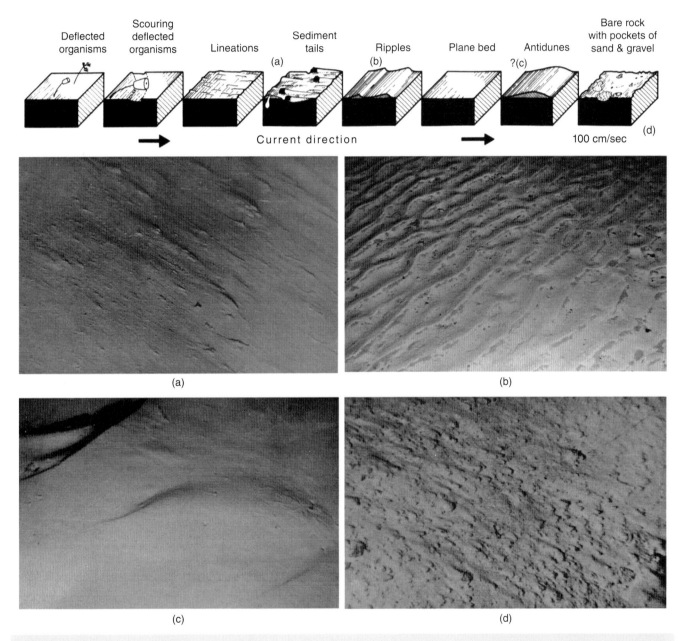

Fig. 6.15 Inferred progression of bedforms beneath abyssal bottom currents of increasing strength (top) together with bottom photographs (bottom) to illustrate typical seabed morphology. (a) Lineations and sediment tails, lower continental rise off New York under the Western Boundary Undercurrent, 5010 m. From Schneider et al. (1967). (b) Ripples beneath Antarctic Circumpolar Current, 3157 m. Courtesy of United States Antarctic Research Program. (c) Fluted clay, 5202 m, Bermuda Rise. Courtesy of Duke University Marine Laboratory. (d) Lineated erosion surface littered with gravel clasts in the Mauritius Trench, 4909 m. From Heezen and Hollister (1964).

bioturbation beneath the oxygenated waters that characterise contour currents (Stow *et al.* 2002a; Stow & Faugères 2008). This bioturbation leads to destruction of most tractional structures, corroborated by cores from sediment drifts that consist of repeated, sometimes subtle, coarsening → fining → coarsening cycles, each ~50–100 cm thick, that are strongly burrowed (Fig. 1.15; Faugères *et al.* 1984; Gonthier *et al.* 1984; Wynn & Stow 2002; Flood & Giosan 2002). Table 6.3 summarises the essential characteristics of muddy and sandy contourites, and reworked turbidites. A convincing demonstration of the lack of tractional structures comes from recent ODP drilling of sediment waves on the Blake–Bahama Outer Ridge. Leg 172 recovered long sections through individual sediment waves, with the facies consisting almost entirely of bioturbated mixtures of nannofossils and terrigenous clay in beds ~0.2–2 m thick. Bed contacts are gradational and burrowed. In the geological record, an exception to such pervasive bioturbation characterises sedimentary rocks of Precambrian age, when for practical purposes burrowing organisms did not exist; in these rocks, tractional structures would likely abound (*c.f.*, Fig. 2.32a).

In the Faro Drift, Gulf of Cadiz (Fig. 6.1), muddy contourites are typically homogeneous and bioturbated, with the preservation of only thin, irregular winnowed silt concentrations and rare silt laminae (Fig. 6.16a; Gonthier *et al.* 1984). In contrast, the so-called sandy contourites of Gonthier *et al.* (1984) show better preservation of primary sedimentary structures, including thin lag concentrations of coarse sand or shell fragments, laminated and cross-laminated sands and silts (Fig. 6.16b). Sand only forms ~5% of these deposits, however, so the term 'sandy contourite' is not entirely appropriate (Shanmugam 2000). Bioturbation is still common, but not as pervasive as in the muddy contourites, presumably reflecting higher energy conditions and/or decreased flux of organic matter. Although the 'average' deposit conforms to the model of Figure 1.15, individual vertical sequences may be thick (Fig. 6.17 KC 8221) or thin with sharp erosional boundaries (Fig. 6.17 KC 8220 and KC 8217), reflecting slower and faster temporal changes in current strength, respectively. Other patterns include the rapid and sudden occupation of a site, followed by a gradual abandonment (Fig. 6.17 KC 8226). Such temporal changes are related to variations of current strength as a result of climatic or oceanographic factors, or the lateral migration of the axis of the bottom current through time (Gonthier *et al.* 1984; *cf.* Ledbetter 1984).

There appears to be a typical textural signal in fine-grained contourites. McCave *et al.* (1995) and McCave (2008) correlate the percentage and mean size of the 'sortable silt' fraction of contourites with the intensity of the current. Sortable silt consists of coarse to medium silt in the size range 63–10 µm that is cohesionless and most affected by low- to moderate-velocity currents. Tracking the amount and character of sortable silt in muddy marine sediments is a possible way to monitor the history of bottom-current activity, and to distinguish contourite mud from turbidite mud and hemipelagic mud.

Where deep currents are stronger than those which form fine-grained sediment waves and sediment drifts, and if sand is available at the seabed, coarse-grained sediment waves resembling large barchan dunes may develop (Kenyon *et al.* 2002; Wynn *et al.* 2002a). Wynn *et al.* (2002a) describe barchan dunes from ~1150 m water depth in the Faroe–Shetland Channel. The barchan dunes have horn-to-horn dimensions of as much as 120 m, heights of <1 m, linguoid current ripples on their backs and presently experience peak flow velocities of 50–60 cm s⁻¹. An 800 m-long transect crossed a total of 22 barchans, spaced about 10–50 m apart, with horn-to-horn width of 10–30 m and estimated height <50 cm. They are aligned transverse to the measured bottom current, with the horns pointing downcurrent. These large bedforms cover a very small fraction of the seabed (Fig. 6.18), and it is not clear that they would leave any widespread deposit. Any deposit would likely consist of thin sheets of cross-bedded sand left behind after migration of a barchan dune across the seabed. In areas of slow deposition, even these sands would likely be disturbed by burrowing.

This book promotes the idea that both muddy and sandy contourites tend to be pervasively bioturbated, based on the results of coring of many modern deposits (Stow & Faugères 2008). A few sedimentologists strongly oppose this viewpoint (e.g., Shanmugam *et al.* 1993a, b; Shanmugam 2008; Martín-Chivelet *et al.* 2008), and interpret stratified and cross-stratified deposits including very sand-prone successions in the ancient record and some oilfields as bottom-current-reworked deposits. Shanmugam (2000) goes further and interprets all beds in ancient deposits lacking a graded Bouma division T_a but consisting of Bouma divisions T_b and T_c, as bottom-current deposits, not the deposits of waning turbidity currents. This latter proposal has the same weaknesses inherent in other interpretations of time-varying bottom-current reworking, as discussed in Section 6.5. Stated briefly, we cannot believe (i) that bottom currents sufficiently strong to rework turbidite sand beds exist in all basins where T_b and T_c divisions occur, and (ii) that after reworking a turbidite top in a developing succession of sand–mud couplets (e.g., Facies Group C2), the bottom current then declines sufficiently in strength to allow accumulation of a mud layer (either turbidite mud or hemipelagic mud) before accelerating just in time to rework the top of the next sandy turbidite, repeating this improbable swing in velocity over and over, in lock-step with the localised triggering and arrival of turbidity currents. In our experience, most beds beginning with Bouma division T_b or T_c have sharp bases and are size graded between these two divisions and into the overlying mud (commonly through a T_d division). Both this size grading and the upward change from planar lamination to ripple-scale cross-lamination (in some cases as climbing ripples) are consistent with waning flow velocity during a relatively short pulse of deposition in an otherwise quiescent environment. Turbidity currents provide the best explanation for such deposits.

Why are we so convinced that most contourites are strongly burrowed, even in sand-prone successions? The primary reason is that these sediments accumulate very gradually and slowly, unlike turbidites and other event-deposits. On the Blake–Bahama Outer Ridge, for example, accumulation rates are ~25 cm kyr⁻¹, and along the Pacific continental margin of the Antarctic Peninsula a rate of ~11 kyr⁻¹ is documented for 122 Ka (Venuti *et al.* 2011). In areas of stronger bottom currents, net accumulation rates may be even lower, or seabed erosion may occur to produce abyssal unconformities (Kennett *et al.* 1975). At such low rates, burrowers are quite capable of repeatedly tilling the uppermost sediments so that they become thoroughly mixed before passing out of the active surface layer and into the geological record. Exceptional deposits that might resist such mixing include gravelly lags developed under strong, episodically erosive bottom currents, and thicker gravity-current deposits that might be intercalated in the contourite record (e.g., Shipboard Scientific Party 1998: p. 167).

A special problem of interpretation can result if strong bottom currents pass across an active submarine fan, or other gravity-flow system. In that case, contourites could result from (i) the reworking of older turbidites or (ii) the 'piracy' of active downslope-moving fine-grained turbidity currents by along-slope bottom currents (e.g., Hill 1984a, b). Potentially diagnostic features of sandy contourites (Lovell & Stow 1981; Stow *et al.* 2002a, Table 6.3) include: coarse lag

Table 6.3 Main characteristics of terrigenous or biogenic muddy contourites, sandy contourites and bottom-current-reworked turbidites (from Stow *et al.* 1998b)

	Muddy contourites	Sandy contourites	Reworked turbidites
Occurrence	Thick uniform sequences of fine-grained sediment in deep-water settings Interbedded with turbidites and other resedimented facies on inferred continental margins	Thin to medium beds in muddy contourite sequences, rarely thick to very thick units Reworked tops of sandy turbidites in interbedded sequences Coarse lag in deep-sea channels and straits	In any normal turbidite setting where strong, permanent bottom currents have been active
Structure	Dominantly homogeneous, bedding not sharply defined, but cyclicity common Bioturbational mottling common to dominant Distinct burrows (typical deep-water assemblage) present in many places Coarse lag concentrations (especially biogenic) reflect composition of coarse fraction in mud Primary silt/mud lamination – rare, but no regular sequence as in turbidites Sharp and erosive contacts locally commom	Generally bioturbated and burrowed throughout with little primary structure remaining Parallel- and cross-lamination more rarely preserved (often with bioturbation) No regular structural sequence as in turbidites May show reverse grading near top, with sharp/erosive contacts common	Lower divisions of turbidite may be preserved, with the upper divisions either removed completely or modified by reworking Bioturbation/burrowing common through reworked-top reverse grading and irregular lag concentrations Bi-directional cross-lamination, may be clean micro-cross-laminated silts with bioturbation Sharp erosive contacts may occur within turbidite sequence
Texture	Dominantly silty mud Frequently 0–15% sand-sized biogenic tests in terrigenous contourites Moderate to poorly sorted, ungraded, no offshore textural trends May show marked textural difference from inter-bedded turbidites if transport distances are different	Silt- to sand-sized, more rarely gravel May be relatively free of mud and well sorted Tendency to low or negative skewness No offshore trends	Removed/non-deposition of fines Significant textural differences from underlying turbidite (e.g. cleaner, better sorted, reverse grading+ lag, negative skewness
Fabric	Mud fabric – typically more parallel alignment of clays than for turbidites, but not well preserved in fossil contourites Primary silt laminae or coarse lag deposits show grain orientation parallel to the current (along-slope)	Indication of grain orientation parallel to the bottom current (along-slope) or more randomised by bioturbation Other features (e.g structures) also indicate along-slope flow, where preserved	Interbedded, reworked turbidite layers may show widely bimodal grain orientations or a more random polymodal fabric
Composition	Mixed contourites have combination of biogenic and terrigenous material (may be distinct from interbedded turbidites) Terrigenous material dominantly reflects nearby land/shelf source with some along-slope mixing and small amount of far-travelled material (no down-slope trends) Biogenic material from pelagic, benthic and resedimented sources, typically fragmented and iron-stained Organic-carbon content very low	Mixed biogenic/terrigenous composition typical Terrigenous composition dependent on local source	Composition entirely reflects that of turbidite, with part of fine fraction removed Long exposure and winnowing may lead to chemo-genic precipitation (probably rare) Organic-carbon content very low
Sequence	Typically arranged in decimetric cycles of grain-size and/or compositional variation with sandy contourites See model (Fig. 1.15) – partial sequences common	Typically arranged in decimetric cycles of grain-size and/or compositional variation with muddy contourites See model (Fig. 1.15) – partial sequences common	Presents a typical turbidite sequence (i.e. top-absent or top reworked) Does not occur within standard contourite sequence of Figure 1.15

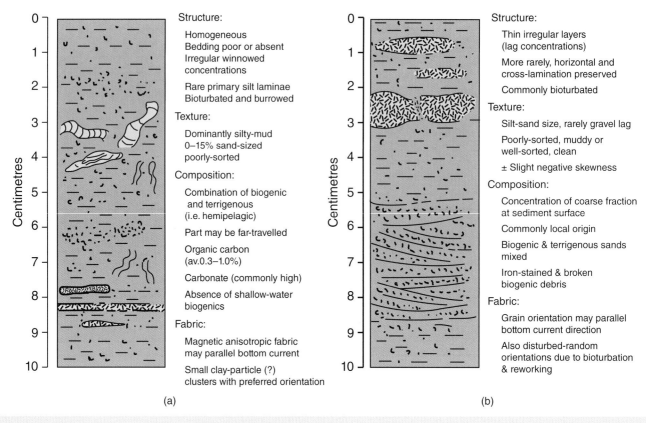

Fig. 6.16   Sedimentary characteristics of (a) muddy contourites and (b) sandy contourites. Redrawn from Stow and Holbrook (1984).

concentrations, especially those with a largely biogenic component (commonly planktonic foraminiferal tests) (Carter *et al.* 1979); inverse grading near the tops of beds and a sharp upper contact (associated with the development of lags) (Carter *et al.* 1979; Lovell & Stow 1981; Shanmugam 2008); and grain fabric patterns, particularly those showing along-slope trends (Ledbetter & Ellwood 1980). Conversely, Piper and Stow (1991) identified a number of features that are present in turbidites, but that should not form by slow deposition under bottom currents (Fig. 1.16). To this, we add the following observations.

- Contour currents tend to be active for long periods of time, certainly hundreds of thousands of years or even millions of years. This is because they represent quasi-stable modes of ocean circulation. Pleistocene currents have been weaker during glacial times (Fagel *et al.* 1996, 1997, 2001; Hanebuth *et al.* 2015), but never ceased to be active. During the rest of Phanerozoic Earth history (greenhouse and 'doubthouse' conditions) there is no reason to suspect abrupt changes in bottom-current configuration, just gradual weakening and strengthening (or shifting up and down the basin slope). For this reason, bottom-current transport should be evident throughout a considerable thickness of a turbidite succession if abyssal circulation is strong, not just in particular beds.
- If bottom currents are sufficiently strong to form lags and winnowed tops to sand beds, then they should be able to successfully prevent deposition of the mud load of turbidity currents (Table 6.3), instead capturing that mud into the nepheloid layer. Because the sand-bed tops would then be exposed at the seabed, they should be much more prone to burrowing than sand beds overlain by many centimetres of turbidite mud.

- The accumulation rate of the contourite component of the stratigraphic succession should be low, perhaps 10% or less of the accumulation rate for the turbidite component. Likewise, bottom currents could probably not erode and entirely rework a significant amount of sediment (particularly sand) during the interval of time between the emplacement of successive turbidites, unless the terrigenous input became very low (e.g., at a sea-level highstand; *cf.* Kenyon *et al.* 2002). It has indeed been proposed that in deep-marine siliciclastic systems, during periods of high sea level, the 'continuous' accumulation of contourites and hemipelagic drapes dominates, whereas during lowstands the 'discontinuous' processes of mass wasting and gravity-flow transport predominate and overwhelm the effects of contour currents (Emery & Uchupi 1972; Gorsline 1980; Sheridan 1986). Modern submarine fans are mostly active during lowstands when thermohaline circulation, at least in the icehouse climatic mode, should be at its weakest.
- It is unlikely that sediment waves would be recognisable in ancient successions, due to insufficient outcrop scale. However, such waveforms might account for the lateral variations seen in sand:mud or silt:mud ratios in some deep-marine sediments. Because sediment waves are mostly deposited on oceanic crust, they are only likely to be preserved in the rock record after uplift and deformation in a mountain belt, making their later recognition difficult to impossible.

To conclude, all but the coarsest Phanerozoic contourite deposits are likely to be heavily burrowed because of slow deposition, and therefore are not likely to be confused with SGF deposits because the

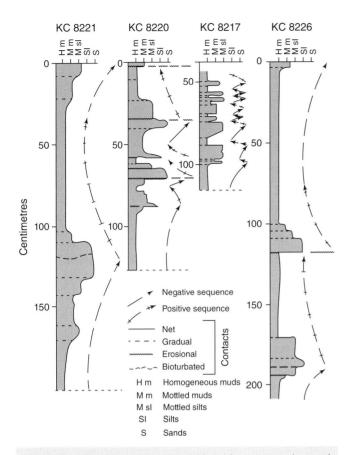

Fig. 6.17 Sketches of core sections through contourite facies of the Faro Drift (Fig. 6.1). There are a number of upward coarsening (negative) and upward fining (positive) trends. All facies are burrow mottled. Redrawn from Gonthier *et al.* (1984).

Fig. 6.18 TOBI side-scan sonar image showing large, isolated barchan dunes adjacent to an area of erosional furrows in the Faroe–Shetland Channel. Note the very wide spacing of the dunes. Horns of the barchans point to the left in the direction of current flow and roughly parallel to furrow orientation. Dark tones = low backscatter, light tones = high backscatter. From Wynn *et al.* (2002a).

latter accumulate very quickly and so tend to preserve their physical and tractional structures. In exceptional cases, strong bottom currents (or deeply penetrating surface currents like the Loop Current in the Gulf of Mexico) are apparently capable of winnowing and transporting sand to form sheets or barchan-shaped abyssal dunes, but these deposits should not be interbedded with mud(stones) because bottom currents are persistent, and so do not turn on and off with the high frequency characteristic of gravity flows. Table 6.3 and Figure 1.16 provide reliable criteria for the distinction of turbidites from contourites.

6.4   Seismic facies of contourites

Other parts of this chapter deal with the morphologies of bottom-current deposits, including large sediment drifts and regularly spaced sediment waves that locally have the geometry of climbing bedforms in seismic profiles. Here, we emphasise stratigraphic variations in reflection amplitude that have been attributed to long-term variations in current strength. Stow *et al.* (2002a) correlate the amplitude, continuity, and shape of seismic reflections with bottom-current intensity (Fig. 6.19). Further, they point to cycles of variable reflection amplitude in seismic data (Fig. 6.20) as evidence for alternating periods of stronger currents (more sand and silt; more hiatuses and condensed intervals; greater seismic amplitude) and weaker currents (more mud; more continuous deposition; low-amplitude to transparent seismic signature). This cyclicity is at a much larger scale than the high-amplitude textural variations that can be seen in outcrops or cores (Fig. 1.15), and suggests that textural trends in bottom-current deposits might be self-similar (fractal) over several length scales.

Although seismic profiles across sediment drifts and sediment waves show parallel to undulatory reflections (Nielsen *et al.* 2008), the deposits recovered in cores are largely unbedded and bioturbated (Stow & Faugères 2008). Clearly, the impedance contrasts responsible for the seismic reflectivity are an integration of lithologic properties over many metres to tens of metres of section, and do not imply that the sediments are either laminated or have sharp bed boundaries. For further information on the criteria for the recognition of contourite versus SGF deposits based on the analysis of many seismic profiles, the reader is referred to Faugères *et al.* (1999).

6.5   The debate concerning bottom-current reworking of sandy fan sediments

Several highly experienced researchers have proposed that certain cross-laminated and cross-bedded deep-marine sandstones in the rock record formed as a result of vigorous reworking of turbidites by strong bottom currents (Mutti *et al.* 1980; Mutti 1992; Shanmugam *et al.* 1993a, b; Shanmugam 1997, 2008; Stanley 1987, 1988). Criteria that have been used to support such interpretations include improved sorting in the proposed bottom-current sands, sharp sand-bed tops, structures that resemble flaser and wavy bedding, and divergent palaeocurrents between gravity-current deposits and interbedded bottom-current deposits. In all cases, the depositional rates are controlled by the rates of turbidite aggradation, and are therefore much higher than rates of accumulation of sediment drifts in the modern oceans, perhaps ten times higher or more.

We have selected the Caledonia Formation (Cretaceous, Virgin Islands; Stanley 1987, 1988) to focus a discussion of bottom-current

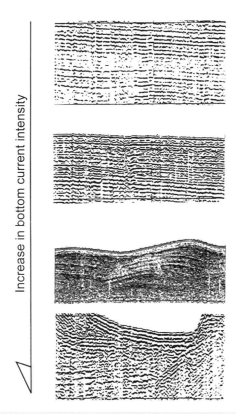

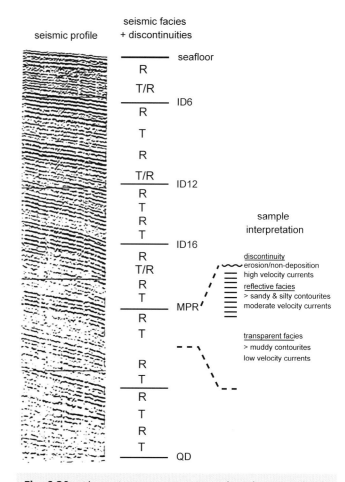

Fig. 6.19 Principal seismic facies found in contourite drifts as a function of bottom-current velocity, according to Stow *et al.* (2002a). From top downward, (i) semi-transparent, reflection-free intervals, (ii) continuous, sub-parallel, moderate- to low-amplitude reflections, (iii) regular, migrating-wave, moderate- to low-amplitude reflections, and (iv) irregular, wavy to discontinuous, moderate-amplitude reflections. Not shown is (v) an irregular, continuous, single high-amplitude reflection caused by a hard seabed. From Stow *et al.* (2002a).

Fig. 6.20 Alternating transparent (T) and moderate-amplitude continuous-reflection (R) seismic facies attributed by Stow *et al.* (2002a) to slower and faster flow velocities, respectively. The T facies are inferred to be mud-prone. The R facies likely contain more sand/silt, hiatuses and condensed sections. From Stow *et al.* (2002a).

reworking. This unit is ~3 km thick and forms part of a ~9 km-thick Upper Cretaceous succession. Accumulation rates (as rock) might therefore be estimated conservatively at ~300 m 10^{-6} years. Because mudstones account for more than 66% of the Caledonia Formation, the accumulation rate as unconsolidated sediment would have been perhaps double this estimate. In any case, the accumulation rate was at least an order of magnitude greater than for contourite drifts in the modern oceans.

Stanley (1987, 1988) very carefully described and interpreted deposits of the Caledonia Formation, and concluded that many of the sandstone beds are Bouma-type turbidites with bottom-current-reworked tops (Fig. 6.21). The sediment in the reworked tops is better sorted than underlying sandstone, has heavy-mineral-rich laminae, forms lenticular deposits, and may be cross-laminated to cross-bedded throughout. Palaeocurrent measurements from the cross-laminated bed tops trend ~90° from sole markings on the turbidites (Stanley 1988).

In spite of apparently strong evidence for reworking after the deposition of each turbidite, there are difficulties with the hypothesis that persistent bottom currents were active throughout deposition of the Caledonia Formation. Stanley (1987) does not discuss the mudstones that overlie the inferred reworked bed tops, nor the large proportion of the formation that consists entirely of fine-grained

rocks. As an example, Stanley's depositional model (Fig. 6.21) hypothesises currents strong enough to erode a muddy bed top (T_e) and several of the Bouma divisions below, eventually in some cases converting the original deposit into isolated ripple lenses. Such strong currents, if persistent like modern bottom currents, would surely have prevented any accumulation of mud (Table 6.3 criteria for reworked turbidites), leading to a stack of amalgamated sand beds. Instead, however, the Caledonia Formation is replete with mudstone interbeds (Stanley 1987), leading to the conclusion that the bottom currents were *not* persistent, but instead were somehow turned on and off at approximately the same frequency as the turbidites were emplaced. Otherwise, the normal sequence of events could not be (i) deposition of a turbidite → (ii) vigorous reworking including erosion → (iii) deposition of mud under tranquil conditions, as advocated by Stanley (1987, 1988). The only alternative we can imagine that could be consistent with the interbedded nature of the formation *and* deposition beneath a persistent thermohaline current would be to suggest that the bottom current periodically received, from an upflow area, a quantity of mud that was spread as a blanket over the sandy substrate. In that case, however, the interbedded muddy contourites would likely have accumulated at rates no faster than those on modern sediment drifts near continental margins, and it

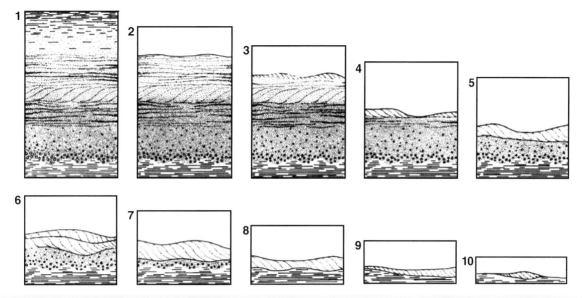

Fig. 6.21   Possible continuum of structures that Stanley (1987) postulated might result from progressive reworking of a turbidite (column 1) by bottom currents. As the reworking continues (from 1–10), the sand bed becomes progressively thinner and is eventually a single wavy to lenticular horizon of cross-laminated sand (columns 9 & 10). From Stanley (1987).

would not have been possible to deposit the amount of mud that is observed between individual sandstone beds in the time available.

Burrows are not common in either the sandstones or mudstones of the Caledonia Formation, although Stanley (1988) reports a greater amount of local bioturbation in his inferred bottom-current-reworked deposits. More commonly, however, the accumulation rate was apparently too high for infaunal colonisation, unlike the situation on modern sediment drifts.

We appreciate the dilemma faced by Stanley (1988) in trying to interpret the local sandstone bed-top erosion and rippling in the Caledonia Formation. However, there are too many discrepancies with the features predicted, or known, for either sandy contourites or bottom-current-reworked SGF deposits (Table 6.3). Most damaging to a bottom-current interpretation is the presence of interbedded mudstones indicating tranquil conditions between the emplacement of the sandstone beds. The same inconsistency permeates the depositional motifs that Shanmugam (2008: his figure 5.9) attributes to bottom-current reworking of sand beds. In our view, the known persistence of bottom currents over at least tens of thousands of years is incompatible with Stanley's (1988) proposed high-frequency alternation between erosion/reworking and mud deposition. Instead, bypassing turbidity currents (steady or waxing; Fig. 1.40) might better account for the modified sand-bed tops. These hypothetical bypassing flows could have occurred either during, or after, the passage of the main depositing flow. Between turbidity-current events, the area would have been sufficiently tranquil to allow the deposition of mud from either the dilute tails of the passing flows, or from the hemipelagic 'rain' of clay and silt grains.

This short discussion of the Caledonia Formation serves to illustrate how difficult it can be to separate the effects of gravity-flow deposition from bottom-current reworking in the geological record. Rigorous application of the criteria in Table 6.3 is essential. Also, geologists need to pay careful attention to the implications of the very different instantaneous and long-term accumulation rates of most turbidites compared with most contourites. High instantaneous deposition rates for turbidites produce wet-sediment deformation, climbing ripple

lamination and prevent bioturbation (Fig. 1.16). Low long-term deposition rates for contourites and other bottom-current deposits ensure thorough bioturbation (unless bottom waters are hostile to life). Future coring of modern sandy contourites and barchan dunes will help improve interpretations, but to date all evidence seems to suggest that (i) sandy contourites are almost everywhere bioturbated, and (ii) bottom currents sufficiently strong to transport sand do not episodically turn off so as to permit the deposition of mud interbeds in an otherwise sandy succession.

6.6   Ancient contourites

The number of convincing ancient sediment-drift or contourite deposits is very small, and many of these have interpretation problems (Hüneke & Stow 2008). This may partly reflect the fact that modern sediment drifts mostly accumulate on transitional crust or oceanic crust, so would be difficult to preserve in the ancient record in an undeformed state.

Stow *et al.* (1998), Luo *et al.* (2002) and Hüneke and Stow (2008) emphasise that ancient contourite drifts must fit criteria at three different scales in order to be unambiguously identified. At the large scale, the palaeogeographic setting of the deposit must be in an area of likely thermohaline circulation, for example along an ancient continental margin on the western side of an ocean basin, or in an area where currents might have been focused through narrow marine straits. At the medium scale, the deposit should be found in a deep-water succession and have mounded morphology or palaeocurrent indicators that support along-slope distribution of the sediment. At the small scale, facies and bedding should be consistent with those features found in modern contour-current deposits, including coarsening–fining cycles and bioturbation.

When considering large-scale criteria, past differences in the driving factors behind bottom-current circulation need to be considered. Thermohaline circulation in modern oceans is dominated by cold dense water masses formed at high latitudes. The geometry of north–south ocean basins that deepen toward the equator is partic- ularly conducive to this type of circulation. During many times in the past, however, the bottom-water circulation was halokinetic, driven by warm and saline bottom waters descending from shallow shelf seas that bordered the eastern sides of ocean basins in arid zones (Brass *et al.* 1982; Hay 1983b; Oberhansli & Hsu 1986). This non-actualistic palaeo-circulation was mainly the result of a different arrangement of continental plates, giving different climatic conditions than those found today. Halokinetic circulation was predominant during the Cretaceous, Palaeocene and Early Holocene. Deep thermohaline circulation was established by the Palaeogene, but was still controlled more by salinity than by temperature (Oberhansli & Hsu 1986). Another major feature of Mesozoic ocean circulation was commu- nication with the Tethys Sea. The Callovian North Atlantic had an east–west equatorial orientation, was narrow (~500 km wide) and deepened by ~300 m due to Callovian subsidence. Sluggish bottom currents gave rise to sediment drifts, with most bottom water derived from Tethys. Trade winds favoured extensive upwelling along the southern African margin (Robertson & Ogg 1986). The effect of the Tethys bottom current extended through the Palaeogene.

The more convincing ancient contourite drifts are developed in carbonate facies (Hüneke & Stow 2008), and are thus of mostly pelagic origin. Examples come from the Ordovician of China (Luo *et al.* 2002) and the Oligocene of Cyprus (Kahler & Stow 1998; Stow *et al.* 2002c). The Cretaceous Talme Yafe Formation of Israel (Bein & Weiler 1976) most successfully demonstrates the matching of defining characteristics at large to small scales.

6.6.1 Talme Yafe Formation

The Cretaceous Talme Yafe Formation, Israel, consists of a huge (>3000 m thick, 20 km wide, >150 km long) prism of calcareous detritus that accumulated on the northwest continental margin of the Arabian craton (Fig. 6.22; Bein & Weiler 1976). The depositional history of this region was influenced mainly by tectonics within the Arabo-Nubian Massif and sedimentation and oceanographic history of the Tethys Sea. The major tectonic element that influenced sedimentation was a north–south 'hinge-line' separating an eastern, shallow, restricted epeiric sea on the Arabo-Nubian craton from an open seaway to the northwest. The Talme Yafe Formation is interpreted as a deposit on the continental rise, from mid-slope to base-of-slope settings. The regional palaeo-circulation within the Tethys Sea at this time was from east to west along its southern margin, but palaeo-Israel is a northerly extension of the Arabian Massif, and as such is interpreted to have had southerly directed contour currents (Fig. 6.23; Bein & Weiler 1976).

In the early Albian, calcareous detritus was derived from the continental shelf and transported downslope by various mass-wasting processes. This material was redeposited on the slope by contour-following currents. Within the whole succession, planktonic fauna are ubiquitous and especially prevalent in the marls to the west, which are interpreted as pelagic deposits (Fig. 6.23). Fine-grained sediment was transported from the shelf and onto the slope as lutite flows (Section 1.2.1).

The Talme Yafe deposit consists of calcilaminite (Facies Class D), calcilutite (Facies Class E), calcarenite (Facies Class B), calcirudite (Facies Class F) and marl (Facies Class G). The bulk of the formation (perhaps a few hundred metres) consists of calcilaminite, which is found in most of the drill holes and along 200 m-thick coastal outcrop sections. The calcilaminite shows a frequent alternation of calcilutite and calcarenite (Fig. 6.24). The laminae are parallel, with

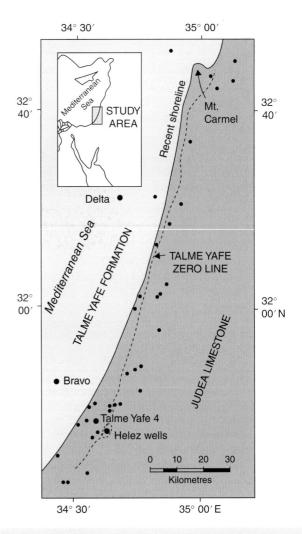

Fig. 6.22 Map of the coastal plain of Israel and its adjacent shelf showing the location of drill holes (dots). From Bein and Weiler (1976).

sharp boundaries; micro-scale cross-lamination is less common. The calcilutite laminae are ungraded, and consist of fine to coarse alter- nations with a small proportion of dispersed sand grains (Fig. 6.25). The calcarenite laminae are formed of fine to coarse grained, poorly sorted sand. Small flute marks, cut-and-fill structures, intraclasts and micro-load structures occur. Some of the units are extensively bioturbated. Erosional channels, up to 20 m wide, occur within the calcilaminite facies.

Calcarenite facies occur in the Talme Yafe deposits west of the Helez oil field (Fig. 6.22). Calcarenites mainly comprise unsorted, randomly oriented skeletal material, up to 0.5 mm in diameter. Structures are limited to flasers and intraclasts. Calcirudites (Facies F1.l) occur locally near the base of the Talme Yafe Formation. The marl facies occurs only in the western Bravo and Delta drill holes (Fig. 6.22) and is mostly structureless. The marl comprises fine lutite, with about 25% clay and 5–10% coarse calcilutite and calcarenite. Planktonic foraminifera and nannoplankton are abundant. It is this facies which has the shape of a mounded drift (Fig. 6.26), and appropriate textural characteristics to justify recognition as a contourite deposit.

The following major features characterise the Talme Yafe deposit: (i) a predominance of shelf-derived calcareous detritus; (ii) textural

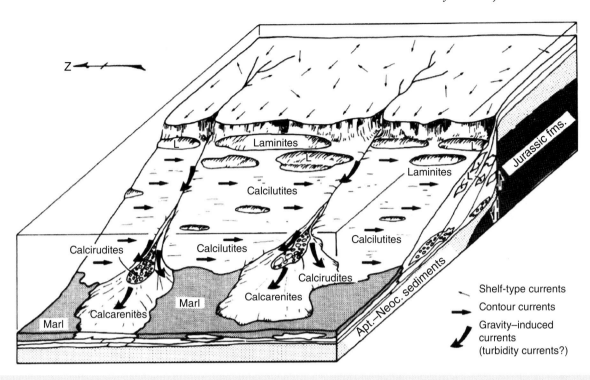

Fig. 6.23   Schematic three-dimensional model of palaeoceanography for the Talme Yafe Formation, Israeli continental shelf. From Bein and Weiler (1976).

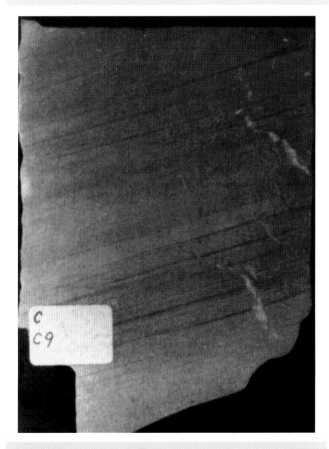

Fig. 6.24   Calcilaminate (Facies Class D) showing alternating calcilutite (light) and calcarenite (dark). Lamina boundaries are sharp and no graded laminae were observed. Length of label = 19 mm. From Bein and Weiler (1976).

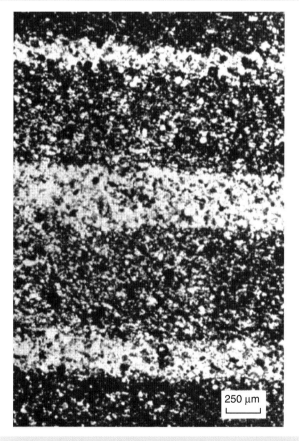

Fig. 6.25   Calcilaminate (Facies Class D), showing detail of Figure 6.24. Note that no clear grain-size sorting exists. Lutite is present in the calcarenite laminae and vice versa. Maximum grain size is 85 μm. From Bein and Weiler (1976).

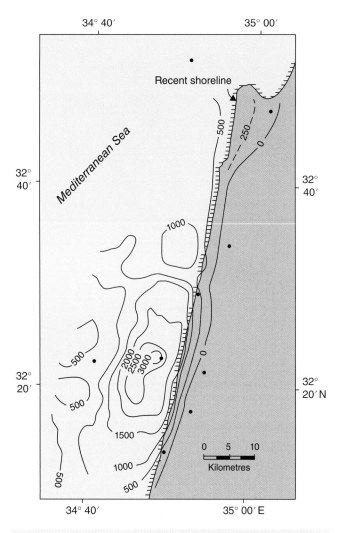

Fig. 6.26 Isopach map of the Talme Yafe Formation from borehole data (Fig. 6.22) and correlated seismic profiles. From Bein and Weiler (1976).

evidence for periodic alternation of strong and weak currents; (iii) a small proportion of turbidites; (iv) an elongated, asymmetric prism, in which the isopach contours parallel the trend of the continental rise (Fig. 6.26); and (v) a stratigraphic position between shallow neritic carbonate-platform deposits and deep-marine pelagic marls.

The overall shape of the formation on isopach maps, palaeogeographic setting, and facies all point clearly to deposition as a sediment drift on the lower continental slope/rise. In this Miocene example, the source of clastic material was from epicontinental platform carbonates located to the east. Downslope movement from the shelf break was by lutite flows for the fine-grained detritus, and more concentrated SGFs, including debris flows, for the coarser grained material. Coarse-load SGFs, which introduced the carbonate sand to the slope and rise, were confined to submarine canyons that dissected the continental slope, and to small submarine fans at the base-of-slope. The main currents that subsequently dispersed this sediment and shaped the continental rise were contour currents. The contour currents emplaced much of the calcilutite as muddy contourites, either laminated or bioturbated. The flaser-bedded and intraclast-strewn calcarenites are interpreted as sandy contourites that likely accumulated beneath the axis of the current. Marls are pelagic–hemipelagic sediments that accumulated in areas more removed from the contour-current influence. Although apparently structureless, these marls might be thoroughly bioturbated.

6.7   Facies model for sediment drifts

In modern settings there is a hierarchy from small to large bedforms: barchan and longitudinal ripples → erosional furrows → muddy sediment waves → drifts. In zones of maximum velocity (i.e., along the axis of a contour current) there may be erosion and non-deposition, with the development of sediment lags, condensed sections and unconformities. These condensed sections or hiatuses may be the result of meandering of the axes of thermohaline currents through time and space.

Possible lateral trends away from the axis of the current can be hypothesised for sediment drifts, based on results from modern studies. An example of this approach can be summarised from data obtained in Orphan Basin, where the WBUC impinges on the lower slope and rise at ~2700 m water depth (Carter *et al.* 1979; Schafer & Asprey 1982) (Fig. 6.27). The trends depend upon the type of sediment source, the direction of supply (downslope versus along-slope), the relative flow velocity from the core to the margins of the current, organic flux to the bottom, the carbonate compensation depth, range of water depths and the bottom gradient, among other factors. For the proposed summary sequence, water depths are 300–3000 m; current strengths are 0–20 cm s^{-1}; a zone of upwelling occurs just upslope of the sediment-drift system; and sediment supply is ice-rafted debris from floating ice and resedimentation of shelf sediment by mass-wasting processes downslope (Aksu & Hiscott 1992; Hiscott & Aksu 1996). Accumulation rates are high on the flanks of the WBUC, diminishing towards the axis of the current. Sand percentages vary from 10% on the upslope flank, to 65% beneath the axis of the current, to 35–40% on the downslope flank. Beneath the axis of the WBUC, sediment is a thin sandy gravel lag, with Fe-Mn coatings on pebbles. Organic-carbon content is low, bioturbation intensity low, and calcium carbonate content is high, reflecting a high diversity of benthic deep-water calcareous foraminifera.

Upslope–downslope trends in sediment character shown in Figure 6.27 may vary both spatially and temporally, depending upon the rate of lateral shifting of the path of the contour current axis up and down the slope (Ledbetter 1979). These shifts generate fining → coarsening → fining sequences. Changes in the bottom-sediment trends across sediment drifts are isochronous, occurring across the entire feature whenever flow configuration is reorganised. The identification of events that led to distinct growth phases in ancient sediment drifts requires stratigraphic preservation of long, correlative along-strike sections.

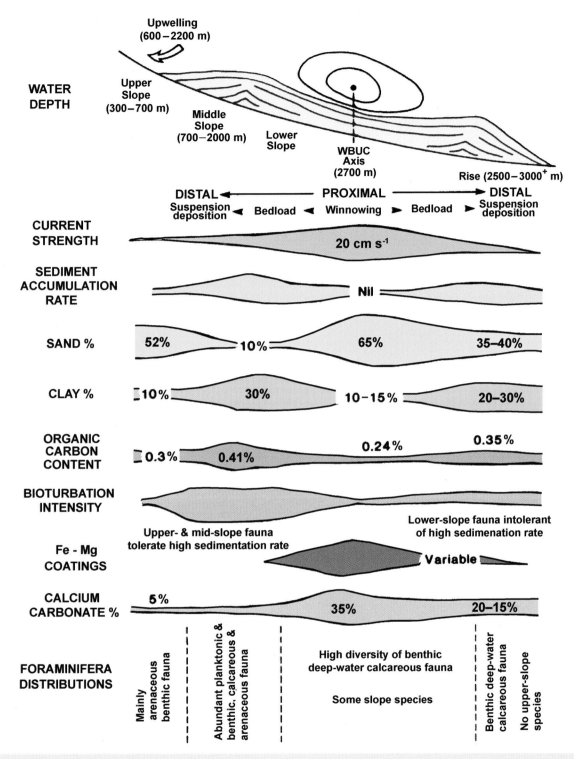

Fig. 6.27 Schematic representation of distal–proximal–distal trends (left to right) in the style of sediment accumulation in the vicinity of a strong contour current flowing along a continental slope and rise, with the development of a double mound. Based on data gathered by Carter *et al.* (1979) and Schafer and Asprey (1982) beneath the Western Boundary Undercurrent near Orphan Knoll, Labrador Slope, eastern Canada. From Pickering *et al.* (1989).

CHAPTER SEVEN

Submarine fans and related depositional systems: modern

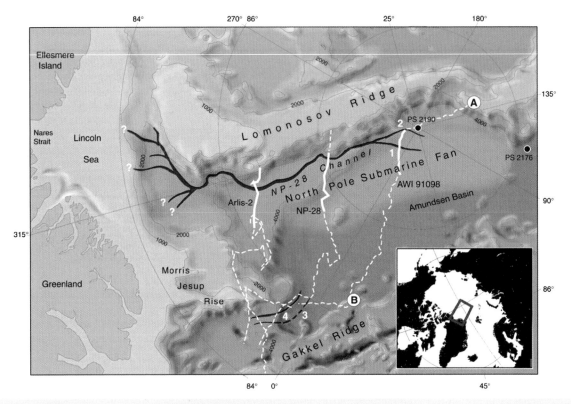

Map of the Amundsen Basin, Arctic Ocean, with the locations of seismic profiles and observed crossings of submarine channels on the North Pole Submarine Fan. Major sediment pathways indicated by a bold dark red line and the outline of fan by dark brown colour. Tributary paths from the Canadian/Greenland continental slope inferred from the position of bathymetric downslope depressions. After Kristoffersen *et al.* (2004).

7.1 Introduction

A *submarine fan* is a cone of sediment at the base of a submarine slope, consisting mostly of the deposits of turbidity currents that travelled into deep water through a submarine canyon from an adjacent shelf area. This definition of a fan excludes the 'debris' fans found at high latitudes seaward of glacially excavated, shelf-crossing troughs (e.g., Vorren & Laberg 1997; Armishaw *et al.* 2000). These 'debris' fans consist instead of stacked successions of glacigenic debrites and hemipelagic sediments, including ice-rafted *Heinrich layers* rich in detrital carbonate that was originally eroded from Palaeozoic bedrock in Arctic regions and delivered to the deep sea by melting icebergs.

Submarine canyons are features with complex erosional and depositional histories that indent the edges of continental shelves, but that can extend almost to the shoreline in the case of narrow shelves along active continental margins. Canyons, as suggested by their name, are commonly hundreds of metres deep with steep rocky walls. *Submarine ramps* consist of a set of laterally coalesced submarine fans that form a slope-parallel wedge, tapering towards the basin floor (Heller & Dickinson 1985; Richards *et al.* 1998). The component parts of submarine ramps and submarine fans are identical, so they are both considered in this chapter.

In our 1989 book on deep-marine deposits (Pickering *et al.* 1989), the only datasets routinely available for the study of submarine

Deep Marine Systems: Processes, Deposits, Environments, Tectonics and Sedimentation, First Edition. Kevin T. Pickering and Richard N. Hiscott.
© 2016 Kevin T. Pickering and Richard N. Hiscott. Published 2016 by John Wiley & Sons, Ltd.
Companion Website: www.wiley.com/go/pickering/marinesystems

End-member submarine-fan models

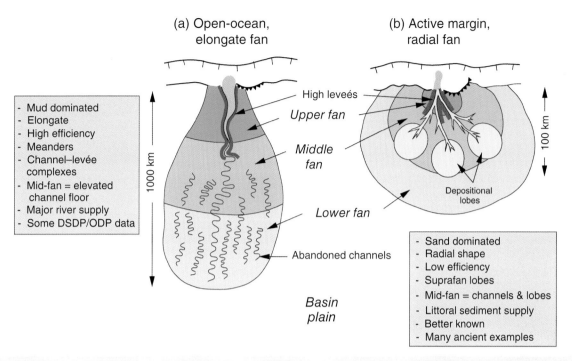

Fig. 7.1 End-member submarine-fan plan-view models, styled after the summary models of Shanmugam and Moiola (1988), who called (a) a 'mature passive-margin fan' and (b) a 'mature active-margin fan'. Open-ocean submarine fans accumulate mainly on oceanic crust, whereas active-margin radial fans accumulate in strike-slip basins, foreland basins and rifts. The *lower fan* and *basin plain* are both underlain by thin-bedded, sheet-like turbidites, so are indistinguishable in ancient deposits. In natural settings, the shapes of submarine fans are rarely this simple because of their impingement on irregular basin margins and bathymetric highs.

fans were outcrops of ancient deposits on land, and marine seismic profiles and shallow piston cores from modern fans (Normark 1978; Walker 1978). This led to a heavy dependence on ancient successions for the understanding of submarine-fan deposits and stratigraphy (Chapter 8). Today, these types of data have been supplemented by 3D seismic attribute maps and horizon maps published by oil explorationists (e.g., Sikkema & Wojcik 2000; Wonham *et al.* 2000), by high-resolution multibeam bathymetry and side-scan sonar images and by long scientific cores from modern fans provided by DSDP (Leg 96) and ODP (Legs 116, 155).

It is common practice in the geoscientific literature to distinguish between so-called *modern fans* (this chapter) and *ancient fans* (Chapter 8). Undisputed ancient fans are those that are now exposed on land because of post-depositional uplift and basin inversion. Undisputed modern fans are those which are linked to an upslope morphological canyon and that have been built by the accumulation of SGF deposits through the Pleistocene and Holocene. There are many submarine-fan deposits, however, which are now buried deeply in continental-margin successions. If these are so deeply buried that their location or size are incompatible with the modern continental drainage or oceanographic setting, or if their deposits cannot be reliably inferred from seismic attributes because of the coarse resolution of industry-scale seismic data, then we consider these to be ancient rather than modern submarine fans. A useful working definition that we adhere to in this book is that modern submarine fans have a bathymetric expression on the modern seafloor and have surface depositional features like channels and levées that formed in Pleistocene or younger times. Their deposits can be imaged using a high-resolution

boomer or equivalent acoustic sources so that details of facies architecture can be readily ascertained. Characteristic deposits can be cored with free-fall corers (e.g., piston corers), although high accumulation rates commonly ensure that all parts of the modern fan cannot be cored except by technologies like the Advanced Piston Corer of the Integrated Ocean Drilling Program (IODP) and its precursors. In contrast, ancient submarine fans are either exposed on land because of uplift, or can occur in the subsurface of offshore areas wherever the deposits are sufficiently deeply buried that they can only be imaged by conventional 2D or 3D seismic systems used for hydrocarbon exploration. Such deposits can only be sampled by rotary coring because they are partially or entirely lithified.

Sophisticated studies of modern fans using deeply towed instruments, including side-scan sonar, began with the work of Normark (1970), with similar techniques subsequently applied to numerous modern fans worldwide (Barnes & Normark 1984). These fans can, in most cases, be divided into (i) an *upper fan* just downslope from the feeder channel, or canyon, and characterised by a single deep channel with levées, (ii) a *middle fan* characterised by updip multiple distributary channels (most inactive) and downdip relatively smooth depositional lobes with shallow channels and relatively high sand content and (iii) a *lower fan* with few if any channels and a smooth surface with low gradient (Fig. 7.1). We consider the Normark (1970) synthesis to be a first-generation model for the development of modern submarine fans.

Geoscientists now appreciate the variability in size, shape, physiography and facies composition of modern fans (Normark *et al.* 1993b), and much current research is focussed on evaluating the

Table 7.1 Major characteristics of submarine fans (this page) and ramps (facing page), based on grain size and sediment source. From Reading & Richards (1994)

Feeder system type	Single Point-Source Submarine Fans			
Dominant grain size	Mud	Mud/Sand	Sand	Gravel
Size	Large	Large-moderate	Moderate	Small
Slope gradient	Low	Low-moderate	Moderate	High
Units = m/km	0.20–18	2.5–18	2.5–36	20–250
Shape	Elongate	Lobate	Radial/lobate	Radial
Radius or Length (km)	100–3000	10–450	10–100	1–50
Source Area				
Size	Large	Moderate	Moderate-small	Small
Gradient	Low	Moderate	Moderate	High
Distance	Distant	Moderate	Close	Close
Feeding Systems	Large, mud-rich river delta	Large, mixed-load river delta and/or its downdip canyon	Shelf failure or shelf canyon	Fan delta or alluvial cone
Sand Percentage	≤30%	30–70%	≥70%	5–50%; >50% gravel
Supply Mechanism	Infrequent slumps and slump-initiated turbidity currents; contour currents	Mainly turbidity currents and concentrated density flows	Reworking or direct access to shelf clastics; low-efficiency turbidity currents and concentrated density flows	Frequent inflated sand/gravel flows; hyperpycnal flows
Size of SGFs	Very large	Moderate	Moderate-small	Very small
Channel System	Large, meandering to straight with well-developed stable levées system	Moderate scale, meandering to braided systems, laterally migrating with levées	Braided to low-sinuosity, ephemeral channels and chutes, rapid lateral migration	Braided, small ephemeral chutes
Distal Slope/Lower Fan Sediments	Thin, sheet-like flows form interbedded sands, silts and muds; coarse intervals form thin clastic sheets	Mixed-load turbidity currents form lobes of interbedded sands and muds	Sand-rich concentrated density flows and turbidity currents form low-relief lobes and sand sheets	Turbidity currents form thin distal beds
Principal Basin Plain Deposits	Turbidites >Hemipelagites	Hemipelagites >Turbidites	Hemipelagites	Hemipelagites
Architectural Elements				
Proximal Area	Channel levées	Channel levées	Channels	Wedges
Distal Area	Sheets	Lobes	Channelised lobes	Sheets
Sandbody Geometry	Large, lenticular channels with multiple, variable-scale sand, silt and mud fills; high degree of heterogeneity; distal fan dominated by thin sand, silt and mud sheets	Lenticular channels dominated by sand or mud fill; downdip lobes formed of interbedded and alternating sands, silts and muds	Broad, sheet-like to low-relief lobate sandbody geometries dominated internally by channelised sand units	Irregular interconnected gravels; proximal areas dominated by gravels and angular clasts; sands dominant within medial to distal parts of the system
Sandbody Communication				
Vertical	Poor	Moderate	Good	Good
Lateral	Poor	Poor	Good	Poor

relative importance of the controlling variables such as tectonics, sea-level changes and climate (e.g., Normark *et al.* 2006; Knudson & Hendy 2009). Reading and Richards (1994) and Richards *et al.* (1998) integrated the complexity of sediment sources and grain size into a second-generation classification of submarine fans (Table 7.1). Mutti and Normark (1987), and later Pickering *et al.* (1995b), took a different approach, similar to the architectural methods suggested for fluvial systems (*cf.* Miall 1986). These authors proposed a hierarchy

of depositional scales, and discussed some of the major lithofacies assemblages or 'elements' that constitute sediment-gravity-flow (SGF) systems in general, and submarine fans in particular. We follow this approach in Section 7.4, below. The approaches of Mutti and Normark (1987), Pickering *et al.* (1995b) and Reading and Richards (1994) represent contrasting (and complementary) attempts to incorporate new data into what we consider to be second-generation submarine-fan classification schemes, focussing primarily on the identification and

Table 7.1 (continued)

Feeder system type			Multiple-Source	Submarine Ramps
Dominant grain size	**Mud**	**Mud/Sand**	**Sand**	**Gravel**
Size	Moderate	Moderate	Moderate	Small
Slope gradient	Moderate	Moderate	Moderate	High
Units = m/km	2.5–25	7–35	>35	20–250
Shape	Lobate	Lobate	Linear-belt	Linear-belt
Radius or Length (km)	50–200	5–75	1–50	1–10
Source Area				
Size	Moderate	Moderate-small	Small	Very small
Gradient	Low	Moderate	Moderate	High
Distance	Distant	Moderate	Close	Close
Feeding Systems	Large, mud-rich river delta	Mixed-load delta, linear shoreline	Sand-rich clastic shoreline/shelf	Alluvial fan or braid plain or fan delta
Sand Percentage	≤30%	30–70%	≥70%	5–50%; >50% gravel
Supply Mechanism	Infrequent slumps and slump-initiated turbidity currents; contour currents	Mainly turbidity currents and concentrated density flows	Reworking or direct access to shelf clastics; low-efficiency turbidity currents and concentrated density flows	Frequent inflated sand/gravel flows; hyperpycnal flows
Size of SGFs	Large	Moderate	Moderate-small	Small
Channel System	Moderate sizes as channel-levée systems	Multiple, levéed channels with meandering to straight planform	Multiple, laterally migrating braided to low-sinuosity channels	Small ephemeral chutes
Distal Slope/Lower Fan Sediments	Thin, sheet-like flows form interbedded sands, silts and muds	Mixed-load turbidity currents form lobes of interbedded sands and muds	Sand-rich concentrated density flows and turbidity currents form low-relief lobes and sand sheets	Turbidity currents form thin distal beds
Principal Basin Plain Deposits	Turbidites >Hemipelagites	Hemipelagites >Turbidites	Hemipelagites	Hemipelagites
Architectural Elements				
Proximal Area	Channel levées	Channel levées	Channels	Wedges
Distal Area	Sheets	Lobes	Channelised lobes	Sheets
Sandbody Geometry	Moderate-scale sand bodies within overall large channel form; sands commonly isolated in both downdip and updip directions	Offset-stacked, lenticular channel sand bodies bounded by fines; levée passing downdip into offset-stacked lobate sand bodies formed of sands and muds	Broad, sheet-like to low-relief lobate sandbody geometries dominated internally by channelised sand units	Irregular interconnected gravels; proximal areas dominated by gravels and angular clasts; sands dominant within medial to distal parts of the system
Sandbody Communication				
Vertical	Poor-moderate	Moderate	Good	Good
Lateral	Moderate	Moderate	Good	Moderate

description of morphological components (architectural elements) and on the range of grain sizes and compositions, respectively. These second-generation schemes attempt to unravel the inherent complexity in deep-marine clastic systems, both in order to better describe and predict depositional patterns, but perhaps more importantly to attempt to relate cause and effect and the magnitude and frequency of natural processes.

This chapter builds on ideas presented in Chapter 4 regarding the sequence stratigraphy of deep-marine systems, including submarine fans. We classify depositional models that incorporate sequence-stratigraphic concepts as third-generation models. These models try to differentiate base-level controls from autocyclic controls on the growth and stacking pattern of SGF facies in submarine fans and related environments. We begin by reviewing the major controls on submarine-fan size, facies, architecture and longevity. The feeder channels, or canyons, are then described. We then describe and explain the origin of the main architectural elements that form the building blocks of modern submarine fans. Modern non-fan dispersal systems that share some attributes with submarine fans are treated briefly at the end of the chapter. An excellent compilation of the attributes of 21 modern fans is presented by Barnes and Normark (1984). Table 7.2 is a summary of some of these data.

Table 7.2 Dimensions, setting and texture of selected submarine fans in modern ocean basins, sorted by area (A) and volume (V). Fans matching the end members in Figure 7.1 are preceded by the appropriate letter, a or b. Margin types are passive (P), active (A) and transform (T). Note that shape is not diagnostic

	Name	Region	Margin type	L (km)	W (km)	A (km²)	Maximum thickness (m)	V (km³)	Shape	Grain size maximum	Grain size average
a	Bengal Fan	Indian Ocean	P/A	2800	1100	3×10^6	>5000	4×10^6	elongate	medium sand	mud
a	Indus Fan	Indian Ocean	P/A	1500	960	1.1×10^6	>3000	10^6	fan	sand	mud
a	Laurentian Fan	eastern Canada	P	1500	400	4×10^5	2000	10^5	elongate	gravel	v. fine sand
a	Amazon Fan	Brazil	P	700	250–700	3.3×10^5	4200	$>7 \times 10^5$	fan	pebbles	mud
a	Mississippi Fan	Gulf of Mexico	P	540	570	$>3 \times 10^5$	4000	3×10^5	conical	gravel	silty mud
a	Nile Fan	Mediterranean	P	280	500	7×10^5	>3000	$>1.4 \times 10^5$	fan	sand	silty mud
	Rhône Fan	Mediterranean	T	440	210	7×10^4	1500	1.2×10^4	***	fine sand	clayey silt
	Ebro Fan	eastern Spain	P	100	50	5×10^3	370	1.7×10^3	oval	medium sand	fine sand
b	La Jolla Fan	California	T	40	50	~1200	1600	~1175	pear	gravel	fine sand
b	Navy Fan	California	T	25	25	560	900	75	triangular	gravel	sandy silt
b	Crati Fan	southern Italy	A	16	5	60	30	0.9	elongate	medium sand	mud

7.2 Major controls on submarine fans

7.2.1 Sediment type

Submarine fans (and ramps) span a spectrum from sand/gravel-rich, to mixed sand–mud systems, to mud-rich systems (Reading & Richards 1994; Richards *et al.* 1998; Table 7.1). Sand/gravel-rich and mixed sand–mud fans are widely recognised in ancient successions, including oil fields of the North Sea (several papers in Doré & Vining 2005), Cenozoic basins of California (tabulated by Mattern 2005) and other intracratonic and rift basins. Modern fans of this type have been studied in great detail in the California Continental Borderland (Normark 1970, 1978; Normark & Piper 1972, 1984; Piper & Normark 1983; Normark *et al.* 1979, 1998, 2006; Piper *et al.* 1999; Bowen *et al.* 1984) and the Mediterranean Sea (Malinverno *et al.* 1988; Piper & Savoye 1993; Savoye *et al.* 1993; Mulder *et al.* 1997; Wynn *et al.* 2002b). These are small, typically radial fans sourced from either relatively small, steep rivers, or from littoral currents that transport sand and silt along an adjacent shelf and into the heads of submarine canyons. For fans supplied by littoral longshore currents, a paucity of mud at the source ensures that sediment gravity flows (SGFs) reaching the fan lack a quasi-continuous fluid phase that is significantly more dense than seawater, and therefore decelerate more rapidly as sand load is lost through deposition. The SGFs are, therefore, less efficient in transporting sand and silt load away from the canyon than they would be if they contained more suspended mud. For this reason, sand-rich fans have been called *low-efficiency fans* by Mutti (1979).

In contrast, mud-rich submarine fans tend to be large, elongate tongues of sediment extending well out onto oceanic crust beyond continental margins, and fed by major rivers draining areas of high sediment yield, particularly areas with a humid climate or active continental glaciation, both conducive to the production of large quantities of mud (Wetzel 1993). Modern examples include the Bengal, Indus, Mississippi, Amazon, Zaire, Laurentian and Rhône Fans. These are the *high-efficiency fans* of Mutti (1979). Experimental results (Salaheldin *et al.* 2000) confirm that the sand-carrying capacity of turbidity currents is significantly increased in flows with a greater mud load, particularly for higher ratios of clay/silt. The mud-rich fans do not possess the distinct suprafan lobes of the Normark (1970) fan model (Fig. 7.2). They can only be divided into upper, middle and lower segments using the characteristics of the highly sinuous and

widespread channel systems on the fan surface. For example, Pirmez and Flood (1995) recognise an *upper fan* where there is a feeder channel with high levées that is cut below the surrounding fan surface, and that is continually occupied during a single lowstand in relative sea level, a *middle fan* where a network of successively active and avulsing channels are perched in prominent levées so that the channel floor is higher than the adjacent fan surface, and a *lower fan* where levées are more subdued (and eventually absent toward the toe of the fan) and provide weak confinement for turbidity currents travelling through channels that are again cut below the level of the surrounding sandy fan surface. Caution is required, however, because other researchers subdivide mud-rich fans in entirely different ways. For example, Curray *et al.* (2003) place the boundary between the upper and middle Bengal Fan at the point where the floor of the most recently active valley ceases to lie at a higher elevation than the surrounding fan surface, and the active valley on the Zaire Fan is everywhere incised into the fan surface (Babonneau *et al.* 2002), so neither of these fans can be subdivided using the criteria of Pirmez and Flood (1995).

The remarkable sinuous channels on mud-rich fans (Figs 7.3 and 7.4) are centred on broad gull-wing-shaped levées, together forming what is called a *channel–levée unit*, or *channel–levée system*. The cross-sectional appearance of channel–levée units is commonly vertically exaggerated in seismic profiles, potentially leading to an incorrect understanding of their shape (Fig. 7.5). Furthermore, gull-wing levées are likely to compact more than channel sands during deep burial (Mutti 1992), potentially diminishing the topographic expression of the levées even more than what is seen on the modern sea floor. However, examples of gull-wing-shaped levées interpreted from industry 3D seismic profiles indicate that this compactional thinning is not necessarily pronounced (Fig. 7.6).

7.2.2 Tectonic setting and activity

Tectonic setting directly affects basin size and shape, bottom gradients, type and rate of sediment supply and longevity of an individual fan. On mature passive continental margins (Chapter 9), the main tectonic process is slow thermal subsidence, which has little effect on the development of large, mature fans with low surface gradients. Many of these large, generally elongate and mud-dominated fans derive their sediment from orogenic belts on the opposite side of the continent. The major rivers that supply the fans flow away from these

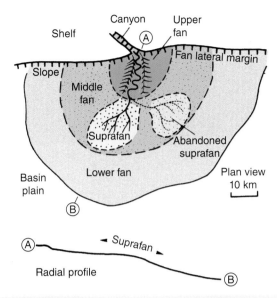

Fig. 7.2 Submarine fan model of Normark (1970, 1978) emphasising growth by successive addition of suprafan lobes on the middle fan (stippled). The 'fan lateral margin' of Pickering (1983a) was not recognised by Normark.

orogens down a topographic gradient toward the distant passive margins (Potter 1978). Examples are the Amazon Fan which is sourced in the Andes, and the Mississippi Fan which is sourced in the Cordillera of western North America.

At active margins, which include convergent (Chapters 10 and 11), oblique-slip (Chapter 12) and young rifted margins (Section 9.3) with rapidly tilting fault blocks (e.g., Surlyk 1978; Normark *et al.* 1998; Nakajima *et al.* 1998), depositional basins may be small and irregular in shape, so that fans, particularly if they are of the delta-fed high-efficiency type, will be constrained by the basin geometry and will not be able to develop the elongate shape typical of unconfined mud-rich fans (*cf.* Pickering 1982c). Fans at active margins are also often transient, because active vertical and horizontal movements of the crust along faults may cut off sediment supply. Escanaba Trough, offshore northern California, provides an example of the intricate routes that some sediment gravity flows take because of the complex seabed bathymetry at active margins. ODP Site 1037 in Escanaba Trough has received thick sand deposits from flows that were sourced from the Columbia River area of Washington State, transited the ~450 km-long Cascadia Channel (bypassing Astoria Fan), then flowed ~120 km west along the Blanco Fracture Zone, ~350 km south across the Tufts Abyssal Plain, ~130 km east along the Mendocino Fracture Zone and finally ~50 km north into Escanaba Trough (Zuffa *et al.* 2000). The total journey is >1100 km, and is remarkably circuitous because of bathymetric barriers. Also, fan sediments at active margins may be deformed and uplifted even as fan growth continues (e.g., Nicobar Fan, Bay of Bengal; Bowles *et al.* 1978). Other unexpected geometries may result from basinal constrictions. For example, the Mozambique Fan has been structurally constrained to develop *parallel* to both the African coastline of Mozambique and the Madagascar Ridge (*cf.* Kolla *et al.* 1980a,b; Droz & Mougenot 1987).

Along the steep, terrestrial to marine (or lacustrine) flanks of oceanic islands or fault-bounded basins, fan deltas may pass gradationally seaward into small, coarse-grained subaqueous fans or slope aprons (Wescott & Ethridge 1982; Postma 1984; Soreghan *et al.* 1999;

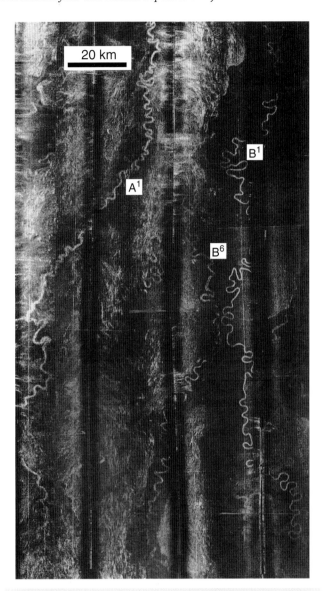

Fig. 7.3 GLORIA side-scan sonar mosaic of part of the middle Indus Fan showing highly sinuous channels. Water depths at the top of the figure are ~3500 m and at the bottom are ~3800 m. The side of the image is oriented ~N15°E. The superscripted letters identify the hierarchy of channels, with lower letters and numbers being younger. A is the youngest channel complex; B the next youngest. From Kenyon *et al.* (1995a).

Wells *et al.* 1999; Sohn 2000). The distal fan deltas and slope aprons are formed of facies similar to those in the proximal parts of small, coarse-grained, fully marine (or lacustrine) fans.

7.2.3 Sea-level fluctuations

Sea-level variations may be (i) eustatic (global) or (ii) regional or relatively local events caused by tectonic processes (Sections 4.3, 4.4 and 4.9). Regardless of cause, relative highstands of sea level generally result in reduced supply of terrigenous sediment to fans due to preferential deposition in coastal systems (e.g., estuarine embayments) and on a relatively broad shelf. Major fan processes during the highstand and early subsequent falling stage then become: (i)

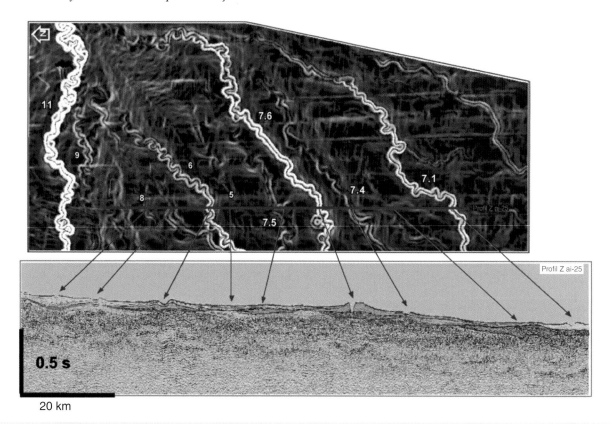

Fig. 7.4 Sinuous channels on the Congo Fan in plan view (top, with land toward the top) and cross-section (bottom). Channel-levée units are laterally and vertically shingled through time as the active channel shifts across the fan surface. From Vittori *et al.* (2000).

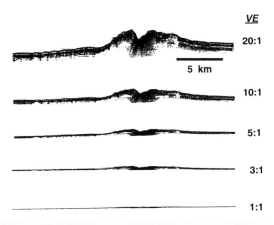

Fig. 7.5 3.5 kHz profile of a channel–levée unit at 3200 m water depth on the Amazon Fan, at various vertical exaggerations. From Damuth *et al.* (1995).

slow deposition of hemipelagites (Facies Class E, e.g., Amazon Fan, Flood & Piper 1997); (ii) reworking of surface sediments by bottom currents (e.g., distal Mississippi Fan, Kenyon *et al.* 2002a) and (iii) mass wasting, producing chaotic deposits of Facies Class F (e.g., Amazon Fan, Piper *et al.* 1997; Rhône Fan, Normark *et al.* 1984). There are notable exceptions like the Zaire Fan, which is active today with episodic sand-load turbidity currents (Khripounoff *et al.* 2003) resulting locally in >10 m of sedimentation during the Holocene (Droz *et al.* 1996), fans sourced from the Himalayas that have been actively

receiving silty/sandy continental detritus during the Holocene (Kolla & Macurda 1988; Weber *et al.* 1997) and California-borderland fans supplied even at highstands by littoral drift into canyons extending nearly to the highstand shoreline (Covault *et al.* 2007). Relative lowstands in sea level produce deposits that are generally much thicker and coarser than their highstand counterparts, particularly if the rate of sea-level fall is sufficiently high to promote river entrenchment right to the shelf edge (Normark *et al.* 2006). A fall in relative sea level leads to the effective narrowing of shallow-marine shelves and forces fluvial/coastal systems to shift toward the shelf-slope break. Additionally, under conditions of lowered sea level, shallow-marine processes, such as severe storm-wave pounding, are more likely to enhance sediment failure along basin slopes and, therefore, further increase sediment delivery rates into deep-marine environments. Exceptions to the rule that higher fluxes of sediment are delivered to the deep sea during lowstands occur in carbonate-sourced systems (Section 4.10), and in rare cases along active margins where tectonic uplift, climatic changes or enhanced littoral supply to canyon heads may contribute to highstand fluxes of coarser detritus (see Sections 4.2 and 4.5).

In seismic data, alternating highstands and lowstands result in packaging of the imaged succession into distinct seismic units bounded by prominent reflections. Three examples are provided here.

(1) The upper Rhône Fan is characterised by a stack of eight lenticular acoustic units (Fig. 7.7a), each about 100 m thick and about 70 km wide. Each acoustic unit consists of a central chaotic facies grading laterally into converging reflections or transparent facies. These are interpreted as channel–levée complexes,

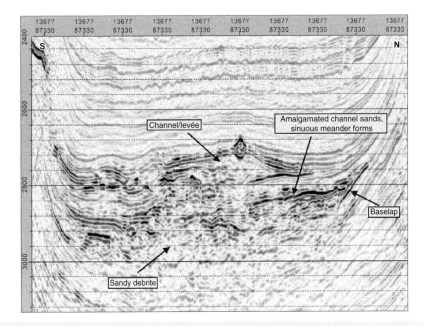

Fig. 7.6 Industry seismic section (from 3D volume offshore Angola) showing several deep-water facies including a gull-wing channel–levée system at the top of a channel complex. From Sikkema and Wojcik (2000).

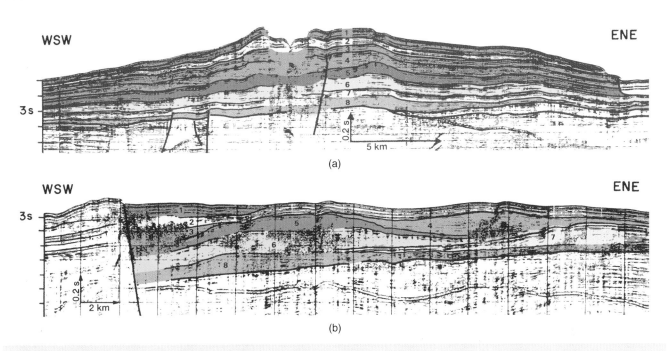

Fig. 7.7 Airgun seismic profiles from the upper (a) and middle (b) Rhône Fan with the same eight lenticular acoustic units numbered on each profile. From Droz and Bellaiche (1985).

each deposited during a lowstand of sea level. The middle fan is also characterised by lenticular acoustic units, but in this case they are not stacked vertically and are instead displaced laterally from one another through time (Fig. 7.7b; see also Fig. 7.4). Each unit is somewhat thicker than beneath the upper fan (maximum 150 m), but lateral shingling leads to an aggregate thickness of only 500 m. The lower fan is also formed of shingled and stacked lenticular acoustic units <70 km wide and ~80 m thick, but

these cannot be correlated with those higher on the fan because of disruption of seismic markers by salt domes and associated faults. In the upper fan, lenticular acoustic units beneath the middle and lower parts of the fan are each interpreted as the depositional record of a lowstand of sea level (Droz & Bellaiche 1985).

(2) The Amazon Fan (Fig. 7.8) consists, in its upper part, of a number of channel–levée complexes (Fig. 7.9). Each channel–levée

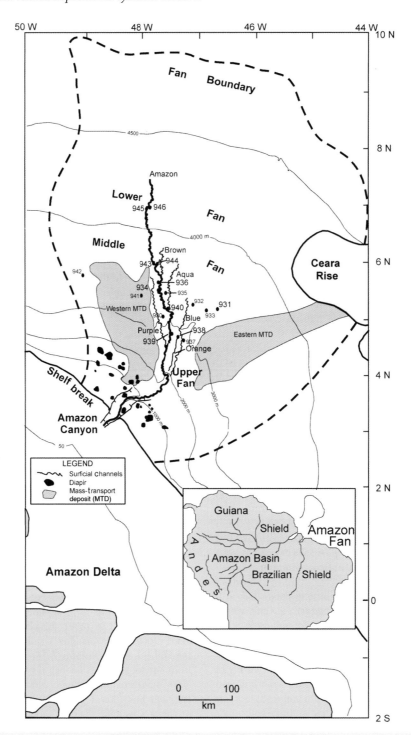

Fig. 7.8 Generalised map of the Amazon Fan showing the location of ODP Leg 155 drill sites in relation to surficial channels and mass-transport complexes (MTCs). From Flood *et al.* (1995) and Piper and Normark (2001).

complex is formed of laterally shingled individual channel–levée units separated by what are called high-amplitude reflection packets (HARPs). Drilling during ODP Leg 155 demonstrated that each channel–levée complex formed during a single low-stand of sea level, and that switching between individual channel–levée units resulted from autocyclic avulsions along what was probably a single active channel (Fig. 7.10). HARPs are

stacked alternations of sand-rich packets (Facies C2.1 and C2.2) and thin-bedded turbidites (Facies C2.3 and D2.1) strikingly like many ancient outcrop successions interpreted as lobe and lobe-fringe deposits (Fig. 7.11).

(3) The Pliocene–Pleistocene Mississippi Fan sediments comprise a stack of 17 discrete sequences (Fig. 7.12), equivalent to the 'turbidite systems' of Mutti and Normark (1987). These have

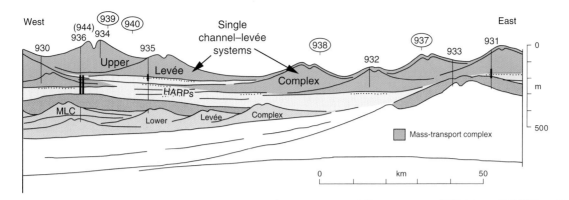

Fig. 7.9 Schematic strike-oriented cross-section through the Amazon middle fan showing HARPs and laterally shingled channel–levée systems of the Upper Levée Complex. This levée complex and its underlying HARPs accumulated during the last lowstand of sea level (~50–12 ka). ODP Leg 155 sites are uncircled if actually in this profile, are circled if projected into the profile from farther up-fan, and are in brackets if projected from farther down-fan. Sites are located on Fig. 7.8. From Shipboard Scientific Party (1995a).

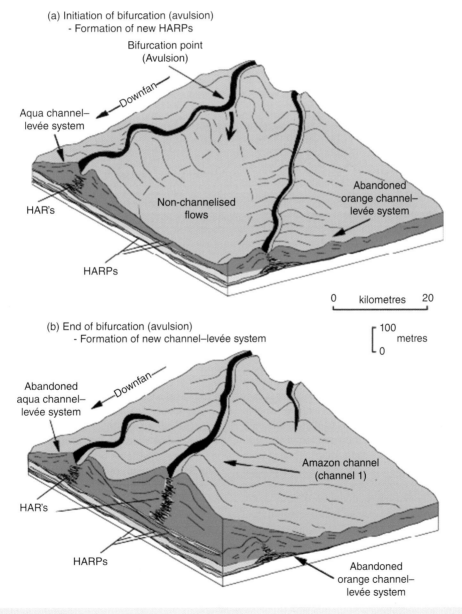

Fig. 7.10 Schematic block diagrams showing the development of channel–levée systems and HARPs following avulsion. From Flood *et al.* (1991).

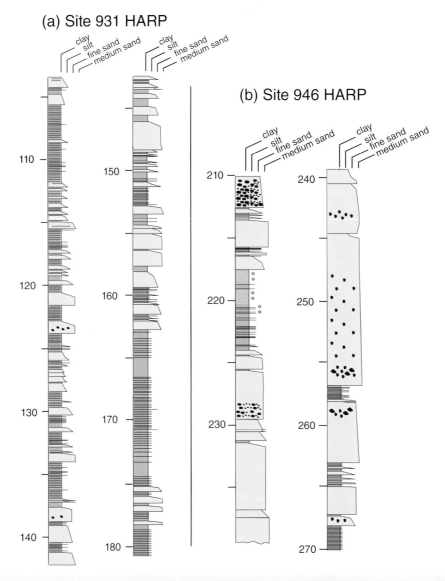

Fig. 7.11 Examples of HARP successions on the Amazon Fan. All depths are in metres below the seafloor. Yellow denotes sand (with local large mud clasts); green denotes turbidite muds, and grey denotes highstand (interglacial), burrow-mottled, bio-calcareous muds. The logs are based on both cores and microresistivity (FMS) logs. The Site 931 HARP is associated with the development of 'Channel 5' of the Upper Levée Complex (last glacial), whereas the Site 946 HARP is from an older glacial epoch, perhaps Stage 12 (Shipboard Scientific Party 1995a). From Hiscott *et al.* (1997c).

been identified from seismic reflection profiles (Weimer & Buffler 1988; Weimer 1991). The closely spaced network of seismic survey lines has allowed individual channel distributary systems from these sequences to be mapped out (Fig. 7.13). There are a series of general trends that can be identified from the sequence of channels, and these are believed to be most strongly controlled by Pliocene–Pleistocene eustacy (Bouma *et al.* 1984; Feeley 1984; Weimer & Buffler 1988; Weimer 1991). During periods of low sea level, canyon backcutting into the slope results in increased sediment supply, and the formation of channel–levée systems (Steffens 1986). When sea level rises, the backstepping of the deltas across the shelf disconnects the direct sediment supply routes, resulting in the gradual infilling of the channel–levée systems. Decreasing sediment grain size throughout the development of individual channel–levée

systems is inferred by Weimer (1991) based on the decreasing strength of the levée reflection character towards the top of each sequence. The decreasing volume and grain size of flows during channel–levée evolution may have caused changes in channel sinuosity and levée relief (Weimer 1991). Indeed, Pickering *et al.* (1986b) attributed the overall upward fining in the Mississippi Channel at DSDP Sites 621 and 622 (Fig. 5.17) to the glacio-eustatic rise in sea level from the Last Glacial Maximum into the Holocene.

The conclusion to be drawn from the Rhône, Amazon and Mississippi systems is that individual lowstands result in the deposition of ~100 m or more of strata, including one or more channel–levée unit and associated HARPs (Flood & Piper 1997; Hiscott *et al.* 1997c; Piper & Normark 2001). HARPs also characterise the Bengal Fan

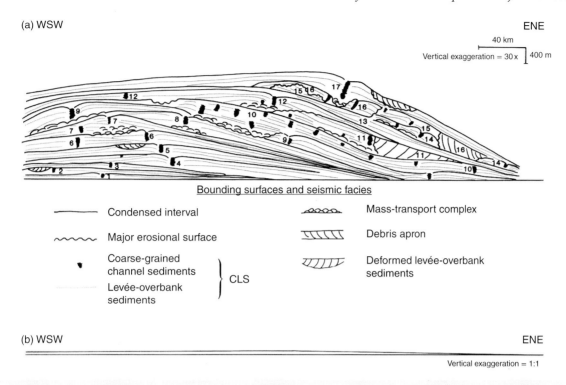

Fig. 7.12 Norxtheast–southwest schematic strike-oriented cross-section across the Mississippi Fan showing the distribution of seismic sequences and depositional environments. The extreme vertical exaggeration in (a) is removed in (b). From Weimer (1995).

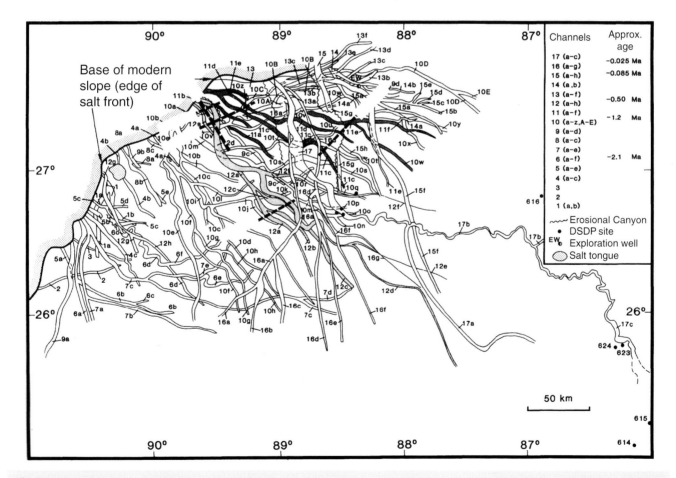

Fig. 7.13 Location of channels associated with the 17 different channel–levée systems on the Mississippi Fan. Within any one channel—levée system, the channels are lettered from oldest to youngest (a, b, etc.). Approximate ages of some sequence boundaries are from Weimer (1990). Heavy dashed lines locate seismic profiles presented in Weimer (1995), the source of this figure.

(Curray *et al.* 2003), the Congo Fan (Vittori *et al.* 2000) and parts of the Indus Fan (as identified by Kenyon *et al.* 1995a, but not identified by McHargue 1991). The great thickness of the lowstand deposits, including stacked mud-prone and sand-prone intervals (packets) runs contrary to the suggestion of Mutti (1985) that 3–15-m-thick alternations of thick-bedded and thin-bedded lobe and lobe-fringe deposits so commonly seen in ancient fan deposits result from a succession of short-term fluctuations of sea level (see also Section 4.2 and Chapter 8). For medium to large submarine fans, the packeting of strata into sand-prone and mud-prone intervals on a scale of ~20 m or less is repeated many times within a single HARP that was deposited during a single lowstand of sea level. We therefore interpret the 3–15 m cyclicity described by Mutti (1985) as a record of autocyclic processes that induce facies shifts at a higher frequency than contemporary sea-level fluctuations (Hiscott *et al.* 1997c).

7.3 Submarine canyons

Submarine canyons are common features along continental margins, and may cut into crystalline rock (around Sri Lanka and the tip of Baja California), consolidated and under-consolidated sediments (the eastern continental margin of the USA), and even evaporites (Congo Canyon). Canyons may be tens of metres to hundreds of metres deep from rim to floor, have widths varying from tens of metres to tens of kilometres, lengths from kilometres to hundreds of kilometres, and show considerable variability in cross-sectional profile. This chapter focusses on those canyons that are linked to downslope submarine fans. Other erosional troughs and smaller erosional shelf-edge gullies are treated in the context of particular tectonic settings in Chapters 9–12.

The study of submarine canyons began in earnest about 60 years ago, accelerated by technological advances in deep-marine surveying apparatus and techniques, for example seismic profiling and side-scan sonar. Much of this research was initially concerned with the preparation of detailed bathymetric charts and explanations for the origin of canyons. Seminal work was done offshore California by Francis P. Shepard and co-workers (e.g., Shepard 1951, 1955, 1963, 1966, 1975, 1976, 1977 & 1981; Shepard & Marshall 1969, 1973a, b, 1978; Shepard *et al.* 1979). Canyons were explained as (i) formerly subaerial river valleys now submerged by elevated sea level (Spencer 1903; Bourcart 1938); (ii) glaciated valleys now below sea level (Shepard 1933); (iii) marine erosional features cut by turbidity currents (Daly 1936); (iv) dissolution features caused by circulation of underground water at continental margins (Johnson 1939, 1967); (v) tsunami-cut depressions (Bucher 1940); and (vi) structurally weak areas like faults that are exploited by erosive processes (Kenyon *et al.* 1978; Picha 1979; Berryhill 1981; Ediger *et al.* 1993; Harris *et al.* 2013). It has become clear that each canyon has a complex and partially unique origin, so a single genetic model is inappropriate (e.g., Mountjoy *et al.* 2009). The main overarching process in all canyons is erosion and deposition by sediment gravity flows, modulated by rising and falling relative sea level (Moore 1965; Mulder *et al.* 1997; von Rad & Tahir 1997; Wonham *et al.* 2000; Propescu *et al.* 2004). Episodic triggering of sediment gravity flows provides a mechanism for the emptying of canyon heads. Bio-erosion and biological weakening of canyon walls by boring organisms are important to the development and maintenance of some canyons (Dill 1964). Bio-erosion is reported from Barrow Canyon in the Beaufort Sea, Arctic Ocean (Eittreim *et al.* 1982), where the ocean-circulation patterns on the shelf and at the shelf-break are believed to cause occasional upwelling events that bring nutrient-rich water up the canyon along the western wall. The upwelling has led to a large population of burrowing and boring organisms with corresponding fast rates of bio-erosion.

Dill (1964) believed that one of the major factors governing the deepening of canyons was sand creep occurring over long time periods. This hypothesis was supported in an experiment where stakes were placed in a line across canyon floors and the central or 'axial' stakes were observed to have moved farther downslope than the 'lateral' stakes. Although sand creep may be important, more catastrophic gravity flows are suggested by the disappearance of some stakes, large concrete blocks and a car body (Dill 1964). Likewise, when a large turbidity current was generated in the Var Canyon by the 1979 collapse of parts of the Nice airport, a 1-m-sized fragment of a bulldozer lost during the slide was found on a terrace at 1400 m water depth, 30 m above the floor of the canyon (Wynn *et al.* 2002b)!

In recent years, research has emphasised the manner in which sediment reaches the deep sea through canyons, including the detailed analysis of events that have culminated in the triggering of large turbidity currents and concentrated density flows in canyons. As examples, Paull *et al.* (2002) used a novel approach to trace fine-grained sediments downslope through the Monterey Canyon. They recognised that the pesticide DDT, only used in California since 1945, would be an excellent tracer for sediment that was introduced into the marine environment by runoff from agricultural lands. DDT concentration along the axis of Monterey Canyon to water depths of >3 km, without significant dilution relative to shelf concentrations, provides strong evidence for down-canyon transport of muds. Paull *et al.* (2003) confirmed this process by documenting a December 2001 inferred turbidity current that carried their current-meter 550 m down-canyon. In the Eel Canyon off northern California, Puig *et al.* (2003) have documented a strong tendency for suspended sediments from the Eel River to be exported from the shelf into the canyon and subsequently down-canyon during storms. For coarser sediments, Khripounoff *et al.* (2003) used current-meters and sediment traps to document the passage of an energetic turbidity current or concentrated density flow through the Zaire Canyon/Valley. Their mooring in the valley at 4000 m water depth only remained on the seabed for a short period following the passage of the flow, because its cable was quickly severed. It nevertheless recorded velocities of >120 cm/s 150 m above the valley floor, and coarse sand and plant debris were collected in a sediment trap 40 m above the floor. Significant amounts of fine sediment reached a second current-meter station 13 km away from the valley axis. Finally, Mulder *et al.* (1997) reconstructed the events which led to transformation of a debris flow to a sand-load, concentrated density flow in the Var Canyon in 1979. They used modelling to predict 6–11 m of erosion of the canyon floor.

Most submarine canyons show the following features (Shepard 1977): (i) sinuous courses, partly true meanders; (ii) canyon floors deepen quite consistently seaward; (iii) canyons lose their deep V-shaped profiles at about the same distance from the shore as the adjacent continental slope shows a marked decrease in slope gradient, transitioning to the continental rise; (iv) canyons are rarely found where the continental slopes are gentle; (v) canyons are cut into rocks of all degrees of hardness as well as into unconsolidated sediments; and (vi) they have tributaries, better developed at the canyon heads than in their lower reaches. Many submarine canyons show an extensive dendritic drainage pattern especially near the canyon heads. McGregor *et al.* (1982) describe gullies arranged in a pinnate pattern around the axes of South Wilmington and North Heyes Canyons, occurring where steep slopes show seaward gradients

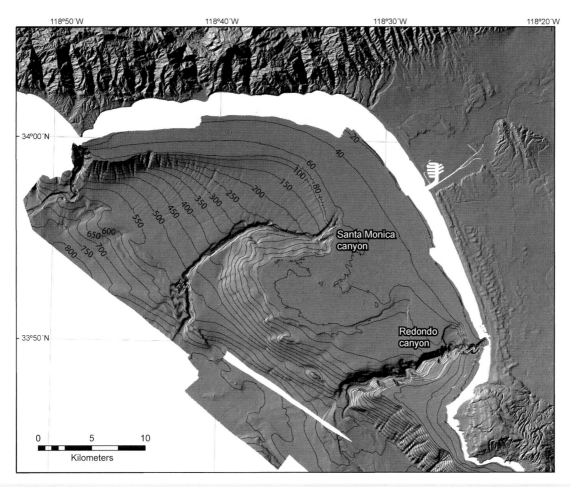

Fig. 7.14 Shaded relief and bathymetry of Santa Monica and Redondo Canyons, offshore California. Contours at a 20 m interval above 100 m depth and a 50 m interval in deeper water. From Gardner *et al.* (2003a).

of up to about 20°. These gullies typically have widths of 75–250 m, depths of 10–20 m and lengths of the order of 1 km. The physiography of modern submarine canyons shows considerable variability, and in general most canyons are cut into the upper part of basin slopes or the continental slope.

Most of the world's large submarine canyons are associated with major rivers, and are of enormous scale. For example, the Zaire Canyon heads 30 km up the Zaire River estuary and incises the continental shelf and slope generating relief of >1.2 km (Babonneau *et al.* 2002). It is as wide as 15 km at the shelf edge. The Swatch of No Ground (Bengal Fan Canyon) has a rim-to-rim width of 20 km on the middle shelf, a floor that is 8 km wide, and depth of 862 m (Curray *et al.* 2003), giving a cross-sectional area of ~11 200 000 m².

The smaller canyons, such as those along the continental shelf and slope off California and Baja California, commonly incise the shelf edge and trap sediment moving along the shelf in longshore current systems. Redondo Canyon (Fig. 7.14) typifies these smaller canyons: the canyon heads in about 15 m of water approximately 300 m from the shoreline, extending seaward for ~15 km to the apex of a submarine fan. The canyon is up to 1.6 km wide with a maximum recorded depth of 395 m, and shows a longitudinal decrease of gradient from about 15° at the head to less than 2° at the mouth. The southern wall is steeper, straighter and higher than the north wall. The position of Redondo Canyon is structurally controlled: the southern wall of the canyon is the surface expression of the Redondo Canyon

Fault. It has been suggested that the ancient 'Gardena' river (no longer extant) excavated a course along the fault trace that initiated development of the canyon (Yerkes *et al.* 1967).

Canyons along mature passive continental margins tend to develop longitudinal and cross-sectional profiles that change relatively consistently downslope, perpendicular to the slope contours. Canyons along active, convergent, destructive or strike-slip margins are more variable. Underwood and Bachman (1982) describe canyons from the Middle America Trench off Mexico, where the largest canyons trend downslope into the trench, but smaller canyons die out or are confined to 'perched basins' in the trench slope (Fig. 7.15). Thus, the smaller canyons are tectonically dammed by either the trench-slope break or, if the forearc basin is filled, by a tectonically formed ridge on the lower slope. Using data from the same area, Farre *et al.* (1983) developed a dynamic working hypothesis to explain the evolutionary phases in canyon development (Fig. 7.16). The starting point is localised slope failure and the headward growth of erosional scours and scars by a variety of mechanisms. Changing physical and/or environmental conditions may modify or halt the initial phase of potential canyon development. Once the valley head breaches the shelf-break, then slope-bypassing of shelf sediments to the deeper water environments may occur, and canyon erosion may be accelerated because of the increased frequency of sediment transport within the canyon.

The distribution of canyons along continental margins, in some cases, has been shown to be related to slope gradient. For

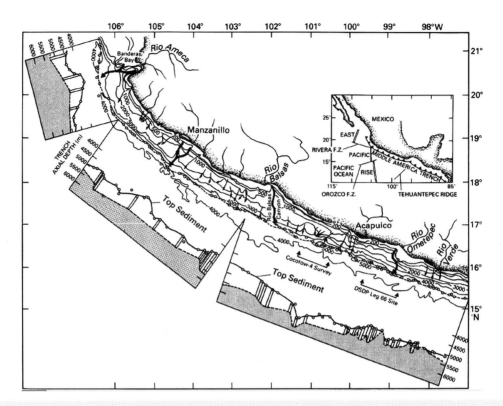

Fig. 7.15   Simplified bathymetric map of the Middle America Trench showing the location and spacing of large and small canyons. Sediment thickness was calculated using an acoustic velocity of 2.0 km/s. From Underwood and Bachman (1982).

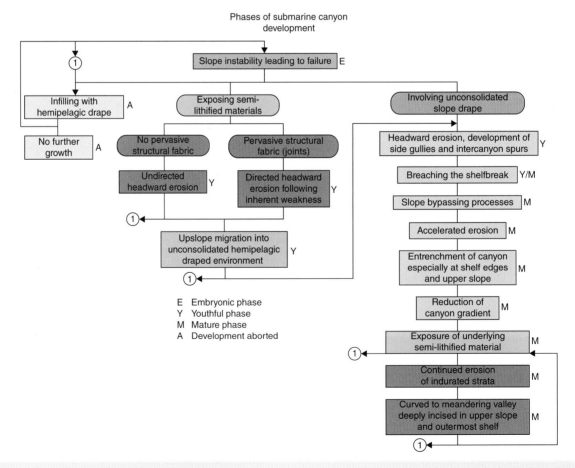

Fig. 7.16   Flow diagram used by Farre *et al.* (1983) to explain evolution of submarine canyons along the Middle America continental margin (Fig. 7.15). The number one, circled, indicates infilling with hemipelagic drape, no further growth, development aborted.

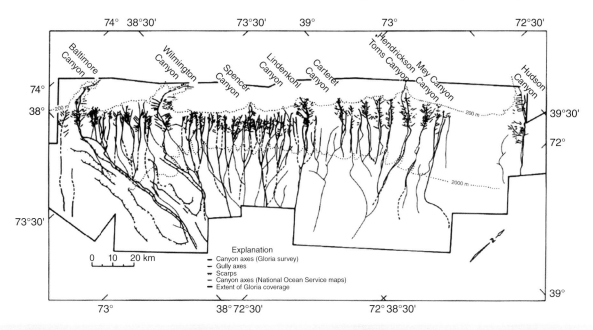

Fig. 7.17   Location of submarine canyons and gullies along the eastern US continental margin. Boxed area was studied using GLORIA side-scan sonar. From Twichell and Roberts (1982).

example, Twichell and Roberts (1982) studied the continental slope between Hudson and Baltimore Canyons (Fig. 7.17) and found that: (i) canyons are absent where the gradient is less than 3°; (ii) canyons are spaced 2–10 km apart where gradients range from 3–5° and (iii) where gradients are greater than 6°, canyons occur ~1.5 km apart.

7.3.1   Shifting locus of coarse-grained clastic input

An important factor in canyon development is the change in sedimentation patterns with time. Felix and Gorsline (1971) have shown just how important this can be in the case of Newport Canyon, California. The origin of Newport Canyon has been ascribed to lateral shifts in: (i) the position of the major clastic source, the Santa Ana River, during the Pleistocene and Holocene and (ii) the point of intersection of longshore drift currents on a narrow shelf. Sediment gravity flows, generated from the spillover of shelf sands onto the continental slope, are believed to have excavated channels that developed into the present-day Newport Canyon. Changes in the location of the river mouth have shifted the point of longshore drift convergence and sand concentration about 1 km northwest of the present canyon head, such that only fine-grained, organic-rich sediments are being deposited in the head. Sand that would have entered the canyon, prior to the relocation of the river-input point, is now accumulating as a lobate shelf deposit north of Newport Canyon, and is beginning to spill over the shelf-break, possibly to initiate the excavation of a new canyon.

Normark *et al.* (1998) and Morris and Normark (2000) describe the effects of westward switching between a set of at least three canyons upslope from Hueneme Fan, Santa Monica Basin, California (Fig. 7.18). With each switch in canyon position, the upper-fan channel (e.g., the older 'channel 2' in Fig. 7.19) shifted to the back side of the former *western* levée, effectively transforming that levée into an *eastern* levée for the new channel (e.g., the younger 'channel 3' in Fig. 7.19, and later 'channel 4'). Once activated, each successively younger channel only had to construct a western levée because

an eastern levée already existed, being inherited from the former channel position.

7.4   Architectural elements of submarine-fan systems

Deep-water deposits may be defined in terms of *architectural elements* (*cf.* fluvial models of Miall 1985), each characterised by a particular facies association and three-dimensional geometry (including orientation). The identification of depositional geometry requires the formalisation of a hierarchy of bounding surfaces, an approach now widely accepted for fluvial and eolian deposits. The concept of *elements* in turbidite systems was introduced by Mutti and Normark (1987) and is equally viable for modern and ancient settings: the elements recognised include large-scale features only, such as channels, overbank deposits and lobes, together with scours which range across several orders of magnitude in scale. These correspond to facies associations and can be placed within a hierarchy of events. Smaller scale depositional and erosional features may also be identified within turbidite systems, but are not discernable in most seismic datasets (Prather *et al.* 2000).

Architectural geometry is defined so as to be independent of scale and facies, at least on scales greater than metres. Even though the deposits themselves are three-dimensional, we have developed two parallel classifications of two-dimensional architectural geometries, in plan view and section view, because the majority of subsurface and seafloor observations are recorded in one or the other of these views. This classification was first published by Pickering *et al.* (1995b) and Clark and Pickering (1996). A strictly three-dimensional classification would have limited value, since very few studies gather data of such quality. However, without the third dimension some features cannot be unambiguously identified. For example, a large scour may have aspect ratios and a sedimentary fill identical to those of a relatively small channel, and cannot be properly interpreted without knowledge of the third dimension.

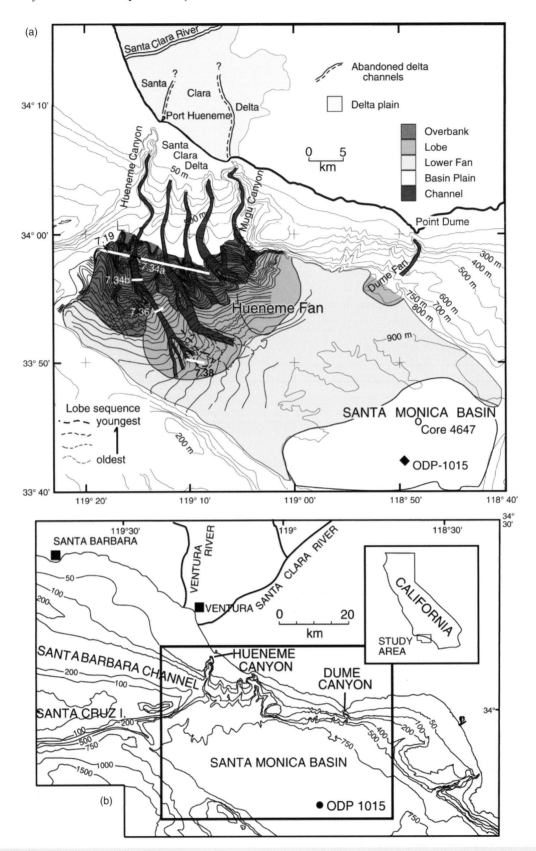

Fig. 7.18   (a) Bathymetry and facies distribution on Hueneme Fan, Santa Monica Basin, California (see map (b) for surroundings). Between 700–900 m water depth, the detailed contour interval in the coloured part of the map is 5 m. The sequence of stacked sand beds on the middle fan around the Fig. 7.38 profile line (oldest to youngest under the heading 'Lobe sequence') is interpreted from boomer profiles that have a vertical resolution of a few tens of centimetres. Fig. 7.19 is the long white strike line; Fig. 7.34a is a shorter part of this line marked in black. Subsequent figures in this chapter are labelled in (a). From Piper and Normark (2001).

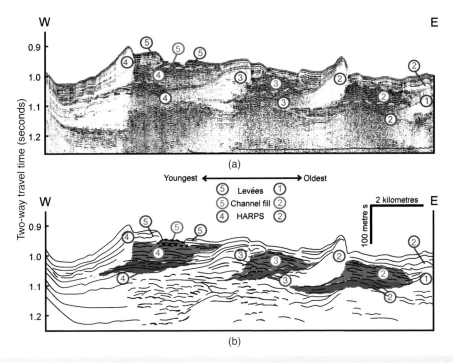

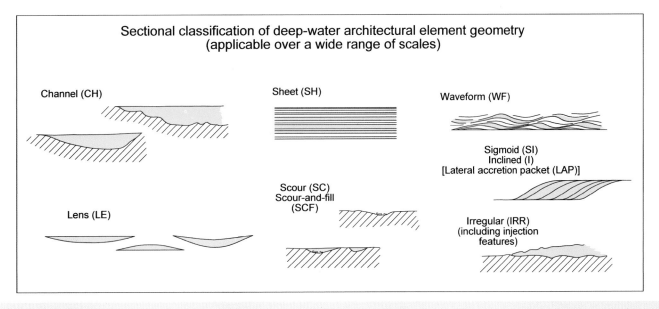

Fig. 7.19 (a) Strike seismic profile of the upper Hueneme Fan, California, obtained using a 655 cm³ sleeve gun, showing a succession of leveed channels with HARP deposits below channel-fill sands, overbank and related facies. (b) Line-drawing interpretation of Morris and Normark (2000). Channels 2, 3 and 4 appear to be largely aggradational channels that show a westward migration, whereas channel 5 is an underfit channel that has backfilled the larger channel 4 as a result of Holocene sea-level rise.

Fig. 7.20 Cross-sectional classification of architectural elements found in submarine-fan successions. From Pickering *et al.* (1995b).

Figures 7.20 and 7.21 show classifications of depositional-erosional two-dimensional architectural geometry both in section and in plan (planform), respectively. These are, in section: channel, sheet, lens, sigmoid/inclined, irregular, waveform and scour/scour-and-fill. In plan, the following elements are recognised: channel, nested channels, mound/lobes, nested mounds, irregular, bedform field

and scour/scour field. It should be noted that the 'channel' is an architectural geometry within which all the other geometries may be found (e.g., Gamberi *et al.* 2013).

Bounding surfaces can be classified by type and order, both in cross-section and in plan view. Figure 7.22 shows schematic sections of submarine-fan deposits and the bounding surface hierarchies, and

Planform classification of deep-water architectural geometry
(applicable over a wide range of scales)

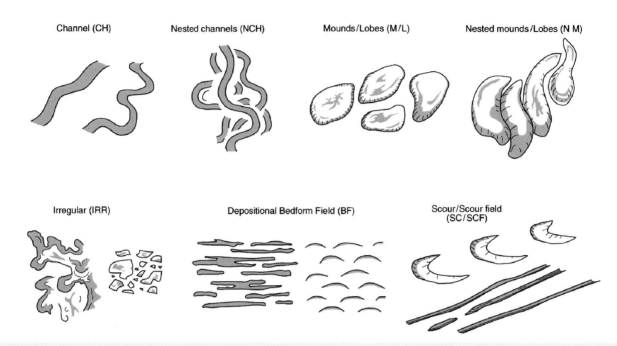

Channel (CH)   **Nested channels (NCH)**   **Mounds/Lobes (M/L)**   **Nested mounds/Lobes (N M)**

Irregular (IRR)   **Depositional Bedform Field (BF)**   **Scour/Scour field (SC/SCF)**

Fig. 7.21   Planform classification of architectural elements found in submarine-fan successions, from Pickering *et al.* (1995b).

illustrates how the classification of these surfaces can easily be applied to sectional geometries of deep-water deposits. Bounding surfaces seen in plan view are less easily defined by current classification schemes, but should nevertheless be used to help characterise architectural elements. Identification of such bounding surfaces, and bounding surface hierarchy, requires some interpretation of the element that is delineated by the bounding surface. For example, a major channel element seen on side-scan images will be defined by third-order bounding surfaces, using diagnostic criteria similar to those adopted for defining sectional bounding surfaces.

In the following sections, we describe with examples the main architectural elements of modern submarine fans. Natural systems are clearly much more complex than the simple sketches in Figures 7.20 and 7.21 would suggest, so a thorough understanding of the range of variation is important.

7.4.1   Channels and channel–levée systems

7.4.1.1   Channel dimensions and classification

Modern deep-marine channel dimensions generally range from ≤ 10 km wide and ~100 m deep (e.g., Upper Indus Fan channel, Kolla & Coumes 1987), to small distal-lobe channels <75 m wide and with depths that are irresolvable on high frequency seismic profiles (<2 m), for example, channels on the outer Mississippi Fan (Twichell *et al.* 1991). Channel relief is a function of the balance between the amount of deposition in the channel axis and on the levées. It can, therefore, be assumed that channel relief is controlled by the hydrodynamics of the channelised flows (Nelson & Kulm 1973) and the texture of the sediment load of transiting sediment gravity

flows. In many cases, submarine channels consist of a major valley with first-order levées, within which there may be secondary levées associated with smaller channel cuts (Piper *et al.* 1999; Kolla *et al.* 2007; Gamberi *et al.* 2013). The inner channel may have considerably higher sinuosity than the enclosing valley (e.g., incised slope channels described by Gee *et al.* 2007). Konsoer *et al.* (2013) compared the dimensions of 177 submarine channels to those of 231 river channels, and determined that submarine channels tend to be wider, deeper and steeper than river channels. The largest submarine channels are an order of magnitude larger than the largest river channel. For submarine and river channels of comparable size, the former are steeper by as much as two orders of magnitude.

The dimensions of selected upper-fan channels are provided in Table 7.3. Farther down the depositional pathway on the middle fan, channel depths and widths decrease and eventually SGFs become unconfined on the outer fan. Unlike rivers, the cross-sectional area of the primary fan channel(s) declines significantly downflow because of a reduction in the volume of turbidity currents through overbank spill and deposition (Pirmez & Flood 1995). Channels can be classified as aggradational, erosional and mixed erosional-depositional (Fig. 7.23). Aggradational channels are associated with high channel sinuosity and lower slopes (Clark *et al.* 1992). These channels tend to be linked to large terrestrial drainage basin areas (e.g., the Amazon, Mississippi and Indus channels; Kenyon 1992), and tend to have finer-grained sediment that ensures greater long-distance transport of sand to the distal part of the fan (i.e., greater efficiency). Such channels commonly have well-developed channel levées, formed from relatively cohesive fine-grained material. Aggradational channels grow upward because sedimentation in the channel axis occurs contemporaneously with deposition of levées, resulting in vertical

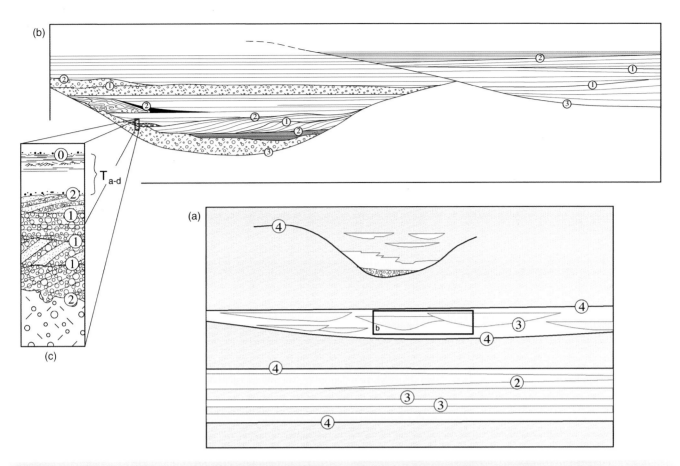

Fig. 7.22 Schematic illustrations of bounding surface hierarchy in a succession of submarine channels (a), the central part of one channel complex (b), and at the scale of individual beds and bedsets (c). The normal concordant bedding contacts between strata and laminae form zero-order bounding surfaces. First-order surfaces bound, for example, packages of cross-bedding sets or concordant bedding. Typically, these surfaces are erosional (concordant or discordant). Second-order surfaces bound units delineated by first-order surfaces to form distinct sedimentary complexes of genetically related facies, possibly with a consistent palaeocurrent trend distinct from surrounding deposits. These complexes are equivalent to the 'storeys' of Friend *et al.* (1979). Third-order surfaces are major erosional features dividing groupings of complexes (as delineated by second-order bounding surfaces). These units are commonly informally referred to in the literature as depositional bodies. Fourth-order surfaces have been added to express erosional contacts that can range up to a basin-wide scale and define, for example, groups of channels and palaeovalleys; they are equivalent to the surfaces that separate Mutti and Normark's (1987) 'stages of growth' within an individual deep-marine system (their third-order of physical scale). Mappable stratigraphic units, such as members or sub-members, are bounded by these fourth-order surfaces (Miall 1989). Fifth-order surfaces (not shown) define individual fan systems. Finally, sixth-order surfaces (not shown) delineate basin-fill sequences and supergroups. From Pickering *et al.* (1995b).

aggradation of the channel–levée complex. Because the muddy levées are difficult to erode, channel growth is mainly vertical once the levées are well established (e.g., the Indus channels). The aggradation of large channel–levée complexes eventually results in major avulsion from one part of the fan to another (Flood *et al.* 1991; Pirmez & Flood 1995; Babonneau *et al.* 2002; Curray *et al.* 2003).

Lateral migration of even the most sinuous submarine channels can be very limited (Kane *et al.* 2008) (Fig. 7.24). Kolla *et al.* (2007) demonstrate, however, that lateral channel migration and point-bar deposition often characterise the early stages of channel development when the channel deposits are more sand-prone (Fig. 7.25). Later in the history of these same channels, the channel profile tends to aggrade with little lateral shifting, and the deposits in the channel are more mud-prone. Janocko *et al.* (2013a, b) assign a greater importance to point-bar migration in deep-marine channels, although the lateral-accretion packages (LAPs) may be confined to

an erosional valley (Fig. 7.26); this limits their ability to sweep across the basin margin in the way that fluvial meanders sweep across many floodplains. Janocko *et al.* (2013b) recognise six features that are common to many submarine valley-fills:

- An erosional base marking deep incision by gravity flows.
- Basal deposits indicative of considerable sediment bypass with little or no aggradation.
- Common occurrences of mass-transport deposits in the lower part.
- An aggradational succession of levéed sinuous channel belts, multi-storey to isolated.
- Non-aggradational meandering channel belts that may occur in the basal and/or top part of a valley-fill or occasionally also in the middle part.
- An overall upward fining with a decreasing proportion of sand.

Table 7.3 Dimensions of selected modern upper-fan channels and abyssal channels arranged by decreasing maximum depth, from Carter (1988), Clark and Pickering (1996), Pirmez and Flood (1995), papers in Pickering *et al.* (1995a) and author unpublished database. Widths are at bank tops; depths at centreline

Name	Depth (m)	Width (km)
Submarine Fan Channels		
Hudson Upper Fan	65–550	1–3.5
Rhône Upper Fan	60–500	2–5
Laurentian Upper Fan	80–410	7–22
Indus Upper Fan	70–410	1–11
Wilmington Upper Fan	30–300	1–6.5
Amazon Upper Fan	70–200	2–4
Monterey Upper Fan	50–175	1–4
Astoria Upper Fan	30–165	2–3
Abyssal Channels		
Bounty Channel	150–650	5–7
Surveyor Channel	100–450	5–8
Valencia Channel	200–350	5–10
Cascadia Channel	40–320	4–7
Maury Channel	~100–300	5–15
Porcupine Channel	120–250	0.75–1.5
Northwest Atlantic Mid-Ocean Ch.	100–200	6–16

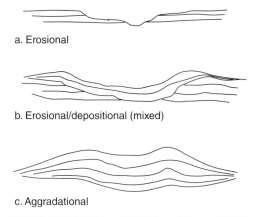

a. Erosional

b. Erosional/depositional (mixed)

c. Aggradational

Fig. 7.23 Three main channel geometries as recognised by Normark (1970). From Pickering *et al.* (1995b).

Modern aggradational channels possess the following features, seen on high-resolution seismic profiles (e.g., Hamilton 1967):

- Apparently continuous reflectors in the levées.
- The channel floor is elevated above the level of the surrounding fan and consists of flat lying high-amplitude reflectors (HARs).
- Levées are always present on both sides–in the northern hemisphere, the right-hand-side levée (looking down-channel) is commonly higher and wider.
- Reflection geometry in the northern hemisphere examples may suggest a shift in channel axis with time to the left-hand side (looking down channel)–lateral migration to this side is favoured by the lower height of the left-hand levée.

Erosional channels are associated with low sinuosity and steep slopes (Clark *et al.* 1992). They typically contain a coarse-grained fill deposited from concentrated density flows, and are sourced from small terrestrial drainage basins (Kenyon 1992). Such channels

commonly show suites of erosional architectural elements (e.g., scours, cut-downs and the deposition of residual facies). Low sinuosity channels may also show braiding patterns with associated extensive channel-bar development (e.g., Haner 1971). Because of the relative coarse-grained nature of the channel deposits, channel levées tend to be poorly developed (and have a lower preservation potential), allowing frequent channel migration, leading to a high connectivity of channel facies in the subsurface.

Mixed erosional-depositional channels are the most common channel type. They apparently form initially under depositional conditions that later are replaced by erosional conditions, promoting down-cutting through older deposits (Heezen *et al.* 1969; Normark 1970; Embley *et al.* 1970; Griggs & Kulm 1973). These channels have the following characteristics:

- Levées are present on both sides; one levée may be higher than the other (the right-hand-side levée looking down channel in the northern hemisphere).
- Reflection geometry may suggest a shift in channel axis with time to the left-hand side (looking down-channel in the northern hemisphere).
- Truncated beds may appear in both channel walls and in the levées.
- The channel floor may be higher or lower than the level of the surrounding seafloor.

Channel–levée systems have remarkably variable lengths (Table 7.2). The longest continuous fan channels are on the largest high-efficiency fans, like the Amazon, Bengal, Zaire and Indus Fans. They have much greater centreline lengths than straight-line lengths because of high sinuosities. In contrast, small low-efficiency fans like those in the California Borderland (e.g., Hueneme Fan, Normark *et al.* 1998) have upper fan channels with high levées (Fig. 7.19) that quickly taper and end in the down-fan direction, over distances of only a few tens of kilometres (Fig. 7.18). Such channel–levée segments are too short to show a sinuous planform shape, but instead may simply have one or two prominent bends in the channel before

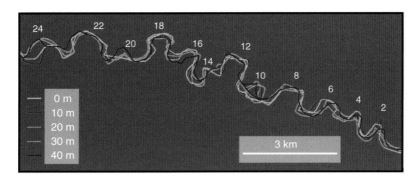

Fig. 7.24 Bend development in a subsurface example of a sinuous submarine channel. This is the uppermost section (~40 m) of a 120-m-thick (±20 m) aggradational channel–levée stack that infills the upper part of a submarine-fan valley on the West African margin. Successive positions of the channel thalweg are shown at aggradation intervals of 10 m, and were picked from amplitudes on successive horizon slices from a 3D seismic volume. Thalweg width is ~60 m, channel width is ~400 m. Flow is from right to left. Note the insignificant amount of lateral and down-fan channel migration. From Peakall *et al.* (2000b).

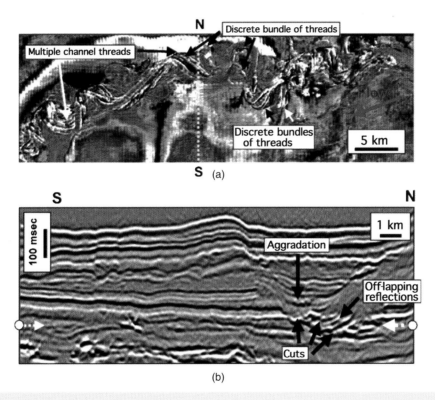

Fig. 7.25 (a) Horizon amplitude slice from the lower part of a Pleistocene deep-water sinuous channel complex, offshore Nigeria. (b) A profile across one loop (marked in part (a)) displays unconformable cuts corresponding to discrete channel shifts. White dashed arrows on the profile indicate the location of the horizon amplitude slice in part (a). Lateral accretion in the lower part of the channel fill gives way to vertical aggradation where marked. From Kolla *et al.* (2007).

the levées merge with the middle fan surface (e.g., Var Fan, Savoye *et al.* 1993).

7.4.1.2 Levée development

Many submarine-fan channels are confined by prominent levées that are initiated and then grow during the advancement of a channel mouth down the fan surface. Menard (1955) introduced the concept of 'bankfull flow' in levéed submarine channels, and

explained how the Coriolis Effect creates asymmetric levées. Coriolis deflection results in levées being higher on the right-hand side looking down-channel in the northern hemisphere (opposite in the southern hemisphere). The amount of cross-flow gradient produced by the Coriolis Effect also depends on the type of turbidity current – the greatest deflection is felt by slow, rather dilute and mud-laden flows (Normark & Piper 1991). Levées may be up to 50 km wide and 300 m above the surrounding fan surface (e.g., Amazon Fan, Damuth *et al.* 1988). Levée deposits, as described or illustrated by many authors

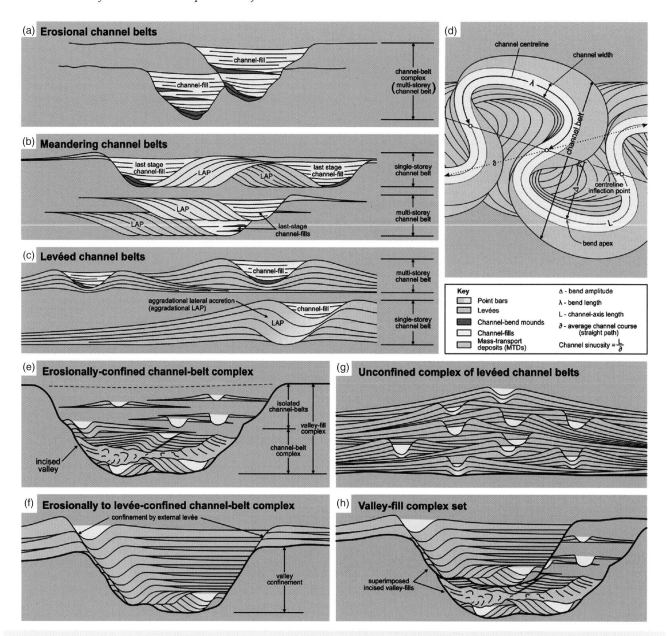

Fig. 7.26   (a–h) Schematic classification of submarine channel belts, in cross-section. Note the common presence of point-bar deposits (lateral-accretion packages) and their tendency to occur toward the base of the channel fills. Part (d) defines the geometrical parameters used to describe any sinuous channel planform. From Janocko *et al.* (2013b).

(e.g., Normark *et al.* 1980; Damuth & Flood 1985; Kolla & Coumes 1987; Piper & Deptuck 1997; Mulder & Alexander 2001), typically show rhythmic alternations of laminated silty beds up to ~3 cm thick, with some medium- to fine-grained sandy beds and lenses up to several centimetres thick. In general, the higher the levée, the more muddy its crest will be because coarser suspended load is not able to escape via overspill. As levée height decreases down-fan, silt and sand content increases because more and more of the lower part of each flow is able to overtop the levées (Fig. 7.27).

The initiation and growth of levées was explained with reference to the Amazon Fan by Flood *et al.* (1991) and was confirmed and clarified by drilling during ODP Leg 155. On the Amazon Fan, new channel–levée segments are initiated by an up-fan avulsion that

breaches the levée of the pre-existing channel (Fig. 7.10). During the late Pleistocene lowstand of marine isotopic stages 4–2, avulsions along the Amazon Channel occurred approximately every 3000 to 10 000 years (Flood & Piper 1997). After the levée break, the first deposits along the new flow pathway are sheet-like sand-rich units correspond to *High Amplitude Reflection Packets* (HARPs). These include both parallel and subtly lens-shaped internal reflections, the latter suggesting the presence of small, unlevéed channels (Pirmez *et al.* 1997). Pirmez (1994) and Pirmez and Flood (1995) present evidence for significant entrenchment of the pre-existing channel at the avulsion site, both around and upslope of the knick point produced at the site of channel bifurcation. Entrenchment along part of the pre-avulsion channel taps previously deposited channel-floor

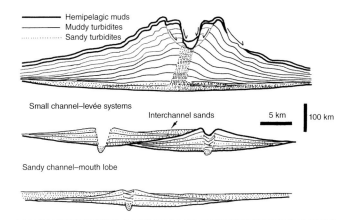

Fig. 7.27 Down-fan change from high muddy levées (top) with sands restricted to the channel floor and the HARP unit under the channel–levée wedge, to laterally shingled low levées containing sandy spillover turbidites (middle), to distal sandy channel–mouth lobes (bottom). Only the high levée is entirely aggradational, whereas the others have a central channel that is excavated below the base of the levée wedge. Note the high vertical exaggeration. Based on seismic profiles across the Indus Fan. From Kenyon *et al.* (1995a).

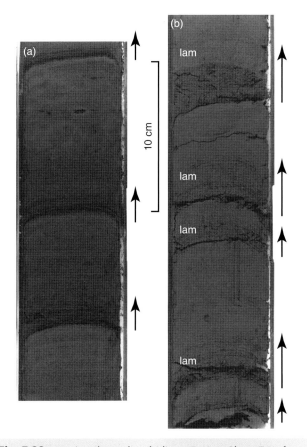

Fig. 7.28 Levée silt–mud turbidites (Facies Class D2) from Amazon Fan, ODP Leg 155. Each sharp-based graded bed is marked with an arrow. Alternating silt–mud laminae are marked 'lam'. (a) 155-935A-8H-2, 5–25 cm (Shipboard Scientific Party 1995f). (b) 155-931A-4H-3, 87–112 cm (Normark *et al.* 1997).

sand and redistributes it farther down-fan by a variety of gravity-flow processes into the new sheet-like HARP unit.

Thick bedded to very thick bedded intervals of sand, 5–25 m thick, characterise HARPs (Fig. 7.11). Mud clasts are common in the thicker sand beds, which are in some cases more than 10 m thick. These sands were probably emplaced by inflated sand flows (sandy debris flows of Hiscott & Middleton 1979 and Shanmugam 1996). The HARPs onlap an angular-erosional unconformity at the base of each channel–levée system. Most sands in the HARP units are bounded by thin-bedded, bioturbated, silty clays below, and show an abrupt decrease in bed thickness and grain size above. The sand beds are organised into either a single group of thick beds, or into a number of packets (i.e., bed clusters), each a few tens of metres thick. The packets generally have sharp basal contacts and fairly sharp tops; locally, the bases of packets are gradational. Some packets contain almost no mud interbeds, whereas others have numerous mud beds ≤ 50 cm thick. Statistical analysis (Section 5.3; Chen & Hiscott 1999a) reveals no significant bed-thickness trends in the sand packets.

The overlying overbank deposits of the levées are characterised by colour-banded muds and thin-bedded turbidites composed primarily of silt and clay. Most of the silt beds, and the less common beds of silty sand, are thinner than ~5 cm, have sharp bases and rapidly gradational tops (Fig. 7.28a). A few beds are thicker, the thickest being ~20 cm. Although lamination was not noted in the shipboard description of most of these beds (they were commonly expanded by bedding-parallel gas voids), X-radiographs show that ~70% of the beds are parallel laminated and ~5% are ripple cross-laminated. Some beds contain 1–2 mm-thick alternations of silt and mud, at the top of the turbidite (Figs 2.33b,c and 7.28b). This is a type of T_d division that is attributed by Hiscott *et al.* (1997b) to overspill from turbidity currents carrying a mixed silt and mud load, which were only just capable of overtopping the levées of the Amazon Channel. Fluctuations of flow depth in the main channel, perhaps as a result of passage of interfacial Helmholtz waves down the channel, explain the well-developed silt–mud laminae. At times when the flow-top was at a Helmholtz wave crest, overspill of silt load would be accentuated; conversely, when an interfacial-wave trough passed in the adjacent main channel, overspill of silt would cease, allowing deposition of a mud lamina. The levée turbidites on the Amazon Fan and other fans are interpreted as the deposits of hyperpycnal turbidity currents with low density but long duration (Mulder *et al.* 2001; Mulder & Alexander 2001).

Thin-bedded silt and silty sand turbidites, interbedded with unburrowed to slightly burrowed mud, constitute the base of a levée succession that is established on top of a HARP unit as the new channel becomes established and its levée 'tips' advance basinward (Fig. 7.10). These turbidites represent the overspill record from the advancing channel (Hiscott *et al.* 1997b). Average sedimentation rates during active growth phases of the Amazon Channel levées are extremely high at 1–2.5 cm/year (Shipboard Scientific Party 1995a), suggesting a recurrence interval for overspilling turbidity currents of perhaps one to a few years during the Pleistocene. The levée successions fine upward, although this trend may be modulated by fluctuations between more silt-rich and more mud-rich bedsets deposited over time scales of a few hundreds of years. Finally, each levée sequence has a 10–25 m-thick, fine-grained cap of bioturbated mud. After each avulsion event, levées apparently build rapidly down-fan from the avulsion point, much as the levées of the distributaries of river-dominated deltas advance seaward (Wright 1977). Once the levée tips reach and then pass over a site, the lower parts of the mixed-load turbidity currents become increasingly confined to the channel axis, and spillover deposits become finer grained until few silts and mostly mud characterise the crests of high levées (Shipboard Scientific Party 1995a; Hiscott *et al.* 1997c).

If the Amazon Fan levées are typical of the levées of mixed sand–mud-load channels, then the diagnostic facies are thin-bedded, fine-grained turbidites, mostly parallel laminated T_{de} beds. There is no evidence for dominance in the levées of ccc-turbidites (ccc = clasts, convolution, climbing ripples) as predicted by Walker (1985). The backslopes of the Amazon Fan levées average about 3–5°, sufficient to initiate localised slumping.

7.4.1.3   Channel gradients and planform shape

The planform geometry of modern submarine-fan channel systems spans the range from straight (sinuosity <1.1), sinuous (sinuosity =1.1–1.5), meandering (sinuosity >1.5) and braided channels. Sinuosity is defined as the distance along the channel centreline divided by the straight-line distance between two points that are more than one meander wavelength apart (Fig. 7.26d). Many submarine fan channels on high-efficiency, mud-rich fans are highly sinuous (Fig. 7.4), with evidence of breached levées on sharp bends (e.g., the Indus, Mississippi, Bengal, Zaire and Amazon Fan channels). Spillover points develop where flows breach levées on sharp bends, and where channel gradients decrease to cause an increase in flow thickness. If the spillover flow is sandy (promoting a large breach of the levée), then it may accelerate and erode a new channel (e.g., the Var Fan, Normark & Piper 1991). If the flow is muddy, however, then the spillover flow will decelerate, resulting in more sheet-like deposition of muddy deposits on the levée and across the adjacent inter-channel area (e.g., turbidity current II on the Navy Fan, Piper & Normark 1983).

The variation of channel sinuosity with increasing gradient is an important relation that has been observed in rivers and in flume-tank experiments (Schumm & Khan 1972; Schumm *et al.* 1972). These authors conclude that as slope increases, channel sinuosity will increase in such a way as to maintain an optimum centreline gradient suitable to accommodate the volume of flow and sediment load in the channel. This process will continue until a 'threshold slope' is reached, after which the channel will seek a more direct course downslope through avulsion, resulting in a rapid decrease in sinuosity. For submarine-fan channels, there is likewise a good correlation between sinuosity and channel slope (Fig. 7.29), with a best-fit line that either fits the plots for river and flume-tank data (Clark *et al.* 1992; Clark and Pickering 1996), or that is parallel to but slightly offset from those plots (Wonham *et al.* 2000). There are different 'threshold' sinuosity and gradient values in different fan systems (Fig. 7.30), suggesting a general tendency for the most sinuous systems (e.g., the Indus Fan channels) to have their 'peak' sinuosity on relatively low gradients (about 1 : 400) and the least sinuous systems (e.g., the Porcupine Seabight channel system) to reach their 'peak' sinuosity on steeper slopes (~1 : 80). For a coarser sediment load, channels tend toward a lower sinuosity 'peak' value than for mud-load (high-efficiency) systems.

Like mature rivers, the extended channels on large mud-rich submarine fans tend toward a 'graded', or equilibrium, channel profile with exponentially decreasing gradient down the channel centreline (Pirmez & Flood 1995; Pirmez *et al.* 2000; Georgiopoulis & Cartwright 2013). Exceptions occur near points of avulsion where the centreline slope is temporarily steepened. However, adjustments in channel depth around the knick point eventually return the channel centreline to a graded longitudinal profile. For example, the gradients along the centrelines of Amazon Channel and the Bengal Fan Channel both decrease from ~8 m/km near the canyon–channel

transition, to ~1 m/km near the distal end of the channel (Pirmez 1994; Curray *et al.* 2003). These channels have concave upward, graded longitudinal profiles like mature rivers, and are most sinuous where the regional slope-to-basin gradient is in the range 3–7 m/km (Pirmez *et al.* 2000). Along this part of a channel, sinuosity is directly proportional to regional gradient, with the result that tighter meandering tends to reduce the effects of steeper regional gradients and smooth fluctuations in channel gradient along the centerline. On the Amazon Fan, the highest channel sinuosity of 1.5–3.0 is found 350–700 km along the channel centreline (Pirmez *et al.* 2000).

A particularly important distinction between these long levéed channels and river channels is a pronounced down-channel decrease in cross-sectional area and levée height (Pirmez 1994; Curray *et al.* 2003). For example, the cross-sectional area of the Amazon Channel decreases by a factor of about 30 from the upper through the middle fan. Because the centreline slope shows a parallel decrease, the flow velocity and therefore absolute discharge within the channel must decrease by a larger factor, perhaps 50–100, over the same distance. This enormous decrease in discharge through the channel cross-section is attributed directly to loss of flow volume and sediment load by overbank spill across the levées and into the inter-channel depressions. The residual turbidity currents that eventually reach the lower fan are able to maintain their density and are enriched in sand (Damuth & Kumar 1975) because finer particles are concentrated in the low-density, overspilling tops of the currents and are therefore preferentially left up-fan, contributing to levée growth. Of course, turbidity currents do not simply lose volume by overspill along the levéed channel – they also gain volume by entrainment of seawater along their upper interface. Clearly, the counteracting effects of turbidity-current inflation by seawater entrainment and continual loss of suspension by overspill across gradually diminishing levées are fundamental controls on levée facies, texture of spillover turbidites and channel cross-sectional area. Hiscott *et al.* (1997b) note that a turbidity current transiting the Amazon Channel would be moving more slowly near its front than near its tail because of the gradually decreasing centreline gradient of the channel. This might allow the flow to effectively replenish its suspended load and maintain itself by becoming shorter, as its tail catches up with its decelerating head.

A challenging question for all students of submarine fans is how highly sinuous channels with prominent levées develop in the first place, given that there is little evidence for significant lateral channel migration beyond the early stages of channel development (Peakall *et al.* 2000a, b; Kolla *et al.* 2007; Fig. 7.31). A sinuous planform shape might be imposed right from the outset, when the levée tips of a new and advancing channel–levée system prograde down the fan surface. The initial imposition of sinuosity could be a consequence of a so-called 'random-walk' process that is commonly used to explain other sinuous geomorphic features (Leopold & Langbein 1962). Alternatively (and speculatively), the initial sinuosity might reflect the fact that the first turbidity currents exiting through a levée breach after an avulsion event, and arriving in the elongate depression between older levées of former channels, might be directed obliquely across the regional gradient, causing those gravity flows to take a series of side-to-side swings in their path down the fan surface. If a number of flows take the same pathway, a preferred route (perhaps a shallow erosional depression) might be established that would 'lead' the levée tips down the fan as they rapidly extend basinward. On the Amazon Fan each channel–levée segment (formed between avulsion events) is ~200 km long along its centreline (Pirmez & Flood 1995), and formed in perhaps 5000 years, so the rate of levée-tip advance is very

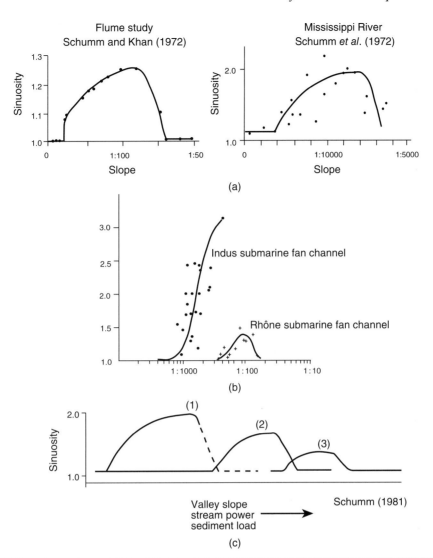

Fig. 7.29 (a) Relation between sinuosity and slope for flume study and Mississippi River. (b) Selected sets of submarine channel data, Indus and Rhône channels, showing relatively good data fit to similar trends found for fluvial data. (c) Effect of valley slope, stream power, or sediment load on river pattern thresholds, for (1) suspended-load channels, (2) mixed-load channels and (3) bed-load channels. From Clark *et al.* (1992).

fast, averaging ~40 m/yr. This implies that single flows must have contributed many metres to the extension of the levée tips.

Some of the sinuosity displayed by mature deep-sea channels can be explained by a phase of early, relatively active lateral channel migration that occurs before channel growth becomes mainly aggradational (Kolla *et al.* 2007; Fig. 7.25). Once aggradation takes over, the pre-existing sinuous planform persists as a relict feature and only minor changes in sinuosity occur between the cross-over points that separate successive bends; the locations of the cross-overs stay essentially fixed (Kolla *et al.* 2001). These minor changes result in an increase in the so-called 'swing' of the meanders (Fig. 7.31). However, even minor increases in sinuosity eventually cease when the levées aggrade well above the surrounding seabed. Instead of shifting across a floodplain like meandering rivers, the sinuous channels of submarine fans reach a mature, static planform shape and any further shifting takes place by wholesale avulsion of the distributary channel to a new location on the fan surface.

The enormous lengths of deep-marine sinuous channels are potentially a consequence of the sinuosity itself. Straub *et al.* (2011) propose that secondary circulation cells at channel bends, and the reversal in flow rotation from one channel bend to the next, vertically mix the suspended load of each transiting turbidity current so that settling and deposition are limited, and the current is able to maintain is density and forward momentum for hundreds to even thousands of kilometres.

7.4.1.4 Channel-floor facies

Facies deposited within modern submarine-fan channels are poorly known because of technical problems taking cores in sandy materials. Besides cores, an alternative strategy is to use ultra-high resolution deeply towed boomer records (e.g., HUNTEC Deep Tow System) with vertical resolution of better than 50 cm to constrain facies characteristics. Below, we select three examples of channel deposits

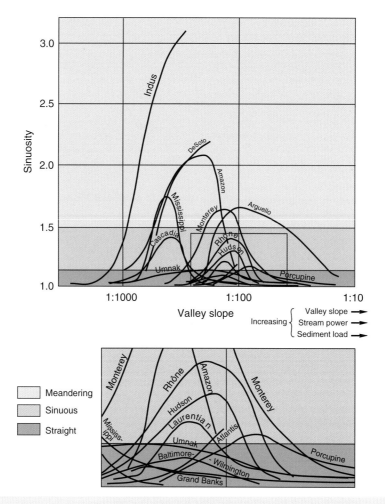

Fig. 7.30   Sinuosity and valley slope relations for 15 submarine fan channels showing trends similar to those of flume and fluvial data shown in Figure 7.29. Lower panel is an enlargement of the box in the upper panel. From Clark *et al.* (1992).

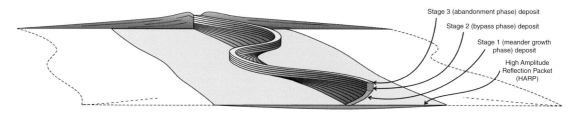

Fig. 7.31   3D block model of bend evolution and sedimentary architecture in a typical aggradational, high-sinuosity submarine channel (not to scale). Through time, the 'swing' of the meanders increases, but the cross-overs between successive bends stay more-or-less fixed. From Peakall *et al.* (2000b).

from modern fans that have been reasonably sampled, photographed, or imaged: the Amazon Channel (ODP Sites 934, 943 & 945, Leg 155); Laurentian Fan Channel, eastern Canada; and the channels of Hueneme Fan, California Borderland.

Three channel axes were drilled on ODP Leg 155, Amazon Fan. Locations are shown in Figure 7.8. Middle fan Site 934 sampled an abandoned meander where the sediments at core depths of ~65–97 m are attributed to deposition on the floor of an active channel. These deposits are overlain by a slump that presumably contributed to abandonment of this channel segment. Site 943 is located in the axis

of the Amazon Channel and recovered <50% of the sandy channel fill. Site 945 is located considerably farther basinward, and only the upper ~10 m at that locality is interpreted as channel fill – below that level the deposits are part of a broad, unconfined lobe that was incised by the channel when it switched to this part of the fan. The facies recovered by hydraulic piston coring are similar at all three sites, and mainly consist of medium to very thick beds of fine to coarse sand, local granules and pebbles, and coarse silt (Fig. 7.32). The coarser deposits are probably somewhat disturbed by coring. The thickest structureless sand units contain 60–70% scattered mud clasts as large

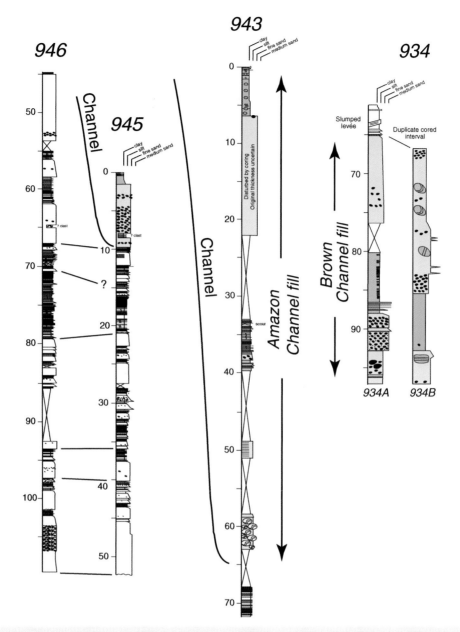

Fig. 7.32 Cored channel-fill deposits from ODP Sites 934 (water depth 3422 m), 943 (water depth 3739 m) and 945 (water depth 4136 m) on the Amazon Fan. Site 946 is ~1.5 km east of Site 945 and the sheet-like deposits of the HARP succession beneath these sites is easily correlated between the drill sites. HARP and levée deposits are not coloured. In the channel deposits, yellow denotes sand (with local large mud clasts); green denotes turbidite muds, and grey denotes highstand (interglacial), burrow-mottled, biocalcareous muds. Each interval with no recovery is marked with a large X.

as ~35 cm in diameter (Fig. 7.33) that presumably were derived from undercutting of levées along the channel. According to Shipboard Scientific Party (1995b, c), these thick sands probably each represent the deposit of one or more sandy debris flow or liquefied flow (inflated sand flow in the classification of this book), whereas the other sand and silt beds are interpreted as the deposits of turbidity currents and concentrated density flows.

Valleys of the Laurentian Fan contain a remarkable suite of coarse-grained deposits. Using side-scan sonar imagery and observations from submersible dives, Piper *et al.* (1985, 1988), Hughes Clarke *et al.* (1990) and Shor *et al.* (1990) have described gravel waves

in water depths of 1600–4700 m. The waveforms tend to occur in lanes of alternating larger and smaller waves (Fig. 1.35). At ~3500 m water depth, wavelength is ~70 m. Waveform heights are ~5–10 m. Downslope faces of the asymmetric waveforms dip at 25–45°. The sinuous crestlines are transverse to flow direction. However, there is no discernable internal stratification. Instead, natural erosional cuts through a number of bedforms indicate clast-supported gravel up to 3 m thick, with mud clasts, which grades upward into sand. Boulders and cobbles are present on the stoss side of the asymmetric bedforms. Sand ribbons overlie the coarse-grained waves in shallow elongate depressions between the lanes. At the distal end of the fan valleys, the

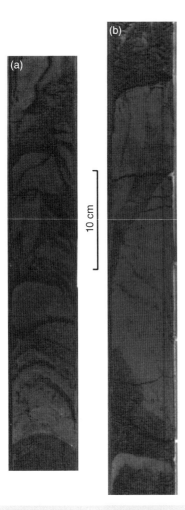

Fig. 7.33 Large mud clasts in medium-grained sand matrix in the channel deposits at ODP Sites 945 and 934. (a) 155-945A-1H-4, 42–80 cm (Shipboard Scientific Party 1995c). (b) 155-934A-11H-4, 47–90 cm (Shipboard Scientific Party 1995b).

Table 7.4 Acoustic facies recognised on the Hueneme Fan by Piper *et al.* (1999)

Facies	Brief description of reflections	Interpretation
I	Parallel, low amplitude	Mud, a few silt beds or laminae
II	Parallel, moderate amplitude	Mud, <20% sand or silt beds
III	Parallel, closely spaced, high amplitude	Mud, >30% sand or silt beds
IV	Discontinuous, subparallel intervals alternating with incoherent intervals	Mud with silt laminae with a few thicker sand beds
V	Discontinuous, irregular, high amplitude	Mud with discontinuous sand beds
VI	Mostly incoherent, high backscatter; some continuous high amplitude reflections	Thick sand units between muddier intervals
VII	Incoherent, high backscatter	Channel-filling sand
VIII	Incoherent, low backscatter	Muddy debrites

gravel waves are progressively blanketed by sand. Piper *et al.* (1988) initially interpreted the gravel waves as the first deposits of an exceptional turbidity current generated by the 1929 Grand Banks earthquake (magnitude 7.2). However, Wynn *et al.* (2002b) withdrew this interpretation and instead implicate a late Wisconsinan glacial outburst flood (jökulhlaup). The sand ribbons and sand sheet, however, are still attributed to the 1929 event, and are interpreted as detritus winnowed from the top of the older gravel by the 1929 sediment gravity flow. This implies that the moulding of the gravel surface might have occurred in 1929, but an older origin cannot be ruled out. Although the features in the Laurentian Fan valley are only known to depths of a few metres, it is clear that a range of graded to disorganised gravels and medium to coarse sands contribute to the fill of these upper fan channels.

Normark *et al.* (1998), Piper *et al.* (1999), Piper and Normark (2001) and Morris and Normark (2000), used ultra high-resolution boomer records to evaluate outcrop-scale facies distribution on Hueneme Fan, Santa Monica Basin, California (Fig. 7.18). No cores are available to ground-truth the interpretation of acoustic facies in the Hueneme inner-fan channel. However, experience gained from the interpretation of sandy and thin-bedded silt/mud successions elsewhere permits a reliable assessment of the texture of the channel deposits. Sandy deposits are characterised by high backscatter and little internal reflectivity, whereas laminated/bedded finer-grained deposits permit deep acoustic penetration and are characterised by continuous, weak to moderate reflections. Table 7.4 indicates the characteristics of eight acoustic facies in the Hueneme Fan. Table 7.5 partitions these acoustic facies among the architectural elements of the fan (overbank element, channel-fill element, lobe element, lower-fan element and basin-plain element). Table 7.6 relates the elements and sub-elements, and the acoustic facies, to the facies classes of Chapter 2.

Channels on the Hueneme Fan are characterised by high acoustic backscatter. As a consequence of partial backfilling of the main inner-fan channel during Holocene sea-level rise, there is a small 'underfit' thalweg channel within the main channel flanked by low secondary levées (or terraces) that contain amorphous lenses (Figs 7.19, 7.34). The channel-floor deposits have the acoustic properties of amalgamated sands/gravels, whereas the inner secondary levées are mixed silts, muds and sand lenses (Fig. 7.35). At the lower end of the fan valley, at the transition to the middle fan, the channel floor is interpreted to be underlain by amalgamated sands of Facies Class B, and the adjacent low levées are interpreted to consist of thickly bedded sandy turbidites (± concentrated density-flow deposits) of Facies C2.1 which pass laterally into thinly bedded sand-mud couplets of Facies C2.3 (Fig. 7.36).

7.4.2   Waveforms (sediment waves)

7.4.2.1   Muddy waveforms in overbank areas

Large asymmetric sediment waves are widely distributed on the levées of modern submarine fans (Normark *et al.* 2002). These are formed of fine-grained turbidites identical to those outside the sediment-wave fields, except that each turbidite is somewhat thicker on the upflow

Table 7.5 Classification and characteristics of Santa Monica Basin fan and slope elements and sub-elements, from Piper *et al.* (1999)

Element/Sub-element	Acoustic Facies	Shape	Characteristic Dimensions	
			horizontal (km)	vertical (m)
Overbank element				
Well stratified levée sub-element	**I, II**	Very large lenses	1–6 wide, 5–9 long	30–100
Confined-levée sub-element	**I, IV**	Small lenses	0.5–2 wide, 5–10 long	5–30
Sediment wave sub-element	**II, III**	Large waveforms	2–4	1–25
Sandy overbank element	**IV, VI**	Lenses	1–2	3–15
Channel fill element				
Upper fan valley-fill sub-element	**VII**	Very large channel	1–4 wide, 7–11 long	20–70
Middle-fan channel-fill sub-element	**VII, I**	Large channel	0.1–0.5 wide, <15 long	3–20
Levée channel-fill sub-element	**VII, VI**	Small lenses	0.05–0.1 wide, ~2 long	2–5
Slope-channel-fill sub-element	**VII,VIII,I,II**	Large channel	0.5–1.5 wide, 5–9 long	2–10
Lobe element				
Channel-termination lobe sub-element	**VII**	Lensoid, wedging	1–?5	≥15
Low-gradient lobe sub-element	**VI**	Tabular, gently lensing	1–7	2–15
High-gradient lobe sub-element (*Dume Fan only*)	**VI**	Wedges	<5	2–10
Scoured lobe sub-element	**V**	Extensive drape, irregular bedding	<2	<5
Lower fan element	**III**	Parallel, subparallel beds	10–20	>100
Gently lensing sub-element	**III**	Large subtle lenses	1–2	<2
Shallow channel-fill sub-element	**I, VII**	Broad, very shallow channel	0.1–0.7	~1
Mounded unstratified sub-element	**VIII**	Erosional base, mounded top	0.5–1.5	<5
Basin-plain element	**III**	Parallel beds	20–30	>100
Basin-slope element				
Slope drape sub-element	**I**	Extensive drape	—	—
Delta slope sub-element	**I, II**	Localised drape	—	—
Blocky sub-element	**VIII**	Irregular blocky masses	0.5–1.5	<10
Mounded unstratified sub-element	**VIII**	Erosional base, mounded top	0.5–1.5	2–10

Table 7.6 Interpreted correspondence between acoustic facies of Piper *et al.* (1999) and the *facies groups* and *facies classes* used in this book

Facies group or class		Acoustic facies	Sub-elements
B	Sands (>80% sand grade) (may include Class A, gravels and pebbly sands)	**VII**	Upper fan valley-fill Middle-fan channel-fill Slope channel-fill Levée channel-fill Channel-termination lobe Lower fan shallow channel-fill
C2	Organised sand–mud couplets		
	C2.1 Very thick- and thick-bedded	**VI**	Low- and high-gradient lobe Sandy overbank Confined levée
	C2.2 Medium-bedded	**III, IV, V**	Scoured lobe Lower fan element Basin-plain element
	C2.3 Thin-bedded	**II, III**	Basin-plain element (locally) Well-stratified levée (in part) Confined levée Sediment-wave (in part)
D2	Organised silts and muds	**I, II, IV**	Overbank element Middle-fan channel fill Basin-slope (lower part)
E	Muds	**I**	Basin-slope (lower part)
F2	Contorted/disturbed mud	**VIII**	Slope channel-fill Mounded unstratified (lower fan)
G2	Biogenic muds	**I**	Slope-drape

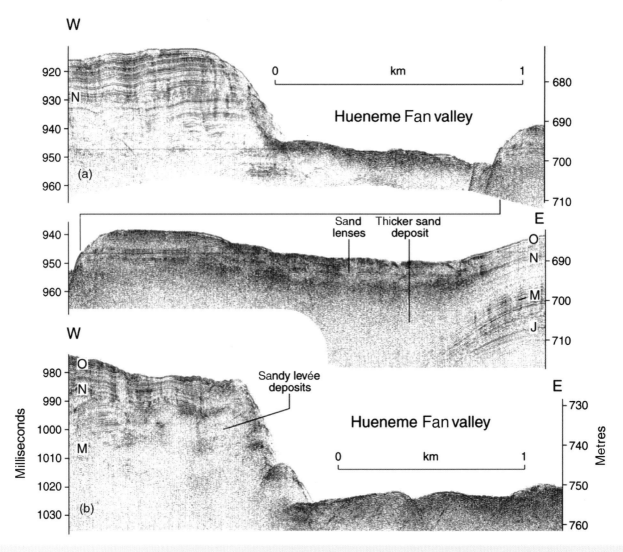

Fig. 7.34 Two boomer profiles across Hueneme fan valley, located in Fig. 7.18a. (a) A pair of contiguous sections through the inner right-hand levée, the inner channel (upper panel), and the low left-hand inner levée. (b) Section showing the inner channel and inner right-hand levée. From Piper *et al.* (1999).

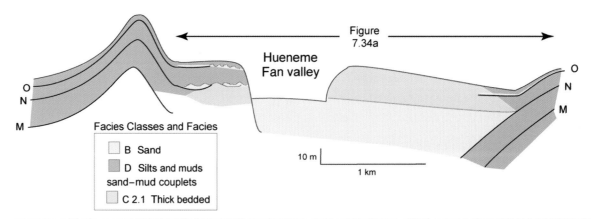

Fig. 7.35 Facies interpretation of boomer profile shown in Fig. 7.34a. The context of this valley is shown in Figs 7.18a and 7.19. From Piper *et al.* (1999).

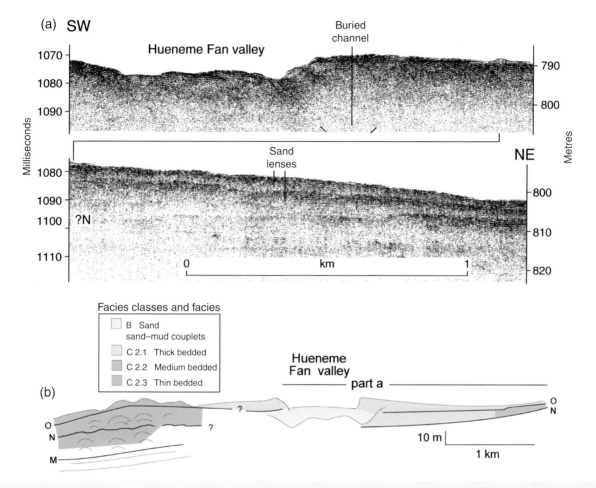

Fig. 7.36 Cross-section of Hueneme Fan valley in the transition between the upper and middle fan, located on Fig. 7.18a. (a) Boomer profile split into two parts to permit greater enlargement. (b) Facies interpretation of boomer profile shown in part (a). From Piper *et al.* (1999).

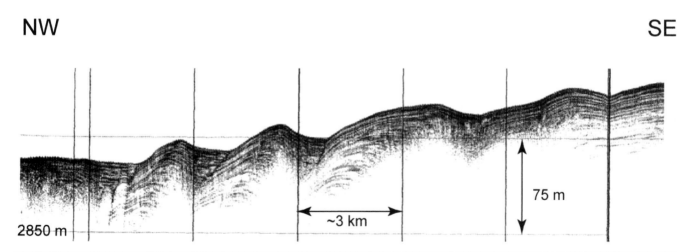

Fig. 7.37 Sediment waves with a height of about 20 m and an apparent wavelength of 3–4 km migrating up the backside of a levée of the Hikurangi Channel at one of its tight right-hand bends. The channel, to the southeast of this image, is ~5 km wide. From Lewis and Pantin (2002).

flank of the sediment wave than on the downflow flank. This leads to climbing of the waves toward the channel responsible for the overspill (Fig. 7.37). It is critical to understand that differential deposition on the upflow side, and not waveform-scale bedload transport of sediment, accounts for the development and maintenance of these sediment waves (Migeon *et al.* 2001).

The sediment waves that characterise levées have wavelengths of ~0.5–5 km and heights of ~3–20 m, although larger exceptions occur (Normark *et al.* 2002; Wynn & Stow 2002). Wave fields occur primarily on bottom slopes in a narrow range from ~0.1–0.7°. Waveform dimensions tend to increase upslope toward the levée crest. The crestlines of the waves are either parallel to channel margins or moderately oblique, depending on the direction of overspill from the adjacent channel (McHugh & Ryan 2000). The waves are interpreted to be the product of deposition under dilute turbidity currents many tens to hundreds of metres thick. Because this process is not unique to submarine fans, similar sediment waves occur in a variety of submarine slope environments where they have in some cases been confused with folded and faulted sediment slides (Lee *et al.* 2002).

Two explanations have been offered for the formation of sediment waves on muddy submarine-fan levées. They are either antidunes generated under slow-moving but very dilute and very thick overspilling flows (Normark *et al.* 1980) or they are deposited under 'lee waves' formed behind an obstacle (Flood 1978, 1988). In this case the obstacle is the levée crest (initially) and subsequently one or more of the early-formed waveforms. Flood (1988) proposes that a stratified flow (for example a density current beneath ambient fluid) is perturbed by the bottom topography leading to a change in the velocity structure of the flow. The turbidity current then decelerates on the upstream flank of the waveform and accelerates on the downstream flank. This results in different accumulation rates from one flank to the other, and enhancement of the waviness on the seabed (Fig. 6.13).

The dimensions of the sediment waves can be used to provide an estimate of depth of the flow beneath which they formed. According to Lewis and Pantin (2002), the wavelength (L) of sediment waves is related to flow thickness (H) by the relationship $L \approx 2\pi H$. Hence sediment waves spaced 0.5–5 km apart are inferred to form under flows ~80–800 m thick.

Many authors point out that sediment waves, once initiated and sufficiently large, may persist long after the conditions for their formation have ceased to exist, simply by the draping of an inherited topography by subsequent deposits.

Utilising the Autonomous Underwater Vehicle technology developed by the Monterey Bay Aquarium Research Institute, Covault *et al.* (2014) studied the San Mateo canyon-channel system, offshore southern California, and interpreted a series of within-channel crescent-shaped bedforms in its thalweg. Numerical modelling, seafloor observations and shallow subsurface stratigraphic imagery suggest that these bedforms are likely to be *cyclic steps*, that is a regular series of long-wave, upstream-migrating bedforms. They interpret each bedform crest to lie between a series of hydraulic jumps in an overriding turbidity current, which is Froude-supercritical over the lee side of the bedform and Froude-subcritical over the stoss side. Numerical modelling and seismic-reflection imagery support an interpretation of weakly asymmetrical to near symmetrical aggradation of predominantly fine-grained, net-depositional cyclic steps. Covault *et al.* (2014) suggest that the interaction between turbidity-current processes and seafloor perturbations appears to be fundamentally important to channel initiation, particularly in high-gradient systems.

Table 7.7 Characteristics of coarse-grained sediment waves in channels and channel-lobe transition zones, from Wynn *et al.* (2002b)

Location	Wave height (m)	Wavelength (m)	Sediment texture
Channels			
El Julan, Canary Islands	6	400–1200	Coarse?
Icod, Canary Islands	?	600–1500	Coarse?
Stromboli Canyon	2–4	20–200	Gravel/sand
Monterey Fan	?	100	Coarse?
Valencia Channel	?	80	Coarse?
Laurentian Fan	1–10	30–100	Gravel/coarse sand
Var Fan	1.5–5	35–100	Gravel
Corinth Graben (ancient)	2–6	50–100	Gravel
Lago Sofia, Chile (ancient)	4	?	Gravel
Channel-lobe transition zones			
Laurentian Fan	4	300	Coarse sand?
Agadir Canyon	?	?	Coarse?
Lisbon Canyon	?	500–2000	Coarse?
Valencia Fan	?	70	Coarse?

7.4.2.2 Sandy/gravelly waveforms in channels and channel-lobe transitions

Wynn *et al.* (2002b) describe coarse-grained sand and gravel waves from submarine fan channels and the areas of transition between channels and lobes. The dimensions of these waveforms are generally smaller than those of muddy overbank sediment waves (Table 7.7). They are typically in the range of a few metres high and tens to hundreds of metres in wavelength (Fig. 1.35). Wave crests are always aligned roughly perpendicular to the dominant flow direction. These waveforms are often developed on erosional surfaces and, when seen in cross-section in ancient examples, typically consist of a thick massive gravel/coarse sand unit capped by a finer-grained, normally-graded unit (*cf.* Ito & Saito 2006). Wynn *et al.* (2002b) interpret the coarse-grained sediment waves as the deposits of concentrated density flows that either acquired a wavy top during deposition, or become wavy after reworking by subsequent turbidity currents.

7.4.3 Lobes

We use the word *lobe* to describe a tongue-shaped deposit of mostly deep-water sands which accumulates beyond the termination of a submarine-fan (or submarine-ramp) channel. Deposition is promoted because of flow expansion as turbidity currents and concentrated density flows exit the channel. Where unconstrained by topography, lobes are broad lenses in cross-sectional view. They include the suprafan lobes of Normark (1970) and the depositional lobes of Mutti and Ricci Lucchi (1972), but with important scale differences that we outline below. We do not support the assertion of Shanmugam and Moiola (1991), re-stated by Shanmugam (2000), that a sediment body cannot be called a lobe unless it shows an upward thickening pattern in its sand beds. Terminology evolves. It is clear that neither Emiliano Mutti nor Franco Ricci Lucchi intended that asymmetric trends be a part of the definition of a 'lobe' – otherwise Ricci Lucchi (1975b) would not have been content to announce that 37% of unchannelised cycles (i.e., lobes) in the Marnoso arenacea do *not* show upward thickening. If Ricci Lucchi believed that a lobe *must* consist only of beds that thicken upward, then such a conclusion would have been nonsense. The early belief that upward thickening is a common characteristic of these deposits had not been tested in 1972, and since then has been seriously challenged (Anderton 1995; Chen & Hiscott 1999a), if not abandoned, by many researchers (Section 5.3).

On small-radius low-efficiency fans, lobes are assigned to the middle fan (Normark 1978). On large high-efficiency fans like the Amazon Fan, lobes are less evident, although parts of the HARP sandy successions that underlie each channel–levée system may consist of lobate sandy sediment bodies of much smaller scale than the other architectural elements of those fans. It is our preference to describe these HARP units under 'sheets' (below).

For land geologists reading this book, it is critical to distinguish between features from modern fans that have been called 'suprafan lobes', and the individual 5–20-m-thick sandstone packets so common in outcrop successions. This is because of a significant mismatch in scale. Modern suprafan lobes are of the order of 100–400 m thick (Emmel & Curray 1981; Garrison *et al.* 1982; Bouma *et al.* 1984; Nelson *et al.* 1984), so they are believed instead to correspond to a multistorey body formed of a number of sandstone packets and intervening finer grained deposits.

The lobes on modern fans consist of channelised and non-channelised parts, for example, those documented from offshore western Corsica and Sardinia by Kenyon *et al.* (2002b). The channels are distributaries of the upper-fan channel, so they are multiple in number, although only one of a set of middle-fan channels is likely the principal sediment conduit at any one time. In contrast to the upper-fan area, the middle-fan channels lack muddy levées. The channels may be only a few metres deep and a few hundreds of metres across. On Navy Fan, offshore Baja California, the middle-fan channels are partly erosional, as shown by abrupt steps and terraces with relief of at most a few metres that cross the channel floors. These are interpreted by Normark *et al.* (1979) to result from differential erosion of sandy and muddy beds in the walls of the channels, but some of the terraces might be depositional in origin, like the terraces described from outcrop studies by Hein and Walker (1982). In spite of evidence for some erosion into older deposits, the smooth surfaces of abandoned lobes on the Navy Fan indicate that channels do not gradually prograde across their depositional lobes (Normark *et al.* 1979). Instead, lobes appear to be aggradational features that are periodically abandoned when deposition shifts to a nearby depression on the middle-fan surface. Normark *et al.* (1998) also concluded that there is a predominance of vertical aggradation, rather than progradation, on Hueneme Fan, Santa Monica Basin.

The transition zone between mid-fan channels (partly confined flows) and the more distal smooth sediment lobes (unconfined flows) is characterised in many cases by coarse-grained sediment waves (Table 7.7). These have been attributed by Mutti and Normark (1987) to increased turbulence that results from flow expansion during a hydraulic jump in the vicinity of the channel exit. In modern systems, the wavelengths of such waves are tens to hundreds of metres.

Over broad areas of modern middle-fan lobes, beyond channel terminations, the depositional surface is smooth at the scale of side-scan sonar imagery. In the shallow subsurface, ultra high-resolution boomer profiles show medium to thick beds (Fig. 7.38), probably sand, with dimensions of 3–7 km (Piper *et al.* 1999). These beds tend to pinch out down-fan, and shingle laterally against one another so that any depressions created by deposition of older beds are filled in and smoothed by deposition of younger beds. This is called a *compensation* effect (Mutti & Sonnino 1981), because the younger beds compensate for the irregularities created by previous

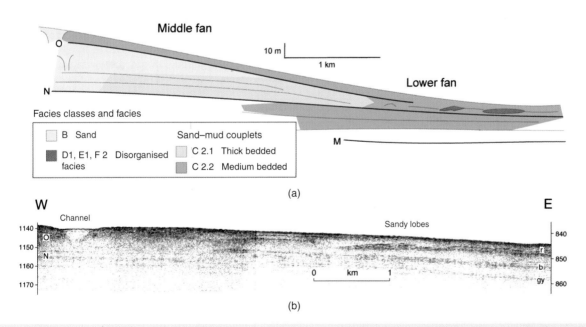

Fig. 7.38 (a) Cartoon showing interpreted facies distribution across the non-channelised parts of the sandy lobes of Hueneme Fan, Santa Monica Basin. (b) Boomer profile, located in Fig. 7.18, showing inferred sand beds (lacking coherent internal reflections) separated by reflections r (red), b (brown) and gy (grey). This profile is entirely within the middle fan and is oblique to the down-fan (dip) orientation of part (a). From Piper *et al.* (1999).

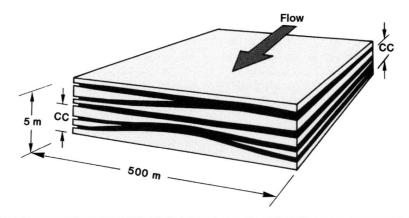

Fig. 7.39 Block diagram illustrating the origin of compensation cycles (CC) by lateral shifts in the thickest parts of successive turbidites, resulting in a smoothing of bottom topography and in formation of upward thickening cycles according to Mutti and Sonnino (1981).

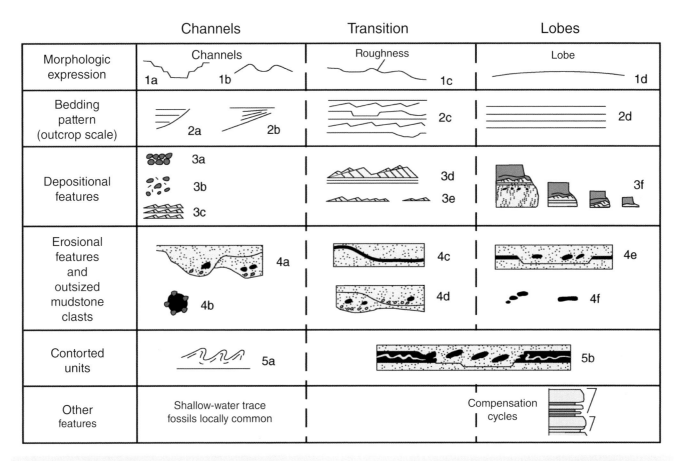

Fig. 7.40 The essential field characteristics of channel, channel-lobe transition and lobe deposits according to Mutti and Normark (1987). 1a = erosional channel; 1b = depositional channel; 1c = zone of roughness; 1d = lobe relief; 2a = beds truncate against channel margin; 2b = beds converge toward channel edge; 2c = bedding irregularity resulting from scours and large-scale bedforms; 2d = parallel bedding planes; 3a = clast-supported lag conglomerate; 3b = mud-supported conglomerate (debrite); 3c = thin-bedded overbank deposits; 3d & 3e = coarse-grained, internally stratified sandstones; 3f = complete and base-missing Bouma sequences; 4a = deep and relatively narrow scours associated with bedrock clasts; 4b = armoured mud clasts; 4c = mud-draped scours; 4d = broad scours locally associated with mud clasts; 4e = tabular scours invariably associated with mud rip-up clasts from underlying deposits; 4f = 'nests' of mud clasts commonly showing inverse grading and evidence of being plucked from the substrate; 5a = slump units; 5b = scours and deformation produced by clast impacts.

deposition (Fig. 7.39). Toward the edge of the lobe, the inferred sand beds taper to a thickness less than ~10 cm and then persist out onto what is called the lobe fringe, near the upslope limit of the lower fan.

The smooth surface of the middle-fan lobes and the gradual tapering of the sand beds are both attributed to progressive deposition from unconfined turbidity currents and concentrated density flows as they spread out, thin, and decelerate after exiting updip channels. A current passing from point *x* to point *y* will undergo a spatial deceleration (Section 1.4.6), whereas an observer stationed at point *x* will eventually see the current slow as first the head and then the body passes the observation site (temporal deceleration). It is this temporal deceleration or a combination of temporal deceleration and lateral size grading in the current (Fig. 1.31) that explains graded bedding in the deposits of middle-fan lobes, whereas it is mainly the spatial deceleration and resultant loss of capacity that drives the deposition (Hiscott 1994a).

7.4.3.1   Lobe facies

Although some inferences can be made from acoustic properties (Table 7.6 and Fig. 7.38), sampling and precise knowledge of the facies beneath modern lobes is meagre because of the resistance of thick sands to gravity and piston coring. We expect considerable similarity to the alternating sand-prone and silt-prone packets of the HARPs recovered during coring of Amazon Fan (Section 7.4.4), but this hypothesis needs testing.

Mutti and Normark (1987) proposed field criteria that can be used to distinguish lobe deposits from channel deposits (Fig. 7.40). These criteria assign particular importance to the greater amount of erosion and bed amalgamation in channel deposits. Erosion produces irregular bed boundaries, bed amalgamation and large intraformational mud clasts. Mutti and Normark (1987) also indicated how the geometry of beds and the proportions of facies likely change across the non-channelised outer parts of depositional lobes (Fig. 7.41). We recommend these field-based criteria as much better tools for the distinction of channel deposits from lobe deposits than subjective attempts to recognise upward-thinning (fining) and upward-thickening (coarsening) cycles. To reiterate from Section 5.3, the number of asymmetric cyles that can be statistically demonstrated in a wide survey of clearly non-channelised turbidite successions, including the modern Amazon Fan, is no greater than the number of such cycles that result from a random draw (e.g., from the series of numbers in a telephone book; Chen & Hiscott 1999a).

Chen and Hiscott (1999b) proposed a new criterion for the differentiation of channel deposits from the deposits of unconfined flows (lobes or sheets). They used the Hurst statistic, *K*, to quantify the degree of facies clustering in submarine-fan and associated deposits. This method is also called *rescaled range analysis* (Feder 1988). Deposits with a high degree of clustering, and a high value of *K*, have groups of thick (or coarse) sand and gravel beds, commonly amalgamated, separated by groups of thin (or fine-grained) beds. We describe such deposits as being highly 'packeted'. In mathematical terms, such deposits possess *serial dependence*, expressed as

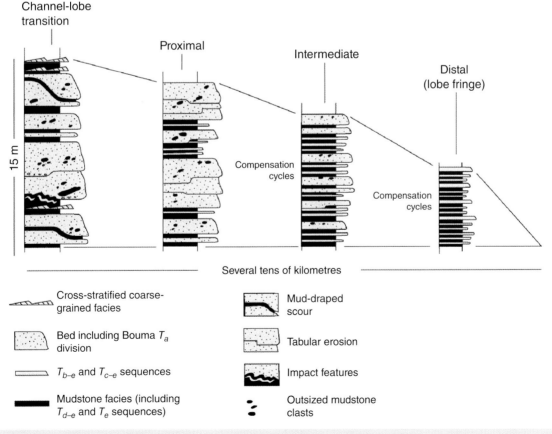

Fig. 7.41   Down-fan facies changes within lobe deposits, based on the Eocene Hecho Group of northern Spain but potentially observable in high-resolution seismic data from modern fans. From Mutti and Normark (1987).

a tendency for there to be irregular groups in which high or low values of some variable predominate. The relationship deduced by Hurst (1951, 1956) is:

$$R/S \sim N^h \tag{7.1}$$

where R is the maximum range of cumulative departures from the mean bed thickness (or grain size), S is the standard deviation of bed thickness (or grain size), N is the number of beds and h is a coefficient which is ~0.5 for random processes and >0.72 for natural phenomena characterised by clustering. Hurst approximated the coefficient h by K, where

$$K = \log (R/S)/\log_{10}(N/2) \tag{7.2}$$

(Equation 7.2, with K replacing h, is easily derived from Equation 7.1 by introducing 2^{-h} as a constant of proportionality on the right-hand side of the first relationship.) To calculate K for a series of N observations, each measurement is first stripped of its units (e.g., cm, φ units) and then replaced by its logarithm, to base 10. This is done because many geological datasets (e.g., bed thicknesses) are approximately log-normally distributed (Drummond & Wilkinson 1996). The sample standard deviation is calculated for these converted data, and R is measured directly from a plot of cumulative departures from the mean over the N sequential observations (Fig. 7.42).

Chen and Hiscott (1999b) considered 19 widely acknowledged submarine-fan successions that span a broad range of geological time, tectonic settings, facies characteristics and depositional environments (Table 5.3). The results of this analysis are included here rather than in Chapter 8 because one of the tested successions is from a modern fan (Amazon Fan), and because sand-prone and mud-prone packets are well described from coring and seismic profiles on modern fans, so deserve some discussion in this chapter.

Three bed-by-bed variables were considered by Chen and Hiscott (1999b): SGF coarse-division thickness (i.e., net thickness, in a

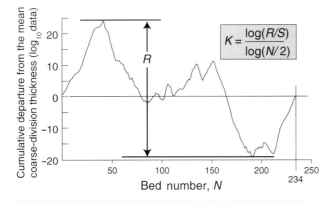

Fig. 7.42 Definition diagram for variables used in the calculation of Hurst's K. R is the maximum range in cumulative departures from the mean over N observations. This plot is for a series of 234 bed-by-bed measurements of coarse-division thicknesses in the Tourelle Formation at Cap Ste-Anne, Québec, Canada. Log values of the bed thicknesses, originally measured in centimetres, were used to calculate the cumulative departures from the mean. Where the departures from the mean are predominantly positive (e.g., rising part of the graph on the left), most beds are thicker than the mean, signifying a thick-bedded 'packet' of sandstones. From Chen and Hiscott (1999b).

single event-bed, of conglomerate + sandstone + siltstone), grain size and coarse-division thickness percentage (relative to thickness of overlying mudstone caps). Sixteen of the 19 analysed turbidite sections (84.2%) show the Hurst phenomenon, that is, irregular, long-term clustering of high and low values of the three bed-by-bed variables. Chen and Hiscott (1999b) then used Monte Carlo simulation to compare the field-based series of bed measurements with 300 randomised series generated from the same population of beds. They discovered that channelised deposits plot separately from lobe deposits and sheet-like deposits on a graph of Hurst K versus the number of standard deviations that the Hurst K for the field section is offset from the mean Hurst K for the 300 randomly shuffled sets of bed-by-bed data (Fig. 7.43).

Chen and Hiscott (1999b) concluded that (i) alternate stacking of channel and levée deposits results in a high Hurst K and a strong departure from the mean value of K expected for random sequences; (ii) lobe and interlobe deposits tend to have an intermediate K and a moderate departure from the mean value of K for random sequences and (iii) basin-floor sheet sand systems have a lower K and a weaker departure from the mean value of K for random sequences. The utility of the Hurst statistic has been confirmed by Felletti (2004) and Felletti *et al.* (2010), providing an additional tool for the recognition of various architectural elements including lobes.

7.4.4   Sheets

A sheet-like deposit is recognised if individual beds in the deep-water succession can be traced for many tens of kilometres with no perceptible change in average bed thickness. Ideally, individual event-beds would persist over the entire area of the sheet, but it is more likely that bed x is widely present in one part of the sheet and is replaced by bed y in a different area, even though the total sheet thickness is approximately constant. Submarine-fan deposits only have a sheet-like geometry in the lower-fan region of small-radius, low-efficiency fans, and in broad inter-levée depressions and the lower fans of large high-efficiency fans. Of course, basin-plain turbidites of abyssal plains, trenches and foreland basins also possess a sheet geometry at the scale of the basin in which they are found (Chapters 10 and 11).

The surface 500–800 m of the Amazon Fan consists of three extensive levée complexes (Fig. 7.9), each formed of a number of laterally shingled channel–levée systems. The youngest of these, the Upper Levée Complex, formed entirely during the last glacial interval of the latest Pleistocene (Flood & Piper 1997). Beneath, and interleaved with channel–levée deposits of the Upper Levée Complex, are widespread, medium- to coarse-grained, sheet-like sand-prone units. In seismic data (Flood *et al.* 1991, 1995; Damuth *et al.* 1995), the sheet-like units correspond to HARPs, which include both parallel and subtly lens-shaped internal reflections, the latter suggesting the presence of small, unlevéed channels. These deposits are the only extensive record from modern fans of the sheet element, so will be described in some detail. Based on several penetrations of HARP units, the Amazon Fan contains widespread sheets with high sand : mud ratios (Fig. 7.11). The sheets are tens of metres thick and several kilometres to several tens of kilometres in lateral extent.

The HARPs are broadly contemporaneous with a set of channel–levée units on the upper and middle Amazon Fan (Pirmez 1994), and with lower fan sandy lobes (Flood & Piper 1997). Hence, the Amazon Fan does *not* show a separation in time between what might be called a *basin-floor fan* (the sandy lobes) and a *slope fan* (the channel–levée complex). Instead, both types of deposit formed

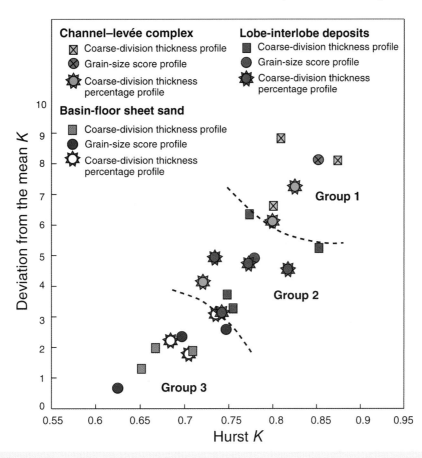

Fig. 7.43 Graph of values of Hurst K for channel–levée successions, lobe-interlobe successions, and sheet-like successions, plotted against the deviation of the Hurst K value from the mean K that was determined from 300 randomly shuffled series of bed-by-bed data. The *y*-axis shows the number of standard deviations that separate the Hurst K value from the mean K for the 300 random trials. For example, Hurst K for one channel–levée succession is ~9 standard deviations away from the mean value of K calculated for 300 randomised series of beds, shuffled from the original field succession. In each case, the standard deviation used to scale the *y*-axis was calculated using the same 300 values of K that were used to calculate the mean. See Chen and Hiscott (1999b) for details.

together during a single sea-level lowstand (Hiscott *et al.* 1997c; Nelson & Damuth 2003). To characterise these sheet sands, Pirmez *et al.* (1997) generated five detailed bed-by-bed sections for Sites 931, 935, 936, 944 and 946 using shipboard core descriptions and interpretations of down-hole logs and borehole images obtained using the Formation MicroScanner™ tool (FMS). Examples are shown in Figure 7.11.

In the middle fan, HARP units associated with individual avulsion events form 10–30 m-thick sand packets (i.e., bed clusters) that alternate with thin-bedded overbank deposits. The packets generally have sharp basal contacts and fairly sharp tops; locally, the bases of packets are gradational. On the lower fan (ODP Sites 945 and 946), similar packets appear to be stacked upon each other with little, if any, intervening overbank deposits, forming sand sheets up to 100 m thick. Where such sandbodies are overlain by an aggrading channel-axis deposit (HAR), total sand thickness may be even greater. Mud clasts are common in the thicker sand beds, which are in some cases more than 10 m thick. Some packets contain almost no mud interbeds, whereas others have numerous mud beds $\leq$ 50 cm thick.

Other large submarine fans like the Indus Fan (Kolla & Macurda 1988; McHargue 1991), the Bengal Fan (Droz & Bellaiche 1991; Thomas *et al.* 2012; France-Lanord, Spiess, Klaus & Expedition 354 Scientists 2015), the Mississippi Fan (Weimer 1991) and the

Rhône Fan (Droz & Bellaiche 1985; d'Heilly *et al.* 1988) also contain onlapping units at the base of channel–levée systems, with both high- and low-amplitude seismic character. By analogy with the Amazon Fan, these and other large fans might contain sand-prone sheets associated with avulsion events and with the distal ends of channels. The sand tends to occur in packets either because mobilisation from the up-fan avulsion site is episodic as a consequence of piecemeal headward erosion and entrenchment of channel segments undergoing gradient changes, or because of autocyclic shifts in the position of focussed sand deposition in the inter-levée low (i.e., switching between shallow feeder channels).

7.4.5 Scours and megaflutes

Normark *et al.* (1979) provided the first detailed description of large asymmetric scours from the surface of the Navy Fan, California Borderland. Inter-channel areas on the middle fan are characterised by an abundance of mesotopographic features, mainly scours and hummocky topography. Scours with widths of 50 to >500 m are most common on levée crests at sharp channel bends where the upper parts of turbidity currents fail to turn the corner and escape the confines of the channel entirely, in a process called *flow stripping* (Fig. 7.44; Piper & Normark 1983).

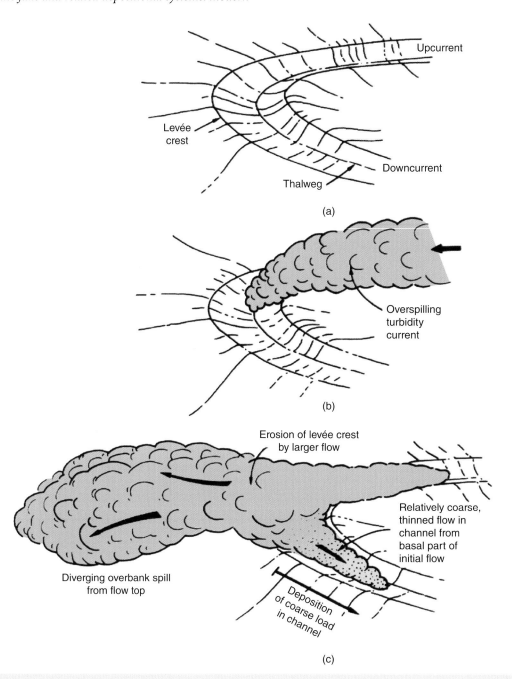

Fig. 7.44 Illustration of the flow-stripping hypothesis of Piper and Normark (1983). Channel curvature (a) causes eventual splitting of the initial flow into two parts (c). Loss of momentum by overbank spill (b) results in deposition of sand just beyond the channel bend. From Pickering *et al.* (1989).

Subsequently, large flute-like scours, called *megaflutes*, have been described from many modern and a few ancient deposits (Shor *et al.* 1990; Masson *et al.* 1995; Kenyon *et al.* 1995b; Vicente Bravo & Robles 1995; Elliott 2000a). The scours commonly have a mud rather than sand drape. They are inferred to characterise levée backsides and channel-lobe transitions where supercritical turbidity currents and concentrated density flows expand and hypothetically undergo a hydraulic jump that induces extreme turbulence and therefore seabed erosion (Mutti & Normark 1987, 1991).

Masson *et al.* (1995) described groups of enormous scours from the Monterey Fan channel that are concentrated at outside bends in the sinuous channel (Fig. 7.45). Scours are strongly asymmetric and from a few tens of metres to ~1 km in diameter; some scours are >10 m deep. Kenyon *et al.* (1995b) describe scours in the channel-lobe transition of the Rhône Fan. These scours are up to 20 m deep and 1 km across (Fig. 7.46). They are crescentic in plan view and have steeper upslope walls. The discovery of such large scours on modern fans signals that a high level of caution is needed in interpreting

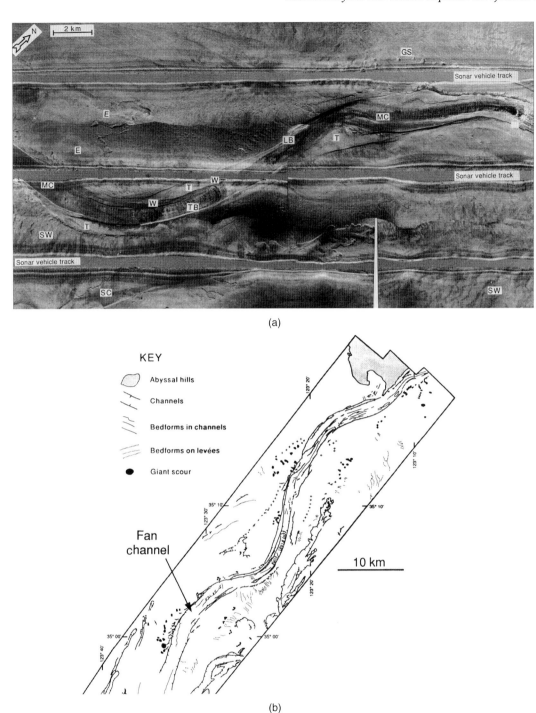

(a)

(b)

Fig. 7.45 (a) Digitally processed mosaic of three TOBI side-scan sonar swathes showing the major features of the Monterey Submarine Channel (MC). T = terraces; W= waterfalls; LB = longitudinal bedforms; TB = transverse bedforms; SW = sediment waves; GS = giant scours; E = aerially-extensive sites of erosion. Light and dark tones are areas of high backscatter and low backscatter, respectively. (b) Line drawing of the Monterey Fan channel based on TOBI 30 kHz deep-towed side-scan sonar and 7 kHz profiler. Note the location of fields of giant scours on the outside bends of meander loops. From Masson *et al.* (1995).

erosional cuts in two-dimensional outcrops of ancient successions, where a large megaflute might be mistaken for a channel. It is not known from modern settings what might constitute the fill of a large asymmetric scour, so good ancient outcrop examples must be relied on for facies information (Vicente Bravo and Robles 1995; Elliott 2000a). The widespread occurrence of large scours and megaflutes on modern submarine fans attests to the passage of large turbulent sediment gravity flows.

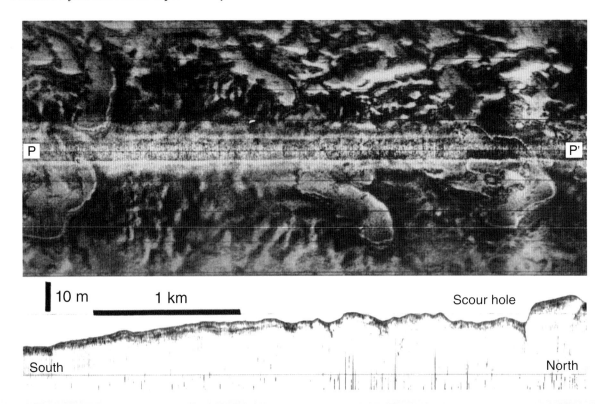

Fig. 7.46 Asymmetric scours on the distal part of the Rhône 'neofan', imaged using a 30 kHz side-scan sonar system and 5 kHz sub-bottom profiler. Scours elsewhere on this submarine fan are even larger. From Kenyon *et al.* (1995b).

7.4.6 Mass-transport complexes

Large parts of the surfaces of many modern fans are covered by *mass-transport complexes* (Walker & Massingill 1970; Damuth & Embley 1981; Twichell *et al.* 1991; Piper *et al.* 1997; Gamberi *et al.* 2010); these are defined as blocky, irregular and mixed units emplaced by one or more cohesive flow, slump or sediment slide. Essentially, these are submarine avalanche deposits (Hampton *et al.* 1996). Seismic imaging and drilling on the Amazon Fan has revealed that these chaotic deposits are also part of the older stratigraphy of the fan

(Piper *et al.* 1997). Mass-transport complexes (MTCs) form ~20% of the fan stratigraphy in the upper few hundred metres, and cover ~40% of the fan surface in water depths less than 4000 m (Fig. 7.8). Individual mass-transport units are of the order of 100–120 m thick where cored. They locally show erosional truncation or disruption of underlying units, particularly adjacent to topographically high levées.

The coring technology used by ODP to recover sediments more deeply buried than ~100 m involves rotation of the drill string and drill bit, which invariably leads to drilling-induced deformation unless the strata are lithified (Shipboard Scientific Party 1995e). It is

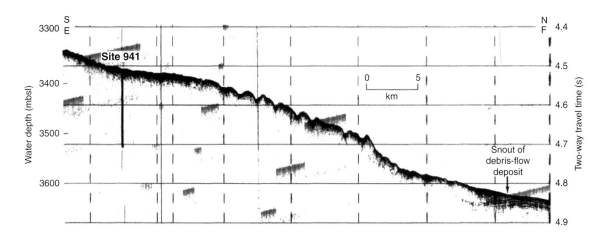

Fig. 7.47 A 3.5 kHz profile of the snout of the MTC cored at ODP Site 941. The MTC is underlain by stratified deposits which can be seen in the vicinity of the snout. mbsl = metres below sea level. From Shipboard Scientific Party (1995e).

therefore fortunate that the Holocene Western MTC on the Amazon Fan (Fig. 7.8; Damuth & Embley 1981) is shallowly buried. The downslope snout of this deposit is abrupt (Fig. 7.47). At ODP Site 941, the upper 85 m of the 125 m-thick Western MTC was cored without drillstring rotation, using a hydraulic piston corer that recovers sediments intact. Although only a single example, the MTC at ODP Site 641 is described here because no other comparable deposits have been recovered, undisturbed, from a modern submarine fan.

Virtually all cores from the MTC have been affected by soft-sediment failure, deformation and redeposition recorded by abrupt changes in lithology and colour, structures indicative of deformation such as folds and faults, discordant stratal relationships and the occurrence of clasts of various sizes (Fig. 7.48, Shipboard Scientific Party, 1995e). Mud clasts range in size from centimetre-scale clasts to metre-scale or larger blocks that span more than a single 1.5 m-long core section. Dips range from very slight to nearly 90°, and dips of 50–80° are common. In some cores, beds dip at different attitudes and directions indicating the occurrence of discordant stratal relationships, in some cases marked by distinct faults. There are folds of various scales. Only a few centimetre-scale folds are

visible in the cores. In one 130 cm-thick cored interval, beds dip in one direction at the top and the opposing direction at the bottom. Palaeomagnetic analysis indicated a change in polarity between these segments with opposing dips. Thus, this interval appears to contain a large fold. Further evidence to suggest folding of the sediment, as well as faulting or shearing, is the deformed nature of colour bands, laminae and beds, most of which appear squeezed or stretched and have irregular boundaries.

Displaced fossils and wood are found in the MTC at Site 941, including echinoderms, gastropods, pteropods, ostracodes, fine disseminated plant material and a 3 cm-across piece of wood. Clasts of lithified sediment and bedrock occur rarely. Softer sedimentary clasts and blocks are commonly deformed with 'crenulated' and serrated boundaries that indicate they are microfaulted or sheared.

The degree of deformation in this MTC from the Amazon Fan is considerable. If this deposit was exposed in outcrop, the observer would see a chaotic jumble of large blocks, some of which would be folded and/or faulted. Matrix would be minor, consisting of mud contributed by disintegration of the margins of clasts and by mobilisation of the surficial sediments at the original failure site.

7.5 The distribution of architectural elements in modern submarine fans

To this point, little attention has been paid to the volumetric importance of each of the architectural elements in modern fans. In Figure 7.49, we present a conceptual view of the contribution that each architectural element makes to modern submarine fans. Mass-transport complexes are not included because they often consist of remobilised older fan sediments that are accounted for separately on the diagram. Levées often are formed by migrating sediment waves. Giant scours are not shown because their record is erosional rather than depositional. The giant open-ocean fans are dominated by channel–levée complexes underlain by sand-prone HARPs. Although the percentage of sandy facies is lower in these fans than in small-radius fans of active-margin basins, the sheer size of these open-ocean fans makes them the largest repositories of terrigenous sand in the ocean basins.

As submarine fans become smaller, the relative volumetric importance of levée deposits decreases at the fastest rate. Medium- to small-radius fans are instead dominated by sand-prone lobes fed by sand/gravel-prone channels. The lower fan is more fine grained and consists of interbedded Bouma-type turbidites (Facies C2.2, C2.3, D2.1) and muds.

The proportions shown in Figure 7.49 are designed to be an integration over the entire volume of selected submarine fans, so do not distinguish between upper, middle and lower fan segments. Piper and Normark (2001) estimated the proportions of architectural elements and their associated facies on the Amazon Fan for five cross-sections between the upper fan and the adjacent abyssal plain (Fig. 7.50). The highest proportion of sand is in the lower fan, where SGFs that were confined to levéed channels higher on the fan are able to spread out, decelerate and deposit the coarser suspended load that is carried in the lower part of each flow.

7.6 Modern non-fan dispersal systems

Sediment dispersal processes either carry detritus from shallow-marine sources (e.g., shelves, flanks of volcanoes) into

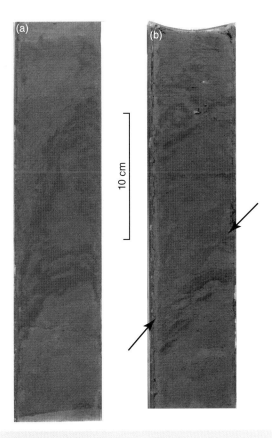

Fig. 7.48 Deformation and large clasts in the Western MTC on the Amazon Fan, cored at ODP Site 941 (located in Fig. 7.8). Both images are photographs of the split faces of 6.6 cm-diameter cores recovered with a hydraulic piston corer, from Shipboard Scientific Party (1995e). (a) Steeply dipping, 'crenulated' colour bands and beds which might be contained within a large clast (ODP reference 155-941A-3H-7, 44–68 cm). (b) Sharply defined margin of a large clast (arrows). The remainder of this interval is believed to be matrix of the MTC, and contains scattered pteropod shells and small clasts (ODP reference 155-941A-4H-2, 3–27 cm).

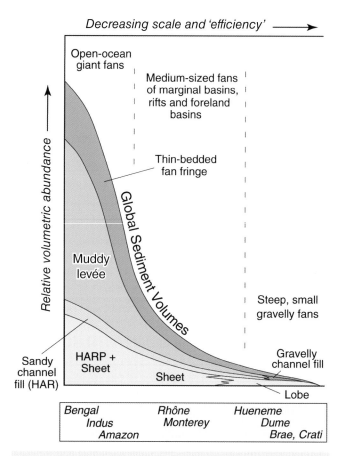

Fig. 7.49 Conceptual overview of the distribution of architectural elements in submarine fans across a wide range of scales. The vertical (*y*-axis) scale shows approximate (and relative) proportions but is not meant to be strictly quantitative. The cumulative thickness of the architectural elements at any point on the diagram represents our assessment of the relative volume of fans of this scale in the modern oceans. The principal message is that a few giant fans contain more sediment than all other fans combined. The relative volumetric importance of each depositional element is represented by the thickness assigned to that element on the diagram; these proportions vary with the scale of fan. Each example (e.g., Hueneme Fan) is believed to have the mix of facies shown vertically above its name. MTCs are not included because they are commonly formed of contorted material that originally accumulated as levées or other elements in the diagram. Highstand 'condensed' hemipelagic deposits found in many modern fans are too thin to represent at this scale. This figure is styled after a plot of depositional environments versus textural maturity in Folk (1974).

the deep sea, or redistribute material by moving it from place to place on the deep-ocean floor. Shelf-to-basin transport is almost entirely attributable to sediment gravity flows (SGFs, see Section 1.4) and gravitational sliding. Redistribution at bathyal and abyssal depths is almost entirely the work of thermohaline currents (e.g., contour currents). Thermohaline currents and the facies they produce are discussed in Chapter 6.

Exceptions to the generalisations in the preceding paragraph are provided by tidal currents in submarine canyons, which promote the basinward transport of fine-grained sediment including a certain amount of sand (Section 1.2.2; Shanmugam 2003), and by surface

currents that may penetrate sufficiently deeply to redistribute sand and mud at water depths of many hundreds to thousands of metres (Kenyon *et al.* 2002; Shanmugam *et al.* 1993a).

Away from submarine-fan systems, shelf-to-basin transport may be facilitated by a network of gullies and canyons which feed a broad *slope apron* rather than a cone-shaped deep-water fan. Stow (1986) summarised three composite slope-apron models: normal (clastic), faulted and carbonate. He also proposed at least ten different types of slope aprons depending on their primary morphotectonic setting. The main subsea landforms recognised by Stow (1986) include: an abrupt shelf break, fault-scarp and reef-talus wedges, slump and slide scars, irregular slump and debris-flow masses (mass-transport complexes), small channels and gullies, larger more complex dendritic canyons, isolated lobes, mounds and drifts, and broad smooth or current-moulded surfaces. Stow (1986) further recognised differences in facies associations between upper and lower slope-apron muddy facies. He proposed that the upper slope apron is marked by more abundant 'higher energy' features, including more resedimented facies, slide scars, erosional gullies and channels; in contrast, the lower slope apron is characterised by more 'lower energy' features, with greater frequency of fine-grained turbidites, isolated submarine channel fills, debrites, slide and slump scars and local interfingering with contourite drifts and/or submarine-fan deposits. Recently, the northwest African continental margin has been promoted as an instructive modern example of a siliciclastic slope apron (Wynn *et al.* 2000). The reader is directed to Section 9.4.3 where this continental margin and its facies are described and discussed.

Large, isolated abyssal channels deliver sediments directly to some of the world's abyssal plains without forming morphological submarine fans along their route. Examples are the Northwest Atlantic Mid-Ocean Channel (NAMOC) and associated braid-plain in the Labrador Sea (Hesse 1995a; Hesse & Chough 1980; Hesse *et al.* 1987, 1990, 1996, 2001; Hesse & Klaucke 1995; Klaucke & Hesse 1996; Klaucke *et al.* 1998a, b), and a set of abyssal channels draining south and north from the Aleutian Islands into, respectively, the Aleutian Abyssal Plain (Grim & Naugler 1969; Mammerickx 1970; Hamilton 1973) and the Bering Sea basin (Carlson & Karl 1988; Kenyon & Millington 1995).

There are three prominent architectural elements in the Labrador Sea basin: (i) a strongly dissected (gullied) continental slope seaward of a plethora of glaciated fjords along the indented coastline (Hesse & Klaucke 1995); (ii) a sinuous trunk channel (NAMOC) which is ~4000 km long (Chough & Hesse 1976) and which has levées built of fine-grained sediment sourced from the slope gullies and (iii) a sandy braid-plain to the northeast of NAMOC which has been attributed to sedimentation from hyperpycnal flows formed during glacial outburst events (jökulhlaups – Hesse *et al.* 2001). Some of the sand beneath the braid plain is coarser than 2 mm. There is a streaky surface pattern in side-scan sonar images which Hesse *et al.* (2001) attribute to both erosional markings (furrows and ridges) and depositional elements (channels, bars, lobes). Locally, energetic flows that transported sand to the braid plain eroded the left-hand levée of NAMOC and thereby gained access to the channel thalweg.

Acoustic properties of the NAMOC levées and a large number of cores indicate that these levées are formed of laminated silt and mud (with minor sand), very much like the levées of large open-ocean submarine fans typified by the Amazon Fan (Fig. 7.1a). Chough and Hesse (1980) attributed textural variations in levée turbidites to alternating head spill and body spill from turbidity currents transiting the channel. The channel floor deposits are sandy and gravelly.

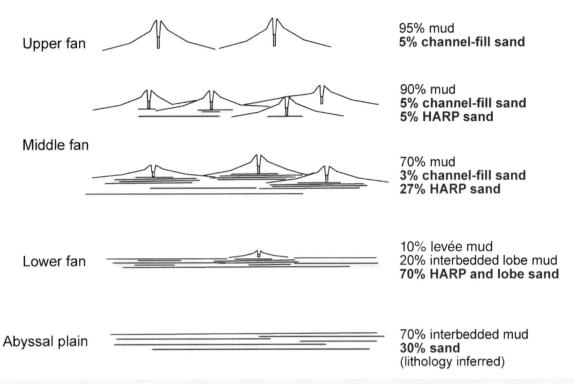

Fig. 7.50 Schematic cross-sections of the youngest levée complex on the Amazon Fan (dating from the last glacial cycle) showing the proportions of muddy and sandy sediment as estimated by Piper and Normark (2001).

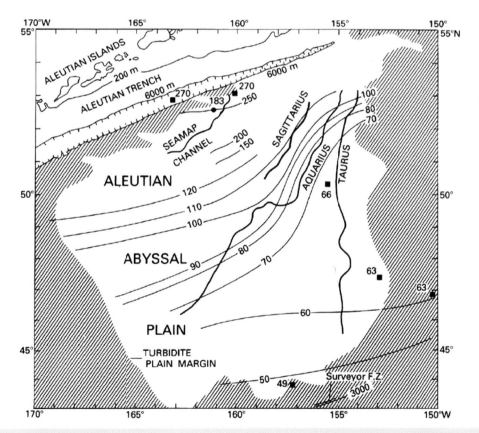

Fig. 7.51 Aleutian Abyssal Plain showing isopachs (contoured in metres) of the pelagic sediment that overlies buried turbidite systems. Deep-marine channels are labelled. Isolated numbers indicate measured thickness of pelagic sediments. From Hamilton (1973).

Fig. 7.52 GLORIA side-scan image of the Umnak distal sand sheet, with braid-like bedforms elongated down the transport path. Flow direction is from upper left to lower right. The broad, high-backscatter strip (Sh) consists of a sheet of sediment up to 8 m thick. From Kenyon and Millington (1995).

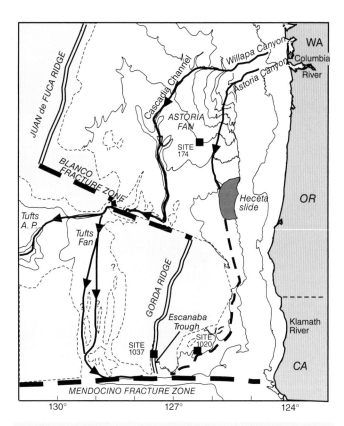

Fig. 7.53 Map showing the circuitous transport path of turbidity currents (arrowed lines) from the Cascadia Channel to ODP Site 1037 in the Escanaba Trough. From Piper and Normark (2001).

The Aleutian Abyssal Plain lies south of the Aleutian Trench, north of the Surveyor Fracture Zone, and is bounded to the east and west by abyssal hills and seamounts (Fig. 7.51). Turbidite deposition began in the middle Eocene and effectively ended in the Oligocene (Scholl & Creager 1971) – the plain now being covered by pelagic sediments. Turbidites were supplied to the plain from a number of levéed channels, especially Sagittarius, Aquarius and Taurus Channels (Mammerickx 1970) and Seamap Channel (Grim & Naugler 1969). A number of additional buried channels have been described by Hamilton (1973). The channels are aggradational and lie along low ridges, with the western levées being higher and broader than those on the east due to the Coriolis Effect. Sediment thicknesses under the western levées range from ~250–800 m (Hamilton 1973).

Kenyon and Millington (1995) described the Umnak depositional system north of the Aleutian chain. The seabed is characterised by elongate bedforms, crescentic scours, braid bars and distal sand sheets. Most striking is the braided channel pattern in the distal part of the dispersal system (Fig. 7.52), several hundreds of kilometres from the upslope terminations of the tributary channels.

Finally, we would like to highlight the role of lineaments and structural depressions in the oceanic crust that serve to constrain the pathways of SGFs away from submarine fans. For example, turbidity currents transiting the Cascadia Channel, offshore Oregon, are first captured by the Blanco Fracture Zone, and later by the Mendocino Fracture Zone as they follow a circuitous path to the Escanaba Trough (Fig. 7.53; Zuffa *et al.* 1997). The slot-like confinement by the relief of the fracture zones prevents spreading of the turbidity currents

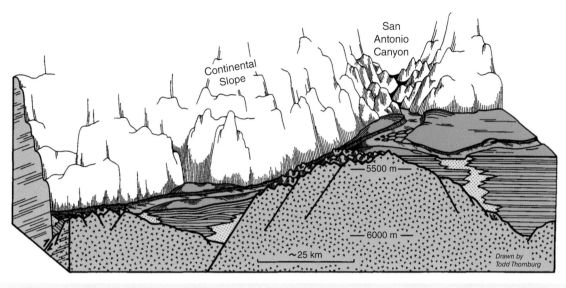

Fig. 7.54 Confluence of the San Antonio Canyon with sheet-like deposits of the Chile Trench (right). There is an axial sandy lobe in the trench axis north (left) of the confluence. Coarse-grained deposits are stippled. From Thornburg and Kulm (1987).

to form fan-shaped bodies of sediment. A similar situation exists in oceanic trenches, where linear rather than fan-shaped dispersal systems predominate (Fig. 7.54; Thornburg & Kulm 1987).

7.7 Concluding remarks

Submarine fans are important targets for hydrocarbon exploration, in particular sand-prone channel fills, lobes, sheet-like deposits (e.g., HARP units) and injection complexes (Section 2.10). In the 1970s and 1980s there were several attempts to simplify and distill submarine-fan (and -ramp) successions into one or two facies models using both real and hypothetical links between vertical successions (from rock outcrops and cores) and the plan-view distribution of sub-environments on modern fans. More recently, it has become apparent that modern (and ancient – see Chapter 8) submarine fans are highly variable and complex, with a range of intrinsic and extrinsic factors controlling the stratigraphic record. There was a phase in the 1980s and 1990s of over-emphasis of the imprint of base-level (e.g., sea-level) changes on sand content, particularly at the scale of 10–30 m-thick clusters of sand-prone versus mud-prone deposits. Examples in this book show, however, that numerous alternations of sand-prone and mud-prone facies clusters can accumulate during single lowstands of sea level – they owe their origin to intrinsic processes like channel switching (avulsion) rather than base-level fluctuations. The nature of sand supply (e.g., river input versus longshore transport into canyon heads) can lead to quite different granulometry of deep-water facies during highstands of sea level and, in the case of riverine supply, the climate in the drainage basin can have a strong influence on sand delivery to the associated submarine fan(s).

Calciclastic submarine fans appear to be rare in the stratigraphic record and no bona fide present-day examples have been described, possibly because they have been largely overlooked by the academic and industrial communities (Payros & Pujalte 2008). From a literature review and their own work, Payros and Pujalte (2008) provide a synthesis of the characteristics of calciclastic submarine fans. They argue that calciclastic submarine fans range in length from a few to >100 km, and recognise three types: (i) coarse-grained, small-sized (<10 km) fans, withr elatively long levéed channels, small radial lobes and abundant calcirudites with little mud; (ii) medium-grained, medium-sized fans, 10–35 km in length, with a tributary network of slope gullies merging to form a levéed channel that opens to the main depositional site, characterised by extensive lobes and/or sheets, which pass into basinal deposits through a narrow fan-fringe area, and typified by abundant calcarenites and lesser amounts of calcirudites and mud and (iii) fine-grained, large fans, with lengths between 50–100 km, containing wide and long slope channels that feed very extensive calciturbiditic sheets, and are rich in calcarenites and mud, but poor in calcirudites. Payros and Pujalte (2008) also propose that in terms of grain-size distribution, the three fan types can be compared with sand/gravel-rich, mud/sand-rich and mud-rich siliciclastic submarine fans, respectively. However, all three calciclastic submarine fans types show notable differences in terms of size and sedimentary architecture, reflecting the different behaviour of their respective SGFs. Most calciclastic submarine fans formed on low-angle slopes and were sourced from distally steepened carbonate ramps subjected to high-energy currents. Under such conditions, shallow-water loose grainy sediments were transferred to the ramp slope and eventually funnelled to the submarine fan by a range of SGFs.

In our view, the most reliable way to study submarine-fan deposits is to match features in the field (or subsurface) to key criteria which have been proposed for the recognition of the major architectural elements (listed in Table 7.1 as levées, channels, lobes, channelised lobes, sheets and wedges). Many of these criteria are summarised in Figures 7.40 and 7.41. Extreme caution must be used in evaluating the presence and significance of asymmetric trends in bed thickness or grain size – many subjectively recognised trends fail the scrutiny of statistical tests, and base-level changes are only one option to explain facies or textural cycles.

Once the primary architectural elements (i.e., building blocks) have been recognised in an ancient or subsurface example, then the type of fan and the likely areal extent and continuity of sediment bodies can be hypothesised by comparing the volumetric abundance of the elements to a summary based on modern fans (Fig. 7.49). Alternatively, if a sufficiently large area has been studied, then palaeocurrent maps and facies maps might improve the chances of selecting appropriate modern analogues amongst the wide range of submarine fans described in the literature. First dissecting, then reconstructing, the spatial distribution of diagnostic facies is a challenge in itself. More difficult, in many cases, is the subsequent determination of the relative importance of external and internal controls on development of a particular submarine fan (or ramp). Perusal of Chapters 4 and 8 will quickly demonstrate that there are many alternatives and complications that can lead to incorrect interpretations when only partial datasets are available to the researcher. Experience, intimate familiarity with the literature, and an open mind are key assets needed to arrive at the most robust interpretations.

CHAPTER EIGHT

Submarine fans and related depositional systems: ancient

Panorama looking north into the Middle Eocene deep-marine Ainsa Basin, Spanish Pyrenees. The basin-bounding Mediano Anticline, showing folded lower Eocene shallow-marine carbonates, is on the eastern (right-hand) side of the Mediano reservoir. The western side of the basin is defined by the Boltaña Anticline, seen left of centre in the background foothills. In this perspective the deep-marine sandy, mainly channelised, submarine fans are all dipping west and defined by ridges in the landscape, for example, left of centre.

8.1 Introduction

The earliest recognition of ancient submarine fans was based on facies associations, presence of channels and palaeocurrent maps in outcrop examples (Sullwold 1960; Jacka *et al.* 1968). In general, ancient submarine fans cannot be recognised on the basis of radial palaeocurrents and shape, but instead must have their origin deduced from facies associations and architecture, as first proposed by Mutti and Ghibaudo (1972) and Mutti and Ricci Lucchi (1972). According to the latter authors, submarine fans consist of (i) an *inner fan* characterised by conglomerate and coarse sandstone facies (our Facies Classes A and B) in large channels cut into fine-grained deposits (our Facies Class E); (ii) a *middle fan* consisting of packets of sandstone and minor amounts of conglomerate (Facies Classes B and A) in thinning-and-fining-upward trends, alternating with packets dominated by Facies Classes C, D and E and (iii) an *outer fan* with few or no channels, and parallel-sided SGF deposits arranged

in thickening-and-coarsening-upward cycles that in the more distal deposits were subsequently called *compensation cycles* (Mutti & Sonnino 1981; Mutti 1984; Ricci Lucchi 1984).

With increased mapping and high-resolution imagery from modern fans and a plethora of detailed outcrop studies in recent years, the tendency to use separate fan models for ancient *versus* modern deposits has all but disappeared. The terminology originally proposed for ancient fans (inner, middle and outer fan) has largely been abandoned to avoid confusion resulting from the fact that the middle and outer fan regions as originally defined by Mutti and Ricci Lucchi (1972) both fall within the middle fan as defined by Normark (1970); the lower fan of Normark (1970) would be classified as part of the basin plain in the scheme of Mutti and Ricci Lucchi (1972). In ancient deposits, the criteria that are necessary to distinguish upper, middle and lower fan segments in a way that would be consistent with the subdivision of modern fans are unlikely to be observable; instead, it is necessary to evaluate the amount of overspill

Deep Marine Systems: Processes, Deposits, Environments, Tectonics and Sedimentation, First Edition. Kevin T. Pickering and Richard N. Hiscott.
© 2016 Kevin T. Pickering and Richard N. Hiscott. Published 2016 by John Wiley & Sons, Ltd.
Companion Website: www.wiley.com/go/pickering/marinesystems

in the succession, which is a function of levée height, amount of confinement and distance down the fan system (Hiscott *et al.* 1997b). The simple, all-purpose submarine-fan model presented by Mutti and Ricci Lucchi (1972), Ricci Lucchi (1975b) and Walker (1978) served the geoscience community well during the 1970s and early 1980s, but is now regarded as an over-simplification. Classifications of fans based on shape (e.g., radial versus elongate, as in Pickering 1982c and Stow 1985) are also simplistic because shape is a dependent variable, reflecting the tectonic setting of the basin (Pickering 1982c; Shanmugam & Moiola 1985).

Since the pioneering work of Mutti and Ricci Lucchi (1972), a number of significant discoveries have been made that require modifications to their depositional model. Here we highlight four of these discoveries. These and other issues will be developed further in subsequent sections. First, it has become apparent that levéed channels are much more common even in small submarine fans than previously thought (e.g., Piper *et al.* 1999). Second, large scours and megaflutes which might be confused with small channels have now been described from ancient systems (e.g., Vicente-Bravo & Robles 1995; Elliott 2000a,b) following their initial discovery on modern fans. Such scours are likely to be widespread in ancient successions but good 3D exposures are needed for their recognition. Third, it has been recognised that calciclastic submarine fans require depositional models that differ from siliciclastic submarine fans (e.g., Payros & Pujalte 2008, and references therein: see Chapter 7).

Fourth, it has been demonstrated through statistical analysis of a large number of sections in classic areas that asymmetric trends of bed thickness and grain size (e.g., thickening/coarsening-upward or thinning/fining-upward trends) are quite uncommon, to the point that they cannot reliably be used to interpret or recognise fan subenvironments (Section 5.3; Harper 1998; Chen & Hiscott 1999a; see also comments of Anderton 1995). This fourth point has created some difficulty in writing this chapter, because many of the references we have used assert the presence of, for example, thinning-upward or fining-upward trends, but without providing the raw data that would be necessary to complete statistical tests to confirm (or refute) such claims (see equations in Section 5.3). In many such cases, the perceived asymmetry is a subjective conclusion stemming from field panoramas, and is not based on the quantitative analysis of bed-by-bed measurements. For this reason, the reader should retain some scepticism toward claims of asymmetric trends that are advanced in this chapter based solely on primary literature sources without recourse to statistical tests.

A major difficulty in correlating attributes of ancient deep-marine fans with processes observed on modern fans is the difference in scale of observation (Normark *et al.* 1979; Wynn *et al.* 2002b), along with the fact that most modern fans have been essentially inactive since the last rise in sea level and are blanketed by hemipelagic mud (Facies Class E). Features observed in the best outcrops are, in all but exceptional circumstances, smaller than the best resolution that can be

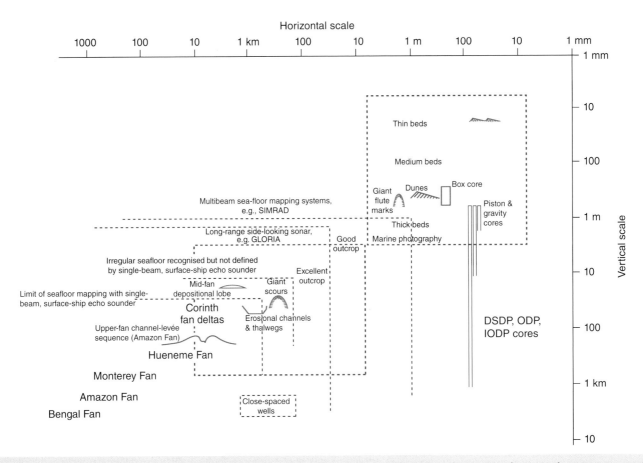

Fig. 8.1   Horizontal and vertical scales of observation employed in studies of modern and ancient submarine fans. Most outcrop observations are at scales within the box in the upper right. Most marine observations are at scales outside this box. From Piper and Normark (2001), updated from Normark *et al.* (1979).

attained with available shipboard deep-sea imagery (Fig. 8.1). Vertical resolution of ultra-high resolution seismic systems has improved to the point that features several tens of centimetres thick can be recognised routinely (Piper *et al.* 1999), but the seabed 'footprint' of side-scan sonar and multi-beam profilers is ~5–10 m across, preventing recognition in all but bottom photographs of outcrop- and core-scale sedimentary structures (e.g., Wynn *et al.* 2002a).

Facies sequences reflect a complex interaction between sediment texture, sediment flux, tectonic uplift/subsidence, eustatic sea-level change and global climate change (at a wide range of temporal scales), together with internal fan processes like channel switching, mass wasting and so on (see Chapter 4). Many ancient fan deposits do not show the abundance of thinning- and thickening-upward trends implicit in published models (Hiscott 1981; Chan & Dott 1983; McLean & Howell 1984; Chen & Hiscott 1999a: see above). There is now some consensus that those thickening-upward trends that do occur might not be the result of lobe progradation, but instead could be generated by subtle lateral shifts in the site of SGF deposition to form compensation cycles (Mutti & Sonnino 1981; Mutti 1984, Ricci Lucchi 1984). When channels are exposed at outcrop, the fill may consist of very fine-grained facies (e.g., Garcia-Mondéjar *et al.* 1985: p. 321) unlike those predicted by extant fan models, although seismic data from some modern fans (e.g., Indus Fan, Kolla & Coumes 1987) do suggest common fining-upward trends within the filled channels of upper and middle fan segments.

Barnes and Normark (1984) summarised the attributes of ten ancient fans, but many more have been described in the literature. It should be stressed that most well-documented ancient fans appear to have been deposited at active continental margins, including foreland basins. Except for the upper Precambrian Kongsfjord Formation of northern Norway (Pickering 1981a,b, 1982a,b, 1983a, 1985; Drinkwater & Pickering 2001, Roberts & Siedlecka 2012), the Neoproterozoic Windermere Group of British Columbia (Ross 1991; Ross & Arnott 2006; Arnott 2007; Schwarz & Arnott 2007; Terlaky *et al.* 2010; Khan & Arnott 2011) and, perhaps, Lower Palaeozoic fans of North Greenland (Surlyk & Hurst 1984), the literature documents very few well-exposed, large-scale, passive-margin fans, probably because such fans (e.g., the Mississippi, Amazon and Laurentian Fans) can only become subaerially exposed through orogenic processes, leading to severe dismemberment as the sediments are thrust away from the oceanic or transitional crust on which they originally accumulated. Remnants of such passive-margin fans in orogenic belts cannot normally be satisfactorily reassembled to permit description of the entire fan.

An interesting facies association of the Kongsfjord Formation that deserves special mention, however, is one that formed at the lateral margin of a submarine fan. The sediments are called *fan lateral-margin deposits* (Macpherson 1978; Pickering 1983a), and are characterised by (i) a relatively high proportion of fine-grained sandstone and siltstone SGF deposits (Facies C2.3, D2.1), (ii) relatively small channels developed at various angles to the regional basin slope direction, (iii) lobes associated with the channels and (iv) abundant clastic dykes and other wet-sediment deformation. The dykes originated in channel sandstones, and are preferentially located near the margins of overlying channel sandstones (*cf.* Hiscott 1979).

8.2   Ancient submarine canyons

Canyon fill has been described almost exclusively from ancient examples (e.g., Whitaker 1974; Carter & Lindqvist 1977; Almgren

1978; Picha, 1979; Arnott & Hein 1986; Bruhn & Walker 1995, 1997; Millington & Clark 1995a; May & Warme 2000; Wonham *et al.* 2000; Anderson *et al.* 2006; Ito & Saito 2006). One well-documented in situ canyon fill of Miocene age, offshore Gabon, has been exquisitely imaged using amplitude maps generated from 3D seismic data (Wonham *et al.* 2000). The fill consists of four 'stratigraphic cycles', each consisting of more than one channel complex (Fig. 8.2). The canyon was 4 km wide and the cumulative fill ~400 m thick. Two of the cycles were studied in detail. They each show a fining-upward trend, but are characterised by different styles of sand accumulation. In the Baliste 1 cycle, there are four fining-upward sandstone-shale cycles, each ~40 m thick. The sandbodies are sheet-like, pinch out toward the canyon margins and are more laterally restricted as they become younger. A 46 m-thick cored interval comprises graded pebbly sandstone with extraformational and rip-up clasts (Facies A2.7), disorganised coarse-grained muddy sandstone with large mud clasts (Facies B1.1 and C1.1) and disorganised cobble conglomerate with a muddy sandstone matrix and large intraformational deformed clasts (Facies A1.1 and A1.2). The organisation into four stacked channel complexes suggests periodic cessation of sediment supply to the canyon, possibly during relative sea-level highstands, or minor baselevel changes culminating in a full highstand consistent with the progressive lateral restriction of component sandbodies. In the Baliste 2 cycle, the canyon is filled by a succession of highly sinuous levéed channels (Fig. 8.3), each of which accumulated in an erosional lane cut in the canyon during a time of bypass, and filled by the levéed channel deposits during times of higher sea level. The canyon fill consists of alternating sand-rich and mud-rich intervals.

Bruhn and Walker (1997) describe an Eocene canyon-fill succession from coastal Brazil that is very similar to the Gabon example of Wonham *et al.* (2000), in that the fill consists of a large number of stacked and amalgamated levéed channels, totalling 38. The Regência Canyon is <6 km wide, ~15 km long and the fill is locally ~1000 m thick. The stratigraphic context is given in Fig. 4.14 (interval labelled MT). Depositional water depth was 200–500 m. Facies are mostly very coarse grained, consisting of (i) unstratified, bouldery to pebbly conglomerate and very coarse-grained sandstone (Facies Classes A and B), (ii) unstratified, coarse-grained sandstone and parallel-stratified medium- to fine-grain sandstone (Facies Classes B and C), (iii) interbedded, bioturbated mudstone and thin-bedded sandstone (Facies Classes D and E) and (iv) mudstone (Facies Class E). The stratified sandstone beds locally contain the trace fossils *Thalassinoides*, *Ophiomorpha*, *Planolites* and *Helminthopsis*. The 38 channels belong to three channel complexes. Channel fills become narrower, thinner and finer-grained upward under the influence of rising relative base level.

Recognition of ancient submarine canyons at outcrop relies upon good lateral and vertical continuity of exposure, together with a tightly constrained stratigraphy. Early well-documented examples of ancient submarine canyons include the six Silurian (Ludlow) canyons of the English Welsh Borderland in which inferred synsedimentary faults controlled the location of some canyons (Whitaker 1962), and the Nesvacilka and Vranovice canyons of the ancient Tethyan margin in Czechoslovakia described by Picha (1979). The Nesvacilka and Vranovice canyons were traced in the subsurface for more than 30 km, with NW–SE-trending axes. Towards the southeast, they merge into a single large canyon. In their upper reaches, the canyons have widths of 1–3 km with slopes down their walls (i.e., transverse to the canyon axes) of 30–35°. Farther downslope the widths increase to ~10 km in Nesvacilka Canyon. The sedimentary fill comprises mainly sandstones and conglomerates of Facies Classes A, B and

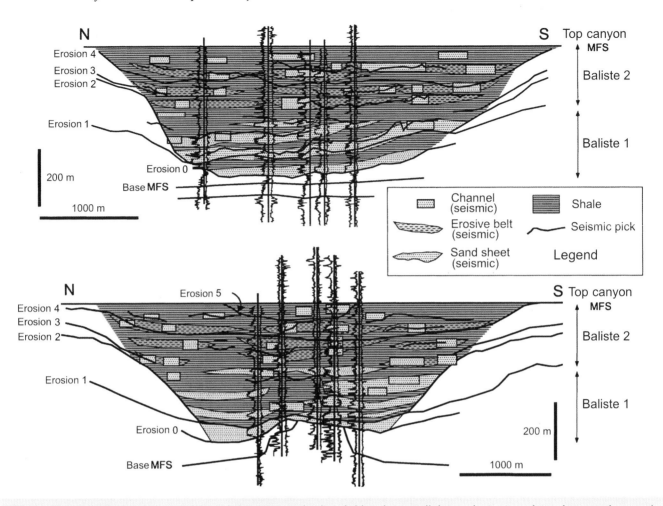

Fig. 8.2 Two cross-sections through the Baudroie Marine and Baliste fields utilising well data and seismic surfaces, from Wonham *et al.* (2000). Six erosion surfaces (numbered 0–5) and two sedimentary packets (Baliste 1 and 2) are shown. The height of the transition from Baliste 1 to Baliste 2, shown at the right, is projected from a position in the axis of the canyon above the fourth sandbody.

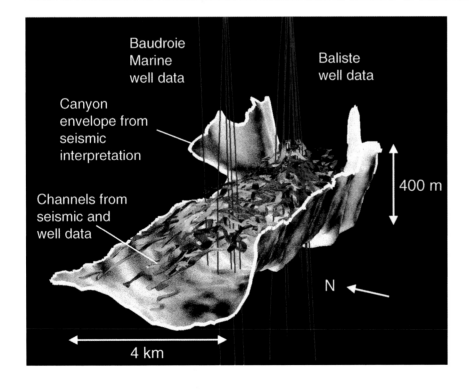

Fig. 8.3 3D model showing distribution of sinuous channels recognised from 3D seismic within the Baliste-Crécerelle Canyon. The image emphasises the labyrinthine complexity of these deposits. From Wonham *et al.* (2000).

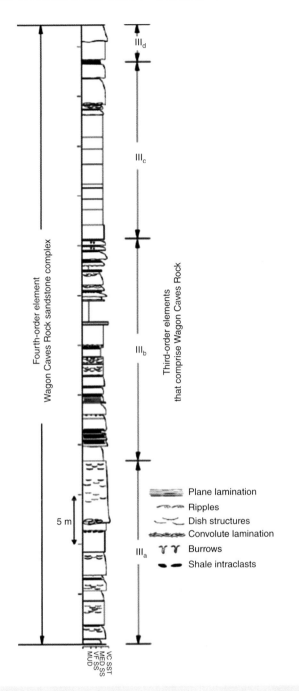

Fig. 8.4 Measured section through the canyon fill at Wagon Caves Rock (WCR), California. This interval represents a significant portion of a fourth-order sandstone complex in the Palaeocene part of the canyon fill. Third-order architectural elements within the WCR outcrop (IIIa–IIId) are packets of sandstone beds characterised by similar bed thickness, bed geometry, grain size and bounding surfaces. See Section 7.4 for a discussion of architectural elements. From Anderson *et al.* (2006).

C toward the base and silty mudstones (Facies Classes D and E) toward the top. The margins show pebbly mudstones, sediment slides (Facies Class F) and other sediment gravity flow (SGF) deposits. The mudstones contain 1–9% organic matter. Three distinct foraminiferal assemblages are recognised within the canyon deposits: (i) indigenous canyon-dwellers; (ii) synsedimentary faunas flushed from estuaries

and the open shelf and (iii) faunas reworked from older sediments. In the Nesvacilka Canyon, the sedimentary fill increases in thickness from 800 m to 1060 m over a distance of 5.5 km along the canyon axis. The time taken to infill the canyons is estimated at 10–12 Myr.

Anderson *et al.* (2006) describe a Cretaceous to Palaeocene canyon from coastal California that has dimensions similar to the nearby modern Monterey Canyon. This ancient canyon is up to 15 km wide and 2 km deep. Parts of the canyon fill are conglomeratic, but medium- and coarse-grained sandstones of Facies Class B constitute more than 97% of the ∼60 m of strata which Anderson *et al.* (2006) studied at a locality called Wagon Caves Rock (Fig. 8.4). There, grain sizes range from silt to boulders. Many SGF deposits contain intraformational mud clasts, either randomly scattered or in distinct horizons. There are scattered subrounded to subangular granitic and high-grade metamorphic basement cobbles and boulders 0.5–3 m in diameter that are interpreted to have tumbled into the canyon from its steep walls.

Ito and Saito (2006) describe a conglomeratic to sandy canyon fill from the Pleistocene Higashihigasa Formation of the Boso Peninsula, central Japan (Fig. 8.5). The canyon walls are locally steep with step-like terraces. This canyon was ∼8 km long, ∼1 km at its widest point, ∼100 m deep and had a sinuosity of ∼1.1. Normally graded depositional units 2–5 m-thick dominate the fill with cross-stratified basal conglomerates (Facies Group A2) overlain sequentially by pebbly sandstones (Facies Class A) and structureless sandstones (Facies Group B1) with locally developed dish structures (Fig. 8.6).

Where sufficiently well exposed, the cross-stratified conglomerates of the Higashihigasa Formation exhibit wave-like, asymmetrically undulated upper surfaces with wavelengths of 10–63 m and relief of 0.6–2.2 m (Fig. 8.6). Climbing forms are locally present. These conglomerates are interpreted as the deposits of gravel waves like those described from some modern inner fan valleys (Fig. 1.35 and Section 7.4.2.2).

8.3 Ancient submarine channels

8.3.1 Channel scale, architecture and stacking patterns

The many studies of ancient (and modern) submarine channels have led to their characterisation as erosional, erosional-depositional or aggradational (Fig. 7.23), and with two end-member sedimentary models; that is, relatively coarse-grained low-sinuosity, to finer-grained high-sinuosity channels (Fig. 8.7; Clark & Pickering 1996a,b). The low-sinuosity, composite-channel model is based mainly on the work of Watson (1981; Fig. 8.8). Examples of simple generic models for the lateral changes in facies and facies associations in channel fills were published by Mutti (1977), Tokuhashi (1979) and Pickering (1982b; Fig. 8.9).

A synthesis of modern, ancient and subsurface industry data by Clark and Pickering (1995, 1996a,b) suggests that the width/depth ratios show considerable spread, but appear to cluster between 1 : 10 and 1 : 100 (Fig. 8.10). Here, channel scale and architecture are considered together as many studies use different definitions of what constitutes a discrete channel *versus* a channel complex. Additionally, many case studies lack the necessary 3D data at a sufficiently high resolution to identify discrete channels or nested channels. Channel stacking patterns are then considered, again acknowledging that many researchers consider together the stacking of discrete channels or channel elements without considering scale.

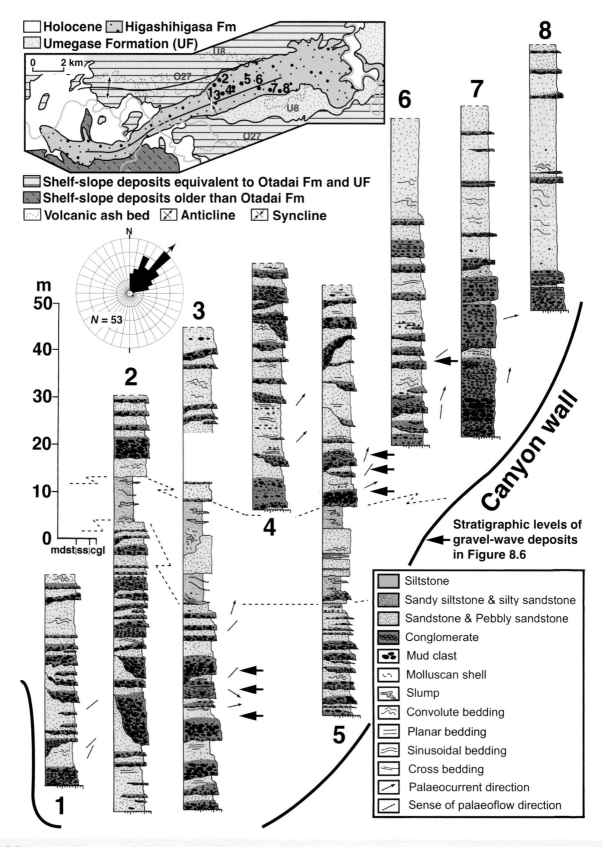

Fig. 8.5 Measured sections of palaeo-canyon-fill deposits of the Higashihigasa Formation. Inset map shows locations of measured sections and distribution of palaeo-canyon-fill and adjacent deposits. Thin dashed lines indicate distribution of slump deposits. Rose diagram indicates palaeocurrent directions measured from cross-stratification. A solid arrow indicates the mean palaeocurrent direction, and *N* is the number of measurements. From Ito and Saito (2006).

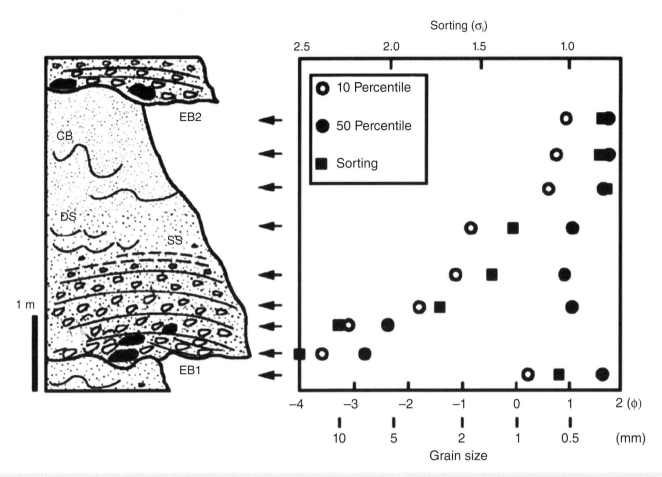

Fig. 8.6 Textural features of a typical couplet of gravel-wave deposits and overlying pebbly sandstones and sandstones from the Higashihigasa Formation, Japan. EB1 and EB2 = erosional bases of single gravity-flow deposits; GW = gravel wave; SS = sinusoidal stratification; DS = dish structure; CB = convolute lamination. Solid arrows on the right side of the column indicate positions of samples. Grain sizes were analysed by standard sieves. From Ito and Saito (2006).

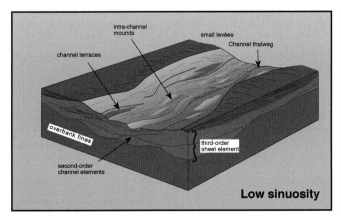

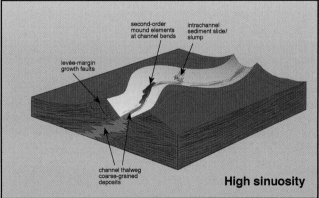

Fig. 8.7 Summary block diagrams for low- and high-sinuosity submarine channels. From Clark and Pickering (1996a,b).

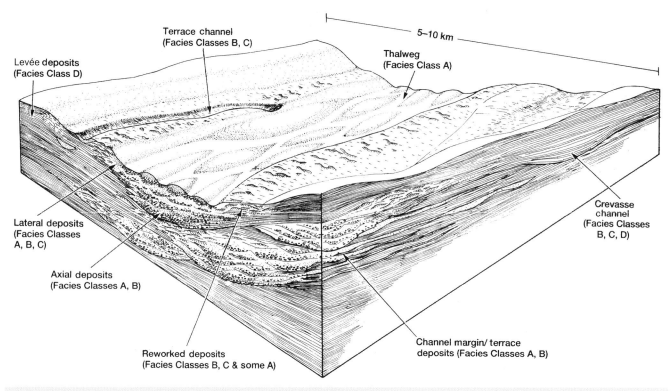

Fig. 8.8   Summary block diagram showing the interpretation by Watson (1981) for inferred upper-fan channel (complex) deposits of the Silurian Milliners Arm Formation, New World Island, Newfoundland, Canada.

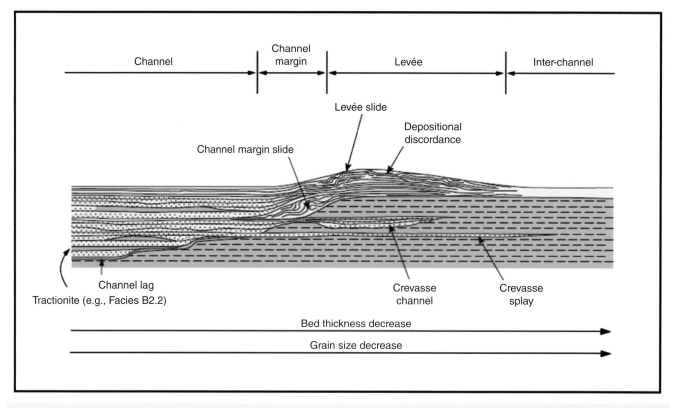

Fig. 8.9   Summary block diagram showing the typical channel fill based on a study of submarine channels in the upper Precambrian Kongsfjord Formation, Finnmark, Arctic Norway. From Pickering (1982b).

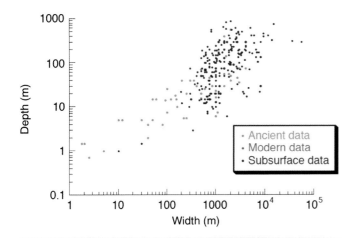

Fig. 8.10 Width/depth of ancient, modern, and subsurface submarine channels. From Clark and Pickering (1996a,b).

In the exploration, appraisal and development of hydrocarbon reservoirs, deep-water channels generally show considerable complexity (e.g., Mayall *et al.* 2006). Mayall *et al.* (2006) utilised oil company subsurface seismic data from West Africa (e.g., Navarre *et al.* 2002), together with well log and core data. They emphasised the complexity and variability of channel fills and, therefore, the limitations of applying single or even multiple depositional models to all the channels even within a given system. They described large erosionally-confined 3rd-order channels (1–2 Myr duration), typically 1–3 km wide and 50–200 m thick, and identified four principal variables controlling the nature of channel fills: (i) channel sinuosity; (ii) constituent facies (including facies-associations as defined in this book); (iii) recognition of repeated cutting and filling episodes and (iv) stacking patterns (Fig. 8.11). Mayall *et al.* (2006) recognised the common recurrence of four main seismic facies (lithostratigraphic facies associations using our terminology): (i) a basal channel lag (Facies Class A); (ii) sediment slide/slump deposits (Facies Class F) and debris-flow deposits (Facies Classes A); (ii) high net:gross stacked channel sands (mainly Facies Classes A, B and C) and (iv) low net:gross channel levées (mainly Facies Classes D and E). According to Mayall *et al.* (2006), most channels contain all of these seismic facies but in widely varying proportions, with repeated cutting and filling as a feature of almost all the channels that were studied. More recent studies by other workers suggest that similar channel (or channel-complex) architecture can accumulate considerably faster than the 1–2 Myr proposed by Mayall *et al.* (2006) (e.g., Pickering & Bayliss 2009; Cantalejo & Pickering 2014, 2015; Scotchman *et al.* 2015b).

In recent years, there has been a proliferation of broadly similar depositional models for channel fills and the most common channel stacking patterns, from generic models (e.g., Mayall *et al.* 2006; McHargue *et al.* 2011; Figs 8.11, 8.12, respectively), to those that are derived from a particular system (e.g., Forties Field, North Sea, Ahmadi *et al.* 2003; Fig. 8.13). Although scales are generally given, the models tend to be presented as widely applicable at any scale. Below, we show several alternative depositional models. While the model of Mayall *et al.* (2006) could be applied to both large inner-fan submarine channels and to many canyon fills, the absence of any 'overspill' deposits (i.e., all deposition is confined within a large 3rd order erosional surface) suggests that it is probably most applicable to submarine canyons and particularly deep upper-fan valleys.

The depositional models proposed by McHargue *et al.* (2011) hypothesise that high abandonment relief can strongly influence the location of subsequent channel elements and will result in an organised channel-stacking pattern in which the path of a younger channel element approximates the path of the former element (e.g., Fig. 7.19). McHargue *et al.* (2011) show models for under-filled and filled channels with respect to coarse-clastic content (Fig. 8.12). They explain the channel fills (and associated stacking patterns) as being dependent on granulometry and aggradation rate. The range of scenarios controlling channel facies are: (i) erosion and sediment bypass; (ii) amalgamation of channel elements associated with a low rate of aggradation; (iii) a disorganised stacking pattern of channel elements associated with a moderate rate of aggradation and (iv)

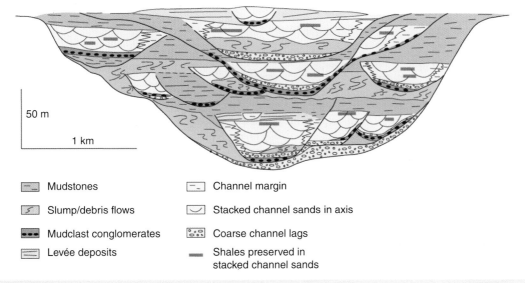

50 m

1 km

	Mudstones		Channel margin
	Slump/debris flows		Stacked channel sands in axis
	Mudclast conglomerates		Coarse channel lags
	Levée deposits		Shales preserved in stacked channel sands

Fig. 8.11 Summary model, mainly based on oil company subsurface seismic data from West Africa (e.g., Navarre *et al.* 2002), together with well log and core data, showing the potential reservoir distribution and heterogeneity patterns in a large 3rd-order erosional channel. Note the repeated cutting and filling. Compare with Fig, 8.2. From Mayall *et al.* (2006).

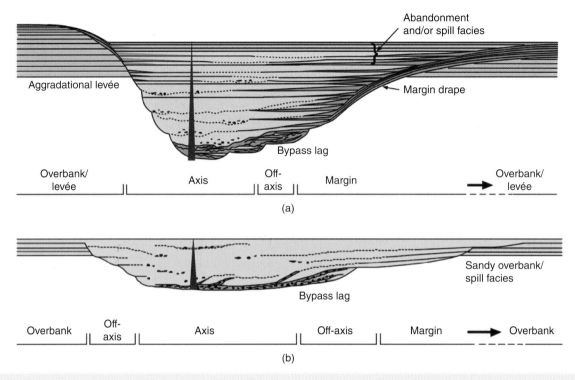

Fig. 8.12 Schematic representations of common facies stacking of under-filled and filled channel elements. Although the contrasting characteristics in (a) and (b) are common for these two channel types, variation is considerable. (a) Under-filled channel element with moderate to high rate of overbank aggradation, semi-amalgamated highly heterolithic fill, common shale/silt drapes, and capped by fining-upward trend, interpreted as an abandonment-fill facies association. (b) Filled channel element with low rate of overbank aggradation, amalgamated and less heterolithic fill, and rare shale/silt drapes. Fining-upward trend, interpreted as an abandonment-fill facies association is thin or absent. If the channel element is over-filled, sandy overbank deposits may be present. Yellow = sand-rich channel-fill sediments, green = mud-rich channel-fill sediments, brown = mud-clast-rich channel-fill sediments, grey = mud-rich pre-existing sediments. From McHargue *et al.* (2011).

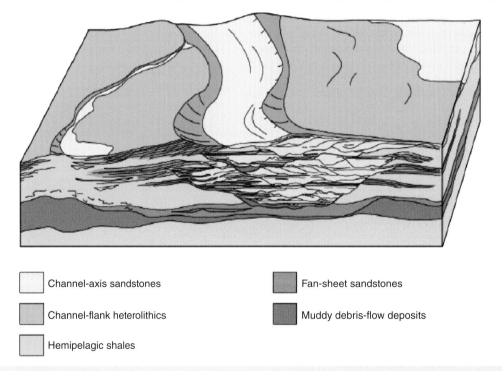

☐ Channel-axis sandstones ▨ Fan-sheet sandstones

▨ Channel-flank heterolithics ▨ Muddy debris-flow deposits

☐ Hemipelagic shales

Fig. 8.13 Schematic representations of the depositional setting for the Palaeocene Forties Field, UK sector, North Sea. From Ahmadi *et al.* (2003).

an organised stacking pattern of channel elements associated with a high rate of aggradation. Scenarios 1 and 2 may be absent or minor in mud-rich systems but prominent in sand-rich systems. Conversely, scenario 4 may be prominent in mud-rich systems but absent in sand-rich systems. Narrow and elongate basins, for example, half-grabens, submarine trenches, foreland basins, thrust-top basins, tend to only have one active axial channel at any time (e.g., as described for the Nankai Trough by Shimamura 1989). Such axial channels tend to occupy the deepest part of the basin near the most active fault-controlled subsidence.

Although most of the published depositional models for channel fills only show a single section perpendicular to palaeoflow, 3D diagrams have been produced for the low-sinuosity, coarser-grained Silurian Milliners Arm Formation, north-central Newfoundland (Watson 1981; Fig. 8.8), and the high-sinuosity Palaeocene Forties Field, UK sector of the North Sea (Ahmadi *et al.* 2003; Fig. 8.13).

The Oligocene Hackberry Formation, southwestern Louisiana, contains hydrocarbon plays within submarine channel reservoirs (Cossey & Jacobs 1992). From accurately controlled well and seismic data, the sinuous course of the sand-filled channels can be mapped out. The production reservoirs within the central channel are chiefly located at sites of channel bends, although not at every channel bend. For example, one occurs immediately downstream from the channel confluence.

Almost all the depositional models for channel fills tend to have the following features in common:

(1) Sharp, in some cases stepped, erosional base into finer-grained non-channelised deposits, possibly with a pre-channel mass-transport complex/deposit (MTC/MTD) and/or relatively unconfined channel mouth-bar deposits.

(2) Overall thinning-and-fining-upward trend, particularly in the younger part of a fill.

(3) Early channel deposits as Facies Classes A and B, with common scour-and-fill, multiple erosional surfaces, common bed amalgamation and poor lateral continuity of beds over tens of metres.

(4) Off-axis finer-grained facies associations (Facies Classes C, D and E), interpreted as either inner or outer levée-overbank deposits.

(5) Channel stacking, shown as major reactivation surfaces (e.g., laterally offset-stacked, vertically-stacked, or more irregularly stacked), above which features 1–4 are repeated several times.

In the Middle Eocene Ainsa Basin, Spanish Pyrenees, Pickering and Corregidor (2005), modified by Pickering and Bayliss (2009), proposed a model to explain such repetitive channel-filling processes within a sequence stratigraphic framework (Fig. 4.36).

The interpretation of many ancient submarine channels is based on the recognition of only one erosional-depositional margin (e.g., Figs 8.14–8.16), although in the largest outcrops it is possible to observe both margins, as in the Permian Brushy Canyon Formation of west Texas (Figs 8.18–8.20), the Permian Karoo System of South Africa (Fig. 8.21) and the Permian East Ford field in the Delaware basin, Texas (Dutton *et al.* 2003). The principal reason for description of only one channel margin is not only a function of the amount of exposure, but also because channel axes commonly tend to sequentially jump, thereby preferentially eroding one margin whilst preserving the other. Many ancient submarine-channel deposits appear to be characterised by an overall fining-upward trend, at least in the younger (upper) part of their fill (Mutti & Ricci Lucchi 1972; Mutti 1977; Pickering 1982b; Chen & Hiscott 1999a,b; Pickering & Corregidor 2005), and

Fig. 8.14   Sandstone-filled margin of submarine channel only preserving the western (right-hand) side, upper Precambrian Kongsfjord Formation, Risfjord, Finnmark, Arctic Norway. The channel sandstones with sequential offset-stacking of successive margins (mainly Facies Classes B and C) are ~15 m thick.

(a)

(b)

Fig. 8.15 General (a) and close-up view (b) of sandstone-filled margin of submarine channel cut into amalgamated sandstones (mainly Facies Classes B and C), Les Scaffarels cliff section, Annot, Oligocene Grès d'Annot Formation, Haute Provençe, southeast France. The steep-sided erosional surface (b) seen cutting down 15–20 m towards the west (left) above the group of geologists on the underlying footpath, is associated with a disorganised pebble- and mud-clast conglomerate of Facies Group A1.

(a)

(b)

Fig. 8.16 (a) Outcrop in submarine channel deposits, Oligocene Sealers Bay, Miocene, New Zealand. Most of the channel deposits are Facies Classes A, B and C. (b) Close-up of an erosional channel margin seen at the extreme left of (a). Note the abundance of mudstone clasts, many of which were probably locally derived from the bedded deposits seen to left. See Fig. 8.17 for line interpretation showing multiple erosional surfaces.

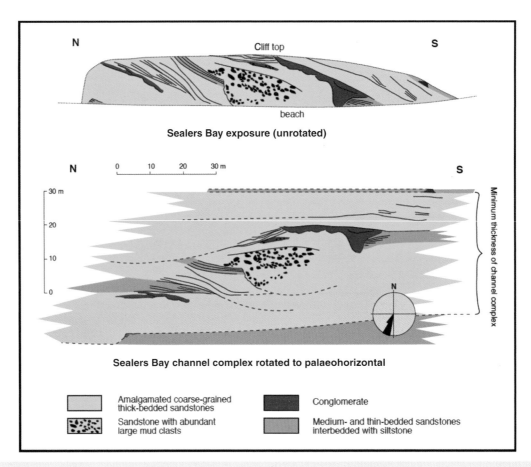

Fig. 8.17   Interpretation of Oligocene Sealers Bay submarine channel complex shown in Fig. 8.16b, Miocene, New Zealand. From Clark and Pickering (1996b).

Fig. 8.18   Northwest (left) to southeast (right) oriented photograph to show sandstone-filled submarine channel in middle field of view, Popo Channel area, Permian, Brushy Canyon, West Texas, USA. See Fig. 8.19 for interpretation.

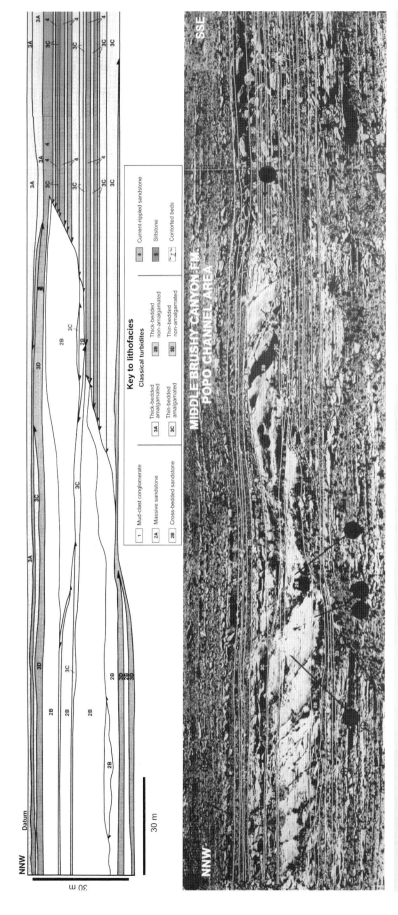

Fig. 8.19 Interpreted photograph of part of Permian middle Brushy Canyon Formation (West Texas, USA), Popo Channel area. Stratal packets labelled 4A comprise laterally continuous, current-rippled sandstones (mainly Facies Class D). From Zelt and Rossen (1995) (*cf.* Blikeng & Fugelli 2000,b).

Fig. 8.20 (a) Interpreted field photograph and (b) line drawing of a datumed cross-section of Permian upper Brushy Canyon Formation, Guadalupe Canyon, West Texas, USA, showing large, vertically-stacked channel fills that are cut into thin-bedded SGF deposits and laminated siltstones (Facies Class D deposits). Most of the channels are filled with thick-bedded SGF deposits and many of the channel margins comprise multiple, stacked, erosional surfaces. Siltstone drapes and conformable bedsets of thin-bedded, non-amalgamated SGF deposits (Facies Group C and Facies Class D deposits) are common along channel margins. Overbank deposits comprise principally fine-grained, non-amalgamated SGF deposits, laminated siltstone and current-rippled sandstone (Facies Groups C and Facies Class D deposits). From Zelt and Rossen (1995).

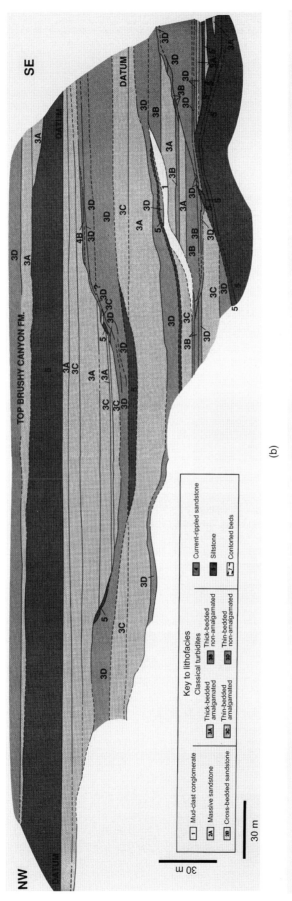

Fig. 8.20 *(continued)*

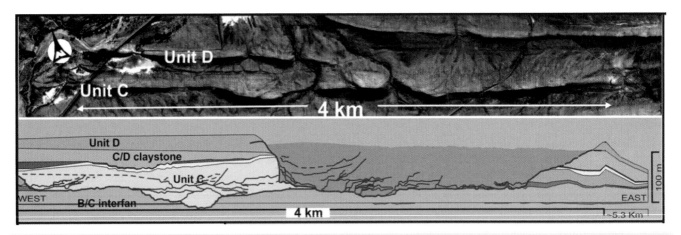

Fig. 8.21 Summary characteristics of Units C and D (so-called lower-middle slope system), a succession dominated by slope channel complexes and exposed around the town of Laingsburg, Permian Karoo Basin, South Africa. Aerial photograph and cartoon correlation panel (based on 200 measured sections) of Units B/C, C and D along the CD Ridge. Note the transect in magenta that includes both margins of an erosively-based channel complex. From Flint *et al.* (2011; *cf.* Hodgson *et al.* 2011).

also by a basal debris-flow/slide deposit, commonly referred to as a mass-transport deposit (MTD) or complex (MTC) (e.g., Fig. 4.36). Readers should exercise caution with these generalisations, because most claims of upward fining are based on subjective field impressions rather than statistical testing (Section 5.3).

The Oligocene Sealers Bay Channel (Figs 8.16, 8.17), Balleny Group, Chalky Island, southwestern New Zealand, is interpreted as a set of middle-fan channels, with a width of ~1500 m for the channel complex and ~50–100 m for the erosive channel elements, a depth of ~40 m (for the channel complex) and ~10 m (single channel elements), an aspect ratio of 37.5 (minimum for the channel complex) and 5–10 (single channel elements) (Clark & Pickering 1996b). The channels are inferred to be sinuous. The Balleny Group represents deep-marine siliciclastics that accumulated along the western side of the New Zealand continental block (northern end of the Solander Basin) and shows a general transgressive sequence (Carter & Lindqvist 1975). The underlying Nuggets Formation consists of mainly marine breccias, interpreted as deposits emplaced by debris flows in a submarine canyon (called 'inertia flows' by Carter & Lindqvist 1975). The overlying Sealers and Munida formations consist of sandstones and conglomerates that have been interpreted as submarine fan sediments, and are overlain by the Chalky Island Formation chalk marls and thin-bedded SGF deposits that represent distal fan and associated abyssal seafloor deposits. The entire succession has been compared to the facies and depositional environments in the Rio Balsas Canyon System and associated littoral facies off western Mexico (Carter & Lindqvist 1975; see paper by Reimnitz & Gutierrez-Estrada 1970).

The Sealers Bay channel-fill contains coarse-grained indistinctly-bedded sandstones (Facies Classes B and C), breccia and conglomerate lenses with large (up to 150 cm) intraformational mud clasts (Facies Class A) (Figs 8.16, 8.17). The channel is cut into similar facies with relatively steep margins. Carter and Lindqvist (1975) describe the channel as being one of five amalgamated channel features in the continuous cliff exposure, which may therefore represent a channel complex at least 40 m thick (Fig. 8.17). This interpretation is supported by the overall thinning-and-fining-upward trends subjectively recognised in the outcrop. Carter and Lindqvist (1975) interpret many of the channel-fill facies as being the deposits

of 'flow-inertia' processes, including sand-flows (inflated sand flows of Section 1.4.1) and slide-creep, giving rise to diffuse contacts between sandstone and conglomerate deposits (Fig. 8.16). The Sealers Bay Channel is interpreted as an erosional middle fan channel, due to the absence of well-developed levée deposits that would be expected in an upper fan environment. Palaeoflow direction is from the north towards the south and south-southeast. The channel is inferred to be relatively close to the sediment source, due to the high-energy nature of the flows (as inferred from grain sizes and poor internal organisation) and the mineralogical immaturity of the deposits (Carter & Lindqvist 1975).

Although there are many outcrop studies of ancient submarine channels, it is commonly difficult to appreciate their 3D geometry, particularly their sinuosity. Amongst the rare exceptions is the Tabernas 'Solitary Channel', from the Late Miocene (Tortonian) Tabernas–Sorbas Basin, southeastern Spain. Within the Tabernas Basin, five channelised sandy systems (erosively-based conglomerate/sandstones units within marlstone successions) are recognised (Hodgson & Pickering, in prep.), making the term 'Solitary Channel' somewhat of a misnomer, although for convenience the term is maintained here. Pickering *et al.* (2001) showed that the Solitary Channel was created by SGFs from the west and not from the north and east as previously interpreted by Kleverlaan (1989a,b, 1984; Cossey & Kleveraan 1995). The dominant palaeoflow in the five systems suggests flow to the east, with only the most northerly (and probably youngest) sandbody differing significantly with flow to the southwest. The third youngest sandbody contains the so-called Solitary Channel. Intrabasinal tectonic growth-folds in the Tabernas–Sorbas basins (trending southwest–northeast) appear to have controlled the depositional architecture of a channelised submarine fan system (the Sandy System). The Sandy System was fed by a series of axial (west-to-east) slope-channel complexes. The presence of seafloor growth anticlines can be identified by the uneven thickness of underlying stratigraphy across fold structures, multiple examples of onlap onto dipping strata indicating intrabasinal bathymetry, and palaeocurrent dispersal patterns. Changes in seabed bathymetry (slope angle and aspect) influenced both the routing and deposition from SGFs, and therefore, the distribution of sedimentary facies and depositional architecture of the confined fan system.

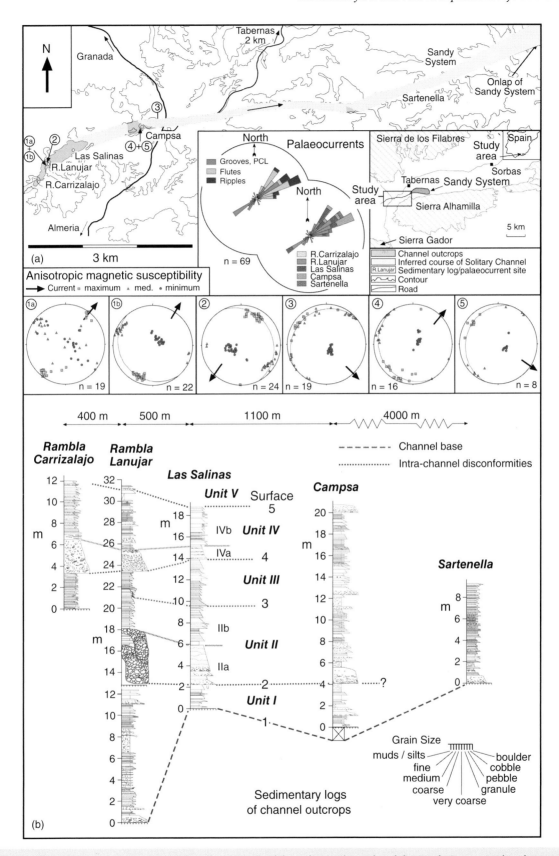

Fig. 8.22 (a) Map to show the outcrop and inferred continuity of the Solitary Channel and the Sandy System (with onlap marked in the Barranco Cerro). Rose diagrams summarising palaeocurrent data from the channel-fill. Note variety of data formats including magnetic susceptibility indicating depositional fabric from long, intermediate and short axes; the *ab* plane defining the up-flow dipping grain imbrication. Each stereonet (Lambert equal area) shows the collation of the data from all samples taken from an individual core set. The data has been corrected for bedding dip. (b) Sedimentary logs from the so-called 'Solitary Channel', Late Miocene Tabernas Basin, southeast Spain, showing lateral correlation of channel units I to V. Down-channel is towards right of figure. Modified after Pickering *et al.* (2001).

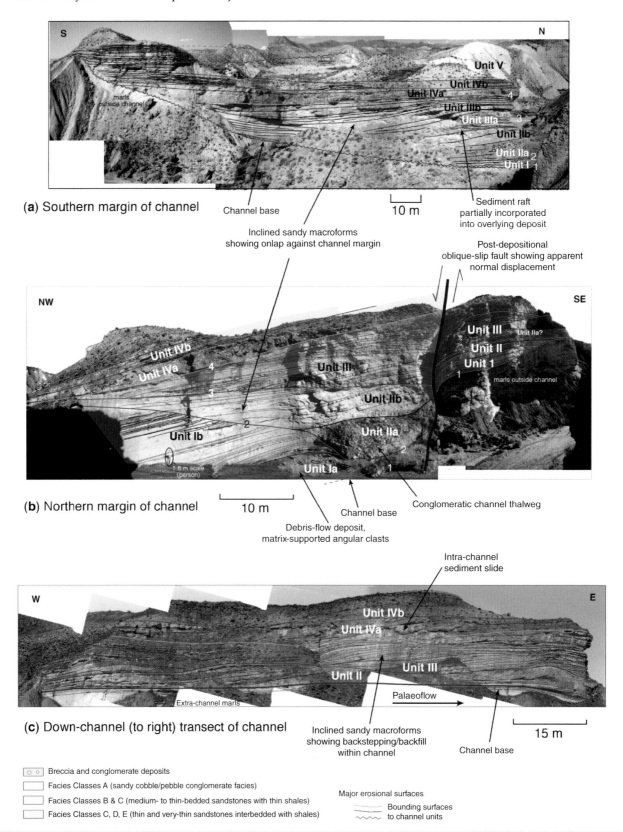

(a) Southern margin of channel

Channel base

Inclined sandy macroforms
showing onlap against channel margin

10 m

Sediment raft
partially incorporated
into overlying deposit

Post-depositional
oblique-slip fault showing apparent
normal displacement

(b) Northern margin of channel

10 m

Channel base

Debris-flow deposit,
matrix-supported angular clasts

Conglomeratic channel thalweg

Intra-channel
sediment slide

(c) Down-channel (to right) transect of channel

Inclined sandy macroforms
showing backstepping/backfill
within channel

Channel base

Palaeoflow

15 m

Breccia and conglomerate deposits

Facies Classes A (sandy cobble/pebble conglomerate facies)

Facies Classes B & C (medium- to thin-bedded sandstones with thin shales)

Facies Classes C, D, E (thin and very-thin sandstones interbedded with shales)

Major erosional surfaces

Bounding surfaces
to channel units

Fig. 8.23 (a and b) Channel margins of the 'Solitary Channel' within the Late Miocene (Tortonian) Tabernas–Sorbas Basin, southeast Spain, with abrupt onlap to south and north, Rambla de Lanujar, constraining channel width to only 200 m in this area. (c) East–west-oriented outcrop of large-scale inclined macroforms (bedforms) within the channel, Las Salinas. Inclined macroform elements are common within the western exposures of Units II, III and IV. The inclined bedding surfaces of the sandy macroforms (second-order bounding surfaces; terminology after Pickering *et al.* 1995a and Fig. 7.22) are observed in sections parallel and perpendicular to flow. See text for explanation of the various units. Modified after Pickering *et al.* (2001). Alternative explanations for these structures include perspective issues for essentially horizontal bedding (Arbués *et al.* 2013), and supercritical flow-regime upstream-inclined bedding (Postma *et al.* 2014).

Mapping of the Solitary Channel and palaeocurrent analysis by Pickering *et al.* (2001) showed that it is an example of a low-sinuosity channel on a relatively steep submarine slope (Figs 8.22, 8.23). They interpreted stratified deposits as '*low-angle sandy macroforms*' (Fig. 8.23). These were subsequently first re-interpreted as lateral-accretion deposits by Abreu *et al.* (2003), and subsequently as upstream-inclined bedding created by supercritical flows by Postma *et al.* (2014). Arbués *et al.* (2013) used LIDAR data to argue that the structures are actually horizontal bedding onlapping channel erosional surfaces, with the horizontal bedding misinterpreted to have a primary dip due to outcrop perspectives. We cite these controversial features here because they are as-yet poorly understood surfaces formed within a channel, albeit with dips of no more than a few degrees.

The Solitary Channel contains cobble/pebble lag deposits (Facies Class A), including breccias, associated with erosional phases with substantial sediment bypass, and a later infill mainly by sandstones (Facies Class B and C deposits), interpreted as channel backfill deposits. The Solitary Channel is abruptly overlain by ∼200 m of marls and then heterolithic sheet-like SGF deposits typical of a confined basin-floor setting. This change in depositional style represents a response to a significant overall decrease in basin-floor gradient, in which there was a change in base level, shown by the coeval development of a major angular unconformity farther east (Sorbas area).

In the western exposures ∼500 m apart (Lanujar and Las Salinas), the 'Solitary Channel' architecture can be divided into five distinct, laterally traceable units (Units I to V, Figs 8.22b and 8.23a). The oldest interval, Unit I, is best exposed at Lanujar, where it comprises a matrix-supported cobble conglomerate (commonly fossiliferous dolomitic limestone, probably derived from the Sierra Gador to the west and southwest) with wet-sediment deformation features such as flame structures. The base of Unit II is marked by a strongly erosional surface forming an angular unconformity (Fig. 8.22b). At Lanujar, the oldest deposits of Unit II (IIa) comprise up to ∼3 m of clast-supported, rounded to well-rounded, cobble/pebble conglomerates as scour infills (Fig. 8.23b). Units IIb and IIIb are characterised by apparently inclined sandy macroform elements, interpreted as backfill deposits. Unit III, showing the same sedimentary motif as in Unit II, also has a basal conglomeratic interval overlain by coarse-grained sandy SGF deposits (Fig. 8.22a). At Las Salinas (Fig. 8.22), Unit IV begins with the same matrix-supported pebble conglomerate observed at Lanujar, overlain by sandy SGFs (probably deposits from concentrated density flows) finer grained than those seen in Units II and III. At Las Salinas, Unit IV contains a large-scale sediment slide of sandy SGFs (Fig. 8.23c). Unit IV is abruptly overlain by thin- and very thin-bedded sandy SGF deposits and marls of Unit V. Locally, Unit V shows evidence for reactivation of a minor channel, as thin- to medium-bedded coarse-grained sandy SGF deposits of Facies Class C (Fig. 8.22a). Units I to V cannot be traced to the adjacent, down-channel outcrop (farther to the east), where the channel system begins to become less confined and the channel architecture is more sheet-like. Within the marls immediately overlying the coarse-grained channel fill, there is a discordance of ∼5°. This truncation surface is poorly exposed, and is either a large-scale early slide scar surface or a later fault.

8.3.2 Channel stacking

Channel stacking patterns can vary from isolated to those showing strong vertical and/or lateral migration (Fig. 8.24). Lateral offset stacking in a consistent direction (e.g., Walker 1975c; Browne & Slatt 2002; Kane *et al.* 2010) may be driven by differential tectonic uplift/subsidence at basin margins (e.g., Clark & Pickering 1996a,b;

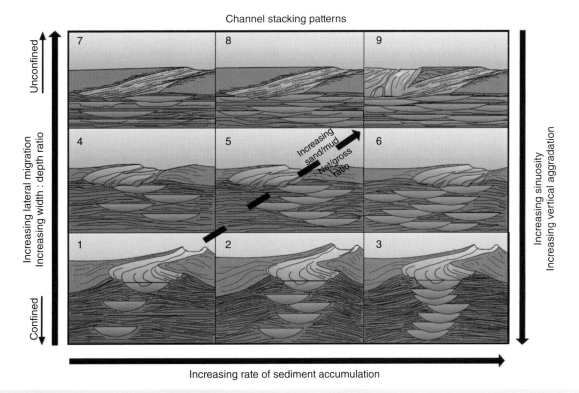

Fig. 8.24 Stacking patterns in submarine channels. From Clark and Pickering (1996a,b).

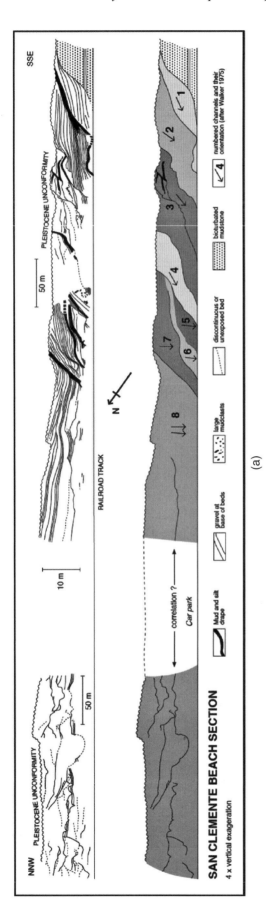

Fig. 8.25 (a) Outcrop sketch of eight nested, laterally offset-stacked, submarine channels in the Upper Miocene Capistrano Formation, San Clemente, California, USA. (b) see p. 331.

(b) see p. 331.

Fig. 8.26 Close-up of channel margins in nested, laterally offset-stacked, submarine channels in the Upper Miocene Capistrano Formation, San Clemente, California, USA.

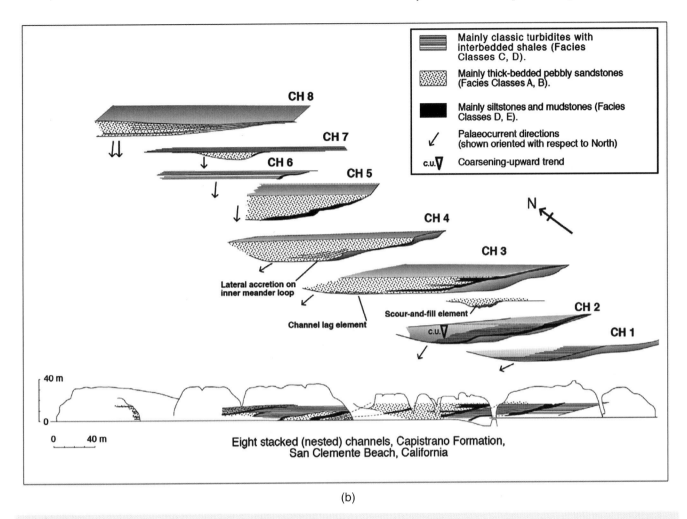

Fig. 8.25 (b) Exploded panel to emphasise the constituent architectural elements. Redrawn and modified after Walker (1975c) in Clark and Pickering (1996b).

McCaffrey *et al.* 2002; Crane & Lowe 2008; Pickering & Bayliss 2009), including that linked to salt or shale diapirism, repeated sediment failure from one margin forcing channels to relocate away from the lower-slope and adjacent basin floor, or the intrinsic tendency for the 'swing' of meanders to increase in sinuous submarine channels (Fig. 7.31; Peakall *et al.* 2000a,b, Kolla et al. 2007). A useful distinction between two factors that can force channel relocation was proposed by Clark and Cartwright (2009). They defined 'channel deflection' as 'a progressive shift in channel position away from the axis of uplift of an adjacent growing structure, causing a shift in channel position to occupy the newly forming topographic low point.' They also defined 'channel diversion' as 'a change in channel course resulting from a pre-existing structure (or series of structures) obstructing the flow pathway of the channel by modifying the slope gradient.'

The upper Miocene San Clemente submarine deposits of the Capistrano Formation (Walker 1975c; Campion *et al.* 2000; Bouroullec & Pyles 2010) provide a well-documented example of the sequential lateral shifting of channelised deposits. The channel-fill facies comprise a range of SGF deposits (Facies Class C), structureless sandstones (Facies Class B), pebbly sandstones (Facies Class A) and mudstone drapes (Facies Class E) (Figs 8.25, 8.26). The eight nested submarine channels described and numbered by Walker (1975c)

can be classified using the architectural element scheme of this book (Fig. 7.22) as third-order sand, pebbly sand and mud-filled channel elements. The sedimentary fill of each third-order channel element, however, may be characterised by lower-order elements of smaller scale; for example, Channel 1 contains a second-order sheet sand architectural element, and the second-order sandy inclined geometry of the lowermost exposed fill of Channel 4 can be interpreted as a lateral accretion element. Fig. 8.25b shows the San Clemente section redrawn using the conventions for naming architectural elements proposed in this book, together with the interpreted depositional history (see Fig. 8.26 for outcrop photographs). Third-order channel elements showing sequential offset stacking relationships, such as the eight nested channels of the Capistrano Formation (Walker 1975c), indicate progressive lateral shifting in the channel axis, due to either channel migration or avulsion. The underlying cause for these shifts, however, remains uncertain.

The Ainsa II and III Fan channels of the Eocene Hecho Supergroup, in the south-central Pyrenees, provide another example showing third-order offset stacked channel sandbodies (Fig. 8.27) that show increasing vertical aggradation with time (Fig. 8.28a,b). These submarine channels developed in a foreland (thrust-top or piggyback) basin with tectonic control on accommodation space

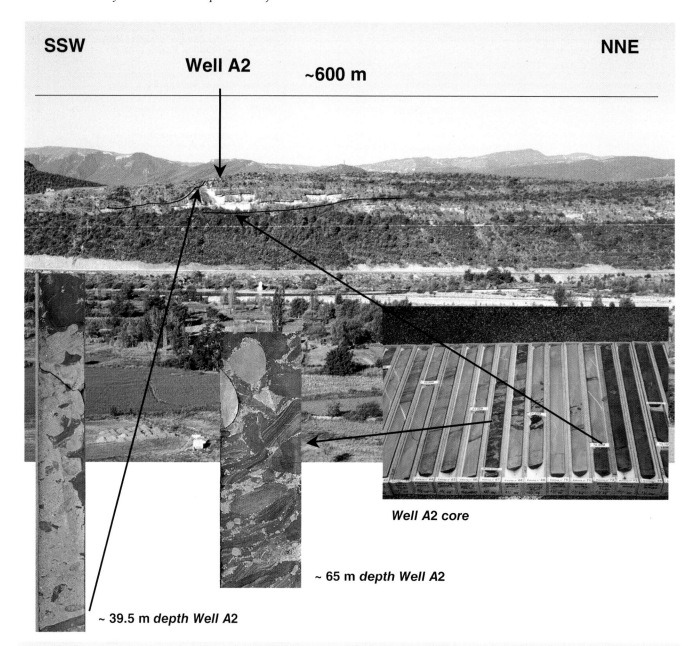

Fig. 8.27 Outcrop in Ainsa II Fan channels immediately north of Ainsa, showing major erosional surfaces associated with the stepwise west-southwest migration of channels, Middle Eocene Ainsa Basin, Spanish Pyrenees. Projection of Well A2 (~80 m behind outcrop) onto cliff face is shown. Example at left shows only basal (pebbly sandstone) part of a mass-transport deposit (MTD). Palaeoflow, towards ~320° and into the cliff face, is very oblique to this outcrop, at ~020°. These intrafan erosional surfaces are defined by type III MTDs, with two examples shown here; 6.5 cm-wide drill cores young to the left in continuous ~1 m lengths. Note basal Ainsa II sands (start of light-coloured split core), above type Ia MTD, with type III MTD ~4 m above the first sands. Inset core image (~65 m depth) shows detail from this type III MTD. The steep-sided, compound erosional surface at the left (both in the main photograph and as an inset from the core, 39.5 m depth) represents a slide scar created by type II MTD/MTCs and subsequent infill by decimetre-thick gravels, pebbly mudstones, pebbly sandstones, sandstones, and marlstones: this surface locally defines the base of the Ainsa III Fan. Modified after Pickering and Corregidor (2005) (*cf.* Dakin *et al.* 2013).

(Section 4.8), and a likely global climatic control on the delivery of voluminous coarse clastics to the deep-marine basin (Pickering & Corregidor 2005; Pickering & Bayliss 2009; Pickering *et al.* 2015).

Second-order offset-stacked channel elements are seen in turbidite channels in the upper Precambrian Kongsfjord Formation, northern Norway (Pickering 1982b) (Fig. 8.14). Detailed correlations of the Llandovery Caban Conglomerate and Ystrad Meurig Grits

Formation, south central Wales, constrained by reliable graptolite biostatigraphic subdivisions (Davies & Waters 1995; Smith & Joseph 2004), demonstrate vertical stacking architecture of the channel conglomerate elements, controlled by synsedimentary faulting (Fig. 8.29a,b). Other examples of vertical stacking of submarine channels have been described by Wild *et al.* (2005) from the Permian Tanqua depocentre, Karoo Basin, South Africa.

In many ancient submarine channel deposits it is difficult to recognise individual, discrete channels; instead, only channel complexes can be confidently defined. One of the main reasons for this is that channels tend to have complex fills, including thalweg channels within larger-scale features, and may show lateral offset-stacking, thereby preferentially removing one channel margin whilst preserving the other. Also, in the absence of exceptional outcrops, facies associations and lateral correlations permit the assignment of the deposits to a channel complex but do not permit a finer level of division into discrete channels or channel elements. An example of channel complexes (but not necessarily individual channels) showing lateral offset-stacking is exemplified by the Middle Eocene Banastón sandy system, Ainsa Basin, Spanish Pyrenees (Fig. 8.30). A wide range of facies are recognised in the Banastón System, particularly deposits of Facies Classes A, C, D and E. Facies Classes C, D and E represent the greatest number of beds observed by Bayliss and Pickering (2015a), whereas Facies Classes A and C represent the highest total thickness of all beds measured. In the Banastón System, six major sandbodies (channelised submarine fans) are recognised. The cumulative thickness of the Banastón System ranges from ~510 m in the Banastón–Usana area, to ~700 m in the San Vicente–Boltaña area (Fig. 8.30). The six channelised fans are recognised by a combination of: (i) a mappable erosional base, usually overlain by (ii) a MTC (typically comprising type II MTCs of Pickering & Corregidor 2005), then (iii) a packet of sandstones and/or heterolithic strata, with (iv) a subtle, lateral displacement in the depositional axis relative to the succeeding sandy fan in the Boltaña–Usana area and (v) a significant west-southwest shift in the depositional axis for the succeeding sandy fan, in the Boltaña–San Vicente area.

Other features observed in many ancient submarine channels are intrachannel sediment slide and slump deposits (e.g., Facies Class F), commonly associated with sedimentary dykes and sills (e.g., Smith & Spalletti 1995). In some cases, such deposits show rotational slide blocks (e.g., Vigorito *et al.* 2005). Vigorito *et al.* (2005) describe a Miocene submarine channel system (and related fan) in a foramol/rhodalgal carbonate setting from the Isili area in the syn-rift Sardinia Basin, Italy. Foramol/rhodalgal carbonate 'factories' developed on submerged structural highs. These carbonate factories were periodically stripped of their unconsolidated veneer mainly during falling sea-level cycles, causing the sediments to be redeposited into the basin via a complex submarine channel network that included a tributary belt, one main channel (Isili Channel) and a related submarine fan. The Isili Channel is up to 1 km wide, 60–100 m deep, and includes two stacked channel complexes each constructed from several smaller-scale channel elements. Complex stratal geometries are observed in the Isili Channel (e.g., overbank, levée, margin and channel thalweg) which also include up to 15 m-high bedforms. The coarsest facies associations include sandy to cobble-sized SGF deposits of Facies Classes A and B, and megabreccias characterised by displaced and tilted blocks (Fig. 8.31) which resulted from major channel-margin collapses (Facies Class F).

8.3.3   Case study: Milliners Arm Formation, New World Island, Newfoundland

Although we do not provide many case studies in the book, here we summarise the work of Watson (1981) who documented in detail ancient examples of low-sinuosity channels from the Llandovery Milliners Arm Formation, New World Island, north-central Newfoundland. This case study shows how detailed field mapping, sedimentary logging and facies analysis led to the depositional model for low-sinuosity channels shown in Figure 8.8.

The Milliners Arm Formation (MAF) is a 4.5 km-thick, progradational fan succession of Late Ordovician to Early Silurian age in the Dunnage Zone of the Appalachian Orogen in Newfoundland (Watson 1981; Arnott *et al.* 1985). The sedimentological data and interpretations that follow are entirely from Watson (1981). The MAF can be conveniently divided into lower and upper parts. The lower part (3.1 km thick) is characterised by sandstone packets, each ~10–20 m thick, separated by units of thin-bedded SGF deposits and shales from ~10 m thick up to ~250 m thick in those parts of the succession that lack sandstone beds. The upper part of the formation (1.4 km thick) is characterised by channelised conglomerate and sandstone bodies (Figs 8.32–8.34). Average palaeoflow is from north to south, but with considerable dispersion about the mean, particularly in facies associations that are non-channelised. The best sections through the lower part of the MAF are narrow coastal exposures that generally do not allow tracing of beds for more than a few tens of metres along strike. In exceptional circumstances, however, lateral tracing of packets for up to 600 m is possible. The upper part of the formation crops out in long strike sections that allow detailed bed-by-bed and packet-by-packet correlations up to 3 km.

The lower part of the MAF can be divided into five facies associations: three non-channelised associations and two channelised associations (Table 8.1). The non-channelised and channelised associations alternate throughout the lower part of the formation.

In non-channelised intervals, the preferred sequence of facies associations is NCF1 ⇒ NCF2 ⇒ NCF3. NCF1 constitutes 72% of all non-channelised deposits in the lower part of the formation, and consists predominantly of thin and very thin-bedded siltstone SGF deposits, thin-bedded sandstone SGF deposits, and structureless mudstone. Individual beds are laterally continuous for >250 m without changes in either bed thickness or internal sedimentary structures. Where asymmetric cycles were subjectively recognised, they most commonly are interpreted to thicken and coarsen upward. NCF2 forms only 11% of the thickness of the lower part of the MAF. Maximum thickness of the association is 10 m, with most occurrences being 3–4 m thick. This association is more sand rich and thicker bedded than NCF1, and is heterogeneous, with interbedded thin and medium beds. Packets of NCF2 are continuous for hundreds of metres along strike, although individual sandstone beds may be lenticular or wedge-shaped. NCF3 forms 17% of non-channelised deposits, and is the thickest bedded and coarsest grained of the three non-channelised associations. Sandstone beds are amalgamated, and sand : shale approaches 100%. Sandstone packets vary in thickness from 2–25 m, with the thickest packets having the most coarse-grained facies. The bases of packets are relatively sharp, but there is no evidence of significant erosion. Asymmetric cycles were rarely identified. Instead, packets are in many cases coarsest in the centre, and finer both at the top and the base. The coarse central parts (Fig. 8.34) consist of lenticular bodies of pebbly sandstone that contrast in their shape with the essentially sheet-like nature of the packets themselves. In one example, the coarsest parts of successive beds are offset from one another, producing a shingling effect.

The NCF3 sandstone packets are interpreted as non-channelised depositional lobes that grade distally into lobe-fringe (NCF2) and fan-fringe (NCF1) deposits. The interlayering of packets of the three facies associations, with the trend NCF1 ⇒ NCF2 ⇒ NCF3, indicates that the dominant growth pattern of the fan at some times was repeated progradation of sandy lobes. Parts of the sections that are characterised by alternations of NCF1 and NCF2, or NCF2 and

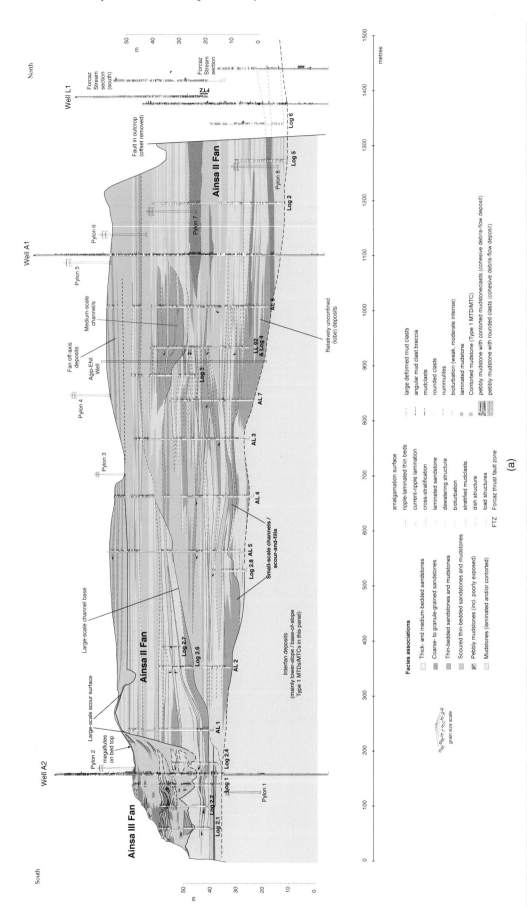

Fig. 8.28 (a) Outcrop in Ainsa II Fan immediately north of Ainsa, along road to Labuerda, showing major erosional surfaces associated with the stepwise overall west-southwest migration of large-scale channels. Projection of Well A2 (~80 m behind outcrop) onto cliff face is shown. Palaeoflow, towards ~320° and into the cliff face, is very oblique to this outcrop, at ~20°. Note the major ~30 m-deep erosion surface that excised large parts of the Ainsa II Fan, and which defines the base of the Ainsa III Fan. Also note the early Ainsa II Fan non-channelised relatively unconfined sandy deposits, interpreted as proximal lobe and channel mouth-bar environments, cut into by small (shallow) channels and scours, then eroded by larger channels and a final thinning-and-fining-upward trend as the Ainsa II Fan was abandoned. (b) Wireline log correlation of seven wells drilled behind outcrops of the Ainsa I, II and III fans, Middle Eocene Ainsa Basin, Spanish Pyrenees. Wireline logs include a filtered gamma-ray log (WIRE.GR_FLTR2_1), neutron log (WIRE.LNN_1) and porosity log (WIRE.NPHI_1). The gamma curve has been coloured with an automated lithofacies interpretation based on V-Shale cut-offs derived from the wire-line logs. A thrust fault identified at outcrop and in the Labuerda-1 Well has resulted in ~34 m of repeated stratigraphy which has been accounted for by splicing the well-log at the fault and showing the repeated section side-by-side.

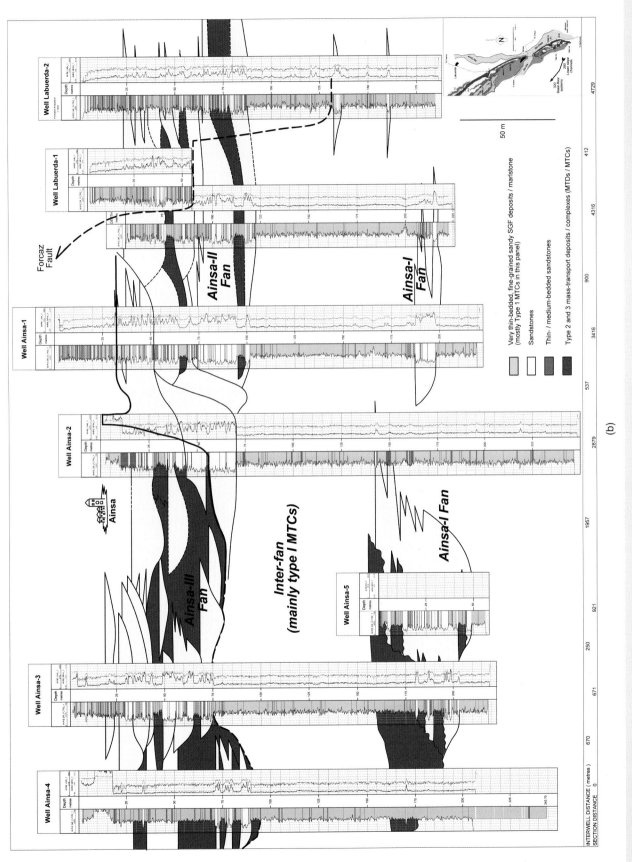

Fig. 8.28 (b) Correlations between wells are interpretive, but are based on the frequency, size, facies distribution and stacking pattern of channel elements and mass transport deposits observed in the outcrop face. The correlation highlights the stratigraphic complexity of these channelised systems and the difficulty in correlating depositional elements between wells only short distances apart. Aside from some aggradational stacking (as seen in the late-stage Ainsa III Fan channels), most channels show incremental west-southwest 'deflection' in response to the growth of the eastern anticline that defined the lateral basin slope outboard of the rising fold-and-thrust belt. From Pickering *et al.* 2015.

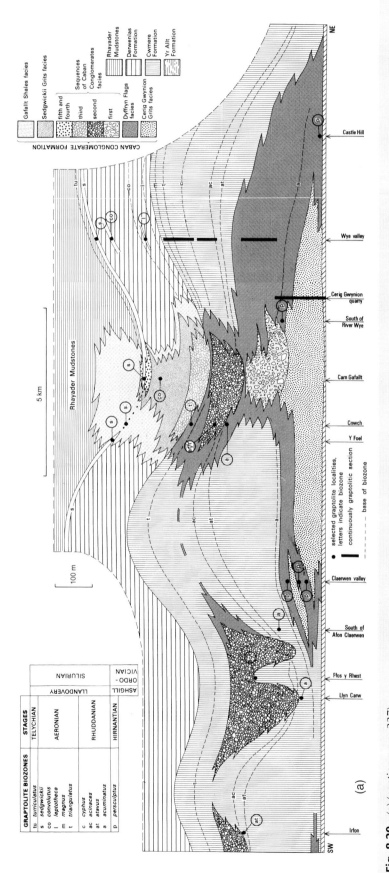

Fig. 8.29   (a) (caption on p. 337)

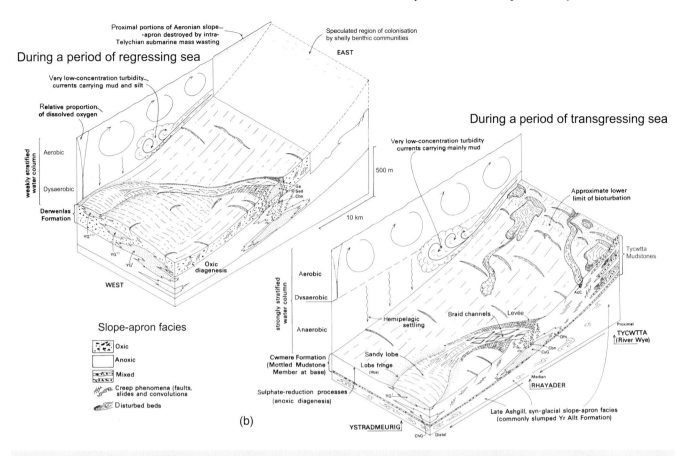

Fig. 8.29 (a; graphic p. 336) Architecture of the Silurian Caban Conglomerate Formation, Welsh Borderland, showing the fault-controlled vertical stacking pattern of the channels. (b) Depositional model for latest Ashgill to Llandovery slope-apron, nested channels and lobes between the Ganth Fault and village of Ystradmeurig in the graptolite *cyphus* (a) and *convolutes* (b) biozones. AcC = Allt-y-clych Conglomerate; Cbn = Caban Coch Conglomerate; Dfn = Dyffryn Flags; Ga = Gafallt Shales; Hbs = Henblas deposits; Sed = Sedgwickii Grits; YG, YG', YG'' = 1st, 2nd and 3rd depositional units of the Ystrad Meurig Grits Formation. From Davies and Waters (1995; *cf.*Smith *et al.* 1991). For detailed descriptions of the facies, the reader is referred to Davies and Waters (1995).

NCF3, probably developed by rapid lateral switching of lobes, with no systematic progradation.

The channelised facies associations are more-or-less equally abundant (CF1 : CF2 = 55 : 45). CF1 consists of strongly lenticular, amalgamated structureless and graded sandstones and pebbly sandstones with parallel- and cross-stratification. Pebbly sandstones may form 50% of CF1 units. The bases of the units are sharp and erosive, being characterised by either abundant rip-up clasts forming a basal lag, or step-wise downcutting to minimum depths of 2–3 m. The tops of CF1 packets tend to show either a step-like thinning upward into overlying fine-grained deposits of CF2, or are abrupt. This upper transition zone is characterised, in about 1/3 of all packets, by Facies B2.2 reworked, cross-stratified, lenticular units to 4 m thick. This facies is almost exclusively confined to the top of CF1 units. CF2 consists of thin-bedded and very thin-bedded siltstone SGF deposits, isolated siltstone ripple trains, and minor thin-bedded sandstone SGF deposits, in units 2–35 m thick. These facies are organised into well defined, subjectively identified thinning-upward trends several metres thick. Palaeoflow data are more variable than for CF1.

Watson (1981) interpreted associations CF1 and CF2 as channel and interchannel deposits, respectively. The interpreted thinning-and-fining-upward trends that characterise the top of CF1 units are believed to indicate progressive channel abandonment (Mutti

& Ghibaudo 1972) and/or shingling of the thickest parts of beds in a wide channel ('compensation cycles' of Mutti and Sonnino 1981; Fig. 7.39). The reworked deposits of Facies B2.2 that occur at the top of sandstone packets are interpreted as channel-margin sediments, formed beneath non-depositing flows that spilled over the channel levees. The specific process involved might have been flow stripping (Piper & Normark 1983). Channel migration led to the superposition of channel-margin deposits on top of channel deposits. Channelised and non-channelised facies associations appear to alternate throughout the lower part of the MAF, and are interpreted as middle and lower fan deposits. Watson (1981) made these channel versus non-channel interpretations from the inferred presence of asymmetric cycles, something that has been challenged as a reliable technique by Chen & Hiscott (1989a,b) in Section 5.3 of this book.

The upper part of the MAF is ~1.4 km thick, and consists predominantly of thick units of conglomerate and pebbly sandstone of Facies Class A (e.g., Fig. 2.17b). Five facies associations are recognised (Table 8.2), the first two being mainly coarse-grained and the last three being finer grained. Abundant channelling and the coarse nature of the deposits led Watson (1981) to interpret all these sediments as upper fan deposits.

Facies association IF1 constitutes 12–70% of measured sections, and is dominated by Facies Class A clast-supported conglomerates.

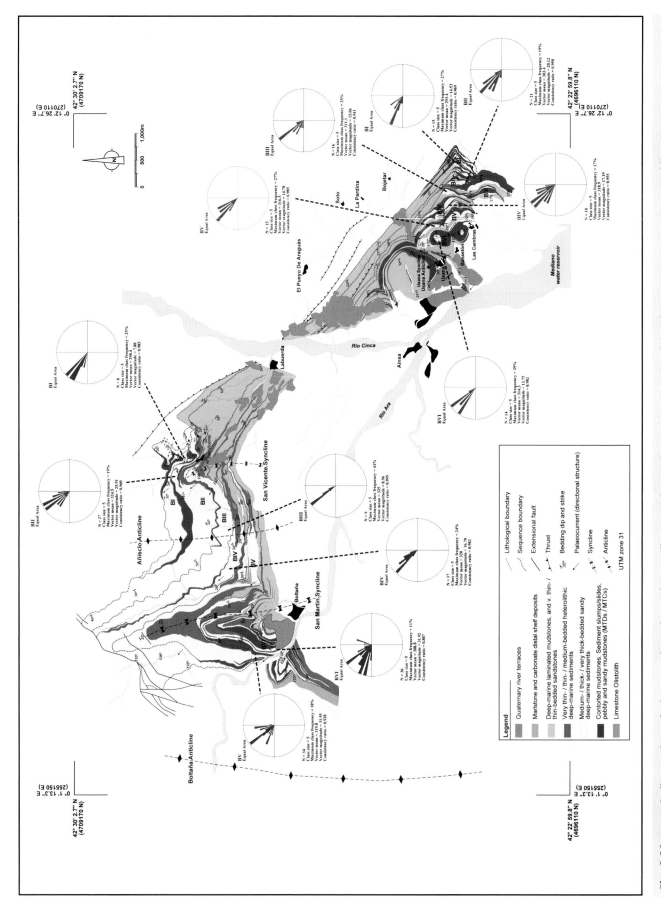

Fig. 8.30   Lateral offset-stacking of submarine channel complexes (yellow = sandy SGF deposits; green = heterolithic deposits), Banastón System, Middle Eocene, Spanish Pyrenees. From Bayliss and Pickering (2015a). See text for explanation.

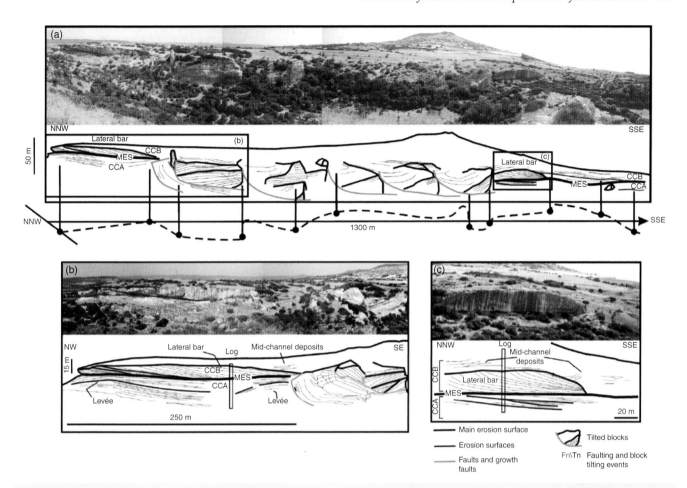

Fig. 8.31 (a) Isili Channel longitudinal section, Miocene, Sardinia Basin, Italy, showing exposures of Channel Complex A (CCA) and B (CCB) along the modern Riu Corrigas Canyon (window of observation) which intersect at different angles the Miocene Isili Channel allowing some 3D control. Trend of the exposures (dashed line) in relation to the Isili Channel trend is shown. Note the multiple tilted fault-blocks (Isili Megabreccias). (b) Lateral bar, up to 15 m high (CCB), prograding over previous channel margin/levée complexes (CCA). On the right side of the photo there is a large, tilted fault-block and related growth faults (numbers indicate the chronology of block tilting events and related faulting). South Riu Corrigas section. (c) Lateral bar downlapping surface labelled MES. Note the wedge-like geometry of the underlying Channel Complex A deposits, which are interpreted to correspond to the transition from the channel margin to the channel axis. From Vigorito *et al.* (2005).

Fig. 8.32 Submarine channel deposits (mainly Facies Classes A and B), with basal stepped erosion surface into Facies Classes C, D and E deposits, Silurian Milliners Arm Formation, New World Island, Newfoundland, Canada. Beds young to right.

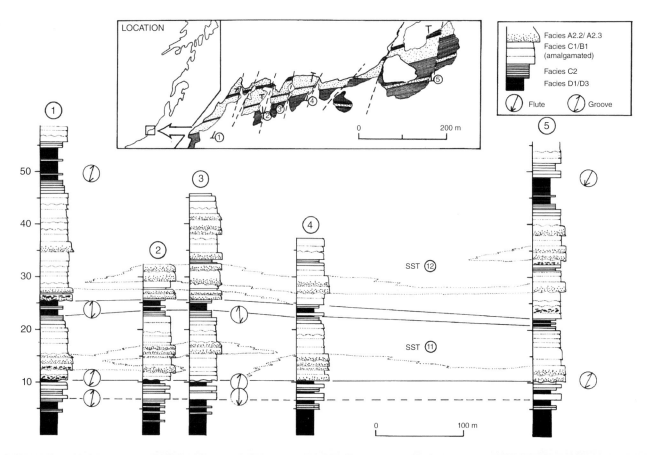

Fig. 8.33 Lateral variation in NCF3 packets over a lateral distance of ~600 m. Whereas entire packets maintain a fairly constant thickness, pebbly sandstone facies are lenticular. Palaeocurrent measurements are plotted with north at the top of the diagram. Silurian Milliners Arm Formation, New World Island, Newfoundland, Canada. From Watson (1981).

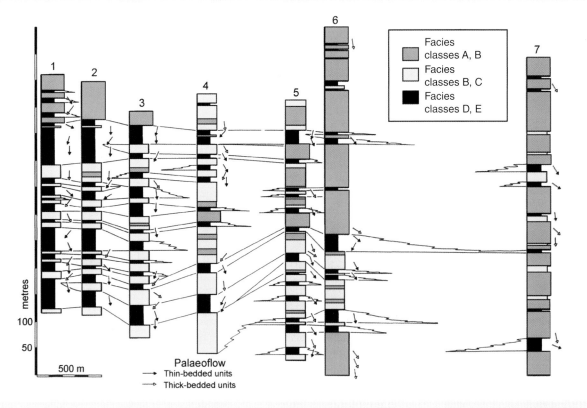

Fig. 8.34 Simplified correlated sections in the upper part of the Silurian Milliners Arm Formation, New World Island, Newfoundland, Canada, showing a general eastward decrease in finer-grained facies, and the importance of channels. From Watson (1981).

Table 8.1 Composition of Non-channelised Fan (NCF) and Channelised Fan (CF) facies associations, Lower Milliners Arm Formation. From Watson (1981)

Association	Percent by facies													
	A1.4	A2.7	A2.8	Bl.l	B2.1	B2.2	Cl.l	C2.1	C2.2	C2.3	D2.1	D2.2	D2.3	El.l
NCF1	—	—	—	—	—	—	3	—	−8	21	40	—	25	1
NCF2	—	—	—	—	—	—	—	11	24	32	28	—	5	—
NCF3	—	10	4	23	—	—	—	45	15	3	—	—	—	—
CF1	3	22	10	27	1	3	—	28	4	2	—	—	—	—
CF2	—	—	—	—	—	—	2	4	—	9	30	13	40	2

Table 8.2 Composition of Inner Fan (IF) facies associations, Upper Milliners Arm Formation. From Watson (1981)

Association	Percent by facies																
	A1.1	A1.4	A2.2	A2.3	A2.4	A2.7	A2.8	B1.1	B1.2	B2.1	B2.2	C2.1	C2.2	C2.3	D2.1	D2.2	D2.3
IF1	13	11	23	27	4	12	8	2	—	—	—	—	—	—	—	—	—
IF2	1	5	—	4	3	23	15	18	—	3	—	25	3	—	—	—	—
IF3	—	—	—	—	—	—	—	—	17	—	65	—	—	5	8	3	2
IF4	—	—	—	—	—	—	—	4	—	1	—	22	30	26	15	—	2
IF5	—	—	—	—	—	—	—	—	—	—	—	—	1	12	61	6	19

Amalgamated units average ~20 m thick, and are interlayered with packets of IF2 to form multiple stacked units up to 130 m thick. Association IF2 forms 20–40% of measured sections, and is dominated by graded and stratified pebbly and non-pebbly sandstones in units 2–18 m thick. The composite IF1/IF2 units have erosive bases, with demonstrable erosion of up to 18 m into underlying thin-bedded units. The erosion is accomplished by a series of steep steps separated by flat terraces (Fig. 8.32). Internal erosion is also characteristic of the IF1/IF2 units.

Association IF3 is volumetrically minor, and comprises very lenticular units, <7.5 m thick, of relatively well-sorted, cross-bedded sandstones (Facies B2.2) and gravel lags (Facies B1.2). These deposits almost always overlie composite IF1/IF2 units and underlie IF4/IF5 units. Cross-bed azimuths differ from sole markings in both underlying and overlying units by 30–60°.

Association IF4 forms 3–33% of measured sections, in units 1–20 m thick, and consists predominantly of Facies Groups C2 and D2. Where units can be traced along strike for distances of several tens of metres, they are seen to have a wedge shape that results from gradual thinning (or thickening) of all sandstone beds in the unit. Both grain size and the proportion of beds starting with Bouma T_a and T_b divisions decrease in the direction of thinning. Palaeoflow for IF4 units differs by up to 50° from the mean direction in IF1/IF2 units.

Facies association IF5 accounts for 30% of sections through the western part of the upper MAF, but forms only 2% of more easterly sections (Fig. 8.35). Siltstone SGF deposits of Facies Group D2 form 85% of this association (Table. 8.2). Packets of IF5 are 4–40 m thick, and commonly gradationally overlie units of IF4. Contorted bedding and truncation surfaces attest to gravitationally-induced sliding. Palaeoflow is broadly the same as for IF1/IF2 units.

In vertical sections through the upper part of the MAF, Watson (1981) suggested a tendency for facies associations to succeed one another in the order IF1 ⇒ IF2 ⇒ IF3 ⇒ IF4 ⇒ IF5 (Fig. 8.35). Partial sequences occur as a result of erosion beneath IF1/IF2 conglomerates and pebbly sandstones. Fully developed sequences from IF1 to IF5 are most easily explained by channel-migration processes, with IF1 deposits corresponding to the channel axis and IF5 deposits to levée and overbank sediments. The reworked deposits of IF3 are assigned to the channel margin, and are believed to have formed during spillover of non-depositing flows, a suggestion supported by the palaeoflow divergence between axial deposits and cross-beds of IF3. Association IF2 was deposited in the channel system, and is intimately associated with conglomerates of IF1. This lateral segregation of conglomerates from pebbly sandstones in the channel suggests a system of thalweg channels (IF1) flanked and separated by terraces (IF2) (Fig. 8.8), much like the depositional model proposed by Hein and Walker (1982) for similar deposits in Québec, Canada.

The entire Milliners Arm Formation is the deposit of a prograding submarine fan system. Oscillations between channelised and non-channelised lobe and lobe-fringe deposits in the lower part of the formation, however, indicate that fan growth was irregular, with frequent lobe abandonment. The upper fan valley was of the order of 5–10 km across, a size that corresponds to valleys of modern fans with a radius of about 50–100 km (Normark 1978).

8.3.4 Levées

The architecture and facies of modern and ancient levée deposits have been described by many authors (e.g., Mutti 1977; Normark *et al.* 1980; Pickering 1982b; Damuth & Flood 1984; Pickering *et al.* 1986b; Kolla & Coumes 1987; Piper & Deptuck 1997). Typically, levée deposits show large-scale bundling of beds, up to many metres thick, into laterally-pinching packets of beds (Fig. 8.36), wedging out over hundreds of metres to a few kilometres. Levée sediments generally show rhythmic occurrences of laminated silty turbidite beds on the order of 1–5 cm thick, with some thicker, medium- to fine-grained sandy beds; bed geometry is commonly very irregular on a scale of metres. Inclined, contorted and folded laminae are common, resulting from wet-sediment sliding within gravitationally unstable accumulations of levée deposits. In some cases, levée SGF deposits may be interbedded with debrites and crevasse sands.

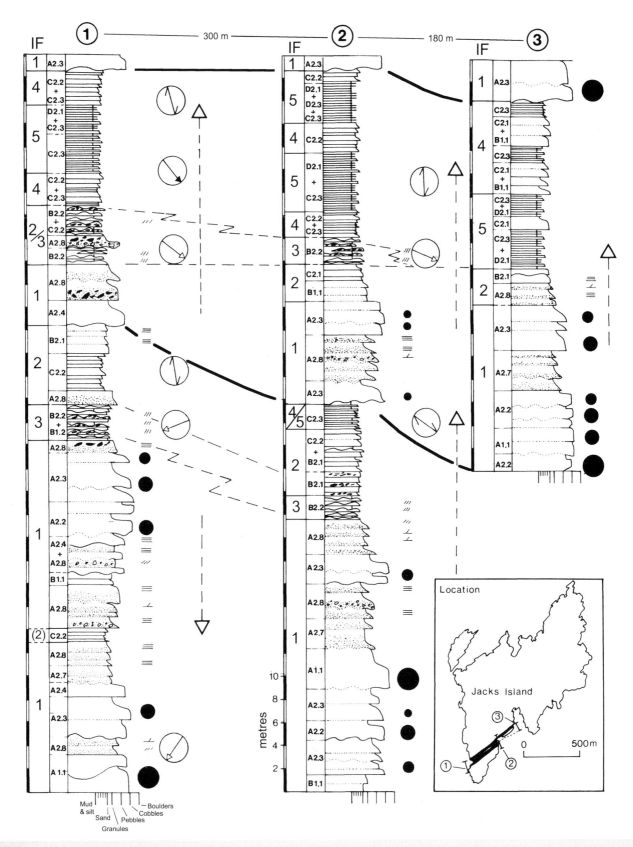

Fig. 8.35   Correlation of detailed sections in the upper part of the Silurian Milliners Arm Formation, New World Island, Newfoundland, Canada, showing a tendency for fining-upward trends with partial or complete cycles of IF1 ⇒ IF2 ⇒ IF3 ⇒ IF4 ⇒ IF5. Coarser grained packets tend to be lenticular. From Watson (1981).

Fig. 8.36 Thin-bedded, fine-grained Facies Classes C and D deposits in a channel off-axis location, interpreted as levée–overbank deposits. Permian Brushy Canyon Formation, West Texas, USA. Note the thinning and wedging out of both individual beds and packets of beds against metre-scale discontinuities or unconformities.

A commonly cited problem with many ancient channel–levée–overbank complexes (generally interpreted by analogy with modern middle-fan channels), is the inability to 'walk-out' beds from within a channel into the levée or overbank deposits. Some rare exposures of ancient channels, however, do permit this, where sandstone beds appear to have onlapped high up the inner levée flanks, and possibly even extend out of the channel into interchannel areas. Figure 8.9 shows the architectural elements of the Hamningberg Channel, Kongsfjord Formation, North Norway (Pickering 1982b). The architecture of this third-order channel contains both axial-fill elements in lateral continuity with channel margin and levée elements. The axial-fill element (a second-order, small pebble-granule/very coarse-grained sandstone and conglomerate channel element) passes laterally into channel margin and levée architectural elements (e.g., a second-order slide element on the channel margin and a thickening-and-coarsening-upward trend), with no observed break in bed continuity. The medium-grained, thick-bedded, sheet element, characterised by palaeocurrents diverging away from the main channel current direction, may be a clastic sill/dike. The continuity of coarser-grained channel facies with overbank sands is consistent with outer-bank spill of density-stratified gravity currents from submarine channel bends characterised by strong run-up (Straub *et al.* 2008) and/or secondary flow reversed from that typical of river channels (Parsons *et al.* 2010).

In contrast to the interpreted thickening- and coarsening-upward trend at the eastern channel margin, the axial-fill element in the western part of the channel fill is interpreted to show a thinning-and-fining-upward trend. A possible explanation is that

the gradual infilling of the channel by vertical accretion resulted in the lateral progradation of the intrachannel deposits, away from the channel axis and onto the channel banks. This process generated a thickening-and-coarsening-upward trend at the channel margin/levée. Thus the two trends can be interpreted as complementary, representing differing responses to one process, that is, 'sympathetic sedimentation' (Pickering 1982b).

Following early studies that recognised overbank–levée deposits (Mutti 1977; Pickering 1982b), there are increasingly more detailed studies of ancient levee–overbank deposits (e.g., Kane *et al.* 2007, 2009; Beaubouef 2004) or off-axis deposits that appear similar to levée deposits (e.g., Figs 8.37. 8.38, 8.39). '*Inner levées*', shown as a component of the channel margin deposits by Pickering (1982b, Fig. 8.9), that form within the confines of large channel or channel-belt conduits, have been shown to be important depositional elements in many deep-water systems (e.g., Hubscher *et al.* 1997; Piper *et al.* 1999; Deptuck *et al.* 2003; Hubbard *et al.* 2008).

Beaubouef (2004) describes migrating, levéed-channel complexes in the Cerro Toro Formation in the Torres del Paine National Park, southern Chile. Collectively, the channel-fill units in his study area form a belt ~5 km wide and several hundred metres thick. Four sets of channel complexes are identified. The channels are filled by bedded conglomerate and amalgamated sandstones interpreted to represent the deposits of various SGFs. Large-scale cross-beds in some of the pebble to cobble conglomerates indicate significant bed-load transport forming bars in the channels. From the channel-axis to the channel-margin, facies change from clast-supported conglomerate to either (i) thick-bedded sandstone or (ii) matrix-supported

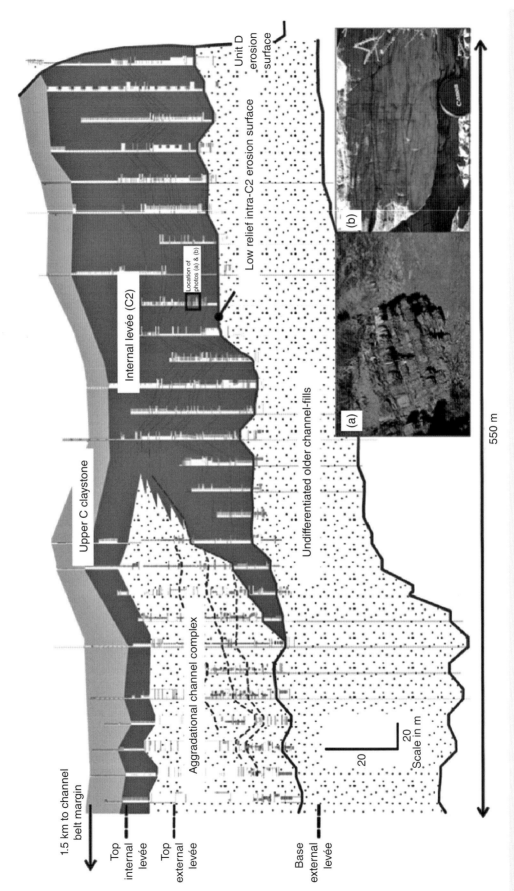

Fig. 8.37 Correlation panel of the eastern so-called 'internal levée' (i.e., deposits that accumulated off-axis and at the channel margin) of Unit C in the C/D ridge, that overlies a widespread low relief erosional surface of the Permian Fort Brown Formation, Karoo Basin, South Africa. Individual correlated beds show the degree of lateral continuity. The top of the aggradational channel complex and the internal levées are palaeo-topographically higher than the external levée, which suggests that the internal and external levées merge. The internal levée is cut by the Unit D erosion surface to the east. Inset photographs illustrate (a) a lack of clear bed stacking and (b) a sandstone bed with climbing ripple lamination. From Kane and Hodgson (2011).

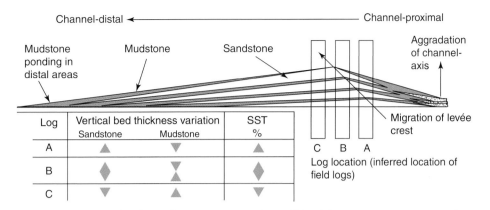

Fig. 8.38 Schematic model of levée growth in the Upper Cretaceous slope sediments, Rosario Formation, Baja California, Mexico, accounting for thinning-upwards sandstone trends in the most proximal Log A, thickening- to thinning-upwards trends in Log B and a well-developed thickening upwards trend in Log C and more distally (in unlabeled Logs D–J). As the levée aggraded vertically, the crest migrated in a channel-distal direction; this scenario requires increasing magnitude of flows (and possibly increasing frequency of flows) combined with channel-floor aggradation maintaining levée relief. Note that the position of the thickest sandstone is always just beyond the levée crest. The model is compiled from sandstone and mudstone thickness data shown in Fig. 8.39. From Kane *et al.* (2007).

conglomerate (*cf.* Jobe *et al.* 2010). Channel-fill facies lie on erosional surfaces that cut into adjacent and subjacent inter-channel facies. Beds thin and onlap these surfaces toward the channel margins. Shale or siltstone drapes of the channel cuts are uncommon and laterally discontinuous. Bed continuity between the channel and adjacent, inter-channel facies is not observed. The inter-channel strata are interpreted to represent levée successions that bound the channels. Strata in the levée units consist of (i) basal, sandy lobe deposits comprised of medium- to thick-bedded SGF deposits and concentrated density-flow deposits and (ii) overbank facies consisting primarily of packets of thinning-and-fining-upward, fine-grained, thin-bedded SGF deposits. The vertical transition between these facies is gradual. Distal levée facies include mudstones with thin-bedded, laterally continuous sandstones. Proximal levée facies include mudstones with both thin- and thick-bedded sandstones; however, the thick-bedded sandstones have lower lateral continuity. The proximal levée facies have a higher sandstone percentage than the distal levée, but also have greater depositional and post-depositional complexity, with sand-filled crevasses, erosional truncation and slumped beds. Field observations suggest that these levéed channels formed in stages that are represented by depositional and/or erosional events. In chronological order, these are (i) an initial stage of relatively unconfined, sand-rich deposition; (ii) aggradation of a mud-rich, confining levée system resulting from overbank deposition as turbidity currents bypassed the area; (iii) erosion as the channel became entrenched or as the channel migrated and (iv) filling of the channel-margin relief by onlap of channel-fill sediments.

Hodgson *et al.* (2011) and Kane and Hodgson (2011) describe two seismic-scale submarine channel–levée systems from the Karoo Basin, South Africa: one a levée-confined channel system (Unit C) and the other an entrenched channel system (Unit D) (Fig. 8.37). They show that in both cases, channel evolution followed a common stacking pattern, with initial horizontal stacking (lateral migration) followed by vertical stacking (aggradation). They interpret these changes in architecture as a response to an equilibrium profile shift from low available accommodation space (slope degradation, composite erosion surface formation, external levée development, sediment bypass) through at-grade conditions (horizontal stacking

and widening) to a time of high available accommodation space (slope aggradation, vertical stacking, internal levée development).

Kane *et al.* (2007) document an Upper Cretaceous submarine channel–levée complex from the Rosario Formation, Baja California, Mexico, which provides constraints on the lithofacies and ichnofacies distribution, and levée depositional thickness decay along transects perpendicular to the channel axis (Fig 8.38, 8.39). Within the levée, they report that both sandstone layer thickness and the overall proportion of sandstone decrease away from the channel axis according to a power law function (*cf.* Dykstra *et al.* 2012). In channel-proximal locations, structureless sands, parallel lamination, overturned ripples and ripple cross-lamination (including climbing ripple cross-lamination) are common, whereas in channel-distal localities, starved ripples are abundant. Sandstone bed thicknesses generally increase upward within the levée succession, which is interpreted to indicate increasing turbidity current strength and/or contemporaneous channel floor aggradation reducing relative levée relief so that the sand-prone lower parts of individual currents could more readily spill across the levées. However, in the most channel-proximal location sandstone bed thicknesses decrease with height; combined with evidence from both facies and palaeocurrent analysis this allows the position of the levée crest to be inferred. The thickest beds occur at higher levels with increasing distance from the channel axis. Using this observation, Kane *et al.* (2007) developed a model for levée growth and migration of the crest (Fig. 8.38).

Ancient levée deposits possess a high potential for subsequent wet-sediment modification, for example, differential compaction, growth faulting and inter-/intra-stratal slip, and are hard to distinguish from other basin-slope deposits. Because it is virtually impossible to preserve and recognise the positive morphological relief of levées in ancient channel–levée complexes, the deposits of ancient levées are incorporated into the category "overbank" deposits, as defined by Mutti and Normark (1987). Overbank facies comprise fine-grained and thin-bedded, current-laminated SGF deposits, commonly interbedded with hemipelagic/pelagic mudstones. Levée architectural elements may be identified within overbank deposits, but it is generally difficult to apply the architectural element scheme to such sequences.

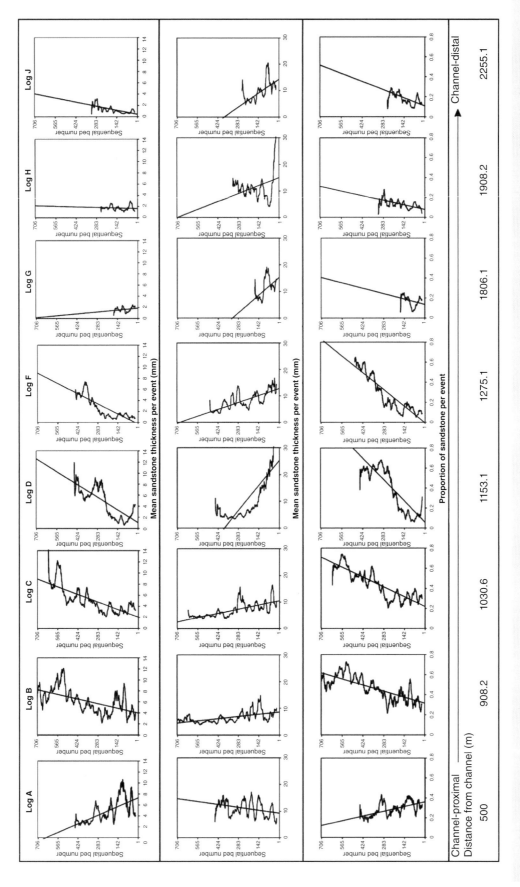

Fig. 8.39 Data from the Upper Cretaceous slope sediments, Rosario Formation, Baja California, Mexico, displayed for all logs with a significant numbers of beds (Logs E and I are omitted for this reason): (a) sandstone thickness per event (i.e., the coarse-division thickness in each sand–mud couplet resulting from a single overspill event); (b) mudstone thickness per event and (c) sandstone proportion (i.e., the ratio of coarse-division thickness to SGF thickness; see Fig. 5.1 for definitions). A 20-bed moving average is employed to smooth the data, allowing qualitative interpretation; a linear trend-line is also used to give an impression of the overall trend. Note that Log A generally shows the opposite trend to other logs and is interpreted to represent inner levée deposition (Fig. 8.36). See text for discussion. From Kane *et al.* (2007).

8.3.5   Lateral accretion deposits (LAPs)

Examples of candidate lateral-accretion bedding (also referred to as *Lateral Accretion Packets* or LAPs) in ancient deep-marine successions are relatively rare, and their nature and origin remain controversial. Documented examples include those from: the Upper Cretaceous Rosario Formation, Baja California (Kneller 2003; Dykstra & Kneller 2009) (Fig. 8.40); the Upper Carboniferous Ross Formation (Elliott 2000a; Lien *et al.* 2003) and Gull Island Formation (Martinsen *et al.* 2000), western Ireland (Fig. 8.41); the middle Miocene Mangarara Formation, Taranaki Basin, New Zealand (Puga-Bernabéu *et al.* 2009: their fig. 9); seismic interpretation from offshore eastern Borneo, Kalimantan, Indonesia (Posamentier & Kolla 2003; Kolla *et al.* 2007) (Fig. 7.25), the Neoproterozoic Castle Creek area of the Windermere Supergroup, Canadian Cordillera (Arnott 2007) (Fig. 8.42); the Permian Beacon Channel Complex, Brushy Canyon Formation, west Texas (Pyles *et al.* 2012), and seismic examples from the Lower Miocene Green Channel Complex, Dalia M9 Upper Field, Block 17, offshore Angola (Abreu *et al.* 2003). Candidate LAPs, although with different nomenclature and varying interpretations, include the so-called 'large-scale oblique laminasets and low-angle accretionary bedsets' in the Oligocene Grès d'Annot, southeast France, described by Guillocheau *et al.* (2004: their fig. 13). LAPs tend to be described from towards the base of evolving channel fills (Kolla *et al.* 2007) and from shallow channels with little confinement of transitting SGFs. Without excellent 3D data, it can be difficult to differentiate between large-scale cross-stratification and LAPs.

Arnott (2007) describes LAPs – referred to in his study as 'lateral accretion deposits' or LADs – from the 2.5 km-thick Neoproterozoic Castle Creek area of the Windermere Supergroup, Canadian Cordillera, where a vertically-dipping succession of basin-floor to base-of-slope deposits is well exposed. These deposits accumulated along the western continental margin of Laurentia following the breakup of the supercontinent of Rodinia (Ross 1991). Arnott (2007) interprets these LAPs as having formed at the inner-bend of sinuous submarine channels (akin to fluvial point-bar deposits); that is, sharp-based, laterally-accreting sinuous channels, of which one is described in detail – the 'Island Creek channel' or IC2.2 (Fig. 8.42).

IC2.2 is up to 13 m thick and extends laterally for at least 400 m. The LAPs are inclined at 7–12° toward the channel base and are ~120–140 m long. There are negligible grain-size changes obliquely-upward along an individual LAP, or vertically upward through the channel fill. LAPs comprise two repeating and interstratified kinds: (i) coarse-grained LAPs consisting of strata up to granule-grade conglomerate (Facies Classes A and B), and (ii) fine-grained LAPs composed of thin- to medium-bedded finer-grained SGF deposits (Facies Class C). In the lower part of the channel fill, strata consist only of amalgamated coarse-grained LAPs as decimetre-thick beds that are typically very coarse sandstone/granule conglomerate grading upward into medium-grained sandstone. Tractional sedimentary structures are absent and fine-grained strata, specifically mudstone, occur only as isolated patches of intraclast breccia. In the upper part of the channel fill, however, LAPs comprise a rhythmic interfingering of coarse- and fine-grained LAPs. Coarse-grained LAPs consist of 2–3 bed-thick packets that are separated and then pinch-out rapidly into fine-grained LAPs. Close to their updip pinch-out these coarse strata commonly comprise poorly-sorted, ungraded very coarse sandstone/granule conglomerate overlain abruptly by planar-laminated or medium-scale (dune) cross-stratified, medium-grained sandstone (Facies B2.1 and B2.2). Fine-grained LAPs are composed of mudstone

interbedded with thin- and medium-bedded T_{bcd} and T_{cd} turbidites that obliquely-downward become truncated as the super- and subjacent coarse-grained LAPs amalgamate.

Arnott (2007) interpreted the rhythmic intercalation of coarse- and fine-grained LAPs to be related to temporal changes in the nature of sediment deposition along the point-bar of a deep-marine sinuous channel. Following failure along the cut-bank margin (outer bend), deposition of coarse-grained sediment on the point bar (inner bend) occurred in order to re-establish an equilibrium channel geometry, and thereby equilibrium sediment transport conditions (i.e., sediment bypass). Once equilibrium was re-established, deposition of finer, thinner-bedded strata of the succeeding fine LAP resumed. These strata represent deposition from the dilute tail region of flows that for the most part had already transitted that particular channel bend and transported the bulk of their coarse sediment farther downdip. This history of alternating coarse- and fine-grained sedimentation was repeated several times in the channel bend as it migrated laterally. Moreover, in coarse LAPs, the restricted occurrence of tractional sedimentary structures close to their up-dip pinch-out suggests that although suspension deposition may have dominated over much of the lateral accretion surface, it was succeeded, at least on the upper part of the lateral accretion surface, by sediment reworking and bed-load transport, possibly related to elevated turbulent stresses caused by mixing along the sharp density interface in a strongly stratified turbulent flow. Although seemingly similar to lateral accretion deposits reported from fluvial point-bars, deep-marine LAPs of the Windermere exhibit many important differences. Some of these differences are likely related to differences in the mode of sand (and coarser) sediment transport in deep-marine versus non-marine environments, specifically suspension versus bedload, respectively. In addition, as discussed by Peakall *et al.* (2000a), Parsons *et al.* (2010) and Pyles *et al.* (2012), fundamental differences in flow structure between subaqueous suspension currents and open-channel flows most probably exert an additional first-order control contributing to these differences.

Another example of slope channels that contain candidate LAPs sits between overlying fluvio-deltaic deposits and underlying basin-floor SGF deposits of the Lower Carboniferous (Namurian) Central Pennine Basin of north England (Walker 1966a,b; Collinson 1968, 1970). This system, within the Namurian *Reticuloceras* zone, shows a general regressive sequence through the Edale shales (basin-floor siltstones and mudstones of Facies Classes D and E), Mam Tor Sandstones (distal SGF deposits of Facies Classes C and D), Shale Grit formation (proximal-fan sandy SGF deposits and concentrated density-flow deposits of Facies Class C with minor Class B), Grindslow Shales (slope environment with Facies Class B and C sandstone-filled channels) and Kinderscout Grit (delta-top and fluviatile channels). Walker (1966b) describes submarine channels from both the Shale Grit Formation and the Grindslow Shales. One of the best-exposed channel outcrops is along the eastern banks of the River Alport in crags known as Alport Castles (Fig. 8.43). The exposed sections are in cliffs up to 50 m high and almost parallel to palaeoflow direction. The facies are predominantly medium- and thick-bedded sandstones, showing a lack of grading (mainly Facies Class B and C). Beds are commonly amalgamated and show common erosive bases. Using the architectural element scheme of Figure 7.22, the outcrop can be divided into 10 distinct second-order architectural elements, containing packets of beds, separated by thin siltstones or connected by downcutting erosive surfaces (Fig. 8.43). The entire cliff section may therefore represent the fill of a large-scale third-order channel or canyon, cut into fine-grained slope sediments, and contains

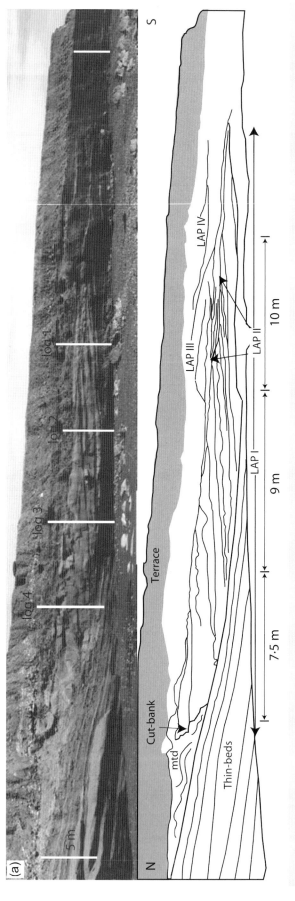

Fig. 8.40 (a) Photomosaic and line-drawing interpretation of the Pelican Point south outcrop in the Upper Cretaceous Rosario Formation, Baja California, Mexico, showing the underlying thin beds that were used to rotate the section to palaeo-horizontal, the four lateral accretion sets and the cut-bank. (b) Sedimentary logs through the lateral accretion sets showing bed correlations, lateral and vertical changes in facies and grain size, and changes in the bedding orientations and palaeocurrent directions. From Dykstra and Kneller (2009).

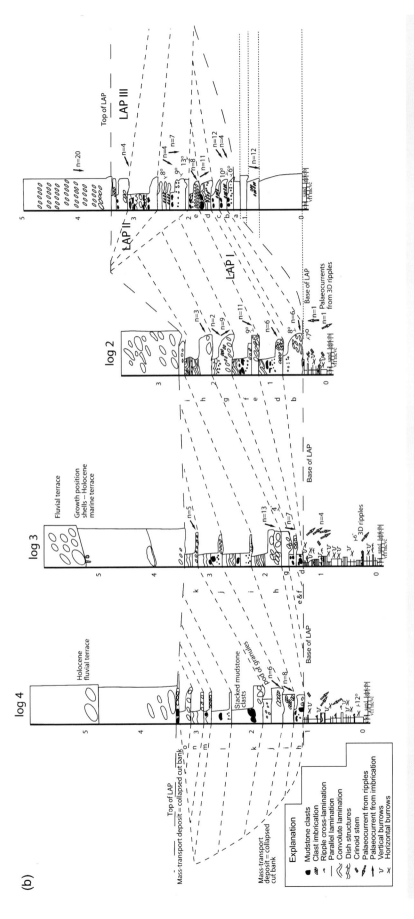

Fig. 8.40 (*continued*)

Fig. 8.41   Lateral accretion bedding (also referred to as a lateral accretion packets or LAP) in Cliffs of Rehy, Namurian Ross Formation, Clare Basin, western Ireland, described by Elliott (2000a) and Lien *et al.* (2003). The base of the channel fill is the downlap surface. The LAP is ~5 m thick and the channel migration is inferred to have been to the west (to the left). Palaeoflow is towards the northeast into the cliff face.

second-order sandstone-filled channel and sheet architectural elements. One such channel element (#2) shows well-developed terraced margins cutting into thin-bedded sandstones and siltstones (Fig. 8.43). The channel-fill is made up of relatively few beds deposited from concentrated density flows (five maximum), and is slightly asymmetrical (Fig. 8.43). The top of the channel-fill ends with an abrupt change from sandstones to laminated siltstones and mudstones. The channels have a width of ~80–100 m (channel element), a depth of ~50 m for the channel complex and ~7 m for discrete channel elements: their aspect ratio is 11–14 (channel element). Based on the presence of consistently-dipping accretion surfaces in a particular channel element (e.g., northwards in channel element #3 and southward in channel element #10/#11) they are inferred to be sinuous (Clark & Pickering 1996b) (Fig. 8.43).

The Middle Eocene Morillo System, Ainsa Basin, Spanish Pyrenees, contains many sandy macroforms and scours, including gravelly macroforms (Fig. 8.44). These 'inclined sandy/gravelly macroforms', or LAPs, have been interpreted as lateral accretion surfaces in high-sinuosity submarine channels within the Morillo System (Abreu *et al.* 2003; Labourdette *et al.* 2008). Possible interpretations of the features shown in Figure 8.44 include lateral accretion surfaces within high- to moderate-sinuosity channels (including thalweg channels), or within-channel accretion surfaces within low-sinuosity channels. The gravelly macroforms may represent the thalweg in larger and straight to low-sinuosity channels. Whilst experimental and theoretical arguments suggest that, compared with river channels, many submarine channels undergo much slower growth and show greater aggradation to produce isolated ribbons of thalweg

deposits (Peakall *et al.* 2000a), the Morillo channels were highly confined by submarine growth anticlines and probably developed on relatively steep seafloor gradients. It therefore seems unlikely that the inclined sandy macroforms are an expression of high-sinuosity channel systems but, rather, some type of side-attached or within-channel barforms, possibly linked with complex and reversed secondary flow around bends (i.e., toward the outer bank; *cf.* Corney *et al.* 2006; Keevil *et al.* 2006, 2007) in a thalweg or local rugose seafloor topography. Also, where steeper slopes exist (as inferred for much of the Morillo System), Georgio Serchi *et al.* (2011) show that reversed secondary flows are more likely.

Mapping, sedimentary logging and lateral correlations suggest that the Morillo channel complexes had depths that were probably in the range 20–60 m and widths of 800–1200 m (Bayliss & Pickering 2015b). The dimensions of individual channels must have been considerably less, but the outcrops and lateral correlations preclude any refinement of these dimensions. Any levée–overbank dimensions are likewise impossible to resolve, a problem commonly encountered with many outcrops in ancient channel systems. However, irrespective of any more precise width/depth measurements, it appears that the inclined sandy macroforms are thin relative to channel depth (also proposed by Labourdette *et al.* 2008), with the height of the inclined sandy macroforms typically in the range 1–4 m (likely about 10–20% of the channel depth), an observation that fits with experimental work (Das *et al.* 2004; Darby & Peakall 2012) suggesting that macroforms in submarine channels are much thinner than channel depth, unlike in rivers (see also Wynn *et al.* 2007 and references therein).

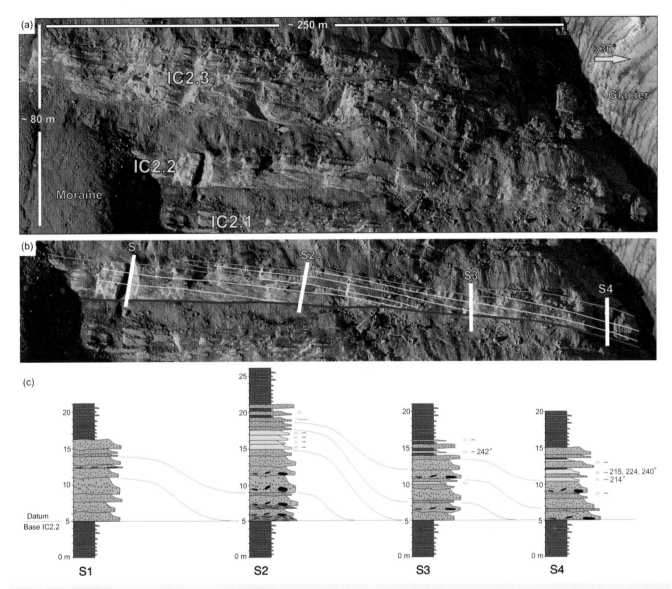

Fig. 8.42 Aerial photo of Isaac Channel Complex 2, Castle Creek south, Windermere Supergroup, Canadian Cordillera. Note that strata are dipping vertically into the photo. (a) Uninterpreted photo showing the three channel fills (IC2.2, IC2.2, IC2.3) that comprise the channel complex 2 between the moraine and the glacier. (b) Interpreted photo of IC2.2. Flat base of IC2.2 is indicated by the red line. Yellow lines represent surfaces that in the field and also on the aerial photos could be traced laterally from the top obliquely downward to the base of the channel fill (from left to right). These surfaces are interpreted to be part of coarse-grained lateral accretion packets, or LAPs, formed by the lateral migration of a deep-marine sinuous channel. The serrated blue line indicates the interfingering of coarse- and fine-grained LAPs at the top of the channel fill. Fine LAPs grade rapidly laterally (toward the left) into contemporaneous fine-grained inner-bend-levée deposits. (c) Line diagrams of bed-by-bed measured stratigraphic sections (S1-S4). Location of each section is shown in (b). From Arnott (2007).

Whilst Bayliss and Pickering (2015b) tend to favour lower-sinuosity, braided channels (relatively steep seafloor, coarse grain size and low-dispersion in palaeocurrents), it should be emphasised that braid bars are in many ways geometrically like back-to-back point bars. Therefore, with only partial outcrop, as is the case for the Morillo System, the inclined sandy macroforms might look very similar to point bars (*cf.* Peakall *et al.* 2007), but could actually be components of partially exposed braid bars. Citing recent sea-floor observations, Bayliss and Pickering (2015b) speculated that the presence of gravel-filled scours within thick MTCs at the base of Morilo I and Morillo II fans (see Fig. 4.34 for map of Morillo System

showing three fans) may represent the down-channel, discontinuous, fill of scours/flute-like structures as imaged by new high-resolution Autonomous Underwater Vehicle (AUV) seafloor images with 1 m lateral resolution and 0.3 m vertical resolution, Lucia Chica, offshore central California (Maier *et al.* 2011). Prior to these sea-floor observations, continuous channel thalwegs were interpreted incorrectly from lower resolution images (e.g., Wynn *et al.* 2007). The apparent scale of the Morillo examples is directly comparable to those observed in the Lucia Chica System (i.e., within-channel low-relief erosional features that are <10 m deep and ~100 m across, including scours and pockmarks; Maier *et al.* 2011).

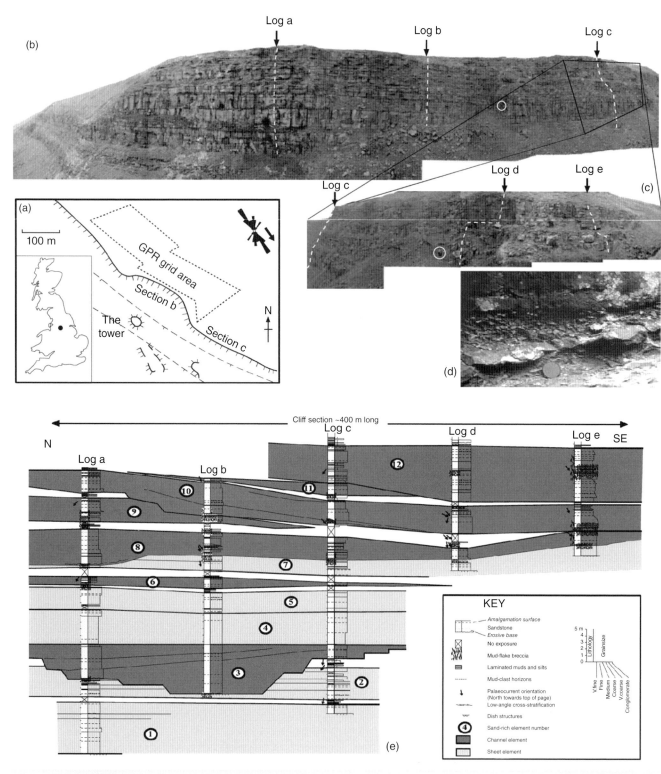

Fig. 8.43   (a) Location map of Alport Castles, Peak District National Park, Derbyshire, Central England, UK. (b, c) Photomosaics of outcrop cliff faces (see (a) for location). Sedimentary log sections are marked. (d) Close-up of intrachannel, mud-flake breccia; palaeocurrent data are shown in (a). Human scales are ringed. (e) Fence diagram interpretation of cliff sections A and B at Alport Castles (modified from Clark & Pickering 1996b). The sinuous cliff (orientated almost perpendicular to palaeocurrent) means that only an approximate horizontal scale can be applied to the correlated sections. The top bed of channel element 12 is used as a datum. From Pringle *et al.* (2004).

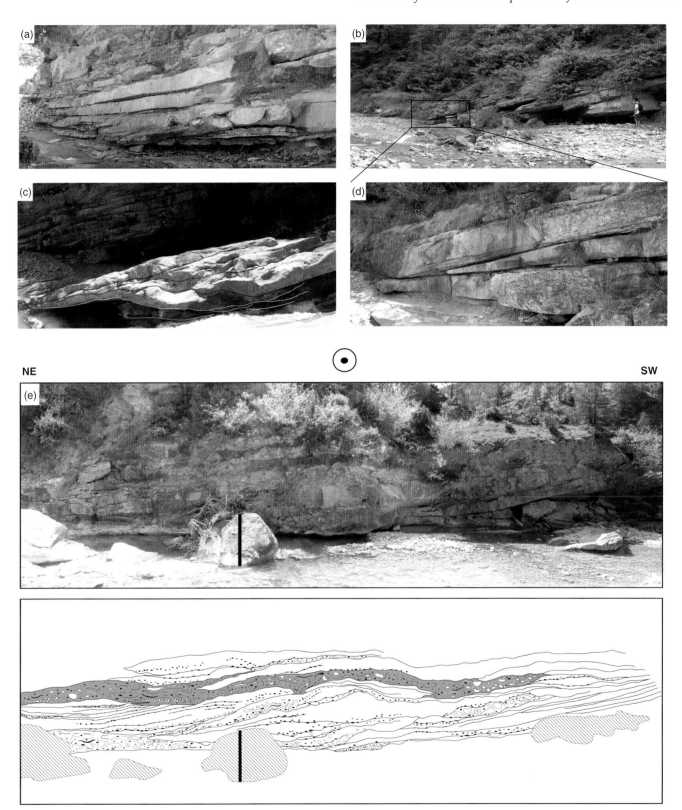

NE   SW

Fig. 8.44   Range of sandy macroforms in the Morillo I Fan, Ainsa Basin, Spanish Pyrenees. See Fig. 4.34 for map of the Morillo System with locations. These examples comprise inclined surfaces that are approximately perpendicular to local palaeoflow from local groove and flute casts. (a) Rucksack scale. (b) Small-scale example, with inclined surfaces dipping to left above a scour with ~1 m maximum depth. Human scale. (c) Human scale. (d) Enlargement of (c) to emphasise wedging out of sandstones towards the base of this ~4 m-thick sandy macroform. (e) Gravel bar form with photo-interpretation, Morillo III Fan. Scale bar is 2 m. The darker layer is a pebble-rich muddy debrite (Facies A1), contrasting with the underlying gravelly sandstones (Facies A). From Bayliss and Pickering (2015b).

8.3.6   Post-depositional modification of channel fills

Post-depositional modification of channel deposits can exert an important influence on the appearance of ancient channels as well as the position and course of subsequent channels. Post-depositional features include clastic injections as dykes and sills (including the formation of brecciated beds), faults (including growth faults) and differential compaction.

The most common post-depositional modification to channel geometry is compaction. There are many examples documented from industry data; for example, from the Gulf of Mexico (Posamentier 2003) and the western Niger delta slope (Deptuck *et al.* 2007). Differential compaction probably accounts for the super-elevated or inverted aspect of many channel fills, suggesting that this fill was substantially less compactable and therefore more sand-prone than the adjacent overbank deposits (Posamentier 2003). Differential compaction tends to create many lens-shaped channels, with the margins at a lower elevation than the more axial fill; for example, as documented by Gee and Gawthorpe (2006) from offshore Angola (see also seismic example in Mayall *et al.* 2006, their fig. 14). Differential compaction of ribbon-like channel sandbodies and their associated overbank–levée deposits may induce growth faults along the margins of the buried channels, and even the injection of sandy clastic dykes along such faults (Fig. 8.45). Differential compaction is also a major feature associated with MTDs/MTCs; for example, as documented from offshore Brazil and the northern slope of the Gulf of Mexico by Alves (2010).

Fluidisation and liquefaction features, such as clastic dykes and sills, are documented from many ancient, including industry subsurface, deep-water systems; for example, slope sandstone gully fills (Surlyk

1987) and inferred submarine-fan channel fills (Hiscott 1979; Pickering 1981b, 1983a; Guy 1992; Kane *et al.* 2009). Although many injections appear to involve sand-grade material, there are examples of conglomeratic injections sourced from submarine channels; for example, the large-scale conglomeratic intrusions sourced from deep-water channel deposits in the Cretaceous Cerro Toro Formation, Magallanes Basin, southern Chile (Hubbard *et al.* 2007).

In the subsurface, large-scale sand-rich injection structures are documented from the Palaeogene of the northern North Sea. These include sand dykes up to many metres wide intruding vertically up through tens of metres of sediments in the Balder Formation, as described by Jenssen *et al.* (1993), and on a decimetre scale in core from the Gryphon oilfield (Newman *et al.* 1993). Industry seismic and borehole evidence have shown that large clastic injections can originate from donor sands in adjacent submarine channel deposits (e.g., Jackson & Sømme 2011). For additional information on sand injections and their relevance to hydrocarbon exploration and production, the reader is referred to the edited volume by Hurst and Cartwright (2007).

A common feature recorded in many industry subsurface cores (e.g., from the Palaeogene of the northern North Sea) is the presence of apparently chaotic mud-flake breccias and conglomerates with a sandy matrix. Although, on the basis of micropalaeontology and composition, including colour, some of these deposits are interpreted as sediment slides and debrites (with extra- and intra-formational clasts), there are many cases where it appears that the deposit was formed by the pervasive injection of semi-consolidated muds. In some field examples the lower part of a sandy SGF deposit has fluidised and pervasively injected the upper muddy part of the same bed, a process here referred to as *autobrecciation*.

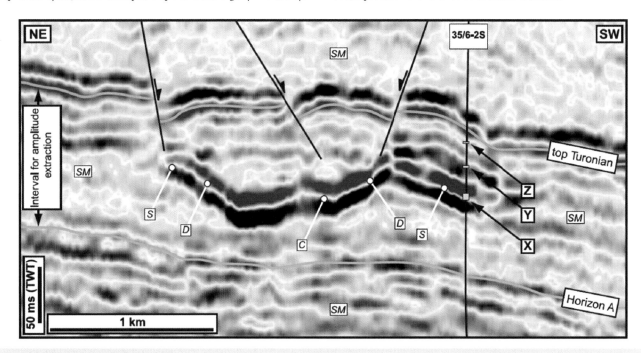

Fig. 8.45 Seismic line across a Late Cretaceous submarine channel, Norwegian sector of the North Sea. TWT = two-way travel time. Here, red reflections are negative (trough), acoustically 'soft' reflection events; black reflections are positive (peak), acoustically 'hard' reflection events. Horizon nomenclature is taken from Jackson *et al.* (2008). The faults above the channel are part of a regional polygonal fault system that is developed in the Upper Cretaceous succession. The location of borehole 35/6-2S is shown; the location of the borehole with respect to the channel-margin amplitude anomaly should be noted. D = clastic dyke; S = clastic sill; C = depositional channel; SM = slope mudstone. Yellow blocks indicate the depth and thickness of sandstones X, Y and Z identified in the borehole, all interpreted as post-depositional injections or sills, From Jackson & Sømme (2011).

Growth faulting, particularly at channel margins associated with differential compaction, may encourage the offset-stacking of younger channelised sandbodies. Growth faults also may act as surfaces and zones of weakness along which large-scale wet-sediment intrusions occur. Growth faults may be reactivated with compaction, even very early after deposition when only perhaps tens of metres of additional strata have accumulated. The location of compaction-related faults can be controlled by buried channel courses, paralleling the channel margins, caused by the differential compaction of channel sands relative to levée and overbank finer grained sediments (see fig. 4 in Jenssen *et al.* 1993; fig. 8 in Newman *et al.* 1993).

8.4 Comparing modern and ancient channels

Many ancient channels, although comparable in scale to modern middle-fan channels (Table 8.3, Fig. 8.10), are erosional and erosional-depositional channel complexes developed within basin-slope sediments, rather than the aggradational (essentially depositional) channel–levée–overbank complexes common to many large-radius, very low-gradient, modern fans. Apart from accretionary prisms, most of the ancient rock record in deep-marine/deep-water systems represents the vestiges of upper continental slope, intra-shelf, or aulacogen-related deposits where basins are up to orders of magnitude smaller than today's continental margins and ocean basins, with few if any essentially flat basin floors, and where seafloor gradients are commonly high. In such rugged areas, it is inevitable that erosional and erosional-depositional, low- to moderate-sinuosity channel systems should prevail, as they do in the ancient record.

For ancient channels, depth estimates rarely include the thickness of significant amounts of fine-grained fill, contrary to the observations in some modern fan channels (e.g., Amazon Fan data in Fig. 7.32; Mississippi Fan data in Fig. 5.17). In ancient outcrops, fine-grained lithologies (muds and silts) are generally poorly exposed at critical outcrops, and/or may have undergone severe post-depositional deformation to obscure original bedding relationships. Such limitations commonly conspire to lead to an under-estimation of true channel dimensions and the interpretation of packets of fine-grained sediments more than a few metres in thickness as 'levée', 'inter-channel', or 'overbank' deposits rather than channel fill.

In many well-documented ancient channel systems that are interpreted as moderate- to high-sinuosity channels (based on the recognition of lateral accretion elements and from 3D seismic mapping), it appears that there may be a predictable evolution from the initial erosional to the final abandonment phase. The establishment of a channel involves an initial erosional phase associated with the deposition of mainly residual deposits and irregular bedding/elements. In many cases, the channel-floor topography probably develops from a number of discrete erosional events in which channel terraces may be formed. After establishment, channels will tend towards an equilibrium profile by becoming more sinuous, unless there are modifying effects that lead to abrupt abandonment of the channel or lead to re-activation of the initial erosional phase (e.g., caused by base-level changes leading to increased rates of sediment supply in flows with greater competence and capacity: *cf.* the observation that at high latitudes, Coriolis forces tend to deflect the downstream velocity core of a turbidity current, and consequently areas of deposition and erosion, to one side of the channel system which, over time may favour the evolution of low-sinuosity submarine channels (Cossu & Wells 2012).

During the development of an increasingly sinuous course, lateral accretion and flow-stripping processes will become significant in the resulting deposits and channel architecture. As the channel segment starts to become abandoned (e.g., as the sediment flux decreases in flows with, on average, decreased volume, competence and capacity), then the channel will tend to backfill and change from offset-stacked inclined macroforms (due to lateral accretion and flow-stripping, or other channel bar forms) to lens-shaped elements, and finally abandonment. It seems reasonable to expect that levée bank-collapse will tend to be more likely between the first two stages as this represents the time when the levée-crest to channel-floor height is greatest and the levées are most unstable. Levées will show their greatest rate of vertical aggradation during the accumulation of the earliest channel deposits since bankfull flow conditions are most likely during the passage of larger sediment gravity flows. The rapidly deposited stage 1 deposits will have relatively high water content and therefore low shear strength.

The evolutionary model outlined here represents an idealised erosional-depositional history and contains a number of implicit assumptions, including: (i) a gradual change between different typical flow conditions within any reach of a channel; (ii) the absence of any major sediment slide/debris-flow event that might leave an anomalously thick deposit within a channel to plug, or partially plug, a reach of the channel and cause channel avulsion and (iii) that the channel is excavated, acts as a conduit for sediment gravity flows, and is filled to abandonment, all within one cycle of falling then rising relative base level, although autocyclic processes like one or more erosionally induced avulsion may be important during a single lowstand (e.g., Pirmez & Flood 1997). Thus, the channel is excavated and acts as a conduit for SGFs during a time of falling and low base level, then is filled to abandonment because of reduced sediment supply either during the next base-level rise or because of avulsion farther up the submarine fan that diverts the coarse basal parts of SGFs elsewhere. It seems reasonable to expect considerable variations in this depositional pattern, both in terms of the thickness and sediment grain sizes associated with these various stages. Additionally, abrupt channel abandonment and deep erosion by subsequent events will lead to incomplete vertical sequences being developed and/or preserved, respectively.

Some of the processes operating in modern submarine channel environments differ from those in fluvial environments. For instance, a sediment gravity flow transitting a submarine channel can overspill the confines of the channel with considerable vertical expansion because of the small density contrast with ambient seawater. The overspill is not necessarily (or generally) fine-grained suspended load from the top of the flow, because of differences in the geometry of secondary flow around channel bends, as compared with rivers. Parsons *et al.* (2010) have shown in field studies of a large natural gravity-flow channel that the sense of spiralling of the secondary flow around channel bends can be opposite to the geometry in fluvial channels, so that coarser suspended load from the deeper parts of stratified SGFs can preferentially spill out of the channel along the outer bends of sinuous channels. Straub *et al.* (2008) attribute such coarse-fraction overspill to enhanced run-up of the basal parts of SGFs because momentum is greatest in that part of the flow. Despite such differences in flow processes, the depositional and erosional geomorphologies of deep-water and fluvial channels are similar. The points of similarity are documented in the literature; for example, meandering channels on the Amazon Fan (Damuth *et al.* 1983), scroll bar features along the Mississippi Fan channel (Pickering *et al.* 1986b) and entrenched thalweg channels on the Rhône Fan

Table 8.3 Dimensions and degree of lateral continuity and vertical connectivity of a selection of channel elements from published literature, including those discussed in previous chapters. Dimensions are measured perpendicular to flow/palaeoflow direction with the exception of the Tabernas Basin Solitary Channel bedforms which are measured in a downcurrent direction. CSA = channel cross-sectional area based on: 1 – author's own work; 2 – Clark (1995); 3 – Cronin (1995); 4 – Damuth *et al.* (1995); 5 – Timbrell (1993); 6 – Vicente Bravo & Robles (1995); 7 – Zelt & Rossen (1995); 8 – Smith *et al.* (1991); 9 – Walker (1975a); 10 – Pickering (1982a); 11– Philips (1987); 12 – Kenyon *et al.* (1995a); 13 – McHargue (1991); 14 – Watson (1981); 15 – Pickering *et al.* (1986a); 16 – Hilton & Pickering (1995); 17 – Masson *et al.* (1995); 18 – Remacha *et al.* (1995); 19 – O'Connell *et al.* (1991); 20 – Kenyon *et al.* (1995a). From Clark and Pickering (1996a).

Element	Width (m)	Thickness (m)	Lateral continuity	Vertical connectivity	CSA/Reference (m²)
Ainsa I 'backfill' element beds'	200	3	65	0.2	6.0×10^2 1
Ainsa I channel	850	30	25	0	2.6×10^4 1
Ainsa I thalweg	230	4	60	0.4	9.2×10^2 1
Ainsa II channels	600	25	24	0.5	1.5×10^4 2
Ainsa II/2 channel/lens elements	50	2	25	0.35	1.0×10^1 2
Ainsa II/3 channel/lens elements	100	3	33	0.2	3.0×10^2 2
Ainsa II/4 axial fill element beds	100	2	50	0.9	2.0×10^2 2
Ainsa II/5 slide elements	100	4	25	0	4.0×10^2 2
Almeria Channel	400	50	8	0	2.0×10^4 3
Almeria Channel thalweg	65	5	13	0	3.3×10^2 3
Amazon-Middle Fan ch.–axis complexes	1250	150	8	0	1.9×10^5 4
Amazon-Middle Fan ch.–levée complexes	30 000	150	200	0.9	4.5×10^6 4
Balder Fm. stacked channels	800	50	16	0.1	4.0×10^4 5
Black Flysch scours	30	5	6	0.2	1.5×10^2 6
Brushy Canyon Fm. 'Salt-flat' Channel	2000	30	66	0	6.0×10^4 7
Brushy Canyon Fm. '100-foot' Channel	400	30	13	0	1.2×10^4 7
Brushy Canyon Fm. Popo channels	160	10	16	0.4	1.6×10^3 7
Brushy Canyon Fm. Brushy Mesa elements	155	15	10	1	2.3×10^3 7
Caban Coch channels	4000	75	50	0.2	3.0×10^5 8
Capistrano channels	250	20	12.5	0.4	5.0×10^3 9
Capistrano lat. accr. elements	50	2	25	1	1.0×10^2 9
Capistrano sheet-fill elements	100	1	100	1	1.0×10^2 9
Hamningberg Channel axial fill	100	10	10	0.3	1.0×10^3 10
Hamningberg channel margin element	75	5	15	0.3	3.8×10^2 10
Indian Draw Field A1 Sandstone elements	400	3	133	0	1.2×10^3 11
Indian Draw Field A3 Sandstone elements	500	5	100	0	2.5×10^3 11
Indian Draw Field channel-fill	1200	30	40	0	3.6×10^4 11
Indus-Middle Fan ch.–axis complexes	1250	75	17	0	9.4×10^4 12
Indus-Middle Fan ch.–axis elements	1000	15	66	0.9	1.5×10^4 12
Indus-Middle Fan ch.–levée complexes	25 000	75	333	0.9	1.9×10^6 12
Indus-Upper Fan ch.–axis complexes	10 000	400	25	0	4.0×10^6 12
Indus-Upper Fan ch.–axis elements	1500	20	75	0.9	3.0×10^4 12
Indus-Upper Fan ch.–levée complexes	50 000	400	125	0.9	2.0×10^7 12
Indus-Upper Fan Channel C axis elements	5000	100	50	0.6	5.0×10^5 13
Indus-Upper Fan Channel Ca axis elements	3000	100	30	0.7	3.0×10^5 13
Indus Upper-Fan Channel Cc axis elements	2500	125	20	0	3.1×10^5 13
Milliners Arm Fm. axial elements	500	20	25	0.3	1.0×10^4 14
Milliners Arm Fm. ch. margin elements	750	15	50	0.4	1.1×10^4 14
Mississippi levées	600	10	60	0.9	6.0×10^3 15
Mississippi migrating thalweg	2000	100	25	0.4	2.0×10^5 15
Montagne de Chalufy turbidite channel	500	50	10	0	2.5×10^4 16
Monterey channel	1500	50	30	0	7.5×10^4 17
Rapitan Channel (RCH-1)	1500	75	20	0.5	1.1×10^5 18
Rapitan channel-fill elements (RCH-2a)	1200	15	80	0.2	1.8×10^4 18
Rapitan channel-fill elements (RCH-3a)	1000	8	125	0.05	8.0×10^3 18
Rhone-channel thalweg	500	130	4	0	6.5×10^4 19
Rhone-Neofan scours	1000	20	50	0.1	2.0×10^4 20
Rhone-Upper Fan levéed valley	1500	100	15	0	1.5×10^5 19
Risfjord channel stacked channels	50	5	10	0.9	2.5×10^2 10
Solitary Channel	200	40	5	0	8.0×10^3 1
Solitary Channel bedforms	150	4	37.5	0.25	6.0×10^2 1

(O'Connell *et al.* 1991). Similarities between the depositional architecture of ancient channel–levée–overbank complexes and those of fluvial deposits are also well documented (e.g., Mutti *et al.* 1985), including lateral accretion deposits (e.g., Phillips 1987), channel lag gravels and channel terraces (e.g., Hein & Walker 1982, Watson 1981), and evidence for channel/canyon sinuosity (e.g., Cossey & Jacobs 1992; von der Borch *et al.* 1985).

8.5 Ancient lobe, lobe-fringe, fan-fringe and distal basin-floor deposits

Although some geophysical and core data are available for modern lobes (e.g., Migeon *et al.* 2010; Mulder *et al.* 2010), the resistance of thick sands to gravity and piston coring means that our knowledge of lobe facies and their variability remains poor (Section 7.4.3.1). Mainly by comparison with modern fans, ancient lobes are inferred to be tongue-shaped deposits of mostly deep-water sandstones which accumulated beyond the termination of a channel on a submarine-fan (or submarine-ramp) (Section 7.4.3). Their geometry or shape is observed to be essentially sheet-like over many tens to hundreds of metres across flow.

Submarine lobes, including their associated high aspect-ratio (width/depth) shallow channels, constitute the bulk of the volume of ancient submarine fans (e.g., Figs 5.19, 8.46, 8.47) and the small-radius modern fans found in tectonically active settings (Fig. 7.49). They generally develop downslope of where levéed channels bifurcate to form a number of distributary channels (Posamentier & Walker 2006). Early studies (e.g., Mutti & Ricci Lucchi 1972; Mutti 1977; Pickering 1981b; Lowe 1982) suggested that lobes comprise sheet-like sandy deposits at the termini of channels and that they

form because decelerating SGFs become unconfined with a corresponding loss of flow competence and flow capacity. More recent studies on modern systems suggest that the internal architecture of lobe deposits is more complex (e.g., Mississippi Fan, Twichell *et al.* 1992[1]; Golo Pleistocene lobe, east Corsica, Gervais *et al.* 2006, Deptuck *et al.* 2008; Amazon Fan, Jegou *et al.* 2008): such complexity is also observed in ancient analogues, for example: the upper Miocene Tabernas basin, southeast Spain (Cossey & Kleverlaan 1995); the Oligocene–Miocene Rochetta Formation, Tertiary Piedmont Basin, northwest Italy (Smith 1995a); the Permian Brushy Canyon Formation, West Texas, (Beauboeuf *et al.* 2000; Carr & Gardner 2000; Gardner *et al.* 2003b, Gardner *et al.* 1985); the Carapebus Formation, Campos Basin, Brazil (Ribeiro Machado *et al.* 2004); the Permian Skoorsteenberg Formation, South Africa (Johnson *et al.* 2001; Sullivan *et al.* 2004; Hodgson *et al.* 2006; Prélat *et al.* 2009), the Carboniferous Ross Formation, western Ireland (Pyles 2008), and the Silurian of the Welsh Basin, UK (Smith 1987, 1995b).

In radial submarine fans where SGFs tend to be relatively unconfined, lobes are more likely to have positive relief above the seafloor (mounded); however, in highly confined settings (e.g., structurally-confined basins and/or deposition on uneven seafloor topography), lobes are likely to onlap basin topography, resulting in subdued relief. If the accommodation space is narrow relative to the natural (unconfined) lobe width, then lobe deposits will tend to stack vertically, with forced progradational and retrogradational stacking (autocyclic processes). In contrast, if the accommodation space is wide relative to the natural (unconfined) lobe width, then lobe deposits will show a greater tendency for lateral offset-stacking with

---

[1] Talling *et al.* (2010) reinterpreted the Twichell *et al.* (1992) deposits as thick bipartite beds, and the fine interbeds as rafts. This new interpretation means that these deposits may not be lobe deposits *sensu stricto*, but rather the products of rare events that buried a lobe.

Fig. 8.46 High aspect-ratio, high-continuity sandbodies interpreted as lobe deposits, Permian Karoo System near Laingsburg, South Africa. Train for scale.

compensation effects, and may be expected to show progradational and retrogradational stacking attributable to allocyclic processes. A consequence of this is that, in the absence of both very high-resolution age dating and excellent 3D reconstruction of basin topography and facies distributions, any attempts to deconvolve tectonic versus climatic controls on deep-marine sedimentation are probably best avoided in lobe, lobe-fringe and fan-fringe deposits.

Outer-fan environments comprise three main facies associations, interpreted primarily by analogy with those described by Mutti and Ricci Lucchi (1975a), Mutti (1977) and Mutti *et al.* (1978) – namely, lobe, lobe-fringe and fan-fringe deposits (Table 8.4; see Figs 7.40 and 7.41 for a schematic summary of lobe deposits). The topographically flattest and smoothest seafloor is referred to as the 'basin-floor environment'. In modern ocean basins, the basin floor is commonly synonymous with the term 'abyssal plain'. As it is extremely difficult in most cases, if not impossible, to recognise true ancient oceanic abyssal-plain deposits, we prefer to use the general terms 'basin-floor' and 'basin-floor deposits' to define the most distal fan and related environments. Perhaps the best indicator of ancient oceanic abyssal plains is a predominance of sheet-like pelagic (including chemogenic) and hemipelagic deposition (Facies Classes E and G), together with diagnostic benthic microfossils. These sedimentary rocks, that likely originally accumulated on oceanic and transitional crust, tend to be severely deformed as slivers of stratigraphy when incorporated into younger orogenic belts.

The facies, facies associations and architecture of basin-floor lobe and related deposits have been studied in some detail from many ancient systems, typically from foreland basins settings

(a)

(b)

Fig. 8.47 Examples of high-continuity sandstone beds and sandy packets interpreted as lobe, lobe-fringe and fan-fringe deposits, Veines section, upper Precambrian Kongsfjord Formation, Arctic Norway. **(a)** Lobe deposits ~15 m thick, as lighter-coloured, medium- to very thick-bedded Facies Classes B and C deposits, surrounded by tens of metres of thinner-bedded, and finer-grained, Facies Classes C, D and E deposits. **(b)** Three stacked lobe deposits (lighter-coloured beds in three distinct packets of mainly Facies Class C deposits). ~50 m of stratigraphy visible, with beds younging towards the left. **(c)** Lobe-fringe and fan-fringe deposits (medium- to thin-bedded Facies Class C, with Facies Classes D and E). ~30 m of stratigraphy visible. See Pickering (1981b) for detailed analysis of these beds.

(c)

Fig. 8.47 *(continued)*

(see Chapter 11). Amongst the most studied examples are the: Miocene–Pliocene Marnoso arenacea (also referred to by some researchers as the Marnoso-arenacea Formation), Italian Apennines (Ricci Lucchi 1975a,b, 1978, 1981, 1984, 1986; Ricci Lucchi & Valmori 1980; Gandolfi *et al.* 1983; Ricci Lucchi & Ori 1984, 1985; Talling 2001; Talling *et al.* 2004; Amy & Talling 2006; Magalhaes & Tinterri 2010; Malgesini *et al.* 2015); Eocene Jaca Basin, Spanish Pyrenees (Mutti 1977, 1985; Labaume *et al.* 1983a,b, 1985, 1987; Mutti *et al.* 1985; Remacha *et al.* 2005); Permian Karoo System, South Africa (Johnson *et al.* 2001; Hodgson 2009; Prélat *et al.* 2009, 2010; Pringle *et al.* 2010; Groenenberg *et al.* 2010; Van der Merwe *et al.* 2010); the Middle Ordovician Cloridorme Formation, Québec Appalachians, Canada (Enos 1969a,b; Hiscott 1984; Pickering & Hiscott 1985; Hiscott *et al* 1986; Edwards *et al.* 1994; Awadallah & Hiscott 2004), and the upper Precambrian Kongsfjord Formation, Finnmark, Arctic Norway (Pickering 1981a,b, 1985; Drinkwater 1995; Drinkwater *et al.* 1996; Drinkwater & Pickering 2001). Figure 8.47 shows some examples of sheet-like deposits from the Kongsfjord Formation interpreted as outer fan lobe and related deposits.

The interpretation of ancient sandy lobe deposits suggests that they are typically formed of packets 5–15 m thick (Mutti 1977; Pickering 1981b; Hadlari *et al.* 2009; Prélat *et al.* 2009). Their internal facies composition and architecture vary in proximal-to-distal, and axial-to-lateral directions, leading to different characteristics for lobe, lobe-fringe and fan-fringe deposits. Whilst distal sandy lobe deposits tend to comprise non-amalgamated beds, the more proximal parts show relatively complex bedding geometry, a wider range of sand/shale ratios (Fig. 7.41) and the presence of shallow and broad scour-and-fill features, together with shallow non-levéed channels (e.g., Fig. 8.48). These features are well illustrated in the upper

Precambrian Kongsfjord Formation, Finnmark, Arctic Norway; a sand-rich, deep-water, turbidite system containing well-defined packets of sandstones and silty mudstones and very minor amounts of hemipelagic deposits (Pickering 1981a,b, 1985).

The western outcrops of the Kongsfjord Formation, east of Berlevåg, comprise two lithostratigraphic members named the Nålneset and Risfjord members (Siedlecki 1980). The older Nålneset Member contains the most coarse-grained facies as packets of sediments 5–20 m thick, consisting of deposits of Facies Classes A and B, but predominantly the latter. Fine-grained deposits within the Nålneset Member also form packets that are typically 1–10 m thick. Sediments within these packets of fine-grained thin-bedded SGF deposits comprise Facies Classes C and D. Virtually all the silty mudstones have been interpreted as SGF deposits (Pickering 1981a). Packets of sediment also characterise the upper part of the formation, the Risfjord Member, but fine-grained, thin-bedded lithologies predominate. Coarser-grained sandstone packets, interpreted as outer-fan lobes, comprise as little as 22% of the total for some sections (Pickering 1981b) and, overall, have a bulk mean grain size finer than the sand-rich packets of the Nålneset Member. Middle-, outer- and transitional-fan environments have been interpreted for this member (Pickering 1981b, 1982b, 1983b).

The bed correlation panels in Figures 8.49a–e are shown to emphasise the variability in sandstone architecture from outcrops; this variability is consistent with proximal lobe deposits as described in the literature (*cf.* Mutti & Ricci Lucchi 1972; Mutti 1977; Pickering 1981b). The more amalgamated facies associations are interpreted to represent channel-lobe transition deposits (Pickering 1983b; Drinkwater 1995; Drinkwater *et al.* 1996; *cf.* channel-lobe transition deposits described by Surlyk 1995), perhaps

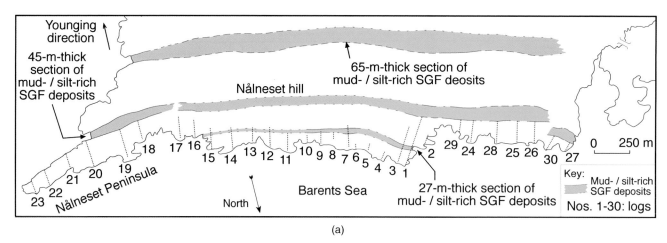

(a)

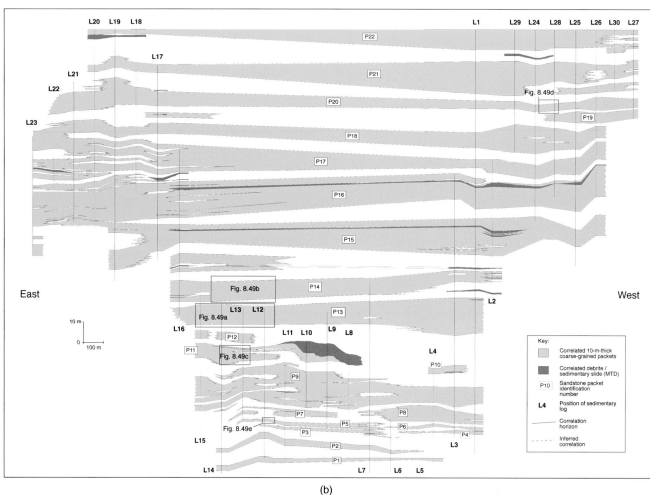

(b)

Fig. 8.48 Map of the Varanger Peninsula, Finnmark, North Norway, north of the Trollfjord-Komalgelv Fault, showing the location of the three main outcrops of the upper Precambrian Kongsfjord Formation (modified from Siedlecka *et al.* 1989). Lower part of figure is a two-dimensional panel from the oldest exposed part of the Nålneset section showing the architecture of 10 m-thick coarse-grained sandstone packets comprising mainly the deposits of concentrated density flows (yellow), interpreted as proximal lobe deposits, between very fine-grained sandstone and siltstone SGF deposits, and mudstones, interpreted as lobe-fringe and fan-fringe deposits (Facies Class C, D and E). Dashed lines show extrapolated sandstone packet continuity. From Drinkwater and Pickering (2001).

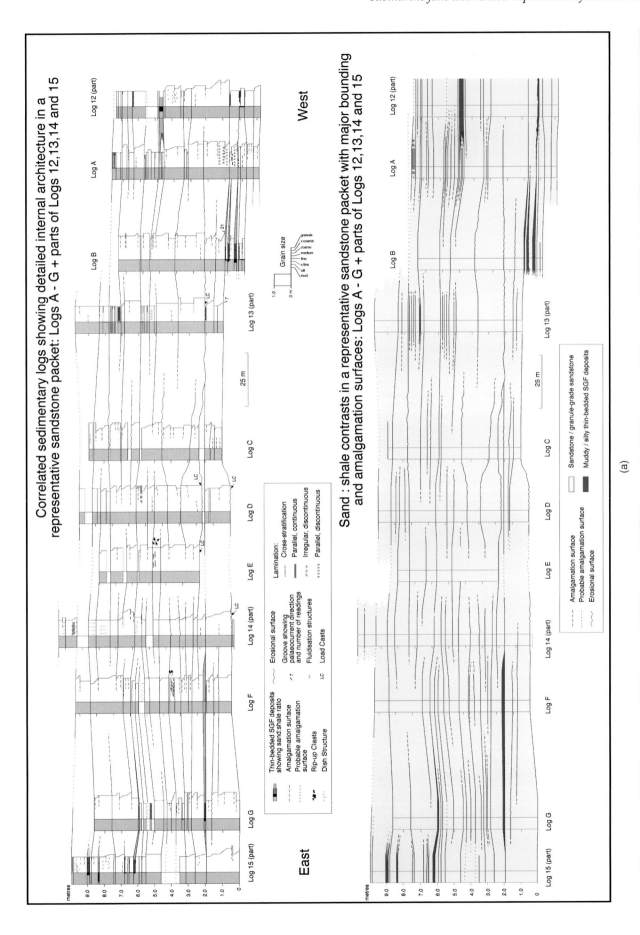

Fig. 8.49 Five representative panels, with varying amounts of bed amalgamation, from the Nålneset section, upper Precambrian Kongsfjord Formation, North Norway, showing the range in architectural elements for sandstone and shale intervals. Panels show correlated sedimentary logs detailing internal architecture (upper) and sand : shale contrasts (lower). (a) Typical sandstone packet with major bounding and amalgamation surfaces (logs A to G and parts of logs 12, 13, 14 and 15). (b) Typical sandstone packet (logs 14A–J). (c) Sandstone packet/thin-bedded turbidite transition zone (logs 13A–F and logs 13 (part) and 14 (part)). (d) Sandstone packet/thin-bedded turbidite transition zone (logs 28A–F). (e) Sandstone packet/thin-bedded turbidite transition zone (logs 12 west A to E). From Drinkwater and Pickering (2001).

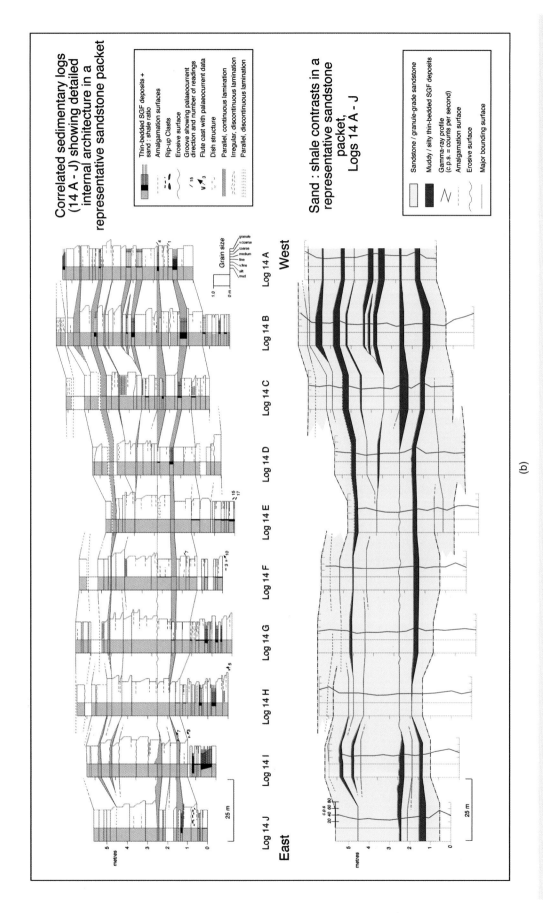

Fig. 8.49   *(continued)*

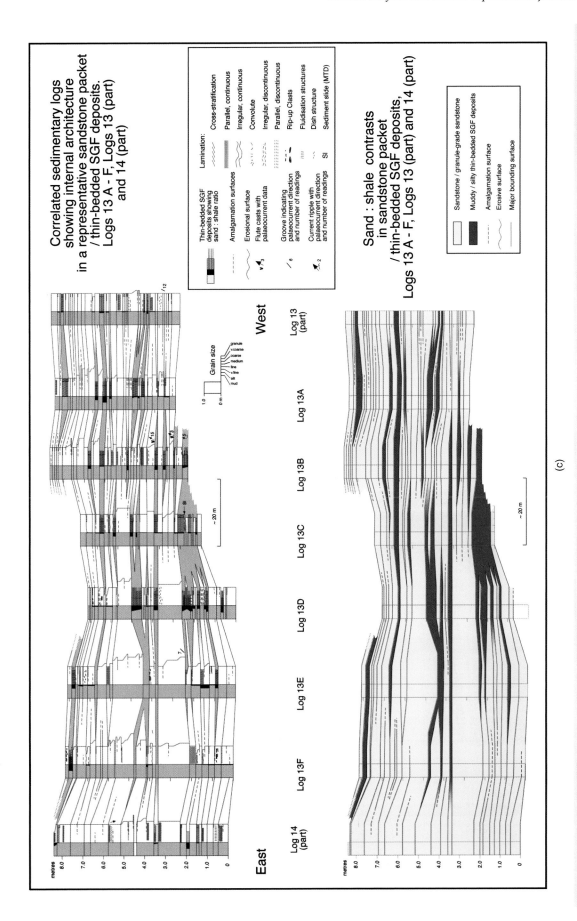

Fig. 8.49 *(continued)*

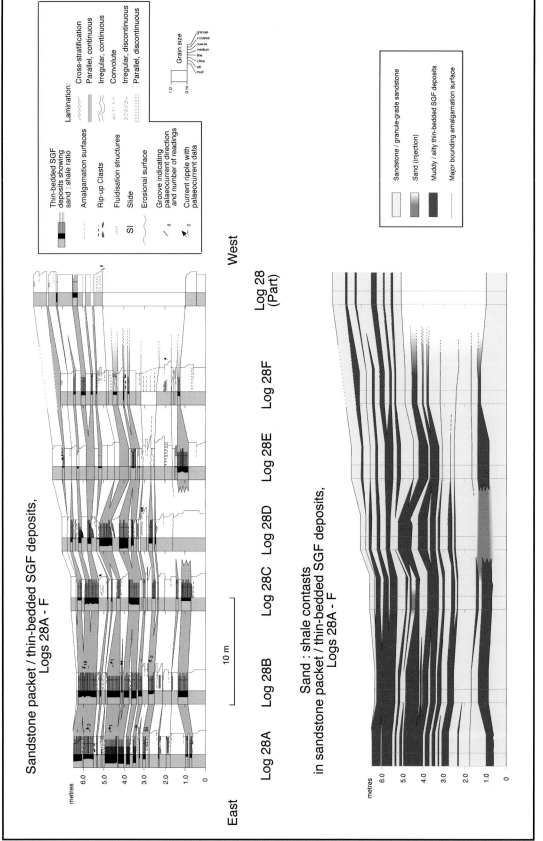

Fig. 8.49 *(continued)*

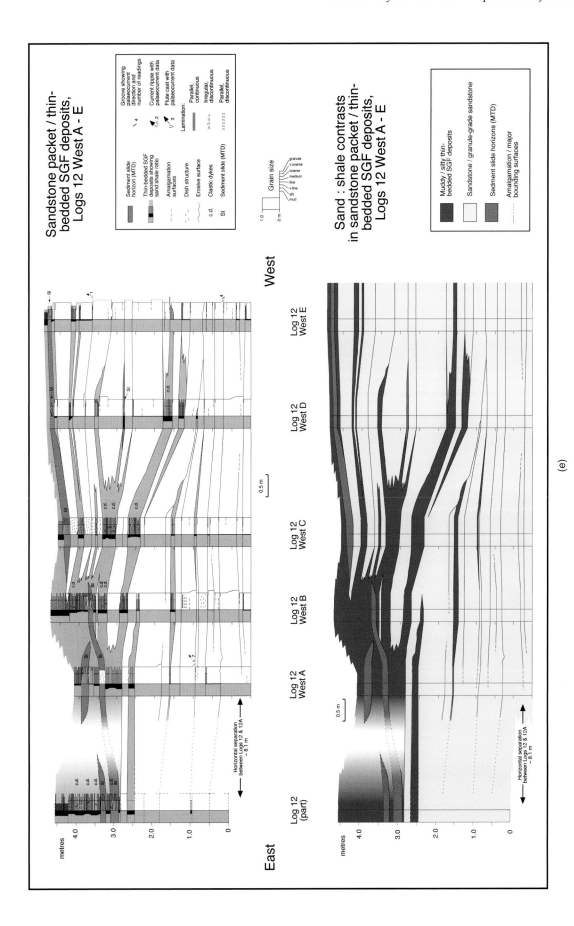

Fig. 8.49 (*continued*)

Table 8.4 Summary of characteristics of lobe, lobe-fringe and fan-fringe deposits. Modified after Pickering (1981a, 1983b).
* denotes information from Mutti, Nilsen and Ricci Lucchi (1978) and Pickering (1981a). Compare with Fig. 7.41

Interpretation	Lobe	Lobe fringe	Fan fringe
Definition	Very thick- to medium-bedded SGF deposits, as sheet-like beds, forming topographic high immediately downfan from a channel mouth (or mouthbar).	Medium- to thin-bedded SGF deposits, peripheral to lobe deposits, as distal and lateral-margin equivalents	Thin- to very thin-bedded SGF deposits in regularly bedded packets, representing the most distal fan deposits.
Bedding pattern	Sheet-like (high-continuity), with localised scour-and-fill (may cut down as packets over many hundreds of metres*).	Sheet-like (high-continuity).	Sheet-like (high-continuity).
Common internal sedimentary structures	T_a to T_c Bouma divisions.	T_b to T_e Bouma divisions.	T_b and T_c to T_e Bouma divisions.
Typical gram size	Very coarse- to medium-grained sandstone.	Fine-grained sandstone.	Very fine-grained sandstone to silts tone.
Estimated % sandstone	>80%	80% to 40%	<40%
Amalgamation	Very common.	Rare.	Absent.
Palaeoflow relative to adjacent environments in *this* table	Similar.	Similar.	Similar.
Other features	Thickening- and/or coarsening-upward trends have been reported (mostly without statistical testing) to be more common than thinning- and/or fining-upward trends (see Section 5.3). Very shallow , high aspect-ratio (width:depth) channels may be present.		Impressive regularity of bedding vertically and laterally. Subtle or absent vertical bedding trends.
Typical facies classes	Classes B, C and D.	Classes C and D.	Classes C, D and E.

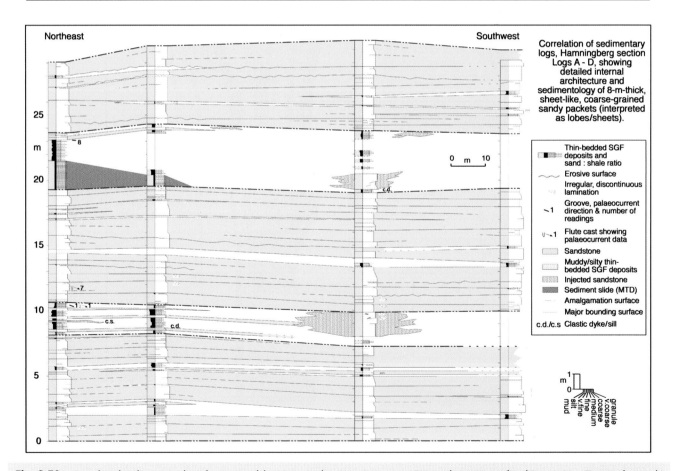

Fig. 8.50 Correlated sedimentary logs from part of the Hamningberg section, upper Precambrian Kongsfjord Formation, Finnmark, North Norway. The panel shows the detailed architecture of 8 m-thick sandstone packets interpreted as depositional lobes. From Drinkwater and Pickering (2001).

more appropriately referred to as *mouthbar deposits*. The Nålneset section is a ~3.5 km-long outcrop oblique to palaeocurrent direction on the central part of the northern coast of the Varanger Peninsula (Figs 8.48, 8.49). The section, which comprises the base of the Nålneset Member, forms the westernmost of the three main Kongsfjord Formation outcrops and is ~60 km west of other extensive Kongsfjord Formation sections at Hamningberg, where lobe, lobe-fringe and fan-fringe deposits are also observed (Fig. 8.50). The Nålneset section is the oldest part of the formation and, therefore, of the Barents Sea Group (Siedlecka 1972; Pickering 1981a). The Nålneset section contains coarse-grained facies (Facies Classes A, B and less commonly C). Shallow channel-form geometry has been described from the upper part of the Nålneset Member (Pickering 1982b, 1985). The oldest part of the member in the Nålneset section, however, comprises essentially non-erosive and laterally continuous (> ~3200 m) sheet-like sandstones with abundant megaripples and decimetre-scale cross-stratification in compositionally mature sets, debrites and scour-and-fill structures with low width/depth ratios. Collectively, these are interpreted to have accumulated in an area dominated by the change from confined (channelised or simply basinally restricted) to relatively unconfined (non-channelised) deposition; for example, the channel-lobe transition (Drinkwater et al. 1996; Drinkwater & Pickering 2001). Three architectural elements were recognised by Drinkwater and Pickering (2001) (data derived from restored, correlated section illustrated in Fig. 8.47): (i) 10 m-thick coarse-grained packets; (ii) 1 m-thick coarse-grained packets and (iii) shale intervals. Five representative detailed panels of individual 10 m-thick coarse-grained packets are shown in Figure 8.49a–e.

The sand-rich elements are labelled *sand-packets* (terminology from Pickering 1981a, 1985 and Drinkwater & Pickering 2001). The packet thickness ranges from 2–16 m. The 1 m-thick packets form the basic building blocks for the 10 m-thick packets. The 1 m-thick packets comprise several composite sand-rich units (CSRUs), which represent the accumulations of individual flow events, identified on textural characteristics such as changes in grain size. Within these 1 m-thick packets, sand-on-sand contacts resulting from bed amalgamation cause some of the internal bounding surfaces of these CSRUs to be unclear to the naked eye. Shales from the Nålneset section are intra-packet shales. In some examples, CSRUs that split shale units thin out completely, and individual shales amalgamate. Subsurface examples of apparently similar deposits to those described from the Kongsfjord Formation are documented from many hydrocarbon basins worldwide, for example, in the Upper Jurassic Angel Formation, Wanea Field, Northwestern Australian shelf (Di Toro 1995).

A detailed study by Amy et al. (2000) of the Peïra Cava sections in the Oligocene Grès d'Annot System, southeastern France, has documented the proximal to distal changes in a basin-floor environment (Figs 8.51a,b). A maximum of ~1200 m stratigraphic thickness is exposed in the Peïra Cava outlier. Amy et al. (2000) studied a 420 m stratigraphic interval that was traced between most measured sections over a downflow distance of ~10 km and a crossflow width of 1–2 km, within which the majority of beds >2 m thick were correlated. They measured ten principal sections at a L : 200 scale, with three 30–50 m stratigraphic intervals logged at a 1 : 20 scale; horizontal distances between neighbouring sections varied from 0.25–4 km. Figures 8.51a and 8.51b show bed correlation for a 170 m-thick and a 70 m-thick stratigraphic interval, respectively, There is a cut-off limit in the confidence of correlation of thinner beds because of the logging scale used and the similar character of beds <0.2 m thick. The principal conclusions of the study of bed properties and downstream sand

bed thickness changes by Amy et al. (2000) were: (i) proximal maximum grain-size, distal bed sand thickness, mud-cap thicknesses and bed sand percentage are not significant predictors of the manner of downstream thickness change; (ii) proximal bed sand thickness and the presence of cross-stratification and erosional surfaces in proximal sections are not by themselves reliable predictors of downstream thickness change and (iii) low thickness to grain-size ratio (<1000) is a good indicator of either less than average downstream thinning or downstream bed thickening in the transect parallel to slope.

Amy et al. (2000) recognised beds having a significant percentage of downstream thickening on the basis of low thickness to grain-size ratio (<1000) and either erosional bases, very coarse-grained basal layers and a sharp upper grain-size break, or the presence of a thin cross-stratified sand interval separated from an overlying normally-graded interval by a sharp grain-size break. The first type has high absolute downstream thickness changes and was inferred to be a good indicator of the bypass of significant sediment volumes. The second shows very high downstream percentage increases in bed sand thickness but only minor absolute thickness increases. These results are in agreement with the analysis of bypass indicators outlined in Mutti's (1992) model. In the Peïra Cava study by Amy et al. (2000), however, only the types of erosional and cross-stratified beds, which they described, appear to be reliable bypass indicators between transects. This is interpreted as the influence of the lateral slope upon deposition. The study by Amy et al. (2000) illustrates that the presence of beds with markedly erosional bases and/or containing cross-stratification are the most reliable indicator of downstream thickening. From a hydrocarbon reservoir perspective, sections containing such beds are also likely to show a greater degree of periodic amalgamation and thus enhanced vertical connectivity.

In many ancient fan systems, it is possible to see an overall vertical progradational succession from more distal (lobe and related) deposits, to more proximal channelised sandbodies, as has been documented by Mulder et al. (2010) from the Lauzanier area in southeast France (Jean et al. 1985) (Fig. 8.52). This area represents the northernmost extension of the Grès d'Annot Formation, where it preserves ~650–900 m of SGF deposits. This basin was active from late Bartonian or early Priabonian to early Rupelian. It comprises two superposed units separated by a major unconformity. The sediment was supplied by confined (channelised) SGFs coming from the south. The size of the particles and the absence of fine-grained sediment suggest transport over a short distance. The 'Lower Unit' is made of coarse-grained tabular sandstone beds, mainly of Facies Classes A and B, interpreted as non-channelised lobe deposits (Mulder et al. 2010). The 'Upper Unit' is made of structureless conglomerates of Facies Class A, interpreted as the channelised part of lobes (Mulder et al. 2010). That these lobe deposits accumulated in a tectonically confined basin is suggested by topographic compensation that occurs from bed scale to unit scale (see also Section 8.6). The abrupt progradation between the lower and the upper unit seems related to a major tectonic uplift in the area. This uplift is also suggested by a change in the mineralogy of the source and an abrupt coarsening of the transported clasts. This field example is relatively unusual in that it documents terminal lobe deposits in a coarse-grained system.

Prélat et al. (2010) describe lobe deposits from fine-grained, sandy basin-floor submarine fans of the Upper Permian Skoorsteenberg Formation that are exposed for ~35 km downdip (S–N) and ~20 km along strike (W–E) in the Tanqua Basin, southwest Karoo Basin, South Africa. The Skoorsteenberg Formation consists of four sandy

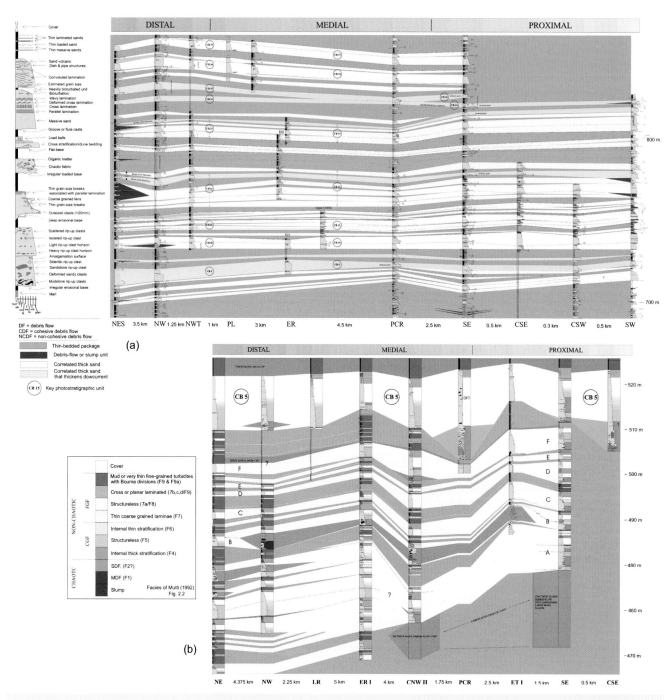

Fig. 8.51   Bed correlation for a 170 m-thick (a) and a 70 m-thick stratigraphic interval (b), respectively. There is a cut-off in confidence of correlation of thinner beds because of the logging scale used and the similar character of beds <0.2 m thick. Genetic units/parasequences of the Grès d'Annot system, southeast France. From Amy *et al.* (2000).

submarine fan systems (Fans 1–4 of Bouma & Wickens 1991; Wickens 1994; Wickens & Bouma 2000; Johnson *et al.* 2001) interpreted as forming the oldest part of a progradational basin-fill succession. These fan systems are overlain by a channelised lower-slope to base-of-slope unit (Unit 5 of Wild *et al.* 2005; Hodgson *et al.* 2006). The stratigraphic and geographic distributions of lithofacies, architectural elements and palaeocurrents were established after integration of research boreholes (Luthi *et al.* 2006) and digital outcrop data (Hodgetts *et al.* 2004), which led to a correlation of several sandy lobes. A common pattern of

fan initiation and growth (basinward movement of successive lobes) followed by fan retreat (landward movement of successive lobes) was identified (Hodgson *et al.* 2006). Prélat *et al.* (2009) showed that the Tanqua submarine fans do not comprise simple sheets in the Skoorsteenberg Formation, even though the sandstone elements commonly appear sheet-like at typical outcrop scales. They identified six sandy lobes (lobes 1–6) separated by thin-bedded silty interlobes (interlobes A–G). A fourfold hierarchy of downstream elongate architectural elements, named *bed, lobe element, lobe* and *lobe complex*

(Fig. 8.53), stack in a pattern that results in intricate depositional geometries and cryptic stratigraphic successions. A quantitative assessment of the data collected is presented by Prélat *et al.* (2009).

Submarine lobe dimensions from six different systems were compared by Prélat *et al.* (2010) (Figs 8.54, 8.55): (i) the exhumed Permian Fan 3 lobe complex of the Permian Tanqua Karoo, South Africa; (ii) the modern Amazon Fan channel-mouth lobe complex, offshore Brazil; (iii) a portion of the modern distal Zaïre Fan, offshore Angola/Congo; (iv) a Pleistocene fan of the Kutai Basin, subsurface offshore Indonesia; (v) the modern Golo System, offshore east Corsica, France and (vi) a shallow subsurface lobe complex, offshore Nigeria. These six systems have significantly different source-to-sink configurations (shelf dimension and slope topography), sediment supply characteristics (available grain size range and supply rate), tectonic settings, (palaeo-) latitude and delivery systems. Despite these differences, lobe deposits appear to share similar geometric and dimensional characteristics. The data led Prélat *et al.* (2010) to group the lobes into two distinct populations of geometries that they believe are related to basin floor topography. The first population corresponds to aerially extensive but thin lobes (average width 14 km, length 35 km and thickness 12 m) that were deposited onto low relief basin-floor areas. Examples of such systems include the Tanqua Karoo, the Amazon and the Zaïre systems. The second population corresponds to aerially smaller but thicker lobes (average width 5 km, length 8 km and thickness 30 m) that were deposited into settings with higher amplitude of relief, like in the Corsican trough, the Kutai Basin and offshore Nigeria. The two populations of lobe types, however, share similar volumes (a narrow range around 1–2 km³), which suggests that there is a control on the total volume of sediment that individual lobes can accommodate before deposition shifts to a new site. This indicates that some extrinsic processes control the number of lobes deposited per unit time rather than their dimensions. Of course lobes experiencing a higher rate of sediment supply will reach their final volume more quickly than lobes in areas of lower sediment supply.

Two alternative hypotheses were presented by Prélat *et al.* (2010) to explain the similarities in lobe volumes calculated from the six very different systems. The first surmises that the wide range in initial flow volumes and grain sizes across all systems is modified to a much narrower range by slope 'filtering' caused by more overspill and intra-channel deposition in larger systems. The second hypothesis proposes a gradual decrease in downstream gradient from the distributive channel base to the lobe top during lobe growth. This decrease in the hydraulic head that drives the flow is not sustainable as the channel will start to aggrade, and when a steeper lateral gradient is available, an avulsion will occur to an adjacent depositional low where subsequent flows will build a new lobe. This analysis of submarine lobe volumes suggests that basin-floor topography influences lobe geometry. However, since lobe volumes appear to have a narrow range there must be a strong influence of intrinsic (i.e., autogenic or autocyclic) processes.

Some researchers, such as Prélat *et al.* (2009: their fig. 16), have suggested that in lobe environments the fine-grained units formed by allogenic and autogenic processes can be distinguished. Specifically, they conclude that fine-grained interlobe deposits will be continuous and will stratigraphically separate the more sandy lobes if the primary control is external (i.e., allogenic) or they will be discontinuous and not fully envelope the more sandy lobes if the controlling process is autogenic. This argument is predicated on the assumption that, if formed by allogenic processes, the fine-grained units represent a decrease in sediment supply to the whole deep-water system, whereas if formed by autogenic processes, the fine-grained units

represent the distal fringes of lobes that thicken toward their axes. Whilst this argument is superficially appealing, the complexity and variability of extrabasinal and intrabasinal processes, the unpredictability of sediment flux to basins, especially for exceptional and rare very large-volume events that may be exceptionally erosive, and topographic complexity, all conspire to make this a very idealised depositional model that is likely to have many exceptions.

8.6   Seafloor topography and onlaps

An appreciation of the nature and range of onlaps in deep water is important for understanding the fluid dynamics and resulting deposition/erosion caused by the interaction between SGFs (e.g., turbidity currents) and basin slopes. Also, it is important for hydrocarbon prospectivity because sand-bed and sand-packet terminations may be associated with poor reservoir sealing and consequent hydrocarbon leakage from stratigraphic traps. The resultant sedimentary characteristics in the vicinity of any particular onlap are a function of (modified after Pickering & Hilton 1998):

(1)   Angle of slope.
(2)   Rate of change of the base-of-slope gradient below decelerating and/or deflecting/reflecting flows.
(3)   Sediment transport process, for example, turbidity current *versus* debris flow.
(4)   Incident angle of flow(s) relative to the maximum slope inclination (i.e., whether or not flows parallel a slope or impinge at a high angle).
(5)   Flow competence and capacity.
(6)   Bed roughness, for example, changing seafloor characteristics, degree of seafloor compaction, lithification and sediment type.
(7)   Height of topography relative to flow height, leading to complete *versus* partial containment of the flow(s), for example, permitting flow stripping (*cf.* Muck & Underwood 1990).

Seafloor topography can be created by the deposition of mounded cohesive debris-flow deposits and sediment slides, for example as described by Caixeta (1998) (Fig. 8.56). This example from a small basin in the Brazilian margin illustrates both intrabasinal and basin-margin onlaps. It shows how cohesive-flow deposits create topography that is subsequently infilled by SGF deposits from non-cohesive and much less cohesive flows.

The Oligocene Grès d'Annot Formation, Haute Provençe, southeast France, crops out in many sub-basins (Fig. 8.57) that reflect the original topography and, therefore, depocentres for deep-marine sediment accumulation (Hilton 1995; Hilton & Pickering 1995; 1998). The Grès d'Annot Formation is a sand-rich deep-marine system deposited in a basin with a complex basin-floor topography subdivided into local topographic highs and lows (e.g., Pickering & Hilton 1998; Apps *et al.* 2004; Salles *et al.* 2014). The topographic lows acted as sites for the preferential accumulation of sediments from SGFs, mainly turbidity currents, concentrated density flows, cohesive debris flows and slides. In many outcrops in the sub-basins, there are excellent examples of onlaps from bed to seismic scale. Within this area, two distinct depositional systems were identified within the much larger Palaeogene Provençal Basin (Pickering & Hilton 1998). One system occurs in the western parts of Haute Provençe and includes the Entrevaux, Annot and Grand Coyer sub-basins, supplied with sediment via the St. Antonin area (fan-deltas), and which form the 'West Basin-Floor System'. In the east, the second system includes the Menton, Contes

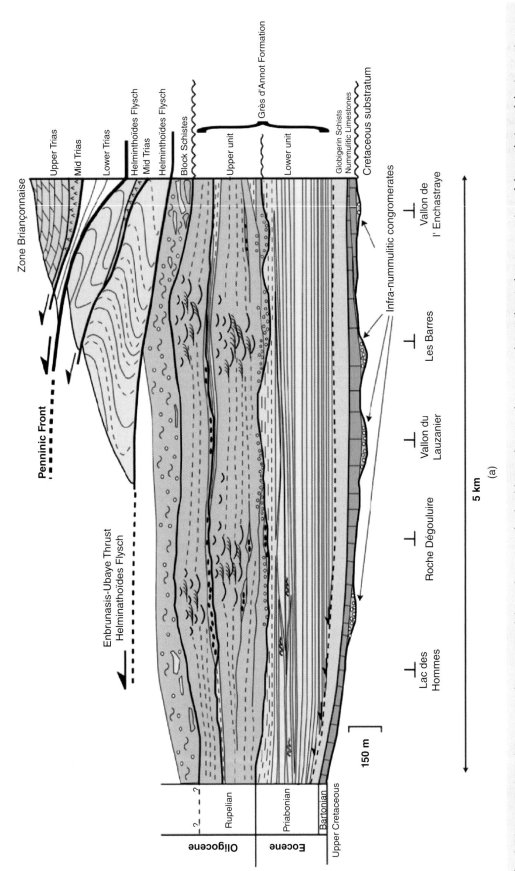

Fig. 8.52 (a) Synthetic lithostratigraphic section of the early Tertiary foreland basin showing the 'Nummulitic Trilogy'. (b) Interpretation of the evolution of the Lauzanier area sub-basin during Priabonian. (A) Deposition of the lower part of Grès d'Annot Formation (sheet sand of the non-channelised lobes over the Nummulitic limestone and Blue Marls. Deposits have an excellent grain-size continuity and little lateral thickness variation. (B) Deposition of the upper part of the Lower Unit of Grès d'Annot Formation. Lateral continuity is less than in the lower part and topographic compensation begins to occur. (C) Deposition of members A and B of the Upper Unit of the Grès d'Annot Formation (conglomerates corresponding to channelised lobes). (D) Deposition of member C and D of the Upper Unit of the Grès d'Annot Formation. Important topographic compensation occurs. (E) Erosion of Grès d'Annot Formation and deposition of the Schistes à Blocs. From Mulder *et al.* (2010).

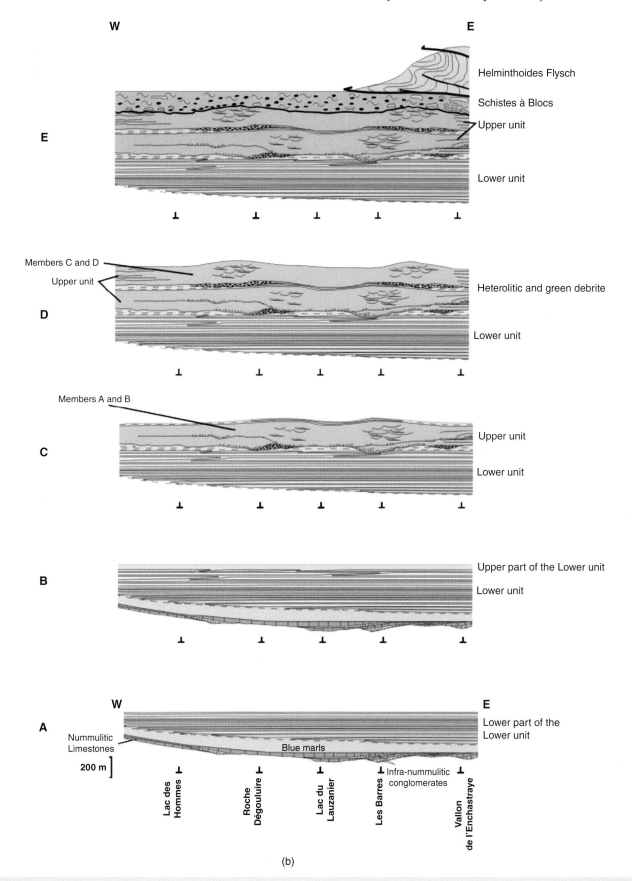

Fig. 8.52 (*continued*)

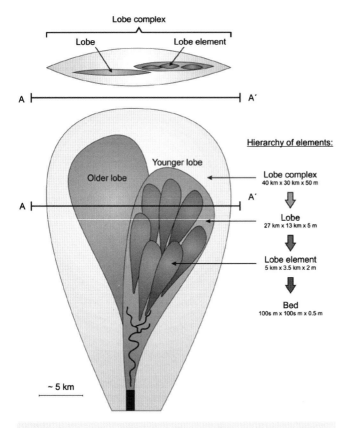

Fig. 8.53 Hierarchical scheme developed from the Fan 3 lobe complex with typical dimensions recorded from the Upper Permian Tanqua System, South Africa. Depositional elements are identified at four scales: bed, lobe element, lobe and lobe complex. Redrawn after Prélat *et al.* (2009) and Groenenberg *et al.* (2010).

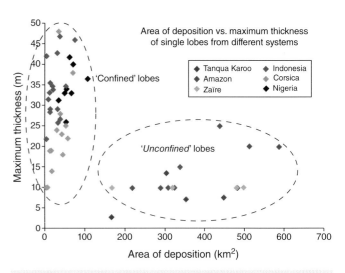

Fig. 8.54 Graph of the area of deposition versus maximum thickness for lobes deposited in the Tanqua Karoo, Amazon Fan, Zaire Fan, the Kutai Basin, Corsican Trough and offshore Nigeria. Two populations are proposed: (i) thin but large lobes, and (ii) thicker but smaller lobes. From Prélat *et al.* (2010).

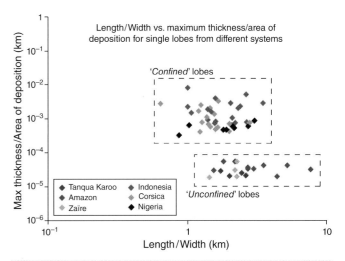

Fig. 8.55 Log-log graph of the length/width versus (maximum thickness)/(area of deposition) for lobes deposited in the Tanqua Karoo, Amazon Fan, Zaire Fan, Kutai Basin, Corsican Trough and offshore Nigeria. All lobes have a similar length to width ratios (1–8) but seem to fall into two clusters with different maximum thickness when normalised to the area of deposition, suggesting two populations of lobes. From Prélat *et al.* (2010).

and Peïra Cava sub-basins that constitute the 'East Basin-Floor System'. Both systems are defined using palaeocurrents and basin-floor morphology. Further information on the Grès d'Annot can be found in the volume edited by Joseph and Lomas (2004a, b).

Two end members in the type of deposition possible at onlaps were recognised by Pickering and Hilton (1998): (i) **Type 1 onlap**: Abrupt edge to the sands which all onlap the same depositional surface, approximating to a near-isochronous surface, and (ii) **Type 2 onlap**: Significant draping of the onlap surface, resulting in an apparent onlap surface which upon closer inspection is in fact diachronous, that is, with multiple stacked onlap surfaces (Fig. 8.58). Type 1 onlaps provide a good reservoir seal unlike Type 2 onlaps. These two types of onlap are exemplified, respectively, in the Grès d'Annot Formation at Chalufy and the lower section at Braux (Pickering & Hilton 1998; *cf.* Smith & Joseph 2004).

Examples of abrupt bed termination against a marl slope can be observed in the Chalufy onlaps in the Trois Evéchés sub-basin (Ghibaudo 1995; Hilton & Pickering 1995; Pickering & Hilton 1998; Joseph *et al.* 2000) (Figs 8.59–8.61). There is a more gradual thinning of beds as a drape onto a marl slope at Braux in the Annot sub-basin (Figs 8.62–8.64) and Tête du Ruch.

At Crête de la Barre, immediately east of the town of Annot (more commonly known as the Braux onlap), a gradual onlap with a progressive thinning of beds laterally (and fining in some beds) has been described in detail (Pickering & Hilton 1998; Kneller &

McCaffrey 1999; McCaffrey & Kneller 2004; Puigdefabregas 2004; Tomasso & Sinclair 2004) (Figs 8.62–8.64). The road section at Braux exposes a sandy packet ~14 m below the main sand-dominated packet that caps the ridge. The character of the onlap of this sand packet is very different to that at Chalufy. Rather than abutting the slope very sharply, as at some of the Chalufy onlap terminations, this particular packet of beds at Braux thins and drapes the slope more gradually (Figs 8.62, 8.63). This section shows both individual bed and packet thinning towards the onlap. In the vicinity of the onlap at Braux, sole marks show a wide dispersion in palaeoflow; which is attributed to abrupt changes in momentum of the sand-laden gravity flows as they rode up and along the local slope. A relatively thick bed was selected from this section to examine in detail the thickness change towards the onlap, and any associated changes in grain size and sedimentary structures. This thick bed is interpreted as the deposit of a concentrated density flow (Section 1.4.1). It shows very little change in grain size, although there is a more gradual thinning towards the onlap. Internally, within the bed, there are small-scale scours and prominent surfaces that can be traced between some logs. There are some very low-angle surfaces which may be the result of low-amplitude bedforms. The common scours and changes from parallel-laminated to ripple-laminated intervals suggest flow unsteadiness during deposition.

Within the Annot sub-basin, east of Annot town, at the Braux onlap, many beds show wedging, typically with decimetre-changes in thickness over a lateral distance of metres (Fig. 8.64a). Many beds also show abundant evidence of flow unsteadiness as alternating Bouma T_b and T_c divisions in what appear to be individual beds and/or genetically-linked events (based on the unusually tight cementation of grouped beds) (Fig. 8.64b, c), wet-sediment deformation as small-scale chaotic horizons, convolute bedding, load structures, flame structures and small-scale injections. In the vicinity of the onlap, sole marks show a large dispersion in palaeocurrents. Such features are predictable where SGFs encounter basin slopes, both approximately parallel to and at oblique angles to the slope (*cf.* Fig. 1.39). Mapping of the Grès d'Annot Formation in the Annot sub-basin shows that the Braux onlap was controlled by syn-sedimentary normal faulting (Fig. 8.65).

In contrast to the Braux onlap, the Montagne de Chalufy outcrop records sand deposition against a generally much steeper slope compared to that of the lower section at Braux. Four sedimentary logs were measured over a lateral distance of 8.2 m, at the termination of the uppermost sandstones at Chalufy, in the oldest sandstone packet that shows a clear thinning-out. Descriptions of this outcrop are provided by Ghibaudo (1995), Hilton and Pickering (1995) and

Smith and Joseph (2004). The logs demonstrate very little change in grain size laterally towards the onlap. More noticeable is the erosive nature of many of the bed bases, and associated thickness changes, suggesting that individual flows did not simply decelerate as a result of the topography but were deflected and locally underwent velocity changes causing both local erosion and deposition: an important factor was almost certainly the sub-parallel orientation of the flows relative to slope contours.

The onlap data were used by Pickering and Hilton (1998) to produce a schematic diagram for the onlap of SGF deposits in the vicinity of a basin slope, together with the generalised internal sedimentary structure divisions of single sandstone SGF deposits within onlapping beds (Fig. 8.58, taken from the Tête du Ruch section). Local topographic highs and palaeoslopes during deposition of the Grès d'Annot Formation are shown. The restored position of the Col de la Cayolle sub-basin is based on Graham (1980). Data for palaeo-highs is also taken from Ghibaudo in Elliott *et al.* (1985) and Hilton (1985).

In terms of reservoir quality and distribution, the situation at Chalufy results in potentially enhanced vertical connectivity of sandbodies at the onlap, because of erosion and the absence of a marked change in grain size towards the onlap. The thinner sands at Braux show a degradation in potential reservoir quality due both to the decreased grain size and bed thinning near the onlap. The issues associated with predicting reservoir quality within the terminal parts of sandy SGF units associated with basinal and intrabasinal slopes requires further research.

The Montagne de Chalufy onlap (Fig. 8.59), Grès d'Annot Formation, southeast France, occurs towards the southern extent of the Trois Evêches sub-basin and is characterised by packets of sandstones and sandstone–mudstone couplets. The Chalufy onlap shows a near continuous exposure of deep-marine basin sediments in a ~32 km long × 5 km wide north-northwest–south-southeast-trending sub-basin. The Trois Evêches

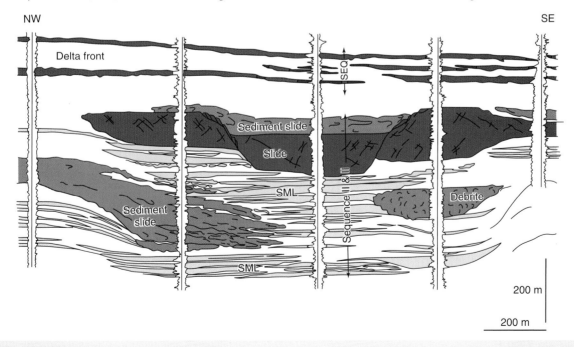

Fig. 8.56 Slope-fronted deltas prograded over deep-water lacustrine rift basins. Jacuipe field, Reconcavo Basin, continental rift megasequence, Brazil margin. Redrawn after Caixeta (1998). This is a good example of the influence of the topography created by mass-transport complexes on subsequent SGF (mainly turbidite) deposition.

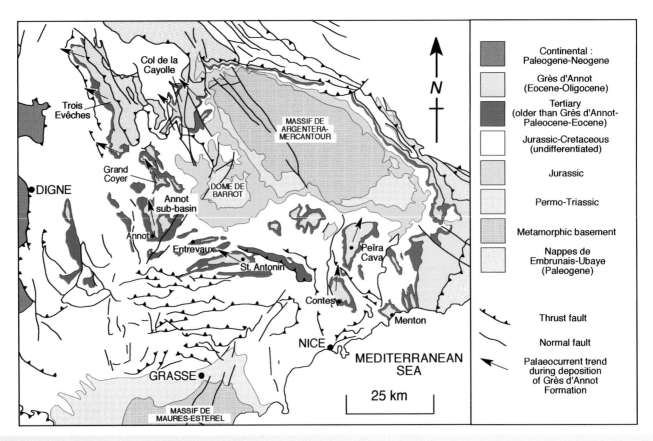

Fig. 8.57 Geological map of the Alpes Maritimes and Haute Provençe regions of southeast France showing Oligocene submarine-fan outcrops of the Grès Armorican or Grès d'Annot Formation. Redrawn with modifications after BRGM Sheet 40 and 45, 1:250,000 Series. From Pickering and Hilton (1998).

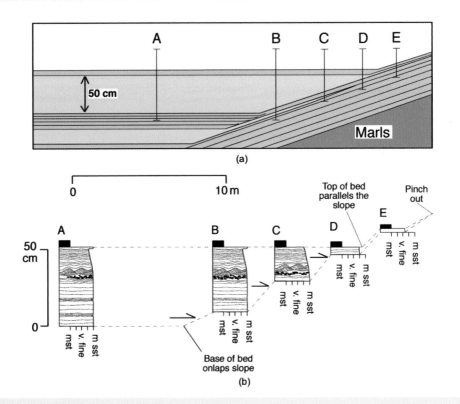

Fig. 8.58 (a) Schematic diagram of SGF deposits (interpreted as mainly turbidites) in the vicinity of a Type 2 onlap against a basin slope. (b) Generalised internal facies organisation within onlapping sandstone turbidite beds. This example is taken from the Tête du Ruch section, Oligocene Grès d'Annot Formation, Haute Provençe. Post-depositional (burial compaction) probably accounts for most of the apparent draping of the onlap surfaces. From Pickering and Hilton (1998).

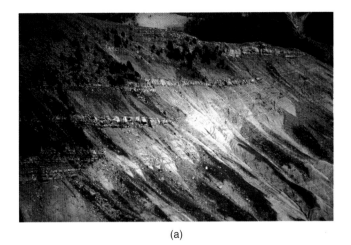

(a)

(b)

(c)

Fig. 8.59 (a) Onlap of Oligocene Grès d'Annot Formation, sandy turbidite packets, Chalufy, Haute Provençe. Four sandy packets can be seen that onlap onto a thin interval of very thin- and thin-bedded sandy SGF deposits immediately overlying the Marnes Bleues Formation (basin-floor and basin-slope marls). Palaeoflow obliquely out of cliff face towards left of view. (b) Close-up of ~35 m-thick sandbody seen at bottom left in (a), to show abrupt termination of sandstones against the basin margin. (c) Close-up of ~15 m-thick sandbody seen in (a), to show termination of sandstones against the basin margin, and large-scale wet-sediment deformation features, that is, folding of sandstone beds. For more detailed descriptions of these outcrops, see Hilton and Pickering (1995), and Pickering and Hilton (1998).

sub-basin locally includes exposures of the Poudingues d'Argens Formation, and near continuous exposure of the Calcaires Nummulitiques, Marnes Bleues and Grès d'Annot formations. Detailed surveys by Le Varlet and Roy (1983) and Inglis *et al.* (1981), summarised by Ravenne *et al.* (1987) show that the base of the Grès d'Annot Formation is characterised by a consistent southerly onlap onto a north-dipping palaeoslope developed in the underlying marls. This north-dipping palaeoslope is exposed at various localities along the length of the sub-basin and the Chalufy onlap represents the southern culmination of these onlaps. The total thickness of the Grès d'Annot Formation at this locality is ~350 m. The sandstone beds and packets onlap a slope in the marls, which locally approached 26° true dip, but was more typically in the order of 12°. The section at Chalufy is aligned at ~40° to the mean palaeocurrent trend. A particularly noteworthy feature of the Chalufy section is the occurrence of a large erosional channel-like feature (Figs 8.60, 8.61). The cross-section through this channel shows both margins and has a width of ~450 m and a depth of fill of ~60 m.

A reconstruction, based on mapping and reconstructing structural contours from the various sub-basins during sand accumulation of the Grès d'Annot Formation is shown as schematic diagrams in Figures 8.66 and 8.67 (fig. 59 of Pickering & Hilton 1998). These sub-basins were probably inter-connected to permit sand-laden SGFs to travel between basins across a submarine ridge. These figures provide a good example of both the 3D complexity of basin-floor topography and their length scales.

Sinclair and Cowie (2003) attempted to isolate the effects of basin-floor topography on SGF bed-thickness distributions using outcrop examples from localities where the ancient sea-floor topography has been reconstructed and is thought to dominate the signal. The Eocene and Oligocene Taveyannaz and Grès d'Annot sandstones of eastern Switzerland and France were deposited in confined intraslope-basin and base-of-slope settings. The deposits of the confined basin record flow ponding and flow stripping; the base-of-slope deposits record the amalgamation of SGFs. Bed thickness data for both the confined basin and proximal base-of-slope settings are best approximated by an exponential distribution; the data from the more distal base-of-slope setting are better described by a power law. Statistical experiments presented here demonstrate that these distributions can be generated by the modification of an input signal with a power-law distribution. In the case of the confined basin, flow ponding causes dramatic thickening of beds. However, flow stripping counteracts this, particularly for the thicker beds, and may account for a very large proportion of the input volume of sediment bypassing the basin even before the basin is filled. For the base-of-slope setting, erosion and nondeposition of beds will result in the preferential preservation of thicker beds; the thick-bed population is also enhanced in the data by potentially unidentifiable amalgamation of beds. Differentiation between distributions that characterise these settings requires careful analysis of the thinnest and thickest portions of the populations and is aided by plotting the data as the log of the cumulative bed number against bed thickness (see also Section 5.1).

In the Grès Annot Formation, no correlation horizons between the various sub-basins have been discovered; therefore, time equivalence between sub-basins remains poorly constrained. Further, there is a paucity of reliable palaeontological evidence to permit detailed correlations between sub-basins. Inferences, therefore, about correlations between the strata within the sub-basins rely solely on similarities in sedimentary facies and the relative positions of sandstone successions.

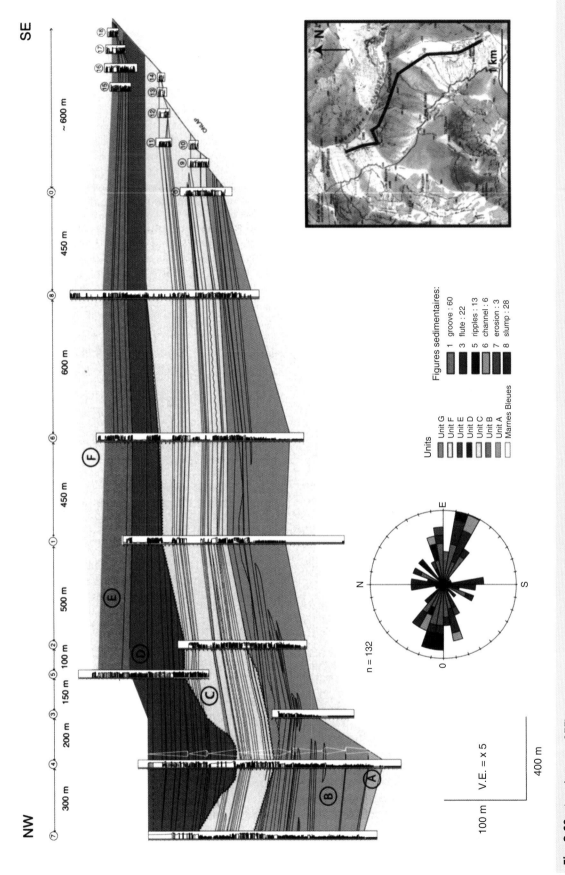

Fig. 8.60 (caption on p.377)

Fig. 8.60 Correlation panel, also marked on an inset topographic map, at Chalufy, Grès d'Annot Formation, Haute Provençe, southeast France. Note the onlap of the beds and sandbodies towards the southeast. See Figs 8.59, 8.61 for outcrop photographs of this basin margin. The Chalufy outcrops are in the southern part of the Trois Evéchés sub-basin and are interpreted by Joseph *et al.* (2000) as the downstream continuation of both the eastern Sanguiniére-Allos sub-basin and the southern Annot-Grand Coyer sub-basin. Nineteen palaeocurrent measurements show a dominant azimuth of 280°, that is, toward the west-northwest. In the Chalufy section, which is continuous over 400 m laterally, five depositional intervals were identified, as follows. Unit B (0–145 m) corresponds to alternations of thin-bedded sandstones and silty-shaly interbeds: sedimentary features, such as parallel lamination, current ripples, climbing ripples and convolutions are well developed. This facies association is interpreted as the distal fringes of lobes. Some sediment slide/slump intervals are identified in this unit. A major unconformity erodes the top of Unit B up to 30 m in some areas, defining the base of Unit C. Benthic foraminiferal associations suggest a palaeo-bathymetry of 200–500 m for the top of the Marnes Bleues and for Unit B. Planktonic foraminiferal associations date the top of Marnes Bleues P16 (late Priabonian) and all the thin levels of Grès d'Annot are dated to P18 (early Rupelian). Unit C (145–205 m) is characterised by the first erosive 'channelised system' and facies correspond to very coarse to granular sandstones (Facies Class A and B). Unit D (205 to 330 m), Unit E (330 to 370 m) and Unit F (370 to 390 m) correspond to apparently structureless sandstone units (Facies Classes B and C). The bottom surface of the channelised Unit D is deeply erosional (up to 100 m), which implies that the thickness of this unit shows considerable lateral thickness changes. The erosion surface is infilled with structureless coarse-grained to pebbly sandbodies (Facies Classes A and B). Units E and D are more tabular. All these units are separated by well-developed heterolithic intervals (10–20 m), which were interpreted by Joseph *et al.* (2000) as corresponding to abandonment episodes of the sand supply (deep-marine equivalent of maximum flooding surfaces). From Fornel *et al.* (2004). See also Ghibaudo (1995: his fig. 34.2) and Joseph *et al.* (2000).

8.7 Scours

Mesotopographic-scale 3D exposures of erosional and depositional features have been documented from the Albian 'Black Flysch', northern Spain, by Vicente Bravo and Robles (1995). They describe large-scale erosional features that include spoon-shaped depressions (flute-like), transverse step-like features and irregular ponded depressions. With limited outcrop and only cross-sectional exposures, the infill of large-scale scours, including megaflutes, are easily mistaken for channel deposits (e.g., Vicente Bravo & Robles 1995; Elliott 2000a,b; Lien *et al.* 2003: cf. modern examples described by Shaw *et al.* 2013).

The large-scale flute-like structures occur as spoon-shaped depressions in plan view and range from 1–5 m deep, and 5–50 m across. They occur as isolated features or are nested together to define fields of megaflutes consistently aligned parallel with the predominant palaeocurrent direction (Fig. 8.68). In transverse section, they appear channel-like, but in longitudinal section the presence of erosional walls normal to the palaeoflow confirm a lack of continuity of these features in the downflow direction (so they are not continuous channels). They also have a distinctive depositional bedform at the downstream end of the spoon-shaped depression that consists of subtle sandy bulges with a hummock-like geometry. Their scour-and-fill geometry suggests that they formed from multiple or unsteady and pulsing events. Vicente Bravo and Robles (1995) interpreted these structures to have been created during flow expansion beneath high-velocity turbulent gravity flows (*cf.* hydraulic jumps, as described by Mutti & Normark 1987; 'defect theory' of Allen 1971).

Fig. 8.61 Onlap of Oligocene Grès d'Annot Formation, sandy packets, Chalufy, Haute Provençe. Palaeoflow obliquely out of cliff face towards left of view. Note the prominent, lowd' aspect-ratio, erosive channel. See Fig. 8.60 for sedimentary logs and scale.

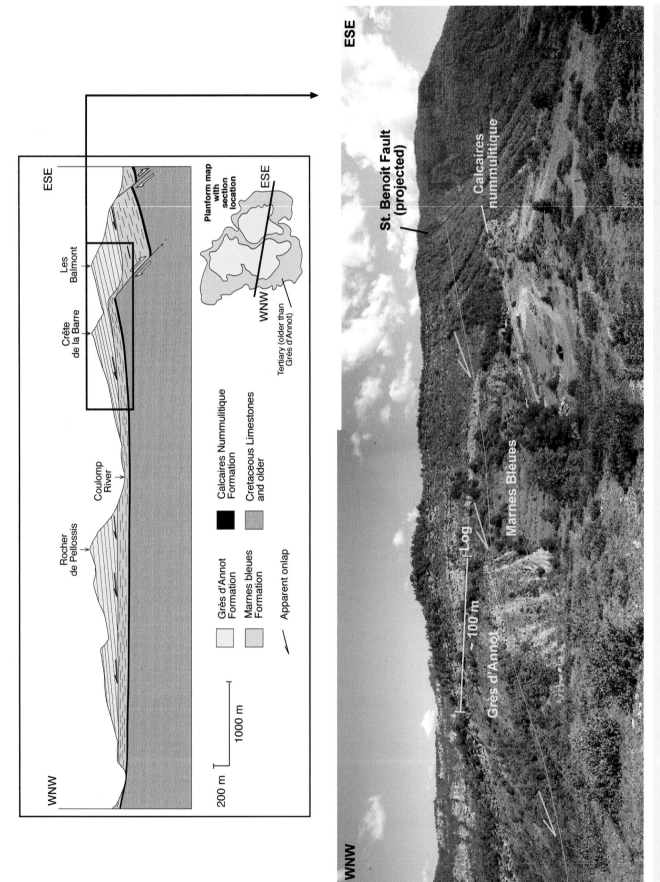

Fig. 8.62 Photo-montage and line interpretation of the onlap of the Grès d'Annot Formation onto the Marnes Bleues marls in the Annot sub-basin at Braux. Location of sedimentary log shown in Fig. 8.63 is marked along the road. Redrawn after Pickering and Hilton (1998).

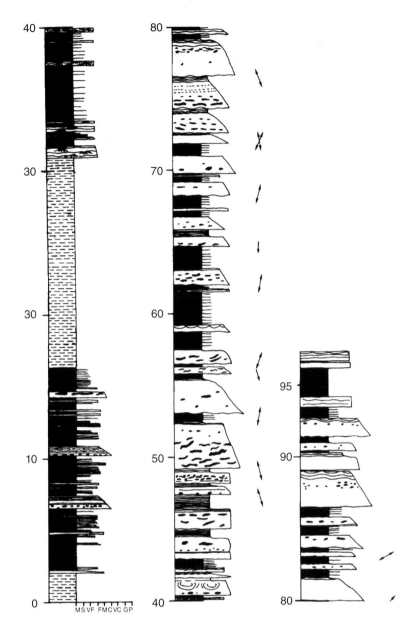

Fig. 8.63 Sedimentary log immediately above the onlap of the Grès d'Annot Formation onto the Marnes Bleues marls in the Annot sub-basin at Braux. Location of sedimentary log shown in Fig. 8.62b along the road. From Pickering and Hilton (1998).

Vicente Bravo and Robles (1995) also described linear step-like features in the Albian Black Flysch. These structures comprise relatively straight erosional crests arranged perpendicular to palaeoflow. Unlike the flute-like structures (above), the linear step-like features do not show a U-shaped headwall, but rather a more rectilinear up-slope edge. The steepest side faces downcurrent, becoming a smoother, near-horizontal or gently upcurrent-dipping erosional surface. Linear step-like features occur either in isolation, or as a bedform field showing a scalloped appearance, with individual bedform heights (depths) ranging from a few decimetres up to 2 m. The genetic processes interpreted for the large-scale flute-like structures were also invoked to explain the origin of the linear step-like structures.

Megaflutes have been documented in the Upper Carboniferous (Namurian) Ross Formation, County Clare, Ireland (also referred to by some researchers as the Ross Sandstone Formation), and their significance and environment of formation has been interpreted in different ways (Chapin *et al.* 1994; Elliott 2000a,b; Lien *et al.* 2003;

Macdonald *et al.* 2011a,b; Haughton & Shannon 2013) (Fig. 8.69). The Clare Basin records ~8 Myr of fine-grained clastic deep-water and fluvio-deltaic sediment accumulation above a foundered carbonate platform succession in an intra-continental basin that was influenced by glacio-eustasy, probably showing 100-kyr Milankovitch cyclicity (Haughton & Shannon 2013). The deep-water stratigraphy comprises up to ~180 m of Clare Shales (Clare Shale Formation of Wignall & Best 2000, 2002), overlain by ~500 m of mainly fine-grained sandstone depositional lobes and shallow channels (Facies Class C and D) in nine cycles (Ross Formation), succeeded by ~550 m of slope deposits with abundant sediment slides and slumps (Facies Class F) (Gull Island Formation): the Ross Formation comprises nine cycles (Haughton & Shannon 2013). Condensed intervals with goniatite horizons provide both age control and correlation horizons throughout the basin.

Based on detailed sedimentary logs and lateral correlations in the Ross Formation, Macdonald *et al.* (2011a) subjectively identified

(a)

(b)

Fig. 8.64 (a, b; caption on facing page)

(c)

Fig. 8.64   (a) Part of road section shown in Fig. 8.62, to show the thinning of individual sandstone beds of the Grès d'Annot Formation onto a sediment-slide deposit (Facies Class F) of Marnes Bleues marlstones above the basal onlap surface in the Annot sub-basin at Braux. West is towards left. (**b, c**) Detail of beds a few metres west of those shown in (a), that might have been emplaced by flows influenced by proximity to the surface in (a), showing abundant evidence of flow unsteadiness as alternating Bouma T_b and T_c divisions in what appear to be individual beds and/or genetically-linked events. Note the convolute bedding in (b), and the flame structures at the base of the Facies C3 bed below centre (c).

thickening-upwards packets interpreted to have resulted from deposition on prograding lobes within a submarine fan. Each thickening-upwards packet is interpreted by those authors to represent one lobe element, and the lobe-elements in turn stack to produce composite lobes. The basal mudstone part of each packet marks a period of lobe-element shutdown, recorded by a thin, fining-upwards trend, and its presence and thickness are dependent upon the magnitude of the avulsion event (lateral distance of the channel shift and time before reintroduction of coarser sediment to the site). Based on these observations, a six-stage model (*t1–t6*) for the development of each lobe-element and its erosive features was proposed (Fig. 8.70), as follows. (*t1*) Initially, in a dominantly depositional distal lobe-element setting, interbedded sands and muds accumulate from successive SGFs, and the bed shear stresses from these expanding and decelerating flows are insufficient to generate any erosional bedforms. (*t2*) As the lobe-element progrades and aggrades, increasing erosion is manifested by the generation of megaflutes. Thicker sands are deposited by higher velocity flows that deposit erosively amalgamated beds. (*t3*) With further progradation and aggradation of the lobe, the frequency of megaflutes becomes greater as the intensity of sediment bypass increases. (*t4*) The period of maximum bypass represents arrival of the distributary channel at a proximal position within the lobe-element; the erosive surface is here termed a proximal lobe bypass surface. At this stage, megaflutes may develop and join laterally or become incorporated into the margin of the bypass surface. (*t5*) Bypass intensity decreases as lobe-element abandonment is initiated; deposition rates increase, resulting in a net accumulation of sediment and the passive filling of the proximal lobe bypass surface. The remaining erosive features become draped or compensationally filled by larger flows. (*t6*) Complete avulsion landward of the site, perhaps linked to the infill of these erosive features, leads to the prograding lobe-element being abandoned. Sediment accumulation at this locality is initially composed of thin sand and mud interbeds and then solely of mud that marks the

beginning of a new thickening-upwards packet and the generation of a new lobe-element.

The model of lobe-element dynamics presented by Macdonald *et al.* (2011a) provides a mechanism to explain some of the thickening-upward described from lobes. This model suggests that a progressive thickening-upwards trend can develop through a combination of locally larger flows and increased bed amalgamation. Macdonald *et al.* (2011a) argued that this produces thickening-upwards trends at all points, with the exception of volumetrically insignificant low-relief channel infills. In contrast, most previous models based on the lateral movement of the thickest parts of deposits in compensation cycles (Fig. 7.39) result in a variety of vertical trends and are unlikely to result in cyclical thickening-upwards packets. The dataset of Macdonald *et al.* (2011a) suggests that progradation can, in certain cases, form thickening-upwards trends of restricted thickness, and furthermore that as a result of bed amalgamation the relative importance of progradation *versus* aggradation does not have to be high.

Macdonald *et al.* (2011a) did not confirm any of these so-called trends with statistical tests, and the proposed asymmetric trends are extremely short (<10 beds) and with steps in bed thickness (*cf.* Fig. 5.18). We would refer to such intervals as step-like trend types which can easily be generated at random. Thus, whilst some of this work is figured and discussed here, we would caution that these so-called trends may in fact be less deterministic than claimed.

Another example of a large-scale scour-and-fill has been documented by Pickering and Hilton (1998) from the Oligocene Grès d'Annot Formation in the Peïra Cava sub-basin, southeast France (Fig. 8.71). The scour is infilled by a very thick Facies A1.1 deposit with extraformational limestone clasts and other pebbles, and intraformational mudstone rip-up clasts. This scour occurred near the base of the Grès d'Annot Formation, and cuts down into the underlying Marnes Bleues Formation.

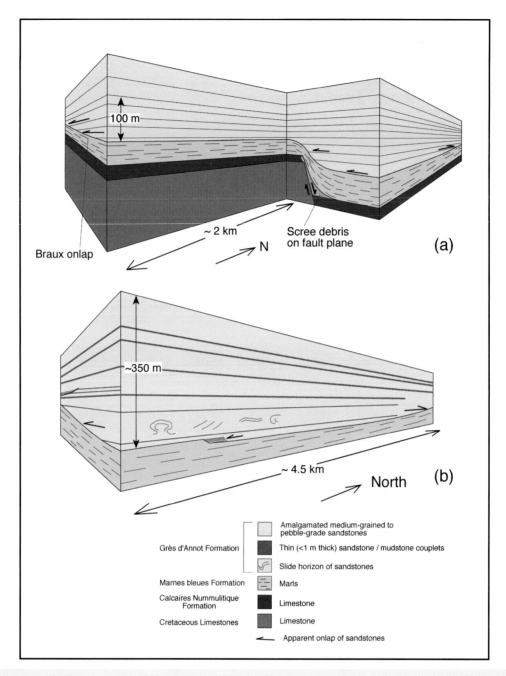

Fig. 8.65   Three-dimensional summary of sandbody architecture and the relationship with the underlying marls. (a) St. Benoit and Braux, and (b) Coulomp Valley sections, Grès d'Annot Formation, Haute Provençe, southeast France. Note the importance of syn-sedimentary normal faulting. From Pickering and Hilton (1998).

8.8   Basin-floor sheet-like systems

Ancient sheet-like basin-floor systems, including abyssal plains, have already been considered in relation to bed-thickness distributions in Section 5.2, as modern systems in Section 7.4.4, and in ancient foreland basins in Chapter 11. Consequently, only a brief discussion of such systems is considered here.

Ancient examples of abyssal plains and other deep-marine plains are difficult to identify, principally because of the lack of extensive outcrop over which bed-by-bed correlations can be made to establish the probable geometry of a basin and its floor. The difficulties are augmented by inadequacies in deducing ancient water depths even from diverse faunal assemblages: for pre-Mesozoic rocks this is effectively impossible. Given the problems in interpretation, the term 'abyssal plain' is less useful for ancient deposits than 'deep-sea plain' or 'basin floor'.

Basin plains that developed on truly oceanic crust have a low preservation potential in the geological record because of subduction processes. Even where incorporated into an orogenic belt as part of an accretionary complex, associated deformation and metamorphism generally are so intense that reconstructing the original depositional environment using palaeocurrent and bed-shape criteria have to be treated with extreme caution. Not surprisingly, some of the best-documented deep-marine sheet systems appear to have formed in foreland basins (e.g., Ricci Lucchi 1975a, 1978; Ricci Lucchi & Valmori 1980; Hiscott *et al.* 1986) (Fig. 8.72). These foreland-basin plains commonly escape the deformation and metamorphism of forearc trench fills, but are not good ancient analogues for the large, modern abyssal plains. Nevertheless, there are many similarities that justify some comparison.

Amongst the best-documented examples are the Lower Palaeozoic Cloridorme Formation, Gaspé Peninsula, Canada (Enos 1969a,b; Pickering & Hiscott 1985; Hiscott *et al.* 1986; Awadallah 2002; Awadallah & Hiscott 2004), the Eocene Jaca and Pamplona basins, Spanish Pyrenees (Seguret *et al.* 1984; Labaume *et al.* 1985; Payros *et al.* 1999; 2007), the upper Oligocene to Pliocene Marnosa-arenacea Formation, Italy (Ricci Lucchi & Pialli 1973; Ricci Lucchi 1975a,b, 1978; Ricci Lucchi & Valmori 1980; Talling 2001; Magalhaes & Tinterri 2010), and the Carpathian Cretaceous to Palaeogene flysch of the East Alps (Hesse 1974, 1975, 1982, 1995b). In this section, a Cretaceous (Carpathians and East Alps), and a Miocene basin-plain (Marnoso-arenacea Formation, Italy) are described as examples of ancient deep-marine plains. Other ancient examples include those described by Scholle (1971), Sagri (1972, 1974), Parea (1975), Hesse and Butt (1976), Robertson (1976), Bouma and Nilsen (1978), Ingersoll (1978a,b), Mutti and Johns (1979) and numerous others.

The Miocene Marnoso-arenacea Formation (Italian Apennines), up to 3500 m thick, crops out for ~200 km in a northwest-southeast belt parallel to the tectonic strike. It is interpreted as an example of deep-sea plain sedimentation (Ricci Lucchi 1975a,b, 1978; Ricci Lucchi & Valmori 1980). The succession has an estimated volume of 28 000 km^3. In 18 logged sections covering an area about 123 × 27 km, Ricci Lucchi and Valmori (1980) found that turbidites and the deposits of concentrated density flows form 80–90% of the volume with hemipelagites and minor debrites (Amy & Talling 2006; Fig. 1.26) accounting for the remainder. From ~14 000 bed-thickness measurements, Ricci Lucchi and Valmori (1980) divided the beds into thick-bedded Facies Class B and C SGFs >40 cm thick, and thin-bedded Facies Class C and D SGF deposits <40 cm thick. Almost 40% of the thick-bedded SGF deposits could be traced over the whole study area, that is, a maximum distance of 123 km.

The thin-bedded SGF deposits tend to consist of T_{cde} turbidite divisions, while the thick-bedded SGF deposits typically include the T_b division, up to 5 m thick in the exceptional Contessa Bed (see below). These beds may show base- and top-absent Bouma sequences. Most SGF deposits are fine-grained and display sheet-like geometry. One of the outstanding features of the Marnoso arenacea is the occurrence of very thick-bedded SGF deposits, or Contessa-like beds (named after the thickest, or Contessa Bed, of Ricci Lucchi & Pialli 1973). These 'megabeds' are characterised by: (i) tabular or sheet-like geometry; (ii) sand : mud ratios less than one; (iii) beds up to several metres thick; (iv) subtle lateral changes in texture and sedimentary structures; (v) generally basin-wide aerial extent and (vi) up-current and basin-margin thinning (Ricci Lucchi & Valmori 1980). The Contessa Bed (Figs 8.73, 8.74), up to 16 m thick, contains a very thick mudstone cap, as do other Contessa-like beds, and appears to have been deposited from a large turbidity current ponded within a confined basin. We ascribe transport to a turbidity current rather than concentrated density flow (Section 1.4.1) because the basal coarse divisions of these beds are stratified throughout, indicating selective,

grain-by-grain deposition and some degree of traction transport throughout the depositional phase.

The Marnoso-arenacea basin was 'over-supplied' in that both normal and exceptional turbidity currents had a large size compared to the area of the receiving basin. Flows were deflected along the basin axis at their entry points – the main sediment source being in the Southern Alps and shifting eastwards with time. While water depth estimates are necessarily crude, Ricci Lucchi (1978) suggests 1000–3000 m from geometrical relationships such as: (i) the inferred distance from the source area *versus* minimum gradients for gravity transport; (ii) ichnofacies and (iii) the paucity of skeletal benthic remains. Sediment accumulation rates averaged 15–45 cm/1000 years, increasing to 75 cm/1000 years in the Tortonian.

The Flysch Zone of the East Alps consists of Cretaceous to Palaeogene trench and arc-related deposits that accumulated over 70 Myr when the subduction zone associated with the trench was dormant (Fig. 8.75). Some of the formations are interpreted as sheet depositional systems, and the evidence for trench sedimentation is shown in Figure 8.75 and Table 8.5 (Hesse 1974, 1982). This 500 km-long and minimum 10 km-wide (possibly 80–100 km-wide) trench was connected to the 1000 km-long Carpathian Trench, the western part of which is not demonstrably associated with subduction until the Oligocene. Although there are many more recent references (e.g, Schnabel 1992; Hesse 1995b; Trautwein *et al.* 2001; Egger *et al.* 2002; Mattern & Wang 2008; Wagreich 2008: see also the earlier volume *Turbiditic Basins of Serbia* edited by Dimitrijevic & Dimitrijevic 1987), the summary presented here remains important for understanding the long-distance correlation of SGF deposits in a basin-floor setting.

In Aptian–Albian times, the trench contained a basin-plain at least 100 km (possibly 200–300 km) in length and below the carbonate compensation depth (CCD). The existence of a basin plain is suggested by the long-distance correlation of individual beds for 50 km in the Campanian, and 115 km in the Aptian to Albian (Hesse 1974). The lack of: (i) syndepositional deformation; (ii) volcanism and (iii) volcaniclastic detritus, suggests that sediments accumulated in a trench that was not associated with active subduction. Major palaeocurrent reversals occur throughout the Cretaceous Flysch Zone; changes in palaeoflow are associated with petrographic changes, that is, different source areas. Also, the palaeocurrents show a remarkable consistency for any single time and suggest very confined axial turbidity currents.

Hesse (1974, 1982), in a study of the lower part of the 1500 m-thick Cretaceous-Palaeogene succession in Bavaria, near the tectonic base of the flysch belt, made a detailed study of the Aptian to Albian Gault Formation (Fig. 8.76). Within the ~200 m-thick Gault Formation, typical thicknesses of the coarse divisions of SGFs (see Fig. 5.1 for definitions) are ~1 m, with some sandstone beds up to 4 m thick. Many beds show a basal Bouma T_b division, so are interpreted here as turbidites; other thicker beds accumulated from concentrated density flows. Sole marks consistently show palaeoflow from west to east throughout Aptian–Albian times. The intercalated packets of claystones average 75 cm with individual green, black and grey layers <20 cm thick. The green claystones are carbonate-free and represent hemipelagites, and the organic-rich, sometimes carbonate-rich, black claystones may represent hemipelagic/pelagic deposition during periods of basin anoxia. A packet of 55 glauconitic sandstone beds has been traced for up to 115 km along strike. Estimated minimum volumes of entrained sediment to form an 'average' 1 m-thick bed, assuming a basin width of 10 km, are of the order of 1 km^3, with possible maximum volumes of about 25 km^3. Sediment supply was axial from west to east with no evidence of lateral supply. The

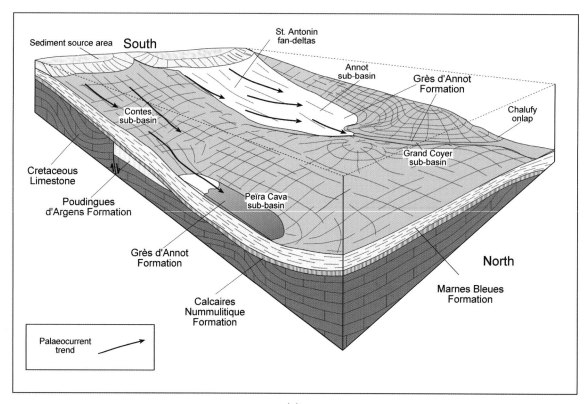

(a)

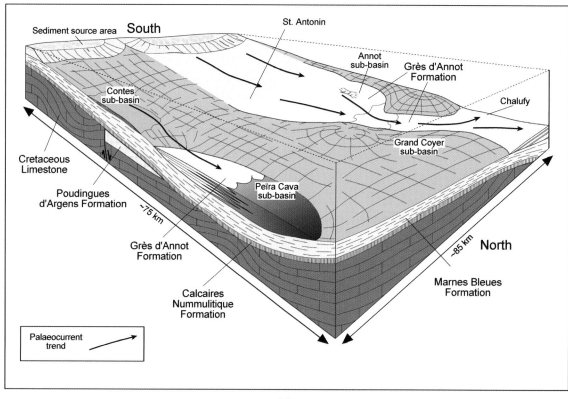

(b)

Fig. 8.66   (a, b; caption on facing page)

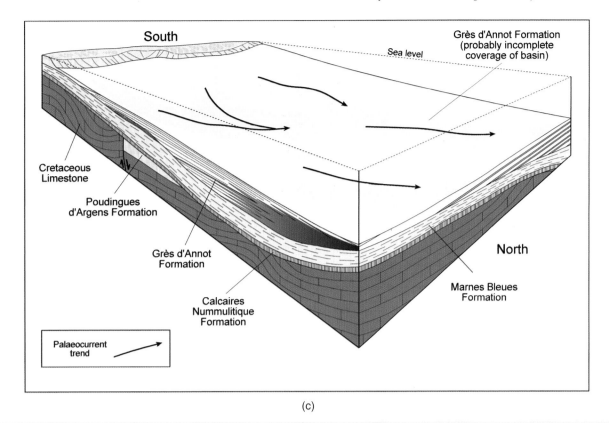

(c)

Fig. 8.66 Palaeogeographic reconstructions for the southern, more proximal, sectors of the Palaeogene Provençal Basin during the deposition of the Oligocene Grès d'Annot Formation. Seafloor topography interpreted from mapping, identifying onlap of the SGF deposits, and structural contours (a) Early stage. (b) Middle stage (main interval for sand accumulation). Sub-basin shown as partially filled, but accurate relative timing of infill is unknown. (c) Late stage. Sub-basins shown as completely filled with blanketing of sub-basins. Southern margin of Peïra Cava sub-basin is not seen and the inferred onlap is conjecture. From Pickering and Hilton (1998).

geometry of the beds suggests a seafloor with negligible relief. Hesse (1974, 1982) envisages an Early Cretaceous abyssal plain or trench floor (Gault Formation) and island arc system associated with the closure of Tethys. Tectonic tilting of the trench axis produced systematic reversals of the dominant palaeocurrent direction. Subsequent sedimentation includes submarine fans and other depositional systems.

Many modern and ancient deep-marine plains (both open-ocean abyssal plains, trenches and arc-related basin floors) tend to have a number of features in common regardless of the specific details of the system: (i) aerially and/or axially extensive, sheet-like SGF deposits; (ii) a high proportion, although variable, of hemipelagites and pelagites compared to other deep-water systems – however trench fills tend to be sandier; (iii) relatively rare wet-sediment sliding; (iv) typical bathyal and abyssal faunas; () possible association with metalliferous, other chemogenic and fine-grained biogenic material and (vi) locally common megabeds (ponded SGF deposits) attributable to major basin-slope failure. These features, together with the overall stratigraphic framework of a succession, may provide the best evidence for the identification of\vadjust{\vfill\eject} ancient sheet depositional systems. The main controls on cyclicity in sheet systems are believed to be: (i) extra-basinal, such as climatically-influenced changes in sediment influx, or those that are tectonically-controlled, and (ii) compensation cycles.

Perhaps the most likely confusion in the interpretation of ancient deep-marine basin plains is their distinction from fan-fringe sedimentation, although clearly the two environments merge with one another. Ricci Lucchi (1978) has suggested the following distinctions: (i) hemipelagites/pelagites are thicker and more regularly interbedded with basin-plain than with outer-fan SGF deposits – this means longer pauses elapsed between successive flows in a plain and only the larger SGFs reached this area; (ii) a higher degree of correlation for basinal beds because of their greater volumes and momentum on reaching plains; (iii) higher mud proportions than for the fan fringe and (iv) the occurrence of ponded flows as megabeds in plains and their absence from fan-fringe deposits.

Finally, many studies of ancient arc-related deep-marine plains (e.g., Ingersoll 1978a,b, Leggett 1980; Nilsen & Zuffa 1982; van der Lingen 1982; Chan & Dott 1983) have shown the complexity of such depositional systems in which preserved vestiges of sheet systems often form only a relatively small part. While this chapter has emphasised the obvious open-ocean abyssal plains, trench floors and arc-related basin floors, it must be stressed that the distinction between sheet and non-sheet systems is not always as sharp as the modern and ancient examples might suggest (e.g., trench systems, Chapter 10). However, whilst accepting that such deep-marine or deep-sea plains are not perfectly smooth horizontal to near-horizontal surfaces, to a first approximation the geometry both of most individual beds and the overall system is sheet-like.

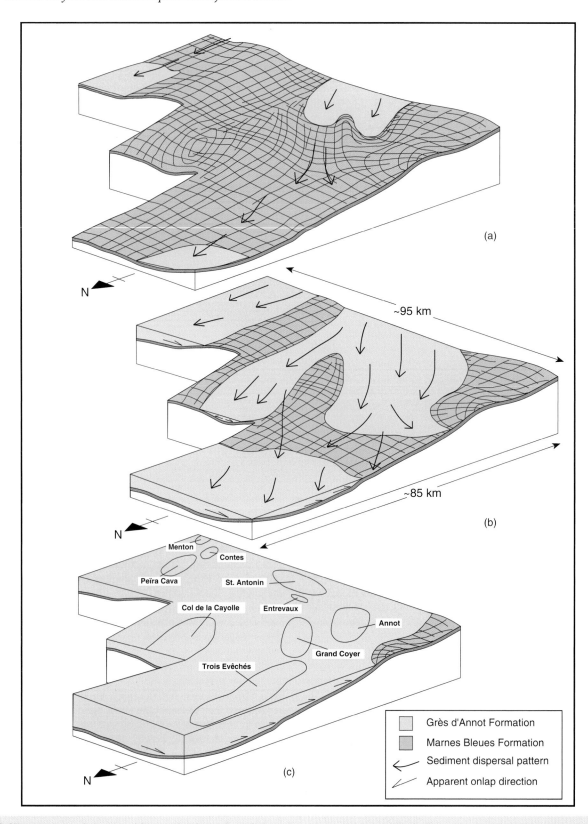

Fig. 8.67 Summary depositional history for the entire Palaeogene Provençal Basin as block diagrams, showing location and extent of the remnants of the deep-marine systems. Sub-basins probably filled in a complex manner rather than being sequentially filled to over-full (over-supply) in a proximal to distal direction (i.e., ideal fill-and-spill; see Section 4.6.2). Age constraints and present correlations have not resolved this problem. Note the inapplicability of the classic (radial or elongate) tri-partite submarine fan model to the Grès d'Annot Formation. From Pickering and Hilton (1998).

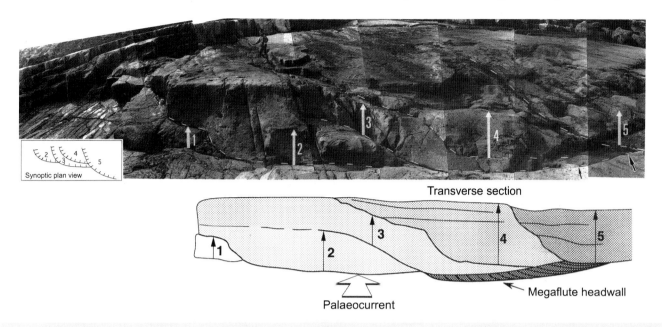

Fig. 8.68 Transverse section through a flute-like scour fill structure from the Matxixako Cape section in the Albian Black Flysch, northern Spain. Photographic montage of the outcrop, with simplified line drawing below. Several episodes of erosion and infilling occur. Although such features may superficially resemble small-scale channels, the presence of erosional walls (black arrow) perpendicular to the mean palaeocurrent direction reveals a lack of continuity of the erosional trough in the downflow direction. The persistence at the same stratigraphic level of scour-and-fills suggests that once a negative-relief erosional asperity or seafloor defect formed, it became a site for multiple erosion by subsequent turbulent SGFs. From Vicente Bravo and Robles (1995).

8.9   Prodeltaic clastic ramps

A particular type of deep-water clastic system that deserves special attention, and for which there are few documented examples, is the so-called prodeltaic clastic ramp, or low-gradient clastic system, that appears to develop at the shallow end of the deep-water spectrum. Whilst many submarine fans appear to contain both well-developed channels and lobes, low-gradient clastic systems appear to have only shallow, high aspect-ratio channels (i.e., high width:depth ratio) and relatively unconfined, laterally-extensive, sandbodies. They also tend to be relatively fine-grained. These systems appear to be linked to shelf-edge deltas (e.g., Mellere *et al.* 2002; Sutcliffe & Pickering 2009).

Perhaps the earliest well-documented example is the Tyee Formation, a 3 km-thick Eocene deposit in Oregon, USA, that consists, in vertical section, of a shallowing succession from a deep-water SGF system, through fan-like and slope deposits, to a mixed delta and shelf succession (Fig. 8.77a). The SGF deposits are sand-rich (Lovell 1969) with many concentrated density-flow and perhaps inflated sandflow deposits (Section 1.4.1), and were fed from multiple feeder channels that crossed the basin-slope from a source area to the south in the Klamath Mountains (Fig. 8.77b). Depositional rates were fast, $\sim$70 cm kyr^{-1} (Chan & Dott 1983) as the delta prograded rapidly from the south (progradation rate of 12.5–25 m kyr^{-1}; Heller & Dickinson 1985).

The Tyee Formation was deposited in a small, tectonically active forearc basin (Chan & Dott 1983) at a time of inferred globally low sea level (Heller & Dickinson 1985). Multiple shallow feeder channels generated an unconfined, sheet-like deposit that shows little of the facies organisation implicit in the Normark (1970) and Mutti and Ricci Lucchi (1972) fan models. Instead, thick, sheet-like, commonly amalgamated sandstones of Facies Group B1 (sand:shale > 9:1) pass basinward into thinner bedded sandstones of Facies Group C2 with a gradual increase in shale content (down to sand:shale = 2:1). There is no differentiation into channel and interchannel deposits, precluding assignment of even the more proximal deposits to an upper or middle-fan setting (Chan & Dott 1983; Heller & Dickinson 1985). This was clearly an oversupplied sandy system, and conforms closely with the Type I systems of Mutti (1985). Lack of a single feeder channel induced Heller and Dickinson (1985) to propose a new name, *submarine ramp*, for the depositional setting of the Tyee Formation, but many researchers still prefer to describe these rocks as submarine fan deposits, recognising that deposit shape is in most cases not that of a simple fan radiating from a point source. In terms of geometry alone, the Tyee 'Fan' is an elongate system (*cf.* Reading & Richards 1994; Richards *et al.* 1988; Table 7.1).

The upper-fan region of Chan and Dott (1983), described as basin slope by Heller and Dickinson (1985), is a muddy succession characterised by a series of channels, up to 350 m wide and 40 m deep, filled with either (i) essentially 100% Facies Class B (mostly Group B1) sandstones (Fig. 8.78), or (ii) thin-bedded fine-grained sandstone, siltstone and mudstone (Facies Groups C2, D2 and E2) identical to the facies into which the channels are cut. According to Heller and Dickinson (1985), the mud-filled channels tend to occur stratigraphically above the sand-filled channels. They attribute the sands to backfilling of the channels at the base of the slope; channels higher on the slope received no sand fill. It is also possible that some of the mud-filled 'channels' are really draped slide scars or megascours (*cf.* Normark *et al.* 1979).

Because of the lack of clear distinguishing features of middle- and lower-fan environments, we will describe the Tyee 'Fan' in terms of proximal and distal segments corresponding to the proximal

(a)

(b)

Fig. 8.69   (a) Exhumed megaflute in the Upper Carboniferous (Namurian) Ross Formation, Clare Basin, western Ireland. Palaeoflow is obliquely from top right to the lower left, with ~2 m-high human scale standing near the separation point that generated the scour when this surface was being swept by a turbulent SGF during the Namurian. These outcrops are described by Elliott (2000a,b). (b) Flat-based scour-and-fill structure infilled mainly with mudstones and siltstones, lower Ross Formation. The observed cut down is ~3 m. Mean palaeoflow is towards the northeast and into the cliff face. This scour may be an oblique section through a megaflute, a linear large-scale scour feature or a more continuous shallow channel.

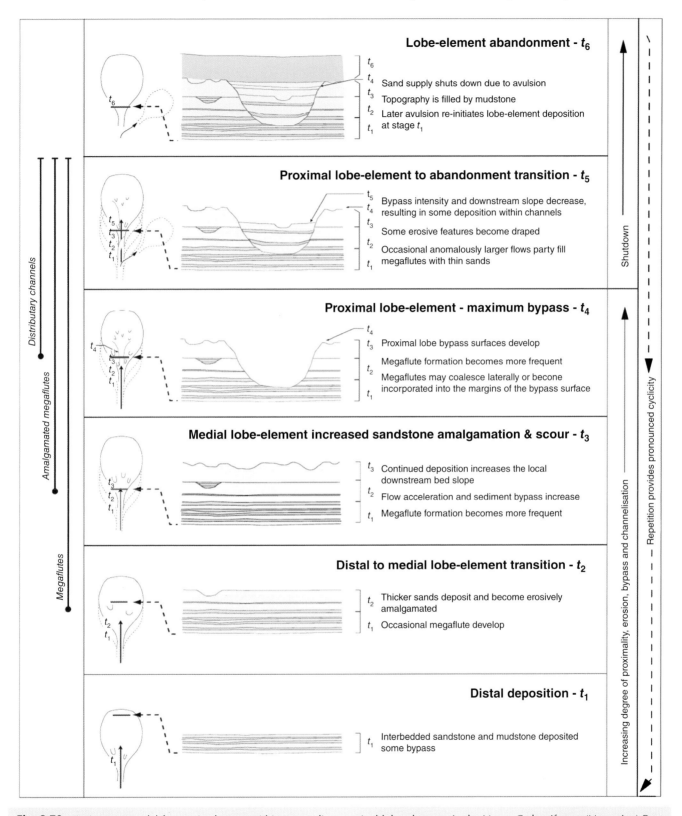

Fig. 8.70 A six-stage model for erosive bypass within prograding terminal lobe-elements in the Upper Carboniferous (Namurian) Ross Formation, County Clare, western Ireland. The schematic lobe-elements are prograding relative to the fixed-position red line. Cross-sections depict interbedded sandstone (yellow) and mudstone (grey), and illustrate the deposits associated with each stage at the fixed-position red line. From Macdonald *et al.* (2011a).

Fig. 8.71   Scour-and-fill in the Oligocene Grès d'Annot Formation, Peïra Cava sub-basin, Haute Provençe, southeast France, as described by Pickering and Hilton (1998). Note the ~2 m-high human scale immediately below the scour towards lower centre of field of view.

ramp facies and distal ramp facies of Heller and Dickinson (1985), and to the mid-fan and lower-fan/basin-plain facies of Chan and Dott (1983). The proximal deposits have sand : shale from 2 : 1–9 : 1. Most sandstone beds are sheet-like, 1–3 m thick, poorly graded, amalgamated, and belong to Facies Class B. Some beds are separated by thin mudstone layers. Only a few of the thickest beds have Bouma T_b, T_c and T_d divisions at the very top. Hence these are mostly interpreted as the deposits of concentrated density flows (Section 1.3.1). Asymmetrical cycles cannot be recognised in the field (Chan & Dott 1983) or through application of statistical techniques (Heller & Dickinson 1985). Only at the inferred base of the slope can a few broad, flat-based channels be recognised. Elsewhere, all bedding contacts are essentially flat.

The distal fan deposits consist of sandstone beds 0.1–1.0 m thick, interbedded with mudstone layers up to 1 m thick. Sand : shale ranges from about 7:1 to about 12:1, a range not unlike that for the proximal deposits, although the distal deposits are on average more mud rich. The sandstones belong to Facies Group C2, and contain various combinations of Bouma divisions, while the thicker beds contain all the divisions (Chan & Dott 1983). The fan deposits of the Tyee Formation are ~1 km thick, while the overlying slope sequence is only 500 m thick. Heller and Dickinson (1985) believe that unconfined, sand-rich fans with multiple feeder channels appear to be associated with relatively thin slope deposits because of the rapid rate of fan (ramp) aggradation in the basin. There is ultimately little bathymetric relief between the edge of the shelf (delta front) and the basin.

Sutcliffe and Pickering (2009) describe a Middle Eocene structurally confined low-gradient clastic system (Fig. 8.79) that can be considered a prodeltaic clastic ramp. The ~300 m-thick Guaso System is the youngest part of the deep-marine Ainsa Basin, Spanish Pyrenees. It is overlain by 150–200 m of fine-grained slope and prodelta deposits, succeeded by ~500 m of (fluvio-) deltaic sediments (Dreyer *et al.* 1999). The Guaso sandbodies represent laterally extensive deposition, built by lateral switching of 3–10 m deep erosional channels, and confined only by basin structure during deposition (Fig. 8.80a). The final tectonic event in this area to influence the deep-marine deposition was a phase of differential tectonic uplift above the underlying (Boltaña) thrust, creating a narrower and shallower basin morphology, thus creating a low-gradient clastic system. Then, the next fall in relative base level was insufficient to permit the cutting of canyons or deeply incised slope-channels, as had been the case earlier when the topographic relief between shelf and basin was at least several hundred metres greater.

Sutcliffe and Pickering (2009) divided the Guaso System into two major depositional units, designated Guaso I and Guaso II (GI and GII, respectively). Each unit comprises a major sandbody and associated multi-event mass transport deposits or complexes (mainly Type I MTCs as defined by Pickering & Corregidor 2005) and minor, typically heterolithic sandbodies. The Guaso deposits (including MTCs) differ from older systems in the Ainsa Basin in that there is very little gravel-grade material. Similarities with the other sandy systems exist, however, such as the overall westward shift in the

Fig. 8.72 Palaeocene Zumaya flysch, northwest Pyrenees, Spain. Beds, mainly of Facies Group C3 and Facies Classes D and E, are near vertical and young to left.

Fig. 8.73 Tabular (sheet-like) deposits of sandstones and shales (Marnoso arenacea, Italian Apennines). Note Facies C2.4 megaturbidites with very thick sandstone basal parts and medium-gray, structureless, silty mudstone caps (e.g., at the level of the white arrow, which points to the cap of the "Contessa" megabed portrayed in Fig. 8.74).

axis of sand accumulation throughout the system and the apparent fining-upward trend of the sandbodies. The facies associations for the Guaso I unit are divided into a lower (GIa) and upper (GIb) part. Above any 'basal' MTC, there are thick-bedded, medium- to coarse-grained, amalgamated sandstones (concentrated density-flow deposits), some of which have sole marks such as flutes and grooves. Locally, particularly in the older parts of the sandbodies, there are abundant angular mud clasts, with very rare small sub-rounded pebbles. A few beds are traceable over >100 m. Others are cut by broad erosional channel-like features ~1–5 m deep and up to ~150 m wide (but typically ~10–20 m wide) that occur in the middle and upper parts of the GI and GII sandbodies (Fig. 8.79). These are typically filled with thick- to medium-bedded, medium-grained, laminated, sandstone SGF deposits (Facies C2.2 and C2.3). Ripple, climbing-ripple and convolute lamination are common throughout the middle and upper parts of the sandbodies where bulk mean grain-size is typically medium to very fine-grained. The younger part of the GI sandbody is characterised by the appearance of medium- to thin-bedded, medium- to fine-grained sandstone SGF deposits with metre-scale relatively shallow erosional scour-and-fill or channel-like features. Beds in the GIb sandbody tend to possess relatively poor lateral continuity, commonly thinning out over tens of metres laterally. The youngest deposits in GIb tend to be thin-bedded and fine-grained, with abundant bioturbation, and show an overall thinning-and-fining-upward trend (interpreted as an 'abandonment stage'). Climbing-ripple lamination and convolute bedding are common throughout the GI sandbody, as Facies Classes C and D. In axial sites, the GII sandbody is erosive into the GI sandbody. Off-axis, however, these sandbodies are separated by a ~50 m-thick interval

of fine-grained, thin- to very thin-bedded and marly laminated sediments. The facies in the Guaso II sandbody overall are coarser grained than those observed in the GI sandbody. The lower parts of the GII sandbody contain extensive Type II MTCs as defined by Pickering and Corregidor (2005), represented by marly sediments with small amounts of extraformational material as small pebbles. Above the MTCs, there are thick- to medium-bedded, coarse- to medium-grained sandstones with rare pebbles. As in the GI sandbody, a variety of similar laminated sediments is present, that is, T_{bcd}, T_b, T_c and T_{cd} turbidites (Facies C2.2, C2.3, D2.2 and D2.3). As for the GI sandbody, the GII sandbody is capped by an overall thinning-and-fining-upward trend with abundant current-ripple lamination and bioturbation. The Guaso sandbodies show abundant evidence for deposition from supercritical SGFs, such as backset bedding, wavy and irregular bedding, discontinuous and irregular decimetre thick sandy bedforms, and local inverse grading within sandstone beds.

Stratigraphically above the Guaso System, there is an upper basin slope and delta system (the Sobrarbe Delta; Dreyer *et al.* 1999). The upper slope consists of marly mudstones and bioturbated, mainly very thin-bedded and thin-bedded sandstones and siltstones, with sandy turbidite-filled channels (Fig. 8.80b). These small-scale upper-slope and prodelta channels are up to >100 m wide, up to >300 m in flow-parallel length (in the three-dimensional outcrops), typically contain 3–4 m of mainly medium-grained sandy SGF deposits (*cf.* features described from the Turronian Ferron Sandstone, Utah, by Fielding 2015), and show palaeoflow that is consistent with sediment input from both a southeasterly and more easterly direction, as for the rest of the Ainsa Basin. In the southwestern part of the study area, eight upper-slope channels have been mapped (Fig. 8.79). The absence of deep erosional features is inconsistent with a canyon-like system, but rather signifies relatively unconfined and laterally-extensive sandbodies. When plotted with equal vertical and horizontal scales, the basal surfaces of the Guaso sandbodies are at a very low-angle to the underlying marls, that is, only up to ~1°. Overall, the Guaso System shows lateral structural confinement, with the sandbodies onlapping the confining basin slopes created by the Boltaña and Mediano anticlines.

Measured sections through the axis of the Guaso System in the Buil Syncline are thicker than off-axis sections at the margins, due to the sandbodies thinning onto the basin margins formed by the syn-depositional topographical highs of the confining growth-anticlines. The map and panel (Fig. 8.79a,b) also show lateral-offset stacking away from the submarine growth structure of the Mediano Anticline (to the east), as the axis of the second Guaso sandbody (GII) shows a westward shift compared to the first Guaso sandbody (GI). The sediments above the Guaso System include nummulitic sandstones and marls.

The off-axis thinning of the Guaso System onto the basin margins, together with the absence of large channel–levée–overbank complexes, precludes a typical middle submarine fan as a likely depositional setting. The pinch-out of sandbodies is by onlap against the basin lateral margins, not due to erosion. Given the scale of the Ainsa Basin, and constraints on palaeogeographic reconstructions, there is no space, even in the subsurface, for there to be the remnants of submarine canyons and deep erosional channels that might have fed the exposed system. As the GII sandbody is overall coarser grained than the older GI sandbody, this might be interpreted as a more proximal deposit, possibly the result of progradation of the depositional system. However, the GII sandbody shows an overall thinning-and-fining-upwards trend and it is overlain by ~200 m of typical fine-grained, marly slope sediments. Neither of these characteristics suggest progradation.

The laterally-extensive geometry of the two main Guaso sandbodies, with multiple shallow (and presumably relatively ephemeral) channels, and the paucity of coarse-grained Type II MTCs, is consistent with an interpretation as a low-gradient clastic system, including base-of-slope deposition following an episode of slope regrading (Pickering & Corregidor 2005). It seems likely, given the confined nature of the Ainsa Basin, with axial input of coarse clastics, that any delta feeding the Guaso sandbodies had prograded to the shelf edge, and was feeding the system essentially directly via a broad fairway of shallow, ephemeral, distributary channels.

8.10　Concluding remarks

Our understanding of both modern (Chapter 7) and ancient submarine fans and related systems (this chapter) shows that their 3D complexity and temporal evolution makes over-simplified depositional models unsustainable. For these reasons, frustrating as it may be to readers searching for simple and general models, we do not conclude with a set of distinct depositional models for submarine fans. However, the models that we outline for discrete fan elements, such as channels, levée-overbank and lobes (e.g., Figs 4.47 and 7.41) should prove more useful, as they can be selectively incorporated into any model for a particular depositional system.

Since the late 1990s, our understanding of submarine-fan and related systems has been considerably enhanced by several integrated outcrop-subsurface studies, such as: the Upper Miocene Mount

Messenger Formation, Taranaki, New Zealand (Coleman *et al.* 2000; Browne *et al.* 2000; Browne & Slatt 2002; Johansson 2005; King *et al.* 2011); the Eocene Ainsa System, Spanish Pyrenees (Pickering & Corregidor 2000, 2005), the Permian Karoo System, South Africa (Wild *et al.* 2005; Prélat *et al.* 2009, 2010; Pringle *et al.* 2010; Di Celma *et al.* 2011; Flint *et al.* 2011), and the Carboniferous, western Ireland (Pierce *et al.* 2010; Haughton & Shannon 2013). Such studies circumvent many of the problems associated with the limitations of outcrop position, orientation and scale, and they have provided unprecedented 3D data from ancient systems. Undoubtedly, in the future, we would anticipate the implementation of more studies of this type, something that will complement offshore DSDP/ODP/IODP programmes. Also, in recent years there have been several attempts to integrate outcrop studies with numerical (computer-based) modelling of SGF deposition, for example, for syn-depositional growth structures and their influence on seafloor topography in the Grès d'Annot Formation (~40–32 Ma), southeast France (Salles *et al.* 2011).

Several relatively new concepts are introduced in this chapter, such as lateral accretion packets (LAPs) and/or inclined sandy macroforms, which show considerable variation in their scale and constituent facies. These features remain poorly understood in terms of their geometry, depositional setting and the sedimentary processes that produce them (*cf.* Grosheny *et al.* 2015). The recognition that helical flow around submarine-channel bends may be counter to that in fluvial channels (Peakall *et al.* 2000a, b; Abad *et al.* 2011; Dorrell *et al.* 2013) has led to discussion that remain unresolved about the likely resulting deposits and vertical sequence of beds, including any characteristic bed-thickness trends.

Channel inner levées or internal levées (deposits that accumulate off-axis and towards a channel margin) may not be present in many channel fills but require more research to understand their genesis during channel development, infill and abandonment.

The recognition of knick-points along the thalweg of sinuous submarine channel–levée systems where the seafloor bathymetry is dynamically changing, for example as documented by Heiniö and Davies (2007) on the slope of the western Niger Delta using 3D seismic data, may influence SGFs passing over such features and lead to both local deposition and erosion around such sites. Knick-points form as a result of changes in slope gradient caused by local uplift, such as because of faults intersecting the seafloor. Locally, channel gradient may be lower upstream causing SGFs within a channel to decelerate and deposit the coarsest sediment load. Basinward-dipping slopes may be local steepened, leading to increased flow velocity and turbulence within SGFs, thus enhancing erosion to create a knick-point. If preserved, for example as a result of channel avulsion or abandonment, the deposits upstream of knick-points might contain significant hydrocarbon reservoir elements (Heiniö & Davies 2007). They can, however, also be partially eroded by headward-migrating knick-points, as the channel attempts to regain its equilibrium profile, leaving remnant sand pockets preserved on channel margins (Heiniö & Davies 2007). Such knick-points may also be associated with cyclic steps between regions of supercritical and subcritical flow (Section 4.13).

We have discussed relatively unconfined or sheet-like deposits at some length, mainly because of their internal complexity and variability (*cf.* detailed study to show the internal complexity of sandy lobes in the Grès d'Annot Formation by Mulder *et al.* 2010; Etienne *et al.* 2012). Readers should be aware that many researchers have described lobe and related deposits with erosionally-based elements – at a range of scales – that actually might be mouthbar deposits, discontinuous shallow and ephemeral channels or megaflutes. For example, industrial data commonly show lateral correlations of sandstone stratigraphic intervals that appear lobe-like, yet may have formed in rapidly laterally-shifting shallow channels such as in a submarine braidplain. The resulting amalgamated sandstones may create laterally continuous depositional units that, in the absence of very high-resolution 3D data, are erroneously interpreted as lobe deposits emplaced by sheet-like flows.

Another aspect of ancient fans and related deposits is the importance of syn- and post-depositional liquefaction and fluidisation processes that modify primary sedimentary structures and original bedding relationships. We have documented some of these features when describing facies in Chapter 2 (e.g., Section 2.10), or in conjunction with general descriptions of architectural elements throughout this chapter. This is a topic that will benefit from more research, especially combining experimental-theoretical and field observational work.

Probably the single most important advance in our understanding of submarine fans over the past few decades has been an appreciation of the variety and complexity of flow processes (Chapter 1), their deposits (Chapters 2 and 3) and the statistical properties of SGFs (Chapter 5) linked to sequence stratigraphy (Chapter 4).

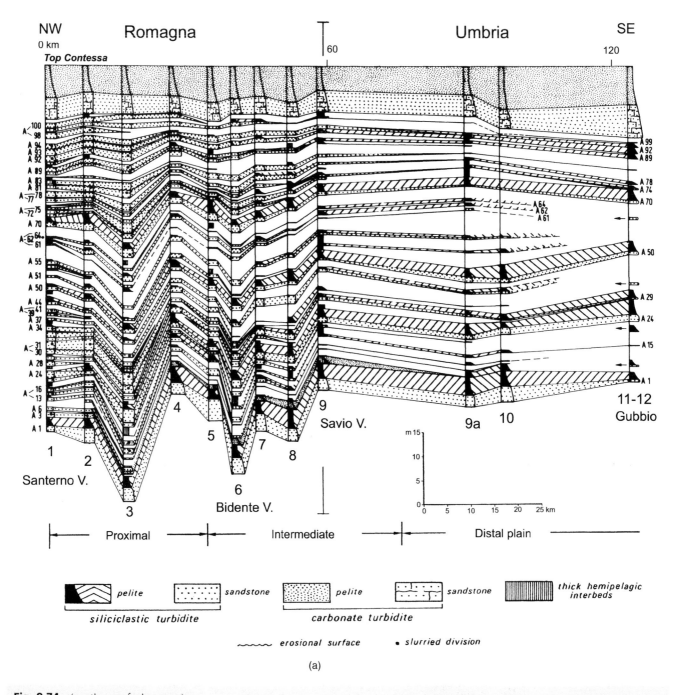

(a)

Fig. 8.74 (caption on facing page)

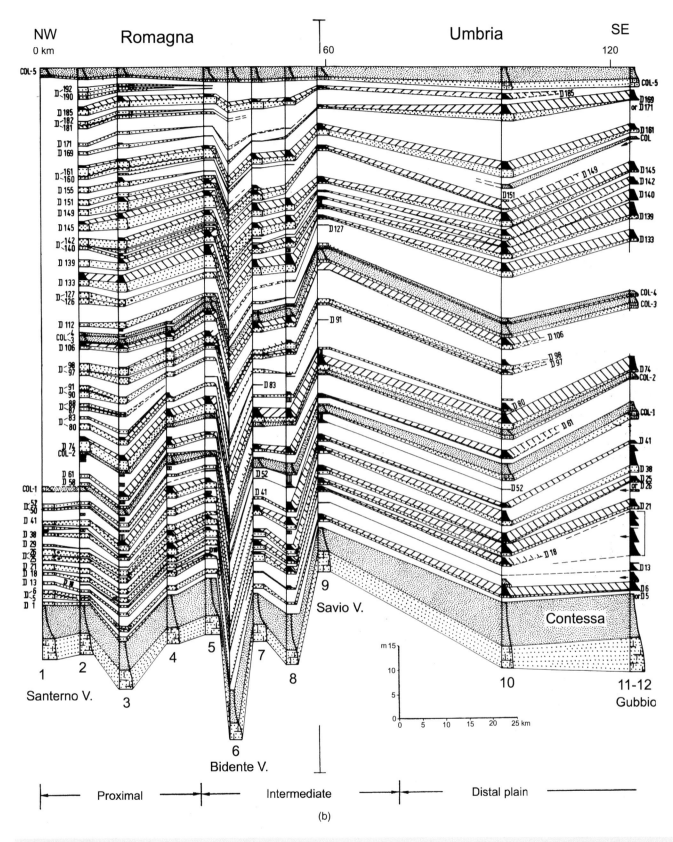

Fig. 8.74 Longitudinal correlation of individual very thick-bedded SGF deposits, pre-Contessa (a) and post-Contessa (b) intervals of the Miocene Marnoso arenacea, northern Italian Apennines. Black parts of columns represent mud-dominated parts, with alternating SGF deposits and hemipalegites. From Ricci Lucchi and Valmori (1980).

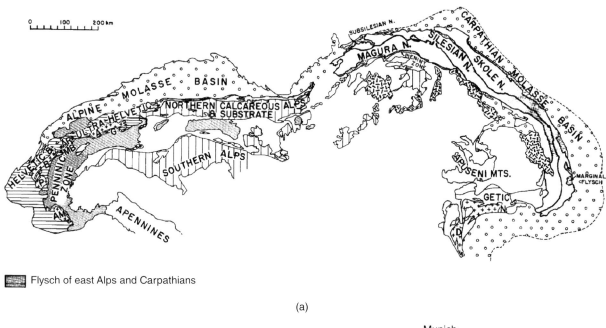

Flysch of east Alps and Carpathians

(a)

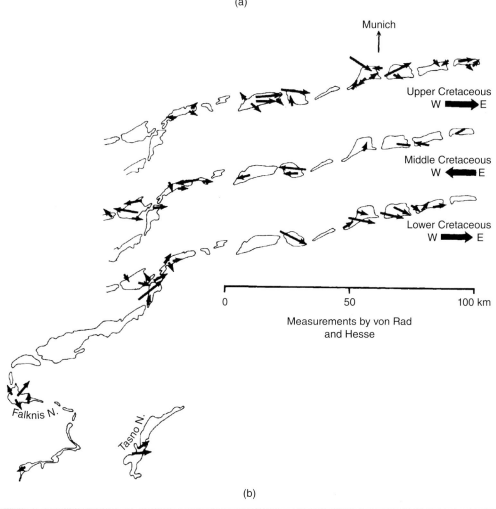

(b)

Fig. 8.75   (a) Flysch Zones of the East Alps and Carpathians, and (b) palaeocurrent pattern in the Flysch Zone for the Lower, Middle and Upper Cretaceous. Note the reversal of axial palaeocurrents with length of arrows proportional to the number of measurements. See text for explanation. From Hesse (1982).

Table 8.5   Summary features of the Flysch Zone of the East Alps, Cretaceous to Palaeocene trench fill. From Hesse (1982)

Evidence for trench:

Dimensions:
    Length: 500 km (without Carpathians).
    Width: 10–50 km (widening toward E: possibly >100 km).
    Depth: below palaeo carbonate compensation level.
       Red claystone as equivalent of brown abyssal clay.

Depositional environments:

(a) Elongate, nearly horizontal basin plain:
    Palaeocurrent directions predominantly parallel to strike (basin axis);
    Repeated reversal of current directions (change in azimuth by 180° in successive SGF deposits)
    Long-distance continuity of individual SGF deposits – Gault Formation, 115 km; Zementmergel
    Formation, 50 km;
    Low gradients of downcurrent change in bed thickness, grain-size, etc along individual beds (high
    gradients perpendicular to flow).
(b) Small to intermediate size deep-sea fan (Falknis-Tasna Nappes);
    Limited number of lateral sediment sources suggests existence of slope basins that intercept
    access routes of turbidity currents.

Lack of contemporaneous subduction:

Lack of volcanism and volcaniclastic detritus.
Continuous sedimentation without major diastrophism for 70 Myr.
Occurrence of slivers of basic and ultrabasic rocks (ophiolite suite) N and S of Flysch Zone.

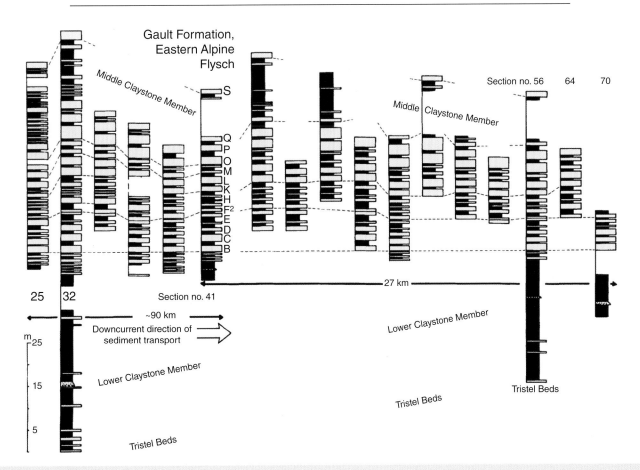

Fig. 8.76   Example of correlation of bed-by-bed sections from the Aptian–Albian Gault Formation, East Alps and Carpathians. Only the lower part of the formation is shown (i.e., Lower Claystone, Lower Greywacke and Middle Claystone Members). Bed-by-bed correlation is nearly perfect except for the most proximal section 25 which contains additional thin beds. Bed F2 is a relatively feldspar-rich marker bed. Letters refer to marker beds. From Hesse (1982).

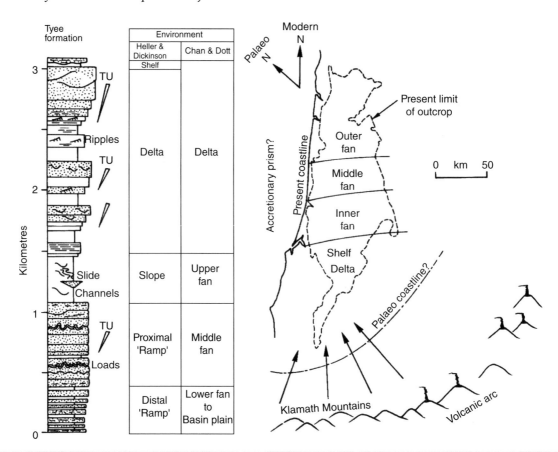

Fig. 8.77 (a) Simplified stratigraphic section through the Tyee Formation on and near the west side of Tyee Mountain, with interpretations of Heller and Dickinson (1985) and Chan and Dott (1983). (b) Palaeogegoraphic sketch map for Tyee Formation deposition. Modern north is at the top of the map. Palaeoflow in the Tyee Formation is from present-day south to north.

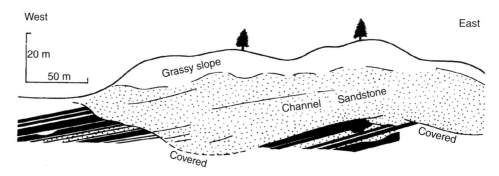

Fig. 8.78 Sketch of channel cut into thin-bedded facies and filled with structureless Facies Class B sandstones. Tyee Formation, Oregon. From Chan and Dott (1983).

Much remains to be learned, especially how the long-term geological evolution of the Earth (e.g., the ocean–atmosphere system with changing seawater chemistry, density and stratification), changing climate (greehouse versus icehouse conditions) and so-called 'extreme events' in Earth history (e.g., leading to the accumulation of organic-rich black shales during ocean anoxic events [OAEs]) influence the spatial and temporal evolution of deep-water systems. This improved understanding will only come about with more detailed, integrated outcrop–subsurface studies of ancient submarine-fan systems based upon high-resolution stratigraphic records and accurate chronology, in order to assess which factors control fan and fan-element deposition across a wide range of time scales. Future research on ancient submarine fans needs to focus more on (a) process-response models than on facies description, (b) the interpretation of single flow events and (c) geometrical analysis of depositional elements.

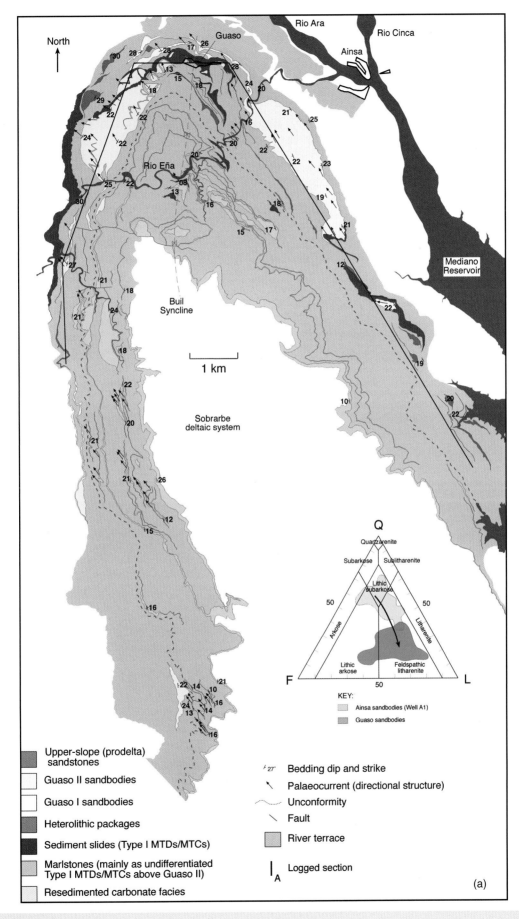

Fig. 8.79 (a) (caption on p. 400)

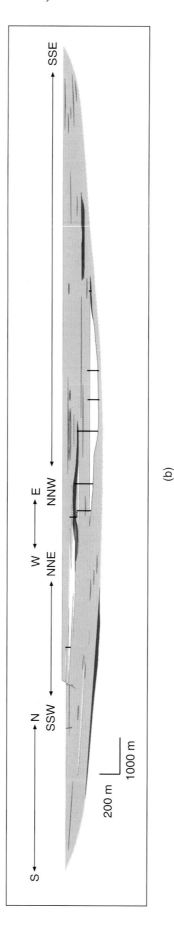

(b)

Fig. 8.79 See page 399 for part (a). (a) Map of the Guaso sandbodies, including directional sedimentary structures. Inset: Quartz- Feldspar-Lithics (QFL) ternary graph to show representative petrography of Guaso sandstones compared to older Ainsa System sandstones, emphasizing the relative compositional immaturity of the Guaso sandstones. (b, graphic p. 400) Correlation panel of the Guaso System along a transect around the Buil Syncline shown in (a). Note that the sandbodies have a high aspect ratio, and without steep erosional margins. Modified after Sutcliffe and Pickering (2009).

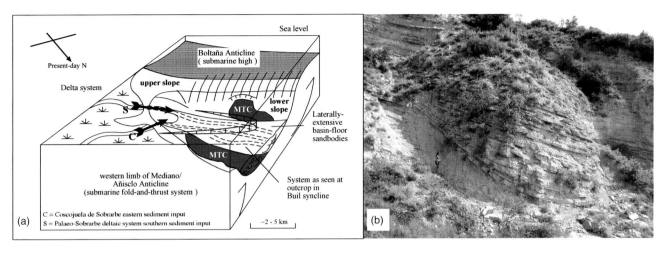

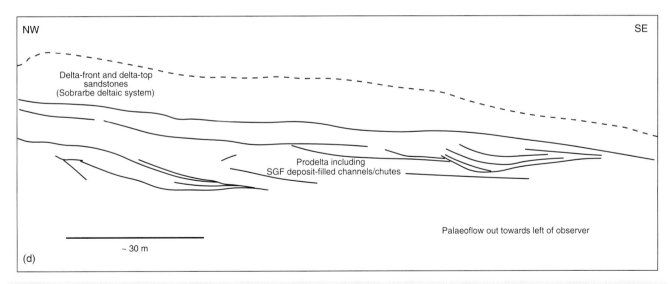

Fig. 8.80 (a) Depositional model for Guaso sandbodies as a generic model for a deep-marine structurally confined low-gradient clastic system. Water depths typically up to maximum several hundred metres. Approximate outcrop pattern shown. Note, likely sediment supply both from south and east via slope gullies, with low-sinuosity channels preserving abundant evidence for supercritical flows as backset bedding, wavy bedding and irregular sandy bedforms. (b) Slope gully filled by thin-bedded sandstone SGF deposits. Person for scale ~1.7 m in height. This gully is stratigraphically above the Guaso System but it is inferred to have been similar, albeit probably on a much smaller scale, to those that are likely to have fed the Guaso System. (c) Photo-montage of prodelta turbidite-filled slope gullies/chutes and overlying Sobrarbe deltaic system. (d) Line interpretation of (c) Modified after Sutcliffe and Pickering (2009).

PART 3

Plate tectonics and sedimentation

CHAPTER NINE
Evolving and mature extensional systems

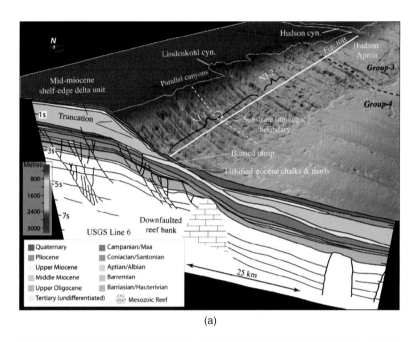

(a)

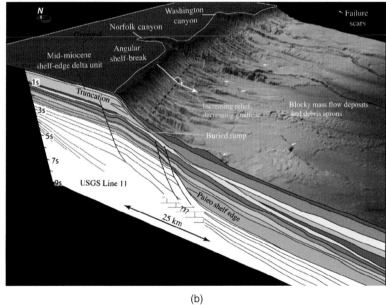

(b)

Merged multibeam bathymetry surface and interpreted 2D seismic reflection profiles, Mid-Atlantic Margin offshore New Jersey (a; USGS Line 6) and offshore Virginia (b; USGS Line 11). (a) Lithologic boundary between the truncated Middle Miocene package and the underlying, indurated Eocene package appears to correspond to a change in gradient along the lower slope. Where the Quaternary unit is thin or missing, slope-confined canyons are parallel, closely spaced and have relatively low relief. Canyon relief is greatest along the upper slope where Quaternary sediment is thicker. Note the increasing relief downslope in (b). From Brothers *et al.* (2013).

Deep Marine Systems: Processes, Deposits, Environments, Tectonics and Sedimentation, First Edition. Kevin T. Pickering and Richard N. Hiscott.
© 2016 Kevin T. Pickering and Richard N. Hiscott. Published 2016 by John Wiley & Sons, Ltd.
Companion Website: www.wiley.com/go/pickering/marinesystems

9.1   Introduction

We recognise five stages in the evolution of extensional continental margins (Table 9.1). *Pre-rift* sediments are deposited on a relatively stable craton, or are the product of an older, unrelated tectonic framework. The *rift* to *breakup* to *drift* stages of extensional-margin development provide several opportunities for the deposition of deep-water sediments. During rifting, narrow and steep-sided basins are sites for the formation of small, sand- and gravel-rich submarine fans. Also, given favourable climatic conditions, organic-rich black shales can accumulate in these narrow rift basins (e.g., Jenkyns 1986; Meyers *et al.* 1987; Chang *et al.* 1992; Cainelli & Mohriak 1999; Strogen *et al.* 2005; Rosas *et al.* 2007; Duarte *et al.* 2010; Gertsch *et al.* 2010; Kreuser & Woldu 2010; Berlinger *et al.* 2012). Following continental breakup, basin slopes may prograde, or be locally eroded and variably moulded by gravitational failures and thermohaline currents. Bottom currents tend to intensify through time as wider ocean basins permit greater access for far-travelled currents.

Rifting may involve several episodes of lithospheric stretching and thermal subsidence before continental breakup. In some cases, protracted stretching never leads to formation of oceanic crust in the rift zone (e.g., North Sea). In other cases, continental crust eventually separates to make way for a new ocean basin. The two continental edges that face this ocean basin are subsequently referred to as *passive margins* because there is no plate boundary at the junction between the continental and oceanic domains, and hence little or no seismicity or volcanism. The longer the time of pre-breakup rifting and the greater the number of associated extensional pulses, the greater will be the proportion of syn-rift sediments in the passive-margin record (Cochran 1983).

Once breakup has occurred, passive margins provide an important repository for shallow- to deep-water successions until the margin eventually is destroyed by ocean closure and subduction. Bradley (2008) has reviewed the chronology of 85 ancient passive margins, 49 of which are sufficiently well dated to assess their longevity (Table 9.2). For these passive margins, the average life span is ~155 Myr, and the average length is ~1115 km.

Excellent examples of industry regional-scale seismic sections and interpretations associated with deep-marine depositional environments from rift-related and passive continental margins and basins are described from the South Atlantic, including the Braziliian continental margin (Niemi *et al.* 2000; Richardson & Underhill 2002; Henry *et al.* 2011; Mann 2013; Saunders *et al.* 2013; Pérez *et al.* 2015; Sachse *et al.* 2015); Barents Sea (Blaich & Ersdal 2011a, b; Abrahamson 2013); West African margin (Martin *et al.* 2009; Greenhalgh & Whaley 2012; Jarsve *et al.* 2012; Wells *et al.* 2012); Gulf of Mexico (Mohn & Bowen 2012; Pindell *et al.* 2011; Radovich *et al.* 2011); North Sea (Jameson *et al.* 2011; Sakariassen *et al.* 2012; Duval 2013; Reiser & Bird 2013); offshore western Ireland (Davison *et al.* 2010); offshore east and northeast Greenland margin (Dinkelman *et al.* 2010; Jackson *et al.* 2012); offshore Florida (Mohn & Bowen 2011, 2012); offshore west Greenland (Bradbury & Woodburn 2011); northwest

Table 9.1   Evolution of passive continental margins (based partly on Brice *et al.* 1983)

Phase	Tectonic signature	Structural style	Seismic signature	Environments and lithological signature
Pre-rift	Stable craton?	Internally concordant Gentle dips Little faulting	Concordant events Strong, low frequency events	Variable shallow-marine to non-marine sediments
Rift or syn-rift	Lithospheric extension	Block faulting (listric or planar) Angular unconformities Steeply dipping strata	Weak, discontinuous reflections Angular reflection patterns Truncated sequences	Shaly and sandy siliciclastics (marine to non-marine), local high organic-matter concentrations, local deep lakes/narrow deep-marine troughs/hypersaline basins
Breakup	Initiation of seafloor spreading Major attenuation of lower crust	Internally concordant Gentle dips Little faulting ± halokinesis with associated complex faulting	Concordant events Strong continuous reflections grading into transparent zones ± complex patterns with halokinesis	Sandy siliciclastics, carbonates
Post-breakup or early drift	Rapid thermal subsidence and deepening	Internally concordant ± deep faulting and shallow folding with halokinesis or shale diapirism	Weak concordant reflections ± complex patterns with strong reflections at base with halokinesis or shale diapirism	Marine siliciclastics and carbonates Transgressive sequence
Mature margin	Slowing thermal subsidence and increasing flexural rigidity of lithosphere	Internally concordant Diminishing faulting/folding due to halokinesis or shale diapirism ± major sediment slides	Weak discontinuous reflections having progradational offlap patterns ± cut and-fill patterns with erosion, chaotic patterns with sediment slides	Shaly siliciclastics, turbidites ± slide sheets, debrites, canyon fills Regressive sequences Sea-level variation increasingly influential

Table 9.2 Start ('breakup' age) and end (tectonic destruction) of 49 ancient passive margins with well-constrained chronology (Bradley 2008). Margins are sorted according to lifespan. A map showing the location of each ancient margin is available in Bradley (2008: fig. 6)

Margin and orogen	Location	Start (Ma)	End (Ma)	Lifespan (Myr)	Length (km)
Baltic craton, E side, Uralian orogen, Phase 1	Russia	1000	620	380	1000
Laurentian craton, W side, northern sector, Antler orogen	Canada	710	385	325	1560
São Francisco craton, E side, Aracauai-Ribeira orogen	Brazil	900	590	310	750
Superior craton, E side, New Quebec orogen ("Labrador Trough")	Canada	2135	1890	245	960
Pilbara Craton, S margin, Ophthalmian orogen	Australia	2685	2445	240	580
Wyoming Craton, S side, Medicine Bow orogen	USA	2000	1780	220	310
Indian craton, N side, Himalayan orogen, Phase 2	India, Nepal	271	52	219	2460
Laurentian craton, S side, Ouachita orogen	USA	520	310	210	1720
Siberian craton, N side, Taymyr, Phase 2	Russia	530	325	205	1380
Sierra de la Ventana (a), Cape (b), and Ellsworth Mtns. (c) Argentina,	S. Africa, Antarctica	500	300	200	2170
Congo Craton, W side, Kaoko Belt (N Coastal Branch) of Damara orogen	Namibia	780	580	200	2175
Kalahari Craton, W side, Gariep Belt, (S Coastal Branch) of Damara orogen	Namibia	735	535	200	590
Hearn craton, SE side	Canada	2070	1880	190	820
South China craton, SE side. Nanling orogen	China	635	445	190	1730
Laurentian craton, W side, southern sector, Antler orogen	USA, Canada	542	357	185	1870
Superior craton, S side, Animike margin, Penokean orogen	Canada, USA	2065	1880	185	1170
Arabia, NE margin, Oman-Zagros orogen	Oman, Iran	272	87	185	2300
Arctic Alaska microcontinent, S side, Brookian orogen	Alaska, Russia	350	170	180	1230
Isparta angle, eastern margin	Turkey	227	53	174	260
Kola craton, S side, Kola suture belt	Russia	1970	1800	170	690
Apulian microcontinent, Pindos ocean	Greece	230	60	170	630
Alborz orogen	Iran	390	210	170	550
Tarim microcontinent, S side, Kunlun orogen	China	600	430	170	950
Isparta angle, western margin	Turkey	227	60	167	280
Australian craton, NE side, New Guinea orogen	New Guinea, Irian Jaya	180	26	154	1380
Australian craton, NW side, Timor orogen	Indonesia	151	4	147	1530
Indian craton, N side, Himalayan orogen, Phase 1	India, Nepal	635	502	133	2460
Slave craton, W side, Wopmay orogen	Canada	2015	1883	132	560
Dzabkhan block	Mongolia	710	580	130	200
S margin of Europe, Alpine orogen	Switzerland	170	43	127	790
Superior Craton, N side, Cape Smith and Trans-Hudson orogens	Canada	2000	1875	125	370
S. American Craton, N side, Venezuela margin	Venezuela	159	34	125	1080
Kalahari Craton, N side, Inland Branch, Damara orogen	Namibia	670	550	120	540
Congo Craton, S side, Inland Branch, Damara orogen	Namibia	670	555	115	1800
São Francisco craton, W side, Brasiliano orogen	Brazil	745	640	105	1080
Baltic Craton, E side, Uralian orogen, Phase 2	Russia	477	376	101	3130
Baltic Craton, Wside, Scandinavian Caledonide orogen	Norway	605	505	100	1550
Saxo-Thuringian block	Germany	444	344	100	400
NW Iberia, Variscan orogen	Spain	475	385	90	210
Guaniguanico terrane	Cuba	159	80	79	150
Laurentian craton, E side, Appalachian margin, Taconic orogen (a) and NW Scotland (b) USA, Canada		540	465	75	3320
Karakorum block	Pakistan	268	193	75	500
South China Craton, NW side, Longmen Shan orogen, Phase 2	China	300	228	72	480
Australia, E side, Tasman orogen	Australia	590	520	70	1670
Baltic craton, S side, Variscan orogen	Ireland to Poland	407	347	60	1080
Amazon craton, SE side, Araras margin, Paraguay orogen	Brazil	640	580	60	470
Cuyania terrane, Argentine Precordillera, E side	Argentina	530	473	57	640
Pyrenean-Biscay margin of Iberia	Spain	115	70	45	390
China, E side, Taiwan orogen	Taiwan	28	5	23	500

Australian margin (Cameron 2010; Silva-Gonzalez 2012; Grahame & Silva-Gonzalez 2013); offshore Lebanon (Peace 2011; Hodgson 2013; Lie *et al.* 2013); offshore East Africa (Danforth *et al.* 2012), and the Seychelles (Morrison 2011). Whilst these papers provide a good reference source for students who wish to appreciate the large-scale structure and stratigraphy of evolving and mature extensional systems, they do not contain in-depth sedimentological data.

This chapter provides examples of typical styles of deep-water sedimentation along passive continental margins and in precursor rift basins, including failed rift arms called *aulacogens*. Unlike our earlier book (Pickering *et al.* 1989), emphasis is placed on basic principles rather than case studies. We begin by briefly outlining the tectonic development of extensional systems from the time of the onset of rifting. Particular emphasis is given to subsidence history because of the role that subsidence plays in creating deep-water depocentres.

9.2 Models for lithospheric extension

Rifting and subsequent continental breakup to form oceanic crust have been modelled by McKenzie (1978) under a regime of pure shear, with crust and lithosphere both thinning as a result of extension (Fig. 9.1a). An alternative model was provided by Wernicke and Burchfiel (1982; also Wernicke 1985), who suggested that rifting under simple shear better explains the geometry of many rift systems. Simple shear results in asymmetric crustal thinning and asymmetric half-graben geometry about the rift axis. With simple shear, continental failure ultimately takes place along a low-angle detachment fault that reaches Earth's surface as a 'breakaway fault' situated to one side of the rift zone (Fig. 9.1b). Toward the end of the rifting period, detachment faults may extend downward into the mantle (Fig. 9.1c), providing a mechanism to 'pull' what was originally sub-continental mantle up to the seafloor as peridotite ridges (Hölker *et al.* 2003). The Wernicke model has been applied to the crustal structure of the Viking Graben by Beach (1986), and the Grand Banks–western Iberia rift zone by Boillot *et al.* (1987) and Tankard and Welsink (1987). Whatever model is valid, perhaps a combination of both, crustal thinning to form oceanic crust has important implications for the stratigraphy of passive margins.

Rifting can be accompanied by either subsidence or uplift. If the site of rifting is over a mantle plume (e.g., Afar region of Ethiopia), then the thermal anomaly creates an updoming of the lithosphere and formation of three rift arms. In such cases, only two of the rift arms ultimately develop into a widening ocean basin, and the third is abandoned after an initial phase of fault-generated subsidence and sediment infilling. Such abandoned rift arms are called *aulacogens* (Burke 1978). Away from strong thermal anomalies, rifting begins by stretching of the lithosphere. As the lithosphere is thinned, hot asthenosphere rises to fill the newly created space. If the crust extends a greater amount than the lithospheric mantle, subsidence will not be preceded by regional uplift because the rising asthenosphere is denser than the crust that it replaces. If a greater amount of stretching occurs in the lithospheric mantle than in the crust, then an initial phase of uplift is predicted because the rising asthenosphere is less dense than the lithospheric mantle that it replaces (Royden & Keen 1980). The stretched lithosphere, although warmer than its surroundings, retains a certain amount of flexural rigidity, so that the margins of the rift may bend upward (uplift) or downward (subsidence) depending on whether the lithosphere 'necks' in its lower or upper part, respectively. Finally, unstretched lithosphere immediately adjacent to the rift may be warmed by lateral escape of heat, causing these marginal areas to

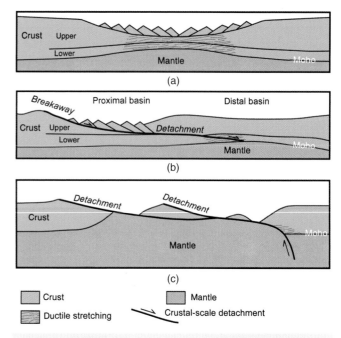

Fig. 9.1 Models for the formation of rift basins. (a) Pure shear model with symmetrical basin developed over thinned crust (from McKenzie 1978). (b) Simple shear model with asymmetric basin developed over a crustal-scale detachment (from Wernicke & Burchfiel 1982). (c) Model involving a concave-downward detachment along which subcontinental mantle can be pulled to the seabed and exhumed (after Hölker *et al.* 2003). Rift-phase and younger sediments are not shown, but would partially or totally fill depressions in each panel, depending on the quantity of detritus available. The sediment load would isostatically depress the lithosphere according to sediment thickness, creating even more room for sedimentary fill.

undergo thermal upwarping that might promote the development of unconformities.

The message to take from the preceding paragraph is that rifting of continental lithosphere is a complicated process that can induce either syn-rift uplift or subsidence, and which can affect rocks outside the rift zone. Extension in rift basins generally involves several episodes (or pulses) of lithospheric stretching separated by less active periods characterised by thermal subsidence (Chadwick 1986; Hiscott *et al.* 1990). On a local scale within the rift zone, brittle extension of the upper continental crust causes rapid subsidence of the hanging walls of normal faults, and concomitant rift-flank uplift of the footwalls as they are unloaded. Later, following (i) the cessation of a pulse of rifting, or (ii) continental breakup and the onset of seafloor spreading, the lithosphere begins to cool and contract, resulting in an extended period of exponentially decreasing thermal subsidence that lasts on the order of 100 Myr (McKenzie 1978). In detail, rift basins marginal to the final zone of continental breakup can experience contraction during the time of breakup because of outwardly directed in-plane stresses originating at the new plate boundary, as spreading is initiated (Withjack *et al.* 1998). Along the central Atlantic portion of the Atlantic rift zone, this contraction resulted in inversion and uplift approaching ~2 km as rift-phase normal faults changed into reverse faults (Withjack *et al.* 1998).

The models described above are only appropriate for what are called 'non-volcanic rifted margins'. In the case of so-called 'volcanic rifted margins', rising asthenosphere in the rift zone undergoes significant partial melting, and large quantities of basalt are extruded in the basins of the rift zone (Nielsen *et al.* 2002). The thinned continental crust is invaded by mafic magma and therefore becomes more dense once this magma solidifies. Deep-water syn-rift sediments are not well described from this setting. After continental breakup, thermal subsidence ensues in the same manner as for non-volcanic margins.

Geodynamic models like the McKenzie (1978) model do not consider changes in basin geometry and fault motions caused by differential sediment loading and salt tectonics. These processes, however, can be of fundamental importance to facies distribution and stratigraphic architecture. For example, in the salt province of the Gulf of Mexico, west of the Mississippi Delta, SGF deposits preferentially accumulate in intraslope basins created by salt intrusion and flow (Beauboeuf & Friedmann 2000). Offshore Angola, high sediment accumulation rates associated with the Congo River system have loaded and reactivated extensional faults and have mobilised salt. As a result, listric faulting has continued long after continental breakup and SGF deposits are in many cases confined to the more subsiding hanging walls of these faults (Anderson *et al.* 2000; Gee *et al.* 2007). In order to distribute sand farther basinward, submarine channels must incise topographic highs, including growing anticlines. This leads to a stratigraphic architecture characterised by deep slot-like channels nested in overbank deposits of the continental slope (Fig. 9.2).

Although geodynamically relevant to other plate-tectonic settings, the development of rift basins is dominated by extensional normal faulting where faults grow, propagate and link to other faults. These kinematic processes mean that basin-floor topography and subsidence patterns can be complicated.

Morley (2002) explains that the linkage and propagation of early extensional faults to form basin-scale boundary faults can occur (i) prior to significant basin formation, (ii) after minor faulting has created an extensive area of subsidence or (iii) during basin development. In the latter case, early subsidence and initial sedimentary thicknesses on the hanging wall of the boundary fault are controlled by the spatial distribution of the precursor short extensional fault segments, each having created a local depocentre midway along the particular fault segment. As the initially short fault segments link

into a through-going boundary fault, the separate depocentres overlap and coalesce. At discontinuities between the original short fault segments, misalignments or en echelon offsets between the segments are accommodated by oblique linkage faults (transfer zones), above which the floor of the basin may be flexed upward to create a transverse anticline on the hanging wall block. The age of the sedimentary successions in the hanging wall can be different from one side of a transverse anticline to the other (Morley 2002), because the original short fault segments might not develop synchronously, and one part of the boundary fault might experience greater dip slip than an adjacent part — later, activity on what had been the less active part of the fault might gain prominence. In some cases, initial sites with least subsidence can become the sites of maximum subsidence later on (Kim & Sanderson 2005: fig. 8). In particular, sites that began as transfer zones between smaller fault segments, and that might have acted as routes for sediment transfer from shallower into deeper water, may then be associated with maximum subsidence on the hanging wall, becoming important depocentres for SGF deposits in a deep-water basin. Such differential subsidence is known to occur in many deep-marine rift basins such as the Palaeocene of the North Sea, and may even offer an explanation for some of the mounded stratigraphy that is observed on seismic profiles (e.g., Kosa 2007).

Gawthorpe *et al.* (1997) consider another important aspect in the structural evolution of normal faults that impacts the stratigraphic architecture of any associated depositional sequence. An an initial growth-fold stage occurs when the fault is a buried structure. During the growth-fold stage, strata thin and become truncated toward the fault zone and are rotated and diverge away from the buried fault into growth synclines. Once the fault breaks the surface, strata form a divergent wedge, which is rotated and thickens into the fault. Both tectono-stratigraphic styles can occur contemporaneously along the length of a single fault segment (Gawthorpe *et al.* 1997). Growth folding characterises deformation around the ends of fault segments where the fault is blind, whereas the central part of fault segments is characterised by surface faulting. The result of such complexity is that significant along-strike variation in stratal geometry and the stacking patterns of facies associations may occur in depositional sequences in areas of normal faulting.

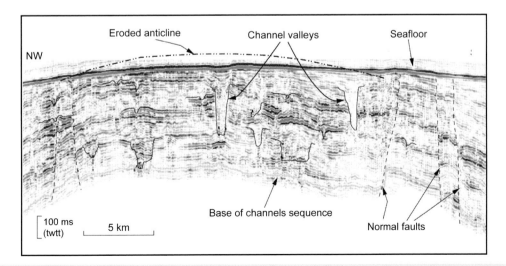

Fig. 9.2 Strike-oriented seismic line from the Angolan continental slope showing cross-sectional geometry of channels, syn-sedimentary normal faults, and a broad, near-surface anticline truncated at the seabed. From Gee *et al.* (2007).

9.3 Subsidence and deep-water facies of rifts and young passive margins

The McKenzie and Wernicke models involve a phase (or phases) of initial fault-controlled subsidence, followed by exponentially decreasing thermally controlled subsidence. On rapidly subsiding hanging walls, deep-marine sediments may be deposited in isolated depocentres soon after rift onset (e.g., Surlyk 1978). Elsewhere, it is only after the beginning of a phase of regional thermal subsidence that large parts of the continental margin descend into deep water. Inboard of the continental edge, rift-stage deep-water sediments that accumulated in hanging wall depressions commonly pass gradually upwards into shallow-marine deposits resulting from slope and shelf progradation. This progradation takes place when sediment supply and accumulation begin to keep pace with, and eventually exceed, the rate of thermal subsidence. The shelf edge can readily advance seaward many tens of kilometres at this time (e.g., Berger & Jokat 2009). Variations in base level (sea level) may complicate this simplified scenario (Section 4.1). Closer to the ocean–continent transition, thermal subsidence takes the seafloor into water so deep (2–4 km) that subsequent sedimentation never leads to significant shallowing. An example comes from the vicinity of the continent-ocean transition in the Newfoundland Basin, probed during ODP Leg 210 (Shipboard Scientific Party 2004a). Breakup along this margin occurred in the Aptian, tectonically exhuming mantle peridotite and creating oceanic crust between the Grand Banks of Newfoundland and Iberia. Leg 210 Site 1276 recovered 615 m of Albian–Cenomanian SGF deposits including beds as thick as 2–3 m with features indicating syndepositional wet-sediment deformation, and interpreted as slurry-flow deposits (Facies C2.5; concentrated density-flow deposits in the classification of Section 1.4.1). The Site 1276 Albian–Cenomanian succession was derived from the west–southwest, with some of the detritus recycled from the European side of the early rift system (Hiscott *et al.* 2008). The volume of sediment in this part of the Cretaceous early-drift succession is $\sim 7 \times 10^4$ km^3, consistent with delivery from a river (or rivers) with about 2/3 the sediment discharge of the modern Po River, Italy (Hiscott *et al.* 2008). Sedimentation rates at Site 1276 were ~ 100 m/Myr in the early to mid Albian, all in deep water because of rapid post-breakup subsidence (Shipboard Scientific Party 2004b). Sedimentation rates declined to 2–4 m/Myr in the Turonian–Santonian and gravity-flow deposits were replaced by bottom-current-reworked silty and sandy facies (Facies C1.2 and D1.3; Fig. 2.30). Slower sedimentation was not the result of shallowing, but rather a dramatically reduced sediment supply.

Thermal-mechanical models for the evolution of passive continental margins take account of the effects of thermal contraction and sediment loading, and the increasing flexural rigidity of the lithosphere as it cools (e.g., Beaumont *et al.* 1982; and references in Coward *et al.* 1987). These models can predict the stratigraphy for an evolving passive continental margin. A comparison of the predicted and observed stratigraphic profiles permits an assessment of the relative importance of the various factors that controlled the development of the particular stratigraphic section (*cf.* Watts & Thorne 1984).

Magnitudes and rates of passive-margin subsidence can be determined by using stratigraphic and facies information, and palaeo sea-level information, to incrementally decompact and strip sediments off a computer model of the margin (e.g., Stam *et al.* 1987; Hegarty *et al.* 1988; Gradstein *et al.* 1989; Hiscott *et al.* 1990). For burial depth, z, in kilometres, and porosity, P, as a per cent of rock volume, Hiscott *et al.* (1990) fitted Mesozoic rift-basin well data with Equation 9.1 for shale and siltstone and Equation 9.2 for sandstone

and limestone:

$$z = 6.02 \, (1 - P)^{6.35} \tag{9.1}$$

$$z = 3.7 \, \ln \, (0.30/P) \tag{9.2}$$

Backstripping involves removing the surface sedimentary layer (layer 1, say) from the model, then adjusting the thickness of deeper layers (2 through n, say) so that their porosity conforms to a new reduced vertical load dictated by the shallower burial depths resulting from the removal of layer 1. Next, an amount of basement uplift is calculated that would be required to bring the palaeo-seabed to the correct water depth (inferred from facies information) relative to the sea level of the day (from global curves). These increments of basement uplift, calculated for a model with time running in reverse (hence *back*stripping), are then plotted, with a change in sign, against absolute time, producing a curve for the history of basement subsidence throughout the depositional phase. User-friendly freeware programs (e.g., OSXBackstrip 2.2) can be downloaded from the internet and used for backstripping analysis.

Backstripping procedures require good estimates, at unit boundaries, of absolute age, palaeo water depth, palaeo sea level and average rock type (essential for assessment of reasonable amounts of compaction). Shallow-marine sediments are ideal for this purpose, because the water depth during deposition can be estimated to the nearest 10 m or so. For deep-water deposits formed during margin subsidence, however, accurate water depths cannot be determined, and subsidence history becomes less well known. Nevertheless, even conservative estimates of water depth in such situations demonstrate that strong pulses of accelerated subsidence occur in rift basins. For example, Hiscott *et al.* (1990) calculated a peak rate of >250 m/Myr for basement subsidence in the early Kimmeridgian Lusitanian Basin of Portugal. This was fault-driven subsidence that depressed the seafloor to water depths >300 m (Wilson *et al.* 1989). Submarine fans of the Abadia Formation accumulated in this extensional depression; this formation is locally >1000 m thick because sediment loading accentuated the initial tectonic subsidence. Upper Jurassic SGF deposits of similar facies occur in the Tempest unit, Jeanne d'Arc Basin of the Canadian Grand Banks of Newfoundland (De Silva 1994). Deposits of Facies Class B dominate the Upper Jurassic deep-water reservoirs of the Buzzard oilfield, UK offshore, where a submarine ramp with stacked amalgamated sand sheets developed in the latest Oxfordian in a basin produced by a pulse of extension-driven subsidence (Doré & Robbins 2005; Fig. 9.3). In a final example from the western Woodlark Basin, offshore Papua New Guinea, lower Pliocene shallow-marine deposits (water depth <20 m) are now buried to a depth of ~ 850 m at ODP Site 1118, where the present water depth is ~ 2300 m. If the ~ 850 m of overburden depressed the continental lithosphere at this site by ~ 280 m (isostatic sediment loading), and if the mid Pliocene sea level was ~ 60 m higher than today, then the amount of extension-driven subsidence in ~ 3.5 Myr has been ~ 2625 m, giving a minimum syn-rift subsidence rate of ~ 750 m/Myr!

Rift sediments (also called 'syn-rift' sediments) are deposited during pre-breakup phases of lithospheric extension and are affected by rotation of upper crustal blocks on listric or planar faults (e.g., Surlyk 1978). Two examples of the facies of syn-rift deep-water successions will be provided to illustrate the style of deposition: (i) Upper Jurassic marine slope deposits of east Greenland (Surlyk 1987), and (ii) Lower Cretaceous deep-lake deposits of eastern Brazil (Bruhn 1998a, b).

The 200–500 m-thick Upper Jurassic Hareelv Formation of east Greenland consists mainly of black shale (6–12% TOC)

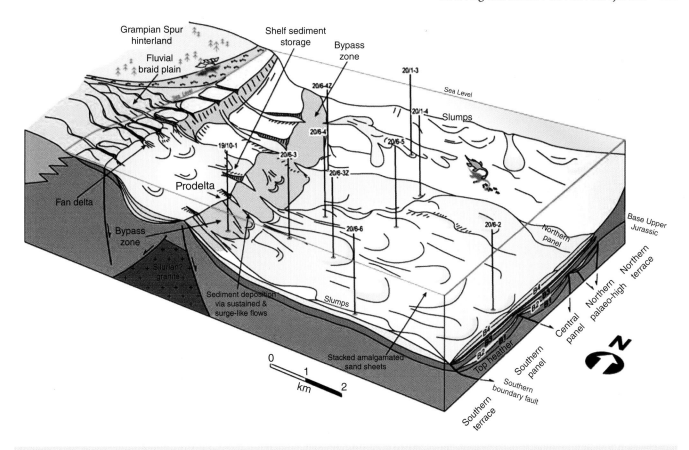

Fig. 9.3 Depositional model for the Upper Jurassic deep-water sandstones in the Buzzard oilfield. Inferred sediment supply was from a broad braid plain with multiple slope feeder channels creating a sand-prone submarine ramp. Vertical bars represent exploration wells. From Doré and Robbins (2005).

and thick-bedded sandstones (Surlyk 1987). The sandstones are interpreted as the deposits of a submarine ramp (Table 7.1) flanking a fault-bounded deep-marine shelf basin. Sediment was supplied along a line source from a sand-dominated shelf. Seaward of the shelf was a narrow, fault-controlled slope, against which slope-apron sediments of the clastic submarine ramp onlapped. Most sandstones were deposited at the base-of-slope and on the basin floor, although some of the outcrop belt may have been deposited directly on the slope. Wet-sediment deformation, including liquefaction structures, clastic dykes and sills, are common within the Hareelv Formation. Some dykes cross-cut bedding for hundreds of metres of vertical section.

Thick-bedded and very thick-bedded (0.5–50+ m) sandstones are ungraded and virtually devoid of primary sedimentary structures (Facies Bl.l), with the exception of faint parallel stratification in some of the very thick beds. These occur as two major types: (i) laterally discontinuous fills of steep-sided (to overhanging) gullies or scours (Fig. 9.4), and (ii) parallel-sided beds that are laterally continuous for hundreds of metres. The gully-fill sandstones comprise up to 50% of the formation and alternate randomly with the laterally-extensive sandstones. Most of the gullies have a sinuous to meandering planform. There is a virtual absence of over-bank deposits along the margins of the gully-fills, and infilling of individual gullies and scours appears to have been accomplished by a few flows.

There are no consistent vertical trends within the measured sections of the Hareelv Formation. The complete lack of apparent organisation and the extremely irregular vertical stacking of

sandstone bodies suggests that these deposits did not accumulate on a submarine fan. The random juxtaposition of gully and sheet-like deposits shows that the depositional system was disorganised, with individual flow events randomly initiated along the fault-scarp line-source and travelling independently for variable distances downslope. Inflated sand flows and concentrated density flows, perhaps triggered by reactivation of basin-margin faults, are thought to have emplaced the sands. The Hareelv Formation is thus interpreted as the product of essentially continuous hemipelagic sedimentation of black shales under anoxic or poorly oxygenated conditions in relatively deep water. This hemipelagic sedimentation was periodically interrupted by gravity flows that transported large volumes of sand into gullies and scours. Flows were triggered along the entire slope by movements along the basin-margin faults. The generation of the flows from a line source resulted in a random distribution of sandstone bodies and precluded the development of a submarine-fan system. Very similar facies associations and comparable interpretations have been made for the rift-related Kimmeridgian strata of northeast Scotland (Pickering 1984b; Wignall & Pickering 1993).

According to Bruhn (1998b), the half-graben rift basins of eastern Brazil are bounded by extensional faults with cumulative throws of as much as 6 km. The basins are mostly a few tens of kilometres wide. During the Berriasian–Valanginian, a series of deep lakes developed because of pronounced extension. The Recôncavo Basin contains >2000 m of mainly dark organic-rich mudstones and sandy SGF

Fig. 9.4 Closely spaced, structureless and massive gully sandstones (Facies B1.1) in partly scree-covered dark mudstone (Facies Class E). Note the lack of organisation of the system. From Surlyk (1987).

deposits that accumulated at rates as high as ~1800 m/Myr. There is a facies change adjacent to the basin-bounding fault, where a thick wedge of gravels and sands was deposited as coalescing fan deltas, and as base-of-slope aprons (Bruhn 1998a). As in the Lusitanian Basin of Portugal (Abadia Formation), the deep-water succession shallows upward into deltaic and fluvial deposits because the rate of subsidence was eventually overwhelmed by a high rate of sediment supply.

The Brazilian lacustrine rifts contain the following depositional elements (Fig. 4.14): gravel/sand-rich channel complexes, sand/mud-rich lobes, gravel/sand-rich aprons, sand-rich density underflow deposits and sand/mud-rich debrites. The aprons along the basin marginal fault form wedges up to 2 km thick, 5–20 km wide, and 5–200 km long. They consist of five facies belonging almost exclusively to Facies Class A:

(1) 0.5–5 m-thick inversely graded beds of unstratified, granular to bouldery conglomerate;

(2) 0.5–3 m-thick normally graded beds of unstratified, bouldery to pebbly conglomerate, granular sandstone, very coarse-grained sandstone and local parallel- or cross-stratified sandstone;

(3) 0.5–12 m-thick normally graded beds of unstratified, intraformational bouldery to granule conglomerate, intraclastic sandstone, very coarse- to coarse-grained sandstone and local parallel-stratified sandstone;

(4) 0.5–10 m-thick beds of disorganised bouldery to pebbly conglomerate; and

(5) 0.5–10 m-thick beds of disorganised bouldery to pebbly mudstones.

The two examples described above suggest that deep-water successions in rift basins are strongly influenced by steep marginal fault(s). This conclusion is supported by observations in the Lusitanian Basin, where the shaly lower part of the deep-water Abadia Formation passes eastward into >2000 m of deep-water arkosic gravels of the Castanheira Member – this member is confined to the vicinity of the eastern basin-bounding fault of the rift basin (Fig. 9.5; Leinfelder & Wilson 1998). The basin margin forms an essentially linear source, promoting the accumulation of ramp or apron deposits rather than point-sourced submarine fans.

The continental breakup phase (Table 9.1) follows major attenuation of the lower crust and the appearance of the first oceanic crust. Breakup is not an instantaneous event along a margin, but may propagate along the rift zone. Even locally, variable stretching rates (β-factors) between segments of the rift zone, each bounded by transfer faults (Gibbs 1984), can lead to a complex and protracted history of breakup. In the South Atlantic, rifting began in the Oxfordian–Kimmeridgian (136–147 Ma), and ended with diachronous breakup from the Hauterivian near the Agulhas Plateau, to early Albian in the vicinity of the Niger Delta (DSDP Leg 75, Hay *et al.* 1984); hence, the rifting phase varied in duration by ~15 Myr.

Some researchers refer to a 'post-rift phase' after continental breakup, but this is an unnecessary and redundant term in the case of passive margins. The rift phase ends as a result of breakup, so the adjective 'post breakup' is preferred for clarity. We reserve the term 'post-rift' for failed rifts in which extension ceases without continental breakup. The term 'post-rift' would therefore be appropriate for aulacogens, or for early failed attempts at continental rupture, for example following the Triassic–earliest Jurassic extensional episode along the North Atlantic borderlands which ended not with breakup but instead with a phase of thermal subsidence (Sinclair 1988; Alves *et al.* 2002).

The post-breakup phase (Table 9.1) involves thermal subsidence, major deepening and the development of onlap sequences. The mature passive margin (Phase 5) is characterised by a prograding slope, and a complex history of both erosion and deposition. The main controls on stratigraphy during the development of a passive continental margin are: (i) thermal contraction of initially warm and thin lithosphere; (ii) sedimentary loading (flexure) of a lithosphere that cools and becomes progressively thicker and more rigid after extension has ceased and (iii) compaction, palaeobathymetry, local erosion and global sea-level changes, all of lesser importance (Watts & Thorne 1984). Global sea-level changes begin to exert a relatively greater influence as tectonic subsidence rate decreases with time after breakup (Thorne & Watts 1984).

Most evolving and mature passive margins show a transition to deep-marine deposition after breakup (Table 9.1). As a general rule, the sediments of the post-breakup, or 'drift' phase tend to become finer grained and more biogenic with time, as continental source

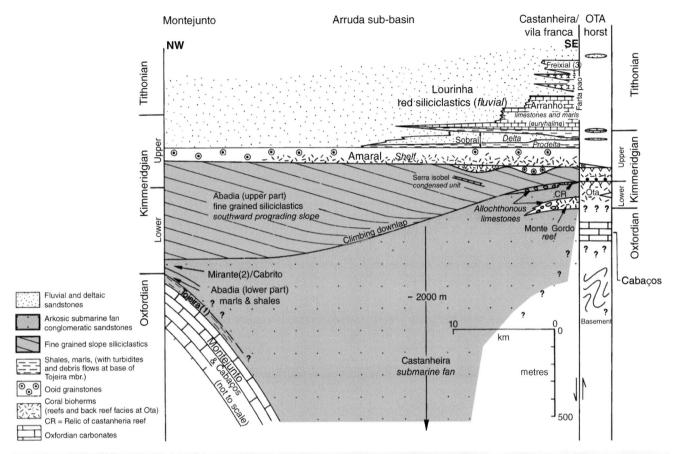

Fig. 9.5 Cross-section of the fill of the Arruda sub-basin, Lusitanian Basin, Portugal. The uppermost Oxfordian to lower Kimmeridgian Castanheira submarine fan/ramp (orange) thins away from the border fault of the basin into a shaly succession of SGF deposits (Tojeira member of the lower Abadia Formation – green). The upper Abadia Formation (also green) is a prograding slope capped by shallow-marine limestones of the Amaral Formation. All other stratigraphic units in the diagram are shallow marine or continental. The deep-water phase resulted from a pulse of rift-induced subsidence. From Leinfelder and Wilson (1998).

areas become relatively less important: coarse-grained deep-marine sediments (Facies Classes A, B, C and F), if present, are overlain mainly by finer grained (Facies Classes D and E) SGF deposits and/or contourites, above which hemipelagic and pelagic deposits (Facies Classes E and G) increase in volume. Submarine canyons and gullies may incise the evolving margin (e.g., Bruhn & Walker 1995), transporting shelf or delta-fed sediments deeper into the basin to form gravity-flow systems such as submarine fans, either within slope basins or on the developing continental rise.

On deep parts of passive continental margins, organic-rich deposits may accumulate if biological productivity is high, for example under zones of upwelling. Enhanced organic-matter deposition may be triggered by sea-level changes. For example, organic-rich black shales accumulated along the Mazagan margin off northwest Africa during the Aptian–Albian and Cenomanian–Turonian *oceanic anoxic events* (OAEs), coincident with times of rising sea level (Fig. 9.6). Furthermore, Leckie (1984) showed that the diversity of planktonic foraminifera decreased during the two OAEs, presumably due to an intensified oxygen minimum zone off central Mazagan in latest Aptian to early Albian, and latest Cenomanian to early Turonian times. Thus global sea-level changes during the evolution of the Mazagan margin exerted a fundamental control on the nature of deep-marine sediment types. A similar correspondence of organic shales with times of maximum flooding is found in the Aptian succession of the Vocontian Basin, France (Friès & Parize 2003).

9.4   The post-breakup architecture of passive margins

Although each evolving passive margin will be influenced by a unique set of oceanographic conditions and a unique sediment-delivery system(s), it is possible to consider a limited number of scenarios, listed below (and treated in the following sections), that represent a majority of modern and ancient post-breakup situations.

(1)   In areas where major rivers debouche onto subsiding passive margins during the drift phase, large open-ocean submarine fans like the Mississippi, Amazon and Congo fans build out onto ocean crust (Chapter 7). Great thicknesses of sediment at such margins load the lithosphere to create additional accommodation space (e.g., Watts 2007: fig. 18).

(2)   In segments of continental margins underlain by mobile syn-rift evaporite deposits (salt), later flowage of the salt under the load of overlying sediments can produce extremely complex seabed

bathymetry. Sediment gravity flows, such as turbidity currents, negotiating the irregular seafloor will produce localised ponded deposits, and some slope minibasins will fill while others are still sediment-starved (e.g., western Gulf of Mexico, Satterfield & Behrens 1990; Beauboeuf & Friedmann 2000; Liu & Bryant 2000).

(3) Along margins with little fluvial/deltaic supply, fine-grained slope aprons are characterised by limited deposition, bypass and gravitational failures. The northwest African margin off the Saharan region offers a prime example of this type of deep-water system (Masson *et al.* 1992; Wynn *et al.* 2000).

(4) Passive margins on the western sides of north–south-oriented ocean basins will be affected by deep thermohaline circulation (Chapter 6) and can experience winnowing, and moulding of silts and muds into contourite drifts. If submarine fans are also encountered by such bottom currents, then the suspended load of sediment gravity flows can be siphoned off into the near-bottom nepheloid layer and advected along the margin. Erosional surfaces and sandy/gravelly lags may develop as bottom currents cut into SGF deposits.

(5) On high-latitude passive margins, grounded continental glaciers tend to excavate shelf-crossing troughs and provide abundant and poorly sorted detritus to the shelf edge, where it episodically fails to form stacked debrites. The result is either a relatively smooth slope because the debris flows fill lows between their predecessors (e.g., Aksu & Hiscott 1992), or the development of a trough-mouth debris fan (e.g., Vorren & Laberg 1997).

(6) In warm tropical areas with limited siliciclastic influx, carbonate platforms may develop sufficiently steep flanks that rubble units and periplatform debrites form a distinctive deep-water succession. If, instead, cool-water grainy carbonates predominate on the shelf, then the adjacent continental slope and rise will resemble siliciclastic systems. An excellent example of this situation is found offshore south Australia (Feary & James 1998; Shipboard Scientific Party 2000).

Continental margins may embrace a number of these 'types' within a relatively restricted geographical area. For example, the Gulf of Mexico (Fig. 9.7) contains carbonate platform-fringed margins (West Florida Slope and Campeche Escarpment), together with a hybrid

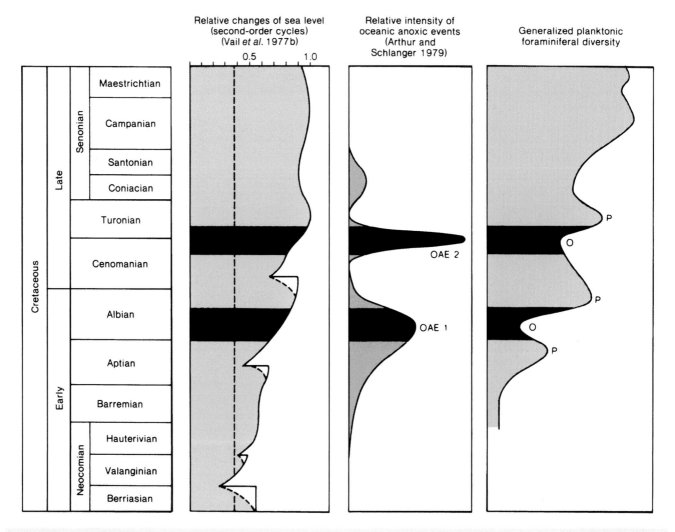

Fig. 9.6 Correlation of global sea-level changes, oceanic anoxic events and diversity trends in planktonic foraminifera. O = oligotaxic; P = polytaxic. From Leckie (1984).

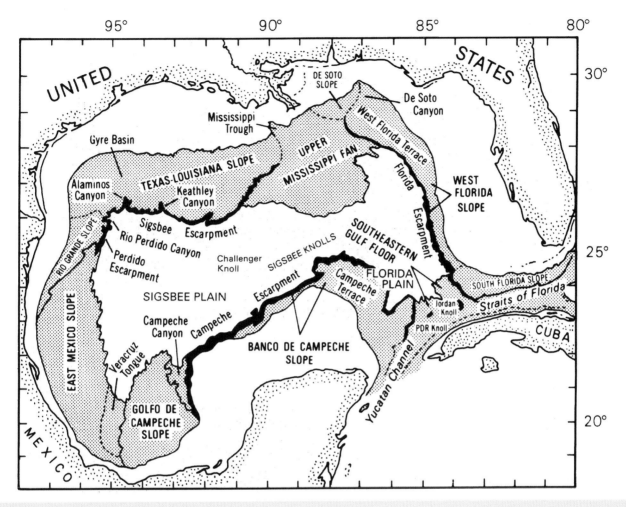

Fig. 9.7 Physiographic provinces of the Gulf of Mexico, including principal submarine canyons, sea valleys and escarpments. From Martin and Bouma (1978).

margin influenced by a major delta and salt diapirs (Texas–Louisiana Slope and Mississippi Fan) (Bouma *et al.* 1978 and references therein).

9.4.1 Passive margins outboard of major deltas

Type 1 passive margins with major river influence are associated with large, open-ocean submarine fans: the Mississippi Fan, Congo Fan, Amazon Fan and Nile Fan. The largest of these fans are not supplied locally, but instead are supplied by orogenic belts on the opposite side of the continent (Potter 1978). According to Wetzel (1993), the lengths of river-fed submarine fans are directly related to the depositional rate (= (fan volume)/age; Fig. 9.8). He also concludes that the amount of detritus transferred to the submarine fan by the river is 6–8 times larger for rivers emptying into estuaries intercepted by deeply incised canyons than it is for rivers terminating in lobate deltas. In the latter case, significant detritus never leaves the shelf or is distributed laterally by longshore currents.

In the major deltas that characterise this type of passive margin, fast and thick accumulation leads to gravitational instability in the form of growth faulting, mud/shale diapirism, wet-sediment sliding and wholesale seaward translation of the deltaic pile along overpressured

decollement surfaces at depth. A classic example of this type of gravitational instability comes from the Niger Delta of west Africa (Damuth 1994). In the adjacent deep-water area, gravity-flow systems are spatially constrained by the complex shape of the deformed seabed (Armentrout *et al.* 2000), which includes piggy-back (thrust-top) basins on the tops of megaslides and depressions along the axes of toe thrusts. Miocene and younger strata off the Niger Delta are folded into synforms separated by structurally complex zones. Channel and lobe architectural elements (Figs 7.20, 7.21) are common in the various slope basins seaward of the Niger Delta.

The Mississippi and Amazon fans are located at passive margins supplied by major rivers. They have both been drilled for scientific purposes (DSDP Leg 96 and ODP Leg 155). Facies and growth history of the Amazon Fan are described in Sections 7.2.3, 7.4.1 and 7.4.4.

9.4.2 Passive margins underlain by mobile salt

Salt diapirism is common in the slope and rise sediments of many passive continental margins, for example off northern Israel (Almagor & Garfunkel 1979; Almagor & Wiseman 1982; Garfunkel 1984; Garfunkel & Almagor 1985), in the northern Gulf of Mexico

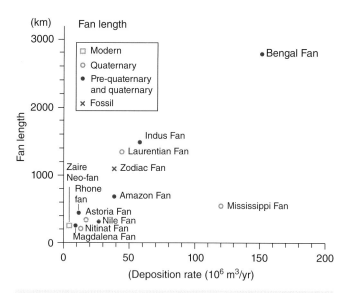

Fig. 9.8 Fan length versus deposition rate for large river fan systems, from Wetzel (1993). According to Wetzel, the amount of material delivered to a submarine fan is a major control on fan size, whereas sea-level fluctuations are of secondary importance and only modulate the delivery of this material to the fan.

(Bouma *et al.* 1978; Humphris 1978; Berryhill 1981; Bouma 1983), off North Carolina, where salt structures extend for 300 km, parallel to and seaward of the base of the continental slope (Cashman & Popenoe 1985), off Angola (Anderson *et al.* 2000; Babonneau *et al.* 2002; Fraser *et al.* 2005; Gee *et al.* 2007) and off Brazil (Adam *et al.* 2012). Diapirism and lateral salt flowage may alter the sediment distribution patterns in deep water by: (i) diverting pre-existing canyons; (ii) creating intraslope basins within which sediment ponding can occur (Satterfield & Behrens 1990; Badalini *et al.* 2000; Beaubouef & Friedmann 2000; Liu & Bryant 2000; Adam *et al.* 2012); (iii) generating slope failure through mass-wasting; (iv) providing a local source of sediment and (v) providing easy-slip horizons for mass-wasting triggered by, for example, earthquakes and other processes (e.g., the Messinian salt horizon in the continental margin off Israel).

The western Gulf of Mexico is a well-studied salt province with exceptionally irregular seabed relief (Fig. 9.9). Gravity flows move through this complex terrain by seeking out the narrow slot-like 'channels' that link the three-dimensional staircase of small basins. These slot-like features are neither depositional nor erosional channels, but are simply tortuous pathways through a complex topography created by differential movement of underlying salt. The mode of deposition has been called 'fill-and-spill', because sediment gravity flows first deposit a sheet-like confined lobe in upslope basin 1, say, then levées are built to create a channel–levée unit which forces subsequent SGFs to eventually bypass basin 1 (confined by levées in that basin) and to enter basin 2, and so on (Section 4.6.2; Fig. 4.29). Each mini-basin is therefore filled with a HARP-like sandbody overlain by a channel–levée unit. In many basins of the northwest Gulf of Mexico, the HARP-like units are underlain by mass-transport complexes (MTCs) of slide and debris-flow deposits (Fig. 9.10). The MTCs appear to coincide with the development of a new transport route from the shelf edge (perhaps by opening of a slide scar tapping into shelf deposits and deltas). Beaubouef and Friedmann (2000) interpret the MTCs to form early during a sea-level lowstand (Fig. 9.11). However, the number of MTCs (and associated cycles) in the last

100 kyr is significantly greater than the number of glacial lowstands, so we instead interpret the fill-and-spill cycles to have a dominant autocyclic control related to shelf-edge instability in an area of high sediment accumulation. Booth *et al.* (2000) and Prather *et al.* (2000, Section 4.7.1) have identified similar fill-and-spill cycles (which they call 'ponded phase' followed by 'spill phase') in Pliocene reservoir units in the northwest Gulf of Mexico (Fig. 9.12). The 'ponded'-phase deposits are sheet-like, and the 'spill'-phase deposits are formed of channelised sandbodies. These are interpreted to have formed in a similar fashion to the cycles described by Beaubouef and Friedmann (2000).

An ancient passive margin influenced by salt tectonics is exposed in the Vocontian Basin, western Alps, France (Friès & Parize 2003). From their abstract:

> key elements of the Vocontian [slope] model are (1) an emphasis on lowstand slope erosion and complex slope morphology controlled by contemporary tectonism and salt diapirism; (2) slope deposition in confined erosional and structurally controlled conduits rather than the build out of slope fans/channel-levée complexes; (3) a dominance of large-volume muddy sediment slide/slump and transitional debris-flow deposits, with subordinate sandy SGF deposits such as turbidites, including significant structureless sandstone facies; (4) common sand injections (sills and dykes) associated with the structureless sandstone facies; and (5) minimal downslope evolution of the flows.

Depositional sequences are defined by major widespread erosional surfaces. Sands were sourced from the adjacent shelf, so are well sorted and accumulated in slope gullies rather than travelling farther into the deep basin; this short transport distance is attributed to low flow efficiency (Section 7.2.1).

9.4.3 Slope apron of the northwest African margin

This section is based on the overview paper of Wynn *et al.* (2000). Because of limited detrital influx from the adjacent arid landmass, sediment supply to the northwest African passive margin is fine grained and hemipelagic. There is some reworking of this material by bottom (contour) currents at depths >3 km. However, the main processes for the delivery of sediment to the deep slopes and adjacent abyssal plains are gravity flows, predominantly debris flows and turbidity currents (Fig. 9.13). The western Saharan shelf has supplied many of the megaturbidites found beneath the Madeira Abyssal Plain (Weaver & Kuijpers 1983; Weaver *et al.* 1986; Weaver & Rothwell 1987; Jones *et al.* 1992; Masson 1994). Weaver *et al.* (1992) demonstrated that the megaturbidites are emplaced primarily as a result of large slope failures that take place during the transitions from glacial to interglacial epochs, and from interglacial to glacial epochs (i.e., at times of both rising and falling relative sea level). This margin was also the source of the spectacular Saharan Debris Flow (Fig. 1.59) with its 'woodgrain' surface textures indicating viscous flow.

Wynn *et al.* (2000) have argued that the northwest African margin is an excellent example of a fine-grained slope apron. It lacks point-source gravity-flow systems like submarine fans, and is instead dominated by bypass and gravitational failures. The slow rate of sediment accumulation on the shelf and upper slope prevents autocyclic processes from controlling sediment delivery to the deep-water areas; instead, long-term controlling factors like base-level (sea-level) variations have a profound effect (Weaver *et al.* 1992).

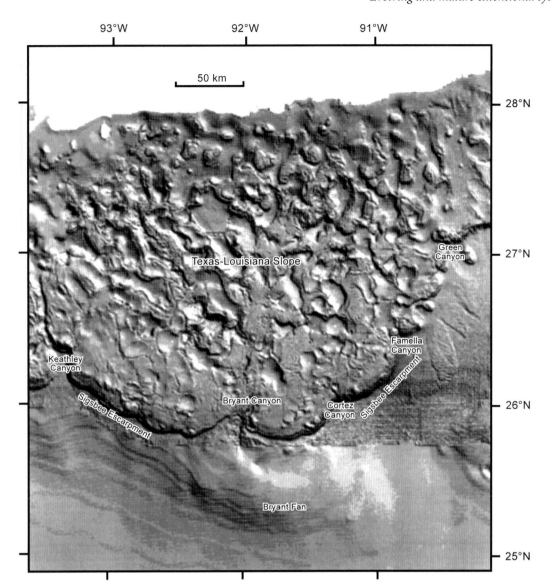

Fig. 9.9 Sun-shaded seafloor relief of the north-central Gulf of Mexico deepwater. The simulated sun illuminates the scene from an azimuth of 45° and an elevation of 45°. The bathymetry becomes smoother south of 26° N because of more meagre multibeam data. From Liu and Bryant (2000).

9.4.4 Passive margins swept by bottom currents

Chapter 6 provides an introduction to the character of sediment drifts along the modern Atlantic Ocean margins (Section 6.2, Fig. 6.1, Table 6.1). Single drifts, paired drifts, or simply a winnowed seafloor at water depths generally exceeding 2500 m characterise this type of margin. Sediment is advected along the margin (Hacquebard *et al.* 1981; Hill 1984a) until a site of reduced flow velocity is encountered, leading to net deposition.

Continental rifting and the development of ocean basins, leading to changes in the distribution of land and sea, are linked to major changes in ocean-current circulation patterns and the global mass balance of water, which, in turn, may be associated with important temporal changes in deep-marine facies. The onset of deep-ocean thermohaline circulation cells may rework passive margin sediments into contourite drifts and other small- to large-scale bedforms.

Neogene–Pleistocene examples of this phenomenon have been documented for the Norwegian–Greenland Sea (DSDP Leg 38, Talwani, Udintsey *et al.* 1976), the Atlantic continental margin off western South Africa (DSDP Leg 75, Hay, Sibuet *et al.* 1984), off the east coast of the USA (DSDP Leg 93, van Hinte, Wise *et al.* 1985a, b; DSDP Leg 95, Poag, Watts *et al.* 1987), and the Gulf of Cadiz as a result of Mediterranean outflow (Faugères *et al.* 1984; Gonthier *et al.* 1984).

Particularly strong bottom currents are responsible for the development of bathyal and abyssal unconformities that are widespread in some parts of the modern oceans (Emery & Uchupi 1984; Fig. 6.7; von Rad & Exon 1983). These unconformities delimit depositional sequences and therefore can logically be called sequence boundaries, but they are genetically and temporally unrelated to sequence boundaries on the adjacent shelf. Thermohaline circulation is apparently stronger during interglacial times (Fagel *et al.* 1996, 1997, 2001), so abyssal unconformities might be

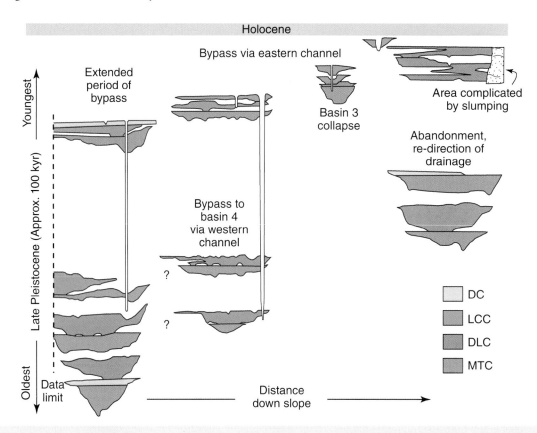

Fig. 9.10 Summary of stratigraphic successions in a linked set of four intraslope basins in water depths of 400–1500 m, northwest Gulf of Mexico. MTC = mass transport complex, DLC = distributary channel–lobe complex, LCC = levéed-channel complex, DC = hemipelagic drape complex. From Beaubouef and Friedmann (2000).

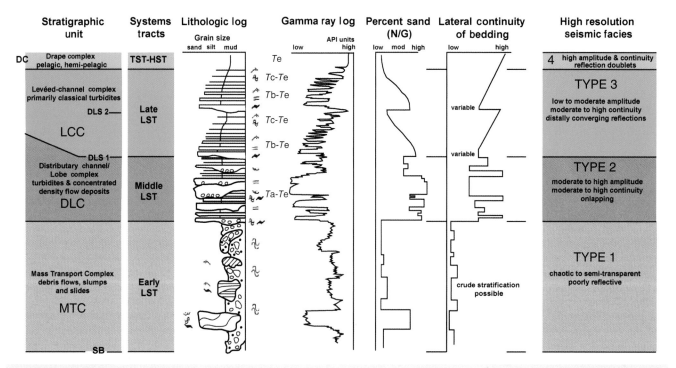

Fig. 9.11 Summary of the interpreted sequence stratigraphy, facies and seismic attributes of a cycle of mini-basin filling, northwest Gulf of Mexico salt province. From Beaubouef and Friedmann (2000).

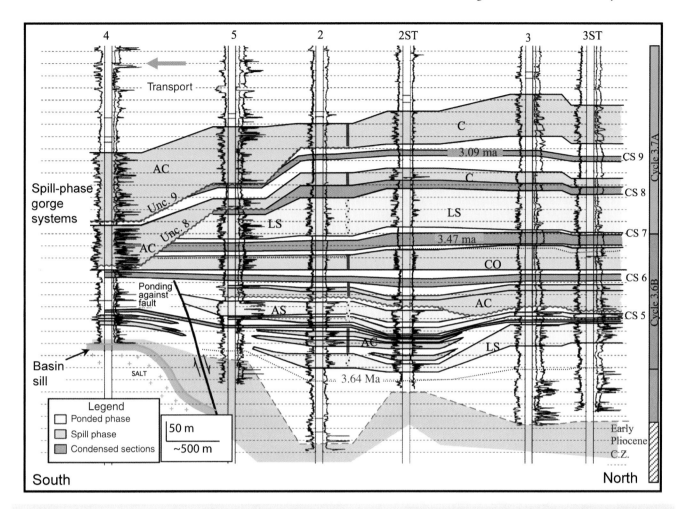

Fig. 9.12 Stratigraphic architecture of the Marconi Field, Gulf of Mexico. The interval below CS 6 contains a number of ponded–spill cycles. The shallower interval above CS 8 is dominated by channels and two large-scale spill-phase 'gorge' systems. Colour bars on right-hand side indicate sea-level cycles. Yellow and red colour bars to the right of well #2 indicate ponded and spill phases, respectively. AS = amalgamated sheet, LS = layered sheet, AC = amalgamated channels, C = isolated channels, CO = channel overbank, DB = debrite/mass flow. From Booth *et al.* (2000).

expected to develop out of phase with shallow-marine, lowstand unconformities.

Beneath the continental slope off New Jersey, there are a number of distinct current-generated unconformities, falling into two groups (Poag & Watts 1987): those that are correlated from basin to basin and appear approximately coincident with the 'global' erosional events postulated by Vail *et al.* (1977b) or Haq *et al.* (1987), and those of a more local extent, restricted to individual basins and slopes. Periods of non-deposition are extremely variable, for example an 11 Myr hiatus between upper Eocene and upper Oligocene strata (observed in the COST B-3 and other wells), and a hiatus of 5 Myr duration between lower and middle Miocene strata. Variations in water depth are recorded across many of these unconformities, for example the 9 Myr gap from the upper Oligocene to middle Miocene strata shows a change from lower to upper bathyal depths in the COST B-3 well, whereas the same hiatus in the COST B-2 well is defined by a change from outer to inner sub-littoral depths, that is, an overall shallowing upwards.

The opening of the Drake Passage in the Antarctic Ocean in the latest Oligocene–early Miocene (25–30 Ma) (*cf.* age ranges from middle Eocene through early Miocene; Ghiglione *et al.* 2008) produced a deep circumpolar current, leading to the thermal isolation of Antarctica and increased global cooling. The cooling is associated with a major turnover in planktonic organisms with significant Neogene extinctions (Keller & Barron 1983). The present circulation in the Antarctic was established 13.5–12.5 Ma.

Although modern thermohaline circulation is initiated by the sinking of denser water in cold polar seas, some ancient bottom currents might have been generated by downwelling of warm, dense, saline waters from restricted low-latitude seaways like the Tethys Ocean (Hay 1983b). Modern bottom currents flow from polar regions across the equator and into the other hemisphere. As a result, passive margins with sediment drifts on the lower slope and continental rise might pass landward into anything from a glaciated shelf (e.g., Labrador Sea, Myers & Piper 1988) to a tropical carbonate platform (e.g., Bahamas).

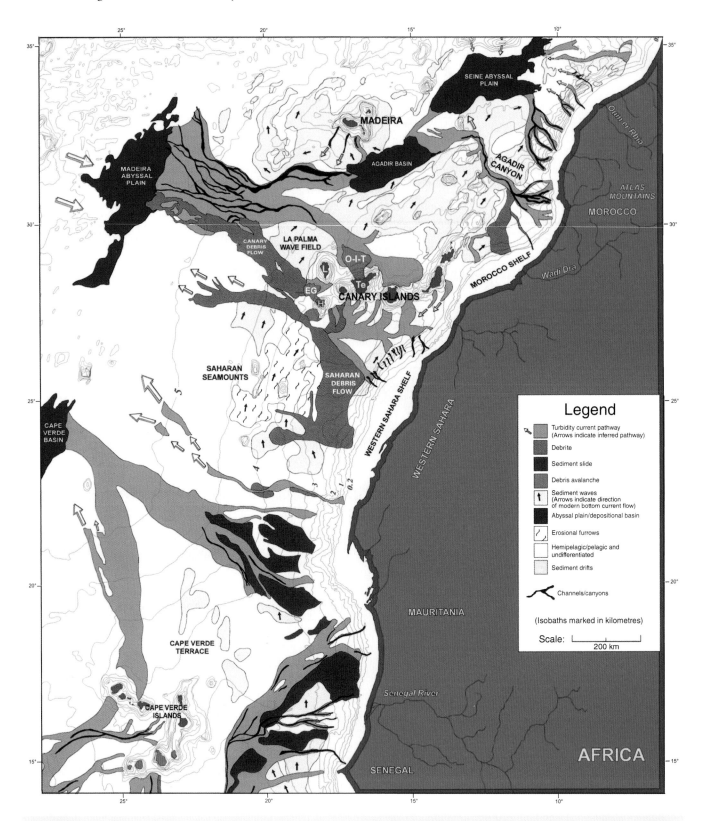

Fig. 9.13 Distribution of deep-water deposits and morphological features along the passive margin of northwest Africa. From Wynn *et al.* (2000).

9.4.5 Glaciated passive margins

Passive continental margins adjacent to continental ice sheets receive enormous volumes of glacial erosion products. Where till and coarse-grained outwash fail close to the shelf edge, disorganised debrites abound. Sandy SGF deposits can also be emplaced by the flows generated by such failures. Subglacial discharge of freshwater heavily freighted with clay- to silt-sized rock flour contributes to a surface plume of suspended material that can rain to the seabed along adjacent slopes to form thick stratified deposits of mostly mud. These plume deposits preferentially accumulate as downslope-trending ridges generally to one side of the freshwater outlet, dependent on deflection by the predominant marine surface current (Hesse *et al.* 1999).

The modern Labrador Sea, Baffin Bay and the Scandinavian off-shore contain many examples of debris fans located at the mouths of over-deepened, shelf-crossing, glacially excavated troughs. Several examples have been described from Scandinavian margins (Eidvin *et al.* 1993; Vorren & Laberg 1997). These deposits, formed during glacial times, are stacked successions of glaciogenic debrites (Fig. 9.14) derived from the failure of till buildups that initially accumulated at the lowstand shelf edge (Aksu & Hiscott 1992; Hiscott & Aksu 1994, 1996). In these areas, stacked debrites alternate with hemipelagic muds, thin-bedded turbidites and ice-rafted 'key' beds

called *Heinrich layers* (Heinrich 1988; Andrews & Tedesco 1992; Broecker *et al.* 1992). On the Northeast Newfoundland Slope (landward of Orphan Knoll, Fig. 6.3), Pleistocene debrite and hemipelagic successions form a seaward-thinning wedge that is ~750 m thick beneath the middle slope (Aksu & Hiscott 1992). This material was delivered through the shelf-crossing Trinity Trough just north of the Grand Banks (Hiscott & Aksu 1996).

Similar trough-mouth fans are found around Antarctica, where they pass basinward into successions influenced by the strong bottom currents that encircle that continent (compare Section 4.3). Passchier *et al.* (2003) describe the ice-marginal passive margin seaward of Prydz Bay, Antarctica, relying on results from ODP Legs 119 and 188 (Figs 9.15 and 9.16). At continental slope Site 1167 (water depth 1640 m), sediments consist of poorly sorted gravelly, muddy sand with 2–5% clasts, poorly sorted muddy sand with dispersed clasts and clast-poor sandy diamicton. In a few intervals the poorly sorted facies are interbedded with coarse sand and granule beds, sandy mud with dispersed clasts, and thinly bedded clay with silt laminae. Passchier *et al.* (2003) interpret these sediments as debrites supplied by an ice-stream in the Prydz Channel, and a mixture of hemipelagic and bottom-current deposits. There are apparently very few turbidites. Oscillations in the texture of the hemipelagic component and the abundance of ice-rafted detritus (IRD) are ascribed by Passchier (2007) to advances and retreats

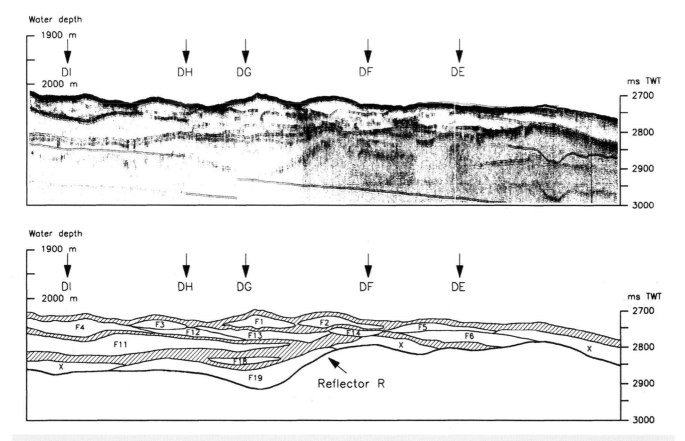

Fig. 9.14 Slope-parallel seismic-reflection profile showing the distribution of debrite lenses (F1–F19 and X) and stratified glacio-marine sediments (cross-hatched pattern) on the northeast Newfoundland Slope, seaward of a glacial outlet called the Trinity Trough. TWT = two-way travel time. Near-horizontal double traces = interference from a Huntec sparker system. Seismic profile is approximately 40 km long. DE-DI = locations of crossing seismic lines. From Aksu and Hiscott (1992).

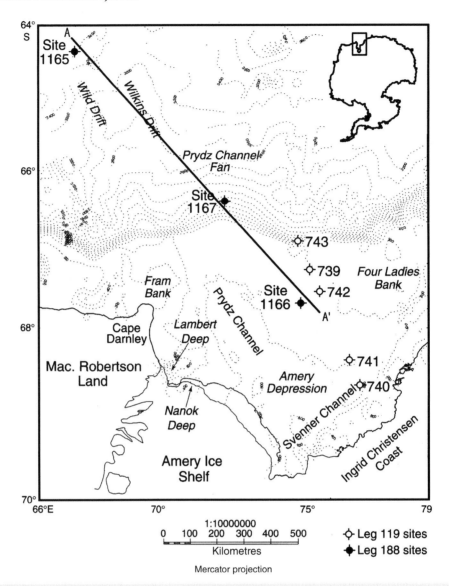

Fig. 9.15 Bathymetric map of the Prydz Bay margin with location of drill holes from Leg 188 (Sites 1165–1167) and Leg 119 (Site 739–742). Note sediment drifts in the northwest of the map. Profile A–A′ is shown in Fig. 9.16. From Passchier *et al.* (2003) after Cooper *et al.* (2001).

of the terminus of the ice-stream, and to large oscillations in sea level caused by far-distant northern-hemisphere glaciations, which resulted in lift-off of Antarctic glaciers from their beds when global sea level was high (i.e., during northern-hemisphere interglacial epochs).

Along the Canadian continental margin of Labrador, Hesse *et al.* (1997, 1999) have described spectacular plume deposits downslope of glacial outlets which existed during the Pleistocene. These form high-relief ridges with weak, continuous acoustic reflections south of the major meltwater outlets (Fig. 9.17), because the buoyant meltwater plumes were deflected southward by the strong palaeo-Labrador Current after entering the marine realm. The deposits themselves are interbedded silts and muds with gradational bases and tops caused by gradual increases and decreases in the velocity and track of the sea-surface plumes (Fig. 9.18). Immediately in front of the Pleistocene

glacial outlets, coarser grained turbidites and debrites are more abundant, and form deposits with relatively subdued seabed relief and better acoustic reflectivity because of strongly contrasting grain sizes between separate beds. The SGFs, including turbidity currents, which were generated by ice-margin failures in this area, also bypassed the slope and followed tributaries to the ~4000 km-long Northwest Atlantic Mid-Ocean Channel (NAMOC), described in Section 7.6.

Just seaward of the primary outlet of the Laurentide Ice Sheet through Hudson Strait, Hesse and Khodabakhsh (2006) describe stacks of ~3 cm-thick graded mud layers reaching composite thicknesses of ~4 m, with scattered ice-rafted dropstones. These are ice-proximal facies of Heinrich layers (Heinrich 1988), and contain >40% detrital carbonate derived by glacial erosion of Palaeozoic carbonate rocks in the Canadian Arctic, and then rafted to the Labrador margin by icebergs. The stacks of graded mud layers are

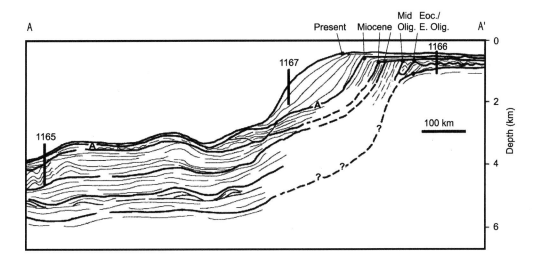

Fig. 9.16 Line interpretation of a seismic profile off Prydz Bay, Antarctica. High volumes of glacially derived detritus form a thick, prograding slope of Oligocene to Modern age. On the slope (Site 1167), debrites and hemipelagic units influenced by bottom currents predominate, whereas in the deep offshore area (Site 1165) there are contourite mounds molded by circum-Antarctic bottom currents. From Passchier *et al.* (2003), after Cooper *et al.* (2001).

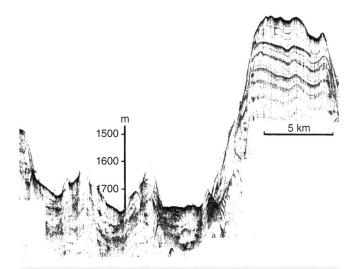

Fig. 9.17 655 cm³ (40 in³) sleeve-gun seismic profile along the strike of the upper Labrador Slope off Hudson Strait showing high-relief ridges formed of plume deposits. There is weak acoustic reflectivity because of rather uniform grain size. Stronger reflections are returned from coarser grained facies. From Hesse and Klaucke (1995).

attributed to *flow lofting*, a process in which a bottom-hugging sediment gravity flow, formed of suspended sediment and essentially freshwater (in this case from glacial meltwater), undergoes a transition from negative buoyancy (the suspension sinks) to positive buoyancy as sediment is deposited (the residual suspension lifts off from the seabed and rises through more dense saline water). The fine-grained residual suspension forms an interflow in the water column and rains mud-sized particles to the seabed over a period of months, during which iceberg-rafted gravel also

Fig. 9.18 X-radiograph of core H92-45-01 (56°09′61′N, 57°07′28″W, 1469 m water depth) showing turbid-surface-plume facies. Core width ~6 cm. Note the blurred boundaries between alternating parallel silt- and clay-rich bands. From Hesse *et al.* (1999).

PERIODS   BUILDUPS   MAJOR BIOTIC ELEMENTS

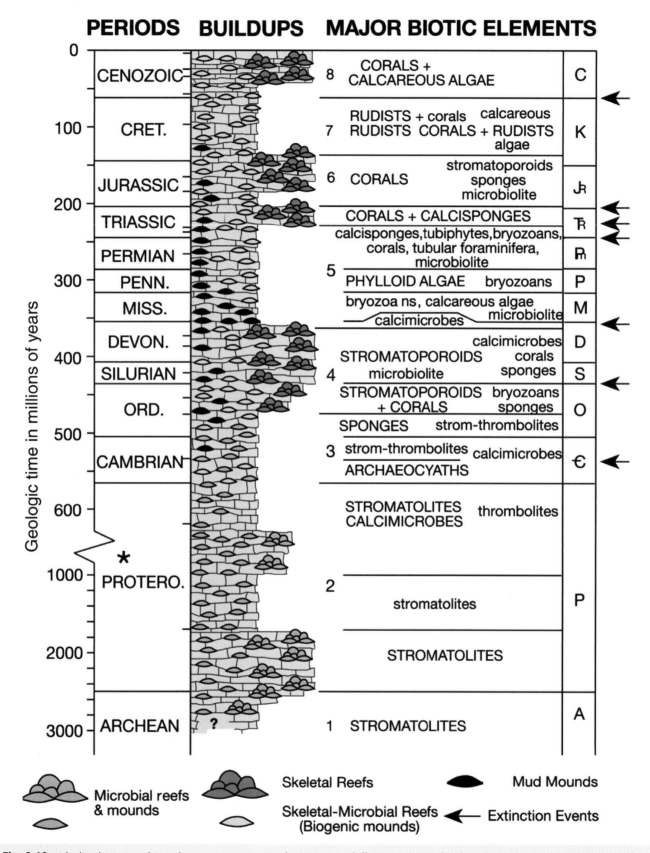

Fig. 9.19 Idealised stratigraphic column representing geologic time and illustrating periods when there were only skeletal–microbial reefs (biogenic mounds), times when there were both skeletal–microbial reefs and skeletal reefs, and periods when mud mounds were important. For example, from the Cambrian through the Early Ordovician, and from the Mississippian through the Early Triassic, no wave-resistant, platform-margin reefs existed on Earth, dramatically limiting the development of steep passive-margin slopes containing bouldery debrites. Numbers indicate different associations of reef- and mound-building biota. Arrows signal major extinction events; * = scale change. From James and Wood (2010).

accumulates. Hence, in this area, the passive margin contains a complex array of glacially influenced deep-marine deposits, from proximal cohesive-flow deposits to sandier SGF deposits, including turbidites, plume deposits and lofted facies, to a distal channel–levée complex forming one of the longest sinuous deep-sea channels on Earth.

9.4.6 Carbonate platforms and ramps

In equatorial latitudes (~30° N to ~30° S), in the absence of significant siliciclastic sediment supply, thick warm-water carbonates can develop on the shelf of a passive continental margin. In Cenozoic oceans, platform-margin reefs have been constructed by scleractinian corals and calcareous red algae. At other times in the geological past, different organisms filled this niche (Fig. 9.19). Exceptions occurred during the Cambrian through Early Ordovician, Carboniferous through Early Triassic, and much of the Cretaceous when platform-margin reef frame builders were absent. Outside the tropics, so-called 'cool-water' carbonate shelves are found wherever siliciclastic input is minor and carbonate particles are produced in sufficient quantities by echinoderms, benthic foraminifera and molluscs. The shelf deposits in these areas are grainy, and the seabed deepens seaward with a profile like that of siliciclastic margins (i.e., there is no platform-margin reef belt or shoals), forming what is called a 'carbonate ramp'. Cenozoic to modern examples are well documented around New Zealand and South Australia (Nelson 1978; James & Bone 1991; Feary & James 1998).

The deep-water environments around rimmed carbonate platforms (i.e., those with fringing reefs) are sites for the deposition of coarse-grained talus, carbonate debrites and associated finer grained sediments. Before the Mesozoic advent of calcareous plankton (e.g., foraminifera, coccolithophorids), the calcareous mud fraction in the deep-water areas consisted of fine material wafted off the platform during storms (Section 1.2.1), and called *periplatform ooze* (Schlager & James 1978). Debrites contain both platform-derived boulders and locally derived clasts of slope limestones (Hiscott & James 1985; Fig. 9.20). In the Cambro-Ordovician examples described by James *et al.* (1989) and Hiscott and James (1985), the largest and most bouldery debrites apparently developed during phases of platform-margin progradation (slowly rising relative sea level: Fig. 9.21), with failure events triggered in the youngest deposits by accelerated faulting and tectonism at the onset of the Taconic Orogeny. The carbonate debrites show all the characteristic features of cohesive flows. They have mounded tops, rapidly tapering marginal 'snouts' (Fig. 9.22), outsized clasts which project from bed tops, and a distinctive clast fabric (Hiscott & James 1985).

The deep-water facies around rimmed carbonate platforms can only be studied effectively in ancient successions, for two reasons: (i) the coarse texture of Holocene debrites with their large lithified blocks precludes sample collection by coring, and (ii) steep slopes and bouldery deposits limit high-resolution seismic profiling close to such platform margins. In ancient successions, the deposits are well exposed and show abundant evidence of gravitational instability (e.g., slide scars and intraformational truncation surfaces, Fig. 9.23). In the Cow Head Group, the maximum amount of section removed at slide scars is 7 m. Other ancient studies document 15–25 m-deep cuts into slope sediments of the Piedmont Basin in Northern Italy (Clari & Ghibaudo 1979) and up to 100–150 m of truncation of Palaeozoic slope carbonates in the Sverdrup Basin (Davies 1977).

The abrupt change in the structural competence of shaly deep-water deposits versus reef-facies limestones, as well as the steep palaeo-relief

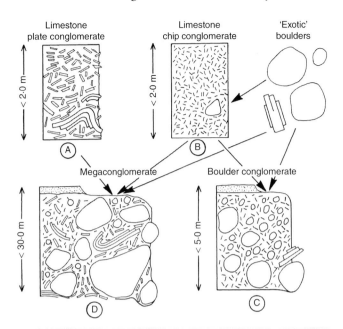

Fig. 9.20 Debrite facies, Cow Head Group, modified from Hiscott and James (1985) and James and Stevens (1986). The three components are 'plates' = cobble- to boulder-sized fragments of thinly bedded slope limestones, 'chips' = pebble- to cobble-sized fragments of slope limestones or shallow-water limestones, and "exotic boulders" = collapsed blocks derived from the Cambro-Ordovician platform margin. Beds 10, 12 and 14 (Fig. 9.21) consist of debrite types C and D.

at the edge of the platform, almost always guarantee that the thrust faults that carry deep-water platform-margin carbonates into their final resting place in orogenic belts will isolate the deep-water successions from their associated platform (e.g., James & Stevens 1986). For this reason, an uninterrupted outcrop traverse from a tropical shelf platform to its contemporary basinal facies is rare, and must be reconstructed instead using high-resolution stratigraphic correlation.

Cool-water carbonates are found outside Earth's tropical belt, in seas where the water temperature is too low for the development of reefs. Shelf deposits tend to be grainy with little carbonate mud (Nelson 1978). Typically, shelf-to-basin transects resemble those along siliciclastic passive margins, with a seaward-deepening shelf and a shelf edge in ~100–200 m of water. These deposits form a type of *carbonate ramp*. This name indicates that there is a progressive deepening away from the shoreline. One of the largest cool-water carbonate systems on Earth extends from outcrops on land to deep basinal areas off the southern coast of Australia. A number of upper-slope sites were drilled in 202–784 m of water during ODP Leg 182, as well as a site in 3875 m of water to evaluate basinal facies (Fig. 9.24). Here we focus on the facies found in the Neogene part of the drilled succession (Miocene and younger) to show the characteristics of this type of passive margin (Fig. 9.25).

A number of facies were recovered from the most seaward to the most landward drill site. The most distal Site 1128 consists of pink to brown, bioturbated nannofossil ooze punctuated by numerous thin glauconitic and planktonic foraminiferal sand calci-turbidites and conglomeratic SGF deposits. There is a chaotic zone of debrites and slumped sediment (54.4–70.0 mbsf). Site 1134 is mostly calcareous nannofossil ooze with varying amounts of planktonic foraminifers (0–33 mbsf), slumped calcareous nannofossil

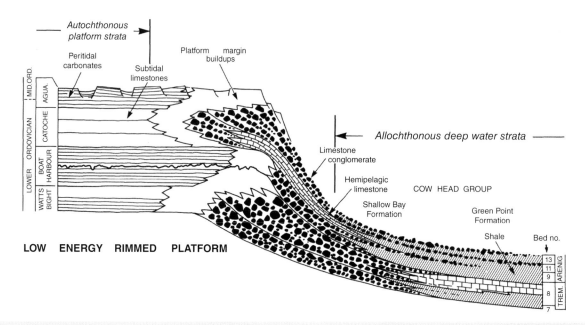

Fig. 9.21 Interpretative cross-section of the restored Early Ordovician rimmed carbonate platform and contemporary deep-water sediments, western Newfoundland, from James *et al.* (1989). The platform margin in the central part of the sketch is not preserved, except as exotic boulders in debrites, so the horizontal scale cannot be quantified. The carbonate-platform succession on the left is ~500 m thick. Even-numbered debrite-rich Beds 10, 12 and 14 are not numbered in the 'Bed no.' column on the right but are shown as trains of boulders (black, filled shapes). They developed during the late Arenig stage of the Early Ordovician as the platform margin prograded and aggraded (i.e., carbonate production >rate of relative sea-level rise). Note that, following mid-20th century usage and consistent with the North American stratigraphic code, Beds in the Cow Head Group are formal stratigraphic units.

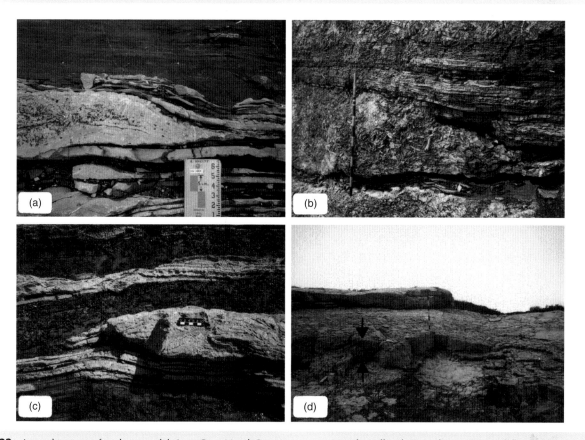

Fig. 9.22 Lateral snouts of carbonate debrites, Cow Head Group, western Newfoundland, Canada. (a) Tapered flow margin at which only a few centimetres of basal erosion can be demonstrated, proving a mounded top. Notebook is ~20 cm long. (b) Tapered debrite margin with onlapping younger ribbon limestones. Scale divisions are 10 cm. (c) Tapered debrite margin with no basal erosion but rather compactional down-bowing of underlying thin beds. Scale divisions are 2 cm. (d) 3D view of a tapered debrite margin. Arrows mark top and base of the debrite. Scale divisions are 10 cm. From Hiscott and James (1985).

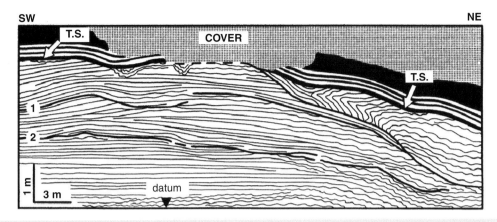

Fig. 9.23 Line drawing of truncation surface (T.S.) interpreted as a slide scar at Green Point, Cow Head Group, western Newfoundland, Canada. '1' and '2' are small-displacement syn-sedimentary faults. Wavy lines toward the right indicate disturbed bedding. Note drag folds in a displaced block just below the main slide scar, which was subsequently buried by little-disturbed sediments. From Coniglio (1986).

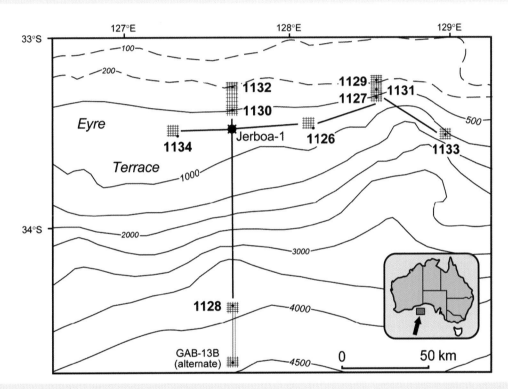

Fig. 9.24 Location of Leg 182 ODP sites on the southern Australian passive margin. From Shipboard Scientific Party (2000).

ooze and calcareous nannofossil foraminifer ooze interleaved with unlithified wackestones, packstones, floatstones and rudstones (33–66 mbsf) – the wackestones to rudstones contain bryozoans, bioclasts, sponge and tunicate spicules, pellets and pebble-sized lumps of calcareous nannofossil ooze interpreted to be reworked clasts – and nannofossil ooze and calcareous nannofossil foraminifer ooze (66–152 mbsf). Site 1126 is similar – mostly nannofossil ooze (Fig. 9.25). Site 1130 occupies the same position as laterally adjacent Site 1127 (Fig. 9.24). At Site 1127, there are very fine to fine-grained, heavily bioturbated, unlithified to partially lithified, greenish grey wackestones to packstones. The succession from ~420–464.5 mbsf has a slumped base with abundant clasts, including bryozoan and large skeletal fragments within an ooze matrix. The bryozoan clasts are derived from nearby biotherms (Fig. 9.26).

As a summary, cool-water carbonates beneath the deep-water parts of the South Australia passive margin are predominantly nannofossil ooze in distal areas and bioclastic packstones and wackestones closer to the shelf edge, with scattered debrite and sediment slide/slump horizons. Lowstand deposits are relatively fine grained and include abundant sponge spicules, whereas transgressive and highstand deposits are coarser grained and contain abundant high-Mg calcite bioclasts (Betzler *et al.* 2005). The highstand deposits are thicker than the lowstand deposits on the slope, consistent with enhanced highstand carbonate production on the shelf and off-shelf transport of this material by storms and shelf-edge failures. Textural cycles

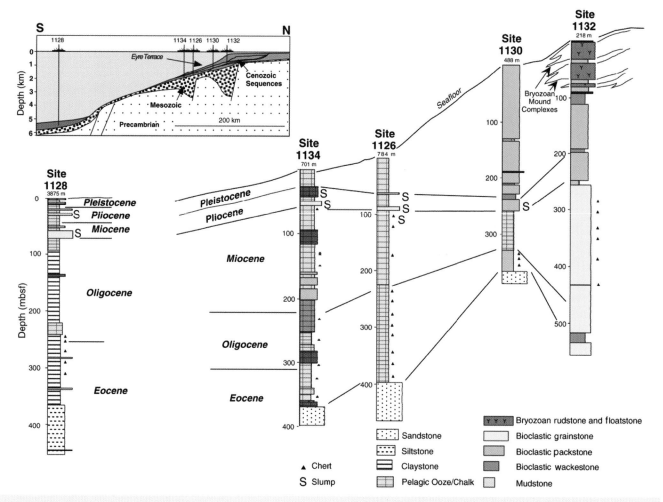

Fig. 9.25 Cross-section showing Neogene facies from the upper slope to deep basin, South Australia cool-water carbonate margin. Water depths at each drill site are indicated below the site name. The inset places the five drill sites within the Cenozoic progradational wedge. Note the widespread top-Miocene through Pliocene slumps (S). From Shipboard Scientific Party (2000).

recovered beneath the slope are several tens of metres thick, and are interpreted to be a response to sea-level fluctuations (Betzler *et al.* 2005). These are essentially hemipelagic deposits advected off the shelf, punctuated by SGF deposits and sediment slides/slumps.

9.5   Failed rift systems

Deep-water sedimentation occurs not only along evolving and mature passive continental margins, but also in 'failed' rift systems (aulacogens), that is, where rifting ceased before any oceanic crust was generated. Aulacogens such as the Precambrian Athapuscow Aulacogen, Great Slave Lake, northern Canada (Hoffman *et al.* 1974), contain deep-water SGF deposits that accumulated during a phase of rapid subsidence of the rift associated with crustal extension. The Benue Trough is a Cretaceous to Cenozoic, 1000 km-long and 100 km-wide aulacogen on the margin of West Africa. At its western end, there are up to 12 km of submarine-fan deposits beneath fluvio-deltaic sediments of the Niger System (Weber 1971; Burke *et al.* 1972; Petters & Ekweozor 1982; Maurin *et al.* 1986; Ofurhie *et al.* 2002; Odigi &

Amajor 2008). The Lower Palaeozoic deep-water sediments of the Ouachita Mountains, Oklahoma, are also interpreted as the infill of a failed rift system associated with the development of the Iapetus Ocean; the arms of the rift system include the South Oklahoma Aulacogen, Reelfoot Rift, early Illinois Basin, Rome Basin and the Ouachita Basin (Lowe 1985). Nearly 12 km cumulative thickness of SGF deposits accumulated in the Ouachita Basin (Lowe 1985).

Because rifting is a protracted and pulsatory process, some rift basins that once formed deep troughs may later be isolated on the margin or interior of a continent if eventual breakup takes place elsewhere. These are not aulacogens, because their axes are parallel to the continental margin, not transverse (Wilson & Williams 1979). The North Sea is a good example. It is a large, failed rift system (Glennie 1986a, b; Brooks & Glennie 1987). Extensive deep-marine siliciclastic systems developed, particularly during the Late Jurassic, Early Cretaceous and Palaeogene (e.g., Stow *et al.* 1982; Bergslien *et al.* 2005; Doré & Robbins 2005; Fitzsimmons *et al.* 2005; Hempton *et al.* 2005; Martinsen *et al.* 2005; Oakman 2005). The North Sea submarine fans typically are small in radius (5–20 km) and sand-rich (Fig. 9.27). Geometry of the deposits is controlled by (i) source area

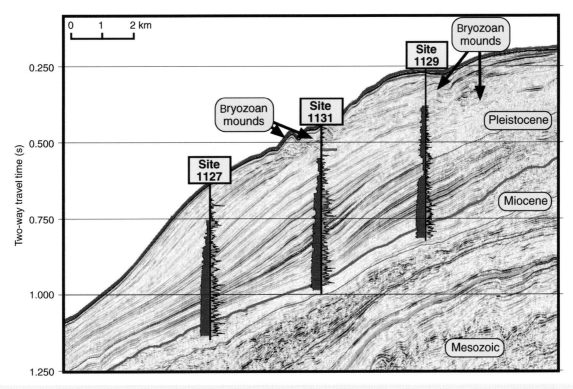

Fig. 9.26 Seismic profile in the vicinity of upper slope ODP Site 1127, at a depth similar to Site 1130 (Fig. 9.25). The prograding deposits are mostly very fine to fine-grained, heavily bioturbated wackestones to packstones. 1.0 s two-way travel time (TWT) = ~1500 m water depth. From Shipboard Scientific Party (2000).

size, (ii) basin and basin-margin physiography and bathymetry, (iii) tectonic history and resulting morphology of drainage and delivery systems of sediments to deep-water areas and (iv) the rate of sediment delivery (Martinsen *et al.* 2005; Fig. 9.28).

The Lusitanian Basin (Fig. 9.5) provides another example of a failed rift basin, now exposed on land in western Portugal because the final continental breakup with what are now the Grand Banks of Newfoundland occurred farther west during the Aptian.

9.6 Fragments of ancient passive margins

Many large fragments of ancient passive margins are documented from orogenic belts. Amongst the thicker successions is the 14-km-thick, upper Precambrian (Riphean–Vendian) Barents Sea Group and Lökvikfjell Group, Finnmark, North Norway (Siedlecka & Siedlecki 1967; Johnson *et al.* 1978; Pickering 1985; Siedlecka 1985). The oldest exposed part of this passive margin consists of at least 3200 m of deep-marine SGF and related deposits (Pickering 1981b, 1985; Drinkwater *et al.* 1996), passing up into 2500–3500 m of upper continental slope and prodelta (Pickering 1981b, 1982a, 1984b), delta-front, delta-top and associated shallow-marine sediments (Siedlecka & Edwards 1980). Up to 1500 m of shallow-marine, intertidal and supratidal carbonates, overlain by 1500 m of essentially fluviatile sediments, form the upper part of the Barents Sea Group (Siedlecka 1978, 1985; Siedlecka *et al.* 1989); 5700 m of upper Precambrian–Eocambrian shelf, marginal-marine and continental sediments of the Lökvikfjell Group (Levell & Roberts 1977; Johnson

et al. 1978; Levell 1980a, b) overlie the Barents Sea Group with disconformity or local angular unconformity. This passive margin stratigraphy, and to the east their along-strike contemporaneous successions in the adjacent Rybachi and Sredni peninsulas, Arctic Russia (Siedlecka *et al.* 1994; Drinkwater *et al.* 1996), developed facing an ocean basin towards the present north–northeast, and was subsequently juxtaposed against the Fennoscandian Shield by major dextral shear between 640–540 Ma (Kjøde *et al.* 1978).

The Aptian Vocontian passive margin is preserved in large thrust slices in southeastern France, having been uplifted during rotation of the Iberian plate as the Atlantic Ocean opened in the Cretaceous (Friès & Parize 2003). Tectonic deformation is minor so these rocks serve as a fine example of the facies in an ancient passive margin. In western Canada, a much older Neoproterozoic passive margin succession is found in the Windermere Supergroup, and contains a number of stacked sand-filled channels, some showing prominent lateral-accretion bedding (Schwarz & Arnott 2007; Fig. 8.42). The basal part of the Windermere turbidite system is formed by the Kaza Group (2–3 km thick), composed of sheet-like sandstone-dominated units and subordinate mudstone packages that accumulated on the basin floor to toe of the slope (Ross *et al.* 1995; Meyer 2004). The Kaza Group is overlain by the Isaac Formation, a mud-dominated unit over 2.5 km thick that consists mostly of thin-bedded turbidites and lenticular, thick-bedded sandstone to conglomerate units (up to 100 m thick). In addition, debrites and slide/slump complexes, some up to 100 m thick, are common.

Many passive margins have been dismembered during later plate convergence and mountain building. A common theme is that the

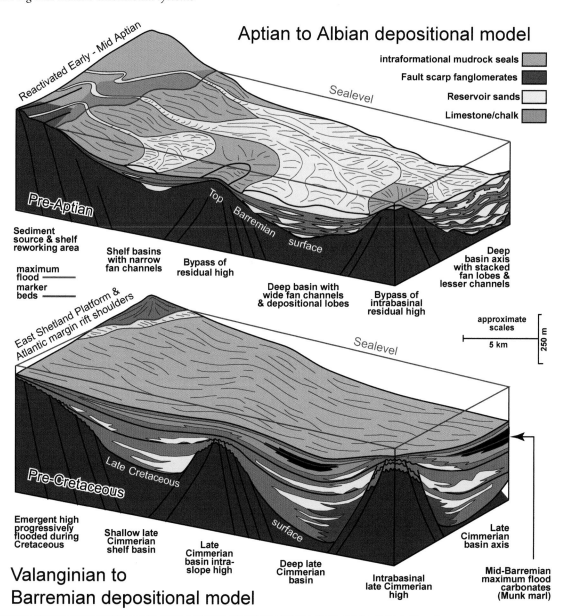

Aptian to Albian depositional model

Reactivated Early - Mid Aptian

Sealevel

Pre-Aptian

Top Barremian surface

intraformational mudrock seals
Fault scarp fanglomerates
Reservoir sands
Limestone/chalk

Sediment source & shelf reworking area

maximum flood marker beds

Shelf basins with narrow fan channels

Bypass of residual high

Deep basin with wide fan channels & depositional lobes

Bypass of intrabasinal residual high

Deep basin axis with stacked fan lobes & lesser channels

East Shetland Platform & Atlantic margin rift shoulders

Sealevel

approximate scales

5 km   250 m

Pre-Cretaceous

Late Cretaceous

surface

Emergent high progressively flooded during Cretaceous

Shallow late Cimmerian shelf basin

Late Cimmerian basin intra-slope high

Deep late Cimmerian basin

Intrabasinal late Cimmerian high

Late Cimmerian basin axis

Mid-Barremian maximum flood carbonates (Munk marl)

Valanginian to Barremian depositional model

Fig. 9.27   Deep-marine depositional models for the Moray Firth basins of the North Sea: Aptian to Albian (top); Valanginian to Barremian (bottom). Note the small dimensions of these depositional systems. From Oakman (2005).

deep-water successions are tectonically transported, as allochthons, so that they rest structurally on top of the coeval and formerly contiguous shelf successions. In western Newfoundland, for example, deposits of a Cambro-Ordovician carbonate slope apron (Cow Head Group) rest structurally on top of contemporaneous intertidal to shallow-marine dolostones and limestones (James & Stevens 1986). The shelf-edge facies of algal boundstones was completely eradicated during Ordovician thrusting, and is only known from large displaced boulders in debrites of the Cow Head Group (James 1981). The large open-ocean submarine fans present in the modern ocean basins lie on oceanic and transitional crust, so cannot survive convergence and mountain building without severe deformation and dismemberment. They are already so large that the component parts (i.e., architectural elements) could never be recognised in even exceptional outcrops. Once penetratively deformed and perhaps considerably

metamorphosed, it is unlikely that reasonable inferences could be made about the primary depositional setting and the scale or geometry of the submarine-fan system.

9.7   Concluding remarks

Passive continental margins develop from intracontinental rift zones. As soon as extension begins, topography is created that provides (i) uplifted sources of detritus and (ii) depressions that may become flooded to form lakes or arms of the sea sufficiently deep to accumulate sediment gravity-flow (SGF) deposits. Although mature passive margins commonly consist of broad monoclinal slopes leading to adjacent abyssal plains, many rifts and young passive margins display ridge-and-trough topography because of fault-block rotation or salt

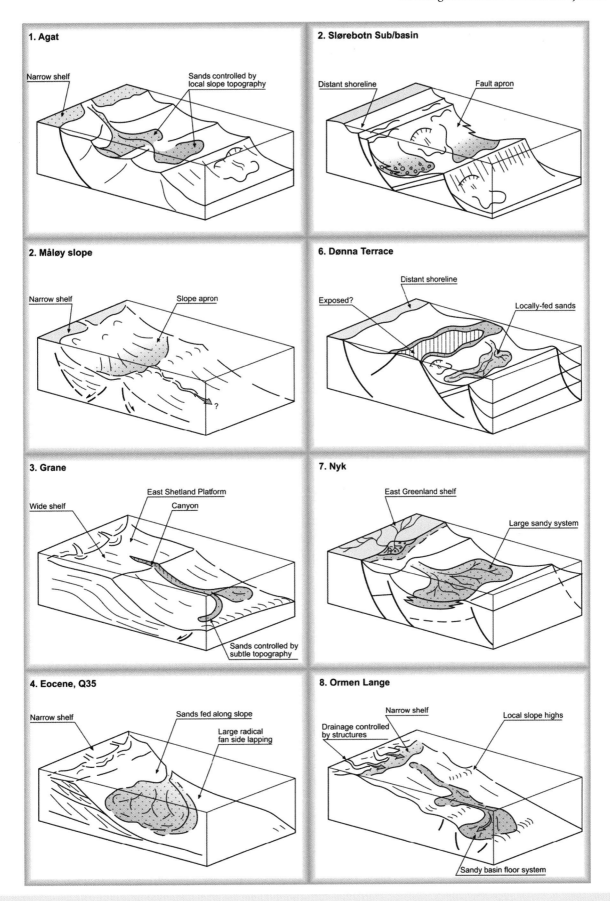

Fig. 9.28 Cartoons summarising the eight different deep-water depositional systems described from basins of the North Sea, Norwegian Sea and east Greenland by Martinsen *et al.* (2005). Outlines of the siliciclastic deep-marine systems are shown to illustrate the difference in shape as a result of the controlling processes. Not to scale.

diapirism. Barriers to SGFs are then able to focus coarser deposits into small slope basins that persist until the barriers are breached by erosion or overtopped when the most landward basins become filled to a spill point. Large sediment slides are common along many passive margins, particularly those outboard of large deltas, glacial troughs, or in areas where major sea-level changes induce slope instability.

The character of sandy SGF deposits will depend on the type of source (e.g., poorly sorted heterolithic delta versus well sorted littoral supply) and the importance of autocyclic (e.g., delta switching) versus allocyclic processes (e.g., prominent sea-level falls, glacial–interglacial climate cycles). Passive-margin successions are typically dismembered during later plate convergence and orogeny, so that their interpretation at outcrop can be difficult. In contrast, deposits that remain buried in a mature passive margin prism (e.g., offshore Brazil) or in failed rifts (e.g., North Sea) are increasingly available for detailed study using cores, 3D seismic surveys and sophisticated processing of these datasets (e.g., amplitude and other horizon maps).

CHAPTER TEN
Subduction margins

(a) (b)

(a) Laminated volcaniclastic siltstones and sandstones showing Facies B2.1 and D2.3 normal grading, plane-parallel lamination, and load structures (interval 322-C0012A-43R-5, 30–56 cm). (b) Facies F2.2 contorted and disturbed strata of (originally laminated) volcaniclastic sandstones involved in a ~50 cm-thick interval (interval 322-C0012A-45R-1, 71–106 cm), showing isoclinal folding, attenuation of beds and sandstone injections). Kashinosaki Knoll (IODP Expedition 322 subduction inputs site) ~10 km south of Nankai Trough, Shikoku Basin, offshore Japan.

10.1   Introduction

This chapter considers forearcs, backarcs and marginal basins with reference to deep-marine sedimentation patterns. The modern examples described in this chapter are chosen to reflect the complex spatial association of such basins associated with active convergent margins. A more wide-ranging treatment of active margins may be found in books written or edited by Burk and Drake (1974), Talwani and Pitman (1977), Watkins *et al.* (1979), Leggett (1982), Kokelaar and Howells (1984), Nasu *et al.* (1985), Allen and Homewood (1986),

Deep Marine Systems: Processes, Deposits, Environments, Tectonics and Sedimentation, First Edition. Kevin T. Pickering and Richard N. Hiscott.
© 2016 Kevin T. Pickering and Richard N. Hiscott. Published 2016 by John Wiley & Sons, Ltd.
Companion Website: www.wiley.com/go/pickering/marinesystems

Coward and Ries (1986), Moore (1986), Busby and Ingersoll (1995), Bebout *et al.* (1996), Dixon and Moore (2007), Draut *et al.* (2008), Brown and Ryan (2011) and Frisch *et al.* (2011). An overview of various aspects of sea-level changes at active margins is provided in the volume edited by Macdonald (1991). Tarney *et al.* (1991) examine the behaviour and influence of fluids in subduction zones. Taylor and Natland (1995) present an interdisciplinary synthesis of results of the Ocean Drilling Program and associated site surveys in the trench–arc–backarc systems in the western Pacific. This edited volume contains 17 multi-authored articles that provide an excellent source of information on the volcanic, fluid, sedimentary and tectonic processes occurring in active margins and marginal basins. Bebout *et al.* (1996) evaluate many aspects of interdisciplinary collaboration in convergent-margin research. This book includes an examination of the critical parameters affecting the dynamics of various subduction zones, the use of shallow and deep earthquake records in studying the evolving structural and mechanical state of subduction zones, and a top-to-bottom consideration of fluxes of energy and matter during subduction. Eiler (2004) examines recent developments in research within the 'subduction factory' using datasets and numerical models. This book contains useful summaries of the geophysics, geochemistry and magmatism at convergent margins. In particular, it reviews the widely studied active convergent margins of the Izu–Bonin–Mariana, Central American and Aleutian arc-related margins. Its perspective tends to focus on the role of water in the Earth's interior, and our understanding of the location and driving force of arc magmatism.

Recent books that examine particular orogenic belts, with their deformed and uplifted deep-marine sedimentary rocks, include the Uralides (Brown *et al.* 2002), the Shimanto Belt, Japan (Taira 1988), the Alaskan orogens (Freymueller *et al.* 2008), the Pacific Rim Kamchatka region (Eichelberger *et al.* 2007), the Himalayas (Treloar & Searle 1995) and active margins and marginal basins of the western Pacific (Taylor & Natland 1995).

A major source of data on modern active margins has come from the Deep Sea Drilling Project (DSDP), the Ocean Drilling Program (ODP), the Integrated Ocean Drilling Program and more recently the International Ocean Discovery Program (IODP). Since publication of our previous book (Pickering *et al.* 1989), the past 25 years have witnessed many ODP and IODP legs that have drilled active convergent margins, and in particular accretionary prisms. Amongst these, the main active margins that have been drilled include: Cascadia margin (ODP Legs 146, 204); Costa Rica margin (ODP Legs 170, 205); Barbados accretionary prism (ODP Legs 110, 156, 171A); Izu–Mariana margin (ODP Legs 126, 185), and Nankai accretionary prism (ODP Legs 131, 190, 196, IODP NanTroSEIZE Expeditions 314, 315, 316, 319, 322, 333, 338 and 348). The Nankai Trough Seismogenic Zone Experiment (NanTroSEIZE) is designed to investigate fault mechanics and seismogenesis along a subduction megathrust, with objectives that include characterising fault slip, strain accumulation, fault and wall rock composition, fault architecture and state variables throughout an active plate boundary system. To achieve these objectives, drilling has been focussed on deepening IODP Site C0002 during Expeditions 338 and 348 located in the Kumano forearc basin (e.g., Moore *et al.* 2009a; Tobi *et al.* 2009; Moore *et al.* 2013a, 2014; Expedition 348 Scientists & Scientific Participants 2014).

Zones of plate convergence tend to be dominated by thrust belt tectonics (*cf.* Boyer & Elliott 1982; Ellis 1988) and sedimentation, together with magmatic arc activity. Where plate convergence is long-lived, orthogonal or oblique, and there is substantial terrigenous sediment supply, then very wide accretionary imbricate systems

may form with widths up to several hundred kilometres. Examples include the Lesser Antilles and Makran accretionary prisms. Active convergent margins can involve: (i) subduction-accretion, with the consumption of oceanic crust to promote arc volcanism, or (ii) continental (or arc) collision-underplating, typically associated with foreland basin thrust systems where neither subduction nor arc volcanism occurs. Examples of the latter include the present-day Timor Trough and Taiwan Foreland Basin (Section 11.2).

There are many examples of basins at convergent margins that are not clearly forearc, backarc/marginal, or foreland basins. For example, the deep eastern Mediterranean, characterised by basin-plain sheet sedimentation, is still tectonically active even though typical subduction processes probably ceased at ~5 Ma with the disappearance of all oceanic crust between Turkey and Africa. Plate convergence continues with only limited under-thrusting of Africa along the Cypriot Arc, but there are regional deformation zones within a possible 300 km-wide band stretching eastward from the Herodotus Abyssal Plain along the northern edge of the African and Arabian Plates (Woodside 1977). Also, halokinesis to form salt pillars and anticlines is an active process in the Herodotus Basin, and the north–south alignment of many diapirs, oblique to the regional trend of the Mediterranean Ridge and the edge of the Nile Cone, suggests basement control, with possible continuation of the north–south structural grain of the North African crystalline basement (Woodside 1977). Thus, while essentially horizontal basin-plain or abyssal-plain sedimentation is occurring, complex compressional tectonics and halokinesis are controlling the stratigraphy.

The Japanese island arc systems, with their range in types of active convergent plate margin, including arc–arc collision and associated backarc/marginal basins, are amongst the most studied worldwide, both onland and offshore (Fig. 10.1). It is for this reason that many of the examples used to illustrate tectonic and sedimentary processes in this chapter are from this geographic region.

Subduction zones are responsible for ~90% of global seismic energy release, generating damaging earthquakes and tsunamis with potentially catastrophic consequences, particularly for densely populated coastal regions (e.g., Lay *et al.* 2005). Understanding the processes that control the nature and distribution of slip along plate boundary faults is essential for evaluating earthquake and tsunami hazards.

NanTroSEIZE is a multi-expedition, multi-stage project focussed on understanding the mechanics of seismogenesis and rupture propagation along plate boundary faults (e.g., Saito *et al.* 2009), these being tectonic processes that are important to understanding the nature of the sediments and stratigraphy of deep-marine deposits that accumulate at active plate boundaries. In recent years, this major scientific programme has been implemented by the Integrated Ocean Drilling Program (IODP) in a coordinated effort to sample and instrument the plate boundary system at several locations offshore from the Kii Peninsula, Japan. Its main objectives are to improve the understanding of the aseismic–seismic transition of the Nankai accretionary-prism megathrust fault system, the mechanics of earthquake and tsunami generation, and the hydrologic behaviour of the plate boundary and subduction margin (Tobin & Kinoshita 2006a, b). This programme involves a combination of riser and riserless drilling, long-term observatories and associated geophysical, laboratory and numerical modeling efforts.

The NanTroSEIZE programme aims to test the following hypotheses (Tobin & Kinoshita 2006a, b; Saito *et al.* 2009): (i) that systematic, progressive material and state changes control the onset of seismogenic behaviour on subduction thrusts; (ii) that subduction

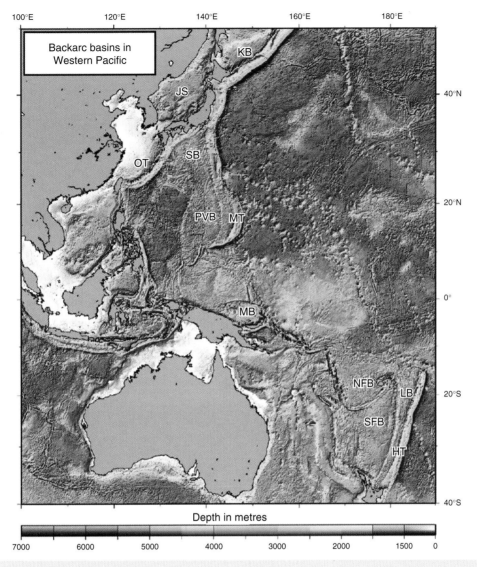

Fig. 10.1 Distribution of backarc basins in the western Pacific. Most backarc basins are located along the western margin of the Pacific Ocean. HT = Havre Trough, JS = Japan Sea, KB = Kurile Basin, LB = Lau Basin, MB = Manus Basin, MT = Mariana Trough, NFB = North Fiji Basin, PVB = Parece Vela Basin, SB = Shikoku Basin, SFB = South Fiji Basin and OT = Okinawa Trough. Basins whose names appear in grey are not currently active rifting/spreading systems. From Martinez *et al.* (2007).

megathrusts are weak faults (i.e., they slip under relatively low stress); (iii) that plate motion is accommodated primarily by co-seismic frictional slip in a concentrated zone; (iv) that physical properties of the plate boundary system change with time during the earthquake cycle (Kimura *et al.* 2011, Strasser *et al.* 2011); (v) that a laterally extensive 'megasplay' fault system slips in discrete events that may include tsunamigenic slip during great earthquakes and (vi) that the fault remains locked during the inter-seismic period and accumulates strain. To test these hypotheses, the initial conditions within subducting sediment and basalt must be known, beginning at 'reference sites' seaward of the deformation front (e.g., IODP Expedition 322; Underwood *et al.* 2009). Although these hypotheses are essentially outside the remit of this book, they involve an understanding of the physical properties of the sediments (that links to depositional processes) and the deep-water stratigraphy. Some of these aspects are explored later in this chapter.

10.2 Modern subduction factories

10.2.1 Forearcs

Wherever plates converge towards subduction zones, a range of deep-marine sedimentary basins may form between the trench and the volcanic arc. Deep-marine forearc basins range from relatively small accretionary-prism slope basins, to large forearc basins such as the Barbados and Tobago basins in the Lesser Antilles forearc, where dimensions are larger than 100 km, and sediment infill >2000 m and 4000 m thick, respectively. Dickinson and Seely (1979) defined the following basins, all of which generally contain substantial thicknesses and volumes of deep-marine sediments: (i) intra-massif basins within and upon basement terranes of the arc massif; (ii) residual basins on oceanic or transitional crust between the arc massif and the site of initial subduction; (iii) accretionary-prism basins upon the accreted elements of the developing subduction–accretion complex and (iv) basins constructed both upon the arc massif and accreted subduction complex. Basins may also develop that are intermediate between types (a)–(d). Dickinson and Seely (1979) fitted these basin types into a range of models for forearc evolution (Fig. 10.2).

The importance of irregular plate boundaries, with promontories and re-entrants in continental crust, has long been recognised as a major control on the type of forearc basin that develops (e.g., Thomas 1977; Seely 1979; Hiscott *et al.* 1986). However, the volume of forearc

sediments, and therefore the size of an accretionary prism, will be influenced largely by factors such as the magnitude of river drainage into the forearc region and the longevity of subduction–accretion processes.

Theoretical models have been proposed for the development of forearcs by frontal accretion and underplating; these take account of the internal deformation in accretionary prisms to explain convergent-extensional wedges (e.g., parts of the Middle America Trench margin) *versus* convergent-compressional wedges (e.g., parts

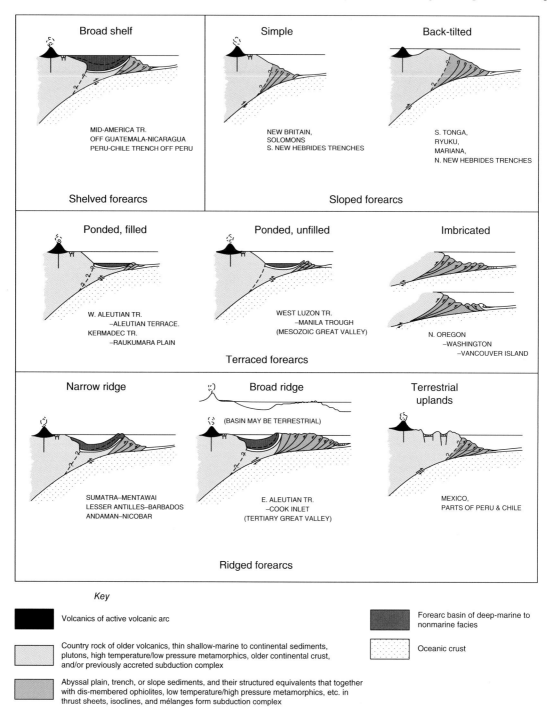

Fig. 10.2 Models of modern forearcs. From Dickinson and Seely (1979). The arc may be submerged below sea level, although in these diagrams it is shown as emergent.

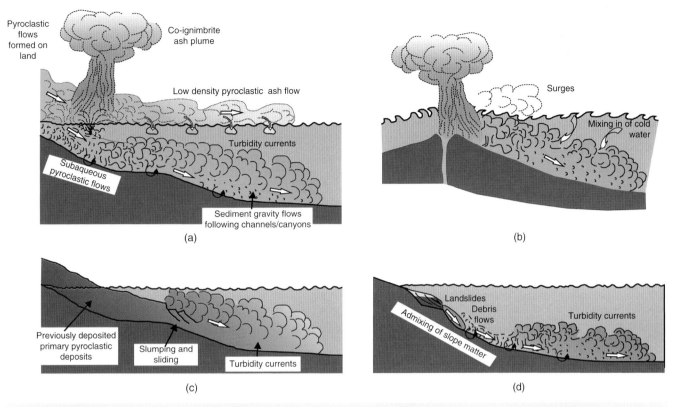

Fig. 10.3 Range of depositional processes that could explain submarine tuffaceous sandstones at IODP Expedition 322 Site C0011. (a) and (b) formation in the course of active eruption will generate deposits predominantly composed of juvenile pyroclastic material as observed in the tuffaceous sandstones at IODP Expedition 322 Site C0011. (c) and (d) post-eruptive reworking most likely creates deposits containing an heterogeneous mixture of volcaniclastic material. (a) Primary pyroclastic flows entering the sea and continuing as SGFs (e.g., turbidity currents) in submarine canyons and channels until the basin plain. (b) Collapse of a subaerial or submarine eruption column from shallow-water vent generating a SGF. (c) Collapse of a subaerial or submarine eruption column from shallow-water vent generating a SGF. (d) SGFs (e.g., turbidity currents) generated by sediment failure due to catastrophic slope-failure processes. From Schindlbeck *et al.* (2013), modified after Saito *et al.* (2010) (*cf.* Freundt 2003; Freundt & Schmincke 1998; Cas & Wright 1991).

of the Nankai Trench prism). For example, the 'critical taper model' of Platt (1986) suggests that if accretionary prisms are disequilibrated by elongation (frontal accretion), then they will tend to an equilibrium geometry and taper by internal folding and thrusting. In contrast, over-shortened prisms will tend to equilibrium by elongation mainly through extensional normal faulting. In the latter case, thicker successions of undeformed sediments might develop upon the accretionary prism, essentially controlled by extensional, normal, growth faulting at shallow depths. Accretionary prisms may grow both by the addition of sediment at the toe of a prism or from 'underplating'. *Underplating* is used to describe two different scales of accretion from below a thrust system at active convergent margins: (i) crustal thickening from continental and/or anomalous crust, including arcs, being accreted along the basal thrust, and (ii) the addition of oceanic-plate sediments from a subducting slab, by being offscraped and attached to the underside of an accretionary prism at depth.

In the following sections we consider the tectonics and sedimentation that occur in the main elements of a forearc; that is, trenches (and the preservation potential of a trench stratigraphy), accretionary prism-top (slope) basins and larger forearc basins. An idealised summary model for forearc sedimentation is presented. There is a section dealing with wet-sediment injection structures in forearcs. There is also a section on backarc/marginal basins. Examples from both modern and ancient convergent margins are considered.

Many of the sediment transport and depositional processes and resulting facies that occur at active convergent margins are described in Chapters 1 and 2. Figure 10.3, however, shows the range of depositional processes that could explain tuffaceous or volcaniclastic sandstones (Schindlbeck *et al.* 2013). There are essentially four main processes: (i) Primary pyroclastic flows entering the sea and then continuing as SGFs in submarine canyons and channels out onto the basin plain; (ii) pyroclastic flows entering the sea, triggering submarine landslides and debris flows, and then continuing as SGFs; (iii) post-eruptive redeposition at some time after an eruption because of catastrophic slope-failure processes and (iv) collapse of a subaerial or submarine eruption column, leading to SGFs. For (ii) and (iii), it is likely that there is a considerable degree of incorporation of previously deposited (terrigenous) sediments, whereas for (i) and (iv), pyroclasts are the predominant component in any deposit.

10.2.2 Trench sedimentation

Using the Middle America Trench as a model, Underwood and Bachman (1982) were amongst the first to describe facies and large variability in depositional style within trench environments (Fig. 10.4). Deep-marine trench sediments may be derived from three separate source areas: (i) oceanic-plate sediments passively conveyed by plate motions into a trench during subduction; (ii) lateral sediment

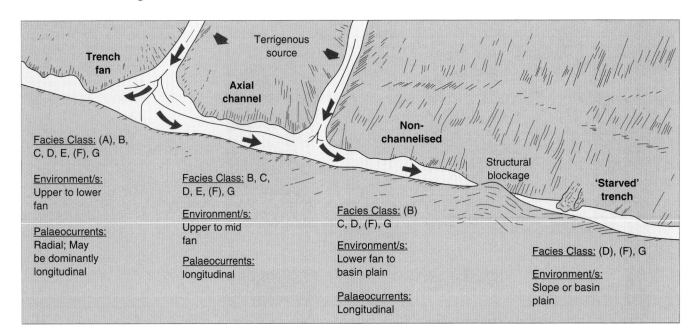

Fig. 10.4 Conceptual diagram modified from Underwood and Bachman (1982) to show the sediment gravity flow (SGF) facies and palaeocurrent patterns predicted for sediment bodies on the trench floor, looking towards the emergent volcanic arc. Blue = SGF pathways. Minor SGF facies are shown in parentheses. Also indicated are submarine-fan facies associations (Mutti & Ricci Lucchi 1972 subdivisions, *cf.* Section 8.1).

input from the forearc and (iii) axial transport of sediments along the trench, possibly from distant areas. Flexure and associated extensional faulting of the subducting oceanic plate will lead to low-angle unconformities, with onlap between oceanic plate sediments and younger strata.

Oceanic-plate sediments tend to occur as a thin veneer of pelagites, hemipelagites and fine-grained turbidites (Facies Classes D, E and G) above oceanic basaltic basement, associated with Facies Group G3 chemogenic sediments such as cherts and manganiferous sediments, on what has been called the 'pelagic plate' (Schweller & Kulm 1978; Lash 1985). Examples of such fine-grained successions, spanning geologically long time intervals, occur on the Pacific and Cocos Plates where sediment accumulation rates are slow, typically from 2–5 mm kyr^{-1} (or 2–5 m Myr^{-1}), and below the carbonate compensation depth (CCD) (Schweller & Kulm 1978). In contrast, in the Indian Ocean, the oceanic plate is buried beneath a thick succession of Bengal Fan sediments that are being conveyed obliquely into the Java Trench. In the north, the Indian Plate is covered by up to several thousand metres of terrigenous SGF deposits (Curray & Moore 1974; Curray *et al.* 1979; Karig *et al.* 1979; Moore *et al.* 1982). An essentially comparable, thick, clastic succession fed northwards from the South American continental margin onto the westward-subducting Atlantic Plate is at least several hundred metres thick over parts of the Lesser Antilles Trench (Westbrook 1982; Biju-Duval *et al.* 1984; Brown & Westbrook 1987; Moore, Mascle *et al.* 1987; Mascle, Moore *et al.* 1988). In the northeast Pacific, coarse clastic sediments, including submarine fans, cover large parts of the Pacific Plate and the Juan de Fuca Plate, and are being subducted below both the Aleutian and Cascadia accretionary complexes, respectively (e.g., the modern Surveyor and ancient Zodiac Fans: Moore *et al.* 1983; Stevenson *et al.* 1983; Harbert 1987; Geist *et al.* 1988; Carlson *et al.* 2006; Jaeger *et al.* 2011; Reece *et al.* 2011). Such thick marine clastic successions on some subducting oceanic plates lead us to prefer the general term 'oceanic plate sediments' rather than 'pelagic plate sediments', because in

many cases the sediments are mainly terrigenous turbidites, not pelagites.

Lateral sediment input from the forearc may produce several contrasting trench-floor deposits: (i) a blocky seafloor, due to slides (Facies Group F2) and debrites (Facies Group A1), either locally derived from the lower slope and inner trench-slope, or from more distant sources such as the upper slope, a forearc ridge and/or shallow-marine shelf areas; (ii) relatively coarse-grained sediments (Facies Classes A, B and C) fed from submarine canyons, channels and gullies to construct trench fans as channel–levée–overbank systems or sheet systems and (iii) smooth-surfaced, relatively fine-grained deposits (Facies Classes D, E and G), for example those supplied by turbidity currents flowing through slope channels, and those deposited from nepheloid layers (see Section 6.2). These latter fine-grained sediments infill and/or mantle uneven topography and tend towards sheet systems with onlap of peripheral slopes.

Large submarine canyons may funnel coarse terrigenous material directly to the trench floor, effectively bypassing the lower slope; examples of such canyons include those documented from the Middle America Trench by Underwood and Bachman (1982) (Figs 7.15 and 10.4). These authors also describe, from the Middle America Trench, tectonic ridges along the trench slope that block and/or deflect smaller canyons so that coarse sediments are trapped within upslope basins. In the Central Aleutian Trench, not associated with well developed canyons, Holocene volcaniclastic sand layers cored in the trench suggest that unconfined, non-channelised, turbidity currents had sufficient velocities to climb the trench-slope break and transport sediment across the lower trench-slope into the trench (Underwood 1986). Thus, although canyons provide a means of effective, efficient, slope bypass for coarse sediments, unconfined flows may also bypass the forearc.

Continent–continent, arc–continent and arc–arc collision will result in trench sedimentation that may incorporate clastic/volcaniclastic material from opposing margins in varying proportions. The submarine flanks of volcanoes will also be associated with clastic aprons (e.g., Expedition 340 Scientists 2012; Le Friant *et al.* 2015; Trofimovs *et al.* 2013; Fig. 10.3). Such laterally supplied sediments may then be funnelled axially along the trench.

Figure 10.5 summarises the eight principal types of trench fill recognised in this book, together with their most characteristic facies classes. The eight types are not mutually exclusive, but serve to emphasise the range of trench fills. Indeed, one trench may contain

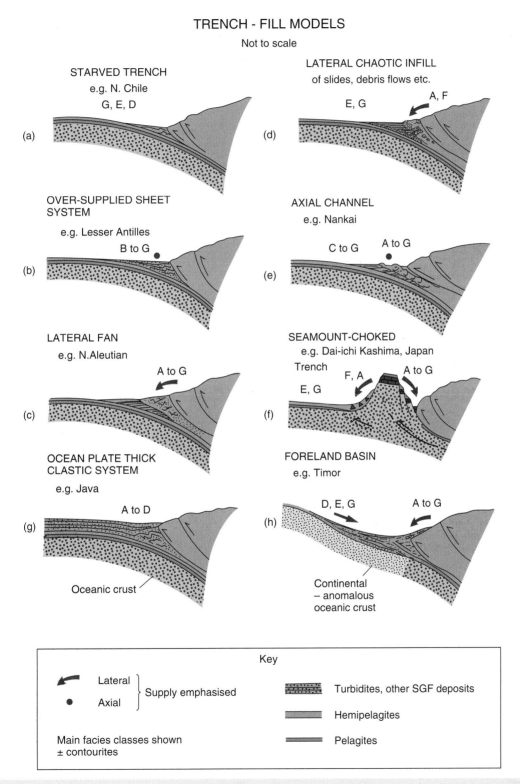

TRENCH - FILL MODELS

Not to scale

(a) STARVED TRENCH
e.g. N. Chile
G, E, D

(b) OVER-SUPPLIED SHEET SYSTEM
e.g. Lesser Antilles
B to G

(c) LATERAL FAN
e.g. N.Aleutian
A to G

(d) LATERAL CHAOTIC INFILL
of slides, debris flows etc.
E, G
A, F

(e) AXIAL CHANNEL
e.g. Nankai
C to G
A to G

(f) SEAMOUNT-CHOKED
e.g. Dai-ichi Kashima, Japan Trench
F, A
A to G
E, G

(g) OCEAN PLATE THICK CLASTIC SYSTEM
e.g. Java
A to D
Oceanic crust

(h) FORELAND BASIN
e.g. Timor
D, E, G
A to G
Continental – anomalous oceanic crust

Key

Lateral
Axial
} Supply emphasised

Main facies classes shown ± contourites

Turbidites, other SGF deposits
Hemipelagites
Pelagites

Fig. 10.5 Trench-fill models to show the main facies classes and predominant transport paths in various settings. The facies classes are listed in order of volumetric importance. See text for explanation.

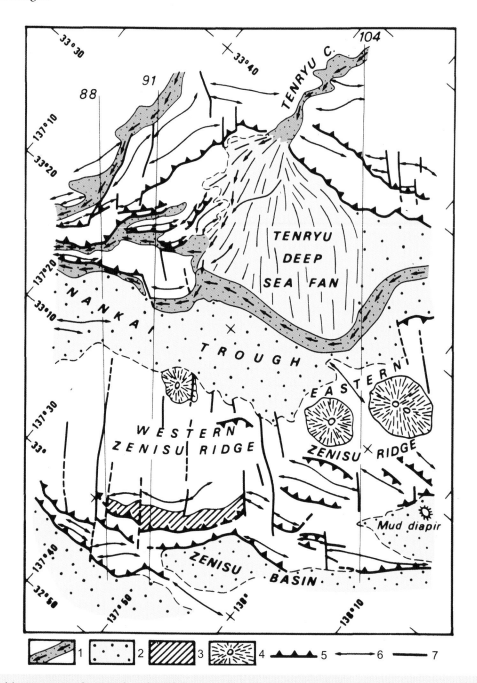

Fig. 10.6 Map of the eastern Nankai Trench to show the main structural and sedimentological features, from Box 5 of the Franco-Japanese KAIKO Project and based on Seabeam bathymetry and single-channel seismic reflection profiles. 1, deep-marine channel; 2, trench fill; 3, probable outcrop of basement; 4, igneous volcano; 5, thrust; 6, fold; 7, unspecified fault. From Le Pichon *et al.* (1987a).

all eight types at various places along the trench axis. The eight end-member types are: (i) starved trench; (ii) sheet system; (iii) trench fan; (iv) chaotic lateral infill from forearc; (v) axial trench channel; (vi) seamount-choked; (vii) subducting oceanic-plate thick clastic system and (h) lateral influx of clastics/volcaniclastics from two opposing continental and/or arc margins in a deep-marine foreland basin (Fig. 10.5).

Axial transport of sediments in trenches may deposit any facies class, although the most common classes are A, B, C, D and E (Table

2.2, Fig. 2.4). Axial channels will tend to develop in topographic lows along the inner trench wall in starved trenches, but on the outer side of a trench if lateral sediment input from the forearc is fast and volumetrically large. Examples of axial channels include the north-ward axial drainage in the Chile Trench (Thornburg & Kulm 1987; Fig. 7.54), and the westward drainage from the triple junction of the Pacific, Philippine and Eurasian plates (mainly from the Fuji River), along the Nankai Trench (Shimamura 1986; Taira & Niitsuma 1986; Le Pichon *et al.* 1987a, b) (Fig. 10.6). Any substantial lateral influx of sediments, such as a trench fan, will tend to divert an axial channel toward the outer (oceanward) floor of the trench – as observed in parts of the Chile Trench by Thornburg and Kulm (1987), and in the Nankai Trench opposite the Tenryu Fan (Le Pichon *et al.* 1987b) (Fig. 10.6). In the Peru–Chile Trench, the BioBio Canyon forms a large regular conical fan that controls the course of the channel that curves around it. The channel has no defined landward wall as fan sediments onlap the channel floor with a low gradient. The fan deposits show a particular reflection pattern characterised by densely spaced hyperbolae, interpreted to reflect small-scale topography such as megaripples, too small to be resolved by the mapping tools available in the 1980s. Undulating microtopography may be due to sediment transport within sediment-laden bottom water. The most prominent feature on the fan is a feeder channel, which forms banks.

Most sediment routing from forearc to trench-floor involves the long-distance axial transport of sediments both within trench axial channels and as less-confined sediment gravity flows. Inferred examples of ancient trench-axis routing include the Andaman Island sandy concentrated density-flow deposits (Fig. 10.7).

Facies Classes A–G occur in all eight trench-fill types (Fig. 10.5), but in varying proportions that reflect the dominant control on sedimentation. For example, starved trenches, where there is little clastic input, will tend to comprise Classes D, E and G, whereas in seamount-choked trenches, near the subducting and/or accreting seamount, Classes A and F may predominate as rockfall, debris-flow and slide deposits. In such cases, the finer grained matrix for any debrites/olistostromes and slides may be derived from pelagites, hemipelagites and fine-grained turbidites that mantled the seamount.

Thermohaline circulation, as ephemeral to semi-permanent ocean currents, may rework trench floor, or forearc slope, sediments into contourites. Thornburg and Kulm (1987) document silt and sand laminae, winnowed from hemipelagic muds and distal turbidites, that are particularly well developed in the sediment starved, constricted parts of the Chile Trench where geostrophic currents accelerate between the steep inner and outer trench walls.

Axial gradients within trenches, combined with the linear basin shape, conspire to favour pronounced asymmetry in axial

sedimentation patterns away from sediment-entry points such as submarine canyons. Thornburg and Kulm (1987), in a study of the Chile Trench, derive proximal to distal trench stratigraphies from 'depositional' to 'erosional' fans, to sheeted then ponded basins (Fig. 10.8). Although we would avoid defining depositional *versus* erosional fans, apparently Thornburg and Kulm (1987) use these terms only to emphasise the predominant interpreted processes operating on such fans. Near sediment entry points, fast rates of sediment accumulation favour fast fan aggradation, together with relatively rapid channel avulsion/migration, and lead to an erosional fan system (Fig. 10.8a, b). More distal trench environments will contain fewer and/or more shallow channel–levée–overbank systems, and will pass into essentially ponded basins, where a sheet system is dominated by fine-grained sediments and, possibly, contourites (Fig. 10.8c, d).

SeaMARC-II side-scan sonar and seismic reflection records from the southern Chile Trench, between 33°S–41°S latitude (Fig. 10.9), image steep erosional escarpments up to 400 m in relief that extend seaward across the trench basin from the mouths of submarine canyons (Thornburg *et al.* 1990). The scarps bisect trench fans into paired lobes of contrasting morphology where sediment gravity flows either follow or oppose the gradient of the axial trough. Fan-lobes are depositional and constructional up-gradient (south) from the canyon mouths, and comprise aggraded channel–levée complexes, smooth and conformable sediment drapes and crescentic levées rimming the headwalls of erosional scarps. Fan-lobes are carved and dissected by erosional processes down-gradient (north) from the canyon mouths. They exhibit amalgamated lag pavements, composite sediment lobes, longitudinal furrows, braided channels and canyon-mouth bars. Thick, structureless-to-laminated sand (Facies Class B) and gravel (Facies Class A) with abundant scour surfaces were observed from the erosional fan-lobes, whereas fine-grained turbidites with expanded hemipelagic intervals typify the depositional fan-lobes.

Fan distributary channels, axial channels, slump scars and erosional gullies are largely localised along structural features (Thornburg *et al.* 1990). Normal faults propagate through the sedimentary

(a)   (b)

Fig. 10.7 Example of trench-axis deposits, as alternating packets of thick-bedded sandy concentrated density-flow deposits and thinner-bedded, finer-grained turbidites in the Andaman Islands, Indian Ocean (a) Beds young to right (east). Human scale on very thick-bedded sandstone bed in centre. Note that the close-up (b) shows thoroughly ripple-laminated fine-grained sandstones with climbing-ripple lamination. The predominant palaeoflow in these Miocene SGF deposits is southwards, away from the Himalayan collision zone.

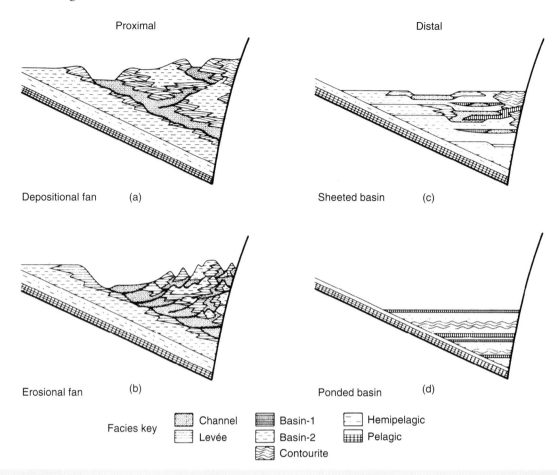

Proximal

Distal

Depositional fan   (a)

Sheeted basin   (c)

Erosional fan   (b)

Ponded basin   (d)

Facies key

Channel	Basin-1	Hemipelagic
Levée	Basin-2	Pelagic
	Contourite	

Fig. 10.8   Diagrams to show the range in trench-fill stratigraphy and environments based on lithologic, morphologic and seismic data from the Chile Trench (from Thornburg & Kulm 1987). The proximal 'depositional' and 'erosional' fans to distal 'sheeted' systems develop along the axis of the trench. In some cases, submarine canyons may debouch sediments directly into sheet systems without the development of a channelised sediment-dispersal system.

cover to create elongate depressions on the seafloor that capture the higher velocity central threads of SGFs. Orthogonal fault sets within the deposits of the trench basin, paralleling the spreading and transform structures of the extinct Pacific-Farallon Rise and the Chile Rise, appear to be reactivated during subduction by flexure of the oceanic basement along the outer wall of the trench basin. Uplifted thrust ridges, generally restricted to a narrow zone along the base of the deformation front, are dissected by distributary channels, and channel courses are locally deflected seaward of these propagating structures. Transform-oriented basement ridges, associated with strike displacements of the axial channel and vertical faults in the trench basin, may accommodate renewed strike-slip motion as the ridges enter the subduction zone and thereby influence the exit points of submarine canyons to the trench floor (Thornburg *et al.* 1990).

Sufficiently fast sediment supply, with relatively stable trench topography and gradients, will allow the axial progradation of clastic systems, potentially generating single or stacked coarsening-upward trends. However, the complexities of changing paths of sediment supply, varying eustatic sea level and tectonic sedimentary processes that alter the shape and gradient of the trench floor and walls, will ensure a far more complicated stratigraphy.

The lateral confinement of SGFs, and in particular the long-distance sediment transport by gravity flows in submarine trenches, permit the more dilute (flow-stripped) part of flows to ride up the basin margins and be deflected/reflected back towards the trench axis. This process was first documented from a modern trench environment (Nankai Trough) by Pickering *et al.* (1992, 1993a, b) (Figs 10.10, 10.11).

Although the spatial variability in trench deposition is now appreciated, many trench-fills appear to be characterised by a first-order coarsening-upward trend within a subduction-related trench-wedge overlying oceanic-plate sediments and capped by the toe of the accretionary prism. This coarsening-upward trend was first described in detail from a modern subduction-accretion system, the Nankai prism, at ODP Leg 131 Site 808 (Pickering *et al.* 1993a, b) (Fig. 10.12). The variability and complexity of the stratigraphy at such sites, created by higher frequency changes in sediment calibre (e.g., driven by global climate change and seismic events), challenges the previous view that a coarsening-upward trend is diagnostic of the vertical succession from subducting oceanic crust, oceanic plate sediments, and overlying trench sediments (Piper *et al.* 1973). Additionally, there is commonly a sharp textural contrast between coarse-grained trench sediments and overlying fine-grained lower-slope deposits,

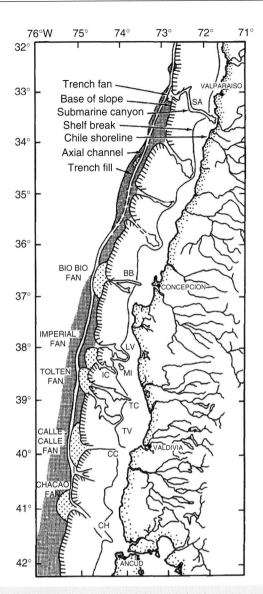

Fig. 10.9 Part of the Peru–Chile Trench showing range of deep-marine environments. North of San Antonio Canyon at 33°S latitude, the volume of the trench basin decreases abruptly, and the sediments are ponded behind the subducting oceanic plate. South of the Strait of Ancud at 42°S latitude, the trench is non-channelised and is filled by relatively unconfined (sheet-like) deposits. Trench sediments are in heavy stipple; trench fans in light stipple; the axial channel is unshaded; BB = Bio Bio Canyon; CC = Callecalle Canyon; CH = Chacao Canyon; IC = Imperial Canyon; LV = Lleulleu Valley; MI = Mocha Island; SA = San Antonio Canyon; TC = Tolten Canyon; TV = Tolten Valley. From Thornburg *et al.* (1990).

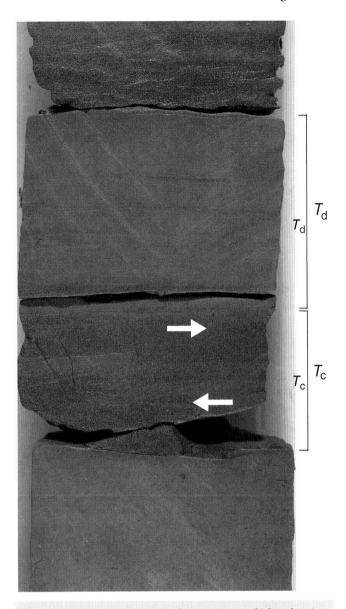

Fig. 10.10 Core photograph of Facies C2.4 turbidite showing opposing flow directions in basal fine-grained sand, and ripple division of bed. Note mud drapes within the T_c division. The overlying silt and silty mudstone are laminated (T_d division) and lack bioturbation, in contrast to surrounding intervals that lack these unusual turbidites. Interval 131-808C-28R, 134–142 cm.

10.2.3 Accretionary prisms

Accretionary prisms develop where sediments on a subducting oceanic lithospheric plate are off-scraped above a detachment or décollement surface in a bulldozer-like manner to create a dynamically deforming fold-and-thrust belt. The plate boundary below a growing accretionary prism comprises both a shallower 'aseismic' (frontal part), and a deeper seismic (or 'seismogenic') component. Frontal accretion processes result in younger sediments defining the outermost part of the accretionary prism, with the oldest sediments forming the innermost parts. Tectonic deformation means that the

a relationship that has been explained in part by the role of submarine canyons in sedimentation along the trench slope. Specifically, canyons cause sediment bypassing and upslope trapping, so that deposits on the lower slope are generally finer grained, consisting largely of hemipelagic muds and lutite turbidites (Underwood & Karig 1980).

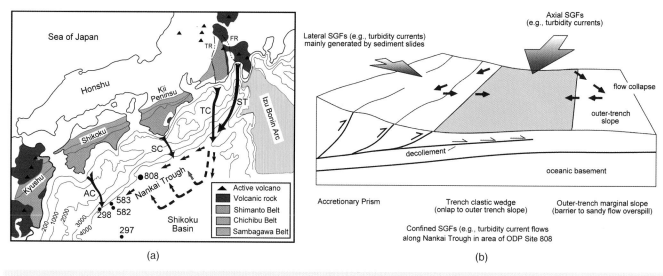

Fig. 10.11 (a) Location map of ODP Site 808 and neighbouring DSDP sites occupied by Legs 37 and 87A, Nankai Trough. Major sediment entry points are as follows: ST = Suruga Trough; TC = Tenryu Canyon; SC = Shiono-Misaki Canyon; SC = Ashizuri Canyon. Bathymetric contours are in metres. FR = Fuji River; TR = Tenryu River. Regions of volcanic rocks include Neogene–Quaternary mafic-intermediate to felsic compositions. (b) Model of confined turbidity-current flow in the vicinity of ODP Site 808. Origin and cause of diverse turbidity-current flow vectors are: (i) lateral input of turbidity currents from upslope on the accretionary prism, followed by high-angle reflection off trench outer slope, and (ii) collapse of axially-flowing and oblique turbidity currents so that the body of the flow on the trench outer slope is deflected downslope towards the trench axis. From Pickering *et al.* (1992, 1993a, b).

older (inner) parts of the accretionary prism are considerably more lithified and tend to be highly deformed with steeper structures compared to the younger (outer) parts. In the basal parts of an accretionary prism, material may be added by a process of underplating. The accretionary prisms in Barbados (Figs 10.13, 10.14), Nankai (Fig. 10.15–10.17), and Cascadia (Fig. 10.18, 10.19) are amongst the best documented worldwide.

The northern Barbados accretionary prism is the leading edge of the Caribbean Plate that is being underthrust by the Atlantic Ocean floor at rates estimated between 20–40 km Myr^{-1} (e.g., Dorel 1981). On the west, the Lesser Antilles defines the volcanic arc, whereas east of the arc the island of Barbados is a high point on the forearc accretionary prism. Frontal structures imaged south of the Tiburon Rise include long-wavelength folds, widely spaced ramping thrust faults and extensive décollement reflections (e.g., Bangs & Westbrook 1991; Westbrook & Smith 1983). North of the Tiburon Rise, trench sediment thickness is much thinner, prism thrusts are more closely spaced (Biju-Duval *et al.* 1982; Westbrook *et al.* 1984), and the accretionary prism reaches at least 10 km thick and 120 km wide in addition to a 50 km-wide forearc basin to the west (Bangs *et al.* 1990; Westbrook *et al.* 1988). Thus, the accretionary prism forms a wide low-taper wedge.

Deep Sea Drilling Project (DSDP) Leg 78A and Ocean Drilling Program (ODP) Legs 110, 156 and 171A all focussed on the northern flank of the Tiburon Rise. Here the décollement is relatively shallow, and the dominantly hemipelagic/pelagic sedimentary section offers good drilling conditions and good biostratigraphic resolution. In this area, previous drilling documented numerous biostratigraphically defined faults of mostly thrust displacement (Brown *et al.* 1990). The décollement zone becomes better defined in a landward direction and is a shear zone up to 40 m thick at Sites 671/948 (Mascle *et al.* 1988; Shipley *et al.* 1995). Anomalies in pore-water chemistry (Gieskes *et al.* 1990) and temperature (Fisher & Hounslow 1990a) indicate focussed

fluid flow along fault zones and in sand layers. Models simulating this fluid expulsion from the prism suggest that the flow is transient (Bekins *et al.* 1995). The faults are characterised by suprahydrostatic, and locally near lithostatic, fluid pressures.

Six sites along two transects across the Nankai Trough accretionary prism were drilled during ODP Leg 190. Two reference sites at the seaward ends of the Muroto Transect (Site 1173) and the Ashizuri Transect (Site 1177) define the stratigraphic framework of the accreting/subducting Shikoku Basin sedimentary section. A thick section of Miocene concentrated density-flow deposits and turbidites (mainly Facies Classes B, C and D) and smectite-rich mudstone (Facies Classes E and G) is present within the subducting section at the Ashizuri site, but is absent in the correlative section at the Muroto site, instead being represented by hemipelagic mudstones (Facies Classes E and G); variations in lithology, mineralogy and hydrologic properties of the incoming sediments probably contribute to the difference in prism wedge taper between the two transects, while possibly controlling the seismic character of the active plate boundary. The décollement in both transects is localised along a stratigraphic unit (~5.9–7 Ma) within the lower Shikoku Basin facies. This horizon can be correlated across both transects through its magnetic susceptibility signature.

A broad low-chloride pore-water anomaly in the lower Shikoku Basin unit, first identified at ODP Site 808, progressively decreases in magnitude from prism to basin along the Muroto Transect. It is unclear whether this landward-freshening trend is due to *in situ* diagenesis, lateral fluid flow, or a combination of the two.

Ideas for the tectonic evolution of the prism in the Muroto Transect changed as a result of ODP Leg 190. Accretion of a Miocene and Pliocene SGF package (mainly Facies Classes B, C and D) formed a large thrust-slice zone (LTSZ). This event was associated with a shift from a transverse sediment transport system that delivered coarse material (mainly Facies Classes B and C) from the arc to the trench,

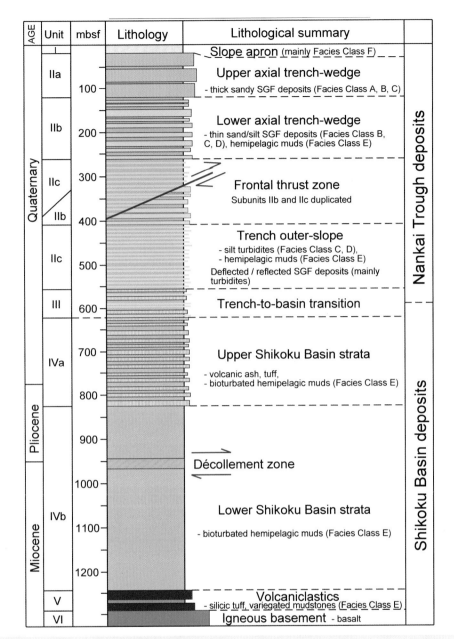

Fig. 10.12 Coarsening-upward trend (Units IIc through IIa) of a subduction-related trench-wedge overlying oceanic-plate sediments and capped by the toe of the accretionary prism, ODP Leg 131 Site 808. From Pickering *et al.* (1993a, b). Many examples, however, do not conform to such a simple coarsening-upward trend (see text for explanation). Note that older parts of this stratigraphic column in the Lower Shikoku Basin deposits are not part of the overall coarsening-upward trend, and that the finer-grained uppermost deposits (Unit I) are lower-slope sediments above the trench stratigraphy.

to an axial transport system that delivered sediment down the trench axis from the Izu collision zone to the east. Growth of the prism from the LTSZ to the toe of the slope (a distance of 40 km) took place rapidly from 2 Ma–Present.

In structural geology, the *vergence* of folds and faults indicates the direction of tectonic transport, hence the direction of overturning in the case of folds and the direction of displacement of the hanging wall slices of thrust faults relative to lower slices. Although the Oregon convergent margin is commonly cited as a type example of a landward-vergent accretionary prism (e.g., Fig. 10.19), this is only true for a relatively small part of the northern Oregon margin.

Generally, the Oregon margin is typified by significant along-strike variability in structural style and prism morphology. In contrast, most of the northern Oregon and Washington accretionary prism is a broad seaward-vergent thrust system with widely spaced folds, and a décollement stepping down to the basement, with virtually all incoming sediment being frontally accreted (Silver 1972; Snavely & McClellan 1987; Mackay *et al.* 1992; Goldfinger 1994; Mackay 1995; Flueh *et al.* 1996), which is a more typical tectonic style in accretionary prisms. This low-taper wedge comprises mainly the Pleistocene Astoria and Nitinat submarine fans (mainly Facies Classes B, C and D) that have been accreting outboard of a narrow, older

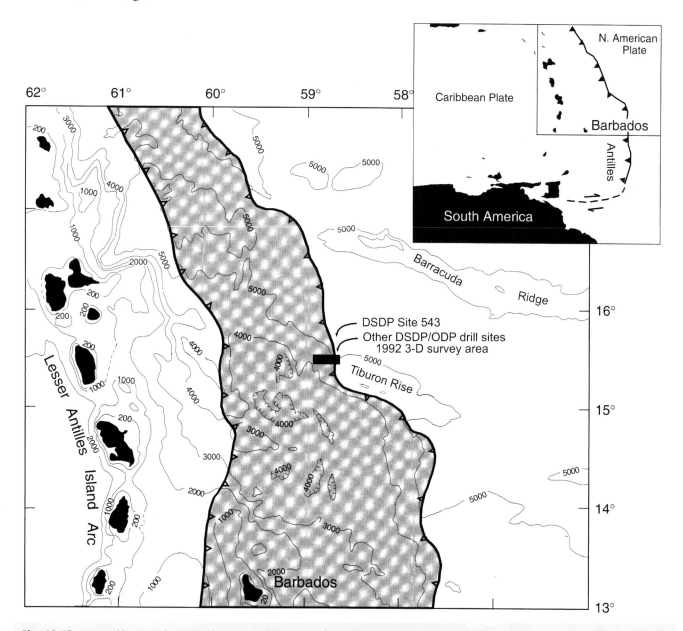

Fig. 10.13 Map of large-scale regional setting and inset map for ODP Leg 171A on the Barbados accretionary prism. Black rectangle locates area of 3D seismic survey and previous drill sites. Shaded region encompasses Barbados accretionary prism with forearc basin to west. Bathymetry is in metres. From Moore *et al.* (2000).

Cenozoic accretionary complex (Goldfinger 2000). This older accretionary complex is situated seaward of an outer arc high and Cenozoic forearc basin. The Eocene Siletzia terrane, an oceanic terrane accreted or perhaps rifted in place during the Eocene (Duncan 1982; Wells *et al.* 1984, 1998; Parsons *et al.* 1999; Haeussler *et al.* 2003), is the basement of the continental forearc, and terminates at the seaward end beneath the forearc basin (Tréhu *et al.* 1994, 1995; Snavely *et al.* 1980). In contrast, the southern Oregon margin is characterised by a steep and narrow chaotic continental slope outboard of the outer arc high and forearc basin, similar to the northern margin, but without the low-taper accretionary prism. Between these distinct provinces, there is a relatively small transitional region typified by the landward-vergent accretionary prism for which the Oregon margin is well known. The steep, narrow, southern margin appears to contain abundant sediment slope failure deposits that dominate the structure and morphology of the continental slope between 42° and 44°N latitude. Arcuate escarpments have been recognised on the Oregon continental slope enclosing regions of hummocky topography, and underlain by detachment surfaces that delineate at least three failure zones encompassing much of the southern Oregon continental slope; beneath the abyssal plain there are widespread subsurface and partially buried debris aprons of Facies Classes A and F (Goldfinger 2000).

Sediment erosion at the toe of the Cascadia accretionary prism was studied by McAdoo *et al.* (1997). Using a combination of geomorphological, seismic reflection, geotechnical data and a series of ALVIN dives in a region south of Astoria Canyon, they investigated the inter-relationship of fluid flow and slope failure in a series of

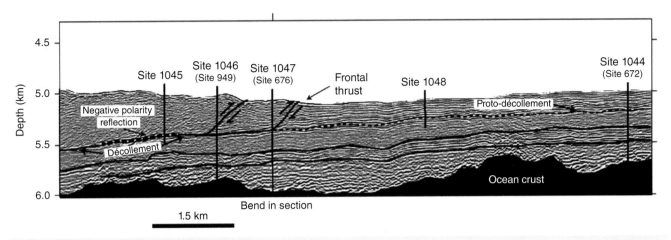

Fig. 10.14 Seismic depth section extending from west of ODP Leg 171A Site 1045 through Sites 676 and 1047 then to Sites 672 and 1044, Barbados accretionary prism. Solid lines below the level of the décollement zone and proto-décollement zone show approximate limits of the sandy underthrust terrigenous sequence. This sandy sequence may be the zone that migrates warm fluids from beneath the accretionary prism, creating a heat-flow anomaly seaward of the deformation front. From Moore *et al.* (2000).

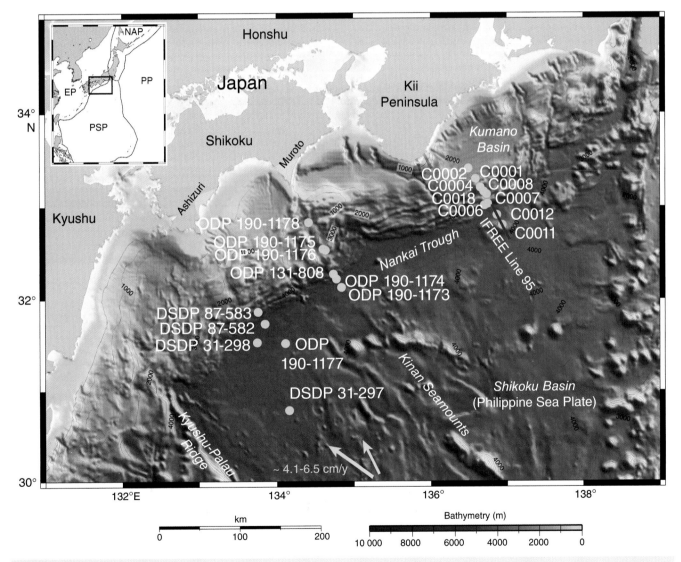

Fig. 10.15 Shaded relief map of the central Nankai Trough showing the regional morphology of the accretionary prism and subducting Shikoku Basin. Modified from Pickering *et al.* (2013).

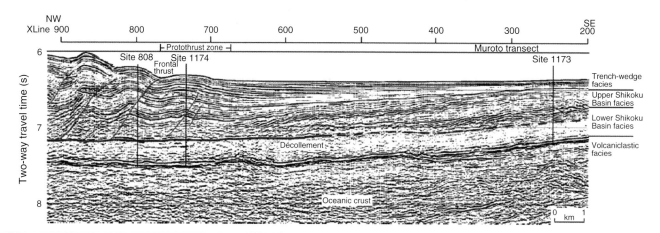

Fig. 10.16 Seismic reflection profile through the Muroto Transect reference (ODP Site 1173) and prism toe sites (ODP Sites 1174 and 808). Correlation of sedimentary facies associations (informally labelled here as facies) to the seismic data is shown on the right. Seismic data are from the 3D seismic survey of Bangs *et al.* (1999) and Moore *et al.* (1999). Xline identifies the crossing line number in the 3D seismic volume. From Moore *et al.* (2001). Most of the Lower Shikoku Basin is characterised by Facies Classes D, E and G, with the addition of Facies Classes B and C in the Upper Shikoku Basin, and Facies Classes A and B in the trench-wedge. Facies Class F is also present in the inner (landward) side of the trench wedge, particularly where slope sediments have been remobilised and deposited within the trench. See Fig. 10.15 for location of ODP sites.

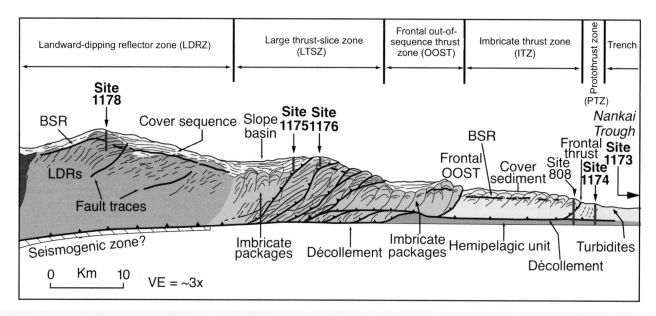

Fig. 10.17 Schematic interpretation of seismic line 141-2D in the Muroto Transect showing tectonic domains and location of Leg 190 drill sites. From Moore, Taira, Klaus *et al.* (2001). See Fig. 10.15 for location of drill sites shown on this transect.

headless submarine canyons. Elevated head gradients at the inflection point of canyons have been inferred to assist in localised failures that feed sediment into a closed slope basin. Measured head gradients are an order of magnitude too low to cause seepage-induced slope failure alone. Therefore, McAdoo *et al.* (1997) proposed transient slope failure mechanisms. Inter-canyon slopes are uniformly unscarred and smooth, although consolidation tests indicated that up to several metres of material may have been removed. A sheet-like failure would remove sediment uniformly, preserving the observed smooth inter-canyon slope. Earthquake-induced liquefaction was also considered a likely trigger for this type of sheet failure, as the slope is too steep and short for sediment flow to organise itself into channels.

Bathymetric and seismic reflection data suggest a local source for sediment in a trench slope-basin between the second and third ridges from the prism toe. A comparison of (a) the amounts of material removed from the slopes and (b) sediment in the basin showed that the amount of material removed from the slopes may slightly exceed the amount of material in the basin, implying that a small amount of sediment has escaped the basin, perhaps when the second ridge was too low to form an effective dam, or through a gap in the second ridge to the south. Regardless, almost 80% of the material shed off the slopes around the basin is deposited locally (mainly as Facies Class C, D and E), whereas the remaining 20% is redeposited on the incoming plate and will be re-accreted.

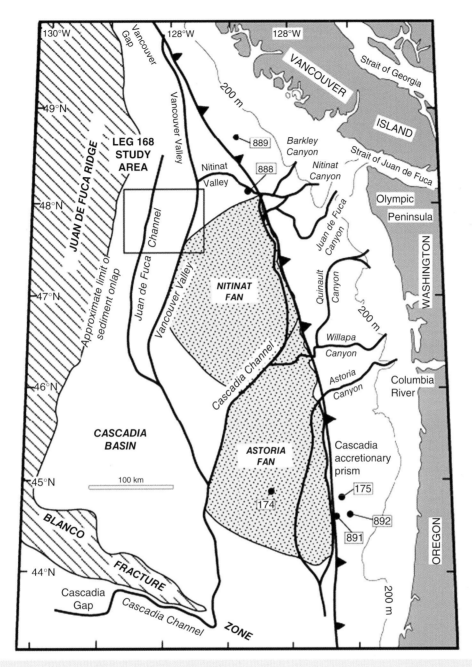

Fig. 10.18 Physiography of the Cascadia Basin region (from Karl *et al.* 1989). Numbers in boxes refer to previous drill sites of the DSDP and ODP. Heavy line with teeth represents approximate position of the Cascadia subduction front and eastern edge of Cascadia Basin. From Riedel, Collett, Malone and Expedition 311 Scientists (2005).

10.2.4   Role of seamounts in subduction factory

Active sediment input from the subducting plate, rather than passive conveyance of open-ocean sediments to the trench, includes seamount-margin sedimentation (slides, debris flows, concentrated density flows, turbidity currents etc.). Such sediments may include carbonate reef clasts, volcanic material and mixed shallow-and deep-marine faunas. The nature of any preserved carbonate may be strongly influenced by the position of the CCD.

Turbidity currents and some concentrated density flows can have sufficient flow thickness and/or momentum to mantle or travel up

onto seamounts. For example, gravity cores from isolated seamounts located within, and rising up to 300 m from, the sediment-filled Peru–Chile Trench off south-central Chile (36° S–39° S) contain abundant SGF deposits that are much coarser (probably Facies Classes B and C) than the hemipelagic (Facies Classes E and G) background sedimentation (Völker *et al.* 2008). Their mineralogical composition suggests that some beds have a mixed provenance and the benthic foraminifera in one of the sand layers show a mixed origin from upper shelf to middle-lower bathyal depths which could have been caused during sediment mixing associated with earthquakes that likely triggered the flows. Völker *et al.* (2008) propose that the SGFs

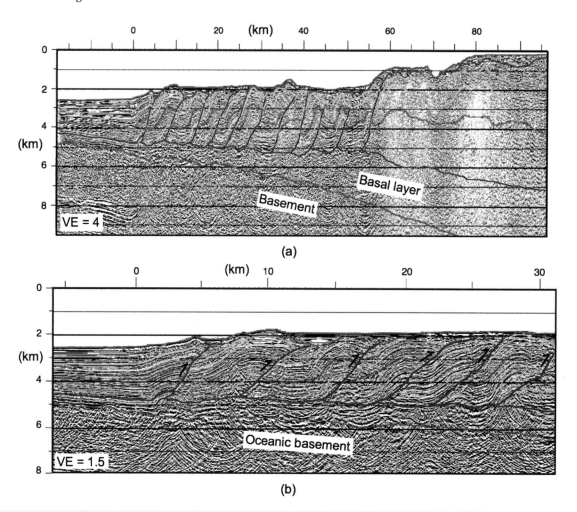

Fig. 10.19 (a) Depth-migrated seismic reflection profile 105 showing 11 landward-vergent thrust slices, Cascadia margin. Thrust faults are red; other colours are correlated reflections. Vertical exaggeration (VE) is 4:1. (b) Close-up of frontal region showing six frontal-most thrust slices. From Gutscher *et al.* (2001). See also, Flueh *et al.* (1998).

rode ~175–450 m higher than the adjacent seafloor, either depositing material by suspension fallout from thick flows or as bedload from relatively unconfined currents that overwhelmed elevated structures: the latter option is considered the least likely.

The drowning of many seamounts, such that their shallow-marine carbonate caps are now in water depths of hundreds to thousands of metres, is generally attributed to a combination of thermal cooling of oceanic lithosphere as volcanic chains are conveyed away from spreading centres, and eustatic sea-level rise. However, at least in some cases the drowning can be attributed to subduction processes. For example, the summit of the carbonate-capped Kikai Seamount, on the northwestern part of the Amami Plateau, northern Philippine Sea, is at 1960 m water depth. Nakazawa *et al.* (2007) interpret its rapid subsidence since the Early Pleistocene as most probably due to collision and subduction of this part of the Amami Plateau. The rapid subsidence may have started when the western corner of the plateau reached the Ryukyu Trench and began subduction beneath the Ryukyu Arc.

Amongst the most studied seamounts associated with the subduction factory are the Dai-ichi Kashima and Erimo seamounts, at the southern and northern ends of the Japan Trench, respectively (Cadet *et al.* 1987; Lallemand & Le Pichon 1987; Pautot *et al.* 1987; Dubois & Deplus 1989; Lallemand *et al.* 1989; Yamazaki & Okamura 1989; Dominguez *et al.* 1998; Nishizawa *et al.* 2009). Wide-angle seismic experiments across the Dai-ichi Kashima and Erimo seamounts show that their maximum crustal thicknesses are 12–17 km greater than typical oceanic crust. The Dai-ichi Kashima seamount (Fig. 10.20a) has been cut into two distinct parts by a large normal fault, resulting in two reef-capped flat tops of the same Lower Cretaceous strata at different water depths: a western (inboard) crest at 5300–5450 m depth, and an eastern (outboard) one at 3880–4000 m depth (Konishi 1989). The present topography of the seamount has resulted from subduction-induced faulting of a once contiguous reef-capped guyot.

Detailed observations by the Franco-Japanese KAIKO Project, near the triple junction of the Izu–Bonin (Ogasawara) and Nankai trenches off southeast central Honshu, have shown the deposits of rockfall, debris flow, slide and other sediment redeposition processes on the disintegrating margins of the Dai-ichi Kashima seamount as it enters the trench (Fig. 10.20b). It is possible that the seamount talus, as an olistostrome, could provide an ideal precursor to mélange during later thrusting and shearing processes as parts of the seamount are incorporated into the accretionary prism.

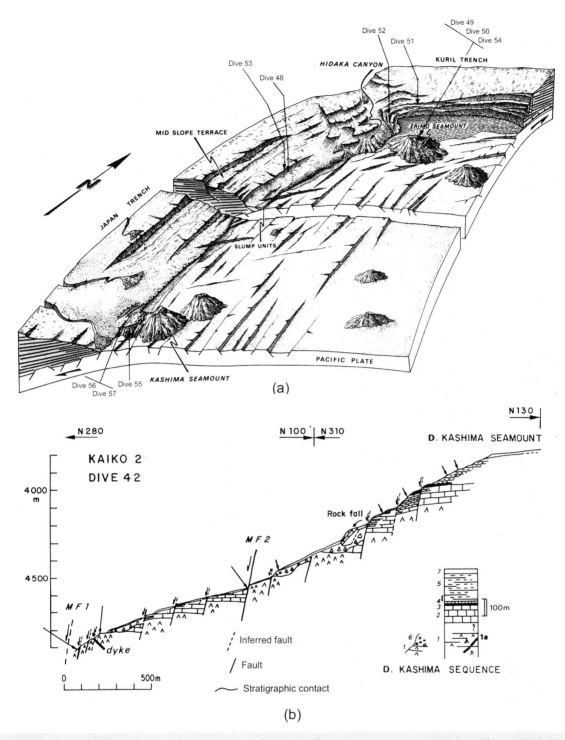

Fig. 10.20 (a) Idealised morphology of the Japan and Kuril trenches based on Seabeam mapping, seismic profiling and submersible dives (locations shown). From Cadet *et al.* (1987). (b) Cross-section of Dai-ichi Kashima Seamount scarp based on dive 42 of the Franco-Japanese KAIKO Project (from Pautot *et al.* 1987). MF1 and MF2 are the main normal faults. 1, basaltic lava (1a = dyke); 2, Lower Cretaceous shallow-marine limestone; 3, brown marls; 4, chalk; 5, upper Miocene–lower Pliocene yellow marl; 6, sedimentary breccia; 7, Recent hemipelagic mud. Sample locations shown by arrows.

The Tonga Trench and forearc have been studied in detail between 14°S and 27°S. Beyond showing many submarine canyons incised in the landward slope, the Capricorn seamount is poised to enter the subduction factory (Wright *et al.* 2000) (Fig. 10.21), to be dismembered like seamounts such as Dai-ichi Kashima.

In many active margins, intense deformation is observed at the front of the overriding plate where seamounts or aseismic ridges subduct and/or accrete. Such deformation appears to be a major tectonic feature of these areas, which influences the morphology and seismicity of a margin (e.g., Scholz & Small 1997; Kodaira

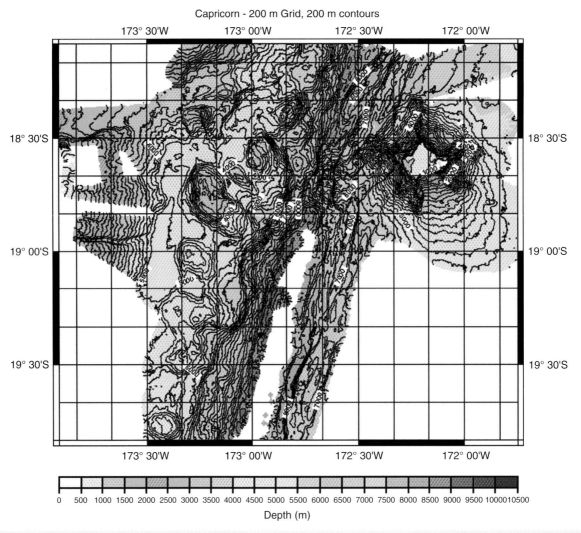

Capricorn - 200 m Grid, 200 m contours

Depth (m)

Fig. 10.21 Bathymetric map (200 m contour interval) of the Tonga Trench and forearc, in the vicinity of the Capricorn seamount. Shading indicates areas deeper than 7000 m. The down-going plate carrying the seamount is to the right (i.e., east). Note the distinct submarine canyon to the west of the trench at ~19°S. Map was created from a 200 × 200 m Seabeam 2000 grid (Boomerang 8), with portions of the trench axis filled in by Seabeam data from Marathon 6. Map projection is Mercator. From Wright *et al.* (2000).

et al. 2000; Tréhu *et al.* 2012; Yang *et al.* 2012; Obana *et al.* 2014; but *cf.* Singh *et al.* 2011; Müller & Landgrebe 2012). Subduction zones respond differently to basement highs (such as seamounts as potential asperities) and smooth basin plains (Cloos 1992; Cloos & Shreve 1996). Dominguez *et al.* (1998) modelled two types of subducting seamounts: relatively small conical seamounts, and larger flat-topped seamounts. In both cases, they found that the indentation of the margin by the seamount inhibits frontal accretion and produces a re-entrant. The margin uplift includes displacement along backthrusts (landward-vergent), which propagate from the base of the seamount, and out-of-sequence forethrusts (seaward-vergent), which define a shadow zone located on the landward flank of the seamount. When the seamount is totally buried beneath the margin, this landward-shielded zone disappears and a larger one is created in the wake of the asperity due to the elevated position of the décollement. As a consequence, a section of the frontal margin of the upper plate follows behind the seamount to greater depth. A 'slip-line' network (a radial array of fractures and strike-slip faults as accommodation

structures) develops concurrently above the subducting seamount flanks from the transtension along the boundaries of the shadow zone. In a final stage, normal faults, controlled by the shape of the seamount, develop in the subsiding wake of the asperity. Swath-bathymetric data from the Costa Rica margin reveal detailed surface deformation of the margin above three subducting seamounts.

The bathymetry and shape, together with sediment transport pathways, in trench-slope basins, including slope basins atop the accretionary prism, are strongly influenced by the 'roughness' of the subducting oceanic plate. ODP Leg 190 yielded important discoveries about the early stages of the tectonic and sedimentary evolution of a trench-slope basin in the Nankai subduction zone (Underwood *et al.* 2003). The lithofacies character, biostratigraphy and seismic-reflection data show that the slope-basin architecture developed during the Early Quaternary by frontal off-scraping of coarse-grained trench-wedge deposits. The clast types in muddy gravel beds (Facies Group A1) suggest that one of the polymictic sources was enriched in low-grade metasedimentary rocks. Outcrops of the Shimanto Belt on the island of Shikoku contain comparable lithologic assemblages, leading Underwood *et al.* (2003) to propose that some of the SGFs were funnelled from that source through a transverse canyon-channel system. Off-scraped trench deposits are mildly deformed and nearly flat, lying beneath the slope basin. Bedding within the basin laps onto a hanging-wall anticline that formed above a major out-of-sequence thrust fault. Rapid uplift brought the substrate above the calcite compensation depth soon after the basin was created. The sediment delivery system probably was re-routed during subduction of the Kiinan seamounts, thereby isolating the juvenile basin from coarse sediment influx. As a consequence, the upper 200 m of basin fill comprises nannofossil-rich hemipelagic mud with sparse beds of volcanic ash and thin silty turbidites (Facies Classes E and D). Intervals of stratal disruption are also common: early sediment folding was caused by north–northeast-directed gravitational failure. The Nankai accretionary prism has grown 40 km in width during the past 1 Myr, and the slope basin is already filled to its spill point on the seaward side. The stratigraphy displays an overall upward thinning-and-fining trend, in contrast to the inferred upward thickening-and-coarsening trend depicted by some conceptual models for slope basins.

In the Chile Trench, the development of a submarine canyon (the San Antonio Canyon) appears to have been controlled by a subducting seamount that formed the San Antonio Re-entrant and warped the middle slope along its landward advancing path (Laursen & Normark 2002). The canyon crosses the forearc slope of the central Chile margin for >150 km before it enters the Chile Trench near 33° S latitude. In its upper reaches, the nearly orthogonal segments of the San Antonio Canyon erode ~1 km into thick sediment following underlying margin-perpendicular basement faults and along the landward side of a prominent margin-parallel thrust ridge on the outer mid-slope. At a breach in the outer ridge, the canyon makes a sharp turn into the San Antonio Re-entrant. The emergence of an obstruction across the head of the San Antonio Re-entrant has trapped sediment in the mid-slope segments of the canyon, such that little sediment presently appears to reach the Chile Trench through the San Antonio Canyon. Laursen and Normark (2002) ascribe the incision of the canyon landward of the outer mid-slope ridge to a combination of headward erosion and entrenchment by captured unconfined turbidity currents, with flushing of the canyon likely enhanced during the lowered sea level of the last glaciation. Where the canyon occupies a triangular embayment of the re-entrant at the base of the slope, sediment has ponded behind a small accretionary ridge (Laursen & Normark 2002). On the trench floor opposite the mouth of the San Antonio Canyon, a 200 m-thick levée–overbank complex has formed on the left side of a distributary channel emanating from a breach in the accretionary ridge. Axial transfer of sediment was inhibited to the north of the San Antonio Canyon mouth, which left the trench to the north starved of sediment. Between ~32° 04'S and 33° 40'S, the Chile Trench axial channel deeply incises the San Antonio distributary complex. This entrenchment may have been initiated when the barrier to northward transport was eliminated.

10.2.5 Very oblique convergence and strike-slip in subduction factory

This section considers an example of very oblique plate convergence (subduction) during the Miocene in the Shikoku Basin that included a phase of strike-slip tectonics. As subduction–accretion processes appear to have predominated along the continental margin during this time interval, this section is placed here rather than in Chapter 12.

An important aspect of IODP NanTroSEIZE Expedition 322 was to characterise the lithologies and physio-chemical properties at reference sites entering the subduction factory – in this case oceanic crust having a rugose surface with many asperities, including ridges, an extinct spreading centre and knolls (Underwood, Saito, Kubo and the Expedition 322 Scientists 2009, 2010) (Fig. 10.15, Figs 10.22–10.24). Site C0011 is located on the northwest flank of a prominent bathymetric high (the Kashinosaki Knoll, a basaltic seamount), whereas Site C0012 is located near the crest of the knoll (Figs 10.15, 10.22). Coring at Site C0012 penetrated ~23 m into igneous basement and recovered the sediment/basalt interface intact at 537.81 mbsf (Fig. 10.24a). The age of the basal sediments (reddish-brown pelagic claystones of Facies G1.1) is >18.9 Ma. Both sites contain tens of metres of deep-marine pyroclastic and volcaniclastic sandstones, mainly of Facies Classes B and C (Figs 10.24b, d). The merger of lithofacies and age-depth models from the two sites (Table 10.1) spans across the Shikoku Basin from an expanded section (Site C0011) to a more condensed section (Site C0012) and captures all of the important ingredients of basin evolution, including a previously unrecognised interval of upper Miocene tuffaceous and volcaniclastic sandstones designated as the middle Shikoku Basin deposits (Fig. 10.23). A stratigraphically lower (lower–middle Miocene) facies-association of sandstones and siltstones (mainly Facies Classes B, C and D) with mixed detrital provenance occurs in the lower Shikoku Basin; this unit may be broadly correlative with superficially similar Miocene SGF deposits on the western side of the basin (Pickering *et al.* 2013). When viewed together, the two sites around the Kashinosaki Knoll demonstrate how basement relief influenced rates of hemipelagic and SGF deposition in the Shikoku Basin. Unlike other so-called reference sites in the Nankai Trough, pore-fluids on top of the basement high show evidence of a seawater-like source, with chlorinity values increasing toward basement because of hydration reactions and diffusion; the fluids are largely unchanged by the effects of focussed flow and/or *in situ* dehydration reactions associated with rapid burial beneath the trench wedge and frontal accretionary prism (Underwood *et al.* 2010).

Seismostratigraphy, coring, and logging-while-drilling during IODP Expeditions 319, 322 and 333 (Sites C0011/C0012) revealed three Miocene submarine fans in the northeast Shikoku Basin, with broadly coeval deposits at ODP Site 1177 and DSDP Site 297 (northwest Shikoku Basin). The oldest, finer-grained fan (Kyushu Fan) has sheet-like geometry; quartz-rich flows were fed mostly from an ancestral landmass in the East China Sea (Pickering *et al.* 2013; *cf.* Clift *et al.* 2013). At that time, the Okinawa Trough would not have blocked sediment delivery into the Shikoku Basin because that bathymetric obstacle did not form until 4–6 Ma (*cf.* Kimura 1996; Miki *et al.* 1990; Lu & Hayashi 2001; Yamaji 2003; Expedition 331 Scientists 2010). Also, within the Xihu depression in the central and northern East China Sea, basin inversion occurred during the late Miocene. The entire Tertiary succession of nearly 10 000 m thickness underwent inversion, with removal of up to 1600 m of sedimentary strata in the north (Yang *et al.* 2011). That erosional event probably contributed sediment influx to the Shikoku Basin.

During prolonged hemipelagic mud deposition at IODP Sites C0011-C0012 (~12.2–9.1 Ma), sand supply continued at Sites ODP 1177 and DSDP 297. Sand delivery to much of the Shikoku Basin halted during a phase of sinistral strike slip to oblique plate motion (and possible 'slow' subduction, after which the Daiichi Zenisu Fan (~9.1–8.0 Ma) was fed by submarine channels. The youngest fan (Daini Zenisu; ~8.0–7.6 Ma) has sheet-like geometry with thick-bedded, coarse-grained pumiceous sandstones (Figs 10.24b, d).

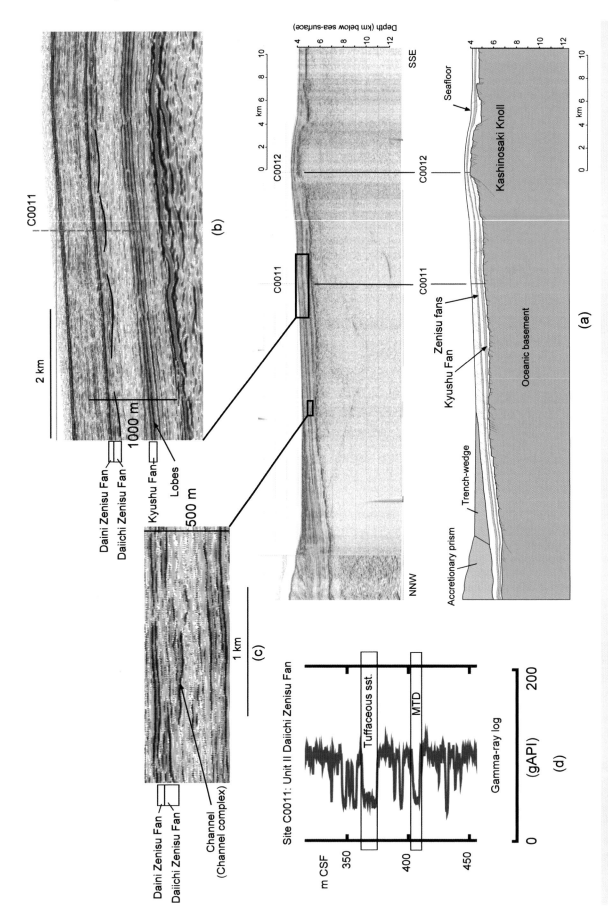

Fig. 10.22 (caption on facing page)

Table 10.1 Lithofacies, lithologic units and age-depth models for IODP Sites C0011 and C0012. Modified from Pickering *et al.* (2013).

Lithologic unit	Unit Name	Depth CSF m*	Thickness (m)*	Age *	Dominant Lithology	Minor Lithology
				Holocene–Late Miocene		
Unit Ia		0.00–251.56	340	0–7.6 Ma	Silty clay(stones) mudstones (Facies Classes E, G)	Ashes
Unit Ib	Upper Shikoku Basin	251.56–347.82	150.86	0–7.8 Ma		
				Late Miocene		Tuffaceous sandstones (Facies Classes B, C)
Unit IIa **Daini Zenisu Fan**	Middle Shikoku Basin	347.82–479.06	139.06	7.6–9.1 Ma	Silty claystones (Facies Classes E, G)	Tuffaceous sandstones Volcaniclastic sandstones (Facies Classes B, C)
Unit IIb **Daiichi Zenisu Fan**	tuffaceous sandstones volcaniclastic sandstones	150.06–219.81	68.95	7.8–9.4 Ma		Clayey siltstones
				Late–Middle Mocene		
Unit III	Lower Shikoku Basin hemipelagites	479.06–673.98 219.81–331.81	194.92 112.00	9.1–12.3 Ma 9.4–12.7 Ma	Silty claystones (Facies Classes E, G)	Calcareous claystones Lime mudstones (Facies G)
				Middle Miocene		
Unit IV **Kyushu Fan**	Lower Shikoku Basin SGF deposits	673.98–849.95	175.97	12.3–13.9 Ma	Tuffaceous silty claystone (Facies Classes E, G)	Silty sandstones (Facies Classes E, G)
		331.81–415.58	83.77	12.7–13.5 Ma	Silty claystones clayey siltstones (Facies Classes E, G)	Siltstones (Facies Classes D, E)
				Middle Miocene		
Unit V	Volcaniclastic–rich	849.95–876.05	>26.10*	>13.9 Ma	Tuffaceous sandy claystones (Facies Classes E, G)	Silty claystones (Facies Classes E, G) Tuff layers
		415.58–528.51	112.93	13.5–>18.9 Ma	Silty claystones (Facies Classes E, G) Sandstones (Facies Classes B, C) (incl.volcaniclastic)	Tuffs –
Unit VI	Pelagic claystone	Not drilled 528.51–537.81	Not drilled 9.3	Early Miocene >18.9 Ma	Red calcareous claystones (Facies Class G)	–
Unit VII	Basement	Not drilled 537.81–576.00	Not drilled 38.2 drilled	Early Miocene >18.9 Ma	Basalt	

*In this table, apart from Unit I (IODP Expedition 333 data), paired depth, thickness and age data are read as: IODP Site C0011 = upper line; IODP Site C0012 = lower line CSF m = Depth in metres cored from seafloor. Modified from Pickering *et al.* (2013).

Fig. 10.22 Facing page. (a) Spliced composite seismic profile of a representative depth section from NanTroSEIZE 3D data volume (Moore *et al.* 2009) and Line 95 from IFREE mini-3D seismic survey (Park *et al.* 2008) that crosses Sites C0011 and C0012 (shown in Fig. 10.15), and interpretation of main lithostratigraphic units. Enlargements of seismic sections show diagnostic depositional environments for the Kyushu and Zenisu submarine fans along this transect, including (b) sheet-like seismic character of Unit IV (Kyushu Fan), interpreted as distal outer-fan lobe deposits, and (b and c) irregular seismic character of Unit IIB (Daiichi Zenisu Fan), interpreted as mainly volcanic sandstone-filled, including MTD, submarine channel deposits, immediately overlain by a continuous, high-amplitude, seismic interval (Unit IIA), interpreted as unconfined, sheet-like volcanic sandstone deposits. (d) Logging while drilling (LWD) data show 10–20 m-thick blocky gamma-ray motif interpreted as an MTD and thick interval of tuffaceous sandstones in the Daini Zenisu Fan. From Pickering *et al.* (2013).

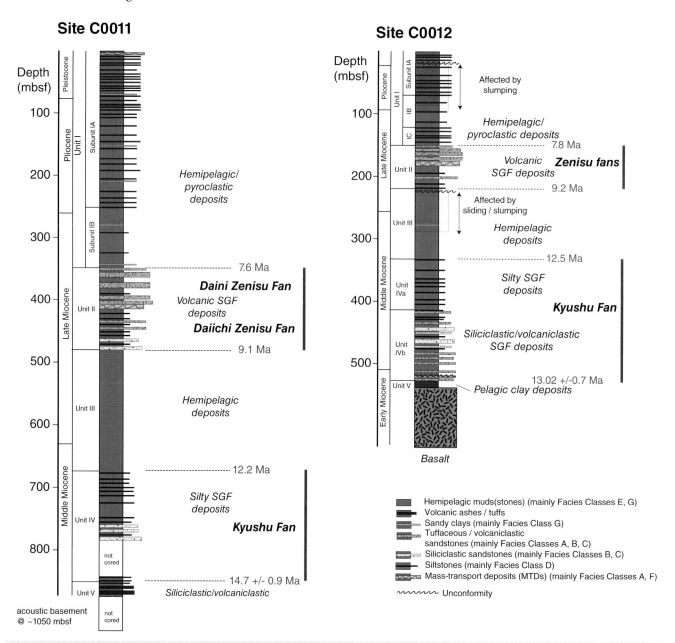

Fig. 10.23 Lithologic columns for IODP Sites C0011 and C0012 drilled in the vicinity of Kashinosaki Knoll, northeast Shikoku Basin, showing IODP Sites C0011 and C0012, based on results from Expeditions 322 and 333 (Underwood *et al.* 2010; Expedition 333 Scientists 2011). See Fig. 10.15 for location of drill sites. See text for explanation of the Kyushu and Zenisu submarine fans. SGF = sediment gravity flow, including turbidity current. Also, see Table 10.1 for summary of lithologic units. Note, in Site C0012, unlike at Site C0011, the two Zenisu fans are not separately defined. Note the overall lateral continuity of the sandy units, interpreted as volcaniclastic submarine fans. Modified from Pickering *et al.* (2013).

The pumice fragments were fed from a mixed provenance that included the collision zone of the Izu–Bonin and Honshu arcs. The shift from channelised to sheet-like (relatively unconfined) SGFs was favoured by renewal of relatively rapid northward subduction, which accentuated the trench as a bathymetric depression. Increased sand supply appears to correlate with long-term eustatic lowstands of sea level (Pickering *et al.* 2013).

In summary, the middle–late Miocene history of the Nankai accretionary prism appears to have been associated with substantial oblique plate convergence/subduction and strike-slip. This resulted in complex and changing deep-marine sediment dispersal patterns in the northern Shikoku Basin. These sediment dispersal patterns have major implications for palaeogeographies in the northern Shikoku Basin at that time, with a changing sediment source from the East China Sea (and Yangtse River drainage basin and its offshore components, e.g., submarine canyons) to sediment dispersal patterns more typical of the present, that is, a substantial input from the Izu Collision Zone (Fig. 10.25).

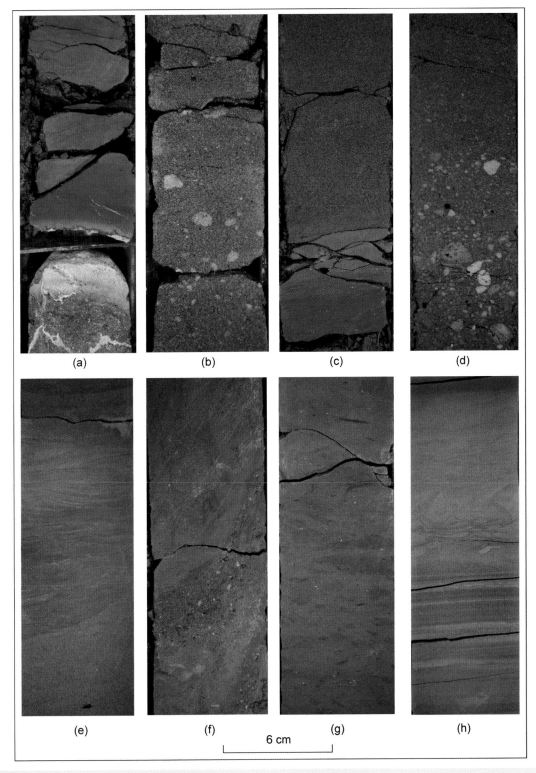

Fig. 10.24 Representative facies in cores from Kashinosaki Knoll (IODP Expedition 322 subduction inputs site) ~10 km south of Nankai Trough, Shikoku Basin, offshore Japan. (a) Sediment/basalt interface showing red Facies Group G1 claysones overlying altered basalt at lithologic Unit VI/VII boundary (interval 322-C0012A-53R-1, 20–35 cm). (b) Pyroclastic sandstone of Facies A1.4 (interval 322-C0011B-3R-3, 28–43 cm). (c) Volcaniclastic sandstone of Facies B1.1, lithologic Unit II (interval 322-C0011B-14R-5, 70–86 cm). (d) Tuffaceous sandstone/gravel of Facies A2.3 with normal grading of subrounded pumice clasts at the base (interval 322-C0011-11R-5, 30–76 cm). (e) Current-rippled silstones (Facies D2.1: interval 322-C0011B-7R-1, 111–126 cm). (f) Folded and attenuated volcaniclastic sandstone layers within silty claystones (Facies F2.1: interval 322-C0011B-8R-7, 24–39 cm). (g) Intensely bioturbated silty claystones with dark green silty claystone layers <0.5 cm thick, upper part of lithologic Unit II, Site C0011 (Facies E1.3: interval 322-C0011B-11R-1, 54–69 cm). (h) Graded and laminated, tuffaceous, very fine-grained silty sandstones showing load and flame structures (Facies D2.1) interbedded with siltstones (Facies D2.3: interval 322-C0011B-58R-1, 30–46 cm).

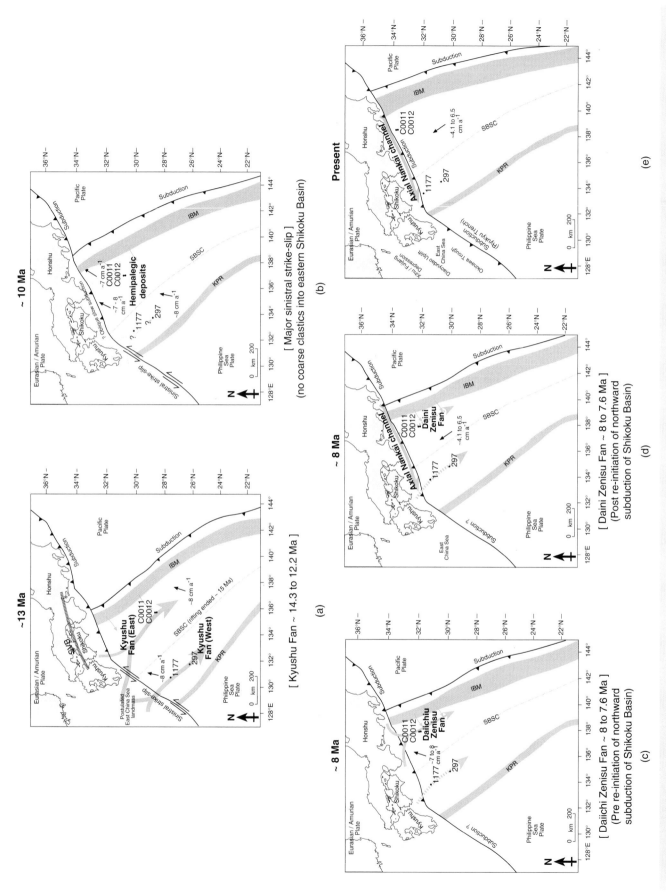

Fig. 10.25 (caption on facing page)

Fig. 10.25 Schematic plate tectonic reconstructions for the Shikoku Basin and adjacent area, redrawn and modified from Mahony *et al.* (2011) to show the dispersal patterns for the coarse sandy sediments into the northeast Shikoku Basin for (a) ~13 Ma, (b) ~10 Ma, (c) ~8 Ma immediately *before* renewed northward subduction, (d) ~8 Ma, immediately *after* renewed northward subduction, and (e) present day. Parenthesis shows more precisely the time corresponding to the displayed palaeogeography. Thickness of sediment dispersal arrows indicates the relative importance of multiple sources. IBM = Izu–Bonin–Mariana Arc; SBSC = Shikoku Basin Spreading Centre; KPR = Kyushu-Palau Ridge. SVB = subduction-related Setouchi Volcanic Belt (12.4–14.8 Ma (Shimoda *et al.* 1998; Kimura *et al.* 2005)). Between 10–6 Ma, there is a lack of subduction-related volcanism (Mahony *et al.* 2011; Tatsumi 2006), with subduction and arc volcanism ceasing 1–2 Myr earlier. Note the change in coarse sediment dispersal at ~8 Ma, from submarine channels supplying sediment far out into the northeast Shikoku Basin to an axial trench (palaeo–Nankai Channel), interpreted as the formation of a topographic trench as northward subduction of the Shikoku Basin oceanic crust resumed, picked up by arc volcanism 1–2 Myr later once the subducted slab had reached appropriate PTT (pressure–temperature–time) conditions. The plate boundary configurations used for the 2–15 Ma reconstructions of southwest Japan utilised published finite rotation poles for the Philippine Sea plate relative to Eurasia from Gaina and Müller (2007). In these reconstructions, Japan is held fixed and the latitude and longitude are for present day Japan. From Pickering *et al.* (2013).

10.2.6 Preservation and recognition of trench stratigraphy

The nature of subduction–accretion processes prevents the common preservation of relatively intact trench stratigraphy in the geological record (e.g., Leverenz 2000). Instead, accretionary processes tend to severely deform the prism sediments by flattening and penetrative simple shear during thrusting and folding. However, some mechanisms of subduction–accretion provide a better chance of accreting a relatively complete trench succession, for example during ocean-ridge and seamount subduction. Basal décollement below the trench fill, for example in oceanic basalts, will greatly increase the probability of preserving a relatively complete trench fill. In many trenches, seismic evidence suggests that the pelagites and hemipelagites immediately overlying oceanic basaltic basement tend to be subducted, whereas it is the trench turbidites, concentrated density-flow deposits and/or oceanic-plate thick clastics that are off-scraped in frontal accretion processes.

Results from the KAIKO Project, in the Nankai Trench, suggest that subduction zones may migrate oceanward in increments, controlled by the position of oceanic basement ridges (Le Pichon *et al.* 1987a, b). In the eastern Nankai Trench, the aseismic Zenisu Ridge is approaching the northward-dipping subduction zone and trench, and multi-channel seismic reflection profiles suggest that the thrust (deformation) front may have jumped to the south and oceanward of the Nankai Trench (Fig. 10.26), due to the mechanical strength of the ridge and upper oceanic crust during flexure. Should such a décollement surface develop into a new subduction zone, thereby de-activating the present trench as the deformation front, then a complete trench-fill stratigraphy floored by oceanic basalts might be accreted. Furthermore, the strength of any accreted basement could shield the overlying strata from the intense internal deformation characteristic of other ocean-facing accretionary prisms. Seamount subduction, with the off-scraping of parts of the basaltic edifice together with the sediment veneer and associated trench sediments, may also preserve an identifiable trench stratigraphy. Such successions may comprise a tectonically-sheared association, or juxtaposition, of open-ocean limestones, cherts (Facies Class G), pelagites (Facies Class E), hemipelagites (Facies Class E) and thin-bedded turbidites (Facies Classes C and D), rockfall (Facies Group F1), olistostromes (Facies Classes F and A) and debrites (Facies Group A1), and relatively coarse-grained, siliciclastic, trench-fill concentrated density-flow deposits (Facies Classes A, B and C). Thornburg and Kuhn (1987) discussed the importance of the nature of trench-fill sediments in determining the location of successive deformation fronts and, therefore, the nature of stratigraphic duplication. They suggested that sediment-starved trenches will tend to decouple within the upper part of the oceanic basaltic layer (although it is hard to see why such décollement would not instead occur immediately above the oceanic crust in trench sediments); also, they suggested that where sediment accumulation rates are fast enough to generate thick trench successions, very little if any oceanic crust is accreted (*cf.* Thornburg & Kuhn 1987: their fig. 18). Where a prominent trench axial channel is developed, they suggested that successive décollement will tend to occur on the oceanward (outer trench) side of such clastics. While the modelling of Thornburg and Kuhn (1987) contains many useful elements, we believe that the controls on preserving certain stratigraphic successions are far more complex and should await more detailed comparative research.

In the absence of any oceanic basalts associated with deep-marine sediments, it is probably impossible to differentiate a forearc basin, accretionary-prism slope basin and trench fill. Apart from the depositional site, there appear to be no unique sedimentary characteristics of trench-fill siliciclastics.

10.2.7 Forearc basins/slope basins

Within a subduction–accretion complex, forearc basins of various sizes develop, controlled by compressional, extensional and in some cases dominated by oblique-slip tectonics. Larger forearc basins tend to develop above older, highly deformed and lithified (metamorphosed) accretionary-prism sediments and associated igneous-volcanic rocks. (e.g., the 20 × 30 km upper Miocene Makara Basin within the North Island oblique-subduction system, New Zealand, that is divided by a narrow zone of deformed Cretaceous–Palaeogene mélange; van der Lingen 1982).

The Western Pacific Ocean, between about 10° and 45°N latitude, contains a number of major arc systems and associated basins (Fig. 10.1). Some aspects of the Japan Trench and Nankai Trench, and forearcs, are considered together with the Ryukyu Arc and associated Okinawa Basin, a marginal basin developed by backarc rifting. Summaries of the tectonic development of this region during the Cenozoic can be found in Letouzey and Kimura (1985, 1986), Taira *et al.* (1989), Pickering and Taira (1994) and in Figure 10.27. The Japanese or Honshu Arc, associated with the triple junction between the Philippine, Eurasian and Pacific plates, contains two discrete trenches – the northeastern Japan Trench and the southwestern Nankai Trench. Along the margin of the Pacific Plate, the Japan Trench passes northward into the Kuril Trench, and southward into the Izu–Bonin (Ogasawara) Trench. The Japan Trench–Honshu Arc is perhaps the most extensively studied Pacific convergent margin.

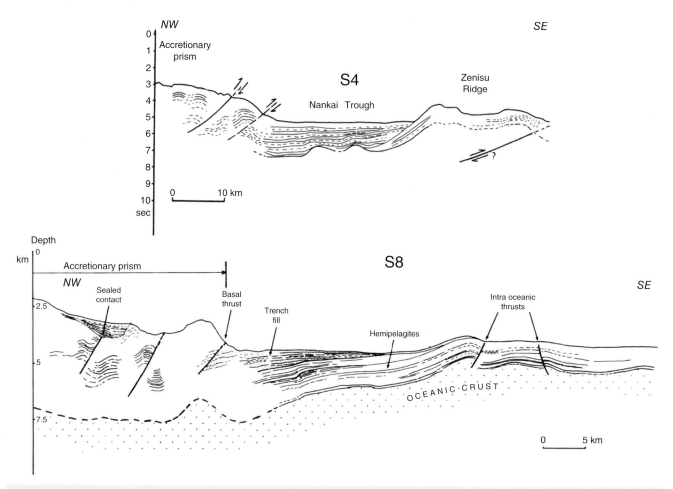

Fig. 10.26 Interpreted multichannel seismic profiles S4 and S8 across the eastern part of the Nankai Trench in Box 5 of the Franco Japanese KAIKO Project. Note the thrusting south of the present Nankai Trench, to the south of the Zenisu Ridge, where a new trench might develop. See text for explanation. From Le Pichon *et al.* (1987b).

Sediment-dominated subduction zones, such as the Nankai margin, are characterised by the repeated occurrence of great earthquakes of ∼M 8.0 (Ruff & Kanamori 1983). Although the causative mechanisms are not well understood (e.g., Byrne *et al.* 1988; Moore & Saffer 2001; Saffer & Marone 2003), the updip limit of the seismogenic zone is thought to correlate with a topographic break along the outer rise of the forearc (e.g., Byrne *et al.* 1988; Wang & Hu 2006). At Nankai, high-resolution seismic reflection profiles clearly document an out-of-sequence thrust or megasplay fault system that branches from the plate boundary (décollement) within the co-seismic rupture zone of the 1944 Tonankai M 8.2 earthquake (Park *et al.* 2002; Ikari & Saffer 2011; Kimura *et al.* 2011; Saffer *et al.* 2013).

All trenches in this region have different sedimentary fills. In the Nankai Trench, large amounts of terrigenous mud and sand have accumulated (Facies Classes A, B, C and D), fed mainly from the Fuji River via submarine canyons into the trench axis to cover the trench floor. Mud diapirism has been reported in this area (KAIKO Project Shipboard Scientific 1985). In the Japan Trench, there is a relatively thin veneer of terrigenous sediments that appears to have been redeposited in the trench by slope instability (mainly sliding) from the forearc lower trench slope (Facies Class F). In contrast, there is a thick sediment cover in the western part of the more northerly Kuril Trench. A very thick, subsiding accumulation of

terrigenous sediments is derived from Boso Canyon and covers the triple junction of the Izu–Bonin (Ogasawara) Trench and Sagami and Japan trenches. The surrounding ocean floors incline towards the centre of the triple-junction triangle. At the triple junction, the rate of subsidence is much faster than the rate of sediment accumulation.

There is considerable variation in the style of accretion in the trenches, with extensional tectonics and uplift also controlling sedimentation. For example, DSDP Leg 87 in the forearc of the Japan Trench has shown that Late Cretaceous and Early Palaeogene uplift created the Oyashio landmass that developed in the outer forearc (Karig, Kagami & DSDP Leg 87 Scientific Party 1983). Subsequent late Oligocene–late Pliocene major subsidence of the forearc led to submergence, with post-early Pleistocene convergence causing renewed uplift and accretion. Thus, a feature of the stratigraphy is the alternation of deep- and shallow-marine sedimentation, associated with local emergence of parts of the forearc.

Subduction style is variable in the 4000–4800 m-deep, 10–20 km-wide, Nankai Trench (Aoki *et al.* 1983). Here, the angle of subduction is steeper along the eastern part of the trench compared to the west. Leggett *et al.* (1985) showed that accretion along the Nankai Trench changes from: (i) subduction–accretion domains with off-scraping of trench deep-marine sediments near the toe

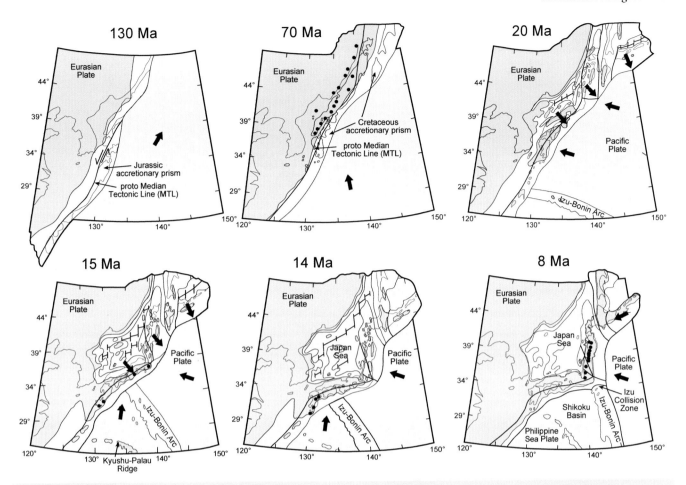

Fig. 10.27 Palaeogeographic reconstruction of the Shimanto Belt and Nankai forearc evolution. Arrows show the direction of convergence. Green = mainly terrestrial environments, blue = oceanic crust and white = transitional (thinned) crust including arc basement. Black dots = volcanic centres. Redrawn and modified from Pickering and Taira (1994) (*cf.* Taira *et al.* 1989, Pickering *et al.* 1993b, Pickering *et al.* 2013).

of the lower trench slope, to (ii) domains of underplating. Where off-scraping and frontal accretion occur, imbricate thrusts are resolvable at intervals of 1–6 km on high resolution, multi-channel, seismic reflection profiles. These thrusts have possible lateral, along-strike continuity up to 70 km. An area of underplating is proposed beneath the growing Minami–Muroto Knoll, an area of uplift associated with contemporaneous extensional growth faults. Upslope from the knoll, an estimated 1200 m of deep-marine sediments have accumulated as a footwall-block succession.

The Japan and Nankai trenches and associated forearc areas demonstrate the complexity and variability in tectonic style associated with subduction–accretion. Not only is there considerable mass wasting of the inner trench slopes by sliding and other processes, but small sedimentary basins have developed in response to both compressional and extensional tectonics: some of these basins having considerable thicknesses of sediment.

Six sites along two transects across the Nankai Trough accretionary prism were drilled during ODP Leg 190, with two reference sites at the seaward ends of the 'Muroto Transect' (ODP Site 1173) and the 'Ashizuri Transect' (ODP Site 1177), to define the stratigraphic framework of the accreting/subducting Shikoku Basin sediments (Figs 10.15–10.17, 10.28–10.31). A thick section of Miocene SGF deposits (mainly of Facies Classes B, C and D) and smectite-rich

mudstone (Facies Classes E and G) occurs within the subducting section at the Ashizuri site. The décollement in both transects is localised along a stratigraphic unit (~5.9–7 Ma) within the lower Shikoku Basin deposits. This horizon can be correlated across both transects through its magnetic susceptibility signature, together with lithostratigraphic correlations to other DSDP and ODP drill sites (Fig. 10.31).

The Middle America and Peruvian convergent margin is an example of a forearc where extensional tectonics are important (*cf.* results of DSDP Leg 66 (Watkins, Moore *et al.* 1982)), Leg 67 (Aubouin *et al.* 1982a, b, c), and Leg 84 (von Huene, Aubouin *et al.* 1985; Bourgois *et al.* 1988)). The margin off Guatemala, containing the Middle America Trench, is one of the classic examples of subduction–accretion, involving the offscraping of sediments and rocks from the downgoing oceanic Cocos Plate. This margin differs from the Lesser Antilles area, because sedimentation is associated with an arc developed on continental crust, with the only substantial deep-marine deposition occurring in the forearc.

Subduction began in the Early Tertiary (von Huene, Aubouin *et al.* 1985). By the Neogene, a well-defined forearc basin had developed in middle to upper bathyal depths with slope and trench sedimentation occurring at abyssal depths >3500 m. Along the axis of the Middle America Trench, hemipelagic silt and clay (essentially

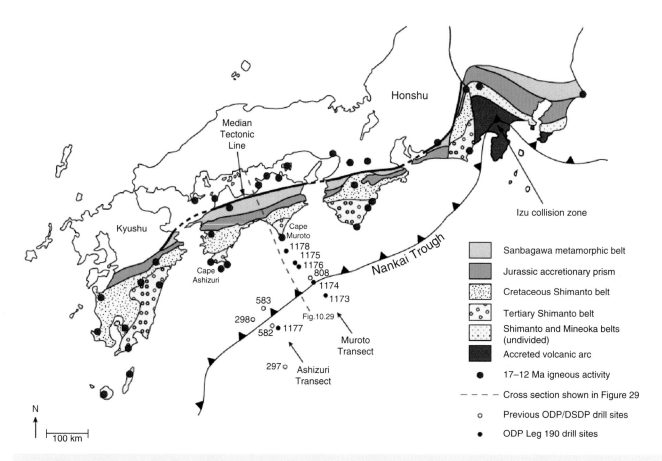

Fig. 10.28 Geologic map of the southwest Japan forearc region and the ODP Leg 190 Muroto and Ashizuri drilling transects. The Shimanto accretionary prism provides a landward analogue of the Nankai accretionary prism. Note the widespread 17–12 Ma igneous activity, probably due to the initial subduction of the young Shikoku Basin oceanic lithosphere. Previous ODP/DSDP drill sites are shown by open circles. The dashed line shows the location of the cross-section shown in Fig. 10.29. From Moore, Taira, Klaus *et al.* (2001).

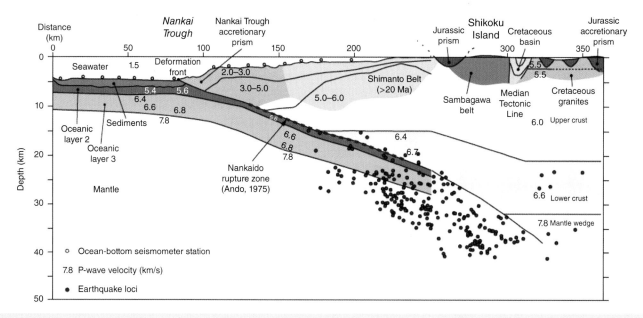

Fig. 10.29 Crustal cross-section of the Nankai Trough forearc (modified from Kodaira *et al.* 2000). Crustal structure, crustal velocities and subducting plate earthquakes are shown. Note that the updip limit of the 1946 Nankaido earthquake rupture zone possibly reaches to the Nankai Trough accretionary prism. From Moore, Taira, Klaus *et al.* (2001).

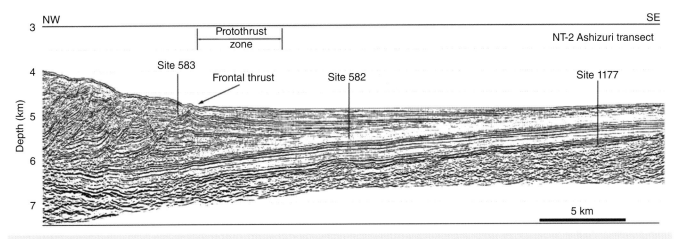

Fig. 10.30 Seismic reflection profile (NT-2) through Ashizuri Transect reference site (ODP Site 1177), trench site (ODP Site 582) and prism toe site (ODP Site 583). From Moore, Taira, Klaus *et al.* (2001). See Fig. 10.15 for location of DSDP and ODP sites.

Facies Classes D and E), together with arc-derived sandy SGF deposits (Facies Classes B, C and D), are ponded on the descending oceanic plate with its veneer of pelagic sediments (Facies Classes E and G). Minor amounts of sand also occur, with displaced shelf and slope faunas. Beneath this trench fill, the Cocos Plate is covered by Miocene, pelagic white foraminiferal/nannoplanktonic chalk and red brown clays, overlain by Pliocene/Quaternary hemipelagic silt and clay, with rare turbidites derived from the inner slope. In contrast, the trench slope, locally incised by submarine canyons, comprises more varied facies, including pebbly mudstones of Facies Group A1. There is evidence to suggest that 50–200 m-thick lobes of slope sediments are subject to non-episodic, geologically continuous, downslope plastic creep processes (Baltuck *et al.* 1985). Sediment accumulation rates are high, for example DSDP Site 565: 165 m Myr^{-1} from 0–80 m sub-bottom, 13 m Myr^{-1} from 80–90 m sub-bottom and 123 m Myr^{-1} from 90–328 m sub-bottom. Despite plate convergence, little compressional deformation is observed off Guatemala in the forearc. Instead, considerable seismic evidence exists for extensional tectonics. A model based on underplating as the main accretionary mechanism can satisfactorily account for the extension. This active margin off Guatemala has been termed a *convergent-extensional active margin* by Aubouin *et al.* (1982a,b, c, 1984), and Bourgeois *et al.* (1988), to contrast it with convergent-compressional margins such as the Lesser Antilles forearc (Biju-Duval, Moore *et al.* 1984). The convergent-extensional active margins appear to contain a thin sediment cover compared to convergent-compressional margins.

In a set of Seabeam, multi- and single-channel seismic reflection surveys of the continental slope off Peru, between 4°–10°S latitude, Bourgeois *et al.* (1988) recognise three distinct morpho-structural domains: (i) an upper slope, to a depth of 2500 m, dipping generally oceanward at ~5° with a slight upward convexity, and cut mainly by straight V-shaped canyons; (ii) a middle slope, characterised by many curvilinear scarps with large offsets of the seafloor of up to 1200 m toward the trench, interpreted as the result of major mass failure and 'tectonic collapse' of the slope, particularly in the upper parts and (iii) a lower slope to the relatively flat trench floor at ~5000 m depth, showing a hummocky topography and interpreted as the deposits of major slope failure from higher up the continental slope. The lower slope shows ridges that are interpreted as the surface expression of major thrusts in the frontal and toe regions of the accretionary prism. The compressional part of the accretionary prism is a minimum of

15 km and maximum of 85 km wide normal to the trench axis; the change from the predominantly compressional to extensional part of the margin is defined at the middle slope–lower slope boundary, at least where it can be recognised in the region of 5°S latitude (Bourgeois *et al.* 1988). The Peruvian margin, like the Middle America forearc studied on DSDP Leg 84 (von Huene, Aubouin *et al.* 1985), comprises a young accretionary complex stacked against continental crust and also represents a 'convergent-extensional margin' (Aubouin *et al.* 1984).

Many of the slope basins that develop above deforming accretionary-prism deposits are controlled by thrust faults, and growing folds in the sediments revealed as ridges at the surface; for example, as seen in the Muroto Transect slope in the Nankai accretionary prism (Fig. 10.32). These basins have dimensions from 3 × 1 km and a few hundred metres of sediment fill, to 21 × 4 km with >600 m of sediment fill (Stevens & Moore 1985). Other good examples revealed on seismic profiles are documented from the Columbia forearc basin, southwest Caribbean Plate (Lu & McMillen 1983), and from the Makran forearc in the Gulf of Oman, northwest Indian Ocean (White & Louden 1983). In the Nias Ridge examples, the thrusts define the arcward side of a basin, while the trenchward margins show progressive onlap by the basin fill. Within individual basins, the development of subsurface folds is contemporaneous with thrusting, and major fold hinges are parallel to oblique to the strike of the forearc slope. The orientation of major structural elements parallel to the strike of the forearc slope tends to favour sediment feeding along the axes of many basins, with geographically restricted lateral input. These slope basins, controlled by compressional tectonics, are also developed in backarc and foreland basins, wherever thrust-imbricate systems form during sedimentation.

In the Columbia Basin, the western (seaward) part of the accretionary prism is smoothed and masked by the Magdalena Fan (e.g., Ercilla *et al.* 2002), with fast rates of sediment accumulation, whereas farther east there is near-surface complex deformation, including a mid-slope structural high and numerous thrust-top basins (Lu & McMillen 1983). Towards the coast of Makran, White and Louden (1983) document slope basins becoming progressively filled, and ridge tops being buried by sediments, thereby smoothing the forearc slope adjacent to areas with fast rates of sediment accumulation. Also, deeply incised canyons and gullies in the Makran upper slope run southwards down the regional slope, but upon reaching exposed

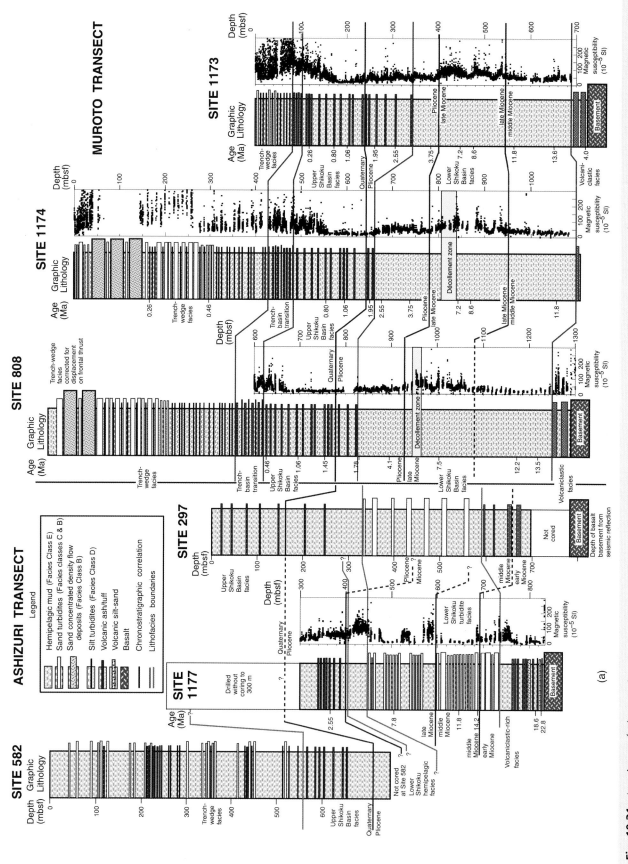

Fig. 10.31 (caption on facing page)

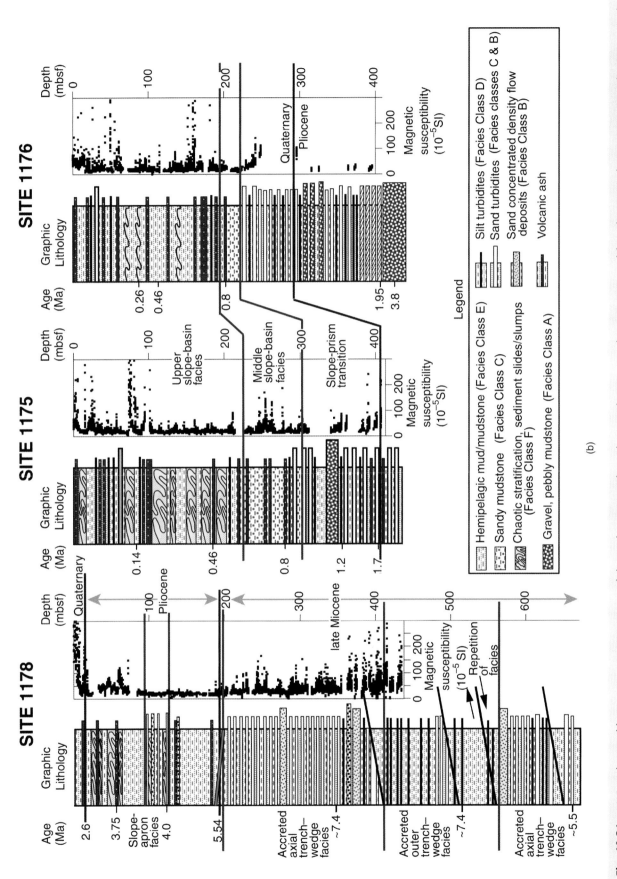

Fig. 10.31 (a) Correlation of facies units, magnetic susceptibility and major time boundaries within stratigraphic successions of the reference and prism toe sites at the Muroto and Ashizuri Transects at the Nankai margin. Time boundaries are in red (solid line). Facies boundaries are in blue (Muroto Transect) and purple (Ashizuri Transect) (patterned lines). Data for DSDP Site 297 are from Shipboard Scientific Party (1975). Data for DSDP Site 582 are from Shipboard Scientific Party (1986). Data for ODP Site 808 are from Shipboard Scientific Party (1991). Note that the effects of sediment imbrication along the frontal thrust of ODP Site 808 have been removed and that the position of the Pliocene/Miocene boundary has been shifted in response to reinterpretation of palaeomagnetic data. From Moore *et al.* (2001). (b) Correlation of facies units, magnetic susceptibility and major time boundaries within stratigraphic successions cored at upslope sites of the Muroto Transect, Nankai margin. Time boundaries are in red (solid line). Facies boundaries are in blue (patterned line). Note that in the original version of this figure at least some concentrated density flow deposits were labelled 'sand turbidites'. From Moore, Taira, Klaus *et al.* (2001).

(b)

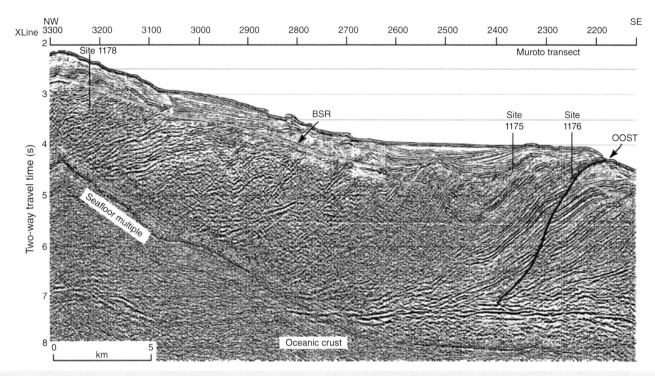

Fig. 10.32 Seismic reflection profile through the Muroto Transect slope ODP Sites 1175, 1176 and 1178. Seismic data are from the 3D seismic survey of Bangs *et al.* (1999) and Moore *et al.* (1999). Xline identifies the crossing line number in the 3D seismic volume. BSR = bottom-simulating reflector related to a gas-hydrate layer; OOST = out-of-sequence thrust. From Moore *et al.* (2001).

fold ridges they are deflected through 90° to run along the axes of the slope terraces/basins.

The clarity of tectonic processes in forearc slope basins generally is best seen where rates of sediment accumulation are relatively slow. For example, Gnibidenko *et al.* (1985) show that the location of many submarine canyons on the forearc upper-slope of the Tonga–Kermadec Trench is strongly influenced by two fault systems, one parallel and the other transverse to the main trend of the trench. Also, horst-and-graben structures, with numerous normal faults, attest to the importance of extensional tectonics at least locally in this forearc where vertical uplift is believed to have been 5000–7000 m (Gnibidenko *et al.* 1985).

During forearc deformation, either compressional or extensional, the rate of back-tilting of the basin floor may be faster than the rate of sediment accumulation and/or slope-incision by canyons, gullies and other channels. In such circumstances, the sediment transport paths may be deflected along the back (arcward) side of these basins until conditions favour re-establishment of regional downslope flow. This process will result in linear facies development along the strike of the forearc slope, and is observed in many slope basins such as the Muroto Basin, a forearc basin off southeast Shikoku, Japan, on the Nankai accretionary prism. Here, the main and tributary submarine canyons have been deflected eastwards along the back of the basin for a distance of ~50 km before turning seaward and down into the Nankai Trench.

The tectonic telescoping of many deep-marine forearcs leads to an overall shallowing-up succession in the basins, ultimately to terrestrial sedimentation unconformably on the vestigial forearc. Such a tectonically controlled shallowing-up will be expressed as a major coarsening-upward trend in the forearc basin, and/or slope-basin fill. These trends are generated because additional off-scraping and frontal accretion of trench sediments causes relatively large lower-slope basins (receiving mainly fine-grained sediments) to be translated gradually toward the arc and into shallower water depths characterised by coarser grained siliciclastics and/or carbonates. An example of such shallowing-up sequences has been described from the Makran accretionary prism by Platt *et al.* (1985), who interpreted a mid-Miocene–Lower Pliocene slope and shelf succession as having been 'deposited directly on abyssal-plain turbidites' (Panigur turbidites) 'without any detectable stratigraphic or structural discordance'. Up to 4 km of shelf deposits (Talar shelf sandstones) pass laterally, southwards, into thin-bedded, fine-grained, deep-marine turbidites over a distance of only a few kilometres. It is hard to see, however, what evidence unequivocally indicates that this supra-accretionary-prism basin developed upon completely undeformed abyssal-plain sediments and not simply upon a relatively unconfined turbidite system of a deep-marine slope basin. Thrust faulting with stratigraphic duplication is common on all scales in the Makran accretionary prism (e.g., duplex structure in thin-bedded turbidites), but apparently does not similarly affect the Panjgur turbidites, contrary to what might be expected if an early compressional deformation had predated the main shallowing-upward sequence in the prism-top basin. For a discussion of this problem, see Platt *et al.* (1985: p. 509).

Figure 10.33 is a conceptual model to show how supra-accretionary-prism basins may develop a shallowing- and coarsening-upward trend as such a basin is translated gradually away from the trench. However, unlike the explanation favoured by Platt *et al.* (1985) for the Makran slope-basin fill (above), our model suggests that: (i) where supra-accretionary-prism basins show large-scale shallowing-upward trends with relatively little internal deformation, it is most likely that they were initiated as slope

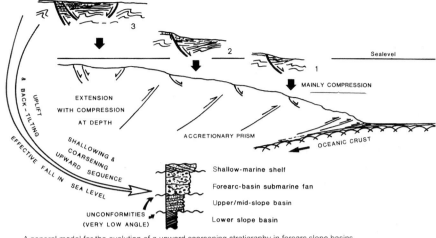

A general model for the evolution of a upward coarsening stratigraphy in forearc slope basins

± Progradation of clastic system at any stage
± Fluctuating eustatic sea level
± Change in nature of source sediments

Fig. 10.33 A general model for the evolution of the stratigraphy in some accretionary-prism-top basins that are progressively uplifted and translated farther back from the trench with time. From Pickering *et al.* (1989).

basins and not trench fills or open-ocean, abyssal-plain, systems; (ii) although difficult to appreciate at outcrop, there may be a series of internal disconformities and/or unconformities, possibly with a consistent overall dip (Fig. 10.33), that reflects accelerated pulses of tectonic activity, such as uplift events and (iii) the internal deformation of such a basin-fill, although gentle, should increase in intensity (and variability?) both with depth and towards the margins of the basin. Naturally, some internal deformation of slope-basin sediments is expected; for example, the margins of these basins, if fault-controlled, are susceptible to tectonic dislocation, but the overall stratigraphy has a high preservation potential, especially in thick successions. High-resolution, multi-channel, seismic reflection profiles through active accretionary prisms may provide the only means of seeing the scale of such subtle features.

In the slope landward of the Middle America Trench prism, off Guatemala, Lundberg (1982) identified a number of approximately 1200 m-thick coarsening-upward trends, but ascribed them to sub-marine fan progradation. However, such trends are of a scale and tectonic location similar to the shallowing-up sequences described above. Clearly, if a slope-basin is translated from the lower to upper slope, but still in deep water, then any shallowing-up may be difficult to identify except on the basis of micropalaeontological evidence and, perhaps, relatively subtle overall grain-size trends. The Great Valley Sequence of California, at present subaerial, evolved from a deep-marine forearc basin that developed during the Mesozoic–Early Cenozoic (Ingersoll 1978a, b), and represents a good example of the shallowing-up (and coarsening-up) trends that can form in forearc basins.

Large-scale coarsening-upward trends generated simply by progradation should not produce unconformities in a forearc basin-fill. However, if the coarsening-upward trends were formed during uplift and back-tilting of forearc slope-basins, then low-angle unconformities should be common, possibly rotated in a consistent direction, and representing phases or pulses of uplift and infill (Fig. 10.33). Unfortunately, the angle of such unconformities may be so small as to escape detection in ancient successions.

Forearc basins may contain relatively coarse-grained deep-marine facies, in part reflecting the proximity to uplifted subaerial sediment sources. The lower–Middle Eocene Scotland Group, exposed in the northeastern part of Barbados, is an example of a forearc basin sitting unconformably on older highly deformed, folded and sheared accretionary prism deposits (Figs 10.34, 10.35). Petrographic and palaeocurrent data suggest that the source for much of the Scotland Group submarine-fan deposits was South America (Pudsey & Reading 1982; Larue & Speed 1983, 1984; Speed 1983; Speed *et al.* 1991, 2005). A possible sediment source was a palaeo Orinoco River, for example as documented for the Orinoco Fan (Deville *et al.* 2003; Callec *et al.* 2010).

Some of the SGF deposits contain multiple Bouma divisions in the same bed, typically with repeated T_b and T_c divisions, interpreted by Larue & Provine (1988) as likely caused by retrogressive slope failure or flow unsteadiness/flow surges: they called these unusual beds 'vacillatory turbidites'.

10.2.8 Fluid flow and plumbing in forearc settings

The subduction factory is associated with some of the world's largest natural disasters linked to tsunamigenic earthquakes (e.g., Tobin & Kinoshita 2006a, b; Moore *et al.* 2007). For example, southeast Japan and the Nankai Trough have a 1300 year documented history of great earthquakes, many of which have generated tsunamis. Recent events of this type include the 1944 Tonankai M 8.2 and 1946 Nankaido M 8.3 earthquakes (Ando 1975; Kodaira *et al.* 2000; Obana *et al.* 2001; Hori *et al.* 2004; Ichinose *et al.* 2003; Kikuchi *et al.* 2003; Baba & Cummins 2005; Satake *et al.* 2005; Baba *et al.* 2006, 2009; Park & Kodaira 2012; Sugioka *et al.* 2012). The role of fluids in the subduction factory plays a key part in understanding the dynamic rheology of accretionary prisms, the development of the seismogenic zone and the internal deformation of the subducting sediments. A detailed treatment of this topic is beyond the scope of this book, but for a more in-depth understanding the reader is referred to Tarney *et al.* (1991), Taira *et al.* (1992) and Moore *et al.* (2001). Figure 10.36 summarises

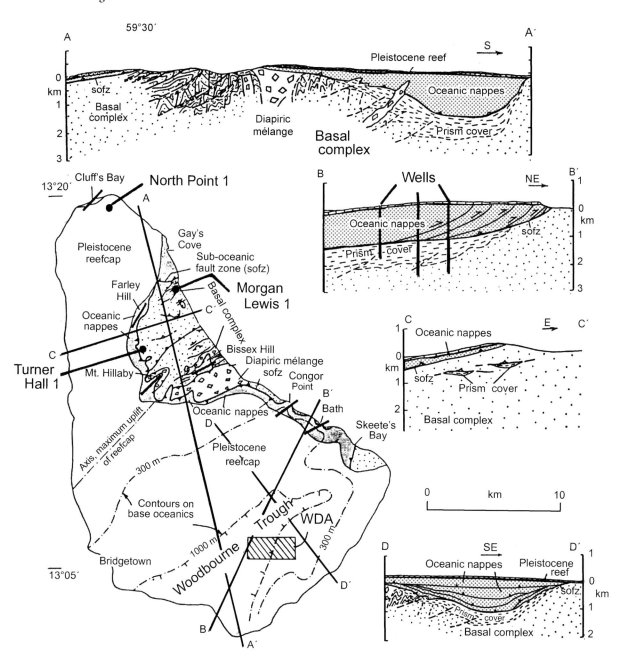

Fig. 10.34 Geological map and structural cross-sections of Barbados. Oceanic nappes are also referred to as oceanic allochthon. The Oceanic nappes contain deep-marine SGFs (see Fig. 10.35) in the lower to middle Eocene Scotland Group as described by Pudsey and Reading (1982). From Speed *et al.* (1991) (*cf.* Donovan 2005; Speed *et al.* 2013).

some of the principal sources and pathways for fluids in accretionary prisms (modified after Tarney *et al.* 1991). Fluid expulsion impacts the chemical and mass budgets at subduction zones, particularly where focussed along faults and stratigraphic conduits.

Geochemical anomalies within and adjacent to décollement zones, such as pore-water freshening (low-chloride anomalies) and the presence of thermogenic hydrocarbons, suggest the long-distance, focussed flow of deeply sourced fluids from farther back in the prism, including the release of structurally bound water during the illitisation of smectites during *in situ* dehydration reactions

(Underwood & Pickering 1996). The décollement is also associated with increased porosities, reduced densities (Figs 10.37, 10.38a, b, 10.39) and reduced velocities. Increased cohesion within sediments due to diagenesis and/or very low-grade metamorphism appear to be important in permitting the transmission of elevated stresses to the updip end of the seismogenic zone within accretionary prisms (e.g., Schumann *et al.* 2014).

Saffer and Screaton (2003) used a simple model that coupled fluid flow and solute transport to evaluate the sharp chemical gradients observed at décollements, and found that the observed geochemical

Fig. 10.35 Pebbly and laminated sandstones of Facies Groups A2, B1, B2 and C2 in the lower to Middle Eocene Scotland Group, eastern Barbados, an example of a forearc slope basin.

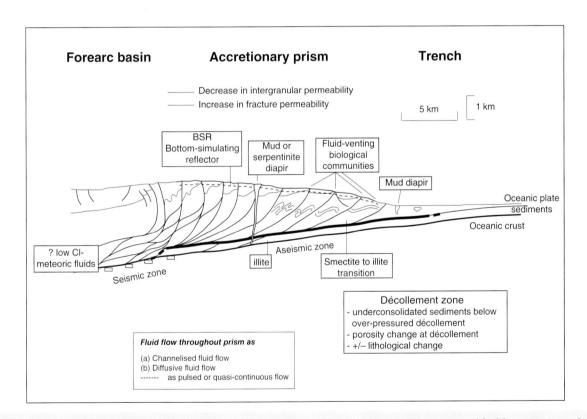

Fig. 10.36 Summary of some of the principal sources and pathways for fluids in accretionary prisms. Modified from Tarney *et al.* (1991).

anomalies at the Northern Barbados and Costa Rican subduction zones could be explained either by recent pulses of flow, or by sustained flow along the décollement zone coupled with modest vertical fluid expulsion from consolidating underthrust sediments. The latter interpretation is consistent with estimates of the upward flow rate at Costa Rica based on estimated pore pressure gradients and measured permeabilities within the underthrust sediments (Saffer

& Screaton 2003). An important implication of this work is that recent pulses of flow along fault conduits may not be required to explain the geochemical anomalies. Additionally, the mixing of locally derived fluids flowing upward from the underthrust sediments and deeply sourced fluids flowing along the décollement zone provides an explanation for the observed changes in pore-water freshening along the décollement at Costa Rica.

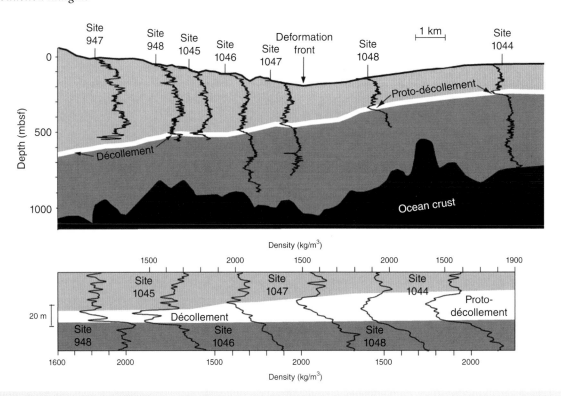

Fig. 10.37 Generalised cross-section through all the ODP Leg 171A sites with LWD density data collected during both ODP Legs 156 and 171A (Bangs *et al.* 1999). The lower panel is an enlargement of the upper panel in the vicinity of the décollement and proto-décollement. The upper lightly shaded area of both panels represents the accretionary prism and the equivalent sediments in the incoming sedimentary section. The lower darker shaded area of both panels indicates the portion of the sedimentary section being underthrust and its equivalent seaward of the deformation front. The white band separating the shaded areas is the low-density interval that is associated with the décollement zone and proto-décollement zone. It is not equivalent in thickness to the structurally defined décollement or proto-décollement but may encompass or be included in them. Depth scales on both upper and lower panels indicate relative, not absolute, depth.

In situ measurement of physical properties in the deep-marine deposits of accretionary prisms permits an evaluation of processes such as consolidation, cementation and dilation during deformation, fluid flow and faulting. Because seismic images are affected by changes in physical properties, their measurement allows for calibration of seismic data as a tool for remotely sensing processes of deformation and fluid flow. Logging While Drilling (LWD) also provides an industry-standard tool for *in situ* evaluation of physical processes, including transient borehole conditions. Such data help with the interpretation of similar, but less active, systems in sedimentary basins elsewhere, thereby contributing to the analysis of groundwater and hydrocarbon migration, and earthquake processes.

Isotopic anomalies within interstitial pore-fluids of forearcs, together with the excursions in element concentrations, show that diagenetic processes operate at depths as shallow as only a few hundred metres, including clay-mineral diagenesis, alteration of tephra, and possible transformation of biogenic silica from abundant diatoms. Dehyle *et al.* (2003) sampled interstitial pore waters from ODP Site 1150, where ~1200 m of sub-seafloor sediment from the upper Japan Trench forearc was recovered. They analysed for elemental concentrations and Cl, Sr and B isotopes. Although chlorinity showed down-hole freshening to values as low as ~310 millimolar (0.55 × seawater) in the deeper part of the claystone-dominated succession, both Sr and B concentrations showed an overall increase. Sr reached concentrations of

up to >250 μM (~3.00 × seawater), whereas B-enrichment was even stronger (3920 μM; i.e., 9.30 × seawater). The variations in concentration correspond to fractionation reactions in the deep, tectonically deformed part of the forearc. The highly fractured portion of ODP Site 1150 (from ~700 m to the total depth of the hole) contains two shear zones that very likely have acted as conduits along which deep-seated fluids were expelled to the seafloor. These fluids not only show the strongest freshening of Cl, but are also characterised by low $\delta^{37}Cl$ measurements (to -1.1‰), the heaviest $\delta^{11}B$ measurements (~40–46‰) and the least radiogenic $^{87}Sr/^{86}Sr$ measurements. Along with the enrichment of some mobile elements (e.g., Sr, B, Li), and the enhanced fluid flow through permeable penetrative faults throughout the forearc (such as along the shear zones at ODP Site 1150), such focussed fluid flow could be an efficient mechanism for reflux of such elements from the deep forearc into seawater.

Logging While Drilling datasets from ODP Legs 170 and 171A enable detailed correlation of sediments between drill sites and estimates of the change of bulk densities (Tokunaga 2000; Saito & Goldberg 2001). The data suggest that the interbedded mudstones in turbidite bedsets have experienced larger volumetric strain while structureless claystones have not. The mudstone beds, sandwiched both above and below by relatively highly permeable sandstones, were interpreted to have undergone dewatering because laterally continuous sandstones acted as effective drains for dewatering. Tokunaga (2000) suggested that the migration of warm fluid along sandstones in

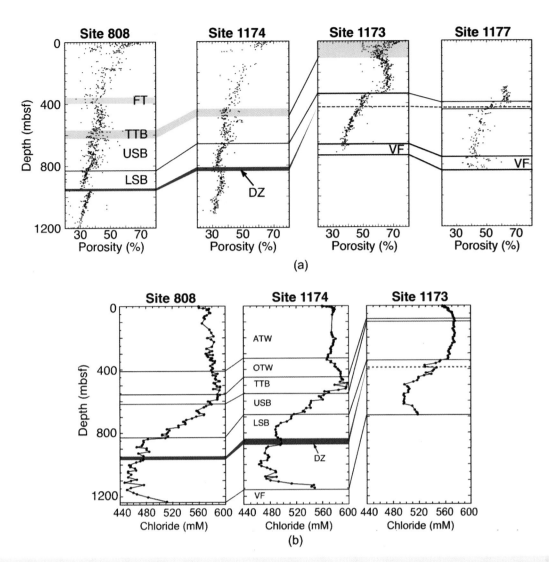

Fig. 10.38 (a) Porosity values across the Muroto Transect (ODP Sites 1173, 1174 and 808) and Ashizuri reference site (Site 1177). Lithologic units and major structural features are shown. The décollement location is shown by red shading where it was observed and by a dashed line at the stratigraphically equivalent depth at the reference sites, which are seaward of the surface trace of the frontal thrust of the accretionary prism. VF, volcaniclastic facies; DZ, décollement zone; LSB, lower Shikoku Basin facies; USB, upper Shikoku Basin facies; TTB, trench-to-basin transition; FT, frontal thrust. From Moore *et al.* (2001). (b) Chloride concentrations in interstitial water samples from the Muroto Transect reference (ODP Site 1173) and prism toe sites (ODP Sites 1174 and 808). VF, volcaniclastic facies; DZ, décollement zone; LSB, lower Shikoku Basin facies; USB, upper Shikoku Basin facies; TTB, trench-to-basin transition; OTW, outer trench wedge facies; ATW, axial trench wedge facies. From Moore *et al.* (2001).

the underthrust package might explain why high heat flow is observed at the ocean floor at Site 672, where sandstones exist, but not at Site 543, where they are absent.

Siliceous layers such as diatomaceous and radiolarian clay tend to be fluid-bearing, and the stratigraphic position of such zones is a critical factor in the fate of the subducted sedimentary section (Saito & Goldberg 2001). On the Costa Rica Margin, Saito and Goldberg (2001) found that the sedimentary section on the Cocos plate is underthrust intact beneath the toe of the Caribbean Plate with no frontal off-scraping where a siliceous fluid-bearing zone is present only in the upper part of the section. On the Barbados margin, a layer of radiolarian clay exists, providing a narrow zone of mechanical weakness and anomalously high dewatering in the middle of the sedimentary section. This layer divides the sediments that are subducted from those that are accreted. Accreted and subducted sediments show different compaction styles. Apparently, the accreted sediments are characterised by rapid compaction with vertical thickening, whereas the subducted sediments are characterised by slow compaction with vertical flattening; the latter being observed in both the Barbados and Cocos margins. Saito & Goldberg (2001) attributed the vertical thickening of the accreted sediments (observed in the Barbados margin) to horizontal tectonic compaction/deformation in the early stages of deformation.

10.2.8.1 Wet-sediment injections

The tectonics and sedimentation patterns of active convergent margins, especially forearcs and accretionary prisms, provide ideal

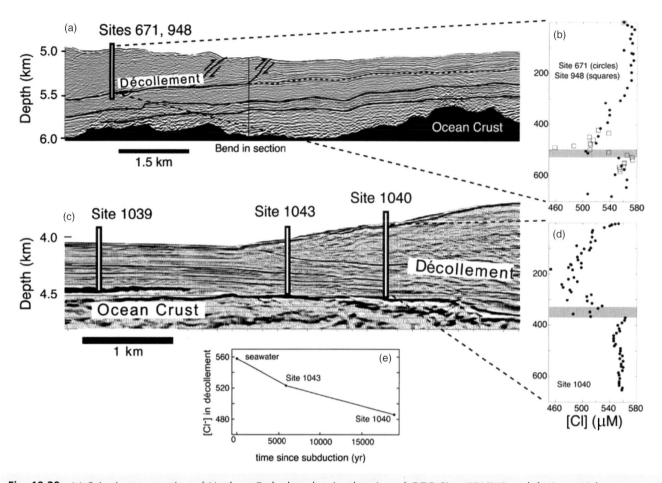

Fig. 10.39 (a) Seismic cross-section of Northern Barbados, showing location of ODP Sites 671/948 and (b) interstitial pore-water chlorinity at ODP Sites 671 (filled circles) and 948 (open squares). (c) Seismic cross-section (line UT-CR20) for Costa Rica, showing location of ODP drill sites, and (d) pore-water chlorinity at ODP Site 1040. Grey-shaded area denotes the décollement location in both geochemical profiles. (e) Minimum chlorinity within the décollement at Costa Rica, shown as a function of time since subduction. From Saffer and Screaton (2003).

environments in which overpressured sediments may become liquefied and fluidised to produce wet-sediment structures and injections such as mud diapirs or seepages. While all scales of sandy injections may occur, the largest diapirs appear to comprise a mud or shale matrix with incorporated blocks of lithified or semi-lithified rock. Examples of mud and shale diapirs (as pipes or ridges) and volcanoes are well documented from the Lesser Antilles accretionary prism (Figs 10.40a, b) (Higgins & Saunders 1967; Biju-Duval *et al.* 1982; Westbrook & Smith 1983; Brown & Westbrook 1987), and from the Timor Trough area (Barber *et al.* 1986; Karig *et al.* 1987). Camerlenghi and Pini (2009) provide an illustrated description and discussion of the possible processes of formation of many circum-Mediterranean mud volcanoes and related deposits/features. Interpretation of a seismic profile across a mud volcano in the Barbados Basin, by Brown and Westbrook (1988), suggests that there are lateral, layer-parallel, injections stretching for more than 2 km from the main feeder pipe. In ancient successions with limited exposure, such chaotic layers, up to hundreds of metres thick, could easily be misinterpreted as debrites, or if severely deformed as entirely 'tectonic' mélanges.

Shale diapirism may develop where sediments become overpressured at depth, leading to intrusion into overlying and surrounding strata in a fluid state. The vertical escape velocities of the mud/water

slurries may reach sufficient magnitudes to hydraulically fracture consolidated and semi-lithified lithologies, incorporate them into the slurry as blocks, and even extrude them from clastic volcanoes on the sea bed, or subaerially on islands: islands formed from mud volcanoes are documented, for example Chatham Island south of Trinidad in the West Indies (Adams 1908; Bower 1951; Arnold & Macready 1956; Birchwood 1965; Higgins & Saunders 1967), and from Timor (Barber *et al.* 1986). An onland eruption of a mud volcano this century occurred near Gisbourne, New Zealand (Strong 1931; Stoneley 1962).

Liquefaction of overpressured sediments at depth and subsequent wet-sediment injection may result from: (i) seismically-induced stresses due to faulting; (ii) loading or overburden stresses induced by tectonic thrust-thickening and/or fast rates of sediment accumulation, including slides; (iii) hydrocarbon build-up, for example methane or gas hydrates, during the decay of organic matter within shales (Hedberg 1974); (iv) escape of deep crustal or upper mantle thermogenic gases and fluids; (v) dehydration reactions during compaction and/or mineralogical transformations in very low-permeability and impermeable clays and (vi) unstable density inversions, possibly caused by thrusting relatively dense lithologies, such as off-scraped oceanic basalts, over less dense, water-saturated muddy sediments.

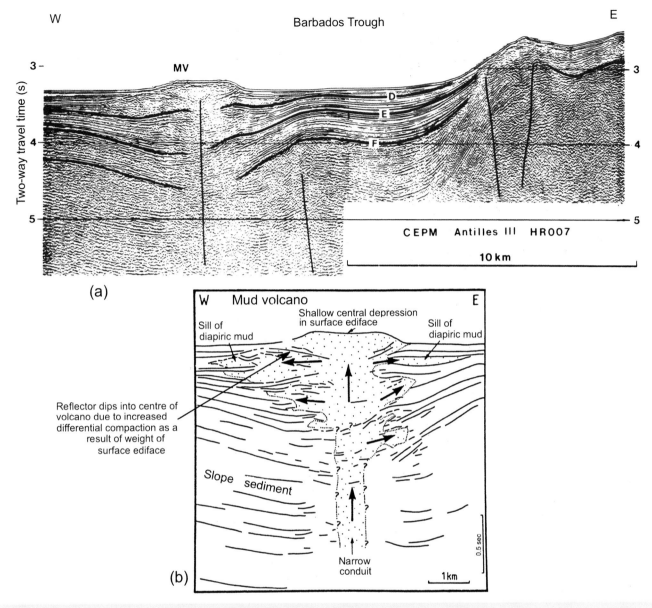

Fig. 10.40 (a) Migrated seismic reflection profile to show a mud diapir from the eastern side of the Barbados Trough, an accretionary-prism-top basin in the Lesser Antilles. The diapir has penetrated the seafloor to form an 80 m high × 2 km wide mud volcano (MV) that has also intruded laterally into the sediments (chaotic reflectors). Uplift of the margin of the basin is shown by the thinning and upwarping of sedimentary horizons D, E and F. From DSDP Leg 78, Biju-Duval *et al.* (1984; *cf.* Westbrook *et al.* 1984). (b) Interpretation of the seismic line in (a) by Brown and Westbrook (1988) to show the shape of the mud diapir and volcano; note the inferred presence of layer-parallel injections >2 km from the main feeder pipe.

Following shale diapirism, the reduced pore pressure at depth causes a volume decrease in the 'donor horizon', with subsidence peripheral to the intrusion, as seen on seismic profiles. Although such intrusions will tend to inject vertically or subvertically, the easiest local escape paths may be horizontal to produce a complex wet-sediment sill and dyke system. In all cases, any obliquely-aligned bedding, layering or foliation immediately adjacent to an intrusion will tend to be deformed into alignment with the margins. Vertical injection through horizontal strata may upturn bedding against a dyke, a feature commonly observed in field exposures.

Mud diapirs may occur as near-circular pipes or as sheets with more irregular geometry. They may: (i) extrude as mud volcanoes and

ridges; (ii) appear related to fault zones and joint patterns or (iii) occur in an apparently random distribution. Wet-sediment extrusion on the seafloor may locally change sediment transport paths and provide a local, exotic, source of sediment whose petrography, faunal content and age are anomalous with respect to the surrounding sediment on the seafloor.

In the Sunda Arc accretionary prism, south of Sumba, Indonesia, Breen *et al.* (1986) used SeaMARC II side-scan sonar mapping to define three distinct structural styles within 15–25 km of the thrust front: (i) lower-slope folds and thrusts, (ii) mud volcanoes and mud ridges paralleling the thrust front and (iii) conjugate strike-slip faults 10–15 km from the thrust front. The mud injections occur in water

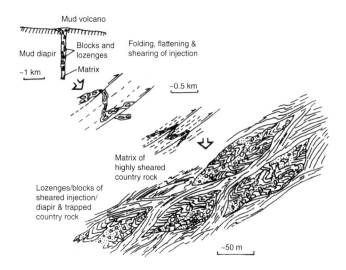

Mud volcano

Mud diapir

Blocks and
lozenges

Folding, flattening &
shearing of injection

~1 km

Matrix

~0.5 km

Matrix of
highly sheared
country rock

Lozenges/blocks of
sheared injection/
diapir & trapped
country rock

~50 m

Fig. 10.41 Conceptual model for the deformation of a mud diapir in a forearc setting by folding, flattening and stretching to produce a mélange. Note that the mélange may only show original wet-sediment deformation features within the 'lozenges' of material, enclosed within entirely tectonically deformed thin-bedded, fine-grained lithologies. Blocks within the mélange may range from igneous and metamorphic to sedimentary, including biogenic, rocks. From Pickering *et al.* (1988a).

depths greater than 4000 m where seafloor gradients range from 2.5°–5.3°. The mud ridges are 200–300 m high, 1–2 km wide, 518 km long, and trend east–west slightly oblique to the thrust front. The mud volcanoes have similar relief above the seafloor but are more symmetrical. Reflections from the mud injections suggest a blocky, rubble-strewn relief. Deformation of the wet sediments in the lower slope occurs by folding and thrusting, wrench faulting and mud intrusion/extrusion. This deformation has altered sediment-transport paths such that submarine feeder channels are sinuous, branching, and locally parallel or oblique to the strike of the slope.

One of the principal objectives of ODP Leg 110 (Mascle *et al.* 1988), in the northern Barbados Ridge, was to investigate the role of pore-fluids in an accretionary prism. Pore-fluid data from the various holes in an east–west transect (along ~15°32'N Latitude), with Site 674 farthest from the deformation front (at ~17 km distance), revealed two distinct fluid realms. The methane-bearing region is restricted to the décollement zone and subjacent underthrust sediment, whereas the methane-free region, with chloride anomalies, occurs in the accretionary prism. Methane content decreases below the décollement zone but increases again towards the permeable sand layer at the base of the hole in Site 671. Similar anomalies are observed at Sites 676 and 674 (see DSDP/ODP/IODP site map in Appendix 1). Carbon isotopic data on methane samples from Leg 110 (Mascle *et al.* 1988) suggests that it has a thermogenic origin. Appropriate physio-chemical conditions are not found in the drilled holes. Negative chloride anomalies occur along faults, for example the chloride minimum at the décollement zone in Site 671 (Figs 10.39a, b). Chloride anomalies are associated with three out of five faults intersected at Site 674. The chloride anomalies may occur because of: (i) ultra filtration processes in the buried clays that act as semipermeable membranes to facilitate the flow of water but inhibit the migration of various ions (*cf.* White 1965), (ii) dehydration of smectites at depth or (iii) melting of gas hydrates as documented from the Middle America Trench by Gieskes *et al.* (1984).

These important results from ODP Leg 110 demonstrate the chemical partitioning that occurs in progressively dewatering sediments due to fluid flow within subduction systems. The stress system within accretionary prisms tends to favour the development of hydrofractures that approximately parallel the major décollement rather than having steep dips. However, the occurrence of mud

volcanoes shows that during times of rapid escape of pore-fluids, subvertical hydrofractures form. Thus, pore-fluid activity, and preferential fluid migration along fault surfaces, provides a fundamental key to understanding wet-sediment deformation within accretionary prisms. Mud diapirism as pipes or sills, seepage and mélange formation, including microstructures such as vein arrays (e.g., Knipe 1986), are all governed by pore-fluid behaviour.

It is important to appreciate that mud or shale diapirism, together with debris flow, slide and rockfall processes, may produce deposits that appear similar to one another in limited outcrops or in cores from the subsurface. Furthermore, the deposits from all these processes may be tectonically sheared to form mélanges (Cowan 1985; Pickering *et al.* 1988a). Thus, the correct interpretation for the origin of chaotic deposits with a mud matrix depends upon very accurate field mapping, detailed small-scale observations and regional considerations. In ancient successions, the deeper part of a mud or shale diapir has a high chance of being preserved whereas the surface and near-surface part probably will be eroded. A corollary of this difference in preservation potential is that while the deeper part of a diapir will tend to be preserved, it is most susceptible to severe deformation. Figure 10.41 shows a likely deformation path for the deeper part of a mud or shale diapir, where folding, flattening and shearing may produce a mélange in which highly sheared country rock encloses lozenges of mud-shale diapir. The original diapiric relationships will only be preserved within the tectonically formed lozenges (Fig. 10.41). Mélanges with these features, and interpreted as highly deformed diapirs, are documented from the Shimanto Belt of southern Japan (Pickering *et al.* 1988a).

10.3 Arc–arc collision zones

Arc–arc collision, unlike processes associated with the subduction of 'normal-thickness' oceanic crust, generates a different tectonic style and corresponding deep-marine basins. In a comparative study of (i) the present collision zone between the Izu–Bonin Ridge (island arc) and mainland Japan (Honshu Arc), and (ii) the late Miocene–Pliocene of onshore southeast Japan, Soh *et al.* (1991) proposed that arc–arc collision processes and the resulting stratigraphic successions may be repetitious and predictable.

In southeast Japan (Fig. 10.42), arc–arc collision has led to the incremental accretion of segments of delaminated Izu–Bonin Arc crust onto the Honshu Arc, associated with the sequential southward migration, as a series of jumps, of the plate boundary and trench (Fig. 10.43). Prior to the accretion of a segment of Izu–Bonin Arc crust, the leading edge undergoes uplift to generate an approximately trench-parallel topographic high, with the Zenisu Ridge (Figs 10.44, 10.45) being the present example and the Hayama–Mineoka uplift zone (HMUZ) a Miocene–Pliocene example (Fig. 10.43) (Soh et al. 1991). The ridge separates a northern trench or trough from a southern intra-oceanic arc basin. During collision-accretion, the trench receives both Honshu Arc-derived, terrigenous, and Izu–Bonin Arc-derived, volcaniclastic, sediments, whereas the arc basin tends to receive only arc detritus. During the final stages of accretion, the arc basin begins to receive ever-increasing volumes of terrigenous, Honshu Arc-derived, detritus fed through basement-controlled canyons. The accretionary process is accompanied by intense deformation and the residual deep-marine basin is then infilled above an angular unconformity.

The Izu Collision Zone has experienced several million years of progressive tectonic deformation, probably initiated in relation to the opening of the Japan Sea ~15 Ma, in the middle Miocene (Itoh 1986), and later changes in plate vectors in the Western Pacific region (*cf.* revised palaeogeographies by Pickering *et al.* 2013; Fig. 10.25). This collision-related deformation resulted in the formation of tectonic segments of up to several tens of kilometres in dimensions, bounded by thrusts, and comprising volcanic and volcaniclastic successions of Izu–Bonin Ridge origin. Relatively short-lived basins filled by an overall coarsening-and-shallowing-upward sedimentary trend, developed along the boundary thrusts. The basin-fill sediments tend to become younger in age in the basins from north–south, and range from middle Miocene to Pleistocene. The thrust movements that generated the basins apparently shift southward with time, which was interpreted by Soh *et al.* (1991) as the incremental migration of the plate boundary caused by the incorporation of the frontal segment of the Izu–Bonin Ridge onto the Honshu Arc side (Taira *et al.* 1989). Collision-related deformation has not been restricted to the Izu Collision Zone but also extends to the nearby Philippine Sea

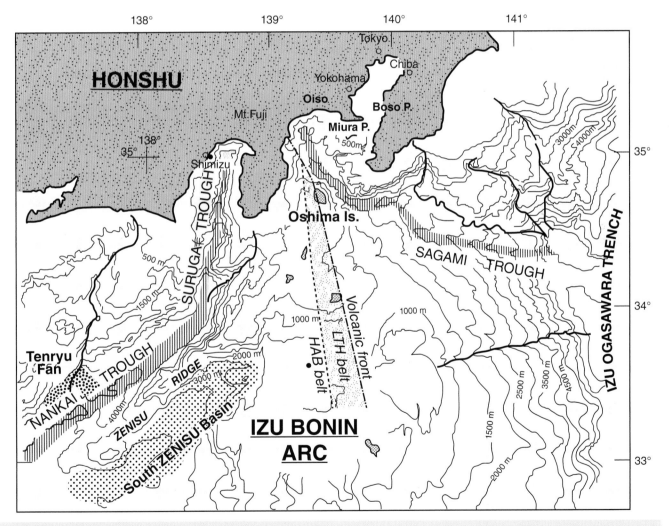

Fig. 10.42 Geomorphological index map around the southeastern part of the Honshu Arc, Central Japan. Note the location of the present volcanic front and geochemical zonation of volcanic rock along the volcanic front (from Takahashi 1986): LTH, low-alkali tholeiite zone; HAB, low-alkali high-alumina basalt zone. From Soh *et al.* (1991).

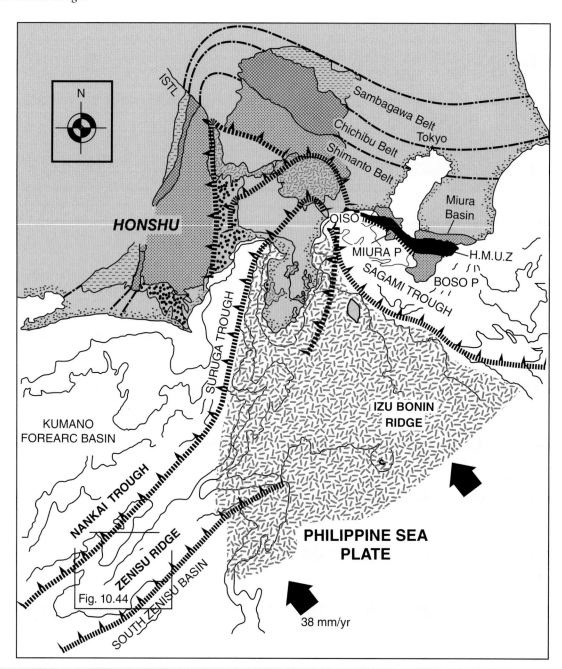

Fig. 10.43 Geotectonic interpretation of the southeastern part of the Honshu Arc (Izu Collision Zone), showing the flexure of geological belts and modern basins or troughs as a result of the collision of the Izu–Bonin Arc with the Honshu Arc. Convergence vector of the Philippine Sea plate from Seno (1989) is shown as arrows. HMUZ, Hayama–Mineoka uplift zone. ISTL = Itoigawa-Shizuoka Tectonic Line. P = Peninsula. From Soh *et al.* (1991).

Plate in the northeastern Shikoku Basin (Chamot-Rooke & Le Pichon 1989). A linear topographic high, the Zenisu Ridge, running from the northeastern Shikoku Basin to the Izu–Bonin Arc (Fig. 10.42), a few tens of kilometres seaward of the Nankai Trough, is a consequence of the rupture and uplift of the Shikoku Basin oceanic crust, due to intraplate compressive tectonics (Le Pichon *et al.* 1987b; Lallemant *et al.* 1987; Taira *et al.* 1989). The probable future result of the interplate deformation is that the intra-oceanic South Zenisu Basin will be activated as the new trench to the south of the Zenisu Ridge. Volcaniclastic material, derived from the Izu–Bonin Ridge to the east,

has been deposited in the intra-oceanic basin. Indeed, Le Pichon *et al.* (1987b) suggested that the thrust marginal to the South Zenisu Basin will change from an intraplate fault to an incipient plate-boundary thrust.

During the development of the Hayama–Mineoka uplift zone, two deep-marine sedimentary basins formed, a Northern and Southern Basin (Figs 10.46, 10.47). The former basin was sited on the plate boundary, and was filled by Honshu Arc-derived detritus and Izu–Bonin Arc-derived material. In contrast, the Southern Basin was an intra-oceanic basin on the forearc of the Izu–Bonin Ridge.

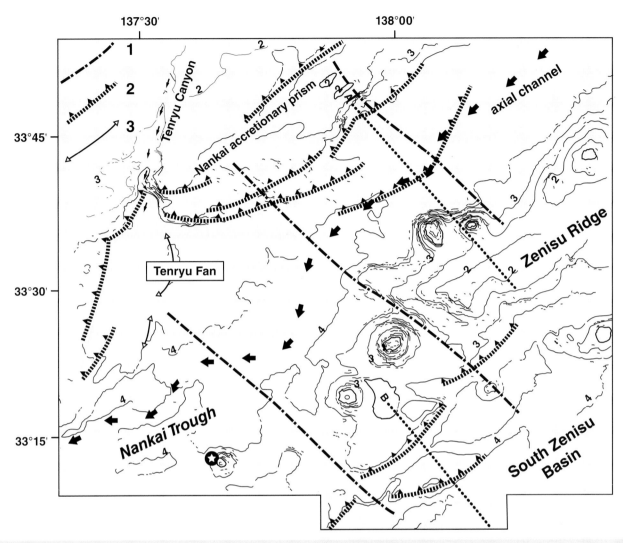

Fig. 10.44 Geomorphological and tectonic map around the Zenisu Ridge region. Bathymetric contours labeled at 2, 3 and 4 km water depth. Note development of thrust and transverse fault systems. Dredged site KH86-5, Hukuho-muru cruise, from where *Calyptogena* and tube-worms were photographed and recovered is shown by a star. 1, strike-slip transverse fault; 2, thrust; 3, anticline with plunge arrows. A and B show seismic lines in Fig. 10.45a, b. From Soh *et al.* (1991).

Lapilli and scoria of low-alkali tholeiite, and tuffaceous mudstones, were deposited in the Southern Basin, without Honshu Arc-derived material. Small graben-like supra-ridge basins in the HMUZ were sites for the accumulation of detritus derived from the underlying Mineoka and Hayama Groups, together with shallow-marine limestone. Ultramafic breccias and chert pebbles, derived from the Mineoka Ophiolite, were deposited in the supra-ridge basins with shallow-marine faunas. Similar lithologic variation can be recognised between the Nankai Trough and the South Zenisu Basin around the present Zenisu Ridge. Lithologic variations in neighbouring tectonic segments with terrigenous detritus-free or detritus-dominant clastics, and ophiolitic (plus island arc) sequences, represent important features in arc–arc collision zones. Finally, intra-plate deformation of a subducting arc can change a plate boundary position, so that intra-oceanic basins become inverted. The dimensions of accreted arc segments are probably governed largely by the spacing of the zones of intra-plate deformation and the position of transform faults cutting a ridge. Apparently, the spacing of such intraplate deformation in the

Izu–Bonin Ridge is narrower than that of the northern Indian Ocean related to the Indo–Asia collision, because of differences in rheological conditions of the colliding crust. Soh *et al.* (1991) speculated that smaller and thinner tectonic segments of volcanic arc origin and foreland basin fill would result from arc–arc collision compared to those of arc–continent and continent–continent collision zones.

The formation of the Hayama–Mineoka uplift belt (HMUZ, Fig. 10.43) as an intra-plate rupture probably coincides chronologically with the transition from the Hayama–Hota Group to the Miura Group (~11–10 Ma), as seen in the synchronous transition from the hemipelagic Hayama–Hota Group to the trench-fill Miura Group in the Northern and Southern basins. The HMUZ was breached before the formation of the Kurotaki Unconformity (~2.5 Ma), because the Hasse Formation in the Southern Basin first received Honshu Arc-derived material at that time. The rheological model of Chamot-Rooke and Le Pichon (1989) suggests continuous uplift at the intra-plate ridge (Zenisu Ridge) and subsidence at the plate boundary basin (Nankai Trough) from the time of initial

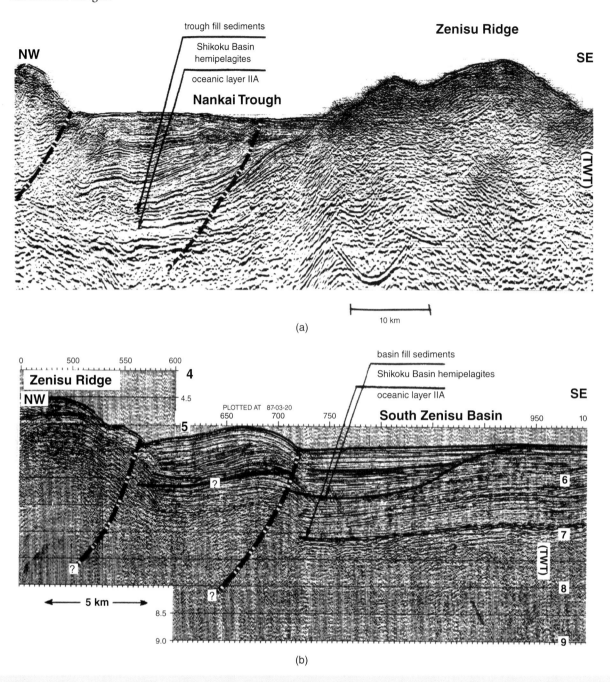

Fig. 10.45 Seismic profile across the Nankai Trough, Zenisu Ridge and South Zenisu Basin. Profile lines are shown in Fig. 10.44. (a) Representative seismic profile across the Nankai Trough and Zenisu Ridge. Note the interpreted level of the boundaries between the trench-fill sediments, hemipelagic sediments and oceanic layer IIA. (b) Seismic profile across the South Zenisu Basin, showing three layers, and deformation due to thrusts. From Soh *et al.* (1991).

lithospheric rupture. As the geohistory and subsidence analysis of the Northern Basin suggests that the basement of the Northern Basin continuously subsided during the development of the HMUZ, the Chamot-Rooke and Le Pichon (1989) model (in terms of compressive mechanical failure of a thin perfectly-elastic plate) appears to best explain the development of the Northern Basin under a compressive, collision-related, stress field. The Miocene–Pliocene HMUZ and the modern Zenisu Ridge demonstrate the repetitious style of tectonic fragmentation of arc crust and basin formation during arc–arc collision.

Soh *et al.* (1991) defined three distinct seismic units in the South Zenisu Basin; 1.6–1.7, 2.3 and 3.5 km s^{-1} (RMS velocity) layers, from bottom to top (Fig. 10.45b). Based on the acoustic characteristics and the samples obtained during the *Tansei Maru* and KAIKO *Nautile* dives (Le Pichon *et al.* 1987b: their table 1). Soh *et al.* (1991) interpreted these as the oceanic basement, the Shikoku Basin hemipelagic cover, and volcaniclastic thin-bedded turbidites, from bottom to top (Fig. 10.45b, 10.47). The seismic profile is consistent with the recognition of the trench wedge shape in the upper volcaniclastic turbidite layer, which is progressively deformed northwestward. The basin fill

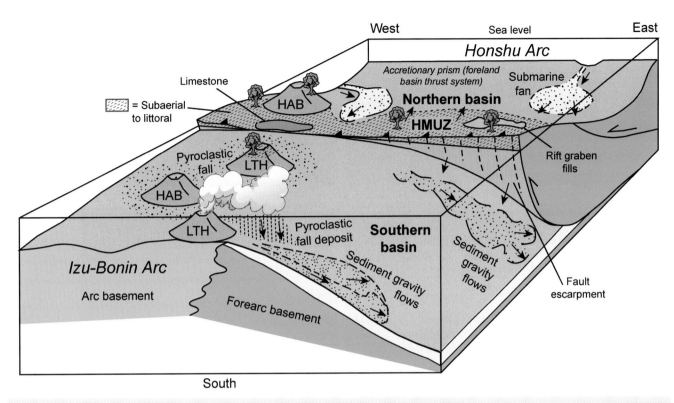

Fig. 10.46 Reconstruction of the sedimentary environments around the Northern and Southern Basins adjacent to the Hayama–Mineoka uplift zone (HMUZ), southeast Japan, ~4–5 Ma. Small arrows show sediment transport paths. Diagonal dashed rule and stylised trees indicate subaerial and infra-littoral environments atop the uplift zone and emergent volcanoes. The "Fault escarpment" is the frontal thrust of the accretionary prism associated with the Honshu Arc, as it bites southward into the subducting Izu-Bonin (Ogasawara) Arc and its forearc basin. HAB = high-alumina basalt; LTH = Low-alkali tholeiite; HMUZ = Hayama–Mineoka uplift zone. From Soh *et al.* (1991).

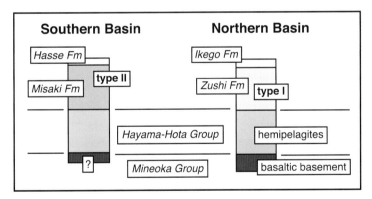

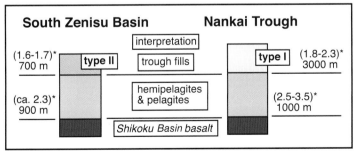

* RMS velocity and layer thickness

type I: trough fills derived from two provinces on different plates

type II: intra-plate trough fills derived from the oceanic plate

Fig. 10.47 Schematic sections of the Miocene–Pliocene Boso–Miura region (Fig. 10.46) and modern Zenisu region. Note the similarities between the Southern Basin and South Zenisu Basin, and between the Northern Basin and Nankai Trough. From Soh *et al.* (1991).

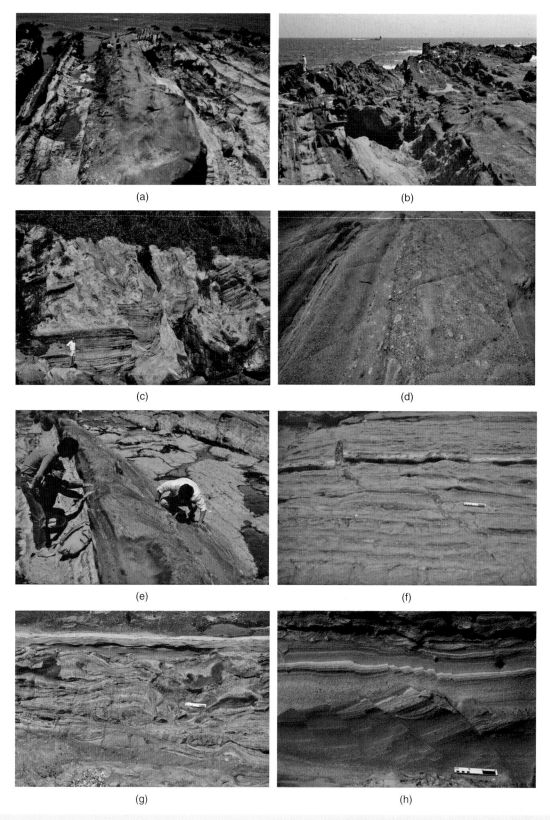

Fig. 10.48   Misaki Formation outcrops on Miura and Boso Peninsulas, southeast Japan. (a) Submarine channel-fill with dark-coloured scoriaceous sediments of Facies Classes A and B. Beds young to left. (b) Submarine channel erosional margin, with infill of Facies Classes A and B cutting into Facies Classes C and D. Beds young to right. (c) Large blocks within 70 m-thick sediment slide of Facies F2.2. (d) Debrite of Facies A1.3 with light-coloured pumiceous clasts. Bed youngs to right. (e) Tripartite volcaniclastic bed of Facies C2.1 (younging to right). (f) Wet-sediment injections infilling fractures. (g) Sediment slide of Facies F2.1 showing irregular upper surface, and draped overlying current-rippled fine-grained sandstones; 15 cm scale bar. (h) Syn-sedimentary normal extensional small-scale faults in Facies B1.2, B2.2 and D2.1; 15 cm scale bar.

is ~1500 m thick: the thickness of the Shikoku Basin hemipelagic sediments does not change laterally from the Zenisu Ridge to the South Zenisu Basin (Fig. 10.47). The upper volcaniclastic sediments were interpreted as derived from the Izu–Bonin (Izu–Ogasawara) Ridge along the axial channel of the elongate basin. Honshu Arc-derived detritus is absent in the South Zenisu Basin. Soh *et al.* (1991) inferred that the South Zenisu Basin, therefore, was likely formed after deposition of the upper volcaniclastic turbidites, as a sedimentary basin due to intra-plate deformation. The sedimentary succession in this basin is comparable to that of the Oligocene–Pliocene Southern Basin, ranging from the Mineoka Group to the Misaki Formation (Fig. 10.47).

The middle Miocene–lower Pliocene (~14–3 Ma) Miura Group rocks (Misaki and Hasse formations, Fig. 10.47) cropping out on the Miura and Boso peninsulas, southeast Japan, comprise a >2 km-thick succession of volcaniclastic sediments that accumulated in a forearc basin on the eastern side of an older segment of the Izu–Bonin Arc or older arc that was sutured with the Izu–Bonin Arc (*cf.* Pickering *et al.* 2013; Fig. 10.25). Later, the Miura block was accreted onto the Honshu Arc through dextral transpressional accretion. The deep-marine Misaki Formation in the southern part of the Miura Peninsula is >850 m thick. In the southern Boso Peninsula, the Misaki Formation is >2000 m thick. The younger Hasse Formation, however, is dominated

by more shallow water, marine currents and storm events and, in part, by fluvial or proximal fan-delta processes (Stow *et al.* 1986b). The basin fill contains a variety of facies deposited from various sediment gravity flows and hemipelagic deposition on a deep-water slope (Fig. 10.48). The main facies groups include dark-coloured, mostly coarse-grained scoriaceous beds, pale-coloured muddy and silty pumiceous bioturbated sediment, yellowish and whitish-coloured tuffaceous horizons, and chaotic sediment slides, slumps, debrites and wet-sediment injections. Many of the beds in the older Misaki Formation have been interpreted to be the deposits of: (i) direct pyroclastic fall through the air and water, for example, with fast suspension fall-out from sinking sediment plumes as described by Fisher and Schmincke (1984) and Cas and Wright (1987); (ii) downslope resedimentation via turbidity currents and related processes, in some cases derived from subaqueous base-surge pyroclastic flow; (iii) hemipelagic settling, commonly under the influence of thermohaline bottom currents. In other cases, composite beds are believed to result from the interaction of these processes in more complex events (Stow *et al.* 1986b).

In the Misaki Formation, three different mechanisms of gravity current generation were proposed by Stow *et al.* (1986b). One involves the slumping of unstable volcaniclastic debris from the flanks of

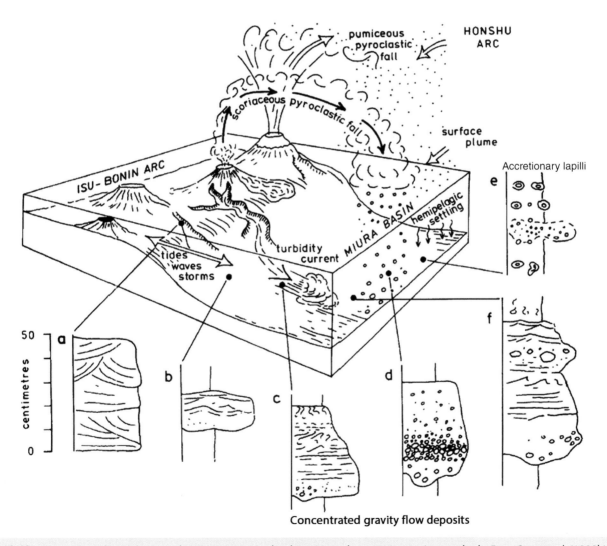

Fig. 10.49 Interaction of pyroclastic and SGF processes in the deposition of composite scoriaceous beds. From Stow *et al.* (1998b).

volcanic seamounts and slope areas surrounding the Miura Basin. The most likely trigger is volcanic-seismic, with the deposits comprising a mix of any material that had accumulated in the transitional source area (e.g., the mixed pumiceous/scoriaceous sandstones). The second mechanism involves the direct transformation of high-concentration subaqueous pyroclastic flows into concentrated density flows (Section 1.4.2). The resulting deposits tend to have a more monomict composition and be less readily distinguished from pyroclastic deposits. The third category involves the transition of sinking dense sediment plumes into downslope SGFs, also resulting in a monomict composition. Many of the acidic tuffaceous beds (Facies Group E) also display fine-grained Bouma turbidite divisions and are therefore interpreted as the deposits of turbidity currents (*cf.* Stow & Piper 1984b).

Stow *et al.* (1986) proposed a composite depositional model for the Miura Basin during the Miocene and early Pliocene (Fig. 10.49). This model should find wide applicability to other slope systems associated with active volcanic island arcs, especially where submarine phreatomagmatic processes operate (*cf.* similar facies described from forearc basins – Van Weering *et al.* 1989; Seyfried *et al.*, 1991; and backarc/marginal basins – Kokelaar & Howells 1984; Pirrie & Riding 1988; MacDonald *et al* 1988).

10.4   Forearc summary model

Active convergent margins evolve through the interaction of sediment flux (commonly including a volumetrically significant contribution by volcaniclastic sediments), eustatic sea-level change and tectonics. The reader is referred to Section 4.4 for a discussion of sea-level change in clastic-dominated deep-marine successions at active plate margins and, therefore, this topic is not considered further here.

The association between forearc basins and slip during subduction-thrust earthquakes led Fuller *et al.* (2006) to propose a link between processes controlling upper plate structure and seismic coupling on the subduction-zone thrust fault. They proposed a mechanism for the formation of forearc basins where sedimentation occurs on landward-dipping segments of the subduction wedge, which itself is actively growing through the accretion of material from the subducting plate. The numerical simulations of Fuller *et al.* (2006) suggest that sedimentation stabilises the underlying subduction wedge, preventing internal deformation beneath the forearc basin. Maximum slip during great-thrust earthquakes tends to occur beneath the sedimentary basins and stabilised subduction wedge. The lack of internal deformation of the wedge in these regions increases the likelihood of thermal pressurisation of the subduction thrust, permits the fault to load more rapidly and promotes enhanced 'healing' of the fault between rupture events. These effects link the degree of deformation of the subduction wedge to the effectiveness of seismic coupling along the subduction thrust.

Figure 10.50 is a summary of the main large-scale sedimentary features found in forearcs. Inevitably, the complex interplay between tectonics and sedimentation may make such environments very difficult to reconstruct accurately from the geological record, especially as the final geological relationships will commonly bear little,

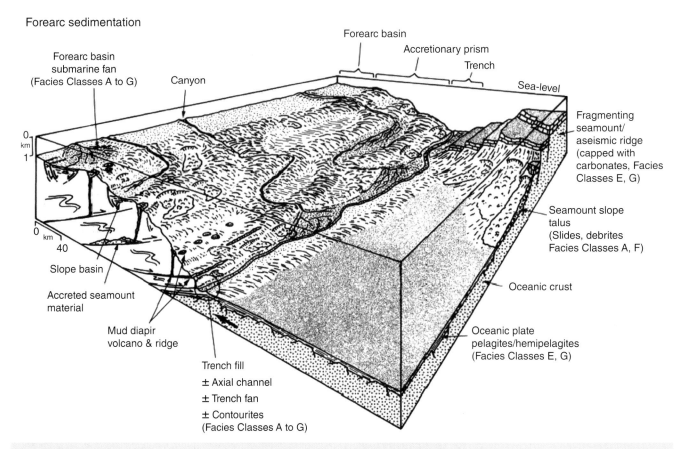

Forearc sedimentation

Fig. 10.50   A general model for the evolution of the stratigraphy in some accretionary-prism-top basins that are progressively uplifted and incorporated farther back from the trench with time. See text for explanation. From Pickering *et al.* (1989).

if any, similarity to the original spatial and temporal association of environments due to subduction-related compression/extension, out-of-sequence thrusting and, possibly, major strike-slip. Furthermore, Figure 10.50 does not distinguish between mud- and sand-rich forearcs, where different processes and deformation styles may prevail. We would expect mud-rich forearcs to contain a greater proportion of mud diapirs, mud ridges and mud volcanoes. At this time, the data-base from active margins precludes a useful comparison between mud- and sand-rich forearcs.

10.5 Marginal/backarc basins

Marginal basins are amongst the more complex of deep-water sedimentary basins (Taylor & Natland 1995, and references therein). Stratigraphy in these basins is extremely variable, and there are few

general models. Marginal basins have been defined as 'relatively small semi-isolated oceanic basins, spatially associated with active or inactive volcanic-arc and trench systems' (Kokelaar & Howells 1984), but that excludes forearc settings. Included in the term 'marginal basin' are those basins that form by extension and thinning of the crust behind magmatic arcs (i.e., backarc basins). A common feature of marginal basins is that their development is controlled by a phase of major crustal extension, involving oceanic, continental or transitional crust. As a general term, 'marginal basin' is preferred for such basins because there is no connotation of an association with an emergent volcanic island arc. The western Pacific has produced >75 % of the marginal basins found on the Earth today (Tamaki & Honza 1991). Figure 10.51 summarises the principal modes of backarc or marginal-basin opening, defined by Martinez *et al.* (2007) as trench roll-back or (subducting) slab 'sea-anchor'. Slab roll-back occurs when the trailing plate is considered fixed (indicated by black

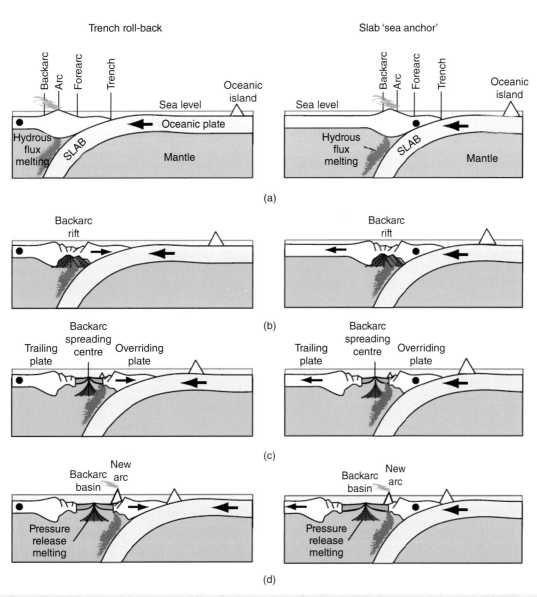

Fig. 10.51 (a–d) Modes of backarc basin opening. The panels from top to bottom show a schematic time sequence of backarc basin opening. Left-hand panels show the case of slab roll-back. In the right-hand panels, the trench hinge is considered to resist motion (black dot) because of a slab 'sea-anchor' force. See text for explanation. From *Martinez et al.* (2007).

dot in Fig. 10.51) and the trench hinge moves relatively seaward (indicated by small right-pointing arrow in Fig. 10.51), breaking off a section of the overriding plate that moves with the trench from a trailing plate that remains fixed. Despite the oceanward motion of the trench, the oceanic plate itself continues to converge with both the trailing and overriding plates as indicated by the relative motion of the oceanic island toward the left (large arrow). In the second scenario for backarc or marginal-basin development, the trench hinge is considered to resist motion (black dot in Fig. 10.51) because of a slab 'sea anchor' force. In Fig. 10.51 panels b–d, what becomes the 'trailing' plate begins to move toward the left (indicated by small arrow in Fig. 10.51), creating a relative opening with respect to the overriding plate. In both cases, mantle melting initially results from hydrous fluxing from fluids released by the slab (pink pattern in Fig. 10.51 panel a). As extension begins (Fig. 10.51 panel b) and the arc plate is rifted and thinned, the underlying mantle begins to rise to fill the space created. This action initiates pressure-release melting (red pattern and dashed

lines indicate mantle advection). Once breakup occurs, the separating overriding and trailing plates continue to drive mantle ascent and pressure-release melting. Note that hydrous flux melting may continue throughout and that arc and spreading-centre melt sources may be initially quite close but more-and-more separated with time (Fig. 10.51 panels b–d).

The nature of the basement crust beneath marginal basins may be oceanic, continental or transitional. Furthermore, large thicknesses of deep-marine sediments can accumulate in such basins, for example in the Sarawak Basin (South China Sea), floored by pre-Oligocene oceanic crust, and presently covered by 8000 m of relatively undisturbed sediments. Also, in marginal basins the progradation of shelf–slope–basin systems can be considerable, as in the Sunda shelf edge that has advanced about 300 km north due to post-Eocene progradation in a marginal basin (Houtz & Hayes 1984).

The initial rifting phase in the development of marginal basins is commonly associated with a large influx of relatively coarse-grained

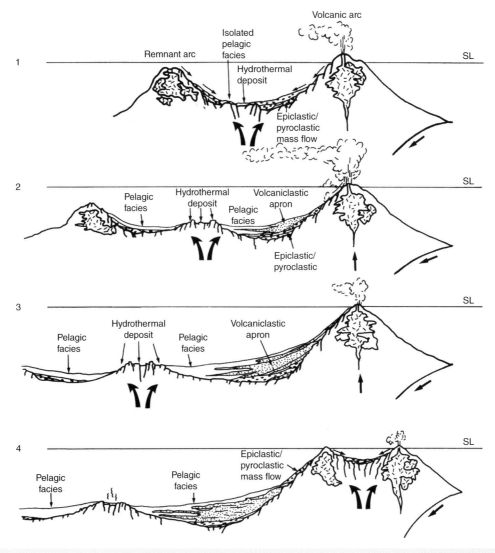

Fig. 10.52 Distribution of volcanigenic sediments in an evolving marginal basin. Stage 1: early rifting and large influx of volcaniclastics. Stage 2: basin widening by backarc spreading with active island arc volcanism. Thick volcaniclastic apron developed marginal to active arc. Stage 3: basin maturity, decreasing arc volcanism favours transgression with finer grained, hemipelagic and pelagic, sedimentation over coarse volcaniclastic apron. Stage 4: basin inactivity, backarc spreading ceases and splitting of volcanic arc initiates a new cycle of basin formation. From Carey and Sigurdsson (1984).

clastics that tends to infill and smooth any irregular basin topography, as both shallow- and deep-water deposits, depending upon the initial bathymetry. As rifting continues, thick successions of relatively coarse-grained deep-water clastics (Facies Classes A, B and C) become restricted to the margins of the basin. Fine-grained sedimentation (Facies Classes D, E and G) dominates the central parts of the basin. Large abyssal plains may eventually develop within 'mature' marginal basins. The final infill of marginal basins is extremely variable and will depend upon the subsidence or uplift history, together with the manner in which the basin is destroyed.

Marginal/backarc-basin sedimentation may include considerable volumes of volcaniclastic deep-water sediments (see Cas & Wright 1987; Taylor & Natland 1995). Carey and Sigurdsson (1984) have related the distribution of volcaniclastic sediments to stages in the evolution of a marginal (backarc) basin (Figure 10.52). They recognise three important contributory sources to the sedimentary apron bordering marginal basins: (i) a primary volcaniclastic influx from subaerial and/or subaqueous arc eruptions; (ii) an epiclastic influx from the erosion of the subaerial and/or subaqueous arc complex and (iii) a continuous accumulation of pelagic biogenic and aeolian particles.

Carey and Sigurdsson (1984) define four stages in the evolution of a backarc (marginal) basin (Fig. 10.52). The first stage represents the initial rifting and development of an intra-arc basin, with steep unstable basin margins typically mantled by various sediment mass-flow deposits (Facies Classes A and F). The second stage involves backarc spreading and island-arc volcanism: initially very unstable faulted basin margins tend to be smoothed by a veneer of

sediment gravity flows (Facies Classes B, C and D). Minor amounts of Facies Classes E and G also accumulate. The third stage, basin maturity, involves a decrease of backarc spreading rates, with increased preservation of fine-grained sediment gravity-flows deposits and hemipelagites/pelagites (Facies Classes D, E and G). Since the rate of aggradation of this apron is directly linked to the magnitude of the arc volcanic activity, an assessment of the rate of sediment accumulation should provide a qualitative estimate of the pace of arc volcanism. Pelagic sediment accumulation rates will depend, amongst other factors, upon the depth of the CCD and the biological productivity in the water column.

As an example of a marginal basin, the Okinawa Basin (behind the Ryukyu Trench) (Fig. 10.1) is considered using data from Letouzey and Kimura (1985) and Kimura (1985). The Okinawa Basin has a width of ~230 km, a length of 1300 km, and an 'inner' graben 50–100 km wide. About 3000–4000 m of mainly deep-marine sediments have accumulated in the south, compared to 7000–8000 m in the north. The history of the Okinawa Basin may be summarised as follows: (i) rifting within the volcanic arc beginning in the Neogene and still active today; (ii) synchronous opening and subsidence of the Okinawa Basin with tilting and subsidence of the forearc terrain – the late Miocene erosional surface now being 4000 m below sea level in the forearc terrace above the trench slope and (iii) a Pliocene, 1.9 Ma, major phase of crustal extension in the southern and central Okinawa Basin. Palaeomagnetic studies suggest a 45°–50° clockwise rotation of the southern Ryukyu Arc since the late Miocene. This rotation was probably related to the collision of Taiwan and the north Luzon Arc

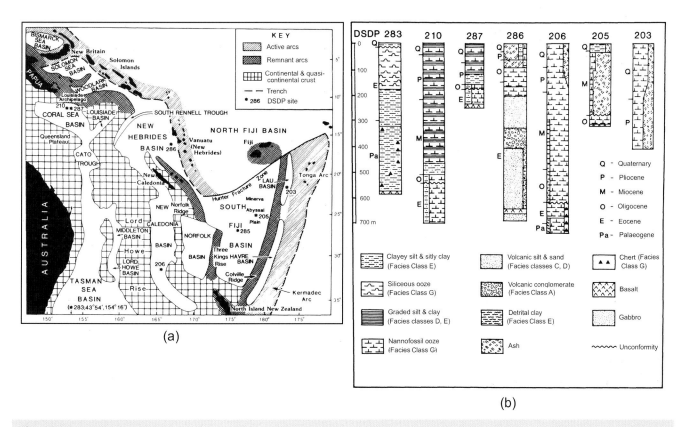

Fig. 10.53 Marginal basins and their stratigraphy. (a) Location of marginal basins in the southwest Pacific; 5° of longitude is ~550 km. (b) DSDP stratigraphies from southwest Pacific marginal basins. Tasman Basin, DSDP Site 283; Coral Sea, DSDP Sites 210 and 287; New Hebrides Basin, DSDP Site 286; New Caledonia Basin, DSDP Site 206; South Fiji Basin, DSDP Site 205; Lau Basin, DSDP Site 203. From Leitch (1984).

with the Chinese mainland margin, provoking the lateral 'extrusion', clockwise rotation and buckling of the southern Ryukyu non-volcanic arc. Crustal extension has produced, on the arc side of the Okinawa Basin, Pleistocene to Recent half-grabens that are 10–80 km wide with floors that dip up to $20°–30°$ to the southeast, away from the arc. Near Taiwan, the slopes are cut by many canyons and channels that obscure the half-graben structure.

Amongst the best-documented marginal basins are those of the southwestern Pacific (*cf.* Leitch 1984) shown in Figure 10.53a. Leitch (1984) summarises some of the main features of the marginal basins of the southwest Pacific as: (i) a spreading history of short duration; (ii) initial rifting of relatively thick crust (the South Rennell Trough being an exception) and (iii) rifting either associated with arc volcanism, as in the southwest Pacific marginal basins (Bismarck, Lau, Havre and North Fiji Basins), or rifting with little or no arc volcanism, as in the northwest Pacific marginal basins. Figure 10.53b shows typical DSDP deep-marine stratigraphy in the southwest Pacific marginal basins. Hundreds of metres of deep-marine, essentially fine-grained, sediments were drilled and show considerable variation, particularly in the Palaeogene–Neogene sediments. Site 206, for example, in the New Caledonia Basin (Fig. 10.53b), comprises mainly siliceous oozes, whereas Site 286 (now a forearc setting), in the New Hebrides Basin, contains abundant volcanigenic conglomerates, sands and silts (Facies Classes A, B, C and D), with relatively little siliceous ooze (Fig. 10.53). Such differences reflect proximity to an active volcanic centre furnishing large volumes of sediments, Site 286 being near the New Hebrides Arc, with Site 206 on oceanic crust hundreds of kilometres from any land.

The Mesozoic–Early Tertiary development of the Black Sea and Caspian Sea provides a useful ancient example of backarc/marginal basins. Zonenshain and Le Pichon (1986) believe that the marginal sea reached its maximum size in the Eocene at ~3000 km in length by 900 km in width, the central part of the basin being subducted during collision between the Arabian Promontory and the Eurasian margin. The large marginal sea consisted of four deep basins floored by oceanic crust, namely the Great Caucasian Sea, South Caspian Sea, and the western and eastern Black Sea basins. An estimated 14–15 km of sediments accumulated in the eastern and western Black Sea basins, upon 5–6 km thick oceanic crust (Zonenshain & Le Pichon 1986).

The interaction of extensional, compressional and oblique-slip tectonics along the perimeters of marginal basins can be extremely complex. In the North Fiji Basin, a major submarine mountain chain formed by two parallel ridges, the d'Entrecasteaux Zone (DEZ) and the West Torres Massif (WTM) of the New Hebrides island arc is causing both a complex plate interaction and indentation of the arc (Collot *et al.* 1985; Fisher *et al.* 1986).

The islands of the New Hebrides form a north-northwest–south-southeast-trending intra-oceanic island arc associated with the eastward subduction of the Indo-Australian Plate beneath the North Fiji Basin; the subduction zone dips toward the Pacific Ocean. The DEZ is an area of high relief, up to 100 km wide and with a mean depth of ~3500 m, composed of MORB affinity basalt. The indentation and oblique subduction $(10°–15°)$ of the DEZ against the New Hebrides, at least since 2 Ma, has led to a mean Quaternary uplift rate of 1 mm yr^{-1} in the arc. This uplift can be explained by radial horizontal stresses resulting from the collision of the DEZ with the New Hebrides. The resultant compressive stresses have modified the width of the arc in this region, with the resulting deformation closely resembling the theoretical stress field produced by the indentation of a long narrow plastic body (arc) by a rigid flat plate (DEZ).

The New Hebrides block is both elastically bent under vertical stresses and plastically deformed and pushed eastwards by the resultant horizontal stresses (Collot *et al.* 1985). The recent uplift of the eastern chain is well explained by this model, with the north and south sections of the arc being influenced mainly by a tensional stress field. Thus, in this case, deep-marine sedimentation is taking place in a range of environments and tectonic regimes from convergence with subduction (New Hebrides forearc), extensional (North Fiji Basin backarc), and oblique slip (north and south of the central block). Furthermore, the rapid Quaternary uplift, due to collision of the DEZ with the New Hebrides Arc, has led to the accumulation of a thick succession of deep-marine sediments near the southern margin of the DEZ, probably eroded from Espiritu Santa Island (Fisher *et al.* 1986).

The Lesser Antilles intra-oceanic arc is situated between the backarc Grenada Basin (or Grenada Trough) to the west, and the westward-subducting Atlantic Ocean Plate to the east (e.g., DSDP Leg 78A (1981), in Biju-Duval, Moore *et al.* (1984), and Brown and Westbrook (1987)) (Fig. 10.54); ocean plate subduction is occurring at about 2 cm yr^{-1}. The Lesser Antilles subduction complex provides a good example of the variation and complexity in tectonics and sedimentation along active convergent plate margins. The forearc of the Lesser Antilles contains the accretionary prism of the Barbados Ridge complex, together with two prominent forearc basins, the Tobago and Barbados basins (Fig. 10.54), the former basin containing >4000 m of sediment fill that becomes progressively more deformed eastwards. The Barbados Ridge Complex is the widest example of an accretionary prism associated with an island arc; even the outer trench rise, defined by a positive isostatic gravity anomaly 150 km seaward of the plate boundary, lies buried beneath the subduction complex. At the deformation front, sediment thicknesses beneath the Atlantic Abyssal Plain vary from >4000 m in the south to ~700 m in the DSDP Leg 78A area.

The Lesser Antilles forearc ranges in width from more than 450 km in the south, to <150 km in the north. The considerable width of the system, together with the very rapid rates of subsidence and fast sediment accumulation rates, at least since the Pliocene, have conspired to mask the trench as a bathymetric feature (Fig. 10.54). Thus, the seaward boundary of the forearc is defined by a deformation front of folding and thrust faulting. In the region of DSDP Leg 78A drilling, the accretionary prism (Barbados Ridge Complex) is ~260 km wide, and to the west of the emergent Barbados Ridge seismic evidence suggests back-thrusting in the deep marine Tobago Basin (Fig. 10.54). Most along-strike structural variation can be related to the influence of basement relief.

The Lesser Antilles volcanic arc has been active since at least the Middle Eocene and possibly since Cretaceous times. North of Martinique opposite the Puerto Rico Basin, the arc divides into an outer arc (Aves Ridge), extinct since the middle Miocene, and an inner arc that has been active from at least about 5 or 6 Ma. In the south, west of Tobago Basin (or Tobago Trough), there is only one arc, probably active since 55 Ma. The crust beneath the arc is ~30 km thick. Between the arc and the subduction zone, there is a 100 km-wide segment of oceanic crust overlain by a thick succession of deep-marine forearc deposits, the Tobago Basin and the Lesser Antilles Basin.

The backarc Grenada Basin, underlain by an anomalously thick oceanic crust (Boynton *et al.* 1979) similar to that of the Venezuela Basin, in part contains >6000 m of essentially arc-derived sheet-like sediments. The Grenada Basin almost certainly had an extensional origin (Fig. 10.52) due to separation of the Lesser Antilles from the Aves Ridge, probably in the Late Cretaceous–Early Tertiary. In a study of 29 piston cores (Pleistocene–Holocene) from the Grenada Basin,

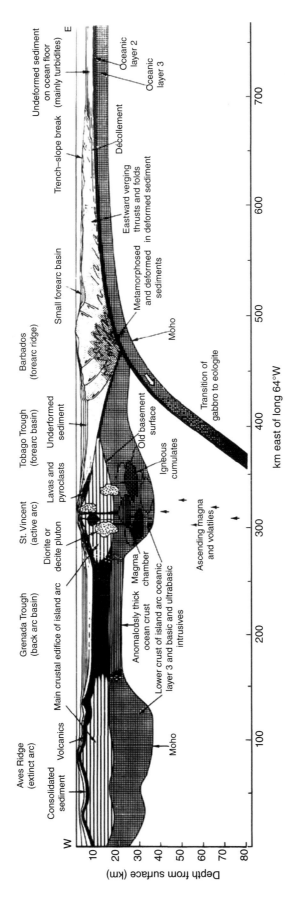

Fig. 10.54 Cross-sectional reconstruction of the Lesser Antilles arc and backarc system through Barbados and St Vincent, derived primarily from gravity and seismic data. From Westbrook *et al.* (1984).

Carey and Sigurdsson (1978), and Sigurdsson *et al.* (1980), describe coarse-grained, volcaniclastic, sediment gravity flow deposits in beds up to 4.5 m thick (Facies Classes A, B and C), interbedded with predominantly green-grey hemipelagic muds and clays (Facies Classes E and G). The upper part of the cores comprise Holocene, brown, pelagic clay of Facies Class G. Their study of the SGF deposits showed that the Bouma T_a division is present in 76% of the beds, T_b in 41%, T_c in 9%, T_d in 6% and T_e in 33%. According to the classification used in this book (Section 1.4.1), most of these beds were emplaced by concentrated density flows. Debrites (Facies Class A or F) contain pumice clasts up to 6.5 cm in diameter in a fine-grained matrix of glass shards, crystals, lithic and biogenic fragments, and clay. Petrographic studies suggest that the volcaniclastic material was derived from multiple sources, including subaerial eruptions on the islands of St Vincent, St Lucia, Martinique, Dominica and Guadeloupe.

Associated with the accretionary prism, there are a number of 'supra-complex' basins, the largest of which is the Barbados Basin (Fig. 10.54; see also Figs 10.34, 10.35) with up to 2000 m of sediment only deformed near widely spaced fault zones. The eastern margin of the Barbados Basin is currently being deformed by east-dipping reverse faults and thrusts with horizontal separations of 10–15 km; the related folds produce ridges 500–600 m above the basin floors (Brown & Westbrook 1987). Locally, such basins have been subject to considerable vertical tectonics with undeformed sediments over 1000 m thick uplifted on ridges between supra-complex basins.

The Tobago Basin is a major forearc basin that was probably formed by the upward growth of the accretionary complex. At the southern end of the basin, its fill oversteps the deformed prism sediments. This stratigraphic relationship could be the result of pulses of deformation at the basin-margin, each followed by relatively stable phases when overstep occurs. However, at the southeastern margin of the Tobago Basin, the overstepping basin sediments have been shielded from compression since the Pliocene by a spur of South American metamorphic basement, also seen on the island of Tobago.

Important, large-scale, forearc mud diapirism has been recognised both from onshore outcrops in Trinidad (Higgins & Saunders 1974), and from seismic profiles and side-scan sonar surveys in the accretionary complex offshore (Fig. 10.40) (Michelson 1976; Biju-Duval *et al.* 1982; Stride *et al.* 1982; Brown & Westbrook 1987). It appears that such diapirs preferentially develop along fault zones, especially those along the margins of sedimentary basins. The mud diapirism appears to be most abundant in the southern part of the accretionary complex where turbidites from the Orinoco submarine fan are accreted rapidly (Brown & Westbrook 1987). The origin of these diapirs, up to 17 km long by 1 km wide, has been ascribed to enhanced pore-fluid pressures created by load-induced stress as the accretionary complex thickens (Westbrook & Smith 1983).

Sediments toward the toe of the Barbados Ridge Complex, drilled on DSDP Leg 78A (Sites 541–543), comprise lower Miocene–Quaternary hemipelagic/pelagic marly calcareous oozes (Facies Group G1), structureless muds (Facies Class E), radiolarian muds (Facies Group G2) and ash bands. These sediments occur at Sites 541 and 543, the latter site also penetrating 44 m of basaltic pillow lavas at 411 m subseafloor. The paucity of terrigenous turbidites at these sites probably reflects deposition on a topographic high. GLORIA long-range side-scan sonar has shown the importance of gravity sliding at the deformation front, opposite the Tiburon Rise (Fig. 10.54) where a debrite and its chute cover an area of 100 km^2.

Sedimentation patterns are principally governed by: (i) long-lived relatively slow subduction–accretion; (ii) a major southern source of terrigenous sediments, axially funnelled from the South American margin (Damuth 1977), although the finer grained silt and clay are carried by the Guiana Current system, and the coarser grained silt and sand transported by turbidity currents down the South American continental slope onto the Atlantic Abyssal Plain; (iii) mass wasting and sediment gravity flows derived from the forearc accretionary prism; (iv) volcanigenic sediments derived from the volcanic island chain and (v) production of pelagic, calcareous, biogenic material, which contribute to the sedimentary record in areas above the CCD.

10.6 Ancient convergent-margin systems

There are many examples of ancient forearc subduction–accretion and backarc systems, interpreted with varying degrees of confidence. Some are very extensive, for example the Cretaceous Chugach Terrane, southern Alaska, with a linear continuity of ~2000 km (Nilsen & Zuffa 1982), whereas other systems have considerably less present-day along-strike continuity such as the Ordovician–Silurian Southern Uplands prism, Scotland, extending for ~120 km (Leggett *et al.* 1982). However, if this system continues into the Northern Appalachians, for example into Central Newfoundland, Canada, as seems likely, then its length is considerably greater. The interpretation of ancient subduction–accretion systems ideally relies upon the recognition of consanguineous arc volcanism with identifiable forearc and backarc tectono-sedimentary environments. If there has been later major oblique-slip tectonics, then reconstructing the ancient arc-related system relies upon correlating far-displaced terranes.

Ideally, in reconstructing ancient subduction–accretion systems, it is important to map out a complete arc–trench system. In the case of the Cretaceous Chugach Terrane, Nilsen and Zuffa (1982) identify: (i) forearc slope deposits, including slope basins, with south-trending submarine canyons; (ii) a major zone of mélange against the flysch zone of the Chugach Terrane, the contact being a major landward-dipping fault zone; (iii) southward-transported forearc-basin deposits forming a belt parallel to other Chugach Terrane tectono-stratigraphic belts and (iv) a major magmatic arc of granitic intrusions bounded by dextral strike-slip fault zones. Also, most fault-bounded blocks dip towards the arc, with a style of faulting, folding and metamorphism typical of modern subduction-accretion systems. Sediment dispersal was mainly along the trench from east to west, with lateral supply from the forearc area to the north.

The deep-water sedimentary fill of arc-related basins can be extremely variable both in its detrital composition and grain size. The lower Eocene Tyee and Flournoy Formations of western Oregon have been interpreted as deep forearc-basin sediments (Chan & Dott 1983), and comprise a 2000-m-thick succession that includes voluminous sands that were transported into deep water along a broad front from the shelf, rather than being fed from point sources such as canyons (Section 8.9). Parts of the sand-rich fan system, however, were fed from complex nested channels, up to 350 m wide and 40 m deep, incised into the shelf and upper slope. Chan and Dott (1983) recognise an overall shallowing-up trend in this forearc basin, with deltaic systems prograding across the narrow shelf and out over the basin-fill in early Eocene times. Minimum sediment accumulation rates were 67 cm kyr^{-1} with 750–1700 years between successive sediment gravity flows (Chan & Dott 1983). As with many forearc basins, submarine fans are unlike those on mature passive margins because of the relatively small, elongate basins; that is, they tend to be more confined at their lateral margins in forearcs.

The template upon which forearc basin sedimentation develops may comprise an accretionary prism and/or volcanic arc basement.

The Cenozoic West Sumatran forearc basin, for example, shows a deep, partially filled basin between the shelf and an outer arc ridge, with flat-lying seismic facies showing onlap over: (i) the arc-massif on the landward side, and (ii) either the subsiding accretionary prism, or attenuated continental crust, on the trench side (Beaudry & Moore 1985). Deposition presently occurs in water depths of 600–1000 m.

The growth of the Japanese arc system has mainly taken place along the continental margin of Asia since the Permian, and is the result of subduction of the ancient Pacific ocean floor. Backarc basin formation in the Tertiary shaped the present-day arc configuration. The Cretaceous–Neogene Shimanto Belt, extending for more than 1800 km from the Nansei Islands in the southwest to Boso Peninsula

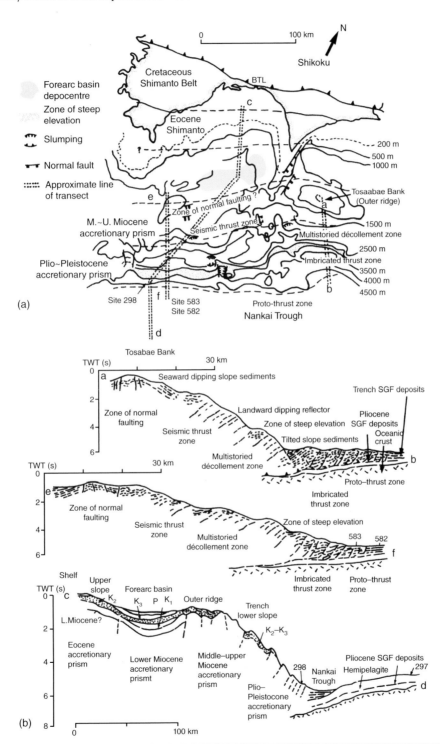

Fig. 10.55 Schematic geological cross-section of southwest Japan, located in (a), to show the long-lived history of Japan as part of an active convergent margin. Main tectono-stratigraphic zones or belts, together with major bounding lineaments (a & b). From Taira (1985). See text for explanation.

southeast of Tokyo (Fig. 10.28), is one of the best studied ancient subduction–accretion systems (Kanmera 1976a, b; Suzuki & Hada 1979; Teraoka 1979; Taira *et al.* 1980, 1982; Tazaki & Inomata. 1980; Sakai & Kanmera 1981; Ogawa 1982, 1985; Taira 1985; Underwood 1993; Taira 2001). In southern Japan, Taira (1985) defined three major tectono-stratigraphic 'terranes' together with several zones of substantial strike-slip displacement: (i) a northern composite terrane of pre-Jurassic rocks accreted to the Asian margin prior to the opening of the Sea of Japan, including the Hida, Sangun and Yamaguchi belts; (ii) a Jurassic subduction complex of the Chyugoku, Chichibu, Tamba, Mino and Ashio belts, and the associated metamorphic Sambagawa and Ryoke belts, and (iii) the Cretaceous–Neogene Shimanto Belt, interpreted as forearc and intra-arc environments. In general, the Shimanto Belt, dislocated by major faults that crop out as northerly-dipping high-angle reverse faults, youngs towards the south. Oblique subduction and strike-slip tectonics have been important in the evolution of Japan since at least Late Palaeozoic times (Taira *et al.* 1983, 1989).

The Shimanto Belt comprises five main sub-belts (Taira 1985) from north–south: (i) extrusive andesites and rhyolites, granitic intrusives and terrestrial sediments of the magmatic arc, with 130–30 Ma ages but clustering between 95–70 Ma; (ii) Upper Cretaceous, mainly deep-marine coarse-grained clastics (Facies Classes A, B and C), fed from east–west into a series of small basins in the forearc; (iii) south of the Median Tectonic Line (MTL), a Carboniferous–Jurassic mostly non-magmatic outer arc of basalts, limestones, cherts, sandstones and mudrocks; (iv) an Early–Late Cretaceous outer-arc shelf-basin of mainly shallow-marine and fluvial to brackish-marine sediments and (v) to the south of the Butsuzo Tectonic Line (BTL), the Cretaceous–Neogene Shimanto subduction–accretion complex (Fig. 10.55). In structural sections, from north–south, across central Honshu and Shikoku to the Nankai Trench, it is possible to appreciate the longevity of Japan as an active convergent margin, with the modern analogue being the present subduction system.

In Shikoku, the Shimanto Belt consists of a northern Cretaceous sub-belt, and a southern Eocene–lower Miocene sub-belt (Taira *et al.*

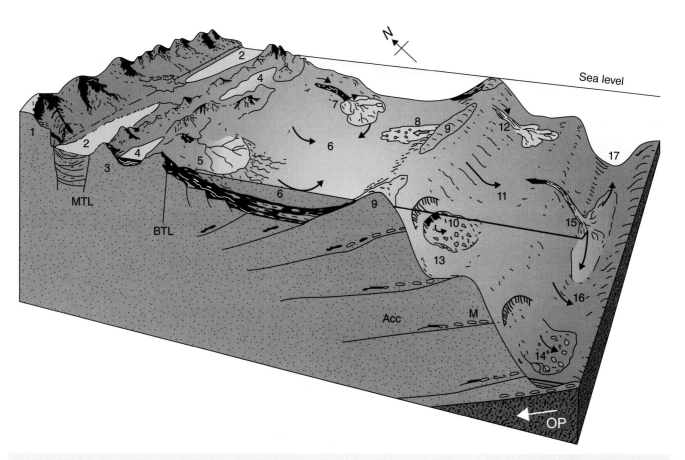

Fig. 10.56 Schematic block diagram showing the depositional environments for the Cretaceous Lower Shimanto Group (not to scale). From Taira *et al.* (1982). 1, seaward limit of Ryoke high P-T belt; 2, intra-arc basins; 3, Sambagawa low P-T and Chichibu belts, with; 4, mainly Lower Cretaceous small, forearc shelf basins with thin fluvial and shallow-marine sands and mudrocks; 5, deltaic deposits; 6, deep-marine, non-fan, turbidite systems (mainly Facies Classes B, C and D); 7 submarine canyon, fan-channel and fan system; 8, slide deposits (Facies Class F), including olistostromes, on upper inner-trench wall; 9, trench-slope break, partly emergent; 10, slide deposits including olistostromes on upper inner-trench wall; 11, non-fan turbidite systems; 12, canyon-fan system contributing sediments to: 13, 'perched' accretionary-prism-top basins; 14, slides including olistostromes on lower inner-trench wall; 15, canyon-fan system; 16, lateral supply of turbidites and related gravity-flow deposits to: 17, trench floor. MTL = Median Tectonic Line; BTL = Butsuzo Tectonic Line; Acc = Accretionary prism wedge separated by thrusts with mélange (M); OP = intermittently subducting oceanic plate. The Palaeogene Shimanto Belt environments are regarded as similar to this Early Cretaceous reconstruction, but with the deformed Lower Shimanto Group being exposed and eroded, and a new accretionary prism and associated environments having developed farther southeast.

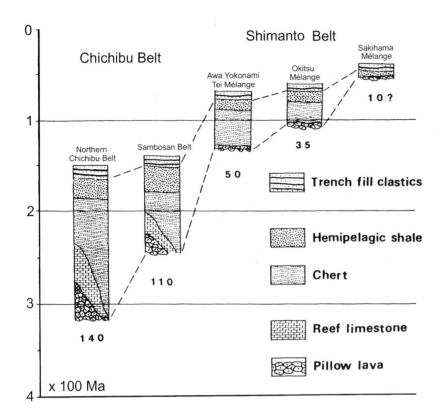

Fig. 10.57 Reconstructed oceanic plate stratigraphy and the ages of high-pressure/temperature (P/T) metamorphism in southwest Japan. From Taira *et al.* (1985).

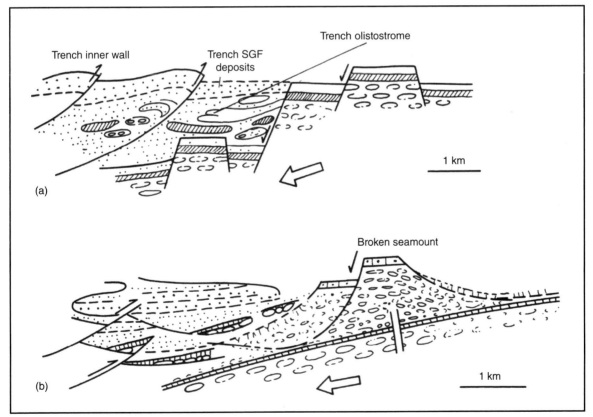

Fig. 10.58 Conceptual model for the accretion and off-scraping of subducting oceanic plate material in a trench, from Ogawa (1985). (a) Formation of trench olistostrome from the normally-faulted oceanic floor, including basaltic material. (b) Off-scraping of parts of a subducting seamount.

1982; Taira 1985). The Cretaceous Shimanto sub-belt is divisible into four major tectono-stratigraphic units bounded by steeply-dipping reverse faults: (i) Neocomian–Coniacian brackish to shallow-marine sediments; (ii) Aptian–Cenomanian turbidites and other sediment gravity-flow deposits (mainly Facies Classes B, C and D); (iii) Coniacian–Campanian turbidites and other sediment gravity-flow deposits, including mélanges (mainly Facies Classes B, C, D and F) and (iv) upper Campanian–Maastrichtian shallow-marine deposits, including sediment slides.

Figure 10.56 (Taira *et al.* 1982) summarises the palaeogeography for the Cretaceous (Lower) Shimanto Group, and emphasises the range of deep-marine environments from forearc-basin submarine fans, to accretionary-prism ridged basins in deeper water, to trench sedimentation. Wet-sediment deformation is recognised, both as surface gravity-slides and as tectonic, thrust-generated, mélange. Work by Pickering *et al.* (1988a) suggests that many of these mélanges may have originated as mud or shale diapirs that were subsequently folded, sheared and flattened.

The Coniacian–Campanian turbidite-mélange unit comprises tectonic slivers of pillow basalts, radiolarian cherts (Facies Class G), hemipelagic mudrocks (Facies Classes E and G), acidic tuffs, sandstone turbidites and concentrated density-flow deposits (Facies Classes B, C and D) in a shale matrix. Studies by Taira *et al.* (1980), Taira (1981) and Kodama *et al.* (1983) suggest that the mélanges were formed in equatorial latitudes, while the overlying Campanian SGF deposits (mainly Facies Classes B, C and D) accumulated essentially at present-day latitudes. There is an approximately 50 Myr age difference between the oceanic plate material (including basalt and limestone) and the matrix of the mélange. The difference in

latitude represents ~3000 km, which can be used with age data to calculate a plate speed of about 6 cm yr^{-1}. The age and composition of the facies within the mélanges show a systematic southward change (Fig. 10.57), thereby allowing successive reconstructions of the trench stratigraphy. They show that progressively younger oceanic crust was being subducted and partially obducted.

The Southern Shimanto Belt comprises Eocene and lower Miocene sub-belts. The lithologies are predominantly Facies Classes B, C and D deep-marine clastics and igneous rocks, interpreted as deposits of forearc accretionary-prism environments (Taira 1985). The offshore geology (Fig. 10.55) reflects modern subduction–accretion processes, emphasising the problem of separating the 'ancient' Upper (Southern) Shimanto Belt from the 'modern' subduction system.

The evolution of the Shimanto Belt has been episodic, as seen by the discrete development of the various 'sub-belts', each during intervals of ~10–20 Myr. This episodic evolution is probably the result of cyclic subduction of the oceanic plate in 10–20 Myr cycles (Taira *et al.* 1980). Structural and stratigraphic data suggest the importance of movement on transform faults during the late Oligocene–early Miocene (Ogawa 1985). For example, the Miura–Boso Terrane of southeast central Honshu was transferred to the Honshu Arc from the Izu–Bonin Arc in this manner. The accreted material comprises: (i) Miocene–Pliocene, volcaniclastic, deep- to shallow-marine sediments; (ii) dismembered ophiolitic rocks and (iii) some rocks of the Shimanto Group. All were accreted under a regime of predominantly dextral transpressive tectonics. This stress regime is believed to have been the result of oblique subduction of the northeast Philippine Plate, with a present average annual relative displacement northwestwards of about 3.5 cm yr^{-1} (Ogawa 1985).

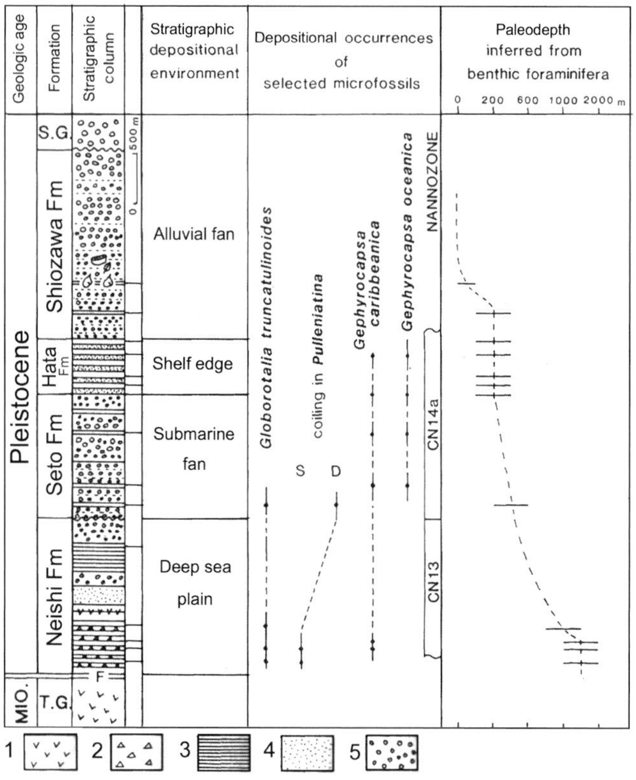

Fig. 10.59 Summary of the stratigraphy and palaeoenvironments of the Pleistocene Ashigara Group, southeast Japan. 1, volcaniclastics; 2, Wapiti tuffs; 3, Facies Class D siltstones; 4, Facies Classes B and C sandstones; 5, Facies Class A conglomerates, S.G. = Suruga gravels, T.G. = Tanzawa Group. From Huchon and Kitazato (1984).

The dextral transpressive shear also is responsible for the Neogene accretion of the Mineoka Ophiolite Belt, a chaotic volcanic and sedimentary succession, including olistostromes, that accumulated in the proto Izu–Bonin Ridge (Izu–Ogasawara Arc) and was later accreted onto the Honshu Arc (Ogawa 1983). The Mineoka Group, probably lower Tertiary–lower Miocene, comprises an ophiolitic and pelagic/terrigenous clastic succession. The lower siliceous and calcareous claystones/mudstones (Facies Classes E and G) are overlain by mainly Facies Classes C and D arkosic turbidites and other sediment gravity-flow deposits, above which there is a shallowing-upward sequence into possible shallow-marine, tidally-dominated clastics (Ogawa 1983); a stratigraphy that may reflect the uplift of the basin during sustained arc–arc collision.

To explain the origin of some mélanges in the Shimanto Belt, Ogawa (1985) proposed a model involving subduction–accretion processes operating in submarine trenches, whereby subducting seamounts are at least partially off-scraped at the leading edge of the accretionary prism (Fig. 10.58). Such off-scraping leads to the incorporation of seamount material, along with oceanic-plate sediments, into the accretionary prism. This model has a superb modern analogue in the Dai-ichi Kashima Seamount (Fig. 10.20) that is currently being dismembered, partially subducted and accreted in the southern part of the Japan Trench (Mogi & Nishizawa 1980).

The Pleistocene Ashigara Group in central Honshu is a further example of a sedimentary basin formed between the colliding Izu–Ogasawara and Honshu arcs (Huchon & Kitazato 1984). Figure 10.59 is a summary of the stratigraphy of the group. The arc–arc collision led to the uplift of a source for the voluminous conglomerates of the Miocene Tanzawa Group. Palaeobathymetric studies, based on benthic foraminifera, suggest that the lower–middle Ashigara Group was deposited in water depths of 1000–2000 m (Fig. 10.59). Above this deep-marine clastic system, there is a coarsening-upward trend, interpreted to reflect a shallowing-up, from the 1500 m-thick Neishi Formation (deep-sea plain) and 1300 m-thick Seto Formation (submarine fan) into the Hata Formation (shelf edge) and Shiozawa Formation (alluvial fan). Soon after the deposition of the upper Ashigara Group, the area was overthrust by the Tanzawa mountains along the Kannawa Fault and folded under northwest–southeast compression. At about 0.3 Ma, the direction of compression changed considerably to north–south or northeast–southwest compression when the colliding Izu–Ogasawara Arc finally locked against central Japan (Huchon & Kitazato 1984).

10.7 Forearc/backarc cycles

Deep-marine backarc and forearc basin stratigraphy may show an overall vertical coarsening-and-thickening-upwards cycle above any igneous/volcanic basement. For example, the igneous/volcanic basement may be immediately overlain by hydrothermal and hydrothermally-altered sediments, in turn covered by pelagic, then hemipelagic and finally terrigenous deposits. A good example of such a cycle occurs in the Lower Palaeozoic of north-central Newfoundland, Canada (Fig. 10.60). Here, in the Ordovician of Notre Dame Bay, interpreted as Early Ordovician (Arenig–Llanvirn) backarc environments, but also associated with major strike-slip plate motions during the closure of the Iapetus Ocean (Pickering 1987a; Pickering & Smith 1995, 1997, 2004; see Figs 12.37–12.39), tholeiitic pillow basalts are locally veneered with red cherty mudstone (jasper),

overlain by red radiolarian-rich (recrystallised) cherts (Fig. 2.43a), then grey and grey-green bioturbated muddy cherts (Fig. 2.43b), then middle Ordovician (late Llandeilo–early Caradoc) graptolite-bearing mudstones, and finally thin-bedded Facies Classes C and D sandy turbidites and other SGF deposits (Fig. 10.61).

10.8 Concluding remarks

Seismic-reflection records and coring results from recent ODP and IODP scientific research have provided a wealth of hitherto unprecedented data on active convergent margins, especially the Nankai Margin through the NanTroSEIZE programme. In general, these data have supported and considerably built upon previous ideas about the structural and tectonic evolution of subduction margins. Such data have also shown the importance of out-of-sequence thrusting in the stratigraphic and tectonic development of both slope basins and forearc basins.

Generic facies models for submarine trench fills generally show a thickening-and-coarsening-upward trend from basaltic basement through open-ocean pelagic oozes and clays (Facies classes E and G), hemipelagic muds (Facies Class E), silty turbidites (Facies classes C and D) and then sandy concentrated density-flow deposits and turbidites (Facies classes A–C), commonly all overlain by sediment slides, slumps (Facies Class F) and debrites from the inner-trench slope (mainly Facies Class A). As a 'first pass' summary of the deep-marine stratigraphic evolution of a trench fill, this vertical trend remains valid.

Studies of modern subduction systems, however, have show the 3D complexity of subduction inputs caused by irregular basement topography associated with seamounts, aseismic ridges and other asperities. Subducting crust can have highly variable thicknesses of siliciclastic sediments, from a thin veneer on sediment-starved plates (e.g., Tonga-Kermadec Trench, South Sandwich Trench linked to the Scotian Island Arc; Heezen & Johnson 1965) that contrasts with thick submarine-fan successions that have completely covered the topographic trench, for example, in the southern part of the Barbados (Biju-Duval *et al.* 1984; Mascle, Moore *et al.* 1988; Moore *et al.* 1990; Moore 2000; Moore, Klaus *et al.* 2000) and the Bengal Fan in the northern Sunda Trench (Curray *et al.* 2003; Weber *et al.* 2003; Schwenk *et al.* 2005). There is an increasing recognition that in forearc regions generally thought to be under compression, such as the Kumano Basin, offshore Japan, they are actually dominated by multiple normal fault populations (Lewis *et al.* 2013; Moore *et al.* 2013b; Sacks *et al.* 2013). This means that sedimentary processes at the seafloor and the stratigraphic architecture can be mainly controlled by extensional normal faulting.

Lithostratigraphic architecture exerts a first-order influence on the material properties and tectonic behaviour of subduction zones, diagenesis and the release of fluid, fluid migration pathways, and the position of the décollement. These aspects still remain relatively poorly understood. Although beyond the scope of this book, the amount and types of clay-size particles affect the coefficient of friction and permeability of any sediment (Underwood 2007). Also, some minerals (e.g., smectite and opal) significantly increase fluid production during diagenesis, commonly with compartments of excess pore-fluid pressure in many depositional systems being set up initially by confined sandbody geometries (Underwood 2007). For more information on the role of clay minerals and clay-mineral

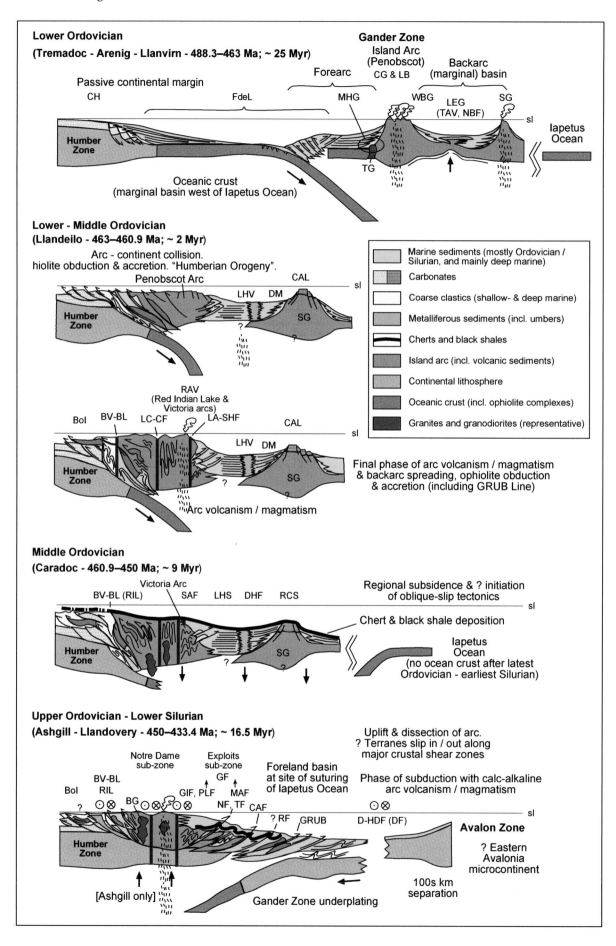

Fig. 10.60 Summary interpretive cross-sections to explain the lower Ordovician-early Silurian (Tremadoc–Llandovery) geological evolution of the Lower Palaeozoic of central Newfoundland (modified from Pickering 1987a, b; Pickering & Smith 1995, 1997, 2004; Waldron *et al.* 2012; Zagorevski *et al.* 2012, and references therein). Stratigraphic thicknesses are highly schematic and have considerable vertical exaggeration. Location of New Bay (including Point Leamington Formation) stratigraphy beneath central part of middle Ordovician to early Silurian (Llandeilo–Llandovery) reconstructions. HZ = Humber Zone; CH = Cow Head Group; FdeL = Fleur de Lys rocks; MHG = Moretons Harbour Group (shown in forearc, but could be intra-arc); TG = Twillingate Granidiorite; CG & LB = Cutwell Group and Lushs Bight Group (incl. Pacquet Harbour Group, Snooks Arm Group); WBG = Wild Bight Group (may incl. Frozen Ocean Group); LEG = Lower Exploits Group, including Tea Arm Volcanics (TAV) and New Bay Formation (NBF); SG = Summerford Group; LHV = Lawrence Head Volcanics; DM = Dunnage Mélange; CAL = Cobbs Arm Limestone; BV–BL = Baie Verte–Brompton Line; RIL = Red Indian Line; LC–CF = Lobster Cove–Chanceport Fault; D–HBF (DF) = Dover–Hermitage Bay Fault (Dover Fault); SAF = Shoal Arm Formation; LHS = Lawrence Harbour Shale; DHF = Dark Hole Formation; RCS = Rogers Cove Shale; RAV = Roberts Arm Volcanic Belt; LA = SHF = Lukes Arm–Sops Head Fault; BG = Burlington Granodiorite; GIF = Gull Island Formation; PLF = Point Leamington Formation; MAF = Milliners Arm Formation (PLF and MAF = part of Badger Group); NF = New Bay Fault; TF = Toogood Fault; CAF = Cobbs Arm Fault; BoI = Bay of Islands and associated ophiolites; GF = Goldson Formation. Mélanges such as the Boons Point Complex are associated with the Lukes Arm–Sops Head Fault, as olistostromes tectonised by the synsedimentary thrust faulting. Subduction directions are shown, together with backarc spreading centres. Arc–continent collision began in the late Llanvirn: this collision involved the SG, DM and CAL, but the 2D sketches do not adequately show this – throughout this time interval, >800 m of tholeiitic basalts (LHV) of the LEG were extruded in a submarine environment, implying long-lived backarc extension. Gander Zone underplating may have occurred from late Ashgill onwards. Sea level shown by lines marked 'sl'. During Caradoc, cessation of arc volcanism may have led to thermal contraction of the lithosphere and associated tectonic subsidence (chert and black shale deposition, possibly with an overall eustatic sea-level rise. Time-scale shown here from Zalasiewicz *et al.* (2009). Ashgill–Llandovery (and Wenlock) history probably involved phases of crustal transtension and associated bimodal igneous activity (Springfield Group and youngest parts of the Topsails igneous complex), but plate-tectonic history was essentially mainly crustal transpression under sinistral shear as Avalonia sutured to the outboard microplates of Laurentia (*cf.* continental reconstructions shown in Pickering & Smith 2004).

diagenesis in subduction zones, the reader is referred to Vrolijk (1990), Underwood (2002), Underwood *et al.* (1993), Underwood and Pickering (1996), Saffer and Marone (2003), Underwood and Fergusson (2005), Saffer *et al.* (2008), Saffer and Tobin (2011), Guo and Underwood (2012) and Guo *et al.* (2014). For information on sand and sandstone petrography/geochemistry in arc and subduction systems, from the voluminous literature, the reader is referred to papers by Dickinson and Suczek (1979), Moore (1979), Maynard *et al.* (1982), Dickinson *et al.* (1983), McLennan *et al.* (1990), Hiscott and Gill (1992), Marsaglia and Ingersoll (1992), Marsaglia *et al.* (1992, 1995), Underwood *et al.* (1993, 1995), Gill *et al.* (1994); Garzanti *et al.* (2007), Hara *et al.* (2012) and Milliken *et al.* (2012). Finally, whilst may researchers have emphasised the undoubted control by seismicity in generating SGF deposits, including turbidites, at subduction margins (for example, Nelson *et al.* 2008; Polonia *et al.* 2013; see also Section 4.7), the relative importance of other processes over long time periods, such as eustatic sea-level change, remains unresolved.

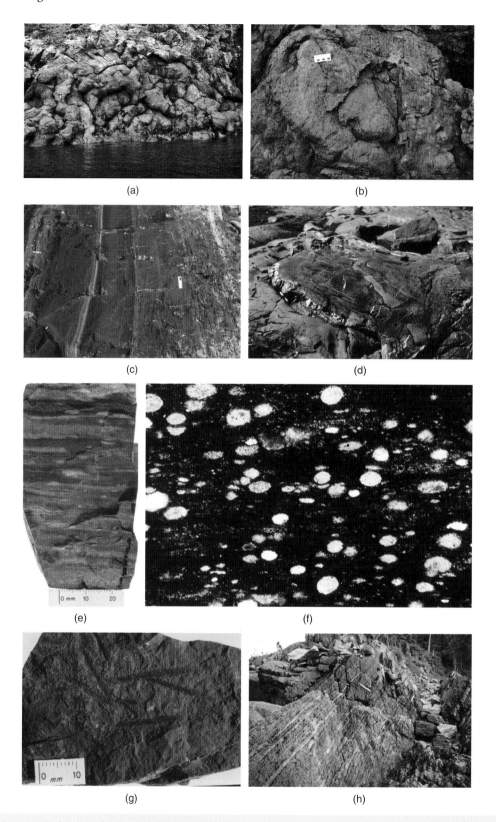

Fig. 10.61 Active-margin deep-marine facies associations in the Lower Palaeozoic (Ordovician) of Notre Dame Bay, Newfoundland. (a) Pillow basalts, Tea Arm Volcanics, Wild Bight Group. (b) Red cherty mudstones (jasper) enveloping pillow basalts, Tea Arm Volcanics, Wild Bight Group. (c) Facies Class E and G multi-coloured cherty mudstones, Wild Bight Group. (d) Basaltic injection into wet sediments of Facies Class E and G, Wild Bight Group. (e) and (f) Outcrop and thin-section of Facies Class G radiolarian cherts between basalts and black shales immediately below Point Leamington Formation.(g) Graptolitic black shales of facies Class G, underlying Point Leamington Formation. (h) Thin-bedded Facies Class D turbidites, Point Leamington Formation.

CHAPTER ELEVEN
Foreland basins

(a)

(b)

(a) Chevron folding in the Upper Carboniferous (Namurian) deep-water Crackington Formation, Millhook Haven, Devon, UK. (b) Flute casts at the base of sandstone bed, Middle Ordovician Cloridorme Formation, Gaspé Peninsula, Quebec Appalachians, Canada. Metre staff divided into 10 cm units.

Deep Marine Systems: Processes, Deposits, Environments, Tectonics and Sedimentation, First Edition. Kevin T. Pickering and Richard N. Hiscott.
© 2016 Kevin T. Pickering and Richard N. Hiscott. Published 2016 by John Wiley & Sons, Ltd.
Companion Website: www.wiley.com/go/pickering/marinesystems

11.1   Introduction

Active convergent plate margins have been divided into those in which oceanic crust is consumed by subduction processes, generally associated with arc magmatism and volcanism, and margins where continents collide with one another or with offshore arcs (e.g., the Palaeozoic Urals; Brown *et al.* 2006), resulting in underplating without subduction, to create 'foreland basins'. Clearly, many remnant sutures mark the sites of complex collisional events that may have initially involved subduction–accretion, finally being replaced by underplating as continent–continent or continent–arc collision locked the system. The term 'foreland basin' was first introduced by Dickinson (1974). Useful summaries of foreland basins and their stratigraphic architecture can be found in Allen and Homewood (1986), Van Wagoner and Bertram (1995), Mascle *et al.* (1998), Joseph and Lomas (2004a,b), Plink-Björklund *et al.* (2001), Lacombe *et al.* (2007), Covault *et al.* (2009), Hubbard *et al.* (2009), Sanchez *et al.* (2011), Brown and Ryan (2011), and Liu *et al.* (2015).

Of the few modern deep-marine foreland basins that have been recognised, the only two that have been described in any detail are: (i) the Timor–Tanimbar foreland basin in the Banda Arc region, formed because of collision between the Banda volcanic arc and the northern Australian continental margin (Audley-Charles 1986a,b; Londoño & Lorenzo 2004), and (ii) the western Taiwan foreland basin that resulted from the collision of the Luzon volcanic arc with mainland China ~4 Ma. The Taiwan foreland basin is 400 km long, 100 km wide and, in the southern part, has water depths up to about 1500 m (Covey 1986). Plate convergence in the Taiwan area is about 7 cm yr^{-1} (70 km Myr^{-1}), shows a diachronous (oblique) north to south closure such that the axis of the Manila Trench dips overall towards the south, involves 160 km of shortening to date, uplift rates of about 5 mm yr^{-1} and denudation rates for the entire island of Taiwan of 5.5 mm yr^{-1} (Covey 1986).

Foreland basins develop as a response to load-induced stresses, for example thrust sheets flexing the lithosphere. Quinlan and Beaumont (1984) considered a continuous visco-elastic plate to be a reliable rheological model of the lithosphere, where the initial elastic response of the lithosphere to loading is independent of any time-dependent assumed rheological properties, and results in a down-warped flexural basin adjacent to the load, and peripheral up-warping along the cratonward edge of the basin (Tankard 1986), producing what is known as the *peripheral bulge* or *forebulge*. If the lithosphere behaved solely as an elastic layer, no further lithospheric deformation would occur if the load remained unchanged. However, for a visco-elastic rheology, relaxation of the plate-bending stress takes place as a result of lithospheric flow. This relaxation results in deepening of the basin as the forebulge rises and migrates towards the load; that is, towards the thrust belt. This process is repeated as each new thrust-sheet advance causes an initial elastic flexural response which is superimposed over the previous relaxation geometry. Quinlan and Beaumont (1984) concluded that a load applied to a thick lithosphere results in a relatively wide and shallow basin, whereas the same load applied to a thin lithosphere generates a relatively narrow and deep basin. Thus, thin lithosphere provides the most favourable template upon which to develop deep-marine foreland basins. The recognition of ancient foreland basin successions relies upon the ability to recognise tectonic loading as the primary control on subsidence.

An important aspect of foreland basin stratigraphy is that subsidence first drowns the ancestral passive margin, leading to the accumulation of mudrocks distal to the rising orogen, and then there is a coarsening-upward trend as basin-plain or basin-floor and later submarine-fan sandy facies prograde toward the foreland. This is the classic flysch to molasse development, as exemplified in the Taconic flysch along the Appalachian–Caledonide System (Section 11.3.3).

An important type of foreland basin, and one for which there are few well-documented case studies, is that which develops in front of obducting ophiolitic basement. A good example is the Upper Cretaceous foreland basin in the Oman Mountains, associated with the obduction of the Semail ophiolite (Robertson 1987). The impingement of the advancing thrust-load on the edge of the Tethys Ocean down-flexed the crust and passive margin in the late Coniacian–Campanian (88.5–73 Ma) to create a foredeep, or foreland basin, that migrated towards the foreland. This foreland basin was partly infilled by terrigenous SGF deposits and olistostromes, in a deep-marine environment below the CCD (Robertson 1987).

Many foreland basins have a complex tectonic history that does not follow a relatively simple and sustained compressional phase to create a fold-and-thrust belt with an essentially consistent sense of fold vergence. Instead, several discrete phases of structural shortening, uplift and erosion, out-of-sequence thrusting, surge zones and lateral ramps, with differential rotations of thrust slices, are commonly documented. An example of such complexity is documented by Ghiglione *et al.* (2010) for the foreland basins of southernmost South America; that is, the Magallanes (Austral) and South Falkland (Malvinas and South Malvinas) sedimentary basins. The tectonic history of these basins involved Triassic–Jurassic crustal stretching, followed by Early Cretaceous thermal subsidence in a backarc setting, then Late Cretaceous–Palaeogene compression, and Neogene–Present deactivation of the fold-and-thrust belt that is now dominated by strike-slip or wrench faulting. Ghiglione *et al.* (2010) also document a Late Cretaceous onset of the foreland basin phase in all three basins, with the main early Palaeocene to early Eocene initial foredeep depocentres being bounded by slivers of basement that are now incorporated into the thin-skinned fold-and-thrust belts. Extensional basins developed in the Magallenes and South Falkland basins during late Palaeocene to early Eocene because of an acceleration in the rate of separation between South America and Antarctica that preceded the initial opening of the Drake Passage (Ghiglione *et al.* 2010). Pervasive normal faults of Palaeocene–Recent age occur throughout the basins, and have been interpreted by Ghiglione *et al.* (2010) to have resulted from flexural bending of the lithosphere.

Both modern and ancient foreland basins, together with other compressional plate-tectonic settings, contain hydrocarbon reserves (Cooper 2007). For example, the Pliocene–Pleistocene deep-marine parts of the Taiwan foreland basin contain gas reserves in combined stratigraphic-structural traps associated with the heads of submarine canyons (Fuh *et al.* 2009). In many of the hydrocarbon fields in geologically older foreland basins, their subtle traps lack surface expression and commonly were discovered by accident while drilling for structural objectives. Advances in seismic imaging, however, have made possible the direct identification of such stratigraphic plays. Over the past decades, unconventional resources (e.g., gas from 'tight' sands, heavy oil and coal-bed gas) have become increasingly important in the reserves mix of well-explored foreland basins. There are many reviews of hydrocarbon accumulations in foreland basins that include deep-water deposits; for example, for central and western China (Song *et al.* 2010), western Europe (Covault & Graham 2008), the Italian Apennines (Carmignani *et al.* 2004; Sani *et al.* 2004; Turini & Rennison 2004; Scrocca *et al.* 2005; Bertello *et al.* 2010; Cazzola *et al.* 2011; Martinelli *et al.* 2012), and the western Carpathians (Nemcok & Henk 2006; Slacka *et al.* 2006; Kotarba 2012; Sandy *et al.* 2012).

11.2 Modern foreland basins

11.2.1 Neogene–Quaternary Taiwan

Oblique arc–continent collision between Taiwan and southeast China created a foreland basin, referred to as the western Taiwan foreland basin (Fig. 11.1) (e.g., Byrne *et al.* 2011, and references therein). This basin evolved into three distinct sub-basins: (i) an over-filled basin proximal to the Taiwan orogen, mainly distributed in the western foothills and coastal plain provinces; (ii) a filled basin occupying the shallow continental shelf in the Taiwan Strait west of the Taiwan orogen and (iii) an under-filled basin distal to the Taiwan orogen in the deep-marine Kaoping Slope offshore southwest Taiwan (Fig. 11.2). Fluvial environments dominate the over-filled depositional phase across structurally controlled piggyback basins. The filled depositional phase in the Taiwan Strait is characterised by shallow-marine environments, and is filled by Pliocene–Quaternary sediments up to 4000 m thick derived from the Taiwan orogen with an asymmetrical and wedge-shaped cross-section. The under-filled depositional phase is characterised by deep-marine environments in a wedge-top basin accompanied by structures formed by active thrust faults and mud diapirs. Sediments derived from the Taiwan orogen have progressively filled the western Taiwan foreland basin across and along the orogen. Sediment dispersal involves orogenic fluvial and shallow-marine sediments mainly being transported perpendicular to hill slopes, and across-strike in the subaerial and shallow-marine environments proximal to the orogen. Fine-grained sediments (mainly Facies Classes D and E), however, are mainly transported longitudinally within the deep-marine environments distal to the orogen.

Chi-Yue Huang *et al.* (2006) document the oblique arc–continent collision in Taiwan that resulted in a four-stage set of tectonic processes. From ~16–15 Ma, subduction of the South China Sea oceanic crust beneath the Philippine Sea plate resulted in volcanism in the Coastal Range in Taiwan and formation of an accretionary prism in the Central Range. During the latest Miocene to earliest Pliocene, subduction was followed by initial arc–continent collision, as suggested

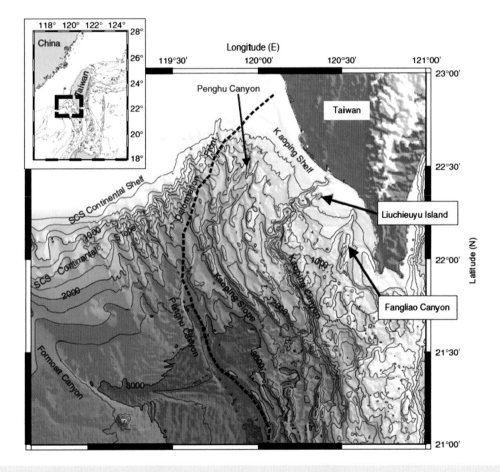

Fig. 11.1 Geomorphologic setting of the area onshore southwest Taiwan. The black dashed line represents the deformation front separating the passive South China Sea (SCS) continental margin from the active submarine Taiwan orogenic wedge. Bathymetric contours are at 250 m intervals. The box inset shows the map area. From Chiu and Liu (2008). The orogenic wedge consists of two broad and deep submarine slopes marked by outwardly bowed bathymetric contours deepening westward and southward. The South China Sea submarine slope on the Chinese cratonic side mainly dips southeast but the Kaoping submarine slope on the Taiwan orogen side dips southwest. These two slopes converge in the north and gradually merge into the Taiwan Strait shallow-marine shelf. The boundary separating these two submarine slopes is the location of the Penghu submarine canyon along the basin axis. The Penghu submarine canyon extends nearly in a north–south direction parallel to the strike of the Taiwan orogen and gradually merges southward into the Manila Trench in the northernmost part of the South China Sea. Numerous gullies and canyons normal or oblique to the shorelines occur on the seafloors of these two slopes.

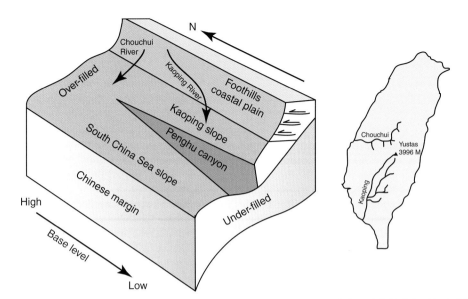

Fig. 11.2 Schematic diagram showing the over-filled and under-filled components of the foreland basin off southwest Taiwan. The Chouchui River transports sediment transversely into the shallow- and deep-marine foreland basin in central Taiwan. The Kaoping River transports sediment longitudinally and most of this is deposited subaerially to form the coastal plain in southwest Taiwan. Sediments derived from Taiwan are transported longitudinally farther into the deep-marine environments, mainly along the basin axis parallel to the strike of the Taiwan orogen. From Yu and Hong (2006), modified from Johnson and Beaumont (1995).

by: (i) unroofing and erosion of the deformed accretionary prism, and deposition of these erosional products in the adjacent accretionary forearc (5 Ma) and slope basins (4 Ma); (ii) waning of volcanism (in the north at ~6–5 Ma, and in the south at ~3.3 Ma); (iii) buildup of fringing coral reefs on previously active volcanoes (in the north at ~5.2 Ma, and in the south at~2.9 Ma); (iv) arc subsidence by strike-slip faulting and the development of pull-apart intra-arc basins (in the north at ~5.2–3.5 Ma, and in the south at ~2.9–1.8 Ma); (v) thrusting of forearc stratigraphy to generate a collision complex starting ~3 Ma and (vi) clockwise rotation of the arc–forearc sequences (in the north at ~2.1–1.7 Ma, and in the south at ~1.4 Ma). The arc–continent collision propagated southward and reached southern Taiwan by ~5 Ma, as recorded by the diachronous and progressive deformation of the associated accretionary wedge. By the earliest Pleistocene, arc–continent collision was at an advanced stage, associated with the westward thrusting and accretion of the Luzon volcanic arc and forearc against the accretionary wedge (in the north at ~1.5 Ma, and in the south at ~1.1 Ma), and exhumation of the underthrust Eurasian continent rocks (in the north at ~2.0–1.0 Ma, and in the south at ~1.0–0.5 Ma). During the past 1 Ma off the northern Coastal Range of Taiwan, arc collapse and subduction recommenced. Overall, the orogen has been associated with the continuous southward migration of tectonic processes and a change in sediment source and structural style.

Cheng-Shing Chiang *et al.* (2004) describe the tectonic influence on the stratigraphy of Taiwan. The wedge-top 'depozone' (Fig. 11.3) in the southern Taiwan foreland basin is confined by the topographic front of the Chaochou Fault to the east and by a submarine deformation front to the west. The non-marine Pingtung plain, the shallow-marine Kaoping shelf and Kaoping deep-marine slope constitute the main morphotectonic components of the wedge-top depozone. On land, alluvial and fluvial sediments accumulate above the frontal parts of the Taiwan orogenic wedge to form the Pingtung plain proximal to high topographic relief. Offshore, fine-grained sediments (Facies

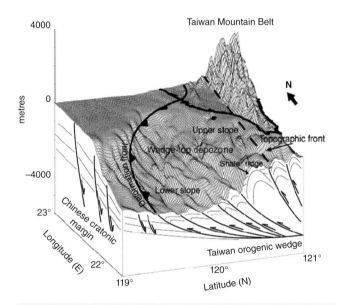

Fig. 11.3 Structural styles of a series of west-verging imbricate thrusts and folds and shale ridges in the wedge-top depozone confined by the topographic front proximal to the Taiwan Orogen and by the distal submarine deformation front at its transition to the passive Chinese margin. From Chiang *et al.* (2004), modified from Wu (1993), Liu *et al.* (1997) and Chiang (1998).

Classes D and E) accumulate on the Kaoping shelf and mainly Facies Class F mass-wasting deposits form the Kaoping slope. Wedge-top sediments are deformed into a series of west-verging imbricate thrusts and folds and associated piggyback basins. A major piggyback basin occurs in the Pingtung plain. Four smaller piggyback basins occur in the shelf-slope region, with many small-sized piggyback basins developed over ramp folds in the lower slope region. Approximately 5000 m

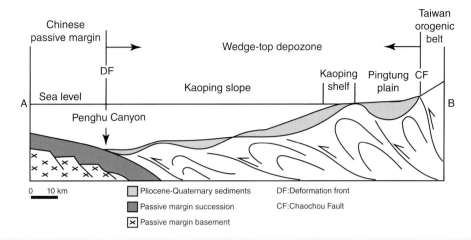

Fig. 11.4 Schematic cross-section, based on seismic data, of the paired mountain belt and under-filled deep-marine foreland basin offshore Taiwan. The foreland basin west of the Taiwan orogen mainly covers the coastal plain and the eastern Taiwan Strait shelf in the north and extends to the deep-marine slope (>3000 m water depth) in the south. The deep-marine foreland basin in southern Taiwan is characterised by accumulation of Pliocene–Quaternary sediments derived from the Taiwan orogen on the Taiwan orogenic wedge. The Kaoping submarine slope onlaps the outer margin of the Chinese craton west of the location of Penghu submarine canyon (shown). From Yu and Hong (2006), modified from Chiang *et al.* (2004).

of Pliocene–Quaternary deep-marine to fluvial sediments have been deposited on top of the frontal orogenic wedge in southern Taiwan. Sedimentary facies shows lateral variations from extremely coarse fluvial conglomerates proximal to the topographic front (Chaochou Fault) to fine-grained deep-marine muds (Facies Class E) close to the deformation front near the base of the slope. The transverse cross-section of the wedge-top depozone in southern Taiwan is a doubly-tapered prism. The northern boundary of the wedge-top depozone in southern Taiwan is placed along the southern limit of the Western Foothills where the frontal orogenic wedge progressively changes southward to a wedge-top depozone (Pingtung Plain), reflecting syn-depositional, southward-moving, oblique collision between the Luzon Arc and mainland China. The wedge-top depozone is bounded to the south by the boundary between continental and oceanic crust.

The sediment dispersal system in the southwestern Taiwan margin consists of two main parts: a subaerial drainage basin and an offshore receiving marine basin (Figs 11.1 to 11.4). In plan view, this sediment dispersal system can be further divided into five geomorphic units (Ho-Shing Yua *et al.* 2008): (i) the Gaoping (formerly spelled Kaoping) River drainage basin, (ii) the Gaoping (Kaoping) shelf, (iii) the Gaoping (Kaoping) submarine slope, (iv) the Gaoping (Kaoping) submarine canyon and (v) the Manila Trench in the northernmost South China Sea. The Gaoping River drainage basin is a small (3250 km²), tectonically active and overfilled part of the foreland basin, which is receiving sediments from denudation of the rising Central Range of Taiwan with a maximum elevation of 3952 m. The Gaoping submarine canyon begins at the mouth of the Gaoping River, crosses the narrow Gaoping shelf (~10 km) and the Gaoping submarine slope, and finally merges into the northern termination of the Manila Trench over a distance of ~260 km.

The southwest Taiwan margin dispersal system is characterised by a direct river-canyon connection with a narrow shelf and frequent episodic sediment discharge events in the canyon head. In a regional source-to-sink scheme, the Gaoping River drainage basin is the primary source area, the Gaoping shelf being the sediment bypass zone, the Gaoping submarine slope being the temporary sink and

the Manila Trench being the ultimate sink of the most far-travelled sediment from the Taiwan orogen. It is inferred from seismic data that the outer shelf and upper slope region can be considered as a line source for mass wasting deposits delivered to the lower Gaoping submarine slope where small depressions between diapiric ridges are partially filled with sediment or are empty. At present, recurrent hyperpycnal flows during the flood seasons are temporarily depositing sediments mainly derived from the Gaoping River in the head of the Gaoping submarine canyon. On decadal and century time scales, sediments temporarily stored in the upper reach are removed, probably by downslope-eroding sediment gravity flows within the canyon. Presently, the Gaoping submarine canyon serves as the major conduit for delivery of sediment from the Taiwan orogen to the marine sink of the Manila Trench. Seismic data indicate that the Gaoping submarine canyon has deeply incised the Gaoping submarine slope, presumably through the action of hyperpycnal flows travelling to the middle and lower reaches of the canyon. The middle reach is a sediment bypass zone whereas the lower reach serves as either a temporary sediment sink or a sediment conduit, depending on the shifting balance between deposition and erosion during canyon evolution.

The axis of the deep-marine, under-filled basin off southwest Taiwan dips southward and parallels the strike of the Taiwan orogen. This is a longitudinal sediment transport route, occupied by the Penghu submarine canyon. The submarine canyon initially developed at the intersection of the upper extremities of the South China Sea and Kaoping submarine slopes. Sediment-gravity flows cut into the slope sediments following regional tilting toward the south, and excavated along the deepest part of the seafloor, forming the present Penghu submarine canyon along the convergent boundary between the frontal Taiwan orogenic wedge and the Chinese cratonic margin. The submarine canyon is ~180 km long and trends north–south. It lies in 240 m of water at its head, increasing to 3200 m at its mouth before merging gradually into the basin floor of the Manila Trench. The Penghu submarine canyon is interpreted as a tectonically controlled canyon rather than a slope incision created by failure events (Ho-Shing Yu & Eason Hong 2005). Tectonic processes control the

Line CPC-1

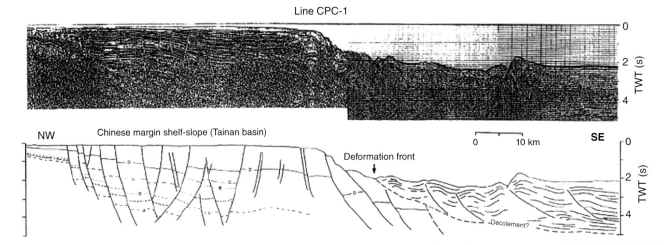

Fig. 11.5 Seismic profile CPC-1 across the submarine deformation front in the Taiwan orogenic wedge, showing the northwestward-advancing submarine deformation front that is overriding the Chinese passive margin. West of the deformation front is the Tainan Basin, which is a Pliocene–Quaternary foredeep characterised by flexural normal faults. Seismic reflectors are labelled: (A) base of Oligocene; (B) top of middle Miocene; (C) top of upper Miocene; and (D) top Pliocene. From Chou (1999), in Chiang *et al.* (2004).

orientation and location of the canyon while sedimentary processes contribute to its excavation and enlargement. The shifting of the axes of Late Pliocene to Pleistocene canyons from onshore southwest Taiwan to the present-day position of the Penghu submarine canyon reflects the evolving foreland basin, with a longitudinal transport system progressively migrating southwestward away from the deformation front.

The western side of the foreland basin is well imaged in seismic profile CPC-1 across the submarine deformation front in the Taiwan orogenic wedge, showing the northwestward-advancing submarine deformation front that is overriding the Chinese passive margin (Fig. 11.5) (Chou 1999; Chiang *et al.* 2004). West of the deformation front is the Tainan Basin, a Pliocene–Quaternary foredeep characterised by flexural normal faults.

11.2.2 Neogene Quaternary Southern Banda Arc

The Indonesian region includes several active volcanic island arcs, and also records Cenozoic volcanic activity caused by subduction of oceanic lithosphere at the margins of southeast Asia. The stratigraphic record in the Indonesian region reflects a complex tectonic history, including collisions, changing plate boundaries, subduction polarity reversals, elimination of volcanic arcs and extension (see summary in Hall & Smythe 2008; Hall 2011). The episodic crustal growth in this region occurred by the addition of ophiolites and continental slivers, and as a result of arc magmatism. In Indonesia, relatively small amounts of material were accreted from the down-going plate during subduction, but there is also little evidence for subduction erosion. The regional high heat flux and associated weak lithosphere in this region mean that the character of sedimentary basins may be unusual, with basins that are characteristically very deep and subside rapidly (Hall & Smythe 2008). Tropical erosion and weathering processes influence the mineralogy and maturity of the sediment, especially the volcanogenic material.

The Java Trench marks the boundary between the northward moving Indian Ocean Plate and the southeastern part of the Asian Plate (Figs 11.6a, b, 11.7, 11.8). At the eastern end of the Java Trench there is northward convergence and underplating of the Australian continental margin beneath the Banda Arc, without any intervening remnant oceanic crust (Jacobson *et al.* 1979; Bowin *et al.* 1980; Audley-Charles 1986b; Karig *et al.* 1987; Fortuin *et al.* 1994; Richardson & Blundell 1996; Hillis *et al.* 2008). Londoño and Lorenzo (2004) infer a maximum subsidence of 3500 m and a maximum width for the basin of ∼470 km; they also conclude that the effective elastic thickness of the Australian lithosphere (∼80–100 km) did not change significantly during basin evolution, as the low curvature imposed on the plate ($\sim 5.1 \times 10^{-8}$ m^{-1}) during bending is too small to weaken the plate. Londoño and Lorenzo (2004) also cite flexural models suggesting that at least 570 km of Australian plate (mostly areas of stretched continental crust) were flexed, primarily by tectonic loading beneath the island of Timor, and that the total amount of subducted plate was at least 100 km during basin evolution.

The non-volcanic Outer Banda Arc can be interpreted as an emergent part of the imbricate thrust system formed in response to the underplating of the Australian margin beneath the Timor–Babar–Tanimbar–Kai island chain. The Timor Trough is a deep-marine foreland basin with its subsidence controlled, at least in part, by loading of the North Australian margin by thrust sheets of the Outer Banda Arc. The Australian shelf represents the distal part of the foreland basin.

Prior to foreland basin development at ∼2.0–2.2 Ma, smaller continental fragments collided with the northern Australian margin (De Smet *et al.* 1990; Snyder *et al.* 1996); subduction–accretion processes similar to those now occurring farther west along the Java Trench are believed to have operated. Farther east, collision between Papua New Guinea and northern Australia occurred at 3.0–3.7 Ma (Abbott *et al.* 1994). The forearc was then converted into a foreland basin as the nature of the crust that was subducted/underplated changed. Before collision and emplacement of the allochthonous nappes in Timor, micropalaeontological studies suggest that sedimentation on both the Asian and Australian margins of the Java Trench, from at least the late Miocene to early Pliocene, occurred in water depths greater than 2000 m.

The curved Banda Arc comprises young oceanic crust enclosed by a volcanic inner arc, outer arc islands and a trough parallel to the Australian continental margin (e.g., Spakman & Hall 2010).

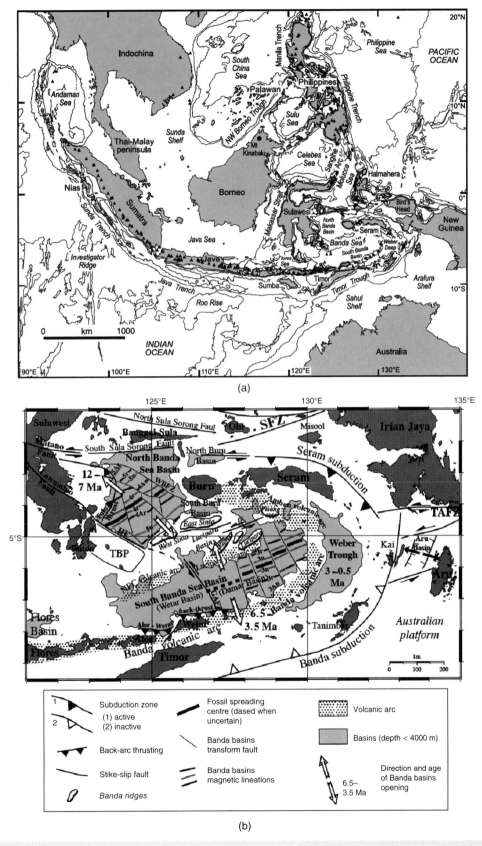

(a)

(b)

Fig. 11.6 (a) Geography of southeast Asia and surrounding regions. Small black filled triangles are volcanoes recognised by the Smithsonian Institution, Global Volcanism Program (Siebert & Simkin 2002). Bathymetry is simplified from the Gebco (2003) digital atlas. Bathymetric contours are at 200 m, 1000 m, 3000 m and 5000 m. From Hall (2011). (b) Map to show the position of the Banda Arc at the intersection of the Pacific, Eurasian and Indo-Australian plates, the location of the inner and outer Banda arcs, the principal tectonic plates and their direction of movement relative to the Eurasian plate, together with the location of deep-marine basins. From Hinschberger *et al.* (2005).

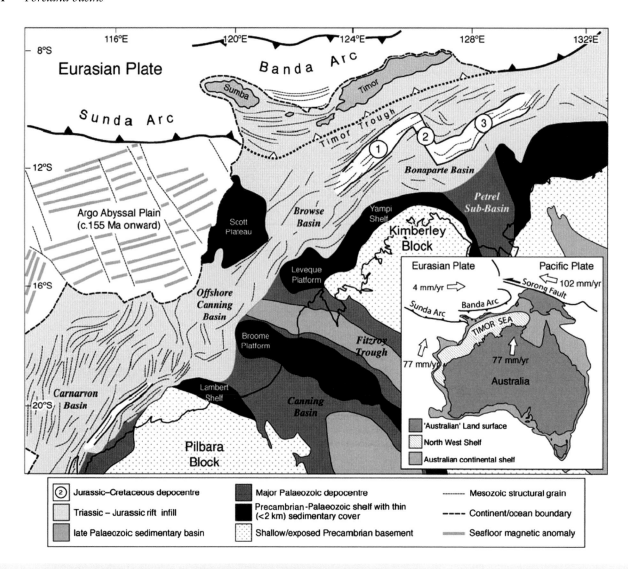

Fig. 11.7   Geological elements of the North West Shelf, after Australian Geological Survey Organisation (AGSO) North West Shelf Study Group (1994), illustrating the submarine extent of the Australian continent and the three major Mesozoic depocentres: 1, the Vulcan Graben; 2, the Sahul Syncline; 3, the Malita Graben. Inset map depicts Neogene plate motions and velocities after Keep *et al.* (1998). From Harrowfield *et al.* (2003).

Strong seismic activity in the upper mantle defines a folded surface, for which there are two contrasting explanations: deformation of a single slab or two separate slabs subducting from the north and south. Spakman and Hall (2010) combined seismic tomography with the plate-tectonic evolution of the region to infer that the Banda Arc results from subduction of a single slab. Their palaeogeographic reconstruction shows that a Jurassic embayment, which consisted of dense oceanic lithosphere enclosed by continental crust, once existed within the Australian plate. Banda subduction began ~15 Ma when active Java subduction relocated eastwards into the embayment. The present morphology of the subducting slab is only partially controlled by the shape of the embayment. As the Australian plate moved northward at a speed of ~7 cm yr⁻¹, the Banda oceanic slab rolled back towards the south–southeast accompanied by active delamination separating the crust from the denser mantle. Increasing resistance of the mantle to plate motion progressively folded the slab and caused strong deformation of the crust.

Micropalaeontological studies in Timor suggest post-collision uplift rates varying from 1.5–3 mm yr⁻¹, associated with crustal shortening by nappe emplacement at rates possibly greater than 62.5–125 mm yr⁻¹ (Audley-Charles 1986a; *cf.* discussion of convergence rates and strike-slip motion along various major faults by McCaffrey & Abers 1991). Tectonic thickening of the Australian continental margin and shelf has occurred by imbrication (mainly by southerly transported thrust sheets) together with other crustal shortening processes. The period of overthrusting has been estimated at 0.4–0.8 Ma (Audley Charles 1986a). Figure 11.9 summarises the stratigraphy on the north Australian passive continental margin (foreland), the colliding island arc complex of Timor, and DSDP Site 262 within the deep-marine foreland basin, or foredeep.

Seismic reflection and SeaMARC II imagery (side-scan and swath bathymetry) reveal a tectonic pattern analogous to that of typical oceanic subduction zones, with a deformation front in the Timor Trough discontinuously advancing southward, as successive

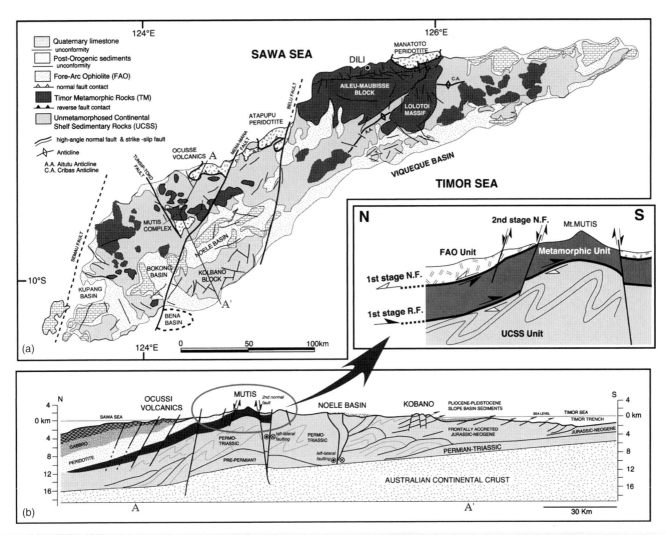

Fig. 11.8 (a) Simplified geologic map of Timor showing the major tectono-stratigraphic divisions. Adapted from Audley-Charles (1968), Rosidi *et al.* (1981), Charlton (1991) and Charlton *et al.* (1991), with additional data from Kenyon (1974), Berry and Grady (1981a,b) and Sopaheluwakan (1990). A–A' is the approximate location of the cross-section shown in B. (b) Cross-section through Mt. Mutis. From Kaneko *et al.* (2007).

thrust sheets of sediment are assembled from the underplating Australian margin (Karig *et al.* 1987). Furthermore, Karig *et al.* (1987) postulate oblique underplating, leading to dextral shear along a northeast-trending fault system that offsets the outer arc between the islands of Savu and Roti. Within Timor, sediments range up to more than 1000 m thick, with rapid lateral thickness changes reflecting the local growth of folds and/or thrusts, at least some of which are controlled by the reactivation of normal faults in the down-bending Australian margin. Within Timor Trough, at DSDP Site 262 (Veevers, Hietzler *et al.* 1974), fine-grained siliciclastics and biogenic carbonates were recovered (Fig. 11.9), suggesting that coarser-grained sediments are trapped on the inner trough slope. Sediment failures on the lower slope provide the main source of material for the trough floor. A large amount of Facies Class F occurs, particularly as sediment slides.

Immediately southwest of the foreland basin, ODP Leg 123 drilled a middle to upper Miocene deep-marine calciclastic depositional system (mainly characterised by SGF deposits) at Site 765 in the southeastern Argo Abyssal Plain, along the northwestern margin of

Australia (Fig. 11.10). The calciclastic deposits consist predominantly of planktonic calcareous components. Textural, mineralogical, bedding and downhole logging trends within the system were used by Simmons (1992) to infer a rapid progradation of the deep-marine system in the middle to late Miocene. The introduction of abundant sediment gravity flows into the area is attributed to changes in oceanographic conditions that accompanied a progressive lowering of sea level associated with major continental ice build-up on Antarctica. These data, in conjunction with seismic stratigraphy, suggest two major depositional pulses. Gradual retrogradation of the system occurred in latest Miocene time, coincident with the subsequent rise in sea level. Retrogradation of the system was overwhelmed by extensive slope failures during the Pliocene, presumably triggered by earthquakes associated with the progressive Miocene–Pliocene collision of the Australian margin and the Sunda–Banda Arc (Simmons 1992).

The overall sheet-like geometry of the foreland basin-floor deep-marine deposits above oceanic basement is revealed in seismic lines, for example the BMR multifold seismic Line 56/22 that

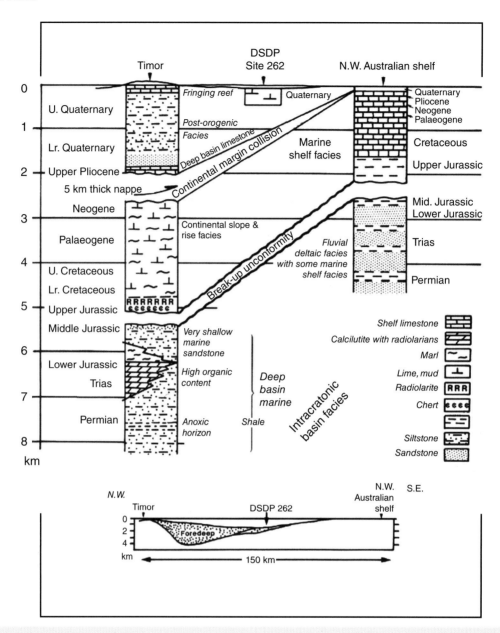

Fig. 11.9 Stratigraphic columns at the margins of the Timor Trough (Timor and the north Australian continental shelf) and within the foreland basin. DSDP Site 262 data from Veevers *et al.* (1978). From Audley-Charles (1986b).

crossed ODP Site 765 (Fig. 11.10b). The southwestern part of this line provides an example of how basement faulting can lead to a relatively restricted depositional site during the early evolution of a foreland basin.

11.3 Ancient deep-marine foreland basins

Ori *et al.* (1986), in a study of the Central Adriatic post-Oligocene foreland basin, make a number of observations that summarise many of the tectono-stratigraphic attributes of foreland-basin development: (i) the foreland basins (foredeeps) migrated in front of the advancing thrust pile, in this case towards the northeast; (ii) thrust highs, with vertical displacements up to 1000 m, can be eroded to produce a local sediment source; (iii) sediment dispersal patterns are mainly

parallel to the long axis of the basin, although lateral supply is at least locally important and (iv) the foredeep was segmented into discrete depocentres as a result of contemporaneous tectonism and diapirism (probably due to basement control). In the Central Adriatic case, the Apennines are a complex thrust belt of tectonic units that are still moving, and have been emplacing thrust sheets since the Oligocene.

Large-scale depositional cycles have been recognised in the Lower Carboniferous Moravian–Silesian Culm Basin (MSCB), the easternmost part of the Rhenohercynian deep-water foreland basin (the so-called 'Culm facies'). The Upper Viséan Moravice Formation (MF) of the MSCB shows a distinct cyclic stratigraphic arrangement, with two major asymmetric megacycles, each ~500–900 m thick (Bábek *et al.* 2004). The megacycles begin with 50–250 m-thick basal units containing erosive channels, overbank and slope-apron deposits interpreted by Bábek

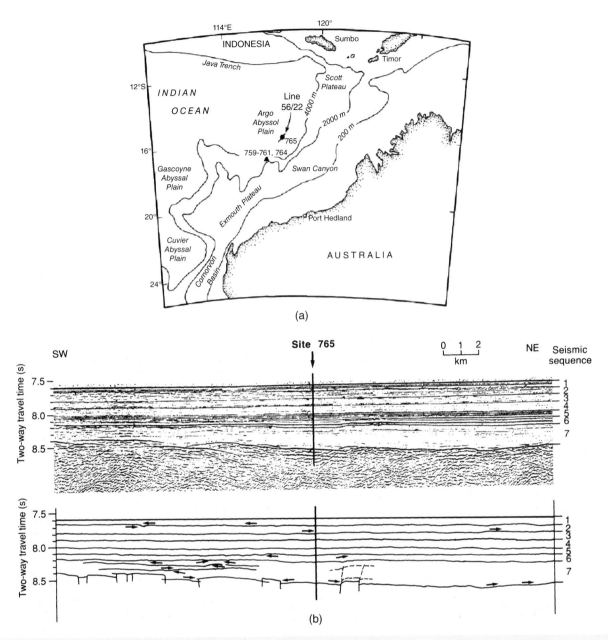

Fig. 11.10 (a) Location of Site 765, Argo Abyssal Plain, and track of BMR seismic Line 56/22. Modified from Ludden *et al.* (1990). (b) BMR multifold seismic Line 56/22 across Site 765. Approximate track location is shown in (a). Line drawing shows interpretation of seismic sequences 1 through 7, including disconformable relationships at sequence boundaries. From Ludden, Gradstein *et al.* (1990).

et al. (2004) as lowstand deep-water systems. Up-section these deposits pass into hundred metre-thick, fine-grained, low-efficiency deep-water systems (defined in Section 7.2.1). Palaeocurrent data show two prominent directions, with: (i) basin-axis-parallel, south–southwest–north-northeast directions, that are common throughout the entire MF, and (ii) basin-axis-perpendicular to oblique, west–east to northwest–southeast directions, which tend to be confined to the basal parts of the megacycles or channel-lobe transition systems in their upper parts. Based on the facies characteristics, palaeocurrent data, sandstone compositional data and trace-fossils, Bábek *et al.* (2004) proposed that periods of increased tectonic activity resulted in slope over-steepening probably combined with increased rates of lateral

west-to-east sediment supply into the basin to produce the basal sequence boundary and the subsequent lowstand deep-water systems. During subsequent periods of tectonic quiescence, the basin was mainly supplied from a more distant southern point source, producing thick, low-efficiency deep-water systems.

11.3.1 Permo–Triassic Karoo foreland basin, South Africa

The Permo–Triassic Karoo foreland basin deposits of South Africa show well-exposed basin-floor and slope deep-water systems. These fine-grained deposits form the basal part of a 400 m-thick progradational succession capped by shallow-marine, deltaic and

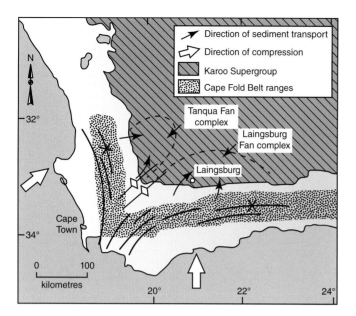

Fig. 11.11 Location of the Tanqua and Laingsburg basin-floor fan-complex in the southwestern corner of the Karoo Basin shown in relation to the two branches of the Cape Fold Belt. From Wickens and Bouma (2000).

fluvial deposits (van der Werff & Johnson 2003a,b; Andersson *et al.* 2004). Many researchers have considered that the Karoo Basin developed as a retroarc foreland basin with subsidence solely due to loading by the fold-and-thrust belt, remnants of which occur along the southern and southwestern margin of the basin (e.g., Johnson 1991; Cole 1992; Visser 1992; Veevers *et al.* 1994). Van der Merwe *et al.* (2010), however, cite petrographic and geochemical studies of the southwest Karoo deep-water deposits (Johnson 1991; Andersson *et al.* 2004; Van Lente 2004) and tectonic models/structural reconstructions of the Cape Fold Belt (CFB) (Tankard *et al.* 2009) that suggest that the thrust-and-fold belt was not emergent at the time of deep-water sand deposition; they, therefore, propose that the main mechanism of early Karoo Basin subsidence may have been dynamic topography effects associated with the subducting slab (Pysklywec & Mitrovica 1999; Tankard *et al.* 2009). Given the uncertainties about the underlying tectonic driver for basin formation, we deal with the Karoo Basin here, accepting that it may not be a typical foreland basin.

In the Tanqua sub-basin or depocentre (Fig. 11.11), the basin-floor fans are relatively undeformed, with an estimated sandstone/shale ratio of ~40–50%. The Tanqua sub-basin submarine-fan complex consists of five north, northeast and east prograding siliciclastic systems (Hodgson *et al.* 2006). Six discrete siliciclastic systems or submarine fans have been recognised, informally called Fans A–F (Sixsmith *et al.* 2004). Each fan has a thickness of 20–60 m, and is separated by a 20–75 m-thick interval of shales and siltstones (Johnson *et al.* 2001).

Within the 'Tanqua depocentre' (Fig. 11.11), Hodgson (2009) documents the stratigraphic and palaeogeographic distribution of so-called 'hybrid event beds' that are interpreted to comprise both debris-flow (cohesive) and turbidity current (non-cohesive) deposits as genetically linked events (compare Section 1.4.2). This was the first study of such beds in a submarine-fan system to combine outcrop and research borehole control. Three types of 0.1–1.0 m-thick hybrid beds are observed, which have a basal, diffusely-graded, fine-grained sandstone, Bouma-type turbidite interval overlain by a division of variable composition that can comprise: (i) poorly-sorted carbonaceous-rich material supported by a mud-rich and micaceous sand-matrix; (ii) poorly-sorted mudstone clasts in a mud-rich sand–silt matrix or (iii)

gravel-grade, rounded mudstone clasts in a well-sorted (mud-poor) sandstone matrix. These upper divisions are interpreted respectively as: (i) the deposit of a debris flow most likely derived from shelf-edge collapse; (ii) the deposit of a debris flow, most likely developed through flow transformation from a turbidity current that eroded a muddy substrate and (iii) the deposit from a turbidity current with mudstone clasts transported towards the rear of the flow. All three hybrid-bed types tend to occur in lobe-fringe environments during fan initiation and growth. The basinward-stepping of successive lobes results in a dominance of hybrid beds toward the base of stratigraphic successions in the middle and outer fan, whereas hybrid beds appear absent in the proximal parts of a fan, and rare and thin in landward-stepping lobes deposited during fan retreat. This distribution was interpreted by Hodgson (2009) to reflect the enhanced amounts of erosion and availability of mud along the transport route during early lowstands of sea level. He concluded that hybrid beds can be used to suggest a fan fringe, to infer lobe stacking patterns, and have sequence stratigraphic significance.

Van der Merwe *et al.* (2010) document the ~380-m-thick fine-grained Vischkuil Formation (~272–262 Ma) that comprises laterally extensive hemipelagic mudstones (Facies Classes E and G), separated by packages of graded sandstone and siltstone SGF deposits (Facies Classes C and D), and volcanic ash beds, as an argillaceous precursor to a 1 km-thick sand-prone basin-floor fan to shelf succession (Laingsburg, Fort Brown and Waterford formations), Karoo Basin, south Africa (Fig. 11.12). Regionally mapped 1–2 m-thick hemipelagic mudstone units (Facies Classes E and G) are interpreted as condensed drapes that represent the starved basin-plain equivalents of transgressive systems tracts and maximum flooding surfaces on the coeval shelf (now removed during later uplift) (Van der Merwe *et al.* 2010). The strata above each mudstone drape comprise siltstone SGF deposits interpreted as highstand systems tract deposits and a surface of regional extent, marked by an abrupt grain-size shift to fine-grained sandstone. They interpret these surfaces as sequence boundaries related to abrupt increases in flow volume and delivery of sand-grade material to the basin plain. The interpreted lowstand systems tract comprises sandstone-dominated SGF deposits and is overlain by another hemipelagic mudstone drape. The upper Vischkuil Formation is marked by three 20–45 m-thick

Fig. 11.12 Field relations in the lower Ecca Group, Geelbek locality. Yellow lines show key mudstone markers. From van der Merwe *et al.* (2010). For a key to the facies in the sedimentary column, see Figure 11.13. See text for discussion of the Vischkuil Formation.

debrites with intraformational sandstone clasts up to 20 cm in diameter. These debrites can be mapped over 3000 km², and induced widespread deformation of the immediately underlying 3–10 m of silty SGF deposits (Facies Class D). A sequence boundary is interpreted at the base of each deformation/debrite package. Van der Merwe *et al.* (2010) recognise six depositional sequences with each successively younger sequence associated with a larger volume of overlying sandstone (Fig. 11.13). The lower two sequences thin to the northwest and show northwest-directed palaeocurrents. The four overlying sequences show a polarity switch in palaeocurrent directions and thinning, to the east and southeast (Fig. 11.13). Sequence 6 is overlain sharply by the 300 m-thick sand-prone Fan A of the Laingsburg Formation. Van der Merwe *et al.* (2010) suggested that the LST debrites might reflect the gradual development of major routing conduits that subsequently fed Fan A. The polarity shift from westward-flowing SGFs to an eastward-prograding deep-water to shelf succession represents the establishment of a long-term feeder system from the west. Van der Merwe *et al.* (2010), therefore, propose that sand supply to the Karoo Basin floor was established in an incremental, stepwise manner. Given the early post-glacial setting of the Karoo deep-water deposits, in an icehouse world, Van der Merwe *et al.* (2010) suggest that glacio-eustatic sea-level changes are likely to have been the main control on sequence development.

The following sections highlight interesting features from a number of ancient deep-water foreland basins. Particular emphasis is placed on the long-distance correlation of beds.

11.3.2 Oligocene–Miocene foreland basin, Italian Apennines

The deep-water Oligocene–Miocene onshore parts of the Periadriatic foreland basin are well documented by Ricci Lucchi (1975a,b; 1981, 1986) and Ricci Lucchi and Ori (1984, 1985). Ricci Lucchi and Ori (1984, 1985) recognise both: (i) major clastic bodies with volumes of 3000–30 000 km³ that represent the progressive infilling of north-easterly migrating foredeeps, together with (ii) minor basin-fills of 50–500 km³. Amongst the smaller basins, thrust-based (thrust-top) basins marginal to the foredeep, or 'piggyback basins' (Ori & Friend 1984), are assigned to a general class of 'satellite basins' (Ricci Lucchi & Ori 1984, 1985). Rates of depocentre (foredeep) shifting from the Oligocene to the present day have been as great as about 7.5 cm yr⁻¹ (Miocene), with fast rates of sediment accumulation tending to occur

during phases of most rapid depocentre shifting, equivalent to periods of greatest tectonic activity; for example, during the fast depocentre migration of the Miocene, sediment accumulation rates reached about 87 cm kyr⁻¹ years (Ricci Lucchi & Ori 1984).

The main style of deep-water sedimentary fill of the Oligocene to Miocene foreland basins described by Ricci Lucchi and others involves essentially sheet systems with many ponded flows, including Facies C2.4 megaturbidites (Ricci Lucchi & Valmori 1980; Gandolfi *et al.* 1983; Talling *et al.* 2004). The basins were effectively over-supplied with respect to their size, a feature that appears common to many foreland-basin successions. Several studies demonstrate long-distance correlations in the Marnoso-arenacea Formation, including the stratigraphic interval containing the so-called Contessa key bed (Ricci Lucchi & Valmori 1980; Amy & Talling 2006; Fig. 1.26). Magalhaes and Tinterri (2010) were the first to show a detailed stratigraphic cross-section, with bed-by-bed correlations of the entire Langhian to Serravallian stratigraphic succession. Their study permits a detailed analysis of the foredeep evolution during the Langhian and Serravallian and a discussion of the facies distribution in relation to thrust propagation.

Magalhaes and Tinterri (2010) present a detailed stratigraphy and facies analysis of an interval of about 2500 m in the Langhian and Serravallian stratigraphic succession of the Marnoso-arenacea Formation (Figs 11.14–11.17). Their high-resolution stratigraphic analysis involved measuring seven stratigraphic logs between the Sillaro and Marecchia lines (60 km apart) with a cumulative vertical stratigraphic thickness of about 6700 m. Their analysis suggests that the stratigraphy and depositional setting of the Marnoso-arenacea Formation was influenced by syn-depositional structural deformation. Magalhaes and Tinterri (2010) subdivided the sections that they studied into five informal stratigraphic units on the basis of the temporal degree of structural control by topographic highs and depocentres, a consequence of thrust propagation. They reconstructed the physiographic changes of the foredeep basin during the progressive appearance and disappearance of thrust-related mass-transport complexes (MTCs) and of five bed types interpreted to be genetically related to structurally controlled basin morphology. Apart from their Bouma-like 'Type-4 beds' (Facies C2.1 to C2.2 in our scheme, Fig. 2.4), 'Type-1 tripartite beds' (Facies C2.5 in our scheme) are characterised by an internal slurry unit, and tend to thicken especially in structurally controlled stratigraphic units where intrabasinal topographic highs and depocentres with variable seabed slopes favoured both mud erosion and decelerations. 'Type-2 beds' (Facies

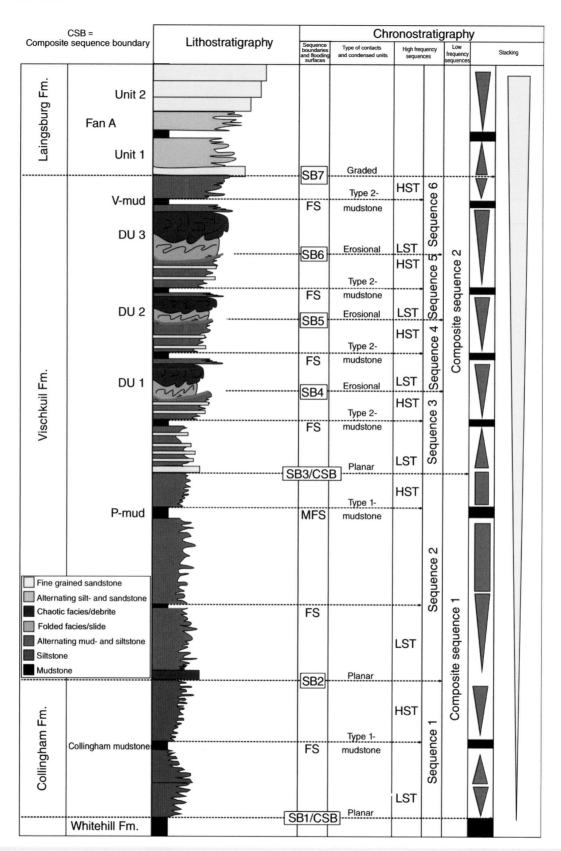

Fig. 11.13 Schematic sequence stratigraphy of the Vischkuil Formation. All major surfaces, sequences and stacking patterns have been mapped regionally. The polarity reversal of palaeocurrents occurs between the oldest two, and four younger sequences at SB 3, which is also interpreted as a composite sequence boundary, with the Vischkuil Formation comprising two composite sequences. Note the upward increase in interpreted flow volume/energy across successive sequence boundaries. From van der Merwe *et al.* (2010).

(a)

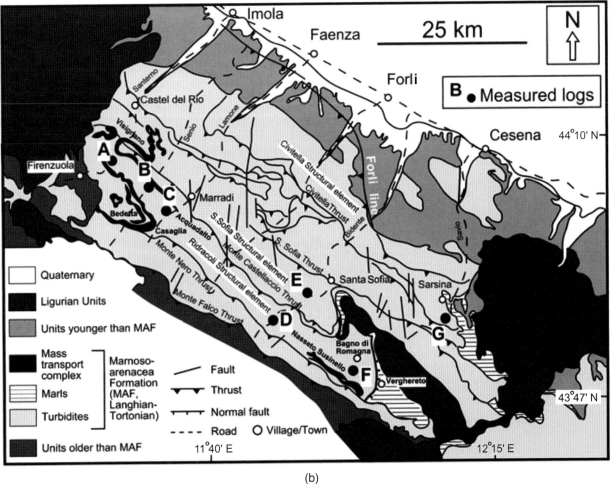

(b)

Fig. 11.14 (a) Location map showing the Marnoso-arenacea Formation outcrops in the northern Apennines. (b) Schematic geological map of the Marnoso-arenacea Formation between the Santerno and Savio Valleys showing the main thrust fronts (modified from Cerrina Feroni *et al.* 2002). The main structural elements are also indicated, whereas the location of the Forli line is taken from Roveri *et al.* (2002, 2003). Capital letters (A, B, C, D, E and F) indicate the location of the seven stratigraphic logs. From Tinterri and Magalhaes (2011).

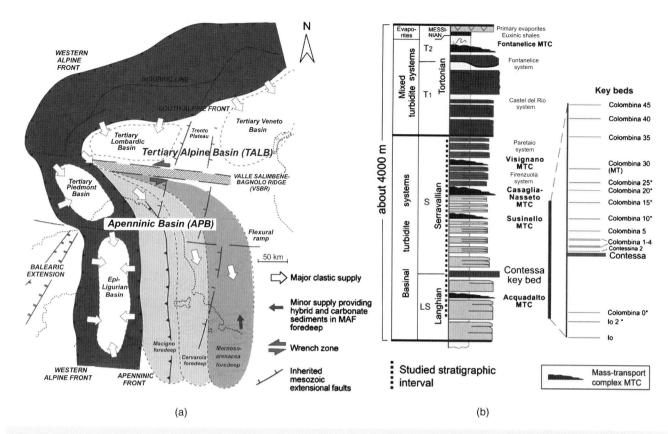

(a)   (b)

Fig. 11.15 (a) Physiographic map showing the inferred configuration and main features of the Proto-Adriatic Basin during the late Oligocene–middle Miocene. (b) Schematic stratigraphic log of the Marnoso-arenacea Formation. The main mass-transport complexes (MTCs) and key beds are also shown. From Tinterri and Magalhaes (2011).

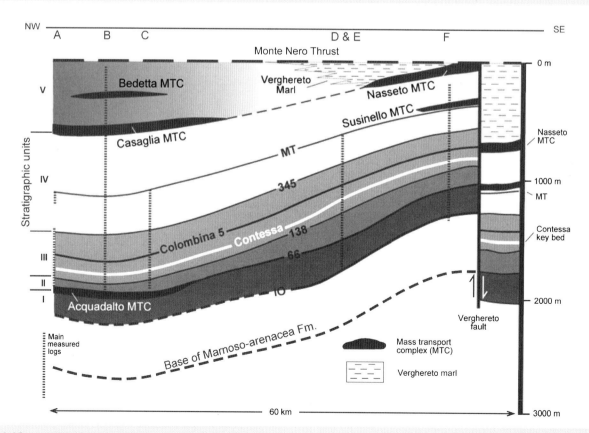

Fig. 11.16 Simplified geological cross-section of the stratigraphic succession studied in the Ridracoli structural element located between the M. Nero and M. Castellaccio thrusts. See Figure 11.14 for the location of the logs. From Magalhaes and Tinterri (2010).

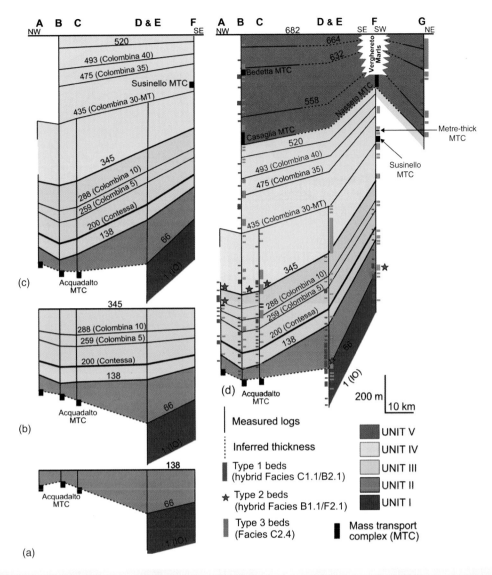

Fig. 11.17 Long-distance correlation of marker beds, and stratigraphic evolution of the Marnoso-arenacea Formation using a progressive flattening approach, with progressively higher units hung on beds that can be traced regionally and that can be linked with physiographic changes in the basin-floor bathymetry/relief. Unit II, for example, is flattened at the top of bed 138; that is, the first bed that can be traced regionally above the Acquadalto MTC (mass-transport complex). See text for an explanation of the three bed types. From Tinterri and Magalhaes (2011).

C1.1 in our scheme), with an internal sediment-slide chaotic unit, characterise the basal boundary of structurally controlled stratigraphic units and are interpreted as indicators of tectonic uplift. 'Type-3 beds' (Facies C2.4 in our scheme) are contained-reflected beds that suggest different degrees of basin confinement, while 'Type-5 deposits' (Facies C2.3 in our scheme) are thin-bedded and fine-grained deposits from dilute reflected turbidity currents that rode up the topographic highs. The vertical and lateral distribution of these deposits was used by Magalhaes and Tinterri (2010) to explain the syn-sedimentary structural control of the studied stratigraphic succession, represented in the Marnoso-arenacea Formation by subtle topographic highs and depocentres created by thrust-propagation folds and emplacements of large mass-transport complexes (MTCs).

11.3.3 Lower Palaeozoic foreland basin, Quebec Appalachians

The Lower Palaeozoic Appalachian foreland basin developed during the closure of the Iapetus Ocean (Thomas 1977, 1985; Hiscott 1984; Hiscott *et al.* 1986; Hatcher 1989; Brett *et al.* 1990; Lehmann *et al.* 1995; Finney *et al.* 1996; Castle 2001; Cawood & Nemchin 2001; Thomas & Becker 2007; Ettensohn 2008; Pinet *et al.* 2010 – see Section 12.6.2). Presently, the deposits of this basin extend for about 2050 km from southern Quebec in Canada to northern Alabama in the USA, covering an area of nearly 536 000 km². During latest Precambrian to Early Ordovician, the southern to southeastern Appalachian margin of the super-continent Laurentia was

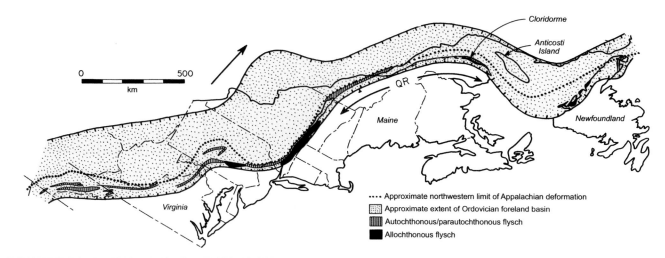

Fig. 11.18 Extent of the Taconic (Ordovician) foreland basin in the Appalachian Orogen, and the location of the Cloridorme Formation in the Quebec Re-entrant (QR). From Hiscott *et al.* (1986).

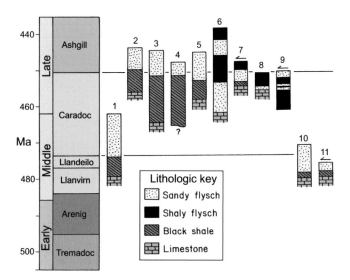

Fig. 11.19 Ages of autochthonous and parautochthonous SGF deposits or flysch, adapted from Barnes *et al.* (1981), Ross *et al.* (1982) and numerous primary references. Faunal zones specified in primary references were related to European series using sheet 1 in Ross *et al.* (1982). 1 = Tellico (Tennessee) and Knobs (Virginia) formations; 2 = Reedsville Formation (Pennsylvania); 3 = Martinsburg Formation (Tennessee, Virginia, West Virginia, Maryland, Pennsylvania, New Jersey); 4 = Shochary Ridge sequence (Pennsylvania); 5 = Schenectady Formation (New York); 6 = Nicolet River Formation (western Quebec); 7 = Beaupré (sandy) and Lotbinière (shaly) formations (Quebec); 9 = Cloridorme Formation (eastern Quebec); 10 = Mainland sandstone (southwest Newfoundland); 11 = Goose Tickle Formation (southwest Newfoundland). Units 7, 9 and 11 are overlain by thrusts. From Hiscott *et al.* (1986).

Fig. 11.20 Simplified model for the collision of the submerged margin of eastern North America with an offshore belt of volcanic arcs. Initial collision was in the north. Arrows summarise palaeoflow data. Palaeoflow was generally away from promontories and into re-entrants. SLP = St. Lawrence Promontory, QR = Quebec Re-entrant, NYP = New York Promontory, PR = Pennsylvania Re-entrant, VP = Virginia Promontory, PT = Piedmont Terrane (micro-continent). The volcanic arc terrane was probably much more complex than shown here, perhaps resembling the complex collage of arcs in the modern western Pacific Ocean. From Hiscott *et al.* (1986).

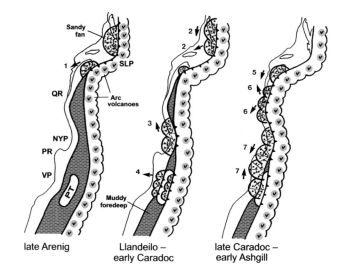

characterised by mainly synrift and postrift, passive-margin environments. By latest Cambrian time, on some of the more distal outboard parts of the Laurentian margin, the initial tectonic reorganisation that would ultimately produce the Appalachian foreland basin had already begun (e.g., Pickering & Smith 1995). The Appalachian foreland basin developed during the Taconic orogeny ~472 Ma (Early–Middle Ordovician).

The Ordovician Cloridorme Formation comprises the Quebec segment of the deep-water foreland-basin succession that formed along the length of the Appalachian Orogen during the Ordovician Taconic Orogeny (Hiscott 1984; Hiscott *et al.* 1986) (Figs 11.18–11.20). The Ordovician Cloridorme Formation is a thick foreland-basin succession of SGF deposits (Fig. 11.21), where the long-distance physical tracing and study of single SGF deposits is possible (Enos 1969a,b; Pickering & Hiscott 1985; Hiscott *et al.* 1986; Slivitzky *et al.* 1991; Kessler *et al.* 1995; Awadallah & Hiscott 2004). The formation is estimated to be ~4 km thick and accumulated at an unrestored rate of ~400 m of rock per million years. Neither its base nor top is exposed. The Cloridorme Formation is deformed by thrusting and folding (Figs 11.22, 11.23), but is interpreted to be parautochthonous and, therefore, little displaced from its site of deposition (Logan 1883; St-Julien & Hubert 1975). The Cloridorme Formation is a true 'flysch' because it consists of a thick succession of SGF deposits derived by erosion of a rising orogenic belt, and shed into an active foreland basin.

Although successions of SGF deposit are characteristically hundreds to thousands of metres thick, it is not common practice to divide them into thinner lithostratigraphic units. This is because of the monotonous stacking of similar facies, and the poor temporal resolution that stems from high sediment accumulation rates of >100 m Myr^{-1}. For example, the Cloridorme Formation equivalents in the USA Appalachians are largely undivided: the Martinsburg Formation has no recognised members (McBride 1960) and the Normanskill Formation consists of only two members (Rickard & Fisher 1973). The Cloridorme Formation has been treated somewhat differently, however, perhaps because it is so well exposed in wave-washed shoreline outcrops along the Gaspé coast. Indeed, the Cloridorme Formation is better exposed than any other Ordovician deep-marine succession in the Appalachians or Caledonides, allowing long-distance correlations of individual beds (Enos 1969b; Ma 1996).

Enos (1969a) divided the formation into informal members α1–α7 and β1–β4 in two different structural blocks, but he was unable to correlate the α and β successions (Fig. 11.24). Pickering and Hiscott (1985) and Hiscott *et al.* (1986) used lithologic characteristics and the biostratigraphy of Riva (1968, 1974) to suggest that α7 = β2; they then renamed the informal members using six local geographic names. Pickering and Hiscott (1985, 1995) used sedimentological observations to show that the lower part of the Cloridorme Formation (their St-Hélier and Pointe-à-la-Frégate members: their fig. 2) includes many widely traceable Facies C2.4 megaturbidite beds that can be used for physical correlation. Because megaturbidites are time markers (*cf.* Ricci Lucchi & Valmori 1980), they also provide a temporal framework for subdivision of the Cloridorme Formation.

Awadallah and Hiscott (2004) advocated abandoning strict lithostratigraphic subdivision in favour of a modified allostratigraphic methodology that permits fine-scale division and long-distance correlation of facies. According to the North American Commission on Stratigraphic Nomenclature (1983), 'An allostratigraphic unit is a mappable stratiform body of sedimentary rock that is defined and identified on the basis of its bounding discontinuities'. The word

'discontinuity' includes various types of unconformities, but Awadallah and Hiscott (2004) elected to use this term as well for unusual and easily traceable event-deposits that interrupt and therefore disrupt the normal depositional record. In the lower Cloridorme Formation, the unusual events that interrupt the stratigraphy are >3–5 m-thick Facies C2.4 megaturbidite beds (Pickering & Hiscott 1985). Previously, only the basin-wide megaturbidites recognised by Skipper and Middleton (1975) and Pickering and Hiscott (1985) were used as time markers, permitting correlation over a lateral distance of ~25 km. Enos (1969a) reported the occurrence of volcanic tuffs, but did not make any detailed study of them. Awadallah and Hiscott (2004) strengthened and revised the stratigraphic framework of the Cloridorme Formation based on megaturbidites by considering nine widely traceable and geochemically finger-printed tuffs (K-bentonites). They traced and correlated 71 Facies C2.4 megaturbidites and nine K-bentonites to erect a high-resolution subdivision of the lower Cloridorme Formation (Fig. 11.26). Particularly thick megaturbidites were used to define the boundaries of three newly defined allostratigraphic members (St-Hélier Allomember, St-Yvon Allomember and Petite-Vallée Allomember). The 71 Facies C2.4 megaturbidites can be correlated on the basis of internal structures and bed thickness, thickness patterns of groups of beds, stratigraphic position in the lower Cloridorme Formation, or a combination of these parameters. Megaturbidites BT-1 through BT-71 vary in thickness from ~1 m to >7 m. Lateral variations in the thickness of individual beds are minor except for beds at one locality, so that bed thickness can be used as a guide to correlation.

Figures 5.15, 5.16, 11.25 and 11.27 show representative correlated sections in the Cloridorme Formation, with long-distance correlation of individual beds and packages of thinner beds (Fig. 11.28). In some cases, sand-prone sets of beds appear to have formed mounds on the seafloor, over which the Facies C2.4 megaturbidites show pronounced thinning (e.g., Figs 5.16, 11.25, 11.27).

11.3.4 South Pyrenean foreland basin and thrust-top/piggyback basins

The Late Cretaceous and Palaeogene, east–west trending, South Pyrenean foreland basin, and associated thrust-top or piggyback basins, are amongst the most studied ancient examples of foreland basins with substantial amounts of deep-marine clastic sediments. The Eocene South Pyrenean foreland basin is not considered here except in passing, because its stratigraphy and sedimentology have been presented in other sections of this book (Sections 4.8, 8.3.2). Here, only the larger scale tectono-stratigraphic and sedimentological aspects, particularly the earlier history of the Pyrenean orogen, are considered.

The South Pyrenean basins developed from latest Cretaceous to Miocene times during north–south compression caused by the collision of the continental crust of Iberia with that of southwestern mainland Europe (Labaume *et al.* 1985; Choukroune *et al.* 1990; Muñoz *et al.* 1992; Sinclair *et al.* 2005; Fig. 11.29). During the Middle Eocene, the foreland basin/thrust-top basin accumulated mainly non-marine/marginal-marine deposits in the east (Tremp-Graus and Ager basins), whilst further west there was an overall change from fluvio-deltaic to deep-marine systems (Ainsa–Jaca Basin), and then the most distal basin-floor deposits in the Pamplona basin (Mutti *et al.* 1988). The earlier Late Cretaceous (~100–75 Ma) phase of subsidence was driven by the creation of the South Pyrenean

Fig. 11.21 (caption on facing page)

Fig. 11.21 Representative plates of SGF deposits in the Cloridorme Formation, Gaspé Peninsula, Quebec Appalachians, Canada. (a) Thin- and very thin-bedded sandstones (Facies Class C and D), with minor amounts of silty mudstones (Facies Class E), Grande-Vallée, Enos (1969a) member β6–β7. Overturned beds young to left. Boulder in foreground left of centre ~1.5 m maximum dimension. (b) Facies Classes B and C sandstones, interpreted as middle-fan deposits, quarry east of Marsoui, Enos (1969a) member γ4. Note, wedging of beds, particularly towards the upper part of the outcrop. Human scale. (c) Heterolithic SGF deposits, Pointe-à-la-Frégate. Overturned beds young to right. Metre staff divided into 10 cm units. (d) Heterolithic SGF deposits, including a prominent Facies C2.4 bed with a light-coloured sandstone basal part and a thick silty mudstone cap on left of field of view near water, Manche d'Epée. Beds young to left. Human scale. (e) ~1.5 m-thick megaturbidite (Facies C2.4) with basal sandy divisions and structureless mudstone cap, St-Yvon section. Beds young to right. Metre staff divided into 10 cm units. (f) Facies C2.4 bed, Pointe-à-la-Frégate. 5 cm scale. Arrows mark the ~180° divergent palaeoflow directions from flutes and ripple lamination. (g) Complexly folded coherent sediment slide deposit (Facies F2.1) within mainly Facies Class D deposits (and overlain by Facies C2.4), St-Yvon. Beds young to right. Metre staff divided into 10 cm units. (h) Packet of amalgamated sandstones of mainly Facies Class C, interpreted as lobe deposits, within finer-grained and thinner-bedded sandstones and siltstones (mainly Facies Group C3 and Facies Class D), interpreted as fan-fringe and basin-floor deposits. Petite-Vallée. Beds young to left. Metre staff divided into 10 cm units. For additional plates of Cloridorme Formation facies, see Figures 2.25f, 2.26b, 2.26c.

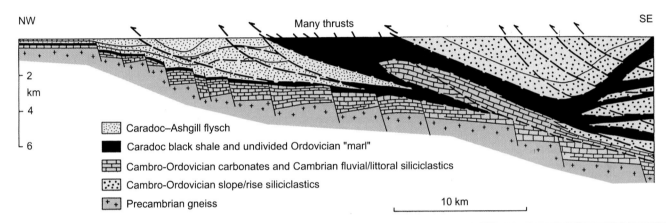

Fig. 11.22 Simplified interpretation of part of the SOQUIP seismic line fully described by Laroche *et al.* (1983). This segment of the line is located towards the western part of the Quebec Re-entrant, labelled QR in Figure 11.20. Foreland-basin SGF deposits or flysch lies stratigraphically above Caradoc black shales, which lie above Cambrian–Ordovician carbonates of a passive-margin sequence. The SGF deposits were subsequently overthrust by nappes containing older rocks originally deposited farther southeast. The Nicolet River and Lotbinière formations form the flysch in this section. From Hiscott *et al.* (1986).

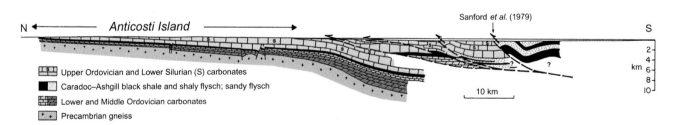

Fig. 11.23 Simplified interpretation of seismic line across Anticosti Island and towards the Gaspé Peninsula coast. The imbricate thrust faults involve Lower Silurian rocks, and therefore are of Acadian age. The southern part of the line crosses part of the Gulf of St Lawrence for which there are no borehole data, so that the interpretation of facies changes at depth is speculative. Maps prepared by Sanford *et al.* (1979) indicate a major change from fault blocks containing Silurian carbonates to parautochthonous Cloridorme Formation SGF deposits (labelled 'flysch') near the southern end of the line. From Hiscott *et al.* (1986).

foreland basin and its basin inversion, as expressed by a regressive megasequence in the Tremp Basin (Tremp Formation). Basin inversion lasted ~10 Myr from ~75–65 Ma (Simó & Puigdefàbregas 1985: their figs 2 and 4). During the Late Cretaceous, the Iberian Plate moved ~400 km southeast with respect to a stable Europe, resulting in sinistral shear along the European–Iberian plate boundary. In the Late Cretaceous to Eocene, the relative plate motions changed to a north-northwest–south-southeast convergence of ~130 km to produce a north–south compressive regime that was active throughout

the Tertiary. Estimated minimum average southward migration of the border of the southern foreland basin is 5 mm yr^{-1}, with the advance of the deformation front from early Eocene to early Miocene at ~3.5 mm yr^{-1}, based on palinspastic reconstructions (Labaume *et al.* 1985).

Within the southern foreland basin, the main transport path for the clastics was axial and from east to west. Today the preserved total basin dimensions are 250 km long and 15–45 km wide, from the Tremp region in the east to the Pamplona region in the west. In the

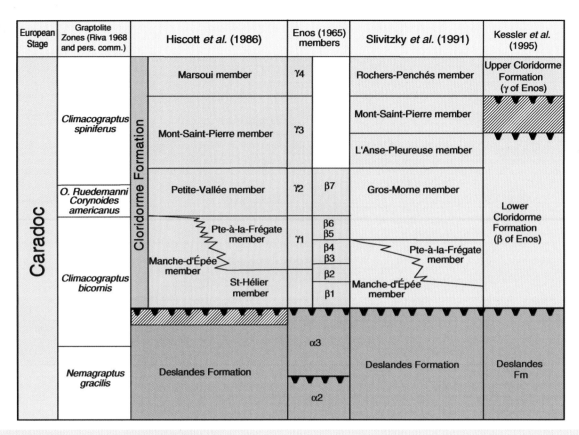

Fig. 11.24   Previously published stratigraphic subdivisions of the Cloridorme Formation. Barbed lines are thrust-fault contacts. Diagonal rule is a hiatus. From Awadallah and Hiscott (2004).

west, a deep-marine basin formed that was of the order of 600–800 m deep and from 15–45 km wide (Mutti *et al.* 1984; Pickering & Corregidor 2005; Pickering & Bayliss 2009). The palaeogeographic boundary between eastern deltas and the deep-marine basin was controlled by a faulted shelf margin coincident with a syn-depositional growing anticline called the Mediano Anticline. Contemporaneous with foreland basin sedimentation, extensive carbonate shelves formed on both the northern and southern basin margins. The northern carbonate shelf is not preserved but has been inferred from palaeocurrents and sediment composition studies.

Along the length of the foreland basin, structural highs, like the Boltaña and Mediano anticlines, were active during sedimentation. These syn-sedimentary highs acted as dams against which deep-marine sediments ponded. The most significant of these structures is the Boltaña Anticline that began to grow in the early Eocene, and divided the Hecho Supergroup into an eastern and western basin (Mutti 1984). The Ainsa Basin siliciclastic deposits occur in two angular-unconformity-bounded units (Muñoz *et al.* 1994, 1998; Fernandez *et al.* 2004; Pickering & Bayliss 2009; Pickering & Cantalejo 2015; Scotchman 2015a), in which the younger unit is both structurally less deformed and shows a southwestward shift in depositional axis (Pickering & Corregidor 2005), demonstrating a first-order tectonic control on accommodation and deposition. Succeeding units are both structurally less deformed and show a west–southwest shift in depositional axis. These two 'tectono-stratigraphic' units contain eight coarse clastic systems, each of the order of 100–200 m thick,

and vertically separated by up to several tens of metres of mainly marls with lesser amounts of thin- to very thin-bedded sandstone SGF deposits. Each system typically contains 2–6 individual sandbodies (amounting to at least 25 throughout the basin), from 30–100 m thick, separated by tens of metres of mainly thin- and very thin-bedded sandstones of Facies Classes C and D with subordinate marls of Facies Class E (Pickering & Bayliss 2009). Tectonics, Milankovitch cyclicity and sub-Milankovitch millennial-scale and sub-millennial-scale cycles, are recognised as having exerted a control on depositional styles within the basin (Section 4.8). The deep-marine deposits are overlain by ~0.5 km of fluvio-deltaic and related sediments fed mainly from the south (Dreyer *et al.* 1999).

A particularly interesting feature of the relatively distal basin-floor deposits in the Jaca basin is the occurrence of 'megaturbidites' in beds up to 200 m thick. The megaturbidites ideally show a five-fold division (Fig. 11.30) and, based on criteria for slope failure and assuming a seismic origin for these beds with earthquakes of at least magnitude 7, Seguret *et al.* (1984) estimated that the sediment gravity flows transported volumes up to 200 km^3. Payros *et al.* (1999, 2007) refer to numerous Eocene carbonate megabreccias intercalated with siliciclastic SGF deposits (and derived by resedimentation of shallow-marine carbonate platforms) as *South Pyrenean Eocene carbonate megabreccias* (SPECM units). Figure 11.31 shows the along-basin, down-system, geometry and internal sedimentary structures of seven of these SPECM units. The SPECM units appear to occur as time-stratigraphic clusters, which can be

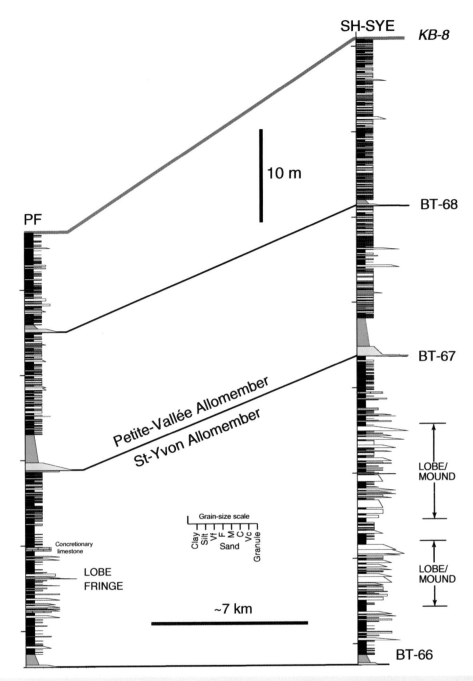

Fig. 11.25 Long-distance correlation of Facies C2.4 megaturbidites (yellow beds with green silty mudstone tops) in the vicinity of the proposed contact between the St-Yvon and Petite-Vallée allomembers of the Cloridorme Formation, Québec. KB-8 is at the top of the correlated sections. The sections are hung on BT-66 because above this point there are two sand-bed clusters at composite section SH-SYE that likely formed a bathymetric high (Awadallah 2002). Thickness variations are greater than in the St-Hélier member (Fig. 5.15) because of the encroachment of sandy submarine fans from the east. See Fig. 11.26 for locality names (e.g., PF). From Awadallah & Hiscott (2004).

correlated with relative sea-level lowstands and linked with phases of tectonic activity (Payros *et al.* 1999). It appears that these megabreccias were derived from a carbonate-platform system along the southern margin of the foreland basin, with episodic instability and mass wasting being triggered by phases of structural steepening (forebulge uplift) accompanied by large-magnitude earthquakes, with the former

causing platform emergence, increased load stresses and excess pore-water pressure in the carbonate ramp (Payros *et al.* 1999). Payros *et al.* (1999) interpret the SPECM deposits to have been emplaced by cohesive debris flows that evolved into concentrated density flows. An ideal SPECM unit comprises: (i) an immature, homogeneous debrite in the proximal part; (ii) a differentiated, bipartite debrite

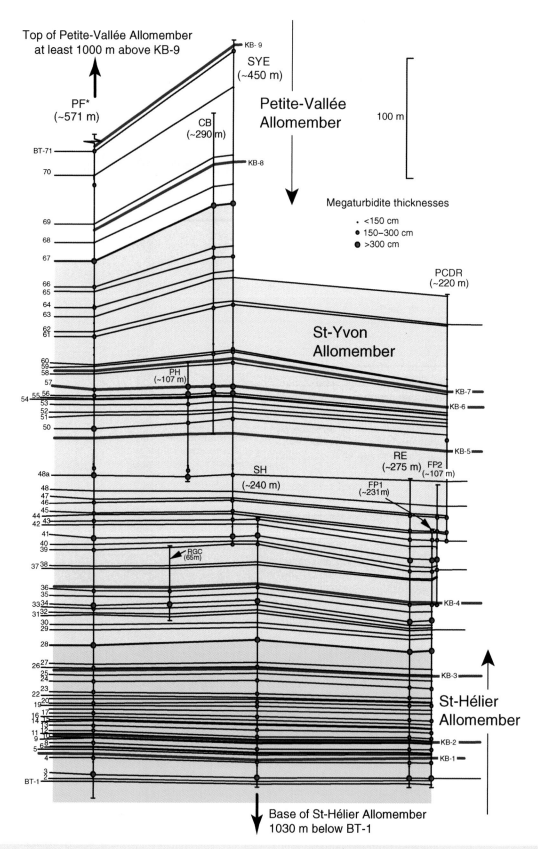

Fig. 11.26 Correlation panel linking ten of thirteen sections through the lower Cloridorme Formation. PF = Pointe à la Frégate; CB = Cap Barré; SYE = St Yvon east; PCDR = Pointe des Canes de Roches; PH = Pointe à Hubert; SH = St Hélier; RGC = Rivière du Grand Cloridorme; RE = Ruisseau à L'Échalote; FP = Fame Point. Tuffs KB-1 through KB-9 are marked by thick grey lines. Facies C2.4 megaturbidites are numbered sequentially from BT-1 to BT-71. The length of each section is indicated in brackets; section PF is limited to the 571 m that can be correlated with other localities, although an additional ~200 m were measured above what is plotted here. See Awadallah and Hiscott (2004) for discussion of allomember definitions, and Awadallah (2002) for details on section and bed thicknesses. From Awadallah and Hiscott (2004).

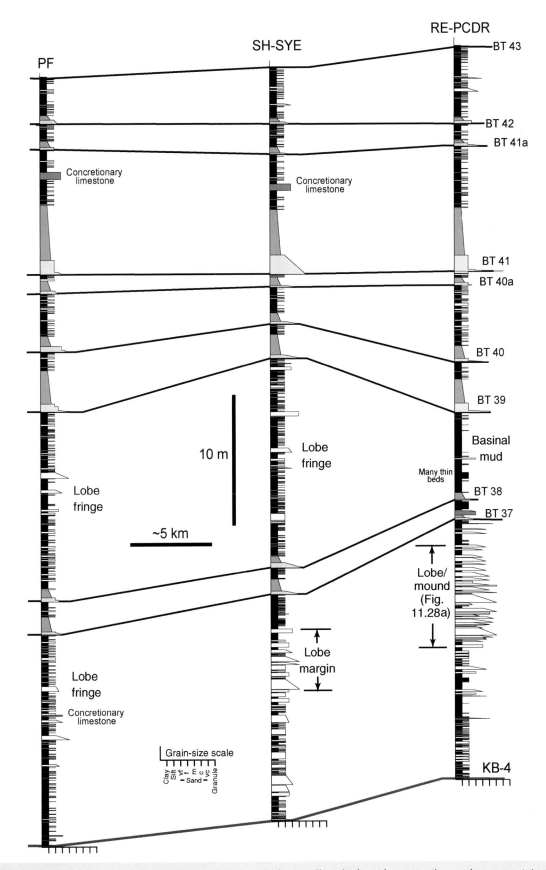

Fig. 11.27 Long-distance correlation of Facies C2.4 megaturbidites (yellow beds with green silty mudstone tops) in the St-Yvon allomember of the Cloridorme Formation, Québec. Bentonite KB-4 is at the base of the correlated sections, which are hung on megaturbidite BT-41. Thickness variations are greater than in the St-Hélier member (Fig. 5.15) because of the encroachment of the distal tips of sandy submarine fans from the east, represented here by interpreted lobe and lobe-fringe deposits. See Fig. 11.26 for locality names (e.g., PF). From Awadallah (2002).

Fig. 11.28 Representative outcrop examples of architectural elements in the Cloridorme Formation, Québec. In all plates, white scale bar is 1 m long. (a) Vertically-dipping sandstone lobe deposits (Facies Class B) in the St-Yvon allomember of the Cloridorme Formation, Québec. Beds young to left. The stratigraphic position is marked in the RE-PCDR column of Fig. 11.27. (b) Lobe-fringe deposits. Beds young to left. (c) Basinal mudstones. Beds young to right. (d) Megaturbidites (Facies C2.4) used for long-distance correlation, bed BT-67 at Pointe à la-Frégate (Fig. 11.25). Beds young to left.

and organised SGF deposit in the medial part; and (iii) an incomplete, base-missing debrite overlain by an organised SGF deposit, or a turbidite or concentrated density-flow deposit alone, in the distal part (vertical sequence shown in Fig. 11.30, with proximal–distal characteristics shown in Fig. 11.31). The debrite component volumetrically predominates in the SPECM units, and the original terms 'megaturbidite' and 'seismoturbidite' thus seem to be inappropriate for these deposits.

Payros *et al.* (2007) reconstructed a transect of a carbonate-ramp slope, using outcrop data from the lower–middle Eocene Anotz Formation in the western Pyrenees (Fig. 11.32). The Anotz Formation contains four calciclastic (mostly bioclastic) members enclosed within hemipelagic marl/limestone alternations, with individual calciclastic members interpreted by those authors as discrete submarine-fan systems. Individual fans comprise a gullied upper

slope, a levéed feeder channel, a channelised lobe area, an unconfined lobe zone and a peripheral lobe-fringe that grades downcurrent into basinal deposits (Fig. 11.32, Fig. 11.33 shows the palaeogeographic context of these submarine fans and related deposits). Quantitative data on the dimensions and degree of lateral continuity and vertical connectivity of the Anotz calciclastic fan elements are presented. These data contribute to a better understanding of the intrinsic nature of calciclastic submarine fans and their reservoir potential. The long-term evolution of the Anotz carbonate slope was generally progradational, as evidenced by the four discrete episodes of calciclastic-fan development. The location of the fans was controlled by the syn-sedimentary tectonic activity of the Pamplona fault, which created a slope valley along which the reworked shallow-water calciclastic sediments were funnelled. In addition, episodic basinward tilting of the shallow-water carbonate ramp, linked to the

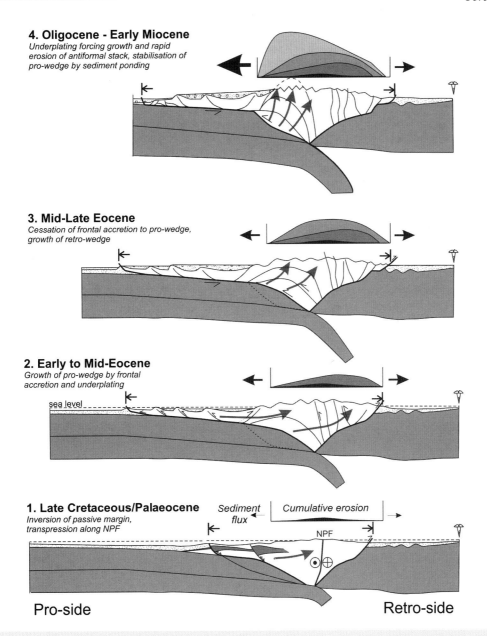

4. Oligocene - Early Miocene
Underplating forcing growth and rapid erosion of antiformal stack, stabilisation of pro-wedge by sediment ponding

3. Mid-Late Eocene
Cessation of frontal accretion to pro-wedge, growth of retro-wedge

2. Early to Mid-Eocene
Growth of pro-wedge by frontal accretion and underplating

sea level

1. Late Cretaceous/Palaeocene
Inversion of passive margin, transpression along NPF

Sediment flux | Cumulative erosion

NPF

Pro-side | Retro-side

Fig. 11.29 Summary of the evolution of the Pyrenean orogen based on fluxes through the system recorded by frontal accretion, underplating, erosion and sedimentation. The total horizontal orogenic shortening is ~165 km, and crustal thickness is ~45 km. 'Pro-side' is Spanish Pyrenees, and 'retro-side' represents the French Pyrenees and Aquitaine Basin. NPF = North Pyrenean Fault. Note overall sinistral transpression during the Late Cretaceous–Palaeocene. From Sinclair *et al.* (2005).

development of the South Pyrenean foreland basin, switched on and off the process of calciclastic resedimentation and determined the growth or abandonment of the fan systems. The Erro Formation, also cropping out in the western Pyrenees, represents the distal part of the Hecho Group SGF deposits (Mutti 1984), a large-scale siliciclastic submarine fan axially fed from the east, and genetically unrelated to the other two formations. The Erro Formation is mainly composed of thin-bedded siliciclastic SGF deposits and hemipelagic marls, but it also includes several large-scale, resedimented carbonate deposits, referred to in the literature as megaturbidites, megabeds or

megabreccias (Labaume *et al.* 1985; Barnolas & Teixell 1994; Payros *et al.* 1999, and references therein).

11.4 Concluding remarks

Deep-marine foreland basins tend to form over 10–20 Myr time spans, and commonly involve several discrete phases of tectonic compression. They may be under- or over-filled. The elongate nature of deep-marine foreland basins (in common with submarine trenches

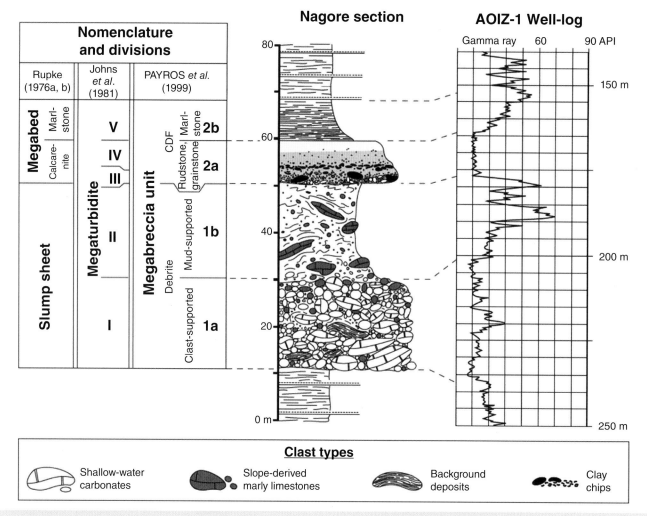

Fig. 11.30 A complete sequence of divisions characteristic of a South Pyrenean Eocene carbonate megabreccia (SPECM unit) using an example from SPECM-b in Nagore section, Urrobi River, and the corresponding wireline log (SPECM-b in Aoiz-1 borehole). Note the deposit nomenclature and divisions used by the previous authors and Payros *et al.* (1999). We have replaced the term 'turbidite' used by Payros *et al.* (1999) by CDF deposit (concentrated density-flow deposit) to be consistent with the process terminology of this book.

formed above subduction zones) means that most submarine fans, including any basin-floor megaturbidites, tend to be linear with strong lateral confinement by the basin slopes. A good example of such confinement has been documented by Bernhardt *et al.* (2011) for the Magallanes foreland basin, Cerro Toro Formation, southern Chile, where they describe submarine channels or channel complexes 700–3500 m wide which occupied a fairway that was 4–5 km wide. Such confinement tends to considerably reduce the potential width of deposits as well as the depositional systems, so that some of the most impressive bed-by-bed correlations have been made in ancient foreland basins (e.g., the Neogene Italian Apennines, Lower Palaeozoic Canadian Appalachians, Late Palaeozoic Karoo and Palaeogene South Pyrenean basins). Syn-sedimentary growth structures, such as anticlines and synclines, are generally important features in controlling the distribution of facies and facies associations, and their architectural elements. Such growth structures tend to force

deposition away from the active fold-and-thrust belt and towards the foreland, resulting in characteristic (but not unique) lateral offset-stacking patterns. Cross-basin topographic highs, commonly linked to basement structures (although these might be sediment dams caused by sediment blocking by MTDs), are generally present and, in some cases, may act as sills that restrict deep-water circulation patterns.

Sediment provenance tends to be point-sourced. Submarine canyons provide lateral input from several locations along a foreland basin, with SGFs abruptly turning along the basin axis. As foreland basins are tectonically very active, repeated and geologically frequent seismicity tends to generate large-volume MTDs as well as megaturbidites. Generally, volcanic activity and, therefore, primary volcanic material, is rare in foreland-basin deposits because arc volcanism ceases or diminishes sharply at the initiation of continent–arc or continent–continent collision.

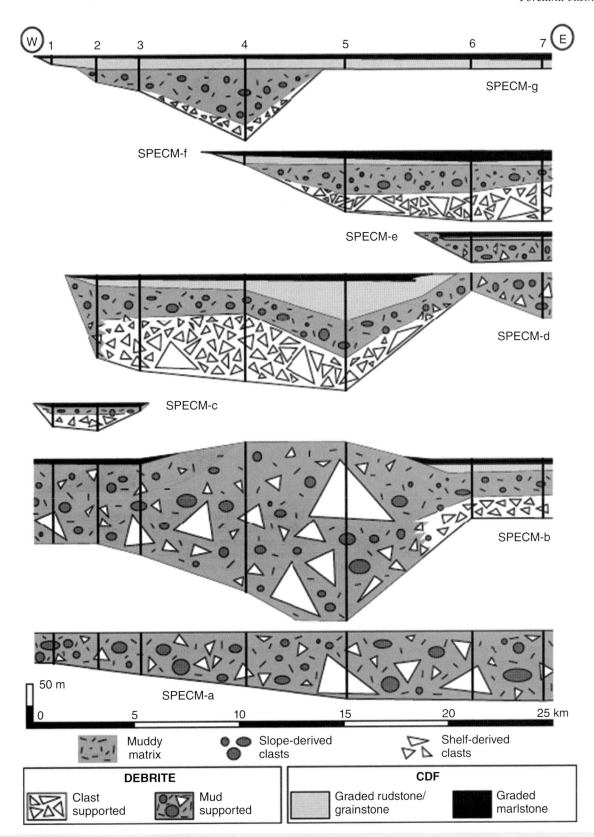

Fig. 11.31 Simplified east–west cross-sections of the South Pyrenean Eocene carbonate megabreccias (SPECM units) in the Pamplona Basin. The sections are transverse (approximately perpendicular to palaeocurrents) and based on seven stratigraphic profiles. From Payros *et al.* (1999). We have replaced the term 'turbidite' used by Payros *et al.* (1999) by CDF deposit (concentrated density-flow deposit) to be consistent with the process terminology of this book.

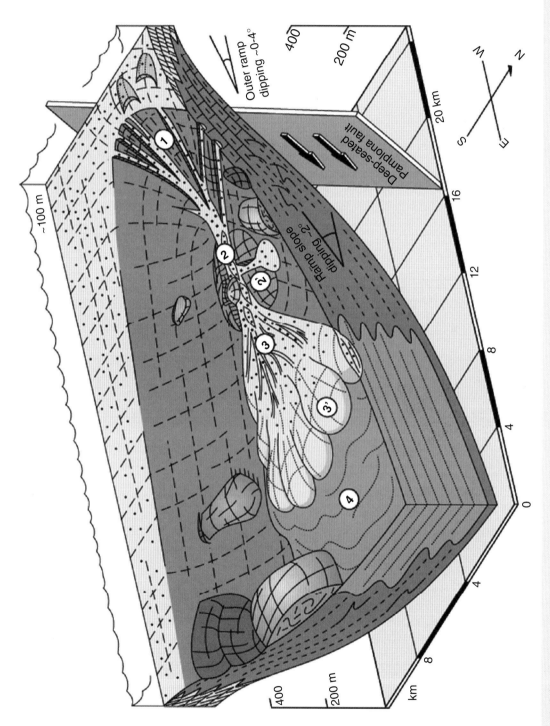

Fig. 11.32 General depositional model for the Anotz CSF systems. 1 = gullied upper slope; 2 = braided submarine channel axis; 2′ = levée/overbank; 3 = proximal lobe deposits; 3′ = distal lobe deposits; 4 = lobe-fringe deposits. From Payros et al. (2007).

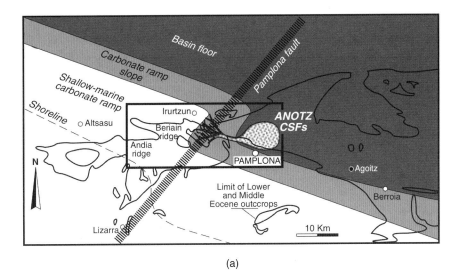

(a)

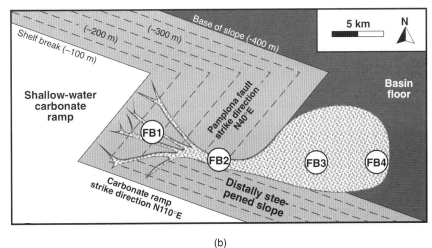

(b)

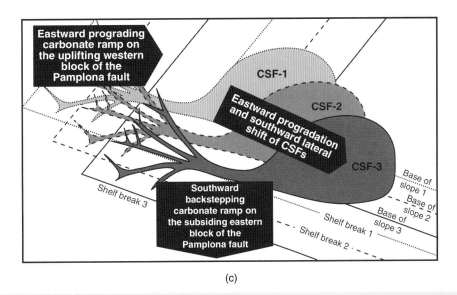

(c)

Fig. 11.33 (a) Palaeogeography of the western Pyrenees during late Ypresian–middle Lutetian times, including the area studied by Payros *et al.* (2007) (boxed). (b) Idealised contour lines in the zone where the Pamplona fault intersected the carbonate slope of the southern margin of the foreland basin. FB1 to 4 correspond to the facies-association belts. (c) Plan view of three successive calciclastic submarine fan systems (CSF-1–3) in the Anotz complex (the scale is approximate), showing their eastward and southward displacement, respectively, attributed to the uplift of the western block of the Pamplona fault and to the southward shift of the depocentre of the foreland basin. From Payros *et al.* (2007).

CHAPTER TWELVE

Strike-slip continental margin basins

Wave-cut platform of sheet-like SGFs, Neogene Aoshima Formation of the Miyazaki Group, eastern Kyushu, Japan (deep-marine distal shelf, fan-delta-linked basin; *cf.* Ishihara *et al.* 2009).

12.1 Introduction

Plate boundaries along which strike-slip (oblique-slip) tectonics predominate may be the site of deep-marine sedimentary basins (Fig. 12.1). While such margins can show tectono-stratigraphic features in common with evolving and mature passive or convergent margins, it is generally the complexity of oblique-slip basins that warrants special consideration (e.g., Busby & Ingersoll 1995). Additionally, basins formed at oblique-slip margins tend to be narrower than many other basins, but with considerable temporal variability. Furthermore, the associated igneous and volcanic stratigraphy appears to be unique to such plate tectonic settings and this latter factor may prove to be one of the best means of interpreting possible ancient oblique-slip margins. The amount of convergence

and divergence of adjacent crustal blocks, the magnitude of such displacements, the rheology of the deforming material, and the nature and position of pre-existing structures, all control the structural patterns in basins formed at oblique-slip margins (Christie-Blick & Biddle 1985). Basin subsidence is driven by crustal thinning, together with any associated thermal and tectonic subsidence, flexural and sedimentary loading. Various theoretical models (e.g., Mann *et al.* 1983) and physical analogue modelling (e.g., Dooley & McClay 1997; Rahe *et al.* 1998; McClay & Bonora 2001) of both transtensional and transpressional basins have been developed to explain their kinematic evolution.

Oblique-slip basins typically show a mismatch in stratigraphy across basin margins, abrupt lateral changes in facies associations and stratal geometry. Unconformities tend not to correlate between

Deep Marine Systems: Processes, Deposits, Environments, Tectonics and Sedimentation, First Edition. Kevin T. Pickering and Richard N. Hiscott.
© 2016 Kevin T. Pickering and Richard N. Hiscott. Published 2016 by John Wiley & Sons, Ltd.
Companion Website: www.wiley.com/go/pickering/marinesystems

adjacent basins. The geometry of such basins (accommodation) also tends to show both lateral and longitudinal asymmetry. Most oblique-slip continental margins record a complex history of extensional (transtensional) and compressional (transpressional) tectonic phases (e.g., Crowell 1974a,b; Reading 1980; Christie-Blick & Biddle 1985; Nilsen & Sylvester 1995; Barnes *et al.* 2001, 2005; Storti *et al.* 2003; Seeber *et al.* 2004; Wakabayashi *et al.* 2004; Zachariasse *et al.* 2008), including alternate left-lateral (sinistral) and right-lateral (dextral) displacements. The Neogene–Holocene development of southern California (Schneider *et al.* 1996; Kellogg & Minor 2005; Ingersoll 2008), and the Sea of Marmara area linked with the North Anatolian Fault Zone (Dewey & Sengör 1979; Sengör 1979; Sengör *et al.* 1985; Aksu *et al.* 2000; Armijo *et al.* 2002; Gürer *et al.* 2003; Okay et al. 1999; Okay & Erdün 2005; Sari & Çagatay 2006; Laigle *et al.* 2008; Elitok & Dolmaz 2011) are good examples of this complexity (Schneider *et al.* 1996; Kellogg & Minor 2005; Ingersoll 2008).

12.2 Kinematic models for strike-slip basins

A simple shear couple (Fig. 12.2) appears to provide an appropriate tectonic model for many cases of basin evolution controlled by oblique-slip (Harding 1974). A more complex modified simple shear model, however, like that described by Aydin and Page (1984) for the San Francisco Bay region, California, may be more useful (Fig. 12.3). Ideally, the first set of strike-slip faults, the R (synthetic) and R' (antithetic) shears, make angles of φ/2 and 90°–φ/2 with the applied shear direction, respectively, where φ, the angle of internal friction, is typically around 30°. With progressive deformation, first the R' and then the R faults are deactivated to become passive strain markers. Eventually, P shears form that are parallel to subparallel with respect to the principal applied shear direction and it is these faults that tend to accommodate the largest displacements.

Oblique-slip tectonics commonly results in complex facies changes over relatively short distances. Synclines and anticlines tend to develop synchronously with sedimentation, to generate elongate sedimentary depocentres that in deep-water examples preferentially concentrate coarser grained clastics as elongate submarine fans and channel systems. The development of syn-sedimentary anticlines can generate intrabasinal highs that may act as local sediment sources, including gravity controlled sediment mass flows. Normal growth faults and thrusts may develop along different margins of the same basin, deforming wet sediment at depth as well as creating areas of local uplift and subsidence. Thus, along the same oblique-slip zone, deep-marine clastic systems can develop under both transpressional and transtensional regimes, the latter producing sag and pull-apart basins. Such pull-apart basins commonly are deep relative to their width, resulting in abrupt lateral facies changes. Pull-apart basins tend to be floored by oceanic crust whereas sag basins have thinned subsided continental crust.

Seismic sections show flower structures as a common feature of oblique-slip zones. *Flower structures* represent the shallow expression of deeper, near-vertical, strike-slip fault zones where either compressive (transpressional) stresses produce splayed reverse faults as a positive flower structure, or tensional forces (transtension) generate curved normal faults as a negative flower structure. Well-illustrated positive flower structures, ranging from relatively shallow features to features incorporating the entire continental crust, occur beneath lower-upper Miocene clastics on the shelf off northwest Palawan, Philippines (Roberts 1983), in the Banggai–Sula–Molucca Sea margin, Indonesia (Watkinson *et al.* 2011), and in the Calabria Arc (Del Ben *et al.* 2008) (Fig. 12.4) as the result of compression with sinistral shear during the development of a thrust imbricate system. Many flower structures appear to be linked to *stepover structures* in strike-slip settings. Where adjacent or nearby fault strands terminate and/or overlap, this results in a wide array of so-called *stepover structures*. Such stepovers are commonly associated with complex fold and fault patterns, and these have been experimentally modelled, for example, by McClay and Bonora (2001) (Figs 12.5, 12.6). In their experiments, they used scaled sandbox models to simulate the geometries and progressive evolution of antiformal pop-up structures developed in a weak sedimentary cover above restraining stepovers in offset sinistral strike-slip fault systems in rigid basement. Pop-ups and transpressional uplifts are an integral part of intraplate and interplate strike-slip fault zones (Sylvester & Smith 1976; Christie-Blick & Biddle 1985; Sylvester 1988; Zolnai 1991), and form at restraining bends or stepovers (e.g., Harding, 1974, 1990; Christie-Blick & Biddle 1985; Harding *et al.* 1985; Lowell 1985).

The models of McClay and Bonora (2001) were run both with and without synkinematic sedimentation, which was added incrementally to cover the growing antiformal structures. Vertical and horizontal sections of the completed models permitted the full 3D structure of the pop-ups to be analysed in detail. Three representative end-member experiments were described: 30° under-lapping restraining stepovers; 90° neutral restraining stepovers; and 150° overlapping restraining stopovers (Fig. 12.5). The experimental pop-ups are typically sigmoidal to lozenge-shaped, antiformal structures having geometries that are dependent on both the stepover angle and stepover width in the underlying basement faults. Although the models generated by McClay and Bonora (2001) do not incorporate plastic or ductile layers designed to simulate weak rocks, such as salt or overpressured shale, they nevertheless provide useful information about the progressive evolution of strike-slip pop-ups as demonstrated by the strong geometric similarities between the models and the natural examples they describe. Thus, the results of experiments like these suggest that deep-water clastic systems in strike-slip basins will tend to have complex geometry and a 3D facies-association distribution that may be less predictable than for many other tectonic settings.

12.3 Suspect terranes

One of the most important aspects of oblique-slip continental margins has been the recognition of 'suspect terranes' or unrelated crustal and geological elements that have been juxtaposed through terrane accretion events (see edited volume by Howell 1985). A suspect terrane is defined as an 'area characterised by an internal continuity of geology, including stratigraphy, faunal provinces, structure, metamorphism, igneous petrology, metallogeny, and palaeomagnetic record, that is distinct from that of neighbouring terranes and cannot be explained by facies changes' (Keppie 1986).

In the Circum-Pacific, palaeobiogeographic and palaeomagnetic studies indicate terrane displacement from a few hundred to >6000 km (Howell *et al.* 1985). Terrane accretion may occur over long time periods, for example China has a complex history of such events from the Late Precambrian, in Mongolia, continuing up until the collision of India with Asia ~45 Ma. Oblique subduction associated with strike-slip, or oblique-slip, has been a feature of terrane accretion in the evolution of Japan, at least throughout the Mesozoic and Cenozoic (*cf.* Taira *et al.* 1983).

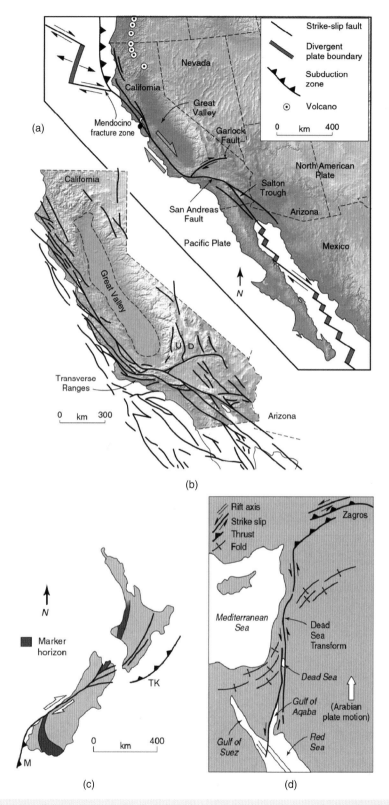

Fig. 12.1 Examples of major strike-slip faults and fault zones. (a) Regional map of the plate boundary between the North American and Pacific plates. The San Andreas Fault is the strike-slip fault zone that defines this boundary in California. (b) An enlargement of California, showing the major strike-slip faults. J & F = Juan de Fuca Plate. (c) The Alpine Fault in New Zealand links the Macquarie Trench (M) with the Tonga–Kermadec Trench (TK). (d) The Dead Sea Transform (DST) runs from the Gulf of Aqaba north to the western end of the Zagros Mountains. It accommodates northward movement of the Arabian Plate. (e) Sketch map of southern Asia, showing the collision of India. Strike-slip faults have developed in several settings here. A boundary transform (the Chaman Fault) delimits the northwestern edge of the Indian subcontinent. Strike-slip faults also form due to oblique collision, oblique convergence, and lateral escape. Small rifts have developed just north of the Himalayas. From van der Pluijm and Marshak (2004).

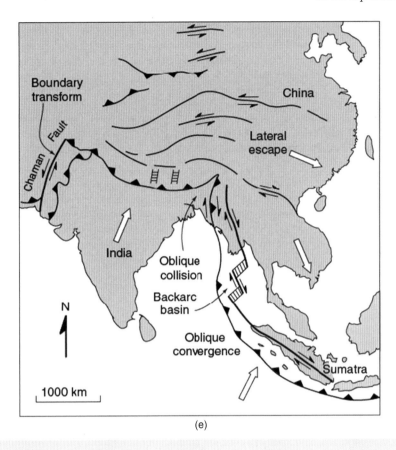

(e)

Fig. 12.1 *(continued)*

The northeastern Pacific continental margin of North America has a long geological history of strike-slip tectonics and terrane accretion, associated with the accumulation of considerable thicknesses and volumes of deep-marine deposits. The suspect terrane of Wrangellia (Fig. 12.7) has been interpreted as a forearc crustal sliver, including a forearc basin, which during its Late Cretaceous accretion and subduction of the Kula plate led to considerable crustal shortening with syntectonic deposition across large parts of southern Alaska (e.g., Nokleberg *et al.* 1994; Plafker & Berg 1994; Trop *et al.* 1999, 2002, 2005; Hults *et al.* 2013; Israel *et al.* 2014; Figs 12.8, 12.9). Up to 400 km of post Late Cretaceous dextral strike slip has occurred along the (backarc) Denali Fault system to the north of Wrangellia (Figs 12.7, 12.8). We recognise, particularly in large slivers of continental crust such as Wrangellia, that the internal stratigraphy of this terrane contains basins that are not strictly speaking formed under a transtensional or transpressional tectonic regime and that might equally be considered in other chapters of this book (i.e., Chapters 9, 10 and 11). However, as Wrangellia is commonly referred to as a suspect terrane, we consider it here.

Eastham and Ridgway (2000) document the Mesozoic stratigraphy of the Kahiltna assemblage in the Alaska Range and Talkeetna Mountains of southern Alaska, where sedimentary basin formation developed along an accretionary convergent margin (Fig. 12.7). The Kahiltna assemblage, containing deep-marine successions, comprises Upper Jurassic–Upper Cretaceous strata that occur between the allochthonous Wrangellia composite terrane and the former Mesozoic continental margin of North America, commonly referred

to as the Yukon–Tanana terrane (Csejtey *et al.* 1982; Jones *et al.* 1982, 1986) (YT in Fig. 12.7).

The Wrangellia composite terrane comprises three tectonostratigraphic terranes: the Wrangellia, Peninsular and Alexander terranes (Plafker & Berg 1994). The thickness of the Kahiltna stratigraphy is unknown but has been estimated at between 4 and 10 km. The only detailed study of the Kahiltna assemblage is from southwestern Alaska, where Wallace *et al.* (1989) showed that the Kahiltna assemblage was derived from the Wrangellia composite terrane and that the basin formed on the suture zone between the Wrangellia composite terrane and former Mesozoic continental margin of North America. The Kahiltna assemblage in south-central Alaska contains up to several thousand metres of intensely deformed and locally highly metamorphosed, Upper Jurassic–Upper Cretaceous deep-marine dark-grey to black argillites (Facies Classes E and G), fine- to coarse-grained lithic greywackes (Facies Classes B, C and D), polymict pebble conglomerates and subordinate black chert-pebble conglomerates (Facies Class A) (Csejtey *et al.* 1992; Eastham & Ridgway 2000). The study by Eastham and Ridgway (2000) focussed mainly on the Kahiltna assemblage exposed in the area between the Denali Fault and the Talkeetna thrust fault (Fig. 12.7). Before their study, few measured stratigraphic sections and compositional data were available from the Kahiltna assemblage in south-central Alaska. They describe the lithofacies, sedimentary structures and provenance of the Kahiltna assemblage in three different areas of the outcrop belt: the northern Talkeetna Mountains, the southern Alaska Range and the Chulitna terrane (Fig. 12.7).

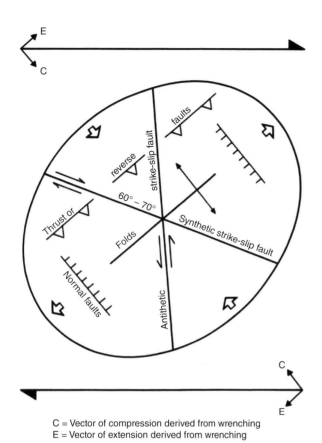

C = Vector of compression derived from wrenching
E = Vector of extension derived from wrenching

Fig. 12.2 Summary diagram to show the structures produced from simple shear under a dextral shear couple. From Harding (1974).

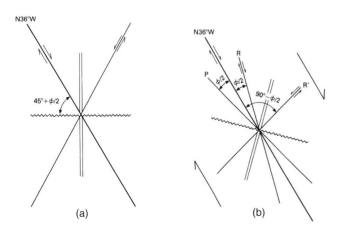

Fig. 12.3 Models for structures produced by a dextral shear couple applied to the San Andreas Fault Zone and associated structures, using (a) Coulomb–Anderson, and (b) Riedel model of simple shear. Wavy lines represent orientation of fold axes, and double parallel line shows orientation of extension. Both cases are drawn with the major dextral strike-slip as N36°W parallel to the Pacific–North American plate boundary and some of the main strike-slip faults. From Aydin and Page (1984). See text for explanation of the P, R and R′ shears.

The Kahiltna assemblage crops out in a northeast–southwest-trending belt that is ~800 km long and ~200 km wide in southern Alaska. The study of the Kahiltna assemblage by Eastham and Ridgway (2000) suggests that the thousands of metres of Upper Jurassic–Upper Cretaceous strata previously grouped together as the Kahiltna assemblage in south-central Alaska may actually represent several different sedimentary basins. Proximal–distal lithofacies trends along the southern part of their traverse in the Clearwater and Talkeetna mountains (Fig. 12.7) suggest deposition in predominately mud-rich submarine fans with mainly northwestward sediment transport. Locally, adjacent to the Wrangellia composite terrane, the Kahiltna assemblage consists of proximal submarine-fan conglomerates and sandstones of Facies Classes A, B and C. Compositional data from proximal conglomerates in this area show a predominance of volcanic (greenstone) and sedimentary clasts that probably reflect unroofing of the Wrangellia composite terrane carried in the hanging wall of the nearby Talkeetna thrust fault. In contrast, proximal-distal lithofacies trends in the Kahiltna assemblage along the northern part of their traverse suggest deposition in gravel- and sand-rich submarine fans with sediment transport mainly towards the southwest. Pebble-boulder conglomerates of the Alaska Range are dominated by sandstone, chert and granitic clasts indicative of a source terrane rich in sedimentary and plutonic rocks. Possible source terranes for the Kahiltna assemblage of the Alaska Range are Palaeozoic continental-margin strata north of the Denali Fault system.

The Kahiltna assemblage of the Alaska Range is separated from the Kahiltna assemblage of the Clearwater and Talkeetna mountains by a large thrust block of Palaeozoic and Mesozoic strata known as the Chulitna terrane (Fig. 12.7). Upper Jurassic–Upper Cretaceous sedimentary strata, also mapped in previous studies as the Kahiltna assemblage, crop out in the Chulitna terrane. These strata are characterised by black shale, chert and in situ fossiliferous limestone that are distinct from lithofacies of the Kahiltna assemblage of the Talkeetna Mountains and Alaska Range. Eastham and Ridgway (2000) interpret the lithofacies of the Chulitna terrane as representing deposition on a bathymetric high relative to submarine-fan lithofacies of the Kahiltna assemblage in the Talkeetna Mountains and Alaska Range.

12.4   Depositional models for strike-slip basins

The tectonic and structural complexity of strike-slip margins militates against simple and universal models for deposition at such margins. Despite the apparently intractable problems for developing such models, there have been attempts to suggest that some aspects of (deep-marine) deposition at strike-slip margins might be predictable. For example, Noda and Toshimitsu (2009) used field surveys and numerical simulations to examine the lithostratigraphic complexity in strike-slip basins. They attempted to bridge the gap between qualitative sedimentary facies analyses and quantitative numerical models in order to better understand the formation of these sedimentary successions. They focussed on the Upper Cretaceous Izumi Group, southwest Japan, which was deposited in an elongate basin (300 km long × 10–20 km wide) along the Median Tectonic Line (Fig. 12.10), which at the time of deposition was a sinistral strike-slip fault related to oblique subduction along a forearc margin. The depositional environments of the group were considered in five lithofacies associations (LAs): submarine channel-fill facies (LA I – mainly Facies Classes A, B and C), proximal facies of lobes or frontal splays (LA II – mainly Facies Classes B and C), distal facies of lobes or frontal

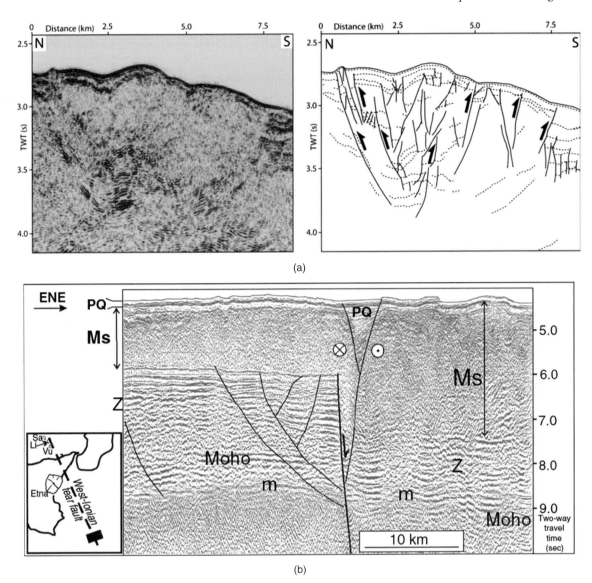

(a)

(b)

Fig. 12.4 Examples of flower structures. (a) Seismic line BS07-36 and interpretation, showing V-shaped reverse faults forming a positive flower-structure along a strike-slip fault, Banggai–Sula–Molucca Sea margin, Indonesia. From Watkinson *et al.* (2011). (b) Seismic profile across the West-Ionian Tear Fault, cutting the Ionian crust, where its eastern sector is deeper and deepening further north-northwestward. It allows easier southward movement and greater thickening of the overlaying sedimentary successions, particularly of the compressed Messinian evaporites (Ms). In the inner sector of the chain, where the vertical displacement of the slab becomes several kilometres, this gives rise to the Vulcano right-lateral strike-slip fault, presently active. PQ = Plio-Quaternary sequence, Z = top of basement, m = sea bottom multiple; volcanic isles in the inset: Vu = Vulcano, Li = Lipari, Sa = Salina. Note, relative sense of displacement shown with conventional symbols. From Del Ben *et al.* (2008).

splays (LA III – mainly Facies Classes C and D), slope-apron facies (LA IV – mainly Facies Classes C, D, E and F), and basin floor facies (LA V – mainly Facies Classes C, D and E) (Fig. 12.11). LAs I–III represent point-sourced channelised submarine-fan successions in the axial facies, with unidirectional palaeocurrent directions from east-northeast to west-southwest, and LAs IV to V are interpreted as the marginal facies, the palaeoslope of which dipped to the south-southwest. Two units of channelised submarine-fan successions are stacked with ~10 km of eastward (backward) shift. Each unit shows a repetitive lithostratigraphy of a stratigraphically-short thickening-and-coarsening-upward trend in the lower part (~350 m thick), overlain by a gradual thinning-and-fining-upward trend

in the upper part (1–3.5 km thick). It is estimated to have taken ~5–7 × 10^5 years for 10 km sinistral offset of correlated markers in each stratigraphic unit.

Although many processes can control the stratigraphic architecture, such as global and local sea-level change, climate and tectonics, the stratigraphic cyclicity observed in the Izumi Group was considered as closely related to the depocentre migration, suggesting that fault movement was the primary control on the stratigraphy. On the assumption that the formation and filling processes of the Izumi sedimentary basin were basically controlled by strike-slip faults, a numerical simulation suggests that episodic changes in fault-slip rate or sediment-supply rate might have controlled the

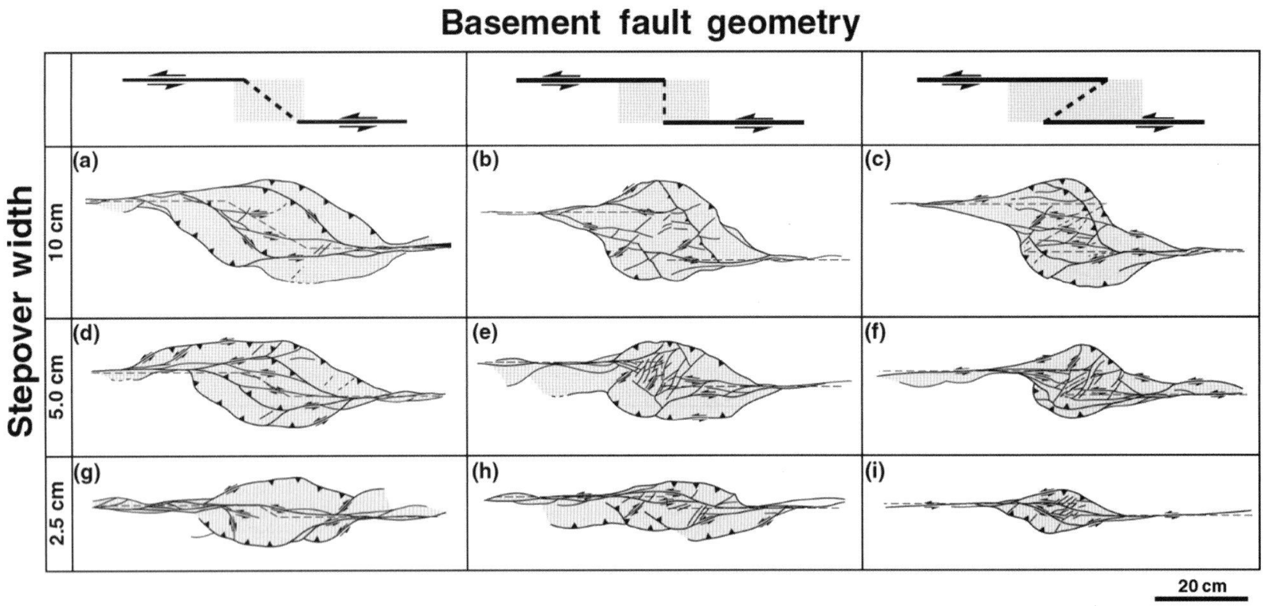

Basement fault geometry

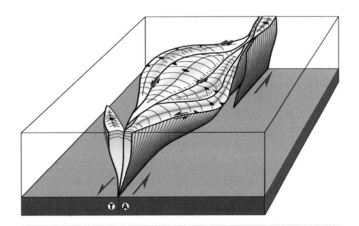

Fig. 12.5 Summary of pop-up structures generated in sandbox experiments for restraining stepover spacing from 10–2.5 cm. In all experiments displacement on the basement master faults was 10 cm. Sandpack thickness 5 cm. From McClay and Bonora (2001).

Fig. 12.6 Synoptic diagram illustrating the 3D geometry of an idealised pop-up structure based on the results of an analogue modelling programme. T = base plate movement toward viewer; A = base plate movement away from viewer. From McClay and Bonora (2001).

stratigraphic cyclicity. Noda and Toshimitsu (2009), therefore, proposed a depositional model where cyclic stratigraphy was ascribed to temporal variations of fault activity controlling accommodation generation, sediment supply and relative sea level, which could generate cyclic stratigraphy associated with depocentre migration in strike-slip basins (Fig. 12.12). Such a depositional model, however, is highly dependent upon the initial assumptions about what is driving stratigraphic architecture, that is, tectonics rather than one or more other factor.

Many palaeogeographic reconstructions of deposition linked to strike-slip faulting incorporate the progressive segmentation and truncation of drainage systems. This has been demonstrated from modern ocean basins, for both individual submarine channels (e.g.,

Appelgate *et al.* 1992) and entire submarine fans (Nagel *et al.* 1986; see details in case study of California Borderland below).

Appelgate *et al.* (1992) mapped the recently active sinistral Wecoma Fault on the floor of Cascadia Basin, offshore Oregon convergent margin, using SeaMARC I side-scan sonar, Seabeam bathymetry and multichannel seismic and magnetic data. This fault intersects the continental slope at 45°10′N and extends northwards (trending 293°) for at least 18.5 km. Its western terminus was not identified, and at its eastern end it splays apart and disrupts the lower continental slope. The fault extends to the base of a 3.5 km-thick sedimentary succession overlying a basement discontinuity. Prominent seafloor features cross-cut by the fault individually indicate sinistral displacements of between 120 and 2500 m. The average slip rate along this fault since 10–24 ka is inferred at 5–12 mm yr⁻¹. Surface structural relationships, combined with the maximum inferred slip rate, suggest that fault activity was initiated at least 210 ka, and that this fault has been active during the Holocene. The Wecoma Fault cross-cuts one of the two principal submarine channels on the Astoria Fan (Nelson *et al.* 1970; Nelson 1976).

An interpretation of one of the SeaMARC I images (2 km swath) between the Wecoma Fault and a slope-base channel shows high backscatter, interpreted as relatively coarse-grained (channel) sediments. The west bank of this channel has been left-laterally offset 120 m along the fault, blocking the channel. (Figs 12.13a,b). Figures 12.13c,d show a submarine knoll in its present configuration (Fig. 12.13c), and after subtracting 350 m of fault motion (Fig. 12.13d), to show the inferred original continuity of an arcuate slump scar (from Appelgate *et al.* 1992). Figure 12.14 shows schematic diagrams of the evolution of a submarine channel on the southern fringes of the Astoria submarine fan that has been displaced by the Wecoma sinistral strike-slip fault. Recent fault activity has blocked the course of this channel to the south of the fault. South of the Wecoma Fault, the west bank of the channel has been offset by several minor faults with individual dextral displacements of 25–50 m.

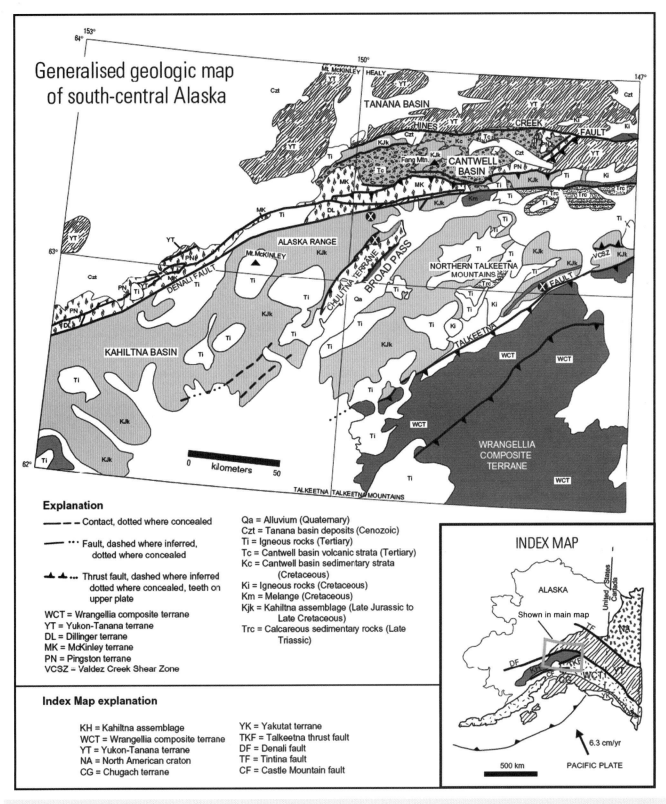

Fig. 12.7 Generalised geologic map of south-central Alaska, showing the suspect terrane of Wrangellia (labelled Wrangellia composite terrane) and the locations of measured stratigraphic sections referred to in the text. Note location of the Kahiltna assemblage (KJk) between the late Mesozoic continental margin of North America (YT) and the Wrangellia composite terrane (WCT). From Eastham and Ridgway (2000).

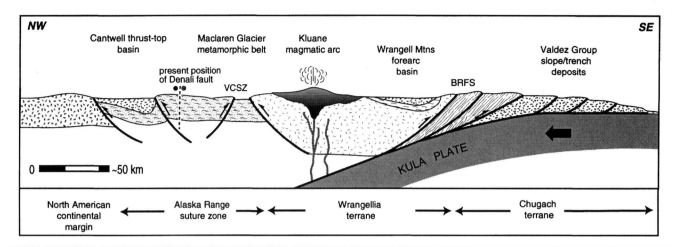

Fig. 12.8 Schematic crustal-scale structural reconstruction across southern Alaska during the Late Cretaceous, based on the regional tectonic models of Nokleberg *et al.* (1994), Plafker and Berg (1994) and Trop *et al.* (1999). During the Late Cretaceous, accretion of the Wrangellia terrane and subduction of the Kula plate resulted in crustal shortening with syntectonic deposition across large parts of southern Alaska. Up to 400 km of post Late Cretaceous dextral strike slip occurred along the Denali Fault System. From Trop *et al.* (1999).

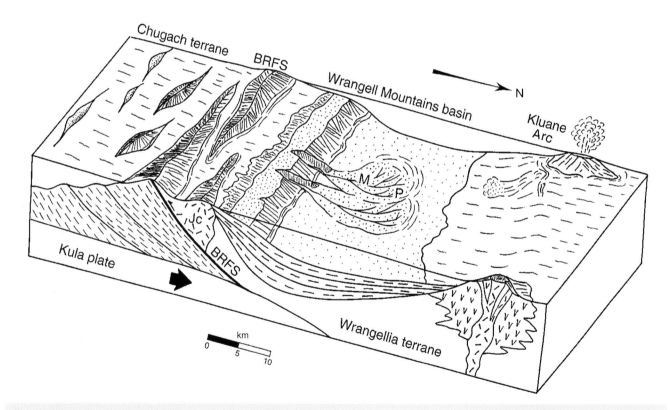

Fig. 12.9 Schematic reconstruction of the Wrangell Mountains basin during the Campanian. The Wrangellia terrane is interpreted as a forearc basin. Coarse-grained submarine fans prograded northward on the hanging wall of the Border Ranges thrust fault. BRFS = Border Ranges fault system; M = MacColl Ridge stratigraphic section; P = Pyramid Peak stratigraphic section; Jc = Chitina Valley Batholith along the BRFS. The BRFS is interpreted as defining the southern margin of the forearc basin, with the Kluane magmatic arc located along the northern basin margin. The abundance of older volcanic and igneous plutonic clastic material, the northward-directed palaeocurrent indicators, and the northward overall fining of the submarine-fan systems, are interpreted to record the uplift and erosion of igneous rocks from the Wrangellia terrane, together with igneous rocks that intruded Wrangellia (e.g., Jc). From Trop *et al.* (1999).

12.5 Modern strike-slip mobile zones

Harding (1983) and Harding *et al.* (1985) document the characteristics of dextral transtension in the Andaman Sea where divergent strike-slip has produced a single major fault bordered by discontinuous oblique faults (Figs 12.15, 12.16). For an introduction to the geology and geophysics of this region, see Malod *et al.* (1995), Susilohadi *et al.* (2005, 2009) and McCaffrey (2009). Water depths in this region are greater than 1000 m. The Andaman Sea is a marginal basin where spreading to produce oceanic crust began ~13.5 Ma, based on the magnetic anomalies. The central spreading axis strikes northeast, and is linked by north–northwest-striking dextral faults (Fig. 12.15). In this region: (i) en échelon folds occur with dextral offsets from 2.4–3.2 km; (ii) abundant normal faults occur, formed contemporaneously with wrench faults that are disposed in a simple en échelon pattern and (iii) there are localised reverse faults near an S-shaped closing or restraining bend at the southern end of the principal fault (Fig. 12.16) (Harding 1983; Harding *et al.* 1985). These observations are consistent with major dextral shear along the West Andaman Fault Zone. The fault density and abundance outside of the principal West Andaman Fault are asymmetric, with more numerous faults to the east (Fig. 12.16). In the southern part of the study area, transtension has produced negative flower structures in Miocene–Pleistocene sediments (Fig. 12.17). Thus, although the deformation patterns are complex and occur with sedimentation, overall there is a predictable geometry consistent with major transtensional dextral shear.

The stratigraphy and sedimentology of New Zealand are dominated by strike-slip tectonics associated with the Alpine Fault Zone, a major dextral fault (Fig. 12.18). Oblique-slip tectonics, therefore, are important in the Hikurangi Trough at the southern end of the Kermadec Trench, off North Island, New Zealand, where the westward-subducting Pacific Plate changes from roughly orthogonal collision with the Indian Plate to approximately transcurrent dextral relative displacement (Fig. 12.19a). Structures in the Hikurangi Trough are consistent with a major dextral shear couple between the Pacific and Indian plates (Fig. 12.19b). The Alpine Fault of South Island, New Zealand, separates blocks of continental crust. Its extension in the Hikurangi Trough separates continental crust to the west from subducting oceanic crust to the east. The Alpine Fault, along the west of South Island, is an intracontinental trench–trench transform that was initiated in the latest Eocene–Oligocene (Norris *et al.* 1978; Lewis *et al.* 1986). In the late Oligocene–early Miocene, a phase of extension was replaced by the present pattern of transpression (Carter & Norris 1976) due to the southward migration of the Indo-Pacific pole of rotation (Walcott 1978). Nicol *et al.* (2007) synthesise the deformation across the active Hikurangi subduction margin, including shortening, extension, vertical-axis rotations, and strike-slip faulting in the upper plate, all of which have been estimated for the last ~24 Myr using margin-normal seismic reflection lines and cross-sections, strike-slip fault displacements, palaeomagnetic declinations, bending of Mesozoic terranes and seafloor spreading information. Post-Oligocene shortening in the upper plate increased southward, reaching a maximum rate of 3–8 mm yr^{-1} in the southern North Island. Upper plate shortening is a small proportion of the rate of plate convergence, most of which (>80%) accrued on the subduction thrust. The uniformity of these shortening rates is consistent with the near-constant rate of displacement transfer (averaged over ≥5 Myr) from the subduction thrust into the upper plate. In contrast, the rates of clockwise vertical-axis rotations of the eastern Hikurangi Margin were temporally variable, with ~3°/Myr since 10 Ma and ~0°-1°/Myr prior to 10 Ma. Post-10 Ma, the rates of rotation decreased westward from the subduction thrust, which resulted in the bending of the North Island about an axis at the southern termination of subduction. With rotation of the margin and southward migration of the Pacific Plate Euler poles, the

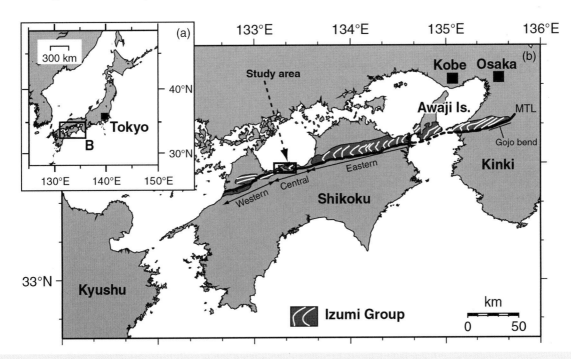

Fig. 12.10 Location of the Cretaceous Izumi Group, Japan. White lines in the Izumi Group show the strike of bedding. Thick black line represents the Median Tectonic Line (MTL). Boxed area is the study area used to develop the palaeogeographic interpretations in Figures 12.11 and 12.12. From Noda and Toshimitsu (2009).

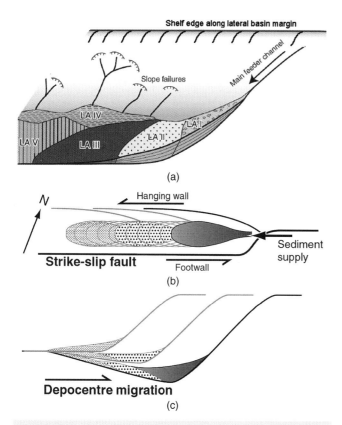

Fig. 12.11 (a) Perspective view of depositional environments of the various lithofacies associations in the Cretaceous Izumi Group, Japan. Plan (b) and cross-sectional (c) models for strike-slip basins characterised by backward shifts of submarine channel-fan successions with progressive retreat/subsidence of the basement. See text for explanation. From Noda and Toshimitsu (2009).

component of the margin-parallel relative plate motion increased to the present. Prior to ~15 Ma, plate convergence dominated the Hikurangi margin, with the rate of margin-parallel motion increasing markedly since 10 Ma. Vertical-axis rotations could accommodate all margin-parallel motion before 1–2 Ma, eliminating the requirement for large strike-slip displacements (for example, >50 km) in the upper plate since the Oligocene.

DSDP Leg 90 (Kennett, von der Borch *et al.* 1986) studied the Hikurangi Trough and Chatham Rise areas within the larger *Eastern New Zealand Ocean Sedimentary System* (ENZOSS; Fig. 12.20a). The net result of the plate motions off northeast New Zealand is that the northwestern edge of Chatham Rise (Fig. 12.19a) is being dragged under the Kaikoura Ranges (Fig. 12.19b) in northeast South Island. Also, the down-warped edge of Chatham Rise can be traced under the southern end of Hikurangi Trough where a thick succession of turbidites has accumulated. The slope between Mernoo Saddle and the southwestern Hikurangi Trough shows evidence of Neogene tectonic activity and superficial slope instability to produce large-scale sedimentary slides of Facies Group F2. The 'tectonic' faults are late Neogene in origin, based on the acoustic units they cut, and generally dip southward; their genesis is ascribed to seismic activity along the boundary, North Mernoo Fault Zone (Lewis *et al.* 1986). Dextral strike-slip tectonics has been important in controlling the development of deep-marine clastics since at least 38 Ma in South Island,

New Zealand (Norris *et al.* 1978); for example, there are hundreds of metres of Oligocene–Miocene calcareous, sand-rich and mud-rich sediments in the Moonlight Zone of southwest South Island. Norris *et al.* (1978) have shown that the Moonlight Zone, with >2000 m of Cenozoic mostly SGF deposits, is an Oligocene–Miocene infill at the northern extension of the north-northeast–south-southwest Solander Trough. In the Oligocene, fault-controlled deep-marine sandy systems (probably small submarine fans) accumulated from sediment sources to the west and east in the approximately north–south basin (Carter & Lindqvist 1975, 1977). By the early Miocene, the dominant transport direction of sandy SGFs had changed from lateral supply to southward, axial, progradation except off parts of the eastern fault-controlled margin, where calcareous sediment was being deposited as small submarine fans (*cf.* Norris *et al.* 1978). By late middle Miocene time, fluvial conglomerates had replaced the deep-marine sedimentation. The critical position of the Moonlight Zone along the boundary of the Indo-Australian and Pacific plates has led to considerable post-Miocene vertical tectonics, with large amounts of uplift, especially in the last few million years. Features that Norris *et al.* (1978) regard as typical of strike-slip tectonics and sedimentation in the Moonlight Zone include: (i) the narrowness of the zone of subsidence, and (ii) the rapidity of lateral and vertical facies changes. Thus, throughout the Cenozoic, southwest South Island and the northern part of the Solander Trough (eastern edge of the Tasman Basin) have been tectonically and sedimentologically a continuous belt of deep-marine basins controlled by oblique-slip processes (Summerhayes 1979).

Although in close proximity to the Alpine Fault segment of the Australian/Pacific plate boundary, the eastern South Island margin to the south of Banks Peninsula is effectively a stable passive margin. The broad shelf experiences high sediment input from the actively rising New Zealand Alps, a strong imprint from eustatic changes in sea level, and a moderately vigorous along-shelf circulation system. The shelf between Clutha and Rangitata is typically 30–80 km wide but reduces to 10 km off the Otago Peninsula. The shelf break is at 125–165 m water depth and is locally indented by the heads of submarine canyons feeding the channel system in Bounty Trough (Fig. 12.20a). Shelf morphology is variable with zones of featureless seabed interspersed with ridge and swale topography, terraces and changes in slope that represent palaeo-shorelines formed at previous still-stands of sea level.

The effects of eustatic sea-level changes on shelf sedimentation have been significant. At the last glacial maximum, sea level was ~118 m lower than present and rivers extended to canyon heads, and discharged directly into the head of Bounty Trough, where SGFs and hemipelagic sedimentation dominated. The subsequent transgression was a series of rapid rises punctuated by still-stands at −89 m (17 ka), −75 m (15 ka) and −55 m (~12 ka), respectively. During these recent lowstand pauses, the palaeo-shoreline and associated longshore transport system were then significantly landward of canyon heads. Bounty Trough was bypassed by the terrigenous supply, which switched from an east-southeast to a northeast route in accord with the developing shelf circulation system. The trough, therefore, changed abruptly from a terrigenous to calcareous bio-pelagic depocentre. The outer shelf and upper slope are strongly affected by northeastward currents associated with the Southland Front, which affects water depths to at least 400 m. Sediment escaping over the shelf-break, therefore, is moved northeast along the margin. Seismic profiles reveal that along-slope transport has created a series of linear sediment drifts with ages back to the late Miocene. Lateral accretion of these slope

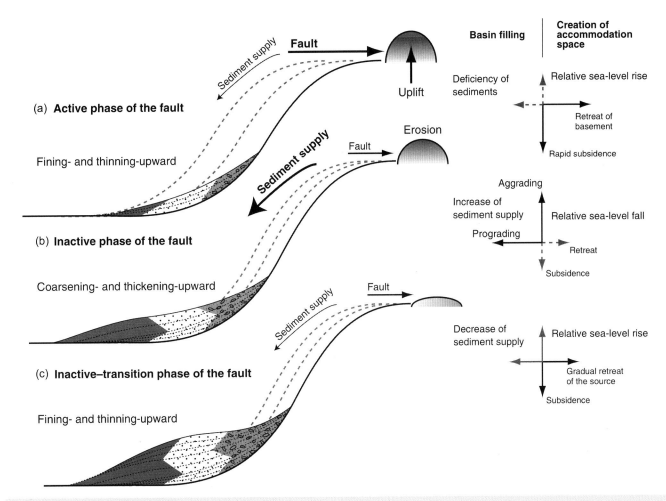

Fig. 12.12 Schematic cross-sectional models for development of cyclic stratigraphy in strike-slip basins, based on a study in the Cretaceous Izumi Group, Japan. (a) In the active phase of the strike-slip fault, rapid retreat and subsidence of the basement make the accommodation space large and create a relative sea-level rise. Sediment input is small in this phase due to a lag between an increase in sediment supply and tectonic uplift, resulting in fining- and thinning-upward trends. (b) Increase of sediment supply in association with enhancement of denudation in the uplifted source region causes progradation/aggradation of delta/submarine fan and fall of the relative sea level and, thus, formation of coarsening-and-thickening-upward successions. (c) Fining- and thinning-upward successions would be developed by gradual decrease of sediment input due to extensive erosion and retreat/subsidence of the basement, leading to relative sea-level rise. From Noda and Toshimitsu (2009).

drifts has caused margin progradation that contrasts with progradation by shelf-normal deposition.

Bounty Trough, a 350 km-wide linear depression between Campbell Plateau and Chatham Rise (Fig. 12.20a), is the main depocentre for the ENZOSS during glacial lowstands. Sediment captured by submarine channels was transported by SGFs, including turbidity currents, 900 km eastward to supply Bounty Fan, which has grown from the trough mouth onto the 4500 m-deep floor of the southwest Pacific Basin (Carter & Carter 1987; Carter *et al.* 2004). Overspill from Bounty Channel (Fig. 12.20a) has created an extensive levée system, laterally constrained by the lower flanks of Bounty Trough. Bounty Fan began in the late Pliocene and is now >400 m thick. The lower fan/channel has extended into the path of the northbound Pacific Deep Western Boundary Current (DWBC), which not only receives channel discharge but also has winnowed the lower fan. Some sediment escapes the ENZOSS to accumulate on sediment drifts formed under the DWBC. The amount of sediment involved is unknown, but it is likely to be small because the fan is not extensively

eroded and sediments are still present on the lower fan. Figure 12.20b summarises the plate-tectonic evolution of New Zealand and the development of the principal deep-marine systems, including the contour currents in relation to the opening of the Tasmanian gateway (after Carter *et al.* 2004).

The central Caribbean region backarc basins, including Cuba and Jamaica, are dominated by sinistral strike-slip (Fig. 12.21a). Mann (2007) and Mann *et al.* (2007) describe the regional fault pattern, geological setting and active fault kinematics of Jamaica, from published geological maps, earthquakes and GPS-based geodesy, to develop a simple tectonic model for both the initial stage of restraining-bend formation and the subsequent stage of bend bypassing (Fig. 12.21b). Restraining-bend formation and widespread uplift in Jamaica began in the Late Miocene, and was probably controlled by the interaction of approximately east–west-trending strike-slip faults with two north-northwest-trending rifts oriented obliquely to the direction of east-northeast-trending, Late Neogene interplate shear. The interaction of the interplate strike-slip fault system (Enriquillo–Plantain

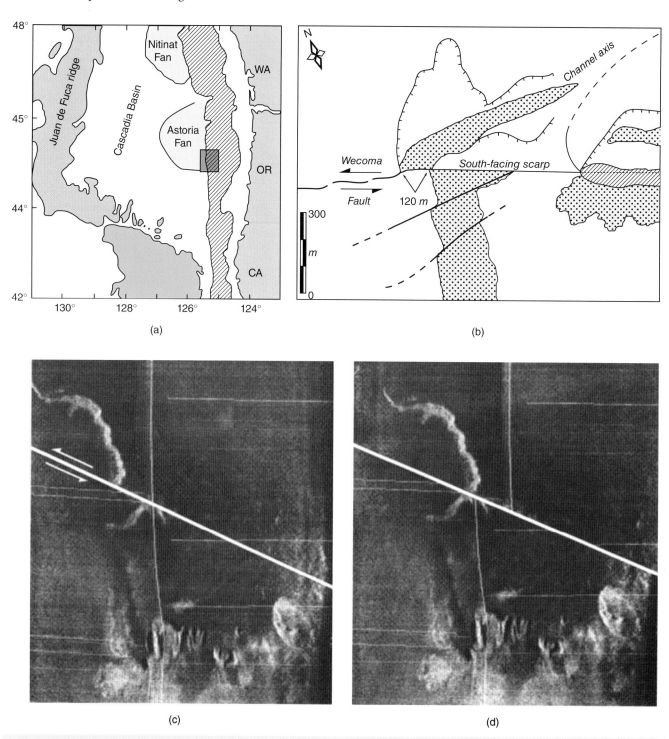

(a)

(b)

(c)

(d)

Fig. 12.13 (a) Study area on Astoria Fan, offshore Oregon convergent margin. WA = Washington State; OR = Oregon; CA = California. Diagonal ruling = continental slope and accretionary complex. (b) Line interpretation of SeaMARC I image (2 km swath) between the Wecoma Fault and a slope-base channel, Astoria Fan. High backscatter, interpreted as relatively coarse-grained (channel) sediments are shown as yellow with stipple. The west bank of this channel has been left-laterally offset 120 m along the fault, blocking the channel. (c, d) View of seafloor, showing a submarine knoll in its present configuration (left), and after subtracting 350 m of fault motion (right), to show the inferred original continuity of an arcuate slump scar. Width of image is ~2.5 km. See text for explanation. From Appelgate *et al.* (1999).

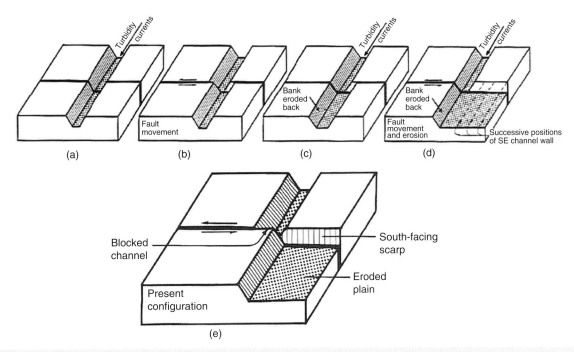

Fig. 12.14 Schematic diagrams showing the evolution of a submarine channel on the southern fringes of the Astoria submarine fan that has been displaced by the Wecoma sinistral strike-slip fault, Cascadia Basin, offshore Oregon. Recent fault activity has blocked the course of this channel. From Appelgate *et al.* (1992).

Garden Fault Zone) and the oblique rifts has shifted the strike-slip fault trace ~50 km to the north and created the 150 km-long by 80 km-wide restraining bend that is now morphologically expressed as the island of Jamaica. The observed GPS velocity field suggests that sinistral shear continues to be transmitted across the Jamaican restraining bend by a series of intervening bend structures, including the Blue Mountain uplift of eastern Jamaica.

Jamaica comprises an emergent Cretaceous-age oceanic volcanic arc and volcanogenic sedimentary rocks, overlain by 5–7 km of Tertiary carbonate rocks (Lewis & Draper 1990). Inversion of the Wagwater Belt has induced folding of the adjacent Mio-Pliocene limestones. The Wagwater Belt is a fault-bounded structural unit in which >3000 m of Early Tertiary sedimentary and volcanic rocks are exposed. Geochemical analyses of the volcanic rocks show that they comprise a bimodal suite of plateau-type tholeiitic basalts and calc-alkaline dacites (Jackson & Smith 1979). The extrusion of these volcanics is associated with the development of an interarc basin, the Wagwater Basin, at the beginning of the Cenozoic. The Wagwater Basin formed as a result of the splitting of a mature Late Cretaceous volcanic arc into a frontal and third arc, represented by the Blue Mountain Massif and the Clarendon Block, respectively. The creation of the Wagwater Basin and the eruption of the basalts are related to the initial opening of the Cayman Trough (Fig. 12.21b). The cessation of dacite volcanic activity in Jamaica signified the separation of the Caribbean Plate from the East Pacific Farallon Plate. The Wagwater Belt contains Middle Eocene SGFs, in the Richmond Formation, now inverted and exposed along the northeastern coast of Jamaica (e.g., Richmond Formation Facies Class A and C deposits, Fig. 12.22). The sedimentology of these Middle Eocene SGFs is described by Wescott and Ethridge (1983), who interpreted them as products of fan-delta and submarine-fan environments. The ichology of the Richmond

Formation is described by Pickerill *et al.* (1993). The deep-marine environments and deposits of the present-day Yallahs fan-delta, on the southeastern part of Jamaica (Wescott & Ethridge 1982), may provide an analogue for the SGFs in the Richmond Formation.

12.5.1 Californian continental margin

The continental margin of both offshore and onshore California is an active transform plate margin dominated by right-lateral or dextral strike-slip motions (Emery 1960; Crouch 1981). The California continental margin (California Borderland) has evolved through the Neogene as part of a broad zone of oblique-slip plate tectonics. Figure 12.23 provides a summary of the geological evolution of this margin since 12.3 Ma (Fletcher *et al.* 2007; *cf.* Saunders *et al.* 1987). Subduction of oceanic lithosphere below western North America occurred from the Cretaceous to Miocene (Atwater & Molnar 1973). The Pacific–Guadalupe ridge intersected the continental margin ~29 Ma to create two triple junctions, the Mendocino transform-transform-trench and the Rivera ridge-trench-transform that migrated towards the north and south respectively. The San Andreas transform fault system developed between the two triple junctions. From ~29–12.5 Ma, the Pacific-Guadalupe spreading centre was progressively consumed southwards at the trench. By ~12 Ma, the entire continental margin as far south as Baja California was converted into a dextral-slip transform system, with the offset of the Magdalena submarine fan and the removal of its sediment source (Fig. 12.23b). In gross terms, the northern sector of the San Andreas Fault system is dominated by transpression in contrast to the Gulf of California where transtension has occurred to generate a narrow ocean basin. There are many small deep-marine basins (Fig. 12.24),

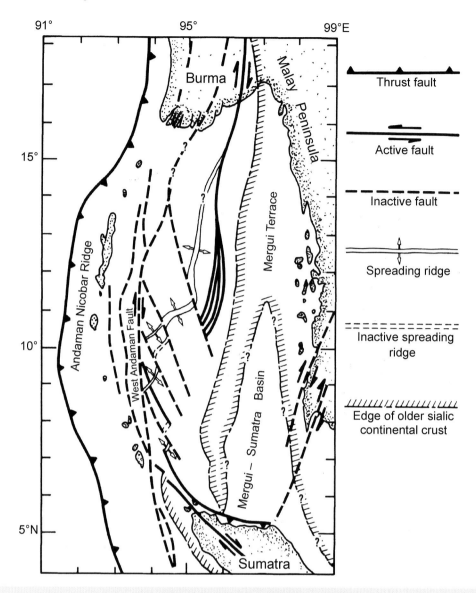

Fig. 12.15 Tectonic framework of the Andaman Sea from Harding *et al.* (1985).

controlled by a complex array of compressive and/or extensional tectonics. Here and in Section 12.5.2, we consider the California margin as two contrasting oblique-slip plate boundaries where deep-marine sedimentation forms a major component of the stratigraphic record.

Legg *et al.* (2007) document exceptional examples of restraining and releasing bend structures along major strike-slip fault zones in the California Borderland. Two large restraining bends with varied structural styles are compared to derive a typical morphology of Borderland restraining bends. A 60 km-long, 15° left bend in the dextral San Clemente Fault has created two primary deformation zones (Fig. 12.24). The southeastern uplift involves unconsolidated to weakly-consolidated SGF deposits (e.g., turbidites) and is expressed as a broad asymmetrical ridge with right-stepping en échelon anticlines and local pull-apart basins at minor releasing stepovers along the fault. The northwest uplift involves more rigid sedimentary and possibly igneous or metamorphic basement rocks creating a steep-sided, narrow and more symmetrical pop-up. The restraining bend terminates in a releasing stepover basin at the northwest end,

but curves gently into a transtensional releasing bend to the southeast. Seismic stratigraphy suggests that the uplift and transpression along this bend occurred within Quaternary times. The 80 km-long, 30–40° left bend in the San Diego Trough–Catalina Fault Zone has created a large pop-up structure that emerges to form Santa Catalina Island. This ridge of igneous and metamorphic basement rocks has steep flanks and a classic 'rhomboid' shape. For both major restraining bends, and most others in the Borderland, the uplift is asymmetrical, with the principal displacement zone lying along one flank of the pop-up. Faults within the pop-up structure are very steep dipping and subvertical for the principal displacement zone. In most cases, a Miocene basin has been structurally inverted by the transpression. Development of major restraining bends offshore of southern California appears to result from reactivation of major transform faults associated with Mid Miocene oblique rifting during the evolution of the Pacific–North America plate boundary. Seismicity offshore of southern California demonstrates that deformation along these major strike-slip fault systems continues today.

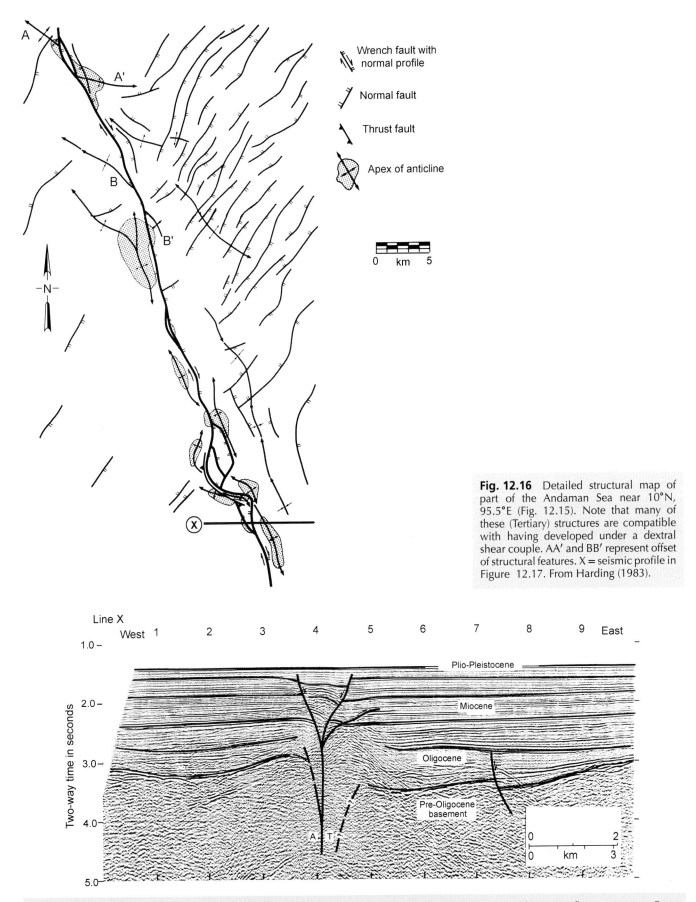

Fig. 12.16 Detailed structural map of part of the Andaman Sea near 10°N, 95.5°E (Fig. 12.15). Note that many of these (Tertiary) structures are compatible with having developed under a dextral shear couple. AA' and BB' represent offset of structural features. X = seismic profile in Figure 12.17. From Harding (1983).

Fig. 12.17 Seismic profile (X in Fig. 12.16) across the Andaman Sea wrench fault to show interpreted negative flower structure. From Harding *et al.* 1985. 'A' = relative motion away, and 'T' towards, observer. These structures typically develop along zones of transtension.

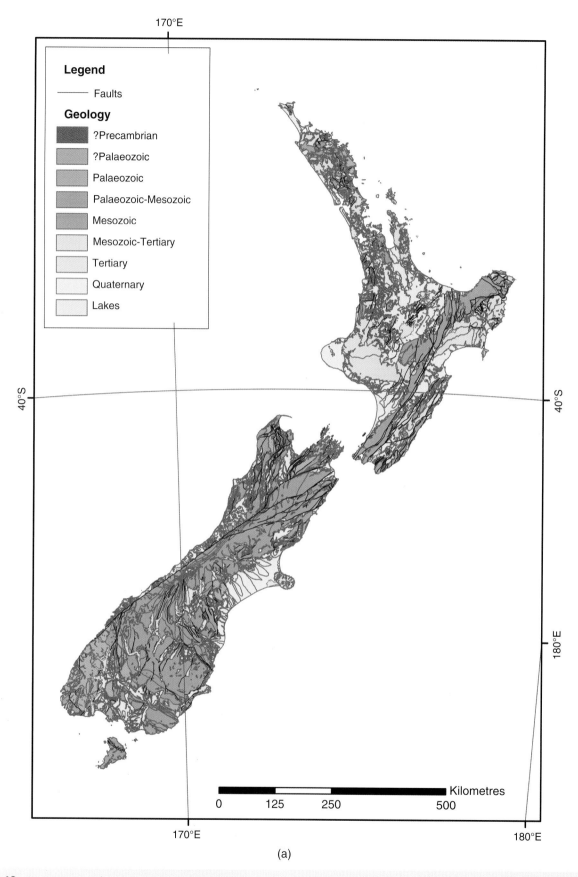

Fig. 12.18   (a) (caption on facing page)

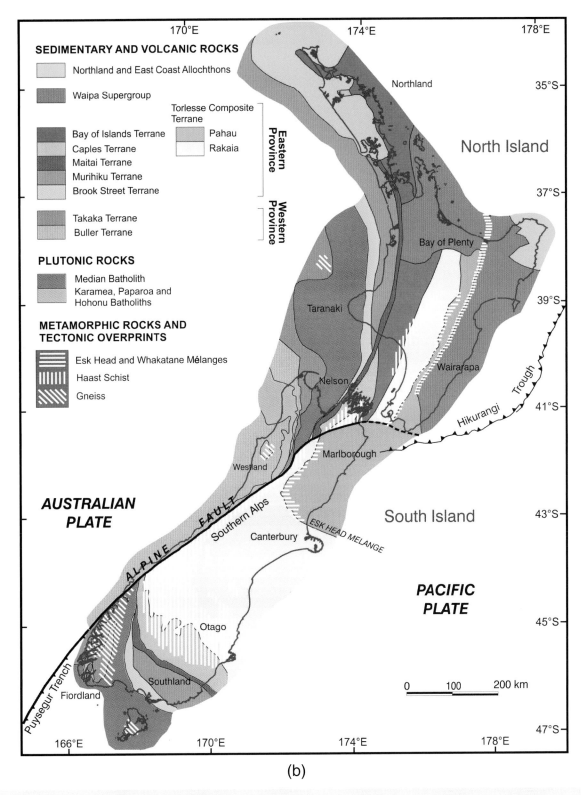

(b)

Fig. 12.18 (a) Geological map of New Zealand, simplified to show the distribution of rocks by stratigraphic age. Note the curvature of the geological units into the major dextral strike-slip Alpine Fault Zone. Courtesy of University of Otago, NZ, website http://www.otago.ac.nz/geology/research/general_geology/maps/nzterranes.html. (b) Geological map of New Zealand to show the tectonostratigraphic terranes that have been identified. Post-110 Ma sedimentary rocks are not shown. The map shows the main terranes that were accreted against the margin of Gondwana during the Palaeozoic and Mesozoic. These were intruded by batholiths, overprinted by metamorphism and deformation, then bent and offset during the Neogene by the Australian-Pacific plate boundary. The Northland and East Coast Allocthons, and Waipa Supergroup, have been overthrust and deposited on Eastern Province terranes, respectively, to form basement in north and east of North Island. Cox & Sutherland (2013). Reproduced with permission of John Wiley & Sons.

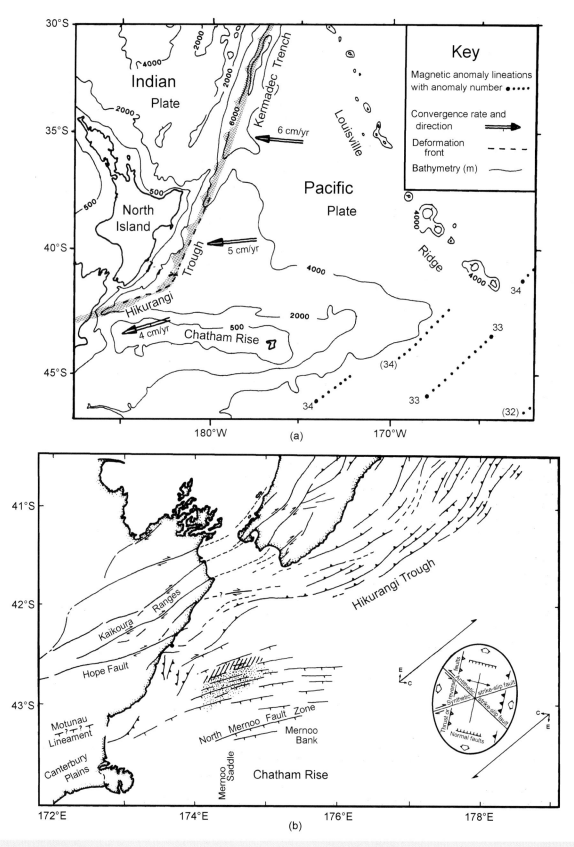

Fig. 12.19 (a) Map of the southwest Pacific, east of North Island New Zealand, showing the main topographic features, boundary between Indian and Pacific plates, direction and rates of plate convergence. From Davey *et al*. (1986). (b) Major faults in the area of the southern Hikurangi Trough (basin) offshore from North and South Islands, New Zealand. Tags on faults show downthrown block; triangles indicate thrust/reverse faults. Inset simple shear strain ellipse in appropriate orientation for major dextral-shear couple between the Indian and Pacific plates. See text. From Lewis *et al*. (1986).

Legg *et al.* (2007) propose a generalised model of the restraining bend pop-up morphology that is derived from interpretations of data along the San Clemente Fault bend region in the Descanso Plain, offshore of northwest Baja California (Fig. 12.26). This model represents a double bend, where the fault bends first to the left and then bends back to the original strike. A transpressional zone forms within the restraining double bend and is manifest as a prominent seafloor uplift that may be a complex structure consisting of right-stepping en échelon anticlines with intervening subsidiary extensional strike-slip basins and diverging oblique-slip fault zones. The uplift is broadly asymmetrical, with the principal displacement zone (PDZ) of the active strike-slip fault located to one side of the axis of uplift. Typically, the PDZ is vertical for well-defined strike-slip faults in the bend. Reverse faults that probably accommodate oblique-slip exist along the flanks of the uplift and trend subparallel to the PDZ. Beyond the ends of the restraining double bend, oblique extension forms strike-slip basins that may occur as pull-apart (stepover) or sag (releasing bend) basins. North-northwest-trending dip-slip (oblique-slip?) faults that diverge from the PDZ could be interpreted as antithetic Riedel shears (Fig. 12.26), but their observed trend is about midway between that predicted for synthetic and antithetic faults (e.g., Wilcox *et al.* 1973). Instead, they trend subparallel to the predicted trend for extension fractures. Although superficially these appear to form graben bounded by normal faults, based on seismic profiles they are high-angle reverse faults (Fig. 12.25b). The high-angle reverse faults elevate the pop-up to create local graben structures where local uplift lags that of adjacent blocks. Some reverse faults in Borderland restraining bends trend east–west as predicted from the wrench fault model, but most trend subparallel to the PDZ. This indicates strain partitioning, with reverse faults accommodating shortening and vertical faults accommodating strike-slip. North–northwest-trending normal-separation (oblique-normal?) faults also exist in the sediments above an acoustic basement block to the west of the PDZ (Fig. 12.26); these result from stretching of the sedimentary cover as the basement pushes upward due to overall transpression. In other areas, such as the Palos Verdes Hills, shallow 'keystone' graben form by extension along the crest of the transpressional uplift (Woodring *et al.* 1946; Francis *et al.* 1999). In contrast, northwest-trending faults that bound the extensional basins beyond the ends of the restraining double bend are subparallel to the PDZ and commonly exist along monoclinal sags that dip inward to the PDZ. This also indicates strain partitioning between normal (oblique?) faults on the flanks of the PDZ that accommodate extension, and a vertical PDZ that accommodates strike-slip. Most faults observed in seismic profiles across Borderland restraining bends have steep dips, measured at 70–80° for secondary oblique-slip (?) faults and vertical for the PDZ.

Phase 1: oblique rifting and formation of transform faults and spreading centres.

The California Continental Borderland formed during the Neogene development of the PAC–NOAM (Pacific–North American) transform plate boundary (Atwater 1970; Legg 1991; Lonsdale 1991; Crouch & Suppe 1993). During mid-Miocene times, oblique rifting of the western Transverse Ranges away from the continental margin formed the Inner Borderland Rift (Legg 1991; Crouch & Suppe 1993). Pre-existing structural fabric from Mesozoic to Early Cenozoic subduction controlled the orientation of the rift, which has a trend of 330° as expressed by the San Diego Trough, roughly parallel to the modern coastline. For a displacement vector about 30° oblique to the rift trend, that is, 290°–308°, a complex pattern of faults formed, including mostly

strike-slip faults, both dextral and sinistral, as well as extensional faults (Fig. 12.27, Phase 1; Withjack & Jamison 1986). New right-slip faults created to accommodate the PAC–NOAM transform motion trend northwest, subparallel to the rift margins. Normal faults formed along a north–south trend. However, when displacement exceeded several kilometres, most of the relative motion became concentrated on a few major right-slip faults that grew parallel to the relative plate-motion vector and linked north-trending continental rift centres. Seafloor spreading, with creation of new oceanic crust along faults orthogonal to the transform faults, did not occur, because the rift initiated in the thickened crust of a former subduction zone accretionary wedge, and extension was insufficient to create zero-thickness lithosphere. Creation of new transform faults in the Inner Borderland Rift was facilitated by thinned continental crust that was also thermally weakened by active volcanism during Mid-Miocene times (Vedder *et al.* 1974; Weigand 1994). The transform-fault trend parallels the displacement vector, whereas synthetic Riedel shears with a more northern trend would become transtensional. Most significantly, a right stepping en échelon pattern of right-slip transform faults linking north-trending pull-apart basins and incipient seafloor-spreading centres was created. This fault pattern resembles the modern Gulf of California transform fault system (Lonsdale 1985) and differs from the classical 'wrench fault tectonics' where dextral strike-slip results in a left-stepping en échelon pattern of synthetic Riedel shears (Wilcox *et al.* 1973; Withjack & Jamison 1986; McClay & White 1995).

Phase 2: clockwise rotation of the relative plate-motion vector, transpression and basin inversion.

In Late Miocene times, the PAC–NOAM relative plate motion vector shifted clockwise (Fig. 12.27, Phase 2; Atwater & Stock 1998). Existing transform faults with the old relative motion trend, to the west or counter clockwise to the new trend, became transpressional. Other right-slip faults with trends more parallel to the new plate-motion vector became purely strike-slip, ceasing transtensional basin formation. Stepover (pull-apart) basins, previously formed between en échelon transform faults, stopped subsiding and became structurally inverted due to transpression. With a rift trend more closely aligned with the relative plate motion vector, formation of new faults would tend to favour strike-slip on synthetic Riedel shears that could also grow into transform faults parallel to the displacement vector. Following the plate-boundary jump inland to the modern San Andreas fault system, at about 6 Ma, creation of the major southern California restraining bend increased the northeast-directed shortening across the Inner Borderland. This latter episode is considered responsible for enhanced restraining bend pop-up formation, such as along the Palos Verdes Hills Fault and at Santa Catalina Island, and may represent a further clockwise rotation of the displacement vector (Fig. 12.27, disp-2b). The Late Quaternary Pasadenan Orogeny (Wright 1991) further increased contractional strain with north-directed shortening between the western Transverse Ranges and the northern Borderland, enhancing transpressional uplift along major Borderland right-slip faults including the Santa Cruz–Catalina Ridge, Palos Verdes and Whittier faults.

Channel–levée systems in the California Borderland basin plains are described by Schwalbach *et al.* (1996). Using long-range large-scale side-scan (GLORIA) data, seismic reflection profiling and sediment cores, Schwalbach *et al.* (1996) describe active

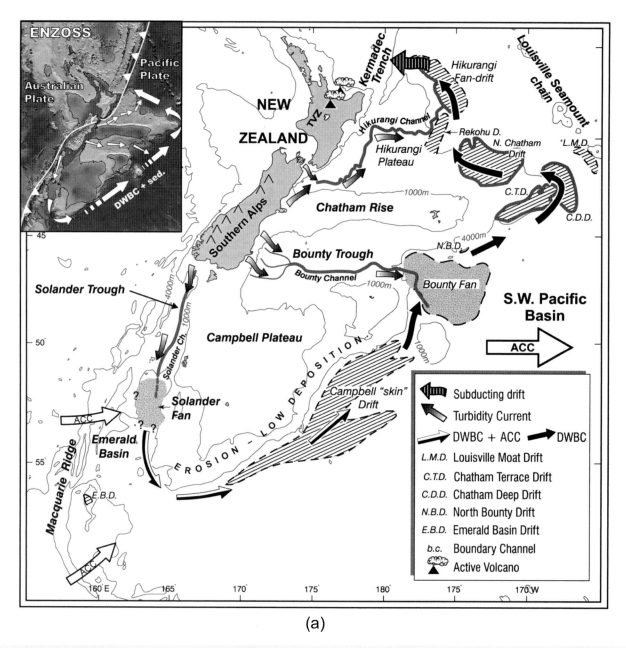

Fig. 12.20   (a) The Eastern New Zealand Oceanic Sedimentary System (ENZOSS) (inset) with the New Zealand source, the main sediment delivery channels (small arrows) and the abyssal transport system (large arrows). The larger map outlines the Solander, Bounty and Hikurangi Channels and the main depocentres beneath the DWBC. The northern limit of ENZOSS is the Hikurangi Fan-drift where it subducts into Kermadec Trench. From Carter *et al.* (2004), modified after Carter and McCave (2002). (b) Generalised plate reconstructions showing key stages in the development of ENZOSS including (30 Ma) the opening of the Tasmanian gateway and inception of regional erosion under the developing Antarctic Circumpolar Current (ACC) and southwest Pacific Deep Western Boundary Current (DWBC), followed by first drift deposition north of Chatham Rise: (20 Ma) expansion of Antarctic ice resulted in stronger ACC and DWBC and widespread erosion/low deposition even north of Chatham Rise, New Zealand plate boundary continued to develop, but with little terrigenous input to deep ENZOSS; (10 Ma) plate boundary established but still no terrigenous input to deep ocean as main supply channels still forming, continued deposition of mainly biogenic-rich drifts north of Rise and ACC-controlled erosion/low deposition to the south, initial offshore deposition of tephra; (5 Ma) continued northward migration of New Zealand reinforced the separation of mainly ACC-erosional and DWBC-drift depositional regimes to the south and north of Chatham Rise respectively, strongly convergent plate boundary supplied terrigenous sediment that accumulated near the continental margin except for localised escape to feed to proto-channel/fan systems, increasing amounts of volcanic ash; (0 Ma) modern ENZOSS fully formed by the Pleistocene with terrigenous sediment injected directly into abyssal flow as (1) Solander Fan/Channel intercepted by a new branch of the ACC, (2) Bounty Fan/Channel extending directly across current's path and (3) Hikurangi Channel diverted from New Zealand into the DWBC, strong orbital control on sediment supply and ACC/DWBC, increased volcanic airfall. From Carter *et al.* (2004).

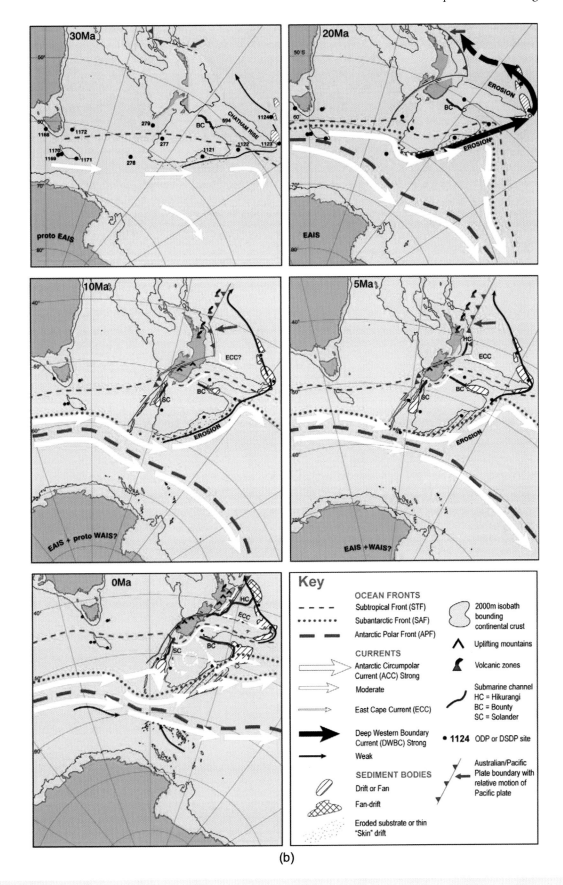

Fig. 12.20 (*continued*)

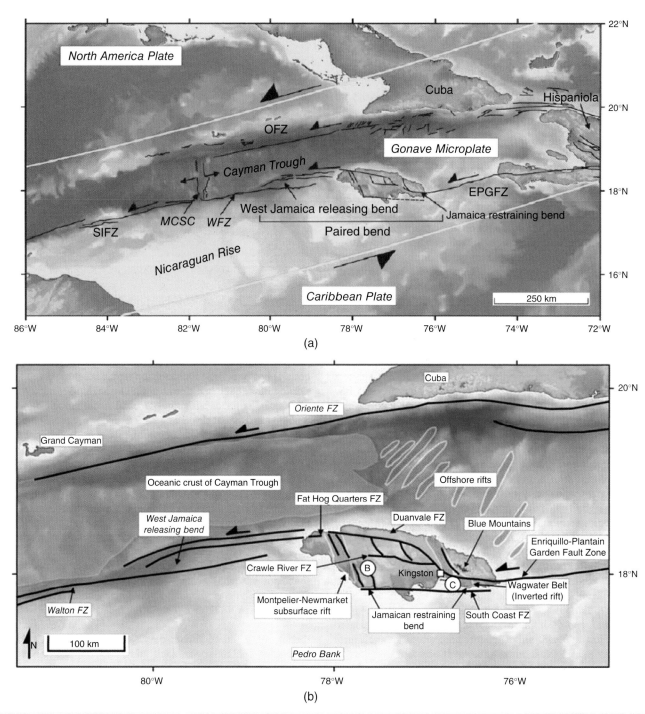

(a)

(b)

Fig. 12.21 (a) Major restraining and releasing fault bends along the active left-lateral strike-slip northern Caribbean fault system separating the North America and Caribbean plates (direction of plate motion is from DeMets *et al.* 2000). The island of Jamaica represents a major restraining bend at a right-step in the Enriquillo–Plantain Garden Fault Zone (EPGFZ) and the Walton Fault Zone (WFZ). The Jamaica restraining bend is flanked to the west by the West Jamaica releasing bend. The adjacent Jamaican restraining bend and the West Jamaica releasing bend are termed a 'paired bend'. A parallel strike-slip fault zone consists of the Oriente Fault Zone (OFZ) and the Swan Islands Fault Zone (SIFZ). The active Mid-Cayman spreading centre (MCSC) is a 100 km-long left-step between the two faults. (b) Topography of Jamaica from the NASA SRTM dataset combined with the GEBCO bathymetry of its offshore area. Labelling identifies the major mapped faults onland (Mann *et al.* 1985) and offshore (Rosencrantz & Mann 1991) in the Jamaican paired-bend system. On- and offshore Palaeocene–Early Eocene rift features in Jamaica associated with the east–west opening of the Early Eocene oceanic Cayman Trough are shown in a pale-brown colour. Onland rifts include the now inverted and highly deformed Wagwater Belt of eastern Jamaica, and the subsurface and undeformed Montpelier-Newmarket Rift of western Jamaica. Oceanic crust of the Cayman Trough is shown by the grey area. From Mann *et al.* (2007).

Fig. 12.22 Middle Eocene SGFs in the Eocene Richmond Formation, Wagwater Belt, Jamaica. (a) Facies Class A deposits. (b) Graded, stratified gravels of Facies A2.1. Camera lens cap for scale.

leveé–channel systems extending basinward from the lower parts of the Hueneme-Mugu and Redondo and Santa Cruz fans in Santa Monica, San Pedro and Santa Cruz Basins, respectively. Channels are incised in the upper to middle fan areas, and become constructional leveéd channels in the lower fan and basin plain as the channel gradient adjusts to maintain a graded profile. SGFs are generally confined to channels in the upper-fan areas, but deposit both channelised and overbank deposits on the lower fan and basin floor (see also Figs 7.18, 7.34–7.36, 7.38; Tables 7.4–7.6). The deposits show that the canyon-fan activity continued during the last phase of rising sea level. Thus, the canyon headward erosion rates must have been equal to or greater than the rate of transgression, with the canyon-fan systems remaining linked with their sediment sources. Although it is likely that the frequency of SGF events was probably higher and more voluminous during glacio-eustatic lowered sea level, flows of sufficient volume to reach the basin floors continue to occur at century or multi-century intervals, an observation also made in many other Borderland basins (Christensen *et al.* 1994; Gorsline 1996; Schwalbach *et al.* 1996; Gorsline *et al.* 2000; Hogan *et al.* 2008). Schwalbach *et al.* (1996) describe these three California Borderland basin fans as probably typical of narrow active margins where the rate of lateral sea-level transgression is less than or equal to the rate of canyon headward erosion. The canyons maintain connections with sediment sources during sea-level rise, and the systems, therefore, are active during an entire sea-level cycle. Sediment supply, therefore, is not a simple function of eustacy, which contrasts with sequence-stratigraphic models developed on passive margins where canyons become inactive as sea level rises.

In a study of the deep-marine siliciclastic systems in the California Borderland basins, Covault and Romans (2009) demonstrated that as these systems infill a basin they show progressively smaller maximum thickness-to-area ratios, that is, system areas increase more than maximum thicknesses during successive growth phases. They interpret

this as most likely a result of progressive SGF deposition 'healing' relatively high-relief bathymetry. They concluded that the growth and morphologies of SGF systems in the relatively confined basins of the California Borderland are similar to systems in the western Gulf of Mexico as a result of similarities between the volumes of sediment supplied and receiving-basin geometries (e.g., fill-and-spill processes, Section 4.6.2).

To demonstrate the complexity of the deep-marine sedimentation in these basins, a brief review of Santa Monica and San Pedro basins is presented, based mainly on Malouta (1981), Nardin (1983) and ODP Leg 167 data (Fisher *et al.* 2003; Normark & McGann 2004). Although not considered in detail here, one of the principal objectives of ODP Leg 167 was to gain a better understanding of millennial-scale variability known to occur along the California margin, with drilling in Santa Barbara Basin revealing major Neogene oceanographic events, with opaline silica burial in the middle and upper Miocene sections showing step-like drops from high opal deposition in the middle Miocene (Lyle *et al.* 2000). Lyle *et al.* (2000) also identified a major drop in opaline silica burial at ~11 Ma (roughly correlative with the eastern equatorial Pacific Miocene carbonate crash), and a second major drop at ~8 Ma, equivalent in age to the top of the Monterey Formation: from ~5-4.2 Ma, the deep-marine sediments are low in all biogenic components, marking the interval separating the Miocene high-opal sediments from upper Pliocene high-carbonate sediments. High $CaCO_3$ deposition occurred all along the entire California margin in the late Pliocene, but $CaCO_3$ burial dropped abruptly with the beginning of Northern Hemisphere glaciation (~2.6 Ma) (Lyle *et al.* 1997, 2000).

Santa Monica and San Pedro basins are northwest–southeast-trending, structurally-controlled depressions with maximum depths of 938 m and 912 m, respectively. These basins subsided during the Pliocene–Quaternary in a predominantly strike-slip tectonic regime (Crouch & Suppe 1993). Except to the southeast, where the Santa

Monica Basin is separated from San Pedro Basin by bedrock of the Redondo Knoll, the basin is bounded by a complex arrangement of strike-slip, reverse, and buried thrust faults (e.g., Nardin & Henyey 1978, Dolan *et al.* 2000, Fisher *et al.* 2003). During the latest Quaternary, the Santa Monica Basin has been filling with sediment at accumulation rates as much as 3 m kyr^{-1} (Shipboard Scientific Party 1997). Dume and Santa Monica submarine canyons provide modest amounts of sediment to construct low-relief submarine fans at the base of the slope, but the dominant source of sediment is the Hueneme Canyon at the western end of the basin (Normark & Piper 1998; Piper *et al.* 1999; Piper & Normark 2001). Hueneme Canyon cuts across the shelf and receives much of the coarse sediment delivered to the coast from the Santa Clara River, from where hyperpycnal flows deliver sediment directly into the canyon and adjacent basin slope (Warrick & Milliman 2003). The Santa Monica Basin receives sediment from at least four submarine canyons that are fed directly from rivers or from sand being transported in the littoral zone, thus the deep-marine deposits in the basin are relatively sand rich (Normark *et al.* 1998; Normark & McGann 2004).

The adjacent shelves and shelf-slopes of Santa Monica Basin are the limbs of large anticlinoria that acted as barriers to sediment transport from the Los Angeles Basin in pre-Late Quaternary times (Emery 1960; Nardin & Henyey 1978). Basin slopes range up to ~18° on the San Pedro escarpment. To the southwest, the basins are bounded by the structurally-controlled Santa Cruz Catalina Ridge. Submarine canyons that head near the shoreline feed sediments, mainly by trapping longshore drift cells, into the basins (Hueneme, Mugu, Dume, Santa Monica, Redondo and San Pedro canyons). Other processes that provide sediment to the basins include shelf bypassing and resuspension (nepheloid layer) of fine-grained sediments, and mass movements due to slope failure. Much of the slope instability and failure is due to both fast rates of sediment accumulation and earthquake ground accelerations (Section 1.2.4). The seismic characteristics of the sediments that are filling the California Borderland basins are shown in Tables 12.1 and 7.4–7.6.

Small radius submarine fans cover large areas of the basins, most of which are well described by the Normark (1970, 1978) 'suprafan model'. Figure 12.28 demonstrates the lateral variation in facies associations over relatively small distances. The geometry of the clastic infills of these basins is primarily governed by local tectonic elements, with developing folds and active faults creating sediment sinks and structural barriers to sediment transport pathways. For example, Santa Monica Canyon now is filled to the crest of a growing (Dume) anticline that has been a sediment dam at the mouth of the canyon (Junger & Wagner 1977; Nardin 1983). A channel now has

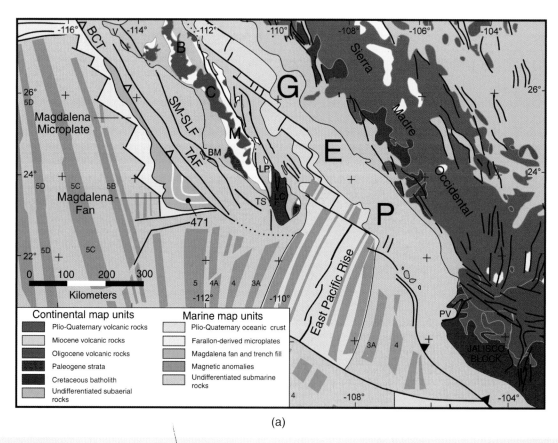

(a)

Fig. 12.23   (a) Tectonic map of the southern Baja California microplate (BCM) and Gulf of California extensional province (GEP). The Magdalena Fan accumulated on oceanic crust of the Farallon-derived Magdalena microplate located west of Baja California. Deep Sea Drilling Project Site 471 is shown as black dot on the Magdalena Fan. BCT = Baja California trench, BM = Bahia Magdalena, LC = Los Cabos block, T = Trinidad block, LP = La Paz, PV = Puerto Vallarta, SMSLF = Santa Margarita–San Lazaro fault, TAF = Tosco–Abreojos fault, TS = Todos Santos, V = Vizcaino Peninsula. Geology is simplified from Muehlberger (1996). Interpretation of marine magnetic anomalies, with numbers denoting the chron of positively magnetised stripes, is from Severinghaus and Atwater (1989) and Lonsdale (1991). From Fletcher *et al.* (2007). (b) Overleaf.

breached this anticline, but as yet has not achieved an equilibrium gradient with the basin floor.

Stable slopes are characterised by onlapping slope reflectors that thin towards the upper slope; locally at the base of the slope such deposits interdigitate with basin-floor sediments, and are locally onlapped by them (see also Piper *et al.* 1999: their figure 14a). Slope sedimentation, in part, has been strongly influenced by the Late Quaternary glacio-eustatic changes in sea level, linked to fluctuating rates of sediment input. Slope instability has created a 140 km² composite slide mass (Facies Class F) between the Hueneme and Mugu fans. The slide mass has controlled fan growth patterns by restricting and deflecting turbidity currents (Nardin 1983). Tectonic deformation also has affected basin-floor/plain deposits (mainly Facies Class E, G). An example where such tectonic effects are well developed is near Avalon Sill, at the southern margin of San Pedro Basin, where late Pleistocene deformation has upturned and faulted onlapping basin floor/margin sediments (Nardin 1983). The San Pedro Basin Fault provides a good example of a transpressional strike-slip fault (with associated flower structure) that has deformed a broad zone within

the San Pedro Basin and the adjacent escarpment, resulting in a complicated assemblage of structures arrayed along the fault's length (Fisher *et al.* 2003). The geologic structure associated with this fault varies over such short distances that even with seismic-reflection lines spaced apart by about 2 km, individual structural elements within the transpressional zone cannot be correlated along strike (Fisher *et al.* 2003).

The detailed analysis of Santa Monica and San Pedro basins led Nardin (1983) to develop a basin-fill model to characterise the sedimentation in the California Borderland basins that includes the following: (i) small fault-controlled basins; (ii) both structural and sedimentological (mass movement) barriers to sediment transport paths; (iii) small radius, generally coarse-grained, submarine fans fed by (iv) canyons that tend to intercept littoral drift currents, to feed sediment into deeper water and (v) fluctuating sea levels exerting considerable influence on the overall growth pattern of the deep-marine clastic systems.

Strike-slip tectonics can cause the truncation of submarine canyon heads. This has been well documented for Ascension Canyon in

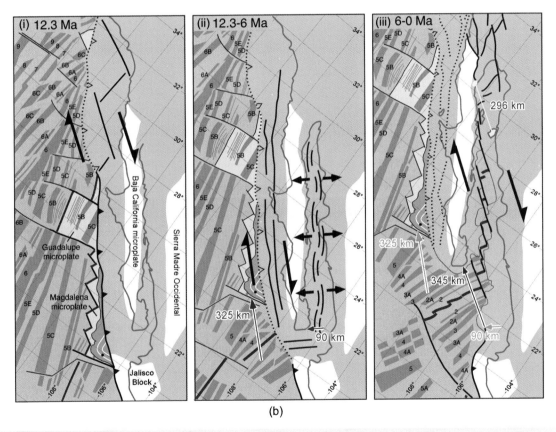

(b)

Fig. 12.23 (*continued*) (b) Revised kinematic model for shearing around the Baja California microplate. The present coastline is shown in blue. A maximum displacement estimate of ~150 km across the Magdalena shelf requires 460 km of shear in the Gulf of California extensional province, which suggests that slip rates gradually decreased in the former and increased in the latter since 12.3 Ma. The assumption is that the Pacific–North American plate motion maintained a constant rate, but became more northward in direction after chron 4 (7.8 Ma). (i) Magdalena microplate restored for chron 5a (12.3 Ma). Spreading ridges west of Baja California just prior to their effective abandonment are shown in pink. (ii) From 12.3 to 7.8 Ma, 75 km and 150 km of integrated transtensional shearing should have accumulated across the Magdalena shelf and Gulf of California extensional province, respectively. (iii) From 7.8 to 0 Ma, 75 km and 310 km of transtensional shearing accumulated across the Magdalena shelf and Gulf of California extensional province, respectively. A new system of en échelon oceanic spreading systems (shown in pink) formed in the southern Gulf of California extensional province with much the same orientation as the one that was abandoned west of Baja California. Maps have Universal Transverse Mercator zone 12 projection with mainland Mexico fixed in present position. From Fletcher *et al.* (2007).

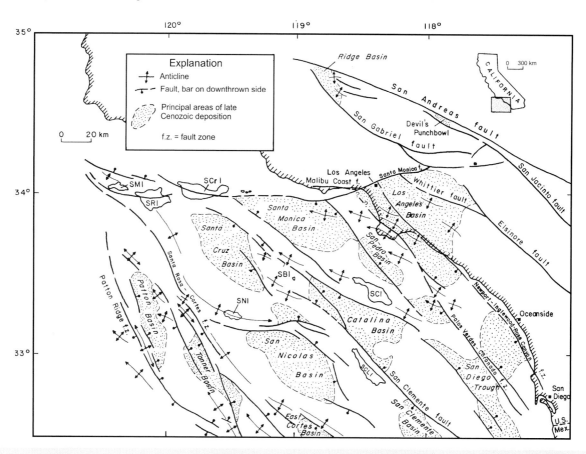

Fig. 12.24 Sedimentary basins and main structural features of the southern California Borderland and immediately adjacent onland areas. SMI, San Miguel Island; SRI, Santa Rosa Island; SCrI, Santa Cruz Island; SCI, Santa Catalina Island; SBI, Santa Barbara Island; SNI, San Nicolas Island; SCLI, San Clemente Island. Approximate thickness of basin fills in kilometres as follows: Ridge Basin, 4.5; Los Angeles Basin, 8.0; Santa Monica Basin, 3.5; San Diego Basin, 2.5; San Pedro Basin, 1.8; Santa Cruz Basin, 1.8; Catalina Basin, 0.8; San Nicolas Basin, 1.4; Patton Basin, 1.7; Tanner Basin, 1.4. Northwest-trending faults are dextral and synthetic to the San Andreas Fault. Santa Monica Fault is sinistral. Anticline trend is mainly west-northwest. From Howell *et al.* (1980).

San Clemente Basin by Nagel *et al.* (1986), on which the following description is largely based. The Outer Santa Cruz Basin, between Monterey Bay and San Francisco, began to form in the Miocene by extension associated with dextral shear along faults of the San Andreas System (Hoskins & Griffiths 1971; Howell *et al.* 1980). This northwest-plunging basin contains >3000 m of Neogene deep-marine sediments (Hoskins & Grifffiths 1971), and is bounded to the northeast by the granitic Farallon Ridge–Pigeon Point High, and to the southwest by Franciscan rocks of the Santa Cruz High (Nagel & Mullins 1983) (Fig. 12.29).

The north–northwest-trending San Gregorio Fault Zone, 1–2 km wide and comprising a complex zone of highly fractured rocks, has controlled the location and offset of canyons. En échelon folds and associated faults are compatible with dextral simple shear. Three canyon cutting episodes are recognised at ~6.6–2.8 Ma, ~750 ka and ~18 ka (Nagel *et al.* 1986). Initial canyon excavation appears to have been associated with relative lowstands of sea level, for example at ~3.8 Ma (Pliocene) the northwest headward region of Ascension Canyon was juxtaposed against the seaward part of Monterey Canyon (Fig. 12.30), coincident with a lowstand of sea level on the Haq *et al.* (1987) curve (Fig. 4.1). Thus, palaeogeographic reconstructions for

this time suggest that Ascension Canyon formed the distal extension of nearby Monterey Canyon, and was subsequently offset along the San Gregorio Fault Zone along which 110 km of dextral displacement have been recorded (Graham & Dickinson 1978). Some of the southeast heads of Ascension Canyon may have been formed during the late Pliocene and early Pleistocene sea-level lowstands at ~2.8 Ma and 1.75 Ma, respectively, and then offset to the north-northwest. There appear to have been at least two canyon-cutting events in the last 750 kyr after the entire Ascension Canyon system migrated to the north-northwest past Monterey Canyon – again controlled by lowstands of sea level.

Nagel and Mullins (1983) estimate ~105 km dextral offset along the San Gregorio Fault Zone since the initial displacement 12 Ma, based on cross-fault offset pairs in upper Miocene Santa Cruz mudstone. Seventy kilometres of this offset occurred since 6.6 Ma (late Miocene), giving an average displacement rate of 1.06 cm yr^{-1}. The last 35 km offset occurred at a slower average rate of 0.65 cm yr^{-1}. Today, Ascension Canyon appears to be an inactive shelf-edge canyon, with the associated fan valley receiving only rare Holocene sand/silt turbidites and mainly hemipelagic muds (Hess & Normark 1976).

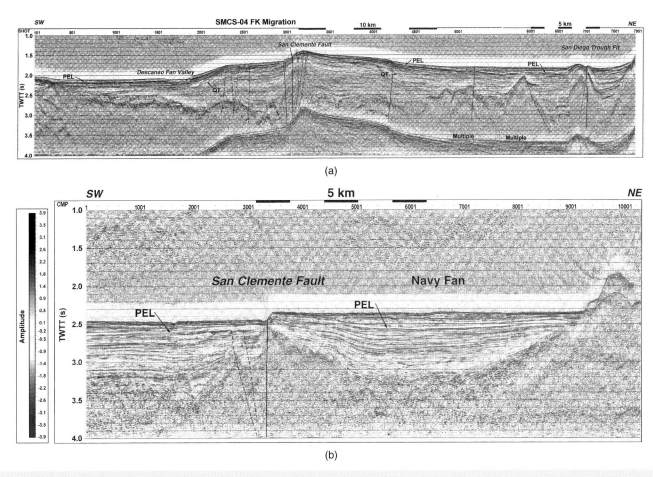

(a)

(b)

Fig. 12.25 (a) High-resolution MCS profiles across the bend region of the San Clemente Fault showing the broad zone of deformation in the pop-up structure. The principal displacement zones for the San Clemente and San Diego Trough faults are bold solid lines; other faults are solid where well-defined, and dashed where buried or less well-defined. Late Quaternary succession PEL is hemipelagic based on its acoustic transparency. Some thin-bedded turbidite sand layers occur in this unit, based on piston-core samples from the nearby San Clemente Basin and Navy Fan (Dunbar 1981). Horizon QT represents the top of the pre-uplift SGF sediments, and records the beginning of transpression. (b) High-resolution multi-channel seismic profile across the north end of the San Clemente Fault bend region. For both panels, data are post-stack migrated, 12-fold with 1.56 m CMP trace spacing. DSV *Alvin* submersible dives observed vertical seafloor scarps, 1–3 m high in mud on the 100 m-high San Clemente Fault escarpment near this profile. Hemipelagic unit PEL is buried beneath young turbidites of the Navy Fan, but may crop out in the escarpment. From Legg *et al.* (2007).

12.5.2 Gulf of California transtensional ocean basin

Seismic crustal-scale imaging from three rift segments in the southern Gulf of California, an oblique extensional (dextral transtensional) paired continental margin, show that over short lateral distances, there are large differences in rifting style and magmatism from wide rifting with minor synchronous magmatism to narrow rifting in magmatically robust segments. (Lizarralde *et al.* 2007). Such studies serve to emphasise the complexity of overlying depositional patterns, both in terms of the routing of sediments into deep-marine settings and any resulting depocentres. Thus, the Gulf of California may be considered both as a young rift system with opposing juvenile passive margins, and as a transtensional oblique-slip zone. Approximately 300 km of dextral slip along the San Andreas and related faults in central and southern California (Ehlig 1981) favours the discussion of this example here rather than in Chapter 9. The Gulf of California can be considered as a deep-marine transtensional

zone located between two triple junctions (Mendocino and Rivera), that passes northwards into a region of predominantly dextral transpressive plate interactions. South of the Rivera triple junction, the plate boundary is characterised by subduction–accretion processes (Fig. 12.31).

The results of DSDP Leg 64 (Curray *et al.* 1982a,b) suggest that at ~5.5 Ma, dextral transform motion between the North American and Pacific plates jumped from offshore to the eastern side of the Peninsular Ranges batholith, thereby initiating: (i) movement along the present San Andreas Fault, and (ii) opening of the Gulf of California. Relative displacement over the last 5.5 Myr, at an estimated rate of 5.6 cm yr^{-1}, is consistent with the matching of geological markers across the plate boundaries and indicates ~300 km offset. By ~3.2 Ma, the first lineated magnetic anomalies and oceanic crust had formed at the mouth to the Gulf. Developing spreading centres erupted basaltic magmas into wet sediments (Einsele *et al.* 1980). Figure 12.31 shows the proposed plate tectonic scenario for the

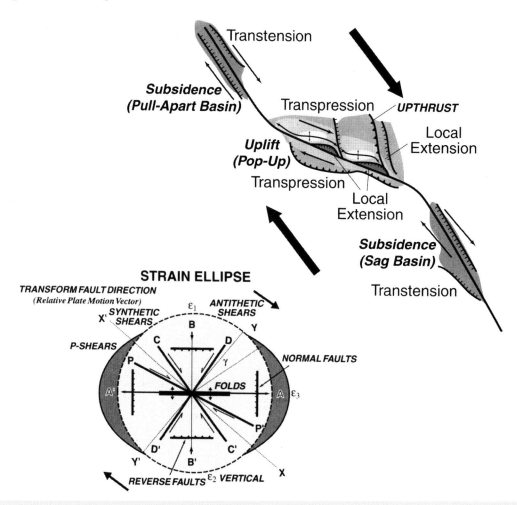

Fig. 12.26 Typical morphology and structure of restraining double bends in southern California, based on the San Clemente Fault bend region. Some features may be missing on other restraining bends, but some features may be more pronounced than in the San Clemente Fault example. Most Borderland restraining double bends have transtensional zones at the ends. A strain ellipse is shown to highlight the expected structural character for different trends in a zone of northwest-directed dextral shear (from Wilcox *et al.* 1973). Contrary to the expected strain patterns, the north-trending faults in the San Clemente Fault bend region are steeply dipping reverse faults, not normal faults. From Legg *et al.* (2007).

evolution of the deep marine Gulf of California. The oldest marine sediments are inferred to be younger than 5.5 Ma.

An intriguing aspect of the palaeogeographic reconstruction, and something that was supported and refined by DSDP Leg 63 results (Yeats, Haq *et al.* 1981) is that at ~13 Ma the rate of sediment accumulation on the Magdalena submarine fan was substantially reduced – a change that appears to correspond to the cessation of delivery of quartzo-feldspathic sandy SGFs to the fan. DSDP Leg 64 results indicate that the source area for the Magdalena Fan was cut off because of movement along a transform, the Tosco–Abriejos Fault, combined with the beginning of substantial crustal extension and associated subsidence causing a regional relative sea-level rise. The ensuing 7 Myr, from 12.5–5.5 Ma, was associated with offshore transform faulting at the plate edge, during which time the triple junction southeast of Cabo San Lucas probably became increasingly unstable because of changes in the plate shear vectors between the North American and Pacific plates (Blake *et al.* 1978; Spencer & Normark 1979).

At ~5.5 Ma, when the transform motion between the North American and Pacific plates jumped towards the northeast, offshore major transform faulting ceased. The magnetic anomalies in the Gulf of California indicate a diachronous transition from extended continental crust to oceanic crust from 4.9–3.2 Ma. The last 3.5 Myr have witnessed crustal thinning by block and listric faulting, together with igneous sill injection about well-defined spreading centres.

DSDP Sites 474, 475 and 476 straddle the boundary from continental to oceanic crust (Fig. 12.32). Sites 475 and 476 penetrated hemipelagic muddy sediments (mainly Facies Group G), a thin interval of sediments containing phosphates and glauconite and a cobble conglomerate of metamorphic clasts, Site 476 ending in weathered granite (Fig. 12.32). Site 474 penetrated mainly muddy turbidites (Facies Groups D and G), ending in middle Pliocene 3 Ma oceanic crust. Depth indicators at these sites suggest water depths of ~1000 m by the time that oceanic crust was first generated.

The range of deep-marine sediments encountered on DSDP Leg 64 is well illustrated by Site 474 (see Table 12.2). Overall, the

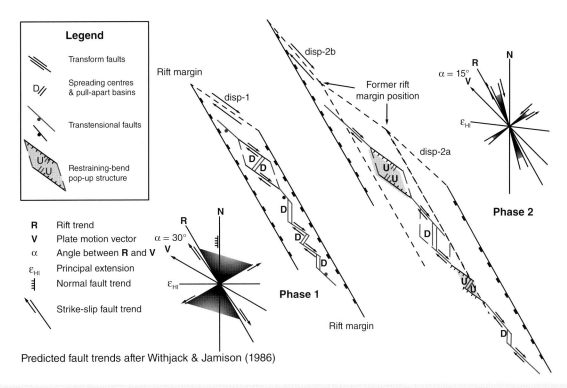

Fig. 12.27 Simplified model for generation of restraining bends in an evolving dextral transform plate boundary. Initial oblique rifting along the transform boundary creates a series of right-stepping en échelon transform faults linked by pull-apart basins. The structural fabric from the prior tectonic boundary style (subduction for California) controls the geometry of the rift formed due to a relative motion vector more westerly than the ancient structural trend. Some new strike-slip faults will have trends oriented clockwise to the transform vector, including synthetic shears. Transtensional basins form along these faults and at the fault stepovers (pull-apart basins). Clockwise rotation of the transform plate-motion vector results in transpression along the original transform faults, and former transtensional faults may become parallel right-slip transform faults. Transtensional basins become structurally inverted and form restraining-bend pop-up structures. From Legg *et al.* (2007).

succession shows a deepening trend from middle-fan to outer-fan and fan-fringe deposits. The upper part of Site 474 is notable for the development of Facies Class F and A deposits, interpreted mainly as deposits of sediment slides and debris flows, respectively. The most rapid sediment accumulation rates are recorded above the Pliocene–Quaternary contact, reaching $\sim$40 cm kyr^{-1}. An important feature of Site 474 is that it represents an entirely deep-marine succession developed on oceanic crust in contrast to Sites 475 and 476 that show an overall deepening stratigraphy from subaerial alluvial plain sediments (e.g., cobble conglomerates) to isolated offshore bank deposition in an oxygen minimum zone (phosphatic sediments with glauconite grains), to deep-marine, diatomaceous, muddy and silty turbidites and hemipelagites.

The second area of investigation on DSDP Leg 64 was in the Guaymas Basin in the central Gulf of California. Sites 477, 478 and 481 in the Guaymas Basin indicate faster rates of sediment accumulation than at the mouth of the Gulf, up to 270 cm kyr^{-1} (van Andel 1964). An interesting aspect of the sedimentology was the identification of hydrothermally altered, basin-plain, mud turbidites in Site 477: the silt/sand turbidites contain an epidote–chlorite–quartz–albite–pyrite–pyrrhotite–dolomite–anhydrite mineral assemblage. Particularly well displayed in Site 480 are Upper Quaternary laminated diatomites (Fig. 12.33) deposited in water depths of $\sim$2000 m. The sediments comprise mainly planktonic

diatoms, with minor silicoflagellates, radiolarians and variable amounts of benthic and planktonic foraminifera; the terrigenous component is olive brown silty clay. Metre-thick intervals of varve-like sediments occur between non-laminated units. The varves are attributed to annual diatom blooms associated with oceanic upwelling events in warm climatic periods, whereas the more homogeneous units are correlated with cooler and drier climatic periods (Kelts & Niemitz 1982).

The Gulf of California drilling results emphasise the complexity of deep-marine sedimentation patterns in a large transtensional ocean basin that forms the southward extension of the San Andreas Fault System. Sustained and protracted dextral-slip plate motions have governed the history of western North America since at least 12.5 Ma.

12.6 Ancient deep-marine oblique-slip mobile zones

The identification of ancient oblique-slip plate margins is one of the most difficult aspects of reconstructing palaeogeographies. By inspection of modern plate margins, some degree of oblique extension or compression is the norm rather than the exception. However,

Table 12.1 Principal seismic facies for Santa Monica and San Pedro basins, California Borderland. From Nardin (1983).

Seismic Facies	Facies parameters			Configuration	Reflection geometry at base	Lateral relationships	Depositional environment
	Amplitude	Frequency	Continuity				
Broad low-relief mound	Variable	Variable	Discontinuous: can be continuous locally	Hummocky, subparallel; on 1-s records relief and gradient decrease from upper to lower mid-fan and reflections become more parallel and even	Onlap toward basin margin and topographic irregularities; downlap locally	Gradational to complex mound, onlap basin fill and onlap slope facies	Mid-fan turbidity currents and concentrated density flows; low-relief channels resolvable on 3-s records; upper and lower mid-fan can be distinguished on 1-s and 0.25-s records and by relief and gradient
High-relief complex mound	Variable	Variable	Discontinuous	Hummocky, subparallel, contorted to chaotic locally; pronounced discordance among some reflections	Onlap toward basin margin and topographic irregularities; downlap	Occurs near submarine canyon and grades to broad low-relief mound facies	Upper fan and canyon; high relief channels and levées
Chaotic mound	Variable	Variable	Discontinuous	Chaotic, distorted or hummocky; high relief surface	Variable	Commonly associated with onlap slope facies and truncated and contorted slope reflections; intercalated or gradational with onlap basin-fill facies	Mass movement–slides, slumps, mass flows; degree of reflection disruption depends on types of mass movement process
Onlap basin fill	Variable on 3-s records continuous reflections tend to be high; discontinuous reflections tend to be low	Tends to be uniform	Mainly continuous on 3-s records, variable on 1-s records	Even, parallel	Onlap toward basin margin and topographic irregularities; concordant at centre of basin	Gradational to broad low-relief mound facies; continuous with onlap slope facies along some horizons	Lower fan and basin plain (can be distinguished on 0.25-s records); relatively low-velocity turbidity currents
Channel fill	Usually high on 1-s records	Variable	Discontinuous on 1-s records	Parallel to hummocky	Concordant or discordant	May be gradational with levees on fan or basin plain if depositional type; onlaps if erosional	Turbidity current channels

Table 12.1 (*continued*)

| Seismic Facies | Amplitude | Facies parameters | | Configuration | Reflection geometry at base | Lateral relationships | Depositional environment |
		Frequency	Continuity				
Onlap slope	Variable but tends to be low on 1-s records	Variable	Continuous	Convergent upslope; may be wavy, disrupted or contorted due to tectonic deformation or mass movement	Onlap	Onlapped by mound or fill facies; gradational with some mound facies; continuous with basin fill along some horizons	Hemipelagic slope deposition; particle settling from low energy turbid layer or nepheloid flows resuspended from shelf and upper slope
Sheet drape	Relatively low on 1-s records	Relatively uniform on 1-s records	Continuous	Parallel and conforms to underlying topography	Concordant on irregular topography; locally onlaps along basin margin	Gradational with onlap slope front fill fades; alternates with onlap basin-fill facies	Basin-plain hemipelagic particle settling from suspended sediment concentrations (nepheloid layers)
Migrating wave	Variable	Relatively high on 0.25-s records	Continuous	Wavy, subparallel; asymmetric; resembles climbing ripples on 0.25-s records; mounded, wavy to hummocky on 3-s records	Concordant	Subfacies with broad, low-relief mound facies (fan); relief decreases and wave length decrease toward margin of fan	Turbidity current phenomenon on large levée complex in mid-fan environment

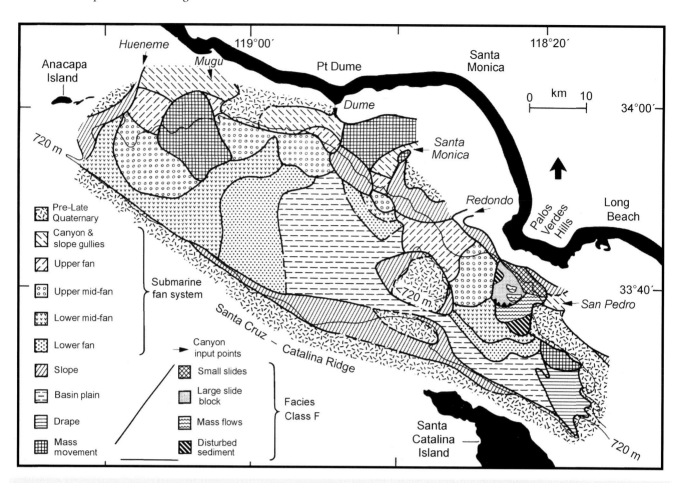

Fig. 12.28 Late Quaternary depositional environments in Santa Monica and San Pedro basins. Note the extreme lateral heterogeneity of deep-marine environments and, therefore, changes in facies. From Nardin (1983).

ancient plate margins where suspect terrain accretion has occurred are most probably associated with considerable strike-slip. Examples include the Cretaceous of Alaska, Wrangellia (Section 12.3), the Lower Palaeozoic northern Appalachians–southern Caledonides in Newfoundland and Britain, and the Mesozoic–Tertiary of the Pyrenees. In all three cases, deep-marine sediments constitute a large part of the stratigraphy, suggesting sedimentation on thin lithosphere subject either to transpression or transtension.

Whilst there may be reasonable arguments for interpreting strike-slip tectonics as important in ancient active plate-margin basins, the polyphase and complex deformation history, commonly lasting several million to many tens of millions of years, together with the preservation of relatively small and fragmented segments of such basins, can make it difficult to generate reliable and robust palaeogeographic reconstructions. For example, the Palaeogene Italian Apennines is commonly interpreted simply in terms of essentially orthogonal convergence between the European and Adriatic plates. However, Marroni and Treves (1998) proposed that a very oblique or transpressional tectonic regime was dominant during the early (pre-collisional) orogenic evolution of the Northern Apennines (Late Cretaceous–early Oligocene). Amongst the arguments for major oblique collision in the Northern Apennines, they cite: (i) the plate tectonic framework, that indicates left-lateral oblique

convergence along the Europe/Adria plate margin; (ii) the lack of a magmatic arc during the entire pre-collisional convergent history of the chain (a time span >45 Myr, from Late Cretaceous–Early Oligocene); (iii) the long (20 Myr) residence time of deep-marine SGF-dominated successions in the trench (the 'dormant' trench); (iv) the multiple source areas of SGFs from both sides of the basin, and the associated coarse-grained deposits; (v) the opposite vergence of the deformation in some oceanic units; (vi) the non-matching stratigraphic features, distinct deformation and metamorphic histories between adjacent overthrust oceanic units (Ligurids), interpreted as tectonostratigraphic terranes. Specific aspects of Apennine stratigraphy and tectonics and the geometry and structure of the contacts between the Ligurid Units led Marroni and Treves (1998) to suggest the existence of a number of terranes juxtaposed by transpression during the early orogenic evolution of the chain.

12.6.1 Mesozoic Pyrenees

The Mesozoic history of the Pyrenees was dominated by seafloor spreading in the Ligurian Tethys, Central and North Atlantic and Bay of Biscay together with the anticlockwise rotation of the Iberian Plate. These plate motions resulted in an overall sinistral shear

Table 12.2 Sedimentary lithological units, DSDP Site 474. From Curray *et al.* (1982a,b)

Unit	Interval	Sub-bottom Depth (m)	lithology	Sedimentary environment	Age	Estimated Sedimentation Rate (m/Myr)	Thickness (m)
I	Cores 474–1 to 474–3	0 0–21.0	dusky yellow green to greyish olive diatomaceous muds to oozes and diatomaceous nannofossil marls with downward increasing clayey silts	hemipelagic/distal fan	to NN21/20	47	305
II	Core 474–4 to Section 474–10, CC	21.0–87.5	olive brown to greyish olive; firm nannofossil diatomaceous muds and a coarse arkose sand to conglomerate	redeposited slump-debrite	(NN19)	very high(')	76.0
III	Sections 474–11–1 to 474A–8, CC	87.5–239.0	clayey silts to silty clays, nannofossil diatomaceous muds, nannofossil marls, and scattered arkosic muds	hemipelagic mud and mud turbidites on outer fan	NN20	395	142.0
IV	Sections 474A–8, CC to 474A–28.CC	239.0–420.0	greyish olive silty claystone to clayey siltstone		early Pleistocene	395–86	181.0
IVA	Sections 474–9–1 to 474A 10, CC	(239.0–258.5)	mostly uniform silty claystones with siliceous fossils and nannofossils	hemipelagic with some mud turbidites	NN19	395	19.5
IVB	Sections 474A–10, CC to 474A–28, CC	(258.5–420.0)	mostly cycles of thick, clayey quartzose siltstone beds and some arkosic sands: mud flows, siliceous fossils diminished	middle fan and mud turbidites		395–86	(161.5)
V	Sections 474A–28, CC to 474A–41–5 (25 cm) and 474A–45–1	420.0–533.0	olive grey clayey siltstone with thick mass flows and cemented arkose, silty claystones between dolerite sills	middle fan, mud turbidites, and hemipelagic mud	late Pliocene	86	(~115)

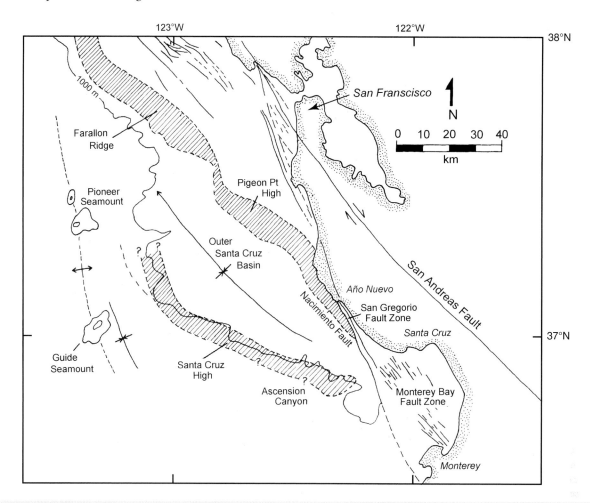

Fig. 12.29 Generalised structural map of the Outer Santa Cruz Basin, also showing the San Gregorio and Monterey fault zones. The area shown in this figure is the reference frame for the palaeogeographies in Figure 12.30. From Nagel *et al.* (1986).

couple between the Iberian and southern European plates (Fig.12.34). Puigdefabregas and Souquet (1986) summarise the transtensional, wrench-faulting phase in the development of the Pyrenees in the following way. In the middle Albian–Cenomanian, transtension was associated with alkaline magmatism, thermal metamorphism and diapirism of the Triassic evaporites. The deeper basins were filled by deep-marine slope-apron and basin SGF deposits of the 'Flysch Noir' or Pyrenean flysch (Souquet *et al.* 1985). Figure 12.35 shows the present-day outcrop of the Flysch Noir, a reconstruction of the lozenge-shaped, deep-marine pull-apart basins, and a schematic cross-section through the basins. An example of slide folds in the Flysch Noir is shown in Figure 2.41b.

During the Cenomanian–middle Santonian, a global rise in sea level effectively widened the basin, and the site of shelf-carbonate sedimentation retreated landward. Turbidity currents from the drowned former shelf transported carbonate detritus into deeper water. In the middle Santonian, there was a relative fall in sea level, together with a tectonic tilting of the margin fault blocks; this resulted in widespread unconformities, and in deeper water the accumulation of slope breccias (Facies Class F) and megaturbidites (Facies Class A and Facies C2.4 and Class F). The late Santonian–Maastrichtian was associated with further normal faulting and subsidence, under sinistral transtension, and the development of a carbonate debris sheet extending

downslope into the basin to interfinger with siliciclastic submarine fans showing axial sediment transport (Van Hoorn 1970). There was a major eustatic rise in sea level in the Santonian to give coastal onlap successions. The development of progressive unconformities (Simó & Puigdefàbregas 1985) shows that sedimentation was synchronous with folding. Sinistral transpression became increasingly more important into the Tertiary. This plate convergence caused uplift in the northeast Pyrenees so that the area became a source for the clastics that were shed into the evolving foreland basins (Section 11.3.4).

Thus, the Mesozoic–Tertiary history of the Pyrenees provides an insight into plate margins that were initially dominated by transtension and then transpression as plate convergence finally closed the remaining seaways between the European and Iberian plates.

12.6.2 Lower Palaeozoic north central Newfoundland and Britain

The large Cambrian Atlantic-type Iapetus Ocean (Wilson 1966) was transformed in the Early Ordovician into a Pacific-type ocean with subduction zones, sites of ophiolite obduction, and complex terrane accretion particularly along the Laurentian margin of North America and its continuation into the British Caledonides (Williams &

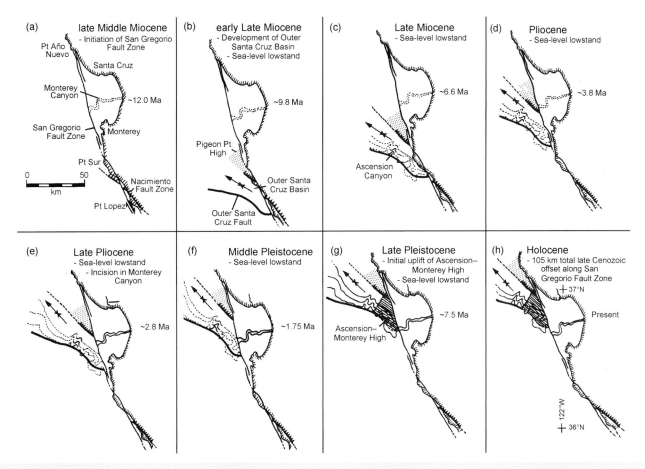

Fig. 12.30 Palinspastic palaeogeographic reconstructions of the south-central California Borderland from the late/middle Miocene–Present (124 Myr). Reference coastline shown in Figure 12.29. Maps show the possible development of the Ascension Submarine Canyon system and adjacent continental shelf in relation to Monterey Canyon. Most tectonic movement is assumed to have occurred along the San Gregorio Fault Zone. See text. From Nagel *et al.* (1986).

Hatcher 1982, 1983; Williams 1984, 1985; Bluck 1985). Terrane accretion occurred progressively outwards from the Laurentian margin (Williams 1985) until the Iapetus Ocean was finally closed in Devonian (Emsian) times (Soper *et al.* 1987). Although the demise of any vestigial ocean did not occur until the Emsian, palaeontologic, tectono-stratigraphic and palaeomagnetic evidence suggests initial closure of the Iapetus Ocean with complete subduction of intervening oceanic crust between Laurentia (North America) in the region of Western Newfoundland and Eastern Avalonia (Britain and Eire south of the Iapetus Suture) probably by the late Ashgill (Pickering 1987a,b; Pickering *et al.* 1988b; Pickering & Smith 1995, 1997, 2004). Western Avalonia (eastern Newfoundland, Acadia and Appalachian Piedmont terrane) was accreted in the Early–Middle Devonian (Keppie 1986). Figure 12.36, from Pickering and Smith (2004) (*cf.* Pickering *et al.* 1988b) summarises the history of the Iapetus Ocean from Arenig–Llandovery times.

Suspect terranes that accreted to the Laurentian margin are widely reported from the Lower Palaeozoic Appalachians and Caledonides (Fig. 12.36) (Williams & Hatcher 1982, 1983; Williams 1984, 1985; Curry *et al.* 1982; Soper & Hutton 1984; Bluck 1985; Keppie 1986; Ziegler 1986a,b,c; Anderson & Oliver 1986; Bluck & Leake 1986; Hutton & Dewey 1986; Elders 1987; Pickering 1987b; Soper *et al.* 1987; Pickering *et al.* 1988b). The Laurentian margin appears to have been a site of complex terrane accretion throughout the Ordovician

until the Middle Devonian, with most evidence supporting terrane docking by sinistral shear (e.g., Soper & Hutton 1984; Bluck 1985; Keppie 1986; Hutton & Dewey 1986; Elders 1987; Pickering 1987a,b; Soper *et al.* 1987; Blewett & Pickering 1988; Pickering *et al.* 1988b).

Although it is difficult to assess the cumulative amount of strike-slip along the Laurentian margin during terrane accretion, the radiometric age range and petrography of granodiorite clasts in the Southern Uplands of Scotland and the Long Range area of Newfoundland, together with palaeocurrent data, suggest that at some time during the Caradoc these terranes may have been juxtaposed (Elders 1987). If so, then at least 1500 km of cumulative sinistral displacement must have occurred from Caradoc–Middle Devonian (Emsian) along various fault zones when the British Isles basement was finally assembled. Both in the British Isles and Newfoundland, deep-marine turbidite systems (mainly slope and slope basin fills) occur throughout the Ordovician and Lower Silurian successions (e.g., Watson 1981; Arnott 1983a,b; Arnott *et al.* 1985; Pickering 1987a), and are overlain by shallow shelf to non-marine deposits.

Regional considerations, together with available palaeomagnetic data, suggest a palaeogeography for the latest Ordovician–earliest Silurian (Fig. 12.36) in which: (i) the once contiguous Iapetus Ocean was finally destroyed by continent–continent collision between Laurentia and Eastern Avalonia to eliminate any intervening oceanic crust; (ii) destruction of oceanic crust continued both to the north and south

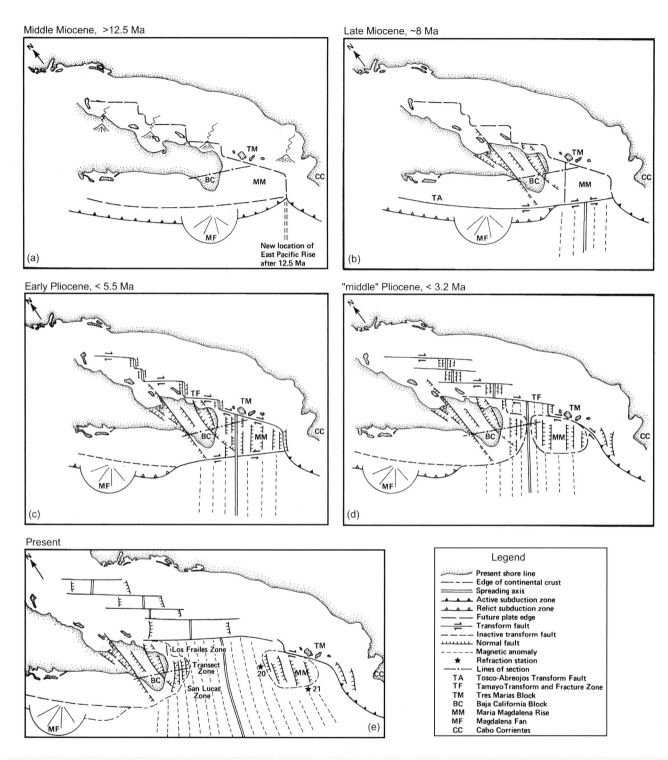

Fig. 12.31 Geological history of the Baja California margin based on the results of DSDP Leg 64. From Curray *et al.* (1982a,b) and various other sources (*cf.* Atwater 1998). See text for explanation.

of this collisional site with westward to northwestward subduction as Baltica and Western Avalonia approached Laurentia; (iii) oblique collision of Eastern Avalonia with Laurentia, probably first occurring opposite western Newfoundland at a promontory in the continental margin, led to sustained sinistral oblique slip, creating transtensional pull-apart, and transpressional thrust-controlled, deep-marine basins.

In north-central Newfoundland, structural and sedimentological studies suggest Ashgill–Wenlock active fault-controlled sedimentation, mainly in relatively small, deep-marine basins controlled by northwest-dipping thrust faults, commonly associated with

tectonised olistostromes (mélanges) and growth faults (Nelson 1981; Arnott 1983a,b; Arnott *et al.* 1985; Pickering 1987a,b). In New World Island, Newfoundland, Arnott (1983a,b) has demonstrated that essentially contemporaneous successions, separated by major synsedimentary faults, probably developed in discrete neighbouring basins. A synsedimentary Late Ordovician–Silurian D1 deformation is recognised both in the Gander and Dunnage terranes. Major tectonic elements associated with the Dunnage–Gander terranes that may have been active, as ancestral structures, during the Late Ordovician–Silurian include the Hermitage Flexure (Brown & Colman-Sadd 1976), the Lobster Cove–Chanceport and Lukes

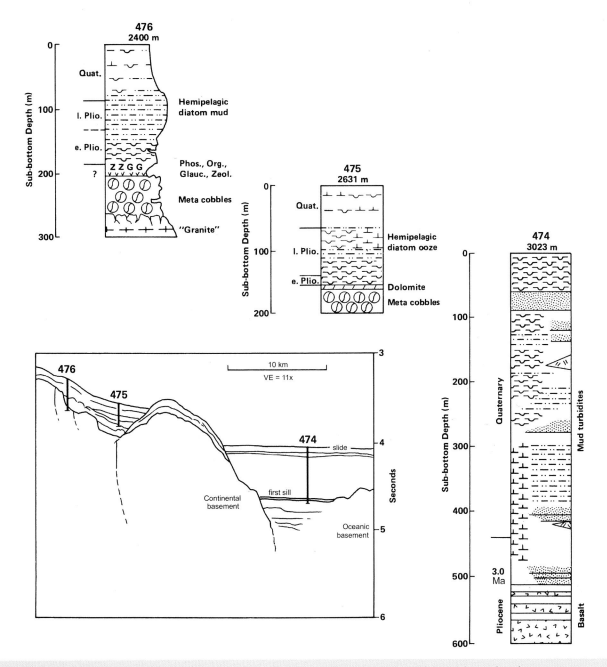

Fig. 12.32 Summary sections through DSDP Leg 64 Sites 474, 475 and 476 in the Gulf of California, located on the inset. For a summary of lithologies, refer to Table 12.2. Note the early Pliocene rapid deepening associated with the opening of this part of the Gulf. From Curray *et al.* (1982a,b).

Centimetres

Fig. 12.33 Organic-rich, laminated, hemipelagic, sediments from DSDP Leg 64, Site 480-29-3, 145–150 cm. These laminites are interpreted as a result of annual phytoplankton and algal blooms. Scale to left graduated in mm and cm. Photograph by K. Kelts.

Arm–Sops Head faults across north-central Newfoundland (Nelson 1981; Watson 1981; Arnott 1983a,b; Arnott *et al.* 1985), with possible northwest-dipping thrust faults including the ancestral Lukes Arm, Toogood, Cobbs Arm, Byrne Cove and Boyds Island faults on New World Island (Nelson 1981; Arnott 1983a,b; Arnott *et al.* 1985), and the ancestral New Bay Fault (Pickering 1987a) (Fig. 12.37).

Figure 12.38 shows a palaeogeographic reconstruction of north-central Newfoundland in the late Caradoc–Ashgill (Blewett *pers. comm.*). This map shows a number of small, fault-controlled, deep-marine basins separated by structural highs that are commonly at an obtuse angle to the main basin-bounding faults. The sediment source for the basins was from the deeply dissected arc terrane to the north, with conglomeratic and sandy facies classes (A, B, C, F) typically occurring in submarine channels and valleys up to several hundred metres deep and a few kilometres wide; the surrounding lithologies mainly being silty or muddy as Facies Classes D and E, with minor G (*cf.* Nelson 1981; Watson 1981; Arnott 1983a,b; Pickering 1987a).

The sedimentology in these deep-marine basins in Newfoundland varies considerably from thick coarse-grained submarine fan deposits, as in the Milliners Arm Formation, New World Island (Watson 1981; Fig. 8.8; Section 8.3.3), to thick, fine-grained, slope-basin fills cut by coarse-grained channel successions, as in the Point Leamington Formation, New Bay area (Pickering

1987a). Furthermore, there is abundant evidence of wet-sediment deformation as sediment slides (Pickering 1987a).

The California Borderland provides a useful modern analogue, with a complex series of fault-controlled basins oceanward of a major zone of strike-slip. A major difference, however, is that during the Silurian, in Newfoundland, there was no major ocean basin to the south/southeast of these marine basins, unlike off California today (see Fig. 12.37). The seaway to the southeast was probably more like the present day Timor Trough. In the example from north-central Newfoundland, the major strike-slip/oblique-slip appears to have occurred under a sinistral shear couple along faults such as the Lobster Cove–Chanceport and Lukes Arm–Sops Head faults (Fig. 12.38). Sinistral shear associated with deep-marine sedimentation appears to have occurred from Late Ordovician to Middle Silurian times, followed by the infilling of remaining marine seaways by shallowing-upward successions. Many of the tectonically-deformed deep-marine sediments show a clockwise transection of cleavage that was imparted during the regional sinistral strike-slip (Fig. 12.39).

12.7   Concluding remarks

Deep-marine basins formed along modern strike-slip continental margins can be three-dimensionally extremely complex, typically with very irregular shapes. Over relatively brief geological time frames, they can undergo reversals in the overall sense of shear, include phases of transtension and transpression. Sediment supply routes can easily become disrupted by faulting, with the truncation of sediment supply from non-marine and shallow-marine into deeper water environments via submarine canyons to feed submarine fans. This is well shown by the Magdalena and Astoria fans in the Californian Borderland. Tectonic processes at such plate margins tend to generate complex and somewhat unpredictable stratigraphic architecture. Abrupt changes in sediment routing can be associated with changing sedimentary petrography in any deep-marine sediments (for example, Dickinson et al. 2005).

Many subduction margins (Chapter 10) and foreland basins (Chapter 11) can be linked to some strike-slip deformation during their evolution, for example the modern Nankai forearc (Section 10.3) and the Lower Palaeozoic of the Caledonide-Appalachian orogen (Sections 10.7 and 12.6.2). Ancient deep-marine basins that form along strike-slip continental margins, therefore, can be difficult to recognise because their constituent facies, facies associations and architectural elements will resemble those formed in other plate-tectonic settings. Without an understanding of the large-scale plate-tectonic framework, it may be very challenging to interpret deep-marine systems as having accumulated at such margins.

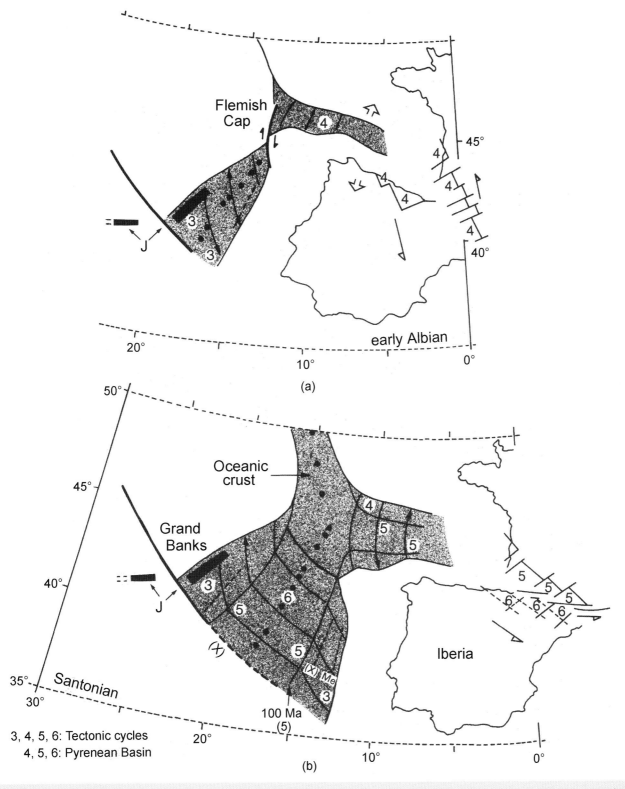

Fig. 12.34 Reconstruction of the southernmost North Atlantic region and Iberia–Armorica in the Early Albian and Santonian to show the development of the Mesozoic–Tertiary deep-marine basins in the Pyrenees. The basins are shown as due to major sinistral transtension along the Iberian and European plate margins. In the middle Albian–Cenomanian, transtension was associated with alkaline magmatism, thermal metamorphism and diapirism of the Triassic evaporites. The deeper basins were filled by deep-marine slope apron and basin turbidites of the 'Flysch Noir' or Pyrenean flysch (Souquet *et al.* 1985). From Puigdefabregas and Souquet (1986). See Figure 12.35 for a more detailed palaeogeography.

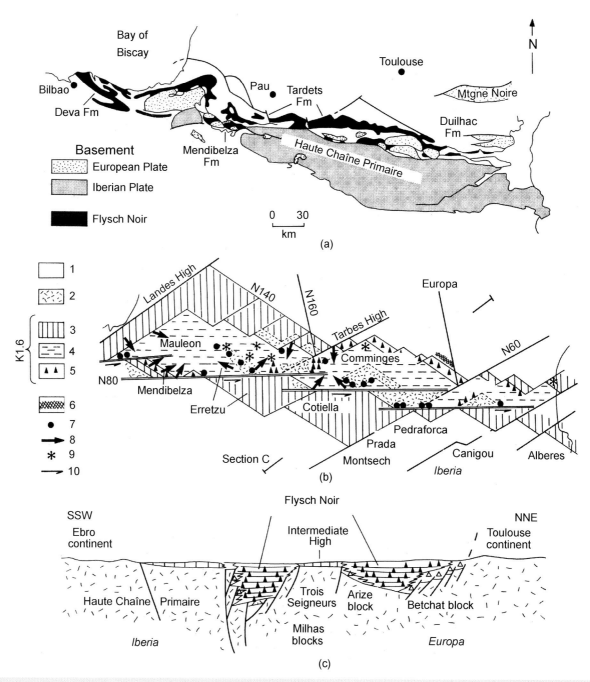

Fig. 12.35 (a) Generalised geological map of the North Pyrenean Fault Zone in the northeastern Pyrenees to show the outcrop of the oldest Pyrenean turbidite successions (Middle Albian–Cenomanian) that infilled the deeper parts of the developing basins, the 'Flysch Noir' Formation. (b) Palaeogeography for the 'Flysch Noir' Formation times, with (c) showing hypothetical cross-section along Section C shown in 12.35b. Key as follows: 1, emergent areas; 2, eroded highs and/or over-riding blocks, 3, shelf deposits, 4, Albian turbidites; 5, fault-controlled debris apron; 6, bauxites; 7, lherzolites; 8, turbidite transport directions; 9, mid-Cretaceous magmatic rocks; 10, inferred strike-slip faults. From Puigdefabregas and Souquet (1986).

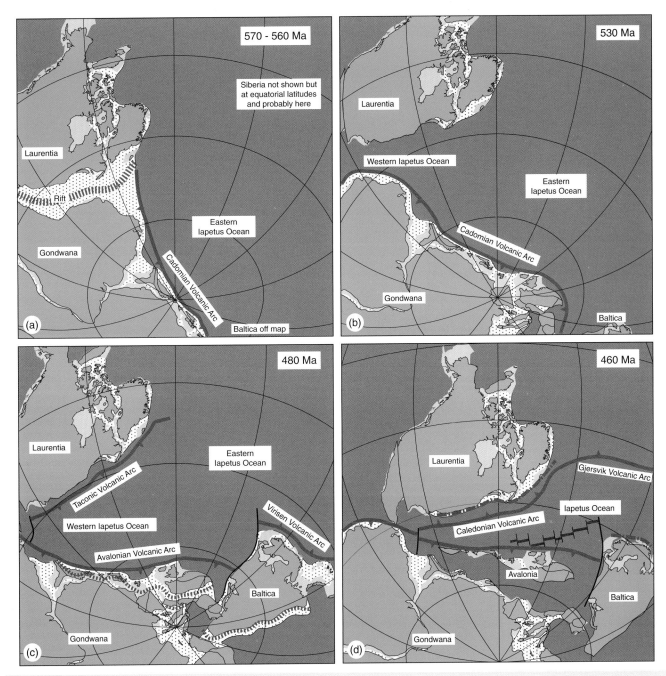

Fig. 12.36 Computer-generated plate reconstructions, based on a synthesis of palaeomagnetic data, for the following time intervals: Late Precambrian (570–560 Ma); Cambrian (530 Ma); Early Ordovician (490–480 Ma); Middle Ordovician (460 Ma); latest Ordovician (440 Ma); Middle Silurian (420 Ma) and early Middle Devonian (390 Ma). Positions of major magmatic arcs are shown schematically. The initial Late Precambrian reassembly joins western South America to eastern Laurentia (present coordinates). Baltica's position is uncertain and is not shown on this map. The positions have been obtained by interpolating between this initial reassembly and positions suggested by Ordovician palaeomagnetic data. The palaeomagnetic evidence suggests that during Cambrian time Gondwana approached Laurentia, and that northwestern South America possibly collided with an oceanic arc or arcs that fringed southeastern Laurentia in Early Ordovician time. Gondwana then rotated anticlockwise in later Ordovician–Early Devonian time, bringing opposite one another those parts of Gondwana and Laurentia (e.g., Florida) that were to collide in the later Palaeozoic. Uncertainties in Baltica's position relative to other continents in Cambrian–Early Ordovician time mean that it has been omitted from maps of 490 Ma and older periods. It is first shown on the 460-Ma map (d) separated from Laurentia by a relatively narrow branch of the Eastern Iapetus Ocean. For compatibility with the events in Svalbard, the distance between the two continents is shown as decreasing between 460 and 440 Ma. Modified from Pickering and Smith (1995, 2004).

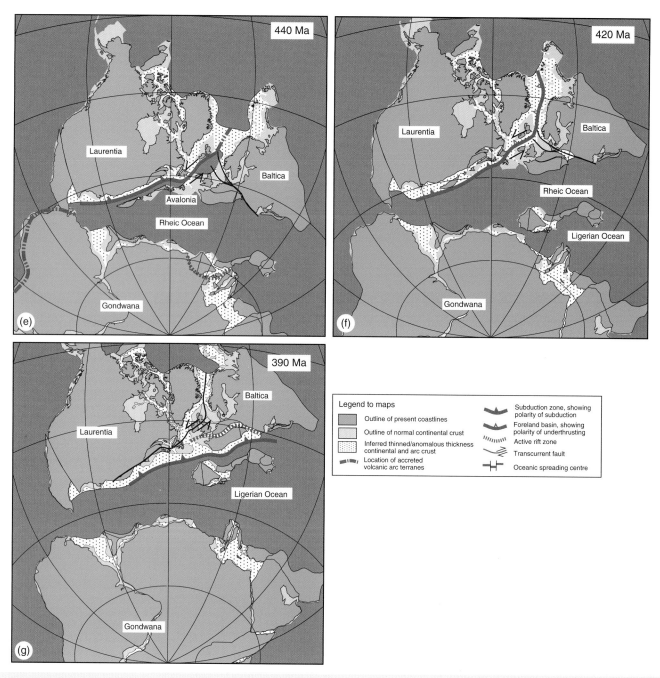

Fig. 12.36 (continued)

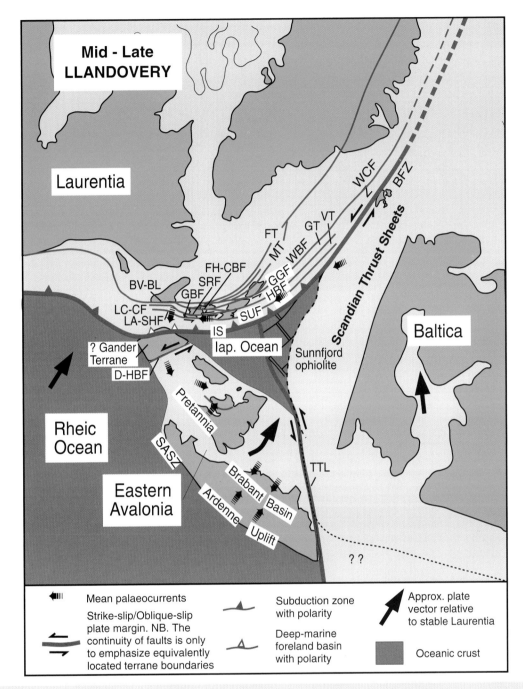

Fig. 12.37 Mid–late Llandovery reconstruction. Plate motion arrows are shown for the Baltica and eastern Avalonia plates relative to a fixed Laurentian (North American) plate. In present-day co-ordinates, Eastern Avalonia approximates to the northern parts of the Avalon Zone in the Appalachians. See Figure 12.36 for outline of Avalonia. Closely spaced stipple outlines parts of Armorica, Britain and the Gander Terrane south of the Iapetus suture, and fragments of western Newfoundland and northwestern Britain with Laurentian crustal affinities prior to ocean closure. Abbreviations for major geologic features: BV–BL = Bay Verte–Brompton Lineament, LC–CF = Lobster Cove–Chanceport Fault, LA–SHF = Lukes Arm–Sops Head Fault, GBF = Galway Bay Fault, SRF = Skerd Rocks Fault, FH–CBF = Fair Head–Clew Bay Fault, SUF = Southern Uplands Fault, HBF = Highland Boundary Fault, GGF = Great Glen Fault, WBF = Walls Boundary Fault, FT = Flannan Thrust, MT = Moine Thrust, IS = Iapetus suture (i.e., Cape Ray-Reach Fault in Newfoundland); BFZ = Billefjorden Fault Zone (Svalbard), WCFZ = Western Central Fault Zone (Svalbard); TTL = Tornquist Teisseyre Lineament; SASZ = South Armorican Shear Zone, D–HBF = Dover–Hermitage Bay Fault. Paired Ordovician arc systems in the northerly parts of the Iapetus Ocean correlated from Newfoundland to Svalbard and separated by ophiolites are as follows: (1) Taconic island arc, Early–Middle Ordovician developed above oceanward-dipping subduction zone, Moreton's Harbour arc (Newfoundland), Lough Nafooey arc (Ireland), Grampian Terrane (Scotland) and Gjersvik arc (Norway), Pearya/Northwestern Svalbard (including Biskayerhalvoya) Terranes; (2) Southern island arc, Ordovician–Silurian arc developed above Laurentia-ward dipping subduction zone, Bronson Hill arc, Robert's Arm arc (Newfoundland), South Connemara arc, Midland Valley terrane (Scotland) Virisen arc (Norway), and western Svalbard. Modified from Pickering and Smith (1995, 2004).

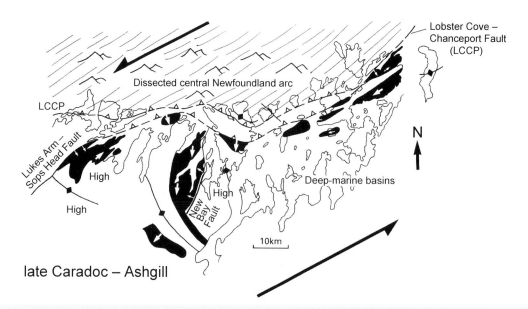

late Caradoc – Ashgill

Fig. 12.38 Late Caradoc–Ashgill palaeogeographic reconstruction for the environments of deposition of the deep-marine successions in north Central Newfoundland, based on unpublished work by K.T. Pickering, R. Blewett and various other sources (*cf.* Arnott 1983a; Pickering *et al.* 1988b). Black legend represents basin depocentres between structural highs; white arrows are schematised palaeocurrent directions; main faults interpreted as active during sedimentation are: LCCP, Lobster Cove–Chanceport Fault; Lukes Arm–Sops Head Fault, both as thrusts/oblique-slip faults. Note the deeply dissected Central Newfoundland Island Arc to the north that collided with Laurentia in the Llandeilo, and provided the main source of sediments for the basins to the south on the plate margin. Also, note the inferred importance of major sinistral shear at that time. By the late Ashgill, Eastern Avalonia probably was sited hundreds of kilometres to the southeast and the intervening environments are interpreted by Pickering *et al.* (1988a) to lie within a major deep-marine foreland basin.

Fig. 12.39 Transecting cleavage in Silurian deep-marine deposits, Northern England, UK. This shows a clockwise overprint or transection of ~10°, consistent with major left-lateral (sinistral) transpression. Similar structural features, also suggesting contemporaneous sinistral transpressional tectonics along strike in Newfoundland, were documented by Blewett and Pickering (1988).

References

Aalto, K.R. 1976. Sedimentology of a mélange: Franciscan of Trinidad, California. *Journal of Sedimentary Petrology*, **46**, 913–929.

Abad, J.D., Sequeiros, O.E., Spinewine, B., Pirmez, C., Garcia, M.H. & Parker, G. 2011. Secondary current of saline underflow in a highly meandering channel: experiments and theory, *Journal of Sedimentary Research*, **81**, 787–813.

Abbate, E., Bortolotti, V. & Passerini, P. 1970. Olistostromes and olistoliths. *Sedimentary Geology*, **4**, 521–557.

Abbott, L.D., Silver, E.A., Thompson, P.R., Filewicz, M.V., Schneider, C. & Abdoerrias 1994. Stratigraphic constraints on the development and timing of arc-continent collision in northern Papua New Guinea. *Journal of Sedimentary Research*, **64**, 169–183.

Abels, H.A., Kraus, M.J. & Gingerich, P.D. 2013. Precession-scale cyclicity in the fluvial lower Eocene Willwood Formation of the Bighorn Basin, Wyoming (USA). *Sedimentology*, **60**, 1467–1483.

Abrahamson, P. 2013. Barents Sea: complete integration of well data and seismic data. *GEO ExPro*, **10**, 60–62.

Abreu, V., Sullivan, M., Pirmez, C. & Mohrig, D. 2003. Lateral accretion packages (LAPs): an important reservoir element in deep water sinuous channels. *Marine and Petroleum Geology*, **20**, 631–648.

Abreu, V., Neal, J.E., Bohacs, K.M. & Kalbas, J.L. (eds) 2010. *Sequence Stratigraphy of Siliciclastic Systems-the Exxon Mobil Methodology: Atlas of Exercises*. Society of Economic Paleontologists and Mineralogists Concepts in Sedimentology and Paleontology, **9**, 226 pp. ISBN: 978-1-56576-288-6.

Adam, J., Ge, Z. & Sanchez, M. 2012. Salt-structural styles and kinematic evolution of the Jequitinhonha deepwater fold belt, central Brazil passive margin. *Marine and Petroleum Geology*, **37**, 101–120.

Adams, J.H. 1908. The eruption of the Waimata mud spring. *New Zealand Mines Record*, **12**, 908–912.

Addy, S.K. & Buffler, R.T. 1984. Seismic stratigraphy of the shelf and slope; northeastern Gulf of Mexico. *American Association of Petroleum Geologists Bulletin*, **68**, 1782–1789.

Addy, S.K. & Kagami, H. 1979. Sedimentation in a closed trough north of the Iberia abyssal plain in Northeast Atlantic. *Sedimentology*, **26**, 561–575.

Ahmadi, Z.M., Sawyers, M., Kenyon-Roberts, S., Stanworth, C.W., Kugler, K.A., Kristensen, J. & Fugelli, E.M.G. 2003. Palaeocene. *In*: Evans, D., Graham, C., Armour, A. & Bathurst, P. (eds), *The Millennium Atlas: Petroleum Geology of the Central and Northern North Sea*. Geological Society, London, 235–259. London: The Geological Society.

Aigner, T., Doyle, M. & Lawrence, D. 1987. Isostatic controls on carbonate platform development. *American Association of Petroleum Geologists Bulletin*, **71**, 524.

Aksu, A.E. 1984. Subaqueous debris flow deposits in Baffin Bay. *Geo-Marine Letters*, **4**, 83–90.

Aksu, A.E. & Hiscott, R.N. 1992. Shingled Upper Quaternary debris flow lenses on the NE Newfoundland slope. *Sedimentology*, **39**, 193–206.

Aksu, A.E., Calon, T.J., Hiscott, R.N. & Yaşar, D. 2000. Anatomy of the North Anatolian Fault Zone in the Marmara Sea, western Turkey: extensional basins above a continental transform. *GSA Today*, **10**, 3–7.

Alexander, J. & Morris, S. 1994. Observations on experimental, non-channelized high-concentration turbidity currents and variations in deposits around obstacles. *Journal of Sedimentary Research*, **A64**, 899–909.

Alexander, J. & Mulder, T. 2002. Experimental quasi-steady density currents. *Marine Geology*, **186**, 195–210.

Allen, J.R.L. 1960. The Mam Tor Sandstones: a 'turbidite' facies of the Namurian deltas of Derbyshire, England. *Journal of Sedimentary Petrology*, **30**, 193–208.

Allen, J.R.L. 1969. Some recent advances in the physics of sedimentation. *Proceedings of the Geologists Association*, **80**, 1–42.

Allen, J.R.L. 1970. The sequence of sedimentary structures in turbidites, with special reference to dunes. *Scottish Journal of Geology*, **6**, 146–161.

Allen, J.R.L. 1971. Instantaneous sediment deposition rates deduced from climbing-ripple cross-lamination. *Journal of the Geological Society (London)*, **127**, 553–561.

Allen, J.R.L. 1982. *Sedimentary Structures: Their Character and Physical Basis. Developments in Sedimentology, 30 (parts I and II)*. Amsterdam: Elsevier.

Allen, J.R.L. 1991, The Bouma division A and the possible duration of turbidity currents: *Journal of Sedimentary Petrology*, **61**, 291–295.

Allen, J.R.L. & Banks, N.L. 1972. An interpretation and analysis of recumbent-folded deformed cross-bedding. *Sedimentology*, **19**, 257–283.

Allen, P.A. & Allen, J.R. 1990. *Basin Analysis: Principles and Applications*. Oxford, England: Blackwell Scientific Publishers, 451 pp.

Allen, P.A. & Densmore, A.L. 2000. Sediment flux from an uplifting fault block. *Basin Research*, **12**, 367–380.

Allen, P.A. & Homewood, P. (eds) 1986. *Foreland Basins*. International Association of Sedimentologists Special Publication, **8**. Oxford: Blackwell Scientific, 453 pp.

Almagor G. & Garfunkel, Z. 1979. Submarine slumping in continental margin of Israel and northern Sinai. *American Association of Petroleum Geologists Bulletin*, **63**, 324–340.

Almagor, G. & Wiseman, G. 1982. Submarine slumping and mass movements on the continental slope of Israel. *In*: Saxov, S. & Nieuwenhuis, J.K. (eds), *Marine Slides and Other Mass Movements*, 95–128. New York: Plenum.

Almgren, A.A. 1978. Timing of Tertiary submarine canyon sand marine cycles of deposition in the southern Sacramento Valley, California. *In*: D.J. Stanley & Kelling, G. (eds), *Sedimentation in Submarine Canyons, Fans, and Trenches*, 276–291. Stroudsburg, PA: Dowden, Hutchinson & Ross.

Deep Marine Systems: Processes, Deposits, Environments, Tectonics and Sedimentation, First Edition. Kevin T. Pickering and Richard N. Hiscott.
© 2016 Kevin T. Pickering and Richard N. Hiscott. Published 2016 by John Wiley & Sons, Ltd.
Companion Website: www.wiley.com/go/pickering/marinesystems

Alpert, S.P. 1974. Systematic review of the genus Skolithos. *Journal of Paleontology*, **48**, 661–669.

Alves, T.M. 2010. 3D Seismic examples of differential compaction in mass-transport deposits and their effect on post-failure strata. *Marine Geology*, **271**, 212–224.

Alves, T.M., Gawthorpe, R.L., Hunt, D.W. & Monteiro, J.H. 2002. Jurassic tectono-sedimentary evolution of the northern Lusitanian Basin (offshore Portugal). *Marine and Petroleum Geology*, **19**, 727–754.

Amy, L.A. & Talling, P.J. 2006. Anatomy of turbidites and linked debrites based on long distance (120 x 30 km) bed correlation, Marnoso Arenacea Formation, northern Apennines, Italy. *Sedimentology*, **53**, 161–212.

Amy, L.A., Kneller, B. & McCaffrey, W. 2000. Evaluating the links between turbidite characteristics and gross system architecture: upscaling insights from the turbidite sheet-system of Peira Cava, SE France. *In*: Weimer, P., Slatt, R.M., Coleman, J., Rosen, N.C., Nelson, H., Bouma, A.H., Styzen, M.J. & Lawrence, D.T. (eds), *Gulf Coast Section-Society of Economic Paleontologists and Mineralogists Foundation 20th Annual Bob F. Hoskins Research Research Conference, Deep-Water Reservoirs of the World*, 1–15. CD-ROM Society of Economic Paleontologists and Mineralogists Special Publications.

Amy, L.A., Kneller, B.C. & McCaffrey, W.D. 2007. Facies architecture of the Grès de Peïra Cava, SE France: landward stacking patterns in ponded turbiditic basins. *Journal of the Geological Society (London)*, **164**, 143–162.

Anastasakis, G.C. & Piper, D.J.W. 2006. The character of seismo-turbidites in the S-1 sapropel, Zakinthos and Strofadhes basins, Greece. *Sedimentology*, **38**, 717–733.

Anderson, J.B., Kurtz, D.D. & Weaver, F.M. 1979. Sedimentation on the Antarctic continental slope. *In*: Doyle, L.J. & Pilkey, O.H. (eds), *Geology of Continental Slopes*, 265–283. Society of Economic Paleontologists and Mineralogists Special Publication, **27**.

Anderson, J.E., Cartwright, J., Drysdall, S.J. & Vivian, N. 2000. Controls on turbidite sand deposition during gravity-driven extension of a passive margin: examples from Miocene sediments in Block 4, Angola. *Marine and Petroleum Geology*, **17**, 1165–1203.

Anderson, K.S., Graham, S.A. & Hubbard, S.M. 2006. Facies, architecture, and origin of a reservoir-scale sand-rich succession within submarine canyon fill: insights from Wagon Caves Rock (Paleocene), Santa Lucia Range, California, U.S.A. *Journal of Sedimentary Research*, **76**, 819–838.

Anderson, L.D. & Ravelo, A.C. 2001. Data report: biogenic opal in Palmer Deep sediments, Site 1098, Leg 178. *In*: Barker, P.F., Camerlenghi, A., Acton, G.D. & Ramsay, A.T.S. (eds), *Proceedings of the Ocean Drilling Program, Scientific Results*, **178**, 1–7. College Station, Texas, USA: Ocean Drilling Program.

Anderson, T.B. & Oliver, G.T.H. 1986. The Orlock Bridge Fault: a major Late Caledonian sinistral fault in the Southern Uplands terrane, British Isles. *Transactions of the Royal Society of Edinburgh, Earth Sciences*, **77**, 203–222.

Andersson, P.O.D., Worden, R.H., Hodgson, D.M. & Flint, S. 2004. Provenance evolution and chemostratigraphy of a Palaeozoic submarine fan-complex: Tanqua Karoo Basin, South Africa. *Marine and Petroleum Geology*, **21**, 555–577.

Anderton, R. 1995. Sequences, cycles and other nonsense: are submarine fan models any use in reservoir geology: *In*: Hartley, A.J. & Prosser, D.J. (eds), *Characterization of Deep Marine Clastic Systems*, Geological Society London Special Publication, **94**, 5–11.

Ando, M. 1975. Source mechanisms and tectonic significance of historical earthquakes along the Nankai Trough, Japan. *Tectonophysics*, **27**, 119–140.

Andresen, N., Reijmer, J.J.G. & Droxler, A.W. 2003. Timing and distribution of calciturbidites around a deeply submerged carbonate platform in a seismically active setting (Pedro Bank, Northern Nicaragua Rise, Caribbean Sea). *International Journal of Earth Sciences (Geologische Rundschau)*, **92**, 573–592.

Andreson, A. & Bjerrum, L. 1967. Slides in subaqueous slopes in loose sand and silt. *In*: Richards, A.F. (ed.), *Marine Geotechnique*, 221–239. Urbana: University of Illinois Press.

Andrews, J.T. & Tedesco, K. 1992. Detrital carbonate-rich sediments, northwestern Labrador Sea: implications for ice-sheet dynamics and iceberg rafting (Heinrich) events in the North Atlantic. *Geology*, **20**, 1087–1090.

Anka, A. & Séranne, M. 2004. Reconnaissance study of the ancient Zaire (Congo) deep-sea fan (ZaiAngo Project). *Marine Geology*, **209**, 223–244.

Anka, Z., Séranne, M., Lopez, M., Scheck-Wenderoth, M. & Savoye, B. 2009. The long-term evolution of the Congo deep-sea fan: a basin-wide view of the interaction between a giant submarine fan and a mature passive margin (ZaiAngo project). *Tectonophysics*, **470**, 42–56.

Anselmetti, F.S., Eberli, G.P. & Ding, Z.-D. 2000. From the Great Bahama Bank into the Straits of Florida: A margin architecture controlled by sea-level fluctuations and ocean currents. *Geological Society of America Bulletin*, **112**, 829–844.

Aoki, Y., Tamano, T. & Kato, S. 1983. Detailed structure of the Nankai Trough from migrated seismic sections. *In*: Watkins, J.S. & Drake, C.L. (eds), *Studies in Continental Margin Geology*, 309–322. American Association of Petroleum Geologists. Memoir, 34.

Appelgate, B., Goldfinger, C., MacKay, M.E., Kulm, L.D., Fox, C.G., Embley, R.W. & Meis, P.J. 1992. A left-lateral strike-slip fault seaward of the Oregon continental margin. *Tectonics*, **11**, 465–477.

Apps, G., Peel, F. & Elliott, T. 2004. The structural setting and palaeogeographical evolution of the Grès d'Annot Basin. *In*: Joseph, P. & Lomas, S.A. (eds), *Deep-Water Sedimentation in the Alpine Basin of SE France: New Perspectives on the Grès d'Annot and Related Systems*. Geological Society, London, Special Publication, **221**, 65–96. London: The Geological Society.

Arbués, P., Granado, P., De Mattheis, M., Cabello, P., Lopez-Blanco, M., Marzo, M., Muñoz, J.A. & Abreu, V. 2013. An integrated outcrop and subsurface study of the Solitary Channel Complex (Tabernas Basin, Spaion). *In*: *30th International Association of Sedimentologists Meeting of Sedimentology, University of Manchester, U.K., 2–5th September 2013, Conference Abstract Volume*. T3S5-O9.

Archer, J.B. 1984. Clastic intrusions in deep-sea fan deposits of the Rosroe formation, lower Ordovician, western Ireland. *Journal of Sedimentary Petrology*, **54**, 1197–1205.

Armentrout, J.M., Kanschat, K.A., Meisling, K.E., Tsakma, J.J., Antrim, L. & McConnell, D.R. 2000. Neogene turbidite systems of the Gulf of Guinea continental margin slope, Offshore Nigeria. *In*: Bouma, A.H. & Stone, C.G. (eds), *Fine-grained Turbidite Systems*, 93–108. American Association of Petroleum Geologists Memoir, **72** & Society of Sedimentary Geologists, Special Publication, **68**. Joint publication, Tulsa, Oklahoma.

Armijo, R., Meyer, B., Navarro, S., King, G. & Barka, J. 2002. Asymmetric slip partitioning in the Sea of Marmara pull-apart: a clue to propagation processes of the North Anatolian Fault? *Terra Nova*, **14**, 80–86.

Armishaw, J.E., Holmes, R.W. & Stow, D.A.V. 2000. The Barra Fan: a bottom-current reworked, glacially-fed submarine fan system. *Marine Geology*, **17**, 219–238.

Armitage, D.A., Piper, D.J.W., McGee, D.T. & Morris, W.R. 2010. Turbidite deposition on the glacially influenced, canyon-dominated Southwest Grand Banks Slope, Canada. *Sedimentology*, **57**, 1387–1408.

Armitage, D.A. & Stright, L. 2010. Modeling and interpreting the seismic-reflection expression of sandstone in an ancient mass-transport deposit dominated deep-water slope environment. *Marine and Petroleum Geology*, **27**, 1–12.

Arnold, R. & Macready, G.A. 1956. Island-forming mud volcano in Trinidad, British West Indies. *American Association of Petroleum Geologists Bulletin*, **40**, 2748–2758.

Arnott, R.J. 1983a. *Sedimentology, structure and stratigraphy of north-east New World Island, Newfoundland*. Unpublished Ph.D. Thesis, Oxford University, UK.

Arnott, R.J. 1983b. Sedimentology of Upper Ordovician – Silurian sequences on New World Island, Newfoundland: separate fault-controlled basins? *Canadian Journal of Earth Sciences*, **20**, 345–354.

Arnott, R.J., McKerrow, W.S. & Cocks, L.R.M. 1985. The tectonics and depositional history of the Ordovician and Silurian rocks of Notre Dame Bay, Newfoundland. *Canadian Journal of Earth Sciences*, **22**, 60–618.

Arnott, R.W. & Hein, F.J. 1986. Submarine canyon fills of the Hector Formation, Lake Louise, Alberta: Late Precambrian syn-rift deposits of the proto-Pacific miogeocline. *Bulletin of Canadian Petroleum Geology*, **34**, 395–407.

Arnott, R.W.C. 2007. Stratal architecture and origin of lateral accretion deposits (LADS) and conterminuous inner-bank levée deposits in a base-of-slope sinuous channel, lower Isaac formation (Neoproterozoic), east-central British Columbia, Canada. *Marine and Petroleum Geology*, **24**, 515–528.

Arnott, R.W.C. & Hand, B.M. 1989. Bedforms, primary structures and grain fabric in the presence of suspended sediment rain. *Journal of Sedimentary Petrology*, **59**, 1062–1069.

Arthur, M.A. & Natland, J.H. 1979. Carbonaceous sediments in North and South Atlantic: the role of salinity in stable stratification of Early Cretaceous basins. *In*: Talwani, M., Hay, M.W. & Ryan, W.B.F. (eds), *Deep Drilling Results in the Atlantic Ocean: Continental Margins and Paleoenvironment*, 375–401. Maurice Ewing Series **3**. Washington: American Geophysical Union.

Arthur, M.A. & Schlanger, S.O. 1979. Cretaceous "Oceanic Anoxic Events" as causal factors in development of reef-reservoired giant oil fields. *The American Association of Petroleum Geologists Bulletin*, **63**, 870–885.

Arthur, M.A., Dean, W.E. & Stow, D.A.V. 1984. Models for the deposition of Mesozoic-Cenozoic fine-grained organic-carbon-rich sediment in the deep sea. *In*: Stow, D.A.V. & Piper, D.J.W. (eds), *Fine-grained Sediments: Deep-water Processes and Facies*, 527–560. Geological Society of London Special Publication, **15**. Oxford: Blackwell Scientific.

Arrhenius, G. 1963. Pelagic sediments. *In*: Hill, M.N. (ed.), *The Sea*, **3**, 655–727. New York: Wiley.

Ashi, J., Lallemant, S., Masago, H. & the Expedition 315 Scientists 2008. NanTroSEIZE Stage 1A: NanTroSEIZE megasplay riser pilot. *IODP Preliminary Report*, **315**. doi:10.2204/iodp.pr.315.2008.

Aswasereelert, W., Meyers, S.R., Carroll, A.R., Peters, S.E., Smith, M.E. & Feigl, K.L. 2013. Basin-scale cyclostratigraphy of the Green River Formation, Wyoming. *Geological Society of America Bulletin*, **125**, 216–228.

Athanasiou-Grivas, D. 1978. Reliability analysis of earth slopes. *Proceedings of the Society of Engineering Sciences, 15th Annual Meeting*, 453–458, Gainesville, University of Florida.

Athmer, W., Gonzalez Uribe, G.A., Luthi, S.M. & Donselaar, M.E. 2011. Tectonic control on the distribution of Palaeocene marine syn-rift deposits in the Fenris Graben, northwestern Vøring Basin, offshore Norway. *Basin Research*, **23**, 361–375.

Atwater, T. 1970. Implications of plate tectonics for the Cenozoic evolution of western North America. *Geological Society of America Bulletin*, **81**, 3515–3536.

Atwater, T.M. 1998. Plate tectonic history of Southern California with emphasis on the Western Transverse Ranges and Santa Rosa Island. *In*: Weigand, P.W. (ed.), *Contributions to the Geology of the Northern Channel Islands, Southern California*. American Association of Petroleum Geologists, Pacific Section, MP **45**, 1–8.

Atwater, T & Molnar, P. 1973. Relative motion of the Pacific and North American Plates deduced from seafloor spreading in the Atlantic, Indian and South Pacific Oceans. *In*: Kovach, R.L. & Nur, A. (eds), *Proceedings of the Conference on Tectonic Problems in the San Andreas Fault System,* 139–148. Stanford University Geological Science Publication, **13**.

Atwater, T. & Stock, J. 1998. Pacific – North America place tectonics of the Neogene southwestern United States; an update. *International Geology Review*, **40**, 375–402.

Aubouin, J., Bourgois, J., von Huene, R. & Azema, J. 1982a. La marge pacifique du Guatemala: un modele de marge extensive en domaine convergent. *Comptes Rendus de l'Académie des Sciences, Paris*, **294**, 607–614.

Aubouin, J., Stephan, J.F., Roump, J. & Renard, V. 1982b. The Middle America Trench as an example of a subduction zone. *Tectonophysics*, **86**, 113–132.

Aubouin, J., von Huene, R., Baltuck, M., Arnott, R., Bourgois, J. *et al.* 1982c. Leg 84 of the Deep Sea Drilling Project, subduction without accretion: Middle America Trench off Guatemala. *Nature*, **297**, 458–460.

Aubouin, J., Bourgois, J. & Azema, J. 1984. A new type of active margin: the convergent extensional margin, as exemplified by the Middle America Trench off Guatemala. *Earth and Planetary Science Letters*, **67**, 211–218.

Audley-Charles, M.G. 1968. *The Geology of Portuguese Timor. Geological Society of London Memoir*, **4**.

Audley-Charles, M.G. 1986a. Rates of Neogene and Quaternary tectonic movements in the Southern Banda Arc based on micropalaeontology. *Journal of the Geological Society, London*, **143**, 161–175.

Audley-Charles, M.G. 1986b. Timor-Tanimbar Trough: the foreland basin of the evolving Banda Orogen. *In*: Allen, P.A. & Homewood, P. (eds), *Foreland Basins*, 91–102. Special Publication International Association of Sedimentologists, **8**. Oxford, UK: Blackwell Publishing Ltd.

Ausich, W.I. & Bottjer, D.J. 1982. Tiering in suspension-feeding communities on soft substrata throughout the Phanerozoic. *Science*, **9**, 173–174.

Australian Geological Survey Organisation (AGSO) North West Shelf Study Group 1994. Deep reflections on the North West Shelf: changing perspectives of basin formation. *In*: Purcell, P.G. & Purcell, R.R. (eds), *The Sedimentary Basins of Western Australia*, 63–74. Perth, Australia: Petroleum Exploration Society of Australia.

Awadallah, S.A.M. 2002. *Architecture and depositional history of the lower Cloridorme Formation, Gaspé Peninsula, Quebec, Canada*. Unpublished PhD Thesis, Memorial University of Newfoundland, St John's, Canada, 376 pp.

Awadallah, S.A.M. & Hiscott, R.N. 2004. High-resolution stratigraphy of the deep-water lower Cloridorme Formation (Ordovician), Gaspé Peninsula, based on K-bentonite and megaturbidite correlations. *Canadian Journal of Earth Sciences*, **41**, 1299–1317.

Awadallah, S.A.M., Hiscott, R.N., Bidgood, M. & Crowther, T.E. 2001. Turbidite facies and bed-thickness characteristics inferred from microresistivity (FMS) images of lower to upper Pliocene rift-basin deposits, Woodlark Basin, offshore Papua New Guinea. *In*: Huchon, P., Taylor, B. & Klaus, A. (eds), *Proceedings of the Ocean Drilling Program, Scientific Results*, **180**, 1–29. College Station, Texas, USA: Ocean Drilling Program.

Azpeitia M.F. 1933. Datos para el studio paleontólogico del Flysch de la Costa Cantábrica y de algunos otros puntos de España. *Boletín. Instituto Geológico y Minero de España*, **53**, 1–65.

Aydin, A. & Page, B.M. 1984. Diverse Pliocene–Quaternary tectonics in a transform environment, San Francisco Bay region, California. *Geological Society of America Bulletin*, **95**, 1303–1317.

Baas, J.H. & Best, J.L. 2002. Turbulence modulation in clay-rich sediment-laden flows and some implications for sediment deposition. *Journal of Sedimentary Research*, **72**, 336–340.

Baas, J.H., Van Dam, R.L. & Storms, J.E.A. 2000. Duration of deposition from decelerating high-density turbidity currents. *Sedimentary Geology*, **136**, 71–88.

Baas, J.H., Best, J.L. & Peakall, J. 2011. Depositional processes, bedform development and hybrid bed formation in rapidly decelerated cohesive (mud-sand) sediment flows. *Sedimentology*, **58**, 1953–1987.

Baba, T. & Cummins, P.R. 2005. Contiguous rupture areas of two Nankai Trough earthquakes revealed by high-resolution tsunami waveform inversion. *Geophysical Research Letters*, **32**, L08305.

Baba, T., Cummins, P.R. & Hori, T. 2005. Compound fault rupture during the 2004 off the Kii Peninsula earthquake (M 7.4) inferred from highly resolved coseismic sea-surface deformation. *Earth, Planets and Space,* **57**, 167–172.

Baba, T., Cummins, P.R., Hori, T. & Kaneda, Y. 2006. High precision slip distribution of the 1944 Tonankai earthquake inferred from tsunami waveforms: possible slip on a splay fault. *Tectonophysics,* **426**, 119–134.

Baba, T., Cummins, T.R., Thio, H.K. & Tsushima, H. 2009. Validation and joint inversion of teleseismic waveforms for earthquake source models using deep ocean bottom pressure records: a case study of the 2006 Kuril megathrust earthquake. *Pure and Applied Geophysics,* **166**, 55–76.

Bábek, O., Mikuláš, R., Zapletal, J. & Lehotský, T. 2004. Combined tectonic-sediment supply-driven cycles in a Lower Carboniferous deep-marine foreland basin, Moravice Formation, Czech Republic. *International Journal of Earth Sciences,* **93**, 241–261.

Babić, L. & Župančić, J. 2008. Evolution of a river-fed foreland basin fill: the North Dalmatian flysch revisited (Eocene, Outer Dinarides). *Nature Croatica,* **17**, 357–374.

Babonneau, N., Savoye, B., Cremer, M. & Klein, B. 2002. Morphology and architecture of the present canyon and channel system of the Zaire deep-sea fan. *Marine and Petroleum Geology,* **17**, 445–467.

Backert, N., Ford, M. & Malartre, M. 2010. Architecture and sedimentology of the Kerinitis Gilbert-type fan delta, Corinth Rift, Greece. *Sedimentology,* **57**, 543–586.

Bacon, C.R. 1983. Eruptive history of Mount Mazama and Crater Lake Caldera, Cascade Range, U.S.A. *Journal of Volcanology & Geothermal Research,* **18**, 57–115.

Badalini, G., Kneller, B. & Winker, C.D. 2000. Architecture and processes in the Late Pleistocene Brazos-Trinity turbidite system, Gulf of Mexico continental slope. *In*: Weimer, P., Slatt, R.M., Coleman, J., Rosen, N.C., Nelson, H., Bouma, A.H., Styzen, M.J. & Lawrence, D.T. (eds), *Gulf Coast Section-Society of Economic Paleontologists and Mineralogists Foundation 20th Annual Bob F. Hoskins Research Conference, Deep-Water Reservoirs of the World,* 16–34. Houston, Texas: Gulf Coast Section, Society of Economic Paleontologists and Mineralogists.

Badescu, M.O., Visser, C.A. & Donselaar, M.E. 2000. Architecture of thick-bedded deep-marine sandstones of the Vocontian Basin, SE France. *In*: Weimer, P., Slatt, R.M., Coleman, J., Rosen, N.C., Nelson, H., Bouma, A.H., Styzen, M.J. & Lawrence, D.T. (eds), *Gulf Coast Section-Society of Economic Paleontologists and Mineralogists Foundation 20th Annual Bob F. Hoskins Research Research Conference, Deep-Water Reservoirs of the World,* 35–39. Houston, Texas: Gulf Coast Section, Society of Economic Paleontologists and Mineralogists.

Bagnold, R.A. 1956. The flow of cohesionless grains in fluids. *Philosophical Transactions of the Royal Society London (A),* **249**, 235–297.

Bagnold, R.A. 1962. Auto-suspension of transported sediment: turbidity currents. *Proceedings of the Royal Society London (A),* **265**, 315–319.

Bagnold, R.A. 1966. An approach to the sediment transport problem from general physics. *U.S. Geological Survey Professional Paper,* 422–I.

Bahk, J.J., Chough, S.K. & Han, S.J. 2000. Origins and paleoceanographic significance of laminated muds from the Ulleung Basin, East Ses (Sea of Japan). *Marine Geology,* **162**, 459–477.

Bak, K. 1995. Trace fossils and ichnofabrics in the Upper Cretaceous red deep-water marly deposits of the Pieniny Klippen Belt, Polish Carpathians. *Annales Societatis Geologorum Poloniae,* **64**, 81–97.

Bailey, E.B. & Weir, J. 1932. Submarine faulting in Kimmeridgian times: east Sutherland. *Transactions of the Royal Society of Edinburgh,* **47**, 431–467.

Bakke, K., Gjelberg, J. & Petersen, S.A. 2008. Compound seismic modeling of the Ainsa II turbidite system, Spain: Application to deep-water channel systems offshore Angola. *Marine and Petroleum Geology,* **25**, 1058–1073.

Baltuck, M., Taylor, E. & McDougall, K. 1985. Mass movement along the inner wall of the Middle America Trench, Costa Rica. *In*: von Huene, R. Aubouin, J. *et al., Initial Reports Deep Sea Drilling Project,* **84**, 551–570. Washington, DC: US Government Printing Office.

Bandy, O.L. 1953. Ecology and paleoecology of some California foraminifera. *Journal of Paleontology,* **27**, 161–203.

Bangs, N.L.B. & Westbrook, G.K. 1991. Seismic modeling of the décollement zone at the base of the Barbados Ridge accretionary complex. *Journal of Geophysical Research,* **96**, 3853–3866.

Bangs, N.L.B., Westbrook, G.K., Ladd, J.W. & Buhl, P. 1990. Seismic velocities from the Barbados Ridge Complex: indicators of high pore fluid pressures in an accretionary complex. *Journal of Geophysical Research,* **95**, 8767–8782.

Bangs, N.L., Shipley, T.H., Moore, J.C. & Moore, G. 1999. Fluid accumulations and channeling along the Northern Barbados Ridge décollement thrust. *Journal of Geophysical Research,* **104**, 20399–20414.

Barber, A.J., Tjokrosapoetro, S. & Charlton, T.R. 1986. Mud Volcanoes, Shale Diapirs, Wrench Faults, and Melanges in Accretionary Complexes, Eastern Indonesia. *American Association of Petroleum Geologists Bulletin,* **70**, 1729–1741.

Barker, P.F. & Camerlenghi, A. 2002. Glacial history of the Antarctic Peninsula from Pacific margin sediments. *In*: Barker, P.F., Camerlenghi, A., Acton, G.D. & Ramsay, A.T.S. (eds), *Proceedings of the Ocean Drilling Program, Scientific Results,* **178**, 1–40. College Station, Texas, USA: Ocean Drilling Program.

Barker, P.F., Camerlenghi, A., Acton, G.D. *et al.* 1999. *Proceedings of the Ocean Drilling Program, Scientific Results,* **178**. College Station, Texas, USA: Ocean Drilling Program.

Barnes, N.E. & Normark, W.R. 1984. Diagnostic parameters for comparing modern submarine fans and ancient turbidite systems. *Geo-Marine Letters,* **3**, map following p. 224.

Barnes, P.M. & Audru, J.-C. 1999. Recognition of active strike-slip faulting from high-resolution marine seismic reflection profiles: Eastern Marlborough fault system, New Zealand. *Geological Society of America Bulletin,* **111**, 538–559.

Barnes, P.M. & Lewis, K.B. 1991. Sheet slides and rotational failures on a convergent margin: the Kidnappers Slide, New Zealand. *Sedimentology,* **38**, 205–221.

Barnes, C.R., Norford, B.S. & Skevington, D. 1981. *The Ordovician System in Canada: Correlation Chart and Explanatory Notes.* International Union of Geological Sciences Publication, **8**. Paris: IUGS Secretariat.

Barnes, P.M., Sutherland, R., Davy, B. & Delteil, J. 2001. Rapid creation and destruction of sedimentary basins on mature strike-slip faults: an example from the offshore Alpine fault, Fiordland, New Zealand. *Journal of Structural Geology,* **23**, 1727–1739.

Barnes, P.M., Sutherland, R. & Delteil, J. 2005. Strike-slip structure and sedimentary basins of the southern Alpine Fault, Fiordland, New Zealand. *Geological Society of America Bulletin,* **117**, 411–435.

Barnolas, A. & Teixell, A. 1994. Platform sedimentation and collapse in a carbonate-dominated margin of a foreland basin (Jaca basin, Eocene, southern Pyrenees). *Geology,* **22**, 1107–1110.

Bartow, J.A. 1966. Deep submarine channel in upper Miocene, Orange County, California. *Journal of Sedimentary Petrology,* **36**, 700–705.

Bassinot, F.C., Labeyrie, L.D., Vincent, E., Quidelleur, X., Shackleton, N.J. & Lancelot, Y. 1994. The astronomical theory of climate and the age of the Bruhnes-Matuyama magnetic reversal. *Earth and Planetary Science Letters,* **126**, 91–108.

Baturin, G.N. 1982. *Phosphorites on the Seafloor: Origin, Composition and Distribution. Developments in Sedimentology,* **33**. Amsterdam: Elsevier.

Bayliss, N.J. & Pickering, K.T. 2015a. Transition from deep-marine lower-slope erosional channels to proximal basin floor stacked channel-levée-overbank deposits, and syn-sedimentary growth structures, Middle Eocene Banastón System, Ainsa basin, Spanish Pyrenees. *Earth-Science Reviews,* **144**, 23–46.

Bayliss, N.J. & Pickering, K.T. 2015b. Deep-marine structurally confined channelised sandy fans: Middle Eocene Morillo System, Ainsa Basin, Spanish Pyrenees. *Earth-Science Reviews,* **144**, 82–106.

Batzle, M. & Gardner, M.H. 2000. Lithology and fluid effects on outcrop seismic models of the Permian Brushy Canyon Formation, Guadalupe Mountains, west Texas. *In*: Bouma, A.H., Stelting, C.E. & Stone, C.G.

(eds), *Fine-Grained Turbidite Systems and Submarine Fans,* American Association of Petroleum Geologists Memoir 72/Society of Economic Paleontologists and Mineralogists Special Publication.

Baztan, J., Berné, S., Olivet, J.-L., Rabineau, M., Aslanian, D., Gaudin, M., Réhault, J.-P. & Canals, M. 2005. Axial incision: the key to understand submarine canyon evolution (in the western Gulf of Lion). *Marine and Petroleum Geology,* **22,** 805–826.

Beach, A. 1986. A deep seismic reflection profile across the northern North Sea. *Nature,* **323,** 53–55.

Beattie, P.D. & Dade, W.B. 1996. Is scaling in turbidite deposition consistent with forcing by earthquakes? *Journal of Sedimentary Research,* **A66,** 909–915.

Beaudry, D. & Moore, G.F. 1985. Seismic stratigraphy and Cenozoic evolution of West Sumatra forearc basin. *American Association of Petroleum Geologists Bulletin,* **69,** 742–759.

Beaubouef, R.T. 2004. Deep-water levéed-channel complexes of the Cerro Toro Formation, Upper Cretaceous, southern Chile. *American Association of Petroleum Geologists Bulletin,* **88,** 1471–1500.

Beaubouef, R.T. & Friedman, S.J. 2000. High resolution seismic/sequence stratigraphic framework for the evolution of Pleistocene intra slope basins, western Gulf of Mexico: depositional models and reservoir analogs. *In:* Weimer, P., Slatt, R.M., Coleman, J., Rosen, N.C., Nelson, H., Bouma, A.H., Styzen, M.J. & Lawrence, D.T. (eds), *Deep-water Reservoirs of the World,* 40–60. Houston, Texas: Gulf Coast Section, Society of Economic Paleontologists and Mineralogists.

Beaubouef, R.T., Rossen, C., Zelt, F.B., Sullivan, M.D., Mohrig, D.C. & Jennette, D.C. 1999. *Deep-water Sand-stones, Brushy Canyon Formation, West Texas: Field Guide.* American Association of Petroleum Geologists Hedberg Field Research Conference.

Beauboeuf, R.T., Rossen, C., Sullivan, M.D., Mohrig, D.C. & Jennette, D.C. 2000. Deep-water sandstones, Brushy Canyon Formation, West Texas. *In: AAPG Hedberg Field Research Conference, American Association of Petroleum Geologists, Studies in Geology,* 1.2–3.9.

Beaubouef, R.T., Abreu, V. & Van Wagoner, J.C. 2003. Basin 4 of the Brazos-Trinity slope system, western Gulf of Mexico: the terminal portion of a late Pleistocene lowstand systems tract. *In:* Roberts, H.H., Rosen, N.C., Fillon, R.H. & Anderson, J.B. (eds), *Gulf Coast Section-SEPM Foundation 23rd Annual Bob F. Perkins Research Conference,* 182–203. Tulsa: Society of Economic Paleontologists and Mineralogists.

Beaumont, C., Keen, C.E. & Bantillier, R. 1982. A comparison of foreland and rift margin sedimentary basins. *In:* Kent, P., Bott, M.H.P., McKenzie, D.P. & Williams, C.A. (eds), *The Evolution of Sedimentary Basins,* 295–318. London: The Royal Society.

Bebout, G.E., Scholl, D.W., Kirby, S.H. & Platt, P. (eds). 1996. *Subduction: Top to Bottom.* American Geophysical Union, Geophysical Monograph, 96, 384 pp. Washington, DC: American Geophysical Union. ISBN: 0-87590-078-X

Bednarz, M. & McIlroy, D. 2009. Three-dimensional reconstruction of "phycosiphoniform" burrows: implications for identification of trace fossils in core. *Palaeontologia Electronica,* **12,** 13A, 15 pp.

Beekman, F., Bull, J.M., Cloetingh, S. & Scrutton, R.A. 1995. Crustal fault reactivation as indicator of lithospheric folding in the Central Indian Ocean. *In:* Nieuwland, D. (ed.), *Modern Examples in Structural Interpretation, Validation and Modelling,* 251–253. Geological Society London, Special Publication, 99.

Beglinger, S.E., Doust, H. & Cloetingh, S. 2012. Relating petroleum system and play development to basin evolution: West African South Atlantic basins. *Marine and Petroleum Geology,* **30,** 1–25.

Bein, A. & Weiler, Y. 1976. The Cretaceous Talme Yafe Formation, a contour current shaped sedimentary prism of carbonate debris at the continental margin of the Arabian craton. *Sedimentology,* **23,** 511–532.

Bekins, B.A., McCaffrey, A.M. & Driess, S.J. 1995. Episodic and constant flow models for the origin of low-chloride waters in a modern accretionary complex. *Water Resource Research,* **31,** 3205–3215.

Belderson, R.H., Kenyon, N.H., Stride, A.H. & Pelton, C.D. 1984. A braided distributary system on the Orinoco deep-sea fan. *Marine Geology,* **56,** 195–206.

Benediktsson, S. 2013. *Paleoenvironmental Reconstruction from Marine Core JM05-30-GC2-1, North West Svalbard Slope.* BS ritgero Jarovisindadeild, Háskóli Íslands, 31 pp.

Benevelli, G., Angella, S., Fava1, L., Rocchini, P. & Valdisturlo, A. 2003. From a Field Based Geological Model to a Seismic Image: The Middle Eocene Ainsa System (South-Central Pyrenees). *American Association of Petroleum Geologists International Conference, Barcelona, Spain, September 21–24, 2003. Programme with Abstracts.*

Bennett, R.H. & Nelson, T.A. 1983. Seafloor characteristics and dynamics affecting geotechnical properties at shelfbreaks. *In:* Stanley, D.J. & Moore, G.T. (eds), *The Shelfbreak: Critical Interface on Continental Margins,* 333–355. Society of Economc Paleontologists and Mineralogists, Special Publication, 33.

Benson, L., Kashgarian, M., Rye, R., Lund, S., Paillet, F., Smoot, J., Kester, C., Mensing, S., Meko, D. & Lindstrom, S. 2002. Holocene multidecadal and multicentennial droughts affecting Northern California and Nevada. *Quaternary Science Reviews,* **21,** 659–682.

Bentham, P.A., Burbank, D.W. & Puigdefabregas, C. 1992. Temporal and spatial controls on the alluvial architecture of an axial drainage system: Late Eocene Escanilla Formation, southern Pyrenean foreland basin: Spain. *Basin Research,* **4,** 335–352.

Bentor, Y.K. (ed.) 1980. *Marine Phosphorites-Geochemistry, Occurrence, Genesis.* Society of Economic Paleontologists and Mineralogists Special Publication, **29.**

Berger, A. & Loutre, M.F. 1991. Insolation values for the climate of the last 10 million of years. *Quaternary Science Reviews,* **10**(4), 297–317.

Berger, A., Loutre, M-F. & Laskar, J. 1992. Stability of the astronomical frequencies over the Earth's history for paleoclimatic studies. *Science,* **255,** 560–566.

Berger, D. & Jokat, W. 2009. Sediment deposition in the northern basins of the North Atlantic and characteristic variaitons in shelf sedimentation along the East Greenland margin. *Marine and Petroleum Geology,* **26,** 1321–1337.

Berger, W.H. 1974. Deep-sea sedimentation. *In:* Burk, C.A, & Drake, C.L. (eds), *The Geology of Continental Margins,* 213–241. New York: Springer.

Berger, W.H. & Piper, D.J.W. 1972. Planktonic foraminifera: differential settling, dissolution and redeposition. *Limnolgy and Oceanography,* **17,** 275–287.

Berger, W.H., Ekdale, A. & Bryant, P.P. 1979. Selective preservation of burrows in deep-sea carbonates. *Marine Geology,* **32,** 205–230.

Bergslien, D., Kyllingstad, G., Solberg, A., Ferguson, I.J. & Pepper, C.F. 2005. Jotun Field reservoir geology and development strategy: pioneering play knowledge, multidisciplinary teams and partner co-operation-key to discovery and successful development. *In:* Doré, A.G. & Vining, B.A. (eds), *Petroleum Geology: North-west Europe and Global Perspectives-Proceedings of the 6th Petroleum Geology Conference, 1,* 99–110. London: The Geological Society.

Bernhard, J.M. 1986. Characteristic assemblages and morphologies of benthic foraminifera from anoxic, organic-rich deposits: Jurassic through Holocene. *Journal of Foraminiferal Research,* **16,** 207–215.

Bernhardt, A., Jobe, Z.R. & Lowe, D.R. 2011. Stratigraphic evolution of a submarine channel–lobe complex system in a narrow fairway within the Magallanes foreland basin, Cerro Toro Formation, southern Chile. *Marine and Petroleum Geology,* **28,** 785–806.

Berry, R.F. & Grady, A.E. 1981a. The age of major orogenesis in Timor. *In:* Barber, A.J. & Wiryosujono, S. (eds), *The Geology and Tectonics of Eastern Indonesia,* 171–181. Geological Research and Development Centre, Republic of Indonesia, Special Publication, **2.**

Berry, R.F. & Grady, A.E. 1981b. Deformation and metamorphism of the Aileu Formation, north coast, East Timor and its tectonic significance. *Journal of Structural Geology,* **3,** 143–167.

Berryhill, Jr., H.L. 1981. Ancient buried submarine trough, northwest Gulf of Mexico. *Geo-Marine Letters,* **1,** 105–110.

Bertello, F., Fantoni, R. Franciosi, R., Gatti, V., Ghielmi, M. & Pugliese, A. 2010. From thrust-and-fold belt to foreland: hydrocarbon occurrences in Italy. *In*: Vining, B.A. & Pickering, S.C. (eds), *Petroleum Geology: From Mature Basins to New Frontiers-Proceedings of the 7th Petroleum Geology Conference,* 113–126 Bath: Geological Society, London.

Bertrand, M. 1897. Structure des alpes français et récurrence de certaines facies sédimentaires. *International Geological Congress, 6th Session, 1984, Comptes Rendus,* 161–177.

Best J.L., Kostaschuk, R.A., Peakall, J., Villard, P.V. & Franklin, M. 2005. Whole flow field dynamics and velocity pulsing within natural sediment-laden underflows. *Geology,* 33, 765–768.

Berthois, L. & Le Calves, Y. 1960. Etude de la vitesse de chute des coquilles de foraminiferes planctoniques dans un fluid com-parativement a celle de grains de quartz. *Institute Scientifique et Technique des Peches Maritimes,* 24, 293–301.

Betzler, C., Saxena, S., Swart, P.K., Isern, A. & James, N.P. 2005. Cool-water carbonate sedimentology and eustasy; Pleistocene upper slope environments, Great Australian Bight (Site 1127, ODP Leg 182). *Sedimentary Geology,* 175, 169–188.

Beverage, J.P. & Culbertson, J.K. 1964. Hyperconcentrations of suspended sediment. *Journal of Hydraulics Division, American Society of Civil. Engineers,* 90, 117–126.

Biju-Duval, B., Le Quellec, P., Mascle, A., Renard, V. & Valery, P. 1982. Multibeam bathymetric survey and high resolution seismic investigations on the Barbados Ridge Complex (Eastern Caribbean): a key to the knowledge and interpretation of an accretionary wedge. *Tectonophysics,* 80, 275–304.

Biju-Duval, B., Moore, J.C. *et al.* 1984. *Initial Reports of the Deep Sea Drilling Program,* 78A. Washington, DC: US Government Printing Office.

Birchwood, K.M. 1965. Mud volcanoes in Trinidad. *Institute of Petroleum Review,* 19, 164–167.

Birkeland, P.W. 1964. Pleistocene glaciation of the northern Sierra Nevada, north of Lake Tahoe, California. *Journal of Geology,* 72, 810–825.

Biscaye, P.E. & Eittreim, S.L. 1977. Suspended particulate loads and transports in the nepheloid layer of the abyssal Atlantic Ocean. *Marine Geology,* 23, 155–172.

Biscaye, P.E., Gardner, W.D., Zaneveld, J.R.V., Pak, H. & Tucholke, B. 1980. Nephels! Have we got nephels! *EOS, Transactions of the American Geophysical Union,* 61, 1014.

Blackbourn, G.A. & Thomson, M.E. 2000. Britannia Field, UK North Sea: petrographic constraints on Lower Cretaceous provenance, facies and the origin of slurry-flow deposits. *Petroleum Geoscience,* 6, 329–343.

Blaich, O.A. & Ersdal, G.A. 2011a. South West Barents Sea: complex structuring and hydrocarbon migration revealed by regional multiclient 3D. *GEO ExPro,* 8, 58–62.

Blaich, O.A. & Ersdal, G.A. 2011b. South-western Barents Sea. *GEO ExPro,* 8, 38–40.

Blake, M.C. Jr, Campbell, R.H., Dibblee, T.W., Howell, D.G., Nilsen, T.H., Normark, W.R., Vedder, J.C. & Silver, E.A. 1978. Neogene basin formation in relation to plate-tectonic evolution of San Andreas Fault system, California. *American Association of Petroleum Geologists Bulletin,* 62, 344–372.

Blatt, H.G., Middleton, G.V. & Murray, R.C. 1980. *Origin of Sedimentary Rocks,* 2nd edn. Englewood Cliffs, New Jersey: Prentice-Hall.

Blewett, R.S. & Pickering, K.T. 1988. Sinistral shear during Acadian deformation in north-central Newfoundland, based on transecting cleavage. *Journal of Structural Geology,* 10, 125–127.

Bluck, B.J. 1985. The Scottish paratectonic Caledonides. *Scottish Journal of Geology,* 21, 437–464.

Bluck, B.J. & Leake, B.E. 1986. Late Ordovician to Early Silurian amalgamation of the Dalradian and adjacent Ordovician rocks in the British Isles. *Geology,* 14, 917–919.

Blikeng, B. & Fugelli, E. 2000. Application of results from outcrops of the deep-water Brushy Canyon Formation, Delaware Basin, as analogues for the deep water exploration targets on the Norwegian shelf. *In*: Weimer, P., Slatt, R.M., Coleman, J., Rosen, N.C., Nelson, H., Bouma, A.H., Styzen, M.J. & Lawrence, D.T. (eds), *Gulf Coast Section-Society of Economic Paleontologists and Mineralogists Foundation 20th Annual Bob F. Hoskins Research Research Conference, Deep-Water Reservoirs of the World,* 61–81. Houston, Texas: Gulf Coast Section, Society of Economic Paleontologists and Mineralogists.

Boccaletti, M., Calamita, F., Deiana, G., Gelati, R., Massari, F., Moratti, G. & Ricci Lucchi, F. 1990. Migrating foredeep-thrust belt system in the northern Apennines and southern Alps. *Palaeogeography, Palaeoclimatology, Palaeoecology,* 77, 3–14.

Bøe, R. 1994. Nature and record of Late Miocene mass-flow deposits from the Lau-Tonga forearc basin, Tongan Plateau (Hole 840B). *In*: Hawkins, J., Parson, L., Allan, J. *et al.*, *Proceedings of the Ocean Drilling Program, Scientific Results,* 135, 87–100. College Station, Texas, USA: Ocean Drilling Program.

Boehm, A. & Moore, C.J. 2002. Fluidized sandstone intrusions as an indicator of Paleostress orientation, Santa Cruz, California. *Geofluids,* 2, 147–161.

Bohaty, S.M. & Zachos, J.C. 2003. Significant Southern Ocean warming event in the late middle Eocene. *Geology,* 11, 1017–1020.

Boillot, G. & Leg 103 Scientific Party 1987. Tectonic denudation of the upper mantle along passive margins: a model based on drilling results (ODP Leg 103, western Galicia margin, Spain). *Tectonophysics,* 132, 335–342.

Boltunov, V.A. 1970. Certain earmarks distinguishing glacial and moraine-like glaciomarine sediments, as in Spitsbergen. *International Geology Review,* 12, 204–211.

Booth, J.R., DuVernay III, A.E., Pfeiffer, D.S. & Styzen, M.J. 2000. Sequence stratigraphic framework, depositional models, and stacking patterns of ponded and slope fan systems in the Auger Basin: central Gulf of Mexico slope. *In*: Weimer, P., Slatt, R.M., Coleman, J., Rosen, N.C., Nelson, H., Bouma, A.H., Styzen, M.J. & Lawrence, D.T. (eds), *Deep-water Reservoirs of the World,* 82–103. Houston, Texas: Gulf Coast Section, Society of Economic Paleontologists and Mineralogists.

Booth, J.S., Sangrey, D.A. & Fugate, J.K. 1985. A nomogram for interpreting slope stability of fine-grained deposits in modern and ancient marine environments. *Journal of Sedimentary Petrology,* 55, 29–36.

Booth, J.S., O'Leary, D.W., Popenoe, P. & Danforth, W.W. 1993. U.S. Atlantic continental slope landslides: their distribution, general attributes, and implications. *In*: Schwab, W.C., Lee, H.J. & Twichell, D.C. (eds), *Submarine Landslides: Selected Studies in the U. S. Exclusive Economic Zone,* 14–22 Bulletin U.S. Geological Survey.

Booth, J.S., DuVernay III, A.E., Pfeiffer, D.S., Styzen, M.J. 2000. Sequence stratigraphic framework, depositional models, and stacking patterns of ponded and slope fan systems in the Auger Basin: central Gulf of Mexico. *In*: Weimer, P., Slatt, R.M., Coleman, J., Rosen, N.C., Nelson, H., Bouma, A.H., Styzen, M.J. & Lawrence, D.T. (eds), *Gulf Coast Section-Society of Economic Paleontologists and Mineralogists Foundation 20th Annual Bob F. Hoskins Research Research Conference, Deep-Water Reservoirs of the World,* 82–103. Houston, Texas: Gulf Coast Section, Society of Economic Paleontologists and Mineralogists.

Bornhold, B.D., Firth, J.V. *et al.* 1998. *Proceedings of the Integrated Ocean Drilling Program, Initial Reports,* 169S, 11–61. College Station, Texas, USA: Ocean Drilling Program.

Bottjer, D.J. & Droser, M.L. 1991. Ichnofabric and basin analysis. *Palaios,* 6, 199–205.

Bouma, A.H. 1962. *Sedimentology of some Flysch Deposits: A Graphic Approach to Facies Interpretation.* Amsterdam: Elsevier.

Bouma, A.H. 1964. Ancient and recent turbidites. *Geologie en Mijnbouw,* 43, 375–379.

Bouma, A.H. 1972. Recent and ancient turbidites and contourites. *Transactions of the Gulf Coast Association of Geological Societies,* 22, 205–221.

Bouma, A.H. 1983. Intraslope basins in northwest Gulf of Mexico: a key to ancient submarine canyons and fans. *In*: Watkins, J.S. & Drake, C.L. (eds), *Studies in Continental Margin Geology*, 567–581. American Association of Petroleum Geologists. Memoir, **34**.

Bouma, A.H. 2000. Fine-grained, mud-rich turbidite systems: model and comparison with coarse-grained sand-rich systems. *In*: Bouma, A.H. & Stone, C.G. (eds), *Fine-grained Turbidite Systems*, 9–20. American Association of Petroleum Geologists. Memoir, **72** & Society of Sedimentary Geologists, Special Publication, **68**. Joint publication, Tulsa, Oklahoma.

Bouma, A.H. & Brouwer, A. (eds) 1964. *Turbidites. Developments in Sedimentology, 3*. Amsterdam: Elsevier.

Bouma, A.H. & Hollister, C.D. 1973. Deep ocean basin sedimentation. *In*: Middleton, G.V. & Bouma, A.H. (eds), *Turbidites and Deep Water Sedimentation*, 79–118. Pacific Section, Society of Economic Paleontologists and Mineralogists Short Course Notes, Anaheim.

Bouma, A.H. & Nilsen, T.H. 1978. Turbidite facies and deep-sea fans-with examples from Kodiak Island, Alaska. *In*: *Proceedings of the 10th Offshore Technology Conference, Houston*, 559–570.

Bouma, A.H. & Ravenne, C. 2004. The Bouma Sequence (1962) and the resurgence of geological interest in the French Maritime Alps (1980s): the influence of the Grès d'Annot in developing ideas of turbidite systems. *In*: Joseph, P. & Lomas, S.A. (eds), *Deep-Water Sedimentation in the Alpine Basin of SE France: New Perspectives on the Grès d'Annot and Related Systems*. Geological Society, London, Special Publication, **221**, 27–38. London: The Geological Society.

Bouma, A.H. & Rozman, D.J. 2000. Characteristics of fine-grained outer fan fringe turbidite systems. *In*: Bouma, A.H. & Stone, C.G. (eds), *Fine-Grained Turbidite Systems*. Society of Economic Paleontologists & Mineralogists, Special Publication, **68**, 291–298.

Bouma, A.H. & Stone, C.G. (eds) 2000. *Fine-grained Turbidite Systems*. American Association of Petroleum Geologists, Memoir, **72** & Society of Sedimentary Geology Special Publication, **68**, Joint publication. Tulsa, Oklahoma.

Bouma, A.H. & Wickens, H.D. 1991. Permian passive margin submarine fan complex, Karoo Basin, South Africa: possible model to Gulf of Mexico. *Transactions of the Gulf Coast Association of Geological Societies*, **41**, 30–42.

Bouma, A.H., Devries, M.B. & Cook, T.W. 1995. Correlation efficiency as a tool to better determine depositional subenvironments in submarine fans. *Gulf Coast Association of Geological Societies Transactions*, **45**, 31–40.

Bouma, A.H., Moore, G.T. & Coleman, J.M. (eds) 1978. Framework, facies, and oil-trapping characteristics of the upper continental margin. *American Association of Petroleum Geologists Studies in Geology*, **7**.

Bouma, A.H., Stelting, C.E. & Coleman, J.M. 1984. Mississippi Fan: internal structure and depositional processes. *Geo-Marine Letters*, **3**, 147–154.

Bouma, A.H., Coleman, J.M. & DSDP Leg 96 Shipboard Scientists 1986. *Initial Reports of the Deep Sea Drilling Project, 96*. Washington, DC: US. Government Printing Office.

Bourcart, J. 1938. La marge continentale: essai sur les regressions et les transgressions marines. *Bulletin Societé Geologie France*, **8**, 393–474.

Bourgeois, A., Joseph, P. & Lecomte, J.C. 2004. Three-dimensional full wave seismic modelling versus one-dimensional convolution: the seismic appearance of the Grès d'Annot turbidite system. *In*: Joseph, P. & Lomas, S.A. (eds), *Deep-Water Sedimentation in the Alpine Basin of SE France: New Perspectives on the Grès d'Annot and Related Systems*. Geological Society, London, Special Publication, **221**, 401–417. London: The Geological Society.

Bourget, J., Zaragosia, S., Garlan, T., Gabelotaud, I., Guyomard, P. Dennielou, B., Ellouz-Zimmermann, N., Schneider, J.L., and the FanIndien 2006 survey crew 2008. Discovery of a giant deep-sea valley in the Indian Ocean, off eastern Africa: The Tanzania channel. *Marine Geology*, **255**, 179–185.

Bourget, J., Zaragosi, S., Mulder, T., Schneider, J.-L., Garlan, T. & van Toer, A. 2010. Hyperpycnal-fed turbidite lobe architecture and recent sedimentary processes: A case study from the Al Batha turbidite system, Oman margin. *Sedimentary Geology*, **229**, 144–159.

Bourget, J., Zaragosi, S., Ellouz-Zimmermann, N., Mouchot, N., Garlan, T., Schneider, J.-C., Lanfumey, V. & Lallemant, S. 2011. Turbidite system architecture and sedimentary processes along topographically complex slopes: the Makran convergent margin. *Sedimentology*, **58**, 376–406.

Bourgois, J., Pautot, G. Bandy, W. Boinet, T. Chotin, P. Huchon, P. Mercier de Lepinay, B. Monge, F. Monlau, J. Pelletier, B., Sosson, M. & von Huene, R. 1988. Seabeam and seismic reflection imaging of the neotectonic regime of the Andean continental margin off Peru (4°S to 10°S). *Earth and Planetary Science Letters*, **87**, 111–126.

Bouroullec, R. & Pyles, D.R. 2010. Sandstone extrusions and slope channel architecture and evolution: Mio-Pliocene Monterey and Capistrano formations, Dana Point harbor, Orange County, California, U.S.A. *Journal of Sedimentary Research*, **80**, 376–392.

Bouroullec, R., Cartwright, J.A., Johnson, H.D., Lansigu, C., Quémener, J.-M. & Savanier, D. 2004. Syndepositional faulting in the Grès d'Annot Formation, SE France: high-resolution kinematic analysis and stratigraphic response to growth faulting. *In*: Joseph, P. & Lomas, S.A. (eds), *Deep-Water Sedimentation in the Alpine Basin of SE France: New Perspectives on the Grès d'Annot and Related Systems*. Geological Society, London, Special Publication, **221**, 241–265. London: The Geological Society.

Bowen, A.J., Normark, W.R. & Piper, D.J.W. 1984. Modelling of turbidity currents on Navy Submarine Fan, California Continental Borderland. *Sedimentology*, **31**, 169–185.

Bower, T.H. 1951. Mudflow occurrence in Trinidad, British West Indies. *American Association of Petroleum Geologists Bulletin*, **35**, 908–912.

Bowin, C., Purdy, G.M., Johnson, C., Shor, G.G., Lawver, L., Hartono, H.M.S. & Jezek, P. 1980. Arc continent collision in the Banda Sea. *American Association of Petroleum Geologists Bulletin*, **64**, 868–915.

Bowles, F.A., Ruddiman, W.F. & Jahn, W.H. 1978. Acoustic stratigraphy, structure, and depositional history of the Nicobar Fan, western Indian Ocean. *Marine Geology*, **26**, 269–288.

Boyer, S.E. & Elliott, D. 1982. Thrust systems. *American Association of Petroleum Geologists Bulletin*, **66**, 1196–1230.

Boynton, C.H., Westbrook, G.K., Bott, M.H.P. & Long, R.E. 1979. A seismic refraction investigation of crustal structure beneath the Lesser Antilles island arc. *Geophysical Journal of the Royal Astronomical Society*, **58**, 371–393.

Braccini, E. de Boer, W., Hurst, A., House, M., Vigorito, M. & Templeton, G. 2008. Sand injectites. *Oilfield Review*, 34–49.

Brachert, T.C., Forst, M.H., Pais, J.J., Legoinha, P. & Reijmer, J.J.G. 2003. Lowstand carbonates, highstand sandstones? *Sedimentary Geology*, **155**, 1–12.

Bradbury, W. & Woodburn, N. 2011. Improved imaging in Baffin Bay. *GEO ExPro*, **8**, 36–40.

Bradley, D.C. 2008. Passive margins through earth history. *Earth-Science Reviews*, **91**, 1–26.

Brady, L.F. 1947. Invertebrate tracks from the Coconino Sandstone of northern Arizona. *Journal of Paleontology*, **21**, 466–472.

Braithwaite, C.J.R. 1973. Settling behaviour related to sieve analysis of skeletal sands. *Sedimentology*, **20**, 251–262.

Bralower, T.J., Thomas, D.J., Zachos, J.C., Hirschmann, M.M., Röhl, U., Sigurdsson, H., Thomas, E. & Whitney, D.L. 1997. High-resolution records of the late Paleocene thermal maximum and circum-Caribbean volcanism: Is there a causal link? *Geology*, **25**, 963–966.

Branney, M.J. & Kokelaar, P. 2002. Pyroclastic density currents and the sedimentation of ignimbrites. *Geological Society of London Memoir*, **27**. London: The Geological Society.

Brass, G.W., Saltzman, E., Sloan II., J.L., Southam, J.R., Hay, W.W., Holser, W.T. & Peterson, W.H. 1982. Ocean circulation, plate tectonics and climate. *In*: *Climate in Earth History, Studies in Geophysics*, Panel on pre-Pleistocene climates, 83–89. Washington, DC: National Academy Press.

Breen, N.A., Silver, E.A. & Hussong, D.M. 1986. Structural styles of an accretionary wedge south of the island of Sumba, Indonesia, revealed by SeaMARC II side scanner. *Geological Society of America Bulletin*, **97**, 1250–1261.

Brett, C.E., Goodman, W.M. & LoDuca, S.T. 1990. Sequences, cycles, and basin dynamics in the Silurian of the Appalachian Foreland Basin. *Sedimentary Geology*, **69**, 191–244.

Breza, J.R. & Wise, S.W.Jr. 1992. Lower Oligocene ice-rafted debris on the Kerguelen Plateau: evidence for East Antarctica continental glaciation. *In*: Wise, S.W.Jr., Schlich, R. *et al.*, *Proceedings of the Ocean Drilling Program, Scientific Results,* **120**, 161–178. College Station, Texas, USA: Ocean Drilling Program.

Brice, S.E., Cochran, M.D., Pardo, G. & Edwards, A.D. 1983. Tectonics and sedimentation of the South Atlantic rift sequence: Cabinda, Angola. *In*: Watkins, J.S. & Drake, C.L. (eds), *Studies in Continental Margin Geology,* 5–18. American Association of Petroleum Geologists Memoir, **34**.

Briedis, N.A., Bergslien, D., Hjellbakk, A., Hill, R.E. & Moir, G.J. 2007. Recognition criteria, significance to field performance, and reservoir modeling of sand injections in the Balder Field, North Sea. *In*: Hurst, A. & Cartwright, J. (eds), *Sand Injectites: Implications for Hydrocarbon Exploration and Production,* 91–102. American Association of Petroleum Geologists Memoir, **87**. Tulsa, Oklahoma.

Broecker, W.S. 1974. *Chemical Oceanography.* New York: Harcourt Brace Javanovich.

Broecker W., Bond, G., Klas, M., Clark, E. & McManus, J. 1992. Origin of the northern Atlantic's Heinrich events. *Climate Dynamics*, **6**, 265–273.

Bromley, R.G. 1996. *Trace Fossils: Biology, Taphonomy and Applications,* 361 pp. London: Chapman & Hall. ISBN 0 412 61480 4.

Bromley, R.G. & Ekdale, A.A. 1986. Composite ichnofabrics and tiering of burrows. *Geological Magazine*, **123**, 59–65.

Bromley, R.G., Pemberton, S.G. & Rahmani, R.A. 1984. A Cretaceous woodground: the Teredolites ichnofacies. *Journal of Paleontology*, **58**, 488–498.

Brooks, J. & Glennie, K.W. (eds) 1987. *Petroleum Geology of North-west Europe.* London: Graham & Trotman.

Brothers, D.A., ten Brink, U.S., Andrews, B.A. & Chaytor, J.D. 2013. Geomorphic characterization of the U.S. Atlantic continental margin. *Marine Geology*, **338**, 46–63.

Broucke, O., Guillocheau, F., Robin, C., Joseph, P. & Calassou, S. 2004. The influence of syndepositional basin floor deformation on the geometry of turbiditic sandstones: a reinterpretation of the Cote de l'Ane area (Sanguinire-Restefonds sub-basin, Grès d'Annot, Late Eocene, France). *In*: Joseph, P. & Lomas, S.A. (eds), *Deep-Water Sedimentation in the Alpine Basin of SE France: New Perspectives on the Grès d'Annot and Related Systems.* Geological Society, London, Special Publication, **221**, 203–222. London: The Geological Society.

Brown, D. & Ryan, P.D. (eds) 2011. *Arc-Continent Collision.* Springer, Frontiers in Earth Sciences, 988 pp. ISBN: 978-3-540-88558-0.

Brown, D., Juhlin, C. & Puchkov, V. 2002. *Mountain Building in the Uralides: Pangea to the Present. American Geophysical Union, Geophysical Monograph Series,* **132**, 288 pp.

Brown, D., Spadea, P., Puchkov, V., Alvarez-Marron, J., Herrington, R., Willner, A.P., Hetzel, R., Gorozhanina, Y. & Juhlin, C. 2006. Arc-continent collision in the Southern Urals. *Earth-Science Reviews*, **79**, 261–287.

Brown, K.M. & Westbrook, G.K. 1987. The tectonic fabric of the Barbados Ridge accretionary complex. *Marine and Petroleum Geology*, **4**, 71–81.

Brown, K.M., Mascle, A. & Behrmann, J.H. 1990. Mechanisms of accretion and subsequent thickening in the Barbados Ridge accretionary complex: balanced cross sections across the wedge toe. *In*: Moore, J.C., Mascle, A. *et al.*, *Proceedings of the Ocean Drilling Program,* **110**, 209–227.

Brown, K.M. & Westbrook, G.K. 1988. Mud diapirism and subcretion in the Barbados Ridge accretionary prism:the role of fluids in accretionary processes. *Tectonics*, **7**, 613–640.

Brown, L.F.Jr. & Fisher, W.L. 1977. Seismic-stratigraphic interpretation of depositional systems: examples from Brazilian rift and pull-apart basin. *In*: Payton, C.E. (ed.), *Seismic Stratigraphy-Applications to Hydrocarbon Exploration,* 213–248. American Association of Petroleum Geologists, Memoir, **26**.

Brown, P.A. & Colman-Sadd, S.P. 1976. Hermitage flexure: figment or fact? *Geology*, **4**, 561–564.

Browne, G.H. & Slatt, R.M. 1997. Thin-bedded slope fan (channel-levée) deposits from New Zealand: an outcrop analog for reservoirs in the Gulf of Mexico. *Gulf Coast Association of Geological Societies Transactions*, **47**, 75–86.

Browne, G.H. Slatt, R.M. & King, P.R. 2000. Contrasting styles of basin-floor fan and slope fan deposition: Mount Messenger Formation, New Zealand. *In*: Bouma, A.H. & Stone, C.G. (eds), *Fine-grained Turbidite Systems,* 142–152. American Association of Petroleum Geologists Memoir, **72** & Society of Sedimentary Geologists, Special Publication, **68**. Joint publication, Tulsa, Oklahoma.

Browne, G.H. & Slatt, R.M. 2002. Outcrop and behind-outcrop characterization of a late Miocene slope fan system, Mt. Messenger Formation, New Zealand. *American Association of Petroleum Geologists Bulletin*, **86**, 841–862.

Browning, J.V., Miller, K.G., McLaughlin, P.P., Kominz, M.A., Sugarman, P.J., Monteverde, D., Feigenson, M.D. & Hernández, J.C. 2006. Quantification of the effects of eustasy, subsidence, and sediment supply on Miocene sequences, mid-Atlantic margin of the United States. *Geological Society of America Bulletin*, **118**, 567–588.

Bruhn, C.H.L. 1994. Sand-rich density underflows from the Early Cretaceous, Recôncavo rift-basin, Brazil; implications for sequence stratigraphic analysis of deep lacustrine successions. *International Sedimentologists Congress, Abstracts*, **14**, S6.1–S6.2.

Bruhn, C.H.L. 1998. Major types of deep-water reservoirs from the eastern Brazillian rift and passive margin basins. *In*: Mello, M.R. & Yilmaz, P.O. (eds), *Extended Abstracts Volume,* American Association of Petroleum Geologists International Conference and Exhibition, Rio de Janeiro, 8-11 November 1998, 14–15.

Bruhn, C.H.L. 1998a. Reservoir architecture of deep-lacustrine sandstones from the Early Cretaceous Recôncavo rift-basin, Brazil. *American Association of Petroleum Geologists Bulletin*, **83**, 1502–1525.

Bruhn, C.H.L. 1998b. Petroleum geology of rift and passive margin turbidite systems: Brazilian and worldwide examples. Part 2: *Deep-water Reservoirs from the Eastern Brazilian Rift and Passive Margin Basins.* Short course 6, American Association of Petroleum Geologists Bulletin. International Conference and Exhibition, Rio de Janeiro, November 12–13.

Bruhn, C.H.L. & Walker, R.G. 1995. High-resolution stratigraphy and sedimentary evolution of coarse-grained canyon-filling turbidites from the Upper Cretaceous transgressive megasequence, Campos Basin, offshore Brazil. *Journal of Sedimentary Research*, **B65**, 426–442.

Bruhn, C.H.L. & Walker, R.G. 1997. Internal architecture and sedimentary evolution of coarse-grained, turbidite channel-levée complexes, early Eocene Regência Canyon, Espírito Santo Basin, Brazil. *Sedimentology*, **44**, 17–46.

Brunt, R.L., Hodgson, D.M., Flint, S.S., Pringle, J.K., Di Celma, C., Prélat, A. & Grecula, M. 2012. Confined to unconfined: Anatomy of a base of slope succession, Karoo Basin, *South Africa. Marine and Petroleum Geology,* doi:10.1016/j.marpetgeo.2012.02.007.

Brunt, R.L., Hodgson, D.M., Flint, S.S., Pringle, J.K., Di Celma, C., Prélat, A. & Grecula, M. 2013a. Confined to unconfined: Anatomy of a base of slope succession, Karoo Basin, South Africa. *Marine and Petroleum Geology*, **41**, 206–221.

Brunt, R.L., Di Celma, C.N., Hodgson, D.M., Flint, S.S., Kavanagh, J.P. & van der Merwe, W.C. 2013b. Driving a channel through a levee when the levee is high: An outcrop example of submarine down-dip entrenchment. *Marine and Petroleum Geology*, **41**, 134–145.

Buatois, L.A., Mangano, M.G. & Sylvester, Z. 2001. A diverse deep-marine Ichnofauna from the Eocene Tarcau Sandstone of the Eastern Carpathians, Romania. *Ichnos*, **8**, 23–62.

Bucher, W.H. 1940. Submarine valleys and related geologic problems of the North Atlantic. *Geological Society of America Bulletin*, **51**, 489–512.

Bugge, T., Befring, S., Belderson, R., Eidvin, T., Jansen, E., Kenyon, N., Holtedahl, H. & Sejrup, H.. 1987. A giant three-stage submarine slide off Norway. *Geo-Marine Letters*, **7**, 191–198.

Bulfinch, D.L. & Ledbetter, M.T. 1984. Deep Western Boundary Undercurrent delineated by sediment texture at base of North American continental rise. *Geo-Marine Letters*, **3**, 31–36.

Bull, S., Cartwright, J. & Huuse, M. 2009. A review of kinematic indicators from mass-transport complexes using 3D seismic data. *Marine and Petroleum Geology*, **26**, 1132–1151.

Burbank, D.W. Puigdefabregas, C. & Muñoz, J.A. 1992. The chronology of the Eocene tectonic and stratigraphic development of the eastern Pyrenean foreland basin, north-east Spain. *Geological Society of America, Bulletin*, **104**, 1101–1120.

Burgess, C.E., Pearson, P.N., Lear, C.H., Morgans, H.E.G., Handley, L., Pancost, R.D.,& Schouten, S. 2008. Middle Eocene climate cyclicity in the southern Pacific: implications for global ice volume. *Geology*, **36**, 651–654.

Burk, C.A. & Drake, C.L. (eds) 1974. *The Geology of Continental Margins*. New York: Springer.

Burke, K. 1978. Evolution of continental rift systems in the light of plate tectonics. *In*: Ramberg, I.B. & Neumann, E.R. (eds), *Tectonics and Geophysics of Continental Rifts*, 1–9. NATO Advanced Study Institute, Series C, Mathematical and Physical Sciences, **37**.

Burke, K.C., Dessauvagie, T.F.J. & Whiteman, A.J. 1972. Geological history of the Benue valley and adjacent areas. *In*: Dessauvagie, T.F.J. & Whiteman, A.J. (eds), *African Geology*, 187–205. Ibadan: Ibadan University Press.

Burne, R.V. 1970. The origin and significance of sand volcanoes in the Bude Formation (Cornwall). *Sedimentology*, **15**, 211–228.

Busby, C.J. & Ingersoll, R.V. 1995. *Tectonics of Sedimentary Basins*. Cambridge, Mass., USA. Oxford, UK: Blackwell Science.

Byers, C.W. 1977. Biofacies patterns in euxinic basins: a general model. *In*: Cook, H.E. & Enos, P. (eds), *Deep-water Carbonate Environments*, 5–17. Society of Economic Paleontologists and Mineralogists Special Publication, **25**.

Byers, C.W. 1982. Geological significance of marine biogenic sedimentary structures. *In*: McCall, P.L. & Tevesz, M.J.S. (eds), *Animal-Sediment Relations*, 221–256. Plenum Press, New York.

Byrne, D.E., Davis, D.M. & Sykes, L.R. 1988. Loci and maximum size of thrust earthquakes and the mechanics of the shallow region of subduction zones. *Tectonics*, **7**, 833–857.

Byrne, T., Chan, Y.-C., Rau, R.-J., Lu, C.-Y., Lee, Y.-H. & Wang, Y.-J. 2011. *In*: Brown, D. & Ryan, P.D. (eds), *Arc-Continent Collision, 213–245*. Springer, Frontiers in Earth Sciences.

Cacchione, D.A. & Drake, D.E. 1986. Nepheloid layers and internal waves over continental shelves and slopes. *Geo-Marine Letters*, **6**, 147–152.

Cacchione, D.A. & Southard, J.B. 1974. Incipient sediment movement by shoaling internal gravity waves. *Journal of Geophysical Research*, **70**, 2237–2242.

Cacchione, D.A., Rowe, G.T. & Malahoff, A. 1978. Submersible investigation of outer Hudson submarine canyon. *In*: Stanley, D.J. & Kelling, G. (eds), *Sedimentation in Submarine Canyons, Fans, and Trenches*, 42–50. Stroudsburg, PA: Dowden, Hutchinson & Ross.

Cadet, J.P., Kobayashi, K., Lallemand, S., Jolivet, L., Auboin, J., Boulégue, J., Dubois, J., Hotta, H., Ishii, T., Konishi, K., Niitsuma, N. & Shimamura, H. 1987. Deep scientific dives in the Japan and Kuril Trenches. *Earth and Planetary Science Letters*, **83**, 313–328.

Cainelli, C. & Mohriak, W.U. 1999. Some remarks on the evolution of sedimentary basins along the eastern Brazilian continental margin. *Episodes*, **22**, 206–216.

Caixeta, J.M. 1998. *Estudo faciologico e caracteristicas de reservatorio dos arenitos produtores de gas do Campo de Jacuipe (Cretaceo inferior), Bacia do Reconcavo, Brasil*. Unpublished M.Sc. Thesis, Universidade Federal de Ouro Preto, Brazil, 131 pp.

Caixeta, J.M., Milhomem, P.S., Witzke, R.E., Dupuy, I.S.S. & Gontijo, G.A. 2007. Bacia de Camamu. *Boletim de Geociências da Petrobrás*, **15**, 455–461.

Callec, Y. 2004. The turbidite fill of the Annot sub-basin (SE France): a sequence-stratigraphy approach. *In*: Joseph, P. & Lomas, S.A. (eds), *Deep-Water Sedimentation in the Alpine Basin of SE France: New Perspectives on the Grès d'Annot and Related Systems*. Geological Society, London, Special Publication, **221**, 111–135. London: The Geological Society.

Callec, Y., Deville, E., Desaubliaux, G., Griboulard, R., Huyghe, P., Mascle, A., Mascle, G., Noble, M., Padron de Carillo, C. & Schmitz, J. 2010. The Orinoco turbidite system: tectonic controls on sea-floor morphology and sedimentation. *American Association of Petroleum Geologists Bulletin*, **94**, 869–887.

Callow, R.H.T. & McIlroy, D. 2011. Ichnofabrics and ichnofabric-forming trace fossils in Phanerozoic turbidites. *Bulletin of Canadian Petroleum Geology*, **59**, 103–111.

Callow, R.H.T., McIlroy, D., Kneller, B. & Dykstra, M. 2013. Integrated Ichnological and sedimentological analysis of a Late Cretaceous submarine channel-levée system: The Rosario Formation, Baja California, Mexico. *Marine and Petroleum Geology*, **41**, 277–294.

Callow, R.H.T., McIlroy, D., Kneller, B. & Dykstra, M. 2012. Integrated Ichnological and sedimentological analysis of a Late Cretaceous submarine channel-levée system: The Rosario Formation, Baja California, *Mexico. Marine and Petroleum Geology*, doi 10.1016/j.marpetgeo.2012.02.001.

Calvert, S.E. 1966. Accumulation of diatomaceous silica in the sediments of the Gulf of California. *Geological Society of America Bulletin*, **77**, 569–596.

Cantero, M.I., Cantelli, A., Pirmez, C., Balachandar, S., Mohrig, D., Hickson, T.A., Yeh, T., Hajime Naruse, H., & Parker, G. 2012. Emplacement of massive turbidites linked to extinction of turbulence in turbidity currents. *Nature Geoscience*, **5**, 42–45.

Camerlenghi, A. & Pini, G.A. 2009. Mud volcanoes, olistostromes and Argille scagliose in the Mediterranean region. *Sedimentology*, **56**, 319–365.

Cameron, B. 2010. Carnarvon Basin. *GEO ExPro*, **7**, 54–56.

Camerlenghi, A. & Pini, G.A. 2009. Mud volcanoes, olistostromes and Argille scagliose in the Mediterranean region. *Sedimentology*, **56**, 319–365.

Campion, K.M., Sprague, A.R., Mohrig, D., Lovell, R.W., Drzewiecki, P.A., Sullivan, M.D., Ardill, J.A., Jensen, G.N. & Sickafoose, D.K. 2000. Outcrop expression of confined channel complexes. *In*: Weimer, P., Slatt, R.M., Coleman, J., Rosen, N.C., Nelson, H., Bouma, A.H., Styzen, M.J. & Lawrence, D.T. (eds), *Gulf Coast Section-Society of Economic Paleontologists and Mineralogists Foundation 20th Annual Bob F. Hoskins Research Research Conference, Deep-Water Reservoirs of the World*, 127–151. CD-ROM Society of Economic Paleontologists and Mineralogists Special Publications.

Campion, K., Sprague, A. & Sullivan M. 2005. *Architecture and Lithofacies of the Capistrano Formation (Miocene-Pliocene), San Clemente, California.Fullerton, California*. Pacific Section SEPM, Society for Sedimentary Geology.

Cantalejo, B. & Pickering, K.T. 2014. Climate forcing of fine-grained deep-marine system in an active tectonic setting: Middle Eocene, Ainsa Basin, Spanish Pyrenees. *Palaeogeography, Palaeoclimatology, Palaeoecology*, **410**, 351–371.

Cantalejo, B. & Pickering, K.T. 2015. Orbital forcing as principal driver for fine-grained deep-marine siliciclastic sediments, Middle-Eocene Ainsa Basin, Spanish Pyrenees. *Palaeogeography, Palaeoclimatology, Palaeoecology*, **421**, 24–47.

Carey, S. & Sigurdsson, H. 1978. Deep-sea evidence for distribution of tephra from the mixed magma eruption of the Soufriere of St Vincent, 1902: ash turbidites and air fall. *Geology*, **6**, 271–274.

Carey, S. & Sigurdsson, H. 1984. A model of volcanogenic sedimentation in marginal basins. *In*: Kokelaar, B.P. & Howells, M.F. (eds), *Marginal Basin Geology*, 37–58. Geological Society, London, Special Publications, **16**. London: The Geological Society.

Carlson, J. 1998. *Analytical and Statistical approaches toward understanding sedimentation in siliciclastic depositional systems*. Unpublished PhD Thesis, Department of Earth, Planetary and Atmospheric Sciences, MIT, Cambridge, MA, 406 pp.

Carlson, J. & Grotzinger, J.P. 2001. Submarine fan environment inferred from turbidite thickness distributions. *Sedimentology*, **48**, 1331–1351.

Carlson, P.R. & Karl, H.A. 1988. Development of large submarine canyons in the Bering Sea, indicated by morphologic, seismic, and sedimentologic characteristics. *Geological Society of America Bulletin*, **100**, 1594–1615.

Carlson, P.R., Stevenson, A.J., Bruns, T.R., Mann, D.M. & Huggett, Q. 2006. Sediment pathways in Gulf of Alaska from beach to abyssal plain. *In*: Gardner, J.V., Field, M.E. & Twichell, D.C. (eds), *Geology of the United States' Seafloor: The View from GLORIA, 255–278*. Cambridge University Press. ISBN: 9780521433105.

Carmignani, L., Cornamusini, C.P. & Meccheri, M. 2004. *The Internal Northern Apennines, The Northern Tyrrhenian Sea and the Sardinia-Corsica Block*. Special Volume of the Italian Geological Society for the IGC 32 Florence-2004, 59–77.

Cantero, M.I., Cantelli, A., Pirmez, C., Balachandar, S., Mohrig, D., Hickson, T.A., Yeh, T., Hajime Naruse, H. & Parker, G. 2012. Emplacement of massive turbidites linked to extinction of turbulence in turbidity currents. *Nature Geoscience*, **5**, 42–45.

Carpenter, G. 1981. Coincident sediment slump/clathrate complexes on the US Atlantic continental slope. *Geo-Marine Letters*, **1**, 29–32.

Carr, M. & Gardner, M.H. 2000. Portrait of a basin-floor fan for sandy deep-water systems, Permian Brushy Canyon Formation. *In*: Bouma, A.H., Stelting, C.E. & Stone, C.G. (eds), *Fine-Grained Turbidite Systems and Submarine Fans*, American Association of Petroleum Geologists Memoir **72**/Society of Economic Paleontologists and Mineralogists Special Publication.

Carson, B., Westbrook, G.K., Musgrave, R.J. & Suess, E. (eds) 1995. *Proceedings. Ocean Drilling Program, Scientific Results*, **146** (Part 1). College Station, Texas, USA: Ocean Drilling Program.

Carter, L. & McCave, I.N. 1994. Development of sediment drifts approaching an active plate margin under the SW Pacific Deep Western Boundary Current. *Paleoceanography*, **9**, 1061–1085. doi:10.1029/94PA01444.

Carter, L. & McCave, I.N. 2002. Eastern New Zealand drifts, Miocene–Recent. *In*: Stow, D.A.V., Pudsey, C.J., Howe, J.A., Faugères, J.-C. & Viana, A.R. (eds), *Deep-water Contourite Systems: Modern Drifts and Ancient Series, Seismic and Sedimentary Characteristics*. Geological Society London Memoir, **22**, 385–407.

Carter, L. & Schafer, C.T. 1983. Interaction of the Western Boundary Undercurrent with the continental margin off Newfoundland. *Sedimentology*, **30**, 751–768.

Carter, L., Carter, R.M. & McCave, I.N. 2004. Evolution of the sedimentary system beneath the deep Pacific inflow off eastern New Zealand. *Marine Geology*, **205**, 9–27.

Carter, L., Schafer, C.T. & Rashid, M.A. 1979. Observations on depositional environments and benthos of the continental slope and rise, east of Newfoundland. *Canadian Journal of Earth Science*, **16**, 831–846.

Carter, L., Milliman, J.D., Talling, P.J., Gavey, R. & Wynn, R.B. 2012. Near-synchronous and delayed initiation of long run-out submarine sediment flows from a record-breaking river flood, offshore Taiwan. *Geophysical Research Letters*, **39**, L12603, 10.1029/2012GL051172.

Carter, L., Gavey, R., Talling, P.J. & Liu, J.T. 2014. Insights into submarine geohazards from breaks in subsea telecommunication cables. *Oceanography*, **27**, 58–67.

Carter, R.M. 1975. A discussion and classification of subaqueous mass-transport with particular application to grain-flow, slurry-flow and fluxoturbidites. *Earth-Science Reviews*, **11**, 145–177.

Carter, R.M. 1988. The nature and evolution of deep-sea channel systems. *Basin Research*, **1**, 41–54.

Cartigny, M., Postma, G., van den Berg, J.H. & Mastbergen, D.R. 2011. A comparative study of sediment waves and cyclic steps based on geometries, internal structures and numerical modeling. *Marine Geology*, **280**, 40–56.

Cartigny, M.J.B., Ventra, D., Postma, G. & van den Berg, J.H. 2014. Morphodynamics and sedimentary structures of bedforms under supercritical-flow conditions: New insights from flume experiments. *Sedimentology*, **61**, 712–748.

Carter, R.M. & Carter, L. 1987. The Bounty channel system: a 55-million-year-old sediment conduit to the deep sea, southwest Pacific Ocean. *Geo-Marine Letters*, **7**, 183–190.

Carter, R.M. & Lindqvist, J.K. 1975. Sealers Bay submarine fan complex, Oligocene, southern New Zealand. *Sedimentology*, **22**, 465–483.

Carter, R.M. & Lindqvist, J.K. 1977. Balleny Group, Chalky Island, southern New Zealand: an inferred Oligocene submarine canyon and fan complex. *Pacific Geology*, **12**, 1–46.

Carter, R.M. & Norris, R.J. 1977. Redeposited conglomerates in Miocene flysch sequence at Blackmount, western Southland, New Zealand. *Sedimentary Geology*, **18**, 289–319.

Carvajal, C.R. & Steel, R.J. 2006. Thick turbidite successions from supply-dominated shelves during sea-level highstand. *Geology*, **34**, 665–668.

Cas, R. 1979. Mass-flow arenites from a Paleozoic interarc basin, New South Wales, Australia: mode and environment of em-placement. *Journal of Sedimentary Petrology*, **49**, 29–44.

Cas, R.A.F. & Wright, J.V. 1987. *Volcanic Successions: Modern and Ancient: A Geological Approach to Processes, Products and Successions*. London: Unwin-Hyman, 528 pp.

Cas, R.A.F. & Wright, J.V. 1991. Subaqueous pyroclastic flows and ignimbrites: an assessment. *Bulletin of Volcanology*, **53**, 357–380.

Cashman, K.V. & Popenoe, P. 1985. Slumping and shallow faulting related to the presence of salt on the continental slope and rise off North Carolina. *Marine and Petroleum Geology*, **2**, 260–271.

Castelltort, S. & Van den Driessche, J. 2003. How plausible are high-frequency sediment supply-driven cyclesin the stratigraphic record? *Sedimentary Geology*, **157**, 3–13.

Castle, J.W. 2001. Foreland-basin sequence response to collisional tectonism. *Geological Society of America Bulletin*, **113**, 801–812.

Cattaneo, A. & Ricci Lucchi, F. 1995. Long-distance correlation of sandy turbidites: a 2.5 km long cross-section of Marnoso arenacea, Santerno Valley, Northern Apennines. *In*: Pickering, K.T., Hiscott, R.N., Kenyon, N.H., Ricci Lucchi, F. & Smith, R.D.A. (eds), *Atlas of Deep Water Environments: Architectural Style in Turbidite Systems*, 303–306. London: Chapman and Hall.

Catuneanu, O. & Zecchin, M. 2012. High-resolution sequence stratigraphy of clastic shelves II: Controls on sequence development. *Marine and Petroleum Geology*, **39**, 25–38.

Cawood, P.A. & Nemchin, A.A. 2001. Source regions for Laurentian margin sediments: constraints from U/Pb dating of detrital zircons in the Newfoundland Appalachians. *Geological Society of America Bulletin*, **113**, 1234–1246.

Cazzola, A., Fantoni, R., Franciosi, R., Gatti, V., Ghielmi, M. & Pugliese, A. 2011. From Thrust and Fold Belt to Foreland Basins: Hydrocarbon Exploration in Italy. *American Association of Petroleum Geologists, Search and Discovery Article #10374*.

Cerrina Feroni, A., Martelli, L., Martelli, P. & Ottria, G. 2002. *Structural-Geologic Map of Emilia Romagna Apennines*, 1:250,000. *Regiona Emilia Romagna e Consiglio Nazionale delle Ricerche (CNR)*, SELCA Firenze.

Chadwick, R.A. 1986. Extension tectonics in the Wessex Basin, southern England. *Journal of the Geological Society (London)*, **143**, 465–488.

Chakraborty, P.P., Mukhopadhyay, B., Pal, T. & Gupta, T.D. 2002. Statistical appraisal of bed thickness patterns in turbidite successions, Andaman Flysch Group, Andaman Islands, India. *Journal of Asian Earth Sciences,* **21**, 189–196.

Chamberlain, C.K. 1971. Morphology and ethology of trace fossils from the Ouachita Mountains, Southeast Oklahoma. *Journal of Paleontology,* **45**, 212–246.

Chambers, J.M., Cleveland, W.S., Kleiner, B. & Tukey, P.A. 1983. *Graphical Methods for Data Analysis.* Belmont, CA: Wadsworth.

Chamot-Rooke, N. & Le Pichon, X. 1989. Zenisu Ridge: mechanical model of formation. *Tectonophysics,* **160**, 175–193.

Champagnac, J.-D., Molnar, P., Sue, S. & Herman, F. 2012. Tectonics, climate, and mountain topography, *Journal of Geophysical Research,* **117**, 1–34. doi:10.1029/2011JB008348.

Chan, M.A. & Dott, R.H. 1983. Shelf and deep sea sedimentation in Eocene forearc basin, western Oregon fan or non fan? *American Association of Petroleum Geologists Bulletin,* **67**, 2100–2116.

Chang, H.K., Kowsmann, R.O., Figueiredo, A.M.F. & Bender, A.A. 1992. Tectonics and stratigraphy of the East Brazil Rift system: an overview *Tectonophysics,* **213**, 97–138.

Charlton, T.R. 1991. Postcollision extension in arc-continent collision zones, eastern Indonesia. *Geology,* **19**, 28–31.

Charlton, T.R., Kaye, S.J., Samodra, H. & Sardjono 1991. Geology of the Kai Islands: implications for the evolution of the Aru Trough and Weber Basin, Banda Arc, Indonesia. *Marine and Petroleum Geology,* **8**, 62–69.

Chapin, M.A., Davies, P., Gibson, J.L. & Pettingill, H.S. 1994. Reservoir architecture of turbidite sheet sandstones in laterally extensive outcrops, Ross Formation, Western Ireland. *In:* Weimer, P., Bouma, A.H. & Perkins, B.F. (eds), *Gulf Coast Section-Society of Economic Paleontologists and Mineralogists Foundation 15th Annual Research Conference, Submarine Fans and Turbidite Systems: Sequence Stratigraphy, Reservoir Architecture and Production Characteristics,* 53–68.

Chen, C. & Hiscott, R.N. 1999a. Statistical analysis of turbidite cycles in submarine fan successions: tests for short-term persistence. *Journal of Sedimentary Research,* **69B**, 486–504.

Chen, C. & Hiscott, R.N. 1999b. Statistical analysis of facies clustering in submarine-fan turbidite successions. *Journal of Sedimentary Research,* **69B**, 505–517.

Cheng-Shing Chiang, Ho-Shing Yu & Ying-Wei Chou 2004. Characteristics of the wedge-top depozone of the southern Taiwan foreland basin system. *Basin Research,* **16**, 65–78.

Chiang, C.S. 1998. *Tectonic features of the Kaoping shelf-slope region off southwestern Taiwan.* PhD Thesis, National Taiwan University.

Chipping, D.H. 1972. Sedimentary structures and environment of some thick sandstone beds of turbidite type. *Journal of Sedimentary Petrology,* **42**, 587–595.

Chiu, J.-K. & Liu, C.-S. 2008. Comparison of sedimentary processes on adjacent passive and active continental margins offshore of SW Taiwan based on echo character studies. *Basin Research,* **20**, 503–518.

Chi-Yue Huang, Peter B. Yuan & Shuh-Jung Tsao 2006. Temporal and spatial records of active arc-continent collision in Taiwan: A synthesis. *Geological Society of America Bulletin,* **118**, 274–288.

Chou, Y.W. 1999. *Tectonic framework, flexural uplift history and structural patterns of flexural extension in western Taiwan foreland basin.* PhD Thesis, National Taiwan University.

Chough, S. & Hesse, R. 1976. Submarine meandering thalweg and turbidity currents flowing for 4000 km in the Northwest Atlantic Mid-Ocean Channel, Labrador Sea. *Geology,* **4**, 529–534.

Chough, S. & Hesse, R. 1980. Northwest Atlantic Mid-Ocean Channel of the Labrador Sea: III. *Head spill vs body spill deposits from turbidity currents on natural levées. Journal of Sedimentary Petrology,* **50**, 227–234.

Chough, S.K. & Hesse, R. 1985. Contourites from Eirik Ridge, south of Greenland. *Sedimentary Geology,* **41**, 185–199.

Choukroune, P., Roure, F., Pinet, B. & ECORS Pyrenees Team 1990. Main results of the ECORS Pyrenees profile. *Tectonophysics,* **173**, 411–423.

Christensen, C.J., Gorsline, D.S., Hammond, D.E. & Lund, S.P. 1994. Non-annual laminations and expansion of anoxic basin-floor conditions in Santa Monica Basin, California Borderland, over the past four centuries. *Marine Geology,* **116**, 399–418.

Christie-Blick, N. & Biddle, K.T. 1985. Deformation and basin formation along strike-slip faults. *Society of Economic Paleontologists and Mineralogists Special Publication,* **37**, 1–34.

Christie-Blick, N. & Driscoll, N.W. 1995. Sequence Stratigraphy. *Annual Reviews in Earth and Planetary Sciences,* **23**, 451–478.

Cita, M.B., Beghi, C., Camerlenghi, A., Karstens, K.A., McCoy, F.W., Nosetto, A., Parisi, E., Scolari, F. & Tomadin, L. 1984. Turbidites and megaturbidites from the Herodotus Abyssal Plain (eastern Mediterranean) unrelated to seismic events. *Marine Geology,* **55**, 79–101.

Clari, P. & Ghibaudo, G. 1979. Multiple slump scars in the Tortonian type area (Piedmont Basin, northwestern Italy). *Sedimentology,* **26**, 719–730.

Clark, I.R. & Cartwright, J.A. 2009. Interactions between submarine channel systems and deformation in deepwater fold belts: Examples from the Levant Basin, Eastern Mediterranean sea. *Marine and Petroleum Geology,* **26**, 1465–1482.

Clark, J.D. 1995. Detailed section across the Ainsa II Channel complex, South Central Pyrenees, Spain. *In:* Pickering, K.T., Hiscott, R.N., Kenyon, N.H., Ricci Lucchi, F. & Smith, R.D.A. (eds). *Atlas of Deep Water Environments: Architectural Style in Turbidite Systems,* 139–144. Chapman and Hall. London.

Clark, J.D. & Pickering, K.T. 1996a. Architectural elements and growth patterns of submarine channels: application to hydrocarbon exploration. *American Association of Petroleum Geologists Bulletin,* **80**, 194–221.

Clark, J.D. & Pickering, K.T. 1996b. *Submarine Channels: Processes and Architecture.* Marketed by: American Association of Petroleum Geologists, and Vallis Press (London), 232 pp. ISBN: 0-9527313-0-4.

Clark, J.D., Kenyon, N.H. & Pickering, K.T. 1992. Quantitative analysis of the geometry of submarine channels: implications for the classification of submarine fans. *Geology,* **20**, 633–636.

Clift, P.D., Carter, A., Nicholson, U. & Masago, H. 2013. Zircon and apatite thermochronology of the Nankai Trough accretionary prism and trench, Japan: Sediment transport in an active and collisional margin setting. *Tectonics,* doi: 10.1002/tect.20033.

Cloos, M. 1992. Thrust-type subduction zone earthquakes and seamount asperities: A physical model for earthquake rupture. *Geology,* **20**, 601–604.

Cloos, M. & Shreve, R. 1996. Shear-zone thickness and the seismicity of Chileanand Marianas-type subduction zones. *Geology,* **24**, 107–110.

Close, D.I., Watts, A.B. & Stagg, H.M.J. 2009. A marine geophysical study of the Wilkes Land rifted continental margin, Antarctica. *Geophysical Journal International,* **177**, 430–450. doi:10.1111/j.1365-246X.2008.04066.x.

Cochran, J.R. 1983. Effects of finite rifting times on the development of sedimentary basins. *Earth and Planetary Science Letters,* **66**, 289–302.

Cofaigh, C.O. & Dowdeswell, J.A. 2001. Laminated sediments in glacimarine environments diagnostic criteria for their interpretation. *Quaternary Science Reviews,* **20**, 1411–1436.

Cole, D.I. 1992. Evolution and development of the Karoo Basin. *In:* De Wit, M.J. & Ransome, I. (eds), *Inversion Tectonics of the Cape Fold Belt, Karoo and Cretaceous Basins of Southern Africa,* 87–99. Rotterdam: Balkema.

Colella, A. & d'Alessandro, A. 1988. Sand waves, Echinocardium traces and their bathyal depositional setting (Monte Torre Palaeostrait, Plio-Pleistocene, southern Italy). *Sedimentology,* **35**, 219–237.

Coleman, Jr., J.L., Browne, G.H., King, P.R., Slatt, R.M., Spang, R.J., Williams, E.T. & Clemenceau, G.R. 2000. The inter-relationships of scales of heterogeneity in subsurface, deep water E&P projects-lessons learned from the Mount Messenger Formation (Miocene), Tartanaki Basin, New Zealand. *In:* Weimer, P., Slatt, R.M., Coleman, J., Rosen, N.C., Nelson, H., Bouma, A.H., Styzen, M.J. & Lawrence, D.T. (eds),

Gulf Coast Section-Society of Economic Paleontologists and Mineralogists Foundation 20th Annual Bob F. Hoskins Research Research Conference, Deep-Water Reservoirs of the World, 263–283. CD-ROM Society of Economic Paleontologists and Mineralogists Special Publications.

Collinson, J.D. 1969. The sedimentology of the Grindslow Shales and the Kinderscout Grit: a deltaic complex in the Namurian of northern England. *Journal of Sedimentary Petrology,* **39,** 194–221.

Collinson, J.D. 1970. Deep channels, massive beds and turbidity current genesis in the Central Pennine Basin. *Proceedings of the Yorkshire Geological Society,* **37,** 495–520.

Collot, J.Y., Daniel, J. & Burne, R.V. 1985. Recent tectonics associated with the subduction/collision of the d'Entrecasteaux Zone in the central New Hebrides. *Tectonophysics,* **112,** 325–356.

Coniglio, M. 1986. Synsedimentary submarine slope failure and tectonic deformation in deep-water carbonates. Cow Head Group, western Newfoundland. *Canadian Journal of Earth Science,* **23,** 476–490.

Cook, H.E. & Enos, P. (eds) 1977. *Deep-water Carbonate Environments.* Society of Economic Paleontologists and Mineralogists Special Publication, **25.**

Cooper, M. 2007. Structural style and hydrocarbon prospectivity in fold and thrust belts: a global review. *In:* Ries, A.C., Butler, R.W.H. & Graham, R.H. (eds), *Deformation of the Continental Crust: The Legacy of Mike Coward,* 447–472. The Geological Society, London, Special Publications, **272.** Bath, UK: The Geological Society, London.

Cooper, A.K., O'Brien, P.E. & ODP Leg 188 Shipboard Scientific Party 2001. Early stages of East Antarctic glaciation-insights from drilling and seismic reflection data in the Prydz Bay region. *In:* Florindo, F. & Cooper, A.K. (eds), *The Geologic Record of the Antarctic Ice Sheet from Drilling, Coring and Seismic Studies,* 41–42. Quaderni di Geofisica, Instituto Nazionale di Geofisica e Vulcanologia, Erice.

Coranumsini, G. 2004. Sand-rich turbidite system of the Late Oligocene Northern Apennines foredeep: physical stratigraphy and architecture of the 'Macigno costiero' (coastal Tuscany, Italy). *In:* Lomas, S.A. & Joseph, P. (eds), *Confined Turbidite Systems,* 261–283. Geological Society, London, Special Publications, **222.** London: The Geological Society.

Corbett, K.D. 1972. Features of thick-bedded sandstones in a proximal flysch sequence, Upper Cambrian, southwest Tasmania. *Sedimentology,* **19,** 99–114.

Corliss, B.H. & Chen, C. 1988. Morphotype patterns of Norwegian Sea deep-sea benthic foraminifera and ecological implications. *Geology,* **16,** 716–719.

Corney, R.K.T., Peakall, J., Parsons, D.R., Elliott, L., Amos, K.J., Best, J.L., Keevil, G.M. & Ingham, D.B. 2006. The orientation of helical flow in curved channels. *Sedimentology,* **53,** 249–257.

Corella, J.P., Arantegui, A., Loizeau, J.L., DelSontro, T., Le Dantec, N., Stark, N., Anselmetti, F.S. & Girardclos, S. 2013. Sediment dynamics in the subaquatic channel of the Rhone delta (Lake Geneva, France/Switzerland). *Aquatic Sciences,* **76,** 73–87.

Cossey, S.P.J. & Jacobs, R.E. 1992. Oligocene Hackberry Formation of southwest Louisiana: sequence stratigraphy, sedimentology, and hydrocarbon potential (1). *American Association of Petroleum Geologists Bulletin,* **76,** 589–606.

Cossey, S.P.J. & Kleverlaan, K. 1995. Heterogeneity within a sand-rich submarine fan, Tabernas Basin, Spain. *In:* Pickering, K.T., Hiscott, R.N., Kenyon, N.H., Ricci Lucchi, F. & Smith, R.D.A. (eds), *Atlas of Deep Water Environments: Architectural Style in Turbidite Systems,* 157–161. London: Chapman and Hall.

Cossu, R. & Wells, M.G. 2012. The evolution of submarine channels under the influence of Coriolis forces: experimental observations of flow structures. *Terra Nova,* **25,** 65–71.

Covault, J.A. & Graham, S. A. 2008. Turbidite architecture in proximal foreland basin-system deep-water depocenters: insights from the Cenozoic of Western Europe. *Australian Journal of Earth Sciences,* **101,** 36–51.

Covault, J.A. & Romans, B.W. 2009. Growth patterns of deep-sea fans revisited: Turbidite-system morphology in confined basins, examples from the California Borderland. *Marine Geology,* **265,** 51–66.

Covault, J.A., Normark, W.R., Romans, B.W. & Graham, S.A. 2007. Highstand fans in the California borderland: The overlooked deep-water depositional systems. *Geology,* **35,** 783–786.

Covault, J.A., Hubbard, S.M., Graham, S.A., Hinsch, R. & Linzer, H.-G. 2009. Turbidite-reservoir architecture in complex foredeep-margin and wedge-top depocenters, Tertiary Molasse foreland basin system, Austria. *Marine and Petroleum Geology,* **26,** 379–396.

Covault, J.A., Kostic, S., Paull, C.K., Ryan, H.F. & Fildani, A. 2014. Submarine channel initiation, filling and maintenance from seafloor geomorphology and morphodynamic modelling of cyclic steps. *Sedimentology,* **61,** 1031–1054.

Covey, M. 1986. The evolution of foreland basins to steady state: evidence from the western Taiwan foreland basin. *In:* Allen, P.A. & Homewood, P. (eds), *Foreland Basins,* 77–90. International Association of Sedimentologists Special Publication, **8.** Oxford: Blackwell Scientific.

Cowan, D.S. 1985. The origin of some common types of melange in the Western Cordillera of North America. *In:* Nasu, N. *et al.* (eds), *Formation of Active Ocean Margins,* 257–272. Tokyo: Terrapub.

Coward, M.P. & A.C. Ries (eds) 1986. *Collision Tectonics.* Geological Society of London Special Publication, **19.** Oxford: Blackwell Scientific.

Coward, M.P., Dewey, J.F. & Hancock, P.L. (eds) 1987. *Continental Extensional Tectonics. Geological Society London Special Publication,* **28.** Oxford: Blackwell Scientific.

Cox, S.C. & Sutherland, R. 2007. A Continental Plate Boundary: Tectonics at South Island, New Zealand. *Geophysical Monograph Series,* **175.** 19–46. American Geophysical Union.

Coxall, H.K., Wilson, P.A., Palike, H., Lear, C.H. & Backman, J. 2005. Rapid stepwise onset of Antarctic glaciation and deeper calcite compensation in the Pacific Ocean. *Nature,* **433,** 53–57.

Crane, W.H. & Lowe, D.R. 2008. Architecture and evolution of the Paine channel complex, Cerro Toro Formation (Upper Cretaceous), Silla Syncline, Magallanes Basin, Chile. *Sedimentology,* **55,** 979–1009.

Cremer, M. 1989. Texrture and microstructure of Neogene-Quaternary sediments, ODP Sites 645 and 646, Baffin Bay and Labrador Sea. *In:* Srivastava, S. P., Arthur, M., Clement, B. *et al., Proceedings of the Ocean Drilling Program, Scientific Results,* **105,** 7–20. College Station, Texas, USA: Ocean Drilling Program.

Crimes, P.T. 1977. Trace fossils of an Eocene deep-sea fan, northern Spain. *In:* Crimes, P.T. & Harper, J.C. (eds). *Trace Fossils 2,* Geological Journal Special Issue, **9,** 71–90.

Crimes, P.T., Goldring, R., Homewood, P., Stuijvenberg, J. & Winkler, W. 1981. Trace fossil assemblages of deep-sea fan deposits, Gurnigel and Schlieren flysch Cretaceous–Eocene, Switzerland. *Eclogae Geologicae Helveticae,* **74,** 953–995.

Cronin, B.T. 1995. Structurally-controlled deep-sea channel course: examples from the Miocene of southeast Spain and the Alboran Sea, southwest Mediterranean. *In:* Hartley, A.J. & Prosser, D.J. (eds), *Characterization of Deep Marine Clastic Systems,* 115–135. Geological Society, London, Special Publication, **94.** Bath: The Geological Society.

Crouch, J.K. 1981. Northwest margin of California Continental Borderland: marine geology and tectonic evolution. *American Association of Petroleum Geologists Bulletin,* **65,** 191–218.

Crouch, J.K. & Suppe, J. 1993. Late Cenozoic tectonic evolution of the Los Angeles basin and inner California borderland – A model for core complex-like crustal extension. *Geological Society of America Bulletin,* **103,** 1415–1434.

Crowell, J.C. 1974a. Origin of Late Cenozoic basins in southern California. *Society of Economic Paleontologists and Mineralogists Special Publication,* **22,** 190–204.

Crowell, J.C. 1974b. Sedimentation along the San Andreas fault, California. *Society of Economic Paleontologists and Mineralogists Special Publication,* **19,** 292–303.

Crowell, J.C. 1957. Origin of pebbly mudstones. *Bulletin of the Geological Society of America,* **68,** 993–1010.

Csejtey, Bela, Jr., Cox, D.P. & Evarts, R.C. 1982. The Cenozoic Denali fault system and the Cretaceous accretionary development of southern Alaska. *Journal of Geophysical Research*, **87**, 3741–3754.

Csejtey, Bela, Jr., Mullen, M.W., Cox, D.P. & Stricker, G.D. 1992. Geology and geochronology of the Healy quadrangle, southcentral Alaska. *U.S. Geological Survey Miscellaneous Investigations Series Map I-1961*, scale 1:250,000, 2 sheets.

Cummings, J.P. & Hodgson, D.M. 2011. Assessing Controls on the Distribution of Ichnotaxa in Submarine Fan Environments, the Basque Basin, Northern Spain. *Sedimentary Geology*, **239**, 162–187.

Curran, K.J., Hill, P.S. & Milligan, T.G. 2002. Fine-grained suspended sediment dynamics in the Eel River flood plume. *Continental Shelf Research*, **22**, 2537–2550.

Curran, K.J., Hill, P.S., Schell, T.M., Milligan, T.G. & Piper, D.J.W. 2004. Inferring the mass fraction of floc-deposited mud: application to fine-grained turbidites. *Sedimentology*, **51**, 927–944.

Curray, J.R. & Moore, D.G. 1974. Sedimentary and tectonic processes in the Bengal deep-sea fan and geosyncline. *In*: Burk, C.A. & Drake, C.L. (eds), *The Geology of Continental Margins*, 617–627. New York: Springer.

Curray, J.R., Moore, D.G., Lawver, L.A., Emmel, F.J., Raitt, R.W., Henry, M. & Kieckhefer, R. 1979. Tectonics of the Andaman Sea and Burma. *In*: Watkins, J.S., Montadert, L. & Dickerson, P.W. (eds), *Geological and Geophysical Investigations of Continental Margins*, 189–198. American Association of Petroleum Geologists Memoir, **29**.

Curray, J.R., Emmel, F.J. & Moore, D.G. 2003. The Bengal Fan: morphology, geometry, stratigraphy, history and processes. *Marine and Petroleum Geology*, **19**, 1191–1223.

Curray, J.R., Moore, D.G. *et al*. 1982a. *Initial Reports Deep Sea Drilling Project, 64*. Washington, DC: US Government Printing Office.

Curray, J.R., Moore, D.G., Kelts, K. & Einsele, G. 1982b. Tectonics and geological history of the passive continental margin at the tip of Baja California. *In*: Curray, J.R., Moore, D.G. *et al*., *Initial Reports Deep Sea Drilling Project, 64*. 1089–1116. Washington, DC: US Government Printing Office.

Curry, G.B., Ingham, J.K., Bluck, B.J. & Williams, A. 1982. The significance of a reliable Ordovician age for some Highland Border rocks in Central Scotland. *Journal of the Geological Society, London*, **139**, 451–454.

Da Silva, J.C.B., New, A.L. & Magalhaes, J.M. 2009. Internal solitary waves in the Mozambique Channel: Observations and interpretation. *Journal of Geophysical Research*, **114**, 1–12. doi:10.1029/2008JC005125.

Dacey, M.F. 1979. Models of bed formation. *Mathematical Geology*, **11**, 655–668.

Dade, W.B. & Friend, P.F. 1998. Grain-size, sediment-transport regime, and channel slope in alluvial rivers. *Journal of Geology*, **106**, 661–675.

Dade, W.B. & Huppert, H.E. 1995. A box model for non-entraining, suspension-driven gravity surges on horizontal surfaces. *Sedimentology*, **42**, 453–471.

Dade, W.B., Lister, J.R. & Huppert, H.E. 1994. Fine-sediment deposition from gravity surges on uniform slopes. *Journal of Sedimentary Research*, **A64**, 423–432.

Dakin, N., Pickering, K.T., Mohrig, D. & Bayliss, N.J. 2013. Channel-like features created by erosive submarine debris flows: Field evidence from the Middle Eocene Ainsa Basin, Spanish Pyrenees. *Marine and Petroleum Geology*, **41**, 62–71.

Daly, R.A. 1936. Origin of submarine canyons. *American Journal of Science*, **31**, 410–420.

Damuth, J.E. 1977. Late Quaternary sedimentation in the western equatorial Atlantic. *Geological Society of America Bulletin*, **88**, 695–710.

Damuth, J.E. 1980. Use of high-frequency (3.5–12 kHz) echograms in the study of near-bottom sedimentation processes in the deep-sea: a review. *Marine Geology*, **38**, 51–75.

Damuth, J.E. 1994. Neogene gravity tectonics and depositional processes on the deep Niger Delta continental margin. *Marine and Petroleum Geology*, **11**, 320–346.

Damuth, J.E. & Embley, R.W. 1981. Mass-transport processes on Amazon Cone: western equatorial Atlantic. *American Association of Petroleum Geologists Bulletin*, **65**, 629–643.

Damuth, J.E. & Flood, R.D. 1984. Morphology, sedimentation processes and growth pattern of the Amazon deep-sea fan. *Geo-Marine Letters*, **3**, 109–117.

Damuth, J.E. & Flood, R.D. 1985. Amazon Fan, Atlantic Ocean. *In*: Bouma, A.H., Normark, W.R. & Barnes, N.E. (eds), *Submarine Fans and Related Turbidite Systems*, 97–106. New York: Springer.

Damuth, J.E. & Kumar, N. 1975a. Amazon Cone: morphology, sediments, age, and growth pattern. *Geological Society of America Bulletin*, **86**, 863–878.

Damuth, J.E. & Kumar, N. 1975b. Late Quaternary depositional processes on the continental rise of the western equatorial Atlantic: comparison with the western North Atlantic and implications for reservoir rock distribution. *American Association of Petroleum Geologists Bulletin*, **59**, 2171–2181.

Damuth, J.E., Kolla, V., Flood, R.D., Kowsmann, R.O., Monteiro, M.C., Gorini, M.A., Palma, J.J.C. & Belderson, R.H. 1983. Distributary channel meandering and bifurcation patterns on the Amazon deep-sea fan as revealed by long-range side-scan sonar (GLORIA). *Geology*, **11**, 94–98.

Damuth, J.E., Flood, R.D., Kowsmann, R.O., Belderson, R.H. & Gorini, M.A.1988. Anatomy and growth pattern of Amazon deep-sea fan as revealed by long-range side-scan sonar (GLORIA) and high-resolution seismic studies. *American Association of Petroleum Geologists Bulletin*, **72**, 885–911.

Damuth, J.E., Flood, R.D., Pirmez, C. & Manley, P.L. 1995. Architectural elements and depositional processes of Amazon Deep-sea Fan imaged by sidescan sonar (GLORIA), bathymetric swath-mapping (SeaBeam), high-resolution seismic, and piston-core data. *In*: Pickering, K.T., Hiscott, R.N., Kenyon, N.H., Ricci Lucchi, F. & Smith, R.D.A. (eds), *Atlas of Deep Water Environments: Architectural Style in Turbidite Systems*, 105–121. London: Chapman and Hall.

Danforth, A., Granath, J.W., Gross, J.S., Horn, B.W., McDonough, K.-J. & Sterne, E.J. 2012. Deepwater fans across a transform margin, offshore East Africa. *GEO ExPro*, **9**, 76–78.

Daniel, W.W. 1978. *Applied Nonparametric Statistics*. Boston: Houghton Mifflin.

Darby, S.E. & Peakall, J. 2012. Modelling the equilibrium bed topography of submarine meanders that exhibit reversed secondary flows. *Geomorphology*, **163–164**, 99–109.

Das Gupta, K. & Pickering, K.T. 2008. Petrography and temporal changes in petrofacies of deep-marine Ainsa-Jaca basin sandstone systems, Early and Middle Eocene, Spanish Pyrenees. *Sedimentology*, **55**, 1083–1114.

Das, H.S., Imran, J. & Mohrig, D. 2004. Numerical modelling of flow and bed evolution in meandering submarine channels. *Journal of Geophysical Research C, Oceans*, **109**, 1–17.

Davey, F.J., Hampton, M., Childs, J., Fisher, M.A., Lewis, K. & Pettinga, J.R. 1986. Structure of a growing accretionary prism, Hikurangi margin, New Zealand. *Geology*, **14**, 663–666.

Davies, I.C. & Walker, R.G. 1974. Transport and deposition of resedimented conglomerates: the Cap Enrage Formation, Cambro-Ordovician, Gaspe, Quebec. *Journal of Sedimentary Petrology*, **44**, 1200–1216.

Davies, G.R. 1977. Turbidites, debris sheets and truncation structures in upper Paleozoic deep-water carbonates of the Svedrup Basin, Arctic Archipelago. *In*: Cook, H.E. & Enos, P. (eds), *Deep-water Carbonate Environments*, 221–249. Society of Economic Paleontologists and Mineralogists, Special Publication, **25**.

Davies, J.R. & Waters, R.A. 1995. The Caban Conglomerate and Ystrad Meurig Grits Formation-nested channels and lobe switching on a mud-dominated latest Ashgill to Llandovery slope apron, Welsh Basin, UK. *In*: Pickering, K.T., Hiscott, R.N., Kenyon, N.H., Ricci Lucchi, F. & Smith, R.D.A. (eds), *Atlas of Deep Water Environments:*

Architectural Style in Turbidite Systems, 184–193. London: Chapman and Hall.

Davison, I., Dinkelman, M.G. & Kool, W. 2010. Two frontier basins come to light. *GEO ExPro*, **7**, 36–40.

De Boer, P.L. & Alexandre, J.T. 2012. Orbitally forced sedimentary rhythms in the stratigraphic record: is there room for tidal forcing? *Sedimentology*, **59**, 379–392.

De Boer, P.L. & Smith, D.G. (eds) 1994. *Orbital Forcing and Cyclic Sequences*. International Association of Sedimentologists Special Publication, **19**.

De Boer, P.L., Pragt, J.S.J. & Oost, A.P. 1991. Vertically persistent sedimentary facies boundaries along growth anticlines and climate-controlled sedimentation in the thrust-sheet-top South Pyrenean Tremp-Graus Foreland Basin. *Basin Research*, **3**, 63–78.

De Quatrefages, M.A. 1849. Note sur la Scolicia prisca (A. De Q.) annélide fossile de la craie. *Annales des Sciences Naturelles, 3 Sèrie, Zoologie*, **12**, 265–266.

De Silva, N.R. 1994. Submarine fans on the northeastern Grand Banks, offshore Newfoundland. *In*: Weimer, P., Bouma, A.H. & Perkins, B.F. (eds), *Gulf Coast Section-Society of Economic Paleontologists and Mineralogists Foundation 15th Annual Research Conference, Submarine Fans and Turbidite Systems: Sequence Stratigraphy, Reservoir Architecture and Production Characteristics*, 95–104.

De Smet, M.E.M., Fortuin, A.R., Troelstra, S.R., Van Marle, L.J., Karmini, M., Tjokosaproetro, S. & Hadiwasastra, S. 1990. Detection of collision related vertical movements in the Outer Banda Arc (Timor, Indonesia), using micropaleontological data. *Journal of SE Asian Earth Sciences*, **4**, 337–356.

DeConto, R.M. & Pollard, D. 2003. Rapid Cenozoic glaciation of Antarctica induced by declining atmospheric CO_2. *Nature*, **421**, 245–249.

Dehyle, A., Kopf, A., Frape, S. & Hesse, R. 2003. Evidence for fluid flow in the Japan Trench forearc using isotope geochemistry (Cl, Sr, B): Results from ODP Site 1150. *The Island Arc*, **13**, 258–270.

Del Ben, A., Barnaba, C. & Taboga, A. 2008. Strike-slip systems as the main tectonic features in the Plio-Quaternary kinematics of the Calabrian Arc. *Marine Geophysical Research*, **29**, 1–12.

DeMets, C. 1992. Oblique convergence and deformation along the Kuril and Japan trenches. *Journal of Geophysical Research*, **97**, 17,615–17,626.

DeMets, C., Gordon, R.G., Argus, D.F. & Stein, S. 1990. Current plate motions. *Geophysical Journal International*, **101**, 425–478.

DeMets, C., Jansma, P.E., Mattioli, G.S., Dixon, T.H., Farina, F., Bilham, R., Calais, E. & Mann, P. 2000. GPS geodetic constraints on Caribbean-North America plate motion. *Geophysical Research Letters*, **27**, 437–440.

Demko, T.M., Fedele, J., Hoyal, D., Pederson, K., Hamilton, P., Abreu, V. & Postma, G. 2014. Bedforms indicative of supercritical flow in steep, sandy submarine afns and fan deltas: Ainsa, Ebro and Tabernas basins, Spain. *Geological Society of America Annual Meeting, Vancouver, British Columbia, 19–22 October*, Abstract *Volume*.

Dennielou, B., Huchona, A., Beaudouinc, C & Bernéa, S. 2006. Vertical grain-size variability within a turbidite levee: autocyclicity or allocyclicity? A case study from the Rhône neofan, Gulf of Lions, Western Mediterranean. *Marine Geology*, **234**, 191–213.

Deptuck, M.E., Steffens, G.S., Barton, M. & Pirmez, C. 2003. Architecture and evolution of upper fan channel belts on the Niger Delta slope and in the Arabian Sea. *Marine and Petroleum Geology*, **20**, 649–676.

Deptuck, M.E., Sylvester, Z., Pirmez, C. & O'Byrne, C. 2007. Migration–aggradation history and 3-D seismic geomorphology of submarine channels in the Pleistocene Benin-major Canyon, western Niger Delta slope. *Marine and Petroleum Geology*, **24**, 406–433.

Deptuck, M.E., Piper, D.J.W., Savoye, B. & Gervais, A. 2008. Dimensions and architecture of late Pleistocene submarine lobes off the northern margin of East Corsica. *Sedimentology*, **55**, 869–898.

Deutsch, C.V. & Journel, A. 1998. *GSLIB: Geostatistical Software Library and User's Guide*, 2nd edn. Oxford University Press, New York 369 pp.

Deutsch, C.V. & Tran, T.T. 2002. FLUVSIM: a program for object-based stochastic modeling of fluvial depositional systems. *Computers & Geosciences*, **28**, 525–535.

Deville, E., Callac, Y., Desaubliaux, G., Mascle, A., Huyghe-Mugnier, P., Griboulard, R. & Noble, M. 2003. Deepwater erosion processes in the Orinoco turbidite system. *Offshore*, 10/01/2003.

Dewey, J.F. 1962. The provenance and emplacement of upper Arenigian turbidites in Co. Mayo, Eire. *Geological Magazine*, **99**, 238–252.

Dewey, J.F. & Sengör, A.M.C. 1979. Aegean and surrounding regions: complex multiplate and continuum tectonics in a convergent zone. *Geological Society of America Bulletin*, **90**, 84–92.

d'Heilly, P., Millot, C., Monaco, A. & Got, H. 1988. Hydrodynamic study of the furrow of the Petit Rhône canyon. *Deep-Sea Research*, **35**, 465–471.

Di Celma, C.N., Brunt, R.L., Hodgson, D.M., Flint, S.S. & Kavanagh, J.P. 2011. Spatial and Temporal evolution of a Permian submarine slope channel-levée system, Karoo Basin, South Africa. *Journal of Sedimentary Research*, **81**, 579–599.

Di Iorio, D., Akal, T., Guerrini, P., Yüce, H., Gezgin, E. & Özsoy, E. 1999. Oceanographic Measurements of the West Black Sea: June 15 to July 5, 1996. *NATO: Report SR-305, SACLANTCEN*, La Spezia, Italy.

Di Toro, G.A.E. 1995. Angel Formation turbidites in the Wanea field area, Dampier sub-basin, North-West shelf, Australia. *In*: Pickering, K.T., Hiscott, R.N., Kenyon, N.H., Ricci Lucchi, F. & Smith, R.D.A. (eds), *Atlas of Deep Water Environments: Architectural Style in Turbidite Systems*, 2260–266. London: Chapman and Hall.

Dickinson, W.R. 1974. Plate tectonics and sedimentation. *In*: Dickinson, W.R. (ed.), *Tectonics and Sedimentation*. Special Publication of the Society of Economic Paleontologists and Mineralogists, **22**, 1–27. Tulsa, Oklahoma.

Dickinson, W.R. & Seely, D.R. 1979. Structure and stratigraphy of forearc regions. *American Association of Petroleum Geologists Bulletin*, **63**, 2–31.

Dickinson, W.R. & Suczek, C. 1979. Plate tectonics and sandstone composition. *American Association of Petroleum Geologists Bulletin*, **63**, 2164–2182.

Dickinson, W.R., Beard, L.S., Brakenridge, G.R., Erjavec, J.L., Ferguson, R.C., Inman, K.F., Knepp, R.A., Lindberg, F.A. & Ryberg, P.T. 1983. Provenance of North American Phanerozoic sandstones in relation to tectonic setting. *Geological Society of American, Bulletin*, **94**, 222–235.

Dickinson, W.R., Ducea, M., Rosenberg, L.I., Greene, H.G., Graham, S.A., Clark, J.C., Weber, G.E., Kidder, S., Ernst, W.G. & Brabb, E.E. 2005. Net dextral slip, Neogene San Gregorio–Hosgri fault zone, coastal California: Geologic evidence and tectonic implications. *Geological Society of America, Special Paper*, **391**.

Dill, R.F. 1964. Sedimentation and erosion in Scripps submarine canyon head. *In*: Miller, R.L. (ed.), *Papers in Marine Geology*, 23–41. New York: McMillan.

Dimberline, A.J. & Woodcock, N.H. 1987. The southeast margin of the Wenlock turbidite system, Mid-Wales. *Geological Journal*, **22**, 61–71.

Dimitrijevic, M.N. & Dimitrijevic, M.D. (eds) 1987. *Turbiditic Basins of Serbia*. Serbian Academy of Sciences and Arts, Monographs, **61**, 304 pp.

Dimitrov, L.I. 2002. Mud volcanoes-the most important pathway for degassing deeply buried sediments. *Earth-Science Reviews*, **59**, 49–76.

Dinkelman, M.G., Granath, J.W. & Whitaker, R. 2010. The NE Greenland continental margin. *GEO ExPro*, **7**, 36–40.

Dixon, J.F., Steel, R.J. & Olariu, C. 2012. River-dominated, shelf-edge deltas: delivery of sand across the shelf break in the absence of slope incision. *Sedimentology*, **59**, 1133–1157.

Dixon, R.J., Schofield, K., Anderton, R., Reynolds, A.D., Alexander, R.W.S., Williams, M.C. & Davies, K.G. 1995. Sandstone diapirism and clastic injection in the Tertiary submarine fans of the Bruce-Beryl embayment, quadrant 9, UKCS. *In*: Hartley, A.J. & Prosser, D.J. (eds), *Characterization of Deep Marine Clastic Systems*, 77–94. Geological Society London Special Publication, **94**.

Dixon, T.H. & Moore, J.C. (eds) 2007. *The Seismogenic Zone of Subduction Thrust Faults.* New York: Columbia University Press, 680 pp.

Dolan, J.F., Beck, C., Ogawa, Y. & Klaus, K. 1990. Eocene–Oligocene sedimentation in the Tiburon Rise/ODP Leg 110 area: an example of significant upslope flow of distal turbidity currents. *In:* Moore, J. C, Mascle, A. *et al., Proceedings of the Ocean Drilling Program, Scientific Results,* **110**, 47–83. College Station, Texas, USA: Ocean Drilling Program.

Dolan, J.F., Sieh, K. & Rockwell, T.K. 2000. Late Quaternary activity and seismic potential of the Santa Monica fault system, Los Angeles, California. *Geological Society of America Bulletin,* **112**, 1559–1581.

Dominguez, S., Lallemand, S.E., Malavieille, J. & von Huene, R. 1998. Upper plate deformation associated with seamount subduction. *Tectonophysics,* **293**, 207–224.

Donovan, S.K. (including a joint contribution with Harper, D.A.T.) 2005. The geology of Barbados: a field guide. *Caribbean Journal of Earth Science,* **38**, 21–33. Geological Society of Jamaica.

Dooley, T. & McClay, K. 1997. Analog modeling of pull-apart basins. *American Association of Petroleum Geologists Bulletin,* **81**, 1804–1826.

Doré, A.G. & Vining, B.A. (eds) 2005. *Petroleum Geology: North-west Europe and Global Perspectives-Proceedings of the 6th Petroleum Geology Conference,* **1**. London: The Geological Society.

Doré, G. & Robbins, J. 2005. The Buzzard Field. *In:* Doré, A.G. & Vining, B.A. (eds), *Petroleum Geology: North-west Europe and Global Perspectives-Proceedings of the 6th Petroleum Geology Conference,* Volume 1, 241–252. London: The Geological Society.

Dott, R.H. Jr. 1961. Squantum "Tillite", Massachusetts: evidence of glaciation or subaqueous mass movements? *Geological Society of America Bulletin,* **72**, 1289–1306.

Dorel, J. 1981. Seismicity and seismic gap in the Lesser Antilles arc and earthquake hazard in Guadeloupe. *Geophysical Journal of the Royal Astronomical Society,* **67**, 679–695.

Dowdeswell, J.A., Whittington, R.J., Marienfeld, P. 1994. The origin of massive diamicton facies by iceberg rafting and scouring, Scoresby Sund, East Greenland. *Sedimentology,* **41**, 21–35.

Downie, N.M. & Heath, R.W. 1983. *Basic Statistical Methods.* New York: Harper & Row.

Doxsee. W.W. 1948. The Grand Banks earthquake of November 18, 1929. *Publications of the Dominion Observatory,* **7**(7), 323–335.

Doyle, L.J. & Pilkey, O.H. (eds) 1979. *Geology of Continental Slopes. Society of Economic Paleontologists and Mineralogists Special Publication,* **27**.

Drake, D.E., Kolpack, R.L. & Fischer, P.J. 1972. Sediment transport on the Santa Barbara-Oxnard shelf, Santa Barbara Channel, California. *In:* Swift, D.J.P., Duane, D.B. & Pilkey, O.H. (eds), *Shelf Sediment Transport,* 307–331. Stroudsburg, PA: Dowden, Hutchinson & Ross.

Drake, D.E., Hatcher, P.G. & Keller, G.H. 1978. Suspended particulate matter and mud deposition in Upper Hudson submarine canyon. *In:* Stanley, D.J. & Kelling, G. (eds), *Sedimentation in Submarine Canyons, Fans, and Trenches,* 33–41. Stroudsburg, PA: Dowden, Hutchinson & Ross.

Draut, A.E., Clift, P.D. & Scholl, D.W. (eds) 2008. *Formation and Application of the Sedimentary Record in Arc Collision Zones.* The Geological Society of America, Special Paper **436**.

Dreyer, T., Corregidor, J., Arbues, P. & Puigdefabregas, C. 1999. Architecture of the tectonically influenced Sobrarbe deltaic complex in the Ainsa Basin, northern Spain. *Sedimentary Geology,* **127**, 127–169.

Drinkwater, N.J. 1995, Sheet-like turbidite system: the Kongsfjord Formation, Finnmark, north Norway. *In:* Pickering, K.T., Hiscott, R.N., Kenyon, N.H., Ricci Lucchi, F. & Smith, R.D.A. (eds), *Atlas of Deep Water Environments: Architectural Style in Turbidite Systems,* 267–275. London: Chapman and Hall.

Drinkwater, N.J. & Pickering, K.T. 2001. Architectural elements in a high-continuity sand-prone turbidite system, late Precambrian Kongsfjord Formation, northern Norway: Application to hydrocarbon reservoir characterization. *American Association of Petroleum Geologists Bulletin,* **85**, 1731–1757.

Drinkwater, N.J, Pickering, K.T & Siedlecka, A. 1996. Deep-water fault-controlled sedimentation, Arctic Norway and Russia: response to Late Proterozoic rifting and the opening of the Iapetus Ocean. *Journal of the Geological Society, London,* **153**, 427–436.

Driscoll, M.L., Tucholke, B.E. & McCave, I.N. 1985. Seafloor zonation in sediment texture on the Nova Scotian lower continental rise. *Marine Geology,* **66**, 25–41.

Dorrell, R.M., Darby, S.E., Peakall, J., Sumner, E.J., Parsons, D.R. & Wynn, R.B. 2013. Superelevation and overspill control secondary flow dynamics in submarine channels. *Journal of Geophysical Research: Oceans,* **118**, 3895–3915.

Droser, M.L. & Bottjer, D.J. 1986. A semiquantitative field classification of ichnofabric. *Journal of Sedimentary Petrology,* **56**, 558–559.

Droser, M.L. & Bottjer, D.J. 1988. Trends in depth and extent of bioturbation in Cambrian carbonate marine environments, western United States. *Geology,* **16**, 233–236.

Droser, M.L. & Bottjer, D.J. 1989. Ichnofabric of sandstone deposited in high-energy nearshore environments: measurement and utilization. *Palaios,* **4**, 598–604.

Droz, L, Rigaut, F., Cochonat, P. & Tofani, R. 1996. Morphology and recent evolution of the Zaire turbidite system (Gulf of Guinea). *Geological Society of America Bulletin,* **108**, 253–269.

Droz, L. & Bellaiche, G. 1985. Rhône deep-sea fan: morphostructure and growth pattern. *American Association of Petroleum Geologists Bulletin,* **69**, 460–479.

Droz, L. & Bellaiche, G. 1991. Seismic facies and geologic evolution of the central portion of the Indus Fan. *In:* Weimer, P. & Link, M.H. (eds), *Seismic Facies and Sedimentary Processes of Submarine Fans and Turbidite Systems,* 383–402. New York: Springer-Verlag.

Droz, L. & Mougenot, D. 1987. Mozambique Upper Fan: origin of depositional units. *American Association of Petroleum Geologists Bulletin,* **71**, 1355–1365.

Drummond, C.N. 1999. Bed-thickness structure of multi-sourced ramp turbidites: Devonian Brallier Formation, central Appalachian basin. *Journal of Sedimentary Research,* **69**, 115–121.

Drummond, C.N. & Wilkinson, B.H. 1996. Stratal thickness frequencies and the prevalence of orderedness in stratigraphic sequences. *Journal of Geology,* **104**, 1–18.

Du Fornel, E., Joseph, P., Desaubliaux, G., Eschard, R., Guillocheau, F., Lerat, O., Muller, C., Ravenne, C. & Sztrakos, K. 2004. The southern Grès d'Annot Outcrops (French Alps): an attempt at regional correlation. *In:* Joseph, P. & Lomas, S.A. (eds), *Deep-Water Sedimentation in the Alpine Basin of SE France: New Perspectives on the Grès d'Annot and Related Systems,* 137–160. Geological Society, London, Special Publication, **221**. London: The Geological Society.

Duarte, L.V., Silva, R.L., Oliveira, L.C.E., Comas-Rengifo, M.J. & Silva, F. 2010. Organic-rich facies in the Sinemurian and Pliensbachian of the Lusitanian Basin, Portugal: Total organic carbon distribution and relation to transgressive-regressive facies cycles. *Geologica Acta,* **8**, 325–340.

Dubois, J. & Deplus, C. 1989. Gravimetry on the Erimo Seamount, Japan. *Tectonophysics,* **160**, 267–275.

Duda, S.J. 1965. Secular seismic energy release in the circum-Pacific belt. *Tectonophysics,* **2**, 409–452.

Dunbar, R.B. 1981. *Sedimentation and the history of upwelling and climate in high fertility areas of the northeastern Pacific Ocean.* Ph.D. thesis San Diego, Scripps Institution of Oceanography.

Dunbar, R.B. & Berger, W.H. 1981. Fecal pellet flux to modern bottom sediment of Santa Barbara Basin (California) based on sediment trapping. *Geological Society of America Bulletin,* **92**, 212–218.

Duncan, R.A. 1982. A captured island chain in the Coast Range of Oregon and Washington. *Journal of Geophysical Research,* **87**, 10,827–10,837.

Dutton, S.P., Flanders, W.A. & Barton, M.D. 2003. Reservoir characterization of a Permian deep-water sandstone, East Ford field, Delaware basin, Texas. *American Association of Petroleum Geologists Bulletin,* **87**, 609–627.

Duval, G. 2013. The UK West Central Graben. *GEO ExPro,* **10**, 58–62.

Dykstra, M. & Kneller, B. 2009. Lateral accretion in a deep-marine channel complex: implications for channellized flow processes in turbidity currents. *Sedimentology*, **56**, 1411–1432.

Dykstra, M., Kneller, B. & Milana, J.-P. 2012. Bed-thickness and grain-size trends in a small-scale proglacial channel-levée system; the Carboniferous Jejenes Formation, Western Argentina: implications for turbidity current flow processes. *Sedimentology*, **59**, 605–622.

Dziadzio, P.S., Enfield, M.A., Watkinson, M.P. & Porebski, S. 2006. The Ciezkowice Sandstone: examples of basin-floor fan-stacking patterns from the Main (Upper Paleocene to Eocene) Reservoir in the Polish Carpathians. *In*: Golonka, J. & Picha, F.J. (eds), *The Carpathians and their Foreland: Geology and Hydrocarbon Resources, 477–496.* American Association of Petroleum Geologists Memoir, **84**.

Dzułynski, S. & Sanders, J.E. 1962. Current marks on firm mud bottoms. *Transactions of the Connecticut Academy of Arts and Sciences*, **42**, 57–96.

Dzułynski, S. & Walton, E.K. 1965. *Sedimentary Features of Flysch and Greywackes. Developments in Sedimentology*, **7**, 274 pp. Amsterdam: Elsevier.

Dzułynski, S., Ksiaskiewicz, M. & Kuenen, Ph.H. 1959. Turbidites in flysch of the Polish Carpathian Mountains. *Bulletin of the Geological Society of America*, **70**, 1089–1118.

Eastham, K.R. & Ridgeway, K.D. 2000. Stratigraphic and provenance data from the Upper Jurassic to Upper Cretaceous Kahiltna assemblage of South-Central Alaska. *Studies by the U.S. Geological Survey in Alaska, U.S. Geological Survey Professional Paper*, **1662**.

Edgers, L. & Karlsrud, K. 1982. Soil flows generated by submarine slides-case studies and consequences. *Norges Geotekniske Instututt*, **143**, 1–11.

Ediger, V., Okyar, M. & Ergin, M. 1993. Seismic stratigraphy of the fault-controlled submarine canyon/valley system on the shelf and upper slope of Anamur Bay, Northeastern Mediterranean Sea. *Marine Geology*, **115**, 129–142.

Edwards, D.A. 1993. *Turbidity currents: dynamics, deposits and reversals. Lecture notes in Earth Sciences, 44*. Berlin: Springer-Verlag. 173 pp.

Edwards, D.A., Leeder, M.R., Best, J.L. & Pantin, H.M. 1994. On experimental reflected density currents and the interpretation of certain turbidites. *Sedimentology*, **41**, 437–461.

Edwards, B.D., Lee, H.J. & Field, M.E. 1995. Mudflow generated by retrogressive slope failure, Santa Barbara Basin, California continental borderland. *Journal of Sedimentary Research*, **65**, 57–68.

Edwards, M.B. 1986. Glacial environments. *In*: Reading, H.G. (ed.), *Sedimentary Environments and Facies,* 2nd edn, 445–470. Oxford: Blackwell Scientific.

Egan, J.A. & Sangrey, D.A. 1978. Critical state model of cyclic load pore pressures. *American Society of Civil Engineers Special Conference, Earthquake Engineering Soil Dynamics*, **1**, 410–424.

Eggenhuisen, J.T. & McCaffrey, W.D. 2012. The vertical turbulence structure of experimental turbidity currents encountering basal obstructions: implications for vertical suspended sediment distribution in non-equilibrium currents. *Sedimentology*, **59**, 1101–1120.

Egger, H., Homayoun, M. & Schnabel, W. 2002. Tectonic and climatic control of Paleogene sedimentation in the Rhenodanubian Flysch basin (Eastern Alps, Austria). *Sedimentary Geology*, **152**, 247–262.

Elders, C. 1987. The provenance of granite boulders in conglomerates of the Northern and Central Belts of the Southern Uplands of Scotland. *Journal of the Geological Society, London*, **144**, 853–863.

Eichelberger, J., Gordeev, E., Kasahara, M., Izbekov, P. & Lees, J. (eds) 2007. *Volcanism and Subduction: The Kamchatka Region. Geophysical Monograph Series*, **172**, 369 pp.

Eidvin, T., Jansen, E. & Riis, F. 1993. Chronology of Tertiary fan deposits off the western Barents Sea: implications for uplift and erosion history of the Barents shelf. *Marine Geology*, **112**, 109–131.

Ehlig, P.L. 1981. Origin and tectonic history of the basement terrane of the San Gabriel Mountains, central Transverse Region. *In*: Ernst, G. (ed.), *The Geotectonic Development of California*, 254–283. New Jersey: Prentice-Hall.

Eiler, J. 2004. *Inside the Subduction Factory. American Geophysical Union, Geophysical Monograph Series*, **138**, 324 pp.

Einsele, G. 1992. *Sedimentary Basins: Evolution, Facies, and Sediment Budget.* Berlin, Germany: Springer-Verlag, 628 pp.

Einsele, G. 1985. Response of sediments to sea-level changes in differing subsiding storm-dominated marginal and epeiric basins. *In*: Bayer, U. & Seilacher, A. (eds), *Sedimentary and Evolutionary Cycles,* 68–97. Lecture Notes in Earth Sciences, **1**.

Einsele, G., Gieskes, J.M., Curray, J.M., Moore, D., Aguayo, E., Aubry, M-P., Fornari, D., Guerrero, J., Kastner, M., Kelts, K., Lyle, M., Matola, M., Molina-Cruz, A., Niemitz, J., Rueda, J., Saunders, A., Schrader, H., Simoniet, B. & Vacquier, V. 1980. Intrusion of basaltic sills into highly porous sediments, and resulting hydrothermal activity. *Nature*, **283**, 441–445.

Eittreim, S., Grantz, A. & Greenberg, J. 1982. Active geologic processes in Barrow Canyon, northeast Chukchi Sea. *Marine Geology*, **50**, 61–76.

Ekdale, A.A. 1980. Graphoglyptid burrows in modern deep-sea sediment. *Science*, **207**, 304–306.

Ekdale, A.A. 1985. Paleoecology of the marine endobenthos. *Palaeogeography, Palaeoclimatology, Palaeoecology*, **50**, 63–81.

Ekdale, A.A. & Berger, W.H. 1978. Deep-sea ichnofacies: modern organism traces on and in pelagic carbonates of the western equatorial Pacific. *Palaeogeography, Palaeoclimatology, Palaeoecology*, **23**, 268–278.

Ekdale, A.A. & Bromley, R.G. 1983. Trace fossils and ichnofabrics in the Kjølby Gaard Marl, uppermost Cretaceous, Denmark. *Bulletin of the Geological Society of Denmark*, **31**, 107–119.

Ekdale, A.A. & Bromley, R.G. 1991. Analysis of composite ichnofabrics: an example in Uppermost Cretaceous Chalk of Denmark. *Palaios*, **6**, 232–249.

Ekdale, A.A. & Mason, T.R. 1988. Characteristic trace-fossil associations in oxygen-poor sedimentary environments. *Geology*, **16**, 720–723.

Ekdale, A.A., Bromley, R.G. & Knaust, D. 2012. The Ichnofabric Concept. *In*: Knaust, D. & Bromley, R.G. (eds), *Trace Fossils as Indicators of Sedimentary Environments,* 139–156. Developments in Sedimentology, **64**. Amsterdam: Elsevier.

Ekdale, A.A., Bromley, R.G. & Pemberton, G. 1985. *Ichnology, Trace Fossils in Sedimentology and Stratigraphy*. Society of Economic Paleontologists & Mineralogists Short Course, **15**.

Elitok, O. & Dolmaz, M.N. 2011. Tectonic escape mechanism in the crustal evolution of Eastern Anatolian region (Turkey) *In*: Schattner, U. (ed.), *New Frontiers in Tectonic Research at the Midst of Plate Convergence*, 289–302. Rijeka, Croatia: Intech.

Elliott, T. 2000a. Megaflute erosion surfaces and the initiation of turbidite channels. *Geology*, **28**, 119–122.

Elliott, T. 2000b. Depositional architecture of a sand-rich, channelized turbidite system: the Upper Carboniferous Ross Sandstone Formation, western Ireland. *In*: Weimer, P., Slatt, R.M., Coleman, J., Rosen, N.C., Nelson, H., Bouma, A.H., Styzen, M.J. & Lawrence, D.T. (eds), *Gulf Coast Section-Society of Economic Paleontologists and Mineralogists Foundation 20th Annual Bob F. Hoskins Research Research Conference, Deep-Water Reservoirs of the World, 342–364.* CD-ROM Society of Economic Paleontologists and Mineralogists Special Publications.

Elliott, T., Apps, G., Davies, H., Evans, M., Ghibaudo, G. & Graham, R.H. 1985. A structural and sedimentological traverse through the Tertiary foreland basin of the external alps of southeast France. *In*: Allen, P.A. & Homewood, P. (eds), *Field Excursions Guidebook for the International Association of Sedimentologists Meeting on Foreland Basins. (Fribourg),* 39–73. International Association of Sedimentologists.

Ellis, M. 1988. Lithospheric strength in compression: initiation of subduction, flake tectonics, foreland migration of thrusting, and an origin of displaced terranes. *Journal of Geology*, **96**, 91–100.

Elmore, R.D., O.H. Pilkey, W.J. Cleary & H.A. Curran 1979. Black Shell turbidite, Hatteras Abyssal Plain, western Atlantic Ocean. *Geological Society of America Bulletin*, **90**, 1165–1176.

Elrick, M. & Snider, A.C. 2002. Deep-water stratigraphic cyclicity and carbonate mud mound development in the Middle Cambrian Marjum Formation, House Range, Utah, USA. *Sedimentology*, **49**, 1021–1047.

Embley, R.W. 1976. New evidence for occurrence of debris flow deposits in the deep sea. *Geology*, **4**, 371–374.

Embley, R.W. 1980. The role of mass transport in the distribution and character of deep-ocean sediments with special refer-ence to the North Atlantic. *Marine Geology*, **38**, 23–50.

Embley, R.W., Ewing, J.I. & Ewing, M. 1970. The Vidal deep-sea channel and its relationship to the Demerara and Barracuda abyssal plain. *Deep-Sea Research*, **17**, 539–552.

Emeis, K.-C. & Weissert, H. 2009. Tethyan-Mediterranean organic carbon-rich sediments from Mesozoic black shales to sapropels. *Sedimentology*, **56**, 247–266.

Emery, K.O. 1960. Basin plains and aprons off southern California. *Journal of Geology*, **68**, 464–479.

Emery, K.O. & Milliman, J.D. 1978. Suspended matter in surface waters: influence of river discharge and of upwelling. *Sedimentology*, **25**, 125–140.

Emery, K.O. & Uchupi, E. 1972. Western North Atlantic Ocean: topography, rocks, structure, water, life and sediments. *American Association of Petroleum Geologists Memoir*, **17**.

Emery, K.O. & Uchupi, E. 1984. *The Geology of the Atlantic Ocean*. New York: Springer-Verlag. 1,050 pp. + 23 oversize charts. ISBN: 3540960325.

Emmel, F.J. & Curray, J.R. 1981. Dynamic events near the upper and mid-fan boundary of the Bengal Fan. *Geo-Marine Letters*, **1**, 201–205.

England, T.D.J. & Hiscott, R.N. 1992. Lithostratigraphy and deep-water setting of the upper Nanaimo Group (Upper Cretaceous), outer Gulf Islands of southwestern British Columbia. *Canadian Journal of Earth Sciences*, **29**, 574–595.

Enos, P. 1969a. *Cloridorme Formation, Middle Ordovcian Flysch, Northern Gaspé Peninsula, Quebec*. Geological Society of America, Special Paper, **117**, 66 pp.

Enos, P. 1969b. Anatomy of a flysch. *Journal of Sedimentary Petrology*, **39**, 680–723.

Enos, P. 1977. Flow regimes in debris flow. *Sedimentology*, **24**, 133–142.

Erba, E. & Silva, I.P. 1994. Orbitally driven cycles in trace-fossil distribution from the Piobbico core late Albian, central Italy. *In*: Deboer, P. & Smith, D (eds), *Orbital Forcing and Cyclic Sequences*, 211–225. International Association of Sedimentologists, Special Publication, **19**.

Erbacher, J., Mosher, D.C., Malone, M.J. *et al.* 2004. *Proceedings of the Ocean Drilling Program, Initial Reports*, **207**. College Station, Texas, USA: Ocean Drilling Program.

Erbacher, J., Friedrich, O., Wilson, P.A., Birch, H. & Mutterlose, J. 2005. Stable organic carbon isotope stratigraphy across oceanic anoxic Event 2 of Demerara rise, western tropical Atlantic. *Geochemistry, Geophysics & Geosystems*, **6**, Q06010. doi:10.1029/2004GC000850.

Ercilla, G., Wynn, R.B., Alonso, B. & Baraza, J. 2002. Initiation and evolution of turbidity current sediment waves in the Magdalena turbidite system. *Marine Geology*, **192**, 153–169.

Ercilla, G., Casas, D., Estrada, F., Vázquez, J.T., Iglesias, J., García, M., Gómez, M., Acosta, J., Gallart, J., Maestro-González, A. & Marconi Team 2008. Morphosedimentary features and recent depositional architectural model of the Cantabrian continental margin. *Marine Geology*, **247**, 61–83.

Ettensohn, F.R. 2008. The Appalachian Foreland Basin in Eastern United States. *In*: *Sedimentary Basins of the World*, **5**, 105–179.

Ettienne, S., Mulder, T., Bez, M., Desaubliaux, G., Kwasniewski, A., Parize, O., Dujoncquoy, E. & Salles, T. 2012. Multiple scale characterization of sand-rich distal lobe deposit variability: examples from the Annot Sandstones Formation, Eocene–Oligocene, SE France. *Sedimentary Geology*, **273–274**, 1–18.

Euzen, T., Joseph, P., Du Fornel, E., Lesur, S., Granjeon, D. & Guillocheau, F. 2004. Three-dimensional stratigraphic modelling of the Grès d'Annot system, Eocene–Oligocene, SE France. *In*: Joseph, P. & Lomas, S.A. (eds), *Deep-Water Sedimentation in the Alpine Basin of*

SE France: New Perspectives on the Grès d'Annot and Related Systems, 161–180. Geological Society, London, Special Publication, **221**. London: The Geological Society.

Evans, M.J., Elliott, T., Apps, G.M. & Mange-Rajetzky, M.A. 2004. The Tertiary Grès de Ville of the Barr6me Basin: feather edge equivalent to the Grès d'Annot? *In*: Joseph, P. & Lomas, S.A. (eds), *Deep-Water Sedimentation in the Alpine Basin of SE France: New Perspectives on the Grès d'Annot and Related Systems*. Geological Society, 97–110. London, Special Publication, **221**. London: The Geological Society.

Ewing, M. & Thorndike, E.M. 1965. Suspended matter in deep ocean water. *Science*, **147**, 1291–1294.

Exon, N.F., Kennett, J.P. & Malone, M.J. 2004 (eds). Leg 189 synthesis: Cretaceous–Holocene history of the Tasmanian gateway. *In*: Exon, N.F., Kennett, J.P. & Malone, M.J. (eds), *Proceedings of the Ocean Drilling Program, Scientific Results*, **189**, 1–37. College Station, Texas, USA: Ocean Drilling Program.

Expedition 317 Scientists 2010. Canterbury Basin Sea Level: Global and local controls on continental margin stratigraphy. *Integrated Ocean Drilling Program, Preliminary Report*, **317**. College Station, Texas, USA: Ocean Drilling Program. doi:10.2204/iodp.pr.317.2010.

Expedition 318 Scientists 2010. Wilkes Land Glacial History: Cenozoic East Antarctic Ice Sheet evolution from Wilkes Land margin sediments. *Integrated Ocean Drilling Program, Preliminary Report*, **318**. College Station, Texas, USA: Ocean Drilling Program. doi:10.2204/iodp.pr.318.2010.

Expedition 331 Scientists 2010. Deep hot biosphere. *Integrated Ocean Drilling Program, Preliminary Report*, **331**. College Station, Texas, USA: Ocean Drilling Program. doi:10.2204/iodp.pr.331.2010.

Expedition 339 Scientists 2012. Mediterranean outflow: environmental significance of the Mediterranean Outflow Water and its global implications. *Integrated Ocean Drilling Program, Preliminary Report*, **339**. College Station, Texas, USA: Ocean Drilling Program. doi:10.2204/iodp.pr.339.2012.

Expedition 340 Scientists 2012. Lesser Antilles volcanism and landslides: implications for hazard assessment and long-term magmatic evolution of the arc. *Integrated Ocean Drilling Program, Preliminary Report*, **340**. doi:10.2204/ iodp.pr.340.2012

Expedition 342 Scientists, 2012. Paleogene Newfoundland sediment drifts. *Integrated Ocean Drilling Program, Preliminary Report*, **342**. doi:10.2204/iodp.pr.342.2012.

Expedition 348 Scientists & Scientific Participants 2014. NanTroSEIZE Stage 3: NanTroSEIZE plate boundary deep riser 3. *Integrated Ocean Drilling Program, Preliminary Report*, **348**. doi:10.2204/iodp.pr.348.2014.

Eyles, N. 1990. Marine debris flows: Late Precambrian "tillites" of the Avalonian-Cadomian orogenic belt. *Palaeogeography, Palaeoclimatology, Palaeoecology*, **79**, 73–98.

Eyles, N. & Eyles, C.H. 1989. Glacially-influenced deep marine sedimentation of the Late Precambrian Gaskiers Formation, Newfoundland, Canada. *Sedimentology*, **36**, 601–620.

Fagel, N., Robert, C. & Hillaire-Marcel, C. 1996. Clay mineral signature of the NW Atlantic Boundary Undercurrent. *Marine Geology*, **130**, 19–28.

Fagel, N., Robert, C. & Hillaire-Marcel, C. 1997. Changes in the Western Boundary Undercurrent outflow since the Last Glacial Maximum, from smectite/illite ratios in deep Labrador Sea sediments. *Paleoceanography*, **12**, 77–96.

Fagel, N., Robert, C., Preda, M. & Thorez, J. 2001. Smectite composition as a tracer of deep circulation: the case of the Northern North Atlantic. *Marine Geology*, **172**, 309–330.

Falcini, F., Marini, M., Milli, S. & Moscatelli, M. 2009. An inverse problem to infer paleoflow conditions from turbidites. *Journal of Geophysical Research*, **114**, doi:10.1029/2009JC005294.

Falivene, O., Arbués, P., Gardiner, A., Pickup, G., Muñoz, J.A. & Cabrera, L. 2006a. Best practice stochastic facies modeling from a channel-fill turbidite sandstone analog (the Quarry outcrop, Eocene Ainsa basin, northeast Spain). *American Association of Petroleum Geologists Bulletin*, **90**, 1003–1029.

Falivene, O., Arbués, P., Howell, J., Muñoz, J.A. & Fernandez, O. 2006b. Hierarchical geocellular facies modelling of a turbidite reservoir analogue from the Eocene of the Ainsa basin, NE Spain. *Marine and Petroleum Geology*, **23**, 679–701.

Farley, K.A. & Eltgroth, S.F. 2003. An alternative age model for the Paleocene–Eocene thermal maximum using extraterrestrial 3He. *Earth and Planetary Science Letters*, **208**, 135–148.

Farre, J.A., McGregor, B.A., Ryan, W.B.F. & Robb, J.M. 1983. Breaching the shelfbreak: passage from youthful to mature phase in submarine canyon evolution. *In*: Stanley, D.J. & Moore, G.T. (eds), *The Shelfbreak: Critical Interface on Continental Margins*, 25–39. Society of Economic Paleontologists and Mineralogists, Special Publication, 33.

Faugères, J.-C. & Mulder, T. 2011. Contour currents and contourite drifts. *In*: Hüneke, H. & Mulder, T. (eds), *Deep-Sea Sediments*. Developments in Sedimentology, **63**, 149–214. Amsterdam: Elsevier.

Faugères, J.-C., Stow, D.A.V. & Gonthier, E. 1984. Contourite drift moulded by deep Mediterranean outflow. *Geology*, **12**, 296–300.

Faugères, J.-C., Stow, D.A.V., Imbert, P. & Viana, A. 1999. Seismic features diagnostic of contourite drifts. *Marine Geology*, **162**, 1–38.

Feary, D.A. & James, N.P. 1998. Seismic stratigraphy and geological evolution of the Cenozoic, coolwater, Eucla Platform, Great Australian Bight. *American Association of Petroleum Geologists Bulletin*, **82**, 792–816.

Feder, J. 1988. *Fractals*. New York: Plenum. 283 pp. ISBN: 0-306-42851-2.

Feeley, M.H. 1984. *Seismic stratigraphic analysis of the Mississippi Fan*. Ph.D. Thesis, Texas &aM University, College Station, Texas.

Felix, D.W. & Gorsline, D.S. 1971. Newport submarine canyon, California; an example of the effects of shifting loci of sand supply upon canyon position. *Marine Geology*, **10**, 177–198.

Felix, M. 2001. A two-dimensional numerical model for a turbidity current. *In*: McCaffrey, W., Kneller, B. & Peakall, J. (eds), *Particulate Gravity Currents*, 71–82. International Association of Sedimentologists, Special Publication, **31**. Oxford: Blackwell Scientific.

Felix, M. 2004. The significance of single value variables in turbidity currents. *Journal of Hydraulic Research*, **42**, 323–330.

Felix, M., Peakall, J. & McCaffrey, W.D. 2006. Relative importance of processes that govern the generation of particulate hyperpycnal flows. *Journal of Sedimentary Research*, **76**, 382–387.

Felletti, F. 2004. Spatial variability of Hurst statistics in the Castagnola Formation, Tertiary Piedmont Basin, northern Italy: discrimination of sub-environments in a confined turbidite system. *In*: Lomas, S.A. & Joseph, P. (eds), *Confined Turbidite Systems*, 285–306. Geological Society London, Special Publication, **222**. London: The Geological Society.

Felletti, F. & Bersezio, R. 2010a. Quantification of the degree of confinement of a turbidite-filled basin: a statistical approach based on bed thickness distribution. *Marine and Petroleum Geology*, **27**, 515–532.

Felletti, F. & Bersezio, R. 2010b. Validation of Hurst statistics, a predictive tool to discriminate turbiditic sub-environments in a confined basin. *Petroleum Geoscience*, **16**, 401–412.

Fernandez, O., Muñoz, J.A., Arbues, P., Falivene, O. & Marzo, M. 2004. Three-dimensional reconstruction of geological surfaces: An example of growth strata and turbidite systems from the Ainsa basin, Pyrenees, Spain. *American Association of Petroleum Geologists Bulletin*, **88**, 1049–1068.

Fielding, C.R. 2015. Anatomy of falling-stage deltas in the Turonian Ferron Sandstone of the western Henry Mountains Syncline, Utah: Growth faults, slope failures and mass transport complexes. *Sedimentology*, **62**, 1–26.

Figuerido, J.J.P., Hodgson, D.M., Flint, S.S. & Kavanagh, J.P. 2010. Depositional environments and sequence stratigraphy of an exhumed Permian mudstone-dominated submarine slope succession, Karoo Basin, South Africa. *Journal of Sedimentary Research*, **80**, 97–118.

Figuerido, J.J.P., Hodgson, D.M., Flint, S.S. & Kavanagh, J.P. 2013. Architecture of a channel complex formed and filled during long-term degradation and entrenchment on the upper submarine slope, Unit F, Fort Brown Fm., SW Karoo Basin, South Africa. *Marine and Petroleum Geology*, **41**, 104–116.

Fildani, A., Normark, W.R., Kostic, S. & Parker, G. 2006. Channel formation by flow stripping: large-scale scour features along the Monterey East Channel and their relation to sediment waves. *Sedimentology*, **53**, 1265–1289.

Fildani, A., Hubbard, S.M., Covault, J.A., Maier, K.L., Romans, B.W., Traer, M. & Rowland, J.C. 2013. Erosion at inception of deep-sea channels. *Marine and Petroleum Geology*, **41**, 48–61.

Fillion, D. & Pickerill, R.K. 1990. Ichnology of the Upper Cambrian? to Lower Ordovician Bell Island and Wabana groups of eastern Newfoundland, Canada. *Canadian Society of Petroleum Geologists*, 7, 119.

Finney, S.C., Grubb, B.J. & Hatcher, Jr., R.D. 1996. Graphic correlation of Middle Ordovician graptolite shale, southern Appalachians: An approach for examining the subsidence and migration of a Taconic foreland basin. *Geological Society of America Bulletin*, **108**, 355–371.

Fischer-Ooster, C. 1858. *Die fossilen Fucoiden der Schweizer Alpen, nebst Erörterungen über deren geologisches Alter*. Huber, Bern. 72 pp.

Fisher, A.T. & Hounslow, M.W. 1990a. Heat flow through the toe of the Barbados accretionary complex. *In*: Moore, J.C., Mascle, A. *et al.*, *Proceedings of the Ocean Drilling Program, Scientific Results*, **110**, 345–363. College Station, TX: Ocean Drilling Program.

Fisher, A.T. & Hounslow, M.W. 1990b. Transient fluid flow through the toe of the Barbados accretionary complex: constraints from Ocean Drilling Program Leg 110 heat flow studies and simple models. *Journal of Geophysical Research*, **95**, 8845–8858.

Fisher, M.A., Collot, J.-Y. & Smith, G.L. 1986. Possible causes for structural variation where the New Hebrides island arc and the d'Entrecasteaux zone collide. *Geology*, **14**, 951–954.

Fisher, M.A., Normark, W.R., Bohannon, R.G., Sliter, R.W. & Calvert, A.J. 2003. Geology of the continental margin beneath Santa Monica Bay, Southern California, from seismic-reflection data. *Bulletin of the Seismological Society of America*, **93**, 1955–1983.

Fisher, R.V. 1983. Flow transformations in sediment gravity flows. *Geology*, **11**, 273–274.

Fisher, R.V. & Schmincke, H.-U. 1984. *Pyroclastic Rocks*. Berlin: Springer-Verlag, 472 pp. ISBN 3-540- 12756-9.

Fisz, M. 1963. *Probability Theory and Mathematical Statistics*. New York: Wiley.

Fitzsimmons, R., Veiberg, D. & Kråkenes, T. 2005. Characterization of the Heimdal Sandstones within Alveim, Quads 24 and 25, Norwegian North Sea. *In*: Doré, A.G. & Vining, B.A. (eds), *Petroleum Geology: North-west Europe and Global Perspectives-Proceedings of the 6th Petroleum Geology Conference*, 123–131. London: The Geological Society.

Flemings, P.B., Behrmann, J., Davies, T., John, C. & Expedition 308 Project Team. 2005. Gulf of Mexico hydrogeology-overpressure and fluid flow processes in the deepwater Gulf of Mexico: slope stability, seeps, and shallow-water flow. *IODP Scientific Prospectus, 308*. College Station, Texas, USA: Ocean Drilling Program.

Fletcher, J.M., Grove, M., Kimbrough, D., Lovera, O. & Gehrels, G.E. 2007. Ridge-trench interactions and the Neogene tectonic evolution of the Magdalena shelf and southern Gulf of California: insights from detrital zircon U-Pb ages from the Magdalena fan and adjacent areas. *Bulletin of the Geological Society of America*, **119**, 1313–1336.

Flint, S.S. & Bryant, I.D. 1993. *The Geological Modelling of Hydrocarbon Reservoirs and Outcrop analogues*. International Association of Sedimentologists, Special Publication, **15**, 269 pp.

Flint, S., Hodgson, D., Sprague, A. & Box, D. 2008. A physical stratigraphic hierarchy for deep-water slope system reservoirs 1: super sequences to complexes. *American Association of Petroleum Geologists International Conference and Exhibition*, Cape Town, South Africa, *Abstracts*.

Flint, S.S., Hodgson, D.M., Sprague, A.R., Brunt, R.L., Van der Merwe, W.C., Figueiredo, J.J.P., Prélat, A., Box, D., Di Celma, C. & Kavanagh, J.P. 2011. Depositional architecture and sequence stratigraphy of the

Karoo basin floor to shelf edge succession, Laingsburg depocentre, South Africa. *Marine and Petroleum Geology*, **28**, 658–674.

Flood, R.D. 1978. *Studies of deep sea sedimentary microtopography in the North Atlantic Ocean*. Ph.D. Thesis. Massachusetts Institute of Technology & Woods Hole Oceanographic Institution, Woods Hole Oceanographic Institution Report WHOI-78-64.

Flood, R.D. 1988. A lee wave model for deep-sea mudwave activity. *Deep-Sea Research*, **A35**, 973–983.

Flood, R.D. & Giosan, L. 2002. Migration history of a fine-grained abyssal sediment wave on the Bahama Outer Ridge. *Marine Geology*, **192**, 259–273.

Flood, R.D. & Piper, D.J.W. 1997. Amazon Fan sedimentation: the relationship to equatorial climate change, continental denudation, and sea-level fluctuations. *In: Proceedings Ocean Drilling Program, Scientific Results*, **155**, 653–675. College Station, Texas, USA: Ocean Drilling Program.

Flood, R.D., Manley, P.C., Kowsman, R.O., Appi, C.J. & Pirmez, C. 1991. Seismic facies and Late Quaternary growth of Amazon Submarine Fan. *In*: Weimer, P. & Link, M.H. (eds), *Seismic Facies and Sedimentary Processes of Submarine Fans and Turbidite Systems*, 415–433. New York: Springer-Verlag.

Flood, R.D., Piper, D.J.W. & Shipboard Scientific Party 1995. Introduction. *In: Proceedings Ocean Drilling Program, Initial Reports*, **155**, 5–16. College Station, Texas, USA: Ocean Drilling Program.

Flood, R.D., Piper, D.J.W., Klaus, A. & Peterson, L.C. (eds) 1997. *Proceedings Ocean Drilling Program, Scientific Results*, **155**. College Station, Texas, USA: Ocean Drilling Program.

Flores, G. 1955. Les résultats des études pour la recherche pétrolifère en Sicile: discussion. *Proceedings, 4th Word Petroleum Congress, Ed. Carlo Colombo, Roma*, 121–122.

Flueh, E.R. & Fisher, M.A. & Cruise Participants 1996. F S Sonne Cruise report SO 108 Orwell. *GEOMAR Report*, **49**, ISSN 0936-5788, GEOMAR, Kiel, Germany, 262 pp.

Flueh, E.R., Fisher, M.A., Bialas, J., Childs, J.R., Klaeschen, D., Kukowski, N., Parsons, T., Scholl, D.W., ten Brink, U., Tréhu, A.M. & Vidal, N. 1998. New seismic images of the Cascadia subduction zone from cruise SO108-ORWELL. *Tectonophysics*, **293**, 69–84.

Folk, R.L. 1974. *Petrology of Sedimentary Rocks*. Austin, Texas: Hemphill Publishing. 182 pp.

Föllmi, K.B. & Grimm, K.A. 1990. Doomed pioneers: gravity-flow deposition and bioturbation in marine oxygen-deficient environments. *Geology*, **18**, 1069–1072.

Ford, M. & Likorish, H. 2004. Foreland basin evolution around the western Alpine Are. *In*: Joseph, P. & Lomas, S.A. (eds), *Deep-Water Sedimentation in the Alpine Basin of SE France: New Perspectives on the Grès d'Annot and Related Systems*. Geological Society, London, Special Publication, **221**, 39–63. London: The Geological Society.

Fortuin, A.R., Roep, Th.B. & Sumosusastro, P.A. 1994. The Neogene sediments of east Sumba, Indonesia products of a lost arc? *Journal of Southeast Asian Earth Sciences*, **9**, 67–79.

France-Lanord, C., Spiess, V., Klaus, A. and the Expedition 354 Scientists 2015. Bengal Fan: Neogene and late Paleogene record of Himalayan orogeny and climate: a transect across the Middle Bengal Fan. *International Ocean Discovery Program Preliminary Report*, **353**. 10.14379/iodp.pr.354.2015.

Francis, R.D., Sigurdson, D.R., Legg, M.R., Grannell, R.B. & Ambos, E.L. 1999. Student participation in an offshore seismic-reflection study of the Palos Verdes Fault, California continental borderland. *Journal of Geo-Science Education*, **47**, 23–30.

Fraser, A.J., Hilkewich, D., Syms, R., Penge, J., Raposo, A. & Simon, G. 2005. Angola Block 18: a deep-water exploration success story. *In*: Doré, A.G. & Vining, B.A. (eds), *Petroleum Geology: North-west Europe and Global Perspectives-Proceedings of the 6th Petroleum Geology Conference*, Volume 1, 1199–1216. London: The Geological Society.

Freund, R. 1974, Kinematics of transform and transcurrent faults. *Tectonophysics*, **21**, 93–134.

Freundt, A. 2003. Entrance of hot pyroclastic flows into the sea: experimental observations. *Bulletin of Volcanology*, **65**, 144–164.

Freundt, A. & Schmincke, H.U. 1998. Emplacement of ash layers related to high-grade ignimbrite P1 in the sea around Gran Canaria. *In*: Weaver, P.P.E., Schmincke, H.-U., Firth, J.V. & Duffield, W.A. (eds.), *Proceedings of the Ocean Drilling Program, Scientific Results*, **157**, 201–218.

Frey, R.W. & Goldring, R. 1992. Marine event beds and recolonization surfaces as revealed by trace fossil analysis. *Geology Magazine*, **129**, 325–335.

Frey, R.W. & Pemberton, S.G. 1985. Biogenic structures in outcrops and cores. 1. Approaches to Ichnology. *Bulletin of Canadian Petroleum Geology*, **33**, 72–115.

Frey, R.W. & Pemberton, S.G. 1987. The Psilonichnus ichnocoenose, and its relationship to adjacent marine and nonmarine ichnocoenoses along the Georgia coast. *Bulletin of Canadian Petroleum Geology*, **35**, 155–158.

Frey, R.W. & Seilacher, A. 1980. Uniformity in marine invertebrate ichnology. *Lethaia*, **13**, 183–207.

Frey, R.W., Howard, J.D. & Hong, J.S. 1987. Prevalent lebensspuren on a modern macrotidal flat, Inchon, Korea: Ethological and environmental significance. *Palaios*, **2**, 517–593.

Frey, R.W., Pemberton, S.G. & Saunders, T.D. 1990. Ichnofacies and bathymetry: A passive relationship. *Journal of Paleontology*, **64**, 155–158.

Freymueller, J.T., Haeussler, P.J., Wesson, R.L. & Ekström G. (eds) 2008. *Active Tectonics and Seismic Potential of Alaska. Geophysical Monograph Series*, **179**, 431 pp.

Friedmann, S.J. & Beaubouef, RT. 1999. Relationships between depositional process, stratigraphy, and salt tectonics in a closed, intraslope basin: E. Breaks area, Gulf of Mexico. *American Association of Petroleum Geologists Annual Meeting, San Antonio, Abstracts with program*.

Friend, P.F., Slater, M.J. & Williams, R.C. 1979. Vertical and lateral building of river sandstone bodies, Ebro Basin, Spain. *Journal of the Geological Society, London*, **136**, 39–46.

Friès, G & Parize, O. 2003. Anatomy of ancient passive margin slope systems: Aptian gravity-driven deposition on the Vocontian palaeomargin, western Alps, south-east France. *Sedimentology*, **50**, 1231–1270.

Frisch, W., Meschede, M. & Blakey, R. 2011. *Plate Tectonics: Continental Drift and Mountain Building*, 217 pp. Berlin: Springer-Verlag. ISBN: 978-3-540-76503-5.

Frost, R.E. & Rose, J.F. 1996. Tectonic quiescence punctuated by strike-slip movement: influences on Late Jurassic sedimentation in the Moray Firth and the North Sea region. *In*: Hurst, A. *et al.* (eds), *Geology of the Humber Group: Central Graben and Moray Firth, UKCS*, 145–162. Geological Society of London, Special Publication, **114**.

Fruth, L.S. Jr 1965. *The 1929 Grand Banks turbidite and the sediments of the Sohm Abyssal Plain*. Ph.D. thesis, Columbia University, New York.

Fu, S. 1991. Funktion, Verhalten und Einteilung fucoider und lophoctenider Lebensspuren. *Courier Forschungs-Institut Senckenberg*, **135**, 1–79.

Fuh, S.-C., Chern, C.-C., Liang, S.-C., Yang, Y.-L., Wu, S.-H., Chang, T.-Y. & Lin, J.-Y. 2009. The biogenic gas potential of the submarine canyon systems of Plio-Peistocene foreland Basin, southwestern Taiwan. *Marine and Petroleum Geology*, **26**, 1087–1099.

Füchs, T. 1895. Studien über fucoiden und hieroglyphen. *Denkschr Akad Wiss Wien Math-Naturwiss Kl*, **62**, 369–448.

Fukushima, Y., Parker, G. & Pantin, H.M. 1985. Prediction of ignitive turbidity currents in Scripps Submarine Canyon. *Marine Geology*, **67**, 55–81.

Fuller, C.W., Willett, S.D. & Brandon, M.T. 2006. Formation of forearc basins and their influence on subduction zone earthquakes. *Geology*, **34**, 65–68.

Fulthorpe, C.S. & Melillo, A.J. 1988. Middle Miocene carbonate gravity flows in the strata of Florida at Site 626. *In*: Austin, J.A., Jr., Schlager,

W. *et al.*, *Proceedings of the Ocean Drilling Program, Scientific Results*, **101**, 179–191. College Station, Texas, USA: Ocean Drilling Program.

Fürsich, F.T., Werner W., Schneider S. & Mäuser M. 2007. Sedimentology, taphonomy, and palaeoecology of a laminated plattenkalk from the Kimmeridgian of the northern Franconian Alb (southern Germany). *Palaeogeography, Palaeoclimatology, Palaeoecology*, **243**, 92–117.

Gabelli, L. De. 1900. Sopra un interessante impronta medusoidae. *Il Pensiero Aristotelico della Scienza Moderna*, **1**, 74–78.

Gaina, C. & Müller, R.D. 2007. Cenozoic tectonic and depth/age evolution of the Indonesian gateway and associated backarc basins. *Earth-Science Reviews*, **83**, 177–203.

Galloway, W.E. 1989a. Genetic stratigraphic sequences in basin analysis; I, Architecture and genesis of flooding-surface bounded depositional units. *American Association of Petroleum Geologists Bulletin*, **73**, 125–142.

Galloway, W.E. 1989b. Genetic stratigraphic sequences in basin analysis; II, Application to Northwest Gulf of Mexico Cenozoic basin. *American Association of Petroleum Geologists Bulletin*, **73**, 143–154.

Galloway, W.E. & Williams, T.A. 1991. Sediment accumulation rates in time and space: Paleogene genetic stratigraphic sequences of the northwestern Gulf of Mexico basin. *Geology*, **19**, 986–989.

Gamberi, F. 2010. Subsurface sediment remobilization as an indicator of regional-scale defluidization within the upper Tortonian Marnoso-arenacea formation (Apenninic foredeep, northern Italy). *Basin Research*, **22**, 562–577.

Gamberi, F. & Rovere, M. 2010. Mud diapirs, mud volcanoes and fluid flowin the rear of the Calabrian Arc OrogenicWedge (southeastern Tyrrhenian sea). *Basin Research*, **22**, 452–464.

Gamberi, F. & Rovere, M. 2011. Architecture of a modern transient slope fan (Villafranca fan, Gioia basin–Southeastern Tyrrhenian Sea). *Sedimentary Geology*, **236**, 211–225.

Gamberi, F., Rovere, M. & Marani, M. 2010. Modern examples of mass-transport complexes, debrite and turbidite associations: geometry, stratigraphic relationships and implications for hydrocarbon trap development. *American Association of Petroleum Geologists Search & Discovery*, **Article #40536**.

Gamberi, F., Rovere, M., Dykstra, M., Kane, I.A. & Kneller, B.C. 2013. Integrating modern seafloor and outcrop data in the analysis of slope channel architecture and fill. *Marine and Petroleum Geology*, **41**, 83–103.

Gandolfi, G., Paganelli, L. & Zuffa, G.G. 1983. Petrology and dispersal directions in the Marnoso Arenacea Formation (Miocene, Northern Apennines). *Journal of Sedimentary Petrology*, **53**, 493–507.

Garcia, D., Joseph, P., Maréchal, P. & Moutte, J. 2004. Patterns of geochemical variability in relation to turbidite facies in the Grès d'Annot Formation. *In*: Joseph, P. & Lomas, S.A. (eds), *Deep-Water Sedimentation in the Alpine Basin of SE France: New Perspectives on the Grès d'Annot and Related Systems*, 349–365. Geological Society, London, Special Publication, **221**. London: The Geological Society.

García, M. & Parker, G. 1989. Experiments on hydraulic jumps in turbidity currents near a canyon-fan transition. *Science*, **245**, 393–396.

Garcia, M., Riquelme, R., Farías, M., Hérail, M. & Reynaldo, C. 2011. Late Miocene–Holocene canyon incision in the western Altiplano, northern Chile: tectonic or climatic forcing? *Journal of the Geological Society, London*, **168**, 1047–1060.

Garcia-Mondejar, J., Hines, F.M., Pujalte, V. & Reading, H.G. 1985. Sedimentation and tectonics in the western Basque-Cantabrian area (northern Spain) during Cretaceous and Tertiary times. *In*: Mila, M.D. & Rosell, J. (eds), *Excursion Guidebook, 6th European Regional Meeting, Lleida*, Spain. International Association of Sedimentologists.

Gardiner, S. & Hiscott, R.N. 1988. Deep-water facies and depositional setting of the lower Conception Group (Hadrynian), southern Avalon Peninsula, Newfoundland. *Canadian Journal of Earth Sciences*, **25**, 1579–1594.

Gardner, J.V., Dartnell, P., Mayer, L.A. & Hughes Clarke, J.E. 2003a. Geomorphology, acoustic backscatter, and processes in Santa Monica Bay from multibeam mapping. *Marine Environmental Research*, **56**, 15–46.

Gardner, M.H., Borer, J. & Johnson, K., 2000. Submarine channel architecture along a slope to basin profile, Permian Brushy Canyon Formation. *In*: Bouma, A.H., Stelting, C.E. & Stone, C.G. (eds), *Fine-Grained Turbidite Systems and Submarine Fans*, 195–215. American Association of Petroleum Geologists Memoir 72/Society of Economic Paleontologists and Mineralogists Special Publication.

Gardner, M.H., Borer, J.M., Melick, J.J., Mavilla, N., Dechesne, M. & Wagerle, R.N. 2003b. Stratigraphic process-response model for submarine channels and related features from studies of Permian Brushy Canyon outcrops, West Texas. *Marine and Petroleum Geology*, **20**, 757–787.

Gardner, W.D., Biscaye, P.E., Zaneveld, J.R.V. & Richardson, M.J. 1985. Calibration and comparison of the LDGO nephelometer and the OSU transmissometer on the Nova Scotian rise. *Marine Geology*, **66**, 323–344.

Garfunkel, Z. 1984. Large-scale submarine rotational slumps and growth faults in the Eastern Mediterranean. *Marine Geology*, **55**, 305–324.

Garfunkel, Z. & Almagor, G. 1985. Geology and structure of the continental margin off northern Israel and the adjacent part of the Levantine Basin. *Marine Geology*, **62**, 105–131.

Garner, J.V., Mayer, L.A. & Hughs Clarke, J.E. 2000. Morphology and processes in Lake Tahoe (California-Nevada). *Geological Society of America Bulletin*, **112**, 736–746.

Garrison, L.E., Kenyon, N.H. & Bouma, A.H. 1982. Channel systems and lobe construction in the Mississippi Fan. *Geo-Marine Letters*, **2**, 31–39.

Garton, M. & McIlroy, D. 2006. Large thin slicing: a new method for the study of fabrics in lithified sediments. *Journal of Sedimentary Research*, **76**, 1252–1256.

Garzanti, E., Doglioni, C., Vezzoli, G. & Andò, S. 2007. Orogenic Belts and Orogenic Sediment Provenance. *The Journal of Geology*, **115**, 315–334.

Gawthorpe, R.L., Sharp, I., Underhill, J.R. & Gupta, S. 1997. Linked sequence stratigraphic and structural evolution of propagating normal faults. *Geology*, **25**, 795–798.

GEBCO 2003. *IHO-UNESCO, General Bathymetric Chart of the Oceans, Digital Edition, 2003*, http://www.gebco.net/.

Gee, M.J.R. & Gawthorpe, R.L. 2006. Submarine channels controlled by salt tectonics: examples from 3D seismic data offshore Angola. *Marine and Petroleum Geology*, **23**, 443–458.

Gee, M.J.R., Gawthorpe, R.L., Friedmann, S.J. 2006. Triggering and evolution of a giant submarine landslide, offshore Angola, revealed by 3D seismic stratigraphy and geomorphology. *Journal of Sedimentary Research*, **76**, 9–19.

Gee, M.J.R., Gawthorpe, R.L., Bakke, K. & Friedmann, S.J. 2007. Seismic geomorphology and evolution of submarine channels from the Angolan continental margin. *Journal of Sedimentary Research*, **77**, 433–446.

Geist, E.L., Childs, J.R. & Scholl, D.W. 1988. The origin of summit basins of the Aleutian Ridge: implications for block rotation of the arc massif. *Tectonics*, **7**, 327–341.

Gennesseaux, M., Guibout, P. & Lacombe, H. 1971. Enregistrement de courants de turbidite dans la vallee sous-marine du Var (Alpes-Maritimes). *Comptes Rendus de l'Académie des Sciences Paris*, **273**, 2456–2459.

Gennesseaux, M., Mauffret, A. & Pautot, G. 1980. Les glissements sous-marins de la pente continentale niçoise et la rupture de cables en mer Ligure (Méditerranée occidentale). *Comptes Rendus de l'Académie des Sciences Paris (D)*, **290**, 959–962.

Georgiopoulou, A. & Cartwright, J.A. 2013. A critical test of the concept of submarine equilibrium profile. *Marine and Petroleum Geology*, **41**, 35–47.

Gersonde, R., Hodell, D.A., Blum, P. *et al.* 1999. *Proceedings of the Ocean Drilling Program, Initial Reports*, **177**. College Station, Texas, USA: Ocean Drilling.

Gertsch, B., Adatte, T., Keller, G., Tantawy, A.A.A.M., Berners, Z., Mort, H.P. & Fleitmann, D. 2010. Middle and late Cenomanian oceanic anoxic events in shallow and deeper shelf environments of western Morocco. *Sedimentology*, doi: 10.1111/j.1365-3091.2010.01151.x.

Gervais, A., Mulder, T., Savoye, B. & Gonthier, E. 2006. Sediment distribution and evolution of sedimentary processes in a small sandy turbidite system (Golo system, Mediterranean Sea): implications for various geometries based on core framework. *Geo-Marine Letters*, **26**, 373–395.

Geyer, W.R., Hill, P.S. & Kineke, G.C. 2004. The transport, transformation and dispersal of sediment by buoyant coastal flows. *Continental Shelf Research*, **24**, 927–949.

Ghadeer, S.G. & Macquaker, J.H.S. 2011. Sediment transport processes in an ancient mud-dominated succession: a comparison of processes operating in marine offshore settings and anoxic basinal environments. *Journal of the Geological Society, London*, **168**, 1121–1132.

Ghibaudo, G. 1980. Deep-sea fan deposits in the Macigno Formation (Middle-Upper Oligocene) of the Gordana Valley, northern Apennines. *Journal of Sedimentary Petrology*, **50**, 723–742.

Ghibaudo, G. 1992. Subaqueous sediment gravity flow deposits: practical criteria for their field description and classification. *Sedimentology*, **39**, 423–454.

Ghibaudo, G. 1995. Sandbody geometries in an onlapping turbiditic basin-fill: Montagne de Chalufy, Alpes des Hautes Provence, SE France. *In*: Pickering, K.T., Hiscott, R.N., Kenyon, N.H., Ricci Lucchi, F. & Smith, R.D.A. (eds), *Atlas of Deep Water Environments: Architectural Style in Turbidite Systems,* 242–243. London: Chapman and Hall.

Ghibaudo, G., Grandesso, P., Massari, F. & Uchman, A. 1996. Use of trace fossils in delineating sequence stratigraphic surfaces, Tertiary Venetian Basin, northeastern Italy. *Palaeogeography, Palaeoclimatology, Palaeoecology*, **120**, 261–279.

Ghiglione, M.C., Quinteros, J., Yagupsky, D., Bonillo-Martínez, P., Hlebszevtich, J., Ramos, V.A., Vergani, G., Figueroa, D., Quesada, S. & Zapata, T. 2010. Structure and tectonic history of the foreland basins of southernmost South America. *Journal of South American Earth Sciences*, **29**, 262–277.

Ghiglione, M.C., Yagupsky, D., Ghidella, M. & Ramos, V.A. 2008. Continental stretching preceding the opening of the Drake Passage: Evidence from Tierra del Fuego. *Geology*, **36**, 643–646.

Gibson, R.E. 1958. The progress of consolidation in a clay layer increasing in thickness with time. *Geotechnique*, **8**, 171–182.

Gibbs, A.D. 1984. Structural evolution of extensional basin margins. *Journal of the Geological Society (London)*, **141**, 609–620.

Gibbs, R.J. 1985a. Estuarine flocs: their size, settling velocity and density. *Journal of Geophysical Research Oceans and Atmospheres*, **90**, 3249–3251.

Gibbs, R.J. 1985b. Settling velocity, diameter, and density of flocs of illite, kaolinite, and montmorillonite. *Journal of Sedimentary Petrology*, **55**, 65–68.

Gibbs, R.J. & Konwar, L. 1986. Coagulation and settling of Amazon River suspended sediment. *Continental Shelf Research*, **6**, 127–149.

Gieskes, J.M., Johnston, K. & Boehm, M. 1984. Appendix. Interstitial water studies, Leg 66. *In*: von Huene, R., Aubouin, J. *et al.*, *Initial Reports of the Deep Sea Drilling Project*, **66**, 961–967. Washington, DC: US Government Printing Office.

Gieskes, J.M., Vrolijk, P. & Blanc, G. 1990. Hydrogeochemistry of the northern Barbados accretionary complex transect: Ocean Drilling Program Leg 110. *Journal of Geophysical Research*, **95**. 8809–8818.

Gilbert, R. 1983. Sediment processes of Canadian arctic fjords. *Sedimentary Geology*, **36**, 147–175.

Gill, J.B., Hiscott, R.N. & Vidal, Ph. 1994. Turbidite geochemistry and evolution of the Izu-Bonin arc and continents. *Lithos*, **33**, 135–168.

Gingras, M.K., Pemberton, S.G., Dashtgard, S. & Dafoe, L. 2008. How fast do marine invertebrates burrow? *Palaeogeography, Palaeoclimatology, Palaeoecology*, **270**, 280–286.

Giorgio Serchi, F., Peakall, J., Ingham, D.B. & Burns, A.D. 2011. A unifying computational fluid dynamics investigation on the river-like to river-reversed secondary circulation in submarine channel bends. *Journal of Geophysical Research-Oceans*, **116**, C06012. doi:10.1029/2010JC006361.

Gladstone, C. & Pritchard, D. 2010. *Patterns of deposition from experimental turbidity currents with reversing buoyancy Sedimentology*, **57**, 53–84.

Gladstone, C. & Sparkes, R.S.J. 2002. The significance of grain-size breaks in turbidites and pyroclastic density current deposits. *Journal of Sedimentary Research*, **72**, 182–191.

Gladstone, C., Ritchie, L.J., Sparks, R.S.J. & Woods, A.W. 2004. An experimental investigation of density-stratified inertial gravity currents. *Sedimentology*, **51**, 767–790.

Glasby, G.P. (ed.) 1977. *Marine Manganese Deposits.* Oceanography Series, **15**. Amsterdam: Elsevier.

Glennie, K.W. (ed.) 1986a. *Introduction to the Petroleum Geology of the North Sea,* 2nd edn. Oxford: Blackwell Scientific.

Glennie, K.W. 1986b. Structural framework and pre-Permian history of the North Sea. *In*: Glennie, K.W. (ed.), *Introduction to the Petroleum Geology of the North Sea*, 853–864. Oxford: Blackwell Scientific.

Gloppen, T. G. & Steel, R.J. 1981. The deposits, internal structure and geometry of six alluvial fan-fan delta bodies (Devonian, Norway) a study in the significance of bedding sequences in conglomerates. *In*: Ethridge, F.G. & Flores, R.M. (eds), *Recent and Ancient Non-marine Depositional Environments: Models for Exploration*, 49–69. Society of Economic Paleontologists and Mineralogists Special Publication, **31**.

Gnibidenko, H.S., Anosov, G.A., Argentov, V.V. & Pushchin, I.K. 1985. Tectonics of the Tonga-Kermadec Trench and Ozbourn Seamount junction area. *Tectonophysics*, **112**, 357–383.

Gold, E. 1929. Notes on the frequency of occurrence of sequence in a series of events of two types. *Royal Meteorological Society Quarterly Journal*, **55**, 307–309.

Goldfinger, C. 1994. *Active deformation of the Cascadia forearc: implications for great earthquake potential in Oregon and Washington.* Oregon State University, Corvallis, Ph.D. Thesis, 202 pp.

Goldfinger, C. 2011. Submarine paleoseismology based on turbidite records. *Annual Reviews of Marine Science*, **3**, 35–66.

Goldfinger, C., Kulm, L.D., Yeats, R.S., Appelgate, B., MacKay, M. & Moore, G.F. 1992. Transverse structural trends along the Oregon convergent margin: Implications for Cascadia earthquake potential. *Geology*, **20**, 141–144.

Goldfinger, C., Kulm, L.D., Yeats, R.S, McNeill, L. 1995. Super-scale slumping of the Southern Oregon Cascadia margin: tsunamis, tectonic erosion, and extension of the forearc. *EOS, Transactions of the American Geophysical Union*, **76**, F361.

Goldfinger, C. Kulm, L.D., McNeill, L.C. & Watts, P. 2000. Super-scale failure of the Southern Oregon Cascadia margin. *Pure & Applied Geophysics*, **157**, 1189–1226.

Goldfinger, C., Nelson, C.H., Johnson, J.E. and the Shipboard Party 2003. Deep-water turbidites as Holocene earthquake proxies: the Cascadia subduction zone and Northern San Andreas Fault systems. *Annals of Geophysics*, **46**, 1169–1194.

Goldfinger, C., Morey, A.E., Nelson, C.H., Gutiérrez-Pastor, J., Johnson, J.E., Karabanov, E., Chaytor, J.D., Eriksson, A. & Shipboard Scientific Party 2007. Rupture lengths and temporal history of significant earthquakes on the offshore and north coast segments of the Northern San Andreas Fault based on turbidite stratigraphy. *Earth and Planetary Science Letters*, **254**, 9–27.

Goldfinger, C., Grijalva, K., Bürgmann, R., Morey, A.E., Johnson, J.E., Nelson, C.H., Gutiérrez-Pastor, J., Eriksson, A., Karabanov, E., Chaytor, J.D., Patton, J. & Gràcia, E. 2008. Late Holocene rupture of the northern San Andreas Fault and possible stress linkage to the Cascadia subduction zone. *Bulletin of the Seismological Society of America*, **98**, 861–889.

Goldfinger, C., Patton, J., Morey, A.E. & Nelson, C.H. 2009. Reply to "Comment on Late Holocene rupture of the northern San Andreas Fault and possible stress linkage to the Cascadia subduction zone". *Bulletin of the Seismological Society of America*, **99**, 2599–2606.

Goldfinger, C., Nelson, C.H., Morey, A.E., Johnson, J.E., Patton, J.R., Karabanov, E., Gutiérrez-Pastor, J., Eriksson, A.T., Gràcia, E., Dunhill, G., Enkin, R.J., Dallimore, A. & Vallier, T. 2012. *Turbidite Event History – Methods and Implications for Holocene Paleoseismicity of the Cascadia Subduction Zone.* USGS Professional Paper 1661–F, 170 pp.

Goldfinger, C., Morey, A.E., Black, B., Beeson, J., Nelson, C.H. & Patton, J. 2013 Spatially limited mud turbidites on the Cascadia margin: segmented earthquake ruptures? *Natural Hazards and Earth System Sciences*, **13**, 2109–2146,

Goldring, R. 1993. Ichnofacies and facies interpretation. *Palaios*, **8**, 403–405.

Gómez-Paccard, M., López-Blanco, M., Costa, E., Garcés, M., Beamud, E. & Larrasoaña, J.C. 2011. Tectonic and climatic controls on the sequential arrangement of an alluvial fan/fan-delta complex (Montserrat, Eocene, Ebro Basin, NE Spain). *Basin Research*, **23**, 1–19.

Gong, G., Wang, Y., Xu, S., Pickering, K.T., Peng, X., Li, W. & Yan, Q. 2015. The northeastern South China Sea margin created by the combined action of down-slope and along-slope processes: processes, products and implications for exploration and paleoceanography. *Marine and Petroleum Geology*, **64**, 233–249.

Gonthier, E.G., Faugères, J.-C. & Stow, D.A.V. 1984. Contourite facies of the Faro Drift, Gulf of Cadiz. *In*: Stow, D.A.V. & Piper, D.J.W. (eds), *Fine-grained Sediments: Deep-water Processes and Facies,* 275–292. Geological Society of London Special Publication, **15**. Oxford: Blackwell Scientific.

Gorsline, D.S. 1980. Deep-water sedimentologic conditions and models. *Marine Geology*, **38**, 1–21.

Gorsline, D.S. 1984. A review of fine-grained sediment origins, characteristics, transport and deposition. *In*: Stow, D.A.V. & Piper, D.J.W. (eds), *Fine-grained Sediments: Deep-water Processes and Facies,* 17–34. Geological Society of London Special Publication, **15**. Oxford: Blackwell Scientific.

Gorsline, D.S. 1996. Depositional events in Santa Monica Basin, California Borderland, over the past five centuries. *Sedimentary Geology*, **104**, 73–88.

Gorsline, D.S., Kolpack, R.L., Karl, H.A., Drake, D.E., Fleischer, P., Thornton, S.E., Schwalbach, J.R. & Svarda, C.E. 1984. Studies of fine-grained sediment transport processes and products in the California Continental Borderland. *In*: Stow, D.A.V. & Piper, D.J.W. (eds), *Fine-grained Sediments: Deep-water Processes and Facies,* 395–415. Geological Society of London Special Publication, **15**. Oxford: Blackwell Scientific.

Gorsline, D.S., De Diegob, T. & Nava-Sanchezc, E.H. 2000. Seismically triggered turbidites in small margin basins: Alfonso Basin, Western Gulf of California and Santa Monica Basin, California Borderland. *Sedimentary Geology*, **135**, 21–35.

Gradstein, F.M., Fearon, J.M. & Huang, Z. 1989. BURSUB and DEPOR version 3.50-two FORTRAN 77 programs for porosity and subsidence analysis. *Geological Survey of Canada Open-File Report*, **1283**, 1–10.

Gradstein, F.M., Ogg, J.G. & Schmitz, M. 2012. *The Geologic Time Scale 2012.* Amsterdam: Elsevier Science & Technology Books.

Graham, R.H. 1978. Wrench faults, arcuate fold patterns and deformation in the southern French Alps. *Proceedings of the Geologists Association*, **89**, 125–142.

Graham, S.A. & Dickinson. W.R. 1978. Evidence for 115 kilometres of right-slip on the San Gregorio-Hosgri fault trend. *Science*, **199**, 179–181.

Grahame, J. & Silva-Gonzalez, P. 2013. Zeus and Zeebries: enhanced exploration potential from new high-quality 3D data acquired by Fugro in the northern Carnarvon Basin. *GEO ExPro*, **10**, 38–40.

Grammer, G.M. & Ginsburg, N. 1992. Highstand versus lowstand deposition on carbonate platform margins: insight from Quaternary foreslopes in the Bahamas. *Marine Geology*, **103**, 125–136.

Grecula, M., Flint, S., Wickens, D. & Potts, G.J. 2003a. Partial ponding of turbidite systems in a basin with subtle growth-fold topography: Laingsburg-Karoo, South Africa. *Journal of Sedimentary Research*, **73**, 603–620.

Grecula, M., Flint, S., Wickens, H. deV. & Johnson, S. 2003b. Upward-thickening patterns and lateral continuity of Permian sand-rich turbidite channel-fills, Laingsburg Karoo, South Africa. *Sedimentology*, **50**, 831–853.

Greene, H.G., Murai, L.Y., Watts, P., Maher, N.A., Fisher, M.A., Paull, C.E. & Eichhubl, P. 2006. Submarine landslides in the Santa Barbara Channel as potential tsunami sources. *Natural Hazards and Earth System Sciences*, **6**, 63–88.

Greenhalgh, J. & Whaley, M. 2012. Exploration potential of the Nigerian transform margin. *GEO ExPro*, **9**, 36–40.

Greenlee, S.M. & Moore, T.C. 1988. Recognition and interpretation of depositional sequences and calculation of sea-level changes from stratigraphic data–offshore New Jersey and Alabama Tertiary. *In*: Wilgus, C.K., Hastings, B.S., Posamentier, H., Van Wagoner, J., Ross, C.A. & Kendall, C.G. St.C. (eds), *Sea-Level Changes: An Integrated Approach.* Society of Economic Paleontologists and Mineralogists, Special Publication, **42**, 329–353.

Griggs, G.B. & Kulm, L.D. 1970. Sedimentation in Cascadia deep-sea channel. *Geological Society of America Bulletin*, **81**, 1361–1384.

Griggs, G.B. & Kulm, L.D. 1973. Origin and development of Cascadia deep-sea channel. *Journal of Geophysical Research*, **78**, 6325–6339.

Grim, P.J. & Naugler, F.P. 1969. Fossil deep-sea channel on the Aleutian Abyssal Plain. *Science*, **163**, 383–386.

Groenenberg, R.M., Hodgson, D.M., Prélat, A., Luthi, S.M. & Flint, S.S. 2010. Flow-deposit interaction in submarine lobes: insights from outcrop observations and realizations of a process-based numerical model. *Journal of Sedimentary Research*, **80**, 252–267.

Grosheny, D., Ferry, S. & Courjault, T. 2015. Progradational patterns at the head of single units of base-of-slope, submarine granular flowdeposits ("Conglomérats des Gâs", Coniacian, SE France). *Sedimentary Geology*, **317**, 102–115.

Grundvåg, S.-A., Johannessen, E.P., Helland-Hansen, W. & Plink-Björklund, P. 2014. Depositional architecture and evolution of progradationally stacked lobe complexes in the Eocene Central Basin of Spitsbergen. *Sedimentology*, **61**, 535–569.

Guillocheau, F., Quéméner, J.-M., Robin, C., Joseph, P. & Broucke, O. 2004. Genetic units/parasequences of the Annot turbidite system, SE France. *In*: Joseph, P. & Lomas, S.A. (eds), *Deep-Water Sedimentation in the Alpine Basin of SE France: New Perspectives on the Grès d'Annot and Related Systems.* Geological Society, London, Special Publication, **221**, 111–135. London: The Geological Society.

Guo, J. & Underwood, M.B. 2012. Data report: clay mineral assemblages from the Nankai Trough accretionary prism and the Kumano Basin, IODP Expeditions 315 and 316, NanTroSEIZE Stage 1. *In*: Kinoshita, M., Tobin, H., Ashi, J., Kimura, G., Lallemant, S., Screaton, E.J., Curewitz, D., Masago, H., Moe, K.T. and the Expedition 314/315/316 Scientists, *Proceedings of the Integrated Ocean Drilling Program*, **314/315/316**. Washington, DC (Integrated Ocean Drilling Program).

Guo, J., Underwood, M.B., Likos, W.J. & Saffer, D.M. 2013. Apparent overconsolidation of mudstones in the Kumano Basin of southwest Japan: Implications for fluid pressure and fluid flow within a forearc setting. *Geochemistry, Geophysics, Geosystems*, **14**, 1023–1038.

Gürer, O.F., Kaymakci, N., Cakir, S. & Ozburan, N. 2003. Neotectonics of the southeast Marmara region, NW Anatolia, Turkey. *Journal of Asian Earth Sciences*, **21**, 1041–1051.

Gutscher, M.-A., Klaeschen, D., Flueh, E. & Malavieille, J. 2001. Non-Coulomb wedges, wrong-way thrusting, and natural hazards in Cascadia. *Geology*, **29**, 379–382.

Guy, M. 1992. Facies analysis of the Kopervik sand interval, Kilda Field, Block 16/26, UK North Sea. *In*: Hardman, R.F.P. (ed.) *Exploration Britain: Geological insights for the next decade.* Geological Society London Special Publication, **67**, 187–220.

Hacquebard, P.A., Buckley, D.E. & Vilks, G. 1981. The importance of detrital particles of coal in tracing the provenance of sedimentary rocks. *Bulletin Des Centres De Recherches Exploration-Production Elf Aquitaine*, **5**, 555–572.

Hadlari, T., Lemieux, Y., Zantvoort, W.G. & Catuneanu, O. 2009. Slope and Submarine Fan Turbidite Facies of the Upper Devonian Imperial Formation, Northern Mackenzie Mountains, NWT. *Bulletin of Canadian Petroleum Geology*, **57**, 192–208.

Haeussler, P.J., Bradley, D.C., Wells, R.E. & Miller, M.L. 2003. Life and death of the Resurrection plate: evidence for its existence and subduction in the northeastern Pacific in Paleocene–Eocene time. *Geological Society of America Bulletin*, **15**, 867–880.

Haines, A.J., Hulme, T. & Yu, J. 2004. General elastic wave scattering problems using an impedance operator approach. 1, Mathematical development. *Geophysical Journal International*, **159**, 643–657.

Haldeman, S.S. 1840. *In*: Paludina & Anculosa. J. Dobson (eds), *Supplement to Number One of "A Monograph of the Limniades, or Freshwater Univalve shells of North America". Containing descriptions of apparently new animals in different classes, and the names and characters of the subgenera*. Philadelphia.

Haldorsen, H.H. & Chang, D.W. 1986. Notes on stochastic shales, from outcrop to simulation model. *In*: Lake, L.W. & Carroll, H.B. (eds), *Reservoir Characterization*, 445–485. London: Academic Press.

Haldorsen, H.H. & Lake L.W. 1984. A new approach to shale management in field-scale model. *Society of Petroleum Engineers Journal April*, 447–457.

Hall, J. 1847. *Paleontology of New York. Volume 1. C*. Albany: Van Benthuysen. 338 pp.

Hall, R. 2011. Australia-SE Asia collision: plate tectonics and crustal flow. *In*: Hall, R., Cottam, M.A. &Wilson, M.E.J. (eds), *The SE Asian Gateway: History and Tectonics of the Australia-Asia Collision, Geological Society, London, Special Publications*, **355**, 75–109. Bath, UK: The Geological Society, London.

Hall, R. & Smyth, H.R. 2008. Cenozoic arc processes in Indonesia: Identification of the key influences on the stratigraphic record in active volcanic arcs. *Geological Society of America, Special Papers*, **436**, 27–54. doi: 10.1130/2008.2436.03.

Hallworth, M.A., Huppert, H.E., Phillips, J.C. & Sparks, R.S.J. 1996. Entrainment into two-dimensional and axisymmetric turbulent gravity currents. *Journal of Fluid Mechanics*, **308**, 289–311.

Hamberg, L., Jepsen, A.M., Borch, N.T., Dam, G., Engkilde, M.K. & Svendsen, J.B. 2007. Mounded Structures of Injected Sandstones in Deep-marine Paleocene Reservoirs, Cecile Field, Denmark. *In*: Hurst, A. & Cartwright, J. (eds), *Sand Injectites: Implications for Hydrocarbon Exploration and Production*, 69–79. American Association of Petroleum Geologists Memoir, **87**. Tulsa, Oklahoma.

Hamilton, E.L. 1967. Marine geology of abyssal plains in the Gulf of Alaska. *Journal of Geophysical Research*, **72**, 4189–4213.

Hamilton, E.L. 1973. Marine geology of the Aleutian Abyssal Plain. *Marine Geology*, **14**, 295–325.

Hamilton, P.B., Strom, K. & Hoyal, D.C.J.D. 2013. Autogenic incision-backfilling cycles and lobe formation during the growth of alluvial fans with supercritical distributaries. *Sedimentology*, **60**, 1498–1525.

Hamilton, P.B., Strom, K. & Hoyal, D.C.J.D. 2015. Hydraulic and sediment transport properties of autogenic avulsion cycles on submarine fans with supercritical distributaries. *Journal of Geophysical Research*, doi:10.1002/2014JF003414.

Hampton, M.A 1975. Competence of fine-grained debris flows. *Journal of Sedimentary Petrology*, **45**, 834–844.

Hampton, M.A. 1972. The role of subaqueous debris flow in generating turbidity currents. *Journal of Sedimentary Petrology*, **42**, 775–793.

Hampton, M.A. 1979. Buoyancy in debris flows. *Journal of Sedimentary Petrology*, **49**, 753–758.

Hampton, M.A., Bouma, A.H., Carlson, P.R., Molnia, B.F., Clukey, E.C. & Sangrey, D.A. 1978. Quantitative study of slope instability in the Gulf of Alaska. *Proceedings of the 10th Offshore Technology Conference OTC3314*, 2307–2318.

Hampton, M.A., Lee, H.J. & Locat, J. 1996. Submarine landslides. *Reviews of Geophysics*, **34**, 33–59.

Hand, B.M. 1997. Inverse grading resulting from coarse-sediment transport lag. *Journal of Sedimentary Research*, **67**, 124–129.

Hand, B.M. & J.B. Ellison, J.B. 1985. *Inverse Grading in Density-current Deposits. Abstracts, 1985 Mid-year meeting, Society of Economic Paleontologists and Mineralogists*, Golden, Colorado.

Hand, B.M., Middleton, G.V. & Skipper, K. 1972. Antidune cross-stratification in a turbidite sequence, Cloridorme Formation, Gaspé, Québec. *Sedimentology* **18**, 135–138.

Hanebuth, T.J.J., Zhang, W., Hoffman, A.L., Löwemark, L.A. & Schwenk, T. 2015. Oceanic density fronts steering bottom-current induced sedimentation deduced from a 50 ka contourite-drift record and numerical modeling (off NW Spain). *Quaternary Science Reviews*, **112**, 207–225.

Haner, B.E. 1971. Morphology and sediments of Redondo submarine fan, southern California. *Geological Society of America Bulletin*, **82**, 2413–2432.

Häntzschel, W. 1975. Trace fossils and problematica. *In*: Teichert, C. (ed.), *Treatise on Invertebrate Paleontology*. Geological Society of America and Kansas University Press.

Haq, B.U. 1981. Paleogene palaeoceanography: Early Cenozoic ocean revisited. *Oceanologica Acta Proceedings of the International Geological Congress*, Geology of Oceans Symposium Paris, 71–82.

Haq, B.U. & Al-Qahtani, A.M. 2005. Phanerozoic cycles of sea-level change on the Arabian Platform. *GeoArabia*, **10**, 127–160.

Haq, B.U. & Schutter, S.R. 2008. A chronology of Paleozoic sea-level changes. *Science*, **322**, 64–68.

Haq, B.U., Hardenbol, J. & Vail, P.R. 1987. Chronology of fluctuating sea levels since the Triassic (250 years ago to present). *Science*, **235**, 1156–1167.

Haq, B.U., Hardenbol, J. & Vail, P.R. 1988. Mesozoic and Cenozoic Chronostratigraphy and Eustatic Cycles. *In*: Wilgus, C.K., Hastings, B.S., Posamentier, H., Van Wagoner, J., Ross, C.A. & Kendall, C.G. St.C. (eds), *Sea-Level Changes: An Integrated Approach*. Society of Economic Paleontologists and Mineralogists, Special Publication, **42**, 71–108.

Hara, H., Kunii, M., Hisada, K., Ueno, K., Kamata, Y., Srichan, W., Charusiri, P., Charoentitirat, T., Watarai, M., Adachi, Y. & Kurihara, T. 2012. Petrography and geochemistry of clastic rocks within the Inthanon zone, northern Thailand: Implications for Paleo-Tethys subduction and convergence. *Journal of Asian Earth Sciences*, **61**, 2–15.

Harbert, W. 1987. New paleomagnetic data from the Aleutian Islands: Implications for terrane migration and deposition of the Zodiac fan. *Tectonics*, **6**, 585–602.

Harding, T.P. 1974. Petroleum traps associated with wrench faults. *American Association of Petroleum Geologists Bulletin*, **58**, 1290–1304.

Harding, T.P. 1976. Tectonic significance and hydrocarbon trapping consequences of sequential folding synchronous with San Andreas faulting, San Joaquin Valley, California. *American Association of Petroleum Geologists Bulletin*, **60**, 356–378.

Harding, T. P. 1983. Divergent wrench fault and negative flower structure, Andaman Sea. *In*: Bally, A.W. (ed.), *Seismic Expression of Structural Styles*, **3**, 4. 2-1–4. 2-8. American Association of Petroleum Geologists, Studies in Geology Series **15**.

Harding, T.P. 1990. Identification of wrench faults using subsurface structural data: criteria and pitfalls. *American Association of Petroleum Geologists Bulletin*, **74**, 1590–1609.

Harding, T.P., Vierbuchen, R.C. & Christie-Blick, N. 1985. Structural styles, plate-tectonic settings, and hydrocarbon traps of divergent (transtensional) wrench faults. *In*: Biddle, K.T. & Christie-Blick, N. (eds), *Strike-slip Deformation, Basin Formation, and Sedimentation*, 51–77. Society of Economic Paleontologists and Mineralogists, Special Publication, 37.

Harland, W.B., Armstrong, R.L., Cox, A.V., Craig, L.E., Smith, A.G. & Smith, D.G.. 1990. *A Geologic Time Scale 1989*. Cambridge: Cambridge University Press.

Harms, J.C. & Fahnestock, R.K. 1965. Stratification, bed forms and flow phenomena (with an example from the Rio Grande). *In*: Middleton, G.V. (ed.), *Primary Sedimentary Structures and their Hydrodynamic Interpretation*, 84–115. Society of Economic Paleontologists and Mineralogists Special Publication, **12**.

Harms, J.C., Tackenberg, P., Pickles, E. & Pollock, R.E. 1981. The Brae oilfield area. *In*: Illing, L.V. & Hobson, G.D. (eds), *Petroleum Geology of the Continental Shelf of North-West Europe*, 352–357. London: Heyden.

Harper, C.W. Jr., 1984. Facies models revisited: an examination of quantitative methods. *Geoscience Canada*, **11**, 203–207.

Harper, C.W., Jr., 1998. Thickening and/or thinning upward patterns in sequences of strata: tests of significance. *Sedimentology*, **45**, 657–696.

Harris, P.T., Barrie, J.V., Conway, K.W. & Greene, H.G. 2013. Hanging canyons of Haida Gwaii, British Columbia, Canada: fault control on submarine canyon geomorphology along active continental margins. *Deep Sea Research Part II: Topical Studies in Oceanography*, doi.org/10.1016/j.dsr2.2013.06.017.

Harrison, C.P. & Graham, S.A. 1999. Upper Miocene Stevens Sandstone, San Joaquin Basin, California: reinterpretation of a petroliferous, sand-rich, deep-sea depositional system. *American Association of Petroleum Geologists Bulletin*, **83**, 898–924.

Harrowfield, M., Cunneen, J., Keep, M. & Crowe, W. 2003. Early-stage orogenesis in the Timor Sea region, NW Australia. *Journal of the Geological Society, London*, **160**, 991–1001.

Hatayama, T., Awaji, T. & Akitomo, K. 1996. Tidal currents in the Indonesian seas and their effect on transport and mixing. *Journal of Geophysical Research, doi*: 101:12,353–12,373.

Hatcher, R.D., Jr., 1989. Tectonic synthesis of the U.S. Appalachians. *In*: Hatcher, R.D., Jr., Thomas, W.A. & Viele, G.W. (eds), *The Appalachian-Ouachita Orogen in the United States:. The Geology of North America*, **F-2**, 511–535.

Haughton, P.D.W. 1994 Deposits of deflected and ponded turbidity currents, Sorbas Basin, southeast Spain. *Journal of Sedimentary Research*, **A64**, 233–246.

Haughton, P.D.W. 2000. Evolving turbidite systems on a deforming basin floor, Tabernas, SE Spain. *Sedimentology*, **47**, 497–518.

Haughton, P.D.W. & Shannon, P. 2013. *Upper Carboniferous Deepwater, Slope and Deltaic Deposits, County Claire, Eire. 30th International Association of Sedimentologists, Manchester, Field Trip Guidebook FTA5.*

Haughton, P., Barker, S.P. & McCaffrey, W.D. 2003. 'Linked' debrites in sand-rich turbidite systems-origin and significance. *Sedimentology*, **50**, 459–482.

Haughton P., Davis, C., McCaffrey, W. & Barker, S. 2009. Hybrid sediment gravity flow deposits–Classification, origin and significance. *Marine and Petroleum Geology*, **26**, 1900–1918.

Haughton, P., Davis, C., McCaffrey, W.D. Barker, S. 2010. Reply to Comment by R. Higgs on 'Hybrid sediment gravity flows-classification, origin and significance'. *Marine and Petroleum Geology*, **27**, 2066–2069.

Hay, A.E. 1983a. On the frontal speeds of internal gravity surges on sloping boundaries. *Journal of Geophysical Research*, **88**, 751–754.

Hay, A.E. 1987a. Turbidity currents and submarine channel formation in Rupert Inlet, British Columbia, Part I: Surge observations. *Journal of Geophysical Research*, **92**, 2875–2882.

Hay, A.E. 1987b. Turbidity currents and submarine channel formation in Rupert Inlet, British Columbia, Part II: the roles of continuous and surge-type flows. *Journal of Geophysical Research*, **92**, 2883–2900.

Hay, A.E., Burling, R.W. & Murray, J.W. 1982. Remote acoustic detection of a turbidity current surge. *Science*, **217**, 833–835.

Hay, W.W. 1983b. The global significance of regional Mediterranean Neogene paleoenvironmental studies. *In*: Meulenkamp, J.E. (ed.), *Reconstruction of marine paleoenvironments*, 9–23. Utrecht Micropaleontology Bulletin, **30**.

Hay, W.W., Sibuet, J.-C. *et al.* 1984. *Initial Reports Deep Sea Drilling Project*, **75**. Washington, DC: US Government Printing Office.

Hay, W.W., DeConto, R., Wold, C.N., Wilson, K.M., Voigt, S., Schulz, M., Wold-Rossby, Dullo, W.C., Ronv, A.B., Baluukhovsky, A.N. & Soeding, E. 1999. Alternative global Cretaceous paleogeography. *In*: Barrera, E. & Johnson, C. (eds), *The Evolution of Cretaceous Ocean/Climate Systems*, 1–47. Geological Society of America Special Paper, **332**.

Hawkins, J., Parson, L., Allan, J. *et al.* 1994. Proceedings of the Ocean Drilling Program, Scientific Results, **135**. College Station, Texas, USA: Ocean Drilling Program.

Heard, T.G. & Pickering, K.T. 2008. Trace fossils as diagnostic indicators of deep-marine environments, Middle Eocene Ainsa-Jaca basin, Spanish Pyrenees. *Sedimentology*, **55**, 809–844.

Heard, T.G., Pickering, K.T. & Robinson, S.A. 2008. Milankovitch forcing of bioturbation intensity in deep-marine thin-bedded siliciclastic turbidites. *Earth and Planetary Science Letters*, **272**, 130–138.

Heard, T.G., Pickering, K.T. & Clark, J.D. 2014. Ichnofabric characterization of a deep-marine clastic system: a subsurface study of the Middle Eocene Ainsa System, Spanish Pyrenees. *Sedimentology*, **61**, 1298–1331.

Heaton, T.H. & Kanamori, H. 1984. Seismic potential associated with subduction in the northwestern United States. *Bulletin of the Seismological Society of America*, **74**, 993–941.

Hedberg, H.D. 1974. Relation of methane generation to under-compacted shale, shale diapirs, and mud volcanoes. *American Association of Petroleum Geologists Bulletin*, **58**, 661–673.

Hedstrom, B.O.A. 1952. Flow of plastic materials in pipes. *Industrial and Engineering Chemistry Research*, **44**, 651–656.

Heer, O. 1877. *Flora fossilis helvetiae. Vorweltliche flora der Schweiz*. J. Wurster and Comp. Zürich. 182 pp.

Heezen, B.C. (ed.) 1977. Influence of abyssal circulation on sedimentary accumulations in space and time. *Marine Geology*, **23** (special issue).

Heezen, B.C. & Ewing, M. 1952. Turbidity currents and submarine slumps and the 1929 Grand Banks earthquake. *American Journal of Science*, **250**, 849–873.

Heezen, B.C. & Hollister, C.D. 1971. *The Face of the Deep*. New York: Oxford University Press.

Heezen, B.C. & Johnson, G.L. 1965. The south sandwich trench. *Deep Sea Research and Oceanographic Abstracts*, **12**, 185–197.

Heezen, B.C., Ericson, D.B. & Ewing, M. 1954. Further evidence for a turbidity current following the 1929 Grand Banks earthquake. *Deep-Sea Research*, **1**, 193–202.

Heezen, B.C., Hollister, C.D. & Ruddiman, W.F. 1966. Shaping of the continental rise by deep geostrophic contour currents. *Science*, **152**, 502–508.

Heezen, B.C., Tharp, M. & Ewing, M. 1959. The floors of the oceans: I. The North Atlantic. *Geological Society of America Special Paper*, **65**.

Heezen, B.C., Menzies, R.J., Schneider, E.D., Ewing, W.M. & Granelli, N.C.L. 1964. Congo submarine canyon. *American Association of Petroleum Geologists Bulletin*, **48**, 1126–1149.

Heezen, B.C., Johnson, G.L. & Hollister, D.C. 1969. The Northwest Atlantic Mid-Ocean Canyon. *Canadian Journal of Earth Sciences*, **6**, 1441–1453.

Hegarty, K.A., Weissel, J.K. & Mutter, J.C. 1988. Subsidence history of Australia's southern margin: constraints on basin models. *American Association of Petroleum Geologists Bulletin*, **72**, 615–633.

Hein, F.J. 1979. *Deep-sea valley-fill sediments, Cap Enragé Formation, Quebec*. PhD Thesis. Hamilton, Ontario, McMaster University.

Hein, F.J. 1982. Depositional mechanisms of deep-sea coarse clastic sediments, Cap Enragé Formation, Québec. *Canadian Journal of Earth Sciences*, **19**, 267–287.

Hein, F.J. & Gorsline, D.S. 1981. Geotechnical aspects of fine grained mass flow deposits: California Continental Borderland. *Geo-Marine Letters*, **1**, 1–5.

Hein, F.J. & Walker, R.G. 1982. The Cambro-Ordovician Cap Enragé Formation, Québec, Canada: conglomeratic deposits of a braided submarine channel with terraces. *Sedimentology*, **29**, 309–329.

Heiniö, P. & Davies, R.J. 2007. Knickpoint migration in submarine channels in response to fold growth, western Niger Delta. *Marine and Petroleum Geology*, 24, 434–449.

Heinrich, H. 1988. Origin and consequences of cyclic ice rafting in the northeast Atlantic Ocean during the past 130,000 years. *Quaternary Research*, 29, 142–152.

Heller, P.L. & Dickinson, W.R. 1985. Submarine ramp facies model for delta-fed, sand-rich turbidite systems. *American Association of Petroleum Geologists Bulletin*, 69, 960–976.

Hempton, M., Marshall, J., Sadler, S., Hogg, N., Charles, R. & Harvey, C. 2005. Turbidite reservoirs of the Sele Formation, central North Sea: geological challenges for improving production. *In*: Doré, A.G. & Vining, B.A. (eds), *Petroleum Geology: North-west Europe and Global Perspectives–Proceedings of the 6th Petroleum Geology Conference*, Volume 1, 449–459. London: The Geological Society.

Hendry, H.E. 1972. Breccias deposited by mass flow in the Breccia Nappe of the French pre-Alps. *Sedimentology*, 8, 277–292.

Hendry, H E. 1973. Sedimentation of deep water conglomerates in Lower Ordovician rocks of Quebec–composite bedding produced by progressive liquefaction of sediment? *Journal of Sedimentary Petrology*, 43, 125–136.

Hendry, H.E. 1978. Cap des Rosiers Formation at Grosses Roches, Quebec-deposits in the mid-fan region on an Ordovician submarine fan. *Canadian Journal of Earth Sciences*, 15, 1472–1488.

Henry, P., Le Pichon, X., Lallement, S., Foucher, J.-P., Westbrook, G. & Hobart, M. 1990. Mud volcano field seaward of the Barbados Accretionary Complex: a deep-towed side scan sonar survey. *Journal of Geophysical Research: Solid Earth*, 95, 8917–8829.

Henry, S., Kumar, N., Danforth, A., Nutall, P. & Venkatraman, S. 2011. Ghana/Sierra Leone lookalike plays in Northern Brazil. *GEO ExPro*, 8, 36–40.

Hernández-Molina, F.J., Llave, E., Preu, B., Ercilla, G., Fontan, A., Bruno, M., Serra, N., Gomiz, J.J., Brackenridge, R.E., Sierro, F.J., Stow, D.A.V., Garcia, M., Juan, C., Sandoval, N. & Amaiz, A. 2014. Contourite processes associated with the Mediterranean Outflow Water after its exit from the Strait of Gibraltar: Global and conceptual implications. *Geology*, 42, 227–230.

Hesse, R. 1965. Herkunfe und Transport der Sedimente im Bayerischen Flyschtrog. *Zeitschrift der Deutschen Gesellschaft für Geowissenschaften*, 116, 147–170.

Hesse, R. 1974. Long-distance continuity of turbidites: possible evidence for an early Cretaceous trench-abyssal plain in the East Alps. *Geological Society of America Bulletin*, 85, 859–870.

Hesse, R. 1975. Turbiditic and non-turbiditic mudstone of Cretaceous flysch sections of the East Alps and other basins. *Sedimentology*, 22, 387–416.

Hesse, R. 1982. Cretaceous–Palaeogene Flysch Zone of the East Alps and Carpathians: identification and plate-tectonic significance of 'dormant' and 'active' deep-sea trenches in the Alpine-Carpathian Arc. *In*: Leggett, J.K. (ed.), *Trench-Forearc Geology*, 471–494. Geological Society of London Special Publication of the Geological Society, London, 10. Oxford: Blackwell Scientific.

Hesse, R. 1995a. Continental slope and basin sedimentation adjacent to an ice-margin: a continuous sleeve-gun profile across the Labrador Slope, Rise and Basin. *In*: Pickering, K.T., Hiscott, R.N., Kenyon, N.H., Ricci Lucchi, F. & Smith, R.D.A. (eds), *Atlas of Deep Water Environments: Architectural Style in Turbidite Systems*, 14–17. London: Chapman and Hall.

Hesse, R. 1995b. Bed-by-bed correlation of trench-plain turbidite sections, Campanian Zementmergel Formation, Rhenodanubian Flysch Zone of the East Alps. *In*: Pickering, K.T., Hiscott, R.N., Kenyon, N.H., Ricci Lucchi, F. & Smith, R.D.A. (eds), *Atlas of Deep Water Environments: Architectural Style in Turbidite Systems*, 307–309. London: Chapman and Hall.

Hesse, R. & Butt, A.A. 1976. Paleobathymetry of Cretaceous turbidite basins of the East Alps relative to the calcite compensation level. *Journal of Geology*, 84, 505–533.

Hesse, R. & Chough, S.K. 1980. The Northwest Atlantic Mid-Ocean Channel of the Labrador Sea: II. Deposition of parallel laminated levée-muds from the viscous sublayer of low density turbidity currents. *Sedimentology*, 27, 697–711.

Hesse, R. & Khodabakhsh, S. 2006. Significance of fine-grained sediment lofting from melt-water generated turbidity currents for the timing of glaciomarine sediment transport into the deep sea. *Sedimentary Geology*, 186, 1–11.

Hesse, R. & Klaucke, I. 1995. A continuous along-slope seismic profile from the upper Labrador Slope. *In*: Pickering, K.T., Hiscott, R.N., Kenyon, N.H., Ricci Lucchi, F. & Smith, R.D.A. (eds), *Atlas of Deep Water Environments: Architectural Style in Turbidite Systems*, 18–22. London: Chapman and Hall.

Hesse, R., Chough, S.K. & Rakofsky, A. 1987. The Northwest Atlantic Mid-Ocean Channel of the Labrador Sea. V. Sedimentology of a giant deep-sea channel. *Canadian Journal of Earth Sciences*, 24, 1595–1624.

Hesse, R., Rakofsky, A. & Chough, S.K. 1990. The central Labrador Sea: facies and dispersal patterns of clastic sediment in a small ocean basin. *Marine and Petroleum Geology*, 7, 13–28.

Hesse, R., Klaucke, I., Ryan, W.B.F., Edwards, M.E. & Piper, D. 1996. Imaging Laurentide ice sheet drainage into the deep sea: impact on sedimentation and bottom water. *GSA Today*, 6(9), 3–9.

Hesse, R., Khodabakhsh, S. Klaucke, I. & Ryan, W.B.F. 1997. Asymmetrical turbid surface-plume deposition near ice-outlets of the Pleistocene Laurentide ice sheet in the Labrador Sea. *Geo-Marine Letters*, 17, 179–187.

Hesse, R., Klaucke, I. Khodabakhsh, S. & Piper, D. 1999. Continental slope sedimentation adjacent to an ice margin. III. The upper Labrador Slope. *Marine Geology*, 155, 249–276.

Hesse, R., Klaucke, I. Khodabakhsh, S., Piper, D.J.W., Ryan, W.B.F. & the NAMOC Study Group 2001. Sandy submarine braid plains: potential deep-water reservoirs. *American Association of Petroleum Geologists Bulletin*, 85, 1499–1521.

Hesselbo, S.P., Jenkyns, H.C., Duarte, L.V. & Oliveira, L.C.V. 2007. Carbon-isotope record of the Early Jurassic (Toarcian) Oceanic Anoxic Event from fossil wood and marine carbonate (Lusitanian Basin, Portugal). *Earth and Planetary Science Letters*, 253, 455–470.

Hicks, D.M. 1981. Deep-sea fan sediments in the Torlesse zone, Lake Ohau, South Canterbury, *New Zealand*. *New Zealand Journal of Geology & Geophysics.*, 24, 209–230.

Higgins, G.E. & Saunders, J.B. 1967. Report on 1964 Chatham Mud Island, Erin Bay, Trinidad, West Indies. *American Association of Petroleum Geologists Bulletin*, 51, 55–64.

Higgins, G.E. & Saunders, J.B. 1974. Mud volcanoes, their nature and origin. *In*: *Contributions to the Geology and Palaeo-biology of the Caribbean and Adjacent Areas*, 84, 101–152. Verhandlungen Naturforschenden Gesellschaft in Basel.

Higgins, J.A. & Schrag, D.P. 2006. Beyond methane: Towards a theory for the Paleocene–Eocene Thermal Maximum *Earth and Planetary Science Letters*, 245, 523–537.

Hill, P.R. 1984a. Facies and sequence analysis of Nova Scotian Slope muds: turbidite vs 'hemipelagic' deposition. *In*: Stow, D.A.V. & Piper, D.J.W. (eds), *Fine-grained Sediments: Deep-water Processes and Facies*, 311–318. Geological Society of London Special Publication, 15.

Hill, P.R. 1984b. Sedimentary facies of the Nova Scotian upper and middle continental slope, offshore eastern Canada. *Sedimentology*, 31, 293–309.

Hill, P.R., Moran, K.M. & Blasco, S.M. 1982. Creep deformation of slope sediments in the Canadian Beaufort Sea. *Geo-Marine Letters*, 2, 163–170.

Hill, P.S. 1998. Controls on floc size in the sea. *Oceanography*, 11(2), 13–18.

Hill, P.S. & McCave, I.N. 2001. Suspended particle transport in benthic boundary layers. *In*: Boudreau, B.P. & Jørgensen, B.B. (eds), *The Benthic Boundary Layer: Transport Processes and Biogeochemistry*, 78–103. New York: Oxford University Press.

Hillenbrand, C.-D. & Fütterer, D.K. 2001. Neogene to Quaternary deposition of opal on the continental rise west of the Antarctic Peninsula, ODP Leg 178, Sites 1095, 1096, and 1101. *In*: Barker, P.F., Camerlenghi, A., Acton, G.D. & Ramsay, A.T.S. (eds), *Proceedings Ocean Drilling Program, Scientific Results*, **178**, 1–33. College Station, Texas, USA: Ocean Drilling Program.

Hillis, R.R., Sandford, M., Reynolds, S.D. & Quigley, M.C. 2008. Present-day stresses, seismicity and Neogene-to-Recent tectonics of Australia's 'passive' margins: intraplate deformation controlled by plate boundary forces. *In*: Johnson, H., Doré, A.G., Gatliff, R.W., Holdsworth, R., Lundin, E. & Ritchie, J.D. (eds) *The Nature and Origin of Compressive Margins*, 71–89. The Geological Society, London, Special Publications, **306**. Bath, UK: The Geological Society, London.

Hilton, V.C. 1995. Sandstone architecture and facies from the Annot Basin of the Tertiary SW Alpine Foreland Basin, SE France. *In*: Pickering, K.T., Hiscott, R.N., Kenyon, N.H., Ricci Lucchi, F. & Smith, R.D.A. (eds), *Atlas of Deep Water Environments: Architectural Style in Turbidite Systems*, 227–235. London: Chapman and Hall.

Hilton, V.C. & Pickering, K.T. 1995. The Montagne de Chalufy turbidite onlap, Eocene–Oligocene turbidite sheet system, Hautes Provence, SE France. *In*: Pickering, K.T., Hiscott, R.N., Kenyon, N.H., Ricci Lucchi, F. & Smith, R.D.A. (eds), *Atlas of Deep Water Environments: Architectural Style in Turbidite Systems*, 236–241. London: Chapman and Hall.

Hinschberger, F., Malod, J.-A., Dyment, J., Honthaas, C., Rehault, J.-P. & Burhanuddin, S. 2001. Magnetic lineations constraints for the back-arc opening of the Late Neogene South Banda Basin (eastern Indonesia). *Tectonophysics*, **333**, 47–59.

Hinschberger, F., Malod, J-A., Réhault, J.P., Villeneuve, M., Royer, J-Y. & Burhanuddin, S. 2005. Late Cenozoic geodynamic evolution of eastern Indonesia. *Tectonophysics*, **404**, 91–118.

Hiscott, R.N. 1979. Clastic sills and dikes associated with deep-water sandstones, Tourelle formation, Ordovician, Quebec. *Journal of Sedimentary Petrology*, **49**, 1–10.

Hiscott, R.N. 1980. Depositional framework of sandy mid-fan complexes of Tourelle Formation, Ordovician, Québec. *American Association of Petroleum Geologists Bulletin*, **64**, 1052–1077.

Hiscott, R.N. 1981. Deep-sea fan deposits in the Macigno Formation (middle-upper Oligocene) of the Gordana Valley, northern Apennines, Italy–Discussion. *Journal of Sedimentary Petrology*, **51**, 1015–1021.

Hiscott, R.N. 1984. Ophiolitic source rocks for Taconic-age flysch: Trace-element evidence. *Geological Society of America Bulletin*, **95**, 1261–1267.

Hiscott, R.N. 1994a. Loss of capacity, not competence, as the fundamental process governing deposition from turbidity currents. *Journal of Sedimentary Research*, **A64**, 209–214.

Hiscott, R.N. 1994b. Traction-carpet stratification in turbidites–fact or fiction? *Journal of Sedimentary Research*, **A64**, 204–208.

Hiscott, R.N. 2001. Depositional sequences controlled by high rates of sediment supply, sea-level variations, and growth faulting: the Quaternary Bartam Delta of northwestern Borneo. *Marine Geology*, **175**, 67–102.

Hiscott, R.N. & Aksu, A.E. 1994. Submarine debris flows and continental slope evolution in front of Quaternary ice sheets, Baffin Bay, Canadian Arctic. *American Association of Petroleum Geologists Bulletin*, **78**, 445–460.

Hiscott, R.N. & Aksu, A.E. 1996. Quaternary sedimentary processes and budgets in Orphan Basin, southwest Labrador Sea. *Quaternary Research*, **45**, 160–175.

Hiscott, R.N. & Devries, M. 1995. Internal characteristics of sandbodies of the Ordovician Tourelle Formation, Quebec, Canada. *In*: Pickering, K.T., Hiscott, R.N., Kenyon, N.H., Ricci Lucchi, F. & Smith, R.D.A. (eds), *Atlas of Deep Water Environments: Architectural Style in Turbidite Systems*, 207–211. London: Chapman and Hall.

Hiscott, R.N. & Gill, J.B. 1992. Major- and trace-element geochemistry of Oligocene to Quaternary volcaniclastic sands and sandstones from the Izu-Bonin Arc. *In*: Taylor, B., Fujioka, K., Janecek, T. *et al.*, Proceedings of the Ocean Drilling Program, Scientific Results, **126**, 467–485.

Hiscott, R.N. & James, N.P. 1985. Carbonate debris flows, Cow Head Group, western Newfoundland. *Journal of Sedimentary Petrology*, **55**, 735–745.

Hiscott, R.N. & Middleton, G.V. 1979. Depositional mechanics of thick-bedded sandstones at the base of a submarine slope, Tourelle Formation (Lower Ordovician), Québec, Canada. *In*: Doyle, L.J. & Pilkey, O.H. (eds), *Geology of Continental Slopes*, 307–326. Society of Economic Paleontologists and Mineralogists Special Publication, **27**.

Hiscott, R.N. & Middleton, G.V. 1980. Fabric of coarse deep-water sandstones, Tourelle Formation, Québec, Canada. *Journal of Sedimentary Petrology*, **50**, 703–722.

Hiscott, R.N. & Pickering, K.T. 1984. Reflected turbidity currents on an Ordovician basin floor, Canadian Appalachians. *Nature*, **311**, 143–145.

Hiscott, R.N., Pickering, K.T. & Beeden, D.R. 1986. Progressive filling of a confined Middle Ordovician foreland basin associated with the Taconic Orogeny, Quebec, Canada. *In*: Allen, P.A. & Homewood, P. (eds), *Foreland Basins*, 309–325. Special Publication of the International Association of Sedimentologists, **8**. Oxford: Blackwell Scientific Publications.

Hiscott, R.N., Cremer, M. & Aksu, A.E. 1989. Evidence from sedimentary structures for processes of sediment transport and deposition during post-Miocene time at Sites 645, 646, and 647, Baffin Bay and the Labrador Sea. *In*: Srivastava, S. P., Arthur, M., Clement, B., *et al.*, *Proceedings of the Ocean Drilling Program, Scientific Results*, **105**, 53–63. College Station, Texas, USA: Ocean Drilling Program.

Hiscott, R.N., Wilson, R.C.L., Gradstein, F.M. Pujalte, V. García-Mondéjar, J., Boudreau, R.R. & Wishart, H.A. 1990. Comparative stratigraphy and subsidence history of Mesozoic rift basins of North Atlantic. *American Association of Petroleum Geologists Bulletin*, **74**, 60–76.

Hiscott, R.N., Colella, A., Pezard, P., Lovell, M.A. & Malinverno, A. 1992. Sedimentology of deep-water volcaniclastics, Oligocene Izu-Bonin forearc basin, based on formation microscanner images. *In*: Taylor, B. & Fujioka, K. (eds), *Proceedings of the Ocean Drilling Program Scientific Results*, **126**, 75–96. College Station, Texas, USA: Ocean Drilling Program.

Hiscott, R.N., Colella, A., Pezard, P., Lovell, M.A. & Malinverno, A. 1993. Basin plain turbidite succession of the Oligocene Izu-Bonin intraoceanic forearc basin. *Marine and Petroleum Geology*, **10**, 450–466.

Hiscott, R.N., Pickering, K.T. Bouma, A.H. Hand, B.M. Kneller, B.C. Postma, G. & Soh, W. 1997a. Basin-floor fans in the North Sea: sequence stratigraphic models vs. sedimentary facies: discussion. *American Association of Petroleum Geologists Bulletin*, **81**, 662–665.

Hiscott, R.N., Hall, F.R. & Pirmez, C. 1997b. Turbidity-current overspill from Amazon Channel: texture of the silt/sand load, paleoflow from anisotropy of magnetic susceptibility, and implications for flow processes. *In*: Flood, R.D., Piper, D.J.W., Klaus, A. & Peterson, L.C. (eds), *Proceedings Ocean Drilling Program, Scientific Results*, **155**, 53–78. College Station, Texas, USA: Ocean Drilling Program.

Hiscott, R.N., Pirmez, C. & Flood, R.D. 1997c. Amazon Submarine Fan drilling: a big step forward for deep-sea fan models. *Geoscience Canada*, **24**, 13–24.

Hiscott, R.N., Marsaglia, K.M., Wilson, R.C.L., Robertson, A.H.F., Karner, G.D., Tucholke, B.E., Pletsch, T., Petschick, R. 2008. Detrital sources and sediment delivery to the early post-rift (Albian–Cenomanian) Newfoundland Basin east of the Grand Banks: results from ODP Leg 210. *Bulletin of Canadian Petroleum Geology*, **56**, 69–92.

Hiscott, R.N., Aksu, A.E., Flood, R.D., Kostylev, V. & Yaşar, D. 2013. Widespread overspill from a saline density-current channel and its interaction with topography on the SW Black Sea shelf. *Sedimentology*, **60**, doi: 10.1111/sed.12071.

Hodgetts, D., Drinkwater, N.J., Hodgson, D.M., Kavanagh, J.P., Flint, S.S., Keogh, K.J. & Howell, J.A. 2004. Three-dimensional geological models from outcrop data using digital data collection techniques: an example from the Tanqua Karoo depocenter, South Africa. *In*: Curtis, A.C. & Wood, R. (eds), *Geological Prior Information: Informing Science and Engineering,* 57–75. Geological Society of London, Special Publication, **239**, London: The Geological Society.

Hodgson, D.M. 2009. Distribution and origin of hybrid beds in sand-rich submarine fans of the Tanqua depocentre, Karoo Basin, South Africa. *Marine and Petroleum Geology*, **26**, 1940–1956.

Hodgson, D.M., Flint, S.S., Hodgetts, D., Drinkwater, N.J., Johannessen, E.P. & Luthi, S.M. 2006. Stratigraphic evolution of fine-grained submarine fan systems, Tanqua depocenter, Karoo Basin, South Africa. *Journal of Sedimentary Research*, **76**, 20–40.

Hodgson, D.M., Di Celma, C.N., Brunt, R.L. & Flint, S.S. 2011. Submarine slope degradation and aggradation and the stratigraphic evolution of channel-levée systems. *Journal of the Geological Society (London)*, **168**, 1–4.

Hodgson, D.M. & Pickering, K.T. *in prep.* Sructurally-confined submarine-fan system the Tabernas-Sorbas Basin, SE Spain.

Hodgson, N. 2013. Power up! Selecting a 3D dataset for the 2013 Licence Round, offshore Lebanon. *GEO ExPro*, **10**, 38–40.

Hodson, J.M. & Alexander, J. 2010. The effects of grain-density variation on turbidity currents and some implications for the deposition of carbonate turbidites. *Journal of Sedimentary Research*, **80**, 515–528.

Hoel, P.G. 1971. *Introduction to Mathematical Statistics,* 4th edn. New York: Wiley.

Hoffert, M. 1980. *Les 'argiles rouges des grands fonds, dans le Pacifique centre-est: authigenese, transport, diagenese'.* Thesis, University Louis Pasteur Strasbourg, Memoir, **61**.

Hoffman, P., Dewey, J.F. & Burke, K. 1974. Aulacogens and their genetic relation to geosynclines, with a Proterozoic example from Great Slave Lake, Canada. *In*: Dott, R.H. Jr & Shaver, R.H. (eds), *Modern and Ancient Geosynclinal Sedimentation,* 38–55. Society of Economic Paleontologists and Mineralogists Special Publication, **19**.

Hogan, P., Lane, A., Hooper, J., Broughton, A. & Romans, B. 2008. Geohazard challenges of the Woodside OceanWay Secure Energy LNG development, offshore Southern California. Offshore Technology Conference, Houston, Texas, 5–6 May 2008, Paper **19563**.

Hogg, N.G. 1983. A note on the deep circulation of the western North Atlantic: its nature and causes. *Deep-Sea Research*, **30**, 945–961.

Hölker, A.B., Manatschal, G., Holliger, K. & Bernoulli, D. 2003. Tectonic nature and seismic response of top-basement detachment faults in magma-poor rifted margins. *Tectonics*, **22**, 1035, doi:10.1029/2001TC001347.

Holl, J.E. & Anastasio, D.J. 1993. Paleomagnetically derived folding rates, Southern Pyrenees, Spain. *Geology*, **13**, 271–274.

Hollister, C.D. & Heezen, B.C. 1972. Geologic effects of ocean bottom currents: western North Atlantic. *In*: Gordon, A.L. (ed.), *Studies in Physical Oceanography, 2*, 37–66. New York: Gordon & Breach.

Hollister, C.D. & McCave, I.N. 1984. Sedimentation under deep-sea storms. *Nature*, **309**, 220–225.

Hollister, C.D. & Nowell, A.R.M. 1991a. Prologue: Abyssal storms as a global geologic process. *Marine Geology*, **99**, 275–280.

Hollister, C.D. & Nowell, A.R.M. 1991b. HEBBLE epilogue. *Marine Geology*, **99**, 445–460.

Hollister, C.D., Southard, J.B., Flood, R.D. & Lonsdale, P.F. 1976b. Flow phenomena in the benthic boundary layer and bed forms beneath deep-current systems. *In*: McCave, I.N. (ed.), *The Benthic Boundary Layer,* 183–204. New York: Plenum.

Hollister, C.D., Craddock, C. et al. 1976a. *Initial Reports Deep Sea Drilling Project, 35.* Washington, DC: US Government Printing Office.

Hopfinger, E.J. 1983. Snow avalanche motion and related phenomena. *Annual Review of Fluid Mechanics*, **15**, 47–76.

Hori, T., Kato, N., Hirahara, K., Baba, T. & Kaneda, Y. 2004. A numerical simulation of earthquake cycles along the Nankai Trough in southwest Japan: lateral variation in frictional property due to the slab geometry controls the nucleation position. *Earth and Planetary Science Letters*, **228**, 215–226.

Horn, D.R. (ed.) 1972. *Ferromanganese Deposits on the Ocean Floor.* New York: Arden House, Harriman and Lamont-Doherty Geological Observatory.

Ho-Shing Yua & Hong, E. 2005. Shifting submarine canyons and development of a foreland basin in SW Taiwan: controls of foreland sedimentation and longitudinal sediment transport. *Journal of Asian Earth Sciences*, **27**, 922–932.

Ho-Shing Yua, Cheng-Shing Chiangb & Su-Min Shen 2008. Tectonically active sediment dispersal system in SW Taiwan margin with emphasis on the Gaoping (Kaoping) Submarine Canyon. *Journal of Marine Systems*, **76**, 369–382.

Hoskin, C.M. & Burrell, D.C. 1972. Sediment transport and accumulation in a fjord basin, Glacier Bay, Alaska. *Journal of Geology*, **80**, 539–551.

Hoskins, E.G. & Griffiths, J.R. 1971. Hydrocarbon potential of northern and central California offshore. *In*: Cram, I.H. (ed.), *Future Petroleum Provinces of the United States – Their Geology and Potential,* 212–228. American Association of Petroleum Geologists Memoir, **15**.

Houseknecht, D.W., Schenk, C.J., Lepain, D.L., Burruss, R.C., Moore, T.E. & Bird, K.J. 1999. *Petroleum Potential of Torok and Nanushuk Depositional Sequences in the National Petroleum Reserve-Alaska (NPRA) and Adjacent Areas Anonymous.* American Association of Petroleum Geologists, Annual Meeting San Antonio, TX, United States, Apr. 11–14, 1999, Expanded Abstracts, **1999**, A62–A63.

Houtz, R.E. & Hayes, D.E. 1984. Seismic refraction data from Sunda shelf. *American Association of Petroleum Geologists Bulletin*, **68**, 1870–1878.

Howard, J.D. & Frey, R.W. 1984. Characteristic trace fossils in nearshore to offshore sequences, Upper Cretaceous of east-central Utah. *Canadian Journal of Earth Sciences*, **21**, 200–219.

Howell, D.G. (ed.) 1985. *Tectono-stratigraphic Terranes of the Circum-Pacific.* Houston Texas: Circum-Pacific Council for Energy & Mineral Resources.

Howell, D.G. & Link, M.H. 1979. Eocene conglomerate sedimentology and basin analysis, San Diego and the southern California Borderland. *Journal of Sedimentary Petrology*, **49**, 517–540.

Howell, D.G., Crouch, J.K., Greene, H.G., McCulloch, D.S. & Vedder, J.G. 1980. Basin development along the Late Mesozoic and Cenozoic California margin: a plate tectonic margin of subduction, oblique subduction, and transform tectonics. *In*: Balance, P.F. & Reading, H.G. (eds), *Sedimentation in Oblique-slip Mobile Zones,* 43–62. International Association of Sedimentologists Special Publication, **4**. Oxford: Blackwell Scientific.

Howell, D.G., Jones, D.L. & Schermer, E.R. 1985. Tectono-stratigraphic terranes of the circum-Pacific region. *In*: Howell, D.G. (ed.), *Tectono-stratigraphic Terranes of the Circum-Pacific,* 3–30. Houston Texas: Circum-Pacific Council for Energy & Mineral Resources.

Hoyal, D., Sheets, B., Wellner, R., Box, D., Sprague, A. & Bloch, R. 2011. Architecture of Froude critical-supercritical submarine fans: tank experiments versus field observations. *American Association of Petroleum Geologists Annual Convention and Exhibition, April 10–13, Datapages/Search and Discovery Article #90124*.

Hoyal, D.C.H., Demko, T., Postma, G., Wellner, R.W., Pederson, K., Abreu, V., Fedele, J.J., Box, D., Sprague, A., Ghayour, K., Strom, K. & Hamilton, P. 2014. Evolution, architecture and stratigraphy of Froude supercritical submarine fans. *American Association of Petroleum Geologists Annual Convention and Exhibition, April 6–9, Datapages/Search and Discovery Article #90189*.

Hsü, K.J. 1974. Mélanges and their distinction from olistostromes. *In*: Dott, R.H.Jr & Shaver, R.H. (eds), *Modern and Ancient Geosynclinal Sedimentation,* 321–333. Society of Economic Paleontologists and Mineralogists Special Publication, **19**.

Hsü, K.J. 1977. Studies of Ventura Field, *California. I: facies geometry and genesis of Lower Pliocene turbidites. American Association of Petroleum Geologists Bulletin*, **61**, 137–168.

Hsü, K.J. & Jenkyns, H.C. (eds) 1974. *Pelagic Sediments: On Land and Under the Sea.* International Association of Sedimentologists Special Publication, **1**. Oxford: Blackwell Scientific.

Huang, H., Imran, J., Pirmez, C., Zhang, Q. & Chen, G. 2009. The critical densimetric Froude number of subaqueous gravity currents can be non-unity or non-existent. *Journal of Sedimentary Research*, **79**, 479–485.

Hubbard, S.M. & Shultz, M.R. 2008. Deep burrows in submarine fan-channel deposits of the Cerro Toro Formation (Cretaceous), Chilean Patagonia: Implications for firmground development and colonization in the deep sea. *Palaios*, **23**, 223–232.

Hubbard, S.M., Romans, B.W. & Graham, S.A. 2007. An outcrop example of large-scale conglomeratic intrusions sourced from deep-water channel deposits, Cerro Toro formation, Magallanes basin, southern Chile. *In*: Hurst, A. & Cartwright, J. (eds), *Sand Injectites: Implications for Hydrocarbon Exploration and Production*, 199–207. American Association of Petroleum Geologists Memoir, **87**. Tulsa, Oklahoma.

Hubbard, S.M., Romans, B.W. & Graham, S.A. 2008. Deep-water foreland basin deposits of the Cerro Toro Formation, Magallanes basin, Chile: architectural elements of a sinuous basin axial channel belt. *Sedimentology*, **55**, 1333–1359.

Hubbard, S.M., de Ruig, M.J. & Graham, S.A. 2009. Confined channel-levee complex development in an elongate depocenter: deep-water Tertiary strata of the Austrian Molasse basin. *Marine and Petroleum Geology*, **26**, 85–112.

Hubbard, S.M., MacEachern, J.A. & Bann, K.L. 2012. Slopes. *In*: Knaust, D. & Bromley, R.G. (eds), *Trace Fossils as Indicators of Sedimentary Environments*, 607–642. Developments in Sedimentology, **64**. Amsterdam: Elsevier.

Hubert, J.F. 1966a. Modification of the model for internal structures in graded beds to include a dune division. *Nature*, **211**, 614–615.

Hubert, J.F. 1966b. Sedimentation history of Upper Ordovician geosynclinal rocks, Girvan, Scotland. *Journal of Sedimentary Petrology*, **36**, 677–699.

Hubert, C., Lajoie, J. & Leonard, M.A. 1970. Deep sea sediments in the Lower Paleozoic Quebec Supergroup. *In*: Lajoie, J. (ed.), *Flysch Sedimentology in North America*, 103–125. Geological Association of Canada Special Paper, **7**. Toronto: Business & Economic Service.

Hubscher, C., Spiesz, V., Breitzke, M. & Weber, M.E. 1997. The youngest channel-levée system of the Bengal Fan: results from digital sediment echosounder data. *Marine Geology*, **21**, 125–145.

Huchon, P. & Kitazato, H. 1984. Collision of the Izu block with central Japan during the Quaternary and geological evolution of the Ashigara area. *Tectonophysics*, **110**, 201–210.

Huchon, P., Taylor, B. & Klaus, A. (eds) 2001. *Proceedings of the Ocean Drilling Program, Scientific Results*, **180**. College Station, Texas, USA: Ocean Drilling Program.

Hughes Clarke, J.E., Shor, A.N., Piper, D.J.W. & Mayer, L.A. 1990. Large-scale current-induced erosion and deposition in the path of the 1929 Grand Banks turbidity current. *Sedimentology*, **37**, 613–629.

Hughes Clarke, J.E., Brucker, S., Muggah, J., Church, I., Cartwright, D. 2011. The Squamish delta repetitive survey program: a simultaneous investigation of prodeltaic sedimentation and integrated system accuracy. *Squamish Repetitive Survey Program, Paper: Mapping-5*. U.S. Hydrographic Conference 2011, 16 pp.

Hughes Clarke, J.E., Brucker, S., Muggah, J., Hamilton, T., Cartwright, D., Church, I. & Kuus, P. 2012a. Temporal progression and spatial extent of mass wasting events on the Squamish prodelta slope. *In*: Eberhardt, E., Froese, C., Turner, K. & Leroueil, S. (eds), *Landslides and Engineered Slopes: Protecting Society Through Improved Understanding*, 1091–1096. London: Taylor & Francis Group, ISBN 978-0-415-62123-6.

Hughes Clarke, J.E., Brucker, S., Muggah, J., Church, I. Cartwright, D., Kuus, P., Hamilton, T., Pratamo, D. & Eisan, B. 2012b. The Squamish prodelta: monitoring active landslides and turbidity currents. *In*: *The Arctic, Old Challenges New*, Niagara Falls, Canada 15–17 May 2012.

Hulme, T., Haines, A.J. & Yu, J. 2004. General elastic wave scattering problems using an impedance operator approach. 2, Two-dimensional isotropic validation and examples. *Geophysical Journal International*, **159**, 658–666.

Hults, C.P., Wilson, F.H., Donelick, R.A. & O'Sullivan, P.B. 2013. Two flysch belts having distinctly different provenance suggest no stratigraphic link between the Wrangellia composite terrane and the paleo-Alaskan margin. *Lithosphere*, **5**, 575–594.

Humphrey, N.F. & Heller, P.L. 1995. Natural oscillations in coupled geomorphic systems: an alternative origin for cyclic sedimentation. *Geology*, **23**, 499–502.

Humphris, C.C. Jr, 1978. Salt movement on the continental slope, northern Gulf of Mexico. *In*: Bouma, A.H., Moore, G.T. & Coleman, J.M. (eds), *Framework, Facies, and Oil-trapping Characteristics of the Upper Continental Margin*, 69–85. American Association of Petroleum Geologists Studies in Geology, **7**.

Hüneke, H. & Mulder, T. (eds) 2011. *Deep-sea Sediments. Developments in Sedimentology, 63, 849 pp.* Amsterdam: Elsevier B.V. ISBN: 978-0-444-53000-4.

Hüneke, H. & Stow, D.A.V. 2008. Identificiation of ancient contourites: problems and palaeoceanographic significance. *In*: Rebesco, M. & Camerlenghi, A. (eds), *Contourites*, 323–344. Developments in Sedimentology, **60**. Amsterdam: Elsevier.

Hurst, A. & Cartwright, J.A. (eds) 2007. *Sand Injectites: Implications for Hydrocarbon Exploration and Production*. American Association of Petroleum Geologists Memoir, **87**. Tulsa, Oklahoma.

Hurst, A., Scott, A. & Vigorito, M. 2011. Physical characteristics of sand injectites. *Earth-Science Reviews*, **106**, 215–246.

Hurst, H.E. 1951. Long term storage capacity of reservoirs. *Transactions of the American Society of Civil Engineers*, **116**, 770–808.

Hurst, H.E. 1956. Methods of using long-term storage in reservoirs. *Proceedings of the Institute of Civil Engineers*, Part I, **5**, 519–590.

Hutton, D.H.W. & Dewey, J.F. 1986. Palaeozoic terrane accretion in the western Irish Caledonides. *Tectonics*, **5**, 1115–1124.

Huyghe, D., Castelltort, S., Mouthereau, F., Serra-Kiel, J., Filleaudeau, P., Emmanuel, L., Berthier, B. & Renard, M. 2012. Large scale facies change in the middle Eocene South-Pyrenean foreland basin: The role of tectonics and prelude to Cenozoic ice-ages. *Sedimentary Geology*, **253–254**, 25–46.

Hyne, N.J. 1969. *Sedimentology and Pleistocene history of Lake Tahoe, California-Nevada*. Los Angeles: University of Southern California, Ph.D. dissertation, 121 pp.

Hyne, N.J., Chelminski, P., Court, J.E., Gorsline, D.S. & Goldman, C.R. 1972. Quaternary history of Lake Tahoe, California-Nevada. *Geological Society of America Bulletin*, **83**, 1435–1448.

Hyne, N.J., Goldman, C.R. & Court, J.E. 1973. Mounds in Lake Tahoe, *California-Nevada: a model for landslide topography in the subaqueous environment: Journal of Geology*, **81**, 176–188.

Ichinose, G.A., Thio, H.K., Sato, T., Ishii, T. & Somerville, P.G. 2003. Rupture process of the 1944 Tonankai earthquake (Ms 8.1) from the inversion of teleseismic and regional seismograms. *Journal of Geophysical Research*, **108**. doi: 10.1029/2003JB002393.

Ikari, M.J. & Saffer, D.M. 2011. Comparison of frictional strength and velocity dependence between fault zones in the Nankai accretionary complex, *Geochemistry, Geophysics, Geosystems*, **12**, doi:10.1029/2010GC003442.

Ikehara, K. 1989. The Kuroshio-generated bedform system in the Osumi Strait, southern Kyushu, Japan. *In*: Taira, A. & Masuda, F. (eds), *Sedimentary Facies in the Active Plate Margin*, 261–273. Tokyo: Terra Scientific Publishing Company.

Ilstad, T., Marr, J.G., Elverhøi, A. & Harbitz, C.B. 2004a. Laboratory studies of subaqueous debris flows by measurements of pore-fluid pressure and total stress. *Marine Geology*, **213**, 403–414.

Ilstad, T., De Blasio, F.V., Elverhøi, A., Hartiz, C.B., Engvik, L., Longva, O. & Marr, J.G. 2004b. On the frontal dynamics and morphology of submarine debris flows. *Marine Geology*, **213**, 481–497.

Imbrie, J. & Imbrie, K.P. 1979. *Ice Ages: Solving the Mystery*. Short Hills New Jersey: Enslow Publications, 224 pp. ISBN: 0333267672.

Imbrie, J. & Imbrie, K.P. 1980. Modelling the climatic response to orbital variations. *Science*, **207**, 943–953.

Imbrie, J., Hays, J.D., Martinson, D.G., McIntyre, A., Mix, A.C., Morley, I.J., Pisias, N.G., Prell, W.L. & Shackleton, N.J. 1984. The orbital theory of Pleistocene climate: support from revised chronology of the marine $\delta^{18}O$ record. *In*: Berger, A.L., Imbrie, J. Hays, J., Kukla, G. & Saltzman, B. (eds), *Milandovitch and Climate, Part I,* 269-305. NATO ASI Series C, 126. Dordrecht: Reidel.

Imran, J., Parker, G., Locat, J. & Lee, H. 2001. 1D numerical model of muddy subaqueous and subaerial debris flows. *Journal of Hydraulic Engineers,* **127,** 959–968.

Ingersoll, R.V. 1978a. Submarine fan facies of the Upper Cretaceous Great Valley Sequence, northern and central California. *Sedimentary Geology,* **21,** 205–230.

Ingersoll, R.V. 1978b. Petrofacies and petrologic evolution of the Late Cretaceous fore-arc basin, northern and central California. *Journal of Geology,* **86,** 335–352.

Ingersoll, R.V. 2008. Reconstructing southern California. *In*: Spencer, J.E. & Titley, S.R. (eds), *Ores and orogenesis: Circum-Pacific tectonics, geologic evolution, and ore deposits. Arizona Geological Society Digest,* **22,** 409–417.

Ingle, J.D., Jr., 1975. Paleobathymetric analysis of sedimentary basins. *In*: Dickinson, W.R. (ed), *Current Concepts of Depositional Systems with Applications for Petroleum Geology, 11-1 to 11-12.* San Joaquin Geological Society, Bakersfield, California.

Ingle, J.C., Jr. 1980. Cenozoic paleobathymetry and depositional history of selected sequences within the southern California continental borderland. *In*: Sliter, W.V. (ed.), *Studies in Marine Micropaleontology and Paleoecology, A mMemorial Volume to Orville L. Bandy,* 163–195. Cushman Foundation for Foraminiferal Research, Special Publication, **19.** Lawrence, Kansas: Allen Press.

Inglis, I., Lepvraud, A., Mousset, E., Salim, A. & Vially, R. 1981. *Étude sédimentologique des Grès d'Annot Région de Colmars les Alpes et du Col de la Cayolle.* ENSPM *Rèf.* **29765.**

Ingram, R.L. 1954. Terminology for the thickness of stratification and parting units in sedimentary rocks. *Geological Society of America Bulletin,* **65,** 937–938.

Inman, D.L., Nordstrom, C.E. & Flick, R.E. 1976. Currents in submarine canyons: an air-sea-land interaction. *Annual Reviews in Fluid Mechanics,* **8,** 275–310.

Isaacs, C.M. 1981. Lithostratigraphy of the Monterey Formation, Coleta to Point Conception, Santa Barbara Coast California. *In*: *American Association of Petroleum Geologists Annual Meeting, Guide* **4,** 9–24.

Isaacs, C.M. 1984. Hemipelagic deposits in a Miocene basin California: toward a model of lithologic variation and sequence. *In*: Stow, D.A.V. & Piper, D.J.W. (eds), *Fine-grained Sediments: Deep-water Processes and Facies,* 481–496. Geological Society of London Special Publication, **15.** Oxford: Blackwell Scientific.

Ishihara, Y., Abe, H. & Oshikawa, M. 2009. Sediment gravity flow deposits and specificity of stratigraphic patterns in the Neogene Aoshima Formation, Miyazaki Group, Nichinan Coast, SW Japan. *Journal of the Sedimentological Society of Japan,* **67,** 65–84.

Ishiwatari, R., Hirakawa, Y., Uzaki, M., Yamada, K. & Yada, T. 1994. Organic geochemistry of the Japan Sea sediments - 1: bulk organic matter and hydrocarbon analyses of core KH-79-3, C-3 from the Oki Ridge for paleoenvironment assessments. *Journal of Oceanography,* **50,** 179–195.

Israel, S., Beranek, L., Friedman, R.M. & Crowley, J.L. 2014. New ties between the Alexander terrane and Wrangellia and implications for North America Cordilleran evolution. *Lithosphere,* doi: 10.1130/L364.1.

Ito, M., Ishikawa, K. & Nishida, N. 2014. Distinctive erosional and depositional structures formed at a canyon mouth: A lower Pleistocene deep-water succession in the Kazusa forearc basin on the Boso Peninsula, Japan. *Sedimentology,* **61,** 2042–2062.

Ito, M., Nishikawa, T. & Sugimoto, H. 1999. Tectonic control of high-frequency depositional sequences with durations shorter than Milankovitch cyclicity: An example from the Pleistocene paleo-Tokyo Bay, Japan. *Geology,* **27,** 763–766.

Ito, M. & Saito, T. 2006. Gravel waves in an ancient canyon: analogous features and formative processes of coarse-grained bedforms in a submarine-fan system, the Lower Pleistocene of the Boso Peninsula, Japan. *Journal of Sedimentary Research,* **76,** 1274–1283.

Ito, Y. 1986. Differential rotation of northeastern part of southwest Japan–paleomagnetism of early to late Miocene rocks from Yatsuo area in Chichibu district. *Journal of Geomagnetism & Geoelectrics,* **38,** 325–334.

Ivanov, V.V., Shapiro, G.I., Huthnance, J.M., Aleynik, D.L. & Golovin, P.N. 2004. Cascades of dense water around the world ocean. *Progress in Oceanography,* **60,** 47–98.

Iverson, R.M. 1997. The physics of debris flows. *Reviews in Geophysics,* **35,** 245–296.

Iwai, I., Osamu, F., Hiroyasu, M., Nozomu, I., Harumasa, K., Motoyoshi, O., Hiromi, M. & Makoto, O. 2004. Holocene seismoturbidites from the Tosabae Trough, a landward slope basin of Nankai Trough off Muroto: Core KR9705P1. *Memoirs of the Geological Society of Japan,* **58,** 137–152.

Jacka, A.D., Beck, R.H., Germain, L.St.C. & Harrison, S.G. 1968. Permian deep-sea fans of the Delaware Mountain Group (Guadalupian), Delaware Basin. *In*: Silver, B.A. (ed.), *Guadalupian Facies, Apache Mountain Area, West Texas,* 49–90. Permian Basin Section, Society of Economic Paleontologists and Mineralogists, Publication 68–11.

Jackson, C. A.-L. & Sømme, T.O. 2011. Borehole evidence for wing-like clastic intrusion complexes on the western Norwegian margin. *Journal of the Geological Society (London),* **168,** 1075–1078.

Jackson, C.A.-L., Barber, G.P. & Martinsen, O.J. 2008. Submarine slope morphology as a control on the development of sand-rich turbidite depositional systems: 3D seismic analysis of the Kyrre Formation (Upper Cretaceous), Maløy Slope, offshore Norway. *Marine and Petroleum Geology,* **25,** 663–680.

Jackson, D., Protacio, A., Silva, M., Helwig, J.A. & Dinkelman, M.G. 2012. The North East Greenland Danmarkshavn Basin. *GEO ExPro,* **9,** 60–62.

Jackson, T.A. & Smith, T.E. 1979. The tectonic significance of basalts and dacites in the Wagwater Belt, Jamaica. *Geological Magazine,* **116,** 365–374.

Jacobs, C.L. 1995. Mass wasting along the Hawaiian Ridge: giant debris avalanches. *In*: Pickering, K.T., Hiscott, R.N., Kenyon, N.H., Ricci Lucchi, F. & Smith, R.D.A. (eds), *Atlas of Deep Water Environments: Architectural Style in Turbidite Systems,* 26–28. London: Chapman and Hall.

Jacobson, R.S., Shor, G.G., Kieckhefer, R.M. & Purdy, G.M. 1979. Seismic refraction and reflection studies in the Timor – Aru Trough system and Australian continental shelf. *In*: Watkins, J.S., Montadert, L. & Dickerson, P.W. (eds), *Geological and Geophysical Investigations of Continental Margins,* 209–222. American Association of Petroleum Geologists Memoir, **29.**

Jaeger, J., Gulick, S., Mix, A. & Petronotis, K. 2011. Southern Alaska margin: interactions of tectonics, climate, and sedimentation. *IODP Scientific Prospectus,* **341.** doi:10.2204/iodp.sp.341.2011.

James, N.P. 1981. Megablocks of calcified algae in the Cow Head Breccia, western Newfoundland; vestiges of a Lower Paleozoic continental margin. *Geological Society of America Bulletin,* **92,** 799–811.

James, N.P. & Bone, Y. 1991. Origin of a cool-water, Oligo-Miocene deep shelf limestone, Eucla Platform, southern Australia. *Sedimentology,* **38,** 323–342.

James, N.P. & Stevens, R.K. 1986. Stratigraphy and correlation of the Cambro-Ordovician Cow Head Group, western Newfoundland. *Bulletin of the Geological Survey of Canada,* **366,** 143 pp.

James, N.P. & Wood, R.A. 2010. Reefs and reef mounds. *In*: James, N.P. & Dalrymple, R.W. (eds), *Facies Models 4,* 421–448. Newfoundland, Canada: Geological Association of Canada.

James, N.P., Stevens, R.K., Barnes, C.R. & Knight, I. 1989. Evolution of a Lower Paleozoic continental-margin carbonate platform, northern Canadian Appalachians. *In*: Crevello, P.D., Wilson, J.J., Sarg, J.F. &

Read, J.F. (eds), *Controls on Carbonate Platform and Basin Development,* 123–146. Society of Economic Paleontologists and Mineralogists Special Publication, **44**, Tulsa, OK.

Jameson, M., Ragbir, S., Loader, C., Bird, T., Smith, C. & Reiser, C. 2011. Central North Sea. *GEO ExPro,* **8**, 58–62.

Janocko, M., Cartigny, M.B.J., Nemec, W. & Hansen, E.W.M. 2013a. Turbidity current hydraulics and sediment deposition in erodible sinuous channels: laboratory experiments and numerical simulations. *Marine and Petroleum Geology,* **41**, 222–249.

Janocko, M., Nemec, W., Henriksen, S. & Warchol, M. 2013b. The diversity of deep-water sinuous channel belts and slope valley-fill complexes. *Marine and Petroleum Geology,* **41**, 7–34.

Jarrard, R.D. 1986. Terrane motion by strike-slip faulting of forearc slivers. *Geology,* **14**, 780–783.

Jarsve, E.M., Thyberg, B.I., Moss, C., Pedley, A. & Berstad, S. 2012. Western margin of UK Central Graben. *GEO ExPro,* **9**, 38–40.

Jipa, D. & Kidd, R.S. 1974. Sedimentation of coarser grained interbeds in Arabian Sea and sedimentation processes of the Indus Cone. *In*: Whitmarsh, R.B., Weser, O.E., Ross, D.A. *et al.*, *Initial Reports Deep Sea Drilling Project,* **23**, 471–495. Washington, DC: US Government Printing Office.

Jean, P.S., Kerckhove, C., Perriaux, J. & Ravenne, C. 1985. Un modéle Paleogéne de bassin á turbidites les Grès d'Annot du NW du Massif de l'Argentera-Mercantour. *Gèologie Alpine,* **61**, 115–143.

Jeffery, G.B. 1922. The motion of ellipsoidal particles immersed in a viscous fluid. *Proceedings of the Royal Society, London, (A),* **102**, 161–179.

Jegou, I., Savoye, B., Pirmez, C. & Droz, L. 2008. Channel-mouth lobe complex of the recent Amazon Fan: the missing piece. *Marine Geology,* **252**, 62–77.

Jenkyns, H.C. 1980. Cretaceous anoxic events: from continents to oceans. *Journal of the Geological Society (London),* **137**, 171–188.

Jenkyns, H.C. 1986. Pelagic environments. *In*: Reading, H.G. (ed.), *Sedimentary Environments and Facies,* 2nd edn, 343–397. Oxford: Blackwell Scientific.

Jenkyns, H.C. 2010. Geochemistry of oceanic anoxic events. *Geochemistry, Geophysics, Geosystems,* **11**, doi: 10.1029/2009GC002788.

Jenssen, A.I., Bergslien, D., Rye-Larsen, M. & Lindholm, R.M. 1991. Origin of complex mound geometry of Paleocene submarine-fan sandstone reservoirs, Balder Field, Norway. *In*: Parker, J.R. (ed.). *Petroleum Geology of Northwest Europe: Proceedings of the 4th Conference,* 135–143. London: Geological Society.

Jobe, Z.R., Bernhardt, A. & Lowe, D.R. 2010. Facies and architectural asymmetry in a conglomerate-rich submarine channel fill, Cerro Toro Formation, Sierra Del Toro, Magallanes Basin, Chile. *Journal of Sedimentary Research,* **80**, 1085–1108.

Jobe, Z.R., Lowe, D.R. & Morris, W.R. 2012. Climbing-ripple successions in turbidite systems: depositional environments, sedimentation rates and accumulation times. *Sedimentology,* **59**, 867–898.

Johansson, M. 2005. High-resolution borehole image analysis in a slope fan setting: examples from the late Miocene Mt. Messenger Formation, New Zrealand. *In*: Hodgson, D.M. & Flint, S.S. (eds), *Submarine Slope Systems: Processes and Products,* 75–88. Geological Society, London, Special Publication, 244.

John, C.M., Karner, G.D., Browning, E., Leckie, R.M., Mateo, Z., Carson, B. & Lowery, C. 2011. Timing and magnitude of Miocene eustasy derived from the mixed siliciclastic-carbonate stratigraphic record of the northeastern Australian margin. *Earth and Planetary Science Letters,* **304**, 455–467.

Johnson, A.M. (with contributions by J. R. Rodine) 1984. Debris flow. *In*: Brunsden, D. & Prior, D.B. (eds), *Slope Instability,* 257–362. New York: Wiley.

Johnson, A.M. 1970. *Physical Processes in Geology.* San Francisco: Freeman, Cooper. ISBN: 0877353204.

Johnson, B.A. & Walker, R.G. 1979. Paleocurrents and depositional environments of deep water conglomerates in the Cambro-Ordovician Cap Enrage Formation, Quebec Appalachians. *Canadian Journal of Earth Sciences,* **16**, 1375–1387.

Johnson, D. 1939. The origin of submarine canyons. *Journal of Geomorphology,* **2**, 42–60, 133–158, 213–236.

Johnson, D.W. 1967 (originally published 1925). *The New England-Acadian Shoreline.* New York: Hafner.

Johnson, D.D. & Beaumont, C. 1995. Preliminary results from a planform kinematic model of orogen evolution, surface processes and the development of clastic foreland basin stratigraphy. *In*: Dorobek, S.L. & Ross, G.M. (eds), *Stratigraphic Evolution of Foreland Basins,* 3–24. Society for Sedimentary Geology Special Publication, **52**.

Johnson, G.L. & Schneider, E.D. 1969. Depositional ridges in the North Atlantic. *Earth and Planetary Science Letters,* **6**, 416–422.

Johnson, H.D., Levell, B.K. & Siedlecki, S. 1978. Late Precambrian sedimentary rocks in East Finnmark, north Norway and their relationship to the Trollfiord-Komagelv fault. *Journal of the Geological Society (London),* **135**, 517–533.

Johnson, S.D., Flint, S., Hinds, D. & Wickens, H. DeV. 2001. Anatomy of basin floor to slope turbidite systems, Tanqua Karoo, South Africa: Sedimentology, sequence stratigraphy and implications for subsurface prediction. *Sedimentology,* **48**, 987–1023.

Johnson, M.R. 1991. Sandstone petrography, provenance and plate tectonic setting in Gondwana context of the southeastern Cape-Karoo Basin. South African *Journal of Geology,* **94**, 137–154.

Jolly, R.J.H. & Lonergan, L. 2002. Mechanisms and control on the formation of sand intrusions. *Journal of the Geological Society, London,* **159**, 605–617.

Jones, D.L., Silberling, N.J., Gilbert, W.G. & Coney, P.J. 1982. Character, distribution, and tectonic significance of accretionary terranes in the central Alaska Range. *Journal of Geophysical Research,* **87**, 3709–3717.

Jones, D.L., Silberling, N.J. & Coney, P.J. 1986. Collision tectonics in the Cordillera of western North America; examples from Alaska. *In*: Coward, M.P. & Ries, A.C. (eds), *Collision Tectonics.* Geological Society of London Special Publication **19**, 367–387.

Jones, K.P.N., McCave, I.N. & Weaver, P.P.E. 1992. Textural and dispersal patterns of thick mud turbidites from the Madeira abyssal plain. *Marine Geology,* **107**, 149–173.

Jones, M.E. & Preston, R.M.F. (eds) 1987. *Deformation of Sediments and Sedimentary Rocks.* Geological Society, London, Special Publication, **29**. Oxford: Blackwell Scientific.

Jones, T.A. 2001. Using flowpaths and vector fields in object-based modeling. *Computers & Geosciences,* **27**, 133–138.

Jonk, R, Duranti, D., Parnell, J., Hurst, A. & Fallick, E. 2003. The structural and diagenetic evolution of injected sandstones: examples from the Kimmeridgian of NE Scotland. *Journal of the Geological Society, London,* **160**, 881–894.

Jopling, A.V. & Walker, R.G. 1968. Morphology and origin of ripple-drift cross-lamination, with examples from the Pleistocene of Massaehusetts. *Journal of Sedimentary Petrology,* **38**, 971–984.

Jordan, D.W., Lowe, D.R., Slatt, R.M., Stone, C.G., D'Agostino, A., Scheihing, M.H. & Gillespie, R.H. 1991. *Scales of Geological Heterogeneity of Pennsylvanian Jackfork Group, Ouachita Mountains, Arkansas: Application to Field Development and Exploration for Deepwater Sandstones.* Dallas Geological Society: Field Trip **3**, Guidebook.

Jordan, D.W., Schultz, D.J. & Cherng, J.A. 1994. Facies architecture and reservoir quality of Miocene Mt. Messenger deep-water deposits, Taranaki Peninsula, New Zealand. *In*: Weimer, P., Bouma, A.H. & Perkins, B.F. (eds), *Gulf Coast Section-Society of Economic Paleontologists and Mineralogists Foundation 15th Annual Research Conference, Submarine Fans and Turbidite Systems: Sequence Stratigraphy, Reservoir Architecture and Production Characteristics,* 151–166.

Jordan, T.E. 1981. Enigmatic deep-water depositional mechanisms, upper part of the Oquirrh Group, Utah. *Journal of Sedimentary Petrology,* **51**, 879–894.

Jorry, S.J., Droxler, A.W. & Francis, J.M. 2010. Deepwater carbonate deposition in response to re-flooding of carbonate bank and atoll-tops at glacial terminations. *Quaternary Science Reviews,* **29**, 2010–2026.

Joseph, P. & Lomas, S.A. (eds) 2004a. *Deep-Water Sedimentation in the Alpine Basin of SE France: New Perspectives on the Grès D'Annot and Related Systems.* Geological Society, London, Special Publication, **221**, 437 pp. London: The Geological Society.

Joseph, P. & Lomas, S.A. (eds) 2004b. Deep-water sedimentation in the Alpine Foreland Basin of SE France: New perspectives on the Grès d'Annot and related systems - an introduction. *In*: Joseph, P. & Lomas, S.A. (eds), *Deep-Water Sedimentation in the Alpine Basin of SE France: New Perspectives on the Grès d'Annot and Related Systems.* Geological Society, London, Special Publication, **221**, 1–16. London: The Geological Society.

Joseph, P., Babonneau, N., Bourgeois, A., Cotteret, G., Eschard, R., Garin, B., Granjeon, D., Lerat, O. & Ravenne, C. 2000. The Annot Sandstone outcrops (French Alps): architecture description as input for quantification and 3D reservoir modeling. *In*: Weimer, P., Slatt, R.M., Coleman, J., Rosen, N.C., Nelson, H., Bouma, A.H., Styzen, M.J. & Lawrence, D.T. (eds), *Gulf Coast Section-Society of Economic Paleontologists and Mineralogists Foundation 20th Annual Bob F. Hoskins Research Research Conference, Deep-Water Reservoirs of the World,* 422–449. CD-ROM Society of Economic Paleontologists and Mineralogists Special Publications.

Junger, A. & Wagner, H.C. 1977. *Geology of the Santa Monica and San Pedro Basins, California Continental Borderland. United States Geological Survey Map* **MF-820**.

Kahler, G. & Stow, D.A.V. 1998. Turbidites and contourites of the Paleogene Lefkara Formation, southern Cyprus. *Sedimentary Geology*, **115**, 215–231.

Kaiho, K. 1991. Global changes of Paleogene aerobic-anaerobic benthic foraminifera and deep sea circulation. *Palaeogeography, Palaeoclimatology, Palaeoecology*, **83**, 65–85.

Kaiho, K. 1994. Benthic foraminiferal dissolved-oxygen index and dissolved-oxygen levels in the modern ocean. *Geology*, **22**, 719–722.

KAIKO II Research Group 1987. *6000 Meters Deep: A Trip to the Japanese Trenches; Photographic Records of the Nautile Dives in the Japanese Subduction Zones,* 104 pp. University of Tokyo Press, IFREMER-CNRS.

KAIKO Project Shipboard Scientific Party 1985. Japanese deep-sea trench survey. *Nature*, **313**, 432–433.

Kaminski, M.A., Aksu, A.E., Hiscott, R.N., Box, M. Al-Salameen, M. & Filipescu, S. 2002. Late glacial to Holocene benthic foraminifera in the Marmara Sea: implications for Black Sea - Mediterranean Sea connections following the last deglaciation. *Marine Geology*, **190**, 165–202.

Kanazawa, T., Sager, W.W., Escutia, C. *et al.* 2001. Site 1179. *Proceedings of the Ocean Drilling Program, Initial Reports,* **191**. College Station, Texas, USA: Ocean Drilling Program.

Kane, I.A. & Hodgson, D.M. 2011. Sedimentological criteria to differentiate submarine channel levée subenvironments: exhumed examples from the Rosario Fm. (Upper Cretaceous) of Baja California, Mexico, and the Fort Brown Formation, (Permian), Karoo Basin, S. Africa. *Marine and Petroleum Geology*, **28**, 807–823.

Kane, I.A., Kneller, B.C., Dykstra, M., Kassem, A. & McCaffrey, W.D. 2007. Anatomy of a submarine channel-levée: An example from Upper Cretaceous slope sediments, Rosario Formation, Baja California, Mexico. *Marine and Petroleum Geology*, **24**, 540–563.

Kane, I.A., McCaffrey, W.D. & Peakall, J. 2008. Controls on sinuosity evolution within submarine channels. *Geology*, **36**, 287–290.

Kane, I.A., Dykstra, M.L., Kneller, B.C., Tremblay, S. & McCaffrey, W.D. 2009. Architecture of a coarse-grained channel-levée system: the Rosario Formation, Baja California, Mexico. *Sedimentology*, **56**, 2207–2234.

Kane, I.A., Catterall, V., McCaffrey, W.D. & Martinsen, O.J. 2010. Submarine channel response to intrabasinal tectonics: The influence of lateral tilt. *American Association of Petroleum Geologists Bulletin*, **94**, 189–219.

Kaneko, Y., Maruyama, S., Kadarusman, A., Ota, T., Ishikawa, M., Tsujimori, T., Ishikawa, A. & Okamoto, K. 2007. On-going orogeny in the outer-arc of the Timor-Tanimbar region, eastern Indonesia. *Gondwana Research*, **11**, 218–233.

Kanmera, K. 1976a. Comparison between past and present geosynclinal sedimentary bodies I. *Kagaku (Science)*, **46**, 284–291.

Kanmera, K. 1976b. Comparison between past and present geosynclinal sedimentary bodies II. *Kagaku (Science)*, **46**, 371–378.

Kao, T.W., Pan, F.-S. & Renouard, D. 1985. Internal solitons on the pycnocline: generation, propagation, and shoaling and breaking over a slope. *Journal of Fluid Mechanics*, **159**, 19–53.

Karig, D.E., Suparka, S., Moore, G.F. & Hehanussa, P.E. 1979. Structure and Cenozoic evolution of the Sunda Arc in the Central Sumatra region. *In*: Watkins, J.S., Montadert, L. & Dickerson, P.W. (eds), *Geological and Geophysical Investigations of Continental Margins,* 223–237. American Association of Petroleum Geologists Memoir, **29**.

Karig, D.E., Kagami, H. & DSDP Leg 87 Scientific Party 1983. Varied response to subduction in Nankai Trough and Japan Trench forearcs. *Nature*, **304**, 148–151.

Karig, D.E., Barber, A.J., Charlton, T.R., Klemperer, S. & Hussong, D.M. 1987. Nature and distribution of deformation across the Banda arc – Australian collision zone at Timor. *Geological Society of America Bulletin*, **98**, 18–32.

Karner, G.D. & Shillington, D.J. 2005, Basalt sills of the U reflector, Newfoundland Basin: A serendipitous dating technique. *Geology*, **33**, 985–988.

Karl, H.A., Hampton, M.A. & Kenyon, N.H. 1989. Lateral migration of Cascadia Channel in response to accretionary tectonics. *Geology*, **17**, 144–147.

Khan, Z.A. & Arnott, R.W.C. 2011. Stratal attributes and evolution of asymmetric inner- and outer-bend levée deposits associated with an ancient deep-water channel-levée complex within the Isaac Formation, southern Canada. *Marine and Petroleum Geology*, **28**, 824–842.

Kikuchi, M., Nakamura, M. & Yoshikawa, K. 2003. Source rupture processes of the 1944 Tonankai earthquake and the 1945 Mikawa earthquake derived from low-gain seismograms. *Earth, Planets and Space*, **55**, 159–172.

Keefer, D.K. 1984. Landslides caused by earthquakes. *Geological Society of America Bulletin*, **95**, 406–421.

Keep, M., Powell, C.McA. & Baillie, P.W. 1998. Neogene deformation of the North West Shelf, Australia. *In*: Purcell, P.G. & Purcell, R.R. (eds), *The Sedimentary Basins of Western Australia,* **2**, 81–91. Perth, Australia: Petroleum Exploration Society of Australia.

Keevil, G.M., Peakall, J., Best, J.L. & Amos, K.J. 2006. Flow structure in sinuous submarine channels: velocity and turbulence structure of an experimental submarine channel. *Marine Geology*, **229**, 241–257.

Keevil, G.M., Peakall, J. & Best, J.L. 2007. The influence of scale, slope and channel geometry on the flow dynamics of submarine channels. *Marine and Petroleum Geology*, **24**, 487–503.

Keigwin, L.D., Rio, D., Acton, G. *et al.* 1998. *Proceedings of the Ocean Drilling Program, Initial Results,* **172**. College Station, Texas, USA: Ocean Drilling Program.

Keith, B.D. & Friedman, G.M. 1977. A slope-fan-basin-plain model, Taconic sequence, New York and Vermont. *Journal of Sedimentary Petrology*, **47**, 1220–1241.

Keller, G.H. 1982. Organic matter and the geotechnical properties of submarine sediments. *Geo-Marine Letters*, **2**, 191–198.

Keller, G. & Barron, J.A. 1983. Paleoceanographic implications of Miocene deep-sea hiatuses. *Geological Society of America Bulletin*, **94**, 590–613.

Kelling, G. 1961. The stratigraphy and structure of the Ordovician rocks of the Rhinns of Galloway. *Geological Society of London Quarterly Journal*, **117**, 37–75.

Kelling, G. & Holroyd, J. 1978. Clast size, shape, and composition in some ancient and modern fan gravels. *In*: Stanley, D.J. & Kelling, G. (eds), *Sedimentation in Submarine Canyons, Fans and Trenches,* 138–159. Stroudsburg, PA: Dowden, Hutchinson & Ross.

Kellogg, K.S. & Minor, S.A. 2005. Pliocene transpressional modification of depositional basins by convergent thrusting adjacent to the 'Big Bend' of the San Andreas fault: an example from Lockwood Valley, southern California. *Tectonics*, **24**, TC1004, 32 pp.

Kelts, K. & Arthur, M.A. 1981. Turbidites after ten years of deep-sea drilling - wringing out the mop? In: Douglas, R.G. & Winterer, E.L. (eds), *The Deep Sea Drilling Project: A Decade of Progress*, 91–127. Society of Economic Paleontologists and Mineralogists Special Publication, **32**.

Kelts, K. & Niemitz, J. 1982. Preliminary sedimentology of Late Quaternary diatomaceous muds from Deep-Sea Drilling Pro-ject Site 480, Guyamas Basin slope, Gulf of California. *In*: Curray, J.R., Moore, D.G. *et al.*, *Initial Reports Deep Sea DrillingProject*, **64**, 1191–210. Washington, DC: US Government Printing Office.

Kendall, M.G. 1969. *Rank Correlation Methods*. London: Charles Griffin & Company.

Kendall, M.G. 1976. *Time Series*, 2nd edn. New York: Hafner Press (Macmillan).

Kendall, M.G. & Gibbons, J.D. 1990. *Rank Correlation Methods*, 5th edn. New York: Oxford University Press.

Kennard, L., Schafer, C. & Carter, L. 1990. Late Cenozoic evolution of Sackville Spur; a sediment drift on the Newfoundland continental slope. *Canadian Journal of Earth Sciences*, **27**, 863–878.

Kennett, J. 1982. *Marine Geology*. Englewood Cliffs, New Jersey: Prentice-Hall. 813 pp. ISBN: 0135569362.

Kennett, J.P., Houtz, R.E., Andrews, P.V., Edwards, A.R., Gostin, V.A., Hajos, N., Hampton, M., Jenkins, D.G., Margolis, S.V., Ovenshine, A.T. & Perch-Nielsen, K. 1975. Cenozoic paleo-oceanography in the southwest Pacific Ocean and the development of the Circumpolar current. *In*: Kennett, J.P., Houtz, R.E. *et al. Initial Reports Deep Sea Drilling Project*, **29**, 1155–1169. Washington, DC: US Government Printing Office.

Kennett, J.P., von der Borch, C.C. *et al.* 1986. *Initial Reports Deep Sea Drilling Project*, **90**, 1325–1337. Washington, DC: US Government Printing Office.

Kenyon, C.S. 1974. *Stratigraphy and sedimentology of the late Miocene to Quaternary deposits of Timor*. PhD Thesis, University College London.

Kenyon, N.H. 1992. Channelised deep-sea siliciclastic systems: a plan view perspective. *In*: *Sequence Stratigraphy of European Basins*, 458–459. Dijon May 18–20 1992. CNRS/Institute français du Petroleum.

Kenyon, N.H. & Millington, J. 1995. Contrasting deep-sea depositional systems in the Bering Sea. *In*: Pickering, K.T., Hiscott, R.N., Kenyon, N.H., Ricci Lucchi, F. & Smith, R.D.A. (eds), *Atlas of Deep Water Environments: Architectural Style in Turbidite Systems*, 196–202. London: Chapman and Hall.

Kenyon, N.H., Belderson, R.H. & Stride, A.H. 1978. Channels canyons and slump folds on the continental slope between south west Ireland and Spain. *Oceanologica Acta*, **1**, 369–380.

Kenyon, N.H., Amir, A. & Cramp, A. 1995a. Geometry of the younger sediment bodies on the Indus Fan. *In*: Pickering, K.T., Hiscott, R.N., Kenyon, N.H., Ricci Lucchi, F. & Smith, R.D.A. (eds), *Atlas of Deep Water Environments: Architectural Style in Turbidite Systems*, 89–93. London: Chapman and Hall.

Kenyon, N.H., Millington, J., Droz, L. & Ivanov, M. K. 1995b. Scour holes in a channel-lobe transition zone on the Rhone Cone. *In*: Pickering, K.T., Hiscott, R.N., Kenyon, N.H., Ricci Lucchi, F. & Smith, R.D.A. (eds), *Atlas of Deep Water Environments: Architectural Style in Turbidite Systems*, 212–215. London: Chapman and Hall.

Kenyon, N.H., Akhmetzhanov, A.M. & Twichell, D.C. 2002a. Sand wave fields beneath the Loop Current, Gulf of Mexico: reworking of fan sands. *Marine Geology*, **192**, 297–307.

Kenyon, N.H., Klaucke, I., Millington, J. & Ivanov, M.K. 2002b. Sandy submarine canyon-mouth lobes on the western margin of Corsica and Sardinia, Mediterranean Sea. *Marine Geology*, **184**, 69–84.

Kern, J.P. 1980. Origin of trace fossils in Polish Carpathian flysch. *Lethaia*, **13**, 347–362.

Keppie, J.D. 1986. The Appalachian collage. *In*: Gee, D.G. & Sturt, B.A. (eds), *The Caledonide Orogen – Scandinavia and Related Areas*, 1217–1226. New York: Wiley.

Kessler II, L. G. & Moorhouse, K. 1984. Depositional processes and fluid mechanics of Upper Jurassic conglomerate accumulations, British North Sea. *In*: Koster, E.H. & Steel, R.J. (eds), *Sedimentology of Gravels and Conglomerates*, 383–397. Calgary Alberta: Canadian Society of Petroleum Geologists Memoir, **10**.

Khripounoff, A., Vangriesheim, A., Babonneau, N., Crassous, P., Dennielou, B. & Savoye, B. 2003. Direct observation of intense turbidity current activity in the Zaire submarine valley at 4000 m water depth. *Marine Geology*, **194**, 151–158.

Kessler, L.G. II,, Prave, A.R., Malo, M. & Bloechl, W.W. 1995. Mid-Upper Ordovician flysch deposition, northern Gaspé Peninsula, Québec; a synthesis with implications for foreland and successor basin evolution in the Northern Appalachian Orogen. *In*: Cooper, J.D., Droser, M.L. & Finney, S.C. (eds), *Ordovician Odyssey; Short Papers for the Seventh International Symposium on the Ordovician System*, 251–255. Pacific Section, Society of Economic Paleontologists and Mineralogists, Field Guide, **77**.

Kidd, R.B. & Hill, P.R. 1987. Sedimentation on Feni and Gardar sediment drifts. *In*: Ruddiman, W.F., Kidd, R.B., Thomas, E. *et al.*, *Initial Reports Deep Sea Drilling Project*, **94**, 1217–1244. Washington DC: US Government Printing Office.

Kidd, R.B., M.B. Cita & W.B.F. Ryan 1978. Stratigraphy of eastern Mediterranean sapropel sequences recovered during DSDP Leg 42A and their paleoenvironmental significance. *In*: Kidd, R.B. & Worstell, P.J. (eds), *Initial Reports Deep Sea Drilling Project*, **42**, 421–443. Washington DC: US Government Printing Office.

Kim, W. & Paola, C. 2007. Long-period cyclic sedimentation with constant tectonic forcing in an experimental relay ramp. *Geology*, **270**, 331–334.

Kim, Y.-S. & Sanderson, D.J. 2005. The relationship between displacement and length of faults: a review. *Earth-Science Reviews*, **68**, 317–334.

Kimura, G., Moore, G.F., Strasser, M., Screaton, E., Curewitz, D., Streiff, C. & Tobin, H. 2011. Spatial and temporal evolution of the megasplay fault in the Nankai Trough. *Geochemistry, Geophysics, Geosystems*, **12**, doi:10.1029/2010GC003335.

Kimura, J.-I., Stern, R.J. & T. Yoshida, T. 2005. Reinitiation of subduction and magmatic responses in SW Japan during Neogene time. *Geological Societ of America Bulletin*, **117**, 969–986.

Kimura, M. 1985. Back-arc rifting in the Okinawa Trough. *Marine and Petroleum Geology*, **2**, 222–240.

Kimura, M. 1996. Active rift system in the Okinawa Trough and its northeastern continuation. *Bulletin of the Disaster Prevention Research Institute*, **45**, 38–27.

Kimura, T. 1966. Thickness distribution of sandstone beds and cyclic sedimentation turbidite sequence at two localities in Japan. *Earthquake Research Institute*, **44**, 561–607.

King, L.H. 1981. Aspects of regional surficial geology related to site investigation requirements; eastern Canadian Shelf. *In*: Ardus, D.A. (ed.), *Offshore Site Investigation*, 37–60. London: Graham & Trotman.

King, P.R., Browne, G.H. & Slatt, R.M. 1994. Sequence architecture of exposed late Miocene basin floor fan and channel-levée complexes (Mount Messenger Formation), Taranaki Basin, New Zealand. *In*: Weimer, P., Bouma, A.H. & Perkins, B.F. (eds), *Gulf Coast Section-Society of Economic Paleontologists and Mineralogists Foundation 15th Annual Research Conference, Submarine Fans and Turbidite Systems: Sequence Stratigraphy, Reservoir Architecture and Production Characteristics*, 177–192.

King, P.R., Ilg, B.R., Arnot, M.J., Browne, G.H., Strachan, L.J., Crundwell, M.P. & Helle, K. 2011. Outcrop and seismic examples of mass-transport deposits from a Late Miocene deep-water succession, Taranaki Basin, New Zealand. *In*: Shipp, R.C., Weimer, P. & Posamentier, H.W. (eds), *Mass-transport Deposits in Deepwater Settings*, 311–348. Special Publication, Society for Sedimentary Geology, **96**. Tulsa, Okla: Society for Sedimentary Geology.

Kjøde, J., Storetvedt, K.H., Roberts, D. & Gidskehaug, A. 1978. Palaeo-magnetic evidence for large-scale dextral displacement along the Trollfiord-Komagelv fault, Finnmark, north Norway. *Physics of Earth and Planetary Interiors*, **16**, 132–144.

Klaucke, I. & Hesse, R. 1996. Fluvial features in the deep-sea: new insights from the glacigenic submarine drainage system of the Northwest Atlantic Mid-Ocean Channel in the Labrador Sea. *Sedimentary Geology*, **106**, 223–234.

Klaucke, I., Hesse, R. & Ryan, W.B.F. 1997. Flow parameters of turbidity currents in a low-sinuosity giant deep-sea channel. *Sedimentology*, **44**, 1093–1102.

Klaucke, I., Hesse, R. & Ryan, W.B.F. 1998a. Seismic stratigraphy of the Northwest Atlantic Mid-Ocean Channel: growth pattern of a mid-ocean channel-levée complex. *Marine and Petroleum Geology*, **15**, 575–585.

Klaucke, I., Hesse, R. & Ryan, W.B.F. 1998b. Morphology and structure of a distal submarine trunk-channel: the Northwest Atlantic Mid-Ocean Channel between 53° and 44°30′N. *Geological Society of America Bulletin*, **110**, 22–34.

Klaucke, I., Masson, D.G., Kenyon, N.H. & Gardner, J.V. 2004. Sedimentary processes of the lower Monterey Fan channel and channel-mouth lobe. *Marine Geology*, **206**, 181–198.

Kleverlaan, K. 1989a. Neogene history of the Tabernas basin (SE Spain) and its Tortonian submarine fan development. *Geologie en Mijnbouw*, **68**, 421–432.

Kleverlaan, K. 1989b. Three distinctive feeder-lobe systems within one time slice of the Tortonian Tabernas fan, SE Spain. *Sedimentology*, **36**, 25–45.

Kleverlaan, K. 1994. Architecture of a sand-rich fan from the Tabernas submarine fan complex, southeast Spain. *In*: Weimer, P., Bouma, A.H. & Perkins, B.F. (eds), *Gulf Coast Section-Society of Economic Paleontologists and Mineralogists Foundation 15th Annual Research Conference, Submarine Fans and Turbidite Systems: Sequence Stratigraphy, Reservoir Architecture and Production Characteristics*, 209–215.

Kleverlaan, K. & Cossey, S.P.J. 1993. Permeability barriers within sand-rich submarine fans: outcrop studies of the Tabernas Basin, SE Spain. *In*: Eschard, R. & Doligez, B. (eds), *Subsurface Reservoir Characterization from Outcrop Observations*, 161–164. Paris, Editions Technip.

Knaust, D. 2009. Characterisation of a Campanian deep-sea fan system in the Norwegian Sea by means of ichnofabrics. *Marine and Petroleum Geology*, **26**, 1199–1211.

Knaust, D. 2012. Trace-fossil systematics *In*: Knaust, D. & Bromley, R.G. (eds), *Trace Fossils as Indicators of Sedimentary Environments*, 79–102. Developments in Sedimentology, **64**. Amsterdam: Elsevier.

Knaust, D. & Bromley, R.G. (eds) 2012. *Trace Fossils as Indicators of Sedimentary Environments*, 960 pp. Amsterdam: Elsevier B.V. ISBN: 9780444538130.

Kneller, B. 1995. Beyond the turbidite paradigm: physical models for deposition of turbidites and their implications for reservoir prediction. *In*: Hartley, A.J. & Prosser, D.J. (eds), *Characterization of Deep Marine Clastic Systems*. Geological Society London Special Publication, **94**, 31–49.

Kneller, B.C. 2003. The influence of flow parameters on turbidite slope channel architecture. *Marine and Petroleum Geology*, **20**, 901–910.

Kneller, B.C. & Branney, M.J. 1995. Sustained high-density turbidity currents and the deposition of thick massive sands. *Sedimentology*, **42**, 607–616.

Kneller, B.C. & Buckee, C. 2000. The structure and fluid mechanics of turbidity currents: a review of some recent studies and their geological implications. *Sedimentology*, **47**, 62–94.

Kneller, B.C. & McCaffrey, W.D. 1999. Depositional effects of flow nonuniformity and stratification within turbidity currents approaching a bounding slope: deflection, reflection and facies variation. *Journal of Sedimentary Research*, **69**, 980–991.

Kneller, B.C. & McCaffrey, W.D. 2003. The interpretation of vertical sequences in turbidite beds: the influence of longitudinal flow structure. *Journal of Sedimentary Research*, **73**, 706–713.

Kneller, B., Edwards, E., McCaffrey, W. & Moore, R. 1991. Oblique reflection of turbidity currents. *Geology*, **19**, 250–252.

Knipe, R.J. 1986. Microstructural evolution of vein arrays preserved in Deep Sea Drilling Project cores from the Japan Trench, Leg 57. *In*: Moore, J.C. (ed.), *Structural Fabric in Deep Sea Drilling Project Cores from Forearcs*, 75–87. Geological Society of America Memoir, **166**.

Knudson, K.P. & Hendy, I.L. 2009. Climatic influences on sediment deposition and turbidite frequency in the Nitinat Fan, British Columbia. *Marine Geology*, **262**, 29–38.

Kodaira, S. Takahashi, N., Nakanishi, A., Miura, S. & Kaneda, Y. 2000. Subducted seamount imaged in the rupture zone of the 1946 Nankaido earthquake. *Science*, **289**, 104–106.

Kodama, K., Taira, A., Okamura, M. & Saito, Y. 1983. Paleomagnetisation of the Shimanto Belt in Shikoku, southwest Japan. *In*: Hashimoto, M. & Uyeda, S, (eds), *Accretion Tectonics in the Circum-Pacific Regions*, 231–241. Tokyo: Terrapub.

Kokelaar, B.P. & Howells, M.F. (eds) 1984. *Marginal Basin Geology: Volcanic and Associated Sedimentary and Tectonic Processes in Modern and Ancient Marginal Basins*. The Geological Society, London, Special Publications, **16**. Oxford: Blackwell Scientific Publications. ISBN 0-632-01073-8.

Kolla, V. & Coumes, F. 1987. Morphology, internal structure, seismic stratigraphy, and sedimentation of Indus Fan. *American Association of Petroleum Geologists Bulletin*, **71**, 650–677.

Kolla, V. & Macurda, D.B. Jr., 1988. Sea-level changes and timing of turbidity-current events in deep-sea fan systems *In*: Wilgus, C.K., Hastings, B.S., Posamentier, H., Van Wagoner, J., Ross, C.A. & Kendall, C.G. St.C. (eds), *Sea-Level Changes: An Integrated Approach*, 381–392. Society of Economic Paleontologists and Mineralogists, Special Publication No. **42**.

Kolla, V., Eittreim, S., Sullivan, L., Kostecki, J.A. & Burckle, L.H. 1980a. Current-controlled, abyssal microtopography and sedimentation in Mozambique Basin, southwest Indian Ocean. *Marine Geology*, **34**, 171–206.

Kolla, V., Kostecki, J.A., Henderson, L. & Hess, L. 1980b. Morphology and Quaternary sedimentation of the Mozambique Fan and environs, southwestern Indian Ocean. *Sedimentology*, **27**, 357–378.

Kolla, V., Bourges, Ph., Urruty, J.-M. & Sufa, P. 2001. Evolution of deep-water Tertiary sinuous channels offshore Angola (west Africa) and implications for reservoir architecture. *American Association of Petroleum Geologists Bulletin*, **85**, 1373–1405.

Kolla, V., Posamentier, H.W. & Wood, L.J. 2007. Deep-water and fluvial sinuous channels - characteristics, similarities and dissimilarities, and modes of formation. *Marine and Petroleum Geology*, **24**, 388–405.

Kolmogorov, A.N. 1951. Solution of a problem in probability theory connected with the problem of the mechanics of stratification. *Transactions of the American Mathematical Society*, **53**, 171–177.

Komar, P.D. 1969. The channelized flow of turbidity currents with application to Monterey deep-sea fan channel. *Journal of Geophysical Research*, **74**, 4544–4558.

Komar, P.D. 1971. Hydraulic jumps in turbidity currents. *Geological Society of America Bulletin*, **82**, 1477–1488.

Komar, P.D. 1977. Computer simulation of turbidity current flow and the study of deep-sea channels and fan sedimentation. *In*: Goldberg, E.D., McCave, I.N., O'Brien, J.J. & Steele, J.H. (eds), *The Sea, Vol. 6, Marine Modelling*, 603–621. New York: Wiley.

Komar, P.D. 1985. The hydraulic interpretation of turbidites from their grain sizes and sedimentary structures. *Sedimentology*, **32**, 395–408.

Kominz, M.A., Browning, J.V., Miller, K.G., Sugarman, P.J., Misintseva, S. & Scotese, C.R. 2008. Late Cretaceous to Miocene sea-level estimates from the New Jersey and Delaware coastal plain coreholes: an error analysis. *Basin Research*, **20**, 211–226.

Konishi, K. 1989. Limestone of the Daiichi Kashima Seamount and the fate of a subducting guyot: fact and speculation from the Kaiko "Nautile" dives. *Tectonophysics*, **160**, 249–265.

Konsoer, K., Zinger, J. & Parker, G. 2013. Bankfull hydraulic geometry of submarine channels created by turbidity currents: relations between bankfull channel characteristics and formative flow discharge. *Journal of Geophysical Research, Earth Surface*, **118**, 216–228.

Kosa, E. 2007. Differential subsidence driving the formation of mounded stratigraphy in deep-water sediments; Palaeocene, central North Sea. *Marine and Petroleum Geology*, **24**, 632–652.

Kotarba, M.J. 2012. Origin of natural gases in the Paleozoic–Mesozoic basement of the Polish Carpathian foredeep. *Geologica Carpathica*, **63**, 307–318.

Kostic, S. & Parker, G. 2007. Conditions under which a supercritical turbidity current traverses an abrupt transition to vanishing bed slope without a hydraulic jump. *Journal of Fluid Mechanics*, **586**, 119–145.

Kranck, K. 1984. Grain-size characteristics of turbidites. *In*: Stow, D.A.V. & Piper, D.J.W. (eds), *Fine-grained Sediments: Deep-water Processes and Facies*, 83–92. Geological Society of London Special Publication, **15**. Oxford: Blackwell Scientific.

Kreuser, T. & Woldu, G. 2010. Formation of euxinic lakes during the deglaciation phase in the Early Permian of East Africa. *Geological Society of America, Special Papers*, **468**, 101–112.

Kreyszig, E. 1967. *Advanced Engineering Mathematics*, 2nd edn. New York: Wiley.

Krissek, L.A. 1984. Continental source area contributions to fine-grained sediments on the Oregon and Washington continental slope. *In*: Stow, D.A.V. & Piper, D.J.W. (eds), *Fine-grained Sediments: Deep-water Processes and Facies*, 363–375. Geological Society of London Special Publication, **15**. Oxford: Blackwell Scientific.

Krissek, L.A. 1989. Late Cenozoic records of ice-raftng at ODP Sites 642, 643 and 644, Norwegian Sea: onset, chronology, and characteristics of glacial/interglacial fluctuations. *In*: Eldholm, O., Thiede, J., Taylor, E. *et al.*, *Proceedings of the Ocean Drilling Program, Scientific Results*, **104**, 61–74. College Station, Texas, USA: Ocean Drilling Program.

Krissek, L.A. 1995. Late Cenozoic ice-rafting records from Leg 145 sites in the North Pacific: Late Miocene onset Late Pliocene intensification, and Pliocene–Pleistocene events. *In*: Rea, D.K., Basov, I.A., Scholl, D.W. & Allan, J.F. (eds), *Proceedings of the Ocean Drilling Program, Scientific Results*, **145**, 179–194. College Station, Texas, USA: Ocean Drilling Program.

Kristoffersen, Y., Sorokin, M.Y., Jokat, W. & Svendsen, O. 2004. A submarine fan in the Amundsen Basin, Arctic Ocean. *Marine Geology*, **204**, 317–324.

Kroon, D., Zachos, J.C. & Leg 208 Scientific Party 2007. Leg 208 synthesis: Cenozoic climate cycles and excursions. *In*: Kroon, D., Zachos, J.C. & Richter, C. (eds), *Proceedings of the Ocean Drilling Program Scientific Results*, **208**, 1–55. College Station, Texas, USA: Ocean Drilling Program.

Książkiewicz, M. 1954. Graded and laminated bedding in the Carpathian flysch. *Annales Societatis Geologorum Poloniae*, **22**, 399–449.

Książkiewicz, M. 1960. Pre-orogenic sedimentation in the Carpathian geosyncline. *Geologisches Rundschau*, **50**, 8–31.

Książkiewicz, M. 1968. O niektórych problematykach z fliszˊ Karpat polskich. Częśc III. (On some problematic organic traces from the flysch of the Polish Carpathians, Part 3). *Rocznik Polskiego Towarzystwa Geologicznego*, **38**, 3–17.

Książkiewicz, M. 1970. Observations on the ichnofauna of the Polish Carpathians. *In*: Crimes, P.T. & Harper, C. (eds). *Trace Fossils*, 283–322. Geological Journal, Special Issue, 3.

Książkiewicz, M. 1977. Trace fossils in the flysch of the Polish Carpathians. *Palaeontologica Polonica*, **36**, 1–208.

Kuenen, Ph.H. 1951. Properties of turbidity currents of high density. *In*: Hough, J.L. (ed.), *Turbidity Currents*, 14–33. Society of Economic Paleontologists & Mineralogists Special Publication, **2**.

Kuenen, Ph.H. 1953. Significant feature of graded bedding. *American Association of Petroleum Geologists Bulletin*, **37**, 1044–1066.

Kuenen, Ph.H. 1964. Deep-sea sands and ancient turbidites. *In*: Bouma, A.H. & Brouwer, A. (eds), *Turbidites*, 3–33. Developments in Sedimentology, **3**. Amsterdam: Elsevier.

Kuenen, Ph.H. 1966. Matrix of turbidites: experimental approach. *Sedimentology*, **7**, 267–297.

Kuenen, Ph.H. & Migliorini, C.I. 1950. Turbidity currents as a cause of graded bedding. *Journal of Geology*, **58**, 91–127.

Kuhn, G. & Meischner, D. 1988. Quaternary and Pliocene turbidites in the Bahamas, Leg 101, Sites 628, 632, and 635. *In*: Austin, J.A., Jr., Schlager, W. *et al.*, *Proceedings of the Ocean Drilling Program, Scientific Results*, **101**, 203–212. College Station, Texas, USA: Ocean Drilling Program.

Kuramoto, S., Ashi, J., Greinert, J., Gulick, S., Ishimura, T., Morita, S., Nakamura, K., Okada, M., Okamoto, T., Rickert, D., Saito, S., Suess, E., Tsunogai, U. & Tomosugi, T. 2001. Surface observations of subduction related mud volcanoes and large thrust sheets in the Nankai subduction margin; Report on YK00-10 and YK01-04 cruises. *JAMSTEC Journal of Deep Sea Research*, **19**, 131–139.

Kurtz, D.D. & Anderson, J.B. 1979. Recognition and sedimentologic description of recent debris flow deposits from the Ross and Weddell Seas, Antarctica. *Journal of Sedimentary Petrology*, **49**, 1159–1169.

Kuribayashi, E. & Tatsuoka, F. 1977. History of earthquake-induced soil liquefaction in Japan. *Public Works Research Bulletin* (Japan Ministry of Construction), **31**, 1–26.

Labaume, P., Mutti, E., Séguret, M. & Rosell, J. 1983a. Mégaturbidites carbonatées du basin turbiditique de l'Eocene inférieur et moyen sud-pyrénéen. *Bulletin de la Société Géologique de France*, 7 (XXV-6), 927–941.

Labaume, P., Mutti, E., Seguret, M. & Rosell, J. 1983b. Mégaturbidites carbonates du basin turbiditique de l'Eocene inférieur et moyen sud-pyrénéen. *Société Géologique de France, Bulletin*, **25**, 927–941.

Labaume, P., Séguret, M. & Seyve, C. 1985. Evolution of a turbiditic foreland basin and analogy with an accretionary prism: Example of the Eocene South-Pyrenean basin. *Tectonics*, **4**, 661–685.

Labaume, P., Mutti, E. & Séguret, M. 1987. Megaturbidites: a depositional model from the Eocene of the SW-Pyrenean Foreland Basin, Spain. *Geo-Marine Letters*, **7**, 91–101.

Labourdette, R., Crumeyrollea, P. & Remacha, E. 2008. Characterisation of dynamic flow patterns in turbidite reservoirs using 3D outcrop analogues: Example of the Eocene Morillo turbidite system (south-central Pyrenees, Spain). *Marine and Petroleum Geology*, **25**, 255–270.

Lacombe, O., Lavé, J., Roure, F.M. & Verges, J. (eds) 2007. *Thrust Belts and Foreland Basins: From Fold Kinematics to Hydrocarbon Systems.* Springer: Frontiers in Earth Sciences, 492 pp. ISBN: 9783540694250.

Laigle, M., Becel, A., De Voogd, B., Alfred Hirn., Taymaz, T., Ozalaybey, S. & Team, Seismarmara Leg 2008. A first deep seismic survey in the Sea of Marmara: Deep basins and whole crust architecture and evolution. *Earth and Planetary Science Letters*, **270**, 168–179.

Laird, M.G. 1968. Rotational slumps and slump scars in Silurian rocks, western Ireland. *Sedimentology*, **10**, 111–120.

Laird, M.G. 1970. Vertical sheet structures - a new indication of sedimentary fabric. *Journal of Sedimentary Petrology*, **40**, 428–434.

Lallemand, S. & Le Pichon, X. 1987. Coulomb wedge model applied to the subduction of seamounts in the Japan Trench. *Geology*, **15**, 1065–1069.

Lallemand, S., Culotta, R. & von Huene, R. 1989. Subduction of the Daiichi Kashima Seamount in the Japan Trench. *Tectonophysics*, **160**, 231–233, 237–247.

Lallemant, H.P., Nakamura, S., Tsunogai, K., Mazzotti, U., Kobayashi, S. & Marine, K. 2002. Surface expression of fluid venting at the toe of the Nankai wedge and implications for flow paths. *Geology*, **187**, 119–143.

Lallemant, S., Chamot-Rooke, N., Le Pichon, X. & Rangin, C. 1987. Zenisu Ridge: a deep intraoceanic thrust related to subduction off Southwest Japan. *Tectonophysics*, **160**, 151–174.

Lamb, M.P., Hickson, T.A., Marr, J.G., Sheets, B., Paola, C. & Parker, G. 2004. Surging versus continuous turbidity currents: flow dynamics and deposits in an experimental intraslope minibasin. *Journal of Sedimentary Research*, **74**, 148–155.

Lambeck, K. & Chappell, J. 2001. Sea level change through the last glacial cycle. *Science*, **292**, 679–686.

Lance, S., Henry, P., Le Pichon, X., Lallemant, S., Chamley, H., Rostek, F., Faugères, J.-C., Gonthier, E. & Olu, K. 1998. Submersible study of mud volcanoes seaward of the Barbados accretionary wedge: sedimentology, structure and rheology. *Marine Geology*, **145**, 55–292.

Lancien, P., Metivier, F., Lajeunesse, E. & Cacas, M. 2004. Simulating submarine channels in flume experiments: aspects of the channel incision dynamic. *American Geophysical Union Fall Meeting, San Francisco*, Abstract #**OS41D-0509**.

Langseth, M.G., Westbrook, G.K. & Hobart, M.A. 1988. Geophysical survey of a mud volcano seaward of the Barbados Ridge Accretionary Complex. *Journal of Geophysical Research: Solid Earth*, **93**, 1049–1061.

Lansigu, C. & Bouroullec, R. 2004. Staircase normal fault geometry in the Grès d'Annot (SE France). *In*: Joseph, P. & Lomas, S.A. (eds), *Deep-Water Sedimentation in the Alpine Basin of SE France: New Perspectives on the Grès d'Annot and Related Systems*. Geological Society, London, Special Publication, **221**, 223–240. London: The Geological Society.

Laroche, P.J. 1983. Appalachians of southern Québec seen through seismic line no. 2001. *In*: Bally, A.W. (ed.), *Seismic Expression of Structural Styles*, volume 3, 3.2.1-7 to 3.2.1-24. American Association of Petroleum Geologists Studies in Geology, **15**.

Larue, D.K. & Provine, K.G. 1988. Vacillatory turbidites, Barbados. *Sedimentary Geology*, **57**, 211–219.

Larue, D.K. & Speed, R.C. 1983. Quartzose turbidites of the accretionary complex of Barbados, I: Chalky Mount succession. *Journal of Sedimentary Petrology*, **53**, 1337–1352.

Larue, D.K. & Speed, R.C. 1984. Structure of the accretionary complex of Barbados, II: Bissex Hill. *Geological Society of America Bulletin*, **95**, 1360–1372.

Lash, G. 1985. Recognition of trench fill in orogenic flysch sequences. *Geology*, **13**, 867–870.

Laskar, J. 1999. The limits of Earth orbital calculations for geological time-scale use. *Philosophical Transactions of the Royal Society, London, A*, **357**, 1735–1759.

Laursen, J. & Normark, W.R. 2002. Late Quaternary evolution of the San Antonio Submarine Canyon in the central Chile forearc (~33°S). *Marine Geology*, **188**, 365–390.

Laval, A., Cremer, M., Beghin, P. & Ravenne, C. 1988. Density surges: two-dimensional experiments. *Sedimentology*, **35**, 73–84.

Lawrence, D.T., Doyle, M., Snelson, S. & Horsfield, W.T. 1987. Stratigraphic modeling of sedimentary basins. *American Association of Petroleum Geologists Bulletin*, **71**, 582.

Lawrence, D.T., Doyle, M. & Aigner, T. 1989. Calibration of stratigraphic forward models in clastic, carbonate and mixed clastic/carbonate regimes. *28th International Geological Conference, Abstracts*, **2**, 264.

Lawrence, D.T. Doyle, M. & Aigner, T. 1990. Stratigraphic simulation of sedimentary basins: concepts and calibration. *American Association of Petroleum Geologists Bulletin*, **74**, 273–295.

Lay, T., Kanamori, H., Ammon, C.J., Nettles, M., Ward, S.N., Aster, R.C., Beck, S.L., Bilek, S.L., Brudzinski, M.R., Butler, R., DeShon, H.R., Ekstrom, G., Satake, K. & Sipkin, S. 2005. The Great Sumatra-Andaman Earthquake of 26 December 2004. *Science*, **308**, 1127–1133.

Le Friant, A., Ishizuka, O., Boudon, G., Palmer, M.R., Talling, P.J., Villemant, B., Adachi, T. Aljahdali, M., Breitkreuz, C., Brunet, M., Caron, B., Coussens, M., Deplus, C., Endo, D., Feuillet, N., Fraas, A.J., Fujinawa, A., Hart, M.B., Hatfield, R.G., Hornbach, M., Jutzeler, M., Kataoka, K.S., Komorowski, J.-C., Lebas, E., Lafuerza, S., Maeon, F., Manga, M., Martinez-Colon, M., McCanta, M., Morgan, S., Saito, T., Slagle, A., Sparks, S., Stinton, A., Stroncik, N., Subramanyam, K.S.V., Tamura, Y., Trofimovs, J., Voight, B., Wall-Palmer, D., Wang, F. & Watt, S.F.L. 2015. Submarine record of volcanic island construction and collapse in the Lesser Antilles arc: first scientific drilling of submarine volcanic island landslides by IODP Expedition 340. *Geochemistry, Geophysics, Geosystems*, **16**, 420–442,

Le Pichon, X., Kobayashi, K., Cadet, J.-P., Iiyama, T., Nakamura, K., Pautot, G., Renard, V. & the Kaiko Scientific Crew 1987a. Project Kaiko - Introduction. *Earth and Planetary Science Letters*, **83**, 183–185.

Le Pichon, X., Iiyama, T., Chamley, H., Charvet, J., Faure, M., Fujimoto, H., Furuta, T., Ida, Y., Kagami, H., Lallemant, S., Leggett, J., Murata, Y., Okada, H., Rangin, C., Renard, V., Taira, A. & Tokuyama, H. 1987b. The eastern and western ends of Nankai Trough: results of Box 5 and Box 7 Kaiko survey. *Earth and Planetary Science Letters*, **83**, 199–213.

Le Varlet, X. & Roy, J.P. 1983. *Étude de la série priabonienne de la région Vallée de l'Ubaye, Les Trois Evêchés*. ENSPM Réf. **32428**.

Leckie, R.M. 1984. Mid-Cretaceous planktonic foraminiferal biostratigraphy offcentral Morocco, Deep Sea Drilling Project Leg 79, Sites 545 and 547. *In*: Hinz, K., Winterer, E.L. *et al.*, *Initial Reports Deep Sea Drilling Project*, **79**, 579–620. Washington, DC: US Government Printing Office.

Ledbetter, M.T. 1979. Fluctuations of Antarctic bottom water velocity in the Vema Channel during the last 160,000 years. *Marine Geology*, **33**, 71–89.

Ledbetter, M.T. 1984. Bottom-current speed in the Vema Channel recorded by particle size of sediment fine-fraction. *Marine Geology*, **58**, 137–149.

Ledbetter, M.T. & Ellwood, B.B. 1980. Spatial and temporal changes in bottom water velocity and direction from analyses of particle size and alignment in deep-sea sediment. *Marine Geology*, **38**, 245–261.

Lee, H.J., Syvitski, J.P.M., Parker, G., Orange, D., Locat, J., Hutton, E.W.H. & Imran, J. 2002. Distinguishing sediment waves from slope failure deposits: field examples, including the 'Humbolt slide', and modelling results. *Marine Geology*, **192**, 79–104.

Lee, S.E., Amy, L.A. & Talling, P.J. 2004. The character and origin of thick base-of-slope sandstone units of the Peira Cava outlier, SE France. *In*: Joseph, P. & Lomas, S.A. (eds), *Deep-Water Sedimentation in the Alpine Basin of SE France: New Perspectives on the Grès d'Annot and Related Systems*. Geological Society, London, Special Publication, **221**, 331–347. London: The Geological Society.

Leeder, M.R. 1983. On the dynamics of sediment suspension by residual Reynolds stresses - confirmation of Bagnold's theory. *Sedimentology*, **30**, 485–492.

Leeder, M.R., Gray, T.E. & Alexander, J. 2005. Sediment suspension dynamics and a new criterion for the maintenance of turbulent suspensions. *Sedimentology*, **52**, 683–691.

Legg, M.R. 1991. Developments in understanding the tectonic evolution of the California Continental Borderland. *In*: Osborne, R.H. (ed.), *From Shoreline to Abyss*, 291–312. Society of Economic Paleontologists and Mineralogists Shepard Commemorative Volume, **46**.

Legg, M.R., Goldfinger, C., Kamerling, M.J., Chaytor, J.D. & Einstein, D.E. 2007. Morphology, structure and evolution of California Continental Borderland restraining bends. *In*: Cunningham, W.D. & Mann, P. (eds), *Tectonics of Strike-Slip Restraining and Releasing Bends*. Geological Society London Special Publication, **290**, 143–168. The Geological Society of London.

Leggett, J.K. 1980. The sedimentological evolution of a Lower Palaeozoic accretionary fore-arc in the Southern Uplands of Scotland. *Sedimentology*, **27**, 401–417.

Leggett, J.K. (ed.) 1982. *Trench-Forearc Geology*. Geological Society of London Special Publication, **10**. Oxford: Blackwell Scientific.

Leggett, J.K. 1985. Deep-sea pelagic sediments and palaeooceanography: a review of recent progress. *In*: Brenchley, P.J. & Williams, B.P.J. (eds), *Sedimentology Recent Developments and Applied Aspects*, 95–121. Geological Society of London Special Publication, **18**. Oxford: Blackwell Scientific.

Leggett, J.K., Aoki, Y. & Toba, T. 1985. Transition from frontal accretion to underplating in a part of the Nankai Trough accretionary complex off Shikoku (SW Japan) and extensional features on the lower trench slope. *Marine and Petroleum Geology*, **2**, 131–141.

Legros, F. 2002. Can dispersive pressure cause inverse grading in grain flows? *Journal of Sedimentary Research*, **72**, 166–170.

Lehmann, D., Brett, C.E., Cole, R. & Baird, G. 1995. Distal sedimentation in a peripheral foreland basin: Ordovician black shales and associated flysch of the western Taconic foreland, New York State and Ontario. *Geological Society of America Bulletin*, **107**, 708–724.

Leinfelder, R.R. & Wilson, R.C.L. 1998. Third-order sequences in an Upper Jurassic rift-related second-order sequence, central Lusitanian Basin, Portugal. *In*: Hardenbol J., Thierry J., Farley, M.B., Jacquin Th., de Graciansky P.-C. & Vail P.R. (eds), *Mesozoic and Cenozoic Sequence Stratigraphy of European Basins*, 509–525. Society of Economic Paleontologists and Mineralogists, Special Publication, **60**.

Leitch, E.C. 1984. Marginal basins of the SW Pacific and the preservation and recognition of their ancient analogues: a review. *In*: Kokelaar, B.P. & Howells, M.F. (eds), *Marginal Basin Geology*, 97–108. Geological Society, London, Special Publications, **16**. London: The Geological Society.

León, R., Medialdea, T., Javier-Gonzalez, Gimenez-Motreno, C.J. & Perez-Lopez, R. 2014. Pockmarks on either side of the Strait of Gibralter: formation from overpressutred shallow contourite gas reservoirs and internal wave action during the last glacial sea-level lowstand? *Geo-Marine Letters*, **34**, 131–151.

Leopold, L.B. & Langbein, W.B. 1962. The concept of entropy in landscape evolution. *U.S. Geological Survey Professional Paper*, **500-A**.

Letouzey, J. & Kimura, M. 1985. Okinawa Trough genesis: structure and evolution of a backarc basin developed in a continent. *Marine and Petroleum Geology*, **2**, 111–130.

Letouzey, J. & Kimura, M. 1986. The Okinawa Trough: genesis of a back-arc basin developing along a continental margin. *Tectonophysics*, **125**, 209–230.

Levell, B.K. 1980a. A late Precambrian tidal shelf deposit, the Lower Sandfiord Formation, Finnmark, North Norway. *Sedimentology*, **27**, 539–557.

Levell, B.K. 1980b. Evidence for currents associated with waves in Late Precambrian shelf deposits from Finnmark, North Norway. *Sedimentology*, **27**, 153–166.

Levell, B.K. & Roberts, D. 1977. A re-interpretation of the geology of north-west Varanger Peninsula, East Finnmark, North Norway. *Norges Geologiske Undersokelse*, **334**, 83–90.

Leverenz, A. 2000. Trench-sedimentation versus accreted submarine fan—an approach to regional-scale facies analysis in a Mesozoic accretionary complex: "Torlesse" terrane, northeastern North Island, New Zealand. *Sedimentary Geology*, **132**, 125–160.

Lewis, D.W. & Ekdale, A.A. 1992. Composite ichnofabric of a Mid-Tertiary unconformity on a pelagic limestone. *Palaios*, **7**, 222–235.

Lewis, J.C., Byrne, T.B. & Kanagawa, K. 2013. Evidence for mechanical decoupling of the upper plate at the Nankai subduction zone: Constraints from core-scale faults at NantroSEIZE Sites C0001 and C0002. *Geochemistry, Geophysics, Geosystems*, **14**, 620–633.

Lewis, J.F., Draper, G., Bourdon, C., Bowin, C., Mattson, P.O., Maurrasse, F., Nagle, F. & Pardo, G. 1990. Geology and tectonic evolution of the northern Caribbean margin. *In*: Dengo, G. & Case, J.E. (eds), *The Caribbean Region, Volume H of the Geology of North America*, 77–140. Boulder Colorado: Geological Society of America.

Lewis, K.B. 1971. Slumping on a continental slope inclined at 1°-4°. *Sedimentology*, **16**, 97–110.

Lewis, K.B. & Pantin, H.M. 2002. Channel-axis, overbank and drift sediment waves in the southern Hikurangi Trough, New Zealand. *Marine Geology*, **192**, 123–151.

Lewis, K.B., Bennett, D.J., Herzer, R.H. & von der Borch, C.C. 1986. Seismic stratigraphy and structure adjacent to an evolving plale boundary, western Chatham Rise, New Zealand. *In*: Kennett, J.P., von der Borch, C.C. *et al.*, Initial Reports Deep Sea Dnlhng Project, **90**, 1325–1337. Washington, DC: US Government Printing Office.

Lie, O., Fürstenau, J. & Comstock, J. 2013. Will Lebanon be the next oil province? *GEO ExPro*, **10**, 36–40.

Lien T., Walker, R.G. & Martinsen, O.J. 2003. Turbidites in the Upper Carboniferous Ross Formation, western Ireland: reconstruction of a channel and spillover system. *Sedimentology*, **50**, 113–148.

Lien, T., Midtbø, R.E., Martinsen, O. 2006. Depositional facies and reservoir quality of deep-marine sandstones in the Norwegian Sea. *Norwegian Journal of Geology*, **86**, 71–92.

Lindsay, J.F. 1968. The development of clast fabric in mudflows. *Journal of Sedimentary Petrology*, **38**, 1242–1253.

Lipman, P.W., Normark, W.R., Moore, J.G., Wilson, J.B. & Gutmacher, C.E. 1988. The giant submarine Alika debris slide, Mauna Loa, Hawaii. *Journal of Geophysical Research*, **93**, 4,279–4,299.

Lisitzin, A.P. (ed.) 1972. *Sedimentation in the World Ocean*. Society of Economic Paleontologists and Mineralogists Special Publication, **17**.

Liu, C-S., Huang, I.L. & Teng, L.S. 1997. Structural features off southwestern Taiwan. *Marine Geology*, **137**, 305–319.

Liu, J.Y. & Bryant, W.R. 2000. Sea floor morphology and sediment paths of the northern Gulf of Mexico deepwater. *In*: Bouma, A.H. & Stone, C.G. (eds), *Fine-grained Turbidite Systems*, 33–46. American Association of Petroleum Geologists Memoir, **72** & Society of Sedimentary Geologists, Special Publication, **68**. Joint publication, Tulsa, Oklahoma.

Liu, S, Qian, T., Li, W., Dou, G. & Wu, P. 2015. Oblique closure of the northeastern Paleo-Tethys in central China. *Tectonics*, **34**, 413–434.

Liu, X. & Galloway, W.E. 1997. Quantitative determination of Tertiary sediment supply to the North Sea basin. *American Association of Petroleum Geologists Bulletin*, **81**, 1482–1509.

Liu, Z., Pagani, M., Zinniker, D., DeConto, R., Huber, M., Brinkhuis, H., Shah, S.R., Leckie, R.M. & Pearson, A. 2009. Global cooling during the Eocene-Oligocene Climate Transition. *Science*, **323**, 1187–1190.

Lizarralde, D., Axen, G.J., Brown, H.E., Fletcher, J.M., Gonzalez-Fernandez, A., Harding, A.J., Holbrook, W.S., Kent, G.M., Paramo, P., Sutherland, F. & Umhoefer, P.J. 2007. Variation in styles of rifting in the Gulf of California. *Nature*, **448**, 466–469.

Locat, J. 1997. Normalized rheological behaviour of fine muds and their flow properties in a pseudoplastic regime. *Proceedings of 1st International Conference, American Society of Civil Engineers, Reston, Virginia*, 260–269.

Locat, J. 2001. Instabilites along ocean margins: a geomorphological and geotechnical perspective. *Marine and Petroleum Geology*, **18**, 508–512.

Logan, W.E. 1883. *Report on the Geology of Canada*. Geological Survey of Canada, Report of Progress to **1863**.

Lonergan, L., Lee, N., Johnson, H.D., Cartwright, J.A. & Jolly, R.J.H. 2000. Remobilization and injection in deepwater depositional systems: implications for reservoir architecture and prediction. *In*: Weimer, P., Slatt, R.M., Coleman, J., Rosen, N.C., Nelson, H., Bouma, A.H., Styzen, M.J. & Lawrence, D.T. (eds), *Gulf Coast Section-Society of Economic Paleontologists and Mineralogists Foundation 20th Annual Bob F. Hoskins Research Research Conference, Deep-Water Reservoirs of the World*, 515–532. CD-ROM Society of Economic Paleontologists and Mineralogists Special Publications.

Londoño, J. & Lorenzo, J.M. 2004. Geodynamics of continental plate collision during late tertiary foreland basin evolution in the Timor Sea: constraints from foreland sequences, elastic flexure and normal faulting. *Tectonophysics*, **392**, 37–54.

Lonergan, J., Borlandelli, C., Taylor, A., Quine, M., Flanagan, K. 2007. The three dimensional geometry of sandstone injection complexes in the Gryphon Field, United Kingdom, North Sea. *In*: Hurst, A. & Cartwright, J. (eds), *Sand Injectites: Implications for Hydrocarbon Exploration and Production*, 103–112. American Association of Petroleum Geologists Memoir, **87**. Tulsa, Oklahoma.

Long, D.G.F. 1977. Resedimented conglomerate of Huronian (Lower Aphebian) age, from the north shore of Lake Huron, Ontario, Canada. *Canadian Journal of Earth Sciences*, **14**, 2495–2509.

Lonsdale, P.F. 1985. A transform continental margin rich in hydrocarbons, Gulf of California. *American Association of Petroleum Geologists Bulletin*, **69**, 1160–1180.

Lonsdale, P. 1991. Structural patterns of the Pacific floor offshore of peninsular California. *In*: Dauphin, J.P. & Simoneit, B.R.T. (eds), *The Gulf and Peninsular Province of the Californias*, 87–125. American Association of Petroleum Geologists Memoir, **47**.

López Cabrera, M.I., Olivero, E.B., Carmona, N.B. & Ponce, J.J. 2008. Cenozoic trace fossils of the Cruziana, Zoophycos, and Nereites ichnofacies from the Fuegian Andes, Argentina. *Ameghiniana*, **45**, 377–392

Lovell, J.P.B. 1969. Tyee Formation: a study of proximality in turbidites. *Journal of Sedimentary Petrology*, **39**, 935–953.

Lovell, J.P.B. & Stow, D.A.V. 1981. Identification of ancient sandy contourites. *Geology*, **9**, 347–349.

Lowe, D.R. 1975. Water escape structures in coarse-grained sediments. *Sedimentology*, **22**, 157–204.

Lowe, D.R. 1976a. Grain flow and grain flow deposits. *Journal of Sedimentary Petrology*, **46**, 188–199.

Lowe, D.R. 1976b. Subaqueous liquefied and fluidized sediment flows and their deposits. *Sedimentology*, **23**, 285–308.

Lowe, D.R. 1982. Sediment gravity flows: II. Depositional models with special reference to the deposits of high-density turbidity currents. Journal of Sedimentary Petrology, **52**, 279–297.

Lowe, D.R. 1985. Ouachita trough: part of a Cambrian failed rift system. *Geology*, **13**, 790–793.

Lowe, D.R. 1988. Suspended-load fallout rate as an independent variable in the analysis of current structuResearch *Sedimentology*, **35**, 765–776.

Lowe, D.R. & Guy, M. 2000. Slurry-flow deposits in the Britannia Formation (Lower Cretaceous), North Sea: a new perspective on the turbidity current and debris flow problem. *Sedimentology*, **47**, 31–70.

Lowe, D.R. & LoPiccolo, R.D. 1974. The characteristics and origins of dish and pillar structure. *Journal of Sedimentary Petrology*, **44**, 484–501.

Lowe, D.R., Guy, M. & Palfrey, A. 2003. Facies of slurry-flow deposits, Britannia Formation (Lower Cretaceous), North Sea: implications for flow evolution and deposit geometry. *Sedimentology*, **50**, 45–80.

Lowell, J.D. 1985. *Structural Styles in Petroleum Exploration*. Tulsa, Oil and Gas Consultants International, 460 pp.

Lowey, G.W. 1992. Variation in bed thickness in a turbidite succession, Dezadeash Formation (Jurassic–Cretaceous), Yukon, Canada: evidence of thinning-upward and thickening-upward cycles. *Sedimentary Geology*, **78**, 217–232.

Lu, H. & Fulthorpe, C.S. 2004. Controls on sequence stratigraphy of a middle Miocene-Holocene, current-swept, passive margin: Offshore Canterbury Basin, New Zealand. *Geological Society of America Bulletin*, **116**, 1345–1366.

Lu, H., Fulthorpe, C.S. & Mann, P. 2003. Three-dimensional architecture of shelf-building sediment drifts in the o!shore Canterbury Basin, New Zealand. *Marine Geology*, **193**, 19–47.

Lu, H. & Hayashi, D. 2001. Genesis of Okinawa Trough and thrust development within accretionary prism by means of 2D finite element method. *Structural Geology* (Journal of Tectonic Research Group Japan), **45**, 47–67.

Lu, N.Z., Suhayda, J.N., Prior, D.B., Bornhold, B.D., Keller, G.H., Wiseman, W.J. Jr, Wright, L.D. & Yang, Z.S. 1991. Sediment thixotropy and submarine mass movement, Huanghe Delta, China. *Geo-Marine Letters*, **11**, 9–15.

Lu, R.S. & McMillen, K.J. 1983. Multichannel seismic survey of the Columbia Basin and adjacent margins. *In*: Watkins, J.S. & Drake, C.L. (eds), *Studies in Continental Margin Geology*, 395–410. American Association of Petroleum Geologists. Memoir, **34**.

Lucchi, R. & Camerlenghi, A. 1993. Upslope turbiditic sedimentation on the southeastern flank of the Mediterranean Ridge. *Bollettino di oceanologia teorica ed applicata*, **11**, 3–25.

Lucchi, R.G., Rebesco, M., Camerlenghi, A., Busetti, M., Tomadin, L., Villa, G., Persico, D., Morigi, C., Bonci, M.C. & Giorgetti, G. 2002. Mid-late Pleistocene glacimarine sedimentary processes of a high-latitude, deep-sea sediment drift (Antarctic Peninsula Pacific margin). *Marine Geology*, **189**, 343–370.

Lucente, C.C. & Pini, G.A. 2003. Anatomy and emplacement mechanism of a large submarine slide within a Miiocene foreddep in the northern Apennines, Italy: a field perspective. *American Journal of Science*, **303**, 565–602.

Ludden, J.N., Gradstein, F.M. *et al.* 1990. *Proceedings of the Ocean Drilling Program, Initial Reports,* **123**. College Station, Texas, USA: Ocean Drilling Program.

Lundberg, N. 1982. Evolution of the slope landward of the Middle America Trench, Nicoya Peninsula, Costa Rica. *In*: Leggett, J.K. (ed.), *Trench-Forearc Geology,* 131–147. The Geological Society of London, Special Publication, **10**. London: The Geological Society.

Lundegard, P.D., Samuels, N.D. & Pryor, W.A. 1980. *Sedimentology, Petrology and Gas Potential of the Brallier Formation-Upper Devonian Turbidite Facies of the Central and Southem Appalachians*. US Department of Energy Report DOE/METC/5201-5.

Lundgren, B. 1891. Studier ofver fossilforande losa block. *Geologiska Föreningens i Stockholm Förhandlingar*, **13**, 111–121.

Luo, S., Z. Gao, He, Y. & Stow, D.A.V. 2002. Ordovician carbonate contourite drifts in Hunan and Gansu Provinces, China. *In*: Stow, D.A.V., Pudsey, C.J., Howe, J.A., Faugères, J.-C. & Viana, A.R. (eds), *Deep-water Contourite Systems: Modern Drifts and Ancient Series, Seismic and Sedimentary Characteristics*, 433–442. Geological Society London Memoir, **22**.

Lüthi, S. 1981. Experiments on non-channelized turbidity currents and their deposits. *Marine Geology*, **40**, M59–M68.

Luthi, S.M., Hodgson, D.M., Geel, C.R., Flint, S.S., Goedbloed, J.W., Drinkwater, N.J. & Johannessen, E.P. 2006. Contribution of research borehole data to modelling fine-grained turbidite reservoir analogues, Permian Tanqua–Karoo basinfloor fans (South Africa). *Petroleum Geosciences*, **12**, 1–16.

Lykousis, V., Sakellariou, D. & Locat, J. (eds) 2007. *Submarine Mass Movements and their Consequences*. Advances in Natural and Technological Hazards Research, **27**, 3rd International Symposium Series. The Netherlands, Springer, 436 pp. ISBN 978-1-4020-6511-8.

Lyle, M. & Wilson, P.A. 2004. Leg 199 synthesis: Evolution of the equatorial Pacific in the early Cenozoic. *In*: Wilson, P.A., Lyle, M. & Firth, J.V. (eds), *Proceedings of the Ocean Drilling Program, Scientific Results,* **199**, 1–39. College Station, Texas, USA: Ocean Drilling Program.

Lyle, M., Koizumi, I., Richter, C. *et al.* 1997. *Proceedings of the Ocean Drilling Program, Initial Results,* **167**. College Station, Texas, USA: Ocean Drilling Program.

Lyle, M., Koizumi, I., Delaney, M.L. & Barron, J.A. 2000. Sedimentary record of the California current system, Middle Mioce to Holocene: a synthesis of Leg 167 results. *In*: Lyle, M., Koizumi, I., Richter, C. & Moore, T.C., Jr., (eds), *Proceedings of the Ocean Drilling Program, Scientific Results,* **167**, 341–376. College Station, Texas, USA: Ocean Drilling Program.

Ma, C. 1996. *Continuity of sandstone beds in the Ordovician Cloridorme Formation, Gaspé Pensula, Québec*. MSc Thesis, Memorial University, St. John's, Newfoundland & Labrador.

Macdonald, D.I.M., Barker, P.F., Garrett, S.W., Ineson, J.R., Pirrie, D., Storey, B.C., Whitham, A.G., Kinghorn, R.R.F. & Marshall, J.E.A. 1988. A preliminary assessment of the hydrocarbon potential of the Larsen Basin, Antarctica. *Marine and Petroleum Geology*, **5**, 34–53.

Macdonald, D.I.M. (ed.) 1991. *Sedimentation, Tectonics and Eustasy: Sea-level Changes at Active Margins*. Special Publication of the International Association of Sedimentologists, **12**, Oxford: Blackwell Scientific Publications, 518 pp.

Macdonald, H.A., Peakall, J., Wignall, P.B. & Best, J. 2011a. Sedimentation in deep-sea lobe-elements: implications for the origin of thickening-upward sequences. *Journal of the Geological Society* (London), **168**, 319–332.

Macdonald, H.A., Wynn, R.B., Huvenne, V.A.I., Peakall, J., Masson, D.G., Weaver, P.P.E. & McPhail, S.D. 2011b. New insights into the morphology, fill, and remarkable longevity (>0.2 m.y.) of modern deep-water erosional scours along the northeast Atlantic margin. *Geosphere*, **7**, 845–867.

MacDonald, G.J. 1990 Role of methane clathrates in past and future climates. *Climate Change*, **16**, 247–281.

Machlus, M.L., Olsen, P.E., Christie-Blick, N. & Hemming, S.R. 2008. Spectral analysis of the lower Eocene Wilkins Peak Member, Green River Formation, Wyoming: support for Milankovitch cyclicity. *Earth and Planetary Science Letters*, **268**, 64–75.

MacKay, M.E. 1995. Structural variation and landward vergence at the toe of the Oregon accretionary prism. *Tectonics*, **14**, 1309–1320.

MacKay, M.E., Moore, G.F., Cochrane, G.R., Moore, J.C. & Kulm, L.D. 1992. Landward vergence and oblique structural trends in the Oregon margin accretionary prism: Implications and effect on fluid flow. *Earth and Planetary Science Letters*, **109**, 477–491.

MacLeay, W.S. 1839. Note on the Annelida. *In*: Murchinson, R.I. (ed.). *The Silurian System, Part II, Organic Remains,* 699–701. J. Murray, London.

Magalhaes, P.M. & Tinterri, R. 2010. Stratigraphy and depositional setting of slurry and contained (reflected) beds in the Marnoso-arenacea Formation (Langhian-Serravallian) Northern Apennines, Italy. *Sedimentology*, **57**, 1685–1720.

Magwood, J.P.A. 1992. Ichnotaxonomy: A burrow by any other name … ? *In*: Maples, C.G. & West, R.R. (eds), *Trace Fossils,* 15–33. Short courses in Paleontology **5**, University of Tennessee, Knoxville.

Mahony, S.H., Wallace, L.M., Miyoshi, M., Villamor, P., Sparks, R.S.J. & Hasenaka, T. 2011. Volcano-tectonic interactions during rapid plate-boundary evolution in the Kyushu region, SW Japan. *Geological Society of America Bulletin*, **123**, 2201–2223.

Maier, K.L., Fildani, A., Paull, C.K., Graham, S.A., McHargue, T.R. Caress, D.W. & McGann, M. 2011. The elusive character of discontinuous deep-water channels: New insights from Lucia Chica channel system, offshore California. *Geology*, **39**, 327–330.

Maier, K.L., Fildani, A., Paull, C.K., McHargue, T.R., Graham, S.A. & Caress, D.W. 2013. Deep-sea channel evolution and stratigraphic architecture from inception to abandonment from high-resolution Autonomous Underwater Vehicle surveys offshore central California. *Sedimentology*, **60**, 935–960.

Maiklem, W.C. 1968. Some hydraulic properties of bioclastic carbonate grains. *Sedimentology*, **10**, 101–109.

Major, J.J. 1997. Depositional processes in large-scale debris-flow experiments. *Journal of Geology*, **105**, 345–366.

Malgesini, G., Talling, P.J., Hogg, A.J., Armitage, D., Goater, A. & Felletti, F. 2015. Quantitative analysis of submarine-flow deposit shape in the Marnoso-arenacea Formation: what is the signature of hindered settling from dense near-bed layers? *Journal of Sedimentary Research*, **85**, 170–191.

Mallarino, G. Droxler, A.W. & Fitton, R. 2005. Timing of turbidite input and Late Quaternary sea level: comparison between siliciclastic (Western Gulf of Mexico), carbonate (Northern Nicaragua Rise), and mixed carbonate-siliciclastic (Pandora Trough, Coral Sea) systems. *American Association of Petroleum Geologists Annual Convention (June 19–22, 2005) Technical Program, Abstract Volume.*

Malinverno, A. 1997. On the power law size distribution of turbidite beds. *Basin Research*, **9**, 263–274.

Malinverno, A., Ryan, W.B.F., Auffret, G. & Pautot, G. 1988. Sonar images of the path of recent failure events on the continental margin off Nice, France. *In*: Clifton, H.E. (ed.), *Sedimentologic Consequences of Convulsive Geologic Events,* 59–75. Geological Society of America Special Paper, **229**.

Malod, J.A., Karta, K., Beslier, M.O. & Zen, Jr.M.T. 1995. From normal to oblique subduction: Tectonic relationships between Java and Sumatra. *Journal of Southeast Asian Earth Sciences*, **12**, 85–93.

Malouta, D.N., Gorsline, D.S. & Thornton, S.E. 1981. Processes and rates of Recent (Holocene) basin filling in an active transform margin: Santa Monica Basin, California Continental Borderland. *Journal of Sedimentary Petrology*, **51**, 1077–1095.

Mammerickx, J. 1970. Morphology of the Aleutian Abyssal Plain. *Geological Society of America Bulletin*, **81**, 3457–3464.

Mancin, N., Di Giulio, A. & Cobianchi, M. 2009. Tectonic vs. climate forcing in the Cenozoic sedimentary evolution of a foreland basin (Eastern South Alpine system, Italy). *Basin Research*, **21**, 799–823.

Manders, A.M.M., Maas, L.R.M. & Gerkema, T. 2004. Observations of internal tides in the Mozambique Channel. *Journal of Geophysical Research*, **109**, 1–9. doi:10.1029/2003JC002187.

Mangano, M.G., Buatois, L.A., Maples, C.G. & West, R.R. 2000. A new Ichnospecies of Nereites from Carboniferous tidal-flat facies of eastern Kansas, USA: implications for the Nereites-Neonereites debate. *Journal of Paleontology*, **74**, 149–157.

Mann, J. 2013. The Santos Basin, Brazil. *GEO ExPro*, **10**, 76–78.

Mann, M.E. & Lees, J.M. 1996. Robust estimation of background noise and signal detection in climatic time series. *Climate Change*, **33**, 409–445.

Mann, P. 2007. Global catalogue, classification and tectonic origins of restraining- and releasing bends on active and ancient strike-slip fault systems. *In*: Cunningham, W.D. & Mann, P. (eds), *Tectonics of Strike-Slip Restraining and Releasing Bends.*Geological Society, London, Special Publications, **290**, 13–142. London: The Geological Society.

Mann, P., Hempton, M.R., Bradley, D.C. & Burke, K. 1983. Development of pull-apart basins. *Journal of Geology*, **91**, 529–554.

Mann, P., Draper, G. & Burke, K. 1985. Neotectonics of a strike-slip restraining bend system, Jamaica. *In*: Biddle, K. & Christie-Blick, N. (eds), *Strike-Slip Deformation, Basin Formation, and Sedimentation.* Society of Economic Paleontologists and Mineralogists, Special Publications, **37**, 211–226.

Mann, P., DeMets, C. & Wiggins-Grandison, M. 2007. Toward a better understanding of the Late Neogene strike-slip restraining bend in Jamaica: geodetic, geological, and seismic constraints. *In*: Cunningham, W.D. & Mann, P. (eds), *Tectonics of Strike-Slip Restraining and Releasing Bends.* Geological Society, London, Special Publications, **290**, 239–253. London: The Geological Society.

Mantyla, A.W. & Reid, J.L. 1983. Abyssal characteristics of the World Ocean waters. *Deep-Sea Research*, **30**, 805–833.

Mariano I. Cantero, M.I., Cantelli, A., Pirmez, C., Balachandar, S., Mohrig, D., Hickson, T.A., Yeh, T., Hajime Naruse, H. & Parker, G. 2012. Emplacement of massive turbidites linked to extinction of turbulence in turbidity currents. *Nature Geoscience*, **5**, 42–45.

Marjanac, T. 1985. Composition and origin of the megabed containing huge clasts, flysch formation, middle Dalmatia, Yugoslavia. *In*: *6th European Regional Meeting, Lleida, Spain, Abstracts and Poster Abstracts Volume,* 270–273. International Association of Sedimentologists.

Marjanač, T. 1990. Reflected sediment gravity flows and their deposits in flysch of Middle Dalmatia, Yugoslavia. *Sedimentology*, **37**, 921–929.

Marr, J.G., Harff, P.A., Shanmugam, G. & Parker, G. 2001. Experiments on subaqueous gravity flows: the role of clay and water content in flow dynamics and depositional structure. *Geological Society of America Bulletin*, **113**, 1377–1386.

Marroni, M. & Treves, B. 1998. Hidden Terranes in the Northern Apennines, Italy: A Record of Late Cretaceous–Oligocene Transpressional Tectonics. *The Journal of Geology*, **106**, 149–162.

Marsaglia, K.M. & Ingersoll, R.V. 1992. Compositional trends in arc-related, deep-marine sand and sandstone: a reassessment of magmatic-arc provenance. *Geological Society of America Bulletin*, **104**, 1637–1649.

Marsaglia, K.M., Ingersoll, R.V. & Packer, B.M. 1992. Tectonic evolution of the Japanese islands as reflected in modal compositions of Cenozoic forearc and backarc sand and sandstone. *Tectonics*, **11**, 028–1044.

Marsaglia, K.M., Torrez, X.V., Padilla, I. & Rimkus, K.C. 1995. Provenance of Pleistocene and Pliocene sand and sandstone, ODP Leg 141, Chile margin. *In*: Lewis, S.D., Behrmann, J.H., Musgrave, R.J. & Cande, S.C. (eds), *Proceedings of the Ocean Drilling Program, Scientific Results,* **141**, 133–151.

Marschalko, R. 1964. Sedimentary structures and paleocurrents in the marginal lithofacies of the central-Carpathian flysch. *In*: Bouma, A.H. & Brouwer, A. (eds), *Turbidites*, 106–126. Developments in Sedimentology, **3**. Amsterdam: Elsevier.

Marschalko, R. 1975. Depositional environment of conglomerate as interpreted from sedimentological studies (Paleogene of Klippen Belt and adjacent tectonic units in East Slovakia). *Nauka o Zemi*, **9**, Veda, Bratislava 1–47 (In Slovak, English summary.)

Martin, J., Toothill, S. & Moussavou, R. 2009. Pre-salt basins identified in Gabon deepwater area. Cameron, B. 2010. Carnarvon Basin. *GEO ExPro*, **6**, 38–41.

Martin, R.G. & A.H. Bouma 1978. Physiography of Gulf of Mexico. *In*: Bouma, A.H., Moore, G.T. & Coleman, J.M. (eds), *Framework, Facies, and Oil-trapping Characteristics of the Upper Continental Margin*, 3–19. American Association of Petroleum Geologists Studies in Geology, **7**.

Martín-Chivelet, J., Fregenal-Martínez & Chacón, B. 2008. Traction structures in contourites. *In*: Rebesco, M. & Camerlenghi, A. (eds), *Contourites*, 159–182. Developments in Sedimentology, **60**. Amsterdam: Elsevier.

Martinez, F., Okino, Y., Ohara, Y., Reysenbach, A.-L. & Goffredi, S.K. 2007. Back-arc basins. *Oceanography*, **20**, 116–127.

Martinelli, G., Cremonini, S. & Samonati, E. 2012. *In*: Al-Megren, H. (ed.), *Geological and Geochemical Setting of Natural Hydrocarbon Emissions in Italy, Advances in Natural Gas Technology*, 79–120. Rijeka, Croatia: IntTech.

Martinsen, O.J., Lien, T. & Walker, R.G. 2000. Upper Carboniferous deep water sediments, western Ireland: analogues for passive margin turbidite plays. *In*: Weimer, P., Slatt, R.M., Coleman, J., Rosen, N.C., Nelson, H., Bouma, A.H., Styzen, M.J. & Lawrence, D.T. (eds), *Gulf Coast Section-Society of Economic Paleontologists and Mineralogists Foundation 20th Annual Bob F. Hoskins Research Research Conference, Deep-Water Reservoirs of the World*, 533–555. CD-ROM Society of Economic Paleontologists and Mineralogists Special Publications.

Martinsen, O.J., Lein, T. & Jackson, C. 2005. Cretaceous and Palaeogene turbidite systems in the North Sea and Norwegian Sea basins: source, staging area and basin physiography controls on reservoir development. *In*: Doré, A.G. & Vining, B.A. (eds), *Petroleum Geology: North-west Europe and Global Perspectives - Proceedings of the 6th Petroleum Geology Conference*, 1, 1147–1167. London: The Geological Society.

Martinsson, A. 1965. Aspects of a Middle Cambrian thanatotope on Öland. *Geologiska Förening in Stockholm Förhandlingar*, **87**, 181–230.

Martinsson, A. 1970. Toponomy of trace fossils. *In*: Crimes, T.P. & Harper, J.C. (eds), *Trace Fossils*, 323–330. Geological Journal Special Issues, **3**.

Mascarelli, A.L. 2009. A sleeping giant? *Nature Reports*, **3**, 46–49.

Mascle, A., Puigde Fabregas, C., Luterbacher, H.P. & Fernandez, M. (eds) 1998. *Cenozoic Foreland Basins of Western Europe. Geological Society Special Publication* **134**, 134–427.

Mascle, A., Moore, J.C. *et al.* 1988. *Proceedings of the Ocean Drilling Program, Initial Reports (Part A)*, **110**. Texas, USA: College Station, (Ocean Drilling Program).

Mascle, J., Zitter, T., Bellaiche, G., Droz, L., Gaullier, V., Loncke, L. & Prismed Scientific Party 2001. The Nile deep sea fan: preliminary results from a swath bathymetry survey. *Marine and Petroleum Geology*, **18**, 471–477.

Maslin, M., Owen, M., Betts, R., Day, S., Dunkley, T. & Ridgwell, A. 2010. Gas hydrates: past and future geohazard? *Philosophical Transactions of the Royal Society A*, **368**, 2369–2393.

Massalongo, A. 1855. *Zoophycos, novum genus Plantarum fossilium. Typis Antonellianis, Veronae*, 45–52.

Masson, D.G. 1994. Late Quaternary turbidity current pathways to the Madeira Abyssal Plain and some constraints on turbidity current mechanisms. *Basin Research*, **6**, 17–33.

Masson, D.G., Kidd, R.B., Gardner, J.V., Huggett, Q. & Weaver, P.P.E. 1992. Saharan continental rise: facies distribution and sediment slides. *In*: Poag, C.W. & de Graciansky, P.C. (eds), *Geological Evolution of Atlantic Continental Rises, 327–343*. New York: van Nostrand Reinhold.

Masson, D.G., Huggett, Q.J. & Brunsden, D. 1993. The surface texture of the Saharan Debris Flow deposit and some speculations on debris flow processes. *Sedimentology*, **40**, 583–598.

Masson, D.G., Kenyon, N.H., Gardner, J.V. & Field, M.E. 1995. Monterey Fan: channel and overbank morphology. *In*: Pickering, K.T., Hiscott, R.N., Kenyon, N.H., Ricci Lucchi, F. & Smith, R.D.A. (eds), *Atlas of Deep Water Environments: Architectural Style in Turbidite Systems*, 74–79. London: Chapman and Hall.

Masson, D.G., Howe, J.A. & Stoker, M.S. 2002. Bottom-current sediment waves, sediment drifts and contourites in the northern Rockall Trough. *Marine Geology*, **192**, 215–237.

Masson, D.G., Herbitz, C.B., Wynn, R.B., Pedersen, G. & Løvholt, F. 2006. Submarine landslides: processes, triggers and hazard prediction. *Philosophical Transactions of the Royal Society, London*, **364**, 2009–2039.

Masson, D.G., Arzola, R.G., Wynn, R.B., Hunt, J.E. & Weaver, P.P.E. 2011a. Seismic triggering of landslides and turbidity currents offshore Portugal. *Geochemistry Geophysics Geosystems*, **12**, 1–19. doi:10.1029/2011GC003839.

Masson, D.G., Huvenne, V.A.I., de Stigter, H.C., Arzola, R.G. & LeBas, T.P. 2011b. Sedimentary processes in the middle Nazaré Canyon. *Deep Sea Research Part II: Topical Studies in Oceanography*, **58**, 2,369–2,387.

Mastbergen, D.R. & Van Den Berg, J.H. 2003. Breaching in fine sands and the generation of sustained turbidity currents in submarine canyons. *Sedimentology*, **50**, 625–638.

Mattern, F. 2005. Ancient sand-rich submarine fans: depositional systems, models, identification, and analysis. *Earth-Science Reviews*, **70**, 167–202.

Mattern, F. & Wang, P. 2008. Out-of-sequence thrusts and paleogeography of the Rhenodanubian Flysch Belt (Eastern Alps) revisited. *International Journal of Earth Sciences*, **97**, 821–833.

Maurer, F., Reijmer, J.J.G. & Schlager, W. 2001. Quantification of input and compositional variations of calciturbidites in a Middle Triassic basinal succession (Seceda, Dolomites, Southern Alps). *International Journal of Earth Sciences* (Geologisches Rundschau), **92**, 593–609.

Maurin, J.C., Benkhelil, J. & Robineau, R. 1986. Fault rocks of the Kaltunga lineament, NE Nigeria, and their relationship with Benue Trough tectonics. *Journal of the Geological Society* (London), **143**, 587–599.

May, J.A. & Warme, J.E. 2000. Bounding surfaces, lithologic variability, and sandstone connectivity within submarine-canyon outcrops, Eocene of San Diego, California. *In*: Weimer, P., Slatt, R.M., Coleman, J., Rosen, N.C., Nelson, H., Bouma, A.H., Styzen, M.J. & Lawrence, D.T. (eds), *Gulf Coast Section-Society of Economic Paleontologists and Mineralogists Foundation 20th Annual Bob F. Hoskins Research Research Conference, Deep-Water Reservoirs of the World*, 556–577. CD-ROM Society of Economic Paleontologists and Mineralogists Special Publications.

Mayall, M., Jones, E. & Casey, M. 2006. Turbidite channel reservoirs - key elements in facies prediction and effective development. *Marine and Petroleum Geology*, **23**, 821–841.

Maynard, J.B., Valloni, R. & Yu, H.-S. 1982. Composition of modern deep-sea sands from arc-related basins. *In*: Leggett, J.K. (ed.), *Trench-Forearc Geology*, 551–561. The Geological Society of London, Special Publication, **10**. London: The Geological Society.

McAdoo, B.G., Orange, D.L., Screaton, E., Lee, H. & Kayen, R. 1997. Slope basins, headless canyons, and submarine palaeoseismology of the Cascadia accretionary complex. *Basin Research*, **9**, 313–324.

McArthur, J.M. 2007. Discussion: Comment on "Carbon-isotope record of the Early Jurassic (Toarcian) Oceanic Anoxic Event from fossil wood and marine carbonate (Lusitanian Basin, Portugal)" by Hesselbo S., Jenkyns H.C., Duarte L.V. & Oliveira L.C.V. *Earth and Planetary Science Letters*, **259**, 634–639.

McBride, E.F. 1960. *Martinsburg flysch of the central Appalachians*. PhD Thesis, Johns Hopkins University, Baltimore, Maryland.

McBride, E.F. 1962. Flysch and associated beds of the Martinsburg Formation (Ordovician), Central Appalachians. *Journal of Sedimentary Petrology*, **32**, 39–91.

McCabe, P.J. 1978. The Kinderscoutian Delta (Carboniferous) of northern England; a slope influenced by density currents. *In*: Stanley, D.J. & Kelling, G. (eds), *Sedimentation in Submarine Canyons, Fans, and Trenches*, 116–126. Stroudsburg Pennsylvania: Dowden, Hutchinson & Ross.

McCaffrey, R. 1992. Oblique plate convergence, slip vectors, and forearc deformation. *Journal of Geophysical Research*, **97**, 8905–8915.

McCaffrey, R. 1993. On the role of the upper plate in great subduction zone earthquakes. *Journal of Geophysical Research*, **98**, 11953–11966.

McCaffrey, R. 2009. The tectonic framework of the Sumatran subduction zone. *The Annual Review of Earth and Planetary Sciences*, **37**, 345–366.

McCaffrey, R. & Abers, G.A. 1991. Orogeny in arc-continent collision: The Banda arc and western New Guinea *Geology*, **19**, 563–566.

McCaffrey, R. & Goldfinger, C. 1995. Forearc deformation and great earthquakes: Implications for Cascadia earthquake potential. *Science*, **267**, 856–859.

McCaffrey, W., Kneller, B. & Peakall, J. (eds) 2001. *Particulate Gravity Currents*. International Association of Sedimentologists, Special Publication, **31**. Oxford: Blackwell Scientific.

McCaffrey, W.D., Gupta, S. & Brunt, R. 2002. Repeated cycles of submarine channel incision, infill and transition to sheet sandstone development in the Alpine Foreland Basin, SE France. *Sedimentology*, **49**, 623–635.

McCaffrey, W.D. & Kneller, B.C. 2004. Scale effects of non-uniformity on deposition from turbidity currents with reference to the Grès d'Annot of SE France. *In*: Joseph, P. & Lomas, S.A. (eds), *Deep-Water Sedimentation in the Alpine Basin of SE France: New Perspectives on the Grès d'Annot and Related Systems*. Geological Society, London, Special Publication, **221**, 301–310. London: The Geological Society.

McCann, T. & Pickerill, R.K. 1988. Flysch trace fossils from the Cretaceous Kodiak Formation of Alaska. *Journal of Paleontology*, **62**, 330–348.

McCave, I.N. 1972. Transport and escape of fine-grained sediment from shelf areas. *In*: Swift, D.J.P., Duane, D.B. & Pilkey, O.H. (eds), *Shelf Sediment Transport: Process and Pattern*, 225–248. Stroudsburg, PA: Hutchinson & Ross.

McCave, I.N. 1982. Erosion and deposition by currents on submarine slopes. *Bulletin d'Institut de Geologie du Bassin d'Aquitaine*, **31**, 47–55.

McCave, 1.N. 1984. Erosion, transport and deposition of fine-grained marine sediments. *In*: Stow, D.A.V. & Piper, D.J.W. (eds), *Fine-grained Sediments: Deep-water Processes and Facies*, 35–69. Special Publication of The Geological Society London, **15**. Oxford: Blackwell Scientific Publications.

McCave, I.N. 2008. Size sorting during transport and deposition of fine sediments: sortable silt and flow speed. *In*: Rebesco, M. & Camerlenghi, A. (eds), *Contourites,* 121–142. Developments in Sedimentology, **60**. Amsterdam: Elsevier.

McCave, I.N. & Hollister, C.D. 1985. Sedimentation under deep-sea current systems: pre-HEBBLE ideas. *Marine Geology*, **66**, 13–24.

McCave, I.N. & Jones, P.N. 1988. Deposition of ungraded muds from high-density non-turbulent turbidity currents. *Nature*, **333**, 250–252.

McCave, I.N. & Tucholke, B.E. 1986. Deep current-controlled sedimentation in the western North Atlantic. *In*: Vogt, P.R. & Tucholke, B.E. (eds), *The Geology of North America. Volume M, the Western North Atlantic Region*, 451–68. Boulder, Colorado: Geological Society of America.

McCave, I.N., Lonsdale, P.F., Hollister, C.D. & Gardner, W.D. 1980. Sediment transport over the Hatton and Gardar contourite drifts. *Journal of Sedimentary Research*, **50**, 1049–1062.

McCave, I.N., Manighetti, B. & Robinson, S.G. 1995. Sortable silt and fine sediment size/composition slicing: parameters for paleocurrent speed and paleoceanography. *Paleoceanography*, **10**, 593–610.

McCave, I.N., Chandler, R.C., Swift, S.A. & Tucholke, B.E. 2002. Contourites of the Nova Scotia continental rise and the HEBBLE area. *In*: Stow, D.A.V., Pudsey, C.J., Howe, J.A., Faugères, J.-C. & Viana,

A.R. (eds), *Deep-water Contourite Systems: Modern Drifts and Ancient Series, Seismic and Sedimentary Characteristics*, 21–38. Geological Society London Memoir, **22**.

McClay, K. & Bonora, M. 2001. Analog models of restraining stopovers in strike-slip fault systems. *American Association of Petroleum Geologists Bulletin*, **85**, 233–260.

McGrail, D.W. & Carnes, M. 1983. Shelf-edge dynamics and the nepheloid layer in the northwestern Gulf of Mexico. *In*: Stanley, D.J. & Moore, G.T. (eds), *The Shelfbreak: Critical Interface on Continental Margins*, 251–264. Society of Economic Paleontologists and Mineralogists, Special Publication, 33.

McClay, K.R. & White, M.J. 1995. Analogue modelling of orthogonal and oblique rifting. *Marine and Petroleum Geology*, **12**, 137–151.

McGregor, B.A., Stubblefield, W.L., Ryan, W.B.F. & Twichell, D.C. 1982. Wilmington submarine canyon: a marine fluvial-like system. *Geology*, **10**, 27–30.

McHargue, T., Pyrcz, M.J., Sullivan, M.D., Clark, J., Fildani, A., Romans, B., Covault, J., Levy, M., Posamentier, H. & Drinkwater, N. 2011. Architecture of turbidite channel systems on the continental slope: patterns and predictions. *Marine and Petroleum Geology*, **28**, 728–743.

McHargue, T.R. 1991. Seismic facies, processes and evolution of Miocene inner fan channels, Indus Submarine Fan. *In*: Weimer, P. & Link, M.H. (eds), *Seismic Facies and Sedimentary Processes of Submarine Fans and Turbidite Systems*, 403–414. New York: Springer.

McHugh, C.M.G. & Ryan, W.B.F. 2000. Sedimentary features associated with channel overbank flow: examples from the Monterey Fan. *Marine Geology*, **163**, 199–215.

McIlroy, D. 2004. Some ichnological concepts, methodologies, applications and frontiers. *In*: McIlroy, D. (ed.), *The Application of Ichnology to Palaeoenvironmental and Stratigraphic Analysis*, 3–27. Geological Society, London, Special Publications, **228**. London: The Geological Society.

McIlroy, D. 2007. Lateral variability in shallow marine ichnofabrics: implications for the ichnofabric analysis method. *Journal of the Geological Society, London*, **164**, 359–369.

McIlroy, D. 2008. Ichnological analysis: The common ground between ichnofacies workers and ichnofabric analysts. *Palaeogeography, Palaeoclimatology, Palaeoecology*, **270**, 332–338.

McKelvey, B.C., Chen, W. & Arculus, R.J. 1995. Provenance of Pliocene–Pleistocene ice-rafted debris, Leg 145, Northern Pacific Ocean. *In*: Rea, D.K., Basov, I.A., Scholl, D.W. & Allan, J.F. (eds), *Proceedings of the Ocean Drilling Program, Scientific Results*, **145**, 195–204. College Station, Texas, USA: Ocean Drilling Program.

McKenzie, D.P. 1978. Some remarks on the development of sedimentary basins. *Earth and Planetary Science Letters*, **40**, 25–32.

McKerrow, W.S., Lambert, R. St-J. & Cocks, L.R.M. 1985. The Ordovician, Silurian and Devonian Periods. *In*: Snelling, N.J. (ed.), *The Chronology of the Geological Record*, 73–80. Geological Society, London, Memoir, **10**.

McLean, H. & Howell, D.G. 1984. Miocene Blanca Fan, northern Channel Islands, California: small fans reflecting tectonism and volcanism. *Geo-Marine Letters*, **3**, 161–166.

McLennan, S.M., Taylor, S.R., McCulloch, M.T. & Maynard, J.B. 1990. Geochemical and Nd-Sr isotopic composition of deep-sea turbidites: crustal evolution and plate tectonic associations. *Geochimica Cosmochimica Acta*, **54**, 2015–2050.

Macpherson, B.A. 1978. Sedimentation and trapping mechanism in upper Miocene Stevens and older turbidite fans of south-eastern San Joaquin Valley, California. *American Association of Petroleum Geologists Bulletin*, **62**, 2243–2274.

McQuillin, R., Bacon, M. & Barclay, W. 1984. *An Introduction to Seismic Interpretation*. Houston: Gulf Publishing Company.

Meacham, I. 1968. Correlation in sequential data—three simple indicators. *Civil Engineering Transactions of the Institution of Engineers (Australia)*, **CE10**, 225–228.

Meadows, A., Meadows, P.S., Wood, D.M. & Murray, J.M.H. 1994. Microbiological effects on slope stability: an experimental analysis. *Sedimentology*, **41**, 423–435.

Melick, J., Cavanna, G., Benevelli, G., Tinterri, R. & Mutti, E. 2004. The Lutetian Ainsa sequence: an example of a small turbidite system deposited in a tectonically controlled basin. *Search and Discovery Article* #**50008**

Mellere, D. Plink-Bjorklund P. & Steel, R. 2002. Anatomy of shelf deltas at the edge of a prograding Eocene shelf margin, Spitsbergen. *Sedimentology*, **49**, 1181–1206.

Menard, H.W. 1955. Deep-sea channels, topography, and sedimentation. *American Association of Petroleum Geologists Bulletin*, **39**, 236–255.

Meneghini, G. 1850. Paleodictyon. *In*: Savi, P. & Meneghini, G. (eds). *Osservazioni stratigrafiche e paleontologicke concernati la geologie della Toscana e dei paesi limitrofi*. Appendix to R.R. Murchinson, *Memoria sulla struttura geologie delle Alpi*, Firenze. pp. 246.

Mensing, S.A., Benson, L.V., Kashgarian, M. & Lund, S. 2004. A Holocene pollen record of persistent droughts from Pyramid Lake, Nevada, USA. *Quaternary Research*, **62**, 29–38.

Métivier, F. 1999. Diffusive-like buffering and saturation of large rivers. *Physical Review E, Statistical Physics, Plasmas, Fluids, and Related Interdisciplinary Topics*, **60**, 5827– 5832.

Métivier, F. & Gaudemer, Y. 1999. Stability of output fluxes of large rivers in South and East Asia during the last 2 million years: implications on floodplain processes. *Basin Research*, **11**, 293– 303.

Meyer, L. 2004. *Internal architecture of an ancient deep-water, passive margin, basin-floor fan system, upper Kaza Group, Windermere Supergroup, Castle Creek, British Columbia [unpublished MS thesis]*. University of Calgary, Alberta, Canada, 175 pp.

Meyers, P.A., Dunham, K.W. & Ho, E.S. 1987. Organic geochemistry of Cretaceous black shales from the Galicia Margin, Ocean Drilling Program Leg 103. *Advances in Organic Geochemistry*, **13**, 89–96.

Miall, A.D. 1985. Architectural-element analysis: a new method of facies analysis applied to fluvial deposits. *Earth-Science Reviews*, **22**, 261–308.

Miall, A.D. 1986. Eustatic sea level changes interpreted from seismic stratigraphy: a critique of the methodology with particular reference to the North Sea Jurassic record. *American Association of Petroleum Geologists Bulletin*, **70**, 131–137.

Miall, A.D. 1989. Architectural elements and bounding surfaces in channelized clastic deposits: notes on comparisons between fluvial and turbidite systems. *In*: Taira, A. & Masuda, F. (eds), *Sedimentary Facies in the Active Plate Margin*, 3–16. Tokyo: Terra Scientific Publishing.

Miall, A.D. 1992a. Alluvial deposits. *In*: Walker, R.G. & James, N.P. (eds), *Facies Models: Response to Sea-Level Change. GeoText*, **1**, 119–143. Geological Association of Canada, St. John's, Newfoundland.

Miall, A.D. 1992b. The Exxon global cycle chart: An event for every occasion? *Geology*, **20**, 787–780.

Michelson, J.E. 1976. Miocene deltaic oil habitat, *Trinidad. Bull. American Association of Petroleum Geologists Bulletin*, **60**, 1502–1519.

Middleton, G.V. 1965. Antidune cross-bedding in a large flume. *Journal of Sedimentary Petrology*, **35**, 922–927.

Middleton, G.V. 1966a. Experiments on density and turbidity currents: I. Motion of the head. *Canadian Journal of Earth Sciences*, **3**, 523–546.

Middleton, G.V. 1966b. Experiments on density and turbidity currents: II Uniform flow of density currents.. *Canadian Journal of Earth Sciences*, **3**, 627–637.

Middleton, G.V. 1966c. Small scale models of turbidity currents and the criterion for auto-suspension. *Journal of Sedimentary Petrology*, **36**, 202–208.

Middleton, G.V. 1967. Experiments on density and turbidity currents: III Deposition of sediment. *Canadian Journal of Earth Sciences*, **4**, 475–505.

Middleton, G.V. 1970. Experimental studies related to problems of flysch sedimentation. *In*: Lajoie, J. (ed.), *Flysch sedimentology in North America*, 253–272. Geological Association of Canada Special Paper, **7**, 405–26.

Middleton, G.V. 1976. Hydraulic interpretation of sand size distributions. *Journal of Geology*, **84**, 405–426.

Middleton, G.V. 1993. Sediment deposition from turbidity currents. *Annual Reviews in Earth and Planetary Sciences*, **21**, 89–114.

Middleton, G.V. & Hampton, M.A. 1973. Sediment gravity flows: mechanics of flow and deposition. *In*: Middleton, G.V. & Bouma, A.H. (eds), *Turbidites and Deep Water Sedimentation*, 1–38. Short course notes, Pacific Section of The Society of Economc Paleontologists and Mineralogists.

Middleton, G.V. & Hampton, M.A. 1976. Subaqueous sediment transport and deposition by sediment gravity flows. *In*: Stanley, D.J. & Swift, D.J.W. (eds), *Marine Sediment Transport and Environmental Management*, 197–218. New York: Wiley.

Middleton, G.V. & Neal, W.J. 1989. Experiments on the thickness of beds deposited by turbidity currents. *Journal of Sedimentary Petrology*, **59**, 297–307.

Middleton, G.V. & Southard, J.B. 1984. *Mechanics of Sediment Transport*, 2nd edn. Society of Economic Paleontologists and Mineralogists Eastern Section Short Course No. 3, Providence.

Middleton, G.V. & Wilcock, P.R. 1994. *Mechanics in the Earth and Environmental Sciences.* Cambridge: Cambridge University Press, 459 pp. ISBN: 0-521-44124–2

Migeon, S., Savoye, B., Faugères, J.-C. 2000. Quaternary development of migrating sediment waves in the Var deep-sea fan: distribution, growth pattern, and implication for levée evolution. *Sedimentary Geology*, **133**, 265–293.

Migeon, S., Savoye, B. Zanella, E. Mulder, T., Faugères, J.-C. Weber, O. 2001. Detailed seismic-reflection and sedimentary study of turbidite sediment waves on the Var Sedimentary Ridge (SE France): significance for sediment transport and deposition and for the mechanisms of sediment-wave construction. *Marine and Petroleum Geology*, **18**, 179–208.

Migeon, S., Ducassou, E., Le Gonidec, Y., Rouillard, P., Mascle, J. & Revel-Rolland, M. 2010. Lobe construction and sand/mud segregation by turbidity currents and debris flows on the western Nile deep-sea fan (Eastern Mediterranean). *Sedimentary Geology*, **229**, 124–143.

Miki, M., Matsuda, T. & Otofuji, Y. 1990. Opening mode of the Okinawa Trough: paleomagnetic evidence from the South Ryukyu Arc. *Tectonophysics*, **175**, 335–347.

Miller, K.G., Fairbanks, R.G. & Mountain, G.S. 1987. Tertiary oxygen isotope synthesis, sea level history, and continental margin erosion. *Paleoceanography*, **2**, 1–19.

Miller, K.G., Wright, J.D. & R.G. Fairbanks, R.G. 1991. Unlocking the ice house: Oligocene–Miocene oxygen isotopes, eustasy, and margin erosion. *Journal of Geophysical Research*, **96**, 6829–6848.

Miller, K.G., Mountain, G.S., the Leg 150 Shipboard Party, and Members of the New Jersey Coastal Plain Drilling Project. 1996. Drilling and dating New Jersey Oligocene–Miocene sequences: ice volume, global sea level, and Exxon records. *Science*, **271**, 1092–1094.

Miller, K.G., Mountain, G.S., Browning, J.V., Kominz, M., Sugarman, P.J., Christie-Blick, N., Katz, M.E. & Wright, J.D. 1998. Cenozoic global sea-level, sequences, and the New Jersey transect: results from coastal plain and slope drilling. *Reviews of Geophysics*, **36**, 569–601.

Miller, K.G., Sugarman, P.H., Browning, J.V., Kominz, M.A., Hernàndez, J.S., Olsson, R.K., Wright, J.D., Feigenson, M.D. & van Sickel, W. 2003. Late Cretaceous chronology of large, rapid sea-level changes: Glacioeustasy during the greenhouse world. *Geology*, **31**, 585–588.

Miller, K.G., Kominz, M.A., Browning, J.V., Wright, J.D., Mountain, G.S., Katz, M.E., Sugarman, P.J., Cramer, B.S., Christie-Blick, N. & Pekar, S.F. 2005a. The Phanerozoic record of global sea-level change. *Science*, **310**, 1,293–1,298.

Miller, K.G., Wright, J.D. & Browning, J.V. 2005b. Visions of ice sheets in a greenhouse world. *Marine Geology*, **217**, 215–231.

Miller, K.G., Mountain, G.S., Wright, J.D. & Browning, J.V. 2011. A 180-million-year record of sea level and ice volume variations from continental margin and deep-sea isotopic records. *Oceanography*, **24**, 40–53. doi:10.5670/oceanog.2011.26.

Miller, M.F. & Smail, S.E. 1997. A semiquantitative field method for evaluating bioturbation on bedding planes. *Palaios*, **12**, 391–396.

Miller, W. III, 1991. Paleoecology of graphoglyptids. *Ichnos*, **1**, 305–312.

Miller, W. III, (ed.) 2007. *Trace Fossils Concepts, Problems, Prospect,* 611 pp. Amsterdam: Elsevier B.V. ISBN: 978-0-444-52949-7.

Milliken, K.L., Comer, E.J. & Marsaglia, K.M. 2012. Modal sand composition at Sites C0004, C0006, C0007, and C0008, IODP Expedition 316, Nankai accretionary prism. *In*: Kinoshita, M., Tobin, H., Ashi, J., Kimura, G., Lallemant, S., Screaton, E.J., Curewitz, D., Masago, H., Moe, K.T. and the Expedition 314/315/316 Scientists. *Proceedings of the Integrated Ocean Drilling Program,* **314/315/316**. 1-17. Washington, DC (Integrated Ocean Drilling Program Management International, Inc.).

Millington, J. & Clark, J.D. 1995a. Submarine canyon and associated base-of-slope sheet system: the Eocene Charo-Arro system, south-central Pyrenees. *In*: Pickering, K.T., Hiscott, R.N., Kenyon, N.H., Ricci Lucchi, F. & Smith, R.D.A. (eds), *Atlas of Deep Water Environments: Architectural Style in Turbidite Systems,* 150–156. London: Chapman and Hall.

Millington J. & Clark J.D. 1995b. The Charo/Arro canyon-mouth sheet system, south-central Pyrenees, Spain: a structurally influenced zone of sediment dispersal. *Journal of Sedimentary Research*, **65**, 443–454.

Minisini, D. & Schwartz, H. 2007. An early Paleocene cold seep system in the Panoche and Tumey Hills, Central California, USA. *In*: Hurst, A. & Cartwright, J. (eds), *Sand Injectites: Implications for Hydrocarbon Exploration and Production,* 185–197. American Association of Petroleum Geologists Memoir, **87**.

Mitchell, N.C. 2006. Morphologies of knickpoints in submarine canyons. *Geological Society of America Bulletin*, **118**, 589–605.

Mitchum, R.M. Jr. 1977. Seismic Stratigraphy and global changes of sea level, part II: Glossary of terms used in seismic stratigraphy. *In*: Payton, C.E. (ed.), *Seismic Stratigraphy - Applications to Hydrocarbon Exploration,* 205–212. American Association of Petroleum Geologists, Memoir **26**.

Mitchum, R.M. Jr. 1985. Seismic stratigraphic expression of submarine fans. *In*: Berg, O.R. & Woolverton, D.G. (eds), *Seismic Stratigraphy II: An Integrated Approach to Hydrocarbon Exploration.* American Association of Petroleum Geologists, Memoir **39**, Tulsa, Oklahoma.

Mitchum, R.M.Jr. & Uliana, M.A. 1985. Seismic stratigraphy of carbonate depositional sequences, Upper Jurassic–Lower Cretaceous, Neuquen Basin, Argentina. *In*: Berg, O.R. & Woolverton, D.G. (eds), *Seismic Stratigraphy II: An Integrated Approach to Hydrocarbon Exploration,* 255–274. American Association of Petroleum Geologists Memoir **39**.

Mitchum, R.M. Jr., Vail, P.R. & Thompson III, S.1977a. Seismic stratigraphy and global changes of sea level, part 2: the depositional sequence as a basic unit for stratigraphic analysis. *In*: Payton, C.E. (ed.), *Seismic Stratigraphy - Applications to Hydrocarbon Exploration,* 53–62. American Association of Petroleum Geologists, Memoir **26**.

Mitchum, R.M. Jr., Vail, P.R. & Sangree, J.B.1977b. Seismic stratigraphy and global changes of sea levels, part 6: stratigraphic interpretation of seismic reflection patterns in depositional sequences. *In*: Payton, C.E. (ed.), *Seismic Stratigraphy - Applications to Hydrocarbon Exploration,* 117–133. American Association of Petroleum Geologists, Memoir **26**.

Mizutani, S. & Hattori, I. 1972. Stochastic analysis of bed-thickness distribution of sediments. *Mathematical Geology*, **4**, 123–146.

Möbius, J., Lahajnar, N. & Emeis, K.-C. 2010. Diagenetic control of nitrogen isotope ratios in Holocene sapropels and recent sediments from the eastern Mediterranean Sea. *Biogeosciences*, **7**, 3901–3914.

Mogi, A. & Nishizawa, K. 1980. Breakdown of a seamount on the slope of the Japan trench. *Proceedings of the Japanese Academy*, **56**, 257–259. doi: 10.2183/pjab.56.257.

Mogi, K. 1990. Seismicity before and after large shallow earthquakes around the Japanese islands. *Tectonophysics*, **175**, 1–34.

Mohn, K. & Bowen, B. 2012. Florida – the next US frontier: revisiting an old exploration region of the Gulf of Mexico. *GEO ExPro*, **9**, 74–78.

Mohn, K. & Bowen, B.E. 2011. Offshore Florida: regional perspective. *GEO ExPro*, **8**, 58–62.

Mohrig, D., Whipple, K.X. Hondzo, M., Ellis, C. & Parker, G. 1998. Hydroplaning of subaqueous debris flows. *Geological Society of America Bulletin*, **110**, 387–394.

Monaco, P. 2008. Taphonomic features of Paleodictyon and other graphoglyptid trace fossils in Oligo-Miocene thin-bedded turbidites, Northern Apennines, Italy. *Palaios*, **23**, 667–682.

Monaco, P., Milighetti, M. & Checconi, A. 2010. Ichnocoenoses in the Oligocene to Miocene foredeep basins (Northern Apennines, central Italy) and their relation to turbidite deposition. *Acta Geologica Polonica*, **60**, 53–70.

Moore, D.G. 1961. Submarine slumps. *Journal of Sedimentary Petrology*, **31**, 343–357.

Moore, D.G. 1965. Erosional channel wall in La Jolla sea-fan valley seen from bathyscope Trieste II. *Geological Society of America Bulletin*, **76**, 385–392.

Moore, G.H. & Wallis, W.A. 1943. Time series tests based on signs-of-differences. *American Statistical Association Journal*, **38**, 153–164.

Moore, G.F. 1979. Petrography of subduction zone sandstones from Nias Island, Indonesia. *Journal of Sedimentary Research*, **49**, 71–84.

Moore, G.F., Curray, J.R. & Emmel, F.J. 1982. Sedimentation in the Sunda Trench and forearc region. *In*: Leggett, J.K. (ed.), *Trench-Forearc Geology,* 245–258. The Geological Society of London, Special Publication, **10**. London: The Geological Society.

Moore, G.F., Taira, A., Kuramoto, S., Shipley, T.H. & Bangs, N.L. 1999. Structural setting of the 1999 U.S.-Japan Nankai Trough 3-D seismic reflection survey. *EOS*, **80**, F569.

Moore, G.F., Taira, A., Klaus, A. *et al.* 2001. *Proceedings of the Ocean Drilling Program, Initial Reports,* **190**: College Station, TX: Ocean Drilling Program.

Moore, G.F., Taira, A., Klaus, A., Becker, L., Boeckel, B., Cragg, B.A., Dean, A., Fergusson, C.L., Henry, P., Hirano, S., Hisamitsu, T., Hunze, S., Kastner, M., Maltman, A.J., Morgan, J.K., Murakami, Y., Saffer, D.M., Sánchez-Gómez, M., Screaton, E.J., Smith, D.C., Spivack, A.J., Steurer, J., Tobin, H.J., Ujiie, K., Underwood, M.B. & Wilson, M. 2001. New insights into deformation and fluid flow processes in the Nankai Trough accretionary prism: results of Ocean Drilling Program Leg 190. *Geochemistry, Geophysics & Geosystems*, **2**, 1058.

Moore, G.F., Bangs, N.L., Taira, A., Kuramoto, S., Pangborn, E. & Tobin, H.J. 2007. Three-dimensional splay fault geometry and implications for tsunami generation. *Science*, **318**, 1128–1131.

Moore, G.F., Park, J.-O., Bangs, N.L., Gulick, S.P., Tobin, H.J., Nakamura, Y., Sato, S., Tsuji, T., Yoro, T., Tanaka, H., Uraki, S., Kido, Y., Sanada, Y., Kuramoto, S. & Taira, A. 2009. Structural and seismic stratigraphic framework of the NanTroSEIZE Stage 1 transect, in Proceedings of the Integrated Ocean Drilling Program, 314/315/316. *In*: Kinoshita, M., Tobin, H., Ashi, J., Kimura, G., Lallemant, S., Screaton, E.J., Curewitz, D., Masago, H., Moe, K.T. & Expedition 314/315/316 Scientists, *Proceedings of the Integrated Ocean Drilling Program,* **314/315/316**, 1–46. Washington, DC: Integrated Ocean Drilling Program Management International Incorporated.

Moore, G.F., Kanagawa, K., Strasser, M., Dugan, B., Maeda, L., Toczko, S. & the Expedition 338 Scientists 2013a. NanTroSEIZE Stage 3: NanTroSEIZE plate boundary deep riser 2. Integrated Ocean Drilling Program Preliminary Report, 338. Washington, DC: Integrated Ocean Drilling Program Management International Incorporated. doi:10.2204/iodp.pr.338.2013.

Moore, G.F., Boston, B.B., Sacks, A.F. & Saffer, D.M. 2013b. Analysis of normal fault populations in the Kumano Forearc Basin, Nankai Trough, Japan: 1. Multiple orientations and generations of faults from 3-D coherency mapping. *Geochemistry, Geophysics, Geosystems*, **114**, 1989–2002.

Moore, G.F., Kanagawa, K., Strasser, M., Dugan, B., Maeda, L., Toczko, S. & the IODP Expedition 338 Scientific Party 2014. IODP Expedition

338: NanTroSEIZE Stage 3: NanTroSEIZE plate boundary deep riser 2. *Scientific Drilling*, **17**, 1–12.

Moore, J.C. 1974. Turbidites and terrigenous muds, DSDP Leg 25. *In*: Simpson, E.S.W., Schlich, R. *et al.*, *Initial Reports Deep Sea Drilling Project*, **25**, 441–479. Washington, DC: US Government Printing Office.

Moore, J.C. (ed.) 1986. *Structural Fabric in Deep Sea Drilling Project Cores from Forearcs*. Geological Society of America Memoir, **166**.

Moore, J.C. 2000. Synthesis of results: logging while drilling, northern Barbados accretionary prism. *In*: Moore, J.C., Klaus, A. *et al.*, *Proceedings of the Ocean Drilling Program, Scientific Results*, **171A**, 1–25. College Station, Texas, USA: Ocean Drilling Program.

Moore, J.C. & Saffer, D.M. 2001. Updip limit of the seismogenic zone beneath the accretionary prism of southwest Japan: An effect of diagenetic to low-grade metamorphic processes and increasing effective stress. *Geology*, **29**, 183–186.

Moore, J.C., Klaus, A. *et al.* 2000. *Proceedings of the Ocean Drilling Program, Scientific Results*, **171A**. College Station, Texas, USA: Ocean Drilling Program.

Moore, J.C., Byrne, T., Plumley, P.W., Reid, M., Gibbons, H. & Coe, R.S. 1983. Paleogene evolution of the Kodiak Islands, Alaska: consequences of ridge-trench interaction in a more southerly latitude. *Tectonics*, **2**, 265–293.

Moore, J.C, Mascle, A. *et al.* 1990. *Proceedings of the Ocean Drilling Program, Scientific Results*, **110**. College Station, Texas, USA: Ocean Drilling Program.

Moore, J.G., Clague, D.A., Holcomb, R.T., Lipman, P.W., Normark, W.R. & Torresan, E. 1989. Prodigious submarine landslides on the Hawaiian Ridge. *Journal of Geophysical Research*, **94**, 17,465–17,484.

Moraes, M.A.S., Blaskovski, P.R. & Joseph, P. 2004. The Grès d'Annot as an analogue for Brazilian Cretaceous sandstone reservoirs: comparing convergent to passive-margin confined turbidites. *In*: Joseph, P. & Lomas, S.A. (eds), *Deep-Water Sedimentation in the Alpine Basin of SE France: New Perspectives on the Grès d'Annot and Related Systems*. Geological Society, London, Special Publication, **221**, 419–436. London: The Geological Society.

Morgan, S.R. & Campion, K.M. 1987. Eustatic controls on stratification and facies associations in deep-water deposits, Great Valley Sequence, Sacramento Valley, California, abstract. *American Association of Petroleum Geologists Bulletin*, **71**, 595.

Morgenstern, N.R. 1967. Submarine slumping and the initiation of turbidity currents. *In*: Richards, A.F. (ed.), *Marine Geotechnique*, 189–220. Urbana: Illinois University Press.

Morley, C.K. 2002. Evolution of Large Normal Faults: Evidence from Seismic Reflection Data. *American Association of Petroleum Geologists Bulletin*, **86**, 961–978.

Morris, R.C. 1971. Classification and interpretation of disturbed bedding types in the Jackfork flysch rocks (Upper Mississippian), Ouachita Mountains, Arkansas. *Journal of Sedimentary Petrology*, **41**, 410–424.

Morris, S.A. & Alexander, J. 2003. Changes in flow direction at a point caused by obstacles during passage of a density current. *Journal of Sedimentary Research*, **73**, 621–629.

Morris, W.R. & Normark, W.R. 2000. Sedimentologic and geometric criteria for comparing modern and ancient sandy turbidite elements. *In*: Weimer, P., Slatt, R.M., Coleman, J., Rosen, N.C., Nelson, H., Bouma, A.H., Styzen, M.J. & Lawrence, D.T. (eds), *Deep-water Reservoirs of the World*, 606–628. Houston, Texas: Gulf Coast Section, Society of Economic Paleontologists and Mineralogists.

Morrison, K. 2011. Unlocking the exploration potential of the Seychelles. *GEO ExPro*, **8**, 58–62.

Mortimer, N. 2004. New Zealand's geological foundations. *Gondwana Research*, **7**, 261–272.

Moscardelli, M. & Wood, L. 2007. Newclassification systemformass transport complexes in offshore Trinidad. *Basin Research*, doi: 10.1111/j.1365-2117.2007.00340.x.

Moscardelli, M., Wood, L. & Mann, P. 2006. Mass-transport complexes and associated processes in the offshore area of Trinidad and

Venezuela. *American Association of Petroleum Geologists Bulletin*, **90**, 1059–1088.

Mosher, D.C., Erbacher, J. & Malone, M.J. (eds) 2007. *Proceedings of the Ocean Drilling Program, Scientific Results*, **207**, 1–26. College Station, Texas, USA (Ocean Drilling Program). College Station, Texas, USA: Ocean Drilling Program.

Mosher, D.C., Moscardelli, L., Shipp, C., Chaytor, J.D., Baxter, C.D.P., Lee, H.J. & Urgeles, R. (eds) 2010. *Submarine Mass Movements and their Consequences*. Advances in Natural and Technological Hazards Research, **28**, 4th International Symposium Series. The Netherlands: Springer, 220 pp. ISBN: 978-90-481-3030-2.

Mouterde, R. 1955. Le Lias de Peniche. *Comunicações Serviços Geológicos de Portugal*, **36**, 87–115.

Mount, J.F. 1993. Formation of fluidization pipes during liquefaction: examples from the Uratanna Formation (Lower Cambrian), South Australia. *Sedimentology*, **40**, 1027–1037.

Mountjoy, J.J., Barnes, P.M. & Pettinga, J.R. 2009. Morphostructure and evolution of submarine canyons across an active margin: Cook Strait sector of the Hikurangi Margin, New Zealand. *Marine Geology*, doi:10.1016/j.margeo.2009.01.006.

Muck, M.T. & Underwood, M.B. 1990. Upslope flow of turbidity currents: a comparison among field observations, theory, and laboratory models. *Geology*, **18**, 54–57.

Muehlberger, W.R. 1996. *Tectonic map of North America*. American Association of Petroleum Geologists, 4 pp., scale 1:5,000,000.

Mueller, C. 1994. Northridge, California, earthquake of January 17, 1994: Ground motion. *Earthquakes and Volcanoes*, **25**, 75–84.

Mulder, T. & Alexander, J. 2001. The physical character of subaqueous sedimentary density flows and their deposits. *Sedimentology*, **48**, 269–299.

Mulder, T. & Cochonat, P. 1996. Classification of offshore mass movements. *Journal of Sedimentary Research*, **66**, 43–57.

Mulder, T. & Syvitski, J.P.M. 1995. Turbidity currents generated at river mouths during exceptional discharge to the world oceans. *Journal of Geology*, **103**, 285–298.

Mulder, T., Savoye, B. & Syvitski, J.P.M. 1997. Numerical modelling of a mid-sized gravity flow: the 1979 Nice turbidity current (dynamics, processes, sediment budget and seafloor impact). *Sedimentology*, **44**, 305–326.

Mulder, T., Migeon, S., Savoye, B. & Faugères, J.-C. 2001. Inversely graded turbidite sequences in the deep Mediterranean: a record of deposits from flood-generated turbidity currents? *Geo-Marine Letters*, **21**, 86–93.

Mulder, T., Syvitski, J.P.M., Migeon, S. Faugères, J.-C. & Savoye, B. 2003. Marine hyperpycnal flows: initiation, behavior and related deposits. A review. *Marine and Petroleum Geology*, **20**, 861–882.

Mulder, T., Callec, Y., Parize, O., Joseph, P., Schneider, J.-L., Robin, C., Dujoncquoy, E., Salles, T., Allard, J., Bonnel, C., Ducassou, E., Etienne, S., Ferger, B., Gaudin, M., Hanquiez, V., Linares, F., Marches, E., Toucanne, S. & Zaragosi, S. 2010. High-resolution analysis of submarine lobes deposits: Seismic-scale outcrops of the Lauzanier area (SE Alps, France). *Sedimentary Geology*, **229**, 160–191.

Müller, R.D. & Landgrebe, T.C.W. 2012. The link between great earthquakes and the subduction of oceanic fracture zones. *Solid Earth*, **3**, 447–465.

Mullins, H.T. 1983. *Modern carbonate slopes and basins of the Bahamas*. Society of Economic Paleontologists & Mineralogists, Short Course Notes, **12**, 4-1–4-138.

Mullins, H.T. & Cook, H.E. 1986. Carbonate apron models: alternatives to the submarine fan model for paleoenvironmental analysis and hydrocarbon exploration. *Sedimentary Geology*, **48**, 37–79.

Mullins, H.T. & Neumann, A.C. 1979. Deep carbonate bank margin structure and sedimentation in the northern Bahamas. *In*: Doyle, L.J. & Pilkey, O.H. (eds), *Geology of Continental Margins*, 165–192. Society of Economic Paleontologists and Mineralogists Special Publication, **27**.

Muñoz, J.A. 1992. Evolution of a continental collision belt: ECORS-Pyrenean crustal balanced section. *In*: McClay, K.R. (ed.), *Thrust Tectonics*, 235–246. New York: Chapman and Hall.

Muñoz, J.A., McClay, K. & Poblet, J. 1992. Synchronous extension and contraction in frontal thrust sheets of the Spanish Pyrenees. *Geology*, **22**, 921–924.

Muñoz, J.A., Arbues, P. & Serra-Kiel, J. 1998. The Ainsa basin and the Sobrarbe oblique thrust system: Sedimentological and tectonic processes controlling slope and platform sequences deposited synchronously with a submarine emergent thrust system. *In*: *International Association of Sedimentologists 15th International Sedimentological Congress, Alicante, Spain, Field Trip Guidebook*, 213–223.

Muñoz, J.A., Beamud, E., Fernández, O., Arbués, P., Dinarès-Turell, J. & Poblet, J. 2013. The Ainsa fold and thrust oblique zone of the central Pyrenees: kinematics of a curved contractional system from paleomagnetic and structural data. *Tectonics*, **32**, 1142–1175.

Münoza, A., J. Cristobo, Rios, P., Druet, M., Polonio, V., Uchupi, E., Acosta, J. & Atlantis Group 2012. Sediment drifts and cold-water coral reefs in the Patagonian upper and middle continental slope. *Marine and Petroleum Geology*, **36**, 70–82.

Murray, C.J., Lowe, D.R., Graham, S.A., Martinez, P.A., Zeng, J., Carroll, A.R., Cox, R., Hendrix, M., Heubeck, C., Miller, D., Moxon, I.W., Sobel, E., Wendebourg, J. & Williams, T. 1996. Statistical analysis of bed-thickness patterns in a turbidite section from the Great Valley Sequence, Cache Creek, northern California. *Journal of Sedimentary Research*, **A66**, 900–908.

Muto, T. 1995. The Kolmogorov model of bed-thickness distribution: an assessment based on numerical simulation and field-data analysis. *Terra Nova*, **7**, 417–423.

Muto, T. & Steel, R.J. 2002. In defense of shelf-edge delta development during falling and lowstand of relative sea level. *Journal of Geology*, **110**, 421–436.

Mutti, E. 1974. Examples of ancient deep-sea fan deposits from circum-Mediterranean geosynclines. *In*: Dott, R.H. Jr. & Shaver, R.H. (eds), *Modern and Ancient Geosynclinal Sedimentation*, 92–105. Society of Economic Paleontologists and Mineralogists Special Publication, **19**.

Mutti, E. 1977. Distinctive thin-bedded turbidite facies and related depositional environments in the Eocene Hecho Group (south-central Pyrenees, Spain). *Sedimentology*, **24**, 107–131.

Mutti, E. 1979. Turbidites et cônes sous-marins profonds. *In*: Homewood, P. (ed.), *Sedimentation détritique (fluviatile, littorale et marine)*, 353–419. Switzerland: Institut Geologique, Université de Fribourg.

Mutti, E. 1984. The Hecho Eocene submarine-fan system, south central Pyrenees, Spain. *Geo-Marine Letters*, **3**, 199–202.

Mutti, E. 1985. Turbidite systems and their relations to depositional sequences. *In*: Zuffa, G.G. (ed.), *Provenance of Arenites*, 65–93. NATO Advanced Scientific Institute. Dordrecht, Holland: D. Reidel.

Mutti, E. 1992. *Turbidite Sandstones*. Parma: Agip and Università di Parma. 275 pp.

Mutti, E. & Ghibaudo, G. 1972. Un esempio di torbiditi di conoide sottomarina esterna: le Arenarie di San Salvatore (Formazione di Bobbio, Miocene) nell'Appennino di Piacenza. *Accademia delle Scienze di Torino, Memorie Classe di scienze fisiche, matematiche e naturali, Serie 4, no. 16*, 40 pp.

Mutti, E. & Johns, D.R. 1979. The role of sedimentary by-passing in the genesis of basin plain and fan fringe turbidites in the Hecho Group System (South-Central Pyrenees). *Societa Geologica Italiana. Memorie*, **18**, 15–22.

Mutti, E. & Normark, W.R. 1987. Comparing examples of modern and ancient turbidite systems: problems and concepts. *In*: Leggett, J.K. & Zuffa, G.G. (eds), *Marine Clastic Sedimentology*, 1–38. London: Graham & Trotman.

Mutti, E. & Normark, W.R. 1991. An integrated approach to the study of turbidite systems. *In*: Weimer, P. & Link, M.H. (eds), *Seismic Facies and Sedimentary Processes of Modern and Ancient Submarine Fans*, 75–106. New York: Springer Verlag.

Mutti, E. & Ricci Lucchi, F. 1972. Le torbiditi dell'Apennino settentrionale: introduzione all'analisi di facies. *Memoirs of the Geological Society, Italy*, **11**, 161–99. (1978 English translation by T.H. Nilsen, *International Geology Review*, **20**, 125–166.)

Mutti, E. & Ricci Lucchi, F. 1974. La signification de certaines unites sequentielles dans les series a turbidites. *Bulletin Geological Society of France*, **16**, 577–582.

Mutti, E. & Ricci Lucchi, F. 1975. Turbidite facies and facies associations. *In*: *Examples of Turbidite Facies and Facies Associations from Selected Formations of the Northern Apennines, Field Trip Guidebook A-ll*, 21–36. IX International Congress of Sedimentologists, Nice, France. International Association of Sedimentologists.

Mutti, E. & Ricci Lucchi, F. 1978. Turbidites of the Northern Apennines: introduction to facies analysis. *International Geology Review*, **20**, 125–166.

Mutti, E. & Sonnino, M. 1981. Compensation cycles: a diagnostic feature of turbidite sandstone lobes. *In*: *International Association of Sedimentologists 2nd European Regional Meeting*, Bologna, Italy, *Abstracts Volume*, 120–123.

Mutti, E., Nilsen, T.H. & Ricci Lucchi, F. 1978. Outer fan depositional lobes of the Laga Formation (upper Miocene and lower Pliocene), east-central Italy. *In*: Stanley, D.J. & Kelling, G. (eds), *Sedimentation in Submarine Canyons, Fans, and Trenches*, 210–223. Stroudsburg, PA: Dowden, Hutchinson & Ross.

Mutti, E., Barros, M., Possato, S., Rumenos, L., 1980. Deep-sea fan turbidite sediments winnowed by bottom currents in the Eocene of the Campos Basin, Brazilian offshore. *In*: *International Association of Sedimentologists First European Regional Meeting, Abstracts Volume*, p. 114.

Mutti, E., Ricci Lucchi, F., Seguret, M. & Zanzucchi, G. 1984. Seismoturbidites: A new group of resedimented deposits. *Marine Geology*, **55**, 103–116.

Mutti, E., Remacha, E., Sgavetti, M., Rosell, J., Valloni, R. & Zamorano, M. 1985. Stratigraphy and facies characteristics of the Eocene Hecho Group turbidite systems, south-central Pyrenees. *In*: Mila, M.D. & Rosell, J. (eds), *International Association of Sedimentologists, 6th European Regional Meeting, Llerida*, Excursion Guidebook, 521–576.

Mutti, E., Seguret, M. & Sgavetti, M. 1988. *Sedimentation and Deformation in the Tertiary Sequences of the Southern Pyrenees. American Association of Petroleum Geologists, Mediterranean Basins Conference, Nice, France, Field Trip 7 Guidebook*. Special Publication of the Institute of Geology of the University of Parma, 169 pp.

Mutti, E., Davoli, G., Mora, S. & Papani, L. 1994. Internal stacking patterns of ancient turbidite systems from collisional basins. *In*: Weimer, P., Bouma, A.H. & Perkins, B.F. (eds), *Gulf Coast Section-Society of Economic Paleontologists and Mineralogists Foundation 15th Annual Research Conference, Submarine Fans and Turbidite systems: Sequence Stratigraphy, Reservoir Architecture and Production Characteristics*, 257–268.

Mutti, E., Tinterri, R., Benevelii, G., di Biase, D. & Cavanna, G. 2003. Deltaic, mixed and turbidite sedimentation of ancient foreland basins. *Marine and Petroleum Geology*, **20**, 733–755.

Myers, R.A. 1986. *Late Cenozoic sedimentation in the Northern Labrador Sea: a seismic-stratigraphic analysis*. MSc. thesis. Halifax: Dalhousie University.

Myers, R.A. & Piper, D.J.W. 1988. Seismic stratigraphy of late Cenozoic sediments in the northern Labrador Sea; a history of bottom circulation and glaciation. *Canadian Journal of Earth Sciences*, **25**, 2059–2074.

Myrow, P.M. & Hiscot, R.N. 1993. Depositional history and sequence stratigraphy of the Precambrian-Cambrian boundary stratotype section, Chapel Island Formation, southeast Newfoundland. *Palaeogeography, Palaeoclimatology, Palaeoecology*, **104**, 13–35.

Nagel, D.K. & Mullins, H.T. 1983. Late Cenozoic offset and uplift along the San Gregorio Fault zone: central California continental margin. *In*: Anderson, D.W. & Rymer, M.J. (eds), *Tectonics and Sedimentation*

Along Faults of the San Andreas System, 91–103. Society of Economic Paleontologists and Mineralogists Symposium, Pacific Section.

Nagel, D.K., Mullins, H.T. & Greene, H.G. 1986. Ascension submarine canyon, California - evolution of a multi-head canyon system along a strike-slip continental margin. *Marine Geology*, 73, 285–310.

Nakajima, T. 2000. Initiation processes of turbidity currents; implications for assessments of recurrence intervals of offshore earthquakes using turbidites. *Bulletin of the Geological Survey of Japan*, 51, 79–87.

Nakajima, T. & Kanai, Y. 2000. Sedimentary features of seismoturbidites triggered by the 1983 and older historical earthquakes in the eastern margin of the Japan Sea. *Sedimentary Geology*, 135, 1–19.

Nakajima, T. & Satoh, M. 2001. The formation of large mudwaves by turbidity currents on the levées of the Toyama deep-sea channel, Japan Sea. *Sedimentology*, 48, 435–463.

Nakajima, T., Satoh, M. & Okamura, Y. 1998. Channel-levée complexes, terminal deep-sea fan and sediment wave fields associated with the Toyama Deep-Sea Channel system in the Japan Sea. *Marine Geology*, 147, 25–41.

Nakazawa, T., Nishimura, A., Iryu, Y., Yamada, T., Shibasaki, H. & Shiokawa, S. 2007. Rapid subsidence of the Kikai Seamount inferred from drowned Pleistocene coral limestone: Implication for subduction of the Amami Plateau, northern Philippine Sea. *Marine Geology*, 247, 35–45.

Nardin, T.R. 1983. Late Quaternary depositional systems and sea level changes – Santa Monica and San Pedro Basins, California Continental Borderland. *American Association Petroleum Geologists Bulletin*, 67, 1104–1124.

Nardin, T.R. & Henyey, T.L. 1978. Pliocene–Pleistocene diastrophism of Santa Monica and San Pedro shelves, California continental borderland. *American Association Petroleum Geologists Bulletin*, 62, 247–272.

Natland, M.L. 1933. Temperature and depth classification of some Recent and fossil foraminifera in the southern California region. *Scripps Institute of Oceanography Bulletin*, Technical Series, 225–230.

Navarre, J.-C., Claude, D., Librelle, F., Safa, P., Villon, G. & Keskes, N. 2002. Deepwater turbidite system analysis, West Africa: sedimentary model and implications for reservoir model construction. *The Leading Edge*, 21, 1132–1139.

Naylor, M.A. 1980. The origin of inverse grading in muddy debris flow deposits – a review. *Journal of Sedimentary Petrology*, 50, 1111–1116.

Naylor, M.A. 1982. The Casanova Complex of the northern Apennines: a mélange formed on a distal passive continental margin. *Journal of Structural Geology*, 4, 1–18.

Naylor, M. & Sinclair, H.D. 2007. Punctuated thrust deformation in the context of doubly vergent thrust wedges: Implications for the localization of uplift and exhumation. *Geology*, 35, 559–562.

Nederbragt, A.J. & Thurow, J.W. 2001. A 6000 yr varve record of Holocene climate in Saanich Inlet, British Columbia, from digital sediment colour analysis of ODP Leg 169S cores. *Marine Geology*, 174, 95–110.

Nederlof, F.H. 1959. Structure and sedimentology of the Upper Carboniferous of the upper Pisuega valleys, Cantabrian Mountains, Spain. *Leidse Geologische Mededelingen*, 24, 603–703.

Nelson, C.H. 1976. Late Pleistocene and Holocene depositional trends, processes, and history of Astoria deep-sea fan, northeast Pacific. *Marine Geology*, 20, 129–173.

Nelson, C.H. & Damuth, J.E. 2003. Myths of turbidite system control: insights provided by modern turbidite studies. *International Conference "Deep Water Processes in Modern and Ancient Environments", Barcelona and Ainsa, Abstracts Volume*, p. 32.

Nelson, A.R., Kelsey, H.M. & Witter, R.C. 2008. Great earthquakes of variable magnitude at the Cascadia subduction zone. *Quaternary Research*, 65, 354–365.

Nelson, C.H. & Kulm, L.D. 1973. Submarine fans and deep-sea channels. *In*: Middleton, G.V. & Bouma, A.H. (eds), *Turbidites and Deep-water Sedimentation*, 39–78. Society of Economic Paleontologists and Mineralogists, Pacific Section Short Course Notes, Anaheim.

Nelson, C.H., Carlson, P.R., Byrne, J.V. & Alpha, T.R. 1970. Development of the astoria canyon-fan physiography and comparison with similar systems. *Marine Geology*, 8, 259–291.

Nelson, C.H., Mutti, E. & Ricci Lucchi, F. 1975. Comparison of proximal and distal thin-bedded turbidites with current-winnowed deep-sea sands. *9th International Congress Sedimentology, Nice, France, Theme 5*, 317–324.

Nelson, C.H., Normark, W.R., Bouma, A.H. & Carlson, P.R. 1978. Thin-bedded turbidites in modern submarine canyons and fans. *In*: Stanley, D.J. & Kelling, G. (eds), *Sedimentation in Suhmanne Canyons, Fans, and Trenches*, 177–189. Stroudsburg, PA: Dowden, Hutchinson & Ross.

Nelson, C.H., Maldonado, A., Coumes, F., Got, H. & Monaco, A. 1984. The Ebro deep-sea fan system. *Geo-Marine Letters*, 3, 125–132.

Nelson, C.H., Twichell, D.C., Schwab, W.C., Lee, H.J. & Kenyon, N.H. 1992 Upper Pleistocene turbidite sand beds and chaotic silt beds in the channelized, distal, outer-fan lobes of the Mississippi fan. *Geology*, 20, 693–696.

Nelson, C.H., Karabanov, E.B., Colman, S.M. & Escutia, C. 1999. Tectonic and sediment supply control of deep rift lake turbidite systems: Lake Baikal, Russia. *Geology*, 27, 163–166.

Nelson, C.S. 1978. Temperate shelf carbonate sediments in the Cenozoic of New Zealand. *Sedimentology*, 25, 737–771.

Nelson, K.D. 1981. Mélange development in the Boones Point Complex, north-central Newfoundland. *Canadian Journal of Earth Sciences*, 18, 433–442.

Nelson, T.A. & Stanley, D.J. 1984. Variable depositional rates on the slope and rise off the Mid-Atlantic states. *Geo-Marine Letters*, 3, 37–42.

Nemcok, M. & Henk, A. 2006. Oil reservoirs in foreland basins charged by thrustbelt source rocks: insights from numerical stress modelling and geometric balancing in the West Carpathians. *In*: Butler, S.J.H. & Schreurs, G. (eds), *Analogue and Numerical Modelling of Crustal-scale Processes*, 253, 415–428. Geological Society, London, Special Publications. Bath: The Geological Society, London.

Nemec, W. 1990. Aspects of sediment movement on steep delta slopes. *In*: Colella, A. & Prior, D.B. (eds), *Coarse-Grained Deltas, 29–73*. International Association of Sedimentologists, Special Publication, 10.

Nemec, W. 1995. The dynamics of deltaic suspension plumes. *In*: Oti, M.N. & Postma, G. (eds), *Geology of Deltas*, 31–93. Rotterdam: A. A. Balkema.

Nemec, W. & Steel R.J. 1984. Alluvial and coastal conglomerates: their significant features and some comments on gravelly mass-flow deposits. *In*: Koster, E.H. & Steel, R.J. (eds), *Sedimentology of Gravels and Conglomerates*, 1–31. Calgary: Canadian Society of Petroleum Geologists Memoir, 10.

Nemec, W., Porebski, S.J. & Steel, R.J. 1980. Texture and structure of resedimented conglomerates - examples from Ksaiz Formation (Famennian-Tournaisian), southwestern Poland. *Sedimentology*, 27, 519–538.

Nemec, W., Steel, R.J., Gjelberg, J., Collinson, J.D. Prestholm, E. & Øxnevad, I.E. 1988. Anatomy of collapsed and re-established delta front in Lower Cretaceous of eastern Spitsbergen: gravitational sliding and sedimentation processes. *American Association of Petroleum Geologists Bulletin*, 72, 454–476.

Newman, M.St.J., Reeder, M.L., Woodruff, A.H.W. & Hatton, I.R. 1993. The geology of the Gryphon Oil Field. *In*: Parker, J.R. (ed.). *Petroleum Geology of Northwest Europe: Proceedings of the 4th Conference*, 123–133. London: Geological Society.

Nichols, R.J. 1995. The liquification and remobilization of sandy sediments. *In*: Hartley, A.J. & Prosser, D.J. (eds), *Characterization of Deep Marine Clastic Systems*, Geological Society London Special Publication, 94, 63–76.

Nichols, R.J., Sparks, R.S.J. & Wilson, C.J.N. 1994. Experimental studies of the fluidization of layered sediments and the formation of fluid escape structures. *Sedimentology*, 41, 233–253.

Nicholson, H.A. 1873. Contributions to the study of the errant annelids of the older Palaeozoic rock. *Proceedings of the Royal Society of London,* **21**, 288–290.

Nicol, A., Mazengarb, C., Chanier, F., Rait, G., Uruski, C. & Wallace, L. 2007. Tectonic evolution of the active Hikurangi subduction margin, New Zealand, since the Oligocene. *Tectonics,* **26**, 24 pp. doi:10.1029/2006TC002090.

Nielsen, T., Knutz, P.C. & Kuijpers, A. 2008. Seismic expression of contourite depositional systems. *In*: Rebesco, M. & Camerlenghi, A. (eds), *Contourites,* 301–321. Developments in Sedimentology, **60**. Amsterdam: Elsevier.

Nielsen, T.K., Christian Larsen, H. & Hopper, J.R. 2002. Contrasting rifted margin style south of Greenland: implications for mantle plume dynamics. *Earth and Planetary Science Letters,* **200**, 271–286.

Niem, A.R. 1976. Patterns of flysch deposition and deep-sea fans in the lower Stanley Group (Mississippian), Ouachita Mountains, Oklahoma and Arkansas. *Journal of Sedimentary Petrology,* **46**, 633–646.

Niemi, T.M., Ben-Avraham, Z., Hartnady, C.J.H. & Reznikov, M. 2000. Post-Eocene seismic stratigraphy of the deep ocean basin adjacent to the southeast African continental margin: a record of geostrophic bottom current systems. *Marine Geology,* **162**, 237–258.

Nijman, W. 1998. Cyclicity and basin axis shift in a piggyback basin: towards modelling of the Eocene Tremp-Ager Basin, South Pyrenees, Spain. *In*: Mascle, A. Puigdefabregas, C., Luterbacher, H.P. & Fernandez, M. (eds), *Cenozoic Foreland Basins of Western Europe,* 135–162. Geological Society of London, Special Publication, 134.

Nijman, W. & Nio, S.D. 1975. The Eocene Montanana Delta (Tremp-Graus Basin, Provinces Lerida and Huesca, Southern Pyrenees, N. Spain). *In*: Puigdefabregas, C. and Rosell, J. (eds), *International Association of Sedimentologists, 9th European Regional Meeting, Nice, Excursion Guidebook,* **19**, Part B. International Association of Sedimentologists.

Nilsen, T.H. 2000. The Hilt Bed, an Upper Cretaceous compound basin-plain seismoturbidite in the Hornbrook forearc basin of southern Oregon and northern California, USA. *Sedimentary Geology,* **135**, 51–63.

Nilsen, T.H. & Simoni, T.R. 1973. Deep-sea fan paleocurrent patterns of the Eocene Butano Sandstone, Santa Cruz Mountains, California. *United States Geological Survey Journal of Research,* **1**, 439–452.

Nilsen, T.H. & Zuffa, G.G. 1982. The Chugach Terrane, a Cretaceous trench-fill deposit, southern Alaska. *In*: Leggett, J.K. (ed.), *Trench-Forearc Geology,* 213–227. The Geological Society of London, Special Publication, **10**. London: The Geological Society.

Nilsen, T.H. & Sylvester, A.G. 1995. Strike-slip basins. *In*: Busby, C.J. & Ingersoll, R.V. (eds), *Tectonics of Sedimentary Basins,* 425–457. Oxford: Blackwell Scientific.

Nisbet, E G. 1992. Sources of atmospheric CH4 in early Postglacial time. *Journal of Geophysical Research,* **97**, 12,859–12,867.

Nishizawa, A., Kaneda, K., Watanabe, N. & Oikawa, M. 2009. Seismic structure of the subducting seamounts on the trench axis: Erimo Seamount and Daiichi-Kashima Seamount, northern and southern ends of the Japan Trench. *Earth, Planets and Space,* **61**, e5–e8.

Nittrouer, C.A., Austin, J.A., Field, M.E., Kravitz, J.H., Syvitski, J.P.M. & Wiberg, P.L. (eds) 2007. *Continental Margin Sedimentation: From Sediment Transport to Sequence Stratigraphy.* International Association of Sedimentologists, Special Publication, **37**, 549 pp. ISBN: 978-1-4051-6934-9.

Noda, A. & Toshimitsu, S. 2009. Backward stacking of submarine channel-fan successions controlled by strike-slip faulting: The Izumi Group (Cretaceous), southwest Japan. *Lithosphere,* **1**, 41–59.

Nokleberg, W.J., Plafker, G. & Wilson, F.H. 1994. Geology of south-central Alaska. *In*: Plafker, G. & Berg, H.C. (eds), *The Geology of Alaska, Volume G-1 of the geology of North America,* 311–366. Boulder Colorado: Geological Society of America.

Normark, W.R. 1970. Growth patterns of deep-sea fans. *American Association of Petroleum Geologists Bulletin,* **54**, 2170–2195.

Normark, W.R. 1978. Fan valleys, channels, and depositional lobes on modern submarine fans: characters for recognition of sandy turbidite environments. *American Association of Petroleum Geologists Bulletin,* **62**, 912–931.

Normark, W.R. 1989. Observed parameters for turbidity-current flow in channels, Reserve Fan, Lake Superior. *Journal of Sedimentary Petrology,* **59**, 423–431.

Normark, W.R. & Dickson, F.H. 1976. Man-made turbidity currents in Lake Superior. *Sedimentology,* **23**, 815–832.

Normark, W.R. & McGann, M. 2004. Late Quaternary Deposition in the Inner Basins of the California Continental Borderland - Part A. Santa Monica Basin. *United States Geological Survey, Scientific Investigations Report,* **2004-5183**, 21 pp. United States Geological Survey.

Normark, W.R. & Piper, D.J.W. 1972. Sediments and growth pattern of Navy deep-sea fan, San Clemente Basin, California Borderland. *Journal of Geology,* **80**, 192–223.

Normark, W.R. & Piper, D.J.W. 1984. Navy Fan, Caiifornia Borderland: growth pattern and depositional processes. *Geo-Marine Letters,* **3**, 101–108.

Normark, W.R. & Piper, D.J.W. 1991. Initiation processes and flow evolution of turbidity currents: implications for the depositional record. *In*: Osborne, R.H. (ed.), *From Shoreline to Abyss: Contributions in Marine Geology in Honor of Francis Parker Shepard,* 207–230. Society of Economic Paleontologists and Mineralogists, Special Publication, **46**.

Normark, W.R. & Piper, D.J.W. 1998. *Preliminary Evaluation of Recent Movement on Structures Within the Santa Monica Basin, Offshore southern California.* U.S. Geological Survey Open-File Report No. **98-518**, 60 pp.

Normark, W.R., Piper, D.J.W. & Hess, G.R. 1979. Distributary channels, sand lobes, and mesotopography of Navy Submarine Fan, California Borderland, with applications to ancient fan sediments. *Sedimentology,* **26**, 749–774.

Normark, W.R., Hess, G.R., Stow, D.A.V. & Bowen, A.J. 1980. Sediment waves on the Monterey Fan levée: a preliminary physical interpretation. *Marine Geology,* **37**, 1–18.

Normark, W.R., Barnes, N.E. & Coumes, F. 1984. Rhône deep sea fan: a review. *Geo-Marine Letters,* **3**, 155–160.

Normark, W.R., Wilde, P., Campbell, J.F., Chase, T.E. & Tsutsui, B. 1993a. Submarine slope failures initiated by Hurricane Iwa, Kahe Point, Oahu, Hawaii. *Bulletin of the US Geological Survey* **2002**, 197–204.

Normark, W.R., Posamentier, H. & Mutti, E. 1993b. Turbidite systems: state of the art and future directions. *Reviews in Geophysics,* **31**, 91–116.

Normark, W.R., Damuth, J.E., Cramp, A., Flood, R.D., Goni, M.A., Hiscott, R.N., Kowsmann, R.O., Lopez, M., Manley, P.L., Nanayama, F., Piper, D.J.W., Pirmez, C. & Schneider, R. 1997. Sedimentary facies and associated depositional elements of Amazon Fan. *In*: Flood, R.D., Piper, D.J.W., Klaus, A. *et al.*, *Proceedings Ocean Drilling Program, Scientific Result,* **155**, 611–651. College Station, Texas, USA: Ocean Drilling Program.

Normark, W.R., Piper, D.J.W. & Hiscott, R.N. 1998. Sea level controls on the textural and depositional architecture of the Hueneme and associated submarine fan systems, Santa Monica Basin, California. *Sedimentology,* **45**, 53–70.

Normark, W.R., Piper, D.J.W., Posamentier, H., Pirmez, C. & Migeon, S. 2002. Variability in form and growth of sediment waves on turbidite channel levées. *Marine Geology,* **192**, 23–58.

Normark, W.R., Piper, D.J.W. & Sliter, R. 2006. Sea-level and tectonic control of middle to late Pleistocene turbidite systems in Santa Monica Basin, offshore California. *Sedimentology,* **53**, 867–897.

Norris, R.J., Carter, R.M. & Turnbull, I.M. 1978. Cainozoic sedimentation in basins adjacent to a major continental transform boundary in southern New Zealand. *Journal of the Geological Society, London,* **135**, 191–205.

North American Commission on Stratigraphic Nomenclature. 1983. North American stratigraphic code. *American Association of Petroleum Geologists Bulletin,* **67**, 841–875.

Nowell, A.R.M. & Hollister, C.D. (eds) 1985. Deep ocean sediment transport – preliminary results of the high energy benthic boundary layer experiment. *Special Issue, Marine Geology*, **66**.

Nummedal, D. 2001. Internal tides and bedforms at the shelf edge. *In: Session No. 67: Dynamics of Sediments and Sedimentary Environments I: A Session in Honor of John B. Southard.* Geological Society of America Annual Meeting, Nov. 5–8, 2001, **Paper No. 67-0**.

Nuñes, F. & Norris, R.D. 2006. Abrupt reversal in ocean overturning during the Palaeocene/Eocene warm period. *Nature*, **439**, 60–63.

O'Connell, S.B. 1990. Sedimentary facies and depositional environment of the Lower Cretaceous East Antarctic margin Sites 692 and 693. *In*: Barker, P.R, Kennett, J.P. *et al.*, *Proceedings of the Ocean Drilling Program, Scientific Results*, **113**, 71–88. College Station, Texas, USA: Ocean Drilling Program.

O'Connell, S., Normark, W.R., Ryan, W.B.F. & Kenyon, N.H. 1991. An entrenched thalweg channel on the Rhône Fan: interpretation from a SEABEAM and SEAMARC I survey. *In*: Osborne, R.H. (ed.), *From Shoreline to Abyss: Contributions in Marine Geology in Honor of Francis Parker Shepard*, 259–270. Society of Economic Paleontologists and Mineralogists Special Publication, **46**. Tulsa, Oklahoma: Society of Economic Paleontologists and Mineralogists.

Oakman, C.D. 2005. The Lower Cretaceous plays of the central and northern North Sea: Atlantean drainage models and enhanced hydrocarbon potential. *In*: Doré, A.G. & Vining, B.A. (eds), *Petroleum Geology: North-west Europe and Global Perspectives - Proceedings of the 6th Petroleum Geology Conference*, **1**, 187–198. London: The Geological Society.

Obana, K., Kodaira, S., Mochizuki, K. & Shinohara, M. 2001. Micro-seismicity around the seaward updip limit of the 1946 Nankai earthquake dislocation area. *Geophysical Research Letters*, **28**, 2333–2336.

Obana, K., Scherwath, M., Yamamoto, Y., Kodaira, S., Wang, K., Spence, G., Riedel, M. & Kao, H. 2014. Earthquake activity in northern Cascadia subduction zone off Vancouver Island revealed by ocean-bottom seismograph observations. *Bulletin of the Seismological Society of America*, doi: 10.1785/0120140095.

Oberhansli, H. & Hsü, K.J. 1986. Paleocene–Eocene paleoceanography. *In*: Hsü, K.J. (ed.), *Mesosoic and Cenozoic Oceans*, 85–100. Geodynamics Series **15**. Boulder, Colorado: Geological Society of America.

O'Brien, P.E., Cooper, A.K., Richter, C. *et al.* 2001. *Proceedings of the Ocean Drilling Program, Initial Reports*, **188**. College Station, Texas, USA: Ocean Drilling Program. doi:10.2973/odp.proc.ir.188.2001.

Odigi, M.I. & Amajor, L.C. 2008. Petrology and geochemistry of sandstones in the southern Benue Trough of Nigeria: Implications for provenance and tectonic setting. *Chinese Journal of Geochemistry*, **27**, 384–394.

Ofurhie, M.A., Agha, G.U., Lufadeju, A.O. & Ineh, G.C. 2002. Turbidite depositional environment in deepwater of Nigeria. *In: Proceedings of the Offshore Technology Conference*, Houston, Texas USA, 6–9 *May 2002*. Paper **OTC 14068**, 10 pp.

Ogawa, Y. 1982. Tectonics of some forearc fold belts in and around the arc-arc crossing area in central Japan. *In*: Leggett, J.K. (ed.), *Trench-Forearc Geology*, 49–61. The Geological Society of London, Special Publication, **10**. London: The Geological Society.

Ogawa, Y. 1983. Mineoka ophiolite belt in the Izu forearc area-Neogene accretion of oceanic and island arc assemblages on the northeastern corner of the Philippine Sea Plate. *In*: Hashimoto, M. & Uyeda, S. (eds), *Accretion Tectonics in the Circum-Pacific Regions*, 245–260. Tokyo: Terrapub.

Ogawa, Y. 1985. Variety of subduction and accretion processes in Cretaceous to Recent plate boundaries around southwest and central Japan. *Tectonophysics*, **112**, 493–518.

Ojakangas, R.W. 1968. Cretaceous sedimentation, Sacramento Valley, California. *Geological Society of America Bulletin*, **79**, 973–1008.

Ojakangas, R.W., Srinivasan, S., Hegde, G.S., Chandrakant, S.M. & Srikantia, S.V. 2014. The Talya Conglomerate: an Archean (~2.7 Ga) Glaciomarine Formation, Western Dharwar Craton, Southern India. *Current Science*, **106**, 387–396.

Okada, H. & Tandon, S.K. 1984. Resedimented conglomerates in a Miocene collision suture, Hokkaido, Japan. *In*: Koster, E.H. & Steel, R.J. (eds), *Sedimentology of Gravels and Conglomerates*, 413–427. Calgary Alberta: Canadian Society of Petroleum Geologists Memoir, **10**.

Okay, A.I., Demirbag, E., Kurt, H., Okay, N. & Kuscu, I. 1999. An active,d eepl narines trike-slipb asina longt he North Anatolian fault in Turkey. *Tectonics*, **18**, 129–147.

Okay, N. & Ergün, B. 2005. Source of the basinal sediments in the Marmara Sea investigated using heavy minerals in the modern beach sands. *Marine Geology*, **216**, 1– 15.

O'Leary, D.W. 1993. Submarine mass movement, a formative process of passive continental margins: the Munson-Nygren Landslide Complex and the Southeast New England Landslide Comples. *In*: Schwab, W.C., Lee, H.J. & Twichell, D.C. (eds), *Submarine Landslides: Selected Studies in the U. S. Exclusive Economic Zone*, 23–39. Bulletin U. S. Geological Survey.

Oliveira, C.M.M., Hodgson, D.M. & Flint, S.S. 2011. Distribution of soft-sediment deformation structures in clinoform successions of the Permian Ecca Group, Karoo Basin, South Africa. *Sedimentary Geology*, **235**, 314–330.

Olivero, E.B., Lopez C, M.I., Malumian, N. & Carbonell, P.J.T. 2010. Eocene graphoglyptids from shallow-marine, high-energy, organic-rich, and bioturbated turbidites, Fuegian Andes, Argentina. *Acta Geologica Polonica*, **60**, 77–91.

Oppenheimer, D., Beroza, G., Carver, G., Dengler, L., Eaton, J., Gee, L., Gonzalez, F., Jayko, A., Li, W.H., Lisowski, M., Magee, M., Marshall, G., Murray, M., McPherson, R., Romanowicz, B., Satake, K., Simpson, R., Somerville, P., Stein, R. & Valentine, D. 1993. The Cape Mendocino, California, earthquakes of April 1992: Subduction at the triple junction. *Science*, **261**, 433–438.

Ori, G.G., Roveri, M. & Vannoni, F. 1986. Plio-Pleistocene sedimentation in the Apenninic-Adriatic foredeep (central Adriatic Sea, Italy). *In*: Allen, P.A. & Homewood, P. (eds), *Foreland Basins*, 183–198. International Association of Sedimentologists Special Publication, **8**. Oxford: Blackwell Scientific.

Orr, P.J. 2001. Colonization of the deep-marine environment during the early Phanerozoic: the ichnofaunal record. *Geological Journal*, **36**, 265–278.

Orr, P.J., Benton, M.J. & Briggs, D.E.G. 2003. Post-Cambrian closure of the deep-water slope-basin taphonomic window. *Geology*, **31**, 769–772.

Osleger, D.A., Heyvaert, A.C., Stoner, J.S. & Verosub, K.L. 2009. Lacustrine turbidites as indicators of Holocene storminess and climate: Lake Tahoe, California and Nevada. *Journal of Paleolimnology*, **42**, 103–122.

Ovenshine, A.T. 1970. Observations of iceberg rafting in Glacier Bay, Alaska, and the identification of ancient ice-rafted deposits. *Bulletin of the Geological Society of America*, **81**, 891–894.

Page, B.M. & Suppe, J. 1981. The Pliocene Lichi melange of Taiwan: its plate-tectonic and olistostromal origin. *American Journal of Science*, **281**, 193–227.

Paillard, D., Labeyrie, L. & Yiou, P. 1996. Macintosh program performs time-series analysis. *Eos, Transactions American Geophysical Union*, **77**, 379.

Palike, H., Shackleton, N.J. & Röhl, U. 2001. Astronomical forcing in Late Eocene marine sediments. *Earth and Planetary Science Letters*, **193**, 589–602.

Pang, X., Chen, C., Peng, D., Zhu, M., Shu, Y., Shen, J. & Liu, B. 2007. Sequence stratigraphy of deep-water fan system of Pearl River, South China Sea. *Earth Science Frontiers*, **14**, 220–229.

Pantin, H.M. 1979. Interaction between velocity and effective density in turbidity flow: phase plane analysis, with criteria for autosuspension. *Marine Geology*, **31**, 59–99.

Pantin, H.M. 2001. Experimental evidence for autosuspension. *In*: McCaffrey, W., Kneller, B. & Peakall, J. (eds), *Particulate Gravity Currents*,

189–206. International Association of Sedimentologists, Special Publication, **31**. Oxford: Blackwell Scientific.

Pantopoulos, G., Vakalas, I., Maravelis, A. & Zelilidis, A. 2013. Statistical analysis of turbidite bed thickness patterns from the Alpine fold and thrust belt of western and southeastern Greece. *Sedimentary Geology*, **294**, 37–57.

Paola, C. 2000. Quantitative models of sedimentary basin filling. *Sedimentology*, **47**, 121–178.

Paola, C. & Southard, J.B. 1983. Autosuspension and the energetics of two-phase flows: reply to comments on 'experimental test of autosuspension' by J.B. Southard & M.E. Mackintosh. *Earth Surface Processes Landforms*, **8**, 273–279.

Paola, C., Heller, P.L. & Angevine, C.L. 1992. The large-scale dynamics of grain-size variation in alluvial basins: I. Theory. *Basin Research*, **4**, 73–90.

Paola, C., Straub, K., Mohrig, D. & Reinhardt, L. 2009. The "unreasonable effectiveness" of stratigraphic and geomorphic experiments. *Earth-Science Reviews*, **97**, 1–43.

Parea, G.C. 1975. The calcareous turbidite formations of the Northern Apennines. *In: Examples of Turbidite Facies and Facies Associations from Selected Formations of the Northern Apennines, Field Trip Guidebook A-ll, 52–62.* IXth International Congress of Sedimentologists, Nice, France. International Association of Sedimentologists.

Park, J.-O. & Kodaira, S. 2012. Seismic reflection and bathymetric evidences for the Nankai earthquake rupture across a stable segment-boundary. *Earth, Planets and Space*, **64**, 299–303.

Park, J.-O., Tsuru, T., Kodaira, S., Cummins, P.R. & Kaneda, Y. 2002. Splay fault branching along the Nankai subduction zone. *Science*, **297**, 1157–1160.

Park, J.-O., Tsuru, T., No, T., Takizawa, K., Sato, S. & Kaneda, Y. 2008. High-resolution 3D seismic reflection survey and prestack depth imaging in the Nankai Trough off southeast Kii Peninsula. *Butsuri Tansa*, **61**, 231–241 (in Japanese, with abstract in English).

Parker, G. 1982. Conditions for the ignition of catastrophically erosive turbidity currents. *Marine Geology*, **46**, 307–327.

Parker, G., Fukushima, Y. & Pantin, H.M. 1986. Self-accelerating turbidity currents. *Journal of Fluid Mechanics*, **171**, 145–181.

Parsons, D.R., Peakall, J., Aksu, A.E., Flood, R.D., Hiscott, R.N., Besiktepe, S. & Mouland, D. 2010. Gravity-driven flow in a submarine channel bend: direct field evidence of helical flow reversal. *Geology*, **38**, 1063–1066.

Parsons, J.D., Bush, J.W.M. & Syvitski, J.P.M. 2001. Hyperpycnal plume formation from riverine outflows with small sediment concentrations. *Sedimentology*, **48**, 465–478.

Parsons, J.D., Friedrichs, C.T., Traykovski, P.A., Mohrig, D., Imran, J., Syvitski, J.P.M., Parker, G., Puig, P., Buttles, J.L. & Garcia, M.H. 2007. The mechanics of marine sediment gravity flows. *In*: Nittrouer, C.A., Austin, J.A., Field, M.E., Kravitz, J.H., Syvitski, J.P.M. & Wiberg, P.L. (eds), *Continental Margin Sedimentation: From Sediment Transport to Sequence Stratigraphy*. 275–337. International Association of Sedimentologists, Special Publication, **37**. ISBN: 978-1-4051-6934-9.

Parsons, T., Wells, R.E., Fisher, M.A., Flueh, E. & ten Brink, U.S. 1999. Three-dimensional velocity structure of Siletzia and other accreted terranes in the Cascadia forearc of Washington. *Journal of Geophysical Research*, **104**, 18015–18039.

Paskevich, V., Twichell, D. & Schwab, W. 2001. SeaMARC1A sidescan mosaic, cores and depositional interpretation of the Mississippi Fan. ArcView GIS Data Release, 2001, *U.S. Geological Survey Open-File Report 00-352*, 1 CD-ROM.

Passchier, S. 2007. East Antarctic ice-sheet dynamics between 5.2 and 0 Ma from a high-resolution terrigenous particle record, ODP Site 1165, Prydz Bay - Cooperation Seain Antarctica. *In*: Cooper, A., Raymond, C. and the 10th ISAES Editorial Team (eds), *Antarctica: A Keystone in a Changing World-Online Proceedings for the 10th International Symposium on Antarctic Earth Sciences, Santa Barbara, California, U.S.A. - August 26 to September 1, 2007*. U.S. Geological Survey and The National Academies; USGS OF-2007-1047, Short Research Paper 043. doi:10.3133/of2007-1047.srp043.

Passchier, S., O'Brien, P.E., Damuth, J.E., Januszczak, N., Handwerger, D.A. & Whitehead, J.M. 2003. Pliocene–Pleistocene glaciomarine sedimentation in eastern Prydz Bay and development of the Prydz trough-mouth fan, ODP Sites 1166 and 1167, East Antarctica. *Marine Geology*, **199**, 279–305.

Patton, J.R., Goldfinger. C., Morey, A.E., Romsos, C., Black, B., Djadjadihardja, Y. & Udrekh, U. 2013. Seismoturbidite record as preserved at core sites at the Cascadia and Sumatra-Andaman subduction zones. *Natural Hazards and Earth System Sciences*, **13**, 1–35.

Paull, C.K., Greene, H.G., Ussler III, W. & Mitts, P.J. 2002. Pesticides as tracers of sediment transport through Monterey Canyon. *Geo-Marine Letters*, **22**, 121–126.

Paull, C.K., Ussler III, W., Greene, H.G., Keaten, R., Mitts, P. & Barry, J. 2003. Caught in the act: the 20 December 2001 gravity flow event in Monterey Canyon. *Geo-Marine Letters*, **22**, 227–232.

Paull, C.K., McGann, M., Sumner, E.J., Barnes, P.M., Lundsten, E.M., Krystle, A., Gwiazda, R., Edwards, B. & Caress, D.W. 2014. Sub-decadal turbidite frequency during the early Holocene: Eel Fan, offshore northern California. *Geology*, doi: 10.1130/G35768.1.

Pautot, G., Nakamura, K., Huchon, P., Angelier, J., Bourgois, J., Fujioka, K., Kanazawa, T., Nakamura, Y., Ogawa, Y., Séguret, M. & Takeuchi, A. 1987. Deep-sea submersible survey in the Suruga, Sagami and Japan Trenches: preliminary results of the 1985 Kaiko cruise, Leg 2. *Earth and Planetary Science Letters*, **83**, 300–312.

Payros, A. & Pujalte, V. 2008. Calciclastic submarine fans: An integrated overview. *Earth-Science Reviews*, **86**, 203–246.

Payros, A., Pujalte, V. & Orue-Etxebarria, X. 1999. The South Pyrenean Eocene carbonate megabreccias revisited: new interpretation based on evidence from the Pamplona Basin. *Sedimentary Geology*, **125**, 165–194.

Payros, A., Pujalte, V. & Orue-Etxebarria, X. 2007. A point-sourced calciclastic submarine fan complex (Eocene Anotz Formation, western Pyrenees): facies architecture, evolution and controlling factors. *Sedimentology*, **54**, 137–168.

Payros, A., Tosquella, J., Bernaola, G., Dinares-Turell, J., Orue-Etxebarria, X. & Pujalte, V. 2009. Filling the North European Early/Middle Eocene (Ypresian/Lutetian) boundary gap: Insights from the Pyrenean continental to deep-marine record. *Palaeogeography, Palaeoclimatology and Palaeoecology*, **280**, 313–332.

Peace, D. 2011. Eastern Mediterranean: the hot new exploration region. *GEO ExPro*, **8**, 38–40.

Peakall, J., McCaffrey, B. & Kneller, B. 2000a. A process model for the evolution, morphology, and architecture of sinuous submarine channels. *Journal of Sedimentary Research*, **70**, 434–448.

Peakall, J., McCaffrey, W.D., Kneller, B.C., Stelting, C.E., McHargue, T.R. & Schweller, W.J. 2000b. A process model for the evolution of submarine fan channels: implications for sedimentary architecture. *In*: Bouma, A.H. & Stone, C.G. (eds), *Fine-grained Turbidite Systems*, 73–88. American Association of Petroleum Geologists. Memoir, **72** & Society of Sedimentary Geologists, Special Publication, **68**. Joint publication, Tulsa, Oklahoma.

Peakall, J., Ashworth, P.J. & Best, J.L. 2007. Meander bend evolution, alluvial architecture, and the role of cohesion in sinuous river channels: a flume study. *Journal of Sedimentary Research*, **77**, 197–212.

Peizhen, Z., Molnar, P. & Downs, W.R. 2001. Increased sedimentation rates and grain sizes 2–4 Myr ago due to the influence of climate change on erosion rates. *Nature*, **410**, 891–897.

Pekar, S.F., Hucks, A., Fuller, M. & Li, S. 2005. Glacioeustatic changes in the early and middle Eocene (51–42 Ma): Shallow-water stratigraphy from Ocean Drilling Program Leg 189 Site 1171 (South Tasman Rise) and deep-sea δ^{18}O records. *Geological Society of America Bulletin*, **117**, 1081–1093.

Pemberton, S.G. & Frey, R.W. 1982. Trace fossil nomenclature and the Planolites-Palaeophycus dilemma. *Journal of Paleontology*, **56**, 843–881.

Pemberton, S.G. & Gingras, M.K. 2005. Classification and characterizations of biogenically enhanced permeability. *American Association of Petroleum Geologists Bulletin*, **89**, 1493–1517.

Pemberton, S.G., MacEachern, J.A. & Frey, R.W. 1992. Trace fossil facies models: environmental and allostratigraphic significance. *In*: Walker, R.G. & James, N.P. (eds). *Facies Models - Response to Sea Level Change*, 47–72. Ottawa: Geological Association of Canada.

Pemberton, S.G. Zhou, Z. & MacEachern, J. 2001. Modern ecological interpretation of opportunistic r-selected trace fossils and equilibrium K-selected trace fossils. *Acta Palaeontologica Sinica*, **40**, 134–142.

Pemberton, S.G., Spila, M.V., Pulham, A.J., Saunders, T., MacEachern, J.A., Robbins, D. & Sinclair, I. 2001. Ichnology and sedimentology of shallow and marginal marine systems: Ben Nevis and Avalon reservoirs, Jeanne D'Arc Basin. *Geological Association of Canada, Short Course Notes*, **15**, 343 pp. St John's, Newfoundland.

Pérez, L.F., Hernández-Molina, F.J., Esteban, F.D., Tassone, A., Piola, A.R., Maldonado, A., Preu, B., Violante, R.A. & Lodolo, E. 2015. Erosional and depositional contourite features at the transition between the western Scotia Sea and southern South Atlantic Ocean: links with regional water-mass circulation since the Middle Miocene. *Geo-Marine Letters*, doi: 10.1007/s00367-015-0406-6.

Petersen, S., Kuhn, K., Kuhn, T., Augustin, N., Hékinian, R., Franz, L. & Borowski, C. 2009. The geological setting of the ultramafic-hosted Logatchev hydrothermal field (14°45′N, Mid-Atlantic Ridge) and its influence on massive sulfide formation. *Lithos*, **112**, 40–56.

Petters, S.W. & Ekweozor, C.M. 1982. Petroleum geology of Benue Trough and southeastern Chad Basin, Nigeria. *American Association of Petroleum Geologists Bulletin*, **66**, 1141–1149.

Philips, S. 1987. Dipmeter interpretation of turbidite-channel reservoir sandstones, Indian Draw Field, New Mexico. *In*: Tillman, R.W. & Weber, K.J. (eds), *Reservoir Sedimentology*, 113–128. Society of Economic Paleontologists and Mineralogists Special Publication, **40**. Tulsa, Oklahoma: Society of Economic Paleontologists and Mineralogists.

Phillips, C., McIlroy, D & Elliott, T. 2011. Ichnological characterization of Eocene/Oligocene turbidites from the Grès d'Annot Basin, French Alps, SE France. *Palaeogeography, Palaeoclimatology, Palaeoecology*, **300**, 67–83.

Pianka, E. R. 1970. On r- and K-selection. *American Naturalist*, **104**, 592–597.

Picha, F. 1979. Ancient submarine canyons of Tethyan continental margins, Czechoslovakia. *American Association of Petroleum Geologists Bulletin*, **63**, 67–86.

Pickerill, R.K., Fillion, D. & Harland, T.L. 1984. Middle Ordovician trace fossils in carbonates of the Trenton Group between Montreal and Quebec City, St. Lawrence Lowland, Eastern Canada. *Journal of Paleontology*, **58**, 416–439.

Pickerill, R.K., Donovan, S.K., Doyle, E.N. & Dixon, H.L. 1993. Ichnology of the Palaeogene Richmond Formation of eastern Jamaica - the final chapter? *Atlantic Geology*, **29**, 61–67.

Pickering, K.T. 1979. Possible retrogressive flow slide deposits from the Kongsfjord Formation: a Precambrian submarine fan, Finnmark, *N. Norway. Sedimentology*, **26**, 295–306.

Pickering, K.T. 1981a, The Kongsfjord Formation - a late Precambrian submarine fan in north-east Finnmark, North Norway. *Norges geologiske Undersøgelse*, **367**, 77–104.

Pickering, K.T. 1981b. Two types of outer fan lobe sequence, from the late Precambrian Kongsfjord Formation Submarine Fan, Finnmark, North Norway. *Journal of Sedimentary Petrology*, **51**, 1277–1286.

Pickering, K.T. 1982a. A Precambrian upper basin-slope and prodelta in northeast Finnmark, North Norway - a possible ancient upper continental slope. *Journal of Sedimentary Petrology*, **52**, 171–186.

Pickering, K.T. 1982b. Middle-fan deposits from the late Precambrian Kongsfjord Formation Submarine Fan, northeast Finnmark, northern Norway. *Sedimentary Geology*, **33**, 79–110.

Pickering, K.T. 1982c. The shape of deep-water siliciclastic systems - a discussion. *Geo-Marine Letters*, **2**, 41–46.

Pickering, K.T. 1983a. Small-scale syn-sedimentary faults in the Upper Jurassic 'Boulder Beds'. *Scottish Journal of Geology*, **19**, 169–181.

Pickering, K.T. 1983b. Transitional submarine fan deposits from the late Precambrian Kongsfjord Formation submarine fan, NE. Finnmark, N. Norway. *Sedimentology*, **30**, 181–199.

Pickering, K.T. 1984a. Facies, facies-associations and sediment transport/deposition processes in a late Precambrian upper basin-slope/prodelta, Finnmark, N. Norway. *In*: Stow, D.A.V. & Piper, D.J.W. (eds), *Fine-grained Sediments: Deep-water Processes and Facies*, 343–362. Special Publication of The Geological Society London, **15**. Oxford: Blackwell Scientific Publications.

Pickering, K.T. 1984b. The Upper Jurassic 'Boulder Beds' and related deposits: a fault-controlled submarine slope, *NE. Scotland. Journal of the Geological Society (London)*, **141**, 357–374.

Pickering, K.T. 1985. Kongsfjord turbidite system, Norway. *In*: Bouma, A.H., Normark, W.R. & Barnes, N.E. (eds), *Submarine Fans and Related Turbidite Systems*, 267–273. New York: Springer.

Pickering, K.T. 1987a. Wet-sediment deformation in the Upper Ordovician Point Leamington Formation: an active thrust-imbricate system during sedimentation, Notre Dame Bay, north-central Newfoundland. *In*: Jones, M.E. & Preston, R.M.F. (eds), *Deformation of Sediments and Sedimentary Rocks*. Special Publication of The Geological Society London, **29**, 213–239. Oxford: Blackwell Scientific Publications.

Pickering, K.T. 1987b. Deep-marine foreland basin and forearc sedimentation: a comparative study from the Lower Palaeozoic northern Appalachians, Quebec and Newfoundland. *In*: Leggett, J.K. & Zuffa, G.G. (eds), *Marine Clastic Sedimentology: New Developments and Concepts*, 190–211. London: Graham & Trottman.

Pickering, K.T. & Bayliss, N.J. 2009. Deconvolving tectono-climatic signals in deep-marine siliciclastics, Eocene Ainsa basin, Spanish Pyrenees: Seesaw tectonics versus eustasy. *Geology*, **37**, 203–206.

Pickering, K.T. & Corregidor, J. 2000. 3D Reservoir scale study of Eocene confined submarine fans, south central Spanish Pyrenees. *In*: Weimer, P., Slatt, R.M., Coleman, J., Rosen, N.C., Nelson, H., Bouma, A.H., Styzen, M.J. & Lawrence, D.T. (eds), *Deep Water Reservoirs of the World*, 776–781. Gulf Coast Section Society of Economic Paleontologists and Mineralogists Foundation 20th Annual Bob F. Perkins Research Conference.

Pickering, K.T. & Corregidor, J. 2005. Mass–transport complexes (MTCS) and tectonic control on basin-floor submarine fans, Middle Eocene, south Spanish Pyrenees. *Journal of Sedimentary Research*, **75**, 761–783.

Pickering, K.T. & Hilton, V.C. 1998. *Turbidite Systems of Southeast France*. Marketed by: American Association of Petroleum Geologists, and Vallis Press (London), 229 pp. ISBN 0-9527313-1-2.

Pickering, K.T. & Hiscott, R.N. 1985. Contained (reflected) turbidity currents from the Middle Ordovician Cloridorme Formation, Quebec, Canada: an alternative to the antidune hypothesis. *Sedimentology*, **32**, 373–394.

Pickering, K.T. & Hiscott, R.N. 1995. Foreland basin-floor turbidite system, Cloridorme Formation, Québec, Canada: long-distance correlation in sheet turbidites. *In*: Pickering, K.T., Hiscott, R.N., Kenyon, N.H., Ricci Lucchi, F. & Smith, R. (eds), *Atlas of Architectural Styles in Turbidite Systems*, 310–316. London: Chapman & Hall.

Pickering, K.T. & Smith, A.G. 1995. Arcs and back-arc basins in the Lower Palaeozoic circum-Atlantic. *The Island Arc*, **4**, 1–67.

Pickering, K.T. & Smith, A.G. 1997. European Caledonides. *In*: Van der Pluijm, B. & Marshak, S. (eds), *Earth Structure: An Introduction to Structural Geology and Tectonics*, 435–444. New York: Wm. C. Brown.

Pickering, K.T. & Smith, A.G. 2004. The Caledonides. *In*: van der Pluijm, B. & Marshak, S. (eds), *Earth Structure: An Introduction to Structural Geology and Tectonics*, 2nd edn, 593–606. New York: W.W. Norton & Company.

Pickering, K.T. & Taira, A. 1994. Tectonosedimentation; with examples from the Tertiary–Recent of southeast Japan. *In*: Hancock, P.L. (ed.), *Continental Deformation*, [Chapter 16] 320–354. Oxford: Pergamon Press.

Pickering, K.T., Stow, D.A.V., Watson, M.P. & Hiscott, R.N. 1986a. Deep-water facies, processes and models: a review and classification scheme for modern and ancient sediments. *Earth-Science Reviews*, **23**, 75–174.

Pickering, K.T., Coleman, J., Cremer, M., Droz, L., Kohl, B., Normark, W., O'Connell, S., Stow, D. & Meyer-Wright, A. 1986b. A high-sinuosity, laterally-migrating submarine fan channel-levée-overbank: results from DSDP Leg 96 on the Mississippi Fan, Gulf of Mexico. *Marine and Petroleum Geology*, **3**, 3–18.

Pickering, K.T., Agar, S.M. & Ogawa, Y. 1988a. Genesis and deformation of mud injections containing chaotic basalt-limestone-chert associations: examples from the southwest Japan forearc. *Geology*, **16**, 881–885.

Pickering, K.T., Bassett, M.G. & Siveter, D.J. 1988b. Late Ordovician–early Silurian destruction of the Iapetus Ocean: Newfoundland, British Isles and Scandinavia—a discussion. *Transactions of the Royal Society of Edinburgh: Earth Sciences*, **79**, 361–382.

Pickering, K.T., Hiscott, R.N. & Hein, F.J. 1989. *Deep Marine Environments: Clastic Sedimentation and Tectonics.* London: Chapman & Hall, 416 pp. ISBN 004-4452012/5511225.

Pickering, K.T., Underwood, M.B. & Taira, A. 1992. Open ocean to trench turbidity-current flow in the Nankai Trough: Flow collapse and reflection. *Geology*, **20**, 1099–1102.

Pickering, K.T., Underwood, M.B. & Taira, A. 1993a. Open-ocean to trench turbidity-current flow in the Nankai Trough: flow collapse and flow reflection. *In*: Taira, A., Hill, I.A.H., Firth, J., Vrolijk, P.J. *et al.* (eds), *Proceedings of the Ocean Drilling Program Leg 131*, 35–43. Texas A&M: Ocean Drilling Program.

Pickering, K.T., Underwood, M.B. & Taira, A. 1993b. Stratigraphic synthesis of the DSDP-ODP sites in the Shikoku Basin, Nankai Trough, and accretionary prism. *In*: Taira, A., Hill, I.A.H., Firth, J., Vrolijk, P.J. *et al.* (eds), *Proceedings of the Ocean Drilling Program Leg 131*, 313–330. College Station, Texas, USA: Ocean Drilling Program.

Pickering, K.T., Hiscott, R.N., Kenyon, N.H., Ricci Lucchi, F. & Smith, R.D.A. 1995a. *Atlas of Deep Water Environments: Architectural Style in Turbidite Systems.* London: Chapman & Hall, 333 pp. ISBN 0-412-56110-7.

Pickering, K.T., Clark, J.D., Smith, R.D.A., Hiscott, R.N., Ricci Lucchi, F. & Kenyon, N.H. 1995b. Architectural element analysis of turbidite systems, and selected topical problems for sand-prone deep-water systems *In*: Pickering, K.T., Hiscott, R.N., Kenyon, N.H., Ricci Lucchi, F. & Smith, R.D.A. (eds), *Atlas of Deep Water Environments: Architectural Style in Turbidite System*, 1–11. London: Chapman & Hall.

Pickering, K.T., Vining, B.A. & Ioannides, N.S. 1997. Core photograph-based study of stratigraphic relationships of some Tertiary lowstand depositional systems in the Central North Sea. *In*: Oakman, C.D., Martin, J.H. & Corbett, W.M. (eds), *Cores from the Northwest European Petroleum Province: An Illustration of Geological Applications from Exploration to Development*, 49–65. Bath: The Geological Society Publishing House.

Pickering, K.T., Souter, C., Oba, T., Taira, A., Schaaf, M. & Platzman, E. 1999. Glacio-eustatic control on deep-marine clastic forearc sedimentation, Pliocene – mid-Pleistocene (c. 1180-600 ka) Kazusa Group, SE Japan. *Journal of the Geological Society* (London), **156**, 125–136.

Pickering, K.T., Hodgson, D.M., Platzman, E., Clark, J.D. & Stephens, C. 2001. A new type of bedform produced by back-filling processes in a submarine channel, Late Miocene, Tabernas-Sorbas basin, SE Spain. *Journal of Sedimentary Research*, **71**, 692–704.

Pickering, K.T., Underwood, M.B., Saito, S., Naruse, H., Kutterolf, S., Scudder, R., Park, J.-O., Moore, G.F. & Slagle, A. 2013. Depositional architecture, provenance, and ectonic/eustatic modulation of Miocene submarine fans in the Shikoku Basin: Results from Nankai Trough Seismogenic Zone Experiment. *Geochemistry, Geophysics, Geosystems*, **14**, 1722–1739.

Pickering, K.T., Corregidor, J. & Clark, J. 2015. Architecture and stacking patterns of lower-slope and proximal basin-floor channelised submarine fans, Middle Eocene Ainsa system, Spanish Pyrenees: an Integrated outcrop–subsurface study. *Earth-Science Reviews*, **144**, 47–81.

Pickett, T.E., Kraft, J.C. & Smith, K. 1971. Cretaceous burrows: Chesapeake and Delaware Canal, Delaware. *Journal of Paleontology*, **45**, 209–211.

Pierce, C., Haughton, P.D.W., Shannon, P.M., Martinsen, O.J., Pulham, A. & Elliott, T. 2010. First results from behind-outcrop boreholes in Clare Basin turbidites, western Ireland. *American Association of Petroleum Geologists 2010 Annual Convention Abstracts Volume*, **90104**.

Pierson, T.C. 1981. Dominant particle support mechanisms in debris flows at Mt. Thomas, New Zealand, and implications for flow mobility. *Sedimentology*, **28**, 49–60.

Pierson, T.C. & Costa, J.E. 1987. A rheological classification of subaerial sediment-water flows. *In*: Costa, J.E. & Wieczorek, G.F. (eds), *Debris Flows/Avalanches: Process, Recognition, and Mitigation*, 1–12. Geological Society of America, Reviews in Engineering Geology, **VII**.

Pindell, J., Radovich, B. & Horn, B.W. 2011. Western Florida: a new exploration frontier in the US Gulf of Mexico. *GEO ExPro*, **8**, 36–40.

Pinet, N., Keating, P., Lavoie, D. & Brouillette, P. 2010. Forward potential-field modelling of the Appalachian orogen in Gaspé Peninsula (Quebec, Canada): implications for the extent of rift magmatism and the geometry of the Taconian orogenic wedge. *American Journal of Science*, **310**, 89–110.

Piper, D.J.W. 1970. A Silurian deep-sea fan deposit in western Ireland and its bearing on the nature of turbidity currents. *Journal of Geology*, **78**, 509–522.

Piper, D.J.W. 1972a. Turbidite origin of some laminated mudstones. *Geological Magazine*, **109**, 115–126.

Piper, D.J.W. 1972b. Sediments of the Middle Cambrian Burgess Shale, Canada. *Lethaia*, **5**, 169–175.

Piper, D.J.W. 1973. The sedimentology of silt turbidites from the Gulf of Alaska. *In*: Kulm, L.D., von Huene, R. *et al.*, *Initial Reports Deep Sea Drilling Project*, **18**, 847–167. Washington, DC: US Government Printing Office.

Piper, D.J.W. 1978. Turbidite muds and silts on deep-sea fans and abyssal plains. *In*: Stanley, D.J. & Kelling, G. (eds), *Sedimentation in Submarine Canyons, Fans, and Trenches*, 163–176. Stroudsburg, PA: Dowden, Hutchinson & Ross.

Piper, D.J.W. & Aksu, A.E. 1987. The source and origin of the 1929 grand banks turbidity current inferred from sediment budgets. *Geo-Marine Letters*, **7**, 177–182.

Piper, D.J.W. & Brisco, D.C. 1975. Deep-water continental-margin sedimentation, DSDP Leg 28, Antarctica. *In*: Hayes, D.E., Frakes, L.A. *et al.*, *Initial Reports Deep Sea Drilling Project*, **28**, 727–755. Washington, DC: US Government Printing Office.

Piper, D.J.W. & Deptuck, M. 1997. Fine-grained turbidites of the Amazon Fan: facies characterization and interpretation. *In*: Flood, R.D., Piper, D.J.W., Klaus, A. *et al.*, *Proceedings Ocean Drilling Program, Scientific Result*, **155**, 79–108. College Station, Texas, USA: Ocean Drilling Program.

Piper, D.J.W. & Fader, G.B. 1990. Acoustic and lithological data. *In*: Keen, M.J. & Williams, G.L. (eds), *Geology of the Continental Margin off Eastern Canada*, 494–497. Geological Survey of Canada, Geology of Canada, no. 2 (also Geological Society of America, The Geology of North America, v. **I-1**).

Piper, D.J.W. & Kontopoulos, N. 1994. Bedforms in submarine channels: comparison of ancient examples from Greece with studies of recent turbidite systems. *Journal of Sedimentary Research*, **64**, 247–252.

Piper, D.J.W. & Normark, W.R. 1983. Turbidite depositional patterns and flow characteristics, Navy Submarine Fan, California Borderland. *Sedimentology*, **30**, 681–694.

Piper, D.J.W. & Normark, W.R. 2001. Sandy fans - from Amazon to Hueneme and beyond. *American Association of Petroleum Geologists Bulletin*, **85**, 1407–1438.

Piper, D.J.W. & Savoye, B. 1993. Processes of late Quaternary turbidity current flow and deposition on the Var deep-sea fan, north-west Mediterranean Sea. *Sedimentology*, **40**, 557–582.

Piper, D.J.W. & Stow, D.A.V. 1991. Fine-grained turbidites. *In*: Einsele, G., Ricken, W. & Seilacher, A. (eds), *Cycles and Events in Stratigraphy*, 360–376. Berlin: Springer-Verlag.

Piper, D.J.W., Normark, W.R. & Ingle, J.C. 1976. The Rio Dell Formation: a Plio-Pleistocene basin slope deposit in northern California. *Sedimentology*, **23**, 309–328.

Piper, D.J.W., Panagos, A.G. & Pe, G.G. 1978. Conglomeratic Miocene flysch, western Greece. *Journal of Sedimentary Research*, **48**, 117–125.

Piper, D.J.W., von Huene, R. & Duncan, J.R. 1973. Late Quaternary Sedimentation in the Active Eastern Aleutian Trench. *Geology*, **1**, 19–22.

Piper, D.J.W., Shor, A.N., Farre, J.A., O'Connell, S. & Jacobi, R. 1985. Sediment slides and turbidity currents on the Laurentian Fan: sidescan sonar investigations near the epicenter of the 1929 Grand Banks earthquake. *Geology*, **13**, 538–541.

Piper, D.J.W., Shor, A.N. & Hughes-Clarke, J.E. 1988. The 1929 "Grand Banks" earthquake, slump, and turbidity current. *In*: Clifton, H.E. (ed.), *Sedimentologic Consequences of Convulsive Geologic Events*, 77–92. Geological Society of America Special Paper, **229**.

Piper, D.J.W., Pirmez, C., Manley, P.L., Long, D., Flood, R.D., Normark, W.R. & Showers, W. 1997. Mass transport deposits of the Amazon Fan. *In*: Proceedings Ocean Drilling Program, *Scientific Results*, **155**, 109–146. College Station, Texas, USA: Ocean Drilling Program.

Piper, D.J.W., Hiscott, R.N. & Normark, W.R. 1999. Outcrop-scale acoustic facies analysis and latest Quaternary development of Hueneme and Dume fans, offshore California. *Sedimentology*, **46**, 47–78.

Piper, D.J.W., Deptuck, M.E., Mosher, D.C., Hughes-Clarke, J. & Migeon, S. 2012. Erosional and depositional features of glacial meltwater discharges on the eastern Canadian continental margin. *In*: Prather, B.E., Deptuck, M.E., Mohrig, D., van Hoorn, B. & Wynn, R.B. (eds), *Application of the Principles of Seismic Geomorphology to Continental Slope and Base-of-slope Systems: Case Studies from Seafloor and Near-seafloor Analogues*, 61–80. SEPM (Society for Sedimentary Geology), **99**.

Pirmez, C. 1994. *Growth of a submarine meandering channel-levée system on Amazon Fan*. Ph.D. Thesis, Columbia University, New York.

Pirmez, C. & Flood, R.D. 1995. Morphology and structure of Amazon Channel. *In*: Proceedings Ocean Drilling Program, *Initial Reports*, **155**, 23–45 College Station, Texas, USA: Ocean Drilling Program.

Pirmez, C., Hiscott, R.N. & Kronen, J.D., Jr., 1997. Sandy turbidite successions at the base of channel-levée systems of the Amazon Fan revealed by FMS logs and cores: unraveling the facies architecture of large submarine fans. *In*: Flood, R.D., Piper, D.J.W., Klaus, A. & Peterson, L.C. (eds), *Proceedings of the Ocean Drilling Program, Scientific Results*, **155**, 7–33. College Station, Texas, USA: Ocean Drilling Program.

Pirmez, C., Beaubouef, R.T., Friedmann, S.J. & Mohrig, D.C. 2000. Equilibrium profile and baselevel in submarine channels: examples from Late Pleistocene systems and implications for the architecture of deepwater reservoirs. *In*: Weimer, P., Slatt, R.M., Coleman, J., Rosen, N.C., Nelson, H., Bouma, A.H., Styzen, M.J. & Lawrence, D.T. (eds), *Deep-water Reservoirs of the World*, 782–805. Houston, Texas: Gulf Coast Section, Society of Economic Paleontologists and Mineralogists.

Pirrie, D. & Riding, J.B. 1988. Sedimentology, palynology and structure of Humps Island, northern Antarctic Peninsula. *British Antarctic Survey Bulletin*, **80**, 1–19.

Pitman, W.C. III, 1978. Relationship between eustasy and stratigraphic sequences of passive margins. *Geological Society of America Bulletin*, **89**, 1389–1403.

Pitman, W.C. III, & Golovchenko, X. 1983. The effect of sea level change on the shelfedge and slope of passive margins. *In*: Stanley, D.J. & Moore, G.T. (eds), *The Shelfbreak: Critical Interface on Continental Margins*. Society of Economic Paleontologists and Mineralogists, Special Publication **33**.

Plafker, G. & Berg, H.C. 1994. Overview of the geology and tectonic evolution of Alaska. *In*: Plafker, G. & Berg, H.C. (eds), *The Geology of North America, G-1 of The Geology of Alaska,* 989–1021. Boulder, Colorado: Geological Society of America.

Platt, J.P. 1986. Dynamics of orogenic wedges and the uplift of high-pressure metamorphic rocks. *Geological Society of America Bulletin*, **97**, 1037–1053.

Platt, J.P., Leggett, J.K., Young, J., Raza, H. & Alam, S. 1985. Large-scale sediment underplating in the Makran accretionary prism, southwest Pakistan. *Geology*, **13**, 507–11.

Plink-Bjorklund P., Mellere, D. & Steel, R.J. 2001. Turbidite variability and architecture of sand-prone, deep-water slopes: Eocene clinoforms in the Central Basin, Spitsbergen. *Journal of Sedimentary Research*, **71**, 895-912.

Plink-Björklund, P. & Steel, R.J. 2004. Initiation of turbidity currents: outcrop evidence for Eocene hyperpycnal flow turbidites. *Sedimentary Geology*, **165**, 29–52.

Plint, A.G. 2009. High-frequency relative sea-level oscillations in Upper Cretaceous shelf clastics of the Alberta Foreland Basin: possible evidence for a glacio-eustatic control? *In*: Macdonald, D.I.M. (ed.), *Sedimentation, Tectonics and Eustasy: Sea-level Changes at Active Margins*, Chapter 22. Oxford, UK: Blackwell Publishing Ltd. doi: 10.1002/9781444303896.

Poag, C.W., Watts, A.B. *et al.* 1987. *Initial Reports Deep Sea Drilling Project*, **95**. Washington, DC: US Government Printing Office.

Poblet, J., Muñoz, J.A., Travé, A. & Serra-Kiel, J. 1998. Quantifying the kinematics of detachment folds using three-dimensional geometry: Application to the Mediano anticline (Pyrenees, Spain). *Geological Society of America Bulletin*, **110**, 111–125.

Polonia, A., Panieri, G., Gasperini, L., Gasparotto, A., Bellucci, L.G. & Torelli, L. 2013. Turbidite paleoseismology in the Calabrian Arc subduction complex (Ionian Sea). *Geochemistry, Geophysics, Geosystems*, **14**, 112–140.

Pond, S. & Picard, G.L. 1978. *Introduction to Dynamic Oceanography*. Oxford: Pergamon.

Porbski, S.J. & Steel, R.J. 2006. Deltas and Sea-Level Change. *Journal of Sedimentary Research*, **76**, 390–403.

Posamentier, H.W. 1988. Fluvial deposition in a sequence stratigraphic framework. *In*: James, D.P. & Leckie, D.A. (eds), *Sequences, Stratigraphy, Sedimentology; Surface and Subsurface*. CSPG Memoir, **15**, 582–583.

Posamentier, H. & Allen, G.P. 2000. *Siliciclastic Sequence Stratigraphy - Concepts and Applications*. SEPM Concepts in Sedimentology and Paleontology Series 7, 204 pp. Tulsa, Oklahoma: Society for Sedimentary Geology (SEPM). ISBN 1-56576-070-0.

Posamentier, H.W. & Jervey, M.T. 1988. Sequence stratigraphy; implications for facies models and reservoir occurrence. *In*: James, D.P. & Leckie, D.A. (eds), *Sequences, Stratigraphy, Sedimentology; Surface and Subsurface*, 1–2. CSPG Memoir, **15**.

Posamentier, H.W. & Kolla, V. 2003. Seismic geomorphology and stratigraphy of depositional elements in deep-water settings. *Journal of Sedimentary Research*, **73**, 367–388.

Posamentier, H.W. & Vail, P.R. 1988a. Eustatic controls on clastic deposition II–sequence and systems tract models. *In*: Wilgus, C.K., Hastings, B.S., Posamentier, H., Van Wagoner, J., Ross, C.A. & Kendall, C.G. St.C. (eds), *Sea-Level Changes: An Integrated Approach*, 125–154. Society of Economic Paleontologists and Mineralogists, Special Publication, 42.

Posamentier, H.W. & Vail, P.R. 1988b. Sequence stratigraphy; sequences and systems tract development. *In*: James, D.P. & Leckie, D.A. (eds), *Sequences, Stratigraphy, Sedimentology; Surface and Subsurface*, 571–572. CSPG Memoir, **15**.

Posamentier, H.W. & Walker, R.G. 2006. Deep-water turbidites and submarine fans. *In*: Posamentier, H.W. & Walker, R.G. (eds), *Facies*

Models Revisited., 397–520. Society of Economic Paleontologists & Mineralogists, Special Publication, **84**.

Posamentier, H.W., Jervey, M.T. & Vail, P.R. 1988. Eustatic Controls on Clastic Deposition I - Conceptual Framework. *In*: Wilgus, C.K., Hastings, B.S., Posamentier, H., Van Wagoner, J., Ross, C.A. & Kendall, C.G. St.C. (eds), *Sea-Level Changes: An Integrated Approach,* 109–124. Society of Economic Paleontologists and Mineralogists, Special Publication No. **42**.

Posamentier, H.W., James, D.P. & Allen, G.P. 1990. Aspects of sequence stratigraphy: recent and ancient examples of forced regressions. *American Association of Petroleum Geologists Bulletin*, **74**, 742.

Posamentier, H.W., Erskine, R.D. & Mitchum, Jr., R.M. 1991. Models for submarine-fan deposition within a sequence-stratigraphic framework. *In*: Weimer, P. & Link, M.H. (eds), *Seismic Facies and Sedimentary Processes of Submarine Fans and Turbidite Systems,* 127–136. New York: Springer.

Posamentier, H.W., Allen, G.P., James, D.P. & Tesson, M. 1992. Forced Regressions in a Sequence Stratigraphic Framework: Concepts, Examples, and Exploration Significance. *American Association of Petroleum Geologists Bulletin*, **76**, 1687–1709.

Posamentier, H.W., Meizarwin, Wisman, P.S. & Plawman, T., 2000, Deep water depositional systems - Ultra-deep Makassar Strait, Indonesia. *In*: Weimer, P., Slatt, R.M., Coleman, J., Rosen, N.C., Nelson, H., Bouma, A.H., Styzen, M.J. & Lawrence, D.T. (eds), *Deep-water Reservoirs of the World,* 806–816. Houston, Texas: Gulf Coast Section, Society of Economic Paleontologists and Mineralogists.

Postma, G. 1984. Slumps and their deposits in fan delta front and slope. *Geology*, **12**, 27–30.

Postma, G. 1986. Classification for sediment gravity-flow deposits based on flow conditions during sedimentation. *Geology*, **14**, 291–294.

Postma, G. & Cartigny, M.J.B. 2014. Supercritical and subcritical turbidity currents and their deposits – a synthesis. *Geology*, **42**, 987–990.

Postma, G., Nemec, W. & Kleinspehn, K.L. 1988. Large floating clasts in turbidites: a mechanism for their emplacement. *Sedimentary Geology*, **58**, 47–61.

Postma, G., Kleinhans, M.G., Meijer, P.-Th. & Eggenhuisen, J.T. 2008. Sediment transport in analogue flume models compared with real-world sedimentary systems: a new look at scaling evolution of sedimentary systems in a flume. *Sedimentology*, **55**, 1541–1557.

Postma, G., Kleverlaan, K. & Cartigny, M.J.B. 2014. Recognition of cyclic steps in sandy and gravelly turbidite sequences, and consequences for the Bouma facies model. *Sedimentology*, doi: 10.1111/sed.12135.

Postma, H. 1969. Suspended matter in the marine environment. *In*: *Morning Review. Lectures of the Second International Oceanographic Congress,* Moscow, 1966, 213–219.

Potter, P.E. 1978. Significance and origin of big rivers. *Journal of Geology*, **86**, 13–33.

Powers, D.W. & Easterling, R.G. 1982. Improved methodology for using embedded Markov chains to describe cyclical sediments. *Journal of Sedimentary Petrology*, **52**, 913–923.

Prather, B.E. 2000. Calibration and visualization of depositional process models for above-grade slopes: a case study from the Gulf of Mexico. *Marine and Petroleum Geology*, **17**, 619–638.

Prather, B.E. 2003. Controls on reservoir distribution, architecture and stratigraphic trapping in slope settings. *Marine and Petroleum Geology*, **20**, 529–545.

Prather, B.E., Booth, J.R., Steffens, G.S. & Craig, P.A. 1998. Classification, lithologic calibration, and stratigraphic succession of seismic facies of intraslope basins, deep-water Gulf of Mexico. *American Association of Petroleum Geologists Bulletin*, **82**, 701–728.

Prather, B.E., Keller, F.B. & Chapin, M.A. 2000. Hierarchy of deep-water architectural elements with reference to seismic resolution: implications for reservoir prediction and modeling. *In*: Weimer, P., Slatt, R.M., Coleman, J., Rosen, N.C., Nelson, H., Bouma, A.H., Styzen, M.J. & Lawrence, D.T. (eds), *Deep-water Reservoirs of the World,* 817–835. Houston, Texas: Gulf Coast Section, Society of Economic Paleontologists and Mineralogists.

Prather, B.E., Pirmez, C. & Winker, C.D. 2012a. Stratigraphy of linked intraslope basins: Brazos-Trinity system, western Gulf of Mexico. *In*: Prather, B.E., Deptuck, M.E., Mohrig, D., van Hoorn, B. & Wynn, R.B. (eds), *Application of the Principles of Seismic Geomorphology to Continental Slope and Base-of-slope Systems: Case Studies from Seafloor and Near-seafloor Analogues,* 83–109. SEPM (Society for Sedimentary Geology), **99**.

Prather, B.E., Pirmez, C. Sylvester, Z. & Prather, D.S. 2012b. Stratigraphy response to evolving geomorphology in a usbmarine apron perchedon the upper Nider Delta slope. *In*: Prather, B.E., Deptuck, M.E., Mohrig, D., van Hoorn, B. & Wynn, R.B. (eds), *Application of the Principles of Seismic Geomorphology to Continental Slope and Base-of-slope Systems: Case Studies from Seafloor and Near-seafloor Analogues,* 145–161. SEPM (Society for Sedimentary Geology), **99**.

Prather, B.E., Deptuck, M.E., Mohrig, D., van Hoorn, B. & Wynn, R.B. (eds), 2012c. *Application of the Principles of Seismic Geomorphology to Continental Slope and Base-of-slope Systems: Case Studies from Seafloor and Near-seafloor Analogues. SEPM (Society for Sedimentary Geology)*, **99**.

Pratson, L.F., Imran, J., Parker, G., Syvitski, J.P.M. & Hutton, E. 2000. Debris flows vs. turbidity currents: a modeling comparison of their dynamics and deposits. *In*: Bouma, A.H. & Stone, C.G. (eds), *Fine-grained Turbidite Systems,* 57–72. American Association of Petroleum Geologists. Memoir, **72** & Society of Sedimentary Geologists, Special Publication, **68**. Joint publication, Tulsa, Oklahoma.

Prave, A.R. & Duke, W.L. 1990. Small-scale hummocky cross-stratification in turbidites: a form of antidune stratification. *Sedimentology*, **37**, 531–539.

Prélat, A., Hodgson, D.M. & Flint, S.S. 2009. Evolution, architecture and hierarchy of distributary deep-water deposits: a high-resolution outcrop investigation of submarine lobe deposits from the Permian Karoo Basin, South Africa. *Sedimentology*, **56**, 2132–2154.

Prélat, A., Covault, J.A., Hodgson, D.M., Fildani, A. & Flint, S.S. 2010. Intrinsic controls on the range of volumes, morphologies, and dimensions of submarine lobes. *Sedimentary Geology*, **232**, 66–76.

Press, W.H., Flannery, B.P., Teukolsky, S.A. & Vetterling, W.T. 1986. *Numerical Recipes: The Art of Scientific Computing.* Cambridge U.K.: Cambridge University Press.

Pringle, J.K., Westerman, A.R., Clark1, J.D., Drinkwater, N.J. & Gardiner, A.R. 2004. 3D high-resolution digital models of outcrop analogue study sites to constrain reservoir model uncertainty: an example from Alport Castles, Derbyshire, UK. *Petroleum Geoscience*, **10**, 343–352.

Pringle, J.K., Brunt, R.L., Hodgson, D.M. & Flint, S.S. 2010. Capturing stratigraphic and sedimentological complexity from submarine channel complex outcrops to digital 3D models, Karoo Basin, South Africa. *Petroleum Geoscience*, **16**, 307–330.

Prior, D.B. & Coleman, J.B. 1982. Active slides and flow in underconsolidated marine sediments on the slopes of the Mississippi Delta. *In*: Saxov, S. & Nieuwenhuis, J.K. (eds), *Submarine Slides and Other Mass Movements,* 21–50. New York: Plenum.

Prior, D.B., Bornhold, B.D. & Johns, M.W. 1984. Depositional characteristics of a submarine debris flow. *Journal of Geology*, **92**, 707–727.

Propescu, I., Lericolais, G., Panin, N. & Normand, A. 2004. The Danube submarine canyon (Black Sea): morphology and sedimentary processes. *Marine Geology*, **206**, 249–265.

Puga-Bernabéu, A., Vonk, A.J., Nelson, C.S. & Kamp, P.J.J. 2009. Mangarara Formation: exhumed remnants of a middle Miocene, temperate carbonate, submarine channel-fan system on the eastern margin of Taranaki Basin, New Zealand. *New Zealand Journal of Geology & Geophysics*, **52**, 73–93.

Puig, P., Ogston, A.S., Mullenbach, B.L., Nittrouer, C.A. & Sternberg, R.W. 2003. Shelf-to-canyon sediment-transport processes on the Eel continental margin (northern California). *Marine Geology*, **193**, 129–149.

Puigdefabregas, C. & Souquet, P. 1986. Tectono-sedimentary cycles and depositional sequences of the Mesozoic and Tertiary from the Pyrenees. *Tectonophysics*, **129**, 173–203.

Pudsey, C.J. & Reading, H.G. 1982. Sedimentology and structure of the Scotland Group, Barbados. *In*: Leggett, J.K. (ed.), *Trench-Forearc Geology*, 197–214. The Geological Society of London, Special Publication, **10**. London: The Geological Society.

Puigdefabregas, C., Gjelberg, J. & Vaksdal, M. 2004. The Grès d'Annot in the Annot syncline: outer basin-margin onlap and associated soft-sediment deformation. *In*: Joseph, P. & Lomas, S.A. (eds), *Deep-Water Sedimentation in the Alpine Basin of SE France: New Perspectives on the Grès d'Annot and Related Systems*. Geological Society, London, Special Publication, **221**, 367–388. London: The Geological Society.

Pugh, F.J. & Wilson, K.C. 1999. Velocity and concentration distributions in sheet flow above plane beds. *Journal of Hydraulic Engineering (ASCE)*, **125**, 117–125.

Püspöki, Z., Tóth-Makk, A., Kozák, M., Dávid, A., McIntosh, R.W., Buday, T., Demeter, G., Kiss, J., Püspöki-Terebesi, M., Barta, K., Csordás, C. & Kiss, J. 2009. Truncated higher order sequences as responses to compressive intraplate tectonic events superimposed on eustatic sea-level rise. *Sedimentary Geology*, **219**, 208–236.

Pyles, D. 2008. Multiscale stratigraphic analysis of a structurally confined submarine fan: Carboniferous Ross Sandstone, Ireland. *American Association of Petroleum Geologists Bulletin*, **92**, 557–587.

Pyles, D.R. & Jennette, D.C. 2009. Geometry and architectural associations of co-genetic debrite-turbidite beds in basin-margin strata, Carboniferous Ross Sandstone (Ireland): Applications to reservoirs located on the margins of structurally confined submarine fans. *Marine and Petroleum Geology*, **26**, 1974–1996.

Pyles, D.R., Syvitski, J.P.M. & Slatt, R.M. 2011. Defining the concept of stratigraphic grade and applying it to stratal (reservoir) architecture and evolution of the slope-to-basin profile: An outcrop perspective. *Marine and Petroleum Geology*, **28**, 675–697.

Pyles, D.R., Tomasso, M. & Jennette, D.C. 2012. Flow processes and sedimentation associated with erosion and filling of sinuous submarine channels. *Geology*, **40**, 143–146.

Pyrcz, M.J., Catuneanu, O. & Deutsch, C.V. 2005. Stochastic surface-based modeling of turbidite lobes. *American Association of Petroleum Geologists Bulletin*, **89**, 177–191.

Pysklywec, R.N. & Mitrovica, J.X. 1999. The role of subduction-induced subsidence in the evolution of the Karoo Basin. *Journal of Geology*, **107**, 155–164.

Quatrefages, M.A. de. 1849. Note sur la Scolicia prisca (A. De Q.) annélide fossile de la craie. Annales des *Sciences Naturelles, 3 Sèrie, Zoologie*, **12**, 265–266.

Quinlan, G.M. & Beaumont, C. 1984. Appalachian thrusting, lithospheric flexure, and the Paleozoic stratigraphy of the Eastern Interior of North America. *Canadian Journal of Earth Sciences*, **21**, 973–996.

Radhakrishnan, S., Srikanth, G. & Mehta, C.H. 1991. Segmentation of well logs by maximum likelihood estimation: the algorithm and Fortran-77 implementation. *Computers and Geosciences*, **17**, 1173–1196.

Radovich, B., Horn, B., Nutall, P. & McGrail, A. 2011. The only complete regional perspective: RTM re-processing gives a new look at the Gulf of Mexico continental margin. *GEO ExPro*, **8**, 38–40.

Rahe, B., Ferrill, D.A. & Morris, A.P. 1998. Physical analog modeling of pull-apart basin evolution. *Tectonophysics*, **285**, 21–40.

Rajchel, J. & Uchman, A. 1998. Ichnological analysis of an Eocene mixed marly-siliciclastic flysch deposits in the Nienadowa Marl Member, Skole Unit, Polish Flysch Carpathians. *Annales Societatis Geologorum Poloniae*, **68**, 61–74.

Ramsay, A.T.S. 1977. Sedimentological clues to palaeo-oceanography. *In*: Ramsay, A.T.S. (ed.), *Oceanic Micropalaeontology*, 1371–1453. London: Academic Press.

Ravenne, C., Vially, R., Riche, P. & Tremolieres, P. 1987. Sédimentation et tectonique dans le bassin marin Eocène supérieur-Oligocène des Alpes du sud. *Revue de l'Institut Français du Pétrole*, **42**, 529–553.

Ray R.D., Egbert, G.D. & Erofeeva, S.Y. 2005. A brief overview of tides in the Indonesian seas. *Oceanography*, **18**, 74–79.

Raymo, M.E., Ruddiman, W.F. & Froelich, P.N. 1988. Influence of late Cenozoic mountain building on ocean geochemical cycles. *Geology*, **16**, 649–653.

Rea, D.K., Basov, I.A., Krissek, L.A. & Leg 145 Scientific Party 1995. Scientific results of drilling the North Pacific transect. *In*: Rea, D.K., Basov, I.A., Scholl, D.W. & Allan, J.F. (eds), *Proceedings of the Ocean Drilling Program, Scientific Results*, **145**, 577–596. College Station, Texas, USA: Ocean Drilling Program.

Reading, H.G. 1980. Characteristics and recognition of strike-slip fault systems. *International Association of Sedimentologists Special Publication*, **4**, 7–26.

Reading, H.G. & Richards, M. 1994. Turbidite systems in deep-water basin margins classified by grain size and feeder system. *American Association of Petroleum Geologists Bulletin*, **78**, 792–822.

Rebesco, M. & Camerelenghi, A. (eds) 2008. *Contourites. Developments in Sedimentology*, **60**, 769pp. Amsterdam: Elsevier B.V. ISBN: 978-0-444-52998-5.

Rebesco, M., Hernández-Molina, F.J., van Rooij, D. & Wåhlin, A. 2014. Contourites and associated sediments controlled by deep-water circulation processes: State-of-the-art and future considerations. *Marine Geology*, **352**, 111–154.

Reece, R.S., Gulick, S.P.S., Horton, B.K., Christeson, G.L. & Worthington, L.L. 2011. Tectonic and climatic influence on the evolution of the Surveyor Fan and channel system, Gulf of Alaska. *Geosphere*, **7**, 830–844.

Rees, A.I. 1968. The production of preferred orientation in a concentrated dispersion of elongated and flattened grains. *Journal of Geology*, **76**, 457–465.

Reijmer, J.J.G., Betzler, C., Kroon, D., Tiedemann, R & Eberli, G.P. 2002. Bahamian carbonate platform development in response to sea-level changes and the closure of the Isthmus of Panama. *International Journal of Earth Sciences (Geologisches Rundschau)*, **91**, 482–489.

Reimnitz, E. 1971. Surf-beat origin for pulsating bottom currents in the Rio Balsas submarine canyon, Mexico. *Geological Society of America Bulletin*, **82**, 81–90.

Reimnitz, E. & Bruder, K.F. 1972. River discharge into an ice covered ocean and related sediment dispersal, Beaufort Sea, coast of Alaska. *Geological Society of America Bulletin*, **83**, 861–866.

Reimnitz, E. & Gutierrez-Estrada, M. 1970. Rapid changes in the head of the Rio Balsas Submarine Canyon system, Mexico. *Marine Geology*, **8**, 245–258.

Reineck, H.E. 1963. Sedimentgefüge im Bereich der südlichen Nordsee. *Abhandlungen der Senckenbergischen Naturforschenden Gesellschaft*, **505**, 1–138.

Reineck, H.E. 1973. Schichtung und Wühlgefüge in Grundproben vor der ostafrikanischen Küste. *Meteor-Forschungs-Ergebnisse, Reihe C*, **16**, 67–81.

Reiser, C. & Bird, T. 2013. Recorded broadband 3D: improving reservoir understanding and characterization with recorded broadband seismic. *GEO ExPro*, **10**, 76–78.

Remacha, E. & Fernández, L.P. 2003. High resolution correlation patterns in the turbiditic systems of the Hecho Group south-central Pyrenees, Spain. *Marine and Petroleum Geology*, **20**, 711–726.

Remacha, E. & Fernandez, L.P. 2005. The Ttransition between sheet-like lobe and basin-plain turbidites in the Hecho Basin (south-central Pyrenees, Spain). *Journal of Sedimentary Research*, **75**, 798–819.

Remacha, E., Pickart, J. & Oms, O. 1991. The Rapitan turbidite channel. *In*: Colombo, F., Ramos-Guerrero, E. & Riera, S. (eds), *1st Congress of the Spanish Group on the Tertiary*, 280–282. Barcelona: University of Barcelona.

Remacha, E., Oms, O. & Coello, J. 1995. The Rapitan turbidite channel and its related eastern levée-overbank deposits, Eocene Hecho group, south-central Pyrenees, Spain. *In*: Pickering, K.T., Hiscott, R.N., Kenyon, N.H., Ricci Lucchi, F. & Smith, R.D.A. (eds), *Atlas of Deep Water Environments: Architectural Style in Turbidite Systems*, 145–149. London: Chapman and Hall.

Remacha, E., Fernández, L.P., Maestro, E., Oms, O., Estrada, R. & Teixell, A. 1998. The Upper Hecho Group turbidites and their vertical evolution to deltas Eocene, south-central Pyrenees. *In: Association of Sedimentologists 15th International Sedimentological Congress Field Trip Guidebook*, University d'Alacant. Alacante, 1–25.

Remacha, E., Oms, O., Gual, G., Bolaño, F., Climent, F., Fernandez, L.P., Crumeyrollle, P., Pettingill, H., Vicente, J.C. & Suarez, J. 2003. Sand-rich turbidite systems of the Hecho Group from slope to basin plain. *Facies, stacking patterns, controlling factors and diagnostic features. Geological Field Trip 12. South-Central Pyrenees. American Association of Petroleum Geologists International Conference and Exhibition*, Barcelona, Spain, September 21–24, 78.

Remacha, E., Fernández, L.P. & Maestro, E. 2005. The transition between sheet-like lobe and basin-plain turbidites in the Hecho Basin south-central Pyrenees, Spain. *Journal of Sedimentary Research*, **75**, 795–819.

Reza, Z.A., Pranter, M.J. & Weimer, P. 2006. ModDRE: A program to model deepwater-reservoir elements using geomorphic and stratigraphic constraints. *Computers & Geosciences*, **32**, 1205–1220.

Rial, J.A. 1999. Pacemaking the Ice Ages by frequency modulation of Earth's orbital eccentricity. *Science*, **285**, 564–568.

Ribeiro Machado, L.C., Kowsmann, R.O., de Almeida, W. Jr, Murakami, C.Y., Schreiner, S., Miller, D.J., Orlando, P. & Piauilino, V. 2004. Geometry of the proximal part of the modern turbidite depositional system of the Carapebus Formation, Campos Basin: a model for reservoir heterogeneities. *Boletim de Geociencias da Petrobras*, **12**, 287–315.

Ricci Lucchi, F. 1969. Channelized deposits in the middle Miocene flysch of Romagna (Italy). *Giornale di Geologia*, **36**, 203–282.

Ricci Lucchi, F. 1975a. Miocene palaeogeography and basin analysis in the Periadriatic Apennines. *In*: Squyres, C. (ed.), *Geology of Italy*, 5–111. Tripoli: Petrolm Exploration Society Libya.

Ricci Lucchi, F. 1975b. Depositional cycles in two turbidite formations of northern Apennines. *Journal of Sedimentary Petrology*, **45**, 1–43.

Ricci Lucchi, F. 1978. Turbidite dispersal in a Miocene deep-sea plain: the Marnoso-arenacea of the Northern Apennines. *Geologie en Mijnbouw*, **57**, 550–576.

Ricci Lucchi, F. 1981. The Marnoso-arenacea: a migrating turbidite basin 'over-supplied, by a highly efficient dispersal svstem'. *In*: Ricci Lucchi, F. (ed.), *Excursion guidebook with Contributions on Sedimentology of some Italian Basins*, 231–275. 2nd European Regional Meeting, Bologna, Italy. International Association of Sedimentologists.

Ricci Lucchi, F. 1984. The deep-sea fan deposits of the Miocene Marnoso-arenacea Formation, northern Apennines. *Geo-Marine Letters*, **3**, 203–210.

Ricci Lucchi, F. 1986. The Oligocene to Recent foreland basins of the northern Apennines. *In*: Allen, P.A. & Homewood, P. (eds). *Foreland Basins*, 105–139. International Association of Sedimentologists Special Publication, **8**. Oxford: Blackwell Scientific.

Ricci Lucchi, F. 1995. *Sedimentographica: A Photographic Atlas of Sedimentary Structures*, 2nd edn. New York: Columbia University Press.

Ricci Lucchi, F. & Ori, G.G. 1984. Orogenic clastic wedges of the Alps and Apennines (abstract). *Geological Society of America Bulletin*, **64**, 798.

Ricci Lucchi, F. & Ori, G.G. 1985, Field excursion D: syn-orogenic deposits of a migrating basin system in the NW Adriatic foreland: examples from Emilia-Romagna region, northern Apennines. *In*: Allen, P.A., Homewood, P. & Williams, G. (eds), *International Symposium on Foreland Basins, Excursion Guidebook*, 137–176. International Association of Sedimentologists. London: CSP Economic Publications Limited.

Ricci Lucchi, F. & Pialli, G. 1973. Apporti secondari nella Marnoso-arenacea; 1. Torbiditi di conoide e di pianura sotto-marine a Est-Nordest di Perugia. *Bulletin of the Geological Society*, Italy, **92**, 669–712.

Ricci Lucchi, F. & Valmori, E. 1980. Basin-wide turbidites in a Miocene, over-supplied deep-sea plain: a geometrical analysis. *Sedimentology*, **27**, 241–270.

Rickard, L.V. & Fisher, D.W. 1973. Middle Ordovician Normanskill Formation, eastern New York: age, stratigraphic and structural position. *American Journal of Science*, **273**, 580–590.

Richards, M., Bowman, M. & Reading, H. 1998. Submarine-fan systems I: characterization and stratigraphic prediction. *Marine and Petroleum Geology*, **15**, 689–717.

Richards, P.C., Ritchie, J.D., Thomson, A.R. 1987. Evolution of deep-water climbing dunes in the Rockall Trough – implications for overflow currents across the Wyville-Thomson Ridge in the (?)Late Miocene. *Marine Geology*, **76**, 177–183.

Richardson, A.N. & Blundell, D.J. 1996. Continental collision in the Banda arc. *In*: Hall, R. & Blundell, D. (eds), *Tectonic Evolution of Southeast Asia*, 47–60. Geological Society of London Special Publication, **106**. Bath: The Geological Society, London.

Richardson, M.J., Wimbush, M. & Mayer, L. 1981. Exceptionally strong near-bottom flows on the continental rise of Nova Scotia. *Science*, **213**, 887–888.

Richardson, N.J. & Underhill, J.R. 2002. Controls on the structural architecture and sedimentary character of syn-rift sequences, North Falkland Basin, South Atlantic. *Marine and Petroleum Geology*, **19**, 417–443.

Ridd, M.F., Barber, A.J. & Crow, M.J. (eds) 2011. *The Geology of Thailand*. London: The Geological Society, 626 pp.

Ridente, D., Tricardi, F & Asioli, A. 2009. The combined effect of sea level and supply during Milankovitch cyclicity: Evidence from shallow-marine $\delta^{18}O$ records and sequence architecture (Adriatic margin). *Geology*, **37**, 1003–1006.

Riedel, M., Collett, T.S., Malone, M.J. & Expedition 311 Scientists 2005. *Proceedings of the Integrated Ocean Drilling Program, 311*. Washington, DC: Integrated Ocean Drilling Program Management International Inc.

Rieth, A. 1932 Neue Funde spongeliomorpher Fucoiden aus dem Jura Schwabens. *Geologische und Palaeontologische Abhandlungen*, **19**, 257–294.

Rigsby, C.A., Zierenberg, R.A. & Baker, P.A. 1994. Sedimentary and diagenetic structures and textures in turbiditic and hemiturbiditic strata as revealed by whole-core X-radiography, Middle Valley, northern Juan de Fuca Ridge. *In*: Mott, M.J., Davis, E.E., Fisher, A.T. & Slack, J.F. (eds), *Proceedings of the Ocean Drilling Program, Scientific Results*, **139**, 105–111. College Station, Texas, USA: Ocean Drilling Program.

Rimoldi, B., Alexander, J. & Morris, S. 1996. Experimental turbidity currents entering density-stratified water: analogues for turbidites in Mediterranean hypersaline basins. *Sedimentology*, **43**, 527–540.

Rindsberg, A.K. 2012. Ichnotaxonomy: finding patterns in a welter of information. *In*: Knaust, D. & Bromley, R.G. (eds), *Trace Fossils as Indicators of Sedimentary Environments*, 45–78. Developments in Sedimentology, **64**. Amsterdam: Elsevier.

Riva, J. 1968. Graptolite faunas from the Middle Ordovician of the Gaspé north shore. *Naturaliste Canadien*, **93**, 1379–1400.

Riva, J. 1974. A revision of some Ordovician graptolites of eastern North America. *Palaeontology*, **17**, 1–40.

Roberts, D.G. & Kidd, R.B. 1979. Abyssal sediment-wave fields on Feni Ridge, Rockall Trough: long range sonar studies. *Marine Geology*, **33**, 175–191.

Roberts, D. & Siedlecka, A. 2012. Provenance and sediment routing of Neoproterozoic formations on the Varanger, Nordkinn, Rybachi and Sredni peninsulas, North Norway and Northwest Russia: a review. *Norges Geologiske Undersøkelse Bulletin*, **452**, 1–19.

Roberts, H.H. & Thayer, D.A. 1985. Petrology of Mississippi fan depositional environments. *In*: Bouma, A.H., Normark, W.R. & Barnes, N.E. (eds), *Submarine Fans and Related Turbidite Systems, Chapter 47*. New York: Springer-Verlag.

Roberts, H.H., Cratsley, D.W. & Whelan, T. 1976. *Stability of Mississippi Delta Sediments as Evaluated by Analysis of Structural Features in Sediment Borings*. Offshore Technical Conference Pap. **OTC2425**.

Roberts, M.T. 1983. Seismic examples of complex faulting from northwest shelf of Palawan, Philippines. *In*: Bally, A.W. (ed.), *Seismic Expression of Structural Styles*, **3**, 4.2-18–4.2-24. American Association of Petroleum Geologists, Studies in Geology, Series 15.

Robertson, A.H.F. 1976. Pelagic chalks and calciturbidites from the Lower Tertiary of the Troodos Massif, Cyprus. *Journal of Sedimentary Petrology*, **46**, 1007–1016.

Robertson, A.H.F. 1984. Origin of varve-type lamination, graded claystones and limestone-shale 'couplets', in the Lower Cretaceous of the western North Atlantic. *In*: Stow, D.A.V. & Piper, D.J.W. (eds), *Fine-grained Sediments: Deep-water Processes and Facies*, 437–452. Geological Society of London Special Publication, **15**. Oxford: Blackwell Scientific.

Robertson, A.H.F. 1987. The transition from a passive margin to an Upper Cretaceous foreland basin related to ophiolite emplacement in the Oman Mountains. *American Association of Petroleum Geologists Bulletin*, **99**, 633–653.

Robertson, A.H.F. & Ogg, J.G. 1986. Palaeoceanographic setting of the Callovian North Atlantic. *In*: Summerhayes, C.P. & Shackleton, N.J. (eds), *North Atlantic Palaeoceanography*, 283–298. Geological Society of London Special Publication, **21**. Oxford: Blackwell Scientific.

Robertson, R. & Ffield, A. 2008. Baroclinic tides in the Indonesian seas: tidal fields and comparisons to observations. *Journal of Geophysical Research*, **113**, C07031, doi:10.1029/2007JC004677.

Rocheleau, M. & Lajoie, J. 1974. Sedimentary structures in resedimented conglomerate of the Cambrian flysch, L'Islet, Quebec Appalachians. *Journal of Sedimentary Petrology*, **44**, 826–836.

Rock, N.M.S. 1988. *Numerical Geology*. Berlin: Springer-Verlag.

Rodine, J.D. & Johnson, A.M. 1976. The ability of debris, heavily freighted with coarse clastic materials, to flow, on gentle slopes. *Sedimentology*, **23**, 213–234.

Röhl, U., Bralower, T.J., Norris, R.D. & Wefer, G. 2000. New chronology for the late Paleocene thermal maximum and its environmental implications. *Geology*, **28**, 927–930.

Röhl, U., Westerfield, T., Bralower, T.J. & Zachos, J.C. 2007. On the duration of the Paleocene–Eocene thermal maximum (PETM). *Geochemistry, Geophysics, Geosystems*, **8**, doi:10.1029/2007GC001784.

Romans, B.W., Fildani, A., Hubbard, S.M., Covault, J.A., Fosdick, J.C. & Graham, S.A. 2011. Evolution of deep-water stratigraphic architecture, Magallanes Basin, Chile. *Marine and Petroleum Geology*, **28**, 612–628.

Rosas, S., Fontboté, L. & Tankard, A. 2007. Tectonic evolution and paleogeography of the Mesozoic Pucara Basin, central Peru. *Journal of South American Earth Sciences*, **24**, 1–24.

Rosencrantz, E. & Mann, P. 1991. SeaMARC II mapping of transform faults in the Cayman Trough, Caribbean Sea. *Geology*, **19**, 690–693.

Rosidi, H.M.D., Suwitodirdjo, K. & Tjokrosapoetro, S., 1981. *Geological Map of the Kupang-Atambua Quadrangles, Timor*, scale 1:250.000. Bandung, Indonesia: Geological Research and Development Centre.

Ross, G.M. 1991. Tectonic setting of the Windermere Supergroup revisited. *Geology*, **19**, 1125–1128.

Ross, G.M. & Arnott, R.W. 2006. Regional geology of the Windermere Supergroup, southern Canadian Cordillera and stratigraphic setting of the Castle Creek study area. *In*: Nilsen, T.H., Shew, R.D., Steffens, G.S.& Studlick, J.R.J. (eds), *Atlas of Deep-water Outcrops*. American Association of Petroleum Geologists, Studies in Geology, **56**.

Ross, G.M., Bloch, J.D. & Krouse, H.R. 1995. Neoproterozoic strata of the southern Canadian Cordillera and the evolution of seawater sulfate: *Precambrian Research*, **73**, 71–99.

Ross, R.J. Jr., and 28 others 1982. *The Ordovician System in the United States: Correlation Chart and Explanatory Notes. International Union of Geological Sciences Publication*, **12**. Paris: IUGS Secretariat.

Ross, W.C. 1989. Modeling base-level dynamics as a control on basin-fill geometries and facies distribution: a conceptual framework. *In*: Cross, T. (ed.), *Quantitative Dynamic Stratigraphy*, 387–399. Englewood Cliffs, NJ: Prentice-Hall.

Ross, W.C., Watts, D.E. & May, J.A. 1995. Insights from stratigraphic modeling: mud limited versus sand-limited depositional systems. *American Association of Petroleum Geologists Bulletin*, **79**, 231–258.

Rossignol-Strick, M. 1985. Mediterranean Quaternary sapropels, an immediate response of the African monsoon to variations of insolation. *Palaeogeography, Palaeoclimatology, Palaeoecology*, **49**, 237–263.

Roth, S. & Reijmer, J.J.G. 2004. Holocene Atlantic climate variations deduced from carbonate periplatform sediments (leeward margin, Great Bahama Bank). *Paleoceanography*, **19**, 1–14.

Rothman, D.H. & Grotzinger, J.P. 1994. Scaling in turbidite deposition. *Journal of Sedimentary Research*, **64**, 59–67.

Rothman, D.H. & Grotzinger, J.P. 1995. Scaling properties of gravity-driven sediments. *Nonlinear Processes in Geophysics*, **2**, 178–185.

Rothman, D.H. & Grotzinger, J.P. 1996. Scaling properties of gravity-driven sediments. *Nonlinear Processes Geophysics*, **2**, 178–185.

Rothman, D.H. & Grotzinger, J.P. 1997. On the power law size distribution of turbidite beds. *Basin Research*, **4**, 263–274.

Rothman, D.H., Grotzinger, J.P. & Flemings, P. 1994. Scaling in turbidite deposition. *Journal of Sedimentary Research*, **64**, 59–67.

Rothwell, R.G., Weaver, P.P.E., Hodkinson, R.A., Prat, C.E., Styzen, M.J. & Higgs, N.C. 1994. Clayey nannofossil ooze turbidites and hemipelagites at Sites 834 and 835 (Lau Basin, southwest Pacific). *In*: Hawkins, J., Parson, L., Allan, J., *et al.*, *Proceedings of the Ocean Drilling Program, Scientific Results*, **135**, 101–130. College Station, Texas, USA: Ocean Drilling Program.

Rothwell, R.G., Thomson, J. & Kahler, G. 1998. Low-sea-level emplacement of a very large Late Pleistocene 'megaturbidite' in the western Mediterranean Sea. *Nature*, **392**, 377–380.

Rouse, H. 1937. Modern conceptions of the mechanics of turbulence. *Transactions of the American Society of Civil Engineers*, **102**, 436–505.

Roveri, M. 2002. Sediment drifts of the Corsica Channel, northern Tyrrhenian Sea. *Geological Society, London, Memoirs*, **22**, 191–208.

Roveri, M., Ricci Lucchi, F., Lucente, C.C., Manzi, V. & Mutti, E. 2002. Stratigraphy, facies and basin fill history of the Marnoso-arenacea Formation. *In*: Mutti, E., Ricci Lucchi, F. & Roveri, M. (eds), *Revisiting Turbidites of the Marnoso-Arenacea Formation and their Basin-Margin Equivalents: Problems with Classic Models*, III-1 to III-15. Excursion Guidebook, Università di Parma and Eni-Agip Division, 64th European Association of Geoscientists & Engineers Conference and Exhibition, Florence (Italy).

Roveri, M., Manzi, V., Ricci Lucchi, F. & Rogledi, S. 2003. Sedimentary and tectonic evolution of the Vena del Gesso basin (northern Apennines, Italy): implications for the onset of the Messinian salinity crisis. *Geological Society of America Bulletin*, **115**, 387–405.

Royden, L. & Keen, C.E. 1980. Rifting process and thermal evolution of the continental margin of eastern Canada determined from subsidence curves. *Earth and Planetary Science Letters*, **51**, 343–361.

Ruff, L. & Kanamori, H. 1993. Seismic coupling and uncoupling at subduction zones. *Tectonophysics*, **99**, 99–117.

Rupke, N.A. 1975. Deposition of fine-grained sediments in the abyssal environment of the Algero-Balearic Basin, western Mediterranean Sea. *Sedimentobgy*, **22**, 95–109.

Rutherford, E., Burke, K. & Lytwyn, J. 2001. Tectonic history of Sumba Island, Indonesia, since the Late Cretaceous and its rapid escape into the forearc in the Miocene. *Journal of Asian Earth Sciences*, **19**, 453–

Ryan, H.F. & Scholl, D.W. 1993. Geologic implications of great interplate earthquakes along the Aleutian arc. *Journal of Geophysical Research*, **98**, 22135–22146.

Ryan, H.F., Draut, A.E., Keranen, K. & Scholl, D.W. 2012. Influence of the Amlia fracture zone on the evolution of the Aleutian Terrace forearc basin, central Aleutian subduction zone. *Geosphere*, **8**, 1254–1273.

Sachse, V.F., Strozyk, F., Anka, Z., Rodriguez, J.F. & di Primio, R. 2015. The tectono-stratigraphic evolution of the Austral Basin and adjacent areas against the background of Andean tectonics, southern Argentina, *South America. Basin Research*, doi: 10.1111/bre.12118.

Sacks, A., Saffer, D.M. & Fisher, D. 2013. Analysis of normal fault populations in the Kumano forearc basin, Nankai Trough, Japan: 2. Principal axes of stress and strain from inversion of fault orientations. *Geochemistry, Geophysics, Geosystems*, **14**, 1973–1988.

Sadler, P.M. 1982. Bed thickness and grain size of turbidites. *Sedimentology*, **29**, 37–51.

Saffer, D.M. & Marone, C. 2003. Comparison of smectite- and illite-rich gouge frictional properties: application to the updip limit of the seismogenic zone along subduction megathrusts. *Earth and Planetary Science Letters*, **215**, 219–235.

Saffer, D.M., Flemings, P.B., Boutt, D., Doan, M.-L., Ito, T., McNeill, L., Byrne, T., Conin, M. Lin, W. Kano, Y. Araki, E. Eguchi, N. & Toczko, S. 2013. In situ stress and pore pressure in the Kumano Forearc Basin, offshore SW Honshu from downhole measurements during riser drilling. *Geochemistry, Geophysics, Geosystems*, **14**, 1454–1470.

Saffer, D.M. & Screaton, E.J. 2003. Fluid flow at the toe of convergent margins: interpretation of sharp pore-water geochemical gradients. *Earth and Planetary Science Letters*, **213**, 261–270.

Saffer, D.M. & Tobin, H.J. 2011. *Hydrogeology and mechanics of subduction zone forearcs: fluid flow and pore pressure. Annual Review of Earth and Planetary Sciences*, **39**, 157–186.

Saffer, D.M., Underwood, M.B. & McKiernan, A.W. 2008. Evaluation of factors controlling smectite transformation and fluid production in subduction zones: Application to the Nankai Trough. *Island Arc*, **17**, 208–230.

Sagri, M. 1972. Rhythmic sedimentation in the turbidite sequences of the northern Apennines (Italy). *In*: *Proceedings of the 24th International Geological Congress*, **6**, 82–87.

Sagri, M. 1974. Rhythmic sedimentation in the deep-sea carbonate turbidites (Monte Antola Formation, northern Apennines). *Bulletin of the Geological Society, Italy*, **93**, 1013–1027.

Sahagian, D., Pinous, O., Olferiev, A. & Zakharov, V. 1996. Eustatic curve for the Middle Jurassic–Cretaceous based on Russian platform and Siberian stratigraphy: Zonal resolution. *American Association of Petroleum Geologists Bulletin*, **80**, 1,433–1,458.

Saito, S. & Goldberg, D. 2001. Compaction and dewatering processes of the oceanic sediments in the Costa Rica and Barbados subduction zones: estimates from in situ physical property measurements. *Earth and Planetary Science Letters*, **191**, 283–293.

Saito, S., Underwood, M.B. & Kubo, Y. 2009. NanTroSEIZE Stage 2: subduction inputs. *International Ocean Drilling Program Scientific Prospectus*, **322**. doi:10.2204/iodp.sp.322.2009.

Saito, S., Underwood, M.B., Kubo, Y. & the Expedition 322 Scientists 2010. *Proceedings of the Integrated Ocean Drilling Program*, **322**. Tokyo (Integrated Ocean Drilling Program Management International, Inc.). doi:10.2204/iodp.proc.322.103.2010.

Sakariassen, R., Dowd, N. & Lowrey, C. 2012. Broadband 3D: building blocks for exploration in a mature setting. *GEO ExPro*, **9**, 36–40.

Salaheldin, T.M., Imran, J., Chaudhry, M.H. & Reed, C. 2000. Role of fine-grained sediment in turbidity current flow dynamics and resulting deposits. *Marine Geology*, **171**, 21–38.

Salles, L., Ford, M., Joseph, P., Le Carlier de Veslud, C. & Le Solleuz, A. 2011. Migration of a synclinal depocentre from turbidite growth strata: the Annot syncline, SE France. *Bulletin de la Société Géologique de France*, **182**, 199–220.

Salles, L., Ford, M. & Joseph, P. 2014. Characteristics of axially-sourced turbidite sedimentation on an active wedge-top basin (Annot Sandstone, SE France). *Marine and Petroleum Geology*, **56**, 305–323.

Samson, T.M., Cruse, A.M. & Paxton, S.T. 2006. Spectral gamma ray logs as paleoenvironmental indicators in Carboniferous black shales. *Geological Society of America Abstracts with Programs*, **38**, 23.

Sanchez, G.J., Baptista, N., Parra, M., Montilla, L., Guzman, O.J. & Finno, A. 2011. The Monagas fold–thrust belt of eastern Venezuela. Part II: structural and palaeo-geographic controls on the turbidite reservoir potential of the middle Miocene foreland sequence. *Marine and Petroleum Geology*, **28**, 70–80.

Sanders, J.E. 1965. Primary sedimentary structures formed by turbidity currents and related resedimentation mechanisms. *In*: Middleton, G.V. (ed.), *Primary Sedimentary Structures and their Hydrodynamic Interpretation*, 192–219. Society of Economic Paleontologists and Mineralogists Special Publication, 12.

Sandy, M.R., Lazăr, I., Peckmann, J., Birgel, D., Stoica, M. & Roban, R.D. 2012. Methane-seep brachiopod fauna within turbidites of the Sinaia Formation, Eastern Carpathian Mountains, Romania. *Palaeogeography, Palaeoclimatology, Palaeoecology*, **323–325**, 42–59.

Sanford, B.V., Grant, A.C., Wade, J.A. & Barss, M.S. 1979. *Geology of Eastern Canada and Adjacent Areas. Geological Survey of Canada Map*, **1401A**.

Sani, F., Ventisette, C. Del, Montanari, D., Coli, M., Nafissi, P., Piazzini, A. 2004. Tectonic evolution of the internal sector of the Central Apennines, Italy. *Marine and Petroleum Geology*, **21**, 1235–1254.

Sari, E. & Çagatay, M.N. 2006. Turbidites and their association with past earthquakes in the deep Çınarcık Basin of the Marmara Sea. *Geo-Marine Letters*, **26**, 69–76.

Satake, K., Baba, T., Hirata, K., Iwasaki, S.-I., Kato, T., Koshimura, S., Takenaka, J. & Terada, Y. 2005. Tsunami source of the 2004 off the Kii Peninsula earthquakes inferred from offshore tsunami and coastal tide gauges. *Earth, Planets and Space*, **57**, 173–178.

Sattar, N., Juhlin, C. & Ahmad, N. 2012. *Seismic Stratigraphic Framework of an Early Cretaceous Sand Lobe at the Slope of Southern Loppa High, Barents Sea*, Norway. American Association of Petroleum Geologists Search and Discovery Article, #30230.

Satterfield, W.M. & Behrens, E.W. 1990. A late Quaternary canyon/channel system, northwest Gulf of Mexico continental slope. *Marine Geology*, **92**, 51–67.

Satur, N., Hurst, A., Cronin, B.T., Kelling, G. & Gürbüz, K. 2000. Sand body geometry in a sand-rich, deep-water clastic system, Miocene Cingöz Formation of southern Turkey. *Marine and Petroleum Geology*, **17**, 239–252.

Saunders, A.D., Rogers, G., Marriner, G.F., Terrell, D.J. & Verma, S.P. 1987. Geochemistry of Cenozoic volcanic rocks, Baja California, Mexico: implications for the petrogenesis of post-subduction magmas. 7. *Volcanology & Geothermal Research*, **32**, 223–245.

Saunders, M., Bowman, S. & Geiger, L. 2013. The Pelotas Basin oil proivince revealed. *GEO ExPro*, **10**, 38–40.

Savage, S.B. 1989. Flow of granular materials. *In*: Germain, P., Piau, M. & Caillerie, D. (eds), *Theoretical and Applied Mechanics*, 241–266. Amsterdam: Elsevier.

Savage, S.B. & Lun, C.K.K. 1988. Particle size segregation in inclined chute flow of dry cohesionless granular solids. *Journal of Fluid Mechanics*, **189**, 311–335.

Savage, S.B. & McKeown, S. 1983. Shear stresses developed during rapid shear of concentrated suspensions of large spherical particles between concentric cylinders. *Journal of Fluid Mechanics*, **127**, 453–472.

Savage, S.B. & Sayed, M. 1984. Stresses developed by dry cohesionless granular materials sheared in an annular shear cell. *Journal of Fluid Mechanics*, **142**, 391–430.

Savoye, B., Piper, D.J.W. & Droz, L. 1993. Plio-Pleistocene evolution of the Var deep-sea fan off the French Riviera. *Marine and Petroleum Geology*, **10**, 550–571.

Savrda, C.E. & Bottjer, D.J. 1986. Trace-fossil model for reconstruction of paleo-oxygenation in bottom waters. *Geology*, **14**, 3–6.

Savrda, C.E. & Bottjer, D.J. 1987. The exaerobic zone, a new oxygen-deficient marine biofacies. *Nature*, **327**, 54–56.

Savrda, C.E. & Bottjer, D.J. 1989. Trace-fossil model for reconstructing oxygenation histories of ancient marine bottom waters: application to Upper Cretaceous Niobrara Formation, Colorado. *Palaeogeography, Palaeoclimatology, Palaeoecology*, **74**, 49–74.

Savrda, C.E. & Bottjer, D.J. 1994. Ichnofossils and ichnofabrics in rhythmically bedded pelagic/hemipelagic carbonates: recognition and evaluation of benthic redox and scour cycles. *In*: Deboer, P. & Smith, D. (eds), *Orbital Forcing and Cyclic Sequences*, 195–210. International Association of Sedimentologists, Special Publication, 19.

Savrda, C.E., Krawinkel, H., McCarthy, F.M.G., McHugh, C.M.G., Olson, H.C. & Mountain, G. 2001. Ichnofabrics of a Pleistocene slope succession, New Jersey margin: relations to climate and sea-level dynamics. *Paleogeography, Paleoclimatology, Paleoecology,* **171**, 41–46.

Saxov, S. & Nieuwenhuis, J.K. (eds) 1982. *Marine Slides and Other Mass Movements.* New York: Plenum.

Severinghaus, J. & Atwater, T. 1989. Cenozoic geometry and thermal condition of the subducting slabs beneath western North America. *In*: Wernicke, B. (ed.), *Basin and Range Extensional Tectonics near the Latitude of Las Vegas,* 1–22. Geological Society of America Memoir, **176**.

Schafer, C.T. & Asprey, K.W. 1982. Significance of some geotechnical properties of continental slope and rise sediments off northeast Newfoundland. *Canadian Journal of Earth Sciences,* **19**, 153–141.

Schäfer, W. 1956. Wirkungen der Benthos-Organismen auf den jungen Schichtverband: Senckengergiana. *Lethaea,* **37**, 183–263.

Scheibner, C., Reijmer, J.J.G., Marzouk, A.M., Speijer, R.P. & Kuss, J. 2003. From platform to basin: the evolution of a Paleocene carbonate margin (Eastern Desert, Egypt). *International Journal of Earth Sciences (Geologisches Rundschau),* **92**, 624–640.

Schindlbeck, J.C., Kutterolf, S., Freundt, A., Scudder, R.P., Pickering, K.T. & Murray, R.W. 2013. Emplacement processes of submarine volcaniclastic deposits (IODP Site C0011, Nankai Trough). *Marine Geology,* **343**, 115–124.

Schlager, W. & James, N.P. 1978, Low-magnesian calcite limestones forming at the deep-sea floor, Tongue of the Ocean, Bahamas. *Sedimentology,* **25**, 675–702.

Schlager. W., Reijmer. J.J.G. & Droxler. A.W. 1994. Highstand shedding of carbonate platforms. *Journal of Sedimentary Research,* **64B**, 270–281.

Schlanger, S.O., Athur, M.A., Jenkyns, H.C. & Scholle, P.A. 1987. The Cenomanian-Turonian Oceanic Anoxic Event, I. Stratigraphy and distribution of organic carbon-rich beds and the marine $\delta^{13}C$ excursion. *In*: Brooks, J. & Fleet, A.J. (eds), *Marine Petroleum Source Rocks,* 371–399. Geological Society, London, Special Publications, **26**. The Geological Society, London.

Schlirf, M. 2000. Upper Jurassic trace fossils from the Boulonnais northern France. *Geologica et Palaeontologica,* **34**, 145–213.

Schlirf, M. & Uchman, A. 2005. Revision of the Ichnogenus Sabellarifix Richter, 1921 and its relationship to Skolithos Haldeman, 1840 and Polykladichnus Fursich, 1981. *Journal of Systematic Palaeontology,* **32**, 115–131.

Schlüter, H.U. & Fritsch J. 1985.Geology and tectonics of the Banda Arc between Tanimbar island and Aru island. *Geologisches Jahrbuch,* **30**, 3–41.

Schnabel, G.W. 1992. New data on the Flysch Zone of the Eastern Alps in the Austrian sector and new aspects concerning the transition to the Flysch Zone of the Carpathians. *Cretaceous Research,* **13**, 405–419.

Schneider, C.L., Hummon, C., Yeats, R.S. & Huftile, G.L. 1996. Structural evolution of the northern Los Angeles basin, California, based on growth strata. *Tectonics,* **15**, 341–355.

Scholl, D.W. & Creager, J.S. 1971. Deep Sea Drilling Project, Leg 19. *Geotimes,* November, 12–15.

Scholle, P.A. 1971. Sedimentology of fine-grained deep-water carbonate turbidites, Monte Antola Flysch (Upper Cretaceous), Northern Apennines, Italy. *Bulletin of the Geological Society of America,* **82**, 629–658.

Scholle, P.A. & Ekdale, A.A. 1983. Pelagic environments. *In*: Scholle, P.A., Bebout, D.G. & Moore, C.H. (eds), *Carbonate Depositional Environments,* 620–691. American Association of Petroleum Geologists, Memoir, **33**.

Scholz, C.H. & Small, C. 1997. The effect of seamount subduction on seismic coupling. *Geology,* **25**, 487–490.

Schrader, H.J. 1971. Fecal pellets: role in sedimentation of pelagic diatoms. *Science,* **174**, 55–57.

Schulz, M. & Mudelsee, M. 2002. REDFIT: estimating red-noise spectra directly from unevenly spaced paleoclimatic time-series. *Computers and Geosciences,* **28**, 421–426.

Schulz, M., Bergmann, M., von Juterzenka, K. & Soltwedel, T. 2010. Colonisation of hard substrata along a channel system in the deep Greenland Sea. *Polar Biology,* **33**,1359–1369.

Schumann, K., Behrmann, J.H., Stipp, M., Yamamoto, Y., Kitamura, Y. & Lempp, C. 2014. Geotechnical behavior of mudstones from the Shimanto and Boso accretionary complexes, and implications for the Nankai accretionary prism. *Earth, Planets and Space,* **66**, doi:10.1186/1880-5981-66-129.

Schumm, S.A. 1981. Evolution and response of the fluvial system, sedimentologic implications. *In*: Ethridge, F.G. (ed.), *Recent and Ancient Nonmarine Depositional Environments; Models for Exploration,* 19–29. Society of Economic Paleontologists and Mineralogists Special Publications, **31**.

Schumm, S.A. & Khan, H.R. 1972. Experimental study of channel patterns. *Bulletin of the Geological Society of America,* **88**, 1755–1770.

Schumm, S.A., Khan, H.R., Winkley, B.R. & Robins, L.G. 1972. Variability of river patterns. *Nature,* **237**, 75–76.

Schuppers, J.D. 1995. *Characterization of deep-marine clastic sediments from foreland basins-outcrop-derived concepts for exploration, production and reservoir modeling.* Ph.D. thesis. Technische Universiteit Delft, Netherlands. 272 pp.

Schwab, W.C., Lee, H.J. & Twichell, D.C. (eds) 1993. *Submarine Landslides: Selected Studies in the U. S. Exclusive Economic Zone.* Bulletin U.S. Geological Survey, 2002.

Schwab, W.C., Lee, H.J., Twichell, D.C., Locat, J., Nelson, C.H., McArthur, W.G. & Kenyon, N. 1996. Sediment mass-flow processes on a depositional lobe, outer Mississippi Fan. *Journal of Sedimentary Research,* **A66**, 916–927.

Schwab, A.M., Cronin, B.T. & Ferreira, H. 2007. Seismic expression of channel outcrops: offset stacked versus amalgamated channel systems. *Marine and Petroleum Geology,* **24**, 504–514.

Schwalbach, J.R., Edwards, B.D. & Gorsline, D.S. 1996. Contemporary channel-levée systems in active borderland basin plains, California Continental Borderland. *Sedimentary Geology,* **104**, 53–72.

Schwarz, E. & Arnott, R.W.C. 2007. Anatomy and evolution of a slope channel-complex set (Neoproterozoic Isaac Formation, Windermere Supergroup, southern Canadian Cordillera): implications for reservoir characterization. *Journal of Sedimentary Research,* **77**, 89–109.

Schweller, W.J. & Kulm, L.D. 1978. Depositional patterns and channelized sedimentation in active eastern Pacific trenches. *In*: Stanley, D.J. & Kelling, G. (eds), *Sedimentation in Submarine Canyons, Fans, and Trenches,* 311–324. Stroudsburg, PA: Dowden, Hutchinson & Ross.

Schwenk, T., Spieß, V., Breitzke, M. & Hübscher, C. 2005. The architecture and evolution of the Middle Bengal Fan in vicinity of the active channel–levee system imaged by high-resolution seismic data. *Marine and Petroleum Geology,* **22**, 637–656.

Scrocca, D. 2005. Deep structure of the southern Apennines, Italy: thin-skinned or thick-skinned? *Tectonics,* **24**, 1–20.

Scotchman, J.I., Bown, P., Pickering, K.T., BouDagher-Fadel, M., Bayliss, N.J. & Robinson, S.A. 2015a. A new age model for the middle Eocene deep-marine Ainsa Basin, Spanish Pyrenees. *Earth-Science Reviews,* **144**, 10–22.

Scotchman, J.I., Pickering, K.T., Sutcliffe, C., Dakin, N. & Armstrong, E. 2015b. Milankovitch cyclicity within the middle Eocene deep-marine Guaso System, Ainsa Basin, Spanish Pyrenees. *Earth-Science Reviews,* **144**, 107–121.

Scott, K.M. 1966. Sedimentology and dispersal pattern of a Cretaceous flysch sequence, Patagonian Andes, southern Chile. *American Association of Petroleum Geologists Bulletin,* **50**, 72–107.

Seeber, L., Emre, O., Cormier, M.-H., Sorlien, C.C., McHugh, C.M.G., Polonia, A., Ozer, N. & Cagatay, N. 2004. Uplift and subsidence from oblique slip: the Ganos-Marmara bend of the North Anatolian transform, western Turkey. *Tectonophysics,* **391**, 239–258.

Seed, H.B. & Lee, K.L. 1966. Liquefaction of saturated sands during cyclic loading. *Journal of Soil Mechanics Found. Division, American Society of Civil Engineers,* **92**, 105–134.

Seely, D.R. 1979. The evolution of structural highs bordering maJor forearc basins. *In*: Watkins, J.S., Montadert, L. & Dickerson, P.W. (eds), *Geological and Geophysical Investigations of Continental Margins*, 245–260. American Association of Petroleum Geologists Memoir, **29**.

Seguret, M., Labaume, P. & Madariago, R. 1984. Eocene seismicity in the Pyrenees from megaturbidites in the South Pyrenean basin (North Spain). *Marine Geology*, **55**, 117–131.

Seilacher, A. 1953a. Studien zur palichnologie. I. Über die methoden der palichnologie. *Neues Jahrbuch für Geologie und Paläontologie*, **96**, 421–452.

Seilacher, A. 1953b. Studien zur palichnologie. II. Die fossilen ruhes-puren *(Cubichnia)*. *Neues Jahrbuch für Geologie und Paläontologie*, **98**, 87–124.

Seilacher, A. 1955. Spuren und fazies im unterkambrium. *In*: Schindewolf, O.H. & Seilacher, A. (eds). Beiträge zur Kenntnis des Kambriums in der Salt Range Pakistan. *Akademie der Wissenschaften und der Literatur, Mainz, Abhandlungen der mathematisch-naturwissenschaftlichen Klasse*, **10**, 261–446.

Seilacher, A. 1964a. Biogenic sedimentary structures. *In*: Imbrie, J. & Newell, N. (eds), *Approaches to Paleoecology*, 296–316. Wiley, New York.

Seilacher, A. 1964b. Sedimentological classification and nomenclature of trace fossils. *Sedimentology*, **3**, 253–256.

Seilacher, A. 1967. Bathymetry of trace fossils. *Marine Geology*, **5**, 413–428.

Seilacher, A. 1974. Flysch trace fossils: evolution of behavioural diversity in the deep-sea. *Neues Jahrbuch für Geologie und Paläontologie, Monatshefte, Jahrgang* **1974**, 233–245.

Seilacher, A. 1977. Pattern analysis of Paleodictyon and related trace fossils. *In*: Crimes, T.P. & Harper, J.C. (eds). *Trace Fossils 2*, 289–334. Geology Journal Special Issue **9**.

Seilacher, A. 2007. *Trace Fossil Analysis*, 226 pp. Heidelberg: Springer Verlag. ISBN 9783 540 47225 4.

Seno, T. 1989. Philippine sea plate kinematics. *Modern Geology*, **14**, 87–97.

Sengör, A M.C. 1979. The North Anatolian transform fault Its age, offset and tectonic significance. *Journal of the Geological Society, London*, **136**, 269–282.

Sengör, A M.C., Gorür, N. & Saroglu, F. 1985. Strike-slip faulting and related basin formation in zones of tectonic escape: Turkey as a case study. *In*: Biddle, K.T. & Christie-Blick, N. (eds), *Strike-slip Deformation, Basin Formation, and Sedimentation*, 227–264. The Society of Economic Paleontologists and Mineralogists (SEPM) Special Publication, **37**.

Sercombe, W.J. & Radford, T.W. 2007. Deep Water Gulf of Mexico High Gamma Ray Shales and their Implications for flooding surfaces Source Rocks and Extinctions. *American Association of Petroleum Geologists/European Region Energy Conference and Exhibition, Technical Program*.

Sestini, G. 1970. Flysch facies and turbidite sedimentology. *Sedimentary Geology*, **4**, 559–597.

Severinghaus, J. & Atwater, T. 1989. Cenozoic geometry and thermal condition of the subducting slabs beneath western North America. *In*: Wernicke, B. (ed.), *Basin and Range Extensional Tectonics Near the Latitude of Las Vegas*, 1–22. Geological Society of America Memoir, **176**.

Sexton, P.F., Wilson, P.A., Norris, R.D. 2006. Testing the Cenozoic multisite composite δ^{18}O and δ^{13}C curves: New monospecific Eocene records from a single locality, Demerara Rise Ocean Drilling Program Leg 207. *Paleoceanography*, **21**, 2019–2036.

Seyfried, H., Astorga, A., Amann, H., Calvo, C., Kolb, W., Schmidt, H. & Winsemann, J. 1991. Anatomy of an evolving island arc: tectonic and eustatic control in the south Central American fore-arc area. *In*: Macdonald, D.I.M. (ed.), *Sedimentation, Tectonics, and EEustasy*, 217–240. Special Publication of the International Association of Sedimentologists, **12**.

Shackleton, N.J., Berger, A. & Peltier, W.R. 1990. An alternative astronomical calibration of the lower Pleistocene timescale based on ODP Site 677. *Transactions of the Royal Society of Edinburgh: Earth Sciences*, **81**, 251–261.

Shanmugam, G. 1980. Rhythms in deep sea, fine-grained turbidite and debris flow sequences, Middle Ordovician, eastern Tennessee. *Sedimentology*, **27**, 419–432.

Shanmugam, G. 1996. High-density turbidity currents: are they sandy debris flows? *Journal of Sedimentary Research*, **A66**, 2–10.

Shanmugam, G. 1997. The Bouma Sequence and the turbidite mind set. *Earth-Science Reviews*, **42**, 201–229.

Shanmugam, G. 2000. 50 years of the turbidite paradigm (1950s–1990s): deep-water processes and facies models - a critical perspective. *Marine and Petroleum Geology*, **17**, 285–342.

Shanmugam, G. 2002. Ten turbidite myths. *Earth-Science Reviews*, **58**, 311–341.

Shanmugam, G. 2003. Deep-marine tidal bottom currents and their reworked sands in modern and ancient submarine canyons. *Marine and Petroleum Geology*, **20**, 471–491.

Shanmugam, G. 2008. Deep-water bottom currents and their deposits. *In*: Rebesco, M. & Camerlenghi, A. (eds), *Contourites*, 59–81. Developments in Sedimentology, **60**. Amsterdam: Elsevier.

Shanmugam, G. 2015. The landslide problem. *Journal of Palaeogeography*, **4**, 109–166.

Shanmugam, G. & Moiola, R.J. 1985. Submarine fan models: problems and solutions. *In*: Bouma, A.H., Normark, W.R. & Barnes, N.E. (eds), *Submarine Fans and Related Turbidite Systems*, 29–34. New York: Springer.

Shanmugam, G. & Moiola, R.J. 1988. Submarine fans: characteristics, models, classification, and reservoir potential. *Earth-Science Reviews*, **24**, 383–428.

Shanmugam, G. & Moiola, R.J. 1991. Types of submarine fan lobes: models and implications. *American Association of Petroleum Geologists Bulletin*, **75**, 156–179.

Shanmugam, G. & Moiola, R.J. 1995. Reinterpretation of depositional processes in a classic flysch sequence (Pennsylvanian Jackfork Group), Ouachita Mountains, Arkansas and Oklahoma. *American Association of Petroleum Geologists Bulletin*, **79**, 672–695.

Shanmugam, G., Spalding, T.D. & Rofheart, D.H. 1993a. Traction structures in deep-marine, bottom-current-reworked sands in the Pliocene and Pleistocene, Gulf of Mexico. *Geology*, **21**, 929–932.

Shanmugam, G., Spalding, T.D. & Rofheart, D.H. 1993b. Process sedimentology and reservoir quality of deep-marine bottom-current-reworked sands (sandy contourites); an example from the Gulf of Mexico. *American Association of Petroleum Geologists Bulletin*, **77**, 1241–1259.

Shanmugam, G, Lehtonen, L.R., Straume, T., Syversten, S.E., Hodgkinson, R.J. & Skibeli, M. 1994. Slump and debris flow dominated upper slope facies in the Cretaceous of the Norwegian and Northern North Seas (61–67°N): implications for sand distribution. *American Association of Petroleum Geologists Bulletin*, **78**, 910–937.

Shanmugam, G., Bloch, R.B., Mitchell, S.M., Beamish, G.W.J., Hodgkinson, R.J., Damuth, J.E., Straume, T., Syvertsen, S.E. & Shields, K.E. 1995. Basin-floor fans in the North Sea: sequence stratigraphic models vs. sedimentary facies. *American Association of Petroleum Geologists Bulletin*, **79**, 477–512.

Shanmugam, G., Bloch, R.B., Damuth, J.E. & Hodgkinson, R.J. 1997. Basin-floor fans in the North Sea; sequence stratigraphic models vs. sedimentary facies; reply. *American Association of Petroleum Geologists Bulletin*, **81**, 666–672.

Shannon, P.M., Stoker, M.S., Praeg, D., van Weering, T.C.E., de Haas, H., Nelsen, T., Dahlgren, K.I.T. & Hjelstuen, B.O. 2005. Sequence stratigraphic analysis in deep-water, underfilled NW European passive margin basins. *Marine and Petroleum Geology*, **22**, 1,185–1,200.

Shaw, J., Puig, P. & Han, G. 2013. Megaflutes in a continental shelf setting, Placentia Bay, Newfoundland. *Geomorphology*, **189**, 12–25.

Shepard, F.P. 1933. Canyons beneath the seas. *Scientific Monthly*, **37**, 31–39.

Shepard, F.P. 1951. Mass movements in submarine canyon heads. *Transactions of the American Geophysical Union*, **32**, 405–418.

Shepard, F.P. 1955. Delta-front valleys bordering the Mississippi distributaries. *Geological Society of America Bulletin*, **66**, 1489–1498.

Shepard, F.P. 1963. *Submarine Geology*, 2nd edn. New York: Harper & Row, xviii + 557 pp.

Shepard, F.P. 1966. Meander in a valley crossing a deep-sea fan. *Science*, **154**, 385–386.

Shepard, F.P. 1975. Progress of internal waves along submarine canyons. *Marine Geology*, **19**, 131–138.

Shepard, F.P. 1976. Tidal components of currents in submarine canyons. *Journal of Geology*, **84**, 343–350.

Shepard, F.P. 1977. *Geological Oceanography: Evolution of Coasts, Continental Margins, and the Deep-sea Floor*. New York: Crane, Russak & Co, 214 pp.

Shepard, F.P. 1981. Submarine canyons: multiple causes and long-time persistence. *American Association of Petroleum Geologists Bulletin*, **65**, 1062–1077.

Shepard, F.P. & Marshall, N.F. 1969. Currents in La Jolla and Scripps submarine canyons. *Science*, **165**, 177–178.

Shepard, F.P. & Marshall, N.F. 1973a. *Storm-generated current in a La Jolla submarine canyon*, California. *Marine Geology*, **15**, Ml9–M24.

Shepard, F.P. & Marshall, N.F. 1973b. Currents along floors of submarine canyons. *Geological Society of America Bulletin*, **57**, 244–264.

Shepard, F.P. & Marshall, N.F. 1978. Currents in submarine canyons and other sea valleys. *In*: D.J Stanley & Kelling, G. (eds), *Sedimentation in Submarine Canyons, Fans, and Trenches*, 3–14. Stroudsburg, PA: Dowden, Hutchinson & Ross.

Shepard, F.P., Marshall, N.F. & McLoughlin, P.A. 1974. Currents in submarine canyons. *Deep-Sea Research*, **21**, 691–706.

Shepard F.P., Marshall, N.F. & McLoughlin, P.A. 1975. Pulsating turbidity currents with relationship to high swell and high tides. *Nature*, **258**, 704–706.

Shepard, F.P., Marshall, N.F., McLoughlin, P.A. & Sullivan, G.G. 1979. Currents in submarine canyons and other sea valleys. *American Association of Petroleum Geologists, Studies in Geology*, **8**, 173 pp.

Sheridan, R.E. 1986. Pulsation tectonics as the control on North Atlantic palaeoceanography. *In*: Summerhayes, C.P. & Shackleton, N.J. (eds), *North Atlantic Palaeoceanography*, 255–275. Geological Society of London Special Publication, **21**. Oxford: Blackwell Scientific.

Sheridan, R.E., Gradstein, F.M. et al. 1983. *Initial Reports of the Deep Sea Drilling Project, 76*. Washington, DC: US Government Printing Office.

Shiki, T., Kumon, F., Inouchi, Y, Kontani, Y., Sakamoto, T., Tateishi, M., Matsubara, H. & Fukuyama, K. 2000. Sedimentary features of the seismo-turbidites, Lake Biwa, Japan. *Sedimentary Geology*, **135**, 37–50.

Shimamura, K. 1986. Topography and geological structure in the bottom of the Suruga Trough: a geological consideration of the subduction zone near the collisional plate boundary. *Journal of Geography, Tokyo Geographical Society*, **95**, 317–338.

Shimamura, K. 1989. Topography and sedimentary facies of the Nankai deep sea channel. *In*: Taira, A. & Masuda, F. (eds), *Sedimentary Facies in the Active Plate Margin*, 529–556. Tokyo: Terra Scientific Publishing Company (TERRAPUB).

Shimoda, G., Tatsumi, Y., Nohda, S., Ishizaka, K. & Jahn, B.M. 1998. Setouchi high-Mg andesites revisited: geochemical evidence for melting of subducting sediments. *Earth and Planetary Science Letters*, **160**, 479–492.

Shipley, T.H., Ogawa, Y., Blum, P. et al. 1995. *Proceedings of the Ocean Drilling Program, Initial Reports, 156*. College Station, TX: Ocean Drilling Program.

Shipp, R.C., Weimer, P. & Posamentier, H.W. 2011. *Mass-Transport Deposits in Deepwater Settings*. SEPM (Society for Sedimentary Geology), **96**, 527 pp. Tulsa, Oklahoma. ISBN: 978-1-56576-286-2.

Shipboard Scientific Party 1975. Site 297. *In*: Karig, D.E., Ingle Jr., J.C. et al., *Initial Reports of the Deep Sea Drilling Project*, **31**, 275–316. Washington, DC: US Government Printing Office.

Shipboard Scientific Party 1986. Site 582. *In*: Kagami, H., Karig, D.E., Coulbourn, W.T. et al., *Initial Reports of the Deep Sea Drilling Project*, **87**, 35–122. Washington, DC: US Government Printing Office.

Shipboard Scientific Party 1990. Sites 792 & 793. *In*: *Proceedings Ocean Drilling Program, Initial Reports*, **126**, 221–403. College Station, Texas, USA: Ocean Drilling Program.

Shipboard Scientific Party 1991. Site 808. *In*: Taira, A., Hill, I., Firth, J.V. et al., *Proceedings Ocean Drilling Program, Initial Reports*, **131**, 71–269. College Station, Texas, USA: Ocean Drilling Program.

Shipboard Scientific Party 1995a. Leg synthesis. *In*: *Proceedings Ocean Drilling Program, Initial Reports*, **155**, 17–21. College Station, Texas, USA: Ocean Drilling Program.

Shipboard Scientific Party 1995b. Site 934. *In*: *Proceedings Ocean Drilling Program, Initial Reports*, **155**, 241–271. College Station, Texas, USA: Ocean Drilling Program.

Shipboard Scientific Party 1995c. Site 945. *In*: *Proceedings Ocean Drilling Program, Initial Reports*, **155**, 635–655. College Station, Texas, USA: Ocean Drilling Program.

Shipboard Scientific Party 1995d. Explanatory notes. *In*: *Proceedings Ocean Drilling Program, Initial Reports*, **155**, 47–81. College Station, Texas, USA: Ocean Drilling Program.

Shipboard Scientific Party 1995e. Site 941. *In*: *Proceedings Ocean Drilling Program, Initial Reports*, **155**, 503–536. College Station, Texas, USA: Ocean Drilling Program.

Shipboard Scientific Party 1995f. Site 935. *In*: *Proceedings Ocean Drilling Program, Initial Reports*, **155**, 273–319. College Station, Texas, USA: Ocean Drilling Program.

Shipboard Scientific Party 1997. Site 1015. *In*: *Proceedings of the Ocean Drilling Program, Initial Reports*, **167**, 223–237. College Station, Texas, USA: Ocean Drilling Program.

Shipboard Scientific Party 1998. Deep Blake-Bahama Outer Ridge, Sites 1060, 1061, and 1062. *In*: *Proceedings Ocean Drilling Program, Initial Reports*, **172**, 157–252. College Station, Texas, USA: Ocean Drilling Program.

Shipboard Scientific Party 2000. Leg 182 Summary: Great Australian Bight - Cenozoic cool-water carbonates. *In*: *Proceedings Ocean Drilling Program, Initial Reports*, **182**, 1–58. College Station, Texas, USA: Ocean Drilling Program.

Shipboard Scientific Party 2004a. Leg 210 Summary. *In*: *Proceedings Ocean Drilling Program, Initial Reports*, **210**, 1–78. College Station, Texas, USA: Ocean Drilling Program.

Shipboard Scientific Party 2004b. Site 1276. *In*: *Proceedings Ocean Drilling Program, Initial Reports*, **210**, 1–358. College Station, Texas, USA: Ocean Drilling Program.

Shor, A.N., Piper, D.J.W., Hughes Clarke, J. & Meyer, L.A. 1990. Giant flute-like scours and other erosional features formed by the 1929 Grand Banks turbidity current. *Sedimentology*, **37**, 631–645.

Shultz, A.W. 1984. Subaerial debris-flow deposition in the Upper Paleozoic Cutter Formation. *Journal of Sedimentary Petrology*, **54**, 759–772.

Siebert, L. & Simkin, T. 2002. *Volcanoes of the World: an Illustrated Catalog of Holocene Volcanoes and their Eruptions*. Smithsonian Institution, Global Volcanism Program Digital Information Series, **GVP-3**, http://www.volcano.si.edu/world/.

Siedlecka, A. 1972. Kongsfjord Formation - a late Precambrian flysch sequence from the Varanger Peninsula, Northern Norway. *Norges Geologiske Undersøkelse*, **278**, 41–80.

Siedlecka, A. 1978. Late Precambrian tidal-flat deposits and algal stromatolites in the Batsfiord Formation, East Finnmark, North Norway. *Sedimentary Geology*, **21**, 177–310.

Siedlecka, A. 1985. Development of the Upper Proterozoic sedimentary basins of the Varanger Peninsula, East Finnmark, North Norway. *Bulletin of the Geological Survey of Finland*, **331**, 175–185.

Siedlecka, A. & Edwards, M.B. 1980. Lithostratigraphy and sedimentation of the Riphean Basnaering Formation, Varanger Peninsula, North Norway. *Norges Geologiske Undersokelse*, **355**, 27–47.

Siedlecka, A. & Siedlecki, S. 1967. Some new aspects of the geology of Varanger peninsula (Northern Norway). *Norges Geologiske Undersokelse*, **247**, 288–306.

Siedlecka, A., Pickering, K.T. & Edwards, M.B. 1989. Upper Proterozoic passive margin deltaic complex, Finnmark, N Norway. *In*: Whateley, M.K.G. & Pickering, K.T. (eds), *Deltas: Sites and Traps for Fossil Fuels, Geological Society of London, Special Publication*, **41**. Oxford: Blackwell Scientific.

Siedlecka, A., Negrutsa, V.Z. & Pickering, K.T. 1994. Upper Proterozoic Turbidite System of the Rybachi Peninsula, northern Russia - a possible stratigraphic counterpart of the Kongsfjord Submarine Fan of the Varanger Peninsula, northern Norway. *In*: *Geology of the Eastern Finnmark - Western Kola Region*, 201–216. Norges Geologiske Undersøkelse. Special Publication No. **7**.

Siegel, S. 1956. *Non-parametric Statistics for the Behavioral Sciences.* New York: McGraw-Hill.

Siemers, C.T., Tillman, R.W. & Williamson, C.R. (eds) 1981. *Deep-water Clastic Sediments: A Core Workshop*. Core workshop 2. Tulsa, Oklahoma. Society of Economic Paleontologists and Mineralogists.

Sigurdsson, H., Sparks, R.S.J., Carey, S. & Huang, T.C. 1980. Volcanogenic sedimentation in the Lesser Antilles Arc. *Journal of Geology*, **88**, 523–540.

Sikkema, W. & Wojcik, K.M. 2000. 3D visualization of turbidite systems, lower Congo Basin, offshore Angola. *In*: Weimer, P., Slatt, R.M., Coleman, J., Rosen, N.C., Nelson, H., Bouma, A.H., Styzen, M.J. & Lawrence, D.T. (eds), *Deep-water Reservoirs of the World*, 928–939. Houston, Texas: Gulf Coast Section, Society of Economic Paleontologists and Mineralogists.

Silva-Gonzalez, P. 2012. Zeebries: new high quality data in Australia's most prospective basin. *GEO ExPro*, **9**, 38–40.

Silver, E.A. 1972. Pleistocene tectonic accretion of the continental slope off Washington. *Marine Geology*, **13**, 239–249.

Simó, A. & Puigdefàbregas, C 1985. Transition from shelf to basin on an active slope, Upper Cretaceous, Tremp area, southern Pyrenees *In*: Mila, M.D. & Rosell, J. (eds), *International Association of Sedimentologists, 6th European Regional Meeting, Llerida, Excursion Guidebook*, 63–108.

Simm, R.W. & Kidd, R.B. 1984. Submarine debris flow deposits detected by long range side-scan sonar 1000 km from source. *Geo-Marine Letters*, **3**, 13–16.

Simm, R.W., Weaver, P.P.E., Kidd, R.B. & Jones, E.J.W. 1991. Late Quaternary mass movement on the lower continental rise and abyssal plain off Western Sahara. *Sedimentology*, **38**, 27–40.

Simmons, J.R. 1992. Evolution of a Miocene calciclastic turbidite depositional system. *In*: Gradstein, F.M., Ludden, J.N. *et al.*, *Proceedings of the Ocean Drilling Program, Scientific Results*, **123**, 151–164. College Station, Texas, USA: Ocean Drilling Program.

Simons, D.B., Richardson, E.V. & Nordin, C.F. Jr., 1965. Sedimentary structures generated by flow in alluvial channels. *In*: Middleton, G.V. (ed.), *Primary Sedimentary Structures and their Hydrodynamic Interpretation*, 34–52. Society of Economic Paleontolgists and Mineralogists Special Publication, **12**.

Simpson, G.G., Roe, A. & Lewontin, R.C. 1960. *Quantitative Zoology*, revised edn. New York: Harcourt Brace.

Simpson, J.E. 1982. Gravity currents in the laboratory, atmosphere, and ocean. *Annual Reviews in Fluid Mechanics*, **14**, 213–234.

Simpson, J.E. 1997. *Gravity Currents:In the Environment and the Laboratory*. Cambridge: Cambridge University Press, 244 pp.

Simpson, S. 1957. On the trace fossil Chondrites. *Quarterly Journal of the Geological Society of London*, **112**, 475–499.

Sinclair, H.D. & Cowie, P.A. 2003. Basin-floor topography and the scaling of turbidites. *The Journal of Geology*, **111**, 277–299.

Sinclair, H.D. & Tomasso, M. 2002. Depositional evolution of confined turbidite flows. *Journal of Sedimentary Research*, **72**, 451–456.

Sinclair, H.D., Gibson, M., Naylor, M. & Morris, R.G. 2005. Asymmetric growth of the Pyrenees revealed through measurement and modelling of orogenic fluxes. *American Journal of Science*, **305**, 369–406.

Sinclair, I.K. 1988. Evolution of Mesozoic–Cenozoid sedimentary basins in the Grand Banks area of Newfoundland and comparison with Falvey's (1974) rift model. *Bulletin Canadian Petroleum Geologists*, **36**, 255–273.

Singh, S.C., Hananto, N., Mukti, M., Robinson, D.P., Das, S., Chauhan, A., Carton, H., Gratacos, B., Midnet, S., Djajadihardja, Y. & Harjono, H. 2011. Aseismic zone and earthquake segmentation associated with a deep subducted seamount in Sumatra. *Nature Geoscience*, **4**, 308–311.

Sixsmith, P.J., Flint, S., Wickens, H deV & Johnson, S. 2004. Anatomy and stratigraphic development of an early foreland basin turbidite system: Laingsburg Formation, Karoo basin, South Africa. *Journal of Sedimentary Research*, **74**, 239–254.

Skene, K.I., Piper, D.J.W. & Hill, P.S. 2002. Quantitative analysis of variations in depositional sequence thickness from submarine channel levees. *Sedimentology*, **49**, 1411–1430.

Skipper, K. 1971. Antidune cross-stratification in a turbidite sequence, Cloridorme Formation, Gaspé, Québec. *Sedimentology*, **17**, 51–68.

Skipper, K. & Middleton, G.V. 1975. The sedimentary structures and depositional mechanics of certain Ordovician turbidites, Cloridorme Formation, Gaspé Peninsula, Quebec. *Canadian Journal of Earth Sciences*, **12**, 1934–1952.

Skipper, K. & Bhattacharjee, S.B. 1978. Backset bedding in turbidites: a further example from the Cloridorme Formation (Middle Ordovician), Gaspé, Québec. *Journal of Sedimentary Petrology*, **48**, 193–202.

Slazca, A. & Walton, E.K. 1992. Flow characteristics of Metresa: an Oligocene seismoturbidite in the Dukla Unit, Polish Carpathians. *Sedimentology*, **39**, 383–392.

Slacka, A., Kruglov, S., Golonka, J., Oszczypko, N. & Popadyuk, I. 2006. Geology and hydrocarbon resources of the Outer Carpathians, Poland, Slovakia, and Ukraine: general geology. *In*: Golonka, J. & Picha, F.J. (eds), *The Carpathians and their Foreland: Geology and Hydrocarbon Resources*, 221–258. American Association of Petroleum Geologists Memoir, **84**.

Sliter, W.V. 1973. Upper Cretaceous foraminifers from the Vancouver Island area, British Columbia, Canada. *Journal of Foraminiferal Research*, **3**, 167–186.

Slivitzky, A., St-Julien, P. & Lachambre, G. 1991. *Synthèse géologique du Cambro-Ordovicien du nord de la Gaspésie*. Ministère de l'Énergie et des Ressources du Québec, **MM-85-04**.

Sloss, L.L. 1979. Global sea level changes: a view from the craton. *In*: Watkins, J.S., Montadert, L. & Dickerson, P.W. (eds), *Geological and Geophysical Investigations of Continental Margins*, 461–467. American Association of Petroleum Geologists Memoir, **29**.

Sluijs, A., Bowen, G.L., Brinkhuis, H., Lourens, L.J. & Thomas, E. 2007. The Palaeocene–Eocene Thermal Maximum super greenhouse: biotic and geochemical signatures, age models and mechanisms of global change. *In*: Williams, M., Haywood, A.M., Gregory, F.J. & Schmidt, D.N. (eds), *Deep-time Perspectives on Climate Change: Marrying the Signal from Computer Models and Biological Proxies*, 323–349. The Micropalaeontological Society, Special Publications. The Geological Society, London.

Sluijs, A., Bijl, P.K. Schouten, S., Röhl, U., Reichart, G.-J. & Brinkhuis, H. 2011. Southern ocean warming, sea level and hydrological change during the Paleocene-Eocene thermal maximum. *Climate of the Past*, **7**, 47–61.

Smith, A.G. & Pickering, K.T. 2003. Oceanic gateways as a critical factor to initiate icehouse Earth. *Journal of the Geological Society, London*, **160**, 337–340.

Smith, R.D.A. 1987. The *griestoniensis* Zone Turbidite System, Welsh Basin. *In*: Leggett, J.K. & Zuffa, G.G. (eds), *Marine Clastic Sedimentology*, 89–107. London: Graham & Trotman.

Smith, R.D.A. 1995a. Complex bedding geometries in proximal deposits of the Castelnuovo Member, Rochetta Formation, Tertiary Piedmont Basin, NW Italy. *In*: Pickering, K.T., Hiscott, R.N., Kenyon, N.H., Ricci Lucchi, F. & Smith, R.D.A. (eds), *Atlas of Deep Water Environments: Architectural Style in Turbidite Systems*, 244–249. London: Chapman and Hall.

Smith, R.D.A. 1995b. Sheet-like and channelized sediment bodies in a Silurian turbidite system, Welsh Basin, UK. *In*: Pickering, K.T., Hiscott, R.N., Kenyon, N.H., Ricci Lucchi, F. & Smith, R.D.A. (eds), *Atlas of Deep Water Environments: Architectural Style in Turbidite Systems*, 250–254. London: Chapman and Hall.

Smith, R.D.A. 1995c. Architecture of turbidite sandstone bodies in a rift-lake setting, Gabon Basin, offshore Gabon. *In*: Pickering, K.T., Hiscott, R.N., Kenyon, N.H., Ricci Lucchi, F. & Smith, R.D.A. (eds), *Atlas of DeepWwater Environments: Architectural Style in Turbidite Systems*, 255–259. London: Chapman and Hall.

Smith, R.D.A. 2004. Silled sub-basins to connected tortuous corridors: sediment distribution systems on topographically complex sub-aqueous slopes. *In*: Lomas, S.A. & Joseph, P. (eds), *Confined Turbidite Systems*, 23–43. Geological Society, London, Special Publication, **222**. London: The Geological Society.

Smith, R.D.A. & Joseph, P. 2004. Onlap stratal architectures in the Grès d'Annot: geometric models and controlling factors. *In*: Joseph, P. & Lomas, S.A. (eds), *Deep-Water Sedimentation in the Alpine Basin of SE France: New Perspectives on the Grès d'Annot and Related Systems*, 389–399. Geological Society, London, Special Publication, **221**. London: The Geological Society.

Smith, R.D.A. & Spalletti, L.A. 1995. Erosional, depositional and post-depositional features of a turbidite channel-fill, Jurassic, Neoquen Basin, Argentina. *In*: Pickering, K.T., Hiscott, R.N., Kenyon, N.H., Ricci Lucchi, F. & Smith, R.D.A. (eds), *Atlas of Deep Water Environments: Architectural Style in Turbidite Systems*, 162–166. London: Chapman and Hall.

Smith, R.D.A., Waters, R.A. & Davies, J.R. 1991. Late Ordovician and early Silurian turbidite systems in the Welsh Basin. *In*: *13th International Sedimentological Congress, Nottingham, Field Guide*, **20**. British Sedimentological Research Group.

Smith, S.B., Karlin, R.E., Kent, G.M., Seitz, G.G. & Driscoll, N.W. 2013. Holocene subaqueous paleoseismology of Lake Tahoe. *Geological Society of America Bulletin*, doi: 10.1130/B30629.1.

Snavely, Jr., P.D., Jr. & McClellan, P.H. 1987. *Preliminary Geologic Interpretation of USGS S.P. Lee Seismic Profile WO 76-7 on the Continental Shelf and Upper Slope, Northwestern Oregon*. U.S. Geological Survey Open File Report, **87-612**, 12 pp.

Snavely, Jr., P.D., Wagner, H.C. & Lander, D.L. 1980. *Geologic Cross Section of the Central Oregon Continental Margin*. Map and Chart Series **MC-28J**, scale 1:250,000. Boulder, Colorado: Geological Society of America.

Snyder, D.B., Prasetyo, H., Blundell, D.J., Pigram, C.J., Barber, A.J., Richardson, A. & Tjokosaproetr, S. 1996. A dual doubly vergent orogen in the Banda Arc continent-arc collision zone as observed on deep seismic reflection profiles. *Tectonics*, **15**, 34–53.

Soh, W. 1987. Transportation mechanism of resedimented conglomerate examined from clast fabric. *Journal of the Geological Society of Japan*, **93**, 909–923.

Soh, W., Pickering, K.T., Taira, A. & Tokuyama, H. 1991. Basin evolution in the arc-arc Izu Collision Zone, Mio-Pliocene Miura Group, central Japan. *Journal of the Geological Society, London*, **148**, 317–330.

Song, Y., Zhao, M., Liu, S., Hong, F. & Fang, S. 2010. Oil and gas accumulations in the foreland basins, central and western China. *Acta Geologica Sinica*, **84**, 382–405.

Sohn, Y.K. 1997. On traction-carpet sedimentation. *Journal of Sedimentary Research*, **67**, 502–509.

Sohn, Y.K. 2000. Depositional processes of submarine debris flows in the Miocene fan deltas, Pohang Basin, SE Korea with special reference to flow transformations. *Journal of Sedimentary Research*, **A70**, 491–503.

Sopaheluwakan, J. 1990. *Ophiolite obduction in the Mutis complex, Timor, eastern Indonesia: an example of inverted, isobaric, medium-high pressure metamorphism*. Ph.D. Thesis, Vrije Universiteit, Amsterdam, the Netherlands.

Soper, N.J. & Hutton, D.H.W. 1984. Late Caledonian sinistral displacements in Britain: implications for a three-plate collision model. *Tectonics*, **3**, 781–794.

Soper, N.J., Webb, B.C. & Woodcock, N.H. 1987. Late Caledonian transpression in north west England: timing, geometry and geotectonic significance. *Proceedings of the Yorkshire Geological Society*, **46**, 175–192.

Soreghan, M.J., Scholz, C.A. & Wells, J.T. 1999. Coarse-grained, deep-water sedimentation along a border fault margin of Lake Malawi, Africa: seismic stratigraphic analysis. *Journal of Sedimentary Research*, **B69**, 832–846.

Southard, J.B. & Mackintosh, M.E. 1981. Experimental test of autosuspension. *Earth Surface Processes Landforms*, **6**, 103–111.

Souquet, P., Debroas, E., Boirie, J.M., Pons, Ph., Fixari, G., Roux, J.C., Dol. J., Thieuloy, J.P., Bonnemaison, M., Manivit, H. & Peybernes, B. 1985. Le Groupe du Flysch Noir (Albo-Cenomanien) dans les Pyrénées. *Bulletin des Centres de Recherches Exploration-Production Elf-Aquitaine*, **9**, 183–252.

Soyinka, O.A. & Slatt, R.M. 2008. Identification and micro-stratigraphy of hyperpycnites and turbidites in Cretaceous Lewis Shale, Wyoming. *Sedimentology*, **55**, 1117–1134.

Spakman, W. & Hall, R. 2010. Surface deformation and slab-mantle interaction during Banda arc subduction rollback. *Nature Geoscience*, **3**, 562–566.

Speed, R.C. 1983. Structure of the accretionary complex of Barbados, I: Chalky Mount. *Geological Society of America Bulletin*, **94**, 92–116.

Speed, R.C. & Larue, D.K. 1982. Barbados: Architecture and implications for accretion. *Journal of Geophysical Research*, **87**, 3633–3643.

Speed, R.C., Barker, L.H. & Payne, P.L.B. 1991. Geologic and hydrocarbon evolution of Barbados. *Journal of Petroleum Geology*, **14**, 323–342.

Speed, R.C., Speed, C. & Sedlock, R. 2013. Geology and geomorphology of Barbados: a companion text to maps with accompanying cross sections, Scale 1:10,000. *Geological Society of America, Special Papers*, **491**, 1–63. doi: 10.1130/2012.2491.

Spencer, J.E. & Normark, W.R. 1979. Tosco-Abreojos fault zone: a Neogene transform plate boundary within the Pacific margin of southern Baja California, Mexico. *Geology*, **7**, 554–557.

Spencer, J.W. 1903. Submarine valleys off the American coast and in the North Atlantic. *Geological Society of America Bulletin*, **14**, 207–226.

Sprague, A.R.G., Sullivan, M.D., Campion, K.M., Jensen, G.N., Goulding, F.J., Garfield, T.R., Sickafoose, D.K., Rossen, C., Jennette, D.C., Beaubouef, R.T., Abreu, V., Ardill, J., Porter, M.L. & Zelt, F.B., 2002. The physical stratigraphy of deep-water strata: a hierarchical approach to the analysis of genetically related stratigraphic elements for improved reservoir prediction (abstract). *American Association of Petroleum Geologists, Annual Meeting, March 10–13, 2002, Houston, Texas, Official Program*, p. A167.

Sprague, A.R.G., Garfield, T.R., Goulding, F.J., Beaubouef, R.T., Sullivan, M.D., Rossen, C., Campion, K., Sickafoose, D.K., Abreu, V., Schellpeper, M.E., Jensen, G.N., Jennette, D.C., Pirmez, C., Dixon, B.T., Ying, D., Mohrig, D.C., Porter, M.L., Farrell, M.E. & Mellere, D. 2005. Integrtaed slope channel depositional models: the key to successful prediction of reservoir presence and quality in offshore West Africa. *CIPM, cuarto EExitep 2005, February 2–23, 2005, Veracruz, Mexico*, 1–13.

Sprague, A., Box, D., Hodgson, D. & Flint, S. 2008. A physical stratigraphic hierarchy for deep-water slope system reservoirs 2: complexes to storeys. *American Association of Petroleum Geologists International Conference and Exhibition, Cape Town, South Africa, Abstracts*.

Squinabol, S. 1890. Alghe a Pseudoalghe fossili italiane. *Atti della Societa Linguistica di Scienze Naturali e Geografiche*, **1**, 29–49, 166–199.

Srivastava, S.P., Arthur, M.A., Clement, B. *et al.* 1987. Site 645. *In*: *Proceedings of the Initial Reports of the Ocean Drilling Program (Pt A)*, **105**, 61–418. College Station, Texas: Ocean Drilling Program.

St. John, K., Leckie, R.M., Pound, K., Jones, M. & Krissek, L. 2012. *Reconstructing Earth's Climate History: Inquiry-Based Exercises for Lab and Class*. John Wiley & Sons, 485 pp. ISBN: 978-0-470-65805-5.

St-Julien, P. & Hubert, C. 1975. Evolution of the Taconian Orogen in the Québec Appalachians. *American Journal of Science*, **275-A**, 337–362.

Stacey, M.W. & Bowen, A.J. 1988. The vertical structure of density and turbidity currents: theory and observations. *Journal of Geophysical Research*, **93**, 3528–3542.

Stam, B., Gradstein, F.M., Lloyd, P. & Gillis, D. 1987. Algorithms for porosity and subsidence history. *Computers Geoscience*, **13**, 317–349.

Stanbrook, D.A. & Clark, J.D. 2004. The Marnes Brunes Inf~rieures in the Grand Coyer remnant: characteristics, structure and relationship to the Grès d'Annot. *In*: Joseph, P. & Lomas, S.A. (eds), *Deep-Water Sedimentation in the Alpine Basin of SE France: New Perspectives on the Grès d'Annot and Related Systems*. Geological Society, London, Special Publication, **221**, 285–300. London: The Geological Society.

Stanley, D.C.A. & Pickerill R.K. 1995. Arenitube, a new name for the trace fossil Ichnogenus Micatuba Chamberlain, 1971. *Journal of Paleontology*, **69**, 612–614.

Stanley, D.J. 1981. Unifites: structureless muds of gravity-flow origin in Mediterranean basins. *Geo-Marine Letters*, **1**, 77–83.

Stanley, D.J. 1987. Turbidite to current-reworked sand continuum in Upper Cretaceous rocks, U.S. Virgin Islands. *Marine Geology*, **78**, 143–151.

Stanley, D.J. 1988. Deep-sea current flow in the Late Cretaceous Caribbean: measurements on St. Croix, U.S. Virgin Islands. *Marine Geology*, **79**, 127–133.

Stanley, D.J. & Kelling, G. (eds) 1978. *Sedimentation in Submarine Canyons, Fans, and Trenches*. Stroudsburg, PA: Dowden, Hutchinson & Ross. 395 pp. ISBN: 0879333138.

Stanley, D.J. & Maldonado, A. 1981. Depositional models for fime-grained sediments in the western Hellenic Trench, eastern Mediterranean. *Sedimentology*, **28**, 273–290.

Stanley, D.J. & Unrug, R. 1972. Submarine channel deposits, fluxoturbidites and other indicators of slope and base-of-slope environments in modern and ancient marine basins. *In*: Rigby, J.K. & Hamblin, W.K. (eds), *Recognition of Ancient Sedimentary Environments*, 287–340. Society of Economic Paleontologists and Mineralogists Special Publication, **16**.

Stanley, D.J., Fenner, P. & Kelling, G. 1972. Currents and sediment transport at Wilmington Canyon shelf-break, as observed by underwater television. *In*: Swift, D.J.P., Duane, D.B. & Pilkey, O.H. (eds), *Shelf Sediment Transport: Process and Pattern*, 630–641. Stroudsburg, PA: Dowden, Hutchinson & Ross.

Stanley, D.J., Palmer, H.D. & Dill, R.F. 1978. Coarse sediment transport by mass flow and turbidity current processes and downslope transformation in Annot Sandstone canyon-fan valley systems. *In*: Stanley, D.J. & Kelling, G. (eds). *Sedimentation in Submarine Canyons, Fans and Trenches*, 85–115. Stroudsburg, PA: Dowden, Hutchinson & Ross.

Stauffer, P.H. 1967. Grain flow deposits and their implications, Santa Ynez Mountains, California. *Journal of Sedimentary Petrology*, **37**, 487–508.

Staukel, C., Lamy, F., Stuut, J.-B.W., Tiedemann, R. & Vogt, C. 2011. Distribution and provenance of wind-blown SE Pacific surface sediments. *Marine Geology*, doi:10.1016/j.margeo.2010.12.006.

Stax, R. & Stein, R. 1994. Quaternary organic carbon cycles in the Japan Sea (ODP-site 798) and their paleoceanographic implications. *Palaeogeography, Palaeoclimatology, Palaeoecology*, **108**, 509–521.

Steckler, M.S. & Watts, A.B. 1978. Subsidence of the Atlantic-type continental margin off New York. *Earth and Planetary Science Letters*, **41**, 1–13.

Stefani, C. De. 1895. Aperçu géologique et description paléontologique de l'île de Karpathos. *In*: Stefani, C. de., Forsyth Major, C.J. & Barbey, W. (eds), *Karpathos. Étude géologique, paléontologique et botanique*, 1–28. Lausanne: G. Bridel.

Steffens, G. 1986. Pleistocene entrenched valley/canyon systems, Gulf of Mexico. *American Association of Petroleum Geologists Bulletin*, **70**, 1189.

Sternberg, G.K. 1833. *Versuch einer geognostisch, botanischen Darstellung der Flora der Vorvwelt. IV Heft*. Brenck, C.E. Regensburg, 48 pp.

Stevens, S.H. & Moore, G.F. 1985. Deformational and sedimentary processes in trench slope basins of the western Sunda Arc, Indonesia. *Marine Geology*, **69**, 93–112.

Stevenson, A.J., Scholl, D.W. & Vallier, T.L. 1983. Tectonic and geologic implications of the Zodiac fan, Aleutian Abyssal Plain, northeast Pacific. *Geologicla Society of America Bulletin*, **94**, 259–273.

Stevenson, A.J., Talling, P.J., Wynn, R.B., Masson, D.G., Hunt, J.E., Frenz, M., Akhmetzhanhov, A. & Cronin, B.T. 2013. The flows that left no trace: Very large-volume turbidity currents that bypassed sediment through submarine channels without eroding the sea floor. *Marine and Petroleum Geology*, **41**, 186–205.

Stevenson, C.J., Talling, P.J., Masson, D.G., Sumner, E.J., Frenz, M. & Wynn, R.B. 2014. The spatial and temporal distribution of grain-size breaks in turbidites. *Sedimentology*, **61**, 1120–1156.

Stommel, H. & Arons, A.B. 1961. On the abyssal circulation of the world ocean. I. Stationary planetary flow patterns on a sphere. *Deep-Sea Research*, **6**, 140–154.

Stoneley, R. 1962. Marl diapirism near Gisbourne, New Zealand. *New Zealand Journal of Geology and Geophysics*, **5**, 630–641.

Storti, F., Holdsworth, R.E. & Salvini, F. (eds) 2003. *Intraplate Strike-Slip Deformation Belts*. Geological Society, London, Special Publications, **210**. London: The Geological Society.

Stow, D.A.V. 1976. Deep water sands and silts on the Nova Scotian continental margin. *Marine Sediments*, **12**, 81–90.

Stow, D.A.V. 1979. Distinguishing between fine-grained turbidites and contourites on the Nova Scotian deep water margin. *Sedimentology*, **26**, 371–387.

Stow, D.A.V. 1981. Laurentian Fan: morphology, sediments, processes and growth patterns. *American Association of Petroleum Geologists Bulletin*, **65**, 375–393.

Stow, D.A.V. 1982. Bottom currents and contourites in the North Atlantic. *Bulletin d'Institut de Geologie du Bassin d'Aquitaine*, **31**, 151–166.

Stow, D.A.V. 1984. Turbidite facies associations and sequences in the southeastern Angola Basin. In: Hay, W.W., Sibuet, J.C. *et al.*, *Initial Reports Deep Sea Drilling Project*, **75**, 785–99. Washington, DC: US Government Printing Office.

Stow, D.A.V. 1985. Deep-sea clastics: where are we and where are we going? *In*: P.J Brenchley & B.P.J. Williams (eds), *Sedimentology: Recent Developments and Applied Aspects*, 67–93. Geological Society, London, Special Publication, **18**. Oxford: Blackwell Scientific.

Stow, D.A.V. 1986. Deep clastic seas. *In*: Reading, H.G. (ed.), *Sedimentary Environments and Facies*, 2nd edn, 399–444. Oxford: Blackwell Scientific.

Stow, D.A.V. & Bowen, A.J. 1980. A physical model for the transport and sorting of fine-grained sediments by turbidity currents. *Sedimentology*, **27**, 31–46.

Stow, D.A.V. & Dean, W.E. 1984. Middle Cretaceous black shales at Site 530 in the south-eastern Angola Basin. *In*: W. W. Hay, W.W., Sibuet, J.C. *et al.*, *Initial reports Deep Sea Drilling Project*, **75**, 809–817. Washington, DC: US Government Printing Office.

Stow, D.A.V. & Faugères, J.-C. 2008. Contourite facies and the facies model. *In*: Rebesco, M. & Camerlenghi, A. (eds), *Contourites*, 223–250. Developments in Sedimentology, **60**. Amsterdam: Elsevier.

Stow, D.A.V. & Holbrook, J.A. 1984. North Atlantic contourites: an overview. *In*: Stow, D.A.V. & Piper, D.J.W. (eds), *Fine-grained Sediments: Deep-water Processes and Facies*, 245–256. Geological Society of London Special Publication, **15**. Oxford: Blackwell Scientific.

Stow, D.A.V. & Johansson, M. 2000. Deep-water massive sands: nature, origin and hydrocarbon implications. *Marine and Petroleum Geology*, **17**, 145–174.

Stow, D.A.V. & Lovell, J.P.B. 1979. Contourites: their recognition in modern and ancient sediments. *Earth-Science Reviews*, **14**, 251–291.

Stow, D.A.V. & Mayall, M. 2000. Deep-water sedimentary systems: new models for the 21st century. *Marine and Petroleum Geology*, **17**, 125–135.

Stow, D.A.V. & D.J.W. Piper (eds) 1984a. *Fine-grained Sediments: Deep-water Processes and Facies*. Geological Society, London, Special Publication, **15**. Oxford: Blackwell Scientific.

Stow, D.A.V. & Piper, D.J.W. 1984b. Deep-water fine-grained sediments: facies models. *In*: Stow, D.A.V. & Piper, D.J.W. (eds), *Fine-grained Sediments: Deep-water Processes and Facies,* 611–646. Geological Society of London Special Publication, **15**. Oxford: Blackwell Scientific.

Stow, D.A.V. & Shanmugam, G. 1980. Sequence of structures in fine-grained turbidites: comparison of recent deep-sea and ancient flysch sediments. *Sedimentary Geology*, **25**, 23–42.

Stow, D.A.V., Bishop, C.D. & Mills, S.J. 1982. Sedimentology of the Brae oilfield, North Sea: fan models and controls. *Journal of Petroleum Geology*, **5**, 129–148.

Stow, D.A.V., Cremer, M., Droz, L., Meyer, A.W., Normark, W.R., O'Connell, S., Pickering, K.T., Stelting, C.E., Angell, S.A. & Chaplin, C. 1986. Facies, composition, and tecture of Missisippi Fan sediments, Deep Sea Drilling Project Leg 96, Gulf of Mexico. *In*: Bouraa, A.H., Coleman, J.M., Meyer, A.W. *et al.*, *Initial Reports of the Deep Sea Drilling Project,* **96**, 475–487. Washington: U.S. Government Printing Office.

Stow, D.A.V., Amano, K., Balson, P.S., Brass, G.W., Corrigan, J., Raman, C.V., Tiercelin, J.-J., Townsend, M. & Wijayananda, N.P. 1990. Sediment facies and processeson the distal Bengal Fan, Leg 116. *In*: Cochran, J.R., Stow, D.A.V. *et al.*, *Proceedings of the Ocean Drilling Program, Scientific Results,* **116**, 377–396. College Station, Texas, USA: Ocean Drilling Program.

Stow, D.A.V., Faugère, J.-C., Viana, A.R. & Gonthier, E. 1998a. Fossil contourites: a critical review. *Sedimentary Geology*, **115**, 3–31.

Stow, D.A.V., Taira, A., Ogawa, Y., Soh, W., Taniguchi, T. & Pickering, K.T. 1998b. Volcaniclastic sediments, process interaction and depositional setting of the Mio-Pliocene Miura Group, SE Japan. *Sedimentary Geology*, **115**, 351–381.

Stow, D.A.V., Faugères, J.-C., Howe, J.A., Pudsey, C.J. & Viana, A.R. 2002a. Bottom currents, contourites and deep-sea sediment drifts: current state-of-the-art. *In*: Stow, D.A.V., Pudsey, C.J., Howe, J.A., Faugères, J.-C. & Viana, A.R. (eds), *Deep-water Contourite Systems: Modern Drifts and Ancient Series, Seismic and Sedimentary Characteristics,* 7–20. Geological Society London Memoir, **22**.

Stow, D.A.V., Pudsey, C.J., Howe, J.A., Faugères, J.-C. & Viana, A.R. (eds), 2002b. *Deep-water Contourite Systems: Modern Drifts and Ancient Series, Seismic and Sedimentary Characteristics.* Geological Society London Memoir, **22**.

Stow, D.A.V., Kahler, G. & Reeder, M. 2002c. Fossil contourites: type example from an Oligocene palaeoslope system, Cyprus. *In*: Stow, D.A.V., Pudsey, C.J., Howe, J.A., Faugères, J.-C. & Viana, A.R. (eds), *Deep-water Contourite Systems: Modern Drifts and Ancient Series, Seismic and Sedimentary Characteristics,* 443–455. Geological Society London Memoir, **22**.

Strachan, L.J. 2008. Flow transformations in slumps: a case study from the Waitemata Basin, New Zealand. *Sedimentology*, **55**, 1311–1332.

Srand, K., Marsaglia, K., Forsythe, R., Kurnosov, V. & Vergara, H. 1995. Outer margin depositional systems near the Chile margin triple junction. *In*: Lewis, S.D., Behrmann, J.H., Musgrave, R.J. & Cande, S.C. (eds.), *Proceedings of the Ocean Drilling Program, Scientific Results,* **141**, 379–397.

Strasser, M., Moore, G.F., Kimura, G., Kopf, A.J., Underwood, M.B., Guo, J.& Screaton, E.J. 2011. Slumping and mass transport deposition in the Nankai fore arc: Evidence from IODP drilling and 3-D reflection seismic data. *Geochemistry, Geophysics, Geosystems,* **12**, doi:10.1029/2010GC003431.

Straub, K.M. 2001. Bagnold revisited: implications for the rapid motion of high-concentration sediment flows. *In*: McCaffrey, W., Kneller, B. & Peakall, J. (eds), *Particulate Gravity Currents,* 91–112. International Association of Sedimentologists, Special Publication, **31**. Oxford: Blackwell Scientific.

Straub, K.M., Mohrig, D., McElroy, B., Buttles, J. & Pirmez, C., 2008. Interactions between turbidity currents and topography in aggrading sinuous submarine channels: a laboratory study. *Geological Society of America Bulletin*, **120**, 368–385.

Straub, K.M., Mohrig, D., Buttles, J., McElroy, B. & Pirmez, C. 2011. Quantifying the influence of channel sinuosity on the depositional mechanics of channelized turbidity currents: a laboratory study. *Marine and Petroleum Geology*, **28**, 744–760.

Straub, K.M. Mohrig, D. & Pirmez, C. 2012. Architecture of an aggradational tributary submarine-channel network on the continental slope offshore Brunei Darussalam. *In*: Prather, B.E., Deptuck, M.E., Mohrig, D., van Hoorn, B. & Wynn, R.B. (eds), *Application of the Principles of Seismic Geomorphology to Continental Slope and Base-of-slope Systems: Case Studies from Seafloor and Near-seafloor Analogues,* 13–30. SEPM (Society for Sedimentary Geology), **99**.

Stride, A.H., Belderson, R.H. & Kenyon, N.H. 1982. Structural grain, mud volcanoes and other features on the Barbados Ridge Complex revealed by GLORIA long-range side-scan sonar. *Marine Geology*, **49**, 187–196.

Stright, L., Stewart, J., Campion, K. & Graham, S. 2014. Geologic and seismic modeling of a coarse-grained deep-water channel reservoir analog (Black's Beach, La Jolla, California). *American Association of Petroleum Geologists Bulletin*, **98**, 695–728.

Strebelle, S., Payrazyan, K. & Caers, J. 2002. Modeling of a deepwater turbidite reservoir conditional to seismic data using multiple-point geostatistics. *In*: *Society of Petroleum Engineers (SPE) Annual Conference and Technical Meeting, SPE 77429, San Antonio, Texas,* 16 pp.

Strogen, D.P., Burwood, R. & Whitham, A.G. 2005. Sedimentology and geochemistry of Late Jurassic organic-rich shelfal mudstones from East Greenland: regional and stratigraphic variations in source-rock quality. *In*: *Petroleum Geology: From Mature Basins to New Frontier–Proceedings of the 7th Petroleum Geology Conference,* **6**, 903–912. London: The Geological Society.

Strong, P.G. & Walker, R.G. 1981. Deposition of the Cambrian continental rise: the St. Roch Formation near St. Jean-Port-Joli, Quebec. *Canadian Journal of Earth Sciences*, **18**, 1320–1335.

Strong, S.W.S. 1931. Ejection of fault breccia in the Waimata survey district, Gisbourne. *New Zealand Journal of Science and Technology*, **12**, 257–267.

Styzen, M.J. (compiler) 1996. *Late Cenozoic Chronostratigraphy of the Gulf of Mexico*. Gulf Coast Society of Economic Paleontologists and Mineralogists Foundation, chart, 2 sheets.

Sugioka, H., Okamoto, T., Nakamura, T., Ishihara, Y., Ito, A., Obana, K., Kinoshita, M., Nakahigashi, K., Shinohara, M. & Fukao, Y. 2012. Tsunamigenic potential of the shallow subduction plate boundary inferred from slow seismic slip. *Nature Geoscience*, **5**, 414–418.

Sullivan, S., Wood, L.J. & Mann, P. 2004. Distribution, Nature and Origin of Mobile Mud Features Offshore Trinidad. *In*: Post, P., Olson, D., Lyons, K., Palmes, S., Harrison, P. & Rosen, N. (eds), *Salt Sediment Interactions and Hydrocarbon Prospectivity: Concepts, Applications, and Case Studies for the 21st Century. 24th GCS-SEPM Annual Proceedings,* **24**, 840–867.

Sullwold, H.H. Jr, 1960. Tarzana fan, deep submarine fan of late Miocene age, Los Angeles County, California. *American Association of Petroleum Geologists Bulletin*, **44**, 433–457.

Sumer, B.M., Kozakiewicz, A., Fredsøe, J. & Deigaard, R. 1996. Velocity and concentration profiles in sheet flow layer of movable bed. *Journal of Hydraulic Engineering (ASCE)*, **122**, 549–558.

Sumner, E.J., Talling, P.J. & Amy, L.A. 2009. Deposits of flows transitional between turbidity current and debris flow. *Geology*, **37**, 991–994.

Summerhayes, C.P. 1979. *Marine Geology of the New Zealand sub- Antarctic Seafloor. New Zealand Department of Scientific and Industrial Research, Bulletin,* **190**.

Surlyk, F. 1978. Submarine fan sedimentatiom along fault scarps on tilted fault blocks (Jurassic–Cretaceous boundary, East Greenland). *Bulletin Grønlands Geologiske Undersøgelse,* **128**.

Surlyk, F. 1984. Fan-delta to submarine fan conglomerates of the Volgian-Valanginian Wollaston Foreland Group, East Greenland. *In*: Koster, E.H. & Steel, R.J. (eds), *Sedimentology of Gravels and Conglomerates,* 359–382. Canadian Society of Petroleum Geologists, Memoir, **10**.

Surlyk, F. 1987. Slope and deep shelf gully sandstones, Upper Jurassic, East Greenland. *American Association of Petroleum Geologists Bulletin*, **71**, 464–475.

Surlyk, F. 1995. Deep-sea fan valleys, channels, lobes and fringes of the Silurian Peary Land Group, North Greenland. *In*: Pickering, K.T., Hiscott, R.N., Kenyon, N.H., Ricci Lucchi, F. & Smith, R.D.A. (eds), *Atlas of Deep Water Environments: Architectural Style in Turbidite Systems,* 124–138. London: Chapman and Hall.

Surlyk, F. & Hurst, J.M. 1984. The evolution of the early Paleozoic deep-water basin of North Greenland. *American Association of Petroleum Geologists Bulletin*, **95**, 131–154.

Surpless, K.D., Ward, R.B. & Graham, S.A. 2009. Evolution and stratigraphic architecture of marine slope gully complexes: Monterey Formation (Miocene), Gaviota Beach, California. *Marine and Petroleum Geology*, **26**, 269–288.

Susilohadi, S., Gaedicke, C. & Ehrhardt, A. 2005. Neogene structures and sedimentation history along the Sunda forearc basins off southwest Sumatra and southwest Java. *Marine Geology*, **219**, 133–154.

Susilohadi, S., Gaedicke, C. & Djajadihardja, Y. 2009. Structures and sedimentary deposition in the Sunda Strait, Indonesia. *Tectonophysics*, **467**, 55–71.

Sutcliffe, C. & Pickering, K.T. 2009. End-signature of deep-marine basin-fill, as a structurally confined low-gradient clastic slope: the Middle Eocene Guaso system, south-central Spanish Pyrenees. *Sedimentology*, **56**, 1670–1689.

Sutherland, R. 1999. Basement geology and tectonic development of the greater New Zealand region: an interpretation from regional magnetic data. *Tectonophysics*, **308**, 341–362.

Suzuki, T. & Hada, S, 1979. Cretaceous tectonic melange of the Shimanto Belt in Shikoku, Japan. *Journal of the Geological Society of Japan*, **85**, 467–479.

Swart, R. 1990. *The sedimentology of the Zerissene turbidite system, Damara Orogen, Namibia*. Ph.D. Thesis, Rhodes University, South Africa.

Swift, D.J.P. 1968. Coastal erosion and transgressive stratigraphy. *Journal of Geology*, **76**, 444–456.

Swift, S.A., Hollister, C.D. & Chandler, R.S. 1985. Close-up stereo photographs of abyssal bedforms on the Nova Scotian continental rise. *Marine Geology*, **66**, 303–322.

Sylvester, A.G. 1988. Strike-slip faults. *Geological Society of America Bulletin*, **100**, 1666–1703.

Sylvester, A.G. & Smith, R.R. 1976. Tectonic transpression and basement-controlled deformation in the San Andreas fault zone, Salton trough, California. *American Association of Petroleum Geologists Bulletin*, **60**, 74–96.

Sylvester, Z. 2007. Turbidite bed thickness distributions: methods and pitfalls of analysis and modelling. *Sedimentology*, **54**, 847–870.

Taira, A. 1981. The Shimanto Belt of southwest Japan and arc-trench sedimentary tectonics. *Recent Progress of Natural Sciences in Japan*, **6**, 147–162.

Taira, A. 1985. Sedimentary evolution of the Shikoku subduction zone: the Shimanto Belt and Nankai Trough. *In*: Nasu, N., Kobayashi, K., Veda, S., Kushiro,. I. & Kagami, H. (eds), *Formation of Active Margins,* 835–851. Tokyo: Terrapub.

Taira, A. 1988. *The Shimanto Belt, Southwest Japan: Studies on the Evolution of an Accretionary Prism. Modern Geology*: Netherlands: Gordon & Breach Science Publishers Ltd. 536 pp. ISBN: 9780677256801.

Taira, A. 2001. Tectonic evolution of the Japanese island arc system. *Annual Reviews in Earth and Planetary Sciences*, **29**, 109–134.

Taira, A. & Niitsuma, N. 1986. Turbidite sedimentation in the Nankai Trough as interpreted from magnetic fabric, grain size, and detrital modal analyses. *In*: Kagami, H., Karig, D.E., Coulbourn, W.T., *et al.*, *Initial Reports Deep Sea Drilling Project*, **87**, 611–632. Washington, DC: US Government Printing Office.

Taira, A., Tashiro, M., Okamura, M. & Katto, J. 1980. The geology of the Shimanto Belt in the Kochi Prefecture, Shikoku, Japan. *In*: Taira, A.

& Tashiro, M. (eds), *Geology and Paleontology of the Shimanto Belt (Cretaceous),* 319–389. Kochi: Rinya-kosaikai Press.

Taira, A., Okada, H., Whitaker, J.H. & Smith, A.J. 1982. The Shimanto Belt of Japan: Cretaceous-lower Miocene active-margin sedimentation. *In*: Leggett, J.K. (ed.), *Trench-Forearc Geology,* 5–26. The Geological Society of London, Special Publication, **10**. London: The Geological Society.

Taira, A., Saito, Y. & Hashimoto, M. 1983. The role of oblique subduction and strike-slip tectonics in the evolution of Japan. *In*: Hilde, T.W.C. & Uyeda, S. (eds), *Geodynamics of the Western Pacific – Indonesian Region*, 303–316. American Geophysical Union Geodynamics Series, **11**.

Taira, A., Tokuyama, H., & Soh. W. 1989, *Accretion tectonics and evolution of Japan. In*: Ben-Avraham, Z. (ed.), *The Evolution of the Pacific Ocean Margins,* 100–123. Oxford Monographs in Geology & Geophysics, **8**. Oxford, UK: Oxford University Press.

Taira, A., Hill, I., Firth, J., Berner, U., Brückmann, W., Byrne, T., Chabernaud, T., Fisher, A., Foucher, J.-P., Gamo, T., Gieskes, J., Hyndman, R., Karig, D., Kastner, M., Kato, Y., Lallement, S., Lu, R., Maltman, A., Moore, G., Moran, K., Olaffson, G., Owens, W., Pickering, K.T., Siena, F., Taylor, E., Underwood, M., Wilkinson, C., Yamano, M. & Zhang, J. 1992. Sediment deformation and hydrogeology of the Nankai Trough accretionary prism: synthesis of shipboard results of Ocean Drilling Program Leg 131. *Earth and Planetary Science Letters*, **109**, 431–450.

Takahashi, T. 1981. Debris flow. *Annual Review of Fluid Mechanics*, **13**, 57–77.

Takano, O. 2002. Changes in depositional systems and sequences in response to basin evolution in a rifted and inverted basin: an example from the Neogene Niigata-Shin'etsubasin, Northern Fossa Magna, central Japan. *Sedimentary Geology*, **152**, 79–97.

Talling, P.J. 2001. On the frequency distribution of turbidite thickness. *Sedimentology*, **48**, 1297–1329.

Talling, P.J. 2014. On the triggers, resulting flow types and frequencies of subaqueous sediment density flows in different settings. *Marine Geology*, **352**, 155–182.

Talling, P.J., Peakall, J., Sparks, R.S.J., Cofaigh, C.Ó., Dowdeswell, J.A., Felix, M., Wynn, R.B., Baas, J.H., Hogg, A.J., Masson, D.G., Taylor, J. & Weaver, P.P.E. 2002. Experimental constraints on shear mixing rates and processes: implications for the dilution of submarine debris flows. *In*: Dowdeswell, J.A. & Cofaigh, C.Ó. (eds), *Glacier-influenced Sedimentation on High-latitude Continental Margins,* 89–103. Geological Society of London Special Publication, **203**.

Talling, P.J., Amy, L.A., Wynn, R.B., Peakall, J. & Robinson, M. 2004. Beds comprising debrite sandwiched within co-genetic turbidite: origin and widespread occurrence in distal depositional environments. *Sedimentology*, **51**, 163–194.

Talling, P.J., Amy, L.A. & Wynn, R.B. 2007. New insights into the evolution of large volume turbidity currents; comparison of turbidite shape and previous modelling results. *Sedimentology*, **54**, 737–769.

Talling, P.J., Wynn, R.B., Rixon, R. & Schmidt, D. 2010. How did submarine flows transport boulder sized mud clasts to the fringes of the Mississippi Fan? *Journal of Sedimentary Research*, **80**, 829–851.

Talling, P.J., Masson, D.G., Sumner, E.J. & Malgesini, G. 2012. Subaqueous sediment density flows: depositional processes and deposit types. *Sedimentology*, **59**, 1937–2003.

Talling, P.J., Malgesini, G. & Felletti, F. 2013. Can liquefied debris flows deposit clean sand over large areas of sea floor? *Field evidence from the Marnoso-arenacea Formation, Italian Apennines Sedimentology*, **60**, 720–762.

Talling, P.J., Joshua Allin, Armitage, D.A., Arnott, R.W.C., Cartigny, M.J.B., Clare, M.A., Felletti, F., Covault, J.A., Girardclos, S., Ernst Hansen, Hill, P.R., Hiscott, R.N., Hogg, A.J., Clarke, J.H., Jobe, Z.R., Malgesini, G., Mozzato, A., Naruse, H., Parkinson, S., Peel, F.J., Piper, D.J.W., Pope, E., Postma, G., Rowley, P., Sguazzini, A., Stevenson, C.J., Sumner, E.J., Sylvester, Z., Watts, C., Xu, J. 2015. Key Future Directions For Research On Turbidity Currents and Their Deposits. *Journal of Sedimentary Research*, **85**, 153–169.

Talwani, K. & Pitman III, W.C. (eds) 1977. *Island Arcs, Deep-sea Trenches and Back-arc Basins.* Maurice Ewing Series **1**. Washington, DC: American Geophysical Union.

Talwani, T., Udinstev, G. *et al.* 1976. *Initial Reports Deep Sea Drilling Project, 38.* Washington, DC: US Government Printing Office.

Tamaki, K. & Honza, E. 1991. Global tectonics and formation of marginal basins: role of the western Pacific. *Episodes*, **14**, 224–230.

Tamburini, F., Adatte, T & Föllmi, K.B. 2003. Origin and nature of green clay layers, ODP Leg 184, South China Sesa. *In*: Prell, W.L., Wang, P., Blum, P., Rea, D.K. & Clemens, S.C. (eds), *Proceedings of the Ocean Drilling Program, Scientific Results, 184*, 1–23. College Station, Texas, USA: Ocean Drilling Program.

Tanaka, K. 1970. Sedimentation of the Cretaceous flysch sequence in the Ikushumbetsu area, Hokkaido, Japan. *Geological Survey of Japan, Report*, **236**, 1–102.

Tankard, A.J. 1986. On the depositional response to thrusting and lithospheric flexure: examples from the Appalachian and Rocky Mountain basins. *In*: Allen, P.A. Homewood, P. (eds), *Foreland Basins*, 369–392. Special Publication of the International Association of Sedimentologists, **8**. Oxford: Blackwell Scientific Publications.

Tankard, A.J. & Welsink, H.J. 1987. Extensional tectonics and stratigraphy of Hibernia oil field, Grand Banks, Newfoundland. *American Association of Petroleum Geologists Bulletin*, **71**, 1210–1232.

Tankard, A., Welsink, H., Aukes, P., Newton, R. & Stettler, E. 2009. Tectonic evolution of the Cape and Karoo basins of South Africa. *Marine and Petroleum Geology*, **26**, 1379–1412.

Tappin, D.R., McNeill, L.C., Henstock, T. & Mosher, D. 2007. Mass wasting processes - offshore Sumatra. *In*: Lykousis, V., Sakellariou, D. & Locat, J. (eds), *Submarine Mass Movements and their Consequences*, 327–336. New York: Springer-Verlag.

Tarlao, A., Tunis, G. & Venturini, S. 2005. Dropstones, pseudoplanktonic forms and deep-water decapod crustaceans within a Lutetian condensed succession of central Istria (Croatia): relation to palaeoenvironmental evolution and palaeogeography. *Palaeogeography, Palaeoclimatology, Palaeoecology*, **218**, 325–345.

Tarney, J., Pickering, K.T., Knipe, R.J. & Dewey, J.F. (eds) 1991. *The Behaviour and Influence of Fluids in Subduction Zones.* London: The Royal Society, 418 pp.

Taylor, A.M. & Gawthorpe, R.L. 1993. Application of sequence stratigraphy and trace fossil analysis to reservoir description: examples from the Jurassic of the North Sea. *In*: Parker, J.R. (ed.). *Petroleum Geology of Northwest Europe: Proceedings of the 4th Conference*, 317–335. London: Geological Society.

Taylor, A.M. & Goldring, R. 1993. Description and analysis of bioturbation and ichnofabric. *Journal of the Geological Society, London*, **150**, 141–148.

Taylor, A.M., Goldring, R. & Gowland, S. 2003. Analysis and application of ichnofabrics. *Earth-Science Reviews*, **60**, 227–225.

Tazaki, K. & Inomata, M. 1980. Umbers in pillow lava from the Mineoka tectonic belt, Boso Peninsula. *Journal of the Geological Society of Japan*, **86**, 413–416.

Taylor, B. & Natland, J. (eds) 1995. *Active Margins and Marginal Basins of the Western Pacific.* American Geophysical Union, Geophysical Monograph Series, **88**, 417 pp.

Tchoumatchenco, P. & Uchman, A. 1999. Lower and Middle Jurassic flysch trace fossils from the eastern Stara Planina Mountains, Bulgaria: A contribution to the evolution of Mesozoic ichnodiversity. *Neues Jahrbuch für Geologie und Paläontologie*, **213**, 169–199.

Tchoumatchenco, P. & Uchman, A. 2001. The oldest deep-sea Ophiomorpha and Scolicia and associated trace fossils from the Upper Jurassic–Lower Cretaceous deep-water turbidite deposits of SW Bulgaria. *Palaeogeography Palaeoclimatology, Palaeoecology*, **169**, 85–99.

Teale, T. 1985. Occurrence and geological significance of olisto-liths from the Longobucco Group, Calabria, southern Italy. *In: 6th European Regional Meeting, Lleida, Spain, International Association of Sedimentologists Abstracts volume*, 457–460.

Teraoka, Y. 1979. Provenance of the Shimanto geosynclinal sediments inferred from sandstone compositions. *Journal of the Geological Society of Japan*, **83**, 795–810.

Terlaky, V. & Arnott, R.W.C. 2014. Matrix-rich and associated matrix-poor sandstones: avulsion splays in slope and basin-floor strata. *Sedimentology*, **61**, 1175–1197.

Terlaky, V., Longuépée, H., Rocheleau, J., Meyer, L., Privett, K., van Hees, G., Cramm, G., Tudor, A. & Arnott, R.W. 2010. Facies, architecture and compartmentalization of basin-floor deposits: Upper and Middle Kaza Groups, British Columbia, Canada. *American Association of Petroleum Geologists Search and Discovery Article*, **#50301**.

Thayer, C.W. 1979. Biological bulldozers and the evolution of marine benthic communities. *Nature*, **203**, 458–461.

Thiede, J. & Myhre, A.M. 1996. Introduction to the North Atlantic-Arctic gateways: plate tectonic-paleoceanographic history and significance. *In*: Thiede, J., Myhre, A.M., Firth, J.V., Johnson, G.L. & Ruddiman, W.F. (eds), *Proceedings of the Ocean Drilling Program, Scientific Results*, **151**, 1–23. College Station, Texas, USA: Ocean Drilling Program.

Thomas, B., Despland, P. & Holmes, L. 2012. Submarine Sediment Distribution Patterns within the Bengal Fan System, Deep Water Bengal Basin, India. *American Association of Petroleum Geologists Search & Discovery*, Article **#50756**.

Thomas, W.A. 1977. Evolution of Appalachian - Ouachita salients and recesses from reentrants and promontories in the continental margin. *American Journal of Science*, **277**, 1233–1278.

Thomas, W.A. 1985. The Appalachian - Ouachita connection: Paleozoic Orogenic Belt at the Southern Margin of North America. *Annual Reviews in Earth and Planetary Science*, **13**, 175–199.

Thomas, W.A. & Becker, T.P. 2007. Crustal recycling in the Appalachian foreland. *Geological Society of America, Memoirs*, **200**, 33–40.

Thompson, A.F. & Thomasson, M.R. 1969. Shallow to deep water facies development in the Dimple Limestone (Lower Pennsylvanian), Marathon region, Texas. *In*: Friedman, G.M. (ed.), *Depositional Environments in Carbonate Rocks*, 57–78. Society of Economic Paleontologists and Mineralogists Special Publication, **14**.

Thompson, B.J., Garrison, R.E. & Moore, C.J., 1999. A late Cenozoic sandstone intrusion west of S. Cruz, California. Fluidized flow of water and hydrocarbon-saturated sediments. *In*: Garrison, R.E., Aiello, I.W. & Moore, J.C. (eds), *Late Cenozoic Fluid Seeps and Tectonics along the San Gregorio Fault Zone in the Monterey Bay Region, California GB-76*, 53–74. Annual Meeting of the Pacific Section. American Association of Petroleum Geologists, Monterey, California.

Thompson, B.J., Garrison, R.E. & Moore, C.J., 2007. A reservoir-scale Miocene injectite near Santa Cruz, California. *In*: Hurst, A. & Cartwright, J. (eds), *Sand Injectites: Implications for Hydrocarbon Exploration and Production*, 151–162. Tulsa, Oklahoma: American Association of Petroleum Geologists Memoir, **87**.

Thornburg, T.M. & Kulm, L.D. 1987. Sedimentation in the Chile Trench: depositional morphologies, lithofacies, and stratigraphy. *Geological Society of America Bulletin*, **98**, 33–52.

Thornburg, T.M., Kulm, L.D. & Hussong, D.M. 1990. Submarine-fan development in the southern Chile Trench: A dynamic interplay of tectonics and sedimentation. *Geological Society of America Bulletin*, **102**, 1658–1680.

Thorne, J. & Watts, A.B. 1984. Seismic reflectors and uncon¬formities at passive continental margins. *Nature*, **311**, 365–368.

Thorne, J.A. & Swift, D.J.P. 1991. Sedimentation on continental margins. *Part II: Application of resin merriamgime concept. International Association of Sedimentologists Special Publication*, **14**, 33–58.

Thornton, S.E. 1981. Suspended sediment transport in surface waters of the California Current off southern California: 1977–1978 floods. *Geo-Marine Letters*, **1**, 23–28.

Thornton, S.E. 1984. Basin model for hemipelagic sedimentation in a tectonically active continental margin: Santa Barbara Basin, California Continental Borderland. *In*: Stow, D.A.V. & Piper, D.J.W. (eds), *Fine-grained Sediments: Deep-water Processes and Facies*, 377–394.

Geological Society of London Special Publication, **15**. Oxford: Blackwell Scientific.

Tillman, R.W. & Ali, S.A. (eds) 1982. *Deep Water Canyons, Fans and Facies: Models for Stratigraphic Trap Exploration. Reprint Series, 26.* American Association of Petroleum Geologists.

Tilman, R.W. & Weber, K.J. (eds) 1987. *Reservoir Sedimentology.* Society of Economic Paleontologists & Mineralogists, Special Publication, **40**. ISBN: 0-918985-69-2.

Timbrell, G. 1993. Sandstone architecture of the Balder Formation depositional system, UK Quadrant 9 and adjacent areas. *In:* Parker, J.R. (ed.), *Petroleum Geology of Northwest Europe: Proceedings of the 4th Conference,* 107–121. London: The Geological Society.

Tinterri, R. & Magalhaes, P.M. 2011. Synsedimentary-structural control on foredeep turbidites: An example from Miocene Marnoso-arenacea Formation, Northern Apennines, Italy. *Marine and Petroleum Geology,* **28**, 629–657.

Tobin, H., Kinoshita, M., Ashi, J., Lallemant, S., Kimura, G., Screaton, E.J., Moe, K.T., Masago, H., Curewitz, D. & the Expedition 314/315/316 Scientists 2009. NanTroSEIZE Stage 1 expeditions: introduction and synthesis of key results. *In:* Kinoshita, M., Tobin, H., Ashi, J., Kimura, G., Lallemant, S., Screaton, E.J., Curewitz, D., Masago, H., Moe, K.T. & the Expedition 314/315/316 Scientists. *Proceedings of the IODP,* **314/315/316**. Washington, DC (Integrated Ocean Drilling Program Management International, Inc.). doi:10.2204/ iodp.proc.314315316.101.2009.

Tobin, H.J. & Kinoshita, M. 2006a. *Investigations of Seismogenesis at the Nankai Trough, Japan.* IODP Sci. Prosp., NanTroSEIZE Stage 1. doi:10.2204/iodp.sp.nantroseize1.2006.

Tobin, H.J. & Kinoshita, M. 2006b. NanTroSEIZE: the IODP Nankai Trough Seismogenic Zone Experiment. *Scientific Drilling,* **2**, 23–27. doi:10.2204/iodp.sd.2.06.2006.

Tokuhashi, S. 1979. Three dimensional analysis of large sandy-flysch body, Mio-Pliocene Kiyosumi Formation, Boso Peninsula, Japan. *Memoirs of the Faculty of Sciences, Kyoto University, Series of Geology & Mineralogy,* **46**, 1–61.

Tokunaga, T. 2000. The role of turbidites on compaction and dewatering of under thrust sediments at the toe of the northern Barbados accretionary prism: new evidence from Logging While Drilling, ODP Leg 171A. *Earth and Planetary Science Letters,* **178**, 385–395.

Tomasso, M. & Sinclair, H.D. 2004. Deep-water sedimentation on an evolving fault-block: the Braux and St Benoit outcrops of the Grès d'Annot. *In:* Joseph, P. & Lomas, S.A. (eds), *Deep-Water Sedimentation in the Alpine Basin of SE France: New Perspectives on the Grès d'Annot and Related Systems.* Geological Society, London, Special Publication, **221**, 267–283. London: The Geological Society.

Toniolo, H., Lamb, M. & Parker, G. 2006. Depositional turbidity currents in diapiric minibasins on the continental slope: formulation and theory. *Journal of Sedimentary Research,* **76**, 783–797.

Tonkin, N.S., McIlroy, D., Meyer, R. & Moore-Turpin, A. 2010. Bioturbation influence on reservoir quality: a case study from the Cretaceous Ben Nevis Formation, Jeanne d'Arc Basin, offshore Newfoundland, Canada. *American Association of Petroleum Geologists Bulletin,* **94**, 1059–1078.

Torell, O.M. 1870. Petrifacta Suecana Formationis Cambricae. *Lunds Universitets Årsskrift,* **6**. 1–14.

Trautwein, B., Dunk, I., Kuhlemann, J. & Frisch, W. 2001. Cretaceous–Tertiary Rhenodanubian flysch wedge (Eastern Alps): clues to sediment supply and basin configuration from zircon fission-track data. *Terra Nova,* **13**, 382–393.

Tréhu, A.M., Asudeh, I., Brocher, T.M., Leutgert, J., Mooney, W.D., Nabelek, J.N. & Nakamura, Y. 1994. Crustal architecture of the Cascadia fore arc. *Science,* **265**, 237–243.

Tréhu, A.M., Lin, G., Maxwell, E. & Goldfinger, C. 1995. A seismic reflection profile across the Cascadia subduction zone offshore central Oregon: New constraints on the deep crustal structure and on the distribution of methane in the accretionary prism. *Journal of Geophysical Research,* **100**, 15,101–15,116.

Tréhu, A.M., Blakely, R.J. &Williams, M.C. 2012. Subducted seamounts and recent earthquakes beneath the central Cascadia forearc. *Geology,* **40**, 103–106.

Treloar, P.J. & Searle, M.P. (eds) 1995. *Himalayan Tectonics.* Geological Society Special Publications, **74**, 630 pp. London: The Geological Society. ISBN 0-903317-92-3.

Tripati, A.K., Eagle, R.A., Morton, A.C., Dowdeswell, J.A., Atkinson, K.L., Bahé, Y., Dawber, C.F., Khadun, E., Shaw, R.M.H., Shorttle, O. & Thanabalasundaram, M.2006. Evidence for glaciation in the Northern Hemisphere back to 44 Ma from ice-rafted debris in the Greenland Sea. *Earth and Planetary Science Letters,* **265**, 112–122.

Trofimovs, J., Talling, P.J., Fisher, J.K., Sparks, R.S.J., Watt, S.F.L., Hart, M.B., Smart, C.W., Le Friant, A., Cassidy, M., Moreton, S.G. & Leng, M.J. 2013. Timing, origin and emplacement dynamics of mass flows offshore of SE Montserrat in the last 110 ka: Implications for landslide and tsunami hazards, eruption history, and volcanic island evolution. *Geochemistry, Geophysics, Geosystems,* **14**, 385–406.

Trop, J.M., Ridgway, K.D., Sweet, A.R. & Layer, P.W. 1999. Submarine fan deposystems and tectonics of a Late Cretaceous forearc basin along an accretionary convergent plate boundary, MacColl Ridge Formation, Wrangell Mountains, Alaska. *Canadian Journal of Earth Sciences,* **36**, 433–458.

Trop, J.M., Ridgway, K.D., Manuszak, J.D. & Layer, P. 2002. Mesozoic sedimentary-basin development on the allochthonous Wrangellia composite terrane, Wrangell Mountains basin, Alaska: a long-term record of terrane migration and arc construction. *Geological Society of America Bulletin,* **114**, 693–717.

Trop, J.M., Szuch, D.A., Rioux, M. & Blodgett, R.B. 2005. Sedimentology and provenance of the Upper Jurassic Naknek Formation, Talkeetna Mountains, Alaska: Bearings on the accretionary tectonic history of the Wrangellia composite terrane. *Geological Society of America Bulletin,* **117**, 570–588.

Tucholke, B.E. 1979. Furrows and focussed echoes on the Blake Outer Ridge. *Marine Geology,* **31**, M13–M20.

Tucholke, B.E., Hollister, C.D., Biscaye, P.E. & Gardner, W.D. 1985. Abyssal current character determined from sediment bedforms on the Nova Scotian continental rise. *Marine Geology,* **66**, 43–57.

Tucker, M.E. & Wright, V.P. 1990. *Carbonate Sedimentology.* Blackwell Scientific Publications (Wiley-Blackwell), 482 pp. ISBN: 978-0-632-01472-9.

Tunis, G. & Uchman, A. 1992. Trace fossils in the "Flysch del Grivó" Lower Tertiary in the Julian Prealps, NE Italy: Preliminary observations. *Gortania,* **14**, 71–104.

Tunis, G. & Uchman, A. 1996a. Trace fossils and changes in Cretaceous–Eocene flysch deposits of the Julian Prealps Italy and Slovenia: consequences of regional and world-wide changes. *Ichnos,* **4**, 169–190.

Tunis, G. & Uchman, A. 1996b. Ichnology of Eocene flysch deposits of the Istria Peninsula, Croatia and Slovenia. *Ichnos,* **5**, 1–22.

Turini, C. & Rennison, P. 2004. Structural style from the Southern Apennines' hydrocarbon province - an integrated view. *In:* McClay, K.R. (ed.), *Thrust Tectonics and Hydrocarbon Systems,* 558–578. American Association of Petroleum Geologists Memoir. **82**.

Turner, C.C., Cohen, J.M., Connell, E.R. & Cooper, D.M. 1987. A depositional model for the South Brae oilfield. *In:* Brooks, J. & Glennie, K.W. (eds), *Petroleum Geology of North-west Europe,* 853–864. London: Graham & Trotman.

Twichell, D.C. & Roberts, D.G. 1982. Morphology, distribution, and development of submarine canyons on the United States Atlantic continental slope between Hudson and Baltimore Canyons. *Geology,* **10**, 408–412.

Twichell, D.C., Kenyon, N.H., Parson, L.M. & McGregor, B.A. 1991. Depositional patterns of the Mississippi Fan surface: Evidence from GLORIA II and high-resolution seismic profiles. *In:* Weimer, P. & Link, M.H. (eds), *Seismic Facies and Sedimentary Processes of Submarine Fans and Turbidite Systems,* 349–364. New York: Springer-Verlag.

Twichell, D.C., Schwab, W.C., Nelson, C.H., Kenyon, N.H. & Lee, H.J. 1992. Characteristics of a sandy depositional lobe on the outer Mississippi Fan from SeaMARC IA sidescan sonar images. *Geology*, **20**, 689–692.

Tyler, J.E. & Woodcock, N.H. 1987. The Bailey Hill Formation: Ludlow Series turbidites in the Welsh Borderland reinterpreted as distal storm deposits. *Geological Journal*, **22**, 73–86.

Uchman, A. 1995. Taxonomy and palaeocology of flysch trace fossils: The Marnoso-arenacea Formation and associated facies, Miocene, Northern Apennines, Italy. *Beringeria*, **15**, 3–115.

Uchman, A. 1998. Taxonomy and ethology of flysch trace fossils: revision of the Marian Książkiewicz collection and studies of complementary material. *Annales Societatis Geologorum Poloniae*, **68**, 105–218.

Uchman, A. 1999. Ichnology of the Rhenodanubian Flysch Lower Cretaceous–Eocene in Austria and Germany. *Beringeria*, **25**, 67–173.

Uchman, A. 2001. Eocene flysch trace fossils from the Hecho Group of the Pyrenees, northern Spain. *Beringeria*, **28**, 3–41.

Uchman, A. 2009. The Ophiomorpha rudis ichnosubfacies of the Nereites ichnofacies: Characteristics and constraints. *Palaeogeography, Palaeoclimatology, Palaeoecology*, **276**, 107–119.

Uchman, A. & Demircan, H. 1999. Trace fossils of Miocene deep-sea fan fringe deposits from the Cingöz Formation, southern Turkey. *Annales Societatis Geologorum Poloniae*, **69**, 125–153.

Uchman, A. & Wetzel, A. 2011. Deep-sea ichnology: the relationships between depositional environment and endobenthic organisms. *In*: Hüneke, H. & Mulder, T. (eds), *Deep-sea Sediments*, 517–556. Developments in Sedimentology, **63**. Amsterdam: Elsevier.

Uchman, A. & Wetzel, A. 2012. Deep-sea fans. *In*: Knaust, D. & Bromley, R.G. (eds), *Trace Fossils as Indicators of Sedimentary Environments*, 643–672. Developments in Sedimentology, **64**. Amsterdam: Elsevier.

Uchman, A., Bubniak, I. & Bubniak, A. 2000. Glossifungites ichnofacies in the area of its nomenclatural archetype, Lviv, Ukraine. *Ichnos*, **7**, 183–193.

Uchman. A., Janbu, N.E. & Nemec, W. 2004. Trace Fossils in the Cretaceous–Eocene flysch of the Sinop-Boyabat Basin, Central Pontides, Turkey. *Annales Societatis Geologorum Poloniae*, **74**, 197–235.

Uenzelmann-Neben, G. & Gohl, K. 2012. Amundsen Sea sediment drifts: Archives of modifications in oceanographic and climatic conditions. *Marine Geology*, **299–302**, 51–62.

Underwood, M.B. 1986. Transverse infilling of the Central Aleutian Trench by unconfined turbidity currents. *Geo-Marine Letters*, **6**, 7–13.

Underwood, M.B. 1993. Thermal Evolution of the Tertiary Shimanto Belt, SW Japan: An Example of Ridge-Trench Interaction. *Geological Society of America, Special Paper*, **273**, 1–172.

Underwood, M.B. 2002. Strike-parallel variations in clay minerals and fault vergence in the Cascadia subduction zone. *Geology*, **30**, 155–158.

Underwood, M.B. 2007. Sediment inputs to subduction zones: why lithostratigraphy and clay mineralogy matter. *In*: Dixon, T.H. & Moore, J.C. (eds), *The Seismogenic Zone of Subduction Thrust Faults*, 42–85. New York: Columbia University Press.

Underwood, M.B. & Bachman, S.B. 1982. Sedimentary facies associations within subduction complexes. *In*: Leggett, J.K. (ed.), *Trench-Forearc Geology*, 537–550. The Geological Society of London, Special Publication, **10**. London: The Geological Society.

Underwood, M.B. & Fergusson, C.L. 2005. Late Cenozoic evolution of the Nankai trench-slope system: evidence from sand petrography and clay mineralogy. *In*: Hodgson, D.M. & Flint, S.S. (eds), *Submarine Slope Systems: Processes and Products*, 113–129. Geological Society, London, Special Publications, **244**. London: The Geological Society.

Underwood, M.B. & Karig, D.E. 1980. Role of submarine canyons in trench and trench-slope sedimentation. *Geology*, **8**, 432–436.

Underwood, M.B. & Norville, C.R. 1986. Deposition of sand in a trench-slope basin by unconfined turbidity currents. *Marine Geology*, **71**, 383–392.

Underwood, M.B. & Pickering, K.T. 1996. Clay-mineral provenance, sediment dispersal patterns, and mudrock diagenesis in the Nankai accretionary prism, southwest Japan. *Clays and Clay Minerals*, **44**, 339–356.

Underwood, M.B. & Steurer, J.F. 2003. Composition and sources of clay from the trench slope and shallow accretionary prism of Nankai Trough. *In*: Mikada, H., Moore, G.F., Taira, A., Becker, K., Moore, J.C. & Klaus, A. (eds), *Proceedings of the Ocean Drilling Program, Scientific Results*, **190/196**. College Station, Texas, USA: Ocean Drilling Program.

Underwood, M.B., Pickering, K.T., Gieskes, J.M., Miriam Kastner, M. & Orr, R. 1993. Sediment geochemistry, clay mineralogy, and diagenesis: a synthesis of data from Leg 131, Nankai Trough. *In*: Taira, A., Hill, I.A.H., Firth, J., Vrolijk, P.J. *et al.* (eds), *Proceedings of the Ocean Drilling Program Leg*, **131**, 343–363. College Station, Texas, USA: Ocean Drilling Program.

Underwood, M.B., Ballance, P.F., Clift, P., Hiscott, R.N., Marsaglia, K.M., Pickering, K.T. & Reid, R.P. 1995. Sedimentation in Forearc Basins, Trenches, and Collision Zones of the Western Pacific: A summary of Results from the Ocean Drilling Program. WPAC. *In*: Taylor, B. & Natland, J. (eds), *Active Margins and Marginal Basins of the Western Pacific*, 315–353. American Geophysical Monograph, **88**. Washington DC: American Geophysical Union.

Underwood, M.B., Moore, G.F., Taira, A., Klaus, A., Wilson, M.E.J., Fergusson, C.L., Hirano, S., Steurer, J. & the LEG 190 Shipboard Scientific Party 2003. Sedimentary and Tectonic Evolution of a Trench-Slope Basin in the Nankai Subduction Zone of Southwest Japan. *Journal of Sedimentary Research*, **73**, 589–602.

Underwood, M.B., Saito, S., Kubo, Y. & the Expedition 322 Scientists 2009. *Integrated Ocean Drilling Program Expedition 322 Preliminary Report NanTroSEIZE Stage 2:Subduction Inputs 1 September–10 October 2009.* Integrated Ocean Drilling Program Preliminary Reports, **322**. doi:10.2204/iodp.pr.322.2009.

Underwood, M.B., Saito, S., Kubo, Y. & the Expedition 322 Scientists 2010. *Exp. 322. Proceedings of the IODP*, **322**. Tokyo: Integrated Ocean Drilling Program. doi:10.2204/iodp.proc.322.101.2010.

Vail, P. R. & Hardenbol, J. 1979. Sea-level changes during the Tertiary. *Oceanus*, **22**, 71–79.

Vail, P.R. 1987. Seismic stratigraphy interpretation using sequence stratigraphy, Part 1: Seismic stratigraphy interpretation procedure. *In*: Bally, A.W. (ed.), *Atlas of Seismic Stratigraphy*. American Association of Petroleum Geologists, Studies in Geology, **27**, 1–10.

Vail, P.R. & Posamentier, H.W. 1988. Principles of sequence stratigraphy. *In*: James, D.P. & Leckie, D.A. (eds), *Sequences, Stratigraphy, Sedimentology; Surface and Subsurface.* CSPG Memoir **15**, 572.

Vail, P.R. & Todd, R.G. 1981. Northern North Sea Jurassic unconformities, chronostratigraphy and sea-level changes from seismic stratigraphy. *In*: Illing, L.V. & Hobson, G.D. (eds), *Petroleum Geology of the Continental Shelf of Northwest Europe, Conference Proceedings*, 216–235. London: Heyden and Son.

Vail, P.R., Mitchum, R.M., Jr., & Thompson, S., III, 1977a. Seismic stratigraphy and global changes of sea level, Part 4, global cycles of relative changes of sea level. *In*: Payton, C.E. (ed.), *Seismic Stratigraphy - Applications to Hydrocarbon Exploration*, 83–89. American Association of Petroleum Geologists, Memoir **26**.

Vail, P.R., Mitchum, R.M., Jr., & Thompson, III, S. 1977b. Seismic stratigraphy and global changes of sea level, Part 3, relative changes of sea level from coastal onlap. *In*: Payton, C.E. (ed.), *Seismic Stratigraphy -Applications to Hydrocarbon Exploration*, 63–82. American Association of Petroleum Geologists, Memoir **26**.

Vail, P.R., Hardenbol, J. & Todd, R.G. 1984. Jurassic unconformities, chronostratigraphy an sea level changes from seismic stratigraphy and biostratigraphy. *In*: Schlee, J.S. (ed.), *Interregional Unconformities and Hydrocarbon Accumulation*, 129–144. American Association of Petroleum Geologists Memoir **36**.

Vail, P.R., Audemard, F., Bowman, S.A., Eisner, P.N. & Perez-Cruz, G. 1991. The stratigraphic signatures of tectonics, eustasy and sedimentation. *In*: Einsele, G. *et al.* (eds), *Cyclic and Events in Stratigraphy*, 617–659. Berlin: Springer-Verlag.

Vakarelov, B.K., Bhattacharya, J.P. & Nebrigic, D.D. 2006. Importance of high-frequency tectonic sequences during greenhouse times of Earth history. *Geology*, **34**, 797–800.

Valle, G.D. & Gamberini, F. 2010. Erosional sculpting of the Caprera confined deep-sea fan as a result of distal basin-spilling processes (eastern Sardinian margin, Tyrrhenian Sea). *Marine Geology*, **268**, 55–66.

Van Andel, Tj.H. 1964. Recent marine sediments of Gulf of California. *In*: Van Andel, Tj.H. & Shor, G.G. (eds), *Marine Geology of the Gulf of California; A Symposium*, 216–310. American Association of Petroleum Geologists Memoir, **3**.

Van Andel, Tj. H. & Komar, P.D. 1969. Ponded sediments of the Mid-Atlantic Ridge between 22° and 23° north latitude. *Geological Society of America Bulletin*, **80**, 1163–1190.

Van den Berg, J.H., van Gelder, A. & Mastbergen, D.R. 2002. The importance of breaching as a mechanism of subaqueous slope failure in fine sand. *Sedimentology*, **49**, 81–96.

Van Daele, M., Cnudde, V., Duyck, P., Pino, M., Urrutia, R. & de Batist, M. 2014. Multidirectional, synchronously-triggered seismo-turbidites and debrites revealed by X-ray computed tomography (CT). *Sedimentology*, **61**, 861–880.

Van Dijk, M., Postma, G. & Kleinhans, M.G. 2009. Autocyclic behaviour of fan deltas: an analogue experimental study. *Sedimentology*, **56**, 1569–1589.

Van der Lingen, G.J. 1969. The turbidite problem. *New Zealand Journal of Geology and Geophysics*, **12**, 7–50.

Van der Lingen, G.J. 1982. Development of the North Island subduction system, New Zealand. *In*: Leggett, J.K. (ed.), *Trench-Forearc Geology*, 259–272. Geological Society of London Special Publication of the Geological Society, London, **10**. Oxford: Blackwell Scientific.

Van der Merwe, W., Flint, S.S. & Hodgson, D.M. 2010. Sequence stratigraphy of an argillaceous, deepwater basin-plain succession: Vischkuil Formation (Permian), Karoo Basin, South Africa. *Marine and Petroleum Geology*, **27**, 321–333.

Van der Pluijm, B.A. & Marshak, S. 2004. *Earth Structure: An Introduction to Structural Geology and Tectonics, 2nd edn, 656 pp*. New York: W.W. Norton & Company.

Van der Werff, W. & Johnson, S. 2003a. High resolution stratigraphic analysis of a turbidite system, Tanqua Karoo Basin, South Africa. *Marine and Petroleum Geology*, **20**, 45–69.

Van der Werff, W. & Johnson, S. 2003b. Deep-sea fan pinch-out geometries and their relationship to fan architecture, Tanqua Karoo basin (South Africa). *International Journal of Earth Sciences (Geologisches Rundschau)*, **92**, 728–742.

Van Hinte, J.E., Wise Jr, S.W, Biart, B.N.M., Covington, J.M., Dunn, D.A., Haggerty, J.A., Johns, M.W. Meyers, P.A., Moullade, M.R., Muza, J.P., Ogg, J.G., Okamura, M. Sarti, M. & von Rad, U. 1985a. DSDP Site 603: first deep (>1000 m) penetration of the continental rise along the passive margin of eastern North America. *Geology*, **13**, 392–396.

Van Hinte, J.E., Wise Jr, S.W, Biart, B.N.M., Covington, J.M. Dunn, D.A. Haggerty, J.A., Johns, M.W., Meyers, P.A., Moullade, M.R., Muza, J.P. Ogg, J.G., Okamura, M. Sarti, M. & von Rad, U. 1985b. Deep-sea drilling on the upper continental rise of New Jersey, DSDP Sites 604 and 605. *Geology*, **13**, 397–400.

Van Hoorn, B. 1970. Sedimentology and paleography of an Upper Cretaceous turbidite basin in the south-central Pyrenees, Spain. *Leidse Geologische Mededelingen*, **45**, 73–154.

Van Lente, B. 2004. *Chemostratigraphic trends and provenance of the Permian Tanqua and Laingsburg Depocentres, southwestern Karoo Basin, South Africa*. Ph.D thesis, University of Stellenbosch, 439 pp.

Van Wagoner, J.C. & Bertram, G.T. 1995. *Sequence Stratigraphy of Foreland Basin Deposits. American Association of Petroleum Geologists, Memoir* **64**, 487 pp. ISBN 0-89181-343-8.

Van Wagoner, J.C., Mitchum Jr., R.M., Posamentier, H.W. & Vail, P.R. 1987. Seismic stratigraphy interpretation using sequence stratigraphy; Part 2, Key definitions of sequence stratigraphy. *In*: Bally, A.W. (ed.), *Atlas of Seismic Stratigraphy*, 11–14. American Association of Petroleum Geologists, Studies in Geology, **27**, volumes **1–3**.

Van Wagoner, J.C., Posamentier, H.W., Mitchum, Jr., R.M., Vail, P.R., Sarg, J.F., Loutit, T.S. & Hardenbol, J. 1988. An overview of the Fundamentals of Sequence Stratigraphy and key definitions. *In*: Wilgus, C.K., Hastings, B.S., Posamentier, H., Van Wagoner, J., Ross, C.A. & Kendall, C.G. St.C. (eds), *Sea-level Changes: An Integrated Approach,* 39–45. Society of Economic Paleontologists and Mineralogists, Special Publication No. **42**.

Van Wagoner, J.C., Mitchum Jr., R.M., Campion, K.M. & Rahmanian, V.D. 1990. Siliciclastic sequence stratigraphy in well logs, core and outcrop: Concepts for high-resolution correlation of time and facies. *American Association of Petroleum Geology, Methods in Exploration*, **7**, 55 pp.

Van Weering, T.C.E., Kridoharto, P., Kusnida, D., Lubis, S., Tjokrosapoetro, S. & Munadi, S. 1989. The seismic structure of the Lombok and Savu forearc basins, Indonesia. *Journal of Sea Research*, **24**, 251–262.

Vandorpe, T., Van Rooij, D. & de Haas, P. 2014. Stratigraphy and paleoceanography of a topography-controlled contourite drift in the Pen Duick area, southern Gulf of Cádiz. *Marine Geology*, **349**, 136–151.

Vanneste, M., Mienert, J. & Bünz, S. 2006. The Hinlopen Slide: A giant, submarine slope failure on the northern Svalbard margin, Arctic Ocean. *Earth and Planetary Science Letters*, **245**, 373–388.

Vossler, S.M. & Pemberton, S.G. 1988. Skolithos in the Upper Cretaceous Cardium Formation: an ichnofossil example of opportunistic ecology. *Lethaia*, **21**, 351–362.

Vassoevitch, N.B. 1948. *Flish i metodika ego izucheniia. Leningrad-Moscow, Vsesoiznyï Nauchno-IssledovateVskii Institut* (also in French translation as *le flysch et les méthodes de son étude*. Bureau de Recherches géologiques et minières, Paris).

Vedder, J.G., Beyer, L.A., Junger, A., Moore, G.W., Roberts, A.E., Taylor, J.C. & Wagner, H.C. 1974. *Preliminary Report on the Geology of the Continental Borderland of Southern California. United States Geological Survey, Miscellaneous Field Studies Map*, **MF-624**.

Veevers, J.J., Hietzler, J.R. *et al.* 1974. *Initial Reports of the Deep Sea Drilling Project*, **27**. Washington, DC: US Government Printing Office.

Veevers, J.J.; Cole, D.I. & Cowan, E.J. 1994. Southern Africa: Karoo Basin and Cape Fold Belt. *In*: Veevers, J.J. & Powell, C. McA. (eds), *Permian-Triassic Pangean Basins and Foldbelts Along the Panthalassan Margin of Gondwanaland*, 223–279. Geological Society of America Memoir, **184**.

Venuti, A., Florindo, F., Caburlotto, A., Hounslow, M.W., Hillenbrand, C.-D., Strada, E., Talarico, F.M. & Cavallo, A. 2011. Late Quaternary sediments from deep-sea sediment drifts on the Antarctic Peninsula Pacific margin: Climatic control on provenance of minerals. *Journal of Geophysical Research*, **116**, B06104, doi:10.1029/2010JB007952.

Vergés, J., Millan, H., Roca, E., Muñoz, J.A., Marzo, M., Cires, J., Denbezemer, T., Zoetemeijer, R. & Cloetingh, S. 1995. Eastern Pyrenees and related foreland basins - Precollisional, syncollisional and postcollisional crustal-scale cross-sections. *Marine and Petroleum Geology*, **12**, 903–915.

Verges, J. Marzo, M. Santaeularia, T. Serra-Kiel, J. Burbank, D.W. Muñoz, J.A. & Gimenez-Montsant, J. 1998. Quantified vertical motions and tectonic evolution of the SE Pyrenean foreland basin. *In*: Mascle, A. Puigdefabregas, C., Luterbacher, H.P. and Fernandez, M. (eds), *Cenozoic Foreland Basins of Western Europe*, 107–134. Geological Society of London, Special Publication, **134**.

Viana, A.R. & Rebesco, M. (eds) 2007. *Economic and Palaeoceanographic Significance of Contourite Deposits*. Geological Society, London, Special Publication, 360 pp. London: The Geological Society.

Vicente Bravo, J.C. & S. Robles 1995, Large-scale mesotopographic bedforms from the Albion Black Flysch, northern Spain: characterization, setting and comparison with recent channel-lobe transition zone analogues. *In*: Pickering, K.T., Hiscott, R.N., Kenyon, N.H., Ricci Lucchi, F. & Smith, R.D.A. (eds), *Atlas of Deep Water Environments: Architectural Style in Turbidite Systems*, 216–226. London: Chapman and Hall.

Vigorito, M. & Hurst, A. 2010. Regional sand injectite architecture as a record of porepressure evolution and sand redistribution in the shallow crust: insights from the Panoche Giant Injection Complex, California. *Journal of the Geological Society, London*, **167**, 889–904.

Vigorito, M. Murru, M. & Simone, L. 2005. Anatomy of a submarine channel system and related fan in a foramol/rhodalgal carbonate sedimentary setting: a case history from the Miocene syn-rift Sardinia Basin, Italy. *Sedimentary Geology*, **174**, 1–30.

Vining, B., Ioannides, N. & Pickering, K.T. 1994. Stratigraphic relationships of some Tertiary lowstand depositional systems in the Central North Sea. *In*: Parker, J.R. (ed.), *Petroleum Geology of Northwest Europe: Proceedings of the 4th Conference*, 17–29. [2 volumes] London: The Geological Society, London.

Violet, J., Sheets, B., Pratson, L., Paola, C., Beaubouef, R. & Parker, G. 2005. Experiment on turbidity currents and their deposits in a model 3D subsiding minibasin. *Journal of Sedimentary Research*, **75**, 820–843.

Visser, J.N.J. 1992. Basin tectonics in southwestern Gondwana during the Carboniferous and Permian. *In*: De Wit, M.J. & Ransome, I. (eds), *Inversion Tectonics of the Cape Fold Belt, Karoo and Cretaceous Basins of Southern Africa*, 109–115. Rotterdam: Balkema.

Vittori, J., Morash, A., Savoye, B., Marsset, T., Lopez, M., Droz, L. & Cremer, M. 2000. The Quaternary Congo deep-sea fan: preliminary results on reservoir complexity in turbiditic systems using 2D high resolution seismic and multibeam data. *In*: Weimer, P., Slatt, R.M., Coleman, J., Rosen, N.C., Nelson, H., Bouma, A.H., Styzen, M.J. & Lawrence, D.T. (eds), *Deep-water Reservoirs of the World*, 1045–1058. Houston, Texas: Gulf Coast Section, Society of Economic Paleontologists and Mineralogists.

Völker, D., Reichel, T., Wiedicke, M. & Heubeck, C. 2008. Turbidites deposited on Southern Central Chilean seamounts: Evidence for energetic turbidity currents. *Marine Geology*, **251**, 15–31.

Von der Borch, C.C., Grady, A.E., Aldam, R., Miller, D., Neumann, R., Rovira, A. & Eickhoff, K. 1985. A large-scale meandering submarine canyon: outcrop example from the Late Proterozoic Adelaide Geosyncline, South Australia. *Sedimentology*, **32**, 507–518.

Von Huene, R., Aubouin, J. *et al.* 1985. *Initial Reports Deep Sea Drilling Project*, **84**. Washington, DC: US Government Printing Office.

Von Lom-Keil, H., Spieß, V. & Hopfauf, V. 2002. Fine-grained sediment waves on the western flank of the Zapiola Drift, Argentine Basin: evidence for variations in Late Quaternary bottom flow activity. *Marine Geology*, **192**, 239–258.

Von Rad, U. & Exon, N.F. 1983. Mesozoic–Cenozoic sedimentary and volcanic evolution of the starved passive continentai margin off northwest Australia. *In*: Watkins, J.S. & Drake, C.L. (eds), *Studies in Continental Margin Geology*, 253–281. American Association of Petroleum Geologists. Memoir, **34**.

Von Rad, U. & Tahir, M. 1997. Late Quaternary sedimentation on the outer Indus shelf and slope (Pakistan): evidence from high-resolution seismic data and coring. *Marine Geology*, **138**, 193–236.

Vorren, T.O. & Laberg, J.S. 1997. Trough mouth fans: paleoclimate and ice-sheet monitors. *Quaternary Science Review*, **16**, 865–881.

Vrolijk, P. 1990. On the mechanical role of smectite in subduction zones. *Geology*, **18**, 703–707.

Wagreich, M. 2008. Lithostratigraphic definition and depositional model of the Hütteldorf Formation (Upper Albian – Turonian, Rhenodanudian Flysch Zone, Austria). *Austrian Journal of Earth Sciences*, **101**, 70–80.

Wakabayashi, J., Hengesh, J.V. & Sawyer, T.L. 2004. Four-dimensional transform fault processes: progressive evolution of step-overs and bends. *Tectonophysics*, **392**, 279–301.

Walcott, R l. 1978. Geodetic strains and large earthquakes in the axial tectonic belt of North Island, New Zealand. *Journal of Geophysical Research*, **83**, 4419–4429.

Walcott, R.I. 1970. Flexural rigidity, thickness and viscosity of the lithosphere. *Journal of Geophysical Research*, **75**, 3941–3954.

Wald, A. & Wolfowitz, J. 1944. An exact test for randomness in the non-parametric case based on serial correlation. *Annals of Mathematical Statistics*, **14**, 378–388.

Waldron, J.W.F. 1987. A statistical test for significance of thinning- and thickening-upward cycles in turbidites. *Sedimentary Geology*, **54**, 137–146.

Waldron, J.W.F., McNicoll, V.J. & van Staal, C.R. 2012. Laurentia-derived detritus in the Badger Group of central Newfoundland: deposition during closing of the Iapetus Ocean. *Canadian Journal of Earth Sciences*, **49**, 207–221.

Walker, J.R. & Massingill, J.V. 1970. Slump features on the Mississippi fan, northeastern Gulf of Mexico. *Geological Society of America Bulletin*, **81**, 3101–3108.

Walker, K.R., Shanmugam, G. & Ruppel, S.C. 1983. A model for carbonate to terrigenous clastic sequences. *Geological Society of America Bulletin*, **94**, 700–712.

Walker, R.G. 1965. The origin and significance of the internal sedimentary structures of turbidites. *Proceedings of the Yorkshire Geological Society*, **35**, 1–32.

Walker, R.G. 1966a. Deep channels in turbidite-bearing formations. *American Association of Petroleum Geologists Bulletin*, **50**, 1899–1917.

Walker, R.G. 1966b. Shale Grit and Grindslow Shales: transition from turbidite to shallow water sediments in the Upper Carboniferous of Northern England. *Journal of Sedimentary Petrology*, **36**, 90–114.

Walker, R.G. 1967a. Upper flow regime bed forms in turbidites of the Hatch Formation, Devonian of New York State. *Journal of Sedimentary Petrology*, **37**, 1052–1058.

Walker, R.G. 1967b. Turbidite sedimentary structures and their relationship to proximal and distal depositional environments. *Journal of Sedimentary Petrology*, **37**, 25–43.

Walker, R.G. 1970. Review of the geometry and facies organisation of turbidites and turbidite-bearing basins. *In*: Lajoie, J. (ed.), *Flysch Sedimentology in North America*, 219–251. Geological Association of Canada Special Paper, **7**. Toronto: Business & Economic Service.

Walker, R.G. 1975a. Generalized facies model for resedimented conglomerates of turbidite association. *Geological Society of America Bulletin*, **86**, 737–748.

Walker, R.G. 1975b. Upper Cretaceous resedimented conglomerates at Wheeler Gorge, California: description and field guide. *Journal of Sedimentary Petrology*, **45**, 105–112.

Walker, R.G. 1975c. Nested submarine-fan channels in the Capistrano Formation, San Clemente, California. *Geological Society of America Bulletin*, **86**, 915–924.

Walker, R.G. 1976. Facies models 2. Turbidites and associated coarse clastic deposits. *Geoscience Canada*, **3**, 25–36.

Walker, R.G. 1977. Deposition of Upper Mesozoic resedimented conglomerates and associated turbidites in southwestern Oregon. *Geological Society of America Bulletin*, **88**, 273–285.

Walker, R.G. 1978. Deep water sandstone facies and ancient submarine fans: models for exploration for stratigraphic traps. *American Association of Petroleum Geologists Bulletin*, **62**, 932–966.

Walker, R.G. 1984. Turbidites and associated coarse clastic deposits. *In*: Walker, R.G. (ed.), *Facies Models*, 2nd edn, 171–188. Geoscience Canada Reprint Series 1. Kitchener, Ontario: Ainsworth Press.

Walker, R.G. 1985. Mudstones and thin-bedded turbidites associated with the Upper Cretaceous Wheeler Gorge conglomerates, California: a possible channel-levée complex. *Journal of Sedimentary Petrology*, **55**, 279–290.

Walker, R.G. 1992. Turbidites and submarine fans. *In*: Walker, R.G. & James, N.P. (eds), *Facies Models: Response to Sea Level Change*, 239–263. St. John's, Newfoundland: Geological Association of Canada.

Walker, R.G. & Mutti, E. 1973. Turbidite facies and facies associations. *In*: Middleton, G.V. & Bouma, A.H. (eds), *Turbidites and Deep-water Sedimentation*, 119–157. Society of Economic Paleontologists and Mineralogists, Pacific Section Short Course Notes, Anaheim.

Wallace W.K., Hanks, C.L. & Rogers, J.F. 1989. The southern Kahiltna terrane; implications for the tectonic evolution of southwestern Alaska. *Geological Society of America Bulletin*, **101**, 1389–1407.

Walton, E.K. 1967. The sequence of internal structures in turbidites. *Scottish Journal of Geology*, **3**, 306–317.

Wan, S., Yu, Z., Clift, P.D., Sun, H., Li, A. & Li, T. 2012. History of Asian eolian input to the West Philippine Sea over the last one million years. *Palaeogeography, Palaeoclimatology, Palaeoecology*, **326–328**, 152–159

Wang, D. & Hesse, R. 1996. Continental slope sedimentation adjacent to an ice-margin. II. Glaciomarine depositional facies on Labrador Slope and glacial cycles. *Marine Geology*, **135**, 65–96.

Wang, K. & Hu, Y. 2006. Accretionary prisms in subduction earthquake cycles: The theory of dynamic Coulomb wedge. *Journal of Geophysical Research*, **111**, B06, B06410, doi:10.1029/2005JB004094.

Warrick, J.A. & Milliman, J.D. 2003. Hyperpycnal sediment discharge from semiarid southern California rivers - Implications for coastal sediment budgets. *Geology*, **31**, 781–784.

Waterston, C.D. 1950. Note on the sandstone injections of Eathie Haven, Cromarty. *Geological Magazine*, **87**, 133–139.

Watkins J.S., Moore J.C. *et al.* 1982. *Initial Report of the Deep Sea Drilling Project, 66*. Washington, DC: US Government Printing Office, Washington.

Watkins, J.S. Montadert, L. & Dickerson, P.W. (eds) 1979. *Geological and Geophysical Investigations of Continental Margins. American Association of Petroleum Geologists Memoir*, **29**.

Watkinson, I.M., Hall, R. & Ferdian, F. 2011. Tectonic re-interpretation of the Banggai-Sula-Molucca Sea margin, Indonesia. *In*: Hall, R., Cottam, M.A. &Wilson, M.E.J. (eds), *The SE Asian Gateway: History and Tectonics of the Australia–Asia Collision, 203–224*. Geological Society, London, Special Publications, **355**.

Watson, M.P. 1981. *Submarine fan deposits of the Upper Ordovician Lower Silurian Milliners Arm Formation, New World Island, Newfoundland.* D. Phil thesis, University of Oxford, UK.

Watts, A.B. 2007. An overview. *In*: Watts, A.B. (ed), *Crust and lithosphere dynamics*, 1–48. Treatise on Geophysics, **6**. Amsterdam: Elsevier.

Watt, S.F.L., Talling, P.J., Vardy, M.E., Heller, V., Hühnerbach, V., Urlaub, M., Sarkar, S., Masson, D.G., Henstock, T.J., Minshull, T.A., Paulatto, M., Le Friant, A., Lebas, E., Berndt, C., Crutchley, G.J., Kartstens, J., Stinton, A.J. & Maeono, F. 2012. Combinations of volcanic-flank and seafloor-sediment failure offshore Montserrat, and their implications for tsunami generation. *Earth and Planetary Science Letters*, **319–320**, 228–240.

Watts, A.B. & Thorne, J. 1984. Tectonic, global changes in sea level, and their relationship to stratigraphical sequences at the US Atlantic continental margin. *Marine and Petroleum Geology*, **1**, 319–339.

Weaver, P.P.E. 1994. Determination of turbidity current erosional characteristics from reworked coccolith assemblages, Canary Basin, north-east Atlantic. *Sedimentology*, **41**, 1025–1038.

Weaver, P.P.E. & Kuijpers, A. 1983. Climatic control of turbidite deposition on the Madeira Abyssal Plain. *Nature*, **306**, 360–363.

Weaver, P.P.E. & Rothwell, R.G. 1987. Sedimentation on the Madeira Abyssal Plain over the last 300 000 years. *In*: Weaver, P.P.E. & Thomson, J. (eds), *Geology and Geochemistry of Abyssal Plains, 71–86*. Geological Society of London Special Publication, **31**. Oxford: Blackwell Scientific.

Weaver, P.P.E., Searle, R.C. & Kuijpers, A. 1986. Turbidite deposition and the origin of the Madeira Abyssal Plain. *In*: Summerhayes, C.P. & Shackleton, N.J. (eds), *North Atlantic Palaeoceanography*, 131–143. Geological Society of London Special Publication, **21**. Oxford: Blackwell Scientific.

Weaver, P.P.E. & Rothwell, R.G. 1987. Sedimentation on the Madeira Abyssal Plain over the last 300 000 years. *In*: Weaver, P.P.E. & Thomson, J. (eds), *Geology and Geochemistry of Abyssal Plains, 71–86*. Geological Society of London Special Publication, **31**. Oxford: Blackwell Scientific.

Weaver, P.P.E., Rothwell, R.G., Ebbing, J., Gunn, D. & Hunter, P.M. 1992. Correlation, frequency of emplacement and source directions of megaturbidites on the Madeira abyssal plain. *Marine Geology*, **109**, 1–20.

Weaver, P.P.E., Masson, D.G., Gunn, D.E., Kidd, R.B., Rothwell, R.G. & Maddison, D.A. 1995. Sediment mass wasting in the Canary Basin. *In*: Pickering, K.T., Hiscott, R.N., Kenyon, N.H., Ricci Lucchi, F. & Smith, R.D.A. (eds), *Atlas of Deep Water Environments: Architectural Style in Turbidite Systems*, 287–296. London: Chapman and Hall.

Weaver, P.P.E., Jarvis, I., Lebreiro, S.M., Alibés, B., Baraza, J., Howe, R. & Rothwell, R.G. 1998. Neogene turbidite sequence on the Madeira Abyssal Plain: basin filling and diagenesis in the deep ocean. *In*: Weaver, P.P.E., Schmincke, H.-U., Firth, J.V. & Duffield, W. (eds), *Proceedings of the Ocean Drilling Program, Scientific Results*, **157**, 619–634. College Station, Texas, USA: Ocean Drilling Program.

Weber, K.J. 1971. Sedimentological aspects of oil fields in the Niger delta. *Geologie en Mijnbouw*, **50**, 559–576.

Weber, M.E., Wiedicke, M.H., Kudrass, H.R., Hübsher, C. & Erlenkeuser, H. 1997. Active growth of the Bengal Fan during sea-level rise and highstand. *Geology*, **25**, 315–318.

Weber, M.E., Wiedicke-Hombach, M., Kudrass & Erlenkeuser, E. 2003. Bengal Fan sediment transport activity and response to climate forcing inferred from sediment physical properties. *Sedimentary Geology*, **155**, 361–381.

Webster, R. 1973. Automatic soil-boundary location from transect data. *Mathematical Geology*, **5**, 27–37.

Webster, R. 1980. Divide: a Fortran IV program for segmenting multivariate one-dimensional spatial series. *Computers and Geosciences*, **6**, 61–68.

Weedon, G. 2005. *Time-series Analysis and Cyclostratigraphy: Examining Stratigraphic Records of Environmental Cycles.* Cambridge: Cambridge University Press.

Weedon, G.P. & McCave, I.N. 1991. Mud turbidites from the Oligocene and Miocene Indus Fan at Sites 722 and 731 on the Owen Ridge. *In*: Prell, W. L., Niitsuma, N. *et al.*, *Proceedings of the Ocean Drilling Program, Scientific Results*, **117**, 215–220. College Station, Texas, USA: Ocean Drilling Program.

Weigand, P.W. 1994. Timing and cause of middle Tertiary magmatism in onshore and offshore southern California (abstract). *American Association of Petroleum Geologists Bulletin*, **78**, 676.

Weimer, P. 1990. Sequence stratigraphy, facies geometries, and depositional history of the Mississippi Fan, Gulf of Mexico. *American Association of Petroleum Geologists Bulletin*, **74**, 425–453.

Weimer, P. 1991. Sesmic facies, characteristics and variations in channel evolution, Mississippi fan (Plio-Pleistocene), Gulf of Mexico. *In*: Weimer, P. & Link, M.H. (eds), *Seismic Facies and Sedimentary Processes of Submarine Fans and Turbidite Systems*, 323–348. New York: Springer-Verlag.

Weimer, P. 1995. Sequence stratigraphy of the Mississippi Fan (late Miocene - Pleistocene), northern deep Gulf of Mexico. *In*: Pickering, K.T., Hiscott, R.N., Kenyon, N.H., Ricci Lucchi, F. & Smith, R.D.A. (eds), *Atlas of Deep Water Environments: Architectural Style in Turbidite Systems*, 94–99. London: Chapman and Hall.

Weimer, P. & Buffler, R.T. 1988. Distribution and seismic facies of Mississippi fan channels. *Geology*, **16**, 900–903.

Weirich, F.H. 1988. Field evidence for hydraulic jumps in subaqueous sediment gravity flows. *Nature*, **332**, 626–629.

Welch, P.D. 1967. The use of the fast Fourier transform for the estimation of power spectra: A method based on time averaging over short, modified periodograms. *IEEE Transactions on Audio and Electroacoustics*, **15**, 70–73.

Weller, S. 1899 Kinderhook faunal studies. I. The fauna of the vermicular sandstone at Northview, Webster County, Missouri. *Transaction of the Academy of Sciences St Louis*, **9**, 9–51.

Wells, J.T., Scholz, C.A. & Soreghan, M.J. 1999. Processes of sedimentation on a lacustrine border-fault margin: interpretation of cores from Lake Malawi, East Africa. *Journal of Sedimentary Research*, **B69**, 816–831.

Wells, R.E., Engebretson, D.C., Snavely Jr., P.D. & Coe, R.S. 1984. Cenozoic plate motions and the volcano-tectonic evolution of western Oregon and Washington. *Tectonics*, **3**, 275–294.

Wells, R.E., Weaver, C.S. & Blakely, R.J. 1998. Fore-arc migration in Cascadia and its neotectonic significance. *Geology*, **26**, 759–762.

Wells, R.E., Blakely, R.J., Sugiyama, Y., Scholl, D.W. & Dinterman, P.A. 2003. Basin-centered asperities in great subduction zone earthquakes: A link between slip, subsidence, and subduction erosion? *Journal of Geophysical Research*, **108**, 2507, doi:10.1029/2002JB002072

Wells, S.W., Warner, M., Greenhalgh, J. & Borsato, R. 2012. Offshore Cote d'Ivoire: a modern exploration frontiere. *GEO ExPro*, **9**, 38–40.

Weltje, G.J. & de Boer, P.L. 1993. Astronomically induced paleoclimatic oscillations reflected in Pliocene turbidite deposits on Corfu (Greece): Implications for the interpretation of higher order cyclicity in ancient turbidite systems. *Geology*, **21**, 307–310.

Weltje, G.J., Assenwoude, V. & De Boer, P.L. 1996. High-frequency detrital signals in Eocene fan-delta sandstones of mixed parentage (south central Pyrenees, Spain): a reconstruction of chemical weathering in transit. *Journal of Sedimentary Research*, **66**, 119–131.

Wentworth, C.M. 1967. Dish structure, a primary sedimentary structure in coarse turbidites (abstract). *American Association of Petroleum Geologists Bulletin*, **51**, 485.

Werner F. & Wetzel, A. 1982. Interpretation of biogenic structures in oceanic sediments. *Bulletin de l'Institut de Géologie du Bassin d'Aquitaine*, **31**, 275–288

Wernicke, B. 1985. Uniform-sense normal simple shear of the continental lithosphere. *Canadian Journal of Earth Sciences*, **22**, 108–125.

Wernicke, B. & Burchfiel, B.C. 1982. Modes of extensional tectonics. *Journal of Structural Geology*, **4**, 105–115.

Wescott, W.A. & Ethridge, F.G. 1982. Bathymetry and sediment dispersal dynamics along the Yallahs fan delta front, Jamaica. *Marine Geology*, **46**, 245–260.

Wescott, W.A. & Ethridge, F.G. 1983. Eocene fan delta/submarine fan deposition in the Wagwater Trough, east-central Jamaica. *Sedimentology*, **30**, 235–247

Westbrook, G.K. 1982. The Barbados Ridge Complex: tectonics of a mature forearc system. *In*: Leggett, J.K. (ed.), *Trench-Forearc Geology*, 275–290. Geological Society of London Special Publication of the Geological Society, London, **10**. Oxford: Blackwell Scientific.

Westbrook, G.K. & Smith, M.J. 1983. Long decollements and mud volcanoes: Evidence from the Barbados Ridge Complex for the role of high pore-fluid pressure in the development of an accretionary complex. *Geology*, **11**, 279–283.

Westbrook, G.K., Mascle, A. & Biju-Duval, B. 1984. Geophysics and the structure of the Lesser Antilles forearc. *In*: Biju-Duval, B., Moore, J.C. *et al*. 1984. *Initial Reports of the Deep Sea Drilling Program*, **78A**, 23–38. Washington, DC: US Government Printing Office.

Westbrook, G.K., Ladd, J.W., Buhl, P., Bangs, N. & Tiley, G.J. 1988. Cross section of an accretionary wedge: Barbados Ridge complex. *Geology*, **16**, 631–635.

Wetzel, A. 1981. Ökologische und stratigraphische Bedeutung biogener Gefüge in quartären Sedimenten am NW-afrikanischen Kontinental-rand. *Meteor Forschungs-Ergebnisse Reihe C*, **34**, 1–47.

Wetzel, A. 1983. Biogenic Sedimentary structures in a modern upwelling region: Northwest African continental margin. *In*: Thiede, J. & Suess, E. (eds). *Coastal Upwelling and its Sediment Record, Part B, Sedimentary Records of Ancient Coastal Upwelling*, 123–144. New York: Plenum.

Wetzel, A. 1984. Bioturbation in deep-sea fine-grained sediments: influence of sediment texture, turbidite frequency and rates of environmental change. *In*: Stow, D.A.V. & Piper, D.J.W. (eds), *Fine-grained Sediments: Deep-water Processes and Facies*, 595–608. Geological Society of London Special Publication, **15**. Oxford: Blackwell Scientific.

Wetzel, A. 1991. Ecologic interpretation of deep-sea trace fossil communities. *Palaeogeography, Palaeoclimatology, Palaeoecology*, **85**, 47–69.

Wetzel, A. 1993. The transfer of river load to deep-sea fans: a quantitative approach. *American Association of Petroleum Geologists Bulletin*, **77**, 1679–1692.

Wetzel, A. & Aigner, T. 1986. Stratigraphic completeness: Tiered trace fossils provide a measuring stick. *Geology*, **14**, 234–237.

Wetzel, A. & Bromley, R.G. 1994. Phycosiphon incertum revisited: *Anconichnus horizontalis* is its junior subjective synonym. *Journal of Paleontology*, **68**, 1396–1402.

Wetzel, A. & Bromley, R.G. 1996. Re-evaluation of ichnogenus Helminthopsis Heer 1877–a new look at the type material. *Palaeontology*, **39**, 1–19.

Wetzel, A. & Uchman, A. 1997. Ichnology of deep-sea fan overbank deposits of the Ganei Slates Eocene, Switzerland - a classical flysch trace fossil locality studied first by Oswald Heer. *Ichnos*, **5**, 139–162.

Wetzel, A. & Uchman, A. 1998a. Deep-sea benthic food content recorded by ichnofabrics: A conceptual model based on observations from Paleogene Flysch, Carpathians, Poland. *Palaios*, **13**, 533–546.

Wetzel, A., & Uchman, A. 1998b. Biogenic sedimentary structures in mudrocks - an overview. *In*: Schieber, J., Zimmerle, W., Sethi, P. (eds). *Shales and Mudrocks. I*, 351–369. Stuttgart: Schweizerbart.

Wetzel, A. & Uchman, A. 2001. Sequential colonization of muddy turbidites: examples from Eocene Beloveza Formation, Carpathians, Poland. *Palaeogeography, Palaeoclimatology, Palaeoecology*, **168**, 171–186.

Wetzel, A. & Uchman, A. 2012. Hemipelagic and pelagic basin plains. *In*: Knaust, D. & Bromley, R.G. (eds), *Trace Fossils as Indicators of Sedimentary Environments*, 673–702. Developments in Sedimentology, **64**. Amsterdam: Elsevier.

Wheeler, H.E. 1958. Time-stratigraphy. *American Association of Petroleum Geologists Bulletin*, **42**, 1047–1063.

Wheeler, H.E. 1959a. Note 24 - Unconformity-bounded units in stratigraphy. *American Association of Petroleum Geologists Bulletin*, **43**, 1975–1977.

Wheeler, H.E. 1959b. Stratigraphic units in time and space. *American Journal of Science*, **257**, 692–706.

Whipple, K.X., Parker, G., Paola, C. & Mohrig, D. 1998. Channel dynamics, sediment transport, and the slope of alluvial fans; experimental study. *Journal of Geology*, **106**, 677–693.

Whitaker, J.H. McD. 1962. The geology of the area around Leintwardine, Herefordshire. *Quarterly Journal of the Geological Society, London*, **118**, 319–351.

Whitaker, J.H. McD. 1974. Ancient submarine canyons and fan valleys. *In*: Dott, R.H. Jr & Shaver, R.H. (eds), *Modern and Ancient Geosynclinal Sedimentation*, 106–125. Society of Economic Paleontologists and Mineralogists Special Publication, **19**.

Whitcar, M.J. & Elvert, M.E. 2000. Organic geochemistry of Saanich Inlet, BC, during the Holocene as revealed by Ocean Drilling Program Leg 169S. *Marine Geology*, **174**, 249–271.

White, D.E. 1965. Saline waters of sedimentary rocks. *In*: Galley, J.E. (ed.), *Fluids in Subsurface Environments*, 342–366. American Association of Petroleum Geologists. Memoir, **4**.

White, R.S. & Louden, K.E. 1983. The Makran continental margin: structure of a thickly sedimented convergent plate boundary. *In*: Watkins, J.S. & Drake, C.L. (eds), *Studies in Continental Margin Geology*, 499–518. American Association of Petroleum Geologists Memoir, **34**.

Wickens, H. deV. 1992. Submarine fans of the Permian Ecca Group in the SW Karoo basin: Their origin and reflection on the tectonic evolution of the basin and its source areas. *In*: de Wit, M.J. & Ransome, I.G.D. (eds), *Inversion Tectonics of the Cape Fold Belt, Karoo and Cretaceous Basins of Southern Africa*, 117–125. Rotterdam: Balkema.

Wickens, H.de.V. 1994. *Basin floor fan building turbidites of the southwestern Karoo Basin, Permian Ecca Group*. Ph.D Thesis, Port Elizabeth University, South Africa, 233 pp.

Wickens, H. deV. & Bouma, A.H. 2000. The Tanqua fan complex, Karoo Basin, South Africa – outcrop analog for fine-grained, deepwater deposits. *In*: Bouma, A.H. & Stone, C.G. (eds), *Fine-grained Turbidite*

Systems, 153–164. American Association of Petroleum Geologists Memoir, **72** & Society of Sedimentary Geologists, Special Publication, **68**. Joint publication, Tulsa, Oklahoma.

Wignall, P.B. 1994. *Black Shales.* Oxford: Oxford Science Publications, 127 pp. ISBN 0-19-854038-8.

Wignall, P. & Pickering, K.T. 1993. Palaeoecology and sedimentology across a Jurassic fault scarp, NE Scotland. *Journal of the Geological Society, London,* **150**, 323–340.

Wignall, P.B. 1991. Dysaerobic Trace Fossils and Ichnofabrics in the Upper Jurassic Kimmeridge Clay of Southern England. *Palaios,* **6**, 264–270.

Wignall, P.B. & Best, J.L. 2000. The Western Irish Namurian Basin reassessed. *Basin Research,* **12**, 59–78.

Wignall, P.B. & Best, J.L. 2002. Reply to: The Western Irish Namurian Basin reassessed – a discussion by O.J. Martinsen & J.D. Collinson. *Basin Research,* **14**, 531–542.

Wilcox, R.E., Harding, T.P. & Seely, D.R. 1973. Basic wrench tectonics. *American Association of Petroleum Geologists Bulletin,* **57**, 74–96.

Wild, R.J., Hodgson, D.M. & Flint, S.S. 2005. Architecture and stratigraphic evolution of multiple, vertically-stacked slope channel complexes, Tanqua depocentre Karoo Basin, South Africa. *In*: Hodgson, D.M. & Flint, S.S. (eds), *Slope Systems, Processes and Products,* 89–111. The Geological Society, London, Special Publication, **244**. London: The Geological Society.

Williams, H. & Hatcher, R.D. Jr., 1982. Suspect terranes and accretionary history of the Appalachian orogen. *Geology,* **10**, 530–536.

Williams, H. & Hatcher, R.D. Jr., 1983. Appalachian suspect terranes. *In*: Hatcher, R.D. Jr., Williams, H. & Zietz, I. (eds), *Contnbutions to the Tectonics and Geophysics of Mountain Chains,* 33–53. Geological Society of America Memoir, **158**.

Wilson, J.T. 1966. Did the Atlantic close and then re-open? *Nature,* **210**, 678–681.

Wilson, P.A. & Roberts, H.H. 1995. Density cascading: off-shelf sediment transport, evidence and implications, Bahama Banks. *Journal of Sedimentary Research,* **A65**, 45–56.

Wilson, R.C.L. & Williams, C.A. 1979. Oceanic transform structures and the development of Atlantic continental margin sedimentary basins: a review. *Journal of the Geological Society (London),* **136**, 311–320.

Wilson, R.C.L., Hiscott, R.N., Willis, M.G. & Gradstein, F.M. 1989. The Lusitanian Basin of west central Portugal: Mesozoic and Tertiary tectonic, stratigraphic and subsidence history. *In*: Tankard, A.J. & Balkwill, H. (eds), *Extensional Tectonics and Stratigraphy of the North Atlantic Margins,* 341–361. American Association of Petroleum Geologists Memoir, **46**.

Winker, C.D. 1993. Levéed slope channels and shelf-margin deltas of the Late Pliocene to Middle Pleistocene Mobile River, NE Gulf of Mexico: comparison with sequence stratigraphic models (abstract). *In*: *American Association of Petroleum Geologists 1993 Annual Convention Program, Abstract Volume,* p. 201.

Winn, R.D. Jr, & Dott, R.H. Jr, 1977. Large-scale traction-produced structures in deep-water fan-channel conglomerates in southern Chile. *Geology,* **5**, 41–44.

Winn, R.D. Jr, & Dott, R.H. Jr, 1978. Submarine-fan turbidites and resedimented conglomerates in a Mesozoic arc-rear marginal basin in southern South America. *In*: Stanley, D.J. & Kelling, G. (eds), *Sedimentation in Submarine Canyons, Fans, and Trenches,* 362–376. Stroudsburg, PA: Dowden, Hutchinson & Ross.

Winn, R.D. Jr, & Dott, R.H. Jr, 1979. Deep-water fan-channel conglomerates of Late Cretaceous age, southern Chile. *Sedimentology,* **26**, 203–228.

Winsemann, J. & Seyfried, H. 2009. Response of deep-water fore-arc systems to sea-level changes, tectonic activity and volcaniclastic input in Central America. *In*: Macdonald, D.I.M. (ed.), *Sedimentation, Tectonics and Eustasy: Sea-level Changes at Active Margins,* Chapter 16. Oxford: Blackwell Publishing Ltd. doi: 10.1002/9781444303896.

Withjack, M.O. & Jamison, W.R. 1986. Deformation produced by oblique rifting. *Tectonophysics,* **126**, 99–124.

Withjack, M.O., Schlische, R.W. & Olsen, P.E. 1998. Diachronous rifting, drifting, and inversion on the passive margin of central eastern North America: an analog for other passive margins. *American Association of Petroleum Geologists Bulletin,* **82**, 817–835.

Wood, A. & Smith, A.J. 1959. The sedimentation and sedimentary history of the Aberystwyth Grits (upper Llandoverian). *Quarterly Journal of the Geological Society, London,* **114**, 163–195.

Wood, A.W. 1981. Extensional tectonics and the birth of the Lagonegro Basin (southern Italian Apennines), *Neues Jahrbuch für Geologie und Paläontologie Abhandlungen,* **161**, 93–131.

Woodcock, N.H. 1976a. Structural style in slump sheets: Ludlow Series, Powys, Wales. *Journal of the Geological Society, London,* **132**, 399–415.

Woodcock, N.H. 1976b. Ludlow Series slumps and turbidites and the form of the Montgomery Trough, Powys, Wales. *Proceedings of the Geologists Association,* **87**, 169–182.

Woodcock, N.H. 1979a. Sizes of submarine slides and their significance. *Journal of Structural Geology,* **1**, 137–142.

Woodcock, N.H. 1979b. The use of slump structures as palaeo-slope orientation estimators. *Sedimentology,* **26**, 83–99.

Wonham, J.P., Jayr, S. Mougamba, R. & Chuilon, P. 2000. 3D sedimentary evolution of a canyon fill (Lower Miocene-age) from the Mandorove Formation, offshore Gabon. *Marine and Petroleum Geology,* **17**, 175–197.

Woodring, W.P., Bramlette, M.N. & Kew, W.S.W. 1946. *Geology and Paleontology of Palos Verdes Hills, California.* United States Geological Survey Professional Paper, **P207**.

Woodside, J.M. 1977. Tectonic elements and crust of the eastern Mediterranean Sea. *Marine Geophysical Research,* **3**, 317–354.

Worthington, L.V. 1976. *On the North American Circulation.* Baltimore: Johns Hopkins University Press.

Wright, D.J., Bloomer, S.H., MacLeod, C.J., Taylor, B. & Goodlife, A.M. 2000. Bathymetry of the Tonga Trench and Forearc: a map series. *Marine Geophysical Researches,* **21**, 489–511.

Wright, L.D. 1977. Sediment transport and deposition at river mouths: a synthesis. *Geological Society of American Bulletin,* **88**, 857–868.

Wright, L.D. 1995. *Morphodynamics of Inner Continental Shelves.* Boca Raton, Florida: CRC Press, 241 pp. ISBN: 084938043X.

Wright, L.D., Wiseman, W.J., Bornhold, B.D., Prior, D.B., Suhayda, J.N., Keller, G.H., Yang, Z.-S. & Fan, Y.B. 1988. Marine dispersal and deposition of Yellow River silts by gravity-driven underflows. *Nature,* **332**, 629–632.

Wright, T.L. 1991. Structural geology and tectonic evolution of the Los Angeles basin. *In*: Biddle, K.T. (ed.), *Active Margin Basins,* 13–134. American Association of Petroleum Geologists Memoir, **52**.

Wu, L.C. 1993. *Sedimentary basin succession of the upper Neogene and Quaternary series in the Chishan area, southern Taiwan and its tectonic evolution.* Ph.D Thesis, National Taiwan University.

Wynn, R.B. & Stow, D.A.V. 2002. Classification and characterisation of deep-water sediment waves. *Marine Geology,* **192**, 7–22.

Wynn, R.B., Masson, D.G., Stow, D.A.V. & Weaver, P.P.E. 2000. The northwest African slope apron: a modern analogue for deep-water systems with complex seafloor topography. *Marine and Petroleum Geology,* **17**, 253–265.

Wynn, R.B., Masson, D.G. & Brett, B.J. 2002a. Hydrodynamic significance of variable ripple morphology across deep-water barchan dunes in the Faroe-Shetland Channel. *Marine Geology,* **192**, 309–319.

Wynn, R.B., Piper, D.J.W. & Gee, M.J.R. 2002b. Generation and migration of coarse-grained sediment waves in turbidity current channels and channel-lobe transition zones. *Marine Geology,* **192**, 59–78.

Wynn, R.B., Cronin, B.T. & Peakall, J. 2007. Sinuous deep-water channels: Genesis, geometry and architecture. *Marine and Petroleum Geology,* **24**, 341–387.

Xie, Y., Deutsch, C.V. & Cullick, S. 2000. Surface geometry and trend modeling for integration of stratigraphic data in reservoir models. *In*: Kleingeld, W.J. & Krige, D.G. (eds), *Geostats 2000,* **1**, 287–295.

Xu, Z., Li, T., Wan, S., Nan, Q., Li, A., Chang, F., Jiang, F. & Tang, Z. 2012. Evolution of East Asian monsoon: clay mineral evidence in the western Philippine Sea over the past 700 kyr, *Journal of Asian Earth Sciences*, doi: 10.1016/j.jseaes.2012.08.018.

Yagishita, K. 1994. Antidunes and traction-carpet deposits in deep-water channel sandstones, Cretaceous, British Columbia, Canada. *Journal of Sedimentary Research*, **A64**, 34–41.

Yagishita, K. & Taira, A. 1989. Grain fabric of a laboratory antidune. *Sedimentology*, **36**, 1001–1005.

Yamada, Y., Kawamura, K., Ikehara, K., Ogawa, Y., Urgeles, R., Mosher, D., Chaytor, J. & Strasser, M. (eds) 2012. *Submarine Mass Movements and their Consequences*. Advances in Natural and Technological Hazards Research, **31**, 5th International Symposium Series, Springer, 756 pp. ISBN 978-94-007-2161-6.

Yeats, R.S., Haq, B.U. *et al.* 1981. *Initial Reports of the Deep Sea Drilling Project*, *63*. Washington, DC: US Government Printing Office.

Yeats, R.S., Yamazaki, H., Taira, A., Goldfinger, C. & Kulm, L.D. 1994. Seismotectonics of the Cascadia and Nankai subduction zones. *Geological Society of America Abstracts with Programs*, **26**, A–456.

Yamaji, A. 2003. Slab rollback suggested by latest Miocene to Pliocene forearc stress and migration of volcanic front in southern Kyushu, northern Ryukyu Arc. *Tectonophysics*, **364**, 9–24.

Yang, F.-L., Xu, X., Zhao, W.-F. & Sun, Z. 2011. Petroleum accumulations and inversion structures in the Xihu depression, East China Sea Basin. *Journal of Petroleum Geology*, **34**, 429–440.

Yang, H., Liu, Y. & Lin, J. 2012. Effects of subducted seamounts on megathrust earthquake nucleation and rupture propagation. *Geophysical Research Letters*, **39**, doi: 10.1029/2012GL053892.

Yamazaki, T. & Okamura, Y. 1989. Subducting seamounts and deformation of overriding forearc wedges around Japan. *Tectonophysics*, **160**, 207–217, 221–229.

Yerkes, R.F., Gorsline, D.S. & Rusnak, G.A. 1967. Origin of Redondo submarine canyon, southern California. *U.S. Geological Survey, Professional Paper*, **575-C**, 97–105.

Yu, Ho-Shin & Hong, E. 2006. Shifting submarine canyons and development of a foreland basin in SW Taiwan: controls of foreland sedimentation and longitudinal sediment transport. *Journal of Asian Earth Sciences*, **27**, 922–932.

Zachariasse, W.J., van Hinsbergen, D.J.J. & Fortuin, A.R. 2008. Mass wasting and uplift on Crete and Karpathos during the early Pliocene related to initiation of south Aegean left-lateral, strike-slip tectonics. *Geological Society of American Bulletin*, **120**, 976–993.

Zachos, J.C., Pagani, M., Sloan, L., Thomas, E & Billups, K. 2001. Trends, rhythms, and aberrations in global climate 65 Ma to present. *Science*, **292**, 689–693.

Zachos, J.C., Dickens, G.R. & Zeebe, R.E. 2008. An early Cenozoic perspective on greenhouse warming and carbon-cycle dynamics. *Nature*, **451**, 279–283.

Zagorevski, A., van Staal, C.R., McNicoll, V.J., Hartree, L. & Rogers, N. 2012. Tectonic evolution of the Dunnage Mélange tract and its significance to the closure of Iapetus. *Tectonophysics*, **568–569**, 371–387.

Zalasiewicz, J.A., Taylor, L., Rushton, A.W.A., Loydell, D.K., Rickards, R.B. & M. Williams, M. 2009. Graptolites in British stratigraphy. *Geological Magazine*, **146**, 785–850.

Zecchin, M. & Catuneanu, O. 2012. High-resolution sequence stratigraphy of clastic shelves I: Units and bounding surfaces. *Marine and Petroleum Geology*, **39**, 1–25.

Zelt, F. & Rossen, C. 1995. Geometry and continuity of deep-water sandstone and siltstones of the Brushy Canyon formation (Permian) Delaware Mountains, Texas. *In*: Pickering, K.T., Hiscott, R.N., Kenyon, N.H., Ricci Lucchi, F. & Smith, R.D.A. (eds), *Atlas of Deep Water Environments: Architectural Style in Turbidite Systems*, 167–183. London: Chapman and Hall.

Zeng, J. & Lowe, D.R. 1997a. Numerical simulation of turbidity current flow and sedimentation: I. Theory. *Sedimentology*, **44**, 67–84.

Zeng, J. & Lowe, D.R. 1997b. Numerical simulation of turbidity current flow and sedimentation: II. *Results and geological applications Sedimentology*, **44**, 85–104.

Zeng, J., Lowe, D.R., Prior, D.B., Wiseman, W.J. Jr., & Bornhold, B. 1991. Flow properties of turbidity currents in Bute Inlet, British Columbia. *Sedimentology*, **38**, 975–996.

Zenk, W. 2008. Abyssal and contour currents. *In*: Rebesco, M. & Camerlenghi, A. (eds), *Contourites*, 37–57. Developments in Sedimentology, **60**. Amsterdam: Elsevier.

Ziegler, P.A. 1986a. Late Caledonian framework of western and central Europe. *In*: Gee, D.G. & Sturt, B.A. (eds), *The Caledonide Orogen–Scandinavia and Related Areas*, 3–18. New York: Wiley.

Ziegler, P.A. 1986b. Caledonian, Acadian-Ligurian, Bretonian, and Variscan orogens – is a clear distinction justified ? *In*: Gee, D.G. & Sturt, B.A. (eds), *The Caledonide Orogen – Scandinavia and Related Areas*, 1241–1248. New York: Wiley.

Ziegler, P.A. 1986c. Geodynamic model for the Palaeozoic crustal consolidation of western and central Europe. *Tectonophysics*, **126**, 303–328.

Zierenberg, R.A., Fouquet, Y., Miller, D.J. & Normark, W.R. (eds) 2000. *Proceedings of the Ocean Driling Program, Scientific Results*, *169*. College Station, Texas, USA: Ocean Drilling Program.

Zolnai, G. 1991. *Continental Wrench-tectonics and Hydrocarbon Habitat*, 2nd edn. American Association of Petroleum Geologists Continuing Education Course Notes, **30**, unpaginated.

Zonenshain, L.P. & Le Pichon, X. 1986. Deep basins of the Black Sea and Caspian Sea as remnants of Mesozoic back-arc basins. *Tectonophysics*, **123**, 181–211.

Zuffa, G.G., de Rosa, R. & Normark, W.R. 1997. Shifting sources and transport paths for the Late Quaternary Escanaba Trough sediment fill (northeast Pacific). *Giornale di Geologia*, **59**, 35–53.

Zuffa, G.G., Normark, W.R., Serra, F. & Brunner, C.A. 2000. Turbidite megabeds in an oceanic rift valley recording jökulhlaups of late Pleistocene glacial lakes of the western United States. *Journal of Geology*, **108**, 253–274.

Index

Deep Marine Systems: Processes, Deposits, Environments, Tectonics and Sedimentation, First Edition. Kevin T. Pickering and Richard N. Hiscott.
© 2016 Kevin T. Pickering and Richard N. Hiscott. Published 2016 by John Wiley & Sons, Ltd.
Companion Website: www.wiley.com/go/pickering/marinesystems